D0468276

AMERICAN HANDBOOK OF PSYCHIATRY

Volume Seven

AMERICAN HANDBOOK OF PSYCHIATRY

Silvano Arieti, EDITOR-IN-CHIEF

Volume One
The Foundations of Psychiatry
EDITED BY SILVANO ARIETI

Volume Two
Child and Adolescent Psychiatry, Sociocultural and Community Psychiatry
EDITED BY GERALD CAPLAN

Volume Three
Adult Clinical Psychiatry
EDITED BY SILVANO ARIETI AND EUGENE B. BRODY

Volume Four
Organic Disorders and Psychosomatic Medicine
EDITED BY MORTON F. REISER

Volume Five
Treatment
EDITED BY DANIEL X. FREEDMAN AND JARL E. DYRUD

Volume Six
New Psychiatric Frontiers
EDITED BY DAVID A. HAMBURG AND H. KEITH H. BRODIE

Volume Seven
Advances and New Directions
EDITED BY SILVANO ARIETI AND H. KEITH H. BRODIE

CONTRIBUTORS

Huda Akil, M.D.
Mental Health Research Institute, Department of Psychiatry, University of Michigan, Ann Arbor.

Lolita O. Ang, M.D.
Clinical Assistant Professor, University of Chicago School of Medicine, Chicago, Illinois.

Silvano Arieti, M.D.
Clinical Professor of Psychiatry, New York Medical College; Training Analyst and Supervisor, William Alanson White Institute of Psychiatry, Psychoanalysis, and Psychology, New York.

Jack D. Barchas, M.D.
Department of Psychiatry and Behavioral Science, Stanford University School of Medicine, Stanford, California.

Jules R. Bemporad, M.D.
Director of Children's Services and Associate Professor of Psychiatry, Harvard Medical School at the Massachusetts Mental Health Center, Boston.

Philip A. Berger, M.D.
Department of Psychiatry and Behavioral Science, Stanford Universtiy School of Medicine, Stanford, California.

Marvin L. Blumberg, M.D., F.A.A.P.
Chairman, Department of Pediatrics, The Jamaica Hospital, Jamaica, New York; Associate Professor of Clinical Pediatrics, School of Medicine, Health and Sciences Center, State University of New York at Stony Brook.

Henry Brill, M.D.
Professor of Clinical Psychiatry, School of Medicine, State University of New York at Stony Brook.

H. Keith H. Brodie, M.D.
Professor and Chairman, Department of Psychiatry, Duke University Medical Center, Durham, North Carolina.

Ewald W. Busse, M.D.
J. P. Gibbons Professor of Psychiatry and Associate Provost and Dean, Medical and Allied Health Education, Duke University Medical Center, Durham, North Carolina.

Justin D. Call, M.D.
Professor and Chief of Child and Adolescent Psychiatry, California College of Medicine, University of California at Irvine; Analyzing and Supervising Instructor, Los Angeles Psychoanalytic Society and Institute, Los Angeles, California.

Robert J. Campbell, M.D.
Director, Gracie Square Hospital; Assistant Clinical Professor of Psychiatry, Columbia University College of Physicians and Surgeons, New York.

Robert Cancro, M.D.
Professor and Chairman, Department of Psychiatry, New York University Medical Center; Director, Bellevue Psychiatric Hospital, New York.

William T. Carpenter, Jr., M.D.
Director, Maryland Psychiatric Research Center, Baltimore; Professor of Psychiatry, University of Maryland School of Medicine, College Park.

John F. Clarkin, Ph.D.
Associate Professor of Clinical Psychology in Psychiatry, Cornell University Medical College; Associate Attending Psychologist, Payne Whitney Clinic, The New York Hospital-Cornell Medical Center, New York.

Jonathan O. Cole, M.D.
Chief, Psychopharmacology Unit, McLean Hospital, Belmont, Massachusetts.

Arnold M. Cooper, M.D.
Professor of Psychiatry, Payne Whitney Psychiatric Clinic, The New York Hospital-Cornell Medical Center, New York.

Raymond R. Crowe, M.D.
Associate Professor, Department of Psychiatry, University of Iowa College of Medicine, Iowa City.

John M. Davis, M.D.
Professor of Psychiatry, University of Chicago, School of Medicine; Director of Research, Illinois State Psychiatric Institute, Chicago.

Tristram H. Engelhardt, Jr., Ph.D., M.D.
Rosemary Kennedy Professor of the Philosophy of Medicine, Kennedy Institute of Ethics, Georgetown University, Washington, D.C.

George Gardos, M.D.
Assistant Professor of Psychiatry, Boston University School of Medicine; Director, Institute for Research and Rehabilitation, Boston, Massachusetts.

Michael S. Gazzaniga, Ph.D.
Division of Cognitive Neuroscience, Department of Neurology, Cornell University Medical College, New York.

Ira D. Glick, M.D.
Professor of Psychiatry, Cornell University Medical College; Associate Medical Director for Inpatient Services, Payne Whitney Clinic, The New York Hospital-Cornell Medical Center, New York.

Karen D. Goodman, M.S., R.M.T.
Music Therapy Training Program, Montclair State College, Upper Montclair, New Jersey; Creative Arts Rehabilitation Center, New York.

Frederick K. Goodwin, M.D.
Chief, Clinical Psychiatry Branch, National Institute of Mental Health, Bethesda, Maryland.

Maurice H. Greenhill, M.D.
Dr. Greenhill died on January 20, 1981. He was Professor of Psychiatry, Albert Einstein College of Medicine, Bronx, New York.

John H. Greist, M.D.
Professor of Psychiatry, University of Wisconsin Medical School, Madison.

Graeme Hanson, M.D.
Assistant Clinical Professor in Psychiatry and Pediatrics, University of California, San Francisco.

Douglas W. Heinrichs, M.D.
Research Associate, Maryland Psychiatric Research Center, Department of Psychiatry, University of Maryland School of Medicine, College Park.

Owen E. Heninger, M.D.
President, Board of Directors, Poetry Therapy Institute, Encino, California; Assistant Clinical Professor and Preceptor in Psychiatry, Department of Community and Family Medicine, University of Southern California School of Medicine, Los Angeles.

Anthony Kales, M.D.
Professor and Chairman, Department of Psychiatry, and Director, Sleep Research and Treatment Center, The Pennsylvania State University College of Medicine, Hershey.

Joyce D. Kales, M.D.
Associate Professor, Department of Psychiatry, and Associate Director, Sleep Research and Treatment Center, The Pennsylvania State University College of Medicine, Hershey.

Helen Singer Kaplan, M.D., Ph.D.
Clinical Professor of Psychiatry, and Director, Human Sexuality Program of The New York Hospital-Cornell Medical Center, New York.

David R. Kessler, M.D.
Associate Clinical Professor of Psychiatry, University of California School of Medicine; Staff Psychiatrist, Langley Porter Psychiatric Institute, San Francisco, California.

Clarice J. Kestenbaum, M.D.
Director, Division of Child and Adolescent Psychiatry, St. Luke's-Roosevelt Hospital Center, New York.

Marjorie H. Klein, Ph.D.
Professor of Psychiatry, University of Wisconsin Medical School, Madison.

Carolyn Refsnes Kniazzeh, M.F.A., A.T.R.
Chairman, Education and Training Board, American Art Therapy Association; practicing and consulting art therapist; exhibiting painter, Gallery NAGA, Boston.

Thomas Kreilkamp, Ph.D.
Staff Psychologist, Harvard Community Health Plan, and Senior Counselor, Harvard College Bureau of Study Counsel; Assistant Attending Child Psychologist, Hall-Mercer Children's Center, McLean Hospital, Belmont, Massachusetts.

Stanley Lesse, M.D., Med. Sc.D.
Editor-in-Chief, American Journal of Psychotherapy.

Laurence B. McCullough, Ph.D.
Associate Director, Division of Health and Humanities, Department of Community and Family Medicine, Georgetown University School of Medicine, Washington, D.C.

Mary C. MacKay, M.D.
Chief-of-Service, Rockland Children's Psychiatric Center; Associate Clinical Professor of Psychiatry, New York Medical College, New York.

CONTENTS

Volume Seven

PART THREE *Adult Clinical Conditions and Therapeutic Techniques*

AMERICAN HANDBOOK OF PSYCHIATRY

SECOND EDITION

Silvano Arieti · Editor-in-Chief

VOLUME SEVEN

Advances and New Directions

SILVANO ARIETI AND H. KEITH H. BRODIE · *Editors*

BASIC BOOKS, INC., PUBLISHERS · NEW YORK

Library of Congress Cataloging in Publication Data (Revised)
Main entry under title:

American handbook of psychiatry.

Includes bibliographies.
1. Psychiatry—Collected works. I. Arieti, Silvano, ed. [DNLM: 1. Mental disorders. WM 100 A503]
RC435.A562 616.89 80-68960
ISBN: 0-465-00157-2

Second Edition

Printed in the United States of America
10 9 8 7 6 5 4 3 2 1

Frank J. Menolascino, M.D.
Professor of Psychiatry and Pediatrics, The University of Nebraska College of Medicine, Omaha.

Carol Nadelson, M.D.
Professor and Vice-Chairman, Department of Psychiatry, Tufts University School of Medicine-New England Medical Center Hospital, Boston, Massachusetts.

Darrel A. Regier, M.D., M.P.H.
Director, Division of Biometry and Epidemiology, National Institute of Mental Health, Rockville, Maryland.

Henry F. Smith, M.D.
Staff Psychiatrist and Clinical Instructor in Psychiatry, Harvard Medical School at the Massachusetts Mental Health Center, Boston.

Constantin R. Soldatos, M.D.
Clinical Associate Professor, Department of Psychiatry, and Associate Director, Sleep Research and Treatment Center, The Pennsylvania State University College of Medicine, Hershey.

Alan A. Stone, M.D.
Professor of Law and Medicine, Faculty of Law and Faculty of Medicine, Harvard University.

Fred D. Strider, Ph.D.
Associate Professor of Medical Psychology, Department of Psychiatry, The University of Nebraska Medical Center, Omaha.

Albert J. Stunkard, M.D.
Professor of Psychiatry, University of Pennsylvania, Philadelphia.

Norman Tabachnick, M.D.
Clinical Professor of Psychiatry, University of California School of Medicine, Los Angeles; Supervising and Training Analyst, Southern California Psychoanalytic Institute, Beverly Hills.

John A. Talbott, M.D.
Professor of Psychiatry, Cornell University Medical College; Associate Medical Director, Payne Whitney Psychiatric Clinic, The New York Hospital-Cornell Medical Center, New York.

Daniel Tarsy, M.D.
Department of Neurology, Boston Veterans' Administration Hospital; Boston University School of Medicine, Division of Neurology, New England Deaconess Hospital; Department of Neurology, Harvard Medical School, Boston, Massachusetts.

Carl A. Taube
Deputy Director, Division of Biometry and Epidemiology, National Institute of Mental Health, Rockville, Maryland.

Pamela J. Trent, Ph.D.
Assistant Medical Research Professor, Department of Psychiatry, Duke University Medical Center, Durham, North Carolina.

Bruce T. Volpe, M.D.
Division of Cognitive Neuroscience, Department of Neurology, Cornell University Medical College, New York.

Stanley J. Watson, Ph.D., M.D.
Mental Health Research Institute, Department of Psychiatry, University of Michigan, Ann Arbor.

Thomas A. Wehr, M.D.
Clinical Psychobiology Branch, National Institute of Mental Health, Bethesda, Maryland.

John S. Werry, M.D.
Professor and Head, Department of Psychiatry, University of Auckland School of Medicine, Auckland, New Zealand.

Redford B. Williams, Jr., M.D.
Professor of Psychiatry and Assistant Professor of Medicine, Duke University Medical Center, Durham, North Carolina.

PART FOUR: *Hospital, Administrative, and Social Psychiatry*

PREFACE

At the end of the sixth volume of this Second Edition of the *American Handbook of Psychiatry,* I, as the editor-in-chief, began my concluding remarks with the following thoughts:

> With this volume the second edition of the *American Handbook of Psychiatry* comes to a conclusion. A reading of its various tables of contents, or even a cursory perusal of a few sections, is sufficient to reveal its scope and magnitude.
>
> If for a few seconds we suspend our scientific judgment and give our human propensity for metaphor free rein, what may come to mind is the building of a Medieval cathedral. In this *Handbook,* as in the cathedral, many people from different fields participated. They were united in the hard and long work by a common vision and aim, which we hope has been achieved, of giving an adequate representation of contemporary psychiatry.

When I wrote those sentences, I felt that the metaphor of the cathedral had ended. But just as many cathedrals have required additional work after they were thought to be completed, so too, I have come to believe, will the *American Handbook of Psychiatry* benefit from modification and improvement over time.

We are participating in an era of great scientific expansion, of biological discoveries, of rapid social changes—all of which affect the individual psyche and group psychology; the influence of these factors on psychiatric conditions has to be constantly reevaluated.

One solution to the problems presented by this uninterrupted growth would be to publish from time to time new editions of the *Handbook.* But the editors and publishers of the *Handbook* feel that this is not the best course of action. Nearly all of what is included in the first six volumes of the Second Edition is still valid and applicable. It would not be fair to encourage readers who have bought the six volumes to make considerable financial investment in a completely new edition.

The solution chosen by the editors and publishers is to include these recent modifications and improvements in this seventh volume, *Advances and New Directions.* It is essential for psychiatrists and those in related fields to familiarize themselves with current developments. Surely all mental health professionals will want to become informed, not purely from the theoretical point of view, but also for a direct approach to clinical practice. Among the topics considered are endorphins and psychosis; implications of split-brain studies; the latest findings on the genetic models of mental illness; psychopharmacotherapy as applied exclusively to children; the child at risk for major psychiatric illness; borderline syndromes in childhood; new changes in the psychoanalytic concepts of narcissism; new concepts of masked depression; biofeedback; the most recent concepts of human sexuality; new management of sleep disorders; the role of adult play in mental health; the prevention and treatment of mental retardation; the roles of computers in psychiatry; changes in law that pertain to psychiatry; current ideas on art, poetry, and music therapy; and the family of the schizophrenic as a participant in the therapeutic task. These are just some of the topics developed in the thirty-nine chapters of this volume.

Whether professional readers are predominantly involved in biological or psychodynamic psychiatry, in work with children, adolescents, or adults, in individual therapy or in hospital, administrative, and social psychiatry, they will find chapters, written by outstanding authorities in the respective fields, that are particularly relevant to the areas of their major interest.

One of the most comprehensive and innovative chapters in this volume is "Liaison Psychiatry," by Dr. Maurice Greenhill. It is with deep sorrow that we must announce that Dr. Greenhill died on January 20, 1981, and was not able to see published the result of many years devoted to this particular study and practice.

With so many new and exciting developments taking place, we cannot know exactly what path tomorrow's psychiatry will take. We can only suspect that the construction of the "cathedral" is far from complete. It is to the careful building of that structure that this seventh volume of the *Handbook* is dedicated.

Silvano Arieti, M.D.
Editor-in-Chief

New York, January 1981

PART ONE

Biological Studies

CHAPTER 1

THE ENDORPHINS AND PSYCHOSIS*

Stanley J. Watson, Huda Akil, Philip A. Berger, and Jack D. Barchas

¶ Introduction

THE LIST of neurotransmitters or neuromodulators has grown enormously in the last few years. Yet, few of these newly discovered brain messengers have so rapidly affected behavioral sciences and biological psychiatry as have the endorphins. Clearly, the thought of having our "own natural opiates" is fascinating and carries more obvious connotations to the psychiatrist than the discovery of other substances with fewer pharmacological ramifications. After all, opiates clearly alter mood and affect, and some of them produce hallucinations and bizarre thought content.[49,74] They have been occasionally used as therapeutic tools in the past, with varying success. Furthermore, addiction to morphine or heroin is a psychiatric and social problem which presents multiple questions as to its psychological versus its physiological roots and manifestations. Finally, morphine, codeine, and other analgesics on the one hand, and heroin and other "street" opiates on the other, are closely associated in many minds with notions of pleasure and pain—notions that lie at the core of many psychological theories of normal and abnormal behavior.

It is fortunate that the endorphins are so intrinsically appealing, because they are also complex, numerous, and sometimes frustrating. In the last few years we have learned a great deal about them and from them. While they have yet to provide a key to understanding psychosis, they have taught us a great deal about brain-pituitary relationships, about the nature of neurotransmission, and peptide biosynthesis. They hold some hope for a better understanding of psychosis, an understanding which should be based on sound knowledge of the underlying physiology.

This chapter begins with a presentation of

*Supported by Scottish Rite Schizophrenia Research Foundation; NIMH Grant No. MH30854; and NIMH Program Project Grant No. MH23861.

the important basic science issues associated with the field of the endorphins, followed by a discussion of the clinical studies on endorphins, and ends with a summary and discussion of the possible relationships between endorphins and psychosis. The goal of the basic science section is to emphasize the complexity and diversity of endorphin systems in the brain. A clear understanding of the basic science issues is necessary before it is possible to appreciate potential etiological and pharmacological relationships between the endorphins and human psychopathology. In the second section, on clinical theory and data, it is hoped that the reader will utilize the perspectives obtained from the basic science discussions and, in doing so, be able to evaluate the many problems clinical investigators face in attempting to apply this information to the study of human psychosis.

Throughout this chapter, we shall use the terms "endorphin" and "opiate peptides" interchangeably to denote the whole class of naturally occurring substances that possess opiate-like properties. Terms such as enkephalin, β–endorphin, α–endorphin, or dynorphin all denote specific peptides with known structures that are members of the "endorphin" class (see table 1–1).

¶ Basic Science Studies On the Endorphins: A Summary

The study of endorphins has been very fast paced and has an extremely broad and complex base. Not only have there been a large number of published papers involving the several opiate peptides, but the types of research carried out have been complex, involving every level from electron microscopy and DNA cloning to clinical studies of schizophrenia. Therefore, this discussion will attempt to compress the salient points, in order to give the reader a basis for following the discussion of the possible role of these substances in psychosis.

Opiate Receptors and Stimulation Produced Analgesia

The advent of the field of endorphins can be traced to two main lines of work which came to fruition in the early 1970s. Classical pharmacology, and with it, the preparation of opiate agonists and antagonists in active and inactive stereoisomer forms, was an enormous impetus to opiate research.[44,46] When this pharmacological armamentarium was combined with the availability of techniques for producing drugs tagged with radioisotopes, it was possible to demonstrate the presence of specific binding proteins ("receptors") in the mammalian brain to both plant and synthetic opiates. These binding sites were characterized by several laboratories[87,98,104] as possessing high affinity for opiates, as being stereospecific, and as being heterogeneously distributed across brain regions and other body tissues. Most importantly, the affinity with which a drug bound to these sites predicted with good accuracy its clinical effectiveness as an opiate.[100] Based on such evidence, it was reasonable to conclude that these membrane sites represented the recognition locus for opiate action and were specific opiate receptors.

At the same time, several investigators at the University of California at Los Angeles were able to demonstrate that electrical stimulation in the brains of rats, and eventually of cats, monkeys, and humans, was capable of producing a substantial degree of analgesia for both acute and chronic pain.[75] This phenomenon is referred to as stimulation-produced analgesia (SPA). These investigators were able to show that the analgesia so produced did not interfere with normal function over a broad set of measures, that it did provide excellent pain relief, and that it was in part reversible by the opiate antagonist naloxone.[2,3] The ability of naloxone to partially reverse stimulation-produced analgesia (SPA) led to the conclusion that the stimulation released an endogenous material that appeared to be acting on the opiate receptor to produce pain control.[3] In effect, the demonstration of the opiate receptor by

TABLE 1–1 **Opioid and Related Peptides**

Methionine-Enkephalin (Met-enkephalin)	Tyr-Gly-Gly-Phe-Met (β-LPH 61–65)
Leucine-Enkephalin (Leu-enkephalin)	Tyr-Gly-Gly-Phe-Leu
β-Endorphin (β-END)	β-LPH 61–91 (See Fig. 1)
β-Lipotropin (β-LPH)	91 Amino Acids—Contains β-END (See Fig. 1)
Adrenocorticotropin (ACTH)	39 Amino Acids (See Fig. 2)
α-Melanocyte Stimulating Hormone (α-MSH)	Ac-Ser-Tyr-Ser-Met-Gly-His-Phe-Arg-Trp-Gly-Lys-Pro-Val-NH (See Fig. 2) Cleaved from ACTH and Processed Further
Corticotropin-Like Intermediate Lobe Peptide (CLIP)	(See Fig. 2) Cleaved from ACTH
Dynorphin	Tyr-Gly-Gly-Phe-Leu-Arg-Arg-Ile-Arg-Pro-Lys-Leu-Lys . . .

binding techniques and the stimulated release of endogenous opiates through electrical means, taken together, were interpreted by several groups as being strong evidence for the existence of an endogenous opiate peptide system in the mammalian brain.

Enkephalins and β-Endorphin

Within a relatively few years after these two discoveries, the pentapeptides methionine- and leucine-enkephalin (met- and leu-enkephalin) were extracted from the brain and sequenced by Hughes, Kosterlitz, and their collaborators[54] (see table 1–1). These two opiate peptides were found to exist in a large number of mammalian brain regions and appeared to agree generally with the distribution of opiate receptors. In their paper describing enkephalin sequences, Hughes and his collaborators[54] pointed out that the structure of met-enkephalin appeared within the longer pituitary peptide β-Lipotropin (β-LPH).[67] Several groups simultaneously recognized that the C-terminal portion of β-LPH, that is β-LPH 61–91 (see figure 1–1),

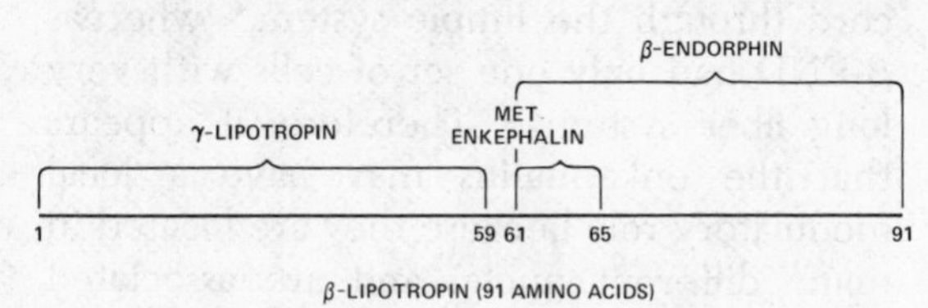

FIGURE 1–1. Structure of β-lipotropin (β-LPH): Contains β-endorphin (β-END 61-91) and has the structure of methionine-enkephalin (61-65). β-LPH 1-59 is known as γ-LPH and is not opiate-like.

which contained met-enkephalin (position 61–65), could very well be an active opiate in its own right.[21,26,47,66] It carried the name C fragment of lipotropin or, eventually, β-Endorphin (β-END). Thus, within a few months we were faced with the existence of not one but three endogenous opiate ligands in mammalian brain (met-enkephalin, leu-enkaphalin and β-endorphin). A large amount of work was carried out to demonstrate the actions and activity of these agents in a wide variety of test systems. The conclusion derived from such pharmacological studies was that these compounds in general had a spectrum of actions very similar to those of the plant alkaloid and synthetic opiates. Like morphine, the endogenous opiates were capable of producing analgesia,[13,70] tolerance and dependence,[129] and positive reinforcement.[12]

But, in early 1976, there developed a sudden embarrassment of riches in having three different opiate compounds. It was not clear at that point whether the shorter peptide (met-enkephalin) was a cleavage product of the longer peptide β-END and was therefore "simply" a metabolite of β-END, or whether β-END was "just" the precursor to enkephalin. The answer to that dilemma came fairly soon with the immunohistochemical work by several groups demonstrating that enkephalin and β-END had distinctly different distributions in the mammalian brain.[20,94,120,124] Both peptides occurred in cells, axons, and terminals, but the two systems exhibited no anatomical relationship to one another. In fact, it was soon concluded that enkephalin could be found in a large number of cell groups throughout the brain, from the spinal cord through the limbic system,* whereas β-END had only one set of cells with very long fiber systems.† Therefore, it appears that the enkephalins may have a local modulatory role because they are located in many different nuclei and are associated with short fiber pathways and local circuit connections. In contrast, β-END is primarily associated with the limbic system; it arises from a single set of cells with a very widespread fiber distribution involving many limbic and brain stem structures. From a physiological point of view, the enkephalins might be thought of as being related to a wide variety of different functions. They are located so that they could be involved in pain perception, control of respiration, motor function, endocrine controls, and even affective states. On the other hand, β-END appears to be much more tightly related to upper brainstem and limbic systems and may be associated with system-wide changes.

Soon after the discovery of the enkephalins, it was possible to demonstrate several important characteristics of their action, suggesting their role as putative neurotransmitters. Not only were they stored intravesicularly with well-demonstrated interaction with specific binding sites, but they could also be shown to be released and degraded rapidly. Release studies were first carried out in humans, since enkephalin-like material could be detected in human lumbar fluid.[6,7,115] The animal work on stimulation-produced analgesia had been extended to the clinical situation, and employed to relieve chronic intractable pain in humans.[90] In neurosurgery, when these patients were electrically stimulated, the concentration of enkephalin-like material in the third ventricular fluid rose significantly.[6] Finally, more recent evidence has suggested the presence of a specific, membrane bound enzyme, termed enkephalinase, which degrades enkephalin with high affinity.[73,103]

β-Endorphin/α-MSH Systems

Studies on the nature of brain β-END have proven to be considerably more complex and have raised more substantial questions than those associated with enkephalins. β-LPH was known as a β-MSH pituitary prohormone for several years prior to the discovery of the enkephalins and β-END.[66] Immunohistochemical studies of β-LPH showed it in the pituitary, in the corticotrophs of the anterior lobe (cells that produce

*See references 36, 51, 56, 97, 107, and 123.
†See references 19, 86, 120, 124, and 130.

adrenocorticotropic hormone [ACTH]), and in all of the cells of the intermediate lobe.[77] When β-END antisera became available, it was possible to demonstrate that β-END had precisely the same localization as did β-LPH, that is the corticotrophs of the anterior lobe and all intermediate lobe cells.[18] Since ACTH was also present within these two cell types, there appeared to be an intimate association between β-END, β-LPH, and ACTH in the pituitary. Pelletier and coworkers[85] and Weber and coworkers,[128] using electron microscopic techniques, were able to demonstrate that β-LPH and ACTH immunoreactivity occurred within precisely the same granules in intermediate lobe and in corticotrophs. By demonstrating the presence of three substances in the same pituitary cells, that is, the corticotroph containing β-LPH, β-END, and ACTH, support was provided for the common biosynthetic origin of ACTH, β-END, and β-LPH. Mains and coworkers[72] and Roberts and Herbert,[92] using a mouse pituitary tumor line, demonstrated that all three substances, ACTH, β-END, and β-LPH, were made from the same precursor molecule—the 31K dalton precursor, also known as pro-opiocortin (see figure 1–2). The 31K molecule contained the three structures mentioned previously, plus a new portion known as the 16K piece, which as yet has no clear physiology associated with it. An elegant and powerful extension of the investigation of this protein chemistry is the work of Nakanishi and associates,[78] which demonstrates the DNA structure for the 31K precursor in the bovine pituitary. Work by Eipper and Mains[35] and Gianoulakis and coworkers[43] went one step further to show that the anterior lobe tended to cleave the 31K precursor down to $ACTH_{1-39}$ and β-LPH (with some β-END), whereas the intermediate lobe of the pituitary took the precursor one step further to produce β-END and γ-LPH (and a small amount of β-LPH), and then cleaved ACTH into two fragments, an 18–39 piece known as corticotropin-like intermediate lobe peptide (CLIP) and α-MSH (N-acetyl ACTH $_{1-13}$ amide or α-Melanocyte Stimulating Hormone). Thus there were two biosynthetic endpoints of the pituitary 31K precursor. Naturally, immunohistochemical studies of the β-END/β-LPH cells in brain were extended to ACTH.[121] Our own work and that of others* demonstrated that the same cells in the brain that contained endorphin/lipotropin contained ACTH as well as the 16K piece. Thus, it appeared that at least three places in the central nervous system contained the genetic machinery for producing 31K precursor: two lobes of the pituitary and the brain arcuate nucleus (see figure 1–3).

As already mentioned, the two pituitary cell types tend to process 31K in different

*See references 16, 80, 83, 101, and 119.

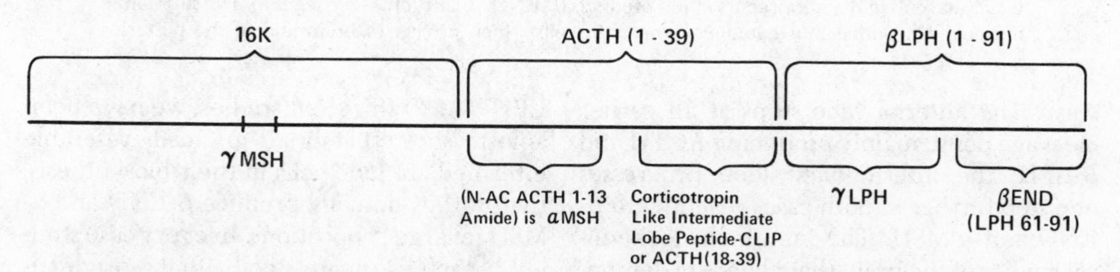

FIGURE 1–2. The 31K precursor for ACTH, β-LPH, and β-END. See Figure 1–1 for β-LPH region model. ACTH 1-39 can be cleaved and modified to produce α-Melanocyte Stimulating Hormone (α-MSH) and Corticotropin-Like Intermediate Lobe Peptide (CLIP). The N Terminus of the 31K precursor is known as the "16K" piece and contains a potentially active sequence similar to ACTH/α-MSH 4-10. This region is known as γ-MSH. The anterior lobe of pituitary mainly produces ACTH 1-39, β-LPH, and some β-END. The intermediate lobe of pituitary and brain go further and produce β-END, γ-LPH, α-MSH, and CLIP. The processing of 16K in these cells is currently under study.
SOURCE: Mains, R.E., Eipper, B.A., and Ling, N. "Common Precursor to Corticotropins and Endorphins," *Proceedings of the National Academy of Science (U.S.A.)*, 74 (1977): 3014–3018; and Eipper, B., and Mains, R. "Existence of a Common Precursor to ACTH and Endorphin in the Anterior and Intermediate Lobes of the Rat Pituitary," *Journal of Supramolecular Structure*, 8 (1978): 247–262.

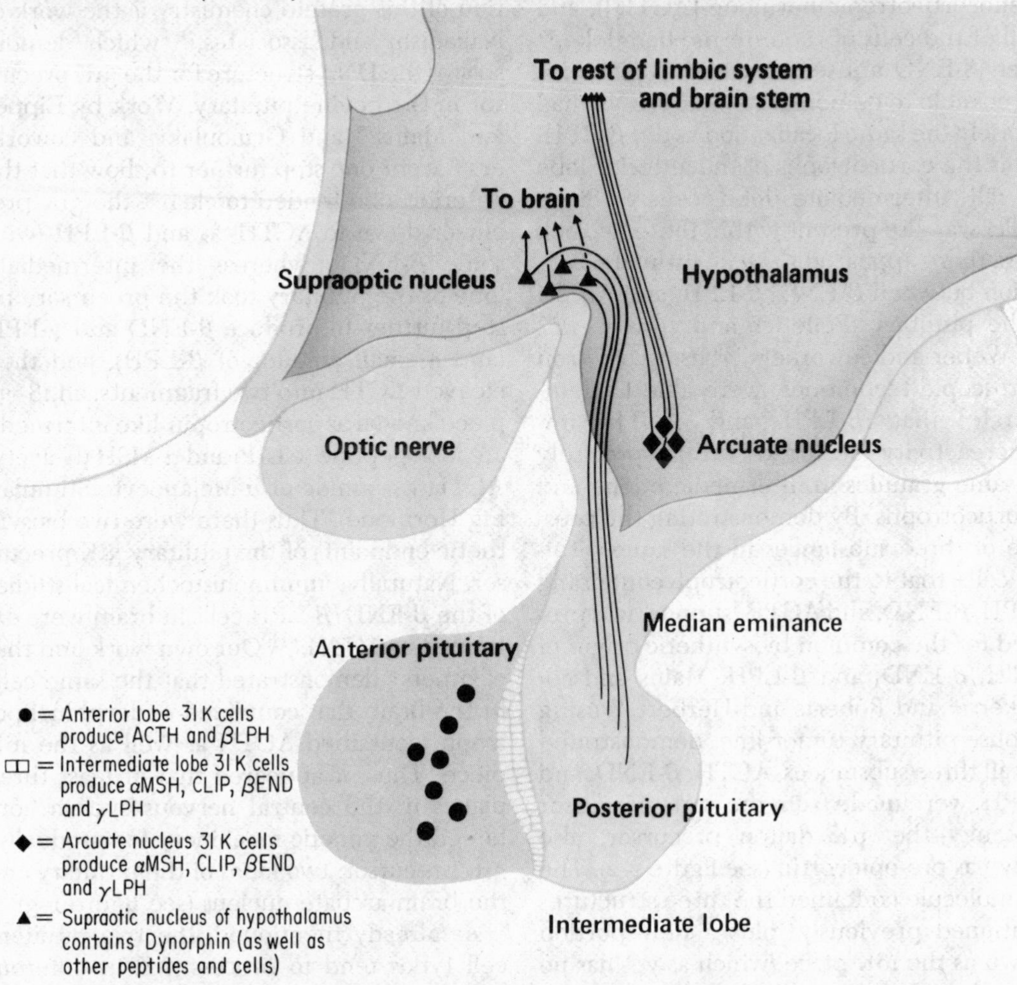

FIGURE 1–3. Schematic of the pituitary and brain cell areas containing β-END and Dynorphin immunoreactivity. The anterior and intermediate lobe of pituitary and arcuate nucleus of hypothalamus contain the 31K precursor and products (β-END, β-LPH etc.—see legend). The supraoptic nucleus of hypothalamus contains Dynorphin cells which project to brain and to the posterior pituitary.

ways. The anterior lobe stops at an earlier cleavage point, mainly producing ACTH and β-LPH, the intermediate lobe progresses one step further in both cases reaching to β-END and α-MSH. The immunohistochemical studies of the brain that allegedly demonstrated "ACTH" in the brain were unclear. It was possible that the antisera employed were actually reacting with γ-MSH and CLIP, rather than full ACTH. Further, there had been several studies demonstrating the presence of α-MSH peptide in the central nervous system and suggesting that this peptide had a distribution similar to β-END/β-LPH.[55,84,109] In recent studies, we have been able to show that brain 31K cells resemble intermediate lobe cells in their biosynthesis, that is, they actually produce β-END and α-MSH in large proportions. In every case studied, each arcuate cell that produces β-END also produces α-MSH and vice versa.[117,118] As an interesting aside, it should be noted that in the process of these studies a second α-MSH system was discovered in the hypothalamus, physically unrelated to the α-MSH/β-END system but positive for α-MSH using a wide variety of α-MSH antibodies.[117,118] (We are once again reminded of the wisdom of nature in using active structures repeatedly: for example, the met-enkephalin structure occurs by itself and also in β-END.) Thus, it appears that α-MSH occurs as a neurally active substance, both as part of the 31K system and independent of it.

In further characterizing the β-END system in the brain, several kinds of studies have been carried out. For example, it has been demonstrated that β-END, when injected intracerebroventricularly, can produce analgesia and tolerance, as one might expect from an opiate.[70,129] Yet, the effects are extremely long lasting, compared to those of enkephalin. The latter peptide is thought to be rapidly degraded—a characteristic of classical neurotransmitters. Furthermore, β-END injections lead to a rigid "catatonic" state in rats, probably due to the production of limbic seizures.[19,20,108] It has been shown that, when one looks at pituitary β-END and studies blood levels, severe stress to the rat can produce a substantial release of β-LPH/β-END in parallel with ACTH.[48] Thus, at least the pituitary system is responsive to stress. It is known that β-END can be detected in the lumbar and third ventricular fluids of normal humans and in pain patients. It has been demonstrated (as mentioned for enkephalin) that when electrodes are placed in the thalamus of humans suffering from chronic pain, electrical stimulation produces pain relief for the patient and a substantial release of β-END immunoreactivity (more dramatic than was observed with enkephalin).[5] In fact, it has been hypothesized that β-END is responsible for the prolonged period of analgesia associated with electrical stimulation. Finally, β-END in a highly specifically labeled form has been made available recently by Dr. C. H. Li,[8,42] and this material also has a high affinity binding site in the brain. We are able to conclude that β-END has many of the characteristics one would expect from an active neuronal substance: It is stored presynaptically, it is synthesized in the cell, it appears to be releasable by electrical and, most recently, by some hormonal means, it appears to produce substantial effects in its own right, and it possesses a binding site in the brain.

To add to the complexity of these systems, one should remember that not only do the β-END cells produce β-END but also produce ACTH in the process of finally synthesizing α-MSH. Thus, α-MSH would appear to be a likely product for release. Not only is α-MSH synthesized in the brain, but it appears to be released into the cerebrospinal fluid (CSF) upon stimulation. In recent studies, the authors and others have been able to demonstrate the presence of α-MSH immunoreactivity in lumbar fluids and its release into ventricular fluids of patients being stimulated for pain relief.[10] The problem of establishing the presence of a receptor for α-MSH and/or ACTH in the central nervous system is a long and complicated one. The fairest summary is to state that many laboratories have tried. Most recently, our laboratory[9] has been successful in producing a small amount of ACTH and α-MSH binding in the brains of several mammalian species, tending to indicate there are relatively few receptor sites available for this material, in contrast to β-END. Further, in very recent studies it has been possible to demonstrate that α-MSH and its deacetylated form are both active in the brain when microinjected into the midbrain periaqueductal gray.[116] It appears that α-MSH is a reasonable candidate for a neurally active substance in that it possesses a binding site, exhibits behavioral effects, and appears to be releasable. However, the reader should be cautioned that these data are preliminary and need further support.

The general principle that a single neuron tends to produce a single active substance has now been brought into question, not only by the work reviewed here but by the elegant work of Hokfelt[50] over several systems, peripherally and centrally. In considering the problem of multiple substances being produced and very likely released by a single neuron, a huge set of complex issues arise. For example, how does the synthesizing neuron control the activity of two potentially active substances? Are there two receptor sites postsynaptically? How do they feed back on the synthesizing neuron? Are they both always released in a fixed ratio? Or does the synthesizing neuron have the capacity to preferentially inhibit one and release the other, or can the neuron make one form relatively more active than the other?

Another issue very relevant to clinical studies is that of the proper way to mimic normal physiology. When one administers an opiate agonist by itself, it is clear that that drug is not mimicking the endogenous β-END/α-MSH system's normal action. It is very likely that the endogenous system is exposed to β–END and α–MSH simultaneously—two different substances for two different receptors. It might be suggested that the proper way to approach such clinical studies would be to administer α-MSH or β-END agonists or antagonists together in some controlled fashion, rather than attempting to separate them, as many studies have inadvertently done.

Dynorphin

Most recently, a new opiate peptide has been discovered by Goldstein and his coworkers at Stanford University.[46] They have called this new peptide dynorphin (from *dynis,* "powerful"). Dynorphin includes the full structure of leu–enkephalin as its first five residues, followed by a highly basic region. This material is extremely potent in the guinea pig ileum; it has yet to be thoroughly studied in the brain. Immunohistochemical studies[127] have revealed that it occurs in the posterior pituitary almost exclusively (unlike β-END which occurs in the intermediate lobe and anterior pituitary). It also has a distribution in the brain with cells in the supraoptic nucleus and perhaps elsewhere and fibers throughout limbic and brain structures. Some fibers have been detected in the guinea pig ileum. While little is known about this newly discovered endorphin, it is clear that its distribution in the brain is different from β-END; its relationship to the enkephalin system is unclear. The cellular overlap relationship between dynorphin and the other posterior lobe peptides (oxytocin and vasopressin) is currently being studied. Certainly, questions of circulating blood levels, CSF levels, or relationship to addiction or psychosis are yet to be addressed, and should prove of great interest.

Summary

In studying the role of endorphins in psychosis, we are faced with three major neuronal systems in the brain, intimately associated with the pain, affect, and endocrine systems. These opiate peptide systems are enormously complex in their own right, and at least one of them appears to contain one other active substance (β-END/α-MSH). Another appears either to occur in the same cells (met– and leu–enkephalin) or to have two different systems, but of extremely similar distribution. The last one, dynorphin, has a third set of brain fibers and major endocrine ramification. By demonstrating the presence of so many systems, all of which would appear to have their own opiate receptors, we are faced with enormous complexity associated with those receptor measurements. It does not seem unlikely that each of these neuronal systems should have a relatively different receptor subtype in order to account for the differences in peptide structure. Of course, the neural connectivity of these systems leads to quite different physiological ramifications.

While this review has focused on the brain and pituitary, it should be noted that opiate peptides exist in the peripheral nervous system—for example, in the gut adrenal.[50,101] These peripheral systems present their own fascinating complexities. For instance, it has recently been discovered that enkephalin and some larger enkephalin-containing peptides occur within the adrenal medulla, are stored within norepinephrine and epinephrine-containing granules, and may be coreleased when those catecholamines are released.[1,53,102,112]

Endorphins are candidates for brain neurotransmitters or neuromodulators, but they are also pituitary, gut, and adrenal hormones. Their peripheral targets are yet to be determined, and their primary brain functions yet to be defined. In retrospect, opiates such as morphine are understandably potent since they access receptors in the brain and in the periphery for numerous endogenous systems with multiple roles. While the complexity of endorphins indicates their importance in behavior, it creates several logistics problems for the clinician. With pharmacological tools, it is difficult to influence one system without affecting the others. Enkephalin analogues are likely to be recognized at the brain or peripheral sites that normally only interact with β-END, and vice versa. Antagonists will block endogenous opiate effects indiscriminately, though possibly with differential efficacy. Thus, interpretation of behavioral and clinical data should proceed cautiously.

¶ Clinical Studies

In the following section we summarize the several investigations of psychosis that have been carried out using opiate pharmacology (or the opiate peptides). In many ways, it is too early to draw conclusions of a valid or lasting nature from this very preliminary set of investigations. Nonetheless, the results to date will probably help shape future research, focusing it in specific directions, while eliminating simplistic approaches or invalid hypotheses.

In general, there have been five main approaches to the problems of relating endorphins to psychiatric illness. These five include studies of cerebrospinal fluid chemistry, the actions of opiate antagonists on psychiatric symptomatology, the actions of opiate agonists (including β-endorphin on psychiatric symptomatology), the attempted removal of endorphins by hemodialysis, and finally, the administration of opiate inactive structural variants of β-END (des-tyrosine-γ-endorphin) to schizophrenic patients. The general thrust in each of these areas has either been aimed at altering symptoms of specific psychiatric illnesses or at correlating the occurrence of the state of the illness with the level of endorphins in CSF, blood, or dialysate.

Cerebrospinal Fluid Studies

Soon after the discoveries of the endogenous opiate peptides, Terenius and associates[105] and Lindstrom and associates[68] carried out a series of clinical studies evaluating the endogenous opiate peptide levels in schizophrenic and manic patients. These investigators did not identify the structure of the peptides they were studying, but only characterized them in terms of their column chromatographic behavior and interactions with opiate receptors. In Terenius' nomenclature, Fraction 1 is a higher molecular weight fraction, but does not appear to contain β-END, whereas Fraction 2 is a lighter fraction and does appear to contain enkephalin-like peptides. When untreated schizophrenics were evaluated, they were found to have elevated levels of Fraction 1 in six out of nine cases; but after treatment, levels returned to normal (less than 2 pmoles per ml) in seven of nine cases. Fraction 1 was further elevated in three out of four cases of mania. However, these same manic subjects also demonstrated highly elevated levels of Fraction 2 during normal mood states. Even though Fraction 1 and Fraction 2 have not been specifically identified, there appears to be significant indication that these opiate peptide fractions may reflect either the effect of treatment, or some aspect of underlying psychopathology in schizophrenic and manic patients. In a replication and extension of this work, Rimon and colleagues[91] have confirmed the elevation of Fraction 1 in first break acute schizophrenics and a few reentering schizophrenics with some normalization after treatment. No pattern to Fraction 2 was detected. Recently, Domschke and colleagues[31] have described studies of spinal fluid β-END in normal and neurological controls and in acute and chronic schizophrenics. They conclude that normal subjects have values of 72 fmoles per ml, whereas neurological controls have values of 92 fmoles per ml. The chronic schizophrenics had only half the normal values (35 fmoles per ml) and acute schizophrenics (definition unclear) had values of 760 fmoles per ml. These normal and neurological control values are in general agreement with the study of Jeffcoate and associates[58] of normal CSF, in which β-END and β-LPH were found in a 60–80 fmoles per ml range. However, both these studies are in striking contrast to the studies from several other laboratories in which the levels of β-END and β-LPH have been found to be much lower. Studies by Akil and colleagues,[10] and Emrich and colleagues[38] generally find normal CSF values to be much lower, in the range of 3–12 fmoles per ml for β-END. Akil and colleagues[10] have studied over sixty spinal taps from chronic schizophrenics and normals and found no difference between them. In a study by Emrich and coworkers,[37] normals and groups of patients with meningitis, disk herniation, lumbago, and schizophrenia all show approximately the same level of CSF endorphins (in the 10–15 fmoles range). Thus, there appears to be substantial disagreement, not only on the levels seen in schizophrenia, but also on the levels seen in normal individuals as well. The studies reporting lower concentrations have used more elaborate biochemical controls and more accurate calibrations of antisera and extraction procedures.

Finally, Dupont and coworkers,[33] in a very interesting study, have approached the problem of peptidase activity in the cerebrospinal fluid of schizophrenics and normals and find that peptidases which are capable of degrading enkephalin are much more active in the spinal fluid of schizophrenics than in that of normal individuals. However, in an attempt at replication, Burbach and associates[23] were unable to reproduce the finding of increased activity of an enkephalin-degrading enzyme when CSF of schizophrenics was contrasted with controls. They were further unable to demonstrate altered β-END breakdown in the same subjects. Thus, the question of altered metabolism of these opiate peptides in the CSF of schizophrenics remains open.

In summary then, investigation of CSF chemistry of the opiate peptides is technically difficult and at an early stage. Studies by Terenius and coworkers[105,106] would appear to be most interesting, as they show consistent pre- and posttreatment differences. The work of other investigators in terms of levels of β-END tends to be inconsistent and falls into two disparate groups.

Opiate Antagonists

Following up on the original results of their studies of elevated endorphin levels in CSF, Gunne, Linstrom, and Terenius[49] attempted to reverse some of the symptoms of schizophrenia by administering the opiate antagonist naloxone. Generally, they reasoned that since elevated levels of endorphins occur during the acute phase of an illness, and decreased levels of endorphins during the recovery phase, it might be concluded that the endorphins were in part responsible for the worsening of psychosis. They speculated that the use of the opiate antagonist naloxone might produce a change in psychotic symptomatology. The initial study of Gunne, Lindstrom, and Terenius[49] was a single-blind study in which a modest dose of naloxone (0.4 mg) was given to six schizophrenic patients. Four of these patients reported significantly decreased auditory hallucinations. This study resulted in several attempts at replications using a double-blind crossover design with basically negative results. Volavka and coworkers[113] used the same dose of naloxone on seven well-chosen schizophrenic subjects and observed no effect. This was a careful study which followed the effects of intravenous administration for over twenty-four hours. Janowsky and coworkers[57] used 1.2 mg of naloxone in studying a rather heterogeneous group of patients for only one hour; they also detected no effect on their eight subjects. Kurland and associates[63] used between 0.4 and 1.2 mg naloxone and found no statistically significant effect on their eight patients. Dysken and Davis,[34] in a single-blind study, evaluated the effects of 20 mg naloxone for only a ten-minute period in a single subject to no avail. Lipinski and associates,[69] in a double-blind crossover, gave 1.6 mg naloxone to schizophrenic patients with no observed effect. Finally, Davis and associates,[28] in a double-blind crossover design, generally using low doses (0.4 mg and in a few instances using 10 mg naloxone), studied fourteen schizophrenic patients, and found a change in "unusual thought content" on the Brief Psychiatric Rating Scale. These same investigators found no change in their "affective" patients. Thus, from the original six studies, only Davis and associates[28] supplied any support for the observations of Gunne and coworkers.[49] However, it should be noted that these were basically low-dose studies using 1.2 mg of naloxone or lower doses.

From basic pharmacological work with the opiate peptides across several test systems, it had become clear that the opiate peptides were relatively resistant to the effects of naloxone, often requiring higher doses, and that naloxone itself, although short-lived in its reversal of morphine, could be detected in its effects on endorphin systems for several hours.[4] Therefore, it seemed logical to attempt to study the effects of naloxone on schizophrenia, using much higher doses and following the patients for longer periods of time. In a study reported by this group,[121] a large number of schizophrenic subjects were prescreened. The authors were able to find patients who met the Research Diagnostic Criteria (RDC) for schizophrenia, were diagnosed either as chronic undifferentiated or paranoid, and were at the same time cooperative, stable on their medication, and noted for their chronic (twice per hour) pattern of hallucinations. In this double-blind crossover design, 10 mg of naloxone was used and the patients were followed for up to two days after the infusion. Under these conditions, naloxone was found to produce a statistically significant effect in reducing the number of hallucinations in these chronic hallucinating schizophrenics at one and one-half to two hours after infusion. Six of the nine patients subjectively reported a clear decrease in auditory hallucinations. It must, of course, be admitted that this is a highly selected subgroup, as over 1,000 schizophrenic patients at the Palo Alto Veteran's Hospital were screened for this study.

Emrich and coworkers[37,38,39,40] report two studies in which they evaluated the effects of naloxone on schizophrenics and other psychotic patients. Their general impression in one study was that naloxone was effective in reducing schizophrenic hallucinations, using between 1.2 and 4 mg of naloxone, at time points between two and seven hours after infusion. In their next study, using much larger doses of naloxone (24.8 mg), although there was a reduction in psychotic symptoms, no reduction in hallucinations were reported in twenty subjects tested. Davis and coworkers[27] in a second double-blind study, using 15 mg of naloxone, found that there was a significant reduction in unusual thought content in their schizophrenic patients and this effect tended to be most pronounced in patients who were maintained on neuroleptics but had been somewhat resistant to their effects. Finally, Lehmann and associates[64] in single- and double-blind paradigms, using 10 mg of naloxone and following the patients for up to three hours, found that three of the six patients reported reduced "tension" and five of five patients demonstrated lessening of their symptoms of thought disturbance and hallucinations.

Thus, when one examines high-dose (between 4 and 25 mg), double-blind, crossover naloxone studies in schizophrenic patients, the results are consistently positive, if somewhat variable. These studies variously report changes in auditory hallucinations, psychotic symptoms, unusual thought content, or tension. From a clinical point of view, all of these symptoms would appear to be essential to the process of psychosis, but their common vulnerability to naloxone is somewhat disconcerting to the clinical investigator, as they are not normally thought of as involving the same underlying variable. Thus, some question should be raised about the nature of the measurement instruments and the consistency of the effects of naloxone. It is of considerable interest, however, that all of the five studies using high-dose naloxone were positive, whereas the majority of the studies using low-dose naloxone were negative. It is of further interest that the effects reported in the high-dose naloxone studies have generally occurred much later than one would expect from the classical pharmacology of that opiate antagonist. This raises an important theoretical question: Is the naloxone effect due to opiate receptor blockade, or is it the result of secondary or tertiary effects resulting from that blockade?

Finally, there has been one study of the effects of naloxone in mania.[60] The study used 20 mg naloxone and reported a reduction of irritability, anger, tension, and hostility in four out of eight manic patients.

The opiate antagonist naltrexone would appear to have several advantages for the study of the effects of antagonists in schizophrenia. It can be given in rather large doses, in an oral form, and has a very long period of action. However, there have been few attempts to study naltrexone in schizophrenia. Milke and Gallant,[76] in an open-label design, gave 250 mg of naltrexone to three schizophrenic patients, but observed no effect. Simpson, Branchly, and Lee[99] studied the use of up to 800 mg naltrexone in a single-blind study of four patients and found no effect.

The authors[125] have evaluated naltrexone in a single-blind paradigm (two subjects) and a double-blind paradigm (two subjects) and found a confusingly mixed picture. Single-blind subjects were improved on 250 mg naltrexone. Of two double-blind subjects, one improved at between 250 and 400 mg, and one worsened, and one subject (one of the single-blind subjects) worsened on 800 mg. The effects of naltrexone would therefore appear to be poorly studied, but also generally negative, although it must be admitted there is the possibility of a therapeutic window in the 250 to 400 mg range.

In an elegant, if indirect, series of studies, Davis and coworkers[29] have been able to show a defect in pain response (flattened evoked response) in their schizophrenic patients. This pattern is similar to that seen in normal patients on morphine. When several schizophrenic subjects were given naltrexone, their evoked responses changed so that they were much more "normal" in appearance. This, again, is suggestive of overactivity in endogenous opioids in some schizophrenics. A broader and more extensive set of studies is needed, however. As Davis and coworkers[29] point out, this paradigm might be useful for selecting individuals with endorphin related psychoses.

The general impression emanating from the study of opiate antagonists in schizophrenia is one of a rather delicate, ephemeral effect with some inconsistency. Certainly, low doses of naloxone can fairly be said to be ineffective in schizophrenia. A large number of excellent and well-qualified laboratories have evaluated its effects and agree on either negative or transient effects. Higher dose naloxone studies are consistent in that there appears to be some effect in each study. However, the nature of that effect tends to change between measurement scales and research groups. Naltrexone does not appear to be particularly effective, although it has not been carefully evaluated.

β-Endorphin Injections

Kline and coworkers[61] originally reported that between 1 and 9 mg of synthetic β-END, when injected intravenously (IV) in an open-label or single-blind fashion, was very effective in altering a wide variety of psychiatric symptoms in schizophrenia and depression. This study was, of course, complicated and difficult to assess in that the patients were well known to the senior investigator, a small amount of β-END was given, and the design was not double-blind. Since the number of subjects carrying the same diagnosis was limited, different drug doses and testing conditions were used. It was not possible to carry out a rigorous statistical analysis of this data. Angst[11] reported that three of six depressed subjects became hypomanic after infusion of

10 mg β-END IV. More recently, the authors[15] have studied the effects of 20 mg of β-END IV in a double-blind crossover design with ten schizophrenic subjects and found a very modest improvement that was not detectable clinically but was seen in the rating scales on days three and five following the infusion. This effect is a very mild one and is smaller than that associated with the first week of the study (an "adaptation" period that was not included in the data analysis). Several technical controls were run to ensure the proper administration of a biologically active compound of the proper specificity. Plasma kinetics by radio immuno assay (RIA) (associated with a rapid rate of infusion and precoating of the syringes and tubing), evaluation of serum prolactin responses, and, in one case, cortical EEG responses were all evaluated and found to be consistent with the injection of an active opiate-like material with the proper molecular weight. Catlin and coworkers[25] have reported, in a very similar design, that β-END is either not effective or mildly agitating to their schizophrenic subjects on the day of infusion. They have also evaluated the compound in depression and found it to be moderately and transiently effective. Finally, Bunney and coworkers[22] have been working with IV β-END in schizophrenic patients and, to date, have seen no reliable effects.

An enkephalin analog FK33–824 synthesized by Sandos has been used by two groups in studies of schizophrenia. Nedopil and Ruther[79] have reported in an open-pilot study that nine schizophrenic patients (receiving 0.5 mg and 1 mg per day for two days) improved significantly. The improvement was reported to last for one to seven days. Another group, Jorgensen and associates,[59] also treated nine chronic psychotic patients in a single-blind fashion and reported striking lessening of their subjects' hallucinations.

The general impression of β-END injections as a mode of treatment for schizophrenia is that of very modest or nonexistent effect. It must be pointed out that a single injection of a neuroleptic would probably be equally ineffective under similar circumstances, especially if the effective dose was not known. Several of the aforementioned studies of β-END or the FK33–824[11,59,61,79] were either single-blind or open-label and therefore susceptible to the criticism of suggestion, subject set, and expectancy by the patient. Other technical problems associated with β-END are rather severe. It is unclear whether β-END is transported into the central nervous system. Recent data suggest that a modest amount is.[88] However, it is not known whether there is substantial breakdown under these conditions or whether the β-END can in fact easily reach the most effective sites. Nor is it clear whether β-END should necessarily be more effective than a synthetic alkaloid such as morphine or levorphanol, unless one invokes notions of multiple opiate receptors,[62,71,74] whose relation to endogenous opioids has yet to be elucidated.

Hemodialysis and Leu-5-β-END

In 1977 Wagemaker and Cade[114] reported that by dialyzing schizophrenic subjects on a weekly basis it was possible to produce a substantial and positive shift in their level of psychopathology. Simultaneously, they and their colleagues[81] reported the existence of a peptide not previously described—that is, a leucine-5-β-END. The existence of leucine in position 5 of β-END had not been reported in any tissue of any species prior to this announcement. They further argued that the effectiveness of dialysis depended on their ability to remove leucine-5-β-END and thereby reduce the amount of this "aberrant" endorphin from the plasma. The hypothesis was that this was an unusual material and that it was producing an unusual effect (that is, psychosis). Thus, several issues were brought together in one very complex package: the question of the clinical efficacy of dialysis, the question of the existence of leucine-5-β-END, and the question of whether removal of that compound or any endorphin by dialysis was an effective means of treating schizophrenia. To address the first issue, there are ongoing studies in several centers using hemodialysis or pressure dialysis. One study by Emrich and associates[39] of three patients reports no benefit from hemodialysis. To date, no other results have been published. There have been many presentations, discussions, and case histories in which most patients are not responsive and some anecdotal reports in which an occasional patient is responsive to dialysis. The type of dialysis membrane, the psychological nature of the setting, and the actual diagnosis of the patient are all major issues. Unfortunately, no conclusion can be reached on the clinical efficacy of hemodialysis as the studies have not been completed. In addition to these preliminary reports, Port and associates[89] have carried out a literature retrospective study of fifty schizophrenics in the Veterans' Administration system who were dialyzed for renal problems. They found that over the time of the dialysis, forty patients were unchanged, eight were improved. They argued that from the multiphasic character of schizophrenia and the heterogeneity of symptoms one would expect this degree of fluctuation among fifty patients. The investigators also contended that the kidney is much more efficient at filtering these molecules from the plasma than is the dialysis machine, and, therefore, they questioned the effectiveness of dialysis as a means of removing from plasma molecules the size of β-END or leucine-endorphin.

On the chemical side, Lewis and coworkers[65] studied the hemofiltrate from dialyzed schizophrenics and could detect no met-5- or leu-5-β-END in the hemofiltrate. It should be pointed out that the level of sensitivity for detecting the levels of met-endorphin (β-END) in normal individuals was quite low in this study. However, the levels reported by Palmour and coworkers[81] are 1,000- to 10,000-fold higher than in normal individuals and should have been detected using the techniques of Lewis and associates.[65]

Ross, Berger, and Goldstein[93] addressed the problem at a somewhat different level. They argued that if schizophrenics had extremely high levels of β-END or leucine endorphin, then this amount of peptide should be obvious in the plasma of these subjects (in contrast to normal individuals) using radioimmunoassays for both met- and leu-endorphin. In ninety-eight patients and forty-two normals, they found an amazingly similar set of levels for endorphin-like immunoreactivity (schizophrenics averaged 2.8 fmoles per ml; normals, 2.4 fmoles per ml). Further, they were unable to detect leu-5-β-END in the dialysate from several schizophrenic patients.

The argument for the existence of a leucine-5-endorphin or elevated levels of endorphins in schizophrenia currently appears to be difficult to justify. Although there have not been a large number of studies, the results of the studies that have been carried out are clear. It is tempting to speculate that because there do not appear to be aberrations of endorphins in schizophrenia, dialysis should not be effective. However, the most productive course, in the long run, would appear to be to wait until the clinical studies are completed and then to address the issue of the mechanism of action of any positive result.

Des-Tyr[1]-Gamma-Endorphin (β-LPH 62-77)

For several years, in a series of pioneering investigations, De Wied and coworkers[30] have been studying the behavioral effects of a large number of neuropeptides. More recently, they have studied the effects of smaller structural variants of β-END, known as α-END and γ-END, and have found that γ-END had effects similar to those of haloperidol, whereas α-END acted more like an amphetamine.[32] This rather extensive set of behavioral and structure activity studies has been most impressive in that they have been confirmed generally by other groups. It has also been possible to alter the molecules so that their basic haloperidol or amphetamine-like effects are still present, but are no longer active in opiate systems (that is, the initial tyrosine from γ-END is removed, resulting in des-tyrosine-γ-endorphin [DTγE]). Following the animal behavioral experiments with γ-END and DTγE, De Wied and coworkers proceeded to test DTγE for its haloperidol-like effects in psychiatric patients. Thus, Verhoeven and coworkers[110] ran a single-blind crossover study, using DT γE in daily 1 mg intramuscular (IM) injections. They reported rapid temporary improvement in schizophrenic symptoms in six out of six patients, with a persistence of that improvement in three of the six patients (all six patients were drug free). They then carried out a double-blind crossover study, using the same compound in the same format.[111] Again, they report substantial improvement in a broad range of schizophrenic symptoms in eight subjects. Several clinical studies by other groups are underway.

There have been studies attempting to investigate whether DTγE acts at the opiate receptor or at the dopamine receptor. Binding in tissue homogenate has not shown any action of this compound at either receptor. However, Pedigo and coworkers[82] have shown that when the compound is administered *in vivo* to rats it decreases the amount of spiroperidol that can bind to the dopamine receptor. Thus, it is conceivable that it has an effect in the live animal that is not seen in *in vitro* assays.

The mode of action of this agent is not clear. Perhaps it acts through the subtle alteration of the dopamine binding previously described. Or perhaps it is an effect on the intermediate metabolism of β-END, acting via negative feedback to alter the amount of neuroactive substance the cell produces. In support of this hypothesis, Burbach and coworkers[24] have reported the existence of DT γE, γ-END, α-END, and DTαE in human lumbar CSF and the formation of α-END, γ-END, DTαE, and DTγE from β-END in brain synaptosomal preparations. Finally, there may be a primary action of the compound on its own receptor in the brain. As yet, none of these possible modes of action has been verified. Certainly, if the agent is demonstrated to be effective in other clinical studies (several studies[14,41,106,126] are currently in the planning stages) then a new approach to the pharmacology of schizophrenia will have been opened. To date, it is reported to be capable of producing antipsychotic effects with no major problems associated with its administration. On a more theoretical level, there are exciting issues associated with the agent in that it might allow a conceptualization of the neuronal chemistry of the psychoses which goes beyond the dopamine theory and the use of antidopaminergic agents.

¶ General Summary

It appears that the CSF studies are promising if somewhat vague and confusing in that there are some endorphin fractions (uncharacterized as yet) that would appear to be very sensitive to the psychotic state of certain individuals. The opiate antagonist naloxone has proven effective in schizophrenia and mania at high doses but not at low doses. Unfortunately, these are rather fleeting effects, difficult to measure, and varying from investigator to investigator. The infusion of β-END itself has been extremely difficult and costly to evaluate. It would appear that a single infusion of β-END is minimally effective as an antipsychotic when administered by some investigators and mildly agitating to the patient when administered by another. In some manic-depressive patients, β-END seems to produce a switch from depression to hypomania. It is reported by some investigators to be an effective antidepressant (if somewhat transient and expensive). The enkephalin analog FK33–824, according to two open- or single-blind reports, seems to be of promise as an antischizophrenic agent. Hemodialysis in schizophrenia is even more complicated, some groups finding it to be effective, others not. The completion of these studies will require some time. However, it seems at this writing that the existence of a leucine-5-β-END is unlikely. Finally, destyrosine-γ-endorphin has been studied fairly extensively in animals, but only in a limited fashion in humans. It does, however, appear to hold great clinical and theoretical promise, should its efficacy be borne out by future studies.

In considering the possibility of an integrated endorphin theory of schizophrenia, which would help make understandable the modes of action of these compounds and their relationship to psychopathology, one is struck with the apparent contradictions of the clinical data and the tremendous diversity of basic information. While it may be possible to construct a hypothesis that integrates the apparent contradictions, the wisest course seems to be to wait until clearer basic science patterns are available and until the various clinical areas are clarified. At that point it may be possible to attain a comprehensive overview that takes into account the basic biology and the psychopathology.

As for the study of affective diseases and endorphins, there are even fewer data. Depression has not been studied with respect to CSF levels of the endorphins, nor have the antagonists been studied for their effects on depression. β-END itself is reported to be moderately effective against depression, but it is not clear that β-END effects in depression surpass those of morphine. There have been no studies with the enkephalin analog, dialysis, or DTγE in depression. In mania, there is preliminary evidence for elevated CSF endorphins (type uncharacterized), and in a single study some effects of opiate antagonists on manic symptoms have been shown. However, there are no other reported studies as of this writing. Therefore, the affective psychoses, which in some ways would appear to be an extremely logical area for study of the effects of opiate agents, have been very poorly examined. One might anticipate a rich reward from careful study of endogenous depression, the endocrinology associated with the effects of opiates, and the mood-altering properties associated with the various types of opiate compounds.

¶ Perspectives and Problems

Some of the issues associated with the study of endorphins and psychosis may be thought of as being related to a lack of precise tools. Certainly, the chronic and persistent complaint that clinical diagnoses are variable and at times unreliable should be borne in mind. But one cannot blame the entire problem in clinical studies on the unreliability of psychiatric diagnoses. The lack of proper design in the use of open-label or single-blind studies has caused problems. Furthermore, the specificity of the pharmacologic agents used, their routes of administration, and the nature of the brain structures affected are open to question. For example, naloxone is not a specific antagonist for enkephalin. It would appear to be active against β-END and dynorphin, as well. We have no specific antagonists aimed at the hypothesized multiple types of opiate receptors. There is a great need for increasing the specificity of opiate agonists and antagonists. These agents should be targeted against specific receptor types in order to alter the physiology of particular systems. For example, it would be most useful to have a specific β-END-like agonist known to cross the blood-brain barrier, so that one could mimic the effects of β-END. However, at the same time one would need a similar compound to mimic the effects of α-MSH, for, as mentioned previously, both β-END and α-MSH are contained within the same neurons and would appear to be coreleased when the neurons fire. Therefore, in order to mimic the normal physiology of that system, it would be necessary to produce actions at the appropriate receptors by using both endorphin and α-MSH analogues. Thus, the pharmacological requirements for improved psychiatric studies are indeed considerable.

In writing this chapter, two different thoughts have come to mind. One is that this is a difficult chapter to write at this juncture. The careful reader has undoubtedly realized by this point that, in many ways, this is a highly premature piece of work. To date, there is no great consistency in the role of endorphins in psychosis. There are trails; there are hints; there is confusion. There may even be some pleasant surprises in the literature, but there certainly emerges no compelling hypothesis. Even tentative conclusions are likely to prove naive and incomplete. The other thought that comes to mind is that this is, in several ways, a rather narrow chapter. It does not take into account an enormous set of recently described substances in the central nervous system. Many of them are peptides and appear to be well located for effects on the limbic system. They represent likely candidates for actions of interest to the psychiatrist. For example, arginine vasopressin has been implicated in memory, opiate addiction, stress, and affect. Cholecystokinin, an active peptide in the brain, has recently been associated with the dopaminergic system, thereby implicating it in many actions classically associated with dopamine. These are but a few of a much larger set of substances, which includes neurotensin, somatostatin, bradykinin, and Substance P. It is clear that the neuropeptides and other neuromodulators, such as the endogenous ligand for diazepam (Valium), are going to mean a very major shift in the role of brain biology for the psychiatrist's understanding of the nature of psychological processes in general and the psychoses in particular.

¶ Bibliography

1. AKIL, H., and HOLZ, R. "Opiate Material is Co-Released with Catecholamines from Bovine Adrenal Chromaffin Granules," in press.
2. AKIL, H., MAYER, D. J., and LIEBESKIND, J. C. "Comparison chez le rat entre l'analgésie induite par stimulation de la substance grise periaqueducale et l'analgésie morphinique," *Comptes Rendus Academie des Sciences,* ser. D. 274 (1972): 3603.
3. ———. "Antagonism of Stimulation Produced Analgesia by Naloxone, a Narcotic Antagonist," *Science,* 191 (1976): 961–962.
4. AKIL, H., et al., "Stress-Induced Increase in Endogenous Opiate Peptides: Concurrent Analgesia and Its Partial Reversal by Naloxone," in Kosterlitz, H. W., ed., *Opiates and Endogenous Opioid Peptides.* Amsterdam: Elsevier/North Holland Press, 1976, pp. 63–70.
5. AKIL, H., et al. "Appearance of β-endorphin-like Immunoreactivity in Human Ventricular Cerebrospinal Fluid Upon Analgesic Electrical Stimulation," *Proceedings of the National Academy of Science (U.S.A.),* 75 (1978): 5170–72.
6. AKIL, H., et al. "Enkephalin-like Material Elevated in Ventricular Cerebrospinal Fluid of Pain Patients After Analgetic Focal Stimulation," *Science,* 201 (1978): 463–465.
7. AKIL, H., et al. "Enkephalin-like Material in Normal Human Cerebrospinal Fluid: Measurement and Levels," *Life Sciences,* 23 (1978): 121–126.
8. AKIL, H., et al. "Binding of ^{3}H-β-endorphin to Rat Brain Membranes: Characterization of Opiate Properties and Interaction with ACTH," *European Journal Pharmacology,* 64 (1980): 1–8.
9. AKIL, H., et al. "^{3}H-α-MSH and ^{3}H-ACTH Binding in Rat Brain," in press.
10. AKIL, H., et al. "β-Endorphin Immunoreactivity in C.S.F. of Schizophrenic Patients," in press.
11. ANGST, J., et al. "Preliminary Results of Treatment with β-Endorphin in Depression," in Usdin, E., and Bunney, W. E., Jr., eds., *Endorphins in Mental Illness.* London: Macmillan Press, 1978, pp. 518–528.
12. BELUZZI, J. D., and STEIN, L. "Enkephalin May Mediate Euphoria and Drive-Reduction Reward," *Nature,* 266 (1977): 556–557.
13. BELUZZI, J. D., et al. "Analgesia Induced *in vivo* by Central Administration of Enkephalin in Rat," *Nature,* 260 (1976): 625–626.
14. BERGER, P. A., et al. Personal communication, 1980.
15. BERGER, P. A., et al. "β-Endorphin and Schizophrenia," *Archives of General Psychiatry,* in press.
16. BLOCH, B., et al. "Immunocytochemical Evidence that the Same Neurons in the Human Infundibular Nucleus are Stained with Anti-Endorphins and Antisera of Other Related Peptides," *Neuroscience Letters,* 10 (1978): 147–152.
17. BLOOM, F. E., et al. "Endorphins: Profound Behavioral Effects in Rats Suggest New Etiological Factors in Mental Illness," *Science,* 194 (1976): 630–632.
18. BLOOM, F. E., et al. "Endorphins are Located in the Intermediate and Anterior

Lobes of the Pituitary Gland, Not in the Neurohypophysis," *Life Sciences,* 20 (1977): 43–48.

19. BLOOM, F. E., et al. "β-Endorphin: Cellular Localization, Electrophysiological and Behavioral effects," in Costa, E., and Trabucchi, M., eds., *The Endorphins: Advances in Biochemical Psychopharmacology,* vol. 18, New York: Raven Press, 1978, pp. 89–109.
20. BLOOM, F. E., et al. "Neurons Containing β-Endorphin in Rat Brain Exist Separately from Those Containing Enkephalin: Immunocytochemical Studies," *Proceedings National Academy of Science (U.S.A.),* 75 (1978): 1591–1595.
21. BRADBURY, A. F., et al. "Liptotropin C-Fragment: An Endogenous Peptide with Potent Analgesic Activity," in Kosterlitz, H. W., ed., *Opiates and Endogenous Opioid Peptides.* Amsterdam: Elsevier/North Holland Press, 1976, pp. 9–17.
22. BUNNEY, W. E., et al. Personal communication, 1980.
23. BURBACH, J. P. H., et al. "Schizophrenia and Degradation of Endorphins in Cerebrospinal Fluid," *Lancet,* 8140 (1979): 480–481.
24. BURBACH, J. P. H., et al. "Selective Conversion of β-Endorphin into Peptides Related to α- and γ-Endorphin," *Nature,* 238 (1980): 96–97.
25. CATLIN, D., et al. "Clinical Effects of β-Endorphin Infusions," in Costa, E., and Trabucchi, E. M., eds., *Advances in Biochemical Pharmacology: Regulation and Function of Neuropeptides.* New York: Raven Press, forthcoming.
26. COX, B. M., GOLDSTEIN, A., and LI, C. H. "Opioid Activity of Peptide (β-LPH 61–91), Derived from β-Lipotropin," *Proceedings of the National Academy of Science (U.S.A.),* 73 (1976): 1821–1823.
27. DAVIS, G. C., et al. "Intravenous Naloxone Administration in Schizophrenia and Affective Illness," *Science,* 197 (1977): 74–77.
28. DAVIS, G. C., et al. "Human Studies of Opioid Antagonist and Endorphins." Presented at the annual meeting of the American College of Neuropharmacology, Maui, Hawaii, December, 1978.
29. DAVIS, G. C., et al. "Analgesia to Pain Stimuli in Schizophrenics and Its Reversal by Naltrexone," *Psychiatric Research,* 1 (1979): 61–69.
30. DE WIED, D. "Peptides and Behavior," *Life Sciences,* 20 (1977): 195–204.
31. DOMSCHKE, W., DICKSCHAS, A., and MITZNEGG, P. "C.S.F. β-Endorphin in Schizophrenia," *Lancet,* 8124 (1979): 1024.
32. DORSA, D. M., VAN REE, J. M., and DE WIED, D. "Effects of [Des-Tyr[1]]-γ-Endorphin and γ-Endorphin on Substantia Nigra Self-Stimulation," *Pharmacology, Biochemistry and Behavior,* 10 (1979): 899–905.
33. DUPONT, A., et al. "Rapid Inactivation of Enkephalin-Like Material by C.S.F. in Chronic Schizophrenia," *Lancet,* 8099 (1978): 1107.
34. DYSKEN, M. W., and DAVIS, J. M. "Naxolone in Amylobarbitone-Responsive Catatonia," *British Journal of Psychiatry,* 133 (1978): 476.
35. EIPPER, B., and MAINS, R., "Existence of a Common Precursor to ACTH and Endorphin in the Anterior and Intermediate Lobes of the Rat Pituitary," *Journal of Supramolecular Structure,* 8 (1978): 247–262.
36. ELDE, R., et al. "Immunohistochemical Studies Using Antibodies to Leucine enkephalin: Initial Observations on the Nervous System of the Rat," *Neuroscience,* 1 (1976): 349–351.
37. EMRICH, H. M., et al. "Indication of an Antipsychotic Action of the Opiate Antagonist Naloxone," *Pharmakopsychiatry,* 10 (1977): 265–270.
38. EMRICH, H. M., et al. "β-Endorphin-Like Immunoreactivity in Cerebrospinal Fluid and Plasma of Patients with Schizophrenia and Other Neuropsychiatric Disorders," *Pharmakopsychiatry,* 12 (1979): 269–276.
39. EMRICH, H. M., et al. "Hemodialysis in Schizophrenia: Three Failures with Chronic Patients," *American Journal of Psychiatry,* 136 (1979): 1095.
40. EMRICH, H. M., et al. "On a Possible Role of Endorphins in Psychiatric Disorders: Actions of Naloxone in Psychiatric Patients," in Obiols, J., et al., eds., *Biological Psychiatry Today.* Amsterdam: Elsevier/North Holland Press, 1979, pp. 798–805.
41. EMRICH, H. M., et al. Personal communication, 1980.
42. FERRARA, P., HOUGHTEN, R., and LI, C.H. "β-Endorphin: Characteristics of Binding

Sites in Rat Brain," *Biochemical and Biophysical Research Communications,* 89 (1979): 786–792.

43. GIANOULAKIS, C., et al. *"In Vitro* Biosynthesis and Chemical Characterization of ACTH and ACTH Fragments by the Rat Pars Intermedia," in Way, E. L., ed., *Endogenous and Exogenous Opiate Agonists and Antagonists.* New York: Pergamon Press, 1980, pp. 289–292.
44. GOLDSTEIN, A. "Opiate Receptors and Opioid Peptides: A Ten-Year Overview," in Upton, M. A., Dimascio, A., and Killam, K. F., eds., *Psychopharmacology: A Generation of Progress.* New York: Raven Press, 1978, pp. 1557–1563.
45. ———, LOWNEY, L. I., and PAL, B. K. "Stereospecific and Non-specific Interactions of the Morphine Congener Levorphanol in Subcellular Fractions of Mouse Brain," *Proceedings National Academy of Science (U.S.A.),* 68 (1971): 1742–1747.
46. GOLDSTEIN, A., et al. "Dynorphin-(1–13), an Extraordinarily Potent Opioid Peptide," *Proceedings National Academy of Science (U.S.A.),* 76 (1979): 6666–6670.
47. GUILLEMIN, R., LING, N., and BURGUS, R. "Endorphins, peptides d'origine hypothalamique and neurohypophysaire d'activité morphinomimétique. Isolement et structure moléculaire d'alpha-endorphine," *Comptes Rendus des Séances de l'Academie des Sciences,* Ser. D. 282 (1976): 783–785.
48. GUILLEMIN, R., et al. "β-Endorphin and Adrenocorticotropin are Secreted Concomitantly by the Pituitary," *Science,* 197 (1977): 1367–1369.
49. GUNNE, L. M., LINDSTROM L., and TERENIUS, L. "Naloxone-Induced Reversal of Schizophrenic Hallucinations," *Journal of Neural Transmission,* 40 (1977): 13–19.
50. HOKFELT, T. "Coexistence of Peptides and Other Transmitters in the Same Neuron," in Costa, E. and Trabucchi, E. M., eds., *Advances in Biochemical Pharmacology: Regulation and Function of Neuropeptides.* New York: Raven Press, forthcoming.
51. ———, et al. "The Distribution of Enkephalin-Immunoreactive Cell Bodies in the Rat Central Nervous System," *Neuroscience Letters,* 5 (1977): 25–31.
52. HÖLLT, V., MULLER, O. A., and FAHLBUSCH, R. "β-Endorphin in Human Plasma: Basal and Pathologically Elevated Levels," *Life Sciences,* 25 (1979): 37–44.
53. HOLZ, R. "Osmotic Lysis of Bovine Chromaffin Granules in Isotomic Solutions of Salts of Organic Acids: Release of Catecholamines, ATP, Dopamine-β-hydroxylase, and Enkephalin-Like Material," *Journal of Biochemistry,* in press.
54. HUGHES, J., et al. "Identification of Two Related Pentapeptides from the Brain with Potent Opiate Agonist Activity," *Nature,* 258 (1975): 577–579.
55. JACOBOWITZ, D. M., and O'DONOHUE, T. L. "α-Melanocyte Stimulating Hormone: Immunohistochemical Identification and Mapping in Neurons of Rat Brain," *Proceedings of the National Academy of Science (U.S.A.),* 75 (1978): 6300–6304.
56. JACOBOWITZ, D. M., SILVER, M. A., and SODEN, W. G. "Mapping of Leu-Enkephalin Containing Axons and Cell Bodies of the Rat Forebrain," in Usdin, E., ed., *Endorphins in Mental Health Research.* New York: Oxford University Press, 1979, pp. 62–74.
57. JANOWSKY, D. S., et al. "Lack of Effect of Naloxone on Schizophrenic Symptoms," *American Journal of Psychiatry,* 134 (1977): 926–927.
58. JEFFCOATE, W. J., et al. "β-Endorphin in Human Cerebrospinal Fluid," *Lancet,* 8081 (1978): 119–121.
59. JORGENSEN, A., FOG, R., and BEILIS, B. "Synthetic Enkephalin Analogue in Treatment of Schizophrenia," *Lancet,* 8122 (1979): 935.
60. JUDD, L. L., et al. "Naloxone Related Attenuation of Manic Symptoms in Certain Bipolar Depressence," in VAN REE, J. and TERENIUS, L., eds., *Characteristics and Functions of Opioids.* Amsterdam: Elsevier/North Holland Press, 1978, pp. 173–174.
61. KLINE, N. S., et al. "β-Endorphin-Induced Changes in Schizophrenic and Depressed Patients," *Archives of General Psychiatry,* 34 (1977): 1111–1113.
62. KOSTERLITZ, H. W. and HUGHES, J., "Development of the Concepts of Opiate Receptors and Their Ligands," in Costa, E., and Trabucchi, M., eds. *The Endorphins: Advances in Biochemical Psycho-*

pharmacology, vol. 18, New York: Raven Press, 1978, pp. 31–44.
63. KURLAND, A. A., et al. "The Treatment of Perceptual Disturbances in Schizophrenia with Naloxone Hydrochloride," *American Journal of Psychiatry,* 134 (1977): 1408–1410.
64. LEHMANN, H., VASAVAN NAIR, N. P., and KLINE, N. S. "β-Endorphin and Naloxone in Psychiatric Patients: Clinical and Biological Effects," *American Journal of Psychiatry,* 136 (1979): 762–766.
65. LEWIS, R. V., et al. "On β_h-Leu^5-endorphin and Schizophrenia," *Archives of General Psychiatry,* 36 (1979): 237–239.
66. LI, C. H., and CHUNG, D. "Isolation and Structure of an Untriakontapeptide with Opiate Activity from Camel Pituitary Glands," *Proceedings National Academy of Science (U.S.A.),* 73 (1976): 1145–1148.
67. LI, C. H., et al. "Isolation and structure of β-LPH from sheep pituitary glands," *Excerpta Medica,* 3 (1965–66): 111–112.
68. LINDSTROM, L. H., et al. "Endorphins in Human Cerebrospinal Fluid: Clinical Correlations to Some Psychotic States," *Acta Psychiatrica Scandinavica,* 57 (1978): 153–164.
69. LIPINSKI, J., et al. "Naloxone in Schizophrenia: Negative Result," *Lancet,* 8129 (1979): 1292–1293.
70. LOH, H. H., et al. "β-Endorphin as a Potent Analgesic Agent," *Proceedings of the National Academy of Science (U.S.A.),* 73 (1976): 2895–2896.
71. LORD, J. A. H., et al. "Multiple Opiate Receptors," in Kosterlitz, H. W., ed., *Opiates and Endogenous Opioid Peptides.* Amsterdam: Elsevier/North Holland Press, 1976, pp. 275–280.
72. MAINS, R. E., EIPPER, B. A., and LING, N. "Common Precursor to Corticotropins and Endorphins," *Proceedings of the National Academy of Science (U.S.A.),* 74 (1977): 3014–3018.
73. MALFROY, B., et al. "High-Affinity Enkephalin-Degrading Peptidase in Brain is Increased after Morphine," *Nature,* 276 (1978): 523–526.
74. MARTIN, W. R., et al. "The Effects of Morphine and Nalorphine-Like Drugs in the Nondependent and Morphine-Dependent Chronic Spinal Dog," *Journal of Pharmacological and Experimental Therapeutics,* 197 (1976): 518–532.
75. MAYER, D. J., et al. "Analgesia from Electrical Stimulation in the Brainstem of the Rat," *Science,* 174 (1971): 1351–1354.
76. MILKE, D. H., and GALLANT, D. M. "An Oral Antagonist in Chronic Schizophrenia: A Pilot Study," *American Journal of Psychiatry,* 134 (1977): 1430–1431.
77. MOON, H. D., LI, C. H., and JENNINGS, B. M. "Immunohistochemical and Histochemical Studies of Pituitary β-Lipotropin," *The Anatomical Record,* 175 (1973): 524–538.
78. NAKANISHI, S., et al. "Nucleotide Sequence of Cloned cDNA for Bovine Corticotropin-β-Lipotropin Precursor," *Nature,* 278 (1979): 423–427.
79. NEDOPIL, N., and RUTHER, E. "Effects of Synthetic Analogue of Methionine Enkephalin FK 33–824 on Psychotic Symptoms," *Pharmakopsychiatry,* 12 (1979): 277–280.
80. NILAVER, G., et al. "Adrenocorticotropin and β-Lipotropin in Hypothalamus," *Journal of Cell Biology,* 81 (1979): 50–58.
81. PALMOUR, R. M., et al. "Characterization of a Peptide Derived from the Serum of Psychiatric Patients," (abstract), Society for Neuroscience, Anaheim, Calif., November 1977, p. 320.
82. PEDIGO, N. W., et al. "Investigation of Des-Tyrosine-γ-Endorphin Activity at Neuroleptic Binding Sites in Rat Brain and of Opiate and Neuroleptic Binding in Human Schizophrenic Brains," Presented at the Annual Meeting of the Society for Neuroscience, Atlanta, Ga., 1979.
83. PELLETIER, G. "Ultrastructural Localization of Neuropeptides with the Post-Embedment Staining Method," Presented at the Annual Meeting of the Histochemical Society, Keystone, Colorado, April, 1979.
84. ———, and DUBE, D. "Electron Microscopic Immunohistochemical Localization of α-MSH in the Rat Brain," *American Journal of Anatomy,* 150 (1977): 201–206.
85. PELLETIER, G., et al. "Immunohistochemical Localization of β-Lipotropin Hormone in the Pituitary Gland," *Endocrinology,* 100 (1977): 770–776.
86. PELLETIER, G., et al. "Immunohistochemical Localization of β-LPH in the Human Hypothalamus," *Life Sciences,* 22 (1978): 1799–1804.
87. PERT, C. B., and SNYDER, S. H. "Opiate Re-

ceptor: Demonstration in Nervous Tissue," *Science,* 179 (1973): 1011–1014.

88. PEZALLA, P. D., et al. "Lipotropin, Melanotropin, and Endorphin: *In vivo* Catabolism and Entry into Cerebrospinal Fluid," *Canadian Journal of Neurological Sciences,* 5 (1978): 183–188.
89. PORT, F. K., KROLL, P. D., and SWARTZ, R. D. "The Effect of Hemodialysis on Schizophrenia: A Survey of Patients with Renal Failure," *American Journal of Psychiatry,* 135 (1978): 743–744.
90. RICHARDSON, D. E., and AKIL, H. "Pain Reduction by Electrical Brain Stimulation in Man. Part 2: Chronic Self-Administration in the Periventricular Gray Matter," *Journal of Neurosurgery,* 47 (1977): 184–194.
91. RIMON, R., TERENIUS, L., and KAMPMAN, R. "Cerebrospinal Fluid Endorphins in Schizophrenia," *Acta Psychiatrica Scandinavica,* in press.
92. ROBERTS, J. L., and HERBERT, E. "Characterization of a Common Precursor to Corticotropin and β-Lipotropin: Identification of β-Lipotropin Peptides and Their Arrangement Relative to Corticotropin in the Precursor Synthesized in a Cell-Free System," *Proceedings National Academy of Science (U.S.A.),* 74 (1977): 5300–5304.
93. ROSS, M., BERGER, P. A., and GOLDSTEIN, A. "Plasma β-Endorphin Immunoreactivity in Schizophrenia," *Science,* 205 (1979): 1163–1164.
94. ROSSIER, J., et al. "Regional Dissociation of β-Endorphin and Enkephalin Contents in Rat Brain and Pituitary," *Proceedings National Academy of Science (U.S.A.),* 74 (1977): 5162–5165.
95. SCHULTZBERG, M., et al. "Enkephalin-Like Immunoreactivity in Gland Cells and Nerve Terminals of the Adrenal Medulla," *Neuroscience,* 3 (1978): 1169–1186.
96. SCHULTZBERG, M., et al. "VIP-, Enkephalin-, Substance P- and Somatostatin-Like Immunoreactivity in Neurons Intrinsic to the Intestine: Immunohistochemical Evidence from Organotypic Tissue Cultures," *Brain Research,* 155 (1978): 239–248.
97. SIMANTOV, R., et al. "Opioid Peptide Enkephalin: Immunohistochemical Mapping in Rat Central Nervous System," *Proceedings National Academy of Science (U.S.A.),* 74 (1977): 2167–2171.
98. SIMON, E. J., HILLER, J. M., and EDELMAN, I. "Stereospecific Binding of the Potent Narcotic Analgesic (^{3}H) Etorphine to Rat-Brain Homogenate," *Proceedings National Academy of Science (U.S.A.),* 70 (1973): 1947–1949.
99. SIMPSON, G. M., BRANCHLY, M. H., and LEE, J. H. "A Trial of Naltrexone in Chronic Schizophrenia," *Current Therapeutic Research,* 22 (1977): 909–913.
100. SNYDER, S. H. "The Opiate Receptor and Morphine-Like Peptides in the Brain," *American Journal of Psychiatry,* 135 (1978): 645–652.
101. SOFRONIEW, M. V. "Immunoreactive β-Endorphin and ACTH in the Same Neurons of the Hypothalamic Arcuate Nucleus in the Rat," *American Journal of Anatomy,* 154 (1979): 283–289.
102. STERN, A. S., et al. "Isolation of the Opioid Heptapeptide Met-Enkephalin [Arg^6, Phe^7] from Bovine Adrenal Medullary Granules and Striatum," *Proceedings National Academy of Science (U.S.A.),* 76 (1979): 6680–6683.
103. SULLIVAN, S., AKIL, H., and BARCHAS, J. D., "*In vitro* Degradation of Enkephalin: Evidence for Cleavage at the Gly-Phe Bond," *Communications in Psychopharmacology,* 2 (1978): 525–531.
104. TERENIUS, L. "Characteristics of the 'Receptor' for Narcotic Analgesics in Synaptic Plasma Membrane Fraction from the Rat Brain," *Acta Pharmacologica Toxicologica,* 33 (1973): 377–384.
105. ———, et al. "Increased Levels of Endorphins in Chronic Psychosis," *Neuroscience Letters,* 3 (1976): 157–162.
106. TERENIUS, L., et al. Personal communication, 1980.
107. UHL, G. R., KUHAR, M. J., and SNYDER, S. H. "Enkephalin Containing Pathway Amygdaloid Efferents in the Stria Terminalis," *Brain Research,* 149 (1978): 223–228.
108. URCA, G., et al. "Morphine and Enkephalin: Analgesic and Epileptic Properties," *Science,* 197 (1977): 83–86.
109. VANLEEUWEN, F. W., et al. "Immunoelectron Microscopical Demonstration of α-Melanocyte-Stimulating Hormone-Like Compound in the Rat Brain," *Journal of Endocrinology,* 80 (1979): 59P–60P.
110. VERHOEVEN, W. M. A., et al. "[Des-Tyr^1]-γ-Endorphin in Schizophrenia," *Lancet,* 8072 (1978): 1046–1047.
111. VERHOEVEN, W. M. A., et al. "Improvement of Schizophrenic Patients Treated

with [Des-Tyr[1]]-γ-Endorphin (DTγE)," *Archives of General Psychiatry,* 36 (1979): 294–298.

112. VIVEROS, O. H., et al. "Opiate-Like Materials in the Adrenal Medulla: Evidence for Storage and Secretion with Catecholamines," *Molecular Pharmacology,* 16 (1979): 1101–1108.

113. VOLAVKA, J., et al. "Naloxone in Chronic Schizophrenia," *Science,* 196 (1977): 1227–1228.

114. WAGEMAKER, H., and CADE, R. "The Use of Hemodialysis in Chronic Schizophrenia," *American Journal of Psychiatry,* 134 (1977): 684–685.

115. WAHLSTROM, A., JOHANSSON, L., and TERENIUS, L. "Characterization of Endorphins in Human CSF and Brain Extracts," in Kosterlitz, H. W., ed., *Opiates and Endogenous Opioid Peptides.* Amsterdam: Elsevier/North Holland Press, 1976, pp. 49–56.

116. WALKER, J. M., AKIL, H., and WATSON, S. J. "Evidence for Homologous Action of Proopiocortin Products," in preparation.

117. WATSON, S. J., and AKIL, H. "The Presence of Two α-MSH Positive Cell Groups in Rat Hypothalamus," *European Journal of Pharmacology,* 58 (1979): 101–103.

118. ———. "α-MSH in Rat Brain: Occurrence Within and Outside Brain β-Endorphin Neurons," *Brain Research,* 182 (1980): 217–223.

119. ———, and BARCHAS, J. D. "Immunohistochemical and Biochemical Studies of the Enkephalins, β-endorphin, and Related Peptides," in Usdin, E., Bunney, W. E., and Kline, N. S., eds., *Endorphins in Mental Health Research.* New York: Macmillan, 1979, pp. 30–44.

120. WATSON, S. J., BARCHAS, J. D., and LI, C. H. "β-Lipotropin: Localization of Cells and Axons in Rat Brain by Immunocytochemistry," *Proceedings National Academy of Science USA,* 74 (1977): 5155–5158.

121. WATSON, S. J., RICHARD, C. W., and BARCHAS, J. D. "Adrenocorticotropin in Rat Brain: Immunocytochemical Localization in Cells and Axons," *Science,* 200 (1978): 1180–1182.

122. WATSON, S. J., et al. "Immunocytochemical Localization of Methionine-Enkephalin: Preliminary Observations," *Life Sciences,* 25 (1977): 733–738.

123. WATSON, S. J., et al. "Effects of Naloxone in Schizophrenia: Reduction in Hallucinations in a Subpopulation of Subjects," *Science,* 201 (1978): 73–76.

124. WATSON, S. J., et al. "Evidence for Two Separate Opiate Peptide Neuronal Systems and the Coexistence of β-Lipotropin, β-Endorphin, and ACTH Immunoreactivities in the same Hypothalamic Neurons," *Nature,* 275 (1978): 226–228.

125. WATSON, S. J., et al. "Some Observations on the Opiate Peptides and Schizophrenia," *Archives of General Psychiatry,* 36 (1979): 35–41.

126. WATSON, S. J., et al. Personal communication, 1980.

127. WATSON, S. J., et al. "Dynorphin Immunocytochemical Localization in Brain and Peripheral Nervous System: Preliminary Studies," *Proceedings of the National Academy of Science (U.S.A.),* in press.

128. WEBER, E., VOIGT, R., and MARTIN, R. "Concomitant Storage of ACTH and Endorphin-Like Immunoreactivity in the Secretory Granule of Anterior Pituitary Corticotrophs," *Brain Research,* 157 (1978): 385–390.

129. WEI, E., and LOH, H. "Physical Dependence on Opiate-Like Peptides," *Science,* (1976): 1262–1263.

130. ZIMMERMAN, E. A., LIOTTA, A., and KRIEGER, D. T. "β-Lipotropin in Brain: Localization in Hypothalamic Neurons by Immunoperoxidase Technique," *Cell Tissue Research,* 186 (1978): 393–398.

CHAPTER 2

SPLIT-BRAIN STUDIES: IMPLICATIONS FOR PSYCHIATRY*

Michael S. Gazzaniga and Bruce T. Volpe

¶ Human Brain Bisection—Present Perspective

BRAIN SCIENCE has for the most part been unable to explain the mechanism through which human beings generate a sense of subjective reality. In the past, most of the energy devoted to the problem was spent on considering whether this question could be reasonably studied. Recently, concerns of a more strategic nature have appeared. The neurobiologist approaches the study of mental processes in a reductionist fashion. As a consequence, current discussion of a mental process such as memory is frequently cast entirely in biochemical terms.[12] Although these studies have begun to elucidate the synaptic and cellular events, it is less clear how they promote an understanding of memory, let alone human memory.

The recurring strategic problem that continually plagues biological approaches to psychological processes is the blurred distinction between levels of analysis. This difficulty becomes apparent when we compare the brain to a computer. There is no way the power of a computer algorithm can be deduced by an analysis of the chemical nature of the individual transistors that subserve those functions. The algorithmic functions are a property of the system resulting from the interaction of elements, and they can only be understood at that level.

In this chapter, the objective is to promote an understanding of conscious processes at the level of human behavior. The approach

*Aided by U.S. Public Health Service Grant no. NS 15053, the McKnight Foundation, and the Alfred P. Sloan Foundation.

is to examine patients who have undergone brain surgery or who have suffered focal brain damage. The experimental data are derived primarily from patients with progressive intractable epilepsy, who have had a surgical procedure in which the largest interhemispheric commissure (the corpus callosum) is sectioned.

First, the early history of human split-brain research will be summarized from the surgical perspective. The neurosurgical procedure is performed only as a final effort to control epilepsy after all drug programs have failed; patients in this group are necessarily few in number. These patients have been followed closely from the therapeutic perspective, but the intensive studies have focused on the cognitive aspects of their course.

Second, it will be demonstrated that the major psychological result of the early studies established that each cerebral hemisphere of the split-brain patient was capable of sustaining autonomous, independent cognitive systems that were outside the realm of awareness of the opposite hemisphere. With language mechanisms generally localized in the left cerebral hemisphere, behavior generated from the right cerebral hemisphere could not, for the most part, be verbalized. Since each hemisphere was ignorant of specific information in the opposite hemisphere, later studies probed the mechanisms by which the integration of these disparate cognitive operations could take place.

Experiments have shown that after a behavior is produced by the nonspeaking (generally right) hemisphere, the subsequent verbal explanation produced by the speaking (generally left) hemisphere delineates an explicit motivation for such activity in spite of the speaking hemisphere having no real prior knowledge of the behavior. Although the model is continuously evolving, it will be argued later in the chapter that an individual is a series of coconscious mental systems each competing for the limited output mechanisms. Of the multiple mental systems present in human beings, usually only one can talk and interpret events linguistically. The view is that the constant flow of emitted behavior is generally interpreted by the verbal system, and provides one with knowledge, opinion, belief about the environment, about oneself, and about one's behavior. By such acts linguistic behavior provides an organizational framework for the individual.

Finally, the split-brain methodology will be modified for clinical neurological studies of patients with focal brain damage. Similar questions about the interaction of verbal and nonverbal mental systems will be discussed. To date, studies suggest that man is not governed by unconscious and generally immutable belief systems, but that his knowledge, opinions, and beliefs about himself and the world arise out of the need to integrate behaviors that are produced from coconscious nonverbal mental systems.

Perspectives in Neurosurgical History

The neurosurgical operations for the control of intractable epilepsy involve the removal of an abnormal cortical area,[17] the removal of a specific lobe (often temporal or frontal),[55,58] or, in certain restricted cases, the removal of a complete cerebral hemisphere.[41] Transection of the corpus callosum for epilepsy control dates to the 1940s when Erickson[15] experimented with the spread of seizure activity in the cerebral hemispheres of monkeys. He suggested that the corpus callosum was the principle pathway for the spread of epileptic discharge from one hemisphere to the other, and that severing this commissural system and other forebrain commissures seemed to prevent that spread. Presently, opposing theoretical viewpoints suggest that the presence of the callosum inhibits the spread of seizure activity, but the issue remains unresolved.[20,42]

At approximately the same time as Erikson, Van Wagenen and Herren[67] independently reached similar conclusions about the importance of the callosum in the spread of the epileptic focus. Their experience was based on clinical observations that epileptic patients who developed tumors of their callosum experienced seizure-free episodes. They took the bold step of performing forebrain commissurotomies on twenty-six patients who suffered intractable epilepsy. Most of the patients underwent partial division of the callosum; only one had the anterior commissure divided. Although the published results of the first ten patients looked optimistic and there was no major change in most patients' behavior, the overall beneficial therapeutic effect was too variable. Nine of the ten patients continued to seize in the first six postoperative months.[67]

An extended follow-up of these patients was undertaken by Akelaitis,[1,2,3] who suggested that there was no apparent decrement in mental functioning. Many of the testing procedures that Akelaitis used were clinical in nature, and they were simply not precise enough to address some of the more subtle issues raised by two disconnected hemispheres in the same cranium. Akelaitis concluded that the great cerebral commissure could be sectioned without apparent clinical consequence. However, more sophisticated techniques later revealed the crisp dissociations of independent cognitive processes.

Since the majority of the fibers in the corpus callosum interconnect homotopic regions in the two hemispheres,* section of the corpus callosum, reserved as it is for the inexorably progressive forms of epilepsy, differs from the cortical removal operations in that the lesion is clearly restricted to these interhemispheric connecting fibers. There is no surround of injury invading adjacent neural areas, such as occurs with cortical ablations. The associated clinically observed deficits are minimal.[9,63,71]

In any case, neurosurgical section of the forebrain commissures was not used again until the 1960s when Bogen and Vogel[9] embarked on a new study. Using similar stringent criteria for selecting patients for operation, two new series were begun. Many of these patients experienced fewer seizures, minimal associated clinical changes, and were managed more successfully on lowered drug dosage.[9,73]

*See references 4, 13, 37, 38, and 39.

The Human Split-Brain Operation in Transition

From the late 1960s through the early 1970s, Donald Wilson and colleagues[72,73] of the Dartmouth Medical School started another series of callosal sectioned patients. Using the accepted criteria for the classification of the epilepsies[19] as well as stringent criteria before accepting the patient for surgery, Wilson sectioned the interhemispheric commissures in several different procedures. In the first series, it was standard practice to open the lateral ventricles and divide the anterior commissure, one fornix, and the corpus callosum.[73] In the second series, Wilson continued to use microsurgical techniques, but he did not enter the ventricles; he divided only the corpus callosum. The most recent group of patients, Wilson's third series, underwent similar microsurgical procedures that were completed in two stages.[74] Several weeks elapsed between the first and the second stage of the commissure section. In this procedure the posterior half of the callosum is sectioned, and several weeks later the remaining callosal fibers are sectioned. There are three patients in this study, each of whom has been tested at each stage of commissurotomy.[61]

¶ The History of Cognition in Each Hemisphere

The initial cognitive studies on Bogen and Vogel's[8,9] first patients were carried out by Gazzaniga and Sperry.* In specific tests, lateralized stimulus information was briefly presented to patients. Most studies involved visual information, although auditory and tactual stimulus presentation modes have also been used. The neural systems that sub-

*See references 21, 24, 25, 26, 27, 62, and 63.

serve these functions are also discretely lateralized. The visual experiments are possible because the retinal-cortical pathways are organized so that tachistoscopic presentation to the left visual field is projected to the right hemisphere, and information presented in the right visual field is projected to the left hemisphere. In general, only stimuli presented in the right visual field or in the right hand can be verbally identified, since these stimuli are discretely projected to the left hemisphere, whereas both hemispheres can respond in a nonverbal fashion.

The early studies demonstrated that information processed by one disconnected hemisphere was not available to the cognitive apparatus of the other hemisphere. Interhemispheric exchange of information was totally disrupted, so that while visual, tactual, and auditory information presented to one hemisphere could be recorded and processed, and a response could be generated by that hemisphere, these activities occurred unknown to the opposite hemisphere until an overt behavior was produced. The data confirmed the experiments of Myers and Sperry[54] in animals, which showed the callosum to be crucial in interhemispheric transfer. The data in humans, however, were more dramatic: Since the left hemisphere controls the language mechanisms in humans, only processes ongoing in the left hemisphere could be verbally described by the patients.

Thus, if a picture of a spoon, for example, was flashed to the right hemisphere, the subject responded by saying "I did not see anything." However, the subject would be able to retrieve the object with the left hand from a series of objects out of vision (see figure 2–1). The right hemisphere could organize the discrete sensorimotor act of the left hand. Further, when this object was held in the left hand out of the patient's view, the response to the experimenter's question, "What are you holding in your left hand?" would persis-

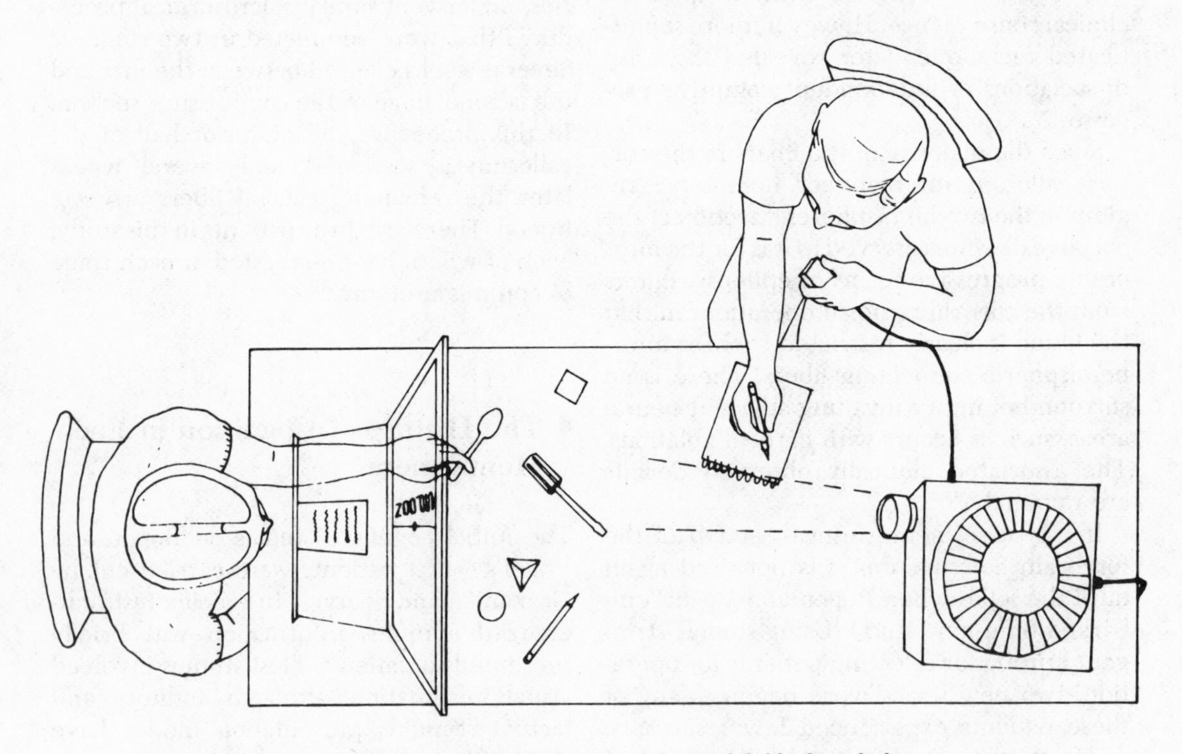

FIGURE 2–1. The word "spoon", lateralized to the left visual field of a split-brain patient, was available only to his right, nonverbal hemisphere. The patient, unable to name it, was able to retrieve the correct object out of view.

tently be "I don't know." In fact, the talking hemisphere behaved as if it did not know what the ipsilateral left hand was holding. It did not see the exposed slide, nor did it have access to the highly refined proprioceptive information from the ipsilateral left side of the body. Clearly, however, the right hemisphere was able to process the projected stimulus and initiate any additional activity necessary to direct the left hand to make a correct choice. Since the right hemisphere in all but a very few split-brain patients is not endowed with formal or sophisticated language mechanisms, the process is distinctly nonverbal.

The studies showed that the left dominant hemisphere was vastly superior to the right in both the production and comprehension of language. At the same time, the right hemisphere possessed superior skills on nonverbal tasks such as drawing and copying designs and in arranging items to construct complex patterns. Although each hemisphere appeared to have some bias for processing certain types of information, the detection of this difference depended critically on the experimental design. LeDoux, Wilson, and Gazzaniga[43] showed, for example, that the right hemisphere advantage on a variety of spatial tasks was dependent on the involvement of manual activities in the perception of spatial relationships. In this study the striking difference in the competence of the right compared to the left hemisphere on tasks of spatial relationships disappeared when use of the hands was prohibited. The extension of the early findings of cerebral lateralization has led to broader claims that argue for the presence of different cognitive styles, each existing exclusively within a cerebral hemisphere.[6,7,47,48] The demonstration of the presence of similar cognitive processes in both hemispheres, however, makes the argument for strict lateralization more apparent than real.

Other investigators[18] have asserted that the isolated right hemisphere is the repository for mental processes that are repressed and "unconscious." They suggest that these right hemisphere processes are congruent with primary-process thinking, and that the right hemisphere is the neural substrate of the unconscious, or the generator of some "preconscious stream." Experiments have demonstrated that the right hemisphere can generate an overt behavior in response to specific and complex stimuli.* While the stimului cannot be verbally described, nor the patients able to evince prior verbal knowledge of this behavior, the ensuing act is appropriate to the stimulus. It is doubtful whether stimulus-appropriate behavior out of verbal awareness can continue to be considered "unconscious."

Some specialists[34] have used the notion of right hemisphere as the neural substrate of the unconscious to support a claim that the split-brain patients do not dream, or at least cannot talk about their dreams. Some years ago a study[33] reported that split-brain patients dream. Over the years these patients continued to report their dreams, revealing a fantasy life that is as full and rich as their peers.

The rubrics of "mind-right"—"mind-left," wholistic mind—analytic mind, intuitive-rational, east-west, and so forth, have all been used to describe the differences in cognitive processing between the two hemispheres. Humans seem to seek dichotomies, yet the appeal of these headings or divisions resulted in an impoverished shorthand that has misrepresented the full story. The taxonomy that has been developed for each hemisphere generally ignores important details and, more specifically, the major issue—the study of integrated behavior.[23]

By demonstrating that information could be accurately processed independently in each hemisphere, the early studies introduced the intriguing question of whether the mechanisms of consciousness were doubly represented following split-brain surgery.† While the conscious properties of the speech-producing hemisphere were apparent, the view that the mute and apparently functionless hemisphere was also "conscious" was

*See references 21, 24, 25, 26, and 27.
†See references 21, 23, 28, 29, 44, and 63.

widely criticized and generally rejected.[14,19] The task, then, was not only to tease out the workings of a speechless hemisphere and recognize its coconscious status, but also to discover the contribution of this hemisphere to the total behavior of the patient.*

¶ Integration of Coconscious Mechanisms in the Split Brain

In recent years this challenge has been answered by designing experiments that focus on the interaction of the two cerebral hemispheres in the split-brain patient. This interaction produces an ever-present sense of unity, even though the experimental evidence clearly shows that the mental phenomena of one hemisphere continue unperceived by the other hemisphere. A series of experimental paradigms address the question of how the dominant left hemisphere deals with the overt and covert behaviors produced by the right hemisphere.

A patient, P, was asked to select from a series of picture cards the one picture that best related to a flashed stimulus. The test picture was flashed tachistoscopically to the right or left visual field and thereby lateralized to the left or right hemisphere. For example, when an "apple" was flashed to a single visual field, the subject was asked to choose from a series of picture cards that might have included a comb, a toaster, and a banana. With the superordinate concept being, in this situation, "fruit," each hemisphere usually made the correct choice. The performance across several superordinate categories was nearly perfect.[29]

To examine how the left, talking hemisphere dealt with behavior produced by the right half of the brain, this experiment was modified slightly: Two pictures were flashed simultaneously, one to each hemisphere. In these critical trials, the patient was again required to point to cards that best related to the flashed stimuli. Only rarely did the response to one of the stimuli, mainly the right visual field-left hemisphere, block a response from the other hemisphere. In general, exposure to both visual fields led to the correct choices. A typical example: A snow scene was exposed to the right hemisphere, and a picture of a chicken claw to the left hemisphere. The best choice for the left hemisphere from the four proffered cards was a chicken, and the best choice for the right hemisphere was a snow shovel. The corresponding choices were made with the hand contralateral to each exposed hemisphere (see figure 2–2).

After the patient had pointed to two out of the eight cards, he was asked to explain the reason for each choice. In the snow scene-chicken claw exposure, the patient explained his choice of a shovel and a chicken by saying, "Oh, that's simple. The chicken claw goes with the chicken, and you need a shovel to clean out the chicken shed."[23]

In test after test, when each hemisphere was given a task to solve requiring an overt and specific response, the left language system behaved as if it viewed the overt behavior of both hemispheres and instantly incorporated that behavior into a general theory of personal motivation. The mode of this incorporation was quite specifically elucidated by the left hemisphere language system. However, the verbal system never admitted to prior knowledge of plans or responses generated by other systems, specifically right hemisphere responses. The striking ease and speed with which a story was completed on this and other occasions demonstrated the need of the organism to establish a framework in which the verbal system defended its sense of conscious unity. Once a behavior was manifest the verbal system explained the external reality.

In other tests, patient J was asked to view two pictures simultaneously exposed, one to each visual field. He then had to choose the identical objects from a box of many objects, all within his field of vision. For example, a spoon was exposed to the right hemisphere, and an apple was exposed to the left, speaking hemisphere. He would pick up the spoon

*See references 23, 28, 29, 44, and 45.

FIGURE 2–2. Two different picture completion tasks were presented simultaneously, one to each hemisphere. The patient was required to point to the appropriate answer, with the hand contralateral to the exposed hemisphere. After both hands moved to complete the task, the patient described the reasons for each choice, even though he could not verbally identify the left visual field.
SOURCE: Gazzaniga, M.S., and LeDoux, J.E. *The Integrated Mind.* New York: Plenum Press, 1978, p. 149.

with his left hand and then say, "This isn't right. I didn't see a spoon, I saw an apple." This said, he easily moved to the apple and picked it up (see figure 2–3). When asked why he picked up the spoon first, he immediately replied, "It was in the way and I wanted to move it so I could pick up the apple." The explanations for his choices took this line throughout the entire experiment. The responses were neither guesses nor the beginning of vivid confabulations; they were statements of fact, a left hemisphere verbal offering to explain a behavior arising from motivations lurking in the right hemisphere. The left hand would always initially pick out the object flashed to the right hemisphere, and the robust left hemisphere verbal system would immediately suggest several reasons for picking an object that the right hemisphere saw but could not describe.

After several trials of this kind, in which the left hemisphere was forced to propose a theory for behaviors produced by independent right hemisphere behaviors, J became agitated. In these particular tests an explicit conflict was induced in the patient, since the right hemisphere was basically listening to a "lie" constructed by the left hemisphere. The right hemisphere knew why the left hand was picking up the spoon. It saw a spoon. However, in J this right hemisphere mental system, as is usually the case, was not capable of speech, and it simply was unable to correct the left hemisphere story.

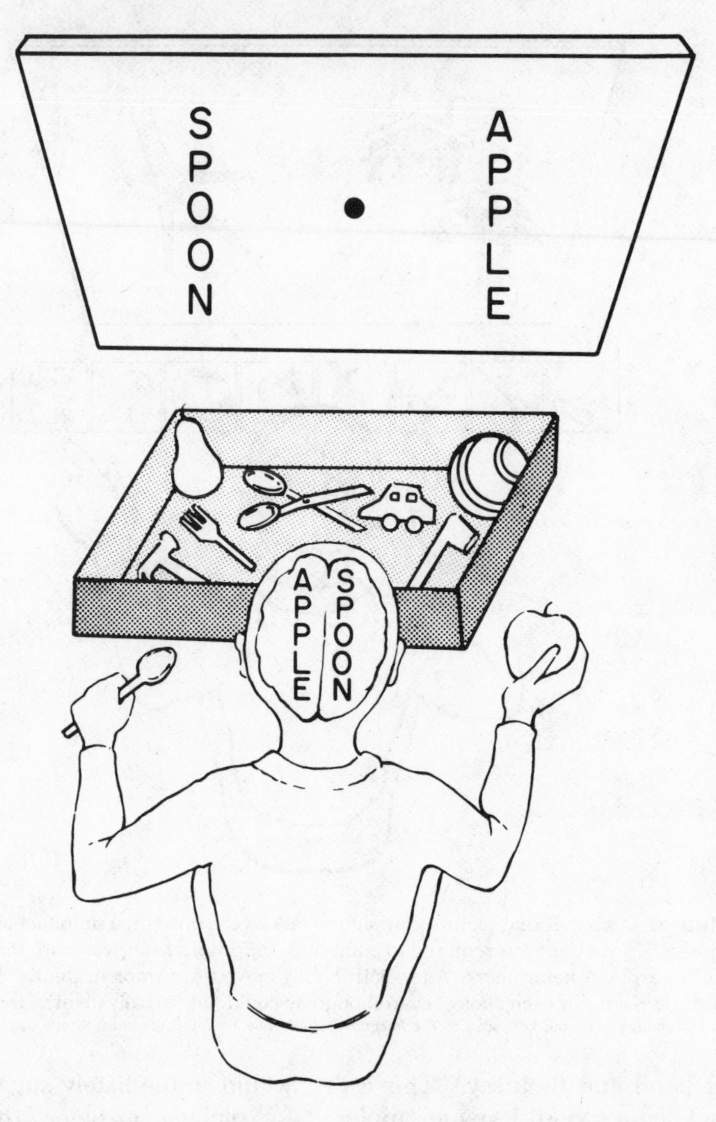

FIGURE 2–3 Two different pictures were presented simultaneously, one to each hemisphere—for example, a "spoon" was exposed on the left visual field and an "apple" was exposed on the right visual field. Although the patient chose the correct objects, the left verbal hemisphere did not know why each object was picked. However, the patient immediately offered an explanation.

A reexplanation of the neurosurgical procedure has always had a calming effect on J. Further, a discussion of the possible reasons for the patient's performance on each of the tasks considerably reduced the patient's anxiety. The patient was always reminded of the artificial system of lateralizing the stimulus; it is a situation that never occurs in daily experience. The investigator might say, "It is our special way of testing that presents pictures to your silent right hemisphere that you simply cannot verbally explain."

Yet on the very next trial, J returned to the typical explanation of the overt behavior produced by the right hemisphere response. He never used his awareness of having a disconnected hemisphere to explain his action. In fact, it can be said that after an action, the immediate drive for consistency and coherence through a verbal description is overwhelming. Patients never use the offered alternative explanation for the overt behaviors that occur outside of verbal awareness. The ability to accept the alternative explanation may require a tolerance of the disparate mental systems that is difficult to acquire. The patient's description of reality seems to arise again and again from considering an overt behavior.

¶ Covert Interactions: Cognitive and Emotional Contributions to Consciousness

The variety of phenomena just described have demonstrated how overt behaviors organized by the right hemisphere were accepted and interpreted by the left hemisphere. Consequently, it is necessary to consider how covert behaviors produced by nonverbal mental systems are interpreted by the verbal system. Under certain conditions behavioral responses have demonstrated insight into cognitive-emotional interactions and reinforced the coconscious multiple-mental-system-interaction model. Before describing split-brain experiments carried out to date, it might be helpful to consider some of the current theories of emotional behavior.

Views of emotional mechanisms and their influence on behavior evolved from a controversy between James[36] and Cannon.[10] James argued that the somatic change that occurred following an exciting stimulus was the emotion. Cannon's refutation was based on the physiology of the peripheral visceral changes as observed in animal experiments. According to Cannon, visceral changes were too slow and too nonspecific to account adequately for emotional change. In fact, in animal experiments, total separation of the viscera from the central nervous system did not alter emotional behavior.[11] Peripheral somatic change caused a general but nonspecific state of arousal.

However, a different and more recent view concerning emotional mechanisms has been constructed by Schachter.[59] After long series of experiments, Schachter maintained that the particular cognitive state of a subject determined the emotional interpretation given to a neutral but arousing physical stimulus. Stated differently, cognitive systems establish dimensions for the crude physiological arousal system that in itself cannot determine positive or negative emotion.

More recent and compelling views stem from the work of Zajonc.[75] In his analysis emotional responses are immediate, precede cognition, and suggest a positive or negative value. These emotional responses, in one experimental setting, were based on stimulus frequency.[53] While being exposed to a random series of words, subjects attached more positive ratings to the more frequently presented words. In this experiment, the words that were repeated with greater frequency were also presented so that they could not be verbally detected. Taken as a whole, this argues for the primacy of affect in cognitive-emotional interactions; it also suggests that people initially assign either positive or negative value to a stimulus and that this judgment takes place independent of cognitive analysis.

Studies carried out on cognitive emotional interactions in the split-brain patient support the latter interpretation and also suggest that a nonverbal mental system making a value judgment about the flashed stimulus can subsequently precipitate an emotional state that the left language system is compelled to interpret. The induced emotional state in the split-brain experiments did not lead to any overt behavior that the left hemisphere could observe and interpret. These conclusions are deduced from the following experiments.

It was known from past observations that the right hemisphere of patient P could perform certain primitive language operations. The development of language skills in the right hemisphere is most unusual, and is the subject of another complete investigation.[30] On a verbal command test, P was instructed to perform the action described by a word flashed before him.[28] His reaction to the left visual field presentation of the word "kiss" proved revealing. Although he could neither describe the word he had seen nor mime the activity, he said, ". . . no way, no way. You're kidding." His smile and nervous laugh on this trial was different from those on other trials. He seemed embarrassed by this flash. On presentation of the word "kiss" to the right visual field (left hemisphere), he would not perform the action and, like the adolescent boy he was, he said, "Kiss . . . No way. Kissin' is not for me." In trials to both hemispheres there was an emotional reaction to the word "kiss." In the latter instance, P could accurately describe the word and the action that he was not going to mime. When the command was exposed to the right hemisphere, he responded with an emotional judgment generated by right hemisphere mechanisms, which he could not describe, but he certainly felt. This independent setting of behavior for an emotionally arousing stimulus has led to a broad exploration of independent hemispheres—specifically, whether each half brain would behave as if it had its own independent system for assigning values to events, setting goals and response priorities.[44]

Since this unusual young man, P, could read in the left visual field, it was possible to pursue the right hemisphere responses that were covertly communicated to both hemispheres and that necessitated verbal interpretation by the left hemisphere. In a series of experiments a dozen words that were known to have positive or negative affective quality were singly presented to P's left visual field. The patient's task was to rank order each word by pointing to one of five ratings: like very much, like, undecided, dislike, and dislike very much. The patient's inability to describe verbally the stimulus lateralized to the left visual field confirmed the notion that the left hemisphere did not have access to the complete critical identification of the information. On specific exposures in the right visual field, the verbal response indicated that the left hemisphere could easily perform the task. However, since the experiment addressed the interaction between the hemispheres caused by the emotional content of the stimuli, the critical exposures were to the left visual field. Once this profile of rank order had been established for the right hemisphere by pointing to rating cards, the words were rearranged and again presented to the right hemisphere. In this series of trials, however, the patient was required to make a verbal response. This verbal response emanated from his left hemisphere and indicated a left hemisphere interpretation of the feeling his right hemisphere had about each stimulus.

As can be seen in figure 2–4, the results under the two test conditions were astonishingly similar. The profile of emotional values that the right hemisphere had independently generated and reported by pointing with the left hand was almost identical to the left hemisphere spoken responses to the same set of left visual field stimuli. The left hemisphere on any particular trial was unable to say what the word had been, although it produced identical rankings. Clearly the emotional dimension was communicated to the left hemisphere.

In the context of interactive coconscious mental systems, these results demonstrated how a nonverbal mental system could precipitate an emotion. Furthermore, it showed that once covert behavior is communicated throughout the brain, it is then incorporated into the ongoing verbal interpretation of the present.[23]

¶ A Model for Anxiety: Further Covert Actions

It is clear from observations made some time ago that each disconnected hemisphere of a split-brain subject can independently express emotion.[26] It would seem possible, therefore, that each hemisphere might possibly evaluate a particular stimulus differently. At a particular moment in time, the left might like a particular idea, concept, or person, while the right might react differently.

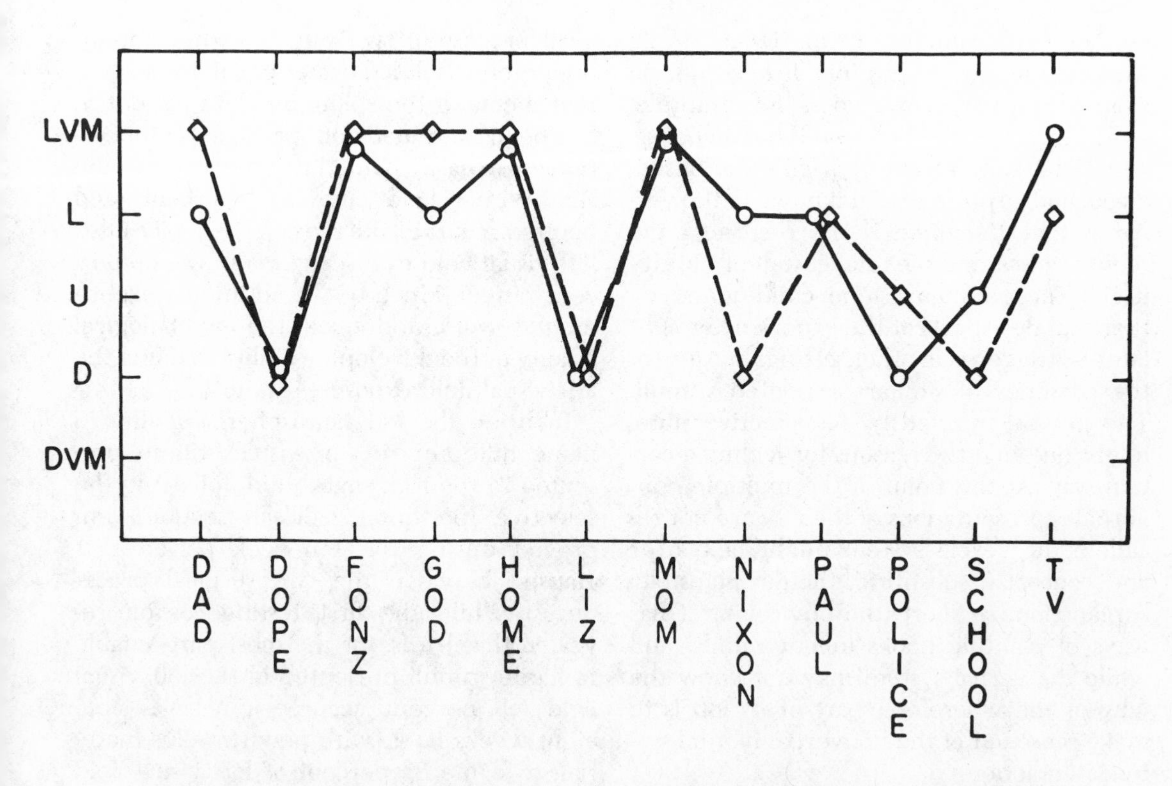

FIGURE 2–4. The left hemisphere and the right hemisphere independently ranked a set of emotional words. These nearly identical rankings suggest that emotional values can be shared by two disconnected hemispheres. The scale consisted of: LVM = like very much; L = like; U = undecided; D = dislike; DVM = dislike very much.
SOURCE: Gazzaniga, M.S., and LeDoux, J.E. *The Integrated Mind.* New York: Plenum Press, 1978, p. 153.

What would be the overall behavioral consequence of this disparate state? Observations of this kind of problem came about on two different test sessions.[44] On a day when P was calm, tractable, and appealing, his left and right hemispheres behaved as if they agreed on, and equally valued, himself, his friends, and other matters (see figure 2–4). Assigning values, generating choices, and making judgments were cognitive tasks easily and independently accomplished by each hemisphere.

At other times, however, there were marked differences between the evaluations made by each hemisphere.[44] Under these circumstances, P behaved in an unusual agitated, aggressive, and restless manner. The right and left hemispheres were producing conflicting evaluations about the same stimuli. It was as if both positive and negative emotional systems were simultaneously active, and the ensuing conflict produced a state of anxiety. In fact, P experienced cold, tremulous extremities, rapid pulse, and dilated pupils—somatic changes frequently associated with anxiety.

This clear example of surgically produced psychological dynamism, seen for the first time in P, raised the question of whether such processes are active in the normal brain. Perhaps most or all episodes of anxiety are the result of discrete mental systems evaluating the same external stimuli or internal thought and assigning different values. Thus, when a nonverbal mental system responds to a particular visual, auditory, sensory, olfactory, or gustatory stimuli, which may or may not enter verbal awareness, it has a pervasive effect on all subsequent processes. These sensations may be conditionally associated with a definite emotional tone so that only a subtle aspect of the experience is necessary to trigger the entire emotional experience. While such conditioning is possible, it need not be available for verbal awareness. For example, in Florence one can be focused on Michelangelo's statue of David and feel so aroused, awed, and inspired that, unknown to the verbal system, the brain is also recording the scents, noises, and the total gestalt of the city itself. The emotional tone conditioned by these subtle aspects of the experience might later be triggered in other settings because of the presence of similar or related stimuli. The person, puzzled by his affective state, might question the reasons for feeling a certain way. At this point, if the multiple nonverbal representations of the city are not recalled, the verbal system might take over and concoct a substitute, though plausible, explanation. In short, the environment has ways of planting hooks in our minds, and while the verbal system may not know the why or the wherefore, part of its job is to make sense out of the nonverbal mental system interaction.

¶ Partial Commissurotomy: Evidence for Multiple Representational Systems

The remarkable split-brain findings of the past twenty years are not apparent in patients who have undergone section of the anterior one-half to two-thirds of the corpus callosum.[29,31,32,65] However, patients with posterior section of the callosum are usually visually split and produce many of the remarkable behaviors already described in patients with complete callosal section. These clinical situations have generally occurred after tumor removal.* This evidence supports a wealth of observations from animal experiments and suggests that disruption of visual communication underlies an important part of the split-brain phenomenon.

Recently, Wilson[71] has carried out the surgical process in two stages, with a patient undergoing isolated posterior callosal section that included the splenium. This patient, J, had been examined both pre- and postoperatively. He easily named the exposures in the right visual field (above 91 percent) and pointed accurately to the correct choice after left visual field exposure (also in the 90 percent range). Much of the additional experimental work addressed the psychological quality of the developing ability to name the left visual field exposures.[61]

In brief, the left hemisphere's ability to name different sets of visual stimuli presented to the right half brain, following the selective posterior callosal section, improved during the ten-week period that elapsed between the two surgical procedures. While the first testing session revealed that J was, for the most part, unable to name stimuli presented in the left visual field (28 percent accuracy), he was able eight weeks later, with new stimulus material, to name 83 percent of left visual field stimuli. At first glance, this kind of result might best be explained by hypothesizing that the stimulus presented to the right hemisphere (left visual field) had been transferred by the remaining commissures to the left hemisphere for analysis and naming. Subsequent careful analysis of each test trial argued against that mechanism. When instructed to name the left visual field, J's behavior was unlike any patient with complete callosal section, particularly because he did not deny having "seen" anything. Also unlike patients who have had complete callosal section but who transfer information via the remaining anterior commissure,[58] he did not name the stimuli immediately.

In fact, his initial response after left visual field exposure was to say that he could "see" a "picture" of the stimulus but he could not name it. The examiner initiated a series of questions whenever J insisted that he had some sense of the left visual field information. This interaction often began with the question, "Is it an object or a living thing?" and continued along these lines. Thus, when

*See references 22, 35, 46, 50, 64, 65, and 69.

a line drawing of a hunter's cap was flashed to the right hemsiphere, J reported that the stimulus was an "object." A number of object classes, such as vehicles, tools, and so on, were then presented. He rejected each, saying "No," until clothing was offered. At this point he responded with an emphatic "Yes." He then recognized that the object was worn by a man, and the particular season in which it was worn. When he recognized the usual (red) color, he quickly exclaimed, "hunting cap." In this manner, J rarely identified the left visual field stimulus immediately. More often he described personally relevant contexts in which the stimulus could be found, yet he guessed infrequently. His choices were precise, and once made, he could not be shaken from his conviction.

It would appear that the stimulus projected to the right hemisphere activated a set of associations that were processed in more anterior regions of the right hemisphere and that were still interconnected by the anterior callosum. Once these attributes were collected, the left verbal system seemed able to deduce what the actual stimulus might have been. This result was more remarkable when particular word-stimuli were considered. To a word exposed in the left visual field—for example, "ship"—he said, "I see a picture of a television show called the *Love Boat,* but it's not boat, ship was flashed." The absence of synonymous substitution errors suggested that the left hemisphere had based the inferential process on more than a pictorial referent.

These and other examples provide converging evidence that the splenium is crucial for interhemispheric visual communication. Moreover, the partial surgical section suggested that the interaction of the verbal and nonverbal mental systems was considerably aided by the construction of complex spatial contexts. The right hemisphere acted to process a visual stimuli not only by pointing to matching choice cards, or completion cards, but also by constructing some representation that became accessible to the verbal system of the left hemisphere.[61]

¶ The Parietal Lobe in Man: Access to the Verbal System After Focal Brain Damage

Although the split-brain patient represents an explicit instance of the interaction of multiple coconscious mental systems, this concept remains to be tested in other situations. The notion of multiple coconscious mental systems can be studied in another clinical neurologic setting. It is generally thought that lesions of the right parieto-occipital cortex in man produce a variety of behavioral disturbances that interfere with the detection of and orientation to external stimuli.[70] A striking example, called "visual extinction," occurs when stimuli are presented simultaneously to both the left and right visual field, and the patient with right parietal damage can identify only the right visual field. Presentation of a single stimulus in any area of the visual field results in accurate detection and description, but simultaneous presentation of two stimuli, one in each field, results in the verbal description of only the stimulus in the right visual field. Although the extinguished stimulus in the left visual field often goes completely unnoticed by these patients, they are able to perform an interfield comparison task between this stimulus, which they cannot name, and the stimulus in the right visual field, which they can name. That is, these patients can make accurate judgments about the similarities or differences between two stimuli, one in each visual field, even though they cannot identify both stimuli and, at times, even deny the presence of the left visual field stimuli.[68]

The results have been documented on seven patients with right parietal damage and on an eighth patient with left parietal damage. Specifically, each patient sat in front of a screen and was required to identify objects or words projected singly to either visual field (see table 2–1). In a second series, stimuli were presented simultaneously to both visual fields, but the response requirements changed. Instead of having to identify both stimuli, the patients were asked to judge whether the stimuli were the same or different. The patients uniformly made accurate judgments when comparing information simultaneously presented to both visual fields, yet on further questioning they were unable to verbally characterize the left visual field information with the same level of accuracy as that of the right visual field (see table 2–2). In fact, during the bilateral simultaneous projections, two patients could not name any of the left visual field stimuli and insisted the task was "absolutely silly," although they continued to make accurate "same/different" judgments. These data bear on the interaction of verbal and nonverbal mental systems. Similar to the split-brain patients, these patients were influenced by information they frequently were unable to verbally identify.

In the trials where the same stimulus was presented to each visual field, the patient concluded, "Well, that was same, and I saw an apple [nodding toward the right visual field] . . . so I guess there was an apple here too [pointing to the left visual field], but I did not see it." On the majority of "different" trials they generally could not name the stimulus in the visual field that projects to the damaged hemisphere.

The stimulus comparison task appears to occur at a postperceptual, preverbal level. Only the comparison and not the specific identification is available to linguistic mechanisms (see figure 2–5). In spite of defective orientation to, and detection of, external stimuli, these patients were able to process stimulus information to permit accurate comparison judgments. Their striking inability, in most instances, to acknowledge the presence of that stimulus supports the notion of coconscious mental systems that function out of verbal awareness.

¶ Further Examination of Information Access to the Verbal System

Intracarotid injections of amobarbital are commonly used to investigate the neural substrate for language and memory mechanisms when neurosurgery on the temporal lobes is contemplated.* This short-acting barbiturate selectively abolishes half-brain

*See references 5, 16, 40, 51, 52, and 60.

TABLE 2–1 **Single Visual Field Naming**

RIGHT PARIETAL LOBE DAMAGE		
	VISUAL FIELD	
PATIENT	LEFT	RIGHT
1	1.00 (15)	1.00 (12)
2	0.94 (16)	0.89 (9)
3	0.86 (14)	1.00 (15)
4	0.91 (33)	0.88 (33)
5	0.77 (26)	0.83 (24)
6	1.00 (30)	1.00 (30)
7	1.00 (10)	0.80 (10)
LEFT PARIETAL LOBE DAMAGE		
PATIENT	LEFT	RIGHT
1	1.00 (10)	0.80 (10)

The proportion of trials that were correctly named by each patient in each visual field is shown. Numbers in parentheses represent total number of trials presented to each visual field. Variability in the numbers of trials among the patients reflects primary concern for their medical care. Performance differences between the visual fields were not significant ($t(7) = 0.833$, p .3).

TABLE 2–2 **"Extinguished" Visual Field Naming After Same/Different Judgments on Double Simultaneous Visual Field Presentation Trials**

RIGHT PARIETAL LOBE DAMAGE		
PATIENT	SAME/DIFFERENT JUDGMENTS	LVF NAMING ON "DIFFERENT" TRIALS
1	1.00 (17)	0.00 (7)
2	0.88 (26)	0.00 (16)
3	0.95 (39)	0.48 (25)
4	0.90 (68)	0.23 (35)
5	0.90 (30)	0.32 (19)
6	0.96 (48)	0.38 (24)
7	0.80 (40)	0.18 (40)

LEFT PARIETAL LOBE DAMAGE		
PATIENT	SAME/DIFFERENT JUDGMENTS	RVF NAMING ON "DIFFERENT" TRIALS
1	0.80 (40)	0.18 (40)

The proportion of correct same/different judgments and proportion of different trials in which the stimulus in the 'extinguished' field was correctly named are shown. Numbers in parenthesis represent the total number of trials. The accuracy of the same/different judgments was significantly greater than the accuracy of naming the extinguished field ($t(7) = 10.935$, $p .005$). LVF = left visual field; RVF = right visual field.

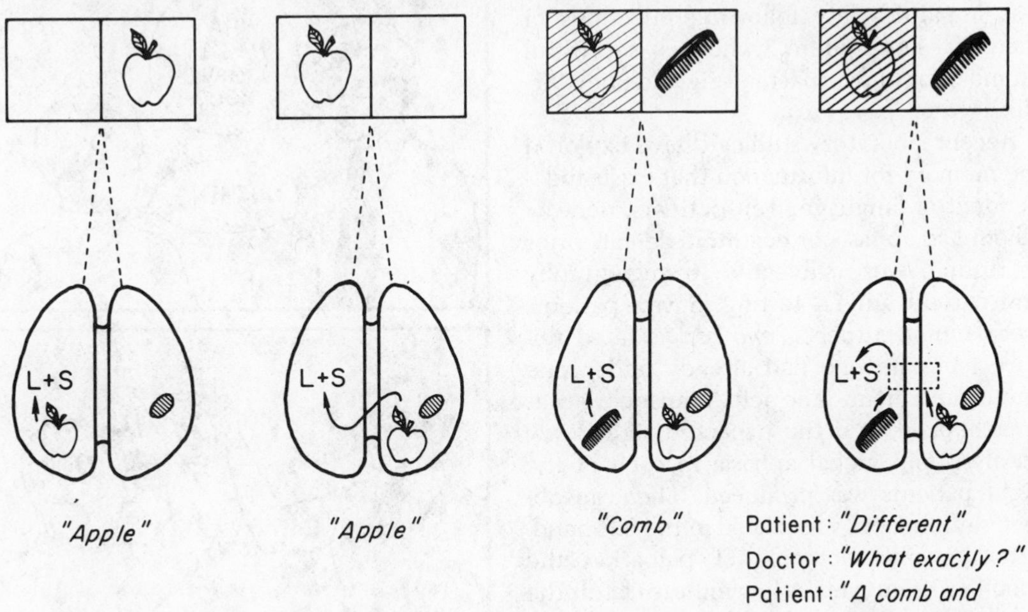

FIGURE 2–5. This picture is a series of typical responses from patients with right parietal lobe damage. Under single-field conditions, objects in each visual field were accurately described, but under bilateral presentation, the left visual field object could not be named even though the patient made an accurate "same/different" judgment.

SOURCE: Volpe, B.T., LeDoux, J.E., and Gazzaniga, M.S. "Information Processing of Visual Stimuli in an 'Extinguished' Field," *Nature*, 282 (1979): 723.

function for a short period of time. In conjunction with this diagnostic effort, experiments have been carried out during the typical test procedures. Prior to the amobarbital test the patient was required to name common pictures and to name objects that were explored tactilely out of the field of vision. When correctly completed these tests signaled adequate processing of visual and tactile information from both visual fields and from both hands. Left visual field information and left-hand information needed further relay to the left hemisphere for naming to occur.

Generally, visual and tactile stimuli were presented to the uninjected and conscious hemisphere during the brief period of depressed functioning of the opposite hemisphere. Reversible paralysis of the contralateral limbs signaled the depressed function of the injected hemisphere. When the effects of the drug had ceased, recall or recognition of the relevant stimuli were assessed. Past studies have demonstrated that memory, specifically for verbal material, was impaired only following injections of the left hemisphere, whereas successful memory for nonverbal material did not lateralize.[5,16,40]

Recent laboratory studies[57] have explored the memory for information that exists independent of language. Ten patients, none of whom had aphasia or cognitive deficits prior to testing, were subjected to angiography and carotid amytal testing. Seven patients were tumor suspects, two had isolated seizure foci, and one had already undergone aneurysm repair. The left hemisphere was anesthetized, and the expected right body paralysis and global aphasia in each of the eight patients was produced. Then, an object, say, a spoon, was placed in the left hand. After several moments of palpation the spoon was removed. A few minutes later, the drug effect dissipated and the patient was completely awake, alert, and had full sensorimotor function. When the patient was asked, "What was placed in your hand?" all eight patients responded, "I don't know," or "Nothing." All patients responded accurately, however, on a test of verbal recall from a period prior to the injection, yet no amount of encouragement or cues prompted a verbal description of the palpated object.

When several choice cards were placed in front of the patients, they pointed, almost instantly, with the left hand to the picture of the correct object (see figure 2–6). This performance reflected the inaccessibility of the nonverbal mental system information to verbal analysis. Six of the eight patients pointed to a picture of an object they had palpated while language mechanisms were selectively

FIGURE 2–6. During a sodium amytal test, the patient's left hemisphere was anesthetized, and the left hand (right hemisphere) was allowed to palpate an object. After the drug effect dissipated, the patient was unable verbally to describe the object, but he was able to retrieve it with the left hand.

depressed, although they could not name this object until after they performed the visual choice task. In a similar fashion, two additional patients were required to palpate objects with the left hand while the left hemisphere (and language function) was temporarily depressed. When the drug effect waned they could not verbally describe the palpated object. In spite of this apparent verbal ignorance, they explored, out of vision, a box full of objects, and with the left hand chose the correct object.

These data suggest that information stored in the absence of language was not easily accessible to language, even when that nonverbal system reemerged and became functional. The tactile information seemed to be represented in a manner that was impenetrable for linguistic analysis. These coexisting systems appeared to be insulated from one another, yet were present within the structure of the brain.

¶ Implications for Psychiatry

The experiments with the split-brain and brain-damaged patients continue to elucidate the heterogenous and autonomous behavioral functions of each hemisphere. Work over two decades has demonstrated that each hemisphere has an independent private complex of cognitive skills that can be mobilized for more than perceptual discriminations. Each hemisphere can autonomously generate opinions, judgments, attitudes, and emotions. Thus, these experiments support a model of interacting coconscious mental systems that directly study the interaction of cognitive processes which must be coordinated for a final behavioral act and which allow controlled observation of the distinctly human behavior of talking about those acts. The data suggest that a dynamic relation holds between nonverbal information processing systems, which can organize, represent, and retrieve information, and the more apparent verbal system. Our approach has been to examine the nature of this interaction in an attempt to gain insights into normal conscious mechanisms.

It should now be clear that by reporting on the interaction of the two hemispheres, we are not simply commenting on the intrinsic or isolated function of the left or the right hemisphere. Split-brain, or two-hemisphere, testing serves as a method for the examination of the interaction of verbal and nonverbal mental systems. Additionally, several of the split-brain testing techniques were modified for use in patients with focal neurologic defects. The experiments with patients who had suffered parietal lobe damage and those undergoing carotid amobarbital injection also demonstrate the interaction between verbal and nonverbal systems.

The recent split-brain experiments explore the extent to which an emotional response of a nonverbal system without access to the verbal system can influence complex behavior. For example, word selection for a simple declarative sentence might be biased by the mood state initiated by another nonverbal system. As a result of this intervention between mental systems, the statement might be at odds with previous attitudes about the point in question and could further lead the person to alter his previous belief about it.

During all of the experiments, the verbal system consistently attempted to incorporate experimentally induced overt behavior produced by nonverbal mental systems into a unifying understanding of personal motivation. For the patients, the language apparatus provided a means to attain a sense of unity. However, in view of the multiple coconscious mental systems that were demonstrated, this unity is an illusion. The ramifications of this notion has particular relevance for forensic psychiatry.

This work is not meant to suggest a specific neural substrate for the basic forces of the psyche as posited by the psychiatric literature. Nor is it meant to provide only a neural model that explains the therapeutic difficulties encountered in traditional psychoanalysis, where conscious experience is assessed exclusively by verbal output. The verbal mechanisms simply do not have access to all the nonverbal specific information systems that may exert crucial influence on behavior. The power and efficiency of verbal communication are not in question. The data do suggest that there are major deficiencies when the verbal system is used to evaluate the behavioral activities of coconscious mental systems.

Experiments with split-brain patients offer a model for behavioral disorders and an insight into the normal conscious mechanisms. A metaphor for the human condition of the duality of human nature, and the struggle with inner conflict, must be expanded to include multiple coconscious mental systems competing for access to the output mechanisms. The primary output mechanism—verbal behavior—may only infrequently have prior knowledge of behavior, as one or another coconscious nonverbal system controls the output mechanism. The language system may, at times, lag behind in organizing the conscious experience. This mental operation is only one of a number of operations, and the study of its interactions with nonverbal systems will continue to capture the imagination.

¶ Bibliography

1. AKELAITIS, A. J. "Studies on Corpus Callosum: Higher Visual Functions in Each Homonymous Field Following Complete Section of Corpus Callosum," *Archives of Neurology and Psychiatry,* 45 (1941): 788–796.
2. ———. "Studies in Corpus Callosum: Study of Language Functions (Tactile and Visual Lexia and Graphia) Unilaterally Following Section of Corpus Callosum," *Journal of Neuropathology and Experimental Neurology,* 2 (1943): 26–32.
3. ———. "A Study of Gnosis, Praxis, and Language Following Section of the Corpus Callosum and Anterior Commissure," *Journal of Neurosurgery,* 1 (1944): 94–103.
4. AKERS, R. M., and KILLACKEY, H. P. "Organization of Corticortical Connections in the Parietal Cortex of the Rat," *Journal of Comparative Neurology,* 181 (1978): 513–538.
5. BLUME, W. T., GRABOW, J. D., and DARLEY, F. L. "Intracarotid Amobarbital Test of Language and Memory before Temporal Lobectomy for Seizure Control," *Neurology,* 23 (1973): 812–819.
6. BOGEN, J. E. "The Other Side of the Brain: II. An Appositional Mind," *Bulletin Los Angeles Neurological Society* 34 (1969): 135–162.
7. ———, and BOGEN, G. "The Other Side of the Brain: III. The Corpus Callosum and Creativity," *Bulletin Los Angeles Neurological Society,* 34 (1969): 191–220.
8. BOGEN, J. E. and VOGEL, P. J. "Cerebral Commissurotomy in Man: Preliminary Case Report," *Bulletin of the Los Angeles Neurological Society,* 27 (1962): 169.
9. ———. "Neurologic Status in the Long Term Following Complete Cerebral Commissurotomy," in Michael, F., and Schott, B., eds., *Les Syndromes de Disconnexion Calleuse chez l'Homme.* Lyons: Hopital Neurologique, 1974, pp. 88–95.
10. CANNON, W. B. "The James Lange Theory of Emotions: A Critical Examination and an Alternative Theory," *American Journal of Psychology,* 39 (1927): 106–124.
11. ———. *Bodily Changes in Pain, Fear, Rage,* 2nd ed. New York: Appleton, 1929.
12. COOPER, J. R., BLOOM, F. E., and ROTH, R. H., *The Biochemical Basis of Neuropharmcology,* 3rd ed. New York: Oxford University Press, 1978.
13. EBNER, F. F., and MYERS, R. E. "Distribution of Corpus Callosum and Anterior Commisure in Cat and Raccoon," *Journal of Comparative Neurology,* 124 (1965): 353–365.
14. ECCLES, J. "The Brain and Unity of Conscious Experience," *19th Arthur Stanley Eddington Memorial Lecture,* Cambridge, Eng.: Cambridge University Press, 1965.
15. ERICKSON, T. C. "Spread of the Epileptic Discharge," *Archives of Neurology,* 43 (1940): 429.
16. FEDIO, P., and WEINBERG, L. K. "Dysnomia and Impairment of Verbal Memory Following Intracarotid Injection of Sodium Amytal," *Brain Research,* 31 (1971): 159–168.
17. FOERSTER, O., and PENFIELD, W. "Struc-

tural Basis of Traumatic Epilepsy and Result of Radical Operation," *Brain,* 53 (1930): 99–120.

18. GALIN, D. "Implications for Psychiatry of Left and Right Cerebral Specialization," *Archives of General Psychiatry,* 31 (1974): 572–583.
19. GASTAUT, H. "Clinical and Electroencephalographical Classification of Epileptic Seizures," *Epilepsia,* 11 (1970): 102.
20. ———, et al. "Lipomas of the Corpus Callosum and Epilepsy," *Neurology,* 30 (1980): 132–138.
21. GAZZANIGA, M. S. *The Bisected Brain.* New York: Appleton-Century-Crofts, 1970.
22. GAZZANIGA, M. S., and FREEDMAN, H. "Observations on Visual Processes After Posterior Callosal Section," *Neurology,* 23 (1973): 1126–1130.
23. GAZZANIGA, M. S., and LEDOUX, J. E. *The Integrated Mind.* New York: Plenum Press, 1978.
24. GAZZANIGA, M. S., BOGEN, J. E., and SPERRY, R. W. "Some Functional Effects of Sectioning the Cerebral Commissures in Man," *Proceedings of the National Academy of Science (U.S.A.),* 48 (1962), 1765–1769.
25. ———. "Laterality Effects in Somesthesis Following Cerebral Commissurotomy in Man," *Neuropsychologia,* 1 (1963): 209–215.
26. ———. "Observations on Visual Perception after Disconnexion of the Cerebral Hemispheres in Man," *Brain,* 88 (1965): 221.
27. ———. "Dyspraxia Following Division of the Cerebral Commissures," *Archives of Neurology,* 16 (1967): 606–612.
28. GAZZANIGA, M. S., LEDOUX, J. E., and WILSON, D. H. "Language, Praxis, and the Right Hemisphere: Clues to Some Mechanisms of Consciousness," *Neurology,* 27 (1977): 1144–1147.
29. GAZZANIGA, M. S., RISSE, G. L., and SPRINGER, S. P. "Psychologic and Neurologic Consequences of Partial and Complete Cerebral Commissurotomy," *Neurology,* 25 (1975): 10.
30. GAZZANIGA, M. S., et al. "Plasticity in Speech Organization Following Commissurotomy," *Brain,* 102 (1979): 805–815.
31. GESCHWIND, N., and KAPLAN, E. "A Human Cerebral Deconnection Syndrome," *Neurology,* 12 (1962): 675–685.
32. GORDON, H. W., BOGEN, J. E., and SPERRY, R. W. "Absence of Deconnexion Syndrome In Two Patients With Partial Section of the Neocommissures," *Brain,* 94 (1971): 327–36.
33. GREENWOOD, P., WILSON, D. H., and GAZZANIGA, M. S. "Dream Report Following Commissurotomy," *Cortex,* 13 (1977): 311–16.
34. HOPPE, K. D. "Split Brains and Psychoanalysis," *Psychoanalytic Quarterly,* 46 (1977): 220–244.
35. IWATA, M., et al. "Etude sur le Syndrome de Disconnexion Visuo-Linguale après la Transection du Splenium du coups Calleux," *Journal of Neurological Sciences,* 23 (1974): 421–432.
36. JAMES, W., and HOLT, H. *The Principles of Psychology.* New York, 1890.
37. JONES, E. G., and POWELL, T. P. S., "The Commissural Connexions of the Somatic Sensory Cortex in the Cat," *Journal of Anatomy* 103 (1968): 433–455.
38. JONES, E. G., BURTON, H., and PORTER, R. "Commissural and Cortico-Cortical Columns in the Somatic Sensory Cortex of Primates," *Science,* 190 (1975): 572–574.
39. JOUANDET, M. L., and GAZZANIGA, M. S. "Cortical Field of Origin of the Anterior Commissure of the Rhesus Monkey," *Experimental Neurology,* 66 (1979): 381–397.
40. KLOVE, H., TRITES, R. L., and GRABOW, J. D. "Intracarotid Sodium Amytal for Evaluating Memory Function," *Electroencephalography and Clinical Neurophysiology,* 28 (1970): 418–419.
41. KRYNAUW, R. A. "Infantile Hemiplegia Treated by Removal of One Cerebral Hemisphere," *Journal of Neurology, Neurosurgery, and Psychiatry,* 13 (1950): 243.
42. KUSSKE, J. A., and RUSH, J. L. "Corpus Callosum and Propagation of Afterdischarge to Contralateral Cortex and Thalamus," *Neurology,* 28 (1978): 905–912.
43. LEDOUX, J. E., WILSON, D. H., and GAZZANIGA, M. S. "Manipulo-spatial Aspects of Cerebral Lateralization: Clues to the Origins of Lateralization," *Neuropsychologia,* 15 (1977): 743–750.
44. ———. "A Divided Mind: Observations on the Conscious Properties of the Separated Hemispheres," *Annals of Neurology,* 2 (1977): 417–21.

45. LEDOUX, J. E., et al. "Cognition and Commissurotomy," *Brain,* 100 (1977): 87–104.
46. LEVINE, D. N., and CALVANIO, R., "Visual Discrimination After Lesion of the Posterior Corpus Callosum," *Neurology,* 30 (1980): 21–30.
47. LEVY, J. "Psychobiological Implications of Bilateral Asymmetry," in DIMOND S. J., and BEAUMONT J. G., eds., *Hemisphere Function in the Human Brain.* New York: Halstead Press, 1974.
48. LEVY-AGRESTI, J., and SPERRY, R. W. "Differential Perceptual Capacities in Major and Minor Hemispheres," *Proceedings of the National Academy of Science (U.S.A.),* 61 (1968): 1151.
49. MACKAY, D. Cited in Gazzaniga, M. S. "One Brain-Two Minds?" *American Scientist,* 60 (1972): 311–317.
50. MASPES, P. E. "Le Syndrome Experimental Chez L'Homme de la Section du Splenium de Coups Calleux Alene Visuelle Pure Hemianopsique," *Revue Neurologie,* 80 (1948): 100–113.
51. MILNER, B. "Psychological Aspects of Focal Epilepsy and its Neurosurgical Management," in Pupura D. P., Penry J. K., and Walter, R. D., eds. *Neurosurgical Management of the Epilepsies,* New York: Raven Press, 1975, pp. 185–210.
52. ———, BRANCH, C., and RASMUSSEN, T., "Study of Short-Term Memory After Intracarotid Injection of Sodium Amytal." *Transactions of the American Neurological Association,* 87 (1962): 224–226.
53. MORELAND, R. L., and ZANJONC, R. B. "Exposure Effects May Not Depend on Stimulus Recognition," *Journal of Personality and Social Psychology,* 37 (1979): 1085–1089.
54. MYERS, R., and SPERRY, R. W. "Interocular Transfer of a Visual Form Discrimination Habit in Cats after Section of the Optic Chiasm and Corpus Callosum," *Anatomical Record,* 115 (1953): 351.
55. RASMUSSEN, T. "Cortical Resection in the Treatment of Focal Epilepsy," in Purpura, D. P., Penry, J. K., and Walter, R. D., eds., *Neurosurgical Management of the Epilepsies,* New York: Raven Press, 1975, pp. 139–154.
56. ———. "Surgery for Epilepsy Arising in Regions Other than the Temporal and Frontal Lobes," in Purpura, D. P., Penry, J. K., and Walter, R. D., eds., *Neurosurgical Management of the Epilepsies,* New York: Raven Press, 1975.
57. RISSE, G. L., and GAZZANIGA, M. S. "Well-Kept Secrets of the Right Hemisphere: Carotid Amytal Study of Restricted Memory Transfer," *Neurology,* 28 (1978): 950–953.
58. RISSE, G. L., et al. "The Anterior Commissure in Man: Functional Variation in a Multisensory System," *Neuropsychologia,* 16 (1978): 23–31.
59. SCHACHTER, S. "Cognition and Peripheralist-Centralist Controversies in Motivation and Emotion," in Gazzaniga, M. S., and Blakemore, C., eds., *Handbook of Psychobiology,* New York: Academic Press, 1975, pp. 529–564.
60. SERAFETINIDES, E.A. "Auditory Recall and Visual Recognition Following Intracarotid Sodium Amytal Injections," *Cortex,* 2 (1966): 367–372.
61. SIDTIS, J. J., et al. "Cognitive Interaction after Staged Callosal Section: Evidence for Transfer of Semantic Activation," *Science,* in press.
62. SPERRY,. R. W. "Mental Unity Following Surgical Disconnection of the Cerebral Hemispheres." *The Harvey Lectures Series,* 62, New York: Academic Press, pp. 293–322.
63. ———, GAZZANIGA, M., and BOGEN, J., "Interhemispheric Relationships: The Neocortical Commissures: Syndromes of Hemispheric Disconnection, in Vinken, P. J., and Bruyn, G. W., eds., *Handbook of Clinical Neurology,* vol 4. Amsterdam: North Holland Publishers, 1969, pp. 273–290.
64. SUGISHITA, M., et al. "Reading of Ideograms and Phonograms in Japanese Patients After Martial Commissurotomy," *Neuropsychologia,* 16 (1978): 417–426.
65. SWEET, W. H. "Seeping Intracranial Aneurysm Simulating Neoplasm, Syndrome of the Corpus Callosum," *Archives of Neurology and Psychiatry,* 45 (1941): 86–104.
66. TRESCHER, J. H., and FORD, F. R. "Colloid Cyst of the Third Ventricle: Report of a Case of Operative Removal with the Section of the Posterior Half of the Corpus Callosum in Man," *Archives of Neurology and Psychiatry,* 37 (1937): 959–973.
67. VAN WAGENEN, W. P., and HERREN, R. Y.

"Surgical Division of Commissural Pathways in the Corpus Callosum," *Acta Neurologica Psychiatrica* 44 (1940): 740–759.

68. VOLPE, B. T., LEDOUX, J. E., and GAZZANIGA, M. S. "Information Processing of Visual Stimuli in an 'Extinguished' Field," *Nature,* 282 (1979): 722–724.
69. WECHSLER, A. F. "Transient Left Hemialexia," *Neurology,* 22 (1972): 628–633.
70. WEINSTEIN, E. A., and FRIEDLAND, R. P., eds., *Hemi-Inattention and Hemisphere Specialization,* New York: Raven Press, 1977,
71. WILSON, D. H., CULVER, C., and WADDINGTON, M. "Disconnection of the Cerebral Hemispheres: An Alternative to Hemispherectomy for the Control of Intractable Seizures," *Neurology,* 24 (1975): 1149.
72. WILSON, D. H., REEVES, A., and GAZZANIGA, M., "Division of the Corpus Callosum for Uncontrollable Epilepsy," *Neurology,* 28 (1978): 649–653.
73. WILSON, D. H., et al., "Cerebral Commissurotomy for the control of intractable seizures. *Neurology,* 27 (1977): 708–715.
74. WILSON, D. H. et al. "Microsurgical, Extraventricular Division of the Corpus Callosum for the Control of Intractable Epilepsy: Evolution of the Operation, Present Technique, and Analysis of Fourteen Consecutive Cases," *Journal of Neurology, Neurosurgery and Psychiatry,* in press.
75. ZAJONC, R. B. "Feeling and Thinking. Preferences Need No Inferences," *American Psychologist,* 35 (1980): 151–175.

CHAPTER 3

BIOLOGICAL RHYTHMS AND PSYCHIATRY

Thomas A. Wehr and Frederick K. Goodwin

Our body is like a clock; if one wheel be amiss, all the rest are disordered, the whole fabric suffers: with such admirable art and harmony is a man composed.

Robert Burton,
The Anatomy of Melancholy (1628)

¶ Introduction

RHYTHMS IN NATURE—the alternation of day and night, the tides and the seasons—govern our lives and structure our experience. The degree to which nature's cycles influence culture is obvious. Less obvious is the fact that they are also impressed upon our genes, for we generate within ourselves days, months, and seasons that mirror and anticipate the rhythmic changes around us. Our biological rhythms make each of us a microcosm of the geophysical world.

Unlike the motions of the planets, biological clocks are imprecise, and their synchronization with external rhythms depends upon their being continually reset. For this purpose we possess special sense organs that lock onto external time cues such as the rising and setting of the sun and that make corresponding adjustments. As we live out our lives, our biological self is always tuned to the rhythms of the world around us, and we are forced to keep time with its march.

Our internal rhythms constitute a kind of temporal anatomy. Each day our body's temperature rises from a predawn nadir to its evening peak. At night the pineal gland secretes a hormone, melatonin. When we fall asleep, growth hormone briefly appears. The adrenals emerge from quiescence abruptly in the middle of sleep and are most active at dawn. And nearly every function of the organism exhibits a 24-hour, or circadian, pattern of variation with its own characteristic timing and waveform. The classical principle of homeostasis must be amended to encompass such physiological variation: cyclic change of the internal milieu is maintained.

Little is known of the functions of our temporal anatomy. We can speculate that biological rhythms help us to adapt to a cyclically changing environment, by modulating a range of specially and sometimes mutually incompatible functions and states that would be impossible in an unchanging organism. A normal temporal anatomy may also be essential to health. Experimental disturbances in circadian rhythms have been shown to adversely affect emotional well-being, mental acuity, longevity, reproductive function, thermoregulation, and restorativeness of sleep.*

Interest in the human circadian system has come late to medical science. Perhaps this is because circadian rhythms exist in a temporal rather than a spatial domain. In addition, there are psychological and technical reasons for the delay: Compared to the contractions of the heart, a rhythm that beats only once in 24 hours is not easily perceived subjectively and is not easily measured objectively. Recently, basic and clinical circadian rhythm research has been greatly stimulated by three developments: (1) the introduction of instruments and techniques for long-term monitoring of physiological and biochemical rhythms in freely moving human subjects;[13] (2) publication of long-term studies of the behavior of circadian rhythms in normal human subjects living in isolation from environmental time cues;[120,121] and (3) identification of the neuroanatomical site in the hypothalamus of a circadian pacemaker and its connections to the eye.[66,70,90]

It is reasonable to suppose that the human circadian system, a central integrative mechanism that programs recurring sequences of physiological and behavioral events and that resides in known neural structures with proven connections to the environment, is subject to disease and disorder. The question of whether there are diseases specific to the human circadian system, however, remains unanswered. It seems likely that such diseases would affect multiple systems, be associated with disruption of function rather than tissue, be of obscure etiology, and have prominent behavioral manifestations, including disturbances of the sleep-wake cycle. In a very general way, this description fits some psychiatric illnesses, especially the affective illnesses—depression and manic-depressive states. In fact, two classical symptoms of depression, early-morning awakening and diurnal variation in mood, implicate circadian rhythms in their pathophysiology and highlight a paradox: The depressive awakens early but arises late. Georgi,[35] writing in 1947, was one of the first to make this connection:

> In the true endogenous depressive we see a shift in the 24-hour rhythm . . . the night becomes day . . . anyone knowing the material would look for the CNS origin in the midbrain, where the entire vegetative nervous system is controlled by a central clock whose rhythmicity . . . regulates and balances the biological system. (Authors' translation)

This chapter will review the evidence that links affective illness to disturbances in the circadian system. It will also describe a circadian theory of depression and mania and possible circadian mechanisms of drugs and procedures that are used in their treatment.

¶ The Human Circadian System

Circadian rhythms are near-24-hour patterns of variation in biological functions. They are ubiquitous in human physiology. Some circadian rhythms, such as the sleep-wake cycle, are familiar to everyone. Others can only be detected with special techniques of measurement. Some of the most striking patterns occur in the various secretions of the endocrine system, but they can only be observed when plasma is sampled every 20 minutes for 24 hours and subsequently analyzed with sensitive radioimmunoassay procedures.

Circadian rhythms are endogenous and self-sustained; they are not simply passive re-

*See references 1, 11, 27, 31, 47, 53, 85, and 103.

sponses to daily changes in the environment, but originate within the organism. In many cases they continue independently of both sleeping and waking. Their endogenous nature can be appreciated when human subjects are deprived of all external time cues, permitting their circadian rhythms to *free-run* in isolation experiments.[121] In this situation the rhythms persist but with a periodicity that is slightly different (usually longer) than 24 hours. The fact that they persist, together with the fact that their period deviates slightly from 24 hours, is evidence that the rhythms are generated by the organism. (If their period continued to be 24 hours, the influence of a subtle environmental factor could not be ruled out.) Wever and Aschoff,[121] German scientists who pioneered long-term temporal isolation experiments, studied hundreds of normal human subjects and found that the average free-running or *intrinsic period* of the human circadian system is about 25 hours.

Since we are able to adhere to a daily schedule that is shorter than the intrinsic 25-hour period of our circadian system, it is obvious that some mechanism continually resets our biological clocks ahead approximately one hour a day. Somehow, physiologically, the human organism "knows" the correct time each day and adjusts accordingly. This ongoing process of *entrainment* of circadian rhythms to the day-night cycle depends upon the capacity of the organism to automatically perceive and respond to stimuli in the environment that have the properties of time cues, or *zeitgebers.* Light-dark transitions (dawn and dusk) are powerful *zeitgebers* for all species, including humans, and light-dark cycles alone can entrain human circadian rhythms.[17] Other *zeitgebers* that may be important for humans include the timing of meals[60] and various social stimuli, which are less well characterized.[121]

The effect of a *zeitgeber* depends upon the phase of the circadian cycle during which it is presented. If, for example, a light stimulus occurs near the expected time of dawn (in an animal whose intrinsic period is longer than 24 hours), circadian rhythms will be shifted earlier, or *phase-advanced;* if the stimulus occurs long after the expected time of dawn, they will be shifted later, or *phase-delayed.* The magnitude and direction of such shifts are a function of how early or late the stimulus occurs. By presenting light stimuli at various phases of the circadian cycle it is possible to define a *phase-response curve* that describes this function.[20] Typically, such curves show an apparent discontinuity, or *switch-over point,* during the circadian phase that corresponds to the middle of the night. If, at this time, a maximally delaying stimulus is shifted a few minutes later, it becomes a maximally advancing stimulus; that is, light at 1 A.M. is perceived as a very late dawning of the previous day and causes a large compensatory phase-delay, whereas light at 2 A.M. is perceived as a very early dawning of the next day and causes a large compensatory phase-advance. The switchover point of the phase-response curve could be an important element in a circadian rhythm theory of affective illness, since the depressive and manic switch processes[7,46,77,79] tend to occur in the middle of the night and appear to be associated with phase-shifts of circadian rhythms.[4,54,77,112]

The human circadian system can be entrained to artificial "days" shorter or longer than 24 hours.[121] The limits of the *range of entrainment* are normally about 21 and 27 hours. The range of entrainment is partly a function of the intrinsic period of the circadian system. Subjects with relatively long free-running circadian periods entrain to long "days" more easily than to short "days"; the converse is true for subjects with relatively short free-running circadian periods. The range of entrainment normally includes the 24-hour period of the day-night cycle, although this might not be the case in diseases of the circadian system.

When a circadian rhythm is entrained to the day-night cycle, it adopts a characteristic timing, or *phase-position,* relative to it. In other words, the peak and trough of the rhythm recur at particular times each day. The phase-position of an entrained circadian rhythm varies slightly from one person to another and is partly a function of the intrinsic period that a person's rhythm would exhibit if it were free-running.[83,121] A person who has a long intrinsic period (for example, 25.5 hours) and thus free-runs relatively slowly will adopt a relatively late phase-position when entrained to the day-night cycle. Someone whose intrinsic period is short (for example, 24.5 hours) and free-runs relatively fast will adopt a relatively early phase-position when entrained to the day-night cycle. For example, the peaks of the rhythms of the two individuals might occur at 5 P.M. and 3 P.M. respectively. Knowledge of this relationship could be useful if circadian rhythm phase-disturbances prove to be important in disease. An abnormally early or late phase-position of a circadian rhythm could indicate that the intrinsic period of that rhythm is abnormally short or long. There is evidence that the phase-position of depressed patients' circadian rhythms is abnormally early relative to the day-night cycle. A possible interpretation is that their intrinsic circadian period is abnormally short.

In the normal situation where the circadian system is entrained to the 24-hour day-night cycle, its intrinsic period is only a latent characteristic—the period its oscillations would exhibit if it were free-running and isolated from external time cues. Obviously, the (latent) intrinsic period is an important characteristic of the circadian system, since it determines whether or not a circadian rhythm can be entrained to a given schedule (the range of entrainment) as well as the phase-position that rhythm will adopt relative to that schedule if entrainment is achieved. The intrinsic period of the circadian system is relatively stable, but it can be altered by certain agents or conditions. The intrinsic period is *history dependent;* for example, if an organism has been entrained to a 22-hour "day," its intrinsic period will gradually become shorter than would be the case if the organism had been entrained to a 26-hour "day."[21] Light intensity, drugs, and hormones can also alter the intrinsic period.* In some species, estrogen and testosterone affect the intrinsic period, which varies with the estrous cycle. These effects of reproductive hormones may be relevant to a circadian rhythm theory of affective illness in light of the increased incidence of depression and mania after puberty, in the puerperium, in association with the menstrual cycle, and during menopause.

*See references 19, 21, 28, 72, and 115.

Many animals, including humans,* exhibit seasonal or annual cycles in behavior and physiology. These cycles regulate levels of sexual, metabolic, and motoric activity related to reproduction, growth, hibernation, and migration. Seasonal rhythms are relevant to the present discussion because they have a circadian basis. Animals use a circadian clock to measure the interval between dawn and dusk (the *photoperiod*).[84,85] Since the photoperiod varies in length during the year, being shortest at the winter solstice and longest at the summer solstice, these measurements can be used to determine the time of year and to trigger appropriate biological and behavioral changes. Seasonal rhythms have been described in humans, and there are reasons to believe that they depend on photoperiodic mechanisms.[17,62] Affective illness shows seasonal patterns of recurrence which might have a circadian basis in the photoperiodic mechanism,[25,111] and at least one type of treatment, partial sleep-deprivation in the second half of the night,[95] artificially lengthens the photoperiod by advancing "dawn."

Most of our knowledge of the human circadian system is derived from the behavior of free-running circadian rhythms in temporal isolation experiments. In general, two types of circadian rhythms have been simultaneously monitored: the rest-activity (sleep-wake) cycle and the circadian temperature rhythm.[121] Subjects in these experiments live in caves, underground bunkers, or aboveground apartments sealed off from the outside world. They wear indwelling rectal temperature probes that are connected to recording equipment. Devices worn on the wrist, or sensors in the floor, detect motion; and the times of switching on and off lights and bedrest is monitored. Sometimes sleep electroencephalogram (EEG) recordings,

*See references 25, 59, 85, 104, and 105.

performance tests, and frequent blood or urine sampling are done. The subjects remain in their quarters from a few weeks to as long as six months, with no knowledge of the time of day or calendar day. They prepare their own meals and select their own times of going to bed and arising. Usually they entertain themselves with books and music and often undertake long-term projects such as the writing of a thesis. When subjects are released into free-running, they typically go to bed and get up about one hour later each day, and the sleep-wake cycle and the circadian temperature rhythm usually oscillate together with the same 25-hour period. However, at some point during free-running their oscillations may spontaneously become dissociated, so that the two rhythms run with different periods and beat in and out of phase with one another every few days.[121] When dissociation of the two rhythms occurs, the subject no longer sleeps regularly at the same phase of his temperature rhythm but goes to sleep and arises at progressively different phases. Sometimes he may sleep while his body temperature is high and be awake while it is low—a reversal of the normal relationship. During this phase-inversion, subjects sometimes report mild psychological and somatic discomforts. Dissociation of the rest-activity cycle and the circadian temperature rhythm, reported by Aschoff, Gerecke, and Wever in 1967,[1] and subsequently observed in many subjects, indicates that the human circadian system consists of at least two potentially separate driving oscillators. Because the two circadian oscillators, which are coupled together in ordinary conditions, sometimes spontaneously dissociate during free-running experiments, it has been possible to learn something of their individual characteristics. Studies by Wever and Aschoff conducted over the past 20 years, and those conducted more recently by Weitzman, Czeisler, Zimmerman, Moore-Ede, and Kronauer[16,20] indicate that the controlling oscillator of the circadian temperature rhythm is a much stronger oscillator than the one which controls the sleep-wake cycle. The intrinsic period of its rhythm is very stable and remains close to 25 hours, even after many months of free-running. Besides body temperature, the strong oscillator also appears to control circadian rhythms in the hypothalamic-pituitary-adrenal axis and in rapid-eye-movement (REM) sleep (see figure 3–1). The weak oscillator, which controls the sleep-wake cycle, however, has a labile intrinsic period that can deviate quite substantially from 24 hours. Sometimes the period of the weak oscillator is much shorter than 24 hours; more often it becomes much longer, reaching 40 or 50 hours per cycle when it dissociates from the strong oscillator during free-running experiments (see figure 3–2). In the latter case, the subject may remain awake for 30 hours or more during each sleep-wake cycle with no perceived need for sleep. At the end of such experiments, subjects are often surprised to learn that many more calendar days have passed than their total number of subjective "days" in isolation.

The different strengths of the two coupled oscillators cause them to respond differently to one another and to external forces acting upon them. For example, because its oscillations are weak, the controlling oscillator of the sleep-wake cycle adjusts more easily to time zone shifts after jet travel than the controlling oscillator of the temperature rhythm. The lag between the time it takes the sleep-wake cycle to shift to a new local time (a day or so) and the time it takes the circadian temperature rhythm to shift (many days) is "jet lag."

Because the forces that tend to couple the two oscillators continue to operate even when they dissociate during free-running, they perturb one another's oscillations as they beat in and out of phase. This means that they tend to keep step with one another for a few cycles, then break away from one another, keep step, then break away, in a kind of recurring tug-of-war in which the slower oscillator acts as a drag on the faster one, and the faster one pulls the slower one ahead of its natural rhythm, until the inherent discrepancy between their respective intrinsic periods wins out and they part com-

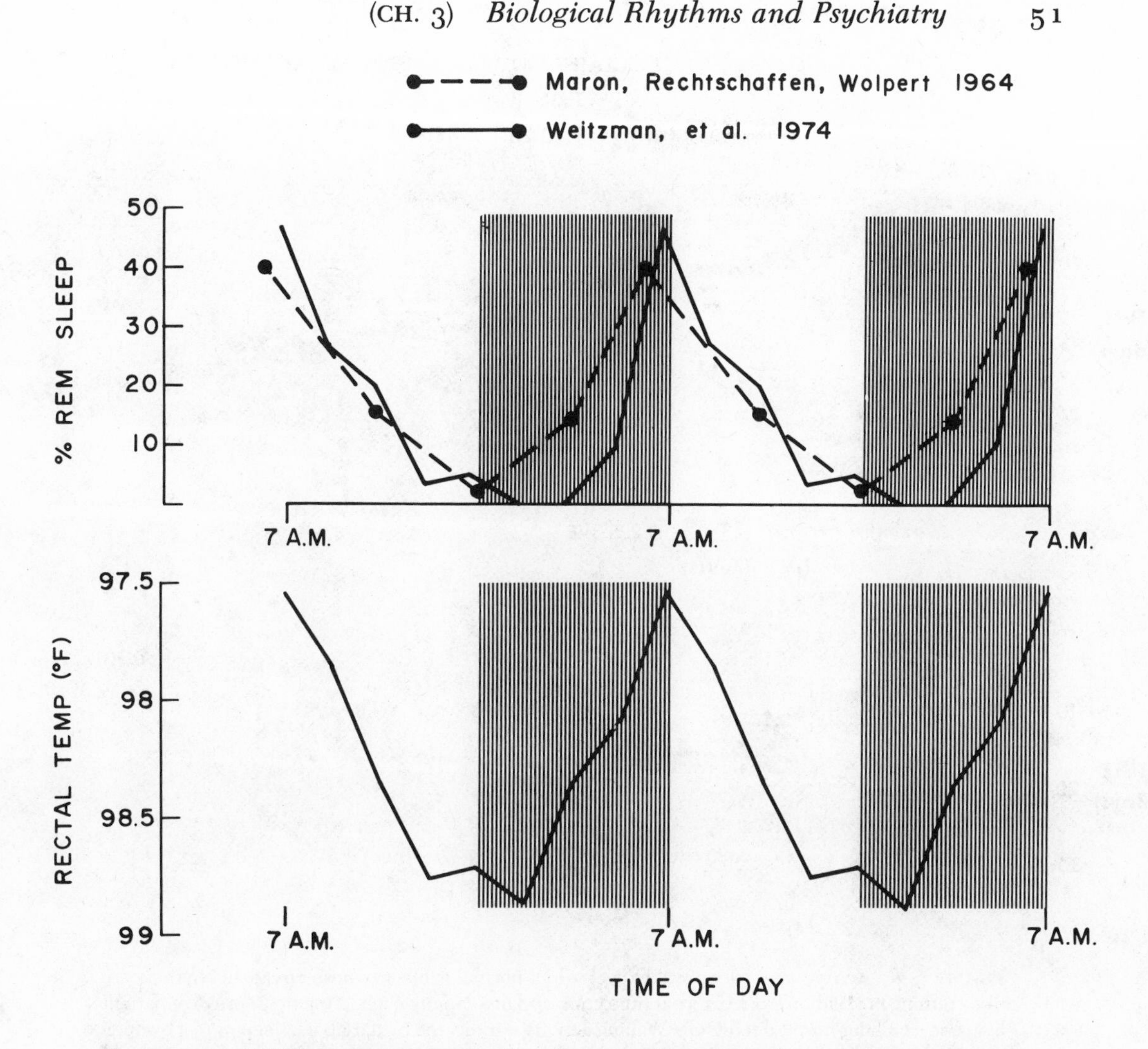

FIGURE 3–1. REM sleep propensity circadian rhythm and its relationship to body temperature circadian rhythm. REM sleep changes in depression may be interpreted as a shift to the left (phase-advance) of the REM rhythm.

NOTE: Maron, L., Rechtschaffen, A., and Wolpert, E.A. "Sleep Cycle During Napping," *Archives of General Psychiatry,* 11 (1964): 503–598. Weitzman, et al. "Effects on Sleep Stage Pattern and Certain Neuroendocrine Rhythms," *Transactions of the American Neurological Association,* 93(1968): 153–157.

pany. This pattern of keeping step, then recurrently breaking out of synchrony has been called *relative coordination* (in contrast to absolute coordination) of two coupled oscillators.[108] At those times when the two relatively coordinated oscillators are temporarily in step with one another, their compromise rhythm is largely determined by the intrinsic period of the stronger, dominating oscillator. Because of this asymmetry, relative coordination can cause the sleep-wake cycle to exhibit runs of short cycles near 25 hours that are periodically interrupted by one or more long cycles 40 or more hours in length, while the circadian temperature rhythm only slightly slows down and speeds up gradually as it breaks in and out of synchrony with the sleep-wake cycle. These features of dissociation and relative coordination of two circadian oscillators may well be connected with circadian rhythm disturbances seen in rapidly cycling manic-depressive patients.

This model of the human circadian system involves a strong and a weak oscillator that normally are coupled bidirectionally to one

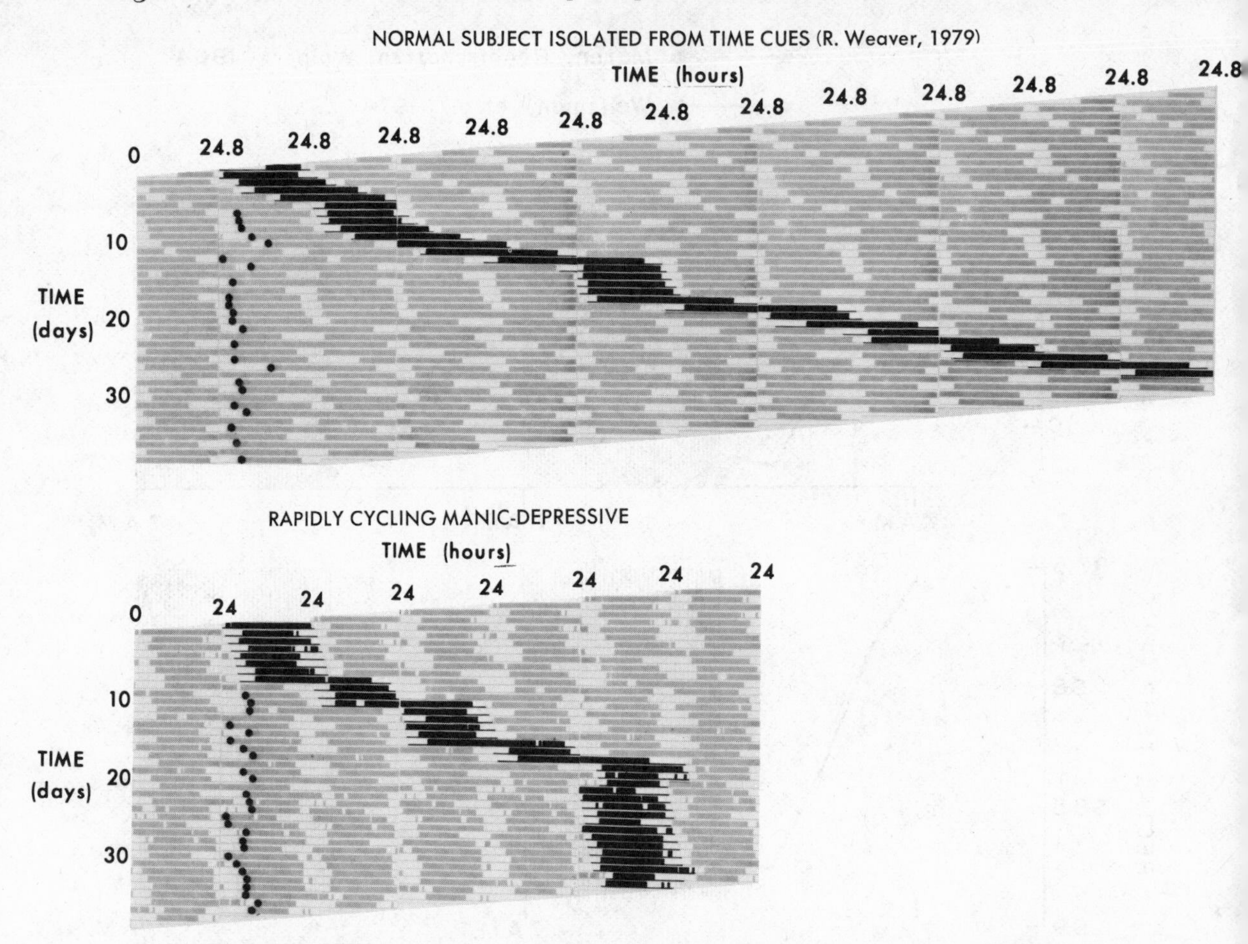

FIGURE 3–2. Activity-rest (sleep-wake) cycles in a normal subject (whose circadian rhythms are free-running in isolation from external time cues) and in a rapidly cycling manic-depressive patient living on a 24-hour hospital schedule. Waking activity is indicated by dark bars, sleep or rest by open spaces between bars. Data are plotted in rasters of consecutive 24.8 hour (top) or 24.0 hour (bottom) segments. Courses of consecutive sleep-wake cycles are emphasized in black. Both the free-running normal subject and the entrained manic-depressive repeatedly exhibit double length (48-hour or circa-bi-dian) sleep-wake cycles, as well as cycles of ordinary length. These may indicate that the driving oscillator of the sleep-wake cycle is escaping from its normal one-to-one mode into a one-to-two mode of coupling to other circadian rhythms and, in the patient, to the external day-night cycle. Black dots indicate that the activity onset times remain highly synchronized with an underlying cyclic process having a period near 24 hours. In the patient, 48-hour sleep-wake cycles occurred in conjunction with switches out of depression into mania.

another and unidirectionally to the day-night cycle. Richard Kronauer,[16] guided by actual data from human free-running experiments, has developed a mathematical model that produces realistic computer simulations of the behavior of this two-oscillator system. A key feature of the model is the absence of direct coupling between the strong oscillator and the external day-night cycle that entrains it. There is a hierarchy of coupling such that the strong oscillator, which controls the circadian temperature, REM propensity, and HPA-axis rhythms, is entrained by the weak oscillator, which controls the sleep-wake cycle and which in turn is entrained directly by the day-night cycle. Thus, the strong oscillator is entrained to the day-night cycle indirectly via the weak oscillator. This hierarchy of coupling is the only model that successfully simulates certain transient behaviors of the system that occur immediately after release into free-running.

This model of the circadian system was developed entirely by analyzing the formal properties of the behavior of circadian rhythms in various conditions. It is remarkable that during the past decade nearly all of the neuroanatomical substrates of the model have been tentatively identified in experimental animals and, to a certain extent, in humans. The weak oscillator, which controls the sleep-wake cycle, appears to reside in neurons of the suprachiasmatic nuclei (SCN) of the hypothalamus.[70] These paired structures are situated in the anterior hypothalamus just above and behind the optic chiasm on either side of the third ventricle. They are innervated by a small group of nerve fibers coming directly from the retina via the retinohypothalamic tract (RHT), which emerges from the chiasm and enters the SCN. It has not been possible to lesion the RHT selectively. However, interruption of optic pathways, which include these fibers, destroys the animal's capacity to entrain to light-dark cycles. Bilateral destruction of the SCN essentially destroys the sleep-wake (or rest-activity) cycle. After lesioning, animals exhibit short bouts of activity and sleep, distributed randomly over time.

It is presently unclear whether SCN lesions destroy the circadian temperature rhythm. Several recent studies[43,125] suggest that the neural substrate of the controlling oscillator of the circadian temperature rhythm, the strong oscillator, may lie elsewhere, possibly in the hippocampus. Two groups[66,70] have recently identified structures corresponding to the SCN in postmortem human brain tissue. Although technical difficulties have prevented identification of the retinohypothalamic tract in humans, two experimental findings suggest strongly that the pathway exists. Czeisler and associates[17] have demonstrated that human circadian rhythms can be entrained to a light-dark cycle alone. Another group[62] has shown that human melatonin secretion by the pineal gland can be suppressed by light (this response is known to be mediated by light acting on the SCN via the RHT in experimental animals). It is therefore likely that a homologous pathway exists in humans.

The neurochemistry, neurophysiology, and neuropharmacology of the SCN are presently the focus of intensive investigation. The visual field of the SCN has been mapped and is very broad; essentially the SCN respond to light impinging on any portion of the retina.[40] The SCN contain the largest concentration of serotonergic nerve terminals in the hypothalamus. These fibers originate in cell bodies in the raphe nuclei. The functions of the serotonergic input to the SCN are unknown; raphe lesions do not alter their rhythmic behavior.[90] Injections of various neuropharmacological agents into the third ventricle adjacent to the SCN indicate that some of the effects of light may be partially mediated by cholinergic mechanisms.[127]

Studies[62] of the effects of light on the putative human RHT-SCN-pineal pathway revealed an interesting difference between humans and other animals (including primates). The threshold for light intensity sufficient to elicit a response is considerably higher in humans. Because of this difference humans are able to discriminate between the brightness of ordinary artificial light and natural light. It is possible that the discovery of fire some 100,000 years ago has shaped human evolution in this respect.

Before concluding this discussion of the human circadian system, an important methodological problem warrants attention. Circadian oscillators are not the only factors that govern body temperature, waking, and sleeping. Changes in our physiological state are also evoked by stress, physical activity, meals, and other variables that strongly reflect the time-structure of our social and physical environment. Evoked physiological responses are superimposed on, and may even obscure, changes related to the circadian rhythm, making it difficult to measure and characterize.

Distortion of the circadian pattern by evoked responses is referred to as *masking.* Another form of masking occurs when one circadian rhythm affects the expression of another. For example, sleep lowers body temperature, and activity raises it; therefore, the timing of the sleep-wake cycle relative to the temperature rhythm will determine the waveform of the rhythm that is being measured. Thus, there are two forms of masking: internal and external. Circadian rhythms in some variables can only be observed in special conditions; for example, the circadian rhythm in REM sleep is expressed only when the subject is asleep. In this case sleep is said to have a *positive masking effect.* The problem of masking highlights the fact that an overt rhythm is only an indirect measure of the oscillator that drives it. In clinical studies of circadian rhythms we must assume that an abnormality in phase-position of a circadian oscillator, if present, will be masked to some extent by sleep schedules and other influences emanating from the temporal structure of the environment.

In summary, the human circadian system is controlled by at least two endogenous, self-sustained, coupled oscillators: a strong one controlling body temperature, REM sleep propensity, and cortisol secretion, and a weak one controlling the sleep-wake cycle and sleep-related neuroendocrine activity. Ordinarily, the longer than 24-hour rhythm of the coupled oscillator system is adjusted to 24 hours exactly by periodic environmental stimuli, such as dawn light, which act on the weaker oscillator through its connections to the eye. Light acting on circadian oscillators is also the basis of seasonal and annual biological rhythms. The normal phase-relationships between the two circadian oscillators and their overt rhythms can be temporarily disturbed by phase-shift experiments and rapid transmeridian travel. Furthermore, the two oscillators may spontaneously dissociate and oscillate with unequal periods when humans are experimentally deprived of external time cues. The circadian oscillators and their connections to the environment have a physical reality in specific nerve cells and tracts in the hypothalamus. The timing of circadian rhythms relative to the day-night cycle and to one another is homeostatistically controlled and partly reflects the period of the intrinsic rhythm of their driving oscillators.

There are several ways in which such a system might be altered by disease and treatment interventions. There could be alterations in coupling between oscillators, coupling of the oscillators to the external day-night cycle, or in the intrinsic periods of the oscillators. These changes could be expected to affect the phase-position of circadian rhythms entrained to the day-night cycle, and may even affect their capacity to be entrained at all. Clinical studies of circadian rhythms are difficult because of the requirement for longitudinal, around-the-clock monitoring of multiple variables and the confounding effect of masking of circadian rhythms by sleep and by exogenous social factors. (In the next section we will examine evidence from many different sources that strongly suggests, but does not yet establish, that affective illness is an expression of pathology in the human circadian system.)

¶ Circadian Rhythm Disturbances in Affective Illness

Historically, four clinical features of depression have stimulated interest in the idea that circadian rhythm disturbances are involved in the disease: early-morning awakening, diurnal variation in symptom severity, seasonality of the illness, and cyclicity of the illness.

Early-morning awakening is one of the hallmarks of endogenomorphic depression. Typically, the depressed patient awakens abruptly at 4 or 5 A.M. and finds he cannot return to sleep, but instead lies in bed wracked with depressive ruminations. Some patients have observed that they are better able to tolerate this problem if they simply accept that they are not going to be able to return to sleep, get out of bed, and begin their day's activities. Early awakening has been regarded as merely one manifestation of a more fundamental insomnia. However, it can also be regarded as a normal event—simply waking up—that occurs at an abnormally early time. In fact, some patients find that they are also able to go to sleep early. The shift in time of awakening has been interpreted to mean that a circadian rhythm is abnormally phase-advanced.

Diurnal variation in mood, with depression worse in the morning, is another hallmark of endogenomorphic depression. Patients awake in the depths of depression. As the day wears on, depressive symptoms abate somewhat, especially after the midafternoon.[109,110] Sometimes in the evening patients feel nearly normal, or even better than normal, but they always dread going to sleep because depression will return at its worst the following morning. This pattern has suggested to some that the depressive process is tied in some way to a circadian rhythm, possibly a normal one. Papoušek,[77] in her elegant and pioneering theoretical paper, interpreted the paradoxical early awakening and late functioning of the depressive as an internal phase-displacement of two circadian rhythms (sleep-wake and "waking-readiness"), one being too early, the other too late.

The seasonality of depression and mania is a striking phenomenon that can be seen in epidemiological studies as well as in individual cases. For depressives, it seems, "April is the cruelest month": Depressive hospital admissions, electroconvulsive treatments, and suicides occur most frequently in the spring.[25] Some studies show secondary peaks in the fall as well. Mania also appears to recur seasonally, although studies differ on whether it is most frequent in the spring, late summer, or fall.[111] Kripke[54] has proposed that seasonal patterns of mania and depression could be atavistic expressions of vestigial seasonal behavioral rhythms based on photoperiodic mechanisms involving the circadian system.

The cyclicity of affective illness need not express itself seasonally. It is characteristic that whenever affective episodes occur, they run their course and remit spontaneously, only to return again at some future time. The recurrent nature of the illness has been repeatedly confirmed by epidemiological studies.[128] Halberg[41] proposed that such long-term cycles of relapse and remission could occur if depression resulted from an abnormal internal phase relationship between two circadian rhythms and if at least one of those rhythms was no longer entrained to the day-night cycle but was free-running and beating slowly in and out of phase with the other rhythm. And Kripke[4,54] has published striking results of longitudinal studies of circadian rhythms in a few rapidly cycling manic-depressive patients that support Halberg's hypothesis.

Some of the first clinical studies[64,75] of circadian rhythms in depression were carried out in England in the 1950s and 1960s. They were inspired by Lewis and Lobban's[61] discovery that the timing of one circadian rhythm relative to another within the same person could be altered experimentally by placing subjects on unusual schedules during the Arctic summer. The clinical studies were designed to explore whether early-morning awakening in depressives is related to an analogous but pathological internal phase-disturbance. Early studies sometimes showed dramatic phase-disturbances in depressives,[64,75] but no consensus about the significance and pattern of these changes emerged, even after thirty years in which more than twenty studies* of circadian rhythms in depressives have been carried out. The results of these studies, to the authors' knowledge, have never been systematically related to one another and have remained on the periphery of psychobiological research in depression. Perhaps this is because there was no context based on normal circadian physiology in which to place their findings. Also, the investigative groups have been widely separated geographically, and the studies widely separated in time. In contrast to circadian rhythms, the sleep of depressives has been intensively studied by several active groups continuously for fifteen

*See references 9, 14, 23, 24, 26, 30, 32, 33, 34, 38, 42, 44, 49, 50, 51, 52, 54, 55, 63, 64, 65, 69, 73, 74, 75, 76, 80, 81, 87, 89, 91, 93, 94, 101, 106, 114, 124, and 126.

years, and a consensus regarding their findings has emerged. Ironically, the results of sleep EEG studies have rekindled interest in a circadian rhythm hypothesis of affective illness, for they can be interpreted as indicating that the circadian rhythm of REM sleep propensity is phase-advanced in depressives.* In light of this finding, the authors retrospectively analyzed published data describing a variety of physiological and biochemical circadian rhythms in depressives as compared with controls (see figure 3-3). These results are compatible with the REM sleep finding and with a *circadian rhythm phase-advance hypothesis of depression.* In the next section this hypothesis and the evidence that supports it will be discussed. In a following section some recent findings about circadian rhythms in mania will be presented.

*See references 77, 107, 112, 116, and 117.

¶ Circadian Rhythm Phase-Advance Hypothesis of Depression

Phase-Advance of the REM Sleep Circadian Rhythm

REM sleep was discovered by Aserinsky and Kleitman in 1953.[3] Episodes of REM sleep lasting 10 to 30 minutes occur about 80 to 90 minutes after sleep onset, then recur cyclically each 90 minutes through the night. It has been called paradoxical sleep because while the EEG shows signs of arousal and the eyes exhibit bursts of vigorous lateral move-

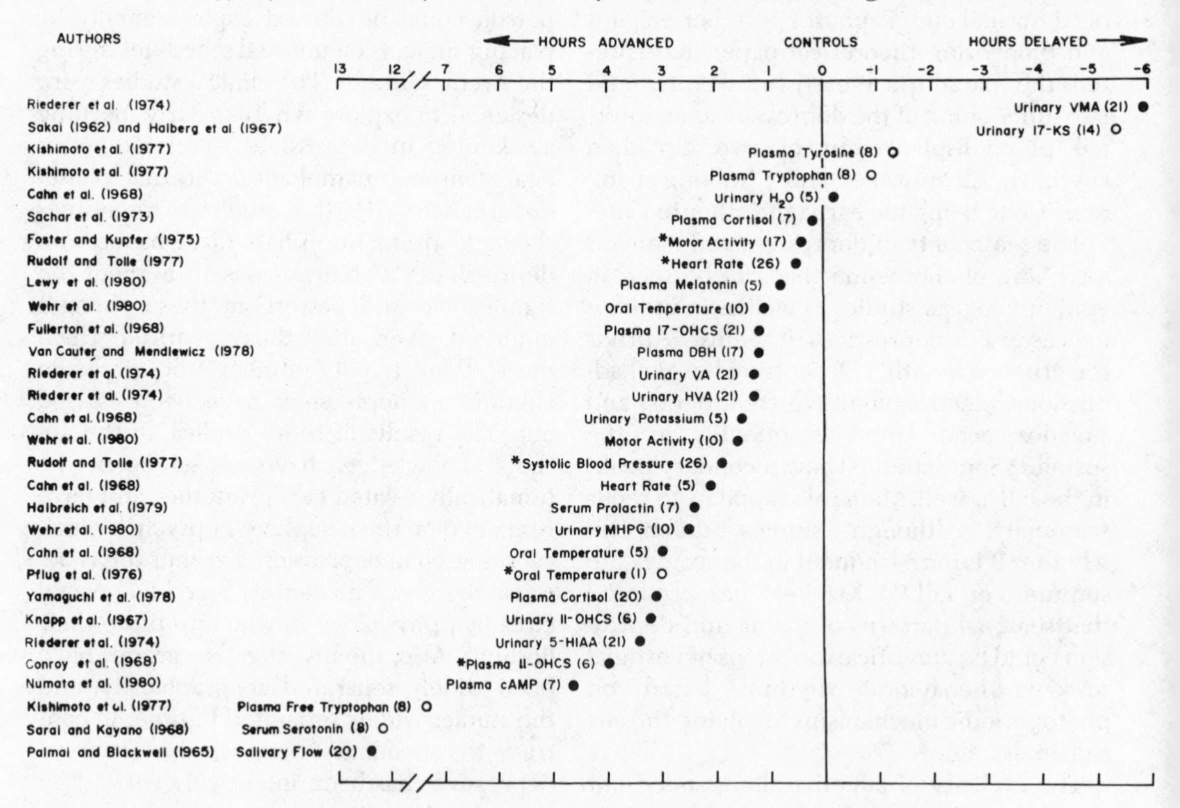

FIGURE 3-3. Early timing (phase-advance) of circadian rhythms in depressives compared with controls. Phase was defined as the time of the peak (acrophase) of a cosine function, $f(t)$ = mean + amplitude $\times$ *cos* (frequency $\times$ t + phase), fitted to published data by the method of least squares. Open circles indicate studies where no data was obtained during sleep. Asterisks indicate studies where control group was not normal.

ments, skeletal muscles are profoundly relaxed. In 1957, Dement and Kleitman[22] observed that the distribution of REM sleep during a night's sleep was skewed, with more REM occurring toward the end of the night than at the beginning. Studying naps in 1965, Maron and colleagues[67] found that afternoon naps contained relatively high amounts of REM while evening maps contained little. Linking their own findings with those of Kleitman and Dement's, they proposed that the propensity for REM sleep is governed by a process that exhibits a circadian rhythm independent of sleep. They also noted that the REM rhythm was more or less inverse to the rhythm of body temperature, and they went on to propose that the two circadian rhythms might be fundamentally related. Subsequent research has borne out their speculations (see figure 3–1). Much about the relationship between the circadian temperature rhythm and REM sleep has been learned from EEG sleep studies of free-running subjects whose sleep-wake cycle and circadian temperature rhythms dissociate and beat in and out of phase with one another.[18,121] In these subjects, the window of sleep through which REM sleep is observed passes repeatedly across all the phases of the temperature rhythm, making it possible to record all 360 degrees of the circadian rhythm of REM sleep. Results of these studies reveal that REM sleep propensity is maximal just after the body temperature falls to its minimum level. When the propensity of REM sleep is high, REM episodes are long and REM latency (the time from sleep onset to REM onset) is short. Also the percentage of sleep that is REM sleep is high. Normally the temperature minimum occurs in the latter half of the sleep period, so that REM sleep is maximal near dawn. The close association between REM sleep, body temperature, and cortisol secretion suggests that their respective circadian rhythms are controlled by the same oscillator—the stronger of the two coupled circadian oscillators.

In one of the first EEG studies of sleep in depressives, Gresham and Agnew[40] found that the normal skewed distribution of REM sleep was altered. Depressives had more REM sleep in the first third of the night and less REM sleep in the last third of the night than did the controls (see figure 3–4). In a way, most subsequent EEG sleep studies of depression have described variations on this theme. Kupfer[57] has emphasized the short REM latency in depressives, and this finding has been extensively replicated. Vogel[107] found that the first REM episode in depressives is sometimes quite long, and that on nights when first REM episodes are long, subsequent episodes are short. All of these changes in the temporal distribution of depressives' REM sleep could result from a phase-advance of the circadian rhythm in the propensity to have REM sleep, such that its maximum, instead of occurring near dawn, occurs nearer to the beginning of sleep. Probably the first person to propose this interpretation was Snyder[118] who, in 1968, noted a similarity between the sleep of normal subjects whose sleep period was experimentally shifted later relative to their REM sleep rhythm and depressed patients whose REM sleep was shifted earlier relative

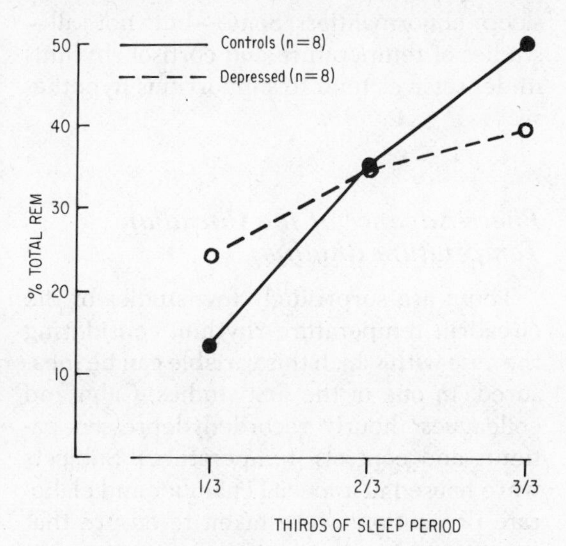

FIGURE 3–4. Early temporal distribution of REM sleep in depressives compared with controls. Many studies indicate that early occurrence of REM sleep in depressives' sleep period is expressed in their short REM latency (time from sleep onset to REM onset) and long first REM episodes.

SOURCE: Gresham, S.C., Agnew, W.F., and Williams, R.L. "The Sleep of Depressed Patients," *Archives of General Psychiatry*, 13 (1965): 503–507.

to their sleep period. Papoušek[77] stated the interpretation more explicitly in 1975.

> Somnopathy in depressive patients is based on an internal and external desynchronization. . . . the reduced REM latency, the relative increase in REM sleep at the beginning of the night and the shortening of the REM cycles . . . find a new interpretation as the expression of a phase displacement of the circadian rhythm of REM activity. (Authors' translation)

Vogel has conducted well-controlled studies of depressive sleep and, working in a different conceptual framework, he arrived at a conclusion that is compatible with Papousek's formulation. Emphasizing the advance of REM sleep within depressives' sleep period, he noted striking similarities between depressive sleep at the beginning of the night and normal sleep in the morning when it is extended past the usual wake-up time.

Because of the close association between circadian rhythms in REM sleep, body temperature, and cortisol secretion, a phase-advance in the latter two rhythms in depressives would tend to support the circadian phase-advance interpretation of the REM sleep abnormalities. Some—but not all—studies of temperature and cortisol rhythms in depressives tend to support this hypothesis.

Phase-Advance of the Circadian Temperature Rhythm

There are surprisingly few studies of the circadian temperature rhythm, considering the ease with which this variable can be measured. In one of the first studies, Cahn and colleagues[9] hourly recorded depressed patients and controls' temperatures. Subjects were housed in a special chamber and elaborate precautions were taken to ensure that measurements were uniformly made and as free as possible of exogenous influences. Five patients were studied. If their results are compared to those of the five controls whose ages are most similar, the depressives show a statistically significant phase-advance (3 hours) in the time of their temperature peak (see figure 3–3).

Nikitoupoulou and Cramer[73] studied manic-depressives in both phases of their illness. In depression, the patients' temperature rhythms became bimodal with two peaks and two troughs per day. This result is difficult to interpret. However, in the authors' studies of normals it was observed that a 180-degree reversal of the timing of sleep relative to the temperature rhythm sometimes results in bimodal patterns; of the two temperature minima, one is related to the true minimum of the circadian temperature rhythm that now occurs late in the waking phase, and the other is related to a lowering of temperature evoked by sleep—a masking effect. If an analogous internal phase-shift is responsible for the patients' bimodal patterns, the phase-displacement of the temperature rhythm would be very great indeed. In the manic phase, the temperature rhythm reverted to its normal unimodal pattern.

In the authors'[116] study of manic-depressive patients during depression it was found that the phase-position of the temperature rhythm was approximately 1 hour advanced compared with controls, although the difference was not statistically significant. Kripke and others[55] reported similar results. Two German groups, Pflug and associates[80,81] and Lund and associates[65] found an early temperature minimum in depression. However, Lund's group observed that, for a given patient, nights when the temperature minimum was particularly early did not necessarily correspond to nights when REM sleep occurred earliest (that is, when REM latency was shortest). Because of the relative ease of measurement, longitudinal studies of the temperature rhythm are possible in individual patients; these are of interest because the patient can serve as his own control and the process of change in circadian phase can be correlated with change in the clinical state. Pflug[80] studied a patient with recurrent depressions for over a year. Marked advances in the phase-position of the circadian temperature rhythm were associated with depressive episodes. Both the San Diego group[4,54] and the authors' group[117] have studied longitudinal changes in temperature rhythms in rapidly cycling manic-depressive patients. In five patients, the point of maximal phase-advance of the circadian temperature rhythm coincided with the switch into the depressive phase of the mood cycles (see figure 3–5). In summary, a small number of preliminary cross-sectional and longitudinal studies of the circadian temperature rhythm tend to support the idea that the early timing of REM sleep in depression is related to a phase-advance of a circadian oscillator.

Phase-Advance of the Circadian Cortisol Rhythm

In 1967, Doig and colleagues[24] noted that the 24-hour pattern of secretion of cortisol was "shifted to the left," that is, phase-advanced in depressives. Subsequent studies of fairly large numbers of patients by Fullerton and associates,[32,33] Conroy and Mills,[14] and Yamaguchi and colleagues[124] have confirmed this finding. (Sachar and coworkers,[92] however, found no shift in the timing of the rhythm.) In the Fullerton study, the degree of phase-advance of the cortisol rhythm was correlated with the severity of the depression. The change in timing of the cortisol rhythm in depressives is accompanied by a change in its waveform. The nadir of the rhythm as well as the time of the daily upsurge in secretion are shifted earlier; the peak of secretion, however, continues to occur near dawn, as in normals. A possible explanation for this change in waveform is that the early hours of sleep suppress cortisol secretion—a masking effect. In other words, the "true" maximum of the circadian rhythm of cortisol excretion may be blunted and obscured by the effects of sleep. An analogus change in waveform can be seen when the cortisol rhythm becomes phase-advanced relative to sleep in normal subjects during free-running experiments[120] (see figure 3–6).

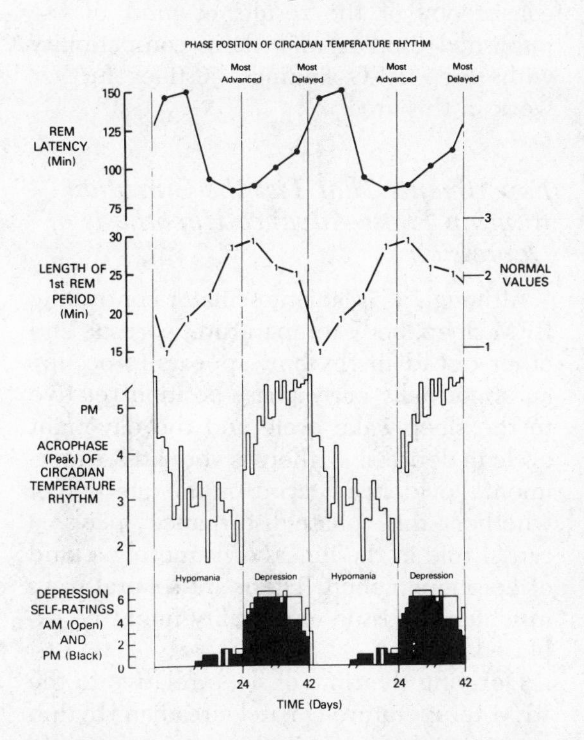

FIGURE 3–5. Phase-advance of circadian temperature rhythm is associated with shortest REM latency (time from sleep onset to REM onset) and longest first REM episodes at switch into depression in a rapidly cycling manic-depressive. Points shown are medians of data obtained in five manic-depressive cycles studied longitudinally.

In most studies of depression there is also evidence of cortisol hypersecretion in depression. Increased activation of the adrenocortical system is also reflected in a tendency of depressives to escape dexamethasone suppression of cortisol secretion.[6,10,96] It is possible that this increase in cortisol secretion in depressives is related to the phase-advance of the circadian rhythm relative to sleep.[15] Analogous shifts in some of the free-running subjects studied by Weitzman and colleagues[120] (see figure 3–6) and by Wever[121] are also associated with increased levels of cortisol secretion. Much more research on the interaction of sleep (or darkness) and the cortisol rhythm is necessary before this explanation of cortisol hypersecretion can be taken seriously. As with temperature, the shift in timing of the daily pattern of cortisol secretion supports a circadian rhythm phase-advance hypothesis of REM sleep changes in depression.

Phase-Advance of Neurotransmitter Metabolite Circadian Rhythms

Alterations of neurotransmitter metabolism have been implicated in the pathophysiology of affective disorders. Circadian rhythms of neurotransmitter metabolites in

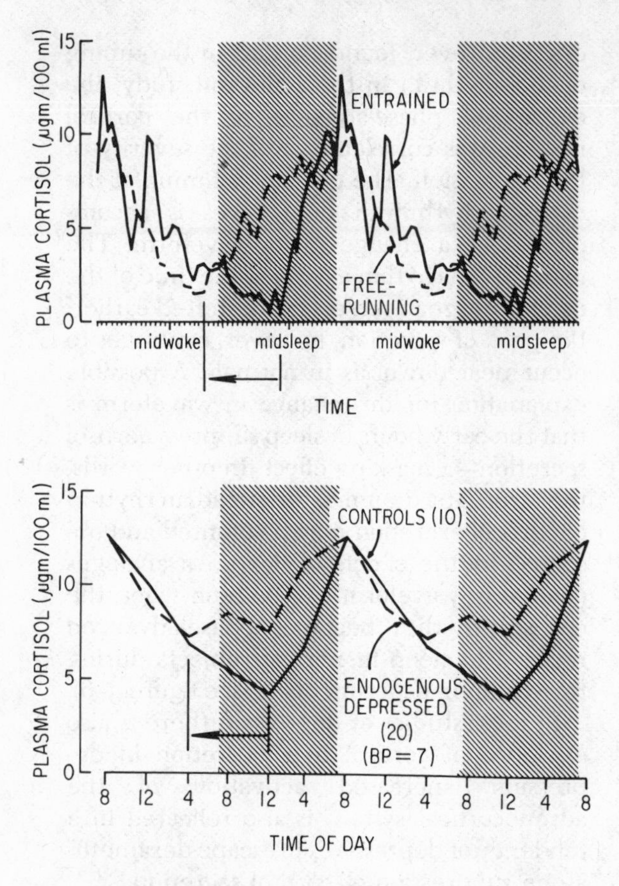

FIGURE 3–6. Change in waveform of circadian rhythm of cortisol secretion associated with its internal phase-advance relative to sleep in a free-running normal subject (top) and in a group of depressives compared with controls (below).

Weitzman, C.D., Czeisler, C.A., and Moore-Ende, M.C. "Sleep-Wake, Neuroendocrine and Body Temperature Circadian Rhythms Under Entrained and Non-Entrained (Free-Running) Conditions in Man," in Suda, M., Hayaishi, O., and Nakagawa, H., eds., *Biological Rhythms and Their Central Mechanism,* Amsterdam: Elsevier/North Holland, 1979, pp. 199–227.

depression are therefore of special interest. Riederer and colleagues[87] studied urinary metabolites of serotonin (5HIAA), dopamine (HVA), and norepinephrine (VMA, VA). An analysis of their results shows that three of the four rhythms are phase-advanced in depressives compared with controls.

The authors'[116] group studied urinary MHPG, a norepinephrine metabolite partially of central origin, and found the circadian rhythm to be approximately 3 hours phase-advanced in depressed manic-depressive patients compared with controls.

Over the years a variety of circadian rhythms have been studied. Taken together, there is a surprising consistency in nearly all of the studies: The phase position of circadian rhythms is abnormally advanced in depression (see figure 3–3).

One's confidence in the validity of this conclusion, however, is limited by methodological deficiencies in the majority of the studies reviewed. No study in which depressed patients were compared with controls adequately controlled for all the possible effects of age, sex, drugs, or experimental conditions. In a number of studies it is not possible to determine even if these factors were examined. Thus, although the research finding of a circadian rhythm phase-advance is consistent with the clinical observation of early morning awakening, our conclusion about the advanced phase-position of circadian rhythms in depressives must remain tentative until additional, better controlled studies are carried out. The consistency of the results of most of the published studies and their compatibility with sleep EEG findings, justifies further work in this area.

Experiments That Test the Circadian Rhythm Phase-Advance Hypothesis of Depression

Although a circadian oscillator controlling REM sleep, body temperature, cortisol, and other circadian rhythms appears to occupy an abnormally early phase-position relative to the sleep-wake cycle and the day-night cycle in depression, there is very little experimental evidence bearing on the question of whether this phase-disturbance plays a causal role in the illness or is merely a kind of epiphenomenon. There are several ways in which the issue of causality might be explored.

Delaying the time of sleep relative to the REM-temperature-cortisol circadian rhythm (causing an advance in the latter relative to the former) would create the depressive-type phase-disturbance and might be expected to precipitate depression in a predisposed individual. Such an experiment may have been unwittingly carried out. Rockwell and coworkers[88] have reported a retrospective analysis of a phase-shift experiment in which four subjects' sleep-wake schedule was delayed 12 hours. In one subject, the circadian temperature rhythm was somewhat phase-advanced relative to the other subjects (evidence of a predisposition?), and failed to adapt to the shifted schedule and instead remained markedly advanced relative to it. This depressive type of phase-disturbance persisted throughout the remainder of the experiment. Two weeks after the experiment the subject committed suicide. Since the psychiatric evaluation was conducted after the fact, it is difficult to know whether the subject was clinically depressed and what role, if any, the phase-shift played in his suicide.

In a person who is already depressed, advancing the time of sleep relative to the REM-temperature-cortisol circadian rhythm (causing a delay in the latter relative to the former) would undo the depressive phase disturbance and might be expected to induce a remission. The authors' group conducted such an experiment with a depressed manic-depressive woman who had a history of prolonged stable depressions.[112] On two separate occasions, when her sleep period was advanced six hours earlier than its usual 11 P.M. to 7 A.M. time, so that she was sleeping from 5 P.M. to 1 A.M. (or 11 A.M. to 7 P.M. after the second shift), she experienced a rapid and complete remission that lasted for almost two weeks. She eventually relapsed each time apparently because the REM-temperature-cortisol circadian rhythm gradually adjusted to the shifted schedule and reestablished the preexisting depressive-type phase disturbance (see figure 3–7). The efficacy of the procedure depended, therefore, on a kind of therapeutic jet lag. Two other patients showed partial and less dramatic responses to the procedure.

Another type of intervention, which has been much more extensively used with depressed patients, may be related to the sleep-wake cycle phase-advance experiment just described. Several studies have demonstrated that waking depressives several hours earlier than usual (for example, 1:30 A.M.) has an antidepressant effect in the majority of cases.[95] In this case, the time of waking is advanced without a corresponding shift in the time of going to sleep. Of course, this procedure results in partial sleep deprivation; however, it is not necessarily sleep deprivation *per se* that induces remission, since sleep deprivation in the first half of the night has no antidepressant effect (see figure 3–8). Total sleep deprivation also induces transient remissions and has been studied much more extensively;[79] its efficacy may also depend on the patient's being awake in the second half of the night (see figure 3–9). Although these procedures confound the variables of phase-shifting and sleep deprivation, they are consistent with a causal role for phase-advanced circadian rhythms in depression.

If being awake in the second half of the night treats depression, then being asleep in the second half of the night presumably sustains it. This apparent interaction of sleep with a specific phase of the circadian system could be a mechanism through which the phase-advance of patients' circadian rhythms could trigger depressive episodes (provided that the causal implications of the hypothesis prove to be correct). In depressives, a phase of the REM-temperature-cortisol circadian rhythm, which is normally associated with the first hours of waking, is advanced into the last hours of sleep (see figure 3–8). Besides the therapeutic efficacy of partial sleep deprivation in the second half of the night, there are other clinical observations that support the idea that this particular consequence of the depressive phase-advance is pathogenic. First, switches into and out of depression tend to occur most often during the night,* that is, near this critical circadian phase. Second, severity of depression is greatest just after and least just before this phase each day (this is just another way of describing depressives' well-known diurnal variation in

*See references 7, 8, 36, 46, 77, 89, 97, and 98.

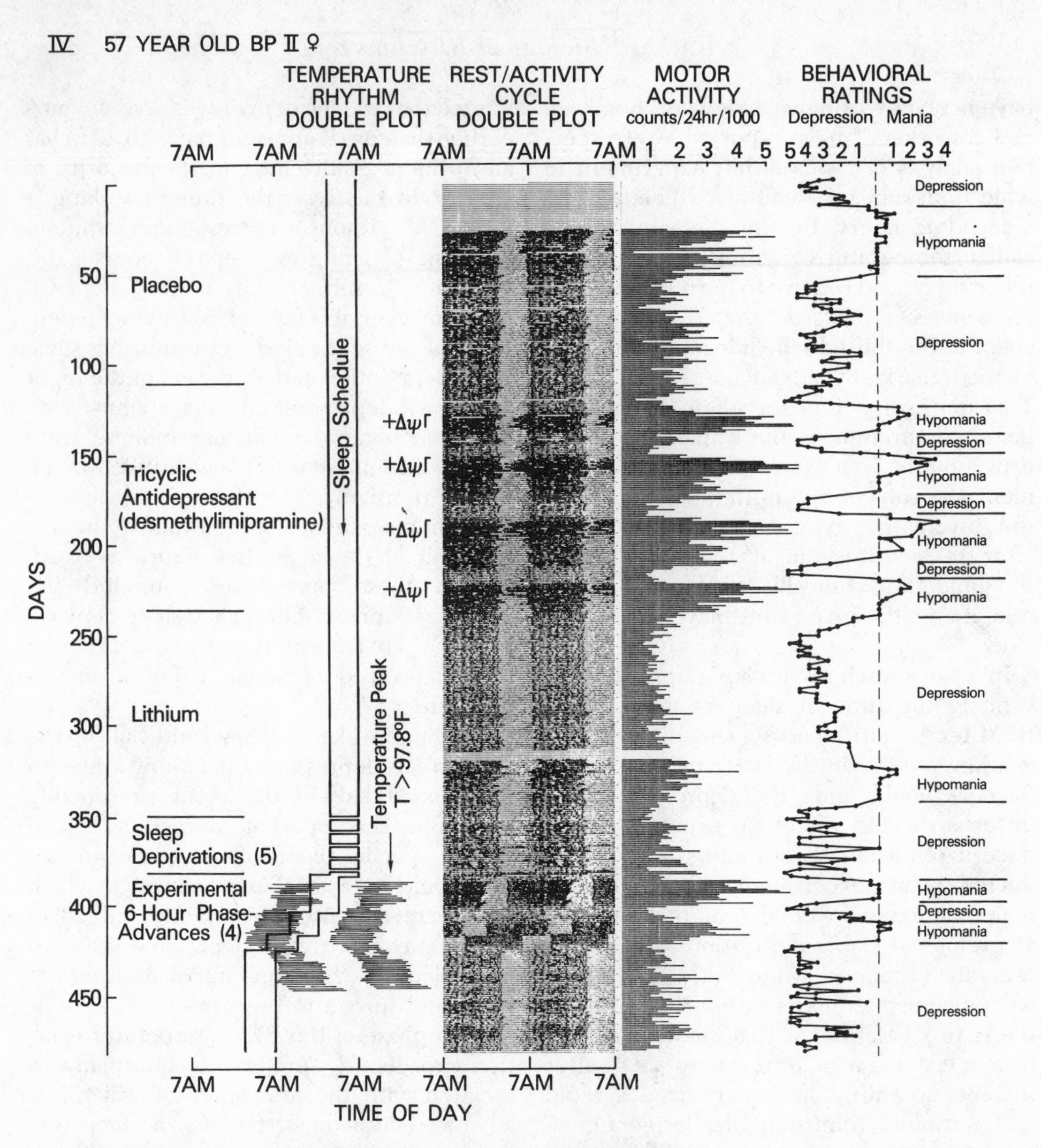

FIGURE 3–7. Longitudinal record of drug and sleep schedule, motor activity, and nurses' behavioral ratings during a phase-shift experiment and the year preceding it. Motor activity was recorded with a small electronic accelerometer worn on the wrist. Movement counts per 15-minute sampling interval were recorded in solid-state memory and later retrieved and analyzed by computer. Activity data are displayed here as actograms similar to those used in animal circadian rhythm studies. Each day's data are plotted as a histogram along a horizontal line beginning at 7 A.M.; consecutive days' data are plotted in sequence beneath each other; the entire display is double-plotted to facilitate visual inspection of changes in the rest-activity cycle. Activity is higher (darker) in hypomania or mania and lower (lighter) in day. Total motor activity (counts per 24 hours divided by 1000) paralleled the clinical state (periods of depression (D) are characterized by low activity). Desmethylimipramine induced rapid cycling between depression and hypomania (H). Drug-induced switches into hypomania are associated with increased motor activity as well as advances of its daily onset time (+ Δψ). Clinical remission induced by 6-hour phase advances, reflected in increased activity, mimics the drug effect. The sleep schedule, which was normal ward routine (11 P.M. to 7 A.M.) for the first year, was interrupted by five weekly deprivations of a single night's sleep (each inducing a day's transient remission), and then followed by the experimental 6-hour phase advances of the sleep schedule. At this time, temperature measurements were taken every 2 hours during waking. The times of day when the longitudinally smoothed temperature curve exceeded 97.8°F are indicated by horizontal lines. The temperature maximum advanced with the first two sleep schedule advances (associated with remission) but broke away after the third phase advance and slowed (associated with unremitting depression).

SOURCE: Wehr, T.A., et al., "Phase Advance of the Sleep-Wake Cycle as an Antidepressant," *Science*, 206 (1979): 710–713.

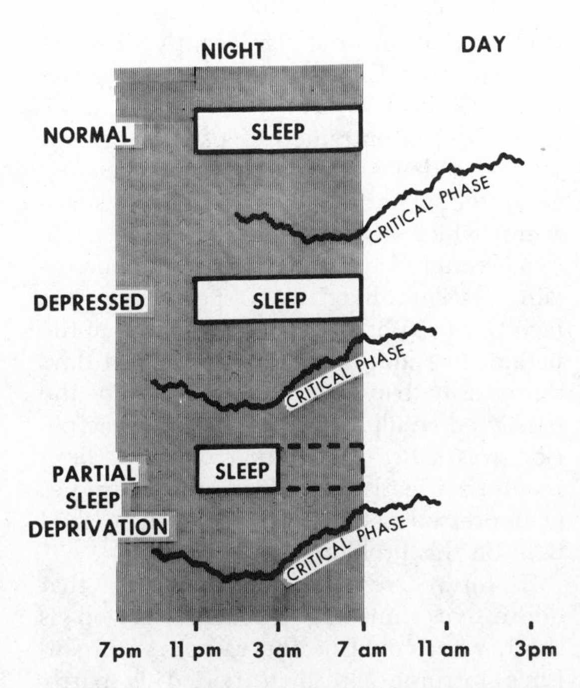

FIGURE 3–8. Hypothesis: Depression occurs in susceptible individuals when sleep interacts with a sleep-sensitive phase of the circadian temperature rhythm that is normally associated with the first hours of waking but becomes advanced into the last hours of sleep. Partial sleep deprivation in the second, but not the first, half of the night induces transient depressive remission. Depression severity is worst just after the sleep-sensitive phase (the classical pattern of diurnal variation in depressive mood) and switches into depression tend to occur during the sensitive phase.

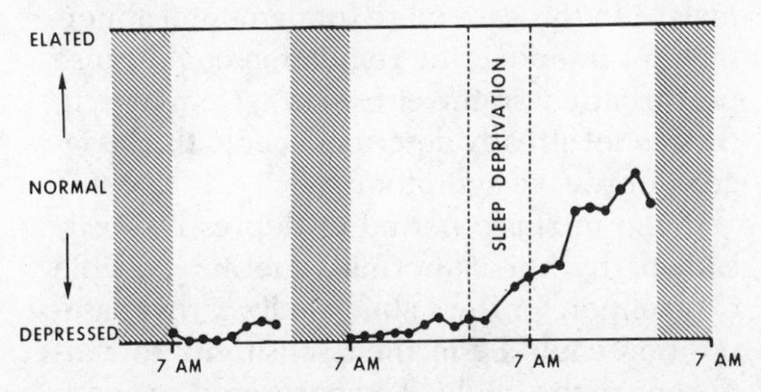

FIGURE 3–9. A depressed woman's mood self-ratings show the typical pattern of depressed patients response to total sleep-deprivation therapy. Switch out of depression occurs during the middle of the night.

mood).* The idea that depression thrives in the dark hours of the morning is not a new one and has been previously discussed by Bunney and associates[7] and Papoušek.[77] The concept of a critical circadian phase that interacts with sleep and wakefulness is similar to photoperiodic mechanisms in which a critical circadian phase interacts with darkness and light. In fact, darkness is a confounding variable in the procedures involving sleep.

Advancing the time of the sleep period relative to the REM-temperature-cortisol circadian rhythm is one of two possible ways of correcting the depressive phase disorder. The other approach would be to delay the phase-position of the REM-temperature-cortisol circadian rhythm relative to the sleep period. The controlling oscillator of this rhythm, it will be recalled, is by far the stronger oscillator and is therefore less easily phase-shifted than the controlling oscillator of the sleep-wake cycle (this difference is the basis of jet lag). As with jet lag, correction of the phase-position of the REM-temperature-cortisol rhythm might require one or two weeks or longer.

The authors' group[115] and others[28,123] have conducted several types of animal experiments that indicate that antidepressants such as monoamine oxidase inhibitors, tricyclic antidepressants, and lithium delay the phase-position of various circadian rhythms and may do so by lengthening the intrinsic period of their driving oscillator. All of these experiments are preliminary; they may not be specific for antidepressant drugs and could be interpreted in other ways. The authors are unaware of any convincing findings about the effects of psychoactive drugs on the human circadian system. Nevertheless, the results of animal studies nearly all point in the direction required of an antidepressant drug by a phase-advance hypothesis of depression.

Selective REM-sleep deprivation has been reported by Vogel[107] to be as effective an antidepressant as imipramine; onset of its therapeutic effects also has a similar time course to that of the drug. In responders, REM sleep deprivation tends to correct their abnormally early temporal distribution of REM sleep. If this normalization of the temporal distribution of REM sleep is due to a phase-delay in the circadian rhythm of REM

*See references 29, 45, 68, 109, and 110.

sleep propensity, then REM sleep deprivation may act on some phase-control mechanism in the circadian system. Since REM sleep is ordinarily maximal near dawn (see figure 3–1), and involves rapid scanninglike movements of the eye, and since light is an important *zeitgeber* (even in humans), it is interesting to speculate about a possible dawn "light sampling" function for REM. In fact, there is some experimental evidence that entrainment of certain circadian rhythms is impaired in animals that are deprived of REM by sectioning of the extraocular muscles.[99]

ANIMAL MODELS OF THE DEPRESSIVE CIRCADIAN PHASE-DISTURBANCE

Under certain conditions there are similarities between circadian rhythms in SCN-lesioned animals and depressed patients. Halberg and associates[43] studied the effects of bilateral lesions of the SCN on the temperature rhythm in rodents living on a standard 24-hour light-dark cycle. An apparent circadian rhythm in temperature persisted in spite of the lesions, although the amplitude was lower. As in depression, the phase-position of the rhythm in lesioned animals was considerably earlier than in controls.

Etiology of the Depressive Circadian Phase-Disturbance

The cause of the abnormally early phase-position of circadian rhythms in depression is presently unknown. One possibility is that the intrinsic period of the circadian rhythms is abnormally short. As indicated, a circadian rhythm that exhibits an unusually short period in free-running conditions will adopt an unusually early phase-position when entrained to the day-night cycle. This abnormality could only be demonstrated by placing depressives in temporal isolation conditions. In the only study of this type known to the authors,[23] a patient with a 48-hour depressive cycle exhibited a free-running period in the cortisol and temperature rhythm that was not significantly different from 24 hours. Either his intrinsic period was unusually short (as predicted by the hypothesis) or he was entrained to a subtle 24-hour *zeitgeber* that somehow was not excluded from the experiment. In another experiment, which may be relevant to the problem, Jenner,[50] together with a 48-hour cycling patient, lived in a special isolation facility on a 21-hour "day." He found that the patient was able to adapt to the short days more easily than he could. This would be the predicted result if the patient's intrinsic period was shorter than that of the experimenter's. Clearly, more free-running studies of depressed patients are required to shed light on this problem.

If future research demonstrates that depressives' intrinsic circadian period is short, what could be the cause? An organism's intrinsic circadian period is partly genetically determined; therefore it might be possible to selectively breed long or short intrinsic circadian periods into certain experimental animals. Of course, a genetic factor is known to be of major importance in affective illness. Another possibility is that endocrine disturbances alter the intrinsic period. For example, estrogen shortens the intrinsic circadian period of certain experimental animals.[72] In this case more fundamental abnormalities in endocrine regulation could cause perturbations in the circadian system that, in the model already described, could then trigger depressive symptoms.

If the intrinsic period of depressives' circadian rhythms is normal, another possible explanation for their abnormally early phase-position could lie in their sensitivity to *zeitgebers,* such as light. The perceived strength of a *zeitgeber* can alter the phase-position of a circadian rhythm entrained to it. In the case of an organism whose intrinsic circadian period is longer than 24 hours, the stronger the *zeitgeber* is, the earlier is the phase-position of the circadian rhythm entrained to it. In a hypersensitive organism the perceived strength of the *zeitgeber* would be abnormally great and the phase-position of circadian rhythms entrained to it would be abnormally early. This hypersensitivity hypothesis could be investigated by testing depressives' responses to stimuli that have the properties of time cues (for example, light).[17,62]

In summary, a reasonably large number of clinical studies have established that a group of circadian rhythms, especially those controlled by the strong circadian oscillator, are abnormally phase-advanced relative to the day-night cycle and the sleep-wake cycle in depressives. A small amount of experimental evidence supports the hypothesis that the phase-disturbance plays a causal role in depressive pathophysiology, and that it may act through a pathogenic mechanism involving the interaction of sleep and waking with a critical circadian phase associated with early morning. The cause of the phase-disturbance is unknown; experiments that might shed light on the problem have been outlined.

¶ Uncoupling of Circadian Oscillators in Mania

Severe disruption of sleep is as characteristic of mania as it is of depression. In contrast to the depressive, however, the manic actually seems to require less sleep. Some nights, manics sleep not at all and yet remain alert and energetic. The authors' group has longitudinally monitored the rest-activity (sleep-wake) cycle in sixteen patients who frequently switched into and out of the manic phase of their illness. With this type of patient it was possible to study manic episodes prospectively. The principal finding was that the majority of patients experienced one or more alternate nights of total insomnia when they switched into mania. When many of these same patients were sleep-deprived for one night during depressive episodes, they experienced transient, or in a few cases sustained, switches into mania or hypomania. Their responses to sleep-deprivation suggest that their nights of spontaneous total insomnia may be causally important in the process responsible for spontaneous switches into mania. In fact, the switch process may depend on the interaction of sleep and wakefulness with a critical circadian phase, as previously discussed in relation to depression.

When patients experienced alternate nights of total insomnia at the beginning of mania, the period of their sleep-wake cycle was, in effect, lengthened to approximately 48 hours (see figure 3–2). These double-length sleep-wake cycles may reflect a corresponding lengthening of the period of their controlling oscillator, a lengthening sufficient to cause it temporarily to escape from its normal one-to-one mode of entrainment to the 24-hour day-night cycle into a one-to-two mode (the secondary mode of entrainment). In the discussion of the normal human circadian system, it was noted that the controlling oscillator of the sleep-wake cycle is a weak oscillator whose period is labile and may lengthen in free-running conditions, escaping from one-to-one coupling with the stronger oscillator, which controls the REM-temperature-cortisol circadian rhythm. In some subjects the period of the sleep-wake cycle may lengthen to such an extent (fifty hours) that it can temporarily recouple with the stronger oscillator in a one-to-two mode[12,16,121] (see figure 3–2). These subjects, like manics, remain energetic and alert without sleep for thirty to forty hours during each sleep-wake cycle. On the basis of these known characteristics of the human circadian system in free-running conditions, it is proposed that the period of the controlling oscillator of the sleep-wake cycle in mania is abnormally long, so that it escapes from its normal mode of entrainment.

If this interpretation is correct and is causally important, then any procedure or agent that lengthens the intrinsic period of circadian rhythms would tend to precipitate mania in susceptible individuals. As noted, evidence from animal studies indicates that some antidepressant drugs may act by lengthening the intrinsic circadian period. By the same mechanism these drugs might be expected to precipitate mania. In fact, antidepressants do tend to precipitate mania in depressed manic-depressives.

In experimental animals, estrogen withdrawal also lengthens the intrinsic circadian period. In this regard, it may be relevant that affective episodes, especially manias, occur more frequently in the postpartum period.[5,86,102]

Animal Models of Circadian Rhythm Disturbances in Mania

Using behavioral stress with primates, Stroebel[100] was able to induce dramatic changes in the pattern of their circadian temperature rhythm. In a subgroup of animals a prominent 48-hour rhythmic component emerged and was superimposed on the basic 24-hour rhythm. This alteration in the temperature rhythm is not unlike that which occurs in free-running human subjects whose sleep-wake cycles reach double length; because of masking effects alternate temperature peaks are blunted by sleep and the intervening peaks are reinforced by wakefulness. The result of this interaction is a waveform with a fundamental near 24-hour component combined with a near 48-hour component. It is not stated whether the 48-hour component was associated with double-length sleep-wake cycles in the primates. The animals developed these patterns while living on a normal light-dark cycle after they were deprived of an important object in their environment. From a psychiatric point of view, the experiment is interesting because it indicates that manialike circadian rhythm disturbances can be precipitated by loss-related stresses.

Rapid Manic-Depressive Cycles as a Beat Phenomenon of Dissociated Circadian Rhythms

In a subgroup of manic-depressive patients who were studied longitudinally,[4,54] it was found that circadian rhythms in temperature and other physiological and behavioral variables were no longer strictly entrained to external time cues, but were free-running with a period *shorter* than 24 hours, and gradually beat in and out of phase with the day-night cycle in such a way, apparently, as to generate the long-term manic-depressive cycles. It is unknown whether Kripke's[54] patients exhibited double-length sleep-wake cycles when they switched into the manic phase of their mood cycles. His data were the first to directly support Halberg's[41] hypothesis that periodic psychiatric disorders might arise as a beat phenomenon of two internally desynchronized circadian subsystems.

In one study,[117] most of the patients who had rapid (one- to six-week) manic-depressive cycles experienced one or more of the double-length sleep-wake cycles at the onset of each manic phase of their mood cycle. It is conceivable that these recurring escapes of the sleep-wake cycle from its primary (one to one) mode to its secondary (one to two) mode of coupling into the day-night cycle and to other circadian rhythms could result from its driving oscillator having an overly long intrinsic period. In this case the manic-depressive cycles could be regarded as a kind of beat phenomenon between the driving oscillator of the sleep-wake cycle and the day-night cycle and other circadian rhythms. Because this oscillator is relatively weak, its oscillations remain relatively well coordinated with the day-night cycle and other circadian rhythms, so that the dissociation of its oscillations is expressed only in the periodic 24-hour phase-jumps associated with the double-length sleep-wake cycles.

It was proposed earlier that antidepressant drugs promote a slowing of the intrinsic rhythm of circadian oscillators, which leads to lengthening of the period of the sleep-wake cycle and its temporary escape from the primary mode of entrainment in mania. A possible endpoint of this drug effect could be a process of frequently recurring escapes and double-length sleep-wake cycles. In this regard, it was observed in a subgroup of manic-depressive patients (all women) that maintenance treatment with antidepressant drugs does in fact induce rapid (three- to six-week) cycling between mania and depression.[56,113] When the drugs are withdrawn, the rapid-cycling stops. Most of these patients experience 48-hour sleep-wake cycles at the beginning of each manic phase. Thus, a drug-induced slowing of the intrinsic rhythm of circadian oscillators, leading to more frequent escapes from the primary mode of entrainment, could be a mechanism underlying the drug-induced rapid manic-depressive cycles. An analogous phenomenon was observed in experimental animals chronically treated with antidepressants.[123] In some cases, the drugs appear to promote dissociation and relative coordination of two oscillatory components of the activity-rest cycle. Although the two components remain synchronized with one another most of the time, the slower component periodically lengthens dramatically and temporarily escapes from one-to-one coupling into the other component. In this way the two components beat in and out of phase every three weeks or so, in a manner similar to the rapidly cycling manic-depressives (see figure 3–10).

¶ Conclusion

This chapter has described the physiology and neuroanatomy of the human circadian system. It has presented evidence from many different sources that disturbances in the circadian system may play an important role in the pathophysiology of affective illness. It is hypothesized that a driving oscillator of the circadian rhythms of REM sleep, body temperature, and cortisol secretion is abnormally phase-advanced in depression, and that in mania the driving oscillator of the sleep-wake cycle is abnormally slow and escapes from its primary mode of entrainment to the day-night cycle. Animal experiments provide only preliminary evidence that antidepressant drugs, by slowing the intrinsic rhythm of circadian oscillators, may counteract depressives' abnormal phase-advance and promote escape from entrainment in mania. Many questions await further study. For example, lithium slows circadian oscillators[28] and treats depression, but unlike tricy-

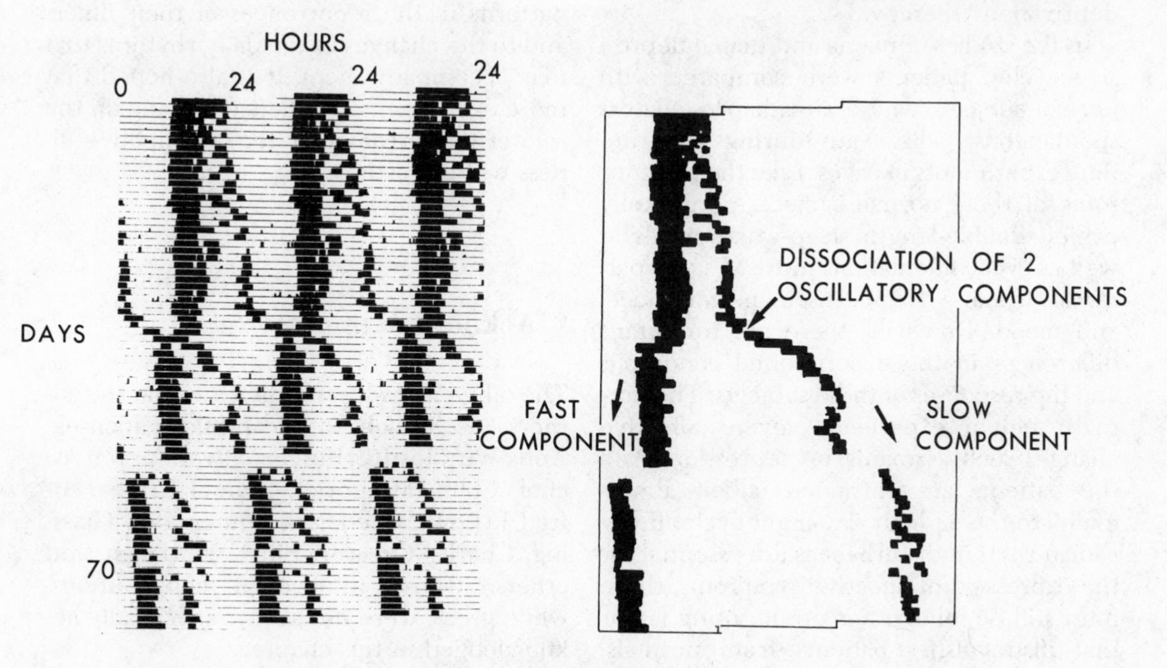

FIGURE 3–10. Uncoupling and relative coordination of the oscillations of two components of a hamster's activity-rest cycle during free-running in isolation from external time cues and after treatment with imipramine. Note three-week cycle of uncouplings. Imipramine sometimes induces three-week manic-depressive cycles in depressed bipolar patients.

clics and monoamine oxidase inhibitors it does not precipitate mania. The possible circadian basis of antidepressant drug action must be further investigated in order to determine whether the circadian rhythm effects are specific for clinically active drugs. Several arguments have been given supporting a hypothesis that the interaction of sleep with a critical circadian phase is a pathogenic mechanism in depression. This hypothesis simply integrates the depressive circadian rhythm phase-advance with the observation that sleep deprivation is an antidepressant. In future depression research, it may be desirable to focus on endocrine (or other) changes that occur during the last hours of sleep (or sleep deprivation) as well as on photoperiodic mechanisms related to the perception of dawn light. The relevance of sleep-deprivation experiments to the pathophysiology of manic-depressive illness is underscored by the fact that 48-hour sleep-wake cycles often occur at the switch out of depression into mania and may result in a kind of spontaneous, endogenous sleep-deprivation "therapy."

In the studies of mania and manic-depressive cycles, patients were compared with normal subjects whose circadian oscillators spontaneously dissociate during free-running experiments in caves. Like the patients, some of these normal subjects periodically exhibit double-length sleep-wake cycles as well as cyclic fluctuations in REM sleep patterns, urinary free-cortisol, performance, and mood. However, there are important differences in the experimental conditions and the responses of these subjects. The normals seldom experience severe affective changes such as mania or depression. And the patients are entrained (albeit abnormally) to the 24-hour day-night cycle. If circadian rhythm disturbances are essential for the expression of affective symptoms, there must still be some other predisposing factor that distinguishes patients from normals. Perhaps the relative chronicity of the rhythm disturbances in patients is important. Or perhaps the interaction of the external day-night cycle with the rhythm disturbances is the critical factor altering their behavior. Although circadian rhythm disturbances may prove to be important pathogenic mechanisms in affective illness, their fundamental cause may lie outside of the circadian system. For example, endocrine disturbances could alter the behavior of an otherwise normal circadian pacemaker.

A circadian theory of affective illness is attractive because it can integrate clinical symptoms (such as early awakening and diurnal variation in mood), epidemiological features (such as seasonality and cyclicity), disturbances in REM sleep, and endocrine function. It also provides a possible way of understanding the effects of psychotropic drugs and procedures such as sleep deprivation. It suggests new animal models of depression, mania, and manic-depressive cycles. Most important, it could point the way to new treatment approaches designed to manipulate the circadian system directly.

For the present, the authors hope that the emerging findings in this area will stimulate clinicians to attend to patients' reports of the patterns in the recurrences of their illness and to the changes in the daily rhythms that may accompany them. It is also hoped that more experimental evidence bearing on the role of the circadian system in affective illness will be forthcoming.

¶ Acknowledgments

The diligent reader will discover that the authors owe much to Mechthild Papoušek, Anna Wirz-Justice, Rutger Wever, Jurgen Aschof, Colin Pittendrigh, Daniel Kripke, Alfred J. Lewy, Frederick Jenner, Rolf Gjessing, Charles Czeisler, Elliot Weitzman, and others, whose intellectual contributions, while great, were not always specifically acknowledged in this chapter.

The authors are also indebted to Eloise Orr, Carolyn Craig, and Marion Webster for their assistance in preparing the manuscript.

¶ Bibliography

1. ASCHOFF, J., GERECKE, U., and WEVER, R. "Desynchronization of Human Circadian Rhythms," *Japanese Journal of Physiology,* 17 (1967): 450–457.
2. ASCHOFF, J., et al. "Re-entrainment of Circadian Rhythms After Phase-Shifts of the Zeitgeber," *Chronobiologia,* 2 (1978): 23–78.
3. ASERINSKY, E., and KLEITMAN, N. "Regularly Occurring Periods of Eye Motility, and Concomitant Phenomena, During Sleep," *Science,* 118 (1953): 273–274.
4. ATKINSON, M., KRIPKE, D. F., and WOLF, S. R. "Autorhythmometry in Manic-Depressives," *Chronobiologia,* 2 (1975): 325–335.
5. BRATFOS, D., and HAUG, J. O. "Puerperal Mental Disorders in Manic-Depressive Females," *Acta Psychiatrica Scandinavica,* 42 (1966): 285–294.
6. BROWN, W. A., JOHNSTON, R., and MAYFIELD, D. "The 24-Hour Dexamethasone Suppression Test in a Clinical Setting: Relationship to Diagnosis, Symptoms and Response to Treatment," *American Journal of Psychiatry,* 136 (1979): 543–547.
7. BUNNEY, W. E., JR., et al. "The Switch Process in Manic-Depressive Illness," *Archives of General Psychiatry,* 27 (1972): 295–302.
8. BURTON, R. *The Anatomy of Melancholy,* vol. 1. New York: Dutton, 1961 (originally published in 1621).
9. CAHN, H. A., POLK, G. E., and HUSTON, P. E. "Age Comparison of Human Day-Night Physiological Differences," *Aerospace Medicine,* 39 (1968): 608–610.
10. CARROL, B. J., CURTIS, G. C., and MENDELS, J. "Neuroendocrine Regulation in Depression: II. Discrimination of Depressed from Nondepressed Patients," *Archives of General Psychiatry,* 33 (1976): 1051–1058.
11. CHERNIK, D. A., and MENDELS, J. "Sleep Reversal: Disturbances in Daytime Sleep Patterns," *Clinical Electroencephalography,* 5 (1974): 143–148.
12. CHOUVET, G., et al. "Periodicite bicircadienne du cycle veille-sommeil dans des conditions hors du temps," *Electroencephalography and Clinical Neurophysiology,* 3 (1974): 367–380.
13. COLBURN, T. R., et al. "An Ambulatory Activity Monitor with Solid-State Memory," *ISA Transactions,* 15 (1976): 149–154.
14. CONROY, R. T. W. L., HUGHES, B. D., and MILLS, J. N. "Circadian Rhythm of Plasma 11-Hydroxy Corticosteroid in Psychiatric Disorders," *British Medical Journal,* 3 (1968): 405–407.
15. CURTIS, G. C. "Psychosomatics and Chronobiology: Possible Implications of Neuroendocrine Rhythms," *Psychosomatic Medicine,* 34 (1972): 235–256.
16. CZEISLER, C. A. "Human Circadian Physiology: Internal Organization of Temperature, Sleep-Wake and Neuroendocrine Rhythms Monitored in an Environment Free of Time Cues," (Ph.D. diss. Stanford University, 1978).
17. ——— et al. "Entrainment of Human Circadian Rhythms by Light-Dark Cycles: A Reassessment," *Photochemistry and Photobiology,* in press.
18. CZEISLER, C. A., et al. "Timing of REM Sleep is Coupled to the Circadian Rhythm of Body Temperature in Man," *Science,* forthcoming.
19. DAAN, S., and PITTENDRIGH, C. S. "Functional Analysis of Circadian Pacemakers in Nocturnal Rodents. II. The Variability of Phase Response Curves," *Journal of Comparative Physiology,* 106 (1976): 253–266.
20. ——— "A Functional Analysis of Circadian Pacemakers in Nocturnal Rodents. III. Heavy Water and Constant Light: Homeostasis of Frequency?" *Journal of Comparative Physiology,* 106 (1976): 267–290.
21. DAAN, S., et al. "An Effect of Castration and Testosterone Replacement on a Circadian Pacemaker in Mice (Mus Musculus)," *Proceedings of the National Academy of Sciences, USA,* 72 (1975): 3744–3747.
22. DEMENT, W. C., and KLEITMAN, N. "Cyclic Variations in EEG During Sleep and Their Relation to Eye Movements, Body Motility, and Dreaming," *Electroencephalography and Clinical Neurophysiology,* 9 (1957): 673–690.
23. DOERR, P., et al. "Relationship Between Mood Changes and Adrenal Cortical Activity in a Patient with 48-Hour Unipolar-Depressive Cycles," *Journal of Affective Disorders,* 1 (1976): 93–104.
24. DOIG, R. J., et al. "Plasma Cortisol Levels in Depression," *British Journal of Psychiatry,* 112 (1966): 1263–1267.

25. EASTWOOD, M. R., and STIASNY, S. "Psychiatric Disorder, Hospital Admission, and Season," *Archives of General Psychiatry,* 35 (1978): 769–771.
26. ELITHORN, A., et al. "Observations on Some Diurnal Rhythms in Depressive Illness," *British Medical Journal,* 2 (1966): 1620–1623.
27. ELLIOTT, J. A., STETSON, M. H., and MENAKER, M. "Regulation of Testis Function in Golden Hamster: A Circadian Clock Measures Photoperiodic Time," *Science,* 178 (1972): 771–773.
28. ENGELMANN, W. "A Slowing Down of Circadian Rhythms by Lithium Ions," *Zeitschrift Für Naturforschung,* 28 (1973): 733–736.
29. FLECK, U., KRAEPELIN, E. "Über die Tagesschwankungen bei Manisch-Depressiven," *Kraepelin's Psychologische Arbeiten,* 7 (1922): 213–253.
30. FOSTER, F. G., and KUPFER, D. J. "Psychomotor Activity as a Correlate of Depression and Sleep in Acutely Disturbed Psychiatric Inpatients," *American Journal of Psychiatry,* 132 (1975): 928–931.
31. FULLER, C. A., SULZMAN, F. M., and MOORE-EDE, M. C. "Thermoregulation Is Impaired in an Environment Without Circadian Time Cues," *Science,* 199 (1978): 794–796.
32. FULLERTON, D. T., et al. "Circadian Rhythm of Adrenal Cortical Activity in Depression: I. A Comparison of Depressed Patients with Normal Subjects," *Archives of General Psychiatry,* 19 (1968): 674–681.
33. ———. "Circadian Rhythm of Adrenal Cortical Activity in Depression: II. A Comparison of Types in Depression," *Archives of General Psychiatry,* 19 (1968): 682–688.
34. GELENBERG, A. J., et al. "Recurrent Unipolar Depressions with a 48-Hour Cycle: Report of a Case," *British Journal of Psychiatry,* 133 (1978): 123–129.
35. GEORGI, F. "Psychophysiologische korrelationen. Psychiatrische Probleme im Lichte der Rhythmus forschung," *Schweizerische Medizinische Wochenschrift,* 49 (1947): 1267–1280.
36. GILLIN, J. C., et al. "An EEG Sleep Study of a Bipolar (Manic-Depressive) Patient with a Nocturnal Switch Process," *Biological Psychiatry,* 12 (1977): 711–718.
37. GILLIN, J. C., et al. "Successful Separation of Depressed, Normal, and Insomniac Subjects by Sleep EEG Data," *Archives of General Psychiatry,* 36 (1979): 85–90.
38. GOODWIN, J. C., et al. "Renal Rhythms in a Patient with a 48-Hour Cycle of Psychosis During a Period of Life on an Abnormal Time Routine," *Journal of Physiology* (London), 176 (1965): 16–17.
39. GRESHAM, S. C., AGNEW, W. F., and WILLIAMS, R. L. "The Sleep of Depressed Patients," *Archives of General Psychiatry,* 13 (1965): 503–507.
40. GROOS, G. A., and MASON, R. "The Visual Properties of Rat and Cat Suprachiasmatic Neurones," *Journal of Comparative Physiology,* 135 (1980): 349–356.
41. HALBERG, F. "Physiologic Considerations Underlying Rhythmometry, with Special Reference to Emotional Illness," in *Symposium Bel-Air III,* Geneva: Masson et Cie, 1968, pp. 73–126.
42. ———, VESTERGAARD, P., and SAKAI, M. "Rhythmometry on Urinary 17-Ketosteroid Excretion by Healthy Men and Women and Patients with Chronic Schizophrenia: Possible Chronopathology in Depressive Illness," *Archives d' Anatomie, d'Histologie, d'Embryologie Normale et Experimentale,* 51 (1968): 301–311.
43. HALBERG, F., et al. "Nomifensine Chronopharmacology, Schedule-Shifts and Circadian Temperature Rhythms in Di-Suprachiasmatically Lesioned Rats—Modeling Emotional Chronopathology and Chronotherapy," *Chronobiologia,* 6 (1979): 405–424.
44. HALBREICH, U., GRUNHAUS, L., and BEN-DAVID, M. "Twenty-four-hour Rhythms of Prolactin in Depressive Patients," *Archives of General Psychiatry,* 36 (1979): 1183–1186.
45. HALL, P., SPEAR, F. G., and STIRLAND, D. "Diurnal Variation of Subjective Mood in Depressive States," *Psychiatric Quarterly,* 38 (1964): 529–536.
46. HARTMANN, E. "Longitudinal Studies of Sleep and Dream Patterns in Manic-Depressive Patients," *Archives of General Psychiatry,* 19 (1968): 312–329.
47. HAUTY, G. T., and ADAMS, T. "Phase Shifts of the Human Circadian System and Performance Deficit During the Periods of Transition: I. East-West Flight. II. West-East Flight," *Aerospace Medicine,* 37 (1966): 668–674.

48. HERTRICH, O. "Beitrag zur Diagnostik und Differentialdiagnostik der leichteren depressiven Zustandsbilder," *Fortschritte Der Neurologie, Psychiatric Und Ihrer Grenzgebiete,* 30 (1962): 237–272.
49. JENNER, F. A., et al. "A Manic-Depressive Psychotic with a Persistent Forty-Eight-Hour Cycle," *British Journal of Psychiatry,* 113 (1967): 895–910.
50. JENNER, F. A., et al. "The Effect of an Altered Time Regime on Biological Rhythms in a 48-Hour Periodic Psychosis," *British Journal of Psychiatry,* 114 (1968): 215–224.
51. KISHIMOTO, H., et al. "The Biochemical Studies of Manic-Depressive Psychosis: I. The Circadian Rhythm of Plasma Tryptophan, Tyrosine and Cortisol, and Its Clinical Significance," *Yokohama Medical Bulletin,* 28 (1977): 23–38.
52. KNAPP, M. W., KEANE, P. M., and WRIGHT, J. G. "Circadian Rhythm of Plasma 11-Hydroxycorticosteroids in Depressive Illness, Congestive Heart Failure and Cushing's Syndrome," *British Medical Journal,* 2 (1967): 27–30.
53. KRIPKE, D. F., COOK, B., and LEWIS, O. F. "Sleep in Night Workers: EEG Sleep Recordings," *Psychophysiology,* 7 (1970): 377–384.
54. KRIPKE, D. F., et al. "Circadian Rhythm Disorders in Manic-Depressives," *Biological Psychiatry,* 13 (1978): 335–350.
55. KRIPKE, D. F., et al. "Circadian Rhythm Phases in Affective Illness," *Chronobiologia,* 6 (1979): 365–375.
56. KUKOPULOS, A., et al. "Course of the Manic-Depressive Cycle and Changes Caused by Treatments," *Pharmakopsychiatrie Neuro-Psychopharmakologie,* forthcoming.
57. KUPFER, D. J. "REM Latency: A Psychobiologic Marker for Primary Depressive Disease," *Biological Psychiatry,* 11 (1977): 159–174.
58. ———, and HENINGER, G. R. "REM Activity as a Correlate of Mood Changes Throughout the Night (EEG Sleep Patterns in a Patient with a 48-Hour Cyclic Mood Disorder)," *Archives of General Psychiatry,* 27 (1972): 368–373.
59. LAGOGUEY, M., and REINBERG, A. "Circannual Rhythms in Plasma LH, FSH and Testosterone and in the Sexual Activity of Healthy Young Parisian Males," *Journal of Physiology,* 257 (1976): 19–20.
60. LEVINE, H., et al. "Changes in Internal Timing of Heart Rate, Blood Pressure and Other Aspects of Physiology and Motor Performance in Presumably Healthy Subjects on Different Meal Schedules," *Proceedings of the 12th International Conference of the International Society for Chronobiology,* Il Ponte, Milan, 1977, pp. 139–148.
61. LEWIS, P. R., and LOBBAN, M. C. "Dissociation of Diurnal Rhythms in Human Subjects Living in Abnormal Time Routines," *Quarterly Journal of Experimental Physiology and Cognate Medical Sciences,* 42 (1957): 371–386.
62. LEWY, A. J., et al. "Light Suppresses Melatonin Secretion in Man," *Science,* forthcoming.
63. LEWY, A. J., et al. "Plasma Melatonin in Manic-Depressive Illness," forthcoming.
64. LOBBAN, M., et al. "Diurnal Rhythms of Electrolyte Excretion in Depressive Illness," *Nature,* 199 (1963): 667–669.
65. LUND, R., and SCHULZ, H. "The Relationship of Disturbed Sleep in Depression to an Early Minimum of the Circadian Temperature Rhythm," *Abstracts of the 12th CINP Congress,* Goteborg, Sweden, 1980, p. 233.
66. LYDIC, R., et al. "Structure Homologous to the Suprachiasmatic Nucleus in Human Brain," *Sleep,* 2 (1980): 355–362.
67. MARON, L., RECHTSCHAFFEN, A., and WOLPERT, E. A. "Sleep Cycle During Napping," *Archives of General Psychiatry,* 11 (1964): 503–598.
68. MIDDLEHOFF, H. D. "Tagesrhythmische Schwankungen bei endogenen Depressionen im symptomfreien Intervall und während der Phase," *Archiv Für Psychiatric und Nervenkrankheiten,* 209 (1967): 315–339.
69. MOODY, J. P., and ALLSOPP, M. N. E. "Circadian Rhythms of Water and Electrolyte Excretion in Manic-Depressive Psychosis," *British Journal of Psychiatry,* 115 (1969): 923–928.
70. MOORE, R. Y. "Central Neural Control of Circadian Rhythms," in GANONG, W. F., and MARTINI, L., eds., *Frontiers in Neuroendocrinology,* vol. 5. New York: Raven Press, 1978, pp. 185–206.
71. ———. Personal communication, 1980.
72. MORIN, L. P., FITZGERALD, F. M., and ZUCKER, I. "Estradiol Shortens the Period

of Hamster Circadian Rhythms," *Science,* 196 (1977): 305–307.

73. NIKITOPOULOU, G., and CRAMMER, J. I. "Change in Diurnal Temperature Rhythm in Manic-Depressive Illness," *British Medical Journal,* 1 (1976): 1311–1314.

74. NUMATA, Y., et al. "Plasma Cyclic AMP in Manic-Depressive Patients: Circadian Variation of Plasma cAMP and the Effect of Sleep Deprivation," *Abstracts of the 12th CINP Congress,* Goteborg, Sweden, 1980, p. 266.

75. PALMAI, G., and BLACKWELL, B. "The Diurnal Pattern of Salivary Flow in Normal and Depressed Patients," *British Journal of Psychiatry,* 111 (1965): 334–338.

76. PALMAI, G., et al. "Patterns of Salivary Flow in Depressive Illness and During Treatment," *British Journal of Psychiatry,* 113 (1967): 1297–1308.

77. PAPOUŠEK, M. "Chronobiologische Aspekte der Zyklothymie," *Fortschritte Der Neurologie, Psychiatrie Und Ihrer Grenzgebiete,* 43 (1975): 381–440.

78. PAYKEL, E. S., et al. "Life Events and Social Support in Puerperal Depression," *British Journal of Psychiatry,* 136 (1980): 339–346.

79. PFLUG, B., and TOLLE, R. "Disturbance of the 24-Hour Rhythm in Endogenous Depression and the Treatment of Endogenous Depression by Sleep Deprivation," *International Pharmacopsychiatry,* 6 (1971): 187–196.

80. PFLUG, B., ERIKSON, R., and JOHNSSON, A. "Depression and Daily Temperature: A Long-Term Study," *Acta Psychiatrica Scandinavica,* 54 (1976): 254–266.

81. PFLUG, B., JOHNSSON, A., and MARTIN, W. "Alterations in the Circadian Temperature Rhythm in Depressed Patients," *Abstracts of the 12th CINP Congress,* Goteborg, Sweden, 1980, p. 280.

82. PITTENDRIGH, C. S., and DAAN, S. "Circadian Oscillations in Rodents: A Systematic Increase of Their Frequency with Age," *Science,* 186 (1974): 548–550.

83. ———. "A Functional Analysis of Circadian Pacemakers in Nocturnal Rodents. I. The Stability and Lability of Spontaneous Frequency," *Journal of Comparative Physiology,* 106 (1976): 223–252.

84. ——— "A Functional Analysis of Circadian Pacemakers in Nocturnal Rode Rodents. IV. Entrainment: Pacemaker as a Clock," *Journal of Comparative Physiology,* 106 (1976): 291–331.

85. ——— "A Functional Analysis of Circadian Pacemakers in Nocturnal Rodents. V. Pacemaker Structure: A Clock for all Seasons," *Journal of Comparative Physiology,* 106 (1976): 333–355.

86. REICH, T., and WINOKUR, G. "Postpartum Psychoses in Patients with Manic-Depressive Disease," *Journal of Nervous and Mental Disease,* 151 (1970): 60–68.

87. RIEDERER, T., et al. "The Daily Rhythm of HVA, VMA, (VA) and 5-HIAA in Depression Syndrome," *Journal of Neural Transmission,* 35 (1974): 23–45.

88. ROCKWELL, D. A., et al. "Biologic Aspects of Suicide—Circadian Disorganization," *Journal of Nervous and Mental Disease,* 166 (1978): 851–858.

89. RUDOLF, G. A. E., and TOLLE, R. "Circadian Rhythm of Circulatory Functions in Depressives and on Sleep Deprivation," *International Pharmacopsychiatry,* 12 (1977): 174–183.

90. RUSAK, B., and ZUCKER, I. "Neural Regulation of Circadian Rhythms," *Physiological Reviews,* 59 (1979): 449–526.

91. SACHAR, E. J. "Twenty-four-hour Cortisol Secretory Patterns in Depressed and Manic Patients," in GISPEN, W. H., et al., eds., *Hormones, Homeostasis and the Brain (Progress in Brain Research, vol. 42),* Amsterdam: Elsevier, 1975, pp. 81–91.

92. ——— et al. "Disrupted 24-hour Patterns of Cortisol Secretion in Psychotic Depression," *Archives of General Psychiatry,* 28 (1973): 19–24.

93. SAKAI, M. "Diurnal Rhythm of 17-Ketosteroid and Diurnal Fluctuation of Depressive Affect," *Yokohama Medical Bulletin,* 11 (1960): 352–367.

94. SARAI, K., and KAYANO, M. "The Level and Diurnal Rhythm of Serum Serotonin in Manic-Depressive Patients," *Folia Psychiatrica et Neurologica Japonica,* 22 (1968): 271–281.

95. SCHILGEN, B., and TOLLE, R. "Partial Sleep Deprivation as Therapy for Depression," *Archives of General Psychiatry,* 37 (1980): 267–271.

96. SCHLESSER, M. A., WINOKUR, G., and SHERMAN, B. M. "Genetic Subtypes of Unipolar Primary Depressive Illness Distinguished by Hypothalamic-Pituitary-Adrenal Axis

Activity," *Lancet,* 1 (April 7, 1979): 739–741.

97. SITARAM, N., GILLIN, J. C., and BUNNEY, W. E., JR. "The Switch Process in Manic-Depressive Illness," *Acta Psychiatrica Scandinavica,* 58 (1978): 267–278.

98. ———. "Circadian Variation in the Time of 'Switch' of a Patient with 48-hour Manic-Depressive Cycles," *Biological Psychiatry,* 13 (1978): 567–574.

99. STEVENS, J. R., and LIVERMORE, A. "Disruption of the Circadian Rhythm of Excitability in the Mesolimbic Catecholamine System by Peripheral Deprivation of Lateral Eye Movements and REM," *Society of Neuroscience Abstracts,* 5 (1979): 347.

100. STROEBEL, C. F. "Biologic Rhythm Correlates of Disturbed Behavior in the Rhesus Monkey," in ROHLES, F. H., ed., *"Circadian Rhythms In Nonhuman Primates," Bibliotheca Primatologica,* 9 (1969), 91–105.

101. SWIFT, N., and ELITHORN, A. "An Experimental Approach to the Problem of Depression," *Proceedings of the 3rd World Conference on Psychiatry,* 2 (1962): 1372–1376.

102. TARGUM, S. D., DAVENPORT, Y. B., and WEBSTER, M. J. "Postpartum Mania in Bipolar Manic-Depressive Patients Withdrawn From Lithium Carbonate," *Journal of Nervous and Mental Disease,* 167 (1979): 572–574.

103. TAUB, J. M., and BERGER, R. S. "Acute Shifts in the Sleep-Wakefulness Cycle: Effects on Performance and Mood," *Psychosomatic Medicine,* 36 (1974): 164–173.

104. Timonen, S., FRANZAS, B., and WICHMANN, K. "Photosensibility of the Human Pituitary," *Annales Chirurgiac et Gynaceologiac Fenniae,* 53 (1964): 165–172.

105. VALSIK, J. A. "The Seasonal Rhythm of Menarche: A Review," *Human Biology,* 37 (1965): 75–90.

106. VAN CAUTER, E., and MENDLEWICZ, J. "24-Hour Dopamine-beta-hydroxylase Pattern: A Possible Biological Index of Manic-Depression," *Life Sciences,* 22 (1978): 147–156.

107. VOGEL, G. W., et al. "Improvement of Depression by REM Sleep Deprivation: New Findings and a Theory," *Archives of General Psychiatry,* 37 (1980): 247–253.

108. VON HOLST, E. "Die relative Koordination als Phanomen und als Methode zentralnervoser Funktionsanalyse," *Ergebnisse der Physiologie,* 42 (1939): 228–306.

109. WALDMANN, H. "Die Tagesschwankung in der Depression als rhythmisches Phanomen," *Fortschritte Der Neurologie, Psychiatrie Und Ihrer Grenzgebiete,* 40 (1972): 83–104.

110. ———. "Schlafdauer und psychopathologische Tagesrhythmik bei Depressiven," in Jovanovic, U. J., ed., *Die Natur des Schlafes,* Stuttgart: G. Fischer, 1973, pp. 184–187.

111. WALTER, S. D. "Seasonality of Mania: A Reappraisal," *British Journal of Psychiatry,* 131 (1977): 345–350.

112. WEHR, T. A., and GOODWIN, F. K. "Rapid Cycling in Manic-Depressives Induced by Tricyclic Antidepressants," *Archives of General Psychiatry,* 36 (1979): 555–559.

113. ———. "Tricyclics Modulate Frequency of Mood Cycles," *Chronobiologia,* 6 (1979), 377–385.

114. WEHR, T. A., MUSCETTOLA, G., and GOODWIN, F. K. "Urinary 3-Methoxy-4-Hydroxyphenylglycol Circadian Rhythm: Early Timing (Phase Advance) in Manic-Depressives Compared with Normal Subjects," *Archives of General Psychiatry,* 37 (1980): 257–263.

115. WEHR, T. A., et al. "Phase Advance of the Sleep-Wake Cycle as an Antidepressant," *Science,* 206 (1979): 710–713.

116. WEHR, T. A., et al. "Clorgyline Slows Circadian Rhythms (I)," *Abstracts of the 12th CINP Congress,* Goteborg, Sweden, June 1980, p. 353.

117. WEHR, T. A., et al. "Uncoupling of Circadian Oscillators in Manic-Depressive Illness," in press.

118. WEITZMAN, E. D., CZEISLER, C. A., and MOORE-EDE, M. C. "Sleep-Wake, Neuroendocrine and Body Temperature Circadian Rhythms Under Entrained and Non-Entrained (Free-Running) Conditions in Man," in SUDA, M., HAYAISHI, O., and NAKAGAWA, H., eds., *Biological Rhythms and Their Central Mechanism,* Amsterdam: Elsevier/North Holland, 1979, pp. 199–227.

119. WEITZMAN, E. D., et al. "Reversal of Sleep-Waking Cycle: Effect on Sleep Stage Pattern and Certain Neuroendocrine

Rhythms," *Transactions of the American Neurological Association,* 93 (1968): 153–157.

120. WEITZMAN, E. D., et al. "Effects of a Prolonged 3-Hour Sleep-Wake Cycle on Sleep Stage, Plasma Cortisol, Growth Hormone and Body Temperature in Man," *Journal of Clinical Endocrinology and Metabolism,* 38 (1974): 1018–1070.

121. WEVER, R. A. *The Circadian System of Man: Results of Experiments Under Temporal Isolation.* New York: Springer-Verlag, 1979.

122. WINOKUR, G., CLAYTON, P. J., and REICH, T. *Manic-Depressive Illness.* St. Louis: C. V. Mosby Co., 1969.

123. WIRZ-JUSTICE, A., et al. "Clorgyline Slows Circadian Rhythms (II)," *Abstracts of the 12th CINP Congress,* Goteborg, Sweden, 1980, p. 361.

124. YAMAGUCHI, N., MAEDA, K., and KUROMARU, S. "The Effects of Sleep Deprivation on the Circadian Rhythm of Plasma Cortisol Levels in Depressive Patients," *Folia Psychiatrica et Neurologica Japonica,* 32 (1978): 479–487.

125. YAMAOKA, S. "Participation of Limbic-Hypothalamic Structures in Circadian Rhythm of Slow Wave Sleep and Paradoxical Sleep in the Rat," *Brain Research,* 151 (1978): 255–268.

126. YAMASHITA, I., et al. "Neuroendocrinological Studies in Mental Disorders and Psychotropic Drugs: Part I: On the Circadian Rhythm of the Plasma Adrenocortical Hormone in Mental Patients and Methamphetamine- and Chlorpromazine-Treated Animals," *Folia Psychiatrica et Neurologica Japonica,* 23 (1969): 143–158.

127. ZATZ, M., and BROWNSTEIN, M. J. "Intraventricular Carbachol Mimics The Effects of Light on the Circadian Rhythm in the Rat Pineal Gland," *Science,* 203 (1979): 558–361.

128. ZIS, A. P., and GOODWIN, F. K. "Major Affective Disorder as a Recurrent Illness," *Archives of General Psychiatry,* 36 (1979): 835–839.

CHAPTER 4

GENETIC MODELS OF MENTAL ILLNESS

Raymond R. Crowe

¶ Introduction

When the long-standing "nature versus nurture" controversy in psychiatry was finally resolved by the adoption studies, demonstrating beyond a reasonable doubt the genetic predisposition to the major mental disorders, attention finally turned to the more productive question of how both genetic and environmental factors are involved in transmitting disease; that is, transmission models.

The earliest of these transmission models were Mendel's laws of inheritance, and early workers in psychiatric genetics naturally attempted to apply these models to the diseases they were studying. However, unlike many medical diseases that fit the Mendelian ratios with such precision that another explanation would be hard to imagine, none of the psychiatric disorders gave an acceptable fit. This led to models modifying Mendel's laws by postulating such variables as incomplete penetrance and additional genes. These explanations were never widely accepted, probably because enough modifications could make the model fit almost any data set and, moreover, no means existed for statistically testing the fit with a probability level.

A major development occurred in the 1960s with the adaptation of the laws of quantitative genetics to the study of discrete traits; that is, the multifactorial threshold model. In the 1970s, this model was expanded to incorporate multiple disease forms, sex differences, and most important, it was rendered statistically testable for goodness of fit. Meanwhile, sophisticated single-locus models that lent themselves to statistical testing were also developed. Finally, the increasing availability of computers made possible analyses that would have been impossible in the early days of model construction.

The term "transmission model" is preferred to "genetic model" because the models do not necessarily require an assumption that the disorder is genetic. Analysis by transmission models is entirely appropriate for conditions of undetermined etiology, even when they may be purely environmental. A transmission model is basically a mathematical hypothesis—a set of predictions based on an assumption about the transmission of the disorder that can be compared with observed data and the fit of model to data tested with a statistic so that the model can be accepted or rejected.

A model is defined by a set of parameters such as the frequency of the trait in the population. The parameters and the assumptions of the model comprise a mathematical formula, or set of formulas that, for any set of parameter values, generate expected values of affected persons among various classes of relatives and the population. Data collected on disease prevalence in the population and in these classes of relatives of affected persons provide the observed values against which the model is tested. The parameters are iterated over a series of values, successive values being substituted for each parameter until the set of parameters giving the best fit of expected to observed values is obtained. The fit can be maximized in either of two ways. First, differences between observed and expected values may be used to compute a chi-square for each parameter set; the parameter set yielding the smallest chi-square is the best fitting one. Alternatively, a likelihood approach may be employed. Under the model, and for each parameter set, the probability of the observed numbers of affected and unaffected persons in each pedigree can be calculated. The product of these probabilities across pedigrees is referred to as the likelihood and is maximized by the best fitting parameter set. Finally, the goodness of fit of the best fitting parameter set is tested statistically to determine whether the model is accepted or rejected. In order to test fit, the degrees of freedom must be determined. They are equal to the number of classes of independent observations minus the number of parameters defining the model. If the number of observations exactly equals the number of parameters, there are 0 degrees of freedom and the model cannot be tested. In the chi-square analysis, the probability level of the smallest chi-square value with its respective degrees of freedom determines whether the model provides an acceptable fit. In the likelihood approach, a likelihood ratio is constructed by the likelihood of the parameter set tested divided by the likelihood of an "unrestricted model" that provides a perfect fit to the data. Twice the logarithm of the likelihood ratio is asymptotically distributed as a chi-square with its respective degrees of freedom.

Slater and Cowie[46,48] proposed a single-gene model of schizophrenia that in some respects represented a forerunner of present transmission models and exemplifies many of these principles. The model proposes a single gene a of frequency p and its normal allele A of frequency $1 - p$. Thus, three genotypes are possible: AA, Aa, and aa, which the law of Hardy-Weinberg equilibrium predicts will occur in the following frequencies respectively: $(1 - p)^2$, $2p(1 - p)$, and p^2, the three frequencies summing to unity. If m is the proportion of heterozygotes (Aa) who develop schizophrenia, then the prevalence of schizophrenia in the population can be expressed by the following formula: $s = 2mp(1 - p) + p^2$. With s fixed at 0.0085, the population prevalence of schizophrenia, m can be calculated from any given value of p. In this way, a wide range of parameter sets of p and m were used to calculate the expected rate of schizophrenia for various classes of relatives of schizophrenics: children, siblings, and second- and third-degree relatives. The expected and observed values were compared by visual inspection rather than a statistical test, and it was concluded that when $p = 0.03$ and $m = 0.13$, a suitable fit was obtained.

Use of Transmission Models

Transmission models have been applied to a number of questions in psychiatry, the problems addressed falling into the following categories.

1. Do the data favor single gene or multifactorial transmission? None of the psychi-

atric disorders fit classic Mendelian segregation ratios, and family studies have been compatible with either a partially penetrant gene or with multifactorial inheritance. With transmission models, the fit between model and data can be expressed quantitatively as a probability level, and in this way different models can be compared and poorly fitting ones rejected.

2. Are the data compatible with a specific genetic or environmental hypothesis? Although the only method for definitively separating heredity from environment is the adoption study, specific hypotheses about genetic and environmental influences can be formulated and tested. For instance, are women less likely to develop antisocial personality or alcoholism because social pressures against this behavior are greater in women or because they are inherently more resistant to the disorder? Each hypothesis predicts different patterns of transmission that may be used to determine which best fits the data.

3. Can two or more conditions be considered subforms of a single disorder? Disorders in psychiatry that may be conceptualized in this way include unipolar and bipolar affective disorder within bipolar families, antisocial personality and hysteria, and male and female alcoholism. Disorders such as these that differ with respect to severity, frequency, or symptomatology but occur together in families can be analyzed to determine whether a single transmission model can account for both or whether separate disorders segregating independently must be assumed.

4. Can mild and atypical forms of a disorder be included as a "spectrum" of that disorder? Examples of such conditions include schizoid traits in relatives of schizophrenics, as well as cyclothymic personality and mild, atypical depressions in relatives of manic-depressives. If these conditions are transmitted independently of the major disorder, the transmission hypothesis may be rejected.

In interpreting the results of studies using transmission models, several points must be kept in mind. First, when a model provides an acceptable fit to the data, this does not mean that it is the only explanation for the observations. Other models not tested, or not yet constructed, may account for the data equally well. Moreover, the disease may subsequently prove to be heterogeneous, with each subtype being transmitted in a manner different from what had originally been thought. Second, if a model does not fit the data, it must be remembered that what has been rejected is that particular model. Thus, rejection of a specific model of environmental transmission does not mean that environmental transmission has been disproven, but only that the formulation of environmental transmission by that model does not account for the data. Finally, data sets in psychiatry often vary greatly from one investigation to another, and models fitting one data set may not fit another of the same disease.

¶ Overview of Transmission Models

This section will present an overview of the transmission models commonly encountered in the psychiatric literature. The following section will review the literature on the actual application of these models. The presentation will be limited to models that have been applied to psychiatric data; models that have been presented in theory only are omitted. The relevant models can be conveniently divided into single-locus and multifactorial ones.

Multifactorial Models

ONE THRESHOLD

The multifactorial threshold model was originally proposed by Carter[3] in 1961, developed into a quantitative mathematical model by Falconer[16] in 1965, and extended to include multiple thresholds by Reich and associates[43] in 1972. This latter development rendered the model testable against data and it has been widely applied in psychiatry since that time. Since the models currently in use in psychiatry were developed by Reich, his notation will be followed.

Basically, the model postulates that multiple genes plus environmental factors contribute to a continuous liability which, although not measurable, is normally distributed in the population, with individuals exceeding a hypothetical threshold manifesting the trait, and those falling short of the threshold liability being unaffected. This formulation is presented diagrammatically in figure 4–1, with the area to the right of the threshold (T) representing the affected portion of the population.

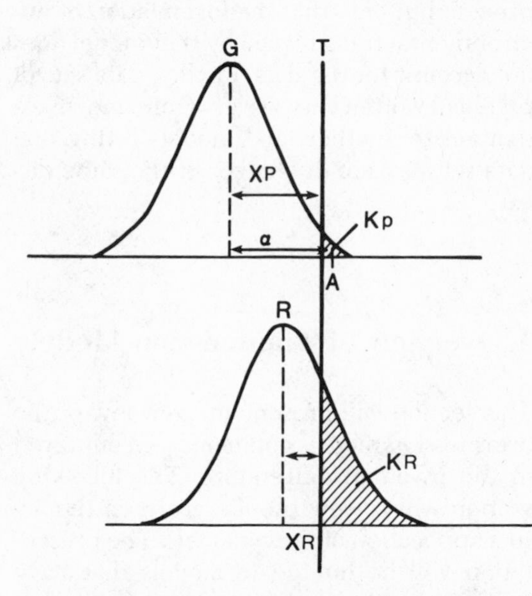

FIGURE 4–1. The Multifactorial Single Threshold Model. The upper distribution represents the liability of the general population and the lower distribution that of the relatives of affected individuals. G and R are the liability distribution means of the general population and of the relatives of affected individuals, respectively. T is the threshold. K_P* and K_R are the prevalences of the trait in the population and among relatives of affected individuals, respectively. X_P and X_R are the deviations of the threshold from the population and relative means respectively. A is the mean liability of affected individuals in the population, and a is its deviation from the general population mean.

*This chapter retains Reich's original notation using an upper case P to designate population parameters. However, in more recent publications a lower case p is used. The two refer to the same parameter ($K_P = K_p$).

SOURCE: Reich, T., Cloninger, C.R., and Guze, S.P. "The Multifactorial Model of Disease Transmission: I. Description of the Model and Its Use in Psychiatry," *British Journal of Psychiatry*, 127 (1975): 2.

Reich and associates[42] have summarized the assumptions on which the model is based: (1) the relevant genetic and environmental factors can be combined into a single continuous liability variable; (2) the liability is normally distributed; (3) a threshold in the liability divides the population into affected and unaffected classes; (4) the relevant genes are each of small effect and act additively; (5) environmental factors are also of small effect and act additively; and (6) since environmental factors may be shared by relatives, the disease may be partly to entirely nongenetic.

In figure 4–1, the upper curve represents the liability distribution of the population, and the lower one, that of first-degree relatives of affected persons. The lower curve is displaced to the right, resulting in a higher prevalence of the trait in the first-degree relatives of affected persons than in the population; that is, the trait is familial. The more strongly familial the disorder is, the farther to the right the relatives' distribution is displaced. This relationship can be expressed as a phenotypic correlation (r). Where K_P and K_R represent the prevalence of the trait in the population and in first-degree relatives of affected persons, respectively, X_P and X_R, the deviation of their respective population means (G and R) from the threshold (T), and a is the mean deviation of the probands from the population mean:

$$r = \frac{X_P - X_R \sqrt{1 - (X_P^2 - X_R^2)(1 - X_P/a)}}{a + X_R^2(a - X_P)} \qquad [4\text{–}1]$$

The values of X_R, X_P, and a can be obtained from tables once the population prevalences (K_P and K_R) are known. The correlation, which can be readily derived from these values, varies from 0 to 1, with 1 indicating that all relatives are affected and 0 indicating that relatives experience the same prevalence as the population.

The model can be fit to data by calculating the heritability (h^2) for correlations between various classes of relatives. This parameter is based on the assumption that the total variance in a trait can be partitioned into genetic and environmental variance ($V_T = V_G + V_E$). The heritability represents the proportion of total variance due to additive genetic variance* ($h^2 = V_A/V_T$). It can be estimated by dividing the correlation between relatives by the coefficient of genetic relationship (1 for monozygote twins, 1/2 for first-degree relatives, 1/4 for second-degree relatives, and so forth). The model predicts that the heritabilities should be consistent across various classes of relatives (for the same trait in the same environment). If this proves to be the case, then the model can be said to fit the data.

The major contribution of this model has been that it greatly modified traditional ways of viewing qualitative traits such as disease. However, it has the disadvantage that with two parameters (K_P and r) and two sets of independent observations (K_P and K_R), there are 0 degrees of freedom remaining for statistical testing. Being untestable, the model has been of mainly heuristic value. The problem of nontestability, however, was circumvented by Reich and coworkers,[43] who extended the model to include multiple thresholds.

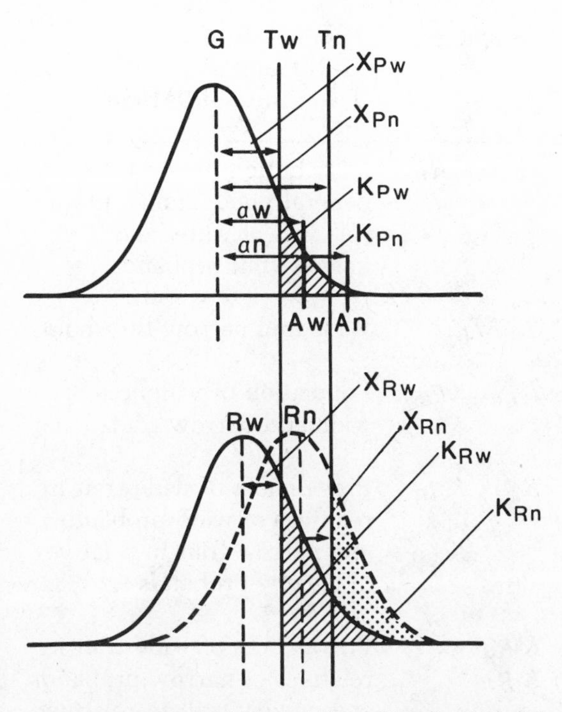

FIGURE 4–2. The Multifactorial Two Threshold Model. The upper distribution represents the liability of the general population, and the lower two distributions that of the relatives of individuals with wide and narrow forms of the trait, respectively. The parameters are listed in table 4–1.

SOURCE: Reich, T., Cloninger, C.R., and Guze, S.P. "The Multifactorial Model of Disease Transmission: I. Description of the Model and Its Use in Psychiatry," *British Journal of Psychiatry*, 127(1975): 5.

The Multifactorial Multiple Threshold Model

Many traits occur in two forms: a less frequent severe form and a more common mild one. For example, in bipolar families, unipolar depression is approximately twice as common as bipolar illness. This situation can be represented by a continuous liability distribution with two thresholds, which is illustrated in figure 4–2. The parameters are defined in table 4–1. This figure is analogous to figure 4–1, with the upper distribution representing the liability distribution of the population and the lower two distributions that of first-degree relatives of persons with either form of the trait and those with only the severe form, respectively. Persons having either form of the trait are designated as having the wide form (w) and are represented by the area to the right of the wide threshold (T_w). Those with only the severe subtype are designated narrow form (n) and are represented by the area under the curve to the right of the narrow threshold (T_n). Those with the mild subtype only are designated wide but not narrow ($w-n$) and are represented by the area between the two thresholds. It can be seen that the liability distribution of relatives of narrow-form probands is displaced to the right of that of the relatives of wide-form probands, resulting in a larger proportion exceeding both thresholds and thus being affected with both forms

*This term refers to genetic variance that can be transmitted to progeny. Siblings can share two genes at a locus through common inheritance, creating "dominance" variance that cannot be transmitted since only one gene is passed to offspring. Thus, $V_G = V_A + V_D$, where V_A and V_D refer to additive and dominance variance, respectively. Heritability based on additive genetic variance ($h^2 = V_A/V_T$) is referred to as heritability in the "narrow" sense and that based on total genetic variance ($h^2 = V_G/V_T$) as heritability in the "broad" sense. (For a complete treatment of this subject, see reference 15.)

TABLE 4–1 **Parameters of the Multifactorial Two-Threshold Model**

G, R_w, R_n	Distribution means for the general population, and for relatives of wide- and narrow-trait probands
T_w, T_n	Wide and narrow thresholds
K_{Pw}, K_{Pn}	Population prevalences of wide and narrow traits
K_{Rw}, K_{Rn}	Prevalences of wide trait in relatives of wide probands and narrow trait in relatives of narrow probands
K'_{Rw}, K'_{Rn}	Prevalences of wide trait in relatives of narrow probands and narrow trait in relatives of wide probands
X_{Pw}, X_{Pn}, X_{Rw}, X_{Rn}, X'_{Rw}, X'_{Rn}	The normal deviate of the respective distribution means from the wide and narrow thresholds respectively
A_w, A_n	Mean liability of wide- and narrow-trait individuals in the general population
a_w, a_n	Deviation of mean liabilities, A_w and A_n, from general population mean

of the trait. The liability distribution of relatives of wide-form probands is in turn displaced to the right of the population distribution, making their risk for both forms greater than the population but less than relatives of narrow-form probands. The model is completely defined by three parameters: the population prevalence of the narrow form (K_{Pn}), the population prevalence of the wide form (K_{Pw}), and the correlation coefficient (r).* The correlation coefficient can be estimated for subjects with the same form of the illness (that is, between wide-form probands and their relatives with respect to wide-form trait, r_{ww}; and between narrow-form probands and their relatives with respect to narrow-form trait, r_{nn}) with formula 4–1. The cross correlations (between wide-form probands and their relatives with respect to narrow-form trait, r_{wn}; and between narrow-form probands and their relatives with respect to wide-form trait, r_{nw}) can be estimated with the following formulas:

$$r = \frac{X_{Pn} - X'_{Rn}\sqrt{1 - (X^2_{Pn} - X'^2_{Rn})(1 - X_{Pw}/a_w)}}{a_w + X'^2_{Rn}(a_w - X_{Pw})} \quad [4\text{–}2]$$

$$r = \frac{X_{Pw} - X'_{Rw}\sqrt{1 - (X^2_{Pw} - X'^2_{Rw})(1 - X_{Pn}/a_n)}}{a_n + X'^2_{Rw}(a_n - X_{Pn})} \quad [4\text{–}3]$$

If the assumptions of the model are correct, and the trait is indeed a unitary one, all four correlations will be equal. Since there are six classes of independent observations (K_{Pn}, K_{Pw}, $K_{Pw} - K_{Pn}$, K_{Rn}, K_{Rw}, $K_{Rw} - K_{Rn}$) and three parameters defining the model (K_{Pn}, K_{Pw}, r) there are three degrees of freedom remaining for testing goodness of fit. Thus, the multiple threshold model represents an important advance because it can be statistically tested and rejected if it does not fit the observations. The usual test statistic is the chi-square goodness-of-fit test.

A simple extension incorporates sex effect into the model.[42] The usual situation is a trait such as unipolar depression in which the prevalence differs between the sexes. This can be represented by a single liability distribution with separate thresholds for the two sexes (T_m and T_f). With the problem set up in this fashion, the prevalence of the less frequently affected sex represents the narrow form and that of the other, the wide form. This model is defined by three parameters: the population prevalence in males (K_m), the population prevalence in females (K_f), and the correlation between relatives (r). There are six classes of independent observations (the male and female population prevalences and the prevalence among male and female relatives of each sex of proband), leaving three degrees of freedom for testing goodness of fit.

*See footnote to figure 4–1 regarding notation.

A further extension of the multifactorial model permits the testing of three hypotheses about subtypes of a trait: (1) the subtypes are different degrees of the same process; (2) they are environmental variants of the same process; or (3) they are transmitted independent of one another.[44] These three hypotheses conform respectively to what is termed (1) the isocorrelational model, (2) the environmental model, and (3) the independent model.

The isocorrelational model assumes that familial transmission factors (genetic and environmental) act equally on all subtypes and that extrafamilial environmental factors likewise affect all subtypes equally. If these assumptions are correct, then all four correlation coefficients should be equal: $r_{ww} = r_{nw} = r_{wn} = r_{nn}$. The model is defined by the three parameters K_{Pw}, K_{Pn}, and r.

In the environmental model, extrafamilial environmental factors are assumed to act preferentially on one subform of the trait so that: $r_{ww} \neq r_{nn}$. In this model the remaining two correlations are equal and are equivalent to the geometric mean of the first two: $r_{wn} = r_{nw} = \sqrt{r_{ww} r_{nn}}$. The model is defined by the four parameters K_{Pw}, K_{Pn}, r_{ww}, and r_{nn}.

In the independent model, the familial factors (genetic and environmental) responsible for transmission are assumed to differ between the subtypes of the trait. Thus, the subtypes are to a greater or lesser extent transmitted independent of one another and each has its own liability distribution with one threshold. The assumption that the correlation in familial factors between subtypes is less than 1 requires that r_{wn} and r_{nw} be significantly less than $\sqrt{r_{ww} r_{nn}}$ as predicted by the environmental model. Thus, the model is defined by five parameters: K_{Pw}, K_{Pn}, r_{ww}, r_{nn}, and r_{wn}. If the model is accepted, the degree of overlap between the two trait forms ($w - n$ and n) can be estimated by the phenotypic correlation

$$r_P = \frac{r_{w-n,\,n}}{\sqrt{r_{w-n,\,w-n,}\, r_{nn}}}$$

which varies from 0 with complete independence to 1 with complete overlap.

The three models are nested within each other: the isocorrelational model represents a special case of the environmental model, which, in turn, is a special case of the independent model. In practice, the isocorrelational model is tested first since it is the most restrictive. If it is rejected the environmental model is tested and, in turn, if it can be rejected, the independent model is tested.

Single-Locus Models

The earliest single-locus model bearing any resemblance to present-day transmission models is the formulation by Slater, which was presented earlier.[46] Since both Slater's single-locus model and Falconer's multifactorial threshold model appeared to fit the data, Slater developed a computational model for determining whether a data set favored single-gene or multifactorial inheritance.[47] Although itself not a transmission model, this method has been widely used in psychiatry to analyze transmission and, therefore, will be included in this review.

Slater reasoned that if a trait is transmitted as a single gene, then ancestral secondary cases should appear predominantly on one side of the pedigree (paternal or maternal). In polygenic inheritance, secondary cases should appear on both sides more frequently than in single-gene transmission. A rigorous solution was not possible, but by using some simplifying assumptions he arrived at the expectation that in polygenic inheritance pairs of ancestral cases should be unilaterally distributed approximately twice as frequently as bilaterally. Any deviation from this expected two to one ratio in the direction of excess unilateral pairs would be evidence for single-gene transmission, and the deviation could be statistically tested with a chi-square test. Slater and Tsuang[49] subsequently introduced a correction factor to allow for the greater weight given to families with large numbers of secondary cases.

As the multifactorial model developed in sophistication, more advanced single-locus models were being developed. Cavalli-Sforza and Kidd[4,29] developed a single-locus threshold model resembling in some respects the multifactorial threshold model. The model, illustrated in figure 4–3, proposes a single locus with two alleles A and a, pro-

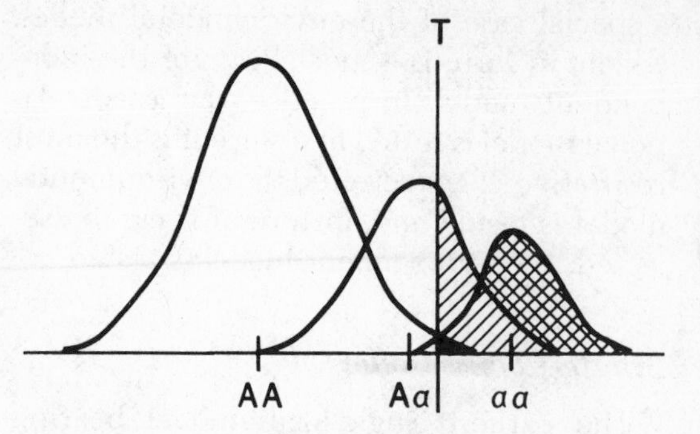

FIGURE 4–3. The Single-Locus Threshold Model. The three distributions represent the liability of *AA, Aa,* and *aa* individuals, respectively. The position of *AA* is arbitrarily set at 0 and that of *aa* at 2. The parameters defining the model are: gene frequency of *a* allele (q); environmental variance (ϵ^2); position of the heterozygote *Aa* (h); and threshold position (T).

SOURCE: Kidd, K.K., and Cavalli-Sforza, L.L. "An Analysis of the Genetics of Schizophrenia," *Social Biology,* 20(1973): 256.

ducing the three genotypes *AA, Aa,* and *aa* with frequencies determined by Hardy-Weinberg equilibrium: $(1 - q)^2$, $2q(1 - q)$, and q^2, respectively where q is the gene frequency of the *a* allele. Each genotype mean is represented on a liability scale, with the *AA* mean arbitrarily set at 0 and the *aa* mean at 2; the *Aa* mean occupies a variable distance between the two, its distance from 0 being represented by h', which can vary from 0 for complete recessiveness (*Aa* = *AA*) to 2 for complete dominance (*Aa* = *aa*). Environmental variance causes the phenotypic liability values of each genotype to vary, forming a distribution of values around each genotype mean. The three distributions are assumed to each be normal with equal variances represented by ϵ^2. The three overlapping liability distributions form a continuous distribution of liability that is divided by a threshold T into affected and unaffected classes. Thus, the model is defined by four parameters: gene frequency (q), dominance (h'), environmental variance (ϵ^2), and threshold position (T).

Expected numbers of affected relatives in various classes can be calculated for any set of parameter values. The model determines the probability of an affected, or unaffected, person being of each genotype. For each genotype, the probability of any class of relative sharing either allele with the proband can be calculated. Finally, for each possible genotype of a relative, the model determines the probability of being affected. A computer program is used to iterate over the parameters, generate the expected values for each parameter set, compare them with the observed ones, and calculate the chi-square for goodness of fit.

Elston and associates[11,12,13] developed models for a limited number of loci (two autosomal and one sex-linked) based on segregation analysis using a likelihood approach. The single locus model proposes two alleles *A* and *a* resulting in the three genotypes *AA, Aa,* and *aa* in Hardy-Weinberg equilibrium with q the frequency of the *A* allele. The probability of being susceptible is represented by λ, which can be made to vary with genotype or to be independent of genotype. Age of onset is considered to be lognormally distributed (the logarithm of the age of onset is normally distributed) with mean μ which may either vary with genotype or be independent of genotype. All age of onset distributions have the same standard deviation σ. The probability of transmitting the *A* allele is represented by τ, which, for the three genotypes, assuming Mendelian inheritance, is $\tau_{AA,A} = 1$, $\tau_{Aa,A} = 1/2$, and $\tau_{aa,A} = 0$. Likewise, there are three complementary probabilities of transmitting the *a* allele. Finally, an ascertainment parameter K is introduced to correct for ascertainment bias.

The model accepts either dichotomous (affected, unaffected) or trichotomous (affected state 1, affected state 2, unaffected) traits. Two hypotheses are tested. The Mendelian hypothesis assumes transmission probabilities ($\tau_{AA,A} = 1$; $\tau_{Aa,A} = 1/2$; $\tau_{aa,A} = 0$) in accordance with Mendel's law. The environmental hypothesis assumes that the probability of transmitting the trait is independent of genotype and, therefore, that the three transmission probabilities are equal ($\tau_{AA,A} = \tau_{Aa,A} = \tau_{aa,A}$). A computer program calculates the likelihood of the data set under each hypothesis and under an unrestricted model, which allows the parameters to vary independently of one another to provide a perfect fit to the data. The Mendelian and environmental hypotheses are tested by means of a likelihood ratio with the unrestricted model. A small likelihood ratio implies a small departure from the unrestricted model and, thus, a good fit to the data and can be tested by the chi-square (2 log [likelihood ratio]).

It can be seen from this brief overview that as transmission models have become more advanced, they have expanded the range of testable hypotheses about disease transmission by incorporating such variables as multiple forms of a trait, environmental effects, and age of onset. Consequently, each transmission model subsumes a number of hypotheses within it. In this sense most of the early hypotheses of single gene (or polygenic) inheritance are tested when one of these broader transmission models is applied. Finally, the single-locus and the multifactorial models represent the extremes of a continuum of potential models from one gene to many. However, until the extremes can be rejected, there is little value in constructing new models with two loci, multiple alleles, and so forth, which, although heuristically less valuable, will fit the data equally well.

TABLE 4–2 **Parameters of the Elston Segregation Analysis Model**

$\tau_{AA,A}$, $\tau_{Aa,A}$, $\tau_{aa,A}$	Probability of each respective genotype transmitting the *A* allele to progeny
q	Gene frequency of the *A* allele
γ	Probability that an individual will develop the trait if he or she lives long enough
μ_{AA}, μ_{Aa}, μ_{aa}	Mean age of onset of each respective genotype on a log scale
σ	Standard deviation of the logarithm of age of onset
K	Ascertainment parameter

¶ Application of Transmission Models

This section will review the application of transmission models to data on schizophrenia, affective disorder, antisocial personality, hysteria, alcoholism, and panic disorder. These studies illustrate the models presented in theory in the last section, as well as their application to the field of psychiatry. The review should also familiarize the reader with recent research in this area. Space does not permit a detailed analysis of each study. However, examples of the major models are reviewed in sufficient detail to provide an understanding in some depth of the use of the models. (For those interested in pursuing the original literature, all of the major studies are reviewed and referenced.)

Schizophrenia

Schizophrenia was a major interest to the early workers in psychiatric genetics who attempted to adapt Mendel's laws to their data. Kallman[28] proposed that an autosomal recessive gene accounted for the inheritance of schizophrenia in his study of a large collection of kindreds. Böök[2] proposed that a single gene with incomplete penetrance accounted for the inheritance of schizophrenia and demonstrated that the model fit his data when the heterozygote penetrance was 0.2, the homozygote penetrance 1, and the gene frequency 0.07. However, because of the unusually high population prevalence in Böök's material, his model would not have fit the data of other workers.

Slater[46,48] further developed the model of a single partially penetrant gene into his formulation (see "Introduction"). He fit the model to data derived from the literature dealing with children of one and two schizophrenic parents, siblings of schizophrenics, and second- and third-degree relatives. Taking 0.0085 as the population prevalence of schizophrenia (S), various values for the gene frequency (p) were substituted into the formula $S = 2mp(1 - p) + p^2$ and values of the heterozygote penetrance (m) calculated. When these parameter sets were used to calculate the expected rates of schizophrenia in the aforementioned classes of relatives, the model seemed to give a good fit at a gene frequency of 0.03 and a penetrance of 0.13. Thus, the model predicted a relatively uncommon gene that was predominantly recessive.

Slater felt that his model accounted for the data as well as the polygenic threshold model and tested the two with his computational model.[49] Nineteen schizophrenic kindreds from the Maudsley Hospital were analyzed, and a unilateral to bilateral ratio of 42 to 11 was found, which was significantly greater than the 2 to 1 ratio predicted by polygenic inheritance. Thus data favored the single-locus model. Since the first study[49] was based on family history, Tsuang[51] applied the model to data obtained from interviews. The twenty-three kindreds in his study revealed a ratio of 43 to 11 unilateral to bilateral pairs, again significantly favoring single-gene transmission.

Gottesman[23] applied Falconer's multifactorial single threshold model to the inheritance of schizophrenia. Data were taken from the literature on monozygotic and dizygotic twins, first-degree and second-degree relatives. Calculations were carried out using a population prevalence of schizophrenia (q) of both 1 percent and 2 percent. Since the model is defined by q and h^2, with q fixed, h^2 can be calculated from the correlation in prevalence between probands and any class of relatives. If the model is to fit, the heritabilities should be consistent across classes of relatives and this was found. At a 1 percent prevalence, h^2 estimates on age-corrected data ranged from 79 percent among first-degree relatives to 106 percent for one set of twins. The consistency was acceptable, especially considering that the data came from five investigations in three countries. The heritabilities indicated a substantial genetic predisposition to schizophrenia. They have been subsequently recalculated on twin data using the tetrachoric correlation, which is more exact than the Falconer method used by Gottesman.[29] Again, both a 1 and 2 percent population prevalence was used in the calculations. At a 1 percent prevalence, the heritabilities ranged from 80 percent to 93 percent, substantiating Gottesman's analysis.

Heston[25] proposed a single-gene model that assumed complete penetrance of the gene. The model made use of the observation that the nonschizophrenic relatives of schizophrenics often manifest other forms of psychopathology, which he termed "schizoidia." If the sum prevalence of schizophrenia and schizoidia was taken, in those studies that recorded this data, the observed proportions of affected relatives came surprisingly close to that predicted by simple autosomal dominance. For children, 49 percent were affected, compared with 50 percent expected, and the respective figures for siblings were 46 percent versus 50 percent; for parents, 44 percent versus 50 percent; for children of two schizophrenic parents, 66 percent versus 75 percent; and for monozygotic co-twins of schizophrenics, 88 percent versus 100 percent.

Using pooled data from the literature on children, siblings, second-degree relatives, and mono- and dizogotic co-twins of schizophrenics, Kidd and Cavalli-Sforza[4,29] and Matthysse and Kidd[32] applied the single-locus threshold model to schizophrenia. They noted considerable heterogeneity in prevalence rates among investigations and dealt with the problem by fitting the model to both a set of "low" and "high" rates. The four parameters of the model, gene frequency (q), environmental variance (ϵ^2), dominance (h'), and threshold position (T), led to a number of parameter sets all fitting the data. For illustrative purposes one of these parameter sets will be discussed in detail: $q = 0.10$, $\epsilon^2 = 0.36$, $h' = 0.25$, and $T = 1.6$. It will be remembered that the AA mean is arbitrarily set at 0 on the liability scale and the *aa* mean at 2.0, with *Aa* falling somewhere in between, its position determined by h′ (0.25). This arbitrary scale can be standardized by using the standard deviation (S.D.) as the unit of measure, which in this case is 0.6 ($\sqrt{0.36}$). The threshold ($T = 1.6$) lies 2.7 S.D. above the *AA* mean and 2.25 S.D. above the *Aa* mean. Thus, relatively few *AA* and *Aa* genotypes will exceed the threshold, and most persons with the disease will be *aa*, the mean of their liability distribution lying 0.67 S.D. above the threshold. Finally, the gene is a common one, with frequency 0.10, meaning 19 percent of the population will carry it [$q^2 + 2q(1 - q)$].

Returning to the overall analysis, all parameter sets predicted a relatively common predominantly recessive gene, as in the illustration. Although persons of the normal *AA* genotype had a small likelihood of developing schizophrenia, 16 to 25 percent of schizophrenics were *AA* depending on the parameter set. (Although proportionally less *AA* are affected, they constitute the majority of the population and, thus, contribute a substantial number of cases.) Thus, the model predicted a sizable proportion of sporadic cases. Another interesting finding was that the expected morbidity risk among siblings was higher than that for parents. This discrepancy, which is seen in most family studies of schizophrenia, is usually considered to be the result of a selection bias, but in fact it is predicted by the single-locus (and polygenic) model. This is because siblings, unlike any other pair of first-degree relatives, can share *both* genes at a locus through common inheritance.

Matthysse and Kidd[32] applied a different single locus model to published schizophrenia data.[26] The parameters of the model are (1) the frequency of the pathogenic *a* allele (q); (2) the probability of a genetically normal *AA* individual becoming schizophrenic (f_0); (3) the probability of the *Aa* heterozygote becoming schizophrenic (f_1); and (4) the probability of the *aa* homozygote becoming schizophrenic (f_2). When the model was applied to the published data, it predicted an unacceptably low morbidity risk for monozygotic co-twins of schizophrenics and for offspring of dual schizophrenic matings (both 19.9 percent). However, for the population prevalence, and for siblings and offspring of schizophrenics, a wide range of parameter sets fit the data. The gene frequency varied between 0.3 percent and 2.2 percent, with the f_0, f_1, and f_2 varying from a high of 0.5 percent, 50.5 percent, and 100.0 percent, to a low of 0.0 percent, 19.4 percent, and 38.9 percent respectively. This model also predicted a high rate of sporadic cases (61.2 percent) with 38.7 percent of schizophrenics being heterozygous and 0.1 percent homozygous.

The same investigators[32] tested a multifactorial model based on the following assumptions: (1) a normally distributed population liability to schizophrenia with a mean of 100 and a standard deviation of 15 arbitrary units, (2) a cumulative normal liability distribution representing the probability that a person with a given liability value will develop schizophrenia. The parameters of the model are the liability value resulting in a 50 percent risk and one resulting in a 99 percent risk. When the model was fitted to the data, liability values of 137 and 148 for the two respective parameters gave a good fit to population and first-degree prevalence, but again the model led to unacceptably high risks for monozygotic co-twins (61 percent) and offspring of dual matings (39 percent).

Elston and associates[14], using Elston's segregation analysis, analyzed a large sample of two-generation kindreds from Kallman's schizophrenia study. These included 178 pedigrees of probands with "nuclear" schizophrenia and 82 pedigrees of probands with paranoid and simple schizophrenia (the "peripheral" group). The data were analyzed first under a model assuming that probability of being susceptible (γ) was the same for all genotypes, but, if susceptible, each genotype was characterized by its unique mean age of onset (μ). The second model assumed that each genotype was characterized by a unique probability of being susceptible (γ), but that age of onset (μ) was the same for all genotypes. Under each model, the Mendelian ($\tau_{AA,A} = 1$, $\tau_{Aa,A} = 1/2$, $\tau_{aa,A} = 0$) and the environmental ($\tau_{AA,A} = \tau_{Aa,A} = \tau_{aa,A}$) hypotheses were tested. A trichotomous classification was used in order to include "schizoidia" as affected state 2. Likelihoods were computed for each set of pedigrees under each hypothesis and under the unrestricted "best fit" hypothesis. The results rejected both the genetic and environmental hypotheses in both data sets. However, the parameters of the unrestricted model were quite similar in the nuclear and "peripheral" groups, indicating a similar pattern of transmission in both subtypes.

This analysis differs from the previous single-locus approaches in using segregation analysis rather than fitting parameters of a model to disease prevalences in various classes of relatives. The former approach is a more powerful tool for detecting lack of fit. (The difference between the two approaches can be illustrated with the following example. Assume two families, each with one affected parent and five children. In the first family, all five children are affected and in the second, none. Taking all children as a group, five of ten are affected, exactly as predicted by autosomal dominance, although in neither family is the segregation ratio close to this expectation.) Perhaps this explains why Elston and Campbell,[10] applying the latter approach to Kallman's data[28] in an earlier analysis, found a good fit to a predominantly recessive single-gene model. The Elston model is a broad one and would subsume most of the early single-gene models (as would the Kidd model[29]), and in addition, his use of "schizoidia" as affected state 2 would subsume Heston's hypothesis[25] as well. Thus, the rejection of the Elston "genetic" hypothesis covers a set of earlier genetic hypotheses. However, it should be remembered that this last analysis was limited to the data from one investigator, and, due to the heterogeneity of the data in this area, the results may not apply to other investigations. It must also be recalled that particular genetic and environmental models were rejected, and the results do not imply that genetics and environment are unimportant in causing schizophrenia.

Affective Disorders

Because of the bipolar-unipolar heterogeneity within affective disorders, early attempts to apply genetic models to affective disorder as a group were doomed to failure. Rosenthal[45] summarized the major hypotheses which included a model with three separate genotypes—two recessive and one dominant and one postulating an autosomal gene for cyclothymia with an *X* chromosome activating factor to explain the greater incidence in women. With the demonstration that bipolar illness segregates independently of unipolar depression,[1,39] renewed interest developed in understanding the genetics of affective disorder, particularly the bipolar form.

Of all the genetic models in psychiatry, Winokur's proposal[54,56] that bipolar illness is transmitted as an *X*-linked dominant trait has created more controversy than any other. This hypothesis was originally suggested by the finding that there was no father-son transmission in sixty-one bipolar families despite frequent occurrences of other types of parent-offspring transmission. Since the *X* chromosome is transmitted from a father to all daughters but not to sons, this hypothesis accounted for the absence of father-son transmission as well as for the excess of affected females usually observed. In two subsequent studies, Winokur and his colleagues[24,53] collected an additional twenty-eight probands and thirty male probands, respectively, again finding virtually no father-son transmission. Other investigators,[22] however, have found frequent instances of father-son transmission in their material, thus contradicting the sex-linkage hypothesis.

A different approach to the question of *X*-linkage is linkage analysis. It is based on the fact that if two genes, a disease gene and a marker gene such as color blindness, lie sufficiently close to one another on the same chromosome, the frequency of recombination will be less than 0.5 and their assortment within families will not be independent of one another. In practice, the pedigree is identified by a proband with one form of each trait, for example, depression and color blindness (each trait exists in two forms: depressed—not depressed and color blind—not color blind). Relatives who have both forms the same as the proband, or neither form the same, are counted as nonrecombinants; those with one form and not the other, as recombinants. For any frequency of recombination (θ), the probability of encountering the observed number of nonrecombinants and recombinants within a family can be calculated from the binomial theorem. The probability at various recombination fractions ($\theta = 0.0, 0.1, 0.2, 0.3, 0.4$) can be compared with the probability under the null hypothesis $\theta = 0.5$ by means of a probability or odds ratio. The odds ratio is usually expressed as the logarithm of the odds (LOD score) and is summed over the pedigrees. If, for any value of θ, the LOD score reaches 3.0, linkage is considered to be established. Likewise, if the LOD score reaches -2.0, linkage can be considered to be ruled out at that recombination fraction.

Winokur's group[56] first suggested genetic linkage of bipolar illness with the genes for deutan and protan color blindness and the *Xg* blood group.[41,54] When Mendlewicz and associates[33] added a number of their own kindreds to this material, ten families informative for linkage with deuteranopia yielded a LOD score of 4.50 at a recombination fraction of 0.07, fifteen informative for protanopia yielded a LOD score of 3.73 at a recombination fraction of 0.10, and twenty-five informative at the *Xg* locus yielded a LOD score of 2.96 at a recombination fraction of 0.19. Thus, linkage was demonstrated at both color blindness loci and strongly suggested at the Xg locus.

These findings have been criticized by Gershon and Bunney,[15] who make the following points regarding linkage work in this area: (1) an association exists in the pedigrees between affective disorder and color blindness that would bias the material toward a finding of linkage; (2) the analytical methods developed for linkage analysis do not allow for such problems as variable age of onset and multiple manifestations of the trait; (3) some of the families are open to alternate interpretations of whether they are informative for linkage; and (4) since the *Xg* and color blindness loci are unlinked, it is unlikely that a third trait would be linked to both. They reanalyzed the data excluding any kindreds they considered to be ambiguous as to informativeness and were unable to support linkage at any of the above three loci. Moreover, Gershon's group has recently studied six new pedigrees informative at the color blindness locus and another six informative at the *Xg* locus. The LOD scores from both these analyses strongly support a verdict of no linkage.[21,31] At the same time, Mendlewicz and associates[36] have published eight new pedigrees which again support linkage between bipolar illness and the color blindness loci (LOD $= 1.55$ at $\theta = 0.15$). Mendlewicz found a significant degree of heterogeneity in his material, with some pedigrees supporting linkage and others not. Indeed, heterogeneity may be the answer to the seemingly endless contradictions in this area.

Since family data from some sources support the *X*-linkage hypothesis and others contradict it, this would appear to be a promising area for the application of genetic models and indeed a number of approaches have been tried.

Several studies have used Slater's computational model to analyze bipolar families. In twenty-six kindreds, Slater and associates[50] found a unilateral to bilateral ratio of affected pairs of relatives of 38 to 30, which was consistent with the polygenic expectation of 2 to 1. In his large family study, Perris[39] examined twenty bipolar and eight unipolar kindreds. The bipolar ratio was 46 to 15 and the unipolar one 12 to 8, both consistent with polygenic transmission. Mendlewicz and associates[35] separated out their bipolar families containing a first-degree relative with bipolar illness, and among these relatives the unilateral to bilateral ratio was 42 to 6, significantly favoring a single gene. The different result may be due to a different method of selecting the pedigrees for analysis. Whatever the reason, this was the only result consistent with a single gene, *X*-linked or not.

Crowe and Smouse[9] performed a pedigree analysis on Winokur's original sixty-one kindreds, which had initiated the sex-linkage hypothesis. An age-correction was introduced and used to calculate the expected numbers of ill relatives under both the sex-linked dominant (SLD) and the autosomal dominant (AD) hypotheses. A likelihood test of fit was used, and both models provided a satisfactory fit, with the SLD hypothesis fitting somewhat better ($p > 0.75$) than the AD ($p > 0.10$). When the models were compared, the sex-linkage hypothesis was favored with an odds ratio of 89 to 1, although the ratio was not statistically significant.

Bipolar illness is suitable for analysis by multiple threshold models, with the bipolar form representing the narrow threshold and the category of bipolar and unipolar illness defining the broad threshold. Gershon's group[18] has analyzed the published data along these lines. In addition to the two thresholds already discussed, they defined a third one to include "related" affective disorders such as mild depressions and cyclothymia. The single-locus model is now defined by the following parameters: (1) gene frequency (q); (2) environmental variance (ϵ^2); (3) dominance (h'); (4) threshold for major affective disorders (T); (5) bipolar threshold (T_{BP}); and (6) in the case of the three-threshold model, threshold for related disorders ($T_{Rel.}$). With two thresholds, there are fourteen independent observations; with three thresholds, twenty, leaving nine and fourteen degrees of freedom to test the respective models.

In the multifactorial model, dominance variance (stronger correlations between siblings than between parents and offspring due to the fact that siblings can share in common both genes at a locus through common inheritance) can be dealt with by computing the parent-offspring and sib-sib correlations separately. If three thresholds are used, this leads to five parameters: three population prevalences and two correlations; in the case of two thresholds, there will be four parameters. With two thresholds, there are fourteen independent observations and with three thresholds, twenty, leaving ten and fifteen degrees of freedom respectively to test each model.

When the four models were applied to Gershon's data, every model gave a satisfactory fit. Taking the multifactorial models first, the two-threshold approximation gave a best-fitting parameter set ($p > 0.3$), which estimated the population prevalence of all affective disorder (bipolar plus unipolar, or $BP + UP$) at 1.8 percent and bipolar illness at 0.4 percent. The sib-sib and parent-offspring correlations were, respectively, 0.37 and 0.31, indicating little or no dominance effect. The three-threshold solution yielded the following population prevalences ($p > .05$): all affective disorder, 3.1 percent; bipolar plus unipolar, 1.6 percent; and bipolar alone, 0.4 percent. The sib-sib and parent-offspring correlations were 0.35 and 0.39, respectively. Thus, both models predicted a relatively common disorder that is strongly familial and no evidence of a dominance effect was found.

Of the two single-locus models, the two-threshold one gave the better fit ($p > 0.5$) and estimated the gene frequency (q) at 0.21 and the variance (ϵ^2) at 0.14, making the standard deviation (ϵ) 0.37. The *a* allele was completely recessive ($h' = 0$), positioning the *Aa* mean at 0 with the *AA* mean. The threshold for major affective disorder (T) was positioned at 2.1 and that for bipolar affective disorder (T_{BP}) at 2.43. These thresholds are 5.7 and 6.6 S.D., respectively, above the *AA* and *Aa* means, making it highly unlikely that a person with either genotype would ever develop affective disorder. Therefore, the model predicts a common completely recessive gene with most affected persons being homozygous recessive but with 33 percent [$2q(1 - q)$] of the population being heterozygous carriers. The three-threshold approximation fit less well ($p > 0.1$) with the following parameter estimates: $q = 0.045$, $\epsilon^2 = 0.28$, $h' = 1.4$, $T_{Rel.} = 1.7$, $T = 1.9$, $T_{BP} = 2.3$. This solution predicts a less frequent gene with moderate dominance such that a substantial portion of the *Aa* liability distribution exceeds the thresholds. Here it is apparent that differences in the beginning assumptions of a model can lead to major differences in what the model predicts about the mode of transmission.

Gershon and associates[20] analyzed the data from Angst's and Perris's studies in a like manner using two categories of affected: bipolar and bipolar plus unipolar. Angst's data fit both models but Perris's rejected both the multifactorial and the single-locus models. For the multifactorial model, the best-fitting parameter set ($p > 0.05$) for Angst's data gave a population prevalence of 0.4 percent for major affective disorder and 0.03 percent for bipolar illness. The respective parent-offspring and sib-sib correlations were 0.43 and 0.47, indicating a strongly familial trait but little or no dominance effect. The single-locus model predicted the following parameter set ($p > 0.05$): $q = 0.06$, $\epsilon^2 = 0.45$, $h' = 0.34$, $T = 1.2$, and $T_{BP} = 2.3$.

Bipolar illness may also be analyzed with respect to sex thresholds, since most studies find females more frequently affected than males, the ratio being approximately 1.5 to 1. Gershon's group[30] applied the single-locus and multifactorial models with sex thresholds (T_m and T_f) to five studies: those of Winokur and associates[56] Mendlewicz and Rainer,[34] Goetzl and associates,[22] James and Chapman,[27] and Gershon and associates.[19] Both the single-locus and the multifactorial models fit the last three studies, but both were rejected by the first two. The three studies fitting the models were then analyzed by the same models without sex thresholds ($T_m = T_f$), and only James's data rejected the models. Because of the question of *X*-linkage, an *X*-chromosome dominant model was tested on the data of Winokur and those of Mendlewicz, the two studies suggestive of *X*-linkage. The model fit Winokur's data but was rejected by those of Mendlewicz. Thus, two studies rejected sex threshold models as an explanation for the sex differences in prevalence but only one of these was compatible with a sex-linkage explanation. Three studies were compatible with sex thresholds but in only one of these were they necessary to account for the observations.

What can be concluded from these studies? Regarding the question of sex linkage, the data that originally suggested the hypothesis have been rigorously tested and continue to support it. However, the majority of family studies do not suggest sex linkage, and unfortunately, in these the multifactorial and single-locus models have been equally satisfactory, with studies that reject one rejecting both. The analyses have demonstrated that a sex-linkage hypothesis is not necessary to account for the sex prevalence differences. Finally, the fact that different data sets lead to very different conclusions, not only in model but also in parameters of the model, speaks for the considerable degree of heterogeneity among studies in this area.

Antisocial Personality and Hysteria

Since there is considerable evidence from adoption studies for a genetic predisposition to antisocial personality, and since antisocial personality and hysteria are typically seen together in families, these disorders provide an ideal situation for analysis by transmission models. The problem requires the use of thresholds for both sex and severity.

The relevant observations are the following: The population prevalence of antisocial personality in males is considerably greater than the female prevalence, and hysteria is found almost exclusively in females. Likewise, among the relatives of male and female antisocials and hysteric women, antisocial personality is found more frequently in males than females and hysteria is found exclusively in females. These observations suggest a model with different thresholds for antisocial personality in males and females, with the female threshold representing a more extreme deviation from the mean. In women, hysteria may be viewed as a milder form of antisocial personality, such that antisocial personality represents the narrow threshold and hysteria plus antisocial personality the broad threshold. Thus, males have a wide threshold for antisocial personality and females a narrow threshold for antisocial personality and a wide threshold for hysteria.

Cloninger and associates[5,6] applied the multifactorial multiple threshold model to their data on antisocial personality and hysteria. The model was first tested on the data on antisocial personality. These were prevalences of antisocial personality in male and female relatives of both male and female antisocials and the population prevalences of both sexes. These six sets of observations were used to obtain the best-fitting set of three parameters: population prevalence in males and females, and correlation between first-degree relatives. The multifactorial model provided a close fit to the data ($p > 0.9$). Among personally interviewed first-degree relatives of white antisocials, the expected population prevalences were 3.6 percent for men and 0.7 percent for women. The first-degree relative correlation of 0.55 indicated strong familial transmission.

When the model is expanded to include hysteria it yields a set of twelve observations: the prevalence of antisocial men, antisocial women, and hysteric women in the general population and in first-degree relatives of each of these three classes of affected subjects. These twelve observations determine a best-fitting set of four parameters: population prevalences of antisocial men, antisocial women, hysteric women, and the correlation between relatives; leaving eight degrees of freedom for testing the minimum chi-square. When the model was compared to the data it provided a close fit ($p > 0.4$), estimating population prevalences of 3.8 percent for antisocial men, 0.5 percent for antisocial women, and 3.0 percent for hysteric women. The correlation between first-degree relatives was 0.54.

The results indicate that the multifactorial threshold model provides a very satisfactory explanation for the data on the familial transmission of antisocial personality and hysteria. Moreover, including hysteria in the analysis leads to an acceptable fit without substantially changing the parameters predicted from antisocial personality alone, providing further evidence that these disorders may be alternate forms of the same process. This was the first example of a genetic model providing a unitary hypothesis explaining the coincidence of two distinct diseases.

The same group used multifactorial models to test hypotheses about sex differences in the prevalence of antisocial personality.[7] The appropriate models are the three modifications of the Reich multifactorial threshold model: the isocorrelational model, the environmental model, and the independent model. The isocorrelational model predicts that extrafamilial factors affecting liability affect the two sexes equally. As a result, all four correlations among relatives (male–male, male–female, female–female, female–male) are equal. The model is tested with six sets of observations: the population prevalence in each sex plus the prevalence in each sex of relative of each sex proband; and is defined by three parameters: male population prevalence, female population prevalence, and correlation among relatives, leaving three degrees of freedom to test the chi-square. Since the model fit the data on antisocial personality very closely ($p > 0.9$), it explained the data without invoking extrafamilial factors that preferentially affect one sex, and testing the environmental or independent models became unnecessary.

Alcoholism

Alcoholism is more prevalent in men than in women and, therefore, may be analyzed with multiple threshold models in the same manner as antisocial personality. Cloninger and associates[7] analyzed a series of pedigrees from their center and found that the isocorrelational model did not lead to a good fit ($p > 0.05$). Thus, it became necessary to test the environmental model. This model predicts that extrafamilial factors that contribute to liability act preferentially on one sex. This is reflected mathematically in the correlation between females being unequal to the correlation between males. Thus, the model is defined by four parameters: the population prevalence in males and in females, the male–male correlation, and the female–

female correlation. Six sets of observations are possible: the population prevalence of each sex, and the first-degree relative prevalence in each sex of each sex proband, leaving two degrees of freedom for testing the best-fitting chi-square. When this model was applied to the data, the fit was very good ($p > 0.8$). The female–female correlation was estimated at 0.18 $\pm$ 0.12, significantly lower than the male–male correlation of 0.53 $\pm$ 0.07, and consistent with the hypothesis of extrafamilial factors acting preferentially in women.

The isocorrelational and environmental models are based on the assumption that the same familial factors are relevant to the etiology of the trait in both sexes. If familial etiologic factors in one sex are only partly correlated with those factors in the other sex, then sex differences occur due to the partial independence of these factors. This is reflected mathematically in a reduced correlation between opposite sexes (male–female, female–male) from the expected geometric mean of the two same-sex correlations. This can be expressed mathematically by the phenotypic correlation (r_p): $r_P = r_{mf}/(r_{mm}r_{ff})^{1/2}$ and $r_P = r_{fm}/(r_{mm}r_{ff})^{1/2}$. If r_P is significantly less than 1, the environmental model is rejected and the independent model is accepted. The data on alcoholism estimated $r_P = 0.94 \pm 0.33$, providing no basis for invoking the independent model.

This set of analyses indicates that the sex differences in alcoholism are compatible with a hypothesis of extrafamilial liability factors acting preferentially on the female but with familial factors being equally important in the two sexes. This set of circumstances might occur, for example, if alcoholism were equally hereditary in both sexes but social pressures made women less likely to drink.

Panic Disorder

The high familial prevalence of panic disorder makes it a good candidate for transmission models. A series of nineteen carefully studied kindreds have recently been analyzed.[37,38] First, the Slater computational model was applied to fifteen informative kindreds with the finding of a unilateral to bilateral ratio of 43 to 4, in contrast to the expected one of 31 to 16. The result was highly significant ($p < 0.001$) in favor of the single-gene hypothesis.

The data were then analyzed by the Elston segregation analysis model with the Mendelian hypothesis providing an acceptable fit ($p > 0.1$). The environmental hypothesis was rejected at a highly significant level ($p < 0.001$). The best-fitting Mendelian hypothesis predicted an *A* allele frequency (q) of 0.014, leading to 4.2 percent of the population having the *A* allele in either homozygous (*AA*) or heterozygous (*Aa*) form $[q^2 + 2q(1 - q)]$. The susceptibility parameter (γ) estimated 75 percent of the population to be susceptible regardless of genotype, but the age of onset distribution of the more frequent *aa* genotype ($\mu = 5.9$, $\gamma = 0.22$) effectively ruled out their ever being affected. The mean age of onset for the *AA* and *Aa* genotypes was twenty-two years with a 2 S.D. range of eighteen to thirty-four. Thus, the model assumes a gene present in 4.2 percent of the population, with 75 percent of the carriers being susceptible and their age of onset distribution being such that 95 percent are affected by age thirty-four.

¶ Conclusion

In conclusion, what have transmission models contributed to the field of psychiatry? It is apparent from the foregoing review that they have not answered the basic question of how any mental illness is inherited. However, in fairness, it is probably asking too much, in our present state of knowledge, to expect this kind of conclusion from them. As long as psychiatric data sets contain the kind of diagnostic heterogeneity that has recently been demonstrated in affective disorders, one can hardly expect firm conclusions about the mode of inheritance. Thus, if the models have not lived up to the promise of clarifying inheritance, this may be because our present mathematical sophistication exceeds our diagnostic sophistication. Nevertheless, transmission models have been influential in modifying traditional ways of thinking about the manner in which genes and environment can cause disease. Modern concepts of disease transmission have come a long way from the simplistic Mendelian concepts of a generation ago. When our diagnostic abilities succeed in rivaling our mathematical ones, the means exist for learning much about disease transmission.

¶ Bibliography

1. ANGST, J. *Zur Atiologie und Nosologie Endogener Depressiver Psychosen.* Berlin: Springer Pub., 1966.
2. BÖÖK, J. A. "Schizophrenia as a Gene Mutation," *Acta Geneticae,* 4 (1953): 133–139.
3. CARTER, C. O. "The Inheritance of Congenital Pyloric Stenosis," *British Medical Bulletin,* 17 (1961): 251–253.
4. CAVALLI-SFORZA, L. L., and KIDD, K. K. "Genetic Models for Schizophrenia," *Neurosciences Research Progress Bulletin,* 10 (1972): 406–419.
5. CLONINGER, C. R., REICH, T. and GUZE, S. B. "The Multifactorial Model of Disease Transmission: II. Sex Differences in the Familial Transmission of Sociopathy (Antisocial Personality)," *British Journal of Psychiatry,* 127 (1975): 11–22.
6. ———. "The Multifactorial Model of Disease Transmission: III. Familial Relationship Between Sociopathy and Hysteria (Briquet's Syndrome)," *British Journal of Psychiatry,* 127 (1975): 23–32.
7. CLONINGER, C. R., et al. "Implications of Sex Differences in the Prevalences of Antisocial Personality, Alcoholism, and Criminality for Familial Transmission," *Archives of General Psychiatry,* 35 (1978): 941–951.
8. ———. "Adoption Studies in Psychiatry," *Biological Psychiatry,* 10 (1975): 353–371.
9. CROWE, R. R., and SMOUSE, P. E. "The Genetic Implications of Age-Dependent Penetrance in Manic-Depressive Illness," *Journal of Psychiatric Research,* 13 (1977): 273–285.
10. ELSTON, R. C., and CAMPBELL, M. A. "Schizophrenia: Evidence for the Major Gene Hypothesis," *Behavior Genetics,* 1 (1970): 3–10.
11. ELSTON, R. C., and RAO, D. C. "Statistical Modeling and Analysis in Human Genetics," *Annual Review of Biophysics and Bioengineering,* 7 (1978): 253–286.
12. ELSTON, R. C., and STEWART, J. "A General Model for the Genetic Analysis of Pedigree Data," *Human Heredity,* 21 (1971): 523–542.
13. ELSTON, R. C., and YELVERTON, K. C. "General Models for Segregation Analysis," *American Journal of Human Genetics,* 27 (1975): 31–45.
14. ELSTON, R. C., et al. "A Genetic Study of Schizophrenia Pedigrees. II. One-locus Hypotheses," *Neuropsychobiology,* 4 (1978): 193–206.
15. FALCONER, D. S. *Introduction to Quantitative Genetics.* New York: Ronald Press, 1960.
16. ———. "The Inheritance of Liability to Certain Diseases, Estimated from the Incidence Among Relatives," *Annals of Human Genetics* (London), 29 (1965): 51–76.
17. GERSHON, E. S., and BUNNEY, W. E. "The Question of X-Linkage in Bipolar Manic-Depressive Illness," *Journal of Psychiatric Research,* 13 (1976): 99–117.
18. GERSHON, E. S., BARON, M., and LECKMAN, J. F. "Genetic Models of the Transmission of Affective Disorders," *Journal of Psychiatric Research,* 12 (1975): 301–317.
19. GERSHON, E. S., et al. "Transmitted Factors in the Morbid Risk of Affective Disorders: A Controlled Study," *Journal of Psychiatric Research,* 12 (1975): 283–299.
20. ———. "The Inheritance of Affective Disorders: A Review of Data and Hypotheses," *Behavior Genetics,* 6 (1976), 227–261.
21. ———. "Color Blindness not Closely Linked to Bipolar Illness," *Archives of General Psychiatry,* 36 (1979): 1423–1430.
22. Goetzl, U., et al. "X-Linkage Revisited: A Further Family Study of Manic-Depressive Illness," *Archives of General Psychiatry,* 31 (1974): 665–672.
23. GOTTESMAN, I. I., and SHIELDS, J. "A Polygenic Theory of Schizophrenia," Proceed-

ings of the National Academy of Science, 58 (1967): 199–205.

24. HELZER, J. E., and WINOKUR, G. "A Family Interview Study of Male Manic-Depressives," *Archives of General Psychiatry,* 31 (1974): 73–77.
25. HESTON, L. L. "The Genetics of Schizophrenic and Schizoid Disease," *Science,* 167 (1970): 249–256.
26. JAMES, J. W. "Frequency in Relatives for an All-Or-None Trait," *Annals of Human Genetics,* 35 (1971): 47–49.
27. JAMES, N., and CHAPMAN, C. J. "A Genetic Study of Bipolar Affective Disorders," *British Journal of Psychiatry,* 126 (1975): 449–456.
28. KALLMAN, F. J. *The Genetics of Schizophrenia.* New York: J. J. Augustin, 1938.
29. KIDD, K. K., and CAVALLI-SFORZA, L. L. "An Analysis of the Genetics of Schizophrenia," *Social Biology,* 3 (1973): 254–265.
30. LECKMAN, J. F., and GERSHON, E. S., "Autosomal Models of Sex Effect in Bipolar-Related Major Affective Illness," *Journal of Psychiatric Res.,* 13 (1977): 237–246.
31. LECKMAN, J. F., et al. "New Data Do Not Suggest Linkage Between the Xg Blood Group and Bipolar Illness," *Archives of General Psychiatry,* 36 (1979): 1435–1441.
32. MATTHYSSE, S. W., and KIDD, K. K. "Estimating the Genetic Contribution to Schizophrenia," *American Journal of Psychiatry,* 133 (1976): 185–191.
33. MENDLEWICZ, J., and FLEISS, J. L. "Linkage Studies with X-Chromosome Markers in Bipolar (Manic-Depressive) and Unipolar (Depressive) Illnesses," *Biological Psychiatry,* 9 (1974): 261–294.
34. MENDLEWICZ, J., and RAINER, J. D. "Morbidity Risk and Genetic Transmission in Manic-Depressive Illness," *American Journal of Human Genetics,* 26 (1974): 692–701.
35. MENDLEWICZ, J., et al. "Affective Disorder on Paternal and Maternal Sides: Observations in Bipolar (Manic-Depressive) Patients With and Without a Family History," British Journal of Psychiatry, 122 (1973): 31–34.
36. ———, et al. "Color Blindness Linkage to Bipolar Manic-Depressive Illness: New Evidence," *Archives of General Psychiatry,* 36 (1979): 1442–1447.
37. PAULS, D. L., CROWE, R. R., and NOYES, R. "Distribution of Ancestral Secondary Cases in Anxiety Neurosis (Panic Disorder)," *Journal of Affective Disorders,* 1 (1979): 287–290.
38. PAULS, D. L., et al. "A Genetic Study of Panic Disorder Pedigrees," *American Journal of Human Genetics,* in press.
39. PERRIS, C. "A Study of Bipolar (Manic-Depressive) and Unipolar Recurrent Depressive Psychoses," *Acta Psychiatrica Scandinavica,* Suppl. (1966): 194.
40. ———. "Abnormality on Paternal and Maternal Sides: Observations in Bipolar (Manic-Depressive) and Unipolar Depressive Psychoses," *British Journal of Psychiatry,* 118 (1971): 207–210.
41. REICH, T., CLAYTON, P. J., and WINOKUR, G. "Family History Studies V: The Genetics of Mania." *American Journal of Psychiatry,* 125 (1969): 1358–1369.
42. REICH, T., CLONINGER C. R., and GUZE, S. B. "The Multifactorial Model of Disease Transmission: I. Description of the Model and Its Use in Psychiatry," *British Journal of Psychiatry,* 127 (1975): 1–10.
43. REICH, T., JAMES, J. W., and MORRIS, C. A. "The Use of Multiple Thresholds in Determining the Mode of Transmission of Semi-Continuous Traits," *Annals of Human Genetics,* London, 36 (1972): 163–184.
44. REICH, T., et al. "The Use of Multiple Thresholds and Segregation Analysis in Analyzing the Phenotypic Heterogeneity of Multifactorial Traits," *Annals of Human Genetics,* London, 42 (1979): 371–390.
45. ROSENTHAL, D. *Genetic Theory and Abnormal Behavior.* New York: McGraw-Hill, 1970.
46. SLATER, E. "The Monogenic Theory of Schizophrenia," *Acta Genetica,* 8 (1958): 50–56.
47. ———. "Expectation of Abnormality on Paternal and Maternal Sides: A Computational Model," *Journal of Medical Genetics,* 3 (1966): 159–161.
48. ———, and COWIE, V. *The Genetics of Mental Disorders.* London: Oxford University Press, 1971.
49. SLATER, E., and TSUANG, M. T. "Abnormality on Paternal and Maternal Sides: Observations in Schizophrenia and Manic-Depression," *Journal of Medical Genetics,* 5 (1968): 197–199.
50. SLATER, E., MAXWELL, J., and PRICE, J. S.

"Distribution of Ancestral Secondary Cases in Bipolar Affective Disorders," *British Journal of Psychiatry,* 118 (1971): 215–218.

51. TSUANG, M. T. "Abnormality on Paternal and Maternal Sides in Chinese Schizophrenics," *British Journal of Psychiatry,* 118 (1971): 211–214.
52. VAN EERDEWEGH, M. R., GERSHON, E. S., and VAN EERDEWEGH, P. "X-Chromosome Threshold Models of Bipolar Illness." Paper presented at the 129th Annual Meeting of the American Psychiatric Association, 1976.
53. WINOKUR, G. "Genetic Findings and Methodological Considerations in Manic Depressive Disease," *British Journal of Psychiatry,* 117 (1970): 267–274.
54. WINOKUR, G., and REICH, T. "Two Genetic Factors in Manic-Depressive Disease," *Comprehensive Psychiatry,* 11 (1970): 93–99.
55. WINOKUR, G., and TANNA, V. L., "Possible Role of X-Linked Dominant Factor in Manic-Depressive Disease." *Diseases of the Nervous System,* 30 (1969): 89–94.
56. WINOKUR, G., CLAYTON, P. J., and REICH, T. *Manic Depressive Illness.* St. Louis: C. V. Mosby, 1969.

CHAPTER 5

ADVANCES IN PSYCHOPHARMACOLOGY

John M. Davis and Lolita O. Ang

THE DISCOVERY of the therapeutic uses of the antipsychotic drugs, the tricyclics, the MAO inhibitors, lithium, and the benzodiazepines have revolutionalized psychiatry.[110] In this chapter, we provide an account of advances in psychopharmacology that have occurred since the appearance of the original section on psychopharmacology in volume V of the *American Handbook of Psychiatry.*

Before we proceed, let us review. Between 1949, when lithium was discovered, and the publication of the most recent chapter on psychopharmacology in the *American Handbook of Psychiatry,* various psychoactive drugs were introduced. The best way to review their status is to give an overall summary of their efficacy. For purposes of comparison and a general presentation, we have compiled summary data to show the efficacy of new treatment in comparison with the older placebo treatment. These are presented in the context of similar data for the classical antibiotics—streptomycin for tuberculosis and penicillin for pneumococcal pneumonia.

Summary: The Power of Psychoactive Drugs

The drug-placebo difference is a meaningful measurement of the overall efficacy of a specific drug administered to patients with a specific disease. In terms of drug-placebo differences, the advance in psychotropic drugs is comparable to major innovations in chemotherapy.

Table 5–1 summarizes data from the National Institute of Mental Health (NIMH) Collaborative Study Number 1 on the efficacy of the treatment of acute schizophrenia with drugs and placebo and from studies by the British Medical Research Council on the efficacy of treatment for tuberculosis with streptomycin. Data on the treatment of pneumococcal pneumonia with penicillin and sulfonilamide and the use of drugs in surgery are also included. The drug-placebo differ-

TABLE 5-1 **Percentage of Patients Who Do Well on Various Treatments**

		WELL	POOR	R
Antipsychotic for treatment of acute schizophrenia	Drug	70%	25%	
	Placebo	30%	75%	.45
Maintenance antipsychotic for prophylaxis	Drug	80%	20%	
	Placebo	47%	53%	.34
Imipramine acute depression	Drug	65%	35%	
	Placebo	32%	68%	.33
Tricyclic prophylaxis of depression	Drug	73%	27%	
	Placebo	48%	52%	.26
Lithium acute mania	Drug	73%	28%	
	Placebo	34%	66%	.38
Lithium prophylaxis of mania depression	Drug	63%	37%	
	Placebo	21%	79%	.43
Streptomycin for tuberculosis	Drug	69%	33%	
	Standard	31%	67%	.36
Penicillin for pneumococcal pneumonia	Penicillin	93%	6%	
	Sulfanilamide	88%	11%	.10
Drugs in surgery—1964–72	New	63%	37%	
	Old	57%	43%	.06

ence is expressed as a product-moment correlation coefficient *R,* in which the higher the correlation, the bigger the difference.

Although we caution against an overly concrete interpretation of such data, it appears that the discovery of effective psychotropic drugs is as much a breakthrough for psychiatry as the discovery of antibiotics was for medicine. Of course, no quantitative comparison can be made on the efficacy of different drugs for different disorders. However, the fourfold drug-placebo differences presented here facilitate qualitative comparisons between diseases treated and drug effects.

There have been many advances in psychopharmacology since the previous volumes of the *American Handbook.* More is known about plasma level and therapeutic efficacy and the dosages. Within the space limitations of this chapter, we shall discuss recent advances in psychopharmacology believed to be particularly significant. In the process of choosing studies to review or topics to cover, we have, of necessity, been selective. There are some important problems that have been omitted because of few significant advances or a lack of firm knowledge.

¶ Antipsychotic Drugs

Since the cause of schizophrenia is unknown, the exact mechanism by which the antipsychotic drugs biologically or psychologically benefit schizophrenics cannot be determined. We have previously examined the different symptoms that are ameliorated by antipsychotic drugs and find that these drugs lessen symptoms typical of schizophrenia, be they fundamental or accessory.[25] If the antipsychotic drugs are, in essence, antianxiety agents, one would expect the greatest effect to be on anxiety and a lesser effect on symptoms more distinctly related to anxiety. This is not the case. All schizophrenic symptoms of abnormality appear to be benefitted by the antipsychotic drugs. Generally, schizophrenia is quantitated by rating scales which, in essence, are Kraepelin in orientation in that they evaluate the severity of symptoms and arrive at a total score as a summation of the severity of each of the individual symptoms. It is of psychological interest to find out whether the antipsychotic drugs relieve the thought disorders of schizophrenia as well as the other symptoms. Although Kraepelin discussed the typical thought disorder that occurs in schizophrenia, Bleuler was responsible for greater focusing and discussion of thought disorders, which may be, in some sense, fundamental to schizophrenia. Holzman and Johnston[86] have developed a psychological instrument to quantitate the degree of schizophrenic thought disorder and to verify that thought is disturbed in schizophrenia. It will be of interest to see if antipsychotic drugs have a beneficial effect on schizophrenic thought disorder, and, if so, whether this improvement is parallel to the extent and rate of the disappearance of the symptoms. Improvement in schizophrenic symptomatology as a function of drug treatment, was further assessed by the authors[37,46] in collaboration with Holzman, Ericksen, and Hurt. Measures were obtained from patients suffering with thought disorder before and after drug administration. This was accomplished by means of a standard rating scale and responses to items from the Wechsler Adult Intelligence Scale and Rorschach cards. Blind assessments of the degree of thought disorder were performed by the psychologists. The most salient findings to emerge were the substantial reduction in psychotic symptoms among the study sample and the observation that the decrease in thought disorder occurred to the same degree and with the same time course as the schizophrenic symptomatology (see figure 5–1).

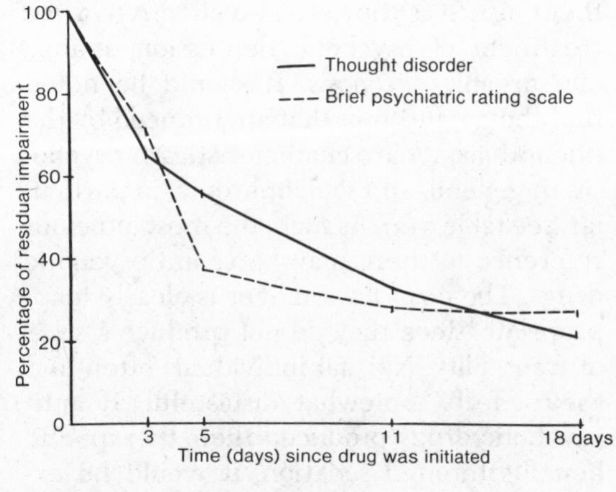

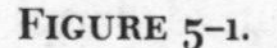
FIGURE 5–1.

Psychologists continue to speculate as to the underlying psychological and functional abnormalities of schizophrenia. It has been suggested that maintenance of a psychological set is problematic for such individuals, that is, they show poor concentration on certain tasks. Spohn and coworkers[191] studied the effects of antipsychotics on the performance of chronic schizophrenic patients. Random assignment to chlorpromazine or placebo was arranged following a six-week washout period. Some of the patients (sixteen out of sixty-three) relapsed during this period, with a relapse rate of 17 percent per month. Forty patients in this study on placebo baseline were examined by means of a psychological test battery during the washout period, and again while on chlorpromazine or placebo at one, four, and eight weeks. Chlorpromazine proved to be more effective than placebo in bringing about improvement, but due to the dropout of a quarter of the samples during the washout period, the drug-placebo differences were probably underestimated. The patients were also given an assortment of protocols aimed at assessing malfunctioning attentional, perceptual, and psychophysiological skills.

It was reported that chlorpromazine reduced overestimation and fixation time on a test of perception and increased verbal accuracy of perceptual judgment. Powers of attention and concentration were improved. The result common to these tests may be the ability to attend appropriately to the task in question. These results accord well with the normalization of symptoms and/or thought disorder just discussed.

In sum, the action of antipsychotic drugs suggests a normalizing effect. They reduce typical hallucinations and delusions. They speed up retarded schizophrenics and slow down the more excited ones. Yet, to classify these agents as antischizophrenic fails to do them justice as they are also effective in the treatment of psychotic depression, mania, and organic psychosis. It should be noted that those symptoms that are reduced by the phenothiazines are characteristics of psychosis, in general, and schizophrenia, in particular (see table 5–2). As such, the most judicious reference to them may be as antipsychotic drugs. The term tranquilizer is clearly inappropriate, since they do not produce a state of tranquility. Normal individuals often find their effects somewhat distasteful. If antipsychotic drugs produced their therapeutic benefits through sedation, it would be expected that sedative drugs would be more efficacious than nonsedative antipsychotics. This is not the case. Furthermore, antipsychotics that are maximally stimulating are as potent as antipsychotics with maximal sedative properties. It is, therefore, an error to conceive of these drugs as a distinctive or special form of sedative.

Does Any Subtype of Schizophrenic Respond Best to Drugs?

Health Collaborative Study No. I has been examined to see which symptoms predict the greatest drug response (drug-placebo difference) and to discover in which subtype of patients (as defined by symptoms) the biggest drug response exists. In essence, there were few significant differences, and it would seem, at least within the limitations of this method of analysis, that there is no marked difference between subtypes of schizophrenics as defined by their response to drugs.[25] Klein and coworkers[95] report a better response to drugs in process schizophrenics. However, Judd and coworkers[90] report that some reactive nonparanoid schizophrenics respond better to placebo. Since these results are apparently contradictory, more work is needed on this question.

Does Failure to Use Antipsychotics Cause Harm?

An important question is whether failure to treat with antipsychotic drugs for an extended period of time will harm the patient permanently. May and Tuma,[121] in an important study, randomly assigned first admission schizophrenics to receive drugs or no drugs and psychotherapy or no psychotherapy. Length of treatment was an essential methodological variable in this study. The experimental design called for six months to one year of treatment, either with or without drugs, which was unusual because most controlled studies last only four to six weeks. Patients who received drugs did substantially better on a number of variables than patients who did not receive drugs (see table 5–3). After the study ended, the patients were followed for the next three to five years. They were able to receive both indicated treatments during this time. It was then possible to determine if failure to treat with drugs resulted in permanent harm to the patient. The authors calculated the number of days in the hospital during the follow-up period, roughly equating many brief hospitalizations with fewer long hospitalizations (see figure 5–2). Patients who initially received drugs did much better in the follow-up period than those who did not. This indicates that withholding antipsychotics during a long hospitalization may result in some sort of permanent harm. The mechanism for this is unknown. Perhaps the disease does not progress so much with medication because the dopamine blockade of the antipsychotics somehow reduces the biological aspects of the psychosis. An alternative explanation is that long hospitalizations permanently sever social ties that are important to a patient's continuous functioning. Whatever the answer to this, it does appear that drugs alter the natural history of schizophrenia.

Role of Psychological Intervention in Acute Treatment

Goldstein and coworkers[62] emphasized a statistically significant effect on the prevention of relapse and rehospitalization due to family therapy. Patients were hospitalized initially for two weeks, at which time outpatient family therapy was introduced for six

TABLE 5–2 **Effect of Phenothiannes on Symptoms in Schizophrenia**[a]

BLEWER'S CLASSIFICATION OF SCHIZOPHRENIC SYMPTOMS	V.A. STUDY NO. 1	V.A. STUDY NO. 3	KURLAND 1962	NIMH-PSC NO. 1	GORNAM AND POKORNY, 1964 VS. GROUP PSYCHOTHERAPY
Fundamental symptoms					
Thought disorder	++	++	++	++	++
Blunted affect-indifference				++	+
Withdrawal-retardation	++	++	o	++	++
Autistic behavior-mannerisms	++	++	o	++	+
Accessory symptoms					
Hallucinations	++	++	+	+	o
Paranoid ideation	o	++	o	+	+
Grandiosity	o	o	o	o	+
Hostility-belligerence	++	++	H.R.	+	+
Resistiveness-uncooperativeness	++	++	H.R.	++	++
Nonschizophrenic symptoms					
Anxiety-tension-agitation	o	o	H.R.	+	o
Guilt-depression	++	o	o	o	o
Disorientation				o	
Somatization					o

[a] ++. Symptom areas showing marked drug-control group differences; +, those showing significant but less striking differences; o, areas not showing differential drug superiority. H.R. heterogeneity of regression found on analysis of covariance of the measures indicated. (This invalidates this particular statistical procedure but does not mean that there was no drug effect [Cole, et al. 1966].)

TABLE 5–3 **Assessment of Outcome in Schizophrenic Patients Treated With and Without Antipsychotic Drugs and Psychotherapy (May, 1978)**

	No Drugs		Drug	
	NO PSYCHOTHERAPY	PSYCHOTHERAPY	NO PSYCHOTHERAPY	PSYCHOTHERAPY
Percent released	58.8	64.4	95.1	96.3
Nurses' rating MACC total	37.7	37.7	47.8	48.1
Menninger nurses' health-sickness rating	26	22.7	28.9	29.8
Nurses' idiosyncratic symptoms (125-X)	37.3	28.8	65.7	74.2
Therapists' rating on symptom rating sheet (50-X)	22.1	20.9	26.4	27.3
Analysis rating of insight	3.4	3.3	3.7	4.1

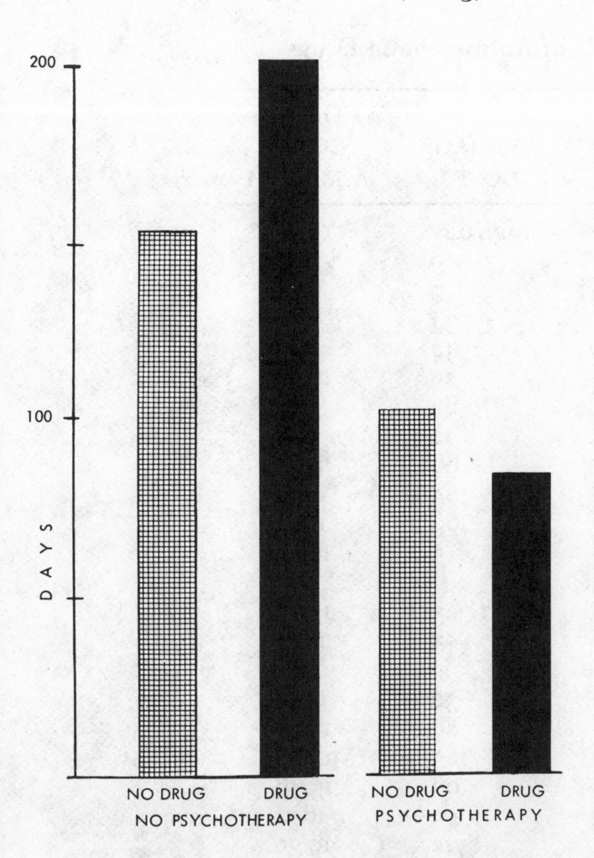

FIGURE 5–2. Three years follow-up after first release.

weeks thereafter. Since this was only a two-week (mean fourteen ± six days) hospitalization, it may be expected that the patient's illness had not remitted completely by the time of discharge, making the psychological support, insight, family care, and patient psychotherapy essential to the treatment program.

Cost of Drug

The continuing rise of the cost of the drugs is an important issue in drug therapy. Cost is not a function of dosage (see table 5–4). For example, the cost of a 25 mg chlorpromazine tablet is comparable to that of a 100 mg tablet. Considerable savings in money and time, however, are to be derived from administering the largest available form once a day. Furthermore, a bedtime or evening dose may be less likely to be forgotten and easier to monitor than a three-times-a-day regimen. Spansules or other delayed release oral forms of some phenothiazines are available by prescription. However, these more expensive preparations have no discernible advantage over the standard tablet form. Since the cost of antipsychotic medication is measured generally in terms of cents per day, it is small in comparison to the amount paid for hospitalization. Costs of medications to the pharmacies are given in table 5–5.

High Dosage Phenothiazine Treatment

A broad range exists between effective dose and toxic overdose of antipsychotic agents. Patients who served as research subjects have been treated safely with ten to one hundred times the agreed-upon therapeutic dose (for example, 1,200 mg fluphenazine). While one should always exercise caution, it is safe to utilize substantially higher doses than those prescribed in the literature. Given reasonable clinical indications, dosage may be increased without great concern.

There is adequate evidence that the antipsychotic drugs help schizophrenia, but little is known about their effective use with respect to optimal load strategies, optimal plasma levels, and so forth. We will briefly review evidence relating both optimal doses and optimal plasma levels. Recently it has been suggested that a faster initial result and perhaps a better ultimate outcome could be achieved by giving a very high dose in the first few days of treatment. Such strategies have been called loading-dose strategies, rapid tranquilization, or high-dose strategies. Two different dimensions are involved. Too often, high doses are said to be better than low doses. We would like to disassociate two concepts, the concept of titration and the concept that a high dose is superior to a low dose. We believe that to say a high dose is better than a low dose is not an accurate representation of the problem. We would rather consider the dosage issue in terms of dose response.

If one is on the linear portion of the dose-response curve, a high dose is better than a

TABLE 5–4 **Comparative Costs of Antipsychotic Drugs**

GENERIC NAME	TRADE NAME	AVERAGE DOSE*	WHOLESALE COST† A MONTH
		mg./day	
Fluphenazine	Permitil	9	$ 4.31
Fluphenazine HCl	Prolixin HCl	9	6.05
Chlorpromazine	Thorazine	734	6.80
Molindone	Moban	44	7.00
Trifluoperazine	Steiazine	20	7.84
Loxapine	Loxitane	64	9.03
Haloperodil	Haldol	12	9.19
Butaperazine	Repoise	66	9.93
Carphenazine	Proketazine	183	11.55
Acetophenazine	Tindal	169	11.93
Fluphenazine decanoate	Prolixin Decanoate	5	12.34
Chlorprothixine	Taractan	323	13.08
Fluphenazine enanthate	Prolixin Enanthate	5	14.18
Thiothixene	Navane	32	14.19
Piperacetazine	Quide	80	14.60
Prochlorperazine	Compazine	103	14.88
Perphenazine	Trilafon	66	15.76
Mesoridazine	Serentil	411	16.21
Thioridazine	Mellaril	712	19.30
Triflupromazine	Vesprin	205	26.03

*Empirically defined average dose for acute treatment (Davis, 1976).

†Cost to retailer for 1-month supply of drug for average acute treatment, based on wholesale price of least expensive (largest) tablet or bottle purchased in largest quantity. Actual cost to the consumer is considerably greater because of physician's ordering of smaller-dose tablets.

TABLE 5–5 **Cost per Milligram for Different Tablet Size as Percentage of Cost of Most Inexpensive Tablet Size**

DRUG	200	100	50	25	16	125	10	8	5	4	25	2	1	0.5
Chlorpromazine	100	180	300	500			1060							
Triflupromazine			100	150			250							
Thioridazine	100	138	231	415			762							
Prochlorperazine				100			210		323					
Perphenazine					100			149		246		358		
Fluphenazine									100		154		270	
Trifluoperazine							100		153			353	553	
Carphenazine		100	165		277									
Butaperazine				100			197		310					
Mesoridazine		100	167	289			545							
Piperacetazine				100			167							
Haloperidol									100			178	238	332
Chlorprothixene		100	165	271			494							
Thiothixene							100		151				291	450

low dose. If one is above the maximal point on the dose-response curve, a high dose is not better than a low dose (see figure 5–3). Indeed, it could be worse. Such patients may suffer more side effects. For theoretical reasons the dose-response curve is a better way to conceptualize the dosage issue than the nonspecific "high-dose-is-good" approach. There is a limited amount of evidence on dose-response relationship in psychiatry, and we will review some of the pertinent studies. In collaboration with Ericksen, Holzman, and Hurt, the authors[37] have performed one study investigating loading doses. The rapid-tranquilization question has also been investigated by Donlon and coworkers.[39] Several groups have studied the use of very high doses in treating acute patients to establish if megadoses achieve a better result than normal doses. With chronic patients or treatment-resistant patients, several groups have investigated whether higher doses are better than lower doses for sustained or maintenance treatment and these studies are also reviewed. It is curious that a certain subgroup of patients displays no significant benefit from any antipsychotic drug medication. The possibility of remission in these patients, had they been treated with higher than normal doses, remains an issue for research. We will, therefore, review the aforementioned studies that utilize a higher than normal dose.

A study of this issue was carried out in our

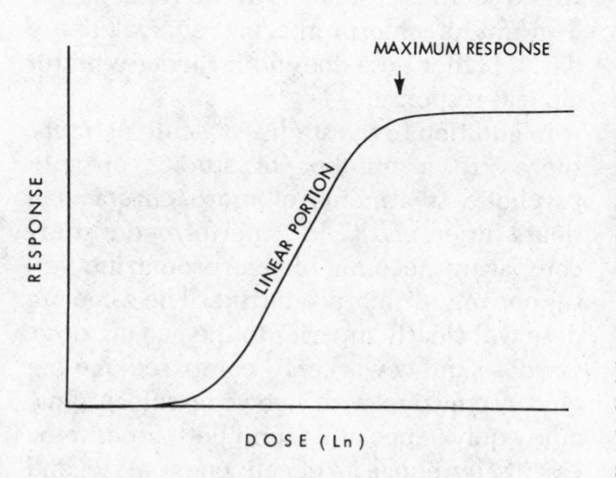

FIGURE 5–3.

laboratory by means of a double-blind design.[46] It was found that a five-day loading dose of 60 mg of haloperidol, administered intramuscularly, was no more effective than 15 mg orally four times a day, both at day five and at three weeks. This raises the question: Does an unusually high loading (digitalizing) dose encourage more rapid improvement during the initial treatment stages than does a normal dose? Given the assumption that this is true, it becomes important to ask if the patients in question maintain a better remission following gradual decrease of the dosage. In other words, is the patient's condition at three weeks more favorable than the condition that would have been achieved if a regular dose had been used? Patients who were acutely decompensated schizophrenics were randomly assigned to one of two groups. Fifteen mg of haloperidol given orally is approximately equivalent to 940 mg of chlorpromazine, a quantity which we consider to be a high-normal dose. Sixty mg given intramuscularly would be the equivalent of 3,600 mg of chlorpromazine, or more, since the bioavailability of intramuscular haloperidol is superior to that which is administered orally. Thus, patients were given massive quantities of haloperidol (five day loading, high-dose group) and were subsequently compared to a group that was given more moderate amounts of the drug (normal dose group). Their medication was reduced to a normal dosage of 15 mg of haloperidol after five days of loading and a few days of tapering, while the normal dosage group received a constant dosage (15 mg) throughout the duration of the study. Between-group evaluation, administered on a double-blind basis, was made on the following measures: global scale, the Brief Psychiatric Rating Scale (BPRS), the New Haven Schizophrenia Index, and the Holzman-Johnston Thought Disorder Ratings. Analysis revealed that the therapeutic outcomes were identical at both points; however, the loading-dose group showed more side effects, in particular, dystonia. Thought disorder improved to the same degree among both patient populations. Such an estimate is imprecise but indicated only as an order of magnitude "guess."

Relevant to this discussion is another investigation, not unlike our own. Donlon and coworkers[39] examined the influence of the rapid treatment of psychosis as compared to a standard dosage of 20 mg of fluphenazine. A maximum loading dosage of 80 mg per day was utilized. The decompensated schizophrenics were followed only a few days—seven days maximum. (During days two through seven of the study, the high-dose group actually underwent recurrent deterioration at a mean of 40 to 74 mg.) In sum, both dose strategies yielded similar results and are, therefore, consistent with our aforementioned findings. Twenty mg of fluphenazine for the standard dose is the equivalent to 1,667 mg chlorpromazine. The high-dose group received approximately 3,000 to 5,000 mg chlorpromazine equivalence.

Furthermore, Wijsenbeek, Steiner, and Goldberg[200] completed a double-blind study in which they assessed the efficacy of 60 to 600 mg of trifluoperazine given to newly admitted schizophrenic patients. Both treatments were found to be equally effective. As such, megadoses do not appear to produce substantially greater improvement than regular doses, at least with newly admitted schizophrenics. One should recall that 60 mg of trifluoperazine is equivalent to approximately 2,000 mg ($60 \times 100/2.8$) of chlorpromazine per day. If a normal dose is equal to 700 mg per day, this amounts to thirty times the normal dose, while 600 mg of trifluoperazine (or 20,000 mg chlorpromazine equivalence) is about thirty times the normal antipsychotic dose. Findings reported by Quitkin, Rifkin, and Klein[145] further substantiate this. They showed that 1,200 mg per day fluphenazine (100,000 mg chlorpromazine equivalent) is no more effective than 30 mg/day (2,500 mg chlorpromazine equivalent).

Goldstein and coworkers[62] divided a group of 196 acute first admission schizophrenics into four subgroups: low dose (6.25 mg q/2 wk) or high dose (25 mg q/2 wk) fluphenazine, family (FT+) therapy or no family (FT−) therapy, for an eight-week trial (two weeks inpatient, six weeks outpatient) using chlorpromazine equivalence of 73 versus 293. These assignments were arranged on a random basis. A clear-cut dose-response relationship emerged. High dose was associated with significantly fewer relapses ($p < .0002$, Fisher Exact Test). Additionally, the family therapy significantly and independently prevented relapse, ($p < .05$). In sum, the numbers of relapses per sample size within the four subgroups were: high dose FT+ = 0/23; high dose FT− = 3/26; low dose FT+ = 2/21; low dose FT− = 5/16. The six-month follow-up showed that both the dose-response drug effect and the family therapy effect endure, and actually become larger; outcome data for these same four groups were 0/23, 5/29, 5/23, and 10/21, respectively. It should be noted that 73 mg per day is less effective than 293 mg per day, given the conversion to chlorpromazine equivalents. This is further evidence that the dose range of 50 to 500 mg is located along the linear portion of the dose-response curve.

Furthermore, for acute patients, it is impossible to bracket precisely the optimal point on the dose-response curve. Very massive doses are no more effective than the equivalent of either 2,000 mg or 2,500 mg of chlorpromazine. A very high loading dose of haloperidol is no more effective than an oral dose of haloperidol, the equivalent of approximately 900 mg of chlorpromazine. It would seem that if one is in the range of 1 or 2 grams of chlorpromazine equivalence a day, a higher dose does not produce a better clinical response.

In addition to the studies of acute patients, there are a number of studies of antipsychotic treatments of more chronic patients. Prien and Cole[142] performed a study comparing 2,000 mg of chlorpromazine versus 300 mg of chlorpromazine. The 2,000 mg dose was clearly superior to the 300 mg dose. Gardose and coworkers[52] compared 440 mg chlorpromazine with 1,760 mg chlorpromazine equivalence and found both doses to be essentially equal in effectiveness. A second study[144] found 15 mg of trifluperazine (equivalent to 535 mg of chlorpromazine) to be as effective as 80 mg of trifluoperazine (2,850 mg of chlorpromazine equivalence). This suggests that 535 mg closely approximates the maximal effective dose, and that an excessively high dose did not lead to additional improvement for this particular population. Clark and coworkers[19] showed, however, that 300 mg or less of the drug is probably not an effective therapeutic dose for chronic patients—600 mg is better than 300, which is better than 150 mg.

Brotman and coworkers[14] initiated a double-blind, noncrossover, random trial which included eighty patients. They were divided into groups receiving either placebo, or 15, 30, or 60 mg per day butaperazine in chlorpromazine equivalence of 167, 333, and 667 mg. The latter group showed a significantly better response on the uncooperative subscale of the BPRS. Analyses of items on the remaining five subscales yielded trends favorable to the high-dose group on the following: emotional withdrawal, tension, mannerism, posture, and flat affect. These differences were small with only one measure achieving statistical significance. However, they afford some limited evidence that 667 mg is better than 333 or less. These results were comparable to those described in the Clark study.[19]

Another study[16] compared 10 mg to 100 mg of trifluoperazine (357 to 3,570 mg chlorpromazine equivalence) using a sample population composed of very chronic patients. No substantial differences emerged upon comparison of these doses. Given the chronic nature of these patients, this study addresses the topic of maintenance; 357 mg of chlorpromazine equivalence would probably constitute an adequate maintenance dose.

These various studies, when examined collectively, reveal considerable variation with regard to the nature of the sample and methodologies employed. However, despite these disparities, an approximate dose response curve can be drawn. Doses approaching either 300 mg or 150 mg of chlorpromazine are rather low for optimal treatment, at least for some patients. Entry of these numbers to the dose-response curve suggests that 150 and 300 mg (chlorpromazine equivalence) are, in a very rough sense, on the linear portion of the curve. More massive quantities are no more effective than those of approximately empirical range 357, 440, 535 mg (chlorpromazine equivalence). Clinically different patients require different doses and, in some cases, higher than normal doses become the treatment of choice. These dose-response considerations are intended as a statement of average dose to acquaint the reader with the nature and implications of the response relationship as a framework in which to evaluate these clinical phenomena. The dose-response curve is shifted to the right for patients who require high doses. Cases of increased sensitivity to the drug would necessitate a shift to the left.

In considering relationships between dosage and therapeutic efficacy, attention should be directed toward the dose-response curve in figure 5–3. It can be observed that dosage increments along the linear portion of the curve are associated with a more favorable response. An inflection point is approached and diminishing clinical returns are apparent upon reaching the top of the linear portion of the dose-response curve. After this, as the dose is increased only a minimal increase is observed in the clinical response, with virtually no increases in clinical response as the dose is further increased. The inflection point, or the place at which the linear portion changes to "diminished returns," is often referred to as the optimal portion of the dose response curve. In essence, all of the clinical response that is potentially achievable occurs at this point. From the foregoing, it may be suggested that a 800 $\pm$ 200 mg chlorpromazine equivalent is located at, or slightly above, the "optimal point."

Distinctions must be drawn between the use of moderately high doses, such as the double normal dose used by Prien and Cole,[138] (not reviewed here) versus the megadose strategy used by Quitkin, Rifkin, and Klein,[145] which was one hundred times the normal dose. As mentioned earlier, one must exercise discretion and guard against an overly enthusiastic attitude with regard to the potential of megadose antipsychotic drug treatment. A more modest increase in dosage may clearly benefit some acute or subacute patients. In selective cases, perhaps, megadose treatment may be tried experimentally. Given the larger clinical literature on megadose fluphenazine therapy, these drugs may be more suitable for high dosage use. Also relevant is the general absence of undue toxicity resulting from larger doses of this drug, so that megatherapy may be introduced without a great deal of concern. Patients who appear to resist treatment may receive at least a trial with a high dose. In sum, there is a grave lack of definitive research on dose levels and dosage response curves for the antipsychotic drugs. Continued efforts should yield information that will assist physicians who must arrive at decisions concerning individual drug treatment programs.

Chlorpromazine Blood Levels

There are reports of patients who respond only to relatively high doses of a given drug. In contrast, there are certain patients for whom only low doses have been beneficial. Studies that attempt to relate blood levels of chlorpromazine to therapeutic improvement and side effects are important in this regard.[31,32] Enormous variability in blood levels may occur with comparable doses. Some individuals receiving a moderate dose of chlorpromazine may show extremely high blood levels with excessive sedation, and striking improvement may be observed only upon dosage reduction. This type of patient may have a defective metabolism and, perhaps, have built up a high toxic level of blood chlorpromazine. It is also possible to find patients with extremely low blood levels, despite their elevated doses. These patients may metabolize chlorpromazine so rapidly that, despite very high doses, the brain is deprived of adequate amounts of chlorpromazine.

A basic assumption underlying plasma level studies is that the rate of metabolism or other factors affect the amount of drug at the receptor site. As such, at a uniform dose, variable amounts of drug reach the receptor site because of individual differences in metabolism. It follows that the plasma level, clinical-response curve is essentially a dose-response curve. A possible plasma level, clinical-response curve is depicted in the classic sigmoid curve in figure 5–4. The first portion of the curve shows the relative lack of clinical response with small doses of drug. The linear portion of the curve represents a more favorable clinical response—when more drug reaches the receptor site. Beyond the linear part of the curve, as larger amounts of the drug are introduced, an area of diminishing returns is observed. A plasma level-toxicity relationship also occurs with various side

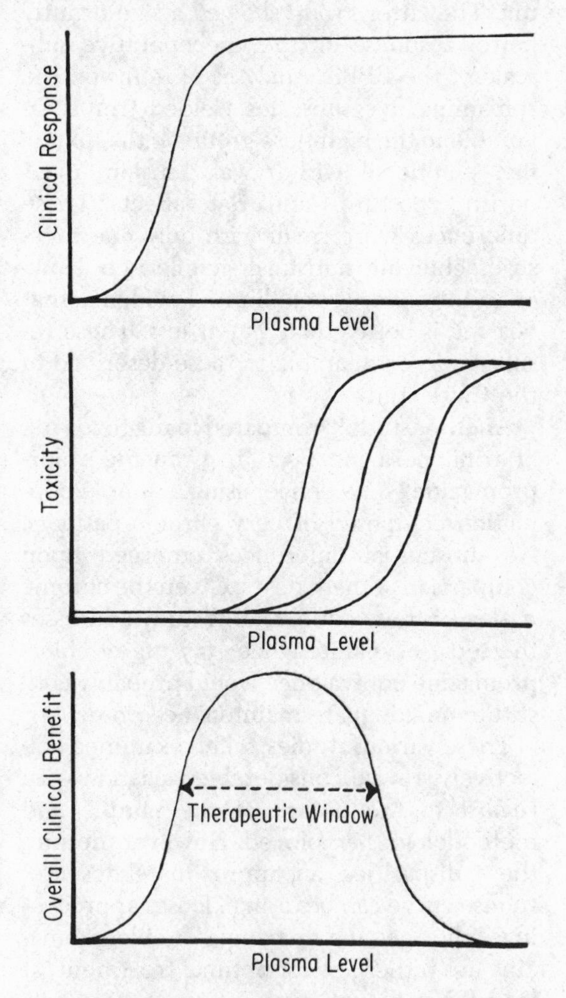

FIGURE 5–4. Theoretical concept of therapeutic window.

effects (middle panel). If the clinical benefit is plotted for the patient against plasma level, it might be anticipated that as plasma levels increase, a better clinical response takes place. Leveling off of the curve indicates the region of optimal benefit. When plasma levels are notably elevated, the resulting detrimental side effects may cancel out any beneficial therapeutic effects. The inverted U-shaped curve, also pictured in figure 5–4, represents the so-called "therapeutic window." In the area below the therapeutic window, insufficient drug reaches the receptor site to produce the desired clinical response. The upper limit can be defined either by toxicity, by a paradoxical pharmacological response, or both. The description of information that indicates whether one or both effects may be occurring is important. Given that various conceptualizations of the upper end of the therapeutic window exist, it is critical that we achieve clarity as to the terms under discussion.

By combining the therapeutic curve and the side effects curve, we can derive the inverted U-shaped curve. Increases in the plasma level, for some drugs, may be associated with a negative therapeutic effect or side effects. Certain agents may stimulate a receptor within a particular range of concentration but, at a higher range, inhibit the receptor. In other words, the therapeutic effectiveness of some drugs may diminish at high plasma levels. As a rule, the majority of drugs (but not all) show increasing side effects with increasing dosage. A drug that displays neither of these effects might be missing the downward portion of the inverted U-shaped curve. Expressed differently, the upper portion of the therapeutic window will be absent.

The concentration of the drug at the particular receptor site is the parameter that, ideally, should be assessed. This is often governed by factors that suggest that concentrations of the drug at these sites may be directly proportional to the plasma levels. As such, plasma levels may constitute valid indices of this event.

Since psychotropic drugs are highly protein bound, it is possible that slight differences in binding could result in different amounts of drug reaching the brain. It is difficult to measure free drug in plasma, but if the same physico-chemical properties govern how much drug passes into red cell that govern how much drug passes into the brain, perhaps red cell levels would be a better correlate to brain than plasma levels.

Data obtained from two controlled studies suggest that some nonresponders have low red blood cell (RBC) levels and may be quick metabolizers (or poor distributors to tissue) and, hence, may show low brain levels of the drug.[17,54] It is possible that such a state of affairs may be responsible, at least in part, for any failure of the drug treatment.

There were not enough subjects in these studies with high levels to prove that there are nonresponders due to excessively high RBC and presumably central nervous system (CNS) levels, but there is some suggestion that it may be true. This aspect is clearly unproven. The majority of the patients did have low RBC levels, and this suggests that the most common clinical pharmacologic reason for nonresponse may be at the lower end of the therapeutic window. The most nonresponsive of all patients are certain patients hospitalized in chronic state hospitals with a long history of failure to respond to drugs. Smith and his coworkers[188] studied a population of extremely poor responders. These patients, treated in an identical protocol, had extremely low plasma and RBC levels. Similarly, patients in an acute hospital who showed no response and were scheduled for transfer to the state hospital or otherwise labeled as extreme nonresponders also showed extremely low red blood cell and plasma levels. There are several studies utilizing low doses of chlorpromazine which found a slight tendency for a worse clinical response in those patients who had the lowest plasma levels.[114,163,172] These differences were modest in direction but consistent with those previously mentioned.

In a pilot study in our laboratory[43], we have also found an inverted U-shaped relationship between plasma fluphenazine levels and clinical response. Although this result was statistically significant, the sample size was relatively small.

We should add that the proper methodology for plasma level studies requires a fixed dose or doses. If dose is varied with clinical response, then the experiment is meaningless. Such studies are difficult to do and there is limited literature on the subject. The limited data just reviewed indicate that variations in rate of metabolism and distribution may be one of the reasons the clinician should adjust the dose to clinical response. The question always arises as to when a research finding is ready for clinical application. At this time, there is not enough evidence relating plasma or red blood cells to therapeutic efficacy to form definitive conclusions. Such conclusions as may be drawn provide preliminary support to the plasma and/or RBC level hypothesis; however, further research is clearly required for verification. The impressive variability in plasma and RBC levels may be one reason why different patients require different doses to achieve similar results. The sensitivity of end organs may also play a role. The clinician should adjust the dosage with regard to the therapeutic responses and side effects.

As an aside, brain sensitivity to drugs may be an inpatient variable. Maxwell and coworkers[120] have illustrated the ability of chlorpromazine to produce central behavioral toxicity in hepatic coma to be an enhanced response of the brain to chlorpromazine's sedative properties. Plasma levels are normal so that such potentiation of hepatic coma does not reflect an impairment in drug metabolism. At present, the methods available for measuring blood levels of antipsychotic drugs remain highly technical and complex. It may be several years before it will be possible for psychiatrists routinely to examine unresponsive patients to make certain that appropriate antipsychotic drug blood levels have been achieved.

Maintenance Treatment with Antipsychotic Medication

The length of time that a patient should be maintained on antipsychotic drug treatment becomes a salient issue due to tardive dyskinesia. Not one single properly controlled double-blind study, among the thirty or so studies carried out, has failed to show that more patients relapsed on placebo than on continuous pharmacotherapy.[33] The difference is significant with a p value of less than 10^{-100} when these studies are combined according to the method of Fleiss.[48] Hogarty and Goldberg[79] performed a particularly impressive study on this question. Three hundred and seventy-four schizophrenic patients who, after recovery, had been discharged from a state hospital were divided into two groups: one group received maintenance chlorpromazine treatment, while the second received placebos. In addition, half of each group received psychotherapy. The major finding was that patients receiving drugs and therapy fared better than those who received only drugs. Very few patients in the placebo group failed to relapse, despite their psychotherapy sessions. Thus, it would appear that maintenance phenothiazines are required for the prevention of relapse in most schizophrenic patients. Psychotherapy did increase the social adjustment, chiefly in the patients who also received a drug.

It is reasonable to ask the question: Are patients especially liable to relapse immediately upon the termination of antipsychotic drugs, or do they tend to relapse at a constant rate with the passage of time? If the latter holds true, then there would be an equal likelihood of patients relapsing during the second month following discontinuance as during the eighth or the fifteenth month. A simple linear plot of patients relapsing is inadequate for the visual display of a constant relapse rate. This is due to the fact that the absolute number of patients relapsing depends upon the number of patients included in the clinical trial; as time goes by, the population in the trial (and at risk for relapse) decreases. To elaborate, we begin with 100 patients and assume a relapse rate of 10 percent per month. During month one, 10 percent of 100, or 10, would have relapsed, leaving 90. During month two, 10 percent of 90, or 9 patients, would have relapsed, leaving 81 patients in the trial. During month eight, 10 per cent of 81 would have relapsed, leaving approximately 72 patients in the trial. During the course of this entire period, the absolute number of patients relapsing per month progressively decreases, given the diminishing number of patients in the trial and the consistency of the relapse rate. These mathematical considerations are identical to those applying to the radioactive half-life ($T_{1/2}$) or the $T_{1/2}$ of drugs in plasma. Questions connected with this issue were first addressed in our laboratory. Data from several large collaborative studies were plotted to illustrate the most suitable fit to an exponential function, relative to a linear function (see figure 5–5). Replotting the data of Hogarty and Goldberg[79] yielded a relapse rate of 10.7 percent for placebo for the first eighteen months. In the study by Caffey and coworkers[15] (not reviewed here), we found a relapse rate of 15.7 per cent. These least squares analyses of the empirical data provide an excellent fit, with r^2 in the vicinity of 0.96. We stress, however, that over an extended trial, some sort of ceiling effect may emerge. In

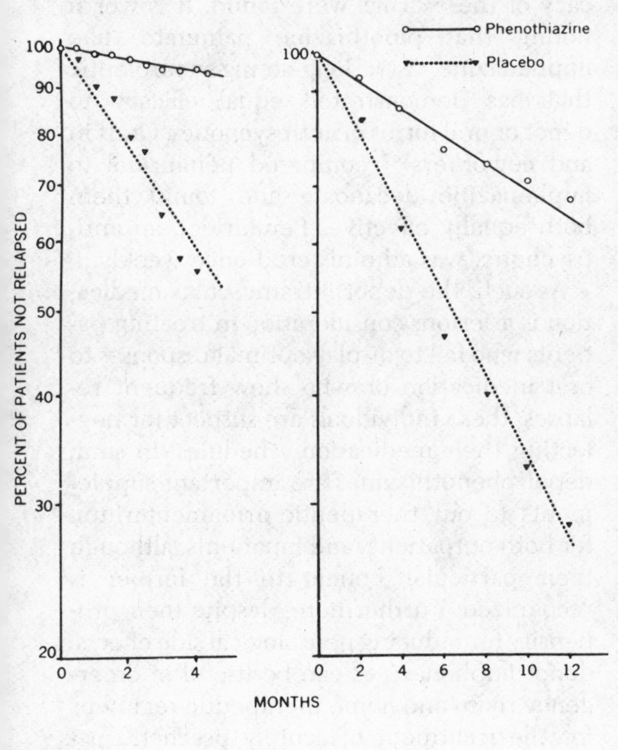

FIGURE 5–5.

other words, all the patients who were at risk for relapse would have relapsed, resulting in attenuation of the progressive nature of the relapse events. If a few patients in the trial do not show the recurrent form of the disease, they may never relapse; hence, they constitute a residual of unrelapsed patients. Empirically, the relapse occurred at a constant rate until about eighteen months in the study. There were so few patients remaining at this point that the empirically observed relapse rate might have been inaccurate; however, there is reason to believe that it may have changed then. Hogarty selected out those patients who had remained in the study for two years or more and examined their relapse rates more closely. Unfortunately, he was able to identify only a few patients in the placebo group who were still unrelapsed, certainly not enough to permit an effective study. In the drug-treated group, however, there remained a sufficient number of subjects available for investigation. Antipsychotic drug treatment had been discontinued for these individuals who did, in fact, relapse in an exponential fashion with a relapse rate similar to that observed initially. It is recognized that while relapse may be checked by drugs for a substantial period (two years), it may occur at approximately the same rate as with patients who are left drug-free after two months of maintenance medication.

Goldberg and his coworkers[61] also examined which schizophrenics had relapses and which did not. Subjects were organized into four groups. It has been documented in the literature that patients with good prognostic signs may not require drugs. However, it was reported that patients with favorable signs appeared to profit the most from drug treatment. Patients who failed to take their medications regularly tended to do poorly. Results were also influenced by psychotherapy. Patients who were asymptomatic with respect to schizophrenic symptomatology seemed to benefit most from psychotherapy, as compared with patients who showed a reasonable degree of psychosis and who seemed to do poorly in the psychotherapy group. It was hypothesized that psychotherapy may pose a stressful situation for patients in borderline compensation, as they fail to deal effectively with encouragement toward social responsibility.

In sum, decisions connected with long-term drug therapy should be derived clinically for each patient, based upon a thorough knowledge of his illness and life situation. It would seem reasonable to maintain the majority of patients on phenothiazines for six months to one year following a psychotic episode; however, over extended time periods, treatment may well require further individual tailoring. Since the so-called "reactive" schizophrenics may experience only a single episode during an entire lifetime, we do not recommend long-term maintenance medication for them.

Obviously, a history of relapse following discontinuation of antipsychotics is an indication for a prolonged period of such treatment. Evidence that antipsychotics may not have helped the patient originally or that the drug's prior discontinuation did not lead to relapse would be indications for the gradual reduction of dosage, leading to the termination of drug treatment. Psychotherapeutic and social interventions during the recovery phase and throughout posthospital care are very important in fostering improved social adjustment and may help to prevent relapse.

Several long-acting antipsychotic agents being studied offer a useful treatment approach for patients who fail to take their oral medication. Fluphenazine enanthate and fluphenazine decanoate are two intramuscular depot forms that are currently used in the United States. Open clinical trials investigating depot medication report evidence of patients who benefitted not only because they had previously failed to take oral medication, but also possibly due to the kinetics of the drugs—factors such as intramuscular versus oral absorption, distribution, and metabolism. Evidence from controlled studies finds depot fluphenazine to be as effective as oral fluphenazine, although the results are somewhat contradictory. Two double-blind studies showed the depot fluphenazine to be superior to the oral formulation, while three other studies found the two preparations to be equally effective.[19] A recent NIMH collaborative double-blind study[168] reported that the depot fluphenazine is essentially comparable to the oral medication. In addition, Rifkin and coworkers[158] have presented data that demonstrate the equivalence of both preparations. Variable patient cooperation in taking the drugs routinely might be invoked as a partial explanation; compliance may have been greater in the latter series of studies. Regardless, it appears certain from anecdotal studies that the depot drug is particularly beneficial for patients who are somewhat negligent about taking their medication.

Simon and coworkers[180] conducted an open maintenance (eighteen month) study in France to investigate the influence of standard neuroleptics, fluphenazine decanoate, and pipothiazine palmitate on chronic schizophrenia. Thirty psychiatrists were employed to assess the effects of these agents, which were administered at random to eighty-one patients from fifteen different wards. No significant differences in the efficacy of these drugs were found. It is worth noting that pipothiazine palmitate, like fluphenazine, is a long-acting neuroleptic that has demonstrated equal efficacy to depot or oral forms of antipsychotics. Quitkin and coworkers[147] compared penfluridol to fluphenazine decanoate and found them both equally effective. Penfluridol, an antipsychotic, was administered once weekly.

As such, the depot intramuscular medication is a serious consideration in treating patients who fail to display optimal responses to oral medication or who show frequent relapses; these individuals are suspect for neglecting their medication schedules. In sum, depot phenothiazines are important supplements to our therapeutic armamentarium, for both outpatients and inpatients, although their particular benefit to the former is recognized. Furthermore, despite their propensity for inducing neurological side effects, depot fluphenazines can be useful in emergency room and home therapeutic regimens for the treatment of acutely psychotic patients, as the psychotic symptoms can be diminished without inpatient admission. This advantage in the emergency situation may outweigh their occasional disposition for inducing neurological side effects.

It is worth determining if one of the two forms of this drug (fluphenazine enanthate and fluphenazine decanoate) may be preferable to the other. Given that these are both long-acting depot fluphenazine, it is anticipated that they would be approximately equal. Empirical studies[94] have, in fact, shown that they are generally equal in potency, efficacy, and side effects. However, several studies[39,195] indicate that fluphenazine decanoate may be slightly more long-acting and may produce slightly fewer extrapyramidal side effects. A number of studies[195] have directly compared fluphenazine decanoate to fluphenazine enanthate. Fluphenazine decanoate appears to be a slightly more potent drug. It requires a lower dosage and less frequent administration, that is, more extended intraindividual dose. Donlon and coworkers[39] found that extrapyramidal side effects may emerge with a slightly higher incidence following treatment with enanthate relative to decanoate. The drugs are extremely similar, yet the five comparison studies have shown that the decanoate form is slightly longer acting and produces slightly fewer side effects. Thus, it is probably the preferable formulation for the administration of long-acting depot fluphenazine.

Van Praag and Dols[195] have described an unusual experimental design for the comparison of these drugs. Thirty patients were randomly assigned to one or another of these drug treatments for the control of an acute schizophrenic episode. Clinicians who were blind to this assignment were permitted to administer supplementary chlorpromazine or antiparkinsonian drug if necessary. One injection for a four-week period was given. The reasoning was that if the influence of the drug diminished during this interval, the patient would then require a greater amount of supplementary chlorpromazine. All patients in the study received placebo throughout so that when chlorpromazine was required on a clinical basis, it replaced the placebo in an identical tablet to maintain the double-blind procedure. The same method was followed if an antiparkinsonian drug was required. Especially during the initial two weeks, it was found that patients receiving enanthate needed more antiparkinsonian drugs. The antipsychotic effect tended to endure longer in the decanoate group, since considerably fewer patients required additional chlorpromazine during the third and fourth week of this trial.

New Antipsychotic Agents

A notably "pure" dopamine antagonistic drug is pimozide. It is this characteristic that makes it a particularly interesting antipsychotic agent. Careful research has documented its antipsychotic properties compared with a placebo among both acute and chronic schizophrenic patients when administered for maintenance treatment. Of additional relevance are data that support the equivalence of maintenance pimozide and standard antipsychotics. Relatively higher doses of the latter are, however, required to achieve an effect that is comparable to the former. Since the efficacy of pimozide has been substantial, it is likely that its approval for release by the Federal Drug Administration is imminent.

Two new agents, molindone (which has an indole structure) and loxapine (which belongs to the dibenzoxapine category), have been clearly shown to have antipsychotic effects (see table 5–6). Six controlled studies[33] involving approximately 200 patients found molindone similar in efficacy to the standard antipsychotic drug: it was slightly superior in one study, not different in three, and slightly inferior in two. Clark and associates[20] conducted a placebo-controlled study and found molindone to be superior to placebo. Qualitatively, molindone improved the same range of schizophrenic symptoms as the other antipsychotic agents and had similar prophylactic effects. Although molindone generally produces the same range of extrapyramidal and autonomic side effects as

TABLE 5–6 **Summary of Studies on the Antipsychotic Efficacy of Loxapine and Molindone in Schizophrenia**

INVESTIGATORS	NO. OF PATIENTS	PATIENT GROUP (SCHIZOPHRENICS)	OUTCOME*
Clark et al. (1975)	37	Newly admitted	lox = tri > pla†
Denber (1970)	31	Newly admitted	lox = tri
Moore (1975)	57	Newly admitted	lox > CPZ
Shopsin et al. (1972)	30	Newly admitted	lox < CPZ
Simpson and Cuculic (1976)	43	Newly admitted	lox = tri
Smith (pers. comm.)	12	Newly admitted	lox = tri
Steinbook et al. (1973)	54	Newly admitted	lox = CPZ‡
Van Der Velde and Kilte (1975)	25	Newly admitted	lox = thi > pla
Charalampous et al. (1974)	54	Subacute	lox = pla < thi
Bishop and Gallant (1970)	24	Chronic	lox = tri
Clark et al. (1972)	50	Chronic	lox = CPZ > pla
Moyano (1975)	48	Chronic	lox = tri
Schiele (1975)	49	Chronic	lox = CPZ
Simpson et al. (1971)	52	Acute	mol = tri
Clark et al. (1970)	43	Chronic	mol = CPZ > pla
Freeman and Frederick (1969)	28	Chronic	mol = tri
Gallant and Bishop (1968)	43	Chronic	mol = tri
Ramsey et al. (1970)	20	Chronic	mol = tri

*Abbreviations: lox, loxapine; mol, molindone; tri, trifluoperazine; thi, thiothixene; CPZ, chlorpromazine; pla, placebo.
†Use of equals sign means any overall difference which might exist could not be detected on basis of number of patients, population, etc.
‡lox > CPZ on some, but not all measures.

the phenothiazines, it does not cause weight gain. Also, since it does not inhibit the noradrenaline (norepinephrine) uptake pump mechanism, it probably does not interfere with the hypotensive action of guanethidine.

Clark and associates[22,23] have provided the best evidence on the efficacy of loxapine. They tested loxapine in carefully executed double-blind studies of hospitalized patients with chronic and acute schizophrenia. In all three studies, loxapine was significantly superior to placebo. Van Der Velde and Kiltie [194] also found loxapine superior to placebo. In another fifteen studies[33] involving about 600 patients, loxapine was found to be indistinguishable from standard antipsychotics. Twelve of those studies[33] showed loxapine to be more effective than thiothixene in younger, but not in older patients, while one group found loxapine less effective than thiothixene and another found loxapine less effective than chlorpromazine. Considered as a group it seems that loxapine is an effective antipsychotic. It has the same range of side effects as the other antipsychotics. Both new antipsychotics are effective drugs and their use is recommended. No occular, liver, blood, or phototoxicity has been reported.

Antiparkinsonian Medications

Prophylactic antiparkinsonian drugs have become a highly controversial issue. Many psychiatrists advocate such treatment for all patients regardless of the specifics of the diagnosis. Others guard against their indication. One research design has been the discontinuation of antiparkinsonian drugs in a group of chronic schizophrenics. Such studies tend to show that between 20 to 50 percent of the subjects will exhibit parkinsonian side effects following withdrawal of the antiparkinsonian drug. Many of the studies did not include a comparison group. Further, it is noteworthy that when a comparison group was used, (antipsychotic + antiparkinsonian drugs versus antipsychotic + placebo) extrapyramidal side effects were detected among some of the group in spite of the drug. Nevertheless, it is clear that when many patients who have received prophylactic antiparkinsonian drugs for extended periods did not relapse, the drug was terminated. Therefore it is reasonable to recommend that after the administration of antiparkinsonian drugs for more than three months, it should be slowly tapered, with eventual discontinuation. Only patients who display parkinsonian symptoms should continue on antiparkinsonian drug treatment. In sum, however, one cannot conclude that antiparkinsonian drugs have no prophylactic efficacy. No doubt, the majority of patients included in these investigations were placed on prophylactic antiparkinsonian drugs. Their failure to show re-emergence of extrapyramidal side effects when their antiparkinsonian drug was discontinued may be accounted for by the large possibility that they never would have had parkinsonian side effects in the absence of prophylactic antiparkinsonian drugs. A more appropriate research design for answering such questions would be to arrange for patients to receive either a prophylactic antiparkinsonian drug or a matched placebo. This was in fact, the methodology employed by Hanlon.[75] Twenty-seven of the patients who had not received antiparkinsonian drugs experienced extrapyramidal side effects, whereas when an antiparkinsonian drug was used, these effects were identified among only 10 percent. This was almost a threefold reduction. A similar study was conducted by Chien and associates.[18]

¶ The Antidepressants

Occasionally, there is a statement in the British literature that the tricyclic drugs are not clinically effective. This is obviously not true. There is overwhelming evidence that all the tricyclic drugs are clinically effective. We, hereby, present an analysis of thirty well-controlled studies of imipramine to demonstrate that if data are properly combined (even in a very crude fashion and based on the simplest dichotomized data), the statistical probability is overwhelming. Substantial evidence for efficacy is found in double-blind studies for all the tricyclics, including the new tricyclics.

Investigators usually express drug effectiveness in one of two ways: (1) by noting the percentages of patients who improve with a drug or placebo or (2) by using a rating scale that demonstrates the mean change in a patient population. Among the forty-four controlled studies[35] comparing imipramine with a placebo, thirty provided data on the percentage of improvement associated with a drug or placebo.

Collectively, the data from these thirty studies yield a total of 1,334 patients treated with imipramine or placebo. Approximately 65 percent of the subjects treated with imipramine showed significant improvement, compared to 30 percent on the placebo. None of the studies in this group demonstrated a greater therapeutic effect from the placebo. The average drug-placebo difference of 35 percent was obtained after subtracting the percentage of improvement with a placebo from the percentage of improvement with imipramine. It is not surprising that since imipramine predicted more favorable improvement than the placebo, there is only an infinitessimal statistical probability that chance alone could lead to these results. The probability of these results, which favor imipramine over a placebo, obtained by chance is 10^{-31}.[48] By enlarging the sample size, one can achieve a highly statistically significant result in a drug that is otherwise of only moderate effectiveness. However, there is an urgent need for sensitive and effective treatment methods since approximately 15 to 30 percent of the depressed patients did not respond to the medication. Covi, and coworkers[30] detected a significant therapeutic effect from imipramine, but were unable to show that group therapy was beneficial. Klerman and coworkers[96] conducted an important controlled study comparing depressed patients who were placed on one of the following treatment modalities—tricyclics, psychotherapy, or combined tricyclics and psychotherapy, as well as a control group. They set up an unusual control group that allowed the patients to request emergency appointments if they felt that they desired assistance. This "demand only" psychotherapy serves as an alternative to "waiting list" controls. The studies demonstrated that tricyclics and psychotherapy, given independently, yield a better therapeutic antidepressant effect than "demand only" psychotherapy. Interestingly, the treatments administered jointly had a significantly better antidepressant effect. This clearly shows that both forms of treatment (chemotherapy together with psychotherapy) do not contradict each other; rather, they tend to complement one another so that the overall therapeutic efficacy is increased.

It is important here to consider what constitutes an adequate "placebo" for psychotherapy, and to remember that patients who were on "waiting list" or "demand only" control were aware that they were not actually receiving psychotherapy. The "demand characteristics" are quite dissimilar from psychotherapy. A highly suitable control experiment would be to compare a given type of standard acceptable "specific" psychotherapy against a "nonspecific, nontherapeutic" type of psychological support, based on an equal number of hours. This control experience will lack the ingredients regarded as specific to psychotherapy. It would also be important to arrange for approximately the same demand characteristics. A key consideration in this type of investigation is that blind assessment may not be blind in actuality. Some patients may be cognizant of their particular psychological intervention and may inadvertently communicate to the blind assessors the "demand characteristics."

Drug Maintenance in Affective Disorders

A major therapeutic question facing the clinician, once the patient is discharged from the hospital, contingent on a good response to treatment with tricyclic indications, is when to discontinue the antidepressant drugs. Since depression is a recurrent disorder, one often wonders whether continued treatment with tricyclic drugs will actually prevent relapse. In an attempt to answer this unsettling question, Mindham and coworkers[124] completed a collaborative study that included thirty-four psychiatrists in Great Britain. Patients studied had suffered a depressive illness and had responded to treatment with either imipramine or amitriptyline in doses of at least 150 mg a day. One group of these patients was continued on treatment with tricyclic drugs in doses of 75 to 100 mg a day, while the other was placed on a placebo for a period of fifteen months. Following this period, relapse was observed among 22 percent of the tricyclic maintenance group, compared to 50 percent of the group treated with a placebo. The relapse rate appears to be linear across time, given that the appropriate correction is made (see "Antipsychotics"). Data are not yet available for predicting which patients needed tricyclic drugs to check relapse. Currently, however, there are numerous studies demonstrating that maintenance tricyclics prevent the recurrence of depression among individuals with multiple relapsing episodes. In two of these studies,[34] patients were first treated successfully with electroconvulsive therapy (ECT), in the five other studies,[34] they were first given tricyclics. Placebo or maintenance tricyclics were administered in a double-blind, random-assignment design. Upon pooling the data from these various studies, there remains no real doubt that maintenance tricyclics are able to prevent relapse (see table 5–7). In cases of multiple relapses, maintenance tricyclics should be considered for the prevention of recurrence. Some patients may have experienced only a single previous depressive episode which might well be their last. Maintenance tricyclics are obviously not recommended for such cases. The decision for prophylaxis should be based on clinical indications, such as the severity of depressions, the frequency of depression, risk of suicide, and so forth. It is interesting to note that maintenance lithium is also used to prevent relapses in patients with recurrent unipolar depression. Lithium is clearly the drug of choice in prevention of relapse in bipolar disease (see table 5–8). The relapse of mania is not prevented by tricyclics, which can, on occasion, precipitate a manic attack.

Klerman and coworkers[96] examined the role of psychotherapy and maintenance treatment. Psychotherapy was not effective in preventing relapse, yet did improve adjustment at the social level. Again, this research describes a qualitatively important role for drugs in the treatment of depressed patients. In sum, both the psychological therapies and drug treatment should be seriously considered.

Plasma Levels

Antidepressant drugs have been shown to assist at least 70 percent of depressed patients. Certainly, some percentage of the patients who have not benefited are suffering from a type of depression that does not respond to this class of drugs. However, others may not respond as a function of clinical and/or pharmacological factors. Plasma levels build up for about two weeks, or until they reach a fairly stable level which is maintained over the course of tricyclic drug administration.

Patient populations yield broad interindividual differences in plasma levels. This raises the question of whether the lack of response can be attributed to an abnormality in the metabolic rate of the tricyclic drug (see table 5–9, figures 5–6, 5–7, and 5–8). Certain individuals might metabolize the drug quickly, fail to build up a sufficient blood level, and, subsequently, show low brain levels. Other patients may have a defective metabolism, so that high plasma brain level may accumulate. Patients may fail to improve clinically either because they are the recipients of toxic doses, or because the drug may lose efficacy at increased levels. It seems, therefore, that there are two possible explanations for nonresponse. Asberg and coworkers[6] examined associations between plasma nortriptyline concentrations and therapeutic response. It was found that some patients who failed to respond had notably reduced levels of blood nortriptyline, while other such patients had relatively higher levels of nortriptyline.

It is predicted that there will be a lower limit to the therapeutic window with virtually all drugs. In other words, if a sufficiently low dose is given, an insufficient quantity of the drug will be available to the receptor site and so fail to produce a therapeutic response. It was reported that the lower level of the therapeutic window was approximately 50 ng/ml because five patients who had levels below this had failed to respond (the precise location of the "low window" derives from data from these five patients).[6] This same study reported that patients who had plasma levels exceeding 140 ng/ml tended to display an unfavorable clinical response. This did not appear to be a result of CNS toxicity; rather, the drug seemed to lose its therapeutic effectiveness when plasma levels were elevated (see figure 5–6).

In view of the tremendous theoretical interest generated by the paradoxical property of the drug, namely, that it seems to diminish in efficacy at high plasma levels, it was studied closely by several investigators. Kraugh-Sorenson[101,102,103] and Montgomery[126] both confirmed this result, in that they noted poor responses with plasma levels above 175 ng/ml and 200 ng/ml, respectively. Montgomery[127] noted that from among a group of sixteen patients with high plasma levels who had had their clinical doses reduced, twelve improved within approximately a week.

TABLE 5–7 **Placebo versus Tricyclics for Prevention of Relapse of Recurrent Depression***

RESEARCHERS		NO. OF PATIENTS WHO RELAPSE OR REMAIN WELL ON PLACEBO OR DRUG	
		PLACEBO	DRUG
Prien et al., 1973	Relapse	24	17
	Well	2	10
Mindham et al., 1975	Relapse	21	11
	Well	21	39
Klerman et al., 1974	Relapse	27	6
	Well	40	33
Coppen et al., 1978	Relapse	5	0
	Well	11	13
Quitkin et al., 1978a	Relapse	6	5
	Well	2	3
Seager and Bird, 1962	Relapse	11	2
	Well	5	10
Kay et al., 1970	Relapse	24	8
	Well	27	26

*$P = 2 \times 10^{-9}$ (Fleiss, 1973).

TABLE 5–8 **Lithium Prevention of Relapse of Unipolar Depression***

RESEARCHERS		NO. OF PATIENTS WHO RELAPSE OR REMAIN WELL	
		PLACEBO	LITHIUM
Baastrup et al. 1970	Relapse	9	0
	Well	8	17
Prien et al., 1973	Relapse	14	13
	Well	2	14
Persson, 1972	Relapse	14	6
	Well	7	15
Coppen et al., 1963	Relapse	12	1
	Well	3	10
Dunner et al., 1976	Relapse	9	8
	Well	5	6

*$p = 3 \times 10^{-8}$

TABLE 5–9 **Plasma Level Studies**

INVESTIGATORS	DOSE	PLASMA LEVELS	NUMBER	STATUS	DRUG*	RESULTS
	mg.	*mg./ml.*				
Asber et al., 1971a	75–225	32–164	29	Inpatients	Nor.	Curvilinear poor results below 50 or above 139 ng./ml.
Kraugh-Sørensen et al., 1973	150	48–238	30	Inpatients	Nor.	Poor results above 175 ng./ml.
Kraugh-Sørensen et al., 1976	Adj.	Adj. > 180 or < 150	24	Inpatients	Nor.	Poor response with plasma level above 180 ng./ml.
Ziegler et al., 1977	Flex.	53–252	19	Outpatients	Nor.	Poor results above 139 ng./ml.
Montgomery et al., 1978	100 m	120–290	18	Inpatients	Nor.	Poor results above 200 ng./ml.
Whyte et al., 1976	40	129–427	28	Inpatients	Prot.	Curvilinear poor response with low (< 14) ng./ml. and poor response above 280 ng./ml.
Biggs et al., 1978	20	22–167	21	Outpatients	Prot.	Poor response in patients with plasma levels under 70 ng./ml.
Khalid et al., 1978	100–150	29–318	15	Outpatients	Dmi.	Linear high plasma level = good response.
Olivier-Martin et al., 1975	150	108–118 Imip. 130–160 DMI	24	Inpatients	Imip.	Good response above Imip. + Dmi. level of 200 ng./ml. in selected endogenous depression.
Reisby et al., 1977	225 nm	58–809	66	Inpatients	Imip.	Linear good response above 240 ng./ml.
Glassman et al., 1975	3.5 mg./kg.	50–1050	42	Inpatients	Imip.	Linear good response above 180 ng./ml.
Muscettola et al., 1978	Flex.	42–432	15	Inpatient	Imip.	Nonsignificant trend for patients over 200 ng./ml. to have better response.
Braithwaite et al., 1972	150	40–313	15	Inpatients and outpatients	Ami.	Good response when level above 90 ng./ml.
Ziegler et al., 1977	Flex. mean	52–318	22	Outpatients	Ami.	Good response above 75 ng./ml.
Kupfer et al., 1977	200 mg.	Mean 275	16	Inpatients	Ami.	Good response above 200 ng./ml.

*Nor. = nortriptyline; Prot. = protriptyline; Dmi. = desipramine; Imip. = imipramine; Ami. = amitriptyline.

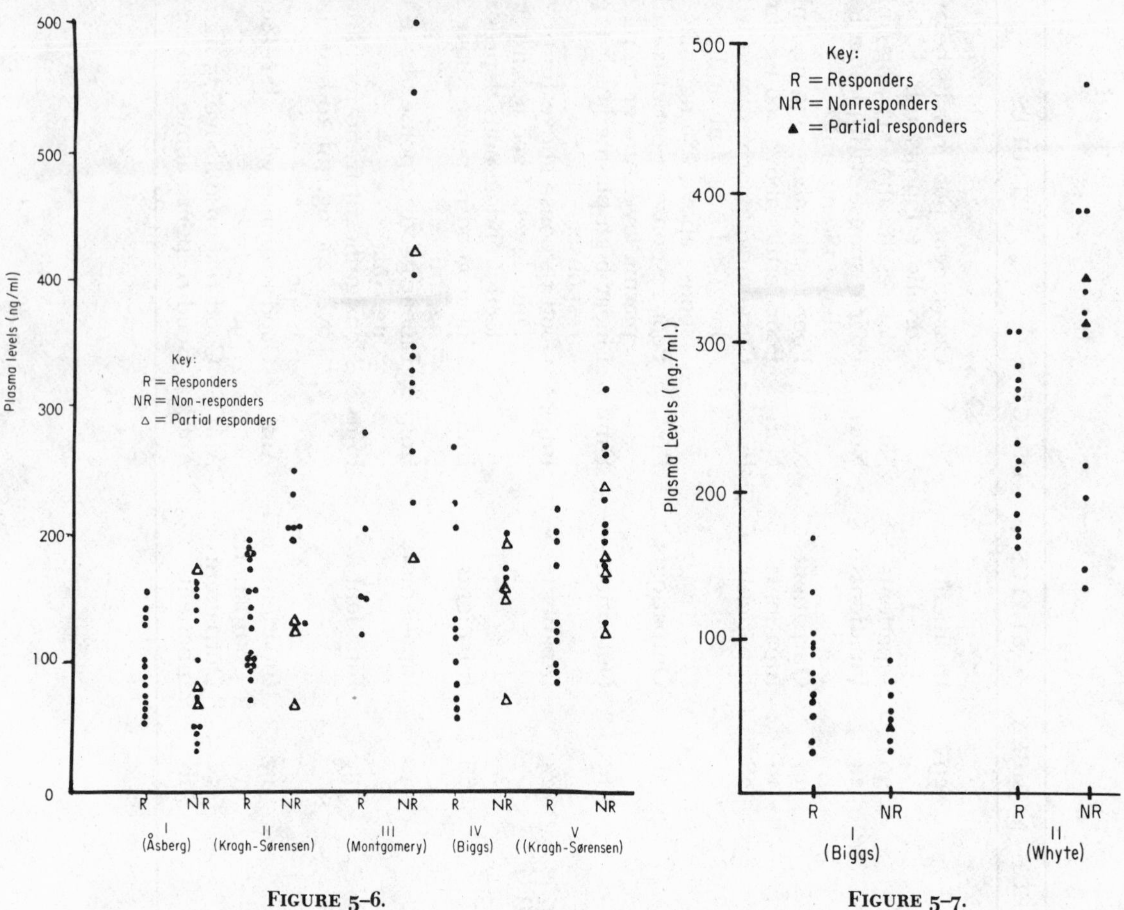

FIGURE 5–6.

FIGURE 5–7.

Kraugh-Sorensen[101,102] examined patients whose doses had been adjusted to either above 180 ng/ml or below 150 ng/ml; a disproportionately large number of subjects in the first group showed a poor clinical response. This study progressed to a second phase in which a subgroup of the original sample was selected at random and the levels of the subgroup of the high plasma level group (180 + ng/ml) were lowered to the therapeutic range. Five out of five patients improved. In contrast, six patients who continued to have high plasma levels had a poor response. Whyte and coworkers[199] suggested an inverted U-shaped relationship with a lower limit to the therapeutic window of below 140 ng/ml and upper above 260 ng/ml using a 40 mg dose. Biggs and Ziegler,[10] using a 20 ng/ml dose, found poor responses in patients with plasma levels under 70 ng. This group used a 50 percent lower dose than did Whyte and coworkers, but observed disproportionately lower plasma levels. Biggs' GCMS method would be specific. The two investigators concur in that there may be a lower limit to the therapeutic window, but disagree as to what this limit would be, that is, 70 or 140 ng/ml.

All three studies of amitriptyline[13,104,202] find a lower limit to the therapeutic window —poor clinical response when plasma levels are below 90 ng/ml, 200 ng/ml, or 75 ng/ml, respectively, but they differ substantially as to the lower-end location. In the well-controlled studies of Glassman[58] and Reisby,[152] patients were administered a fixed dose of imipramine yielding plasma levels from 50 ng/ml to 1,000 ng/ml (see table 5–10). Both workers plus Olivier-Martin[133] reported poor clinical responses with low plasma lev-

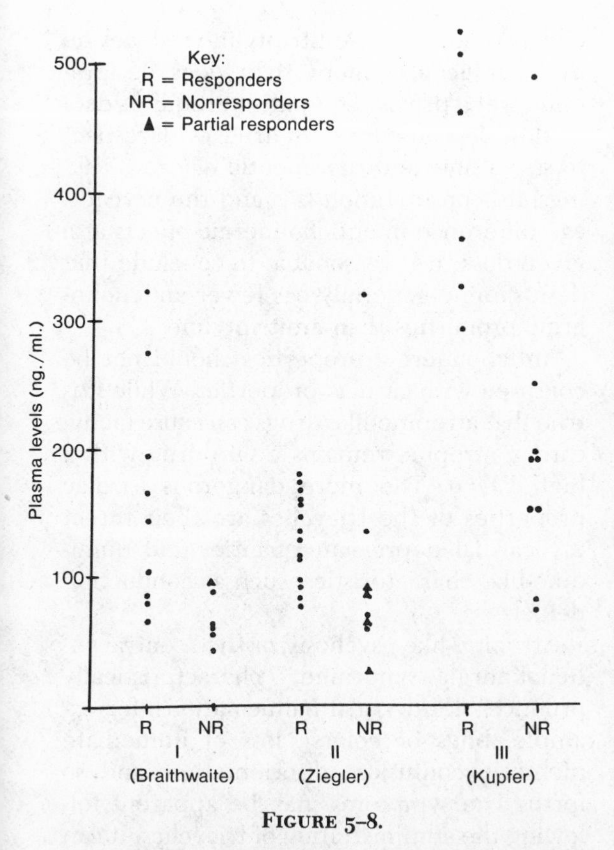

FIGURE 5–8.

els—under 180 ng/ml or 240 ng/ml, respectively. They did not find an upper limit for the therapeutic window in the sense that they did not find a loss in the efficacy of the drug effect with high plasma levels. They did, of course, find toxicity in some patients with very high plasma levels, yet this defines a different type of upper limit to the therapeutic window. Most reviews place absolute limits on the upper and lower limits of the therapeutic window. If the limit of the therapeutic window is from 100 to 200, then a patient with 99 ng/ml is certainly below, and a patient with 201 is certainly above the window. From inspection of figures 5–6, 5–7, and 5–8 it would appear that such an absolute interpretation of the therapeutic window is not supported. This is shown most clearly by protriptyline and amitriptyline. Patients within the therapeutic window in one study would be clearly below the therapeutic window in another study. Different authors measured the tricyclics at different times. For example, Asberg measured tricyclics during the first two weeks; Kraugh-Sorensen at the fourth week. This is a particularly difficult problem with protriptyline due to its very long half life. A method is needed to predict steady state from a test dose or from the first weeks of treatment because values not drawn on steady state are too low for comparison to the norms. Steady state levels should be appropriate levels for normative purposes[2] and plasma levels should be proportioned to doses. We suggest a given value be used to correct the steady state that is relative to the observed normative values. A relatively low dose can be adjusted upward, or a high dose can be adjusted downward, if this makes clinical sense. The most important data for dose adjustment are side effects and clinical response. Plasma levels in the rela-

TABLE 5–10 **Plasma Levels versus Clinical Response of Patients Treated with Imipramine**

Glassman and Perel, 1974	R*	6		19†
	NR	16		1
Reisbey et al., 1977	R	1	1	10
	NR	17	4	4
Martin et al., 1978	R	0	2	2
	NR	3	1	3
Muscettola et al., 1978	R	2	0	3
	NR	4	4	1
Plasma levels in ng./ml.		0–180	180–240	+240

*R = responders; NR = nonresponders. †Author gave plasma levels only as 180 and above.

tive sense can sometimes help in dose adjustment.

More work is needed to define the therapeutic window so that the plasma levels can be clinically useful. Not only must the existence of the therapeutic window be firmly established, but it is also necessary for laboratories to agree on its exact location. These figures will help the reader place a given plasma level within perspective. Plasma levels differ widely among individuals due to differences in the rate of metabolism in the liver. Hence, the clinician must adjust the dose for each patient to achieve maximum benefit with minimal side effects. Accurately determined plasma levels may well aid in dose adjustment when the literature provides enough data to establish the therapeutic window. We believe that the best way to get a feel for plasma levels is not through arbitrarily selected numbers defining the upper or lower end to the therapeutic window, but by placing a given plasma level in the context of the data presented graphically.

Several studies indicate that imipramine or amitriptyline, given at a dose range of 240 to 300 mgs, are far more effective than an average dose of approximately 150 mg.[184]

Tricyclic Drug Toxicity

Tricyclic drugs differ in their anticholinergic properties. A patient who has experienced unfavorable anticholinergic side effects, such as urinary retention, constipation, or the like may be better suited to treatment with a tricyclic that possesses few anticholinergic properties. Snyder and Yamamura[189] find that amitriptyline displays the most potent anticholinergic properties, doxepin has intermediate characteristics, and desipramine is the weakest. Several *in vitro* techniques exist for assessing the anticholinergic properties. One can, for example, measure the binding of atropine-like agents to the muscarinic receptor of the brain, intestine, or other tissue. However, the magnitude of the correlation between the *in vitro* preparations and the *in vivo* effects is not clear. Amitriptyline reduces saliva significantly more than does desipramine, yet a precise comparison requires data on dose response for both drugs with respect to saliva flow and therapeutic efficacy. Taking into consideration this and the percentage difference in anticholinergic effects at a given dose, it is reasonable to conclude that desipramine generally has fewer anticholinergic properties than amitriptyline.

Anticholinergic properties should not be confused with cardiac properties. While it is true that atropine-like drugs can cause tachycardia, atropine remains a safe drug with a high LD_{50}. The more dangerous cardiac properties of the tricyclics are their direct myocardial depressant qualities and quinidine-like characteristics, such as conduction defects.

Atropine-like psychosis, or the "central anticholinergic syndrome," characteristically produces florid visual hallucinations (for example, bugs or colors), loss of immediate memory, confusion, disorientation, and so forth. The symptoms may be apparent following the administration of tricyclics. It can occur with a given tricyclic alone or when the anticholinergic properties of multiple anticholinergics summate. They are reversible by the administration of physostigmine which is an agent that increases brain acetylcholine and pharmacologically overcomes the atropine blockade. The usual clinical treatment is discontinuation of the anticholinergics, which allows the syndrome to subside within a day. In selected cases, physostigmine can be introduced to produce this dramatic reversal. Misdiagnosis or the use of too much physostigmine may produce cholinergic toxicity. The more conservative course of treatment is the withdrawal of the anticholinergics.

Since tricyclics may frequently convert a depression into a mania, one should remain alert to this switch among bipolar patients. Although the relevant control studies are lacking, some clinicians use lithium tricyclic combinations for the treatment of bipolar depressive episodes. Tricyclics do not usually exacerbate schizophrenia, although on occasion they may do so. Mild withdrawal reactions have been observed upon the sudden termination of imipramine, following two months of treatment at a dosage of 300 mg daily. The reactions consist of nausea, vomiting, malaise, and headaches. A gradual decrease in dosage is usually preferred to abrupt withdrawal, so these reactions should not present a serious clinical problem. The sedation produced by the more sedative-type antidepressants may add to the sedation produced by ethyl alcohol. Indeed, empirical studies verify the common sense observation that the sedation caused by the different sedative type of drugs, or by alcohol, can combine to exert a greater effect than either agent acting alone. This need not necessarily apply to all tricyclics, but only to those with significant sedative properties.

Cardiovascular Effects

Tricyclic drugs, administered in accordance with the generally prescribed dosage schedule, may have various side effects. Electrocardiograms have revealed tachycardia, flattened T-waves, prolonged QT intervals, and depressed S-T segments. Most importantly, the tricyclics have quinidine-like properties. Imipramine has been demonstrated to decrease the frequency of premature ventricular contractions as a beneficial effect. Unfortunately, these drugs may prolong conduction, causing a significant problem in patients with a conduction defect.[9] It is imperative that the clinician remain attentive to the evidence of impaired conduction and omit tricyclics from the treatment programs of patients who display conduction defects. These drugs also become arrhythmogenic at elevated plasma levels.

Moir and coworkers[125] have documented cases of cardiovascular upset in patients for whom heart disorders have been previously diagnosed. When cooccurrence with the drug treatment arises by chance, this exacerbates the difficulty associated with distinguishing actual cardiac malfunctioning from an unfavorable event directly caused by introduction of the pharmacological agent. Systematic surveillance of cardiac function, coupled with dosage that is initially lower, is strongly advocated.

Overdosage of Antidepressants

Overdosage of an imipramine-type antidepressant results in a clinical picture marked by temporary agitation, delirium, weakness, increased muscle tone, slurred speech, hyperreflexia, and clonus. The patient then progresses to coma with hypotension. The most important problems are disturbances of cardiac rhythm, such as tachycardia, atrial fibrillation, ventricular flutters, and atrioventricular or intraventricular block. The lethal dose of these drugs is from ten to thirty times the daily therapeutic dose level.

Treatment of tricyclic overdoses should include vomiting or gastric aspiration and lavage with activated charcoal to reduce tricyclic absorption.[38] Tricyclic coma is generally of short duration—less than twenty-five hours. Since death due to cardiac arrhythmia is not uncommon, management of cardiac function is critical. If the patient survives this acute period, recovery without sequelae is probable, and vigorous resuscitative measures (such as cardioversion, continuous electrocardiogram monitoring, and chemotherapy to prevent and manage arrhythmias) should be applied in an intensive care unit. Arrhythmias may be mediated, in part, by the tricyclics, which are direct myocardial depressants and have quinidine-like properties. Detection of conduction defects is particularly important.

Physostigmine has a most dramatic effect in counteracting anticholinergic toxicity or coma produced by the tricyclics. It slows down the atropine-induced tachycardia, thus rousing the patient from atropine coma. Given this unusual action, physostigmine has the risk of overuse. Cholinergic toxicity, such as excess secretions, respiratory depressions, or seizures, can occur when physostigmine is given to a patient who is erroneously diagnosed as having atropine coma. Physostigmine toxicity may also occur if too great a quantity is given to a patient who is in actual atropine-like coma. Generally, the best treatment response to atropine overdose is "benign neglect." This atropine toxicity disappears as the atropine-like agent is metabolized. Physostigmine should be used selectively and judiciously and should be administered only by those familiar with its toxicity.

Bicarbonate is also helpful in preventing arrhythmias. Propranolol is useful in this treatment. It is recommended that patients be placed under medical supervision for several days since cardiac difficulties may occur a few days later, following the regaining of consciousness. Even though the average half-life of tricyclics ranges between approximately sixteen and twenty-four hours, there are many patients in whose case the half-life is significantly longer than the average. Hence, the plasma level can be expected to be high even as late as three to five days after ingestion of tricyclics.

The internist is chiefly responsible for the management of serious overdosage; however, the psychiatrist should be familiar with the gravity of the various difficulties associated with tricyclic overdose (for example seizures and arrhythmias) that are not always recognized. Caution is advised to ensure that the suicidal depressed patient does not gain access to an excessive quantity of antidepressant tablets, since several grams (twenty to forty of the 50 mg tablets) can be fatal.

Overdosage of MAO Inhibitors

Intoxication produced by MAO inhibitors is generally characterized by agitation that progresses to coma, hyperthermia, increased respiratory rate, tachycardia, dilated pupils, and hyperactive deep tendon reflexes. Involuntary movements, particularly of the face and jaw, may be present. The clinician should be aware of the lag period, which is an asymptomatic period lasting between one and six hours after ingestion of the drugs, prior to the general appearance of the symptoms of drug toxicity. Acidification of the urine markedly hastens the excretion of tranylcypromine, phenelzine, and amphetamine.

New Antidepressants

Maprotiline was compared to a standard tricyclic preparation (usually amitriptyline or imipramine) by means of the random assignment of 2,078 patients to one of these drugs. Examination of outcomes by number and percentage of individuals, as well as an analysis of the data by means of the method of Fleiss,[48] revealed no difference in efficacy among the groups. Six hundred and sixty patients (73.6 percent) who had received maprotiline did well, while six hundred and forty (72.6 percent) showed moderate improvement, or better. In contrast, 247 patients on marprotiline and 255 patients on one of the two standard tricyclics demonstrated either marginal improvement, no change, or a less favorable result.

Another heavily investigated drug, whose safety and efficacy have been demonstrated by a large number of double-blind studies, is amoxapine.[77,164] Maprotiline and amoxapine are both norepinephrine uptake inhibitors.

Numerous studies of nomifensine have been conducted in Europe and South America. An impressive number of double-blind studies has compared it to standard antidepressants and/or placebo.[50,71] It has been demonstated that nomifensine is effective as an antidepressant and comparable in efficacy to the standard antidepressants. It is both a safe and nonsedating agent[49,72] and is associated with a relatively reduced incidence of anticholingergic side effects. It has also been shown to inhibit the uptake of norepinephrine and dopamine. It has been extensively investigated and is, at present, under serious consideration by the Food and Drug Administration for release. Mianserin has been shown to be an effective antidepressant.[130] While not an uptake or MAO inhibitor, it does increase the turnover through a presynaptic mechanism. Other experimental antidepressants, such as the NE uptake inhibitor, vilozazine, and the 5HT uptake inhibitor, are now undergoing clinical investigation.[119,140] Trazodone is used in Europe and has been thoroughly investigated in a number of well-controlled double-blind trials in the United States. It is unequivocally an effective antidepressant. Another interesting antidepressant initially developed as a minor tranquilizer is trazolam. Early clinical-drug evaluation study shows it to be promising as an antidepressant.

Benzodiazepines As Antidepressants

Since anxiety is often associated with depression, the logical question that comes to mind is: Are benzodiazepines effective antidepressants? In a review of the relevant literature, Schatzberg and Cole[166] failed to find any study that indicated that benzodiazepines were significantly better than antidepressants. However, benzodiazepines were significantly inferior to the antidepressants in ten studies and there were no significant differences in the other nine studies.[166]

Ives and coworkers[88] divided depressed patients into several groups and treated them with phenelzine plus chlordiazepoxide, phenelzine alone, a placebo alone, and a placebo plus chlordiazepoxide. Chlordiazepoxide did not alleviate depression with the placebo nor with the phenelzine. In separate studies, Kay,[91] Lipman and coworkers,[111,112] and Covi and coworkers[30] found that diazepam was inferior to imipramine in treating depression. The majority of the studies show that benzodiazepines are virtually ineffective in the treatment of depression. There are some investigations, however, that show that minor tranquilizers may temporarily relieve symptoms of depression. More studies of the exact role of benzodiazepines in depression are needed.

Anxiolytic Tricyclic Combinations

A fixed dose preparation that combines chlordiazepoxide and amitriptyline for the treatment of mixed anxiety depression in outpatients has recently been introduced. The effectiveness of this combination was demonstrated by the results from a double-blind, multicenter collaborative study that compared their combination (trade name Limbitrol), which consists of 10 mg chlordiazepoxide plus 25 mg amitriptyline, to amitriptyline alone, chlordiazepoxide alone, and placebo.[47] There were 279 outpatients in this study. The criteria for primary depression were derived from scores obtained on the Hamilton Depression Scale (twenty points or higher), the Beck Depression Scale (fourteen points or higher), and the Covi Anxiety Scale (eight points or higher). Chlordiazepoxide and amitriptyline are both sedative drugs that possess antianxiety properties that are beneficial to insomnia and psychic and somatic anxiety. Due to their combined sedative-anxiety properties, it is not surprising that they contribute to the sedative and antianxiety effects. The combination clearly produces greater improvement on items from the Hamilton Depression Rating Scale such as insomnia, agitation, and somatic and psychic anxiety during week one of treatment. Overall scores on the Global Hamilton Depression Scale and the Beck Inventory demonstrated that the combination was superior to either of the individual drugs by the end of the first week. The superiority of all drugs, relative to placebo, is quite clear. At present, a relevant research question is if the degree of increased improvement observed after one week is a function of the extra antianxiety properties of two sedative drugs, or if it reflects the added therapeutic influence on depression. It is interesting to note that the combination resulted in greater improvement on the Depression Inventory on items such as pessimism, dissatisfaction, and guilt. After several weeks, amitriptyline alone was as effective as the combination on many of the measures, and slightly, but not significantly, exceeded effectiveness of the combination on certain measures, such as the depression factor of the Hopkins Symptom Checklist and the Beck Depression Inventory. The combination was slightly superior to amitriptyline alone by other measures, such as the Hamilton Depression Scale. Unfortunately, the investigators presented a selected finding, rather than a more systematic portrayal of the data. Indeed, it would be desirable to examine a detailed publication of this collaborative study.

Rickels and coworkers[155] conducted a comparable study, which included 243 mild to moderately depressed outpatients suffering from either a reactive neurotic depression or from a mixed anxiety depressive reaction. Owing to the fact that patients were evaluated at the end of two and four weeks, it is impossible to determine if the initial improvement of the combination observed by the Roche collaborative study at one week occurred here as well. This study demonstrated that the drug effects were statistically significant for both the combination and for each of the individual components, in comparison to a placebo. However, it is difficult to interpret this study, due to the failure of the investigators to present statistics on the comparisons between the combination versus each drug separately. It is difficult for the reader to acquire a complete grasp of whether the combination was superior to one or another of the individual drugs.

Hare[76] completed two modest studies that included twenty outpatients with mixed anxiety and depression. He compared the influence of the drug combination against amitriptyline given alone. In the first study, after one week, the former appeared superior on a number of dimensions. By the third week, similarity as to the degree of improvement from both increased, so that the initial difference had been largely observed. In the second study, the effects of the combination were similar to those from the amitriptyline alone. The statistically significant improvement on some measures still persisted for the combination, however. Haider,[73] examining seriously depressed inpatients, noted that the combination was superior to amitryptyline alone at the end of a three week period.

Clearly, the addition of chlordiazepoxide to a tricyclic does not alter its therapeutic action. Some evidence available from the two studies cited here[73,76] indicates that it may be helpful after week one of treatment. However, one study[76] failed to find it to be superior at that time, and another[73] did not investigate efficacy at the close of one week. After three weeks, it is not clear that the combination is superior. Another methodological problem is diagnostic homogeneity. A method is needed to separate out the two populations of patients: the one with pure depression with only secondary anxiety, and the other with pure anxiety accompanied by only secondary depression. Also, note the nonspecific nature of these rating scales: for example, depression scales have anxiety items and vice versa. Furthermore, improvement in anxiety may artificially improve depression due to a sort of halo effect. The evidence of the addition of a minor tranquilizer was summarized here so that the reader can draw his own conclusion. The addition of an antianxiety agent in the first week would do no harm and possibly have some beneficial effect, but more work is needed before forming definitive conclusions.

¶ Minor Tranquilizers

For many years, physicians have treated anxiety with sedative agents—alcohol, barbiturates, meprobamate, chlorpromazine, and benzodiazepines. Do the benzodiazepines offer any advantages over the classical barbiturates?

Comparison With Other Drugs

Several studies have compared the sedative properties of benzodiazepines with those of barbiturates, meprobamate, and a placebo. There is inconclusive evidence that some benzodiazepines may be slightly superior to barbiturates in the treatment for anxiety. Schapira and coworkers[165] found 4.55 mg a day of lorazepam slightly more effective than similar doses of barbiturates. Similarly, in another study 25 mg of diazepam was superior to 463 mg of amytobarbital.[192] In general, benzodiazepines have slightly stronger antianxiety effects than barbiturates at equal sedative doses.[94] One can only make a weak case for this. Overall, there are more visible similarities than differences between benzodiazepines and barbiturates. The therapeutic efficiencies of benzodiazepines, meprobamates, and barbiturates are about equal, although some research indicated that benzodiazepines may be slightly superior.

A substantial number of studies have firmly established that benzodiazepines are more efficient sedative-type antianxiety agents than placebos.[63] The therapeutic efficacy of prazepam was verifed in controlled multiclinical investigations by its sponsor[197] and several academic researchers.[156] The antianxiety effects of chlorazepate have been proved superior to a placebo and comparable to diazepam.[42,117,160] Since both benzodiazepines are precursors of the same active metabolite, desmethyldiazepam, they should be equally efficient as tranquilizers. In a thorough investigation of lorazepam, Richards[153] found the drug more effective than a placebo for treating anxiety neurosis, anxiety manifested in gastrointestinal and cardiovascular disorders and as a component of depression. The study reports on combined data from a large number of controlled trials involving 5,960 patients, of whom 3,520 received lorazepam. Davis and Greenblatt[36] offer a recent review of the treatment of anxiety and personality disorders.

Metabolism

The similarity of the benzodiazepine derivatives can be understood by looking at the active intermediates found in their metabolism. Chlordiazepoxide first metabolized to desmethylchlordiazepoxide, which is in turn converted to demoxepam. Demoxepam is then metabolized to desmethyldiazepam, which is converted to oxazepam. Oxazepam is then conjugated to its glucuronic acid. Probably these four metabolites are all clinically active. The unchanged drug along with the first two metabolites constitutes most of the compounds that are found in the plasma,[70] with their observable levels ranging from approximately 300 ng/ml to 3,000 ng/ml. Desmethyldiazepam and oxazepam are minor metabolites of chlordiazepoxide with a lower plasma level of about 200 ng/ml. Desmethyldiazepam is a major metabolite of diazepam. Direct metabolism of diazepam to desmethyldiazepam is followed by conversion to oxazepam, which, in turn, is metabolized to its inactive conjugate glucuronic acid. Two studies compared the plasma levels of diazepam and desmethyldiazepam. Robin and coworkers[160] found steady state plasma levels after a 5-mg, three-times-a-day dose of diazepam and desmethyldiazepam to be 2,400 ng/ml and 3,500 ng/ml., respectively. Similarly, Tansella and coworkers[192] found plasma levels of diazepam (742 ng/ml) and desmethyldiazepam (762 ng/ml) using a mean dose of 26 mg per day.

Prazepam, a recently introduced benzodiazepine, is essentially a precursor of desmethyldiazepam. It is almost completely metabolized to desmethyldiazepam in a first-pass effect. The possible metabolites, 3-hydroxy prazepam and oxazepam, are not observed in the plasma after administration of prazepam. This suggests that desmethyldiazepam is the active substance responsible for the drug's antianxiety effect.[64,65,66]

Another desmethyldiazepam precursor is chlorazepate, which is completely converted to this metabolite by rapid spontaneous dehydration and decarboxylation occurring in the acidic environment of the stomach. One study found that after administering equal molar doses of chlorazepate and diazepam, the plasma levels of desmethyldiazepam transformed from chlorazepate are essentially equal to the sum of the plasma levels of diazepam and its converted desmethyldiazepam.[160]

Chlorazepate, prazepam, and diazepam have similar clinical effects since each drug is metabolized to the same active substance—desmethyldiazepam. Considering that the half-life of desmethyldiazepam ranges from 30 hours to 200 hours, one daily dose is quite sufficient for all three drugs. Both chlorazepate and prazepam are precursors of desmethyldiazepam since they are nearly completely converted to it during metabolism. Thus, they are essentially identical from a clinical point of view. The two products do differ in the rate in which conversion to desmethyldiazepam occurs. The pharmacokinetics of chlorazepate are rapid; once in the stomach, the drug is immediately converted to desmethyldiazepam and rapidly absorbed. If the stomach is not maintaining its usual low pH level, the rapid conversion rate may not apply. Prazepam, however, results in a slower appearance of desmethyldiazepam due to the first-pass effect in the liver. Its plasma level versus time plot yields a curve similar to the one observed for sustained released preparation. The clinical effects due to the pharmacokinetic differences are most notable when they involve the first dosage of chlorazepate and the subsequent rapid sedative action.[64,65,66] Using multiple dose administrations, the differences between prazepam and chlorazepate are minimal. There may be differences in the pharmacological properties of diazepam and the two products since 40 percent of the compound present in the plasma is unchanged diazepam.

As mentioned earlier, the half-life of desmethyldiazepam can vary greatly. However, several studies show a steady-state half-life of approximately fifty hours.[99,141] Unconverted diazepam exhibits a similar long steady-state half-life of about forty to fifty hours. Chlordiazepoxide, the first of the benzodiazepine derivatives, and its two principal active metabolites, diazepam and desmethyldiazepam, have half-lives in the range of twenty to forty hours.

Each of the four benzodiazepine derivatives discussed achieves a steady state in two to three weeks. However, pharmacological effects should persist for a few days following termination of treatment with chlordiazepoxide, diazepam, and desmethyldiazepam because of their long half-lives.

Lorazepam and oxazepam exhibit pharmacokinetic similarities. Both compounds are metabolized at the 3-hydroxy position by conjugation with glucuronic acid. The resultant conjugate, which is excreted in the urine, is pharmacologically inactive.[106] Compared to other benzodiazepine derivatives, oxazepam and lorazepam have a substantially shorter half-life. Their half-lives range between ten to twenty hours for the majority of patients with oxazepam showing a slightly shorter half-life. Since the B-phase half-life is less than fifteen hours, two- or three-times-a-day dosage schedules are practicable, and steady state levels can be reached within four days after initiating such treatment. However, after suspension of treatment, the pharmacological effects of oxazepam and lorazepam are maintained for a much shorter time than for other benzodiazepines.

Benzodiazepines and Liver Disease

The half-life and volume distribution of some benzodiazepines are markedly greater in patients with liver diseases, such as alcoholic cirrhosis, acute viral hepatitis, and hepatic malignancy. This is caused by the pathway of metabolism through the liver. In the case of diazepam and chlordiazepoxide plasma bindings are reduced, thus increasing the volume of distribution and decreasing clearance (this does not apply in extrahepatic obstructive liver disease). Lorezapam and oxazepam are relatively unaffected by liver disease since they are excreted by a different route. However, in cirrhosis, there is a small increase in the half-life of lorazepam and a significant change in its plasma clearance.

It is well known that the effects of sedatives on the elderly, who are especially sensitive to these agents, suggest that the mechanism of this effect may be increased brain sensitivity to the drug and impaired metabolism resulting in elevated brain drug levels. Specifically, there is a lowered clearance of desmethyldiazepam and chlordiazepoxide in the elderly.[98] The clearances of diazepam, oxazepam, and lorazepam are unaffected by age.

Physical Dependence and Tolerance

Very little research has been directed to documenting the dependence liability of antianxiety drugs. Limited information has suggested that 3,200 mg of meprobomate for forty days or 300 mg of chlordiazepoxide for a month can cause addiction. There is approximately one case of benzodiazepine dependency per 50 million patient months of therapeutic use.[118] Pevnick and coworkers[138] studied the withdrawal symptoms of dependency on diazepam. These were characterized by tremor, dysphoric mood, muscle twitches and cramps, facial numbness, insomnia, anorexia, weakness, nervousness, weight loss, and increased orthostatic pulse. This collective syndrome usually starts on the sixth day of withdrawal, peaks on the seventh day, and disappears by the ninth.

Anxiety-dysphoric symptoms may occur after discontinuation of long-term benzodiazepine treatment, but it has been extremely difficult, if not impossible, to determine whether these symptoms were induced by physical dependence or the re-emergence of the previously suppressed anxiety. Further research on this question is necessary, and physicians should watch patients for any signs of addiction, such as obtaining early refills or consulting several doctors simultaneously for medication.

Cross-tolerance exists among sedative drugs. One can be utilized to treat the withdrawal symptoms caused by addiction to another. As of late, diazepam is replacing pentobarbital as the standard detoxificating agent because it is a longer-acting drug.

¶ Hydergine

Many well-controlled double-blind scientific studies found that Hydergine is effective in treating mental disorders in the elderly. However, it is unclear whether the alleviation of symptoms such as confusion, anxiety, agitation, and irritability result from Hydergine's vasodilating action. Improvement with Hydergine is noted to be slow in onset (about three months), and further research is essential to determine the mechanism by which the drug accomplishes clinical improvement.

¶ Bibliography

1. AGNEW, P. C., et al. "A Clinical Evaluation of Four Antidepressant Drugs (Nardil, Tofranil, Marplan, and Deprol)," *American Journal of Psychiatry,* 118 (1961): 160.
2. ALEXANDERSON, B. "Pharmacokinetics of Nortriptyline in Man After Single and Multiple Oral Doses," *European Journal of Clinical Pharmacology,* 4 (1972): 82.
3. ANANTH, J., and STEEN, N. "A Double-Blind Controlled Comparative Study of Nomifensine in Depression," *Current Therapeutic Research,* 23 (1978): 213.
4. APPLETON, W. S., and DAVIS, J. M. *Practical Clinical Psycho-pharmacology,* 4th ed. New York: Medcomb, 1980.
5. ASBERG, M., PPRICE-EVANS, D., and SJOQUIST, F. "Genetic Control of Nortriptyline Kinetics in Man: A Study of Relatives of Proposits with High Plasma Concentrations," *Journal of Medical Genetics,* 8 (1971): 129.
6. ASBERG, M., et al. "Relationship Between Plasma Levels and Therapeutic Effect of Nortriptyline," *British Medical Journal,* 3 (1971): 331.
7. BAASTRUP, P., POULSEN, K. S., and SCHOU, M. "Propohylactic Lithium: Double-Blind Discontinuation in Manic-Depressive and Recurrent-Depressive Disorders." *Lancet* 2 (1970): 326.
8. BARO, F., et al. "Maintenance Therapy of Chronic Psychotic Patients with a Weekly Oral Dose of R 16341," *Journal of Clinical Pharmacology,* 10 (1970): 330.
9. BIGGER, J. T., JR., et al. "Cardiac Antiarrhythmic Effect of Imipramine Hydrochloride," *New England Journal of Medicine,* 296 (1977): 206.
10. BIGGS, J. T., and ZIEGLER, V. E. "Protriptyline Plasma Levels and Antidepressant Response," *Clinical Pharmacology Therapy,* 22 (1978): 269.
11. BISHOP, M. P., and GALLANT, D. M. "Loxapine: A Controlled Evaluation in Chronic Schizophrenic Patients," *Current Therapeutic Research,* 12 (1970): 594.
12. BLACKWELL, B., et al. "Anticholinergic Activity of Two Tricyclic Antidepressants," *American Journal of Psychiatry,* 135 (1978): 722.
13. BRAITHWAITE, R. A., et al. "Plasma Con-

centration of Amitriptyline and Clinical Response." *Lancet,* 1 (1972): 1297.

14. BROTMAN, R. K., MUZEKARI, L. H., and SHANKEN, P. M. "Butaperazine in Chronic Schizophrenic Patients:" A Double-Blind Study, *Current Therapeutic Research,* 11 (1969): 5.
15. CAFFEY, E. M., et al. "Discontinuation or Reduction of Chemotherapy in Chronic Schizophrenics," *Journal of Chronic Diseases,* 17 (1964): 347.
16. CARSCALLEN, H. B., ROCHMAN, H., and LOVEGROVE, T. D. "High-Dosage Trifluoperazine in Schizophrenia," *Canadian Psychiatric Association Journal,* 13 (1968): 459.
17. CASPER, R. C., et al. "Phenothiazine Levels in Plasma and Red Blood Cells." *Archives of General Psychiatry,* 37 (3) (1980): 301–307.
18. CHIEN, C. P., DIMASCIO, A., and COLE, J. O. "Antiparkinsonian Agents and the Depot Phenothiazine," *American Journal of Psychiatry,* 131 (1974): 86–90.
19. CLARK, M. L., et al. "Chlorpromazine in Chronic Schizophrenia: Behavioral Dose-Response Relationship," *Psychopharmacologia,* 18 (1970): 260.
20. CLARK, M. L., et al. "Molindone in Chronic Schizophrenia," *Clinical Pharmacology Therapy,* (1970): 680.
21. CLARK, M. L., et al. "Chlorpromazine in Chronic Schizophrenia," *Archives of General Psychiatry,* 27 (1972): 479.
22. CLARK, M. L., et al. "Evaluation of Loxapine Succinate in Chronic Schizophrenia," *Diseases of the Nervous System,* 33 (1972): 783.
23. CLARK, M. L., et al. "Loxapine in Newly Admitted Chronic Schizophrenic Patients," *Journal of Clinical Pharmacology,* 15 (1975): 286.
24. COHEN, J., et al. "Diazepam and Phenobarbital in the Treatment of Anxiety. A Controlled Multicenter Study Using Physician and Patient Rating Scales," *Current Therapeutic Research,* 20 (1976): 184.
25. COLE, J. O., GOLDBERG, S. C., and DAVIS, J. M. "Drugs in the Treatment of Psychosis: Controlled Studies," in Solomon, P., ed., *Psychiatric Drugs.* New York: Grune & Stratton, 1966.
26. COLE, J. O., GOLDBERG, S. C., and KLERMAN, G. L. "Phenothiazine Treatment in Acute Schizophrenia," *Archives of General Psychiatry,* 10 (1964): 246.
27. COPPEN, A., GUPTA, R., and MONTGOMERY S. "Miansern Hydrochloride: A Novel Antidepressant," *British Journal of Psychiatry,* 129 (1976): 342.
28. COPPEN, A., et al. "Double-Blind and Open Prospective Studies of Lithium Prophylaxis in Affective Disorders," *Neurologia, Neurochiruygia, Psychiatrica,* 76 (1963): 500.
29. COPPEN, A., et al. "Continuation Therapy with Amitriptyline in Depression," *British Journal of Psychiatry,* 133 (1978): 28.
30. COVI, L., et al. "Drugs and Group Psychotherapy in Neurotic Depression," *American Journal of Psychiatry,* 131 (1974): 191.
31. CURRY, S. H., et al. "Chlorpromazine Plasma Levels and Effects," *Archives of General Psychiatry,* 22 (1970): 289.
32. CURRY, S. H., et al. "Factors Affecting Chlorpromazine Plasma Levels in Psychiatric Patients," *Archives of General Psychiatry,* 22 (1970): 209.
33. DAVIS, J. M. "Overview: Maintenance Therapy in Psychiatry: Schizophrenia," *American Journal of Psychiatry,* 132 (1976): 1237–1245.
34. DAVIS, J. M., "Overview: Maintenance Therapy and Psychiatry: II. Affective Disorders," *American Journal of Psychiatry,* 133 (1976): 1–13.
35. DAVIS, J. M., and ERICKSEN, S. E., "Controlled Trials of Imipramine," *British Journal of Psychiatry,* 129 (1976): 192.
36. DAVIS, J. M., and GREENBLATT, D. *Psychopharmacology Update: New and Neglected Areas.* New York: Grune & Stratton, 1979.
37. DAVIS, J. M. et al. Unpublished paper. 1980.
38. DAWLING, S., CROME, P., and BRAITHWAITE, R. A. "Effect of Delayed Administration of Activated Charcoal on Nortriptyline Absorption," *European Journal of Clinical Pharmacology,* 14 (1978): 445.
39. DONLON, P. T., et al. "Comparison of Depot Fluphenazines: Duration of Action and Incidence of Side Effects," *Comprehensive Psychiatry,* 17 (1976): 369–376.
40. DONLON, P. T. et al. "High vs. Standard Dosage Fluphenazine HCl in Acute Schizophrenia," *Journal of Clinical Psychiatry,* 39 (1978): 800.
41. DUNNER, D. L., STALLONE, F., and FIEVE,

R. R. "Lithium Carbonate and Affective Disorders: V. A. Double-Blind Study of Prophylaxis of Depression in Bipolar Illness," *Archives of General Psychiatry,* 33 (1976): 117.

42. DUREMAN, I., and NORMANN, B. "Clinical and Experimental Comparison of Diazepam, Chlorazepate, and Placebo. *Psychopharmacologia,* 40 (1974): 279.
43. DYSKEN, M. Unpublished paper. 1980.
44. EL-YOUSEF, M. K., et al. "Reversal of Benztropin Mesylate Toxicity by Physostigmine." *Journal of the American Medical Association,* 220 (1972): 125.
45. ———. "Reversal by Physostigmine of Antiparkinsonian Drug Toxicity: A Controlled Study," *American Journal of Psychiatry,* 130 (1973): 141.
46. ERICKSEN, S., et al. "Haloperidol Dose, Plasma Levels and Clinical Response: A Double-Blind Study," *Psychopharmacology Bulletin,* 14 (1978): 15.
47. FEIGHNER, J., et al. "A Placebo-Controlled Multicenter Trial of Limbitrol Versus its Components (Amitriptyline and Chlordiazepoxide) in the Symptomatic Treatment of Depressive Illness" *Psychopharmacology,* 61 (1979): 217.
48. FLEISS, J. L. *Statistical Methods for Rates and Proportions.* New York: John Wiley, 1973.
49. FORREST, A., HEWETT, A., and NICHOLSON, P. "Controlled Randomized Group Comparison of Nomifensine and Imipramine in Depressive Illness," *British Journal of Clinical Pharmacology,* 4 (Suppl. 2) (1977): 215S.
50. GAILANT, D. M., et al. "Amoxapine: A Double-Blind Evaluation of Antidepression Activity," *Current Therapeutic Research,* 15 (1973): 56.
51. GARDOSE, G. "Are Antipsychotic Drugs Interchangeable?" *Journal of Nervous and Mental Disorders,* 159 (1978): 343.
52. GARDOSE, G., et al. "High and Low Dose Thiothixene Treatment in Chronic Schizophrenia," *Diseases of the Nervous System,* 35 (1974): 53.
53. GARVER, D. L., et al. "Pharmacokinetics of Red Blood Cell Phenothiazine and Clinical Effects," *Archives of General Psychiatry,* 33 (1976): 862.
54. GARVER, D. L., et al. "Neuroleptic Drug Levels and Therapeutic Response: Preliminary Observations With Red Blood Cell-Bound Butaperazine," *American Journal of Psychiatry,* 134 (1977): 304.
55. GELENBERG, A. J., and KLERMAN, G. L. "Antidepressants: Their Use in Clinical Practice," *Rationalization of Drug Therapy,* 12 (1978): 1.
57. GLASSMAN, A., KANTOR, S., and SHOSTAK, M. "Depression, Delusions, and Drug Response," *American Journal of Psychiatry,* 132 (1975): 716.
56. GLASSMAN, A. H., and PEREL, J. M. "Plasma Levels and Tricyclic Antidepressants," *Clinical Pharmacology Therapy,* 16 (1974): 198.
58. GLASSMAN, A. H., et al. "Clinical Implications of Imipramine Plasma Levels for Depressive Illness," *Archives of General Psychiatry,* 34 (1977): 197.
59. GOLDBERG, H. L., and FINNERTY, R. J. "A Double-Blind Study of Prazepam vs Placebo in Single Doses in the Treatment of Anxiety," *Comprehensive Psychiatry,* 18 (1977): 147.
60. GOLDBERG, S. C., et al. "Prediction of Response to Phenothiazines in Schizophrenia: A Cross-Validation Study," *Archives of General Psychiatry,* 26 (1972): 367.
61. GOLDBERG, S. C., et al. "Prediction of Relapse in Schizophrenic Outpatients Treated by Drug and Sociotherapy," *Archives of General Psychiatry,* 34 (1977): 171.
62. GOLDSTEIN, M., et al. "Drug and Family Therapy in the Aftercare of Acute Schizophrenics," *Archives of General Psychiatry,* 35 (1978): 1169.
63. GREENBLATT, D. J., and SHADER, R. I. *Benzodiazepines in Clinical Practice.* New York: Raven Press, 1974.
64. ———. "Pharmacotherapy of Anxiety with Benzodiazepines and Beta-Adrenergic Blockers," Lipton, M. A., DiMasacio, A., and Killiam, K. F., ed. *Psychopharmacology: A Generation of Progress.* New York: Raven Press, 1978, p. 1381.
65. ———. "Prazepam and Lorazepam: Two New Benzodiazepines," *New England Journal of Medicine,* 299 (1978): 1342.
66. ———. "Prazepam: A Precursor of Desmethyldiazepam," *Lancet,* 1 (1978): 720.
67. GREENBLATT, D. J., et al. "Influence of Magnesium and Aluminum Hydroxide Mixture on Chlordiazepoxide Absorp-

tion," *Clinical Pharmacology Therapy,* 19 (1976): 234.

68. ———. "Absorption Rate, Blood Concentrations and Early Response to Oral Chlordiazepoxide," *American Journal of Psychiatry,* 134 (1977): 559.

69. GREENBLATT, D. J., et al. "Absorption of Oral and Intramuscular Chlordiazepoxide," *European Journal of Clinical Pharmacology,* 13 (1978): 267.

70. GREENBLATT, D. J., et al. "Clinical Pharmacokinetics of Chlordiazepoxide," *Clinical Pharmacokinetics Journal* 3 (1978): 381.

71. GROF, P., SAXENA, B., and DAIGLE, L. "Dopaminergic Agonist Nomifensine Compared with Amitriptyline: A Double-Blind Clinical Trial in Acute Primary Depressions," *British Journal of Clinical Pharmacology,* 4 (Suppl. 2) (1977): 221S.

72. HABERMAN, W. "A Review of Controlled Studies with Nomifensine Performed Outside the U.K." *British Journal of Clinical Pharmacology,* 4 (Suppl 2) (1977): 237S.

73. HAIDER, I. "A Comparative Trial of Ro 4-6270 and Amitriptyline in Depressive Illness," *British Journal of Psychiatry,* 113 (1967): 993.

74. HAIZLIP, T. M., and EWING, J. A. "Meprobamate Habituation: A Controlled Clinical Study," *New England Journal of Medicine,* 258 (1958): 258.

75. HANLON, T. E., et al. "Perphenazinebenzotropine Mesylate Treatment of Newly Admitted Psychiatric Patients," *Psychopharmacologia,* 9 (1966): 328.

76. HARE, H. P. "Comparison of Chlordiazepoxide-Amitriptyline Combination With Amitriptyline Alone in Anxiety-Depressive States," *Journal of Clinical Pharmacology,* 2 (1971): 456.

77. HEKIMIAN, L., FRIEDHOFF, A. J., and DEEVER, E. "A Comparison of the Onset of Action and Therapeutic Efficacy of Amexapine and Amitriptyline," *Journal of Clinical Psychiatry,* 39 (1978): 633.

78. HIRSCH, S. R., et al. "Outpatient Maintenance of Chronic Schizophrenic Patients with Long-Acting Fluphenazine Double-Blind Placebo Trial," *British Medical Journal,* 1 (1973): 633.

79. HOGARTY, G. E., and GOLDBERG, S. C. Drugs and Sociotherapy in the Aftercare of Schizophrenic Patients," *Archives of General Psychiatry,* 28 (1973): 54.

80. HOGARTY, G. E., et al. "Drugs and Sociotherapy in the Aftercare of Schizophrenic Patients," *Archives of General Psychiatry,* 31 (1974): 603.

81. HOGARTY, G. E., et al. "Drug Discontinuation Among Long-Term Successfully Maintained Schizophrenic Outpatients," *Diseases of the Nervous System,* 37 (1976): 494.

82. HOGARTY, G. E., et al. "Fluphenazine and Social Therapy in the Aftercare of Schizophrenic Patients," *Archives of General Psychiatry,* 36 (1979): 1283.

83. HOLLISTER, L. E. "Tricyclic Antidepressants," *New England Journal of Medicine,* 299 (1978): 1106–1168.

84. ———, and GIAZENER, F. S. "Withdrawal Reactions from Meprobamate Alone and Combined with Promazine: A Controlled Study," *Psychopharmacologia,* 1 (1960): 336.

85. HOLLISTER, L. E., MOTZENBECKER, F. P., and DEGAN, R. O. "Withdrawal Reactions from Chlordiazepoxide (Librium)," *Psychopharmacologia,* 2: (1961): 63.

86. HOLZMAN, P., and JOHNSTON, M. H. *Assessing Schizophrenic Thinking.* San Francisco: Jossey-Bass, 1980.

87. ITIL, T. M., et al. "Clinical and Quantitative EEG Changes at Different Dosage Levels of Fluphenazine Treatment," *Acta Psychiatric Scandanavia,* 47 (1971): 440.

88. IVES, J., et al. "The Ineffectiveness of Chlordiazepoxide in Depression Disorders," *Psychiatric Journal of the University of Ottawa,* 3 (1978): 115.

89. JACOBS, M. A., GLOBUS, G., and HEIM, E. "Reduction in Symptomatology of Ambulatory Patients. The Combined Effects of a Tranquilizer and Psychotherapy," *Archives of General Psychiatry,* 15 (1966): 45.

90. JUDD, L. L., GOLDSTEIN, M. J., and RODNICK, E. H. "Phenothiazine Effects in Good Premorbid Schizophrenics Divided into Paranoid-Nonparanoid Status," *Archives of General Psychiatry,* 29 (1973): 207–221.

91. KAY, D. W. K., FAHY, T., and GARSIDE, R. F. "A Seven-Month Double-Blind Trial of Amitriptyline and Diazepam in ECT-Treated Depressed Patients," *British Journal of Psychiatry,* 117 (1970): 667.

92. KLEIN, D. F. "Delineation of Two-Drug Re-

sponsive Anxiety Syndrome," *Psychopharmacologia,* 5 (1964): 397.

93. ———. "Importance of Psychiatric Diagnosis in Prediction of Clinical Drug Effects," *Archives of General Psychiatry,* 16 (1967): 118.

94. KLEIN, D. F., and DAVIS, J. M. *"Diagnosis and Drug Treatment of Psychiatric Disorders.* Baltimore: Williams & Wilkins, 1969.

95. ———, HOMGFELD, G., and FEIDMAN, S. "Prediction of Drug Effect in Personality Disorders," *Journal of Nervous Mental Disease,* 156 (1973): 183.

96. KLERMAN, G. L., et al. "Treatment of Depression by Drugs and Psychotherapy," *American Journal of Psychiatry,* 131 (1974): 186.

97. KLETT, C. J., and CAFFEY, E. "Evaluating the Long-Term Need for Antiparkinson Drugs by Chronic Schizophrenics," *Archives of General Psychiatry,* 26 (1972): 374.

98. KLOTZ, U., and MULLER-SEYDLITZ, P. "Altered Elimination of Desmethyldiazepam in the Elderly," *British Journal of Clinical Pharmacology,* (1979): 119.

99. KLOTZ, U., ANTONIN, K. L., and BIECK, P. R. "Comparison of the Pharmacokinetics of Diazepam After Single and Subchronic Doses," *European Journal of Clinical Pharmacology,* 10 (1976): 121.

100. KLOTZ, U., et al. "The Effects of Age and Liver Disease on the Disposition and Elimination of Diazepam in Adult Man," *Journal of Clinical Investigation,* 55 (1975): 347.

101. KRAUGH-SORENSON, P., ASBERG, M., and EGGERT-HANSEN, C. "Plasma Nortriptyline Levels in Endogenous Depression," *Lancet,* 1 (1973): 113.

102. KRAUGH-SORENSEN, P., HANSEN, I. E., and ASBERG, M. "Plasma Levels of Nortriptyline in the Treatment of Endogenous Depression," *Acta Psychiatric Scand,* 49 (1973): 445.

103. KRAUGH-SORENSEN, P., et al. Self-Inhibiting Action of Nortriptyline Antidepressive Effect at High Plasma Levels," *Psychopharmacologia,* 45 (1976): 305.

104. KUPFER, D. J., et al. "Amitriptyline Plasma Levels and Clinical Response in Primary Depression," *Clinical Pharmacology Therapy,* 22 (1977): 904.

105. KURKLAND, A. A. and RICHARDSON, J. M. "A Comparative Study of Two Long-Acting Phenothiazine Preparations, Fluphenazine Enanthate and Fluphenazine Decanoate," *Psychopharmacologia,* 9 (1966): 320.

106. KYRIAKOPOULOS, A. A., GREENBLATT, D. J., and SHADER, R. J. "Clinical Pharmacokinetics of Lorazepam: A Review," *Journal of Clinical Psychiatry,* 39 (1978): 16.

107. LAPOLLA, A., and NASH, L. R. "Treatment of Phenothiazine-Induced Parkinsonism with Biperiden," *Current Therapeutic Research,* 7 (1965): 536.

108. LASCELLES, R. G. "Atypical Facial Pain and Depression," *British Journal of Psychiatry,* 112 (1966): 651.

109. LEFF, J. P., and WING, J. K. "Trial of Maintenance Therapy in Schizophrenia, *British Medical Journal,* 3 (1971): 599.

110. LEHMAN, H. E. "Drug Treatment of Schizophrenia," *International Psychiatry Clinics,* 2 (1965) 717–752.

111. LIPMAN, R. S., et al. "Patient Report of Significant Life Situation Events," *Diseases of the Nervous System,* 26 (1965): 586.

112. LIPMAN, R. S., et al. "Medication, Anxiety Reduction and Patient Report of Significant Life Situation Events," *Diseases of the Nervous System,* 32 (1971): 240.

113. LIPSEDGE, M. S., et al. "The Management of Severe Agoraphobia," *Psychopharmacologia,* 32 (1973): 67.

114. LOGA, S., CURRY, S., and LADER, M. "Low Doses and Low Plasma Levels," *British Journal of Clinical Pharmacology,* 2 (1975): 197–208.

115. MCCELLAND, H. A., et al. "Very High-Dose Fluphenazine Decanoate: A Controlled Trial in Chronic Schizophrenia, *Archives of General Psychiatry,* 33 (1976): 435.

116. MACLEOD, S., et al. "Interactions of Disulfiram with Benzodiazepines," *Clinical Pharmacology Therapy,* 24 (1978): 583.

117. MAGNUS, R. V., DEAN, B. C., and CURRY, S. H. "Clorazepate: Double-Blind Crossover Comparison of a Single Nightly Dose with Diazepam Twice Daily in Anxiety, *Diseases of the Nervous System,* 38 (1977): 819.

118. MARKS, J. *The Benzodiazepines: Use, Overuse, Misuse, Abuse.* Lancaster England: University of Cambridge, 1978.

119. MARTIN, L., BAKER, G., and MITCHELL, P. "The Effects of Vilozazine HC1 on the

Transport of Noradrenaline Dopamine and 5-Hydroxytryptamine and Amino Buteric Acid in Rat Brain Tissue, *Neuropharmacology,* 17 (1978): 421.

120. MAXWELL, J. D., et al. "Plasma Disappearance and Cerebral Effects of Chlorpromazine in Cirrhosis," *Clinical Science,* 43 (1972): 143.

121. MAY, P. R. *Treatment of Schizophrenia.* New York: Science House, 1968.

122. ———, TUMA, A. H., and DIXON, W. J. "For Better or For Worse? Outcome Variance with Psychotherapy and Other Treatments for Schizophrenia," *Journal of Nervous Mental Disease,* 165 (1977): 231.

123. MICHAUX, M. H., KURLAND, A. A., and AGALLIANOS, D. "Chlorpromazine-Chlordiazepoxide and Chlorpromazine-Imipramine Treatment of Newly Hospitalized Acutely Ill Psychiatric Patients," *Current Therapeutic Research,* 8 (Suppl.) (1966): 117.

124. MINDHAM, R. H. S., HOWLAND, C., and SHEPHERD, M. "Continuation Therapy with Tricyclic Antidepressants in Depressive Illness," *Lancet,* 2 (1972): 854.

125. MOIR, D. C., et al. "Cardiotoxicity of Tricyclic Antidepressants," *British Journal of Pharmacology,* 44 (1972): 371.

126. MONTGOMERY S. A., BRAITHWAITE, R. A., and CRAMMER, J. L. "Routine Nortriptyline Levels in Treatment of Depression," *British Medical Journal,* 2 (1977): 166.

127. MONTGOMERY, S. A., et al. "High Plasma Nortriptyline Levels in the Treatment of Depression," I. *Clinical Pharmacology Therapeutics,* 23 (1978): 309.

128. MOORE, D. F. "Treatment of Acute Schizophrenia with Loxapine Succinate (Loxitane) in a Controlled Study with Chlorpromazine," *Current Therapeutic Research,* 18 (1975): 172.

129. MOYANO, C. "A Double Blind Comparison of Loxitane Loxapine Succinate and Trifluoperazine Hydrochloride in Chronic Schizophrenic Patients," *Diseases of the Nervous System,* 36 (1975): 301.

130. MURPHY, E. J., DONALD, J. F., and MOLLA, A. L. "Mianserin in the Treatment of Depression in General Practice," *Practitioner,* 217 (1976): 135.

131. MUSCETTOLA, G., et al. "Imipramine and Desipramine in Plasma and Spinal Fluid," *Archives of General Psychiatry,* 35 (1978): 621.

132. NIES, A., et al. "The Efficacy of the Monoamine Oxidase Inhibitor Phenelzine: Dose Effects and Prediction of Response," in Boissier, J. R., Hipptus, H., and Pichot, P., eds., *Neuropsychopharmacology,* vol. 39. Amsterdam: Excerpta Medica, 1975, p. 765.

133. OLIVIER-MARTIN, R., et al. "Concentrations Plasmatiques de l'Imipramine et de la Desmethylimipramine et Effet Antidepresseur au Cours d'un Traitement Controlé," *Psychopharmacologia,* 41 (1975): 187.

134. ORLOV, P., et al. "Withdrawal of Antiparkinsonian Drugs," *Archives of General Psychiatry,* 25 (1971): 410.

135. PECKNOLD, J. C., et al. "Lack of Indication in Use of Antipsychotic Medication," *Diseases Nervous System,* 32 (1971): 538.

136. PERSSON, G. "Lithium Prophylaxis in Affective Disorders: An Open Trial with Matched Controls," *Acta Psychiatry Scandinavia,* 48 (1972): 462.

137. PETERSON, G., et al. "Anticholinergic Activity of the Tricyclic Antidepressants Desipramine and Doxepin in Nondepressed Volunteers," *Communications in Psychopharmacology,* 2 (1975): 145.

138. PEVNICK, J., JASENSKI, D., and HAERTZEN, C. "Abrupt Withdrawal from Therapeutically Administered Diazepam," *Archives of General Psychiatry,* 35 (1978): 995.

139. PIAFSKY, K. M., et al. "Increased Plasma Protein Binding of Propranolol and Chlorpromazine Mediated by Disease-Induced Elevations of Plasma and Acid Glycoprotein," *New England Journal of Medicine,* 299 (1978): 1435.

140. PICHOT, P., GUELT, J., and DREYFUS, J. F. "A Controlled Multicentre Therapeutic Trial of Viloxazine (Vivaian)," *Journal of Internal Medical Research,* 3 (1975): 30.

141. POST, C., et al. "Pharmacokinetics of N-Desmethyldiazepam in Healthy Volunteers After Single Daily Doses of Dipotassium Chlorazepate," *Psychopharmacology,* 5 (1977): 105.

142. PRIEN, R. F., and COLE, J. O. "High-Dose Chlorpromazine Therapy in Chronic Schizophrenia," *Archives of General Psychiatry,* 18 (1968): 482.

143. PRIEN, R. F., CAFFEY, E. M., JR., and

KLETT, C. J. "Lithium Carbonate and Imipramine in Prevention of Affective Episodes," *Archives of General Psychiatry,* 29 (1973): 420.

144. PRIEN, R. F., LEVINE, J., and COLE, J. O. "High-Dose Trifluoperazine Therapy in Chronic Schizophrenia," *American Journal of Psychiatry,* 126 (1969): 305.

145. QUITKIN, F., RIFKIN, A., and KLEIN, D. F. "Very High Dose vs Standard Dosage Fluphenazine in Schizophrenia," *Archives of General Psychiatry,* 32 (1975): 1276.

146. QUITKIN, F., et al. "Phobic Anxiety Syndrome Complicated by Drug Dependence and Addiction," *Archives of General Psychiatry,* 27 (1972): 159.

147. QUITKIN, F., et al. "Long-Acting Oral Versus Injectable Antipsychotic Drugs in Schizophrenics," *Archives of General Psychiatry,* 35 (1978): 389.

148. ———. "Prophylactic Effect of Lithium and Imipramine in Unipolar and Bipolar II Patients," *American Journal of Psychiatry,* 135 (1978): 570.

149. RASKIN, A., et al. "Differential Response to Chlorpromazine, Imipramine and a Placebo: A Study of Hospitalized Depressed Patients," *Archives of General Psychiatry,* 23 (1970): 165.

150. RAVARIS, C. L., et al. "A Multiple-Dose, Controlled Study of Phenelzine in Depression-Anxiety States," *Archives of General Psychiatry,* 33 (1976): 347.

151. REES, L., and DAVIES, B. "A Controlled Trial of Phenelzine (Nardil) in the Treatment of Severe Depressive Illness," *Journal of Mental Science,* 107 (1961): 560.

152. REISBY, N., et al. "Imipramine: Clinical Effects and Pharmacokinetic Variability," *Psychopharmacology,* 54 (1977): 263.

153. RICHARDS, D. J. "Clinical Profile of Lorazepam," *Diseases of the Nervous System,* 39 (1978): 36.

154. RICKELS, K. "Use of Antianxiety Agents in Anxious Outpatients," *Psychopharmacology,* 58 (1978): 1.

155. ———, et al. "Drug Treatment in Depressive Illness," *Diseases of the Nervous System,* 31 (1970): 305.

156. RICKELS, K., et al. "Doxepin and Amitryptyline-Perphenazine in Mixed Anxious-Depressed Neurotic Outpatients: A Collaborative Controlled Study," *Psychopharmacologia,* 18 (1977): 239.

157. RICKELS, K., et al. "Prazepam in Anxiety: A Controlled Clinical Trial," *Comprehensive Psychiatry,* 18 (1977): 239.

158. RIFKIN, A., et al. "Fluphenazine Decanoate Oral Fluphenazine and Placebo in the Treatment of Remitted Schizophrenics. I: Relapse Rates After One Year," *Archives of General Psychiatry,* 34 (1977): 1215.

159. ROBERTS, R. K., et al. "The Effect of Age and Parenchymal Liver Disease in the Distribution and Elimination of Chlordiazepoxide (Librium)," *Gastroenterology,* 75 (1978): 479.

160. ROBIN, A., CURRY, S. H., and WHELPTON, R. "Clinical and Biochemical Comparison of Chlorazepate and Diazepam," *Psychological Medicine,* 4 (1974): 388.

161. ROBINSON, D. S., et al. "The Monoamine Oxidase Inhibitor, Phenelzine, in the Treatment of Deprssive-Anxiety States," *Archives of General Psychiatry,* 29 (1973): 407.

162. ROBINSON, D. S., et al. "Clinical Pharmacology of Phenelzine," *Archives of General Psychiatry,* 35 (1978): 629.

163. SAKALIS, G., et al. "Physiologic and Clinical Effects of Chlorpromazine and their Relationship to Plasma Level," *Clinical Pharmacology and Therapeutics,* 13 (1972): 931–946.

164. SATHANANTHAN, G. L., et al. "Amoxapine and Imipramine: A Double-Blind Study in Depressed Patients," *Current Therapeutic Research,* 15 (1973): 919.

165. SCHAPIRA, K., MCCLELLAND, H., and NEWELL, D. "A Comparison of High and Low Dose Lorazepam with Amylobarbitone in Patients with Anxiety States," *American Journal of Psychiatry,* 134 (1977): 25.

166. SCHATZBERG, A. F., and COLE, J. O. "Benzodiazepines in Depressive Disorders," *Archives of General Psychiatry,* 35 (1978): 1359.

167. SCHIELE, B. C. "Loxapine Succinate: A Controlled Double-Blind Study in Chronic Schizophrenic," *Diseases of the Nervous System,* 18 (1975): 361.

168. SCHOOLER, N. R., and LEVINE, J. "Fluphenazine and Fluphenazine HCı in the Treatment of Schizophrenic Patients," in Deniker, P. Raduc-Thomas, C., and Villeneuve, A., *Proceedings of the Meeting of the Collegium International Neuro-Psy-*

chopharmacologicum, vol. 11. Oxford: Pergamon Press, 1978, p. 418.

169. ———, and SEVERE, J. B. "Depot Fluphenazine in the Prevention of Relapse in Schizophrenia: Evaluation of a Treatment Regimen," *Psychopharmacology Bulletin*, 15 (1979): 44.

170. SCHOU, M., THOMSEN, K., and BAASTRUP, P. C. "Studies on the Course of Recurrent Endogenous Affective Disorders," *International Pharmacopsychiatry*, 5 (1970): 100.

171. SEAGER, C. P., and BIRD, R. L. "Imipramine with Electrical Treatment in Depression Controlled Trial," *Journal of Mental Science*, 108 (1962): 704.

172. SEDVALL, G. "Relationships Among Biochemical, Clinical, and Pharmecokinetic Variables in Neuroleptic-Treated Schizophrenic Patients," in Cattabeni, F., ed., *Long Term Effects of Neuroleptics*. New York: Raven Press, 1980, pp.

173. SELLERS, E. M., et al. "Influence of Disulfiram and Disease on Benzodiazepine Disposition," *Clinical Pharmacology and Therapeutics*, 24 (1978): 583.

174. SHADER, R. I., et al. "Impaired Absorption of Desmethyldiazepam from Clorazepate by Co-Administration of Maalox," *Clinical Pharmacology and Therapeutics*, 24 (1978): 308.

175. SHEEHY, L. M., and MAXMEN, J. "Phenelzine Induced Psychosis," *American Journal of Psychiatry*, 135 (1978): 1422.

176. SHEPHARD, M. "Report to the Medical Research Council by its Clinical Psychiatry Committee: Clinical Trial of the Treatment of Depressive Illness," *British Medical Journal*, 1 (1965): 881.

177. SHOPSIN, B., et al. "A Controlled Double-Blind Comparison Between Loxapine Succinate and Chlorpromazine in Acutely Newly Hospitalized Schizophrenic Patients," *Current Therapeutic Research*, 14 (739): 1972.

178. SHOPSIN, B., et al. "Clozapine Chlorpromazine and Placebo in Newly Hospitalized Acutely Schizophrenic Patients," *Archives of General Psychiatry*, 36 (1979): 36.

179. SHULL, H. J., WILKINSON, G. R., and JOHNSON, R. "Normal Disposition of Oxazepam in Acute Viral Hepatitis and Cyrrhosis," *Annals of Internal Medicine*, 84 (1976): 420.

180. SIMON, P., et al. "Standard and Long-Acting Depot Neuroleptics in Chronic Schizophrenics: An 18-Month Open Multicentric Study," *Archives of General Psychiatry*, 35 (1978): 893.

181. SIMPSON, G. M., AMIN, M., and EDWARDS, J. G. "A Double-Blind Comparison of Molindone and Trifluoperazine in the Treatment of Acute Schizophrenia," *Journal of Clinical Pharmacology*, 11 (1971): 227.

182. SIMPSON, G. M., et al. "Problems in the Evaluation of the Optimal Dose of a Phenothiazine (Butaperazine)," *Diseases of the Nervous System*, 29 (1968): 478.

183. SIMPSON, G. M., et al. "Role of Antidepressants and Neuroleptics in the Treatment of Depression," *Archives of General Psychiatry*, 27 (1972): 337.

184. SIMPSON, G. M., et al. "Two Dosages of Imipramine Hospitalized Endogenous and Neurotic Depressives," *Archives of General Psychiatry*, 33 (1976): 1093.

185. SMITH, R. C. "Amoxapine, Imipramine and Placebo in Depressive Illness," *Current Therapeutic Research*, 18 (1975): 346.

186. ———, TAMMINGA, C., and DAVIS, J. M. "Effects of Apomorphine on Chronic Schizophrenic Symptoms," *Journal of Neurological Transmitters*, 40 (1977): 171.

187. SMITH, R. C., et al. "Plasma Butaperazine Levels in Long-Term Chronic Non-Responding Schizophrenics," *Communications in Psychopharmacology*, 1 (1977): 319.

188. SMITH, R. C., et al. "Blood Levels of Neuroleptic Drugs in Non-Responding Chronic Schizophrenic Patients," *Archives of General Psychiatry*, 36 (1979): 579.

189. SNYDER, S. H., and YAMAMURA, H. I. "Antidepressants and the Muscarinic Acetylcholine Receptor," *Archives of General Psychiatry*, 34 (1977): 326.

190. SOLYOM, L., et al. "Behavior Therapy Versus Drug Therapy in the Treatment of Phobic Neurosis," *Canadian Psychiatric Association Journal*, 18 (1973): 25.

191. SPOHN, H. E., et al. "Phenothiazine Effects on Psychological and Psychophysiological Dysfunction in Chronic Schizophrenics," *Archives of General Psychiatry*, 34 (1977): 633.

192. TANSELLA, C. Z., TANSELLA, M., and LADER, M. "A Comparison of the Clinical and Psychological Effects of Diazepam

and Amylobarbitone in Anxious Patients," *British Journal of Clinical Pharmacology,* 7 (1979): 605.

193. TYRER, P., CONDY, J., and KELLY, D. "Phenelzine in Phobic Anxiety: A Controlled Trial," *Psychological Medicine,* 3 (1973): 120.
194. VAN DER VELDE, C., and KILTIE, H. "Effectiveness of Loxapine Succinate in Acute Schizophrenia: A Comparative Study with Thiothixene," *Current Therapeutic Research,* 17 (1975) 1–11.
195. VAN PRAAG, H., and DOLS, L. C. W. "Fluphenazine Enanthate and Fluphenazine Decanoate: A Comparison of Their Duration of Action and Motor Side Effects," *American Journal of Psychiatry,* 130 (1973): 801.
196. VOGEL, H. P., BENTE, D., and FEDER, J. "Manserin Versus Amitriptyline: A Double-Blind Trial Evaluated by the AMP System," *Internatinal Pharmacopsychiatry,* 11 (1976): 25.
197. WEIR, J. H. "Prazepam in the Treatment of Anxiety," *Journal of Clinical Psychiatry,* 39 (1978): 841.
198. WEISSMAN, M. M., et al. "The Efficacy of Drugs and Psychotherapy in the Treatment of Acute Depressive Episodes," *American Journal of Psychiatry,* 136 (1979): 555.
199. WHYTE, S. F., et al. "Plasma Concentrations of Protriptyline and Clinical Effects in Depressed Women," *British Journal of Psychiatry,* 128 (1976): 394.
200. WIJSENBEEK, H., STEINER, M., and GOLDBERG, S. C. "Trifluoperazine: A Comparison Between Regular and High Doses," *Psychopharmacologia,* 36 (1974): 147.
201. WILKERSON, R., and SANDERS, P. "The Antiarrhythmic Action of Amitriptyline on Arrhythmias Associated with Myocardial Infarction in Drugs," *European Journal of Pharmacology,* 51 (1978): 193.
202. ZIEGLER, V. E., CLAYTON, P. J., and BIGGS, J. T. "A comparison Study of Amitriptyline and Nortriptyline with Plasma Levels," *Archives of General Psychiatry,* 34 (1977): 607.
203. ZINN, C. M., KLEIN, D. F., and WOERNER, M. G. "Behavior Therapy, Supportive Psychotherapy, Imipramine and Phobias," *Archives of General Psychiatry,* 35 (1978): 307.

PART TWO

Childhood and Adolescence

CHAPTER 6

PSYCHOPATHOLOGICAL DEVELOPMENT IN INFANCY

Justin D. Call

BETWEEN 1895 and 1923 Freud postulated that psychoneuroses, sexual aberrations, character defects, and the disguised wishes of anxiety dreams all had their roots in "infantile" experience. Accordingly, many of his followers, along with some educators, believed that the true path to optimum mental health had been found. This theory has been described as "the original sin hypothesis." According to this hypothesis, what was necessary for either the treatment or the prevention of such disorders was the same; that is, the therapist was required to locate and resolve the original infantile source of such difficulty and create for the patient a more loving and less conflictual kind of experience in which the psyche could flourish. This prescription, however, has turned out to be oversimplistic and becomes, in effect, an obfuscating half-truth when applied exclusively to problems of human development.

Freud[24] himself never espoused such a theory or remedy, for he was deeply mindful of the biological underpinnings of all mental functioning, both normal and abnormal. What he meant by "infantile" was often inexact; it could refer to any time from birth through prepuberty. He also recognized that conflict did not exist in isolated psychoneuroses, but was part of human existence itself.[25] (Anna Freud[23] has discussed these misconceptions of psychoanalytic theory.) Freud was successful, however, in convincing most of western society that the infant and young child did possess a psyche in which complex fantasies and feelings were elaborated, even though most of what took place there was not readily accessible to parents, teachers, or other caretakers.

The idea of predicting long-term outcome from infantile experience alone is recognized to be full of so many difficult methodological issues that it is no longer seriously pursued. Benjamin[2] has espoused short-term predictions that utilize outcome measures specified at the time predictions and underlying hypotheses are made. As our knowledge of average-expectable patterns of normal development for the first three years of life has increased over the past twenty-five years,[13] short-term predictions and the theories they are based upon can be more scientifically stated and tested.

The idea that infants may show evidence of serious psychopathology during their development is still surprising and shocking to many people. While it is recognized that human babies are more vulnerable to certain kinds of infections, malignancies, gastrointestinal disorders, metabolic disturbances, dehydration, and electrolite imbalance occasioned by stressful physical events, it is not generally acknowledged that the human infant also is capable of reacting and is vulnerable to the stresses of pain, violent arguments between parents, screaming, changes in caretakers, separation from the primary caretaker, and sudden changes in living circumstances. What constitutes stress for an infant as opposed to an older child or an adult is even less well understood and acknowledged. The many rapid changes in normal development over the first three years of life necessitate examination of these issues as related to infants of a specific age. Age differences must be taken into account in order to accurately evaluate the impact of stress, loss, deprivation, and trauma. Thus, the only way in which an infant can be adequately evaluated is in the context of his own developmental path. Such paths are not the same for all infants. Boys are more vulnerable to caloric deficiencies than girls.[8] Temperamental differences exist.[38] What is stressful for one infant may not be to another of the same age because of these differences.

Many well-defined physical illnesses in infancy reflect disturbances in the caretaking environment; these illnesses respond favorably to improved relations between the infant and his caretakers and to the establishment of a more stable, predictable pattern of caretaking within a socially stable and reliable caretaking relationship.[13] In fact, a number of illnesses previously thought to have specific physical etiology are now known to be associated primarily with disturbances in the psychobiosocial matrix surrounding normal development. Failure to thrive without organic cause is now known to be associated with attachment disorder of infancy.[11] Projectile vomiting associated with pylorospasm improves dramatically with changes in feeding technique.[20] Even so-called congenital pyloric stenosis responds to changes in feeding style,[3] as do rumination[21] and psychogenic megacolon.[14] Psychosocial dwarfism has been described by Kavanaugh and Mattsson.[31]

¶ Some Case Examples

Eczema

A four-month-old boy was admitted to the hospital with oozing eczema from head to foot. He was extremely irritable and cried most of the time. The eczema, at the time of admission to the hospital, had been unresponsive to nonmilk nutritional supplements, such as soybean formula. His mother had a history of eczema and asthma, and eczema had occurred in a less severe form in an older sibling. Breast feeding had failed because of "attachment difficulties" and "inadequate milk supply." Eczema made its first appearance with the change from breast to formula feeding at four weeks, and continued when cow's milk was discontinued at three months. A specific nurse was assigned to provide all regular care. Soybean formula was continued. The nurse held him closely at feeding, played with him, and established well-reciprocated visual and vocal exchanges. At this point the eczema began improving rapidly.

COMMENT

While eczema occurs in babies with a known hereditary allergic diathesis to cow's milk, the onset and severity of the illness is known to be influenced by psychologically stressful situations that occur in the family and in the mother-infant relationship. A total estrangement had developed between this infant and his mother, even before the onset of eczema. Eczema is rarely seen in breast-fed infants. The positive response to exclusive nursing care illustrates the significance and importance of closely coordinated reciprocal exchanges between the mothering figure and the infant in soothing and regulating the baby. Both physiological and emotional reciprocity is required. This usually occurs normally and without conscious thought or instruction. In this case, however, the process was derailed and had to be reconstructed.

Asthma

A nine-month-old boy was hospitalized because of bronchiolitis associated with croup. He clung desperately to his mother, who stayed with him in the croup tent. A year later, he was still clinging to his mother, and asthma had developed. The mother continued watchful anxiousness and close monitoring of breathing for signs of recurrent croup and asthma.

COMMENT

Approximately 50 percent of children who contract bronchiolitis after six months of age develop asthma by the age of two. Asthma is also frequently associated with eczema. When it does occur, the mother-infant relationship is grossly distorted, with the infant showing what has now been described as attachment disorder of the symbiotic type.[11] It is likely that the psychological factors that lead to asthma in these cases are set in motion by the parents' anxiety about their infant's health. Bronchiolitis with croup is a threatening illness that arouses the parents' concern about death. And while the infants recover, the parents often do not fully recover psychologically. They remain traumatized by the experience. This leads to a pathologically close symbiotic relationship with the child and a failure of both parent and child to proceed through the stages of separation-individuation. This psychological constellation renders the child vulnerable to asthmatic attacks precipitated by minor stress. Prevention could be effected by facilitating successful psychological convalescence of both parent and child.

Dog Bite

A twenty-month-old boy was bitten severely on the face by a German shepherd while standing in the driveway of his home with his mother by his side. He still remembered the incident twenty months later when he said, "Canis the dog bit me on the cheek because I was only one year old. He was a mean dog. He doesn't like people. I'm mad at Canis and when I grow up, I'll get a real gun with real bullets and shoot him. I want him to die. He took two bites with his big teeth." He began having nightmares three times a night following the bite. This continued for the next twenty months. He showed stuttering for two months and clung desperately to his mother, showing severe separation anxiety. All of his thinking and play centered around dogs that bit and fear of dogs. He showed a very serious demeanor and seemed to gain very little pleasure in life. His favorite story was about how Louis Pasteur saved a boy bitten by a dog from rabies. Kevin was frightened of the billboard freeways advertising the movie *Jaws.*

COMMENT

Gislason and Call[27] have studied three young children who were bitten severely by dogs when less than three years of age. The march of symptoms and constriction of personality in these children are similar to, but are a more severe form of, what has been described in children who have had tonsillectomies at less than three years of age.[30] These dog-bitten children have been emotionally traumatized and show a severe form of infantile traumatic neurosis. They are highly subject to subsequent traumatic events and a distortion of personality development caused by the initial and subsequent traumas. The close relationship with the parent is caused by the traumas experienced by both child and parent. The child's normal development is distorted because of the extreme difficulty that these traumatized children show in going through the separation-individuation phases of development. Psychiatric treatment using play techniques and psychoeducational approaches with the parent could, in all likelihood, be helpful to such children.

Psychogenic Megacolon

A little girl, eighteen months of age, appeared in the outpatient clinic with a large abdomen. Her mother reported that the difficulty began when she was an infant of only a few months. The mother observed that the child grunted a lot and developed a red face when pushing out a bowel movement. The mother thought her infant was constipated and reported this to the doctor. The doctor prescribed a small suppository. The grunting and red face continued. Enemas were then prescribed, which led to increased withholding of bowel movements. Finally, a chronic state of withholding and constipation developed. No spontaneous bowel movements occurred. Enemas, suppositories, and cathartics were now used regularly to produce bowel movements. Occasionally, a very large fecal mass was produced by the enemas. The child's appetite was poor; she subsisted mainly on milk and soft foods, and she did not use her teeth in biting food. Her nutrition was poor, her arms were spindly, and because of the large abdomen she was short of breath and became sedentary. A developmental history showed that she reached her motor landmarks at the usual time, although she was considered a quiet child by the age of one, and at eighteen months she did not speak. Her eyes were large and observant, and she had an anxious expression on her face. Her abdomen was found to be filled with fecal masses. The remainder of the physical examination was normal. Radiographic and biopsy studies showed normal bowel activity and the presence of ganglion cells in the outer adventitial layers of the rectum and lower colon. The diagnosis was psychogenic megacolon with delayed expressive language function and infantile feeding pattern.

COMMENT

This diagnosis could have been made without radiographic or biopsy studies since the onset of the infant's constipation did not date back to birth but rather a few months after birth. Most healthy infants develop a red face when pushing out a bowel movement, but some parents are particularly anxious about bowel functioning and regard the red face, as this mother did, as an indication of abnormal distress. The mother herself had been constipated and had been subjected to enemas and cathartics all her life, and as an adult had used cathartics regularly. Thus, it is no surprise that she resorted to this method of producing a bowel movement in her infant. Doctors unwittingly collaborated with her. Thus, the mother's anal fixations were passed on to the child with whom she identified. The treatment in these cases requires discontinuation of all anally oriented treatment procedures, including enemas, cathartics, and suppositories. In addition, a reorientation of the parents to the more general needs of the child, together with an attempt to identify and resolve the sources of anal conflict in each of the parents and between them, is necessary since they often collaborate with one another in an anal orientation to the child. In the course of such treatment, children do not get well all at once and continue to show exacerbations and remissions.

Modern Child Abuse

Infants and children throughout history have always been abused. Even in our present enlightened society the rights of infants and young children in regard to such issues as adoption, visitation, and custody are frequently ignored by the courts. The issue of spanking children in the schools is still being hotly debated. Discipline remains confused with punishment. The rediscovery of physical abuse of children at the hands of their parents began in 1946 when a child showing bilateral subdural hematomas of unknown origin was hospitalized at New York Hospital. A radiographic study of the abdomen was also carried out and, quite incidentally, it was observed that one of the arms showed an area of subperiosteal hemorrhage with new bone formation—telltale signs of a healing fracture. Dr. John Caffey[5] suggested the possibility of a previously unrecognized traumatic fracture; and a more careful psychosocial history showed a strong presumption that the fracture had been caused by an abusing mother. This was, however, considered to be a rarity, and it was another fifteen years before the problem of physical abuse of children was clearly recognized as a small epidemic. In 1962, through the work of Dr. Henry Kempe,[32] the problem was placed on the map of admitting rooms and emergency rooms throughout the country. Eight years later another variant of child abuse was described: the shearing of the cerebral veins as they enter the superior longitudinal sinus. This was caused by the violent shaking of the infant by irate parents or babysitters who, when questioned closely, felt that spankings and shakings were justified because of defiance and disobedience by stubborn, frightened children who could not speak. Also included in this group were children who did not perform up to standards expected by their parents or babysitters or who could not or would not explain their misbehavior or stubborn refusal to conform to expectations.

COMMENT

In these cases, a high level of moral accountability is expected by the parents of immature infants and children long before meaningful communication is possible. Studies of the parents by Steele and others[37] have shown that abusing parents are often reenacting their own abuse as children, motivated to do so by a rigid, demanding, and cruel conscience acquired through identification with their own demanding parents. Child abuse is only one of the many parent-child difficulties showing the nongenetic transmission of psychopathological behavior from generation to generation. The passage of other psychopathological child-rearing patterns from one generation to another has also been observed in children showing psychogenic megacolon, attitudes toward separation, giving the child over to the care of others, feeding difficulties, anorexia nervosa, delinquency, and environmental retardation.

Alternative Pathways of Mental Development Demonstrated by Physically Handicapped Children

Until 1941 rubella, or German measles, was considered an unimportant minor illness, significant only as a part of the differential diagnosis for rashes. No significant complications were said to exist with rubella, which often went undiagnosed because of the mild respiratory symptoms and the often unrecognized reddish-brown granular rash appearing on the upper parts of the body and face. This picture for rubella changed dramatically in 1951 when Dr. N. M. Gregg,[29] an Australian family practitioner, observed a much higher incidence of congenital corneal ulcers in newborn infants whose mothers had become pregnant during the rubella epidemic of that year. All of the mothers whose infants showed corneal ulcers had contracted rubella in the first three months of the pregnancy, though only half of these women had been diagnosed as having contracted rubella during the pregnancy. Gregg reasoned that the fetuses were infected in the first three months, but not subsequently. It was later shown that the rubella infants not only showed corneal ulcers, but also were frequently deaf, had other central nervous system deficits causing sensorimotor difficulties, and had a high incidence of congenital heart disease—all occurring in various degrees and combinations. Rubella children were also frequently considered mentally retarded. But the way in which such multihandicapped children developed was not adequately studied until Stella Chess,[15] a child psychiatrist in New York City, began reporting her findings on 235 youngsters, victims of a 1964 worldwide rubella epidemic, whom she began studying when they were two years old. Development showed an overall slowness during the first two years of life, with concomitant slowing of language, motor, and sensorimotor functions. One-third of the children were diagnosed as showing various degrees of mental retardation during the preschool period, while only one-fourth showed evidence of mental retardation at ages eight and nine. Thus, a significant number of such children at all levels of retarded mental development showed considerable improvement and moved into the normal ranges of intellectual functioning. The IQs of the nonretarded children also showed progressive increases as they entered the school-age period. Detailed case studies of a number of the children who showed such improvement demonstrated that they came through a diverse and roundabout pattern to normal school functioning, often pioneering new territory in the acquisition of language, social development, and learning—thereby affirming the inherent plasticity of human brain function in the young child.

All handicaps can be very discouraging and stressful for parents who have both consciously and unconsciously expected to have a perfect child. The marriages in these families and the siblings of families with a severely handicapped child often become seriously stressed. Some siblings, for example, become pseudomature in their development. Even physicians become discouraged in handling multiply handicapped children, and specialized programs with a specially trained staff should be set up to meet these children's needs. However, many parents do not find their way to these programs and many communities do not have specialized rehabilitation programs for such children. When these programs are available, however, such children may have a rather successful convalescence. This acts as an encouragement to the parents to remain involved and committed to the development of their children; and the long-term follow-up of such children is quite good when adequate care has been provided.

COMMENT

This and many other pieces of clinical data suggest that the diagnosis of mental retardation should not be made on very young children showing delayed mental and motor functioning, since such a diagnosis carries with it a very bad prognosis and sets up the expectancy of poor performance, which may then become a self-fulfilling prophesy, the so-called Pygmalion effect. Similar individual-specific roads to cognitive language and social functioning have been demonstrated for (1) children with congential heart disease who had corrective surgery, (2) children who contracted polio before the days of the Salk vaccine; and (3) children with rheumatic fever and with chronic kidney disease. Most recently, Selma Fraiberg's[22] studies of blind children have demonstrated similar plasticity. Eleanor Galenson's[26] and Hilda Schlessinger's[36] studies of deaf children have revealed the importance of the integration of many elements of the total communication system for a child who has a severe sensory defect. Such integration aids in language formation, concept formation, intelligence, social development, and learning skills. Many such children often become late bloomers and manage to make up for what they have missed in earlier years both in educational and social development.

Autistic Behavior in Early Infancy

The normal child of two and one-half years of age begins to show individual-specific personality patterns in his behavior. He becomes a "little character." Such trends are evident by the way he shows his emotions: for example, in his speech, smiling, facial and bodily expression, and bodily reactions. He responds in a comprehensible way to the emotions of others. His interests and curiosity expand as his experience of the world expands. A normal child of two and one-half years expresses his likes and dislikes directly; he utilizes language, affects, mimicry, and bodily posture, all in a smoothly coordinated and usually predictable way to show that he thinks about what is happening to him from within (regarding his wishes, fears, and feelings) and from without. A stage of "normal echolalia" is found in the language development of nonautistic children.

In striking contrast to the normal child is the two and one-half-year-old child who withdraws from others, who fails to establish eye contact, who shows highly ritualistic, unconventional play that is difficult to understand and respond to, who demonstrates both hypersensitivity and insensitivity at various times to visual and auditory stimuli, who relates to his own body and to others in a fragmentary, transient, mechanical way, who uses a few strange, mechanical sounds in place of speech or shows echolalic speech without meaningful contextual reference, and who uses people as if they were mechanical objects. This child does not respond to the usual psychologic testing. He often gives the overall impression of being retarded and sets up a highly controlled, nonhuman environment for himself. He invests this mechanical world with his interest, concern, and love; he becomes attached to things, he identifies parts of himself with inanimate objects and fails to distinguish clearly between his own body boundaries as he relates to others and the inanimate world. Eating and sleeping patterns are highly idiosyncratic, and changes in these idiosyncratic routines or in aspects of the nonhuman environment often induce more severely regressive behaviors. Such children are commonly called "autistic."

COMMENT

It is now possible to partially reconstruct the early developmental history of children showing autistic behavior. Information, for example, can be derived from the developmental histories of such children, from the direct observation of infants showing such behavior at an earlier age, and from a review of family films. The earliest forms of autistic behavior can thus be identified within six different age groups up to the age of two and one-half, taking into account the average-expectable behaviors of those ages.[6] It is clear from observing the natural history of the illness that the earliest signs of autistic behavior are the most subtle. Once the syndrome becomes fully established, however, the serious distortion of development that characterizes autism is so compelling in its organization that all subsequent developmental aspects of the child are crystallized around this central autistic structure. Thus, no matter what original etiologic factors may have played a role in the syndrome, the child's own development and the responses of caretakers are dominantly organized around the pathology rather than the potential for normal development. Treatment of these very young children and prevention has been discussed elsewhere,[6] and there is evidence that early treatment does, in fact, prevent the later sequelae of the illness.[9]

Infantile Substrate of Borderline Functioning

Matthew seemed normal at birth and there were no complications during pregnancy or delivery. At age three and one-half weeks he began turning away from his mother during feeding and making all the positive social responses toward his father. Matthew's mother had a series of psychotic episodes beginning at the time of the breakup of the marriage when the child was thirteen months of age. Consequently, Matthew did not receive a steady home situation until he was about seven, at which time he suffered from nightmares, sleepwalking, enuresis, and retarded psychosocial and intellectual development. His speech was unclear, and he had not learned much in school, having had to repeat kindergarten. However, he responded positively to the new steady home situation, and by young adolescence, judging by his overt behavior, was functioning normally in all areas. His inner fantasy life, however, was still chaotic and he was unable to make friends. He continued appearing well until about age seventeen when his foster father became discouraged with his poor performance in science subjects in high school and presented him with an ultimatum. Matthew responded with regressive, withdrawn behavior. He became zombielike in his attitudes and appearances, he seemed transfixed by the television set and, at times, drifted off into an autistic world of his own. I saw him again at age eighteen and one-half. He responded rapidly to supportive psychotherapy and began doing better in school. However, his inner psychic life was still fluidly organized and he had no interest whatever in social development or in girls.

COMMENT

In light of his earlier history, Matthew was able to function better than might have been expected. However, the underlying vulnerability to psychotic functioning was clearly revealed during his regression in adolescence. He shows what is commonly known as the "borderline syndrome." While the effects of his earlier trauma and deprivation experiences were covered over by acceptable behavior for a good period of time, his inner vulnerability eventually revealed itself. Academic performance in school, conforming behavior, and absence of delinquincy are insufficient criteria to assume healthy psychological functioning.

This rambling visit to the troubled land of infancy as seen through the eyes of the pediatrician and psychiatrist is *meant* to be alarming. Infancy is not synonymous with bliss. Nor does the capacity to impregnate and to conceive bear upon the capacity to be a parent. The either-or cause-and-effect determinism, raging as the nature-nurture controversy, is obsolete and nongenerative when imposed upon the real world of infancy. The environment of the cell is as important as the genetic code in ordering the ontogeny of enzyme systems.[17] Maturation and experience are codetermined. The timing of events or the absence of critical substrate, whether cellular, environmental, or intrapsychic, plays a significant role during the whole course of development from fetal life through old age.

Infants show evidence of suffering from psychic as well as from physical trauma. Why has this lesson been so difficult to comprehend? The onset of language does not mark the onset of the capacity to think and to feel. Distress, neglect, abuse, and fortuitous circumstance all influence the post-natal development of the human fetal brain as well as the growing mind of the child.

I have outlined elsewhere[7] average-expectable normal developmental sequences in infant behavior as distinguished from parental worries, including a description of symptomatic abnormal behavior within seven different age ranges, from birth to age three. These symptomatic patterns may serve as a checklist for evidence of psychological difficulty. However, any symptom must be viewed developmentally and in the context of positive developmental achievements and capacities.

Some Misconceptions and New Findings

Our present knowledge of infant mental development can be divided broadly into two areas: (1) the clarification of misconceptions, and (2) new discoveries. The present excitement and progress in the field began, I believe, with a paper by Dr. Peter Wolff entitled "Observations on Newborn Infants," published in 1959.[41] Wolff demonstrated that newborn behavior made sense, was consistent, replicable, and was thus predictable when the behavior of the infant was cross-correlated with the infant's state of arousal. Wolff defined six separate states: regular sleep, irregular sleep, drowsy, awake active, awake inactive, and crying. Irregular sleep was cross-correlated with rapid eye movement sleep. Within each of these states, various other behaviors were universally and predictably observable; for example, startles occurred only in regular sleep. During the awake attentive state the infant was indeed capable of following objects visually. Most smiling, Emde[19] discovered, occurred during rapid eye movement sleep. The infant's behavioral responses became disorganized only during loud crying, not soft crying. Since 1959, all systematic observations of infants have considered their biobehavioral states as a primary variable. Recently, Wolff[42] and Sander[35] have begun to investigate the way in which behavioral states themselves are organized. Behavioral states observed in the newborn become further differentiated and may be the basis of subsequent moods and emotions in the course of the mothering figure's responses to them. The conception of the infant as a reflexly organized, passive, drive-organized creature with only random movements has given way to a picture of the infant as an active, competent individual, capable of initiating reciprocity with his mother and setting in motion the mother's caretaking responses to the infant. Thus, the infant becomes an active participant in the architecture of his own experience. He is considered capable of inspiring his caretakers to provide those elements of experience required for his own ongoing brain maturation. The human infant is born with a "fetal brain" that continues its rapid development over the first two years of life, and then develops at a somewhat slower rate through the age of twelve. The actual number of neurons doubles in the first two years of life. Thus, brain development itself, by virtue of the infant's dependency during the period of brain maturation, can be significantly influenced by ongoing environmental experience. Neurons can be programmed by postnatal socialization and learning experience.

Another early misconception has now been dispelled. What the infant becomes can no longer be looked upon merely as a result of toxic or deprivatory experiences with deviant parents. Mothers estranged from their infants were, in the past, erroneously considered to be rejecting their infants. We now know that the estrangement which parents, in fact, feel toward their infant may be, to some extent, iatrogenically caused by, for example, hospital routines, illness in the infant, disappointment, inadequate social support, a stressful marriage, or maternal depression. The terms "rejecting mother," "toxic mother," or "depriving mother" no longer do justice to the data.

A third misconception, one that has a tendency to recur in waves with each new generation, is that infants have no psyche until they can speak; that is, that they are incapable of thought and incapable of developing subtle specific affective states. Piaget has become the most widely quoted of all infant development workers.[16,18] His studies, including replication by others, have amply demonstrated how cognitive sequences develop in infancy during the sensorimotor period prior to the establishment of language. Infant mental testing has been shown to be useful during infancy itself. These tests have not been successful in predicting later outcome except at the extremes. Boys are less predictable than girls in such testing. Because of these facts it is no longer appropriate to make even the diagnosis of mental subnormality based on infant tests during the first two years of life. Such a diagnosis must await further longitudinal study of a given infant considered suspect.

A fourth conception about infancy, now seriously undermined by newer findings, is the idea that all infants progress neatly through various stages of mental development and that a disruption of mental development at one period in life will have long-term consequences for future disruption. Studies of blind children, deaf children, and rubella infants have already been cited to cast serious doubt on this idea.

There is another important problem in the study of infant development. Infant development is highly vulnerable to such changes as rapid regressions and progressions. There is no straight-line progression in any function. For example, transformations of hand use from reflexive grasping to intentional manipulation are not always easily discerned, and yet the meaning of each achievement is monumentally different. The same is true in language development.

Newer findings in bioneurological research have shown that in the immature organism with organic brain insult, recovery can be expected if nutritional needs are met and richness of sensorimotor patterning is maintained. Nutritional defects may be overcome by richness of sensory experience. The picture that seems to be emerging is that the brain itself is quite plastic and resilient. Peter Wolff[42] has expressed this as follows:

> Contemporary attempts to predict normal and deviant developmental outcome from neural behavioral status at birth have been generally disappointing. Failure of prediction may be due in part to the inadequate selection of study samples, and the failure to standardize examination procedures in the irrelevance of outcome measures. However, uncontrolled genetic factors, the effect of unsuspected environmental influences and particularly the lack of knowledge concerning developmental transformations in the covert mechanisms that control manifest behavior, severely limit our ability to predict from neonatal behavior to psychological adaptation in the mature person.[p. 304]

Roger Walsh,[40] summarizing experimental data on rats, has commented:

> It is clear that the brain must now be recognized as a plastic organ whose chemical and physiological systems change in an exquisitely sensitive and probably functionally adaptive manner to the environment. The large majority of parameters examined to date have been found responsive and in fact it may well be time for environmental complexity-deprivation studies in particular, and neuroscience in general, to undergo a shift in perspective.

Another area of major development has been our increasing knowledge of the specifics of object orientation, attachment, and anchorage as reflected in the infant-caretaker interaction. We now can understand a whole series of developmental sequences that are set in motion at the time of birth and that show how the human infant becomes attached and anchored to its caretakers and vice versa. As studies indicate,[4, 35, 38] these reciprocal relations between infant and caretaker are quite striking. Recently this author[12] was able to demonstrate how early patterns of both physiological and behavioral reciprocity between mother and infant set in motion the communicative apparatus of the infant and mother, which in turn sets the stage for language development.

Proposed Classification of Maladaptation Syndromes in Infancy

Clinical diagnostic assessment is difficult if not impossible during infancy because (1) the rate of change for all mental functions is rapid; (2) change in a given individual does not take place in a regular line of progression but is characterized by periods of progression and regression; (3) periods of confusion occur during transition; (4) prediction of later outcome from earlier events is more hazardous, since the personality is, in general, more plastic and changeable, as well as more capable of adaptation to change; (5) the brain itself is essentially still fetal in character and does not consolidate its fully functional structure until around the age of five; and (6) assessment itself is complicated by the infant's helplessness and dependency; that is, assessment of the infant must always be qualified by the context in which the assessment is made.

One might consider diagnosing only those conditions in infancy that are significant precursors of diagnostic entities identifiable in later childhood or adult life. However, this approach, which has characterized earlier attitudes toward infant problems, if carried too far, supports the conceptions of the infant as a very young child or embryonic adult. Such an attitude ignores infants as persons with their own strengths, weaknesses, and peculiarities. All of these considerations favor a multiaxial approach to diagnosis. The issues involved in developing a system of diagnostic classification in infancy are sufficiently unique to require a specialized approach that takes into account all of the factors previously mentioned. What follows is a general outline of a group of diagnostic categories that might prove useful.

1. Healthy responses, no maladaptive disturbance. Example: a temporary regression due to illness.
2. Developmental disturbances:
   a. Primary developmental deviation—developmental disturbance without demonstrable deficit in brain or bodily functions. Examples: slow motor and language development, autism, environmental retardation, language delay.
   b. Developmental disturbance associated with impairment of brain structure. Example: Down's syndrome.
   c. Developmental disturbance in association with deficiency of brain function, but without demonstrable structural change. Example: hypsarhythmia.
   d. Developmental disturbance associated with bodily illness or physical handicap without deficit in structure or functioning of the central nervous system. Examples: developmental disturbance associated with blindness, deafness, arthrogryposis, congential heart disease, or kidney disease.
3. Psychophysiologic disturbances. Examples: bronchial asthma, eczema, peptic ulcer, rumination.
4. Attachment disorders of infancy.[11]
   a. Primary attachment failure; that is, failure to thrive without organic cause.
   b. Attachment disorder, anaclitic type; that is, anaclitic depression.
   c. Attachment disorder, symbiotic type; that is, prolongation of symbiosis associated with recent divorce of parents.
5. Disturbed parent-child relationship. Examples: sadomasochistic relationship and dyadic dyssynchrony in infancy, communication impass.
6. Behavioral disturbances of infancy. Examples: attention disorder of infancy, sleep problems, irritable infant syndrome, "colic."
7. Disturbances of the environment (beyond the adaptive capacity of a healthy infant).
   a. Disturbances of fetal environment. Examples: fetal alcohol syndrome, RH or sensitization, rubella infant.
   b. Disturbances of environment during the perinatal period. Examples: anesthesia effect, barbiturate poisoning, acute brain syndrome due to traumatic delivery.
   c. Primary deficit in caretaking after birth. Example: institutionalized infant.
   d. Iatrogenic disturbance.
      (1) Secondary to inappropriate medical care or advice.
      (2) Secondary to psychological care or advice.
      (3) Secondary to bad educational care or advice.
   e. Acute distress syndrome of infancy associated with physical or psychic trauma. Examples: starvation, dog bite, traumatic separation, death of parent.
8. Genetic disturbance.
   a. With phenotypic expression.
   b. Without phenotypic expression.
   c. In family after transmission to child.
9. Language disorder of infancy.

A multiaxial approach to diagnosis of infancy must be considered. Rutter and colleagues[34] proposed five axes that they considered relevant for child psychiatric diagnosis.

Axis I: Clinical psychiatric syndrome.
Axis II: Developmental disorder.
Axis III: Intellectual level.
Axis IV: Associated biological factors.
Axis V: Associated psychosocial factors.

The new *Diagnostic and Statistical Manual of Mental Disorders*[1] axes include:

Axis I: Clinical psychiatric syndrome(s) and other conditions.
Axis II: Personality disorders (adults) and specific developmental disorders (children and adolescents)
Axis III: Nonmental medical disorders.
Axis IV: Severity of psychosocial stressors.
Axis V: Highest level of adaptive functioning past year.

This classification of syndromes in infancy includes developmental consideration and biological factors. Intellectual level in infancy is included as one aspect of development in infancy. Psychosocial stressors are different in infancy and need to be specified. A global rating of severity of disturbance in infancy still needs to be worked out. It is clear from all of these considerations that a great deal of work needs to be done by people dealing with various aspects of infant study. It is especially clear that diagnostic systems worked out for older children and adults are not applicable to the period of infancy.

¶ Bibliography

1. AMERICAN PSYCHIATRIC ASSOCIATION. *Diagnostic and Statistical Manual of Mental Disorders,* 3rd ed. Washington, D.C.: American Psychiatric Association, 1980.
2. BENJAMIN, J. D. "Prediction and Psychopathological Theory," in Jessner, L., and Pavenstedt, E., eds., *Dynamic Psychopathology in Childhood.* New York: Grune & Stratton, 1959, pp. 6–77.
3. BERESFORD, T., and PRUGH, D. "Infantile Hypertrophic Pyloric Stenosis in Denver 1965–1970," unpublished manuscript.
4. BRAZELTON, T. B., LOSLOWSKI, B., and MAIN, M. "The Origins of Reciprocity: The Early Mother-Infant Interaction," in Lewis, M., and Rosenblum, L., eds. *The Effect of the Infant on Its Caregiver.* New York: Wiley-Interscience, 1974.
5. CAFFEY, J. "Infantile Cortical Hyperostosis," *Journal of Pediatrics,* 29 (1946): 541–559.
6. CALL, J. D. "Autistic behavior in infants," in *Brennemann's Practice of Pediatrics,* vol. 1, New York: Harper & Row, 1977.
7. CALL, J. D. "Psychologic and Behavioral Development of Infants and Children," in *Brennemann's Practice of Pediatrics,* vol. I, New York: Harper & Row, 1977.
8. CALL, J. D. "Differential Vulnerability and Coping in Boys and Girls at Birth," in Anthony, E. J., Koupernik, C., and Chiland, C., eds., *The Child in His Family: Vulnerable Children,* vol. 4. New York: John Wiley, 1978, pp. 145–172.
9. CALL, J. D. "Follow-up Study of 18 Infants Showing Autistic Behavior Before Age 3," Presented at the Second Symposium on Infant Psychiatry. New York, 1978.
10. CALL, J. D. "Introduction to Normal Development," in Call, J. D., Noshpitz, J., and Cohen, R. L., eds., *Basic Handbook of Child Psychiatry,* vol. 1. New York: Basic Books, 1979, pp. 3–10.
11. CALL, J. D. "Attachment Disorders of Infancy," in Kaplan, H. L., Freedman, A. M., and Sadock, B. J., eds., *Comprehensive Textbook of Psychiatry,* vol. 3. New York: Williams & Wilkins, 1980.
12. CALL, J. D. "Some Prelinguistic Aspects of Language Development, *Journal of the American Psychoanalytic Association,* 1980, pp. 259–289.
13. CALL, J. D., REISER, D. E., and GISLASON, I. L. "Psychiatric Intervention with Infants," in Harrison, S., and Noshpitz, J., eds., *Basic Handbook of Child Psychiatry,* vol. 3, New York: Basic Books, 1979, pp. 457–484.
14. CALL, J. D., et al. "Psychogenic Megacolon in Three Preschool Boys—A Study of Etiology through Collaborative Treatment of Child and Parents," *American Journal of Orthopsychiatry,* 33 (1963): 923–928.
15. CHESS, S., KORN, S., and FERNANDEZ, P. *Psychiatric Disorders of Children with Congenital Rubella.* New York: Brunner-Mazel, 1971.
16. DECARIE, T. *The Infant's Reaction to Strangers.* New York: International Universities Press, 1974.
17. EIDUSON, S. *Biochemistry and Behavior.* Princeton: Van Nostrand, 1964.
18. ELKIND, D. *Studies in Cognitive Development: Essays in Honor of Jean Piaget.* New York: Oxford University Press, 1969.
19. EMDE, R. N., and HARMON, R. J. "Endogenous and Exogenous Smiling Systems in Early Infancy," *Journal of the American Academy of Child Psychiatry,* 11 (1972): 177–200.
20. FLEISHER, D. R. "Functional Vomiting Syndrome of Infancy-Rumination versus 'Nervous Vomiting,'" unpublished manuscript, 1979.
21. FLEISHER, D. R. "Infant Rumination Syndrome: Report of a Case and Review of the Literature," *American Journal of Diseases of Children,* 133:3 (1979): 266–269.
22. FRAIBERG, S. *Insights from the Blind.* New York: International Universities Press, 1977.
23. FREUD, A. *Normality and Pathology in Childhood.* New York: International Universities Press, 1965.
24. FREUD, S. "Beyond the Pleasure Principle," in Strachey, J., ed., *The Standard Edition of the Complete Psychological Works of Sigmund Freud* (hereafter *The Standard Edition*), vol. 3. London: Hogarth Press, 1955, pp. 3–64 (originally published in 1920).
25. ———. "Analysis Terminable and Interminable," in Strachey, J., ed., *The Standard Edition,* vol. 20. London: Hogarth Press, 1959, pp. 7–175 (originally published in 1937).

26. GALENSON, E. "Assessment of Development in the Deaf Child," *Journal of the American Academy of Child Psychiatry,* 18 (1978): 128–142.
27. GISLASON, I. L., and CALL, J. D. "Dog Bites in Infancy: Trauma and Personality Development," presented at The First World Congress on Infant Psychiatry, Cascais, Portugal, April, 1980.
28. GISLASON, I. L., and CALL, J. D. "The Influence of Trauma on Personality Formation (Three Cases of Severe Dog Bites in Infancy)," Presented at the First World Congress on Infant Psychiatry, Estoril, Portugal, March 30–April 3, 1980.
29. GREGG, N. M. "Rubella During Pregnancy of the Mother, with its Sequelae of Congenital Defects in the Child," *Medical Journal of Australia,* 1 (1945): 313–315.
30. JACKSON, K., et al. "Behavioral Changes Indicating Emotional Trauma in Tonselectomized Children," *Pediatrics,* 12 (1953): 22–29.
31. KAVANAUGH, J. G., Jr., and MATTSSON, A. "Psychophysiologic Disorders," in Noshpitz, J. D., ed., *Basic Handbook of Child Psychiatry.* New York: Basic Books, 1979, pp. 341–380.
32. KEMPE, C. H., et al. "The Battered Child Syndrome," *Journal of the American Medical Association,* 181 (1962): 17–24.
33. PRUGH, D. Personal communication, 1979.
34. RUTTER, M., SHAFFER, D., and SHEPHERD, M. *A Multi-Axial Classification of Child Psychiatric Disorders.* Geneva: World Health Organization, 1975.
35. SANDER, L. W. "Infant State Regulation and the Integration of Action in Early Development," presented at The First Infant Psychiatry Institute, Costa Mesa, Calif., February 24–26, 1978.
36. SCHLESINGER, H. S., and MEADOW, K. *Sound and Sign: Childhood Deafness and Mental Health.* Berkeley: University of California Press, 1972.
37. STEELE, B. "Parental Abuse of Infants and Small Children," in Anthony, E., and Benedek, T., eds., *Parenthood: Its Psychology and Psychopathology.* Boston: Little, Brown, 1970, pp. 449–477.
38. STERN, D. N. *The First Relationship, Infant and Mother.* Cambridge: Harvard University Press, 1977.
39. THOMAS, A., CHESS, S. and BIRCH, H. G. *Temperament and Behavior Disorders in Children.* New York: New York University Press, 1968.
40. WALSH, R. N. "The Ecological Brain: A Review of the Effects of Environmental Complexity and Deprivation on Brain Chemistry and Physiology," *International Journal of Neuroscience,* in press, 1980.
41. WOLFF, P. H. "Observations on Newborn Infants," *Psychosomatic Medicine,* 21 (1959): 110–118.
42. ———. "The Development of Behavior in Human Infants, Premature and Newborns," *Annual Review of Neurosciences,* 2 (1979): 291–307.

CHAPTER 7

THE CHILD AT RISK FOR MAJOR PSYCHIATRIC ILLNESS

Clarice J. Kestenbaum

DURING the past two decades, increasing numbers of research projects have focused attention on those children who have a high probability of developing a major psychiatric disorder in adult life. These studies are subsumed under the rubric "risk research" and have been directed for the most part toward determining which children are most vulnerable to eventual schizophrenic illness.[36] Affective disorders have only recently begun to be studied in similar fashion.

"Risk" refers to the greater likelihood that certain individuals will develop a mental disorder than will others selected at random from the same community.[36] The children of schizophrenics, for example, are at greater than average risk for developing schizophrenia than the general population (10 to 15 percent for the child with one schizophrenic parent compared with approximately 1 percent for a child with nonschizophrenic parents).

"Vulnerability" implies that each individual is endowed with a degree of vulnerability to illness that under certain circumstances will become manifest.[112] In one vulnerability model two major components are described —inborn and acquired. Inborn vulnerability refers to that which is in the genes and is reflected in the internal environment and neurophysiology of the organism. The acquired component is due to the influence of specific diseases, traumas, perinatal complications, family interactions, and life events that either enhance or inhibit the development of future disorder. For an individual vulnerable to schizophrenia, for instance, periods of acute stress may result in a failure of coping mechanisms so that maladaptation results.[60] According to Arieti,[5] Adolph Meyer became convinced that dementia praecox resulted from inefficient coping mechanisms which led to eventual breakdown. Meyer believed there was either a gradual transition to a schizophrenic episode or an acute onset precipitated by a catastrophic event. In contrast to vulnerability, "invulnerability describes the organism's ultimate resilience to the load imposed on it by life event stressors" [p. 114].[110]

¶ Goals of High-Risk Research

Erlenmeyer-Kimling[28] has summarized the goals of high-risk research as follows: (1) development of the natural history of the disorder being studied; (2) identification of the specific members of the high-risk group most likely to become affected; (3) determination of early predictors of the disorder, particularly those indicative of underlying biological deficits; and (4) examination of the various factors, environmental in nature, that increase the likelihood of prepsychotic illness becoming manifest as well as those that may protect a vulnerable individual from becoming ill.

An additional far-reaching goal of high-risk investigations concerns the question of intervention. If one could detect the vulnerable child and provide competence-enhancing interventions, could the future patient be sufficiently protected so that subsequent illness could be prevented? Current research is undertaking the examination of biological, psychological, and sociological factors that may be conducive to the development of schizophrenia and affective disorders.

¶ Children At Risk for Schizophrenia

"Schizophrenia," which Bleuler[10] coined in 1911, is a concept that is still developing and changing today. Bleuler renamed Kraepelin's term, dementia praecox,[58] because he considered "splitting" of the different psychic functions one of the most important characteristics of the condition. Schizophrenics are well known to constitute a heterogeneous group (paranoid, hebephrenic, catatonic, and simple). There are marked differences in terms of genetics, biochemistry, symptom clusters, course, and outcome. Langfeldt[59] tended to lump the schizophrenias into two principal groups, process and reactive (which other investigators have called chronic and acute). Other labels have indicated that an individual may demonstrate some schizophrenic attributes without ever having been psychotic (that is, the nonpsychotic schizophrenic[41] or schizotypal personality).[99] The concept of a schizophrenic spectrum embraces these categories.[89] Such diverse diagnostic considerations have made past research efforts unwieldy, and collaborative efforts difficult. Thus, research diagnostic criteria have evolved that define more sharply the illness under study. Strauss and associates[98] have defined schizophrenia, for example, as "a psychiatric disorder without demonstrable organic etiology that is manifested by such phenomena as delusions, hallucinations, bizarre behavior and thought disorder in which manic or depressive features are not predominant."

Numerous models for the transmission of schizophrenia have been promulgated. At the core of research into high-risk schizophrenia is the question of whether a given deficit in premorbid adjustment is sufficient to cause a schizophrenic breakdown in later life, or whether a particular environmental stress is necessary for the development of the illness.

¶ Etiological Models for the Transmission of Schizophrenia

Psychogenic Models

Various etiological models have been reviewed in the current psychiatric literature.* Proponents of the psychogenic theory of the

*See references 37, 50, 53, 62, and 98.

transmission of schizophrenia contend that abnormal patterns of family interaction are largely responsible for pathological adaptation and subsequent schizophrenic breakdown. Wynne and Singer[108] have viewed the future schizophrenic as an individual highly vulnerable to a disorder in family communication. Bateson and associates[6] have, in fact, labeled that individual chiefly responsible for extreme confusion in the young child as the "double bind" mother. They contend that the double message provided by an extremely ambivalent parent and the subsequent failure in family communication is instrumental in setting the stage for subsequent breakdown in thought and communication in the vulnerable child. Because distortions of reality are so frequent and denial on the part of the ill parent such a prevalent defense, the child subjected to intense and frequent "double bind" communication is unable to acknowledge his own perceptions as being accurate.

Lidz[61] and his collaborators have also been proponents of a psychogenic theory that holds that severe chronic disequilibrium and discord (marital schism) and submission to pathological attributes of the schizophrenic parent (marital skew) are primary etiological factors in later schizophrenic breakdown.

Genetic Models

Genetic approaches to the study of schizophrenia have been in existence for six decades.* Franz Kallmann[44] was the chief proponent of the theory that the more closely related an individual was to a schizophrenic relative, the greater was the likelihood of his becoming schizophrenic. Kallmann, studying monozygotic and dizygotic twins pairs, held that if one monozygotic twin was known to be schizophrenic, the risk to the cotwin also schizophrenic was 86 percent.[45] Recent studies, using sophisticated techniques and "blind raters," achieve a concordance rate of 42 percent for monogygotic twins as against 10 percent for dizygotic twins, approximately the same as sibling rates concordant for schizophrenia.[40]

Since the twins were usually reared together, opponents of the genetic theory argued that environmental influences should not be underestimated. The adoption studies have removed that variable by examining offspring of schizophrenic mothers reared by foster mothers. Heston[43] demonstrated that of forty-seven adult offspring of schizophrenic mothers who were adopted within the first few days of life, 10 percent developed schizophrenia in adult life opposed to none in the control group (children of nonschizophrenic mothers adopted at birth by nonschizophrenic women). When Kety and Rosenthal,[53] using a Danish population, corrobrated Heston's findings, they found that the incidence of schizophrenia among the biological relatives of adopted schizophrenics was six times as high as the incidence noted among the relatives of the control adoptees. Wender and associates[106] moreover, reported a cross-fostering study in which adopted offspring of normal biological parents were reared by schizophrenic adoptive parents. These children were compared with: (1) adoptees of normal biological parents reared by normal adoptive parents; and (2) adoptees of schizophrenic biological parents who were then reared by normal adoptive parents.

Wender and his collaborators noted a significantly higher prevalence of psychopathology among the adopted offspring of schizophrenic parents in contrast to the other two groups.

Summary of the Evidence in Support of Hereditary Factors[26, 39]

1. Schizophrenialike behavior has been reported from antiquity to the present day and occurs in both industrialized and undeveloped societies.
2. The prevalence throughout the world is 0.8 to 1.5 percent, although there exist isolated populations with higher and lower rates.
3. The male-female ratio is one to one.

*See references 27, 41, 45, 55, 68, 87, and 93.

4. In urban communities there is a marked social-class gradient in the prevalence of schizophrenia, usually attributed to a premorbid "downward drift" of predisposed individuals.[55]
5. No environmental causes have been identified.
6. Frequently there is no known precipitating event.
7. Risk figures are compatible with a genetic hypothesis.[40,92]
   a) First degree relatives of schizophrenics are at greater risk of becoming schizophrenic compared to the general population (5.5 percent for parents, 10 percent for siblings, 11 percent for children).
   b) Monozygotic-twin concordance rates for twins reared apart are approximately the same as for twins reared together. They are three times as high as the concordance rates for fraternal twins.
   c) Concordance rates in monozygotic cotwins vary with the degree of severity of illness in proband; high concordance rates are associated with severe illness.
   d) Schizophrenia occurs at the same rate in children of affected fathers as in children of affected mothers.[25]
   e) The risk for children of two schizophrenic parents is three times that for children of one schizophrenic parent, whether the children are reared by their own parents or not.[23]
8. Adoption studies have demonstrated:[43,53,106]
   a) Children of normal parents who are reared by schizophrenic adoptive parents do not have an increased rate of schizophrenia.
   b) Children of schizophrenics adopted by normal families develop schizophrenia at the same rates as those of children reared by their own ill parents.
   c) Unlike the biologic relatives of schizophrenic adoptees, the adoptive relatives do not have elevated rates of schizophrenia.
9. There is no significant increase in schizophrenia risk among children with early deafness, despite the communication difficulties they experience.[1]
10. Family members of individuals who develop schizophreniform psychoses following head injuries have risk figures equal to the general population, unlike the figures for families of "true" schizophrenic probands.[19,110]*

Regarding the strength of the genetic factor in schizophrenia and the mode of transmission, the best available evidence points to the likelihood that a genetic factor constitutes a necessary, though not sufficient, antecedent. Although the predominant mode of transmission may be polygenic,[40] Matthysse and Kidd[67] have shown that a single-major-locus model cannot be discarded. Heterogeneity[27] may also be present in the sense that individuals classed as schizophrenic may be regroupable as belonging to several differing patterns of transmission or to different sets of noxious genes.

Evidence in Support of Environmental Influences (Diathesis-Stressor Framework)[16]

Despite the fact that monozygotic twins share all their genes in common, more than half of the pairs are discordant for schizophrenia. In other words, factors other than heredity are responsible for either protecting a constitutionally vulnerable individual from breakdown or for precipitating a breakdown. According to the diathesis-stressor framework, genetic predisposition, or diathesis, comes under the influence of environmental stressors which eventually result in physiologic and biochemical change in the organism.

Gottesman[39] has stated that to date no environmental causes have been found that will invariably produce genuine schizophrenia in persons unrelated to a schizophrenic. However, the contributions from environmental stressors determine for each individual the point at which a clinical

*The need for careful diagnosis in order to distinguish a schizophrenic genotype from a similar phenocopy is demonstrated by the genetic studies. There are many imitations of schizophrenia, such as temporal lobe epilepsy and manic psychosis (both can resemble an acute schizophrenic reaction). Careful family and developmental history are often helpful in resolving the problem.

threshold is passed. Gottesman stated: "Starting with a subject's genetic liability for schizophrenia at birth, one then adds psychological stress over time that has the effect of adding to his liability" [p. 36]. Stress factors may include chaotic home life, repeated separation and loss, exececessive demands from authority figures, or various organic insults.

In discussing the environmental variables and their impact on a constitutionally vulnerable child, Anthony[3] has noted the effects of the ill parent on the child in a detailed description of parent-child interactions. In his study of the children of psychotic parents, Anthony observed that the sicker schizophrenic parents (diagnosed hebephrenic or catatonic) produced healthier children, while the relatively healthier parents (diagnosed schizoid or borderline) produced more deviant children. He took into consideration the age of the child at the time of the mother's illness and the family support system in terms of other available members with whom the child could identify or become involved. He felt that the most disturbed children were those symbiotically involved with the ill parent and that children under two, still in the phase of separation-individuation, were especially sensitive to the disturbing influence of the chronically ill mother. Whether the child internalizes or externalizes conflict, how compliant or negativistic he is, how prone to a *folie-à-deux*, and how identified with the ill parent, determine his ultimate development. Anthony observed that the more disturbed children suffered a loss of ego-skills by school age and exhibited symptoms such as nightmares, phobias, obsessions, or antisocial behavior. In a similar vein, Kauffman and associates[46] found:

> that the mother's current level of functioning is even more important than the diagnosis of her condition in understanding the impact of her disturbance on the child's later development. Women who are isolated from social contacts and who cannot function effectively in their adult social roles have children with lower competence . . . [p. 1401]

Thus, environmental influences, particularly paucity of supports, produced less competent children.

¶ Research in Early Identification of Schizophrenia

The four methods used for studying the premorbid history are:

1. clinical retrospective (reports of life history, patient or family reminiscences)
2. follow-back study (case records of adult schizophrenics traced back to childhood)
3. follow-up study (life history of the child at-risk who was evaluated in childhood and reevaluated as an adult)
4. prospective longitudinal study (a high-risk group followed into adult life so that incipient pathology may be studied in depth)

The last method has the obvious advantage of a carefully designed methodology and permits the researcher greater objectivity by reliance on a matched control sample, double-blind ratings, and the passage of time.

Impressionistic data derived by the other three methods, however, produced hypotheses about the premorbid state which could then be tested by the prospective longitudinal method.

Example of the Follow-Back Study Method

Watt,[101,102] in a controlled study, examined past cumulative school records of fifty-four hospitalized schizophrenic adults. He found that over one-third were identified by teachers as being deviant in childhood before they showed psychotic disorganization. Patterns of maladjustment for boys and girls differed.

Preschizophrenic boys exhibited primary evidence of unsocialized aggression and secondary evidence of internal conflict, overinhibition, and depression. Preschizophrenic girls exhibited primary evidence of oversensitivity, conformity, and introversion.

Watt found that both sexes had been subjected to heightened frequency of parental deaths and neurological abnormalities compared to controls. He proposed five postdictive indices of schizophrenia: (1) parental death; (2) severe organic handicap; (3) extreme emotional instability; (4) extreme introversion (females); and (5) extreme disagreeableness (males). In a subsequent study, low social competence was associated with withdrawal, thought disorder, and antisocial acting out.[82]

Example of the Follow-up Study Method

Waring and Ricks,[100] using the follow-up method, identified 75 adult schizophrenics selected from the records of 18,000 children seen in a child guidance clinic since 1917. Subjects were divided into two groups: chronic (those whose illness did not remit) and released (those who had been hospitalized intermittently for schizophrenic illness). Differences between schizophrenic patients and normal controls were: (1) developmental history—fewer than 40 percent had normal births; (2) early neurological dysfunction (slower motor development, poorer coordination, unclear speech); and (3) absence of heterosexual experience or extreme homosexual fears during childhood.

Differences between the two groups were also noted. As compared to the released group, the chronic schizophrenics were characterized by: (1) family history of schizophrenia; (2) mothers who were more disturbed (psychotic or functioning on a borderline level); (3) a symbiotic relationship with the ill mother; (4) a schizoid premorbid personality with few close peer relationships; (5) no clear precipitating event necessitating hospitalization; and (6) no history of sociopathy, (truancy, stealing).

Ricks and Berry,[85] discussing children who become schizophrenic, maintained that chronic schizophrenics have biological and social equipment that offers small margin for error in development, and their coping mechanisms are clearly maladaptive. The investigators feel that the IQ, degree of social and vocational success, the home environment, and the presence or absence of biological handicap are all relevant.

The Prospective-Longitudinal Method of High-Risk Research

Pearson and Kley[79] were the first investigators to propose a prospective study of individuals at high statistical risk for schizophrenia. In 1959, nonblind studies were instituted by B. Fish,[32] who subsequently followed up infants born to schizophrenic mothers, and Sobel,[93] who attempted to study infants of two schizophrenic parents. "Pan developmental retardation," Fish[32] suggested, is a transient dysregulation of motor, visual-motor, and physical development noted between birth and two years; this includes erratic and disorganized maturational patterns in activity and alertness, as well as autonomic instability, which, Fish believes, predict vulnerability to later schizophrenia.

In 1962 Mednick and Schulsinger,[69] using more sophisticated methodology, began a longitudinal study of Danish children of schizophrenic mothers. The investigators believed that the offspring of schizophrenic mothers demonstrated a particular vulnerability to schizophrenia that is a joint function of genetic loading and pregnancy and birth complications. This combined liability, they contended, results in an infant who demonstrates a labile pattern of autonomic responsivity.

Neurological "Soft Signs"

There is still controversy about some of the recorded physiological data (i.e. regarding differences in galvanic skin responses between high-risk subjects and normal controls),[99] but there is high agreement among investigators as to the neurological soft signs noted in the offspring of schizophrenics. Such findings have been reported in many studies.* Marcus,[64] in a study undertaken in Israel in 1965, identified a group of offspring born to schizophrenic parents. The children were characterized by soft signs of neurological dysfunction, (deficits in fine motor development, visual-motor coordination, and auditory-visual integration). The finding suggested to Marcus a genetic basis for vulnerability to schizophrenia. In a second study, Marcus and associates[63] were able to identify a subgroup of infants born to schizophrenics who demonstrated the same dysmaturation in motor functions as in perceptual development. The investigators, using a Multidimensional Scalogram Analysis, found that roughly one-half of the high-risk infants exhibited less than optimal functioning. They noted, moreover, lower birth weights in these infants as compared to controls and the other group of high-risk infants whose functioning was optimal. In a subsequent study, Marcus and Mednick[64] examined the data on a subsample of the Danish high-risk, birth-matched subjects and reanalyzed them using a Multidimensional Scalogram Analysis.

When area judgment scores were reexamined, approximately half of the high-risk children could be identified by the repeated findings of less than optimal functioning in motor and coordination tasks, posture, and gait.

Obstetrical Complications

Mednick's[69] original high-risk study demonstrated that the clinically deviant group of adolescent offspring had had a significant increase in anomalous autonomic responses and high rates of obstetrical complications. McNeil and Kaij,[76] in reviewing over eighty papers relating to obstetrical factors in the development of schizophrenia, observed that the outcome of these studies demonstrated few significant differences between the high-risk subjects and their controls in overall obstetrical complications—pregnancy complications (PC), birth complications (BC), neonatal complications (NC), and obstetrical complications (OC).* Two of the studies however, showed significantly more fetal and neonatal deaths in index pregnancies;[65,68] the 1978 report by McNeil and Kaij[76] of a Swedish high-risk sample, revealed increased NC's and OC's corroborating Mednick's 1970 study. Low birth weight was reported in three studies.[65] Marcuse and Cornblatt,[65] in a review of current findings of OC's in relation to soft signs, noted that the relationship between neurological signs in high-risk children is currently unresolved. In their investigation of the relationship between OC's and neurological outcome in a sample of children with schizophrenic parents, they reviewed information in preliminary analysis data on seventeen high-risk children and sixty-eight matched controls. The index cases were selected from the Collaborative Perinatal Project, which recorded obstetrical information on 55,000 pregnancies with follow-up from birth through age five. Positive findings were: (1) male high-risk children had more NC's and total OC's than control cases or high-risk females (The two NC items individually significant were prematurity and low one-minute Apgar scores); and (2) the high-risk group was markedly deficient in the Auditory-Vocal Association Test, block sorting, and school achievement (IQ was a controlled variable). The investigators noted that there was a subgroup of individuals who scored poorly on both OC's and neurological variables, and who had at least one obstetrical indicator and abnormal neurological signs (that is, poor coordination, short attention span, mixed dominance, and other anomalies of lateral dominance). Marcuse and Cornblatt noted that "the consistency of the neurological findings across studies is quite remarkable. Nearly every attempt to examine motor functions and soft signs has demonstrated statistical group differences or associative patterns that distinguish the offspring of schizophrenics from controls."[65]

In another carefully controlled prospec-

*See references 25, 63, 66, 71, 78, and 85.

*Pregnancy Complications = PC; Birth Complications = BC; Neonatal Complications = NC; Obstetrical Complications = OC.

tive study, Hanson and associates[42] collected developmental data on thirty-three children of schizophrenic parents from the Collaborative Perinatal Project, Minnesota Sample. Matched control groups consisted of children of other psychiatric patients and normal parents. Seventeen percent of the high-risk sample had positive scores on the three following indicators: poor motor abilities, schizoid behavior, and marked inconsistencies in academic cognitive achievement. While some of the controls provided positive findings on one or two indices, none provided positive scores on all three. The authors concluded that these specific individuals were especially vulnerable to schizophrenia.

Example of a High-Risk Prospective Study

One of the most extensive of the high-risk prospective studies has been in progress since 1971 at the New York State Psychiatric Institute, where the chief investigator is L. Erlenmeyer-Kimling.[24,28] The risk group includes 205 children who were between the ages of seven and twelve at the time of initial examination (sample A). The 80 high-risk subjects are subdivided into three groups: 44 with schizophrenic mothers, 23 with schizophrenic fathers, and 13 with two schizophrenic parents. The two control groups consist of 25 children of parents with another psychiatric disorder and 100 children whose families do not have psychiatric disorders. The index cases are matched for age, sex, ethnicity, and social class. Strict diagnostic criteria were used in selecting the parents. The chief focus of the study is on psychophysiological, psychiatric, neurological, psychological, and social measures. According to hypotheses based on research with adult schizophrenics, which postulate that schizophrenic individuals may have difficulty in normal processing of stimuli, measures were selected that would be expected to be deviant in preschizophrenic individuals.[57]

Tests were selected that measure attentional dysfunction and distractability. It was felt that genetically vulnerable individuals may have difficulty in "filtering out" background stimuli or may be unable to disengage from a stimulus having once attended to it.[15,66,73]

The subjects are given home interviews and laboratory testing that include structured and semistructured interviews, neurological examination, and psychological tests such as the Wechsler Intelligence Test, Bender-Gestalt, Human Figure Drawing, and projective tests. School records are collected; a variety of cognitive, attentional and distractability measurements are administered, as well as EEG and auditory evoked potentials. A videotaped psychiatric interview using a semistructured interview—the Mental Health Assessment Form[51]—was developed for the project. The interviewer is "blind" as to the child's parental background, as are three raters who independently rate the videotapes for psychopathology and diagnostic impressions. A second group under investigation consisting of 150 subjects, (sample B) has been included as a replication sample (44 children of one or two schizophrenic parents, 40 whose parents have affective disorders, and 66 with normal parents).

Preliminary Results of Group Differences

In addition to corroboration of positive neurological findings in the index cases consistent with those of other investigators, a study of attentional tasks has emerged that differentiates the high-risk group from the controls.[16,28,91] The first sample (sample A) has been tested on three occasions over a nine-year period. Consistent group differences were found on several attentional measures, the Continous Performance Test (CPT)*[91] and the Attention Span Measure,

*The CPT, a measure of sustained attention, involves a succession of playing cards projected onto a screen. The subject's task is to respond when two identical slides appear in sequence, such as the six of clubs followed by the six of clubs. Half the trials are presented without distraction and half are presented in the presence of external auditory distraction (a tape recorded female voice reciting numbers at varying speeds and tones). A false response is considered incorrect (when the subject responds to a six of clubs, for example, followed by a six of hearts).

which requires subjects to recall immediately a series of either three or five letters presented by tape recorder, with or without distraction.

The results on the CPT indicated that the high-risk subjects made significantly fewer correct responses and significantly more random commission errors than did the normal comparison group, with and without distraction. On the Attention Span Measure, high-risk subjects also made significantly fewer correct responses than the normal comparison groups under certain conditions.

NEUROPSYCHOLOGICAL AND NEUROPHYSIOLOGICAL MEASURES

The high-risk subjects scored lower on the tests of neuropsychological development (Bender-Gestalt[24,56] and Human Figure Drawing Test). The finding suggested a developmental lag as described by B. Fish. Some of these subjects showed unusual patterns of auditory-evoked potentials, particularly when attention was required as a task.[34]

CURRENT CLINICAL STATUS

In a preliminary analysis of differences between the index and control children using the videotaped Mental Health Assessment Form, there were significant differences on global assessment of function, anxiety, depression, history of angry feelings, disturbance in relationship with the mother, occurrence of nightmares, and measures of self-esteem.[51]

DEVIANT COGNITIVE PERFORMANCE-DEVIANT-BEHAVIOR OVERLAP[16,28]

Behavioral disturbances were measured according to a five point Behavioral-Global Adjustment Scale that relies heavily on parents' information. "The subgroup identified as deviant within the high-risk group has been found to show an increasing overlap with the subjects showing behavioral problems as they reach adolescence."[27]

Teachers' reports also showed the high-risk subgroup having increasing school difficulty when these children entered adolescence. The current clinical status of the study children, whose mean age is fifteen years, reveals that to date eight subjects of the original 205 (4 percent) have been hospitalized or treated for serious disorders. Five of the eight hospitalized children are from the high-risk group, two are from the psychiatric comparison group, and one from the normal comparison group. All but the child from the normal comparison group had demonstrated dysfunction in the CPT and other attentional measures several years prior to breakdown.

Summary of Findings

The study thus far points to the presence of neurological soft signs in the high-risk group in early childhood. Attentional and cognitive measures as well as attention-related auditory-evoked potentials appear to discriminate between a subgroup of high-risk and control subjects at early ages. These measures are associated with clinical deviance in adolescence, noted by parents, by teachers, and global assessment. The fact that the children with early deficits on laboratory measures become increasingly deviant behaviorally as they get older supports the hypothesis that attentional dysfunctions serve as early predictors of later pathology.

CASE ILLUSTRATION*

Mona, the only child of a middle-aged couple, was referred for psychiatric treatment at age six by her school principal because of extreme shyness, school refusal, inconsistent work habits, and "joylessness." Additional symptoms reported by her parents were: frequent nocturnal, and occasionally diurnal, enuresis; multiple fears (of animals, of the dark, and of being alone); belief in "supernatural powers" (for example, she was convinced that the eyes in photographs or paintings followed her around the room).

Early manifestations of deviance (her mother's pregnancy and delivery having been normal) were hypersensitivity to noise and change (new faces or surroundings), and

*This case is not connected with any of the high-risk projects under discussion.

more-than-expectable separation anxiety. Food "fads" and rejection of all but five or six foods were reported along with a preoccupation with thoughts of vomiting (which had occurred on occasion during febrile illnesses).

Significant traumatic events when Mona was four were the simultaneous death of her grandmother and a month-long hospitalization of her mother for an operable malignancy.

Family history: Mr. S. was a psychiatrically well, successful businessman, "obsessional and quiet," by his own report. Mrs. S. had a history of emotional problems. Her father was alcoholic, suspicious, physically abusive, unable to keep steady employment, and was known in his family as "the crazy one." Never hospitalized for psychiatric illness, he had a brother who died in a mental institution. Mrs. S. left home at age sixteen and worked as a secretary until her marriage at age thirty-five. A "loner," she was in treatment for agoraphobia, which was so incapacitating at times that she could not attend her psychiatric sessions. She had the habit of sending the therapist many pages of "associations" (described as loose and rambling) in lieu of sessions.

Evaluation

Because of Mona's "staring spells," a neurological examination was performed. Results were within normal limits.

A psychological examination revealed a full scale IQ of 115 (WISC) "with higher potential" and subtest scatter (lowered scores for language comprehension and picture arrangement). Language structural skills were poor. Projective tests were characterized by "peculiar percepts: monsters, dragons, skeletons, eyes. . . . Suspicious, phobic, and much like an adult diagnosed paranoid personality," reported the psychologist.

Therapeutic Course

During the initial biweekly psychotherapy sessions, Mona avoided eye contact. Themes of her drawings included fire, explosion, death, and destruction. A well-executed drawing of a smiling girl brought forth the comment, "She's happy because she's eavesdropping on her enemies. Everything about this girl is bad; she has no friends, she is mean, and she hates everyone".

The initial mode of therapy was doll play. Mona reenacted with the dolls events she would not describe in words, giving detailed accounts of what the dolls thought and felt. For example, doll A explained to doll B, "It's dangerous to eat something you don't like; then you'll vomit and your head will fall off and your stomach will burst." Mona was convinced eyes had special controlling powers and thoughts could kill.

Initial therapeutic intervention was in the nature of establishing trust, the therapist serving as a reality-organ, correcting Mona's distortions and, whenever possible, offering interpretations. The parents were counseled not to comply with Mona's every command —for example, to remove the pictures from the wall or allow her to remain home from school. The teachers had a direct line to the therapist as well and instituted special educational therapy to help Mona develop language skills and powers of concentration.

As Mona developed more confidence, she acquired several friends and joined the girl scouts clubs. In therapy, new themes continued to emerge involving her low self-esteem, confused body image, and maladaptive defense mechanisms, chiefly denial and projection. When Mona was eight for example, she enacted the role of eight-year-old Moira "who was born two months ago—she didn't want to come out of her mother's tummy; she has two brains; one which turns itself off when she wants to go into her secret world."

The therapist helped Mona work through her feelings of being "queer," or different from other children, and helped her find better ways of solving problems. In preadolescence, Mona became deeply upset by her budding sexuality. Her thoughts became confused and she experienced episodes of depersonalization.

The bizarre quality of her imagination and fear of pubertal change and bodily damage was exemplified by one of her stories when she was twelve years old: "Moira was bad, went to jail, got pregnant, and had an abortion. She stuck the [fetus] back into her vagina to grow again, but instead, Moira turned herself inside out, upside down, and her ovaries started to walk on two little tube-like legs. She cried, but instead of tears falling, little eggs dropped out."

The therapist again served as "the voice of reality," correcting distortions, and encouraging Mona to use her verbal expressiveness to create stories that could more appropriately be shared with teachers and schoolmates. Mona became editor of the school paper and achieved success and won admiration from her peers for her writing skills. Treatment was terminated at age fourteen when Mona's family moved out of state.

Ten-year follow-up revealed that Mona had made a good adjustment to college life (in a small non-pressured college), and had selected several male teachers as "mentors" to guide her in her literary interests and writing skills. She had one or two friends and found employment as a school librarian. A psychiatric consultant considered her diagnosis to be "schizoid personality" or "schizotypal personality disorder" since she met all the eight inclusion criteria.* The question as yet unanswered is: what part did early intervention play in preventing a schizophrenic breakdown in a vulnerable individual?

¶ The Child at Risk for Manic-Depressive Psychosis: Historical Review and Risk Data[56]

Kraepelin,[58] in 1896, classified manic-depressive psychosis as a unitary form of mental illness distinct from schizophrenia. He noted, in his 1921 monograph, that the strength of the hereditary factor was 75 percent, that 70 percent of his cases were women, and that 25 percent of his manic-depressive patients were alcoholics. According to Rosenthal,[88] there is a high rate of lifetime prevalence of affective disorders in the first and second degree relatives of patients with primary affective disorders as compared with the general population (6 to 24 percent versus 1 to 2 percent). Bipolar manic-depressive psychosis occurs in 0.5 percent of the population as contrasted with the 0.85 to 1.5 percent worldwide incidence of schizophrenia. Risk figures for manic-depressives psychosis in the first degree relatives of manic-depressive index cases from eleven studies from 1921 to 1953 are the following: parents 7.8 percent; siblings 8.8 percent; children 11.2 percent.

A genetic hypothesis has been supported by Zerbin-Rudin[111] in a review of six major studies of twins. Overall concordance rates for monozygotic twins were consistently higher than for dizygotic twins (74 percent versus 19 percent).

Employing strict diagnostic criteria, Bertelsen and associates[69] studied sixty-nine monozygotic probands and determined that the concordance rate was 67 percent versus the corresponding dizygotic twin concordance of 20 percent. (The difference is significant at $p < 0.001$.) A rare case of monozygotic twins reared apart who were concordant for manic-depressive illness was reported by Rosanoff and associates[86] in 1935, and the data reanalyzed by Farber[29] in 1979.

Cadoret[12] found that the incidence of depression was significantly higher in the affect-disordered parent adoptees than in adoptees whose biological parents had other psychiatric conditions or were apparently psychiatrically well. Findings of Mendlewicz and Rainer[71] point to a similar conclusion. Psychopathology in the biological parents was in excess of that found in the adoptive parents of the same manic-depressive offspring.

CASE ILLUSTRATION

Marcy, a fourteen-year-old girl was suspended from school for trauncy and marijuana intoxication. Considered "wild" by

*Odd communication without formal thought disorder, self-referential thinking, suspiciousness, depersonalization, magical thinking, inadequate rapport, hypersensitivity, and social isolation.

parents and teachers, she was subject to intense mood swings. Popular with girls and boys, she had several close friends and did well academically until seventh grade when school performance became inconsistent.

Family background was positive for affective disorder on both sides. The chronically depressed maternal grandfather had been hospitalized once and received electroconvulsive therapy (ECT). The mother, an explosive, volatile woman, had been treated for depression with psychotherapy and antidepressant medication for many years. One of her brothers had undergone repeated hospitalizations for alcoholism, antisocial behavior, and suicide attempts. The father, psychiatrically stable, had a sister who committed suicide at age twenty-five following psychiatric hospitalization for a "nervous breakdown" (diagnosis uncertain). Marcy's sixteen-year-old brother had been hyperactive in childhood and had been expelled from school in early adolescence for stealing from classmates and vandalism.

Psychological tests revealed a 14 point discrepancy between the verbal and performance WISC (114 versus 100). The lower performance score was due to inattentiveness to objective detail on the picture completion test and coding. Projective tests revealed a projective tendency, distrust of authority figures, rage, and an underlying depressive trend.

Psychotherapy was refused. Because of temper tantrums and impulse dyscontrol, Marcy was sent to a boarding school known for its therapeutic milieu. The staff reported extreme lability of mood; she would shift from ebullience to despair within hours. Her dreams were frightening or gloomy ("I dreamed I was dead, I went to my own funeral and saw myself lying in a coffin"). At other times she was overexcited and verbose.

Marcy remained in the school two years. The staff reported that at sixteen Marcy's school work began to deteriorate. Marcy became sexually promiscuous, refused to obey school rules, was often truant, and explosive rage reactions became more frequent. She ran away from school, having stolen a car. Grossly delusional, she was convinced she was about to become a famous Hollywood singer.

A psychiatric consultant recommended hospitalization for manic-depressive psychosis with appropriate medication and psychotherapy following the hospital course.

Discussion

Current research concerning the precursors of schizophrenia and affective disorder is still in its infancy, yet a number of factors are becoming apparent. Certain symptom clusters in genetically predisposed individuals may be predictive of future illness. Specific differences in symptomatology together with a family history positive for either schizophrenia or affective disorder may help in identifying the underlying disturbance.

The case of Mona, for example, demonstrates many of the features noted in children at risk for schizophrenia: positive family history, early signs of hypersensitivity, traumatic separations, illness and death of relatives, and an emotionally labile mother. Investigators have described symptom clusters similar to hers in children who became schizophrenic in adult life[100,101] (extreme shyness and introversion, magical thinking not commensurate with age or intelligence, anhedonia, phobias, attentional problems, and maladaptive defense mechanisms such as projection and denial). Mona exhibited failure of repression and a tendency toward regression reminiscent of the children described by Ekstein and Wright,[22] and Kestenbaum.[49]

The case of Marcy, an example of adolescent manic-depressive disorder, is illustrative in that different symptom clusters are in evidence. The family history of affective disorder and Marcy's early history of temper tantrums, impulsivity, lowered frustration tolerance and dysphoria, behavior problems and lack of judgment point to a manic-depressive diathesis.

An interesting psychometric finding, in contrast to Mona's IQ differential, was Marcy's superior verbal IQ and significantly lower performance scores on the WISC—a finding noted in Anthony's 1960 case report.[4] Low scores on the visuo-constructive tests are reminiscent of the scores obtained by children with certain types of minimal brain damage, according to Gardner.[35] Appreciation of spatial relationships is considered to be primarily a function of the right cerebral hemisphere as is the processing of complex nonverbal sensory input. Right hemispheric dysfunction might explain the inability to perform well on block design, picture arrangement, picture completion, object assembly, and mazes subtests.[90]

Certain investigators have postulated that individuals who have a family history positive for manic-depressive illness may have a greater right than left hemisphere deficit.[7,37] Moreover, they may have a high incidence of disinhibition syndrome which could be the result of subtle frontal brain systems dysfunction, similar to that which characterizes hyperkinetic disorder.

Flor-Henry[33] has formulated a similar hypothesis for adults with manic-depressive disorder. The findings suggest the possibility of a fundamental genetic liability—the lack of some central inhibitory regulating mechanism—that may lead to a manic-depressive illness in later life. This observation is in contrast to the possibility of left-hemispheric dysfunction in schizophrenics, which has been reported in recent neurophysiological investigations.[11,74] At this time, controlled studies are sparse and further investigation is required for better understanding of the neurophysiological basis of mental functioning.

Anthony and Scott,[4] in a review of the literature from 1896 to 1960, concluded that manic-depressive illness in childhood is extremely rare. They contend that the early variety may be due to heavy genetic loading and intense environmental experience, that it may be manifested during childhood under strong physical or psychological pressure, and that it may, under certain circumstances, become clinically recognized as a psychosis. Youngerman and Canino[109] reviewed 190 cases of lithium carbonate in children and adolescents and noted that "many adolescent manic-depressives have histories of behavior and mood disorders often dating back to early childhood. Affective symptoms are mixed and masked in childhood, and it is difficult to elicit reports of sustained mood swings" [p. 223].

Other investigators[2,4,8,30] have described children with a family history positive for bipolar illness, and various symptom complexes including sleep disorder, night terrors, rage attacks, grandiosity, and socially inappropriate behavior. Davis[19] has proposed that there is an identifiable syndrome, which he calls manic-depressive variant syndrome of childhood (MDV), characterized by positive family history of affective disturbance, hyperactivity, temper tantrums, and impairment in personal relationships.

¶ Possible Predictors of Future Affective Disorders

There are few prospective studies as yet of children at high risk for affective disorder. Several studies demonstrate that a significant number of patients hospitalized for depression were parents of children who exhibited episodes of depression.[75,105] There are relatively few studies, however, of the children of bipolar manic-depressive parents.[21,50] Clinical descriptions from actual examination rather than from parents' reports together with psychological test scores are sparse. Kestenbaum[48] has described thirteen children with a family history of bipolar manic-depressive disorder; six of the children (four males, two females) exhibited the following features:

1. Family history positive for bipolar illness;
2. Specific clinical symptomatology including temper tantrums, compulsive rituals, dysphoria, lability, obessional preoccupation, learning disability, hyperactivity, impulsivity;
3. Specific patterns in psychological test scores (WISC) revealing verbal achievement sig-

nificantly greater than performance, with considerable subtest scatter.

Of the remaining seven, three (all females) had psychological test scores that did not follow the pattern described above; four (two males, two females) were not given psychological tests. The presenting symptoms of these seven children were depressed mood (N=5) and behavior problems (N=2). The presenting symptoms of the six children exhibiting the triad of features of the triad mentioned above were learning problems with depressed mood (N=5) and hyperactivity with behavior problems (N=1) [p. 1207].

Genetic Themes of Transmission

Genetic and clinical variables have been noted which differentiate bipolar from unipolar illness.[107] Bipolar probands were observed to have a higher suicide rate, earlier age of onset, and peptic ulcers in greater numbers than unipolar probands. Bipolar females demonstrated heightened vulnerability to postpartum psychosis. Unipolar disorders tend to begin later (forty-three versus thirty-one years), are more frequently females, and are less severe. Winokur[107] has suggested that depressive disorder be divided into autonomous subtypes based on family history: (1) depressive spectrum disease in an individual with a first degree history of alcoholism or antisocial personality; and (2) pure depressive illness in an individual without similarly affected relatives. Thus, family studies have led some investigators to conclude that unipolar and bipolar illnesses are genetically different entities. Theories of genetic transmission of affective disorders postulate: (1) autosomal dominance;[86] (2) genetic heterogeneity;[83] (3) X-linked gene associated with red, green colorblindness and Xga blood type;[13,72] and (4) multiple threshold models.[38] (A variable liability to a disorder is postulated to which genetic and independent factors may contribute. If the net liability crosses a certain threshold, the disorder becomes manifest). Gershon[38] suggested that unipolar and bipolar illnesses represent positions on a continuum of liability.

Greater liability would tend to manifest itself as bipolar illness, lesser liability as a unipolar disorder. Along with Dunner and Goodwin, Gershon contended that the individual with a manic-depressive illness has "an inherited vulnerability to loss, with increased liklihood of development of pathological loss reactions" [p. 8].

¶ Juvenile Manic-Depressive Illness

Serious forms of depression are commonly encountered in childhood.[17,80,81,104] Symptoms include sad affect, social withdrawal, psychomotor retardation, anxiety, school failure, sleep disturbance, feelings of hopelessness and helplessness, suicidal preoccupation, and self-deprecatory ideation. All of the investigators noted that while depressive symptoms are common, mania in childhood is extremely rare.[2,30,103] Recent genetic studies, however, have demonstrated that a first episode of bipolar illness in adolescence is not uncommon.[79] Winokur and associates[107] noted that one-third of their bipolar cases had a first episode occurring between ten and nineteen years of age. Nonetheless, manic episodes in adolescence are frequently misdiagnosed. Carlson and Strober[14] described six cases of adolescents initially diagnosed as schizophrenics who were, at a later admission, rediagnosed manic-depressive. Stone[96] also described the present tendency to label young patients "schizophrenic until proven otherwise" [p. 16].

¶ Conclusion: Prevention and Intervention

Hypotheses about the environmental variables that interact with constitutional factors in the development of schizophrenia or affective disorders include neurophysiological dysfunctions, which can be tested in a laboratory, as well as family interaction patterns and coping skills, which are currently being examined by accurate measurements outside the laboratory. Pooling data from different research projects provides a massive amount of information about the at-risk children so that it should become easier, in time, to isolate those variables that predict eventual breakdown.

Interventions are particularly useful when they are specifically directed toward correcting primary problems.[25] Until more is known about the core psychopathology involved in the major psychoses, primary prevention and therapeutic interventions involve much guesswork.

The evidence gathered thus far indicates that the preschizophrenic child has difficulty filtering stimulus input and has problems in attention that subsequently lead to school difficulties and social problems. The premanic-depressive child may exhibit difficulty with impulse control and regulation of moods as well as with other subtle manifestations of nonverbal learning disability.

Early intervention should include genetic counseling, careful perinatal examination, and frequent pediatric developmental evaluation of at-risk children, with particular focus on language delay. Therapeutic nurseries and language-and-learning therapies might be made available to children who showed signs of early deviance.[47]

> The at-risk child who is "bright but not living up to his potential" should be evaluated as soon as symptoms appear; school failure, attentional problems, social withdrawal, loss of self-esteem should not be left unnoticed until symptoms become fixed. . . . Special school programs for the child with attentional problems should focus on strengthening existing assets which would enhance self-esteem.[49] [p. 174]

Recommendations for family treatment, selection of a proper school or camp, and individual psychotherapy or pharmacotherapy might be made to fit the particular needs of each family. Such environmental interventions may not prevent a psychotic breakdown in adult life, but the coping mechanisms acquired by such intervention techniques may shorten the course of the illness.

An interesting outcome of Anthony's study[3] was his discovery that some of the children of schizophrenic parents (5 to 10 percent) reacted with supernormality, tolerating the family problems with equanimity. These so-called "invulnerable" children demonstrated unusual talents and coping abilities. Other investigators[46] have described such "superkids" who have emerged from sick families as having a wide variety of interests and unusual capabilities. Most of the children were from the schizophrenic group and had mothers who, despite their illness, were warm and supportive. Depressed mothers, unable to respond to their infants' needs, seem to have a more profoundly negative impact on their children.[20] One must assume that an invulnerable child in a sick family is one who is constitutionally better equipped to cope with stress than his less fortunate siblings and who, in addition, has access to beneficial environmental support systems.

The study of children at risk is producing new information about mental illness *in statu nascendi*. Understanding the process involved may provide basic solutions to one of mankind's most pervasive problems.

¶ Bibliography

1. ALTSHULER, K. Z. "Deafness and Schizophrenia: Interrelation of Communication Stress, Maturation Lag, and Schizophrenic Risk," in Kallmann, F. J., et al., eds., *Expanding Goals of Genetics in Psychiatry.* New York: Grune & Stratton, 1962, pp. 52–62.
2. ANNELL, A. "Lithium in the Treatment of Children and Adolescents," *Acta Psychiatrica Scandinavica, Suppl.* 207 (1969): 19–30.
3. ANTHONY, E. J. "A Clinical Evaluation of Children with Psychotic Parents," *Ameri-*

can Journal of Psychiatry, 126:2 (1969): 177–184.

4. ANTHONY, E. J., and SCOTT P. "Manic-Depressive Psychosis in Childhood," *Journal of Child Psychology and Psychiatry,* 1 (1960): 53–72.
5. ARIETI, S. *Interpretation of Schizophrenia.* 2nd ed., New York: Basic Books, 1974.
6. BATESON, G., et al., "Toward a Theory of Schizophrenia," *Behavioral Science.* 1 (1956): 251–264.
7. BEMPORARD, B. "Children of Manic Depressives," unpublished Ph.D. diss., Columbia University, forthcoming.
8. BERG, I., et al. "Bipolar Manic-Depressive Illness in Early Adolescence. A Case Report," *British Journal of Psychiatry,* 125 (1974): 416–417.
9. BERTELSEN, A., HARVALD, B., and HAUGE, M. "A Danish Twin-Study of Manic Depressive Disorder," *British Journal of Psychiatry,* 130 (1977): 330–357.
10. BLEULER, E. *Dementia Praecox or The Group of Schizophrenias.* New York: International Universities Press, 1950.
11. BUCHSBAUM, M. "Average Evoked Response Augmenting/Reducing in Schizophrenia and Affective Disorder," in Freedman, D.X., ed., *Biology of the Major Psychoses.* New York: Raven Press, 1975, pp. 129–142.
12. CADORET, R. J. "Evidence for Genetic Inheritance of Primary Affective Disorder in Adoptees," *Americal Journal of Psychiatry,* 17:1 (1978): 138–153.
13. CADORET, R. J., and WINOKUR, G. "X-Linkage in Manic-Depressive Illness," *American Review of Medicine,* 26 (1975): 21–25.
14. CARLSON, G. A., and STROBER, M. "Manic-Depressive Illness in Early Adolescence," *Journal of American Academy of Child Psychiatry,* 17 (1978): 138–153.
15. CHAPMAN, J., and MCGHIE, A. "A Comparative Study of Disordered Attention in Schizophrenia," *Journal of Mental Science,* 108 (1962): 487–500.
16. CORNBLATT, B., and MARCUSE, Y. "Children at High-Risk for Schizophrenia: From Childhood to Adolescence," in Erlenmeyer-Kimling, L., Dohrenwend, B. S., and Miller, N., eds., *Life Span Research on the Prediction of Psychopathology,* forthcoming.
17. CYTRYN, L., and MCKNEW, D. H. J. "Proposed Classification of Childhood Depression," *American Journal of Psychiatry,* 129 (1972): 149–155.
18. DAVIDSON, K., and BAGLEY, C. R. "Schizophrenia-like Psychoses Associated with Organic Disorders of the Central Nervous System: A Preview of the Literature," *British Journal of Psychology, Special Publication* no. 4, ("Current Problems in Neuropsychiatry"), 1969 p. 113.
19. DAVIS, R. E. "Manic-Depressive Varient Syndrome of Childhood (The M.D.V. Syndrome, A Preliminary Report," *American Journal of Psychiatry,* 136:5 (1979): 702–705.
20. DINCE, P. R. "Maternal Depression and Failures of Individuation During Adolescence," in Musserman, J., ed., *Science and Psychoanalysis,* New York: Grune & Stratton, 1966, pp. 26–37.
21. DYSON, W. L., and BARCAI, A. "Treatment of Children of Lithium-Responding Parents," *Current Therapeutic Research—Clinical and Experimental,* 12 (1970): 286–290.
22. EKSTEIN, R., and WRIGHT, D. C. "The Space Child," *Bulletin of the Menninger Clinic,* 16 (1952): 211–224.
23. ERLENMEYER-KIMLING, L. "Studies on the Offspring of Two Schizophrenic Parents," in Rosenthal, D. and Kety, S. S., eds., *The Transmission of Schizophrenia.* New York: Pergamon Press, 1968, pp. 65–83.
24. ———. "A Prospective Study of Children At-Risk for Schizophrenia: Methodological Considerations and Some Preliminary Findings," in Wirt, R. D., Winokur, G., and Roff, M., eds, *Life History Research in Psychopathology.* vol. 4, Minneapolis: University of Minnesota Press, 1976, pp. 22–46.
25. ———. "Issues Pertaining to Prevention and Intervention in Genetic Disorders Affecting Human Behavior," in Albee, G. W. and Joffe, J. M., eds., *Primary Prevention in Psychopathology.* Hanover, N.H.: University Press of New England, 1977, pp. 68–91.
26. ———. "Genetic Approaches to the Study of Schizophrenia: The Genetic Evidence as a Tool in Research," *Birth Defects* (Original Article Series), 14:5 (1978): 59–74.
27. ERLENMEYER KIMLING, L., and PARADOWSKI, W. "Selection and Schizophrenia," in

Bajema, C. J., ed., *Natural Selection in Human Populations.* New York: John Wiley, 1971, pp. 651–665.

28. ERLENMEYER KIMLING, L., CORNBLATT, B., and FLEISS, J. "High-Risk Research in Schizophrenia," *Psychiatric Annals,* 9(1) (1979): 38–51.
29. FARBER, S. *Identical Twins Reared Apart.* New York: Basic Books, 1981.
30. FEINSTEIN, S., and WOLPERT, E. A. "Juvenile Manic-Depressive Illness Clinical and Therapeutic Considerations," *Journal of American Academy of Child Psychiatry,* 12 (1973): 123–136.
31. FISH, B. "Abnormal States of Consciousness and Muscle Tone in Infants Born to Schizophrenic Mothers," *American Journal of Psychiatry,* 119 (1962): 439–445.
32. ———. "Biologic Antecedents of Psychosis in Children," in Freedman, D.X., ed., *Biology of the Major Psychoses.* vol. 54, New York: Raven Press, 1975, pp. 49–80.
33. FLOR-HENRY, P. "On Certain Aspects of the Localization of the Cerebral Systems Regulating and Determining Emotions," *Biological Psychiatry,* 14:4 (1979): 666–697.
34. FRIEDMAN, D., VAUGHAN, H., and ERLENMEYER-KIMLING, L. "Stimulus and Response Related Components of the Late Positive Complex in Visual Discrimination Tasks," *Electroencephalography and Clinical Neurophysiology,* 45 (1978): 319–330.
35. GARDNER, R. A. *The Objective Diagnosis of Minimal Brain Dysfunction.* Cresskill, N.J.: Creative Therapeutics, 1979.
36. GARMEZY, N. "Children-at-Risk: The Search for The Antecedents of Schizophrenia, Part 1: Conceptual Models and Research Methods," *Schizophrenia Bulletin,* 8 (1974): 14–90.
37. GAZZANIGA, M. S. *The Bisected Brain.* New York: Appleton-Century-Crofts, 1970.
38. GERSHON, E. S., DUNNER, D. L., and GOODWIN, F. K. "Toward a Biology of Affective Disorders," *Archives of General Psychiatry,* 25 (1971): 1–15.
39. GOTTESMAN, I. I. "Schizophrenia and Genetics: Toward Understanding Undertainty," *Psychiatric Annals,* 9 (1) (1979): 26–37.
40. GOTTESMAN, I. I., and SHIELDS, J. *Schizophrenia and Genetics: A Twin Study Vantage Point.* New York: Academic Press, 1972.
41. GRINKER, R. R. Sr., and HOLZMAN, P. S. "Schizophrenic Pathology in Young Adults," *Archives of General Psychiatry* 28 (1973): 168–175.
42. HANSON, D. R., GOTTESMAN, I. I., and HESTON, L. L. "Some Possible Childhood Indicators of Adult Schizophrenia Inferred from Children of Schizophrenics," *British Journal of Psychiatry,* 129 (1976): 142–154.
43. HESTON, L. L. "Psychiatric Disorders in Foster Home Reared Children of Schizophrenic Mothers," *British Journal of Psychiatry,* 112 (1966): 819–825.
44. KALLMANN, F. J. *The Genetics of Schizophrenia.* Locust Valley, New York: J. J. Augustin, 1938.
45. ———. "The Genetic Theory of Schizophrenia: An Analysis of 691 Schizophrenic Twin Index Families," *American Journal of Psychiatry.* 103 (1946): 309–322.
46. KAUFFMAN, C., et al. "Superkids: Competent Children of Psychotic Mothers," *American Journal of Psychiatry,* 136:11 (1979): 1398–1402.
47. KESTENBAUM, C. J. "Psychotherapy of Childhood Schizophrenia," in Wolman, B., Egan, J., and Ross, A., eds., *Handbook of Treatment of Mental Disorders in Childhood and Adolescence.* Englewood Cliffs, N.J.: Prentice Hall, 1977, pp. 354–384.
48. ———. "Children at Risk for Manic-Depressive Illness: Possible Predictors," *American Journal of Psychiatry,* 136:9 (1979): 1206–1208.
49. ———. "Children At-Risk for Schizophrenia," *American Journal of Psychotherapy,* 34:2 (1980): 164–177.
50. ———. "Adolescents At-Risk for Manic-Depressive Illness," in Feinstein, S. C., and Giovacchini, P. L., eds., *Adolescent Psychiatry,* vol. 7, forthcoming.
51. ———, and BIRD, H. "A Reliability Study of the Mental Health Assessment Form for School Age Children," *Journal of American Academy of Child Psychiatry,* 17:2 (1978): 338–347.
52. KETY, S. S. "Studies Designed to Disentangle Genetic and Environmental Variables in Schizophrenia; Some Epistemological Questions and Answers," *American Journal of Psychiatry* 133 (1975): 1134–1137.
53. KETY, S. S., et al. "The Types and Preva-

lence of Mental Illness in the Biological and Adoptive Families of Adoptive Schizophrenics," in Rosenthal, D., and Kety, S. S., eds., *The Transmission of Schizophrenia.* Oxford: Pergamon Press, 1968, pp. 345–362.

54. KRINGLEN, E. "An Epidemiological Clinical Twin Study on Schizophrenia," in Rosenthal, D., and Kety, S. S., eds., *The Transmission of Schizophrenia.* Oxford: Pergamon Press, 1968, pp. 49–63.
55. KOHN, M. "Social Class and Schizophrenia: A Critical Review and A Reformulation," *Schizophrenia Bulletin,* 7 (1973): 60–79.
56. KOPPITZ, E. M. *The Bender Gestalt Test for Young Children.* New York: Grune & Stratton, 1963.
57. KORNESTSKY, C., and MIRSKY, A. "On Certain Psychopharmacological and Physiological Differences Between Schizophrenics and Normal Persons," *Psychopharmacologia,* (Berlin) 8 (1966): 309–318.
58. KRAEPELIN, E., *Manic-Depressive Insanity and Paranoia.* Edinburgh: Livingston, 1921.
59. LANGFELDT, G., "Diagnosis and Prognosis of Schizophrenia, *Proceeds of the Royal Society of Medicine,* 53 (1969): 1047–1052.
60. LAZARUS, R. S., AVERILL, J. R., and OPTON, E. M. "The Psychology of Coping: Issues of Research and Assessment," in Coelho, G. V., Hamburg, D. A., and Adams, J. E., eds., *Coping and Adaptation.* New York: Basic Books, 1974.
61. LIDZ, T. *The Origin and Treatment of Schizophrenic Disorders.* New York: Basic Books, 1973.
62. MARCUS, J. "Cerebral Functioning in Offspring of Schizophrenics: A Possible Genetic Factor," *International Journal of Mental Health* 3:1 (1974): 57–73.
63. ———, et al. "Infants At-Risk for Schizophrenia: The Jerusalem Infant Development Project," *AMA Archives of General Psychiatry,* in press.
64. MARCUS, J., and MEDNICK, S. "A Longitudinal Perspective of Children At Risk for Schizophrenia," unpublished manuscript.
65. MARCUSE, Y., and CORNBLATT, B. "Proceedings of the Society for Life History Research in Psychopathology, 1979," (in press).
66. MATTHYSSE, S. "The Biology of Attention," *Schizophrenia Bulletin,* 3 (1977): 370–72.
67. ———, and KIDD, K. K. "Estimating The Genetic Contribution to Schizophrenia," *American Journal of Psychiatry,* 133 (1976): 185–191.
68. MEDNICK, S. A. "Breakdown In Individuals at High Risk for Schizophrenia: Possible Predispositional Perinatal Factors," *Mental Hygiene,* 54 (1970): 50–63.
69. MEDNICK, S., and SCHULSINGER, F. "Some Premorbid Characteristics Related to Breakdown in Children with Schizophrenic Mothers," in Rosenthal, D. and Kety, S. S., eds., *The Transmission of Schizophrenia.* Oxford: Pergamon Press, 1968, pp. 267–291.
70. ———, et al. "Perinal Perinatal Conditions and Infant Development in Children with Schizophrenic Parents," *Social Biology* 18 (Suppl.) (1971): 103–113.
71. MENDLEWICZ, J. and RAINER, J. "Adoption Study Supporting Genetic Transmission in Manic-Depressive Illness," *Nature,* 268 (1977): 327–329.
72. MENDLEWICZ, J., RAINER, J. D., and FLEISS, J. L. "A Dominant X-linked Factor in Manic Depressive Illness; Studies with Color Blindness," in Fieve, R. R., Rosenthal, A., and Brill, H., eds., *Genetic Research in Psychiatry.* Baltimore: The Johns Hopkins University Press, 1975, pp. 241–255.
73. MCGHIE, A., and CHAPMAN, J. "Disorders of Attention and Perception in Earlier Schizophrenia," *British Journal of Medical Psychology,* 34 (1961): 103–116.
74. MCKINNON, J. A. "Two Semantic Forms, Neuropsychological and Psychoanalytic Descriptions," *Psychoanalysis and Contemporary Thought* 2 (1979): 25–76.
75. MCKNEW, D. W., et al. "Offspring of Patients with Affective Disorders," *British Journal of Psychiatry,* 134 (1979): 148–152.
76. MCNEIL, T. F., and KAIJ, L. "Obstetric Factors in The Development of Schizophrenia: Complications in the Births of Preschizophrenics and in Reproduction by Schizophrenic Parents," in Wynne, L. C., Cromwell, R. L., and Matthysse, S., eds., *The Nature of Schizophrenia.* New York: John Wiley, 1978, pp. 401–429.
77. ORVASCHEL, H., et al. "The Children of Psychiatrically Disturbed Parents: Differences as a Function of the Sex of the Sick

Parent," *Archives of General Psychiatry* 36 (1979): 691–695.

78. PEARSON, J. S., and KLEY, I. B. "On the Application of Genetic Expectancies as Age-Specific Base Rates in the Study of Human Behavior Disorder," *Psychological Bulletin,* 54 (1957): 406–420.
79. PERRIS, C., ed. "A Study of Bipolar (Manic-Depressive) and Unipolar Recurrent Depressive Psychoses," *Acta Psychiatrica Scandinavica,* 42 (Suppl. 194) (1966): 1–189.
80. PETTI, T. A. "Depression in Hospitalized Child Psychiatric Patients: Approaches to Measuring Depression," *American Journal of Child Psychiatry,* 17 (1978): 49–59.
81. POZNANSKI, E. "Natural Course of Childhood Depression," *Acta paedo psychiathca,* (in press).
82. PRENTKY, R. A., WATT, N. F., and FRYER, J. H. "Longitudinal Social Competence and Adult Psychiatric Symptoms at First Hospitalization," *Schizophrenia Bulletin,* 5 2: (1979): 306–312.
83. PRICE, J. S. "Genetic and Phylogenetic Aspects of Mood Variation," *International Journal of Mental Health,* 1 (1972): 124–144.
84. REIDER, R., and NICHOLAS, P. "Offspring of Schizophrenics III: Hyperactivity and Neurological Soft Signs," *Archives of General Psychiatry,* 36 (1979): 665–674.
85. RICKS, D. F., and BERRY, J. C. "Family and Symptom Patterns that Precede Schizophrenia," in Roff, M., and Ricks, D. F., eds., *Life History in Psychopathology.* Minneapolis: University of Minnesota Press, 1970, pp. 31–51.
86. ROSANOFF, A. J., HANDY, L. M., and PLESSET, I. R. "The Etiology of Manic-Depressive Syndromes with Special Reference to Their Occurrence in Twins," *American Journal of Psychiatry,* 91 (1935): 725–726.
87. ROSENTHAL, D. "An Historical and Methodological Review of Genetic Studies of Schizophrenia," in Romano, J., ed., *The Origins of Schizophrenia.* Procedings of the First Rochester International Conference on Schizophrenia, Rochester, N.Y. 1967, pp. 15–26.
88. ———. *Genetic Theory and Abnormal Behavior.* New York: McGraw-Hill, 1970.
89. ———. "The Spectrum Concept in Schizophrenic and Manic-Depressive Disorders," in Friedman, D. X., ed., *Biology of the Major Psychoses.* New York: Raven Press, 1975, pp. 19–25.
90. RUSSELL, E. W., NEURINGER, C., and GOLDSTEIN, G. *Assessment of Brain Damage.* New York: Wiley-Interscience, 1970.
91. RUTSCHMANN, J., CORNBLATT, B., and ERLENMEYER-KIMLING, L. "Sustained Attention in Children At-Risk for Schizophrenia," *Archives of General Psychiatry,* 34 (1977): 571–575.
92. SLATER, E., and COWIE, V. *The Genetics of Mental Disorders.* London: Oxford University Press, 1971.
93. SOBEL, D. E. "Children of Schizophrenic Parents: Preliminary Observations on Early Development," *American Journal of Psychiatry* 118 (1961): 512–519.
94. SPITZER, R. L., ENDICOTT, J., and GIBBON, M. "Crossing the Border into Borderline Personality and Borderline Schizophrenia. The Development of Criteria," *Archives of General Psychiatry* 36 (1979): 17–24.
95. SPRING, B., and ZUBIN, J. "Vulnerability to Schizophrenic Episodes and Their Prevention in Adults," in Albee, G. W., and Joffe, J. W., eds., *Primary Prevention in Psychopathology,* Hanover, N.H.: University Press of New England, 1977, pp. 254–284.
96. STONE, M. H. "Mania—A Guide for the Perplexed," *Psychotherapy and Social Science Review,* 5(10) (1971): 14–18.
97. STONE, M. H. "Etiological Factors in Schizophrenia: A Reevaluation in the Light of Contemporary Research," *Psychiatric Quarterly,* 50:2 (1978): 83–119.
98. STRAUSS, J., et al. "Premorbid Adjustment in Schizophrenia: Concepts, Measures, and Implications," *Schizophrenia Bulletin,* 3:2 (1977): 182–258.
99. VENABLES, P. H. "The Electrodermal Psychophysiology of Schizophrenics and Children at Risk for Schizophrenia: Controversies and Developments," *Schizophrenia Bulletin* 3 (1977): 28–48.
100. WARING, N., and RICKS, D. F. "Family Patterns of Children who Become Adult Schizophrenics," *Journal of Nervous and Mental Disease,* 140 (1965): 351–364.
101. WATT, N. F. "Patterns of Childhood Social Development in Adult Schizophrenics," *Archives of General Psychiatry,* 35 (1972): 160–165.
102. ———. "Longitudinal Changes in the Social

Behavior of Children Hospitalized for Schizophrenia as Adults," *Journal of Nervous and Mental Disease,* 155 (1972): 42–54.

103. WEINBERG, W. A., and BRUMBACK, R. A. "Mania in Childhood: Case Studies and Literature Review," *American Journal of Disease of Children,* 130 (1976): 380–385.
104. WEINBERG, W. A., et al. "Depression in Children Referred to an Educational Diagnostic Center: Diagnosis and Treatment," *Journal of Pediatrics,* 83 (1973): 1065–72.
105. WELNER, Z., et al. "Psychopathology in Children of Inpatients with Depression: A Controlled Study," *Journal of Nervous and Mental Disease,* 164:6 (1977): 408–413.
106. WENDER, P. H., et al. "Crossfostering: A Research Strategy for Clarifying the Role of Genetic and Experiential Factors in the Etiology of Schizophrenia," *Archives of General Psychiatry,* 30 (1974): 121–129.
107. WINOKUR, G., CLAYTON, P. J., and REICH, T. *Manic Depressive Illness.* St. Louis: C.V. Mosby, 1969.
108. WYNNE, L. C., and SINGER, M. T. "Principles for Scoring Communication Defects and Deviances of Parents of Schizophrenics in Psychological Test Transactions," *NIMH Annual Report,* M.H. Intramural Research Programs, 1972, pp. 11–13.
109. YOUNGEMAN, J., and CANINO, I. "Lithium Carbonate Use in Children and Adolescents," *Archives of General Psychiatry,* 35 (1978): 216–224.
110. ZERBIN-RUDIN, E. "Schizophrenic Head Injured Persons and Their Families," Abstract 356, Report of 3rd International Congress on Human Genetics, Chicago, 1966.
111. ———. "The Genetics of Schizophrenia: An International Survey," *Psychiatric Quarterly,* 46 (1972): 371–383.
112. ZUBIN, J., and SPRING, B. "Vulnerability—A New View of Schizophrenia," *Journal of Abnormal Psychology,* 86:2 (1977): 103–126.

CHAPTER 8

CHILD ABUSE AND NEGLECT

Marvin L. Blumberg

DURING this century, tremendous progress has been made in the development of scientific and medical methods for improving the health, nutrition, and general development of children. Yet, on the other hand, the incidence of child abuse and neglect has increased greatly. Perhaps this has been in part a result of the violent and frenetic state of the world, or of universal economic uncertainties, or of the beaurocratic depersonalization of the individual. In any event, the abuse of its offspring is almost exclusively a human phenomenon. Man with his superior intelligence has evolved a complex societal environment and he reacts to its pressures with human emotions, which often lead to character disorders.

From time immemorial children have been considered to be expendable and replaceable, with their parents having unquestioned authority over them. Children were frequently sacrificed to pagan gods, such as the Canaanite deity Baal. The biblical patriarch Abraham was prepared to sacrifice his beloved son, Isaac, at the Lord's request. Through the ages, it was believed that the rights of parents and the will of a deity condoned child abuse and murder. In the agricultural society of the last century, there was more concern for necessary domestic animals than for children. However, in 1875 a societal superego emerged and the first Society for the Prevention of Cruelty to Children was formed in New York City. It was almost one century later before the states began passing laws defining child abuse and neglect and mandating the reporting of cases. More recently, faith in the myth of the maternal instinct and the widespread concept that everyone loves and protects innocent children has tended to obscure the extent of child abuse.

The legal definition of child abuse refers to inflicted wounds or sexual molestation of a child under sixteen years of age. Child neglect involves the deprivation of food, clothing, shelter, and medical care for a person under eighteen years of age. The objective medical view of abuse or neglect relates to the nature and severity of the wounds or the neglected nutritional and chemical aspects of the child's body. The psychiatric concern is for the emotional effects of the maltreatment of the child at the present time and in the future. The ideal approach to the management of child abuse and neglect should encompass all three concepts—legal, medical, and psychiatric—and should include the child, the abusing parent, and the total family.

It is a curious fact that for a long time the psychiatric conceptualization of the mechanisms and the nature of human behavior included little recognition of the emotional effects of ambience on children's personality, character disorders, and future behavioral patterns as adults. Sigmund Freud regarded children's behavior as being based largely on fantasy and derived much of his thinking in this regard from his psychoanalytic interpretation of adult abreaction of childhood physical and sexual abuse as largely fantasy and unresolved Oedipus complex. In 1919, Freud[10] published a clinical study in the *Zeitschrift für Psychoanalyse* entitled "A Child Is Being Beaten." This was a detailed study of an erotic fantasy involving shame, guilt, and relief by masturbation. Alfred Adler was one of the earlier psychiatrists to recognize the influence of environment on a child's behavior. Then later the concept of a child as a person was further developed by the studies of Jean Piaget, Anna Freud, and Melanie Klein, among others, who observed children psychoanalytically during their developmental stages. Only lately has affective disturbance of infants and young children, such as depression, begun to be recognized and described.

Four popular misconceptions must be dispelled at the outset of any discussion of child abuse.[2] First, maternal instinct is an illusion. There is no inherent mother love that automatically invests a biologic parent with positive cathexis for her child. At least 70 percent of cases of serious child abuse are attributable to the mothers and most of the victims are under three years of age. Parenthood is biological; parenting is emotional and, in a practical sense, a skill. Second, psychosis is rarely a factor in child abuse. It is unusual for children to be seriously harmed or killed by schizophrenic or otherwise psychotic parents. The abusing parent is almost always aware of the nature, if not the immediate reason, for the deed. Third, aggression and violence are not instinctive. They are reactions that are learned from culturally determined practices, incited by ambient pressures, and influenced by exposure to the brutality that is portrayed by the public media. Fourth, every mother is not potentially a child abuser who only needs sufficient provocation to trigger such a reaction. There is a qualitative distinction between discipline and punishment on the one hand and abuse on the other. The former may be rationalized as being a deterrent or corrective action intended for the child's benefit. Abuse, however, is indefensible and inexcusable on any grounds. It is true that there are instances when a normal child may provoke a normal parent to an impulse toward harmful retaliation. The thought itself serves as a trial action whereby the person realizes the nature of the conceived deed and is enabled to discharge some of the anger. The normal ego mechanism of reality testing checks the potential act of violence. The child abuser lacks psychological restraints and harms the child by impulsive violence, often because of anger over unrelated matters.

Behavior is essentially the result of the sequence of action, reaction, and interaction within the circumstances of the family and the societal existence of the individual. This basic concept is operative in the triad of factors that lead to child abuse: the early personality development of the parent, the provocative characteristics of the child and of the family, and societal influences.

¶ Personality of the Abusing Parent

It is a generally accepted fact that the fundamental structure of an individual's personality is set during the first few formative years of life. At a time when a child's world is largely presented, represented, and interpreted by his parents, his impressions, attitudes, and reactions develop as a result of the nature of the parenting and nurturing that he receives. Later, circumstances and relationships may modify the basic personality construct, depending on many factors. Emotions are not instinctive. Love and trust, or hate and mistrust, that lead to violence or to personal withdrawal are learned reactions.

Although there is no complete correlation, most abusing parents were themselves abused, neglected, and deprived of love and proper nurture in their early years.[2] Such persons are usually narcissistic, immature, have poor ego control, and seek love for themselves. They have a poor self-image and low self-esteem. Their threshold for frustration tolerance is low so that they find it difficult to accept criticism or to face adversity. When they are faced with troubles this characterologic construct causes them to react with impulsive violence, often against their own infants. In effect, child abuse frequently appears to be a generational perpetuation.

The abuse-prone parent is usually a mistrusting loner without available extended family and without friends. The young woman, married or single, may have become pregnant and may have had a child with the desire and expectation that the infant would offer her love and gratification for her dependency needs. Since the baby cannot furnish active love and she is incapable of deriving passive satisfaction from the child, her mechanisms of denial and projection are activated and she perceives the child as manifesting her own negative traits. As she rejects and mistreats him, the infant becomes irritable and more demanding. The mother interprets this as rejection and punishes the child accordingly.

An interesting phenomenon that may be observed in the behavior of the abusing parent has been termed role reversal.[13,21,32] When ambient pressures upset the unstable equilibrium of her narcissistic psyche, in her unconscious she identifies her child with herself and herself with her own cruel, rejecting mother. She thus externalizes and projects her aggressive hostility against her child. In another manifestation of role substitution, the young insecure mother may feel jealousy and competition with her young child for her husband's love and, therefore, identify herself, in her unconscious, as her child's sibling and out of resentment, then beat him.

Some authors, such as Steele and Pollock,[31] express the opinion that only a small number of child abusers are psychopaths or sociopaths. This appears to be a matter of semantics rather than demonstrable fact. The psychological symptoms of their character disorders and their resulting behavior toward others certainly indicate aberrant personalities that reflect psychopathy.

¶ Physical Neglect

Physical neglect of a young child is a course of action that may represent an active anticathexis or a total lack of cathexis on the part of the mother toward the child. A rejecting mother, for whatever motivation, whose limited superego prevents her from battering her child may elect instead to ignore him by withholding proper nourishment, merely thrusting a bottle of milk into his mouth to silence his crying. Refraining from proper cleansing, dressing, fondling, and the offering of other comforts usually accompanies this form of neglect. A similar situation may occur in a relationship in which the mother, because of her youth, her unmarried status, and her lack of experience or instruction, is totally incompetent and unable to cope with a dependent infant.

Physical neglect is generally more apparent in the case of the infant than in that of the older child. Although there may be no obvious wounds or scars, the emaciated and often dehydrated condition of the infant and the poor state of his skin are sufficient indications of his lack of attention and nurture. The child is irritable and poorly fed, his sleep is disturbed, and he is curiously resistant to cuddling. The fact is that physical neglect of an infant or toddler may be as devastating to his body and his psyche as severe battering. Occasionally it may be as fatal.

Role of the Child

Although the adult is the perpetrator, the child is often the unwitting, or even the deliberate instigator, of the abuse. Among the factors that play a role is the child's own organic and behavioral construct. As in many other aspects of living, there are some individuals who create adversity and problem situations; others accept difficulties passively with resignation, and some cope with their problems more or less successfully.

Some children are in danger of abuse only because of the time and circumstances of their birth.[2,3] The child in a second marriage, who was born to a now divorced or deceased parent, may be only tolerated, if at all, by the stepparent and thus may be a scapegoat for punishment for any unfortunate home or family occurrence. There are children who are conceived and born without being wanted. The "accidental" pregnancy of the unwed young woman that is carried to term because of moral conviction or other reasons may be regretted later, and the child may then suffer. The infant who is had in the hopes of repairing and preserving a failing marriage rarely, if ever, serves the purpose and is, therefore, an undesirable person. It is an interesting fact that the abusing mother who has several children will almost always single out one, for actual or fancied reasons, as the subject of her maltreatment.

A child with a physical or mental disability is particularly at risk for neglect or abuse by a parent predisposed in that direction. Many normal parents accept the unfortunate burden and cope with the situation properly with love and concern. Yet, while some parents are overwhelmed with pity or guilt feelings and devote to the child more of their physical and emotional resources than is necessary, other parents project their guilt feelings onto the disabled youngster, blaming him for the problems that he presents and, consequently, mistreat him. Superstitious belief that the child's disability is the result of retribution from God for sinful transgressions by either parent, or even by the child, may provide the parent with motivation for punishing the child.

There are studies that suggest that prematurely born, low birth weight, and sick neonates are at risk for maternal neglect and later abuse.[17,18] The reason for this is considered to be the failure of bonding or attachment of the mother to her child because of the prolonged separation while the infant remains in the intensive care unit. Of course, this rejection does not always take place. It is dependent to a large extent on the personality of the mother.

After the true nature and extent of child abuse reached professional attention and then public recognition, concern became focused mainly upon the aberrant parent, with the aim of either punishment or treatment. Subsequently, psychiatrists have emphasized the collateral or even the primary role of the child as the provoker in many instances of his own abuse. In opinions reminiscent of Ivan Pavlov's and Sigmund Freud's conclusions, recent investigators believe that the child has basic inherent qualities, such as temperament. In 1927, Pavlov[23] wrote of the existence of congenitally determined types of nervous systems as basic to the course of subsequent behavioral development and not influenced by postnatal experiences. In 1937, Freud[9] stated that "each individual ego is endowed from the beginning with its peculiar dispositions and tendencies." Significant individual differences in the behavioral characteristics of infants, even in the first few weeks of life, have been noted by Gesell and Ames[12] and by Thomas and Chess.[33,34]

Temperament is defined as the way and how an individual behaves rather than what he does or why he does it. Chess and Hassibi[6] conceptualize temperament as innate and as existing in three major characterologic types, the difficult child, the "slow-to-warm-up" child, and the easy child. The first type exhibits biologic irregularity and a predominantly negative mood, making him difficult to nurture and prone to develop behavior disorders. The second type is shy, anxious, difficult to cuddle, and slow to adapt to surroundings or to situations. The easy child generally accepts nurturing, adapts well, and is least likely to be maltreated.

Of course, parents are not responsible for all behavioral deviations in their children. Yet, sometimes in considering the abuse situation, the etiological discussion may be almost as philosophical as the old dilemma of what came first, the chicken or the egg. Whatever the nature of the young child and however provocative he may be, the fact remains that it requires an action to stimulate a reaction, and the two to create an interaction. An adult with competent ego mechanisms and controls can avoid inhumane responses to a young child's provocative behavior.

There are times when the critical assault represents the culmination of an escalating negative relationship between the child and the abuse-prone parent. Conscious and subconscious mechanisms incite the youngster to retaliate against repeated punishment and emotional abuse with irritating bad behavior that incurs more abuse. The child may even exhibit counteraggression against the parent, a younger sibling, or a household pet, thereby inviting further and more severe punishment. Thus, there develops an escalating cycle of abuse-retaliation-abuse.[2]

¶ Extrinsic Influencing Factors

In addition to the interaction of individuals with each other, the ecological influences of their family attitudes and of their broader societal environment affect their behavior in the matter of what they do and why they do it. Marital incompatibilities or other difficulties often involve a clash of aggressive and passive personalities that may lead to an exaggerated dominant-submissive relationship of the parents. The hostility that the weaker male feels toward the other but is afraid to express is projected onto the child. In a situation where the husband or boyfriend deserts the home, the mother with a weak character and with no confidant may react to her frustrations by beating her child.[2]

Some fathers have backgrounds of deficient rearing and present character disorders just as do abuse-prone mothers. They too have the potential of mistreatment and do mistreat children when stress occurs while young victims are at hand. Mothers, however, who must spend entire days and nights with the infants rather than only intermittent hours, are, of course, more likely and more frequently the abusers. It should be noted, therefore, that it is mainly because of their greater proximity and not because of their greater propensity that the word mother and feminine pronouns are more often employed in the context of child abuse than the term father and masculine pronouns.

Although child abuse does occur at all socioeconomic and educational levels and in most cultural groups, the causes are more prevalent in the poor and less educated societal segments. (The effects of child abuse are more easily concealed by the more affluent.) Poverty and unemployment contribute to parent tensions. With the rent unpaid, the utilities disconnected, and food supplies at a bare subsistence level, one childish misdeed or accident, or an infant crying irritably because of the mother's tense handling, can unleash anger and cause impulsive damage for the child. For a person already endowed with a poor self-image and low self-esteem, the depersonalization that has become characteristic of contemporary beaurocratic society tends to reduce further any vestige of ego control.

Exposure to violence, either actual or fictional, has long been recognized by psychiatrists and psychologists as the instructional model and the incentive for potential criminals to commit antisocial acts. Unfortunately, the state of the world today is one of repeated, if not continuous, violence of minor or major proportions. The parent with abusive tendencies who is a loner watching fictional violence on television or reading about serious crimes in the newspaper will often be influenced to beat her child in response to some slight provocation.

Immorality and crime, alcoholism and drug addiction also, of course, play a significant role in child abuse and neglect. The mother may resort to alcohol or tranquilizers to calm her tensions and feelings of inadequacy. Under their influence she is more likely to neglect her child than to beat him. If her mate is an alcoholic or a drug addict, he may fail to give her emotional support. Occasionally when he is drunk or in need of his narcotic, he may become vicious to the point of beating the child or the mother. Because of her fear of the man, the mother projects her anger upon the child and beats him.

¶ Sexual Abuse

Within the purview of this discussion, the concept of sexual abuse will be limited to intrafamily or in-household occurrences. Violent rape by outside criminal psychopaths has other connotations, both psychiatric and legalistic. Sexual misuse, or even the legal euphemism, sexual molestation, is perhaps a more applicable term than sexual abuse for the carnal relationship between an adult and a child who is acquainted with him. The issue of sexual abuse of children is much more complicated than that of physical abuse in terms of its emotional and social aspects. In a sense it is a paradox of adult behavior. It is not impulsive, no harm is intended, and it is usually not forcibly aggressive so that rarely is physical injury involved.[4] In fact, the older girl and the adolescent may, in some cases, acquiesce willingly or may even be the seducers. For these reasons and for others that will be discussed, sexual abuse remains covert far more often than physical abuse.

Long before historical times children had been sexually misused in every conceivable fashion, heterosexually and homosexually. Early psychoanalysts persistently regarded child sexuality as mainly fantasies of incest and the Oedipus complex. Since later investigators have recognized the actual sexual abuse of children by adults, there have been contradictory opinions concerning the motivations of the sexual abusers, the roles of the child victims, and the emotional effects on the children. Bender and Grugett,[1] and Yorukoglu and Kemph[38] expressed the opinion that incest might not be emotionally traumatic or interfere appreciably with the child's emotional development and later adult sexual functioning. While Ferenczi[7] and Panucz[22] insisted that the sexually abused child is always the innocent victim of the adult perpetrator, Revitch and Weiss[27] decided that most pedophiles are harmless but that their supposed victims are usually agressive and seductive. Of course, it is a mistake to be dogmatic either way on such a complex subject. Conjectures must be supported by statistics and mature investigation, and these vary with the family, the ethnic, and the sociocultural contexts.

Sexual misuse is defined by Brant and Tisza[5] from the psychiatrist's point of view as exposure of a child to sexual stimulation inappropriate for the child's age, level of psychosexual development, and role in the family. The law in many states defines sexual misuse as a crime involving carnal knowledge, digital interference, manipulation of the genitals, masturbation, fellatio, sodomy, and indecent exposure, all structurally noted with no recognition of the emotional aspects. It has been estimated that a parent or a parent surrogate is involved in over 70 percent of cases, either by perpetration or by intentionally ignoring the act. While the father, another male relative, or the mother's lover may be the one who commits the deed, the mother may contribute by affording tacit permission, by denial, or by failure to report the offense. Although this may be difficult to believe, it does sometimes occur for one of two reasons. An uncaring mother may silently approve of her adolescent daughter's incestuous involvement in order to free herself from what she considers a burdensome sexual role. In other instances, a mother, although distressed over the occurrence, may be afraid to report her guilty husband to the authorities lest he be jailed and, consequently, deprive the family of financial support.

Incest has been generally defined as sexual intercourse between two closely related blood relatives. It has a history of taboo stretching back into antiquity. The ancient taboo was more sociological than ethical for it was condoned among royalty, though not for the common man. Anthropological explanations of reasons for the insistence on mating outside of the family vary from the desire to expand wealth and power, to the young person's need to escape from the immediate family control. It is difficult to decide when and how incest became an emotional and moral prohibition.

Currently the psychological concept and even the legal definition of incest have been broadened to include any genital contact, such as oral or anal. Rosenfeld and associates[28] note that actual coitus is rare with prepubertal girls. In nonviolent sexual misuse, the adult and the child usually behave sexually in a manner appropriate to the age of the child. Although the majority of sexual victims are girls, boys are not infrequently the objects of offense. Homosexuality with young boys as participants is not new to the world and is still practiced widely. Boys who have been molested sexually are just as traumatized emotionally and with equally serious long-range consequences as sexually misused girls.

It is a myth that sexual abuse occurs only in lower socioeconomic families. Careful investigation[35] has revealed that incest occurs at all levels of society and primarily in unbroken families. Brother and sister incest is probably the more common form in upper- and middle-class homes. Father and daughter incest is next in frequency and more common in lower-class homes. Mother and son, mother and daughter, and father and son relations are less frequent occurrences. It should be noted that no one type of incest is limited exclusively to any one class of homes. Even some professionals, clerics, and other educated, socially respected adults engage in incestuous activities.

Personality of the Offenders

Sex drive and the desire for libidinous gratification are not usually the motivations for sexual maltreatment of children. In fact many forms of child molestation, such as indecent exposure and manipulation of the child's genitals, do not offer physical satisfaction. Anyone craving nonmarital or extramarital sexual relations can almost always find a willing adult partner or a prostitute. There must, then, be some basic emotional aberration that motivates the sexual abuser of children. Brant and Tisza[5] maintain that this abnormal propensity is often common to adults who were themselves sexually abused during childhood.

Child molesters generally fit into a rather typical characterological construct. They are almost all men and usually have normal intelligence. They are neither psychotic nor do they manifest overt psychopathology. They generally have passive-aggressive personalities and strong dependency needs, which often render them ineffective in their jobs and in social relationships. Kolb[19] regards an adult's sexual interest in children as a variant of homosexuality. Sometimes pedophiles are impotent or they function on an immature psychosexual level. They expect failure or rejection in attempts at adult heterosexual relationships so they seek children as sex partners. Occasionally a man with a latent tendency toward pedophilia who has been repeatedly denied sexual gratification by his wife and perhaps has been constantly belittled by her may resort to sexual gratification with his pubertal daughter.

A particularly profligate form of sexual abuse of children is the subjecting of young girls and boys to acts of coitus, sodomy, and fellatio for the purpose of recording these in photographs or motion pictures.[4] This most unconscionable enterprise is operated for profit by individuals who are not necessarily child molesters themselves. They serve the vicarious pleasures of other individuals.

Role of the Child

Aside from the paradox of the behavior of the adult sexual molester, an equally strange incongruity is the role of the involved child who participates without reporting the occurrence, despite the fact that neither the use nor the threat of force takes place. The child usually knows and trusts the adult, believing that he can do no wrong. The man may bribe the youngster with toys or baubles and play upon her sense of loyalty to gain a promise of secrecy. The young child lacks a concept of sexuality and is, consequently, unable to make decisions concerning her or his actions. The older pubertal girl with some sexual orientation may feel flattered by the attention and excited by the stimulation, and thus be a willing if not an active partner. A mentally retarded girl may be an easy subject for sexual misuse because of her lack of judgment and comprehension.

There are some more profound psychological reasons that may motivate the young participant. Rosenfeld and associates[28] and Weitzel and associates[36] indicate that caring and warmth are more likely the desires of the girl participant rather than genital stimulation. An older girl feeling unloved by her family may accept the sexual advances of her father or another adult male in the household in her longing for affection. Her acquiescense may be abetted by libidinous Oedipus yearning. Another incitement may be a desire for revenge against a depriving or physically abusing mother. This might be further enhanced by the girl's sympathy for a weak, deprived father. The sexual misuse of a particular child is usually a repeated rather than a singular occurrence, frequently with the same offender. Mainly this is the result, not so much of the girl's enjoyment, as her feelings of guilt over her participation and her fear of punishment if this is revealed.

¶ Emotional Abuse and Neglect

Parents who abuse or neglect their children emotionally rather than physically have some abreactive mechanisms in common with those who mistreat their children physically. Having been emotionally abused themselves as children, they repeat the pattern with their own youngsters. Sometimes, adults who were battered as children do develop sufficient ego controls to prevent them from physically injuring their children, but they may still abuse or neglect them emotionally.

Emotional abuse usually takes the form of constantly belittling and denigrating a child. He or she is often singled out among the siblings as the constant object of unfavorable criticism and comparison. The parents frequently shame him or her in public. A parent or an older sibling is often presented as a role model for the child, not in a positive sense but for unsatisfactory contrast and further degradation. Frequently the youngster is threatened with punishment or denial of privileges for failure to achieve the goals set by the parent. Consequently, the child is quiet, fearful, and withdrawn in the classroom. Receiving a low grade on an examination is a frightful experience. A successful accomplishment, however, affords only relief rather than pleasure.

Emotional neglect is more often an act of omission rather than of commission. A busy, often incompetent mother, overwhelmed by household work, care of other children, and lack of help from a weak or absent mate, may neglect the young child emotionally and offer no social stimuli, while feeding and dressing him almost mechanically. There is another type of parents who neglect their children inadvertently. These are the mothers and fathers who are constantly busy attending social parties or professional meetings, and who leave the older children to their own devices. These parents may attempt, usually unsuccessfully, to compensate for their neglect by lavishing gifts or money on the children. The personality of these parents generally precludes their paying more than superficial attention to their children even when they are home with them.

Psychologists have concluded that mild to borderline mental retardation is almost always the result of severe continued deprivation of psychosocial stimuli for the developing, otherwise normal, child during the first few years of his existence. This is different in etiology from the more severe types of mental retardation that accompany genetic diseases, gestational pathology, birth injuries, and early postnatal central nervous system diseases. Mild mental retardation that is apparent in some children in the early school grades is observed mainly in the inner-city slum areas. Poor socioeconomic conditions, low educational levels, and ethnic alienation may tend to encourage emotional neglect and are thus more likely to forestall proper psychosocial stimulus input for youngsters than a better socioeconomic and a better educational environment.

Psychosocial Dwarfism

A most interesting phenomenon that demonstrates the intricate interrelationship of emotional stress and physiological function has been termed psychosocial dwarfism.[29] Severe emotional deprivation and rejection of children in the first two or three years of life has been observed to result, in some extreme cases, in marked stunting of linear growth and retardation of bone age. The children also show severe wasting of body tissues in spite of voracious appetites and abnormal eating habits, accompanied by frequent vomiting. Their behavior is further characterized by depression, temper tantrums, and poor toilet habits. At this stage, psychological testing indicates their average IQ to be less than 90. If these children are relocated in a favorable, attentive home environment, there will usually be rapid improvement in the form of accelerated linear bone growth, increased height and weight, and advancement of their IQs. Their behavior disorders may improve more slowly. Progress in this area is dependent upon the severity and duration of the aberrations, the youngsters' ages at the time of removal to their new homes, and the kind of nurture that they receive. These facts present strong evidence that psychosocial dwarfism and its accompanying behavioral deviations are the result of environmental rather than intrinsic factors.

The mechanism of this phenomenon involves a complicated neurophysiological and hormonal interaction. The hypothalamus mediates emotional reactions and also affects the function of the pituitary gland that secretes growth hormone and regulates thyroid gland action. Severe emotional stress for the infant or toddler causes the hypothalamus to suppress growth hormone production from the pituitary gland. The suppressed pituitary gland then affects thyroid hormone release and, consequently, the metabolism in body tissues. If the emotional environment of the distressed youngster is improved early and sufficiently, he is able to regain normal homeostasis.

¶ Long-Range Effects of Abuse

The immediate effects of child abuse are usually fairly obvious. The physical damage is visible. The behavior disorders indicate the presence of emotional disturbance. Injuries that have plainly been inflicted, such as contusions, hematomas, lacerations, and fractures, in areas and in appearance that belie accidental occurrence, unlikely accounts of the incident, and parental attitudes should all lead to the consideration of the possibility of child abuse. It is important to educate and to sensitize professionals, day-care personnel, teachers, and others who are frequently in contact with children to be highly suspicious of any unusual appearance in a child.

Unless the physical injury has produced an irreparable deformity or permanent neurological damage, the bodily wounds will heal. It is the emotional trauma and the resulting personality impairment that cannot be cured with medications. Aside from the immediate treatment of an abused child and hopefully the prevention of recurrent episodes, the more important considerations are the long-range emotional effects. These have implications for the individual in terms of lasting character disorders, and for the society in which he will live.

As a preschooler, the abused child's social contacts outside the home may be limited, if there are any at all. His first actual experience with social interaction occurs with the advent of school attendance. The abused child often adjusts poorly in school. Depending on his home background, he may be either fearful, withdrawn, inattentive, and an underachiever, or a negativistic aggressive disrupter of the class. He may act out his home frustrations by fighting. With a sadomasochistic attitude, he may deliberately misbehave in order to invite punishment. He is labeled a bad child and continues to play the role.

This type of school misfit is often set on a course for the next phase of deviance as he progresses to upper grades. Truancy is more often a deliberate, spiteful, and aggressive act than one of apathy or discouragement. It is frequently a projection of the child's resentment of the rejecting or abusing parent onto authority that at this age and stage of his development is represented by the teacher or the principal. Truancy is but a short step away from school dropout as the individual advances to high school grades and the adolescent years. Apart from other implications, leaving school before graduation greatly minimizes the youth's future chances of suitable employment.

Adolescence is ordinarily a period of transition that arouses doubts and feelings of insecurity for a normal individual facing a life of adult responsibilities and self-reliance and leaving behind the comfort and security of a parent protected childhood. Separation from parental controls, seeking one's own identity while identifying with a peer group, strengthening one's ego, and furthering a superego are all parts of adolescence. This is the time that a youth with personality deficits and a character disorder as the result of having been an abused child may become an antisocial being and a burden for society. The less serious acts of juvenile delinquency, such as truancy and dropping out of school, may advance to more serious juvenile crimes ranging from misdemeanors of property damage and petty thefts to felonies of mugging, sex crimes, arson, and murder.

Life in the stressful environment of an inner-city slum area is arduous and often brutal. Many youngsters from loving, intact families in these neighborhoods have proper guidance and good example from earnest parents. These boys and girls are imbued with benign aggression that enables them to resist destructive peer pressures and to maintain their determination to be good. On the other hand, teenagers who have a home background of abuse and poor parenting lack self-esteem, have a poor self-image, and succumb easily to evil peer pressures. In an attempt to establish their identities, these youths join the indigenous street gangs. The gang leaders are usually power-wielding sociopaths with no social conscience. The members are mainly adolescents, who seek approval and gratification from their leaders and associates by engaging in antisocial exploits. Their passive-aggressive personalities, under stress, convert benign aggression into the malignant aggression of criminal behavior.

Alcohol and drug abuse have penetrated into the younger age groups. It is not unusual to see ten-year-old boys and girls, who had started smoking cigarettes, now smoking marijuana, drinking wine and beer, and even experimenting in the use of some drugs. Some teenagers use narcotics and strong sedatives, stimulants, and hallucinogens. The motivations for these decadent activities are peer example or peer pressure, being able to boast of one's prowess, and, sadly, the need to escape from pressures of home and street. A most distressing fact is the recent large increase in suicides among youths, not only as accidents of drug overdosages but premeditatedly as a result of their awareness of the futility of their existence and of their bleak and hopeless future.

Running away from home usually represents a rebellion against overstrict, demanding parents. It may also be an escape from a seriously disturbed family environment. Runaways do not, as a rule, become involved in antisocial acts and are often depressed loners. Girls are more likely than boys to become runaways. Some leave home to escape from incestuous attempts or actual molestation by their fathers.

The complexity of the world and the disenchantment of many persons with the "establishment," in society, government, and religion, have produced a flourishing of cults. Most of these are not religious in the sense of worshipping a deity, nor do they embrace a system of ethics and morals. They offer a life style of group identity, social contact, and simple doctrines of set routines. This attracts the nonviolent youths from deficient family backgrounds who have poor self-image and who are seeking ego identity. They are the disillusioned dropouts from school, from home, and from society. Interestingly, these youths who resent and defy authority otherwise now submit to the authoritarianism of the cult leaders, who demand constant reaffirmation of their commitment. The reason for this compliance is probably that the leaders condemn the society and the family structure from which the young persons have escaped. Far from improving their self-image and strengthening their ego structure as a step towards reentering and coping with society, these young men and women totally submerge their individuality in an unmotivated mass of communal clones.

Sexual abuse of children may have serious and persisting effects on the victims.[4] Molesting of young children may produce disturbances of feeding and sleep, irritability, enuresis, and phobias. In the school-age child, sexual misuse may cause the interruption of normal development of the concept of sexuality. Adolescent boys who are subjected to homosexual acts may become homosexuals or may develop psychological impotence. As previously noted, some adolescent girls who participate willingly in incestuous sexual intercourse may suffer no emotional trauma and no later sexual inhibitions. Others, however, and especially those girls who were coerced, are very apt to develop emotional and personality problems. Their anxiety and guilt feelings about their participation may drive them to a masochistic continuation of promiscuous sexual experiences in an attempt to master the earlier emotional trauma. Those who decide to avoid further illicit sexual activity may, in some cases, develop psychosomatic gynecologic disturbances, such as dysmenorrhea and amenorrhea. Later, they may suffer frigidity and dyspareunia that will create marital difficulties.

Teenage pregnancies have increased markedly and present problems for the girls' families, the medical profession, the legislature, and society in general. The subject of abortion with its medical, legal, and ethical implications is a thesis in itself. The option of the unwed gravida to carry the pregnancy to term and to deliver the baby brings an infant into the world who may be at risk for being neglected or abused. Sometimes the pregnant girl is persuaded or even forced to marry the father of her child or another man chosen by her family, in order to avoid social disgrace. Needless to say, the child born under these circumstances is also a prime subject for neglect or abuse.

¶ Management and Treatment

Child abuse and neglect cannot be prevented or cured by legislative fiat or by medical measures. Their occurrence can be lessened and the effects can be mitigated by the proper and timely identification of situations, by preventive measures, and by a multidisciplinary approach to the management of the subject. Pediatrician, psychiatrist, social worker, and often the court judge must all work cooperatively toward a resolution of the problem and the rehabilitation of the family in crisis.

The first step in management is recognition of the situation. While sometimes police are summoned to the scene of violent abuse by a neighbor who is alerted by the screaming, most abused children are brought to medical attention, usually to the hospital emergency room, by the abusing mother herself. Often she will lie or conceal the truth concerning the injury. Thus, the professionals must be educated to recognize cases of abuse. Reporting of cases to official agencies, as mandated by law, is not so much intended for punishment of the offender as for investigation of the circumstances with a view to helping the child and the parent.

Hospitalization of the injured child may be necessary for medical or surgical treatment. Furthermore, until all facts are investigated and the situation is under supervision, the child must be removed from the home for his safety. Preferably, separation from the parents should be only a temporary measure if rehabilitation of the family is at all possible. For the young child, his mother is the person with whom he is most familiar. Even though she has hurt him on occasions, she has always been with him and may have provided some nurture. As confused and as emotionally hurt as he is by her actions, prolonged separation from her can have more serious effects on him than the physical damage. Of course, in those situations where the mother and the circumstances of the home are judged to be disastrous and uncorrectable, a suitable foster home becomes the only alternative.

When the abusing mother of a young infant appears to be contrite and yet incapable of proper nurturing because of her own deficient background, there is an effective method of treatment. The mother and the child may be admitted to a live-in facility where both she and her baby can be nurtured by trained personnel. Now she will learn how to be a parent by being parented herself. This method has proven successful in more than 60 percent of cases where it has been applied.

A toddler or a preschool child is unable to respond to direct psychotherapy. However, his management in a hospital, a residential institution, or a foster home might well be directed by a child psychiatrist. An abused school-age child can benefit from psychotherapy that is instituted to help him resolve his emotional dilemmas and resentments before a developing character disorder becomes firmly entrenched.

A child who has been sexually abused must be treated according to her or his age and the nature of the physical acts. The young child who has been subjected to acts other than coitus or sodomy may be helped sufficiently by simple explanations and discussions of sex commensurate with her or his age, accompanied by psychotherapy to overcome fears or other negative emotions. For the older girl who has been involved in sexual intercourse, therapy may uncover a number of conflicts in her psyche including fear, guilt feelings, and sexual repugnance. Intense, prolonged therapy will be required in such cases. A similar situation may hold true for the adolescent boy who has been subjected to homosexual practices. The sooner psychotherapy is instituted for sexually abused children and adolescents, the better will be the prognosis for normal psychosexual development and for normal adult sexual adjustment.

Psychotherapy for the physically abusing mother is not a simple procedure because a number of factors must be considered.[3] The psychiatrist must remain objective lest any personal emotions concerning child abuse create within him a negative countertransference toward his patient. Therapy must focus primarily on the parent's intrapsychic conflicts and not on the parent-child interaction. If the latter is emphasized too early in therapy, a narcissistic parent will become resentful and feel competitive with the child for the therapist's attention. This can create for the mother a negative transference toward the psychiatrist. Because of the early origins and the extent of the subconscious mechanisms behind the patient's character disorder, psychoanalytically oriented psychotherapy would seem to be an effective course of treatment.

As important as treatment is for the child and the abusing parent, if a husband is in the picture, he must also be involved in treatment if the situation is to improve. He may be a dominant, aggressive individual who has been stressing his wife to the point of causing her to vent her anger on the child. He may be a weak person who lacks purpose, a sense of self-worth, and the capability of furnishing his wife with the emotional support that her deficient personality requires. Both husband and wife must, therefore, be treated in order for them to understand their own and their mutual problems and to accept each other. This can contribute a good deal toward rehabilitating the family and helping them to learn better child-rearing practices.

Many abusing parents have never been able to relate to or to communicate with authoritative figures whom they identify with their own feared and hated parents. For these persons, an individual therapeutic relationship is impossible at the outset. An alternative course for them is the so-called parents anonymous groups. Here group sessions are held at regular intervals under professional supervision. At such sessions, the parents can confess, ventilate, and share their problems with peers who have similar troubles and anxieties. Many parents experience considerable relief from their tensions and continue with this form of group therapy. Others will proceed to individual psychotherapy to achieve a more complete resolution of their personality difficulties.

One place where a parent may be recognized early as having a possible potential for abusing or neglecting her newborn infant is the hospital maternity division.[13] A new mother or father may exhibit rejecting behavior or distress concerning the infant. Trained hospital nurses can recognize and report this to the physician. A social worker may uncover emotional, social, or economic factors that are troubling the parent, and proper support and/or therapeutic measures can be instituted. Early intervention and assistance can lead to resolution of the problem and develop a satisfactory family situation.

Some adjunctive helpful measures to calm the abuser and to assist her in readjusting her life are day-care centers for the infants and toddlers and nursery schools for the preschool children. These can offer a mother hours of relief and an opportunity for employment with consequent greater economic sufficiency and a feeling of increased self-worth. Homemaker assistants can help to train the incompetent or poorly coping mother. Altogether though, the entire program of rehabilitation must be coordinated and long-term follow-up must be maintained by social agencies.

Studies in various areas of the country have shown that, with proper programs geared to help rather than to punish, a majority of abusing parents can be rehabilitated and their critical family situations can be improved. Of course, a case of serious or permanent damage to a child, or the murder of a youngster by an adult, must be adjudicated by the court and punished as a felony. Otherwise, if circumstances permit, the emphasis must be on restoration. A rehabilitated mother can best rear her own child on a course of normal personality development.

¶ Bibliography

1. BENDER, L., and GRUGETT, A. E. "A Follow-up Report on Children Who Had Atypical Sexual Experience," *American Journal of Orthopsychiatry,* 22 (1952): 825–837.
2. BLUMBERG, M. L. "Psychopathology of the Abusing Parent," *American Journal of Psychotherapy,* 28 (1974): 21–29.
3. ———. "Treatment of the Abused Child and the Child Abuser," *American Journal of Psychotherapy,* 31 (1977): 204–215.
4. ———. "Child Sexual Abuse: Ultimate in Maltreatment Syndrome," *New York State Journal of Medicine,* 78 (1978): 612–616.
5. BRANT, R. S., and TISZA, V. B. "The Sexually Misused Child," *American Journal of Orthopsychiatry,* 47 (1977): 80–90.
6. CHESS, S., and HASSIBI, M. *Principles and Practice of Child Psychiatry.* New York: Plenum Press, 1978.
7. FERENCZI, S. "Confusion of Tongues Between Adults and the Child," *International Journal of Psychoanalysis,* 30 (1949): 225–230.

8. FONTANA, V. J., and BESHAROV, D. J. *The Maltreated Child,* 4th ed. Springfield, Ill.: Charles C Thomas, 1978.
9. FREUD, S. "Analysis, Terminable and Interminable," in Strachey, J., ed., *Collected Papers,* vol. 5. London: Hogarth Press, 1950.
10. FREUD, S. GESAMMELTE WERKE, vol. 12. London: Imago, 1952.
11. FRIEDRICH, W. N., and BORISKIN, J. A. "The Role of the Child in Abuse: A Review of the Literature," *American Journal of Orthopsychiatry,* 46 (1976): 580–590.
12. GESELL, A., and AMES, L. B. "Early Evidences of Individuality in the Human Infant," *Journal of General Psychology,* 47 (1937): 339.
13. GREEN, A. H. "A Psychodynamic Approach to the Study and Treatment of Child-Abusing Parents," *Journal of the American Academy of Child Psychiatry,* 15 (1976): 414–429.
14. HELFER, R. E. "Early Identification and Prevention of Unusual Child Rearing Practices," *Pediatric Annals,* 5 (1976): 91–105.
15. HUNTER, R. S., et al. "Antecedents of Child Abuse and Neglect in Premature Infants: A Prospective Study in a Newborn Intensive Care Unit," *Pediatrics,* 61 (1978): 629–635.
16. KEMPE, C. H., and HELFER, R. E., eds., *Helping the Battered Child and His Family.* Philadelphia: J. B. Lippincott, 1972.
17. KLAUS, M. H., and KENNEL, J. H. *Maternal-Infant Bonding.* St. Louis: C.V. Mosby Co., 1976.
18. KLEIN, M., and STERN, L. "Low Birth Weight and the Battered Child Syndrome," *American Journal of Diseases of Children,* 122 (1971):15–18.
19. KOLB, L. C. *Clinical Psychiatry,* 9th ed. Philadelphia: W.B. Saunders, 1977, p. 62.
20. MORRIS, M. G., and GOULD, R. W. *Role Reversal: A Concept in Dealing With the Neglected/Battered Child Syndrome.* New York: Child Welfare League of America, 1963, pp. 29–49.
21. ———. "Role Reversal: A Necessary Concept in Dealing with the Battered Child Syndrome," *American Journal of Orthopsychiatry,* 33 (1963): 298–299.
22. PANUCZ, A. "The Concept of Adult Libido and Lear Complex," *American Journal of Psychotherapy,* 5 (1951): 187–195.
23. PAVLOV, I. P. *Conditioned Reflexes: An Investigation of the Physiological Activity of the Cerebral Cortex.* London: Oxford University Press, 1927.
24. PETERS, J. J. "Children Who Are Victims of Sexual Assault and the Psychology of Offenders," *American Journal of Psychotherapy,* 30 (1976): 398–421.
25. PETERSON, G. H., and MEHL, L. E. "Some Determinants of Maternal Attachment," *American Journal of Psychiatry,* 135 (1978): 1168–1173.
26. POLLOCK, C., and STEELE, B. "A Therapeutic Approach to Parents," in Kempe, C. H., and Helfer, R. E., eds., *Helping the Battered Child and His Family.* Philadelphia: J. B. Lippincott, 1972, pp. 3–21.
27. REVITCH, F., and WEISS, R. "The Pedophiliac Offender," *Diseases of the Nervous System,* 23 (1962): 73–78.
28. ROSENFELD, A. A., et al. "Incest and Sexual Abuse of Children," *Journal of the American Academy of Child Psychiatry,* 16 (1977): 327–339.
29. RUDOLPH, A. M., BARNETT, H. L., and EINHORN, A. H. *Pediatrics,* 16th ed. New York: Appleton-Century-Crofts, 1977, pp. 1611–1612.
30. SMOLLER, B., and LEWIS, A. B., JR. "A Psychological Theory of Child Abuse," *Psychiatric Quarterly,* 49 (1977): 38–44.
31. STEELE, B., and POLLOCK, C. "A Psychiatric Study of Parents Who Abuse Infants and Small Children," in Helfer R. E. and Kempe, C. H., eds., *The Battered Child.* Chicago: University of Chicago Press, 1974.
32. TERR, L. C. "A Family Study of Child Abuse," *American Journal of Psychiatry,* 127 (1970): 665–671.
33. THOMAS, A., and CHESS, S. *Temperament and Development.* New York: Brunner/Mazel, 1977.
34. ———, and BIRCH H. *Temperament and Behavior Disorders in Children.* New York: New York University Press, 1968.
35. WEEKS, R. B. "The Sexually Exploited Child," *Southern Medical Journal,* 69 (1976): 848–852.
36. WEITZEL, W. D., POWELL, B. J., and PENICK, E. C. "Clinical Management of Father-Daughter Incest," *American Journal of Diseases of Children,* 132 (1978): 127–130.

37. WERTHAM, F. "Battered Children and Baffled Adults," *Bulletin of the New York Academy of Medicine,* 48 (1972): 887–898.
38. YORUKOGLU, A., and KEMPH, J. P. "Children Not Severely Damaged by Incest with a Parent," *Journal of the American Academy of Child Psychiatry,* 5 (1966): 111–124.

CHAPTER 9

SPECIAL SYMPTOMS OF CHILDHOOD

Mary C. MacKay

THE SIX SYMPTOMS covered in this chapter—enuresis, encopresis, tics, stuttering, thumb-sucking, and nail-biting—are special in the sense that each may occur as an isolated symptom, not necessarily part of any syndrome. When first diagnosed as a symptom, the response occurs at a specific stage of development and is the result of stress related to that particular stage. Beginning as a bodily response to some inner and/or outer conflict, it may eventually become a habit with its own functional autonomy.

There are other special symptoms of children, including fire setting, hair pulling, and various eating and sleeping disorders. These particular six were chosen, however, not because they are any more troublesome than others, but because they are more likely to occur as isolated symptoms and to be treatable as such.

Since these symptoms are all developmentally determined, they may cluster together in various combinations to produce age-related syndromes which are then more difficult to treat. Since they may occur as part of normal development, it is important to make an early differential diagnosis, to treat only if necessary, and then in the least obtrusive manner.

There is controversy as to the exact etiology of these symptoms, with the spectrum ranging from simple maturational factors through elaborate psychoanalytic constructs. Most psychiatrists today, including this author, are of the opinion that, for the most part, they are multifaceted in origin.

¶ Enuresis

Diagnoses

DEFINITION

In the United States, if a child continues to suffer urinary incontinence diurnally past the age of three years or nocturnally past the age of five years, he is usually considered to be enuretic. Dividing the enuretics into primary (those who have never stopped wetting) and secondary (those who have stopped wetting and started again after at least a six-month period of dryness) has proved to be a necessary division when attempting to diagnose this symptom. It is normal for a child to be incontinent on occasions during the transition from first having an autonomous bladder to the developing of regular urinary control.

PREVALENCE

Enuresis is found in all parts ofthe world, but the prevalence figures for specific ages are different in various countries, probably due to the difference in child-rearing practices. For example, Sweden reports only 8 percent of the children with enuresis at five years, while for the United States, overall estimates range from 10 to 15 percent at this age, with a drop to 5 percent at eleven years. In general, enuresis occurs more frequently in boys than in girls. At ages five or six, however, there is little sex difference in the prevalence figures, while at age eleven there are twice as many boy enuretics as there are girls. The ratio of primary to secondary enuretics is about four to one.

EVALUATION

When a psychiatrist begins treatment of an enuretic child, he must be sure that the child has had a thorough physical examination, a urinalysis, and, particularly in the case of girls, a urine culture. An adequate medical history should be taken, including the type, frequency, and place of the wetting. A family history should be obtained, in which all previous efforts to cope with the wetting should be explored. It is important to rule out any nonfunctional reason for the urinary incontinence, such as obstructive uropathy, infection, foreign body, sickle-cell anemia, or diabetes insipidus. About 10 percent of the children who suffer urinary incontinence do so because of organic reasons.

Etiology

NEUROPHYSIOLOGICAL

From a neurophysiological point of view, there is still much controversy about the etiology of enuresis. Although Hallgren[22] considered it to be an inborn developmental deviation, which was, for the most part, genetically determined, Esman[15] claims that enuresis is a maturational disorder, associated with the child's pathological awakening from Stage 4 sleep.

Much information has recently been provided by sleep researchers on the possible connection between nocturnal enuresis and sleep disturbances. Most researchers feel that children who are deep sleepers and consequently in Stage 4 for longer periods of time will be unaware of the messages coming from their full bladders.

Esman[15] has written about the maturational lag theory—as to the cause of enuresis in children. These lags show up as immature functioning in other areas, from reading disabilities to delayed dentition. Among them, and often picked as a prime suspect for the cause of the enuresis, is an immature bladder with less than age-appropriate capacity. This belief has to be weighed carefully against the findings of those who study bladder capacity and who have demonstrated that an improvement in bladder functioning alone does not correct bed-wetting. In addition, Dische[13] found that children who wet by night and day are more likely to be disturbed emotionally than those who wet by night alone, which suggests that the cause of bed-wetting is more than simply neurophysiologically determined.

INTERPERSONAL

Continuing conflicts in the home, in school, and with peers, resulting in depression, regression, or aggression, have in the past been considered by many psychiatrists to be an important cause of primary enuresis. More recently, however, these conflicts have been thought to be the result of the wetting rather than the cause of it.

INTRAPSYCHIC

In a study of seventy-two enuretics, Gerard[18] found that most of the 90 percent left after organicity had been ruled out suffered from oedipal level conflicts. The personality traits exhibited were aggressive or passive-aggressive in nature, and the conflicts revolved around sibling rivalry or strong identification with the parent of the opposite sex.

Both Katan[28] and Sperling[48] have described the fantasies of genital damage expressed by their enuretic patients. Katan[28] reported that frequently a traumatic experience had preceded the appearance of the symptom. She listed the nature of the traumas as follows: (1) separation from a beloved person who had trained the child; (2) jealousy aroused by the birth of a sibling; (3) the discovery of the differences between the sexes by viewing another child; or (4) viewing an adult; and (5) an operation that was experienced as a castration.

Treatment

BIOLOGICAL THERAPY

Because of the large developmental factor in the etiology of enuresis, most children eventually stop bed-wetting spontaneously, with a 12 to 15 percent remission rate yearly. In the meantime, however, since the incidence of secondary emotional and behavioral problems is so high, Dische[2] considers it most beneficial to eliminate the symptom when possible.

The tricyclic antidepressant, imipramine, is the drug most commonly used in treating enuresis. While 80 percent of the children reduced their wetting and 50 percent had achieved total dryness when placed on imipramine, Shaffer and associates[46] reported that the drug started to help immediately but that the wetting resumed when the drug was discontinued. Imipramine can occasionally have the worrisome side effect of postural hypertension and of possible heart block in patients with preexisting cardiac condition defects. In an effort to discover if the tricyclic worked for enuretics by its alpha-adrenolytic action on the neurotransmitters, Shaffer and associates[46] gave the nontricyclic, indoramin, (which mimics the tricyclics in the above effect on the neurotransmitters) to fourteen enuretic children (aged four to twelve years), but the drug had no significant effect on the bed-wetting.

Ritvo and associates,[43] in studing imipramine's effect on the sleeping patterns of enuretics, found that the wetting occurred in all stages of sleep, rather than only in Stage 4. They postulated that the imipramine reduced the amount of REM sleep. Kales and associates[26] concluded that imipramine's effect was more related to the time of night than to the stage of sleep. They felt that when the child was in the deepest sleep early in the night, imipramine reduced the bladder's excitation and/or increased the child's bladder capacity. (Later in the night, in light sleep, the child presumably is more responsive to the bladder's signals.)

BEHAVIORAL THERAPY

A conditioning approach in the treatment of enuresis was first used by Mowrer and Mowrer[37] in 1938. However, it is only recently that the positive effects of a mechanical device have overcome the reluctance of pediatricians to use it with children who, the pediatricians felt, might be psychologically traumatized by it. In 1960, Eysenck[16] wrote that since enuresis represented a failure of the child to learn to gain urinary continence, a conditioning device should work. He felt that symptom substitution could not follow since the enuresis, as a symptom by itself, had no underlying conflict. By 1966, Werry[51] reported that in studies involving more than 1,000 children a 75 percent success rate with the bell-pad mechanical device was achieved with no increase of psychiatric problems. He advised that the bell-pad device be considered a possible treatment for any enuretic child of six years or above who is disturbed by his symptom and who has failed to respond to simpler measures. Presently quoted figures for the bell-pad device are usually 60 to 75 percent success with about a 25 percent relapse rate. A relapse simply leads to another period of treatment, which usually meets with success. Some earlier problems with conditioning device treatment have been resolved, as the bell-pad machines are now easier to get and safer to use. If they are used as the only treatment, they may still present problems since the noise created by the machine often wakes the entire household. The child may allow the noise to continue as a means of getting the attention he unconsciously seeks, in spite of the nuisance to himself. Usually this secondary gain can be avoided by using some other type of behavioral modification as well, such as offering the child a reward if he achieves dryness.

The assumptionis that the bladder of an enuretic child is smaller than the age-appropriate size. Yet in a study of eighteen enuretic children and controls, Harris and Purohit[24] proved this assumption unfounded. They conducted bladder training with the enuretics for thirty-five days. The children drank water, practiced holding, and were rewarded for retaining progressively larger volumes of fluid before voiding. Although the bladders did become larger, the wetting didn't decrease. Some authorities claim, however, that because it is urodynamically a rational idea and puts the responsibility on the child, bladder training is worth trying if other methods of treatment have not worked. Controlled breathing exercises have also been suggested as a means of helping the child hold the urine longer.

Brazelton[9] outlined a child-oriented approach to bladder training that may begin when the child is around two years of age, and that is geared to each child's developmental capacities. Of 1,170 children trained in this manner, he found that only 1.5 percent were wetting by five years of age.

Although waking a child to prevent wetting at night was a favorite method of behavioral therapy in the past, it is felt to have many drawbacks. It is often difficult to waken the child since he is in deep sleep, and also, it puts the responsibility on the one who wakens him, rather than on the child.

Behavioral therapy will only be successful when the child is ready to be trained and only if the parents have the motivation and ability to help in the training. If there is more than one nocturnal wetter in the family, it is best to start with the oldest first, as there is a possibility that the others may identify with his success and stop without further help. The reward system usually works better than the punitive one. Since the reward system most often used is the operant conditioning model, with success or failure so dependent on the person (usually the mother) setting up the guidelines and consistently following through on them, behavioral therapy by itself is seldom totally successful. With the adjunctive use either of medication, or of a conditioning device, it has a better chance to succeed.

Psychotherapy

Although only about 2 percent of enuretics (the secondary type) will need psychotherapy for either underlying intrapsychic or interpersonal conflicts, it is very important for them to receive it. Since all these children will have had at least a six-month period of dryness, they may resume wetting by day as well as by night. Because of the nature of the conflicts usually involved, it is best initially to see the mother and child together. Later it may be necessary to see the child alone in either psychoanalytically-orientated or supportive psychotherapy, according to the severity of the conflicts.

At times, the primary enuretic will also need some supportive psychotherapy for emotional problems related to the enuresis, such as a poor self-image and a disturbed relationship with family and peers. The type of therapy used here should be most often supportive in nature.

Hypnotherapy has been reported by Olness and Gardner[38] to be a successful type of treatment, but controlled studies of its efficacy in comparison to other existing treatment modalities have yet to be performed. The therapist's thorough familiarity with the techniques of hypnosis is clearly required if this modality is to be used either independently or in combination with other methods of treatment.

Prognosis

Most primary enuretics will usually show satisfactory progress without treatment if they have no associated neurophysiological immaturities. Many do relatively well with environmental supports alone even when associated immaturities, such as learning disabilities or poor motor skills, are present.

When the enuresis is associated with the more severe immaturities, such as encopresis and stuttering, the child will usually develop psychiatric symptoms of a characterological nature, if not given adequate treatment.

Most secondary enuretics who receive no treatment will develop psychiatric problems of a neurotic nature.

¶ Encopresis

Diagnoses

Definition

If a child continues to have fecal soiling past the age of two and one-half to three years, and if a trial of toilet training has been unsuccessful, he is usually considered to be a primary encopretic. If he reverts to soiling again after a six-month period of fecal continence, he is considered to be a secondary encopretic.

Prevalence

Encopresis occurs in 1.2 to 1.5 percent of children, at a ratio of about five boys to each girl. About one-half of the encopretics are of the primary type and one-half are of the secondary type. According to Bellman,[5] encopresis, in general, declines spontaneously at a rate of 28 percent per year and has almost disappeared by sixteen years.

Evaluation

Richmond and associates[42] noted that since the constipation that often accompanies the soiling may be associated with a psychogenic megacolon, Hirschsprungs disease, as well as other organic problems, must be ruled out. A physical examination should be performed whenever the diagnosis is in question. Fecal soiling is also seen in the child with mental retardation or infantile autism, and as such, the soiling should be understood in a different manner diagnostically than in the child without these diagnoses. It is essential to get an adequate psychiatric history at the first meeting with the encopretic child and his parents. Whether the soiling is continuous from infancy or whether it is contiguous with some important event in the child's life are factors to be explored, as well as where and in what manner it occurs. The relationship between the parents and the child and between the child and his siblings and peers is also vital information. Although enuresis may often occur in the child who is encopretic, encopresis as a symptom is more troublesome than enuresis and, in turn, is usually indicative of a more troubled child. Psychological testing can be helpful in delineating associated psychopathology.

Etiology

Neurophysiological

Since many children who are encopretic have soft neurological signs, language disorders, poor coordination, and other stigmata associated with the syndrome of minimal cerebral dysfunction, it is logical that one of the first special assessments made of the the child should be along neurodevelopmental lines. Bellman[5] reported that although there is a 15 percent incidence of childhood encopresis in the fathers of encopretics, genetics do not appear to be involved in the etiology of it. Among the many neurophysiological immaturities often found in the encopretic child is an immaturity of the musculature of the bowel that probably interferes with toilet training. Nevertheless, many children who have immature musculature of the bowel and who continue soiling past the usual age of achieving bowel autonomy do not go on to become full-blown encopretics. Therefore, further causative factors must be suspected.

INTERPERSONAL

A mother who is adequate in other areas may have conflict over toilet training her child. When she tries to help the child who appears to be having a difficult time achieving bowel control, she may become engaged in a power struggle involving the child's conflict as to who or what has control over his bowel functioning—himself, his mother, or his bowels. Since this is the first instance of the child's giving up internal function to satisfy parental love and demands, he presumably finds it developmentally difficult. Prugh[41] found that most encopretics had a history of early and often harsh toilet training. Bemporad[6] believes it takes a particular kind of mother to produce a chronic primary or secondary encopretic. The mother who is unable to extricate herself from a power struggle is described by him as being erratic, emotionally inappropriate, and distant. Added to this, the father is often himself troubled, depressed, and sometimes schizoid. He is also frequently away from home. Being without the support of the father, both child and mother stay locked in a power struggle, in which both can only lose. On the other hand, a child's too close identification with the father may also cause encopresis as a result of an oedipal conflict regression to an anal struggle. Because the youngster eventually gains some gratification in winning over his mother or his father, he may fail to notice the effect his encopresis has on siblings, peers, and teachers. When, and if, he finally recognizes it, he may feel helpless and eventually become solitary and depressed. Anthony[2] noted that the child's fears at school, in sports, and in social situations may precipitate one stress after another. Eventually his encopresis may be used as a means of keeping people away from him, so that his maladaptive mode of interacting with his mother also becomes characteristic of his relations with others.

INTRAPSYCHIC

In distinguishing between primary and secondary encopresis, Easson[14] wrote of the secondary type as being of psychogenic origin. He suggested that some of the precipitating emotional stresses may include the birth of a sibling, the loss of a parent, anxieties over sexual feelings, poor peer relations, or scholastic difficulties. Because of the conflicts aroused by these stresses, the child's growing independence is threatened so that he may regress to an earlier and more dependent type of behavior, which will include fecal soiling.

Treatment

BIOLOGICAL THERAPY

When first seen, many encopretics are constipated and may even be obstipated. Hence, the first task is to clear the bowel and then to help the child achieve and maintain a normal functioning bowel until, with the help of other forms of therapy, he can regulate himself. Halpern[23] suggests the use of a laxative suppository only, while Hein[25] uses stool softeners and enemas as well. Hein, a pediatrician, discovered that the amount of help produced by the child's success in having the fecal soiling under control, even if it wasn't completely his own control, is worth the effort put into the early physiological engineering. Because of the child's need to escape the long ongoing power struggle with his mother over control of his bowels, it is best to limit the amount of this type of engineering to the least possible amount that will achieve success in the least obtrusive manner.

BEHAVIORAL THERAPY

Since the encopretic child has not learned to respond to rectal cues, and either holds back and eventually becomes obstipated with leakage occurring or gives the feces up in an inappropriate place, he might be considered for behavioral training as soon as undue resistance to toilet training begins. Since so many encopretic children are late developers, however, it is important to distinguish between the children who are difficult to toilet train because of immaturity and those who are resisting for other reasons. If the former are forced to participate in a training program before they are ready, they may well become encopretics. Behavioral therapy should be attempted once the child has been declared by the pediatrician to be physiologically able to be toilet trained.

Wright and Walker[52] claimed 100 percent success with 100 children over a five-year period, with an average duration of treatment of about four months. Once the behavioral program was set up and explained to the child and the mother, the physician remained remote except for letters or phone calls. Carefully monitored charts and rewards were used.

PSYCHOTHERAPY

In a child who has been conflicted over a long period of time as a result of fixation or regression, neurotic defenses will not necessarily disappear as a result of the symptom being eliminated. It is therefore helpful to provide supportive psychotherapy if associated psychopathology warrants. If the encopresis is treated before the child has developed ingrained neurotic defenses, formal psychotherapy may not be necessary after the symptom has cleared.

The most important and often the most difficult part of the treatment plan is the counseling of the parents. Because the parents are usually ashamed of the symptom, the child will often not be brought for help until teachers or relatives complain. By this time, there is often much hostility between the parents and the child and between the parents themselves. Because of their own personality problems and/or because of their conflicts over trying to manage the child's encopresis, it may be that each parent should be seen separately at first. Eventually they must be seen together, as their approach to the child and his problem must be consistent.

Because of the encopretic child's frequently associated immaturities, it may be necessary to help him achieve better academically, motorically, and socially. This will often involve special remedial help for him, as well as counseling by the school. Once the actual fecal soiling is stopped, however, he will be more acceptable to school personnel as well as to peers. In turn, this may help build his ego enough so that he will become more responsive to other types of therapy.

Bemporad[6] noted that fathers were often more difficult to engage in therapy than the mothers. In some cases, this may possibly be due to the presence of childhood encopresis in the history of the fathers, as noted previously. In many situations, it may be necessary to involve parents in therapy for their own emotional problems.

Prognosis

If the primary encopretic has a supportive environment and does not have other associated neurophysiological immaturities, he usually does well with minimal professional help.

The secondary encopretic, whose symptom usually begins with some traumatic incident, may, if not treated, develop escalating psychiatric symptoms of a neurotic nature. These symptoms may continue, even if the encopresis phases out by itself at a particular developmental stage such as puberty.

Because of the difficulties inherent in the family structure, as well as the child's own neurophysiological immaturities, the chronic (primary or secondary) encopretic may well go on to have emotional problems of a characterological nature.

¶ Tics

Diagnoses

DEFINITION

Tics are recurrent, involuntary spasms of specific skeletal muscle groups. They develop most frequently from ages four to ten. There are two types of psychogenic tics: first, the simple type, which usually involves a single set of muscles, such as those used in eye-blinking; and second, the more complex type, involving many sets of muscles, including those used in producing vocal sounds. Shapiro and associates[47] differentiate those complex tics that involve the production of coprolalic sounds, which are usually referred to as Gilles de la Tourette's disease, from those which are noncoprolalic. He suggests that the former are neurogenic in origin and that psychiatric problems associated with them are secondary in nature.

PREVALENCE

The overall prevalence rate of tics in childhood is about 5 percent, Safer[44] states, and by the age of twelve years, the rate has decreased to 1 percent, with little difference in the ratio between boys and girls for the simple tics. The more complex tics appear more frequently in boys.[8]

EVALUATION

In taking the psychiatric history of a ticquer, it is important to distinguish between a compulsive act and a tic. The ticquer, by definition, does not know when or how often he performs his tic, whereas the compulsive patient does know when he performs his act and may even try to stop himself from doing it. It is useful to obtain information on the motor patterns of the child, as well as on the prevalence of tics in the family background. Since severe tics often occur with the post encephalitic syndrome, obvious organicity must be ruled out.

Etiology

NEUROPHYSIOLOGICAL

Children who develop simple tics may be hyperactive and are often accident prone, which contributes to their underlying concern about body hurt. Usually, however, they have good motor skills, particularly gross motor ones. In contrast, children who develop multiple tics, in addition to being hyperactive, very often have poor gross and fine motor skills. A family history of tics is often found with the latter type. Although there are rarely any hard signs of organicity, there are usually several soft signs.

INTERPERSONAL

Both Levy[32] and Mahler[33] noted similarities in the parents of ticquers. They believed the parents overprotected and infantalized their children but expected them to perform well, particularly academically. Anthony[3] also described the parents, particularly the mother, as being inclined to restrict and pressure the child, without being interested in the child's locomotion, athletic ability, or independence. The child's motor development may have been restricted because of the parents' fear that the child's experimentation with his body might cause him physical harm. In some cases, the child may have had a physical illness that restricted his activity.

In a three-year follow-up study of 615 childhood ticquers at a clinic in Japan, Abe and Oda[1] suggested that many, especially those who also had a family history of tics, may have developed their own tics by imitating family members. They also believed that for the tic to have become successfully established in the child, there had to have been neurophysiological proneness present.

INTRAPSYCHIC

The triggering emotional stress that produces the simple tic is usually posited by psychoanalytically-oriented writers to be oedipal in nature. In her study of ten children with simple tics, Gerard[19] noted that in each case the symptom began following a traumatic incident that aroused fear of being injured. At any age, the child has a great emotional investment in his body, but at the oedipal stage he has specific fears of bodily damage, particularly of the genital area. For example, if he has seen something that he feels he should not have seen, an eye-blinking tic may develop, with the unconscious purpose of protecting him from punishment. By alternately seeing and not seeing, he can go on doing and undoing the punishable act, without fear of being punished but at the expense of not resolving the conflict. In Mahler's words, "tics are an attempted drainage of a chronic state of emotional tension and also the physiological accompaniment of a chronic affective attitude."[33] Eventually the tic, if untreated, may become a way of responding to any anxiety-producing event. In other words, the tic, which had become a defense response appropriate at the moment of the trauma, becomes inappropriate when further used in response to a different trauma.

The triggering incidents of the more complex, multiple tic syndromes are usually anal and oral in nature, with the conflict being aggressive rather than erotic. The child who has usually had many previous emotional problems presents with a history of the gradual emergence of several tics following a series of traumatic instances. The underlying concern can be one of injurying rather than being injured. The child is therefore both infantile and grandiose.

Treatment

Biological Therapy

Because of their muscle-relaxant qualities as well as their anxiolytic qualities, promethazine or diazapam may be useful as short-term adjunctive pharmacotherapy when the child may be having a difficult time in psychotherapy for a simple tic.

Haloperidol is the drug of choice for the more complex tics. In a recent study of its use, Bruun and associates[10] report impressive results. Haloperidol's effectiveness, however, may eventually decrease with use.

Behavioral Therapy

For some of the children who have simple tics and for most of those who have the more complex tics, a conditioning type of behavioral therapy can sometimes alleviate the underlying fears or anxieties. Nevertheless, some of the behavioral therapy described by Yates[54] has not fulfilled the expectations aroused by early reports.

Psychotherapy

Psychotherapy will only be successful when the tic develops abruptly after a traumatic event. Since the traumatic event may be hard to track down because of the child's need to defend against the anxiety associated with it, parents may have to be involved in the therapy at first. Eventually, however, the child must be seen alone, and, in many cases, in psychoanalytically-oriented psychotherapy.

If the tic has become chronic and is no longer directly associated with the original traumatic event but is now associated with any anxiety-producing event, other types of treatment may have to be employed in addition to the psychotherapy.

Parent counseling may deemphasize attempts to alleviate the symptom by simply telling the child to stop it. Trying to get the child to stop the tic without recognizing that he is defending himself against some fear or anxiety only makes him more aware of the tic; he will then become more anxious and the tic may be increased.

In counseling, the parents should be urged to exert less pressure on the child. If the child has been overprotected either because of an illness, or because of parental concerns over bodily hurt, the parents should be encouraged to help him establish more appropriate body boundaries for himself so that he can become more independent.

Prognosis

Many simple tics will disappear without treatment by the time the child has reached puberty. Since the tic appears as the result of some specific conflict related to the emotional stage of development that the child has reached, often it will no longer be needed when he moves on to the next stage. At puberty, many old defenses are given up as new ones develop. Should the tic continue past puberty, it is unlikely to disappear without treatment.

The complex tics have a poor prognosis. Instead of disappearing as the child reaches puberty, they may become more severe. Even when they disappear with the use of medication, the underlying psychiatric symptoms, which are usually characterological in nature, respond poorly to psychotherapy.

¶ Stuttering

Diagnoses

DEFINITION

Stuttering, a disturbance of the flow of speech, is usually first noticed between the ages of two and four. It is important to make a distinction between the nonpathological and the pathological type of stuttering. The nonpathological type is a simple repetition of certain sounds and words and does not appear to trouble a child emotionally. The pathological stutterers are very aware of their impairment and often have blocking and avoidance rituals as a result of the tension produced by the stuttering.

PREVALENCE

Safer[44] has stated that the prevalence of stuttering in preschoolers is about 4 percent. This figure drops to about 2 percent in school-age children, as approximately one-half of the early stuttering is nonpathological in nature. Three-quarters of the children who are still stuttering by age ten will have become pathological stutterers. The symptom occurs more frequently in boys than in girls, at an average ratio of six to one.[29]

EVALUATION

In addition to obtaining an adequate psychosocial history, it is necessary to make a speech and language appraisal. Freeman and Ushijima[17] suggested that the stuttering appears to occur in the attempted production of a stressed vowel. The child will stop on the consonant immediately before the stressed vowel and will continue to repeat it with a muscular spasm that involves breathing, speech, and articulation.

No significant differences in IQ have been reported. Although it is always tempting to have projective testing done in an effort to track down the origins of the child's particular conflicts, caution must be observed while attempting to test a stuttering child because of the child's anxiety associated with attempting to participate verbally in the test.

Etiology

NEUROPHYSIOLOGICAL

About half a century ago, Orton[39] presented the theory that when one hemisphere of the brain is not sufficiently dominant, both will function independently and the resulting two parts of the speech musculature will be poorly integrated and thus produce verbal difficulty. More recently, Moore and Lang[36] found a reduction in alpha waves over the left hemisphere in nonstutterers and reduction in the right hemisphere for stutterers.

K. de Hirsch[12] states that the stuttering is usually first noted when the child, in a short span of eighteen months, passes from a primitive to a highly integrated form of language organization. She feels it occurs when the child is beginning to think faster than he can verbalize, but that it will usually disappear, if the child is left alone, as soon as his neurophysiological development catches up with his cognitive development. Kolb[30] and others remind us that, since stuttering often runs in families, many of the nonpathological stutterers can be made into potentially pathological stutterers by anxious parents or relatives trying to help them, either by slowing them down in speaking or by saying the words for them.

Almost all current researchers in this area, including this author, accept the fact that there is a constitutional predisposition to stuttering, and that the origin of the strong familial history of stuttering is more likely the result of a neurophysiological inheritance than an imitative habit.

INTERPERSONAL

Wyatt and Herzan[53] believe that stuttering is due to a disturbance in the mother-child relationship when the child is first learning to talk. Kessler[29] questions the significance of this factor, noting that many parents fail to respond to their child on his own verbal level without necessarily causing stuttering in the child.

Whether caused in any way by the relationship between the child and significant others, stuttering can certainly be made worse by their influence. In many cases parents and/or teachers may be made uncomfortable by stuttering, and a rejecting attitude may develop that may make the already anxious child even more anxious. Peers and siblings may begin to shun the child, and the child, in turn, may isolate himself from them. It is possible that as the child becomes aware of being rejected, panic sets in, causing his articulative muscles to go into spasm. Thus, stutterers who are often thought to be nonassertive, tense, and insecure, may indeed be struggling with underlying aggressive feelings towards others. The expectation of failure to speak fluently may be self-fulfilling.

Treatment

BIOLOGICAL THERAPY

Burns and associates,[11] who believe that there is a problem with the central dopamingeric system in the pathogenesis of stuttering, found haloperidol to produce a 40 percent improvement in speech by blocking the dopaminergic system postsynaptically. This was in contrast to apomorphine, which had no significant effect on the speech. Although it too acted as a blockade, it did so presynaptically. Haloperidol does not appear to improve the speech by reducing anxiety as other anxiety-reducing drugs do not help stutterers.

BEHAVIORAL THERAPY

As yet there has been no reliable attempt to evaluate mechanical devices for stuttering. Some devices tried are designed to prevent physical action, such as clamping of teeth, improper movement of the tongue, and improper breathing. Others have been designed to help certain muscles function more appropriately.

For those who believe that stuttering is a learned response, speech therapy often appears to be the treatment of choice. The aim is to improve the speech as early as possible, so the attitude of others toward the child will change, in turn improving his image of himself. However, most authors have found that although operant conditioning may modify speech habits, to cure the stuttering child one has to use a diagnostic and treatment approach beyond the resources of the general theory of learning.

Timmons and Boudreau[49] believe that auditory feedback therapy can be useful if it is not too rigorously administered. Brady[8] has had success with a mechanical device worn as a hearing aid that helps the patient maintain normal rhythm of speech through a hidden metronome, the rate of which can be controlled by the patient.

PSYCHOTHERAPY

If psychotherapy is undertaken early, it will be more effective if the mother is involved in the therapy as well as in separate counseling. There has been little success with analytically-oriented psychotherapy. Kaplan[27] suggests that the only way to have the child released from the image he usually has of himself is to use a Gestalt approach, helping the child in gradual steps to be more aware of himself, and in doing so, making fluency become part of himself. If the child does begin to become fluent while undergoing psychotherapy, speech therapy will not then be necessary. Most children who have not received help until latency age will require both psychotherapy and speech therapy. Although hypnotherapy has been used successfully with some stuttering adults, there has been little effort made in this direction with children.

Prognosis

When the stuttering has been of the nonpathological types, and recognized early as such, the child will rarely resume stuttering at a later date.

Because of the difficulties presented to persons involved with the stuttering child, there are few, if any, reports of the natural course of the untreated pathological stutterer. Intervention of one kind or another will have been introduced even before the stuttering has been declared pathological.

¶ Thumb-Sucking

Diagnoses

DEFINITION

The age at which thumb-sucking is defined as a problem depends on whether one is viewing it strictly from a dentitional point of view or from an emotional point of view. Dentitionally, if the child has given the habit up by age four years, most of the dental pathology will be self-corrective. From a physiological point of view, most of the current literature quotes one year as being the age at which normal habitual thumb-sucking should cease, but from an emotional point of view it can still be considered normal up to the age of three years if it is connected with sleep or stress.

PREVALENCE

Thumb-sucking beyond age three is very common in many parts of the world. An overall prevalence rate for preschoolers is 61 to 87 percent. However, by age six the overall rate is down to about 20 percent. There is very little difference in the sex ratio of thumb-suckers.

EVALUATION

A thorough psychosocial history should be taken and a dental examination given before thumb-sucking is diagnosed as a symptom. If it is diagnosed too early as a dentitional problem and if efforts are made to have the child give it up before he is emotionally ready, the habit may only become more entrenched.

Etiology

NEUROPHYSIOLOGICAL

Although a review of the etiology of pathological thumb-sucking from a neurophysiological point of view produces contradictory and frequently inconclusive results, the etiology of normal sucking as a basic physiological need for general survival as well as for the development of face and jaws is noncontroversial. Ozturk and Ozturk[40] feel that persistent thumb-sucking is specifically related to the immature position in which a child falls asleep. Others feel it is associated with the more general neurophysiological immaturities that often accompany special symptoms of childhood.

According to the psychoanalytic theorists, thumb-sucking, when it first appears, is an auto-erotic mechanism that the child employs when he is anxious. The anxiety develops because of the child's insecurity regarding his relationship with the feeding person, whether the feeding is by breast, bottle, or cup. Later the thumb-sucking may be used as a masturbatory equivalent.

INTERPERSONAL

Sears and Wise[45] felt that although a premature emotional withdrawal of the mothering person could be the triggering mechanism for thumb-sucking, there must be other factors that keep it going, since there was no significant difference in his study between the persistent and the non-persistent suckers in regard to such a factor. On the other hand, Massler[34] says that in Africa where breast feeding is prolonged, no habitual thumb-sucking occurs.

Golden[20] compared studies of latency-age children in the Kibbutzim in Israel where the percentage of habitual thumb-suckers was 56 with a similar age group of children in Switzerland where the percentage was 59. It was thought that perhaps the statistics were so high in the Kibbutz study, because of the separation of the child from his parents, but the group of children who spent the night with their parents produced as many thumb-suckers as the group who lived in the homes for children only. The high prevalence in the Swiss population was ascribed to strictness in the home, but in the Kibbutz, where the prevalence is almost as high, childrearing is noted for its leniency.

INTRAPSYCHIC

Levy[31] suggested that thumb-sucking occurred because some children have such a strong oral drive that it cannot be satiated even by generous breast or bottle feeding. Kessler[29] indicated that some children rely on sucking as a substitute for other pleasures, both social and physical. She says that the child can learn how to retreat from both his own feelings and difficult situations by sucking in a trancelike state. But, she adds, it will serve as a successful defense against anxiety only until the child begins to feel guilt and/or shame about the symptom.

Treatment

BIOLOGICAL THERAPY

If the child's persistent thumb-sucking appears to be associated with generalized stress, a mild anxiolytic medication, such as promethazine, may be used for a short time. If the sucking is accompanied by other immaturities, then the medication used, if any, should be related to the specific immaturity. For instance, methylphenidate may be used for certain types of learning disorders.

BEHAVIORAL THERAPY

Only mechanical devices that are not painful should be used and then only in a supportive manner. Pacifiers can be used during the day and a dental device at night. In general, behavioral therapy by itself does little to alleviate thumb-sucking, as drawing attention to the symptom may only produce more denial, at which the youngster is already very adept. But as an adjunct to other therapy, and if the child becomes interested in eliminating the symptom, an operant conditioning type of behavioral therapy with some type of reward system may be successful.

PSYCHOTHERAPY

Psychotherapy is rarely necessary for thumb-sucking alone. If thumb-sucking continues past early childhood and into latency, it is usually indicative of some emotional conflicts, and supportive psychotherapy might be required to uncover and resolve the conflicts. It is important to determine whether the habit is continued because there has been no effort to eliminate it by the counseling of the parents, with or without behavioral therapy, or whether the child is continuing to use it as a defense against anxiety or fear.

If an adequate psychosocial history is obtained, it will not be difficult to counsel the parents. The mother should be helped to distinguish problems the child may be having because of his own continuing immaturities from those caused by difficulties in the mother-child relationship. At times it may be necessary to help institute environmental changes.

Prognosis

The prognosis for thumb-sucking by itself is very good. If untreated, not more than a third go beyond nine years. By the early teenage years, there are very few thumb-suckers. Because of the threat to the dentition and the youngsters' gradually developing poor image of themselves if thumb-sucking continues, it is important to get help as soon as it becomes a problem, in spite of the benign nature of the symptom. If it is accompanied by an underlying emotional disturbance, then this should be treated rather than the thumb-sucking itself.

¶ Nail-Biting

Diagnoses

DEFINITION

Nail-biting, which may occur at any age, usually begins at about five years. It is rarely defined as a symptom, unless it is accompanied by other more troublesome symptoms.

PREVALENCE

Although nail-biting rarely begins before age five years, the peaking of prevalence does not occur until ages twelve to fourteen years. Most authors found that the sex ratio remained the same until about that time, when girls begin to bite their nails less frequently. In a study conducted in Yorkshire, England, Birch[7] found an overall prevalence of 51 percent in 4,223 children.

EVALUATION

Since many normal children bite their nails when under stress, it is important to study the frequency, the pattern, and the cause of nail-biting before diagnosing it as a problem. If the habit begins to become so compulsive that it interferes with the child's other activities, if it grows more aggressively vigorous, or if it is accompanied by other special symptoms such as enuresis and stuttering, the child should be evaluated psychiatrically.

Etiology

NEUROPHYSIOLOGICAL

Nail-biting is associated with general body tension, which at times may be accompanied by a tensing of the musculature. However, Massler and Malone[35] felt that nail-biting alone is a simple tension-reducing mechanism, particularly under conditions of situational stress. When it occurs with other special symptoms of childhood, it is quite possible that there may be some unevenness of maturation that interferes with the child's functioning either at home, with peers, or at school.

INTERPERSONAL

The origin of nail-biting has sometimes been ascribed to the child's feelings of rejection, since high statistics are found among institutionalized children as compared with those in family placement. It is possible, however, that many of the children who have remained institutionalized are children who have severe underlying problems that have kept them from being placed in foster homes. Therefore, as Goldfarb[21] suggests, the rejection may not have been the actual cause of the nail-biting but, instead, the nail-biting and the feelings of rejection may both be associated with the internal tensions created by being in the institution.

INTRAPSYCHIC

In psychoanalytic terms, nail-biting is viewed as an oral, sadistic habit that begins when the child is going through the oedipal stage of development. As with thumb-sucking, it is an autoerotic mechanism but at a higher level of functioning. At times it, too, may be a masturbatory equivalent.

Since many children go through this oedipal stage without nail-biting, one would have to suppose that the nail-biting child is in a more hostile battle with a parent than is normal. Since he is still dependent on the parent, the child can express his aggression by nail-biting without harming the the object of his aggression. In addition, he can relieve the guilt caused by such hostility by hurting himself.

Nail-biting rarely occurs with thumb-sucking and a thumb-sucker rarely, if ever, becomes a nail-biter. Unlike the thumb-sucker, whose anxieties appear to be held at bay by the sucking, the nail-biter, whose anxieties appear to be near the surface, is a more outwardly troubled child.

If the symptom occurs by itself, it may be that the child is going through a period when his body, which had served him well up to this stage, no longer pleases him. If he eventually accepts his body, either through identification with the parent of the same sex or, possibly in some cases, through identification with his parent of the opposite sex, he may outgrow the habit. However, if the nail-biting is accompanied by stuttering and/or enuresis, it may be that because of uneven maturation his body has not served him well during the earlier stages and is still not serving him well.

Treatment

Biological Therapy

There are no known mechanical devices that work. Trying to keep the child's hands from his mouth by immobilizing them only makes the problem worse. The child must be encouraged to use all his body parts, and particularly his hands, in vigorous gross motor activity, so that he learns that he has a usable and reliable body.

Behavioral Therapy

Negative reinforcement conditioning should be avoided. Since the child is already in a power struggle with himself, the significant people in his environment should set up a conditioning plan only if the child is in charge of it. He must be helped to feel that he is capable of setting up a self-help program, and following through on it, because of something that is in it for him.

Psychotherapy

Since the child who is nail-biting only is seldom seen in referral, when he is seen it is usually when the parents and child have not been able to resolve the child's oedipal strivings or the child has other symptoms as well as the nail-biting. In the first case, counseling of the parents is usually sufficient. The parents often need help in being more supportive of the child's wish for greater control of his world, while at the same time setting appropriate limits for him. Some environmental changes must at times be made if the child's conflicts, which basically revolve around his self-image, cannot be resolved otherwise.

If the child has other symptoms as well, he will probably require psychotherapy in a setting where the parent(s) can be involved if necessary.

Prognosis

In his study of adult nail-biters, in which he found an overall prevalence of 24 percent, Ballinger[4] did not find untreated nail-biting to be an important psychiatric symptom. This finding was corroborated in another study of adults conducted by Walker and Ziskind,[50] in which the prevalence of nail-biting by normals, mental defectives, and those who were emotionally disturbed was much the same. They did find, however, that the prevalence of nail-biting among sociopaths was higher than among normals. As indicated by the authors, this finding does not suggest that nail-biters have a tendency toward sociopathy, but rather suggests that sociopaths may experience more anxiety than is generally attributed to them.

¶ Bibliography

1. Abe, K., and Oda, N. "Follow-up Study of Children of Childhood Ticquers," *Biological Psychiatry,* 13 (1978): 629–630.
2. Anthony, J. "An Experimental Approach

to the Psychopathology of Childhood: Encopresis," *British Journal of Medical Pathology,* 30 (1959): 156–174.

3. ANTHONY, J. "Neuroses in Childhood," in Freedman, A., Kaplan H., and Saddock, B., eds., *Comprehensive Textbook of Psychiatry,* Baltimore: Williams & Wilkins, 1975, pp. 2143–2160.
4. BALLINGER, B. "The Prevalence of Nail-biting in Normal and Abnormal Populations," *British Journal of Psychiatry,* 117 (1970): 445–446.
5. BELLMAN, M. "Studies on Encopresis," *Acta Paediatrica Scandinavica,* supp., 170 (1966): 1–151.
6. BEMPORAD, J. "Characteristics of Encopretic Patients and their Families," *Journal of the Academy of Child Psychiatry,* 10 (1971): 272–292.
7. BIRCH, L. "The Incidence of Nail-biting Among School-children, *British Journal of Education and Psychology,* 25 (1955): 123–128.
8. BRADY, J. P. "Metronome Conditioned Speech Retraining for Stuttering," *Behavioral Therapy,* 2 (1977): 129–150.
9. BRAZELTON, T. B. "A Child-Oriented Approach to Toilet Training," *Pediatrics,* 29 (1962): 121–128.
10. BRUUN, R., et al. "A Follow-up of 78 Patients with Gilles de la Tourette's Syndrome," *American Journal of Psychiatry,* 133 (1976): 944–947.
11. BURNS, D., BRADY, J. P., and KURUVILLA, K. "The Acute Effects of Haloperidol and Apomorphine on the Severity of Stutterers," *Biological Psychiatry,* 13 (1978): 255–264.
12. DE HIRSCH, K. "Language Disturbances," in Freedman, A., Kaplan, H., and Saddock, B., eds., *The Comprehensive Textbook of Psychiatry.* Baltimore: Williams & Wilkins, 1975, pp. 2108–2116.
13. DISCHE, S. "Childhood Enuresis. A Family Problem," *Practitioner,* 221 (1978): 323–330.
14. EASSON, W. "Encopresis," *Canadian Medical Association Journal,* 82 (1960): 624–628.
15. ESMAN, A. "Nocturnal Enuresis. Some Current Concepts," *Journal of the Academy of Child Psychiatry,* 16 (1977): 150–158.
16. EYSENCK, H. J. "Learning Theory and Behavior Therapy," in *Behavior Therapy and the Neurosis.* London: Pergamon Press, 1960: 4–21.
17. FREEMAN, F. J., and USHIJIMA, T. "Laryngeal Muscle Activity During Stuttering," *Journal of Speech and Hearing Research,* 21 (1978): 538–563.
18. GERARD, M. "Enuresis: A Study in Etiology," *American Journal of Orthopsychiatry,* 9 (1939): 48–58.
19. GERARD, M. "The Psychogenic Tic in Ego Development," in Essler, R. L. et al., eds., *The Psychoanalytic Study of the Child.* vol. 2, New York: International Universities Press 1946, pp. 133–162.
20. GOLDEN, A. "Patterns of Child Rearing in Relation to Thumb-Sucking," *British Journal of Orthodontics,* 5 (1978): 81–88.
21. GOLDFARB, W. "Infant Rearing Problem Behavior," *American Journal of Orthopsychiatry,* 13 (1943): 249–265.
22. HALLGREN, B. "Enuresis," *Acta Psychiatrica Scandinavica,* Supp., 114 (1957): 1–139.
23. HALPERN, W. I. "The Treatment of Encopretic Children," *Journal of the American Academy of Child Psychiatry,* 17 (1977): 478–516.
24. HARRIS, L., and PUROHIT, A. "Bladder Training and Enuresis: A Controlled Study," *Behavioral Research Therapy,* 15 (1977): 485–490.
25. HEIN, H. "Who Should Accept Primary Responsibility for the Encopretic Child?" *Clinical Pediatrics,* 17 (1978): 67–70.
26. KALES, A., et al. "Enuretic Frequency and Sleep Stages," *Pediatrics,* 60 (1977): 431–436.
27. KAPLAN, N. R., and KAPLAN, M. "The Gestalt Approach to Stuttering," *Journal of Communication Disorders,* 11 (1978): 1–9.
28. KATAN, A. "Experiences with Enuretics," in Eissler, R. L., et al., eds., *The Psychoanalytic Study of the Child.* vol. 2, New York: International Universities Press, 1946, pp. 241–255.
29. KESSLER, J. W. *Psychopathology of Childhood.* New York: Prentice-Hall, 1966.
30. KOLB, L. C. *Modern Clinical Psychiatry,* Philadelphia: W. B. Saunders, 1977.
31. LEVY, D. M. "Finger-sucking and Accessory Movements in Early Infancy: An Etiological Study," *American Journal of Psychiatry,* 7 (1928): 881–918.
32. LEVY, D. "On the Problem of Movement Restraint," *American Journal of Orthopsychiatry,* 14 (1944): 644–671.
33. MAHLER, M. "A Psychoanalytic Evaluation of Tic in Psychopathology of Children:

Symptomatic and Tic Syndrome," in Eissler, R. L., et al., eds., *The Psychoanalytic Study of the Child.* vol. 3/4, New York: International Universities Press, 1949, pp. 279–310.

34. MASSLER, M. "Oral Habits: Origin, Evaluation and Current Concepts in Management," *Alpha Omegan,* 56 (1963): 127–134.
35. MASSLER, M., and MALONE, A. "Nail-biting, A Review," *Journal of Pediatrics,* 36 (1950): 523–531.
36. MOORE, W. H., JR., and LANG, M. K. "Alpha Asymmetry Over the Right and Left Hemispheres of Stutterers," *Perceptual Motor Skills,* 44 (1977): 223–230.
37. MOWRER, O. H., and MOWRER, W. A. "Enuresis: A Method for its Study and Treatment," *American Journal of Orthopsychiatry,* 8 (1938): 436–459.
38. OLNESS, K., and GARDNER, G. "Some Guidelines for Uses of Hypnotherapy in Pediatrics," *Pediatrics,* 62 (1978): 228–233.
39. ORTON, S. *Reading, Writing and Speech Disorders in Children.* New York: W. W. Norton, 1937.
40. OZTURK, M., and OZTURK, O. M. "Thumb-sucking and Falling Asleep," *British Journal of Medical Psychology,* 50 (1977): 95–103.
41. PRUGH, D. "Child Experience and Colonic Disorders," *Annals of the New York Academy of Science,* 58 (1954): 355–376.
42. RICHMOND, J., EDDY, E., and GARRARD, S. "The Syndrome of Fecal Soiling and Megacolon," *American Journal of Orthopsychiatry,* 24 (1954): 391–401.
43. RITVO, E., et al. "Arousal and Non-Arousal Types of Enuresis," *American Journal of Psychiatry,* 126 (1969): 77–84.
44. SAFER, D. "Guiding Parents of Children with Habit Symptoms," in Arnold, L. E., ed., *Helping Parents Help Their Children.* New York: Brunner/Mazel, 1978, pp. 232–243.
45. SEARS, R. R., and WISE, G. W. "Relation of Cup Feeding in Infancy to Thumb-sucking and The Oral Drive," *American Journal of Orthopsychiatry,* 20 (1950): 123–139.
46. SHAFFER, D., HEDGE, B., and STEPHENSON, J. "Trial of an Alpha-adrenolytic Drug (Indoramin)," *Developmental Medicine and Child Neurology,* 20 (1978): 183–188.
47. SHAPIRO, A., et al., *Gilles de la Tourette Syndrome.* New York: Raven Press, 1977.
48. SPERLING, M. "Dynamic Considerations and Treatment of Enuresis," *Journal of the American Academy of Child Psychiatry,* 4 (1965): 19–31.
49. TIMMONS, B. A., and BOUDREAU, J. P. "Speech Dysfluencies and Delayed Auditory Feedback Reactions," *Perceptual Motor Skills,* 47 (1978): 859–862.
50. WALKER, B., and ZISKIND, E. "Relationship of Nail-biting to Sociopathy," *Journal of Nervous and Mental Disease,* 164/165 (1977): 64–65.
51. WERRY, J. "The Conditioning Treatment of Enuresis," *American Journal of Psychiatry,* 123 (1966): 226–229.
52. WRIGHT, L., and WALKER, C. "Treatment of the Child with Psychogenic Encopresis. An Effective Program of Treatment," *Clinical Pediatrics,* 16 (1977): 1042–1045.
53. WYATT, G. L., and HERZAN, H. M. "Therapy with Stuttering Children and Their Mothers," *American Journal of Orthopsychiatry,* 32 (1962): 645–660.
54. YATES, A. *Theory and Practice in Behavior Therapy.* New York: Wiley-Interscience, 1975.

CHAPTER 10

ALTERNATIVE LIFE-STYLES AND THE MENTAL HEALTH OF CHILDREN

Carol Nadelson

IN RECENT YEARS, social changes have brought about major differences in family life and expectations. There have been increased numbers of women in the work force, especially women with young children; an increased divorce rate has meant that women have frequently become heads of families; changes in custody practice have led to fathers more often being awarded custody of children; a general decline in birth rate has resulted in smaller families; there have been changes in sexual behavior; there has been an alteration of traditional role behaviors; and progressive urbanization has affected family ties.

The ability to control reproduction has had a major impact. Birth control methods have made it possible to limit birth rates with consequent diminution of the impact of pregnancy and childbirth on the lives of women.

An important related change has been in the area of work and careers of women. For some families this change has been enriching, for others it has generated conflicts about achievement, family responsibilities, and role redefinition. Social changes have spurred challenges to the traditional views of male and female development. They have also affected the conditions of growing up and effected the sex role stereotyping of childrearing. Family life-styles and models have begun to include many alternatives to the "traditional" family.

Reference to the work alternative immediately brings up the question, alternative to what? It implies a norm, and thus an alternative to that norm. The obvious response to this question is that the norm is a two-parent family with two children, in which the parents are married to each other, the father is away at work during the day, and the mother is a "housewife."

When we face the reality of the American family, we discover to our surprise that only 6 percent of American families currently fit the model.[63,80] Thus, these "traditional" families have become the deviant ones. Our pluralistic society offers a variety of alternative models, some relating to differences in cultural norms, others occuring because of socioeconomic conditions, and still others, the product of changing mores and values. In this chapter we will review the various alternatives as they currently exist and describe how these affect the rearing and mental health of children.

¶ What Has Changed

One immediately apparent change is that the birth rate has dropped dramatically. Americans are having fewer children than in any time in our history. However, the number of unwed mothers has increased. Census data show that there are fewer marriages and more divorces. In 1975, the census bureau reported a 3 percent drop in the number of marriages. Furthermore, the proportion of women remaining single until they are between the ages of twenty and twenty-four has increased by one-third since 1960, and the number of unmarried women and men living together has increased by 800 percent during the 1960s.[101] The divorce rate in the United States is now at the highest point it has ever been, and it is higher than any other industrialized country. One million divorces occurred in 1976 compared to two million marriages. Two million adults and one million children under the age of eighteen were involved directly in divorce; (including indirectly a multitude of friends, relatives, neighbors and colleagues).[74] At today's rate, one-third of first marriages and close to one-half of second marriages will end in divorce. One-half of all separations lead to reconciliation, but half of these reconciliations are temporary and end ultimately in divorce.[102] A rough projection based on divorce rates is that 30 percent of children growing up in the 1970s will experience parental divorce.[74] Thus, divorce has created many single-parent families.

Within what has been considered the traditional family structure, the situation has also changed. Both partners are more apt to be working even when young children are at home. Approximately one-half of all women sixteen years or older are in the work force or actively seeking employment. Forty-one percent of these working women have children under eighteen years of age; 31 percent of children under six have working mothers. The number of women in the work force has increased by 60 percent in the past decade.[63,80]

While this represents substantial change, it is important to note that the "traditional" pattern itself, with a full-time, life-time mother has actually been traditional for only one or two generations, the result of increased industrialization and relative affluence. Previous to this, women who were in the lower classes had always filled many roles and shared the burden of work in order to sustain their families.

Another change has been in the definition and relationship of family members to each other. The time spent living together in a family unit has come to be only a small part of the life cycle. Recent census data also reveal a rapid expansion of the numbers of people living alone. Almost 20 percent of American households now contain only one person. Most of the people in this age group are either young and not yet married, formerly married people, or older people who have been widowed. While women predominate in one-person households, a considerable number of divorced men now live alone. Furthermore, a young adult without a spouse is more likely to be living alone than with a family.[96]

The past few years have also seen many changes in attitudes about career and family life-styles, especially among young people. In a 1967 study, 50 percent of women college students stated that in addition to being a wife and mother, having a career was important.[55] By 1971, 81 percent of a sample of college students held this view. At the same time, contrary to prevailing misconception, 91 percent of male students expressed interest in marriage to a wife with a career out of the home. Furthermore, 60 percent of male and female students thought that fathers and mothers should spend equal time with children, 44 percent of males felt that men and women should share household responsibility, and 70 percent of females and 40 percent of males felt that both partners should contribute equally to family financing. This latter discrepancy may represent either a realistic evaluation of the present differences in male/female earning power or a reluctance to relinquish the privileges accompanying the role of the major "breadwinner."[55] By 1977, three-fourths of college men said that they expected to spend as much time as their wives in bringing up children.[47]

Potential parents are now more concerned about the high cost, both personally and financially, of having children. People report wanting fewer children, and those who have raised a family often express regrets about their choice.[96]

The law has provided other indicators of change. The 1973 Supreme Court decision permitting legal abortion, the emergence of no-fault divorce, increased recognition of illegitimate children, and joint custody following divorce are among the changes which affect life-style.

Lest we see these as radical departures from tradition, we must remember that many of the so-called radical ideas of this part of the century recapitulate much of the earlier ferment around women's rights, male and female sex roles, sexuality, and the relationships between the individual family and society. The last half of the nineteenth century and the beginning of the 1920s brought a sharp increase in these concerns. The depression of the early thirties, followed by World War II in the forties, interrupted the development of alternatives, and brought forth the era of togetherness, the 1950s. The baby boom and the increased domesticity of women were the aftermaths of this disruption.[91] In a sense then, we are witnessing an evolutionary process.

Since the turn of the century, scholars have debated the impact on the family of these changes. Some see the family as underminded by its loss of economic functions, the movement of populations, the shift of manufacturing from home to industry, and the emergence of public schools and day nurseries. For the first time in history, both men and women can find work and satisfy basic needs without family ties. Thus, while the family no longer may function as a work place, school, or hospital, this does not imply that the basic family functions of nurturance, raising children, and providing companionship and refuge are not important.

Another important change has been the reduction of infant and child mortality and the development of contraceptives. When the life cycle was shorter and the youngest child usually left home close to the end of the average mother's life expectancy, there was less concern about planning for later life. Today the average women can expect several decades without maternal responsibilities.[68]

Rossi[87] estimated that the average woman marries at age twenty-two, has two children two years apart, and dies at seventy-four, nine years after the husband's death at age sixty-seven. She thus has fifty-six years of adulthood (starting at age eighteen): (1) 23 percent (thirteen years) are without a husband; (2) 41 percent (twenty-three years) are with a husband but without children under eighteen; and (3) 36 percent (twenty years) are with a husband and at least one child under eighteen.

If we assume that full-time parenting is required only up to school entry, then only 12 percent of a woman's adulthood (or seven years) consists of full-time mothering of preschool children. While these figures continue to change, it is clear that even the nonworking woman in contemporary society will spend almost two times as many years with neither husband nor dependent children as she does in caring for preschool children. Furthermore, the rising divorce rate makes it a distinct possibility that many women will be alone for even a longer time than in the past, and that they will have to support themselves, and often their children.

¶ The Choice of Singleness

A new style of "singlehood" has emerged. It appears that those who remain single by choice are increasing in numbers. In 1962, a study of unmarried college students reported that only 2 percent of them had little or no interest in future marriage.[9] A decade later, Stein[98] found that 2.7 percent of freshman and 7.7 percent of seniors did not expect to marry. By 1973, however, 40 percent of senior women said they did not know whether they would or would not marry. Further, 39 percent of seniors felt that traditional marriage was becoming obsolete, and 25 percent agreed with the statement that the traditional family structure of mother, father, and children living under one roof was not a viable family organization.[98] Bird[16] believes that changes in sexual mores and sex roles have contributed to this development. Men no longer have to marry to have sex and women no longer marry to get financial support.

In the 1950s, Kuhn[50] studied a group of individuals who had not married. Their reasons for remaining unmarried included: hostility toward marriage or toward members of the opposite sex; homosexuality; emotional involvement with parents; poor health or physical characteristics; unattractiveness; unwillingness to assume responsibility; inability to find the "true love"; social inadequacy; marriage perceived as a threat to career goals; economic problems; and geographic, educational, or occupational isolation that limited the chances of meeting an eligible mate. The main thrust of these findings is that those who did not marry were seen as failing. Stein,[98] in a more recent study, reported positive reasons for remaining single. Individuals spoke of freedom, enjoyment, opportunities to meet people and develop friendships, economic independence, more and better social experience, and opportunities for personal development. Adams,[3] in 1971, suggested that for women to choose singleness they must have economic independence, social and psychological autonomy, and a clear intent to remain single by preference.

Since the dominant value in society is for marriage, individuals who choose to remain single often experience some stress. Some of this is because singleness is considered a threat to married people.[32] Further, some report active discrimination against single people. For example, 80 percent of the companies reporting in Jacoby's 1974 survey, asserted that while marriage was not essential to upward mobility, the majority of executives and members of junior management were married. Over 60 percent felt that single executives tend to make snap judgments, and 25 percent believed that singles were less stable than married people.[44]

There appear to be differences between the life experiences of single men and single women. Single women more often than single men find themselves frustrated in obtaining intimate relationships. Sometimes, intense relationships, either heterosexual or homosexual, may help develop a sense of community or family. Friendships become an important source of support and sustenance. Many single women live together or with their parents or families for emotional or economic support, or to ease the loneliness and isolation that may occur when there is no one to count on.[37] However, marriage clearly does not guarantee reliability and companionship either.

Research data indicate that unmarried women have better physical and psychological health than unmarried men.[13,15] It is not clear why this is true, although people have speculated that men who do not marry may have more serious physical or emotional problems than women who do not marry.

There have been few studies that have examined the reasons for remaining single or the consequences of remaining single for a lifetime. There are, however, some women who make the choice because they do not wish to make unacceptable compromises. Since women have tended to marry up (that is, into a higher social status), and men down, there may be fewer men available for women at the top. Furthermore, physical attractiveness and agreeable personality characteristics have in the past tended to be seen as more important for attracting a partner for women than for men.

There has been even less information gathered on women who remain single because they choose lesbian relationships, since most research on homosexuals focuses on males. Lesbians tend to have long-term stable love relationships. Some lesbians also desire to be parents and may achieve parenthood either by adopting children or by bearing them with or without marriage. Some lesbians are bisexual, and formerly married; others have children by artificial insemination or by finding a man who agrees to be a "donor."[33]

Abbott and Love[1] reported that one-third of the members of the Daughters of Bilitis (a gay women's organization) had children. Those who were divorced often had serious problems about child custody. Some of the court decisions have allowed them to retain custody of their children, as long as their partner did not live in the household. There has been increasing interest in lesbian mothers and the effects of their sexual preference on their children. Although there is little long-term data available, it appears that the children of lesbians, including those who live in lesbian households, grow up with the same range of sexual interests and preferences as do other children.[37]

¶ Why Marry?

Marriage is one of the few institutions which exists in all societies. It has a number of important functions: (1) it allows for regulation and stabilization of sexual relationships; (2) it is a solution to the problem of replacement in society and legitimizes children; and (3) it divides labor. The American perspective has been that marriage occurs for a number of reasons: (1) happiness; (2) political or economic alliance; (3) as a springboard for society; and (4) to have children.[2] For the individual a number of other complex reasons are added, including: societal and family pressures; fantasies about a "perfect" life; concerns about loneliness; and needs for intimacy. Specific choices are often made because of romantic expectations and wishes rather than more practical considerations, such as the similarity of life goals or the ability to work out problems together.[67]

People do consider a number of factors, including the compatibility of ideas, values, expectations, and desires, as well as family background, age, education, financial success, and future goals. Most often, however, romance starts a relationship. While it may persist and even expand, the realities of building a life with another person often present problems that were not initially expected. There may be concrete recognition of disparities, or dissatisfaction may arise from unconscious issues related to past experiences and conflicts. An aspect of the process of selection of a mate, after all, is based on unconscious signals. Marriage also presents a paradox, especially for people who marry early before they have settled their own concerns about identity and independence.[67]

From the perspective of mental health, marriage appears to be a benefit for men, but a stress for women. Married women seek help for emotional problems more often than married men or single women.[14] Marriage may be more stressful for women than men for a variety of reasons:[37]

1. The woman who marries modifies her life more than does her husband and risks more loss of autonomy. Although this pattern is changing, and there are an increasing number of couples who work out alternative life-styles (for example, dual-careers, commuting, and so forth), these patterns are by no means problem-free, nor do they necessarily change some of the pressures.

2. For women who marry and become housewives, the tendency toward lack of role differentiation and diversity may lead to decreased self-esteem. The fact that the housewife role is an ascribed rather than achieved role, and that it is expected that women perform well in it without opportunity for diver-

sification, implies that all women must succeed. In fact, women who do not succeed in this role are frequently viewed as life failures.

3. A loss of status may occur for a woman who has had an active career and then gives it up to marry. She often finds her role as housewife and mother devalued, although lip service is paid to its importance. While fewer women without children give up work than do mothers, many women who marry alter their career or work patterns and take positions with more flexibility and fewer time demands in order to spend more time at home or to be available to travel or entertain.

Recent work in adult development has indicated that important shifts occur throughout life.[53,75,106] The concept of a marital life cycle has also evolved.[12] Stages in the individual life cycle may or may not be paralleled with stages in marriage, and conflicts may result from the fact that the partners are in different stages of development in their lives or because there are shifts in tasks, expectations, and demands which include the birth of children, changes in careers, relationships with relatives and friends and the physical environment. Alternative styles must often be worked out in the course of a marriage since complementary shifts may not occur in both partners. If alternatives are not possible, the balance of the relationship may be disturbed.[71]

¶ Children versus Childlessness

More and more couples today are choosing not to have children. Many decide to postpone childbearing temporarily or indefinitely because of career goals, dislike of children, anticipated problems, or desire for more freedom. Some of these people seek sterilization, others rely on birth control. This is a major social change; it is the first time in human history that the *choice* of childlessness has been safe and possible. The impact of this change is not yet clear, although it appears that the stigma against childlessness has diminished.[68]

Vivers[108] has reported that about 5 percent of all couples voluntarily forgo parenthood. There appear to be two paths. One is an explicitly stated intention, the other is postponement until it becomes too late. Since there is strong cultural pressure to have children, it may be easier to avoid the issue rather than to make a deliberate decision.

Many people who make the initial decision to remain childless, later change their minds. This is indicated by the fact that an increasing number of "older" women (past thirty-five) bearing their first child state that they had decided not to have children earlier in their lives, and by the increasing number of people who have requested reversals of sterilization procedures.[108]

Little long-term data are available about those who have chosen to remain childless. Couples who are childless however do report greater marital satisfaction during the early years of marriage than do those with young children. A national survey reported that 88 percent of childless wives and 73 percent of childless husbands between ages eighteen and twenty-nine felt happy with their lives, compared with 65 percent of husbands and wives with children under six.[105] Contentment dropped and stress increased when couples had their first child. As the children grew older, marital happiness increased, and this pattern continued after the children left home. Over the course of their adult lives, however, women with children have reported greater overall life satisfaction than childless women.[37]

This situation is not entirely due to the conditions of parenthood itself, but also relates to other pressures including economic strain, relationships with family, the need to adjust to differences in backgrounds and personalities, and the expectations of often immature and inexperienced people that they can live together and agree on a number of previously unknown aspects of life.

The decision to bear children is a complex

one. Motivations include a love for children and the feeling that they will offer a full life, that it is "natural" to have children, that children will create a closeness and family atmosphere, that children will please a partner (or parents), and that having children will provide a kind of immortality. People also want children because they are insecure, dependent, or seek love and approval. Having children may also be viewed as a way of finding normality, approval, or reaffirmation of sexuality.[108]

While there is evidence that women who marry but remain childless are more successful in their careers, the higher a woman's educational level, the less likely she is to stop her career when she has children.[6] The stresses of motherhood are high because of beliefs and attitudes about child rearing which place higher demands on mothers than on fathers. Although pregnancy and childbearing do involve women more, mothers also continue to remain the major rearers of children, despite work and careers.

¶ The Impact of the Changing Role of Women

Responses to the changing role of women have varied. There are those who herald a new era of freedom while others cite mothers and wives who are out of the home as the cause of divorce, teenage pregnancy, delinquency, violence, homosexuality, and other "ills" of society. This view reflects both anxiety about change and the inability to assess realistically its impact. In many instances, the assumption that all women have a choice does not take account of the pressures on women to work and the lack of real alternatives.

The working mother has become an increasingly more familiar maternal model, and marital status no longer determines whether a mother works. Although, currently, most working mothers are married, live with their husbands, and have school-age children, a large number (more than 15 million) are single, separated, widowed, or divorced. A substantial proportion of this group have young children. In the past decade, families headed by women have grown in number ten times as rapidly as two-parent families.[80]

Women who are single parents either because of divorce or because they never married suffer significant economic problems. Such women are increasingly likely to be poor. In 1974, one-eighth of all families in the United States were headed by women, and one-third of these had incomes below the poverty level. This figure, which is on the rise, includes women who were formerly considered middle class.[78]

Most women work out of economic necessity; they remain in low-status jobs that provide limited personal goals.[78] Moreover, they add their job responsibilities to their traditional roster of activities and often think of their work as an extension of their nurturant maternal and providing role, rather than as an independent activity about which they may have some choice and from which they obtain personal gratification.[8] These women often experience role strain as a result of the proliferation of areas of responsibilities.[45]

Women who work bear a considerably greater psychological burden than do men in reconciling family and work demands. This burden has been increased by current reemphasis on the importance of early childhood relationships which are still usually thought of as mother-child, rather than parent-child.

Concern and conflict about the care of their children is a prominent feature in working women's lives. If children develop physical or emotional problems, women are usually quick to be blamed and also to blame themselves, although the etiology factors may not be at all clear.[68] The working woman often believes that her work is not in the best interest of her family. She may, in fact, overcompensate by asking for less help from other family members than the nonworking woman.[76]

In assessing the effect of maternal employment, data on maternal separation and maternal deprivation have been confounded with those on employment. Thus, it is often difficult to delineate factors. In addition to the work and child-care situations, one must consider the nature of the early mother-child relationship, the personality characteristics of the mother prior to the birth of the child, her concept of her role, the age of the child and his/her emotional and cognitive state, and a variety of family variables. Furthermore, these factors, in turn, affect the mother's relationship to the child and her decisions about work. Additional evidence suggests there may be sex differences in the responses of children to maternal employment; for example, a boy's development has been reported to be enhanced by a stimulating environment outside the home, whereas a girl may benefit more from close contact with the mother.[118] These findings are also related to the age of the child when the mother is less available and to the role of the father. The interpretation of these data must also address the values involved in these reports. Namely, what is considered a desirable personal outcome for boys as compared with girls?

The impact of sociocultural factors, especially poverty, in the lives of families is, in addition to psychological factors, an important concern. Bronfenbrenner[23] points out that deprivation and poverty interfere with parent-child relationships, as do social forces undermining the confidence and the motivation of people to be good parents. Poverty as well as limitations in real options, outlook, and assessment of opportunities, can contribute to a limited view of life and to depression. Furthermore, deprivation and abandonment are not only the result of the physical absence of a parent, but also derive from preoccupation, depression, and emotional unavailability.[115] If the mother is already depressed, full-time employment can further sap her energy.

Bronfenbrenner,[23] Lamb,[51] and Brown[24] also call attention to the effects on children of the absences of fathers, whether by virtue of working long hours or because of divorce. Absence of adults increases a child's susceptibility to group and peer influences, which may, in some cases, intensify antisocial tendencies.

An additional way in which adults fail to play a meaningful part in children's lives derives from discontinuity between generations and from the instability of many families resulting from physical scattering of family members. Intergenerational bonds are often lacking or diffused, and parents and grandparents do not have opportunities for contact.

¶ Maternal Deprivation and Attachment

There has been a great deal of concern about how to integrate emerging data on the importance of mother-infant attachment, with the fact that a growing number of women are leaving their children, including young infants, in the care of others. The impact of this on children's development is not clear.

In their follow-up study on the effects of close contact between the mother and newborn infant, Klaus and Kennell[49] found that there were positive effects from the intimate contact, such as greater attachment behavior, weight gain, and fewer illnesses in the first year. Also, there was greater language development at five years of age.

While this and other data emphasize the importance of early bonding, the equation of maternal employment with risk of failure of attachment cannot be made for a number of reasons. Many authors[27,59,113] emphasize the necessity of considering multiple variables in context, since to consider the effect of maternal employment alone is simplistic and limited in applicability. Cox,[27] for example, found that the effect of maternal employment was negative in father-absent families but not in families with two parents. Resch,[83] Johnson and Johnson,[45] Murray,[66] Hoffman,[41] and Moss[64] have reported that the distress of children in substitute care arrangements was eliminated, or modified, when there was familiarity with the substitute care setting, consistency and caring in relationships with caretakers and other children, and emerging play interests. Marantz and Mansfield[59] suggest that the cognitive stage of the child and other developmental variables are also important.

In a review of studies of the effect on infants of substitute care, Murray[66] concluded that developmental progress of the infant appears to be related to the strength of the mother-infant attachment and the level of stimulation in the home, regardless of the setting of rearing. Separations do not in themselves appear to be harmful, particularly after the first year, as long as they are accompanied by predictable substitute care. Emotional disturbances are reported when mothers put such young children in unstable care. Moore[62] found that children placed in day care before the age of one year showed more dependent attachment to parents and more fear than those who went into day care later. In comparing all children receiving substitute care with exclusively home-reared infants, he found that home-reared infants were less aggressive, more obedient, more docile, and more concerned with approval.

Much of the literature decrying early mother-child separation is based on the assumption that infants are monotrophic, that is, form attachments with one primary caretaking person.[21] This idea has been widely accepted, despite contradictory evidence.[25,66,90] There is considerable evidence to support the view that multiple attachments can form and that the strength of attachments depends upon the amount and quality of attention received from caretakers. In his review of the literature in this area, Rutter[90] emphasizes that the quality of mothering is the more critical variable.

Moreover, Schaeffer and Emerson[93] report that children may also select their fathers for attachment. They challenge the assumption that the child's preference is inevitably for the mother because of her nurturant relationship with the child. Other researchers agree that multiple caretakers per se are not harmful, provided there is stability and predictability in the child's environment.[25,64,90]

Thus, while there is evidence that the caretaker does indeed have to be available and consistent in the early period, there is no clear evidence to support the conclusion that there are negative effects of sharing the caretaking among consistent responsive individuals. Certainly this is the pattern in many cultures in which a grandmother, older sibling, or other relative or friend regularly performs part of the caretaking.[116]

This raises still another basic issue, that is, how one decides on what constitutes a "good" or "bad" outcome in a child. At this time the connections between early patterns and later developmental outcomes are unclear. Furthermore, there are important cultural variations. Since children are reared to become adaptive adults, there are no absolute answers as to what must be provided in every instance. Many models may be applicable as long as basic needs for stable care and contact are met.

Furthermore, the importance of looking at the development of attachment as a mutual interaction is often overlooked. Studies of constitutional traits have recognized that, in the initial month of a child's life, the child's development involves a mutual interaction with the caretaker.[22,36] An understanding of the reciprocity of the relationship between the infant and caretaker raises doubt about the conventional concept of the mother as the one person responsible for shaping development of a plastic, unformed individual.

¶ Sex Roles—Biology and Communication

While much of the literature considers the specific impact of particular behaviors, such as presence or absence of a mother or the socioeconomic condition of the family, there has been less attention paid to the significance of communications about expectations, particularly with regard to sex roles.

The choices and adaptations of children reared in home environments with differing values and expectations are affected by these communications.

Parents do treat little boys and girls differently.[70] They do this in many subtle ways, consciously and unconsciously expressing their views of sex role differences, for example, the perception by parents of an active little girl as "masculine" and a quiet, reflective little boy as "feminine" occurs frequently and is clearly perceived as approval or criticism by a child.

The evidence that parents have different expectations and, in fact, behave differently toward male and female infants is convincing.[18] Rubin, Provenzano, and Luria[88] found that there were consistent differences in the reports of mothers about the characteristics of their one-day-old infants that were not objectively confirmed. Female infants were seen as significantly softer, finer featured, smaller, and more inattentive than male infants. Furthermore, they reported that from the birth of the child mothers acted upon these perceptions. Studies of older infants continue to reveal mother-child interactional differences. Moss[64] found that mothers tended to soothe girls when they cried by picking them up, smiling, and vocalizing, while boys were more often held, rocked, and aroused.

In another study, when the same six-month-old infant was presented to mothers who were not told the sex of the infant, a differential response was reported depending on which sex they thought the baby was. When the mothers thought that the infant was a girl, "she" was handed a doll more frequently; when they thought it was a boy, a train was given. In addition, the mothers did not recognize that their behaviors differed. They also thought the female was "softer" and so forth, even though the same infant was presented to all mothers.[46]

While there is evidence that biological differences in the development and functioning of the nervous and endocrine system exist between girls and boys, the relevance of these have been difficult to establish. The powerful influence of culture and the enormity of individual variation are so critical and begin to have an impact so early that it becomes difficult to draw conclusions about which factors are innate or fixed, and which derive from postnatal influences. Maccoby and Jacklin,[58] in an exhaustive review of research data and cross cultural observations, point to evidence that a large number of widely held views are not supported by data.

Block[18] emphasizes the importance of differences between mothers and fathers in their interaction with male and female children of different ages. She also notes the relative paucity of data on fathers' interactions with their children. Block's data indicate that both mothers and fathers appear to emphasize achievement and competition more for their sons than for their daughters, and that there is a greater emphasis on control of affect, independence, and assumption of personal responsibility by the parents of males. Block[19] states that age-specific socialization practices exist and are related to the developmental level of the child and to the tolerance of society for impulse expression in particular areas.

Restrictive stereotypes limit the functioning of both men and women and determine many aspects of their behavior. The prohibition against affective expression by men inhibits the development of sensitivities in them, just as the overemotional, dependent model inhibits the expression of activity and independence by women.[18] Block[18,19] feels that because children are socialized early into culturally-defined sex-appropriate roles, introspection and self-evaluation, which are essential for psychological growth, are discouraged. She states that socialization appears to expand the options of the male, who is encouraged to be competent and actively instrumental in tasks, but that its effects on the female are more negative since individuation and self-expressive traits may be discouraged, leading to repression of activity-oriented impulses and renunciation of achievement and autonomy.[18,19]

The potential developmental significance of early communications leads us to consider the implications for psychological development when communications are confused, dissonant, or changing.

¶ Impact of Maternal Employment

We can now redirect ourselves to another aspect of the implications of alternative life-styles, the effect of maternal work as a communication with children. A growing body of data supports the idea that there are benefits for mothers, children, and families when the mother works, even if it is out of necessity rather than desire.* Hoffman[41] finds that in her relationship with her children, she, as a working mother, feels less hostility and more empathy toward her children, expresses more positive affect, and employes less coercive discipline, although she may be somewhat overindulgent. Birnbaum[17] studied the attitudes of professional women toward their children and compared them with nonworking mothers. She found that the professional mother experienced greater pleasure in her children's growing independence, was less inclined to be overprotective, and placed less emphasis on self-sacrifice.

There is considerable additional evidence that a working mother has a positive effect, particularly on her daughters.[76] The daughters of working mothers were found to be more likely to choose their mothers as models and as the people they most admired. Adolescent daughters of working mothers, particularly in middle- and upper-socioeconomic groups, were active and autonomous and admired their mothers, but were not unusually tied to them. Some authors, in fact, have speculated that greater maternal distance may result in greater later achievement by daughters.[8,29,46,95] For girls of all ages, having a working mother contributed to their concept of the female role as less restricted. They usually approved of maternal employment and planned to seek employment when they grew up and became mothers. Unlike daughters of nonworking mothers, they did not assume that women were less competent than men. Vogel and associates[109] and Douvan[31] found that the daughters of working mothers have less traditional sex role attitudes than the daughters of nonworking mothers. Interestingly, these findings were obtained in a period when working mothers were less often found than at present.

Studies of daughters' academic and career achievements, provide additional evidence of the positive effects of a mother with career interests. A number of investigators have found that women who achieve or who aspire to careers, particularly to less conventionally feminine careers, are more likely to be the daughters of educated and employed women.* Data on the husbands of working women indicate that they are more actively involved in the care of the children and that the active involvement of fathers has a positive effect on both male and female children.[52,120] Furthermore, the husbands of professional women are more likely to respect competence and achievement in women.†

¶ Models of Marital Interaction: Dual Workers and Dual Careers

The Rapoports,[81] in 1971, suggested that the continuing increase in dual-career families would require more child care, revisions in sex-role attitudes, and reconsiderations of the organization of productive work and of family life. In their subsequent volume,[82] they pointed out that while the dual-career pattern was being chosen by more people, many of the necessary changes in life-styles and patterns had not occurred because of the inflexibility of social systems and internalized resistance to change.

*See references 4, 42, 43, 66, and 113.

*See references 5, 17, 54, 76, and 100.

†See references 5, 17, 30, 38, 54, 57, 76, 81, and 100.

If both partners in a dual-career marriage have high commitment careers with responsibilities extending beyond the usual eight-hour work day, the commitment demands modification in roles, tasks, and decision making. For example, a husband cannot assume that his wife will take off from work when a child is ill or when household repairs are needed, and a wife cannot count on her husband to be available to repair the car when it breaks down. Couples must work out a variety of strategies to cope with both the ordinary aspects of life and the special circumstances that are created by the lack of availability of one partner who functions as the "wife"—the partner who tackles the chores, arranges child care, schedules social activities, and buffers the other (the husband) from the demands of daily life.

Weingarten[114] describes a typical scene which points to some of the adaptational issues faced by the dual-career couple.

Mr. Jones is in his study finishing a speech he will be delivering in New York the next day, while making calls to the usual network of babysitters to arrange for someone to stay with his children for the evening. Dr. Jones is talking with her answering service on the other line to ascertain how high a fever Billy Smith has, while simultaneously heating up a stew for dinner. In an hour, Mr. Jones will drive to the airport, the babysitter will arrive, and Dr. Jones will meet Billy Smith and his parents at a local hospital emergency room.

Meanwhile, at Billy Smith's home, Mr. Smith is calling his wife at her law office to ask her to stop off on her way home and buy a pizza so that they will not have to prepare dinner in case Dr. Jones wants to see Billy that evening.

Clearly, dual-career marriages compel both partners to make adaptations that may not be required within more conventional marriages, and these have important implications for children. Modifications in many areas of life can be seen in terms of decision making and allocation of responsibilities for family maintenance and care of children. Couples must often redefine gender-oriented activities and adapt emotionally to the stresses of new roles and expectations. Obviously, when both spouses are heavily invested in their careers, giving up one's aspirations for the benefit of the other is likely to be viewed as a loss.

According to Rosen, Jerdee, and Prestwich,[86] it is rare for couples to acknowledge that they have chosen to give the wife's career precedence, because of the conflict with traditionally accepted values. Berger and associates[11] report that even when this was done, the family often maintained the myth that the traditional model still existed. The husband who was a house-husband denied that he was unemployed, even to friends.

Lein and associates[52] found that for spouses who viewed work as a financial necessity rather than a personal commitment, it was easier to be supportive of each other's job-seeking endeavors. Under these circumstances the wife's work was seen as a way of expressing interest in the family, rather than competition with the husband's work. Among dual-career couples, the wife's work was more likely to be thought of as competing with the husband's work.

The dual-career model is relatively recent, and except for a few reports, there are few data on its long-term viability. For example, the widely held myth that marital discord or dissatisfaction may arise from conditions of status inequality in the marriage has not been supported. Richardson[84] was not able to substantiate the presumption of marital troubles arising in dual-career families in which wives were equal to or higher in occupational prestige than their husbands.

With regard to children, there are even fewer clear data, except that derived from the evidence of achievement of the children, noted previously. On the other hand, there is positive developmental value connected with communicating to children that any number of instrumental, cognitive, and emotional functions are not necessarily gender-oriented or rigidly associated with sex role. It seems reasonable to hope that children who grow up in a home where caretaking activities are shared between the father and mother will be able to develop flexible ideas of their personal work role and family role identities. Furthermore, children can develop the idea that both parents can be readily and realistically available for all kinds of problem resolution.

¶ Divorce

An alternative, the single parent or reconstituted family, may result when divorce has occurred. The fact that most divorced people remarry, although men are more likely to do so than women, indicates that there are likely to be significant changes in family structure. This includes the problems and adjustments of reconstituted families and the increased number of single parents as a result of divorce.[7]

People considering divorce often do not expect to be vulnerable to the stress which occurs particularly in the period of time shortly after the divorce. They may be unprepared for the loneliness, the isolation, and the emptiness they often feel. While separation may initially be a relief if a marriage has been filled with tension and conflict, at some point those who divorce usually do experience profound feelings about the loss.

The stress of marital disruption may lead to emotional or physical symptoms. It has been reported that generally, widowed, separated, and divorced persons have higher rates of disability than married or never-married persons. Separated and divorced individuals, particularly women, seem unusually vulnerable to acute ill health. The most vulnerable period appears to occur during the separation phase.[73]

People who are separated or divorced have also been found to be overrepresented among psychiatric patients. Of all the social variables related to the distribution of psychotherapy in the population, none has been more consistently found to be so crucial than marital status.[28]

In both sexes, the automobile-accident fatality has been reported to be higher among the divorced than among any other marital status group.[72] It has been found that the accident rate doubled during the period from six months before to six months after a divorce.[60] It also appears that when there has been a loss of a loved person within the previous year (by death, divorce, or separation), the potential for self-destruction is greater. The suicide rate is higher among the divorced than among people of any other marital status.[94] There is increasing evidence linking human loneliness, or a sense of separateness, to disease and to premature death. "Although marital status is not clearly indicative of the presence or absence of loneliness, a few comparative statistics in the premature death rates of white males reveal strikingly higher death rates among the unmarried." Besides statistics from the suicide and automobile fatalities, data demonstrate increased mortality rates from cirrhosis of the liver, pulmonary carcinoma, gastrointestinal carcinomas, cerebrovascular accidents, and cardiac disease among the divorced and widowed population.[56]

People in the throes of marital crisis may experience lowered self-esteem and may be more vulnerable to symptoms, and this may have a substantial impact of their children, acutely as well as chronically. Statistically the relationship between marital disruption and psychopathology appears to be greater for men than for women.[20] Separated men complain of a sense of rootlessness, a loss of purpose, and a lack of meaning in their lives. They seem to experience, more than women, a sense of personal failure at the ending of a marriage.

Perlin and Johnson[79] found that symptoms of depression were particularly common among the formerly married, especially those with children, those with inadequate financial resources, and those who were socially isolated. The rate of depression among the divorced increased with the number of children at home and was greater when the children were younger.

The demands as well as the anger and anxiety of children may be draining and difficult.

Parents having marital difficulties often feel guilty about the distress they see in their children, and they may be overly sensitive to criticism or disappointment. Even the "normal" problems of growing up can be seen as related to the marital failure. Self-blame and anger at the partner frequently increase the burden for children.

When children are involved, divorce ends the marriage but not the relationship with the children, or usually, with the ex-spouse, thus adding to the parent's experience of stress. Fathers in particular suffer from being displaced. Losing a spouse may be viewed as a blessing, but the absence of children can be extremely upsetting for the noncustodial parent. Men rejected by their wives may fear equal rejection by their children and may then alter their behavior in order to protect themselves from this potential loss. The children, in turn, may lose their father and find themselves with a "buddy" who tries desperately to please them. They may experience their father in a new way—as dependent upon them.

Some parents seek remarriage to provide a home, or to make up for an absent parent, only to find that new problems emerge. Children may reject the new partner because they had hoped for the reunion of their parents, they may see the new spouse as a competitor, or they may not like the new "parent." It is important to recognize that this new "parent" is, in fact, a stepparent and not a replacement. A relationship must develop over time. It is a new relationship with unclear definition for all members of the family. Because a marriage has taken place and the parent and new partner view themselves as a couple, does not mean the children see the situation the same way.[107]

When there are children, divorce should be seen as representing a change in the structure and function of a family, rather than as a dissolution of the family itself, since the children continue to have their original family as well as family that is added. When bitterness is contained, and the best possible decisions are made, adaptation can be easier.

¶ Divorce and Children

It has been estimated that of all children born in the decade of the 1970s, 30 percent will experience parental divorce. As the number of children involved in marital disruption increases, it is necessary to understand the effects of this experience on them. Currently, 60 percent of all divorces involve children under the age of eighteen. In 1976, over one million children were involved in divorce proceedings.[7]

Rutter's work[89] serves to put in perspective the impact of marital disruption. He found that there is an incidence of behavior problems with 5 percent of the children with a good relationship with one or both parents in a stable marriage, 25 percent of the children with a poor relationship with one or both parents in a stable marriage, 40 percent of the children with a good relationship with one or both parents in a conflict-ridden marriage, and 90 percent of the children with a poor relationship with parents in a poor marital relationship. Thus, it appears that parental conflict produces symptoms in children and that children appear to function better in an atmosphere of contentment, whether in a two-parent home or in a single-parent situation.

Wallerstein and Kelly[110,111,112] focused on the child's developmental level when they reported on the impact of divorce on children. They divide children into four groups: (1) preschool ages three to six; (2) early latency ages seven to eight; (3) late latency ages eight to eleven; and (4) adolescence.

They found that preschool children tended to focus explanations on themselves, for example, "Daddy left because I was a bad girl." Self-blame for illogical reasons were seen frequently and were not easily treated. Most of these young children had difficulty expressing their feelings. They also lacked a firm sense of continuing family relationships after the divorce, since their conception of a family was based on those individuals who lived together in the same household. These findings suggest that long-term disruption is more probable in younger children. It has also been noted that a single parent is most vulnerable to depression when his/her children are under the age of five.

By the age of seven, children were more aware of their feelings and better able to admit to sadness, although not so able to acknowledge anger. They tended to blame themselves less, but experienced strong feelings of abandonment and rejection. These were coupled with fear of their mother's anger. By age nine, children rarely felt at fault, but experienced a profound sense of rejection and abandonment by the departing parent. They acknowledged their feelings of anger toward one or both parents, but they attempted to conceal their pain and wanted to appear to be courageous to the outside world. Intrapsychically, they experienced intense loneliness and they were torn by the conflict between anger and loyalty. They felt betrayed by their parents. They were also troubled by the sense that their relationship with one parent was jeopardized because of that parent's withdrawal of emotional investment.

Adolescents seemed painfully aware of their feelings and were frequently able to express the anger, sadness, loss, betrayal, and shame. Generally, they seemed to recognize their parents as individuals and they achieved some insight. They demonstrated an understanding of their parents' incompatibility, however, they were extremely concerned about their own future marriage and the possibility of sustaining relationships. It appeared that those teenagers who were able to detach themselves from their parents' emotional turmoil soon after disruption of the marriage achieved a better long-term adjustment.[97]

Frequently the pain of children is denied by one or both parents, who are struggling with their own emotional responses. Children fear that they will lose the love of their remaining parent if they continue to love the absent one. For youngsters, divorce implies loss, often in the context of unresolved ambivalence about the intense need for the absent parent and the wish to have the conflict between parents ended. While the parents may continue to be available, the noncustodial parent may not be involved as actively. The children's fantasies of reconciliation last for years, even after one parent has remarried. Fox[35] found that mothers of preschool children were more likely than mothers of older children to report their children's jealousy of mother's boyfriends. Mothers of older children were more likely to report that their children encouraged "father shopping."

It is a major psychological accomplishment for a child to complete the task of mourning the loss of the predivorce family and to accept a new family constellation, which includes a new relationship with both the custodial and the visiting parent. The child's daily routine will probably change and he or she will have to adapt to the stress of a new economic and social situation, including the possibility of a new home, new friends, a new school, and living with a parent who will undoubtedly be preoccupied, overburdened, anxious, and maybe depressed.

A variety of symptoms may appear in children, depending on the child's developmental level and previous adaptive capacity. Symptoms include insomnia, nightmares, inability to concentrate, decline in schoolwork, emotional outbursts and regressive behavior, eneuresis, thumb sucking, compulsive masturbation, and behavior that is overtly sexual or aggressive.[92] Children may also report a number of functional complaints, including muscle weakness, fatigue, change in appetite, abdominal pain, headaches, and general anxiety symptoms. Psychosomatic disorders, such as asthma and ulcerative colitis, have also been reported, and pre-existing symptoms may be exacerbated.

Preventive approaches can increase the capacity for adjusting to new situations, by fostering a sense of security and love. Lacking these approaches, children are in danger of losing their developmental stride or of internalizing a sense of low self-esteem.[99]

¶ The Single Parent

In 1976, 85 percent of all white children in this country lived with both parents. The remaining 15 percent, or approximately 7 million children, lived in a single-parent family, largely as result of divorce. Eleven and nine-tenths percent of all children lived with mothers only; 1.2 percent with fathers only; the remainder lived with neither.[103]

When a single parent is the mother, the economic burden becomes awesome. Divorce thrusts most families into a lower socioeconomic situation, especially since a large number of fathers do not provide financial support for the children after divorce, and fathers often also lose active contact with the children. Thus, the earning potential of female heads of families is crucial. The number of divorced mothers in the labor force is higher than that of any other group of mothers. Most of these women are concentrated in low-paying jobs. In 1976, 52 percent of children in female-headed families were in families living below the poverty line.[104]

When there is a divorce, adaptation is required of all family members. Tasks are often reassigned and the children are expected to assume more household and personal care responsibilities. Single-parent families often suffer from a sense of social isolation and from a more erratic life-style.[97] For many children, this includes being shuffled from weekday homes to weekend homes and visiting in artificial settings.

The single mother or father may turn to a child to assume some aspect of the role of the missing spouse, an experience which can be distressing to the child. A father who wins custody of his children may unconsciously set up a new household in which his oldest daughter replaces his wife and becomes the caretaker of the family. She may have to struggle with her own wishes and fears about replacing her mother. Not infrequently a situation such as this leads to adolescent acting out behavior that results in pregnancy.[69] Similarly, the visiting parent may establish an inappropriate relationship with his or her children. For example, the visiting father may treat his teenage daughter more as a date than as his child and he may be surprised when she states that she will only visit him if she brings a friend with her.

Children in postdivorce single-parent families have been reported to seek treatment at a rate of one per 100 population, or more than double the rate for children in two-parent families. Furthermore, the impact on children in single-parent families appeared greatest among the youngest children. Children aged six to seventeen in female-headed families had a treatment-seeking rate more than double that of children of the same age in two-parent families, but children under six years had a rate that was four times as large as that of children under six in two-parent families. While children under six from all family types combined had a lower help-seeking rate than older children, children under six living with separated single-parent mothers had a higher rate than children six to thirteen living with their mothers. This reversal of a positive association of age and utilization rate is cause for concern.[10,40]

¶ Reconstituted Families

The current high divorce rate is coupled with a high remarriage rate. Four out of five divorced individuals remarry, with younger individuals more likely to remarry than older ones, and men more likely to remarry than women. Current statistics show that only about half of these marriages succeed.[39]

By 1975, 15 million children in this country under the age of eighteen were living in stepfamilies, and there were approximately 25 million husbands and wives who were stepparents.[85] There has been to date no clear definition of the role of the stepparent. The only social and legal model has been the natural parent role.

In most cases the stepparent is a newcomer who is not welcomed into the existing system and is experienced by the children as an intruder. Stepfamilies are likened by Visher and Visher[107] to open systems with greater instability and lack of control than nuclear families. They list the popular myths about stepparents: (1) stepfamilies are the same as nuclear families; (2) the death of a spouse makes stepparenting easier; (3) stepchildren are less burdensome when not living at home; (4) love of a partner guarantees love of the partner's children; and (5) every new stepparent will immediately love his or her stepchildren.

Most stepmothers feel pressure to achieve closeness with stepchildren and to avoid being cast in the role of "wicked stepmother." Women tend to experience more guilt than men if they do not succeed in this task. A new stepfamily is plunged into a sea of unresolved feelings toward ex-spouses and lost parents. Anger and competition are evoked. Children often attempt to fragment the adults' relationships in order to reunite the original family.

The first three to four years of a new family unit are the most turbulent as the husband and wife struggle to create a new relationship. There is little privacy, and much need to devote time and energy to children. Often the children attempt to exclude the stepparent from family intimacy.

Remarried women often fear that their own children from the previous marriage will suffer, or that they must favor a stepchild. Stepparents often find it difficult to know how to behave. Remarried men often feel guilty and concerned that their children are being neglected, especially if they have left them, or if there is a new baby by the new wife. Stress may be greater during visits by the children on holidays. Self-doubt, insecurity, and inferiority feelings are frequent reactions to the confusing demands and pressures. Lowered self-esteem, feelings of helplessness, and depression may be the psychological risks of stepparents.[61]

¶ Two-Location Families

From an anthropologic point of view, families do not necessarily form households. The U.S. Census Bureau, however, assumes that married people live together and that those who are unmarried live alone.[102] While there are few data to indicate how many people do live separately, certainly war, immigration, and economic necessity, as well as some jobs (such as salesman) that require frequent travel, have kept families apart. Today, however, a new pattern is emerging: the female-determined two-location marriage. Kirschner and Walum[48] reported a variety of modes of coping with living apart, including the use of the telephone and letters. They found that couples did not waste time together in meaningless activities, but used their time effectively. Often work activity was compressed during the couple's separation to allow for more free time when they were together.

Couples use their separation to maximize their individual career opportunities. Farris[34] reported a high level of education and income in these couples. She also found that dissatisfaction with the arrangement increases with the duration of separation, thus the couple who commute on a weekly basis have less difficulty than those who commute on a monthly basis.

Those couples who did not have children did not tend to view either home as the primary one. If there were children, the primary one was considered the one in which the children lived. The woman-determined two-location family leaves the male either as a new or renewed single in an established network, or in a new location as a married single. In either case, friendship or support systems may be difficult to establish or maintain because of lack of understanding and discomfort experienced by others. The female member finds herself in a similar situation. The extended family may deprecate her, coupled friends may avoid her, and new male friends may try to seduce her. The male culture, furthermore, is not supportive or sympathetic to the male who "lets" his wife go off on her own, and this is especially so if she has left him with the children.

Commuting families with children continue to be rare, and there is little data available on the impact of this life pattern on children.

¶ Other Alternative Styles

Communes and communal living arrangements, which have become common since the 1960s, have long been a tradition in American society. More than 200 were founded in the nineteenth century alone.[65] Those communes that have survived the longest have been groups with a strong unifying drive or theme, often religious, and with a strong, usually patriarchal leader. Communes tend to consist of a group of people, married or single, who decide to live together and share certain aspects of their lives. This may include money, a home, activities, child rearing, or other tasks, as well as life goals. Many communes have evolved into self-contained communities and have been in existence for ten or more years, whereas others are temporary and continue only as long as the members remain in similar positions in life.

Most often, people who live in communes, if they are married, do maintain monogamous life-styles within the structure of the community. This is to be distinguished from group marriage situations which are essentially polygamous and where sexual exchange occurs and women may have children fathered by anyone in the group. Over time, these group marriages have tended to be unsatisfactory to the participants because of the difficulty most people have in sharing all aspects of their lives.[117]

Another alternative, the open marriage, is also defined in different ways, depending upon the participants. Although the couple is formally married open marriage for some implies total freedom of life-style including the possibility of multiple sexual partners; for others, freedom is limited to friendships and activities, but other sexual partners are not expected to be part of the arrangement.

The O'Neills[77] proposed an open marriage based on equal freedom and identity. The premise is that both partners change in a marriage and that each must accept responsibility for himself or herself and grant the same to the partner. They also believe that children need not be required as proof of love. In contrast to the mystique of togetherness, the guidelines in this marriage are equality, which indicates respect for the equal status of the other; role flexibility; open companionship; identity; the development of the individual through the realization of potential and growth toward autonomy; privacy; open, honest communication; living for the present; and trust. This marital pattern is still marriage conceptualized within the framework of a primary relationship between two people.

Another model is the multilateral marriage where at least three individuals live together and have a commitment that is essentially analogous to marriage.[26] Those groups studied revealed that the members were motivated by personal growth opportunities and by interest in having a variety of sexual partners. In these families, sex roles were much less differentiated than in the average nuclear family.

Because they are relatively recent in origin, there have been few careful evaluations of the successes or problems of these alternative life-styles. While there are distinct advantages to sharing goals, values, and property, in many of the alternative living situations jealousy, competitiveness, and communication problems exist and may ultimately disrupt the group or couple when other partners or demands force choices or preferences. It is clear that changes and experiments will continue, and that there are many reasons why men and women seek other sources of support, companionship, and intimacy. In this situation, as well as in traditional marriages, the fairy tale concept of living happily ever after is not necessarily realized. Further, we have few data on the impact of these environments upon children, particularly over time.

¶ Bibliography

1. ABBOTT, S., and LOVE, B. *Sappho Was a Right-On Woman.* New York: Stein & Day, 1972.
2. ABERNATHY, V. "American Marriage in Cross-Cultural Perspective," in Grunebaum, H. and Christ, J., eds., *Contemporary Marriage: Structure, Dynamics and Therapy.* Boston: Little, Brown, 1976, pp. 33–52.
3. ADAMS, M. "The Single Women in Today's Society," *American Journal of Orthopsychiatry,* 41 (1971): 776–786.
4. AL-TIMIMI, S. "Self Concepts of Young Children with Working and Non-Working Mothers," (Ph.D. diss., Peabody College, 1976) unpublished.
5. ALMQUIST, E., and ANGRIST, S. "Role Model Influences on College Women's Career Aspirations," *Merrill Palmer Quarterly of Behavior and Development,* 17 (1971): 263–279.
6. BAILYN, L. "Family Constraints on Women's Work," *Annals of the New York Academy of Sciences,* 208 (1973): 82–90.
7. BANE, M. J. "Marital Disruption and the Lives of Children," in Levinger, G. and Moles, O., eds., *Divorce and Separation.* New York: Basic Books, 1979, pp. 276–286.
8. BARDWICK, J. *Psychology of Women.* New York: Harper & Row, 1971.
9. BELL, R. *Marriage & Family Interaction.* Homewood, Ill.: Dorsey, 1971.
10. BELLE, D. "Who Uses Mental Health Facilities?" in Guttentag, M. and Salasin, S., eds., *Families Abandoned: The Cycle of Depression.* New York: Academic Press, (forthcoming).
11. BERGER, M., FOSTER, M. and STRUDLE-WALLSTON, B. "Finding Two Jobs," in Rapoport, R. and Rapoport, R., eds., *Working Couples.* New York: Harper, Colophon, 1978, pp. 23–35.
12. BERMAN, E. M., and LIEF, H. I. "Marital Therapy From a Psychiatric Perspective: An Overview," *American Journal of Psychiatry,* 132 (1975): 583–592.
13. BERNARD, J. "The Paradox of the Happy Marriage," in Gornick, V. and Moran, B., eds., *Women in a Sexist Society.* New York: Basic Books, 1971, pp. 145–162.
14. ———. *The Future of Marriage.* New York: Bantam Books, 1972.
15. ———. *The Future of Motherhood.* New York: Dial, 1974.
16. BIRD, C. "The Case Against Marriage," in Howe, L. K., ed., *The Future of the Family.* New York: Simon and Schuster, 1972, pp. 341–348.
17. BIRNBAUM, J. "Life Patterns and Self Esteem in Family Oriented and Career Committed Women," in Mednick, M., Tangri, S., and Hoffman, L., eds., *Women and Achievement: Social and Motivational Analysis.* New York: John Wiley, 1975, pp. 396–419.
18. BLOCK, J. "Conceptions of Sex Role: Some Cross-Cultural and Longitudinal Perspectives," *American Psychologist,* 28 (1973): 6.
19. ———. "Another Look at Sex Differentiation in the Socialization Behaviors of Mothers and Fathers," in Sherman, J. A., and Denmark, F. L., eds., *Psychology of Women: Future Directions of Research.* New York: Psychology Dimensions, 1978, pp. 29–87.
20. BLOOM, B. L., ASHER, S. J., and WHITE, S. W. "Marital Disruption as a Stressor: A Review and Analysis," *Psychological Bulletin,* 85 (1978):867–894.
21. BOWLBY, J. *Attachment and Loss: Attachment,* vol. 1. London: Hogarth Press, 1969.
22. BRAZELTON, B., and KEEFER, C. "The Early Mother-Child Relationship," in Nadelson, C. and Notman, M., eds., *The Woman Patient,* vol. II, New York: Plenum Press, (forthcoming).
23. BRONFENBRENNER, U. *Two Worlds of Childhood: U.S. and U.S.S.R.* New York: Simon and Schuster, 1970.
24. BROWN, O. "Macrostructural Influences on Child Development and the Need for Childhood Social Indicators," *American Journal of Orthopsychiatry,* 45 (1975): 4.
25. CASLER, L. "Maternal Deprivation: A Critical Review of the Literature," *Sociological Review Monographs,* 26 (1961): 2.
26. CONSTANTINE, L. I., and CONSTANTINE, J. M. "Group and Multilateral Marriage: De-

partmental Notes, Glossary and Annotated Bibliography," *Family Process,* 10 (1971):-157–176.

27. COX, M. "The Effects of Father Absence and Working Mothers on Children," (Ph.D. diss., University of Virginia, 1975) unpublished.
28. CRAGO, M. A. "Psychopathology in Married Couples," *Psychological Bulletin,* 77 (1972):114–128.
29. CRANDALL, V. "Achievement Behavior in Young Children," *Young Children,* 20 (1964):77–90.
30. DIZARD, J. *Social Change in the Family.* Chicago: University of Chicago Press, 1968.
31. DOUVAN, E. "Employment and the Adolescent," in Nye, F. I. and Hoffman, L., eds., *The Employed Mother in America.* Chicago: Rand McNally, 1963, pp. 142–164.
32. DUBERMAN, L. *Marriage and Its Alternatives.* New York: Praeger, 1974.
33. ESCAMILLA-MONDANARO, J. "Lesbians and Therapy," in Rawlings, E. and Carter, D., eds., *Psychotherapy for Women.* Springfield, Ill.: Charles C. Thomas, 1977, pp. 256–265.
34. FARRIS, A. "Commuting," in Rapoport, R. and Rapoport, R., eds., *Working Couples.* London: Routledge & Kegan Paul, 1978, pp. 100–107.
35. FOX, E., personal communication to Belle, D., 1976. In Heckerman, C., ed., *The Evolving Female.* New York: Human Science Press, 1980, pp. 74–91.
36. FRIES, M. "Longitudinal Study: Prenatal Period and Parenthood," *Journal of the American Psychoanalytic Association*, 25 (1977): 115–132.
37. FRIEZE, I., et al. *Women and Sex Roles.* New York: Norton, 1978.
38. GARLAND, T. N. "The Better Half?: The Male in the Dual Profession Family," in Safilios-Rothschild, C., ed., *Toward a Sociology of Women.* Lexington, Mass.: Xerox College Publishing, 1972, pp. 199–215.
39. GLICK, P. C. "A Demographic Look at American Families," *Journal of Marriage and the Family,* 37 (1975):15–26.
40. GUTTENTAG, M., SALASIN, S., and BELLE, D. "Executive Summary: National Patterns in the Utilization of Mental Health Service," (unpublished paper, Harvard University, 1977).
41. HOFFMAN, L. "Early Childhood Experiences and Women's Achievement Motives," *Social Issues,* 28 (1972):129–55.
42. HOWELL, M. "Effects of Maternal Employment on the Child (II)," *Pediatrics,* 52:3 (1973): 327–343.
43. HOWELL, M. "Employed Mothers and Their Families (I)," *Pediatrics,* 52:2 (1973): 252–263.
44. JACOBY, S. "49-Million Singles Can't Be All Right," *The New York Times Magazine,* 17 February 1974.
45. JOHNSON, F., and JOHNSON, C. "Role Strain in High Commitment Career Women," *Journal of the American Academy of Psychoanalysis,* 4:1 (1976): 13–36.
46. KAGAN, J., and MOSS, H. *Birth to Maturity.* New York: John Wiley, 1962.
47. KATZ, J. "Past and Future of the Undergraduate Woman," presented at Radcliffe College, Cambridge, Mass., April 1978.
48. KIRSCHNER, B. F., and WALUM, L. R. "Two Location Families: Married Singles," *Alternative Life Styles,* 1 (1978):513–525.
49. KLAUS, M., and KENNELL, J. *Maternal Infant Bonding.* St. Louis: C.V. Mosby, 1976.
50. KUHN, M. "How Mates Are Sorted," in Becker, H. and Hill, R., eds., *Family, Marriage and Parenthood.* Boston: D.C. Heath, 1948, pp. 246–275.
51. LAMB, M. "Fathers: Forgotten Contributors to Child Development," *Human Development,* 18 (1975):245–266.
52. LEIN, L., et al. "Final Report: Work and Family Life." National Institute of Education Project #3-3094. Cambridge, Mass.: Center for the Study of Public Policy, 1974.
53. LEVENSON, D. *The Seasons of a Man's Life.* New York: Knopf, 1978.
54. LEVINE, A. G. "Marital and Occupational Plans of Women in Professional Schools," (Ph.D. diss., Yale University, 1968) unpublished.
55. LOZOFF, M. "Changing Life Styles and Role Perceptions of Men and Women Students," presented at *Women: Resource for a Changing World,* Radcliffe College, Cambridge, Mass., April 1972.
56. LYNCH, J. J., and CONVEY, W. H. "Loneliness, Disease and Death: Alternative Approaches," *Psychosomatics,* 20 (1979):702–708.
57. MACCOBY, E. "Sex Differences in Intellectual Functioning," in Maccoby, E. E., ed.,

The Development of Sex Differences. Stanford, Calif.: Stanford University Press, 1966, pp. 25–56.

58. MACCOBY, E., and JACKLIN, E. *Psychology of Sex Differences.* Stanford, Calif.: Stanford University Press, 1974.
59. MCMURRAY, L. "Emotional Stress and Driving Performance: The Effect of Divorce," *Behavioral Research in Highway Safety,* 1 (1970):100–114.
60. MARANTZ, S., and MANSFIELD, A. "Maternal Employment and the Development of Sex Role Stereotyping in Five- to Eleven-Year-Old Girls," *Child Development,* 48 (1977):668–673.
61. MESSINGER, L., WALKER, K. N., and FREEMAN, S. J. J. "Preparation for Re-marriage Following Divorce: The Use of Group Techniques," *American Journal of Orthopsychiatry,* 48 (1978):263–272.
62. MOORE, T. "Children of Working Mothers," in Yudkin, S. and Holme, W., eds., *Working Mothers and Their Children.* London: Michael Joseph, 1963, pp. 105–124.
63. MORONEY, R. "Note from the Editor," *Urban and Social Change Review,* 11 (1978):2.
64. MOSS, H. "Sex, Age, and State as Determinants of Mother-Infant Interaction," *Merrill Palmer Quarterly of Behavior and Development,* 13 (1967):19–36.
65. MUNCY, R. L. *Sex and Marriage in Utopian Communities: 19th Century America.* Bloomington, Ind.: Indiana University Press, 1973.
66. MURRAY, A. "Maternal Employment Reconsidered: Effects on Infants," *American Journal of Orthopsychiatry,* 45 (1975):-773–790.
67. NADELSON, C. "The Treatment of Marital Disorders from a Psychoanalytic Perspective," in Paolino, T. and McCrady, B., eds., *Marriage and the Treatment of Marital Disorders: Psychoanalytic, Behavioral and Systems Theory Perspectives.* New York: Brunner/Mazel, 1978, pp. 101–164.
68. NADELSON, C., and NOTMAN, M. "Medicine: A Career Conflict for Women," *American Journal of Psychiatry,* 130 (1973):1123–1127.
69. NADELSON, C., NOTMAN, M., and GILLON, J. "Adolescent Sexuality & Pregnancy," in Notman, M. and Nadelson, C., eds., *The Woman Patient.* Plenum Press, 1978, pp. 123–130.
70. MILLER, J., et al. "Some Considerations of Self-esteem and Aggression in Women," in Klebanow, S. ed., *Changing Concepts in Psychoanalysis.* New York: Gardner Press, forthcoming.
71. NADELSON, C., POLONSKY, D. C., and MATHEWS, M. A. "Marriage and Midlife: The Impact of Social Change," *Journal of Clinical Psychiatry,* 40 (July 1979):292–298.
72. NATIONAL CENTER FOR HEALTH STATISTICS. "Mortality from Selected Causes by Marital Status," Vital and Health Statistics Series 20, nos. 8a, 8b. Washington, D.C.: U.S. Government Printing Office, 1970.
73. ———. "Differentials in Health Characteristics by Marital Status: United States, 1971–72," Vital and Health Statistics Series no. 104. Washington, D.C.: U.S. Government Printing Office, 1976.
74. ———. "Births, Marriages, Divorces and Deaths for 1976," *Monthly Vital Statistics Report No. 25(2).* Washington, D.C.: U.S. Government Printing Office, 8 March 1977.
75. NEUGARTEN, B. *Middle Age and Aging.* Chicago: University of Chicago Press, 1968.
76. NYE, F. I., and HOFFMAN, L. *The Employed Mother in America.* Chicago: Rand McNally, 1963.
77. O'NEILL, G., and O'NEILL, N. *Open Marriage.* New York: Avon, 1972.
78. PEARCE, D. "The Feminization of Poverty: Women, Work and Welfare," *Urban and Social Change Review,* 11 (1978):28–36.
79. PERLIN, L. I., and JOHNSON, J. S. "Marital Status and Depression," *American Sociological Review,* 42 (1977):704–715.
80. PIFER, A. "Women and Working: Toward a New Society," *Urban and Social Review,* 11 (1978):3–11.
81. RAPOPORT, R., and RAPOPORT, R. *Dual-Career Families.* Middlesex, Eng.: Penguin Books, 1971.
82. ———. *Dual-Career Families Re-examined.* New York: Harper, Colophon, 1976.
83. RESCH, R. "Separation: Natural Observations in the First Three Years of Life in an Infant Day Care Unit," (Ph. D. diss., New York University, 1975) unpublished.
84. RICHARDSON, J. G. "Wife Occupational Su-

periority and Marital Troubles: An Examination of the Hypothesis," *Journal of Marriage and the Family,* 41 (1979):63–72.

85. ROOSEVELT, R., and LOFAS, J. *Living in Step.* New York: Stein and Day, 1976.
86. ROSEN, B., JERDEE, T., and PRESTWICH T. "Dual-Career Marital Adjustment: Potential Effects of Discriminatory Managerial Attitudes," *Journal of Marriage and the Family,* 37:3 (1975):565–572.
87. ROSSI, A. "Family Development in a Changing World," *American Journal of Psychiatry,* 128 (1972):1057–1066.
88. RUBIN, J., PROVENZANO, F., and LURIA, Z. "The Eye of the Beholder: Parents' Views on Sex of Newborns," *American Journal of Orthopsychiatry,* 44 (1974):512–519.
89. RUTTER, M. "Parent-Child Separation: Psychological Effects on the Children," *Journal of Child Psychology and Psychiatry and Allied Disciplines,* 12 (1971):233–260.
90. RUTTER, M. *The Qualities of Mothering: Maternal Deprivation Reassessed.* New York: Jason Aronson, 1974.
91. RYDER, N. B. "The Family in Developed Countries," *Scientific American,* 231(3): September 1974, 123–132.
92. SANTROCK, J. W. "Relation to Type and Onset of Father Absence to Cognitive Development," *Child Development,* 43 (1971):455–469.
93. SCHAEFFER, H., and EMERSON, P. "The Development of Social Attachments in Infancy," *Sociological Review, Monographs,* 29 (1964):94.
94. SCHNEIDMAN, E. S., and FARBEROW, N. L. "Statistical Comparisons Between Attempted and Committed Suicides," in Farberow, N. L. and Schneidman, E. S., eds., *The Cry for Help.* New York: McGraw-Hill, 1961, pp. 19–47.
95. SILVERMAN, J. "Attentional Styles and the Study of Sex Differences," in Mostofsky, D. L. ed., *Attention: Contemporary Theory and Analysis.* New York: Appleton-Century-Crofts, 1970, pp. 61–98.
96. SKOLNICK, A., and SKOLNICK, J. "Introduction," in *Family in Transition.* Boston: Little, Brown, 1977.
97. SOROSKY, A. D. "The Psychological Effects of Divorce in Adolescents," *Adolescence,* 12 (Spring 1977).
98. STEIN, P. *Single in America.* Englewood Cliffs, N.J.: Prentice-Hall, 1976.
99. SVECHIN, T. "Separation and Divorce: Crisis and Development," in Marcotte, D. and Nadelson, C., eds., *Treatment Interventions in Human Sexuality.* New York: Plenum Press, (forthcoming).
100. TANGRI, S. "Role Innovation in Occupational Choice," (unpublished Ph.D. diss., University of Michigan, 1969).
101. U.S. BUREAU OF THE CENSUS. *Population Distribution,* Washington, D.C.: 1975.
102. ———. *Current population reports. Marital Status and Living Arrangements.* March, 1976.
103. ———. *Statistical Abstract of the United States,* 1977.
104. UNIVERSITY OF MICHIGAN INSTITUTE FOR SOCIAL RESEARCH. "National Survey," in *St. Petersbury Times.* 8 December 1974.
105. VAILLANT, G. E. *Adaptation to Life.* Boston: Little, Brown, 1977.
106. VISHER, E. B., and VISHER, J. S. "Common Problems of Stepparents and Their Spouses," *American Journal of Orthopsychiatry,* 48 (1978):252–262.
107. VIVERS, J. "Voluntary Childlessness: A Review of Issues and Evidence," *Marriage and Family Review,* 2 (1979):1–26.
108. VOGEL, S., et al. "Maternal Employment and the Perception of Sex Roles among College Students," *Developmental Psychology,* 3 (1970):384–391.
109. WALLERSTEIN, J. S., and KELLY, J. B. "The Effects of Parental Divorce: The Adolescent Experience," in Anthony, E. J. and Kowpernick, C., eds., *The Child in His Family: Children at Psychiatric Risk,* vol. 3. New York: John Wiley, 1974, pp. 479–505.
110. ———. "The Effects of Parental Divorce: Experiences of the Preschool Child," *Journal of the American Academy of Child Psychiatry,* 14 (August 1975):600–616.
111. ———. "The Effects of Parental Divorce: Experiences of the Child in Later Latency," *American Journal of Orthopsychiatry,* 46 (April 1976):256–269.
112. WARSHAW, R. "The Effects of Working Mothers on Children." (unpublished Ph.D. diss., Adelphi University, 1976).
113. WEINGARTEN, K. "Interdependence," in Rapoport, R. and Rapoport, R., eds., *Working Couples.* New York: Harper & Row, 1978, pp. 147–158.
114. WEISMAN, M., and PAKEL, G. *The De-*

pressed Woman. Chicago: University of Chicago Press, 1974.

115. WHITING, S., and EDWARDS, C. "A Cross Cultural Analysis of Sex Differences in the Behavior of Children Age Three through Eleven," in Chess, S. and Thomas, P., eds., *Annual Progress in Child Psychiatry and Child Development, 1974,* New York: Brunner/Mazel, 1975, pp. 32–49.
116. WILLIAMS, J. *Psychology of Women.* New York: Norton, 1977.
117. YARROW, L. "Separation From Parents During Early Childhood," in Hoffman, M. and Hoffman, L., eds., *Review of Child Development Research.* New York: Rand, Soje, Faindat, 1964, pp. 89–136.
118. YOUNG, S. "Paternal Involvement as Related to Maternal Employment and Attachment Behavior Directed to the Father by the One-Year-Old Infant," (unpublished Ph.D. diss., Ohio State University, 1975).

CHAPTER 11

THE DIAGNOSIS AND TREATMENT OF BORDERLINE SYNDROMES OF CHILDHOOD

Jules R. Bemporad, Graeme Hanson, and Henry F. Smith

¶ Introduction

THE POSSIBILITY of borderline conditions occurring before adolescence is a relatively recent concept in child psychiatry. No such official diagnosis exists for the pediatric population; yet there is a growing body of evidence that a pathological process that is midway between neurosis and psychosis is indeed manifested by child patients. Clinical descriptions of such children have appeared in the literature with growing frequency during the past three to four decades, although, in the older publications, various other diagnostic labels such as "benign psychosis," "severe neurosis," or "atypical child" were used. Perhaps the growing interest in borderline conditions in adults has prompted clinicians to search out similar disorders in children and also to view these seriously disturbed children in new ways. Finally, the extensive work on the borderline syndrome in adults may have given authority to the consideration of an analogous syndrome in younger patients.

In this chapter, the authors will attempt to review the literature on borderline children, present criteria for diagnosis (which due to developmentally different levels of maturity, are different from those of borderline adults), and discuss specific guidelines for treatment.

Finally, this chapter should be seen as a preliminary step in a longer and ongoing exploration of this fascinating syndrome of childhood psychopathology.

¶ Review of the Literature

History of the Borderline Concept in Adults

The concept of a disorder that lies diagnostically between the psychoses and the neuroses or, more precisely, of one in which superficially appropriate functioning masks underlying psychosis, can be traced at least as far back as Bleuler's use of the term "latent schizophrenia."[23] Equally as venerable is the debate over the existence of such disorders. Glover[21] attacked the concept in 1932 and his critique may be the first actual appearance of the term "borderline" in the literature. Glover wrote: "I find the terms 'borderline' or 'pre-psychotic,' as generally used, unsatisfactory. If a psychotic mechanism is present at all, it should be given a definite label."

Actually, the term "borderline state" was given status in 1953 by Robert Knight[27] in his description of certain adult patients for whom accurate diagnosis was difficult, although he too argued against its use as a diagnostic label. Borrowing the familiar metaphor of the retreating army, he wrote: "The superficial clinical picture—hysteria, phobia, obsessions, compulsive rituals—may represent a holding operation in a forward position, while the major portion of the ego has regressed far behind this in varying degrees of disorder."

Knight referred to those authors who had suggested alternative terms, such as the "as if personality" of Helene Deutsch[8] and Hoch and Polatin's "pseudoneurotic schizophrenia."[24] Frosch[15] later described the "psychotic character," and significant contributions and refinements of the borderline concept were presented by Schmideberg,[37] Modell,[33] Zetzel,[42] Grinker,[22] Masterson,[32] and others. In recent years, Kernberg[25,26] and Kohut,[28] although differing, have added to the metapsychological understanding and to the treatment of such patients.

In 1975, Gunderson and Singer[23] reviewed the clinical descriptions of adult borderline patients in the literature and suggested six features on which to base the diagnosis: (1) the presence of intense affect; (2) a history of impulsive behavior; (3) superficially appropriate social adaptiveness; (4) brief psychotic episodes; (5) specific patterns on psychological testing; and (6) disturbed interpersonal relationships.

Introduction to the Borderline Concept in Children

When one attempts to apply this concept to children, one faces further complications. Because the organism is in its most rapid period of change and development, diagnostic criteria are more difficult to establish with children than with adults. There is considerably more controversy even over the use of such terms as "psychosis" and "schizophrenia" as they apply to children. The history of the description of borderline children, nevertheless, roughly parallels that of borderline adults, with similar concepts discussed but with remarkably little dialogue between the two fields. Gradually there has been an attempt to separate out a group of children with disorders distinct from psychosis or neurosis; to define them descriptively, developmentally, and metapsychologically; to characterize their psychological testing and treatment; and to bring some order to the concept of borderline syndrome in childhood.

Early Descriptions

As early as 1942, Bender[2,3] proposed criteria for the diagnosis of schizophrenia in children. She felt that it was a form of encephalopathy with behavioral pathology in every area of functioning of the central nervous system: vegetative, motor, perceptual, intellectual, emotional, and social. Shortly afterward, Geleerd[19] published a description of a somewhat different type of child. Not always considered psychotic, "they behave overtly as if they may be suffering from a milder behavior disorder." Geleerd did consider the disorder to be a psychotic one, "most likely a forerunner of schizophrenia," but, presented in the mid-forties and early fifties, hers was among the first of several detailed clinical descriptions of these puzzling children.

The children Geleerd described demonstrated low frustration tolerance and poor impulse control. Pleasant and intelligent when alone with an adult, they became uncontrollably aggressive or withdrawn in a group. They reacted to frustration with an extremely severe form of temper tantrum. Firm handling of their tantrums, which would be the treatment of choice for neurotic children, led these children into paranoia, panic, and loss of contact. Only a loving, soothing attitude of a familiar, affectionate adult could resolve the tantrum. These children were interested in inanimate objects and animals (although they could be cruel to them) more than in humans. By history, they were deviant in all stages of development, and they presented with a variety of "neurotic" symptoms, such as phobias, nightmares, compulsions, tics, and eating, sleeping, and toileting disturbances.

In this and in a later paper,[20] Geleerd related the psychodynamics of the disorder to the infant-mother relationship and felt the child demonstrated interpersonal skills appropriate to a much earlier period in development, namely extreme dependency, difficulty sharing the love object, fantasies of omnipotence, and loss of contact and withdrawal into fantasy when left alone. She stressed the importance of the mother-child relationship as follows:

> In the case of psychotic and borderline children the feeling of loss of the mother is experienced whenever she absents herself for a brief moment or when the child perceives her as not all-loving. . . . [p. 285] To the borderline child, losing the mother or being absent from her means being attacked and spells annihilation. [p. 287][20]

Additionally, during a temper tantrum, the love object became the attacker and the child increasingly lost contact, left alone with his terrifying fantasies. Regarding the tantrum, Geleerd wrote, "The child was overpowered by his fantasies—by his id—and the ego was completely defenseless. The child can only be described as being in a panic. I believe that this helplessness of the ego is pathognomonic for borderline cases."[20] Only by maintaining a fantasy of omnipotence and control over the love object, Geleerd believed, could the child ward off the threat of attack and abandonment in order to function adequately. Geleerd concluded that borderline psychotic children could be distinguished from neurotic children on the basis of an early disturbance in ego development and an inability to function adequately in the absence of the mother or mother-substitute. Much of her discussion anticipates later theories regarding the disturbances of ego and object relations in the etiology of borderline conditions.

Mahler,[30] in an attempt, in 1948, to separate different categories of psychotic children according to the severity of symptoms, age of onset, and period of developmental arrest, described a "more benign" case of childhood psychosis with "neurosislike defense mechanisms." However, it was not until 1953 that there again appeared in the literature detailed clinical descriptions of children that seem much like those described by Geleerd. In that year, Annemarie Weil[38,39] devoted two separate papers to a discussion of borderline children and, for the first time, compared them to the adult "borderline states" (as described by Greenacre), as well as to adults diagnosed as having early, subclinical, or latent schizophrenia.

A prime feature of the children described by Weil[39] was that they "do not acquire latency characteristics in time." She wrote:[38] "The ego has not reached the consolidation which usually gives the characteristic imprint of reasonableness, attempt at control and integration to children of that age." While not as sick as those commonly called schizophrenic, they differed from neurotic children in their degree of ego disturbance, character pathology, fears, obsessions, and free anxiety, and in their capacity to alternate symptoms within weeks.

The histories of these children revealed disturbances in eating, sleeping, motility, and language, and multiple "neurotic" symptoms. Their play was unproductive. Under- or overimpulsive, they might have become addicted to pretend play or remained listless, concrete, and unimaginative, apparent slaves to reality. There was a general lack of playfulness and a lack of nuance and proportion. Weil wrote:[39] "These children hardly ever hit the middle line. They do react in extremes." Thus they can be oversensitive to others or completely tactless, lacking empathy and discrimination. They may lie excessively or be scrupulously honest. Weil added that in their communications there was only a thin veiling of symbolic material in contrast to neurotic children who do not so readily reveal conflictual material.

Weil described[39] three general presenting complaints:

1. Problems in social-emotional adaptation are common. Borderline children demonstrate extreme ambivalence and may be very aloof or extremely clinging with family members. She coined the phrase "clinging antagonism" to describe the ambivalent attachment and wrote, "unmitigated outbursts of love and hate in a child well in latency age are characteristic, especially when in close alternation." With nonfamily members, they may be either extremely shy or indiscriminately friendly.

2. Problems of management often occur, with severe temper tantrums.

3. "Neurotic" manifestations, such as fears, obsessions, phobias, or fetishes are found.

In another paper in 1953, Weil defined these children's deficits in ego development in terms that anticipate later descriptions of the adult borderline:

> "Their faulty ego development . . . consists in a marked deficiency in the development of object relationship with all its consequences (giving up of omnipotence, of magical thinking, acceptance of the reality principle), in reality testing, in the development of the synthetic function, and in the proper use of age-adequate defenses. Moreover, in many of the children this picture is accompanied by an abundance of diffuse or bound anxiety. [p. 272][38]

She further described their object relations in both papers as imitative and need satisfying. In terms of prognosis, Weil indicated that some proceed to frank psychosis, as Geleerd also found,[20] while others have a better outcome if their symptoms diminish by puberty. Those with a relatively good course may show a rigid obsessional organization and superficial socialization. Weil suspected, as do other writers, that there is a hereditary factor in the etiology of this syndrome.

In 1956, Anna Freud[13] presented a paper on "The Assessment of Borderline Cases," but it was not until 1969 that it was published as part of her collected works. In this paper, she emphasized the necessity of making qualitative, not merely quantitative, distinctions between borderline and neurotic children, focusing on the following differences.

1. Borderline children manifest deeper levels of regression and more massive developmental arrests.

2. They tend to withdraw libido from the object world and attach it to the body of the self. For example, instead of fighting for the right to stay up at night as most children do, they choose to withdraw into bed and sleep, preferring their own company to that of their families.

3. They show an inability to receive comfort from others.

4. They demonstrate a number of ego defects, consisting of (1) unstable ego boundaries and confusions between themselves and others; (2) relative defects in reality testing with difficulty distinguishing fantasy from reality; (3) inadequate synthetic functions; (4) inadequate development of defenses beyond the use of denial, projection, and introjection and little use of repression, reaction formation, and sublimation; (5) primary process thinking; and (6) concretization of thought processes.

Fluctuations in Ego State

Whereas Weil was the first to compare these children to adult borderlines, Ekstein and Wallerstein,[9] in 1954, were the first to call them "borderline children." In "Observations on the Psychology of Borderline and Psychotic Children," they focused on one particular feature, reminiscent of encounters with adult borderline patients, namely, "marked and frequent fluctuation in ego states, visible in the treatment process," which lends a characteristic unpredictability to their behavior. They dramatically described this feature as follows:

> Time and again the child will begin the therapy hour with conversation or play wholly suited to his chronological age, so that the clinical observer may reasonably be led to conjecture the presence of a relatively intact ego, well able to use and to sustain the demands and vicissitudes of classical child therapy and analysis. Yet suddenly and without clearly perceptible stimulus, a dramatic shift may occur: the neurotic defenses crumble precipitously; and the archaic mechanisms of the primary process and the psychotic defenses erupt into view. Then they recede just as rapidly, and the neurotic defenses or perhaps more accurately, the pseudoneurotic defenses, reappear. [p. 345][9]

Ekstein and Wallerstein distinguished the ego of the borderline from that of the neurotic child by its particular vulnerability to regression and compared the ego of the borderline child to "a delicate permeable membrane through which the primary process penetrates with relative ease from within and which external forces puncture easily from without."

They hypothesized three precipitants for such regression: (1) changes in the therapeutic transference; (2) autistically derived changes within the child; and (3) feelings within the child related to "changing introjects." In ways suggestive of the later adult literature on problems in the countertransference and on the difficulty maintaining empathic contact with adult borderline and narcissistic patients, they wrote:

> In reviewing our clinical material we repeatedly found that the ego regression was directly preceded by an inadvertent rebuke or lack of comprehension by the therapist of the child's message, and the return into the secondary process followed directly upon the therapist's retrieving of his error and demonstrating his sympathy and understanding." [p. 350]

Regressions and regressive fantasy occurred in response to some affective threat within the transference and served simultaneously to withdraw from and yet maintain contact with the therapist. They suggested that "every fantasy production carries this double message. It reveals both an attempt to master conflict and a confession of current inability to do so." The most regressive fantasies were described by the authors as oral in character with the specific themes of separation, abandonment, bodily distortion and disintegration, and cannibalism.

Along with the fluctuation in ego state, each child demonstrated considerable fluctuation in the level of interpersonal relatedness. Thus, the relationship with the therapist at times seemed clearly autistic or symbiotic, at other times more characteristically neurotic. Finally, Ekstein and Wallerstein placed borderline children on a continuum between neurotic and schizophrenic children with respect to the degree of conscious control they could exercise over these fluctuations in ego function. Similar characteristics were observed on psychological testing of borderline children. (See "Performance on Psychological Testing.")

Problems in Psychotherapy

In 1956, Ekstein and Wallerstein[10] continued their discussion of these children with a paper on the psychotherapy of the borderline child. In contrast to the psychotherapy of the neurotic child, in which the therapist can address interpretations to the most advanced level of ego organization, such interpretation precipitates panic, withdrawal, regression, or superficial conformity by the borderline child. Furthermore, unlike the neurotic child, the borderline child frequently cannot make use of displacement as a defense against hostile and sexual impulses, which then threaten the relationship with the therapist. Thus, with the borderline children, the ability to displace is often a sign of therapeutic progress. Regarding interpretation, therefore, Ekstein and Wallerstein recommended that the therapist *not undo* whatever displacements the child can manage. Rather, he should frame his interventions within the child's own language, primary process fantasy, and level of communication at any given time. They called this "interpretation within the regression" or "within the metaphor." Thereby, in joining the child at his own stage of ego development, the therapist attempts to maintain the relationship while laying the foundation for more mature development through identification.

In "An Attempt to Formulate the Meaning of the Concept Borderline,"[35] Rosenfeld and Sprince reviewed the psychoanalytic literature, including Weil, Geleerd, Ekstein, and others, and reported on their own clinical observations. They emphasized that the appreciation of ego deviation alone is insufficient for making the diagnosis and recommended an overall metapsychological assessment, pointing out, among other features, the lack of libidinal phase dominance in both latency and adolescent borderline children and the relative preservation of reality testing as compared to psychotic children.

Rosenfeld and Sprince touched briefly on the problem of interpretation in the therapy of these children. Whereas interpretation can decrease the acting out and fantasy production of neurotic children, with borderline children, it appears to increase both and to undermine defensive structure. The authors recommended that the therapist begin by facilitating defensive development and avoid content interpretation until later in therapy.

In "Some Thoughts on the Technical Handling of Borderline Children," Rosenfeld and Sprince[36] expanded on these recommendations, emphasizing the danger of mistimed interpretation, which tends to increase anxiety and aggressive behavior. They recommended that in the beginning of treatment the therapist adopt ego-supportive techniques to facilitate displacement and to "arouse in the child the ambition of giving up primitive gratification or direct discharge."

Rosenfeld and Sprince warned of the particular stress these children impose on the therapist, who must for brief periods allow himself to "meet the child's needs for symbiosis." (p.513) Difficult too is the task of finding the balance between the child's need for physical contact and distance. There is also narcissistic hurt in store for the therapist during the periods when he is "unheard," ignored, or "treated as insane." (p. 514) These are dilemmas reminiscent of the therapy of adult borderline patients.

Later Refinements of the Concept

Based on these fundamental contributions, recent authors have elaborated on specific details of the syndrome. In "Borderline States in Children,"[14] Frijling-Schreuder's major emphases were on the development of language by these children and the quality of their anxiety. The author pointed out that the use of language or of "inner speech," which she called "thinking in words," helps the child to convey secondary process, to master impulses, and to tolerate anxiety, especially the anxiety related to the threat of engulfment. The child's use of language, then, is a major prognostic sign and a feature differentiating borderline from psychotic children.

Whereas Geleerd found these children lacking in signal anxiety and Rosenfeld and Sprince believed they experience signal anxiety itself as a threat leading to fears of destruction and annhilation, Frijling-Schreuder found that their anxiety relates to the fear of becoming someone else and made the useful observation that, in contrast to psychotic children, their anxiety is in fact a measure of relatively advanced ego function, for only the more structured ego can be aware of the threat to its own integrity. This awareness, then, gives rise to their anxiety.

Frijling-Schreuder further pointed out

that these children have a tendency to regress into "micropsychotic" states when stressed. The author warned the therapist, for example, not to push the adolescent into dating, which can be highly threatening. Left to their own devices these children may in time find mutually dependent partners. The author commented in passing on the perverse traits of these children and emphasized most poignantly their extreme loneliness and isolation. "They feel like toddlers whose mothers are permanently out of the room." It is this loneliness that, the author believed, leads to their intense need for conformity. In contrast to the patient with childhood psychosis, Frijling-Schreuder found that the borderline child's improvement may be rapid if he feels understood or, in the case of the young child, if the mother's handling of him improves.

Chethik and Fast[6,12] focused on those features of ego development and object relations that indicate what they called the child's "transition out of narcissism." They placed borderline children between neurotic and psychotic children in terms of the degree of commitment to the "independent reality of the external world" at one extreme and the "narcissistic world of the pleasure ego" at the other.

To Chethik and Fast, narcissistic fantasy serves several purposes for borderline children. First, it is a source of present gratification that wards off pain. Second, it provides a base from which the child can begin to test the external world as a source of pleasure. Third, in therapy it offers a means of expressing fears associated with the child's further investment in the external world. Therefore, they recommend that the therapist actively participate in and encourage the elaboration of the fantasy in order to delineate and work through the fears that block the child's further development in his "transition out of narcissism."

In this endeavor, the therapist serves several functions: first, he provides a "complementary object fragment" or transitional object for the borderline child, who has not developed coherent self- or object-representations; second, he acts as a bridge between the child's world of narcissistic fantasy and the external world; and third, he is a stable object on which further self-object differentiation may be practiced.

In 1974, based on an extensive review of the literature and his own clinical experience, Fred Pine[34] differentiated the borderline syndrome into six clinical subtypes. All six manifest severe primary developmental failures in ego function and object ties, as opposed to secondary regressions following developmental conflict. While this distinguishes them from the neuroses, Pine found no sharp distinction between borderline children and some childhood psychoses.

The first subtype is made up of children with *chronic ego deviance.* These children are similar to those described by Weil and by Rosenfeld and Sprince whose ego deficits are " 'silently' present at all times." These children show a simultaneous mixture of varying levels of ego function, drive level, and object relationship. They do not improve with a change in environment.

Whereas the children of this first group show a mixture of ego strengths and deficits, the second group demonstrates *shifting levels of ego organization.* These are the children described by Ekstein and Wallerstein, who, Pine concurs, demonstrate a true ego organization at two different levels, allowing them to make a total shift from one to the other in order to avoid panic.

The third group of children manifest *internal disorganization in response to external disorganization.* These are children from deprived, impulse-ridden homes, who integrate rapidly in a benign hospital environment. Their borderline symptomatology is reactive, although Pine acknowledged the presence of other psychopathology which is not as easily reversed.

The fourth group is closely related to the third and contains children with *incomplete internalization of psychosis.* They manifest psychoticlike phenomena as a result of their attachment to a psychotic love object, usually a parent. Children in the first two categories have a greater internalized ego defect than in the latter two, who in turn manifest a larger reactive component.

Pine described a fifth group of children with what he called *ego limitation.* These are children with severely stunted ego development in all areas, resulting in what some have called an inadequate personality. Pine described two such children and speculated that the clinical picture had been caused in one case by severe stimulus deprivation and in the other by a combination of intense chronic anxiety and minimal cerebral dysfunction. In both cases the child appeared dull, with below normal intellect and judgment, incapacity for self-care and planning, and limited sense of self.

Pine designated a sixth and final subtype, *schizoid personality in childhood,* a group others have called "isolated personality." These children show constricted affective life, distance in human relationships, and preoccupation with their own fantasy life. Their fantasies serve to ward off panic, permitting them to function, albeit aloofly, in the real world.

Whether or not all six subtypes that Pine has described should be considered "borderline," his is a creative and useful contribution to the diagnostic organization of the complicated group of children who are the subject of this chapter.

As the preceding review of the literature indicates, in the past four decades, investigators working independently have described groups of children with certain shared characteristics, that may represent one or several related syndromes. The symptom clusters and clinical description of these children will be presented in the following section.

¶ Manifest Symptomatology

The borderline child exhibits varying degrees of pathology in all major areas of psychic functioning. No single symptom is pathognomonic, and the entire clinical picture must be considered when entertaining this diagnosis. These children may differ in the severity of their impairments in only certain aspects of the symptom Gestalt. When the total compilation of dysfunction is evaluated, however, a coherent and consistent pattern emerges that may be differentiated from childhood neurosis and psychosis as well as from other established diagnostic entities such as psychopathy or organic syndromes.

Often this overall Gestalt is not apparent at initial clinical contact but may only become manifest after prolonged therapeutic work. Most of the literature on such children, in fact, has been contributed by therapists who observed the symptom picture emerge in the course of psychotherapy. While some borderline children may appear different solely on the basis of observation, others may initially present as neurotic or even normal children and it is only in the course of therapy or in unstructured situations that the full extent of their psychopathology emerges. The symptoms described in the following paragraphs, therefore, may not be observed on a routine or structured interview. Only after the clinician has become familiar with the child and has established a therapeutic relationship can the diagnosis of borderline be made with assurance.

Another important factor in diagnostic assessment is the age or developmental level of the child. Many of the characteristics that are defined as pathological symptoms in borderline children may be found in the behavior of normal but much younger children. The borderline child thus presents overt behavior, as well as deeper modes of psychologic organization, that are grossly immature; a finding that has led some to explain this disorder as resulting from massive developmental arrests. Before school age (or latency period) much of the borderline child's behavior may appear normal. Therefore, it is impossible to make this diagnosis with certainty before this developmental stage.

Fluctuation of Functioning

One of the most frequently mentioned characteristics of the borderline child is a rapid shifting in levels of psychological functioning, from healthy or neurotic organization to psychoticlike states, with intrusion of bizarre thinking, grossly inappropriate behavior, and overwhelming anxiety. This psychological disintegration occurs with extreme rapidity, often followed by an equally rapid reintegration at a healthier level of functioning. As mentioned previously, Ekstein and Wallerstein[9] believe that this fluctuation in ego states occurs in the context of a significant relationship and that deterioration follows a sense of being rebuked or of not being understood by an important adult. Conversely, the return of a feeling of being understood or approved by the adult lessens the fear of loss and allows the child to once again function at a healthier level.

Not all authors would agree with this interpretation of the borderline child's fluctuations in behavior although all of those who have worked with such children have remarked on the child's alternation between a reality-oriented mode of relating and an idiosyncratic universe of fantasy. These rapid fluctuations are one of the major differences between these children and their neurotic or psychotic counterparts. Neurotic children may display episodes of extreme anxiety or behavioral dyscontrol, but their thinking remains in a reality context. At the other extreme, schizophrenic children remain in fantasy for very long periods and, in contrast to borderline children, evolve semifixed psychotic delusions that partly ease their anxiety.

Borderline children appear to express great affect in a manner that is neither as well structured as that of neurotics nor as entirely alienated from others as that of schizophrenics. Their "blow-ups" tend to involve others or even provoke others into an emotional engagement. For example, an eleven-year-old borderline boy reacted to the head nurse paying attention to another child by suddenly running around the ward muttering to himself and bumping into the corridor walls. Then, in full view of a ward counselor, he picked up another child's toy and smashed it, forcing the counselor and others to give him attention. When he had been calmed by the staff, he again became rational and able to relate in an adequate fashion.

Nature and Extent of Anxiety

Another key symptom of borderline children concerns the constant presence of varying degrees of anxiety as well as the inability to control the escalation of anxiety. These children do not seem capable of responding to signal anxiety with adequate defenses or activities so that these minimal increments in anxiety due to a conflictual situation rapidly mount to panic and terror. Some borderline children do have a few maneuvers, albeit inadequate ones, to deal with anxiety, such as calling on trusted adults to alter the situation for them or to give reassurance. Others simply escape physically from conflict situations. They will stay away from any object or experience which might arouse their anxiety, for instance by avoiding knives. However, this only adds to their behavioral peculiarity. Many such children, however, decompensate in the face of anxiety-provoking situations and experience states of panic.

There are also indications that the very nature of the anxiety experienced by borderline children differs from the analogous phenomenon in neurotic or normal children. The anxiety of the borderline child is both more global and overwhelming and appears to derive from a different magnitude of threat to the self. The neurotic child may experience anxiety over an urge to disobey some socially or parentally imposed restriction that he is aware may result in punishment or a loss of esteem. In contrast, the borderline child's anxiety appears to derive from a fear of psychological annihilation, body mutilation, or catastrophic destruction. Rosenfeld and Sprince,[35] for example, reported an eleven-year-old borderline child saying, "You never know what will happen when you sit on that hole—everything may fall out and you'd find yourself being a skeleton with nothing but this bit of skin holding you together." Another child, described by the same authors, was afraid that he would separate from himself, that his mind and body would fall to pieces. It would appear that these fears of destruction are readily aroused in borderline children when they are stressed by frustration, interpersonal difficulties, or other conflicts. However, at more secure and peaceful times such apprehensions appear absent. Also, such fears are not elaborated into the gross distortions of schizophrenic defenses, despite similarities in the nature of the causative anxiety. Frijling-Schreuder[14] spoke to this point, suggesting the borderline child may suffer more anxiety than the blatantly schizophrenic child because he is more aware of the inner threat of psychic disintegration, and is unable to form stable delusions, which, while impairing his relationship to reality, would offer relief from anxiety.

In summary, borderline children share the same nonneurotic form of self-threatening anxiety as schizophrenic children, but only for short periods and in pure form, without psychotic elaborations and defenses.

Thought Content and Processes

From the earliest reports, borderline children have been observed to demonstrate disturbances in the "synthetic functions of the ego" as well as revealing contamination of the "conflict free" spheres of the ego as a result of emotional difficulties. Despite this similarity to psychotic children, the borderline child never seems to lose touch completely with reality or, as stated previously, to create fixed schizophrenic delusions. Borderline children nevertheless show some thought disturbances typical of schizophrenic cognition, although in milder and less extensive forms.

Such children alter otherwise accurate ways of dealing with reality because of their persistent underlying fears of destruction or mutilation. Rosenfeld and Sprince,[35] for example, reported a borderline child who, upon hearing that his school would "break up" for the holidays, envisioned with apprehension the annihilation of the school buildings.

An eleven-year-old borderline boy asked a counselor if another child whom he had befriended would be going home, a possibility that aroused great anxiety. The counselor replied that the decision was "up in the air." The patient seemed suddenly afraid and started looking around the ward, asking "Where?" as if he expected to find the answer literally hanging in midair.

These are illustrations of the process of concretization which, as described by Arieti[1] and others, forms part of the cognitive pathology of schizophrenia, but is also seen in borderline conditions.

Borderline children also manifest bizarre phobias and obsessions that go beyond the neurotic spectrum. Frijling-Schreuder[14] described a child who would not shake hands with his therapist because he believed she had poison glands on her hands. Pine[34] reported a child who thought people came at night to cut open his stomach. The authors have seen a child who refused to enter any large body of water, including swimming pools, because he was sure there were water demons who would drag him down and drown him. This same child experienced panic during rainstorms and required ear plugs to muffle the sound of thunder. Another child, who suffered from constipation, believed that sharp, "thorny things" would be expelled in his stools and cut his intestines and rectum.

Some of the borderline child's inappropriate fantasies need not relate directly to fears of survival and body integrity. Rosenfeld and Sprince[35] described a nine-year-old boy who at times behaved as if he were a female character from a Dickens novel—to the extent that he requested female clothes and wanted to do housework. Later the same child acted as if he were a "lorry" and invented a "lorry language" with which others had to comply. For example, his food had to be called "petrol." In many cases we have seen, the parents appear to condone or even encourage such unrealistic behavior in the child. One eight-year-old boy was regularly sent to school by his father dressed in his younger sister's clothes. The mother of a seven-year-old borderline girl actually believed that her daughter was a witch with special malevolent powers and treated her accordingly.

In borderline children thought content in general flows with excessive fluidity from reality to fantasy and back again. A persistent and objective relationship to reality is not maintained for long periods but follows the vicissitudes of the stresses encountered. Over two decades ago, Geleerd[19] aptly described these children as having a short "reality span." Ultimately, fantasied themes of mutilation, survival, and catastrophe intrude into consciousness, accompanied by massive anxiety. Furthermore, these fluctuations between near normal and near psychotic mentation seem chiefly determined by the ups and downs of relationships with others.

Another peculiar aspect of the borderline child's thought content may be his uneven and impractical store of knowledge. These children are frequently quite intelligent but use their intellect in the pursuit of obscure subjects while remaining blatantly ignorant of basic, everyday commonsense facts. One eight-year-old was conversant with the methods by which the ancient Egyptians mummified bodies and was also an expert on dinosaurs (he even knew which were carniverous and which were herbivorous), but he had no understanding of elementary social interactions. These children appear especially drawn to scientific subjects such as astronomy, the biology of insects, or the classification of reptiles; however, the authors have known some borderline children with interests in artistic crafts such as making jewelry or photography. It may be that these idosyncratic interests represent psychic islands of security and regularity in an unpredictable and fearful world, or possibly that these hobbies may be attempts at controlling fears and threats in a symbolic manner. On the other hand, these interests may serve no real defensive purpose and, by excluding a wider range of interests, may represent merely another area of deviant development.

Finally, borderline children often present with a heterogeneous collection of cognitive defects that are picked up on psychological testing. There is no consistent pattern to these deficiencies and different children may have difficulties with focusing attention, learning, reading, perceptual-motor tasks, or abstract concept formation. These deficiencies may indicate the presence of some neurological defect that contributes to the overall clinical picture.

Relationships to Others

It would be surprising if children who manifest the many difficulties just listed did not also exhibit serious problems in relationships. The deficiencies in this area are, indeed, so profound that some clinicians have considered the interpersonal process the foundation of the borderline syndrome on which the other symptoms are based. Rosenfeld and Sprince[35] postulated that borderline children are caught in a dilemma in which they wish to merge psychologically with a trusted adult but, simultaneously, are terrified by the loss of identity and psychological integrity that would follow from this merger. These authors interpreted the borderline child's symptoms as representing both this desire to merge and his defenses against psychological merger.

Ekstein and Wallerstein[9] described the retreat into fantasy as a means of expressing conflictual material without endangering the relationship with a needed other person and thus also underline the importance of relationships in this condition. Frijling-Schreuder[14] concluded that borderline children are fixated at the symbiotic stage of development, according to Mahler's scheme of psychological maturation. Anna Freud[13] also described these children as exhibiting modes of relating that are appropriate for much younger children, using another person almost exclusively for satisfying everyday needs.

These children do appear to utilize others to fulfill functions that should normally be autonomous or fulfilled by the child himself at his stage of development. They require constant reassurance from others that the environment is safe and that they will be protected. When they feel secure in a relationship, these children may function very well and demonstrate considerable innate talent and intellectual ability. If they sense rejection or criticism, they react wildly with massive anxiety, destructive rage, or bizarre thinking. It is as if their internal psychic order depends on a modulating external source.

Yet, despite this great need for a stable relationship, the borderline child does not seem to form great attachments to any one person and may rapidly substitute one relationship for another as long as he receives the support and reassurance he needs. The other person is valued for the functions he performs rather than for his offer of intimacy or concern. Rosenfeld and Sprince[35] described a borderline child who became very upset when he learned that his headmaster was leaving, but when questioned about his grief, his primary anxiety was that no one would do all the things that the headmaster had done for him. Other children display a lack of differentiation in their expectations from others and appear to believe that all others will expect them to behave in the same manner and that they are to behave toward others in an identical fashion despite differences in situations or familiarity with other people.

For example, an eight-year-old girl would approach a total stranger on the street and treat him in an inappropriately familiar manner. She would behave as if the stranger were an old friend or a member of her family. This superficial intimacy immediately ceased, however, if the stranger did not react in a warm, gratifying manner. The child would not seem upset but would simply leave the person and approach a new stranger. This girl's pattern of superficial relatedness demonstrates the borderline child's indiscriminate use of others (who are often interchangeable) to fulfill inner needs.

Anna Freud[13] also commented on a narcissistic investment in others but interpreted the form of relationship as a regression from "object cathexis" to "primitive identification." Rather than considering another person as an individual who may give pain or pleasure but who remains a separate entity, the borderline child, according to Freud, pathologically merges with the other and by assuming characteristics of the other, attempts to share the other's invulnerability and power. In the search for emotional safety, the borderline child will psychologically become one with the other, according to Freud, and, for example, believe that they experience the world in an identical manner. Freud considered this mode of relating as an indication of an arrest at a very early stage of development and as a grossly primitive defensive maneuver. As mentioned previously, Rosenfeld and Sprince,[35] as well as Frijling-Schreuder,[14] described similar phenomena in their observations of borderline children.

The authors have observed children who exhibit an "as if" quality, who mirror the behavior of others, and appear to have no true stable personality of their own. Sometimes this mirroring behavior becomes manifest only in times of stress. For example, one boy who regularly became upset when his mother had to leave the ward after visiting him, behaved like her for some time after her departure. Other children will show the "merging" type of behavior under similar types of stress. Another boy, for example, became upset by his mother when she visited the ward but felt empty and sad after she left. He became very tense and excitedly walked over to a staff doctor, grabbed his tie, and said "I want your tie." Then the boy rapidly said he wanted the doctor's eyes, nose, mouth, and soon until he wanted "all" of him. He wanted to become one with the doctor to undo his sense of anxiety and depression.

These are descriptions of the relationship of the borderline child with one or a small group of sympathetic adults encountered in a therapeutic setting. The borderline child's reaction to his peers is quite different. They usually do not get along well with other children, who are not as tolerant or as predictable as adults. They bully or torment younger children, by whom they do not feel threatened. At the same time, they are fearful of older children and erroneously expect to be attacked by them. With their own peers, they tend to be withdrawn and jealous of adult attention, although they occasionally fly into rages during which they may attack others or vent their fury on themselves. One consistent finding is that borderline children do not do well in groups with either normal or disturbed children. In such situations they experience chronic anxiety, which they try to relieve either by withdrawing into fantasy or by attempting to monopolize the adult group leader by inappropriate behavior. Classroom teachers frequently report that such children attempt to climb on their laps during class or ask for some other sort of tangible show of emotional support. Borderline children usually do not grasp the nuances of social behavior and misinterpret group situations causing them embarrassment and anxiety. They are often perceived as odd by their peers, who eventually avoid them or make them the butt of jokes.

The sharp contrast between the borderline child's behavior in a group of peers and his behavior in a one-to-one situation with a benevolent adult may lead to conflicting reports being brought to the attention of the clinician. Similarly, if the clinician is supportive and the interview is highly structured and nonthreatening during diagnostic sessions, the child may evidence little pathology, presenting a markedly different picture than that provided by his history. Later on, during the course of therapy, when free play is introduced or nonsupportive interpretations are ventured, the extent of psychopathology may become manifest.

Lack of Control

The final general category of difficulties experienced by borderline children concerns their inability to inhibit impulses, to delay gratification, to control aggressive outbursts, and to suppress frightening fantasies.

During periods of frustration, they may exhibit mounting tension, anxiety, or anger. These experiences syncretically spill over into motor behavior so that they will appear hyperactive and agitated. They may rock back and forth, start running around aimlessly, or talk in a rambling fashion. They may also work themselves into a rage, becoming verbally and physically abusive, and indiscriminately attack other people or objects or even themselves. During these attacks, they appear wild and out of contact but they usually can be calmed down by a supportive adult.

Another aspect of the borderline child's inability to control his inner states can be observed in the escalation of frightening fantasies accompanied by the increasing anxiety, described previously. If allowed to play freely with dolls or to associate freely verbally, without imposed structure, the content of their productions will often begin to move toward themes of destruction and mutilation. These themes then continue to escalate as the child is apparently helpless to stem the flow of one terrifying thought after another (this is especially observed on projective testing—as will be discussed). Unless the therapist intercedes, the child will soon be overwhelmed by panic and will verbalize grotesque visions of world catastrophes or gory details of bodily injury. This lack of control over thought content again illustrates the borderline child's excessive reliance on external structure to regulate his inner world. This deficit may account for some of the pseudoneurotic forms of defensive symptoms encountered in these children. They may develop obsessions as a way of protecting themselves against some dreaded event. These rituals take on a magical quality and are adhered to desperately, suggesting that they represent a basic means of survival to these children. Another crippling means of defense consists of phobias that allow the child to avoid those objects that might elicit the chain of terrifying thoughts. Here again, the phobic object does not symbolize rebelliousness or secretly desired transgression of social taboos but is associated with fantasies of annihilation or horrible injury.

Associated Symptoms

In addition to the pathological manifestations listed previously, some borderline children exhibit additional symptoms which are not easily lumped under one overall area of functioning. Among these symptoms may be listed poor social awareness, lack of concern for bodily safety or personal grooming, and an inconsistent ability to adapt to new situations. Others may show "soft" signs on neurological examination, complain of difficulty sleeping, have difficulty concentrating, or have periods of restlessness. While not a specific symptom, many authors have commented on the unevenness of development in these children. This lack of developmental uniformity can be seen in terms of a mixture of ways of relating (genital and pregenital drives), of utilizing defenses (neurotic and primitive), of using intellectual abilities (gaps in knowledge, great disparity on age appropriate tasks), and of general self-management. These children do present a conglomeration of advanced, normal, and grossly delayed behavior, all at the same time, with fluctuations between high and low levels of functioning in almost all areas.

Performance on Psychological Testing

Since borderline conditions of children represent a controversial diagnostic label as well as a fairly recent addition to the unofficial nomenclature, articles dealing with results of psychological testing with these children are quite rare. Appropriate testing of these children, however, would greatly aid in timely diagnosis and would serve as a fruitful area for future delineation of the borderline syndrome. Two reports on testing of borderline children are summarized in order to broaden the description of the clinical syndrome, as well as to add to the clinical evidence that borderline conditions do form a separate diagnostic category of childhood psychopathology.

Engel[11] described the major psychological themes that emerge from responses of borderline children to projective tests. The first issue mentioned is that of *survival.* Engel noted that fears of annihilation appear in a variety of ways on all of the tests. In the stories of these children, incidental aspects of reality are "woven into the fabric of the dread of death." In contrast to schizophrenic children, Engel found that borderline children can transform these morbid ideas into a realistic narrative and are better able to share them with the examiner. The second issue Engel mentioned was the *struggle for reality control.* This aspect is similar to that described by clinicians as fluctuation in ego states. To quote Engel: "What we see in the test material of borderline children is not the complete disruption of reality contact. Nor do we see the reiteration of concrete, sterile, conventional realities; rather the tests contain illustrations of waxing and waning of reality testing." When terrifying fantasies intrude, the child may say he is just pretending. A ten-year-old boy, for example, did not know whether to interpret a Rorschach card as a monster or a cowboy. He eventually stated that it was really a cowboy and he was just imagining it was a monster. The third finding Engel described concerned the child's attempt to cope with *insurmountable demands.* She found her patients suffer an "ego exhaustion" in their efforts to master frightening forces from without and from within. The child seems to expect no solution to his problems. A fourth manifestation is the selective *distance devices* used by borderline children. In the course of responding to the test material the child will embark on a story that gets out of hand and, by raising frightening material, arouses great anxiety. As mentioned by clinical observers, the child appears to lack the foresight to avoid terrifying material in his productions. Once the child arrives at this anxious point, according to Engel, he utilizes fantastic distortions of time and space in his stories to get back to safer territory. The child distances himself figuratively and concretely to avoid anxiety but at the expense of reality constraints. One borderline boy who was seen in the authors' clinic tried to bite the Rorschach cards and actually pounded them with his fists, apparently in an attempt to control the fantasies that the cards elicited from him.* In an analogous manner, border-

*The authors thank Dr. Aydin Wysocki for contributing this clinical illustration.

line children defend themselves by *pivotal interruptions in testing.* When absorbed in anxiety-arousing material, the children seek relief by asking for a drink of water, sharpening a pencil, or other maneuvers to interrupt the process. Although all children use some tactics to obtain relief from the demands of testing, especially when confronted by possible failure or other embarrassment, the borderline child attempts to interrupt testing when he is about to lose control and be overwhelmed by anxiety. His motor activity escalates and he appears near to panic, and, most telling, he turns to the examiner for help. If the examiner alters his role to that of therapist and offers reassurance, the child appears relieved and is able to quiet down and return to work. Therefore, by seeking help from an adult, the child transforms the testing situation into a therapy session.

In regard to this last observation, Engel made a number of cogent remarks about the effect of such a child on the examiner. Engel wrote, "From the first moment, such children make much more vigorous use of the testing relationship than do others and cast their intrapsychic struggles upon the testing situations in large and bold signals." The examiner is puzzled and intrigued by such children but most of all he is involved with them more closely than he is with aloof schizophrenic youngsters or self-sufficient neurotic patients. The examiner finds himself intuitively responding to the repeated disintegration suffered by the child as well as to the child's attempt to reintegrate on a reality level. The examiner may be bewildered by the fluctuation between harmless stories and themes of disaster and horror. Finally, the examiner is caught in the struggle between his duty to obtain an unbiased record and his wish to reassure the child against the mounting anguish of uncontrolled fantasies.

Wolff and Barlow[41] recently administered a number of cognitive, language, and memory tests, and measured the use of emotional constructs in groups of normal, high functioning autistic and "schizoid" children. Wolff and Barlow borrowed the term "schizoid," from the writings of Asperger, who wrote a monograph on disturbed children in 1944. They share Asperger's belief that the condition described is not an illness with an onset and a cause but a consistent, fairly permanent personality pattern. Wolff and Barlow believe that "schizoid" conveys the meaning of Asperger's diagnosis of autistic personality. At the same time, these authors note the similarity of their sample to "borderline" children and cite the same articles mentioned in this chapter in their description of these children. Finally, many of the clinical features they describe are those of borderline children delineated here. It may therefore be assumed that they are referring to borderline children despite the different diagnostic label.

In contrast to Engel's use of projective material, Wolff and Barlow administered mainly structured tests that provided little opportunity for the elaboration of fantasy material. Their results were numerically tabulated so that the three groups could be compared. What is most pertinent is that the three groups could be differentiated by objective tests, so that the schizoid group appeared to be a distinct sample of disturbed children. In general, the schizoid group scored midway between the normal and autistic groups. For example, they exhibited more scatter than normal children on the subtests of the Wechsler Intelligence Scale for Children but less than the autistic group. Perhaps the most interesting finding was that the schizoid group was more distractable on many measures. Wolff and Barlow believe that this is not the result of an organic attentional defect but rather due to these children being more attuned to their inner world than to the test situations.

Homogeneity and Heterogeneity of Borderline Conditions

In concluding this section on manifest symptoms, it is worthwhile to consider whether borderline syndromes in childhood actually form a separate clinical group. Certainly, children so diagnosed show marked differences in superficial symptomatology;

some may be overtly aggressive and others shy and withdrawn, while still others may present with conspicuous phobias and obsessions. Pine[34] has, in fact, lumped together a variety of subgroups of children with different etiology, course, and response to treatment under the rubric borderline which, he believes, does *not* really define a separate clinical entity but rather a descriptive comment on other diagnoses. Until we have more data on the etiology and course of borderline children, it will be difficult to decide whether they do form a valid clinical group. On the basis of manifest symptoms, however, these children do appear to share a constellation of specific and fundamental areas of psychopathology that sets them apart from other disturbed individuals. These areas have been described above and are listed in Table 11–1. It is this Gestalt rather than any single symptom that is characteristic. The first step in the delineation of any pathologic syndrome is a purely clinical description of symptoms. We are still at this initial state of knowledge in regard to borderline conditions of childhood but it is hoped that this attempt will lead to further definition, understanding, and, ultimately, appropriate treatment for these children.

¶ Etiology, Clinical Course, and Prognosis

There is currently very little certain knowledge regarding the cause of borderline syndromes in childhood, the direction these disorders take throughout development, or the eventual outcome of such children in adult life. The literature on borderline children consists almost entirely of scattered single, or small series of case reports, with a paucity of any overall large and systematic study of this syndrome. Furthermore, these case reports present minimal clinical data on history or outcome, stressing instead diagnostic documentation or metapsychological interpretations of the symptoms. An additional limitation is the lack of extensive neuropsychological testing which might better define possible organic impairments in borderline children. Therefore, our knowledge of borderline children is based primarily on subjective and anecdotal material that variously presents data on antecedent factors and eventual outcome. Yet despite this diversity of reporting from different orientations, the clinical descriptions of borderline children are remarkably similar.

One systematic study on possible etiological factors is a recent report by Bradley[4] that examines early separation experiences of borderline children. Bradley found more frequent separations from mother figures or caretakers during the first five years of life in borderline children than in neurotic, psychotic, or delinquent children. Though a definite step in the right direction, this study is hampered by significant problems. One is that the children were diagnosed by criteria devised by Gunderson and Singer[23] that were intended to describe adult and not child patients. Another difficulty is that the nature and extent of the separations were not specified.

On the basis of a review of the literature and of clinical experience, the authors have formulated a few speculations regarding the causes of this syndrome. These will be presented here with the knowledge that validation awaits more vigorous and extensive investigations. The authors have found evidence of mild to moderate organic impairment in borderline children as compared to either their siblings or children with other forms of psychopathology. This organicity may manifest itself in "soft" neurological signs, disorders of impulse control, or cognitive deficits. These defects are identified on neuropsychological testing but are often missed in a routine physical examination.

Often these children are remembered as difficult infants with poor homeostatic patterning. They were irregular in sleeping, eating, or elimination. As toddlers, some are described as excessively clinging, others as hyperactive, and still others as aloof and withdrawn. All are reported as "different"

TABLE 11–1 **Major Areas of Psychopathology**

- I) Fluctuation of functioning
  - A) Rapid decompensation secondary to objectively minimal emotional stress with rapid reintegration after reassurance from environmental figures
  - B) Brief shifts from neurotic to psychotic ideation
  - C) Recurrent intrusions of bizarre preoccupations and fantasies
  - D) Extreme dependence of level of functioning on environmental support
- II) Nature and extent of anxiety
  - A) Inability to contain anxiety with rapid escalation of anxiety to panic unless helped by environmental figures
  - B) Inability to utilize signal anxiety
  - C) Basis of anxiety residing in fears of destruction, mutilation, and emotional annihilation
  - D) Greater suffering from anxiety due to inadequacy of neurotic defenses and lack of psychotic reconstitutive symptoms
- III) Thought content and processes
  - A) Inadequate "synthetic ego functions" with some gross distortions and concretizations but without stable delusions, hallucinations, or prolonged or profound loss of reality contact
  - B) Excessive fluidity of thought between fantasy and reality with inability to control potentially frightening avenues of association
  - C) Short "reality span" with recurrent but transient intrusion of grotesque and bizarre fantasy themes
  - D) Concern with survival manifested by poorly developed defenses (obsessions, phobias, extreme dependency, merging) to ward off possibility of catastrophic destruction
  - E) Proficiency in obscure areas of knowledge with lack of awareness of practical, everyday matters
  - F) Heterogeneous cognitive defects
- IV) Relationships to others
  - A) Immature attachments to need-fulfilling adults (merging, primitive identification, dependency)
  - B) Excessive reliance on others to maintain inner security, function well with trusted adult
  - C) Poor relationship with peers, inability to utilize intellectual talents in group situations
- V) Lack of control
  - A) Inability to delay gratification or tolerate frustration
  - B) Syncretic expression of anxiety and tension by action and aggression
  - C) Inability to contain inner life so that anxiety leads to action
- VI) Associated symptoms
  - A) Social awkwardness, lack of adaptiveness
  - B) Neurological "soft" signs
  - C) General unevenness in development

and as presenting management problems for various reasons.

Toward middle childhood, problems with peers become apparent as greater socialization is expected. Other difficulties are lack of social and practical judgment, poor coordination, excessive aggression and rage attacks, incipient anxiety attacks with associated phobias and sleep disorders, and precocious intellectual interests.

While the evidence for constitutional factors is still inconclusive, there is little doubt that borderline children come from chaotic and unstable homes. The mothers of these children were found to be disturbed in varying degrees. Most frequently, the mothers exhibit symptoms of the adult borderline syndrome. They are unstable, easily frustrated, quick to anger, unable to sustain an empathic relationship, and likely to distort essential aspects of interpersonal relationships. Many of the mothers of borderline children exhibited poor judgment and lacked common sense in childrearing. Often they would excessively stimulate their children sexually or have the child participate in the acting out of the mother's fantasies. For example, one nine-year-old boy was regularly taken into bed by his mother, where they would play at biting each other. Another mother regularly exposed her breasts to her son in a seductive manner. The mother of an eighteen-year-old patient, who has progressed rather well, still tries to bathe him and dress him despite his mature physical development.

The fathers of these children also manifested instability in emotional control and relationships. Violent scenes and physical fights were common among the parents. Some of the children were "kidnapped" back and forth by combative parents who used the children as pawns in their battles. We also found significant but inconsistent abuse and neglect of the child, depending on parental moods. One child was repeatedly smashed into the wall because her mother believed the child was a witch. Another child, as an infant, was given LSD by his father. Yet another child was routinely beaten by a paranoid, acting out father. The most characteristic quality of childrearing was lack of consistency of care. Many of the parents were quite successful in their professional or adult social lives but were unable to care adequately for their children because of personal psychopathology. Those families who could afford it often turned the care of the children over to a succession of baby sitters or nursemaids. Others utilized older siblings in caretaker roles.

The status of borderline children after they reach adulthood is unknown. Some[20,38] report that they become "odd" adults. Others[30] consider them preschizophrenic. We have followed some children to adolescence and found a heterogeneous outcome. One child, who is now fifteen years old, had progressed well in a special boarding school until he learned his family was moving out of state. Upon hearing this news, he became very anxious, started stealing, initiated sex play with a younger girl, and eventually exhibited psychotic behavior. Another formerly borderline child, now eighteen years old, attends a regular school, excels in electronics, but is socially shy and withdrawn. Another eighteen-year-old former patient is doing well and leads a fairly normal life. However, he tends to lack certain social graces and remains ignorant of social amenities. Those children with the best outcome had minimal organic impairment, took part in prolonged intensive individual psychotherapy, and experienced considerable improvement in their home environment. The authors have found that the greatest impediment to a favorable long-term outcome has been a continued chaotic family situation which all too often rapidly nullifies improvement laboriously obtained in therapy.

¶ Treatment of Borderline Children

It is difficult to make general statements regarding the treatment of borderline children. Since these children present with a wide range of symptoms as well as deviations in their overall development, a broad range of therapeutic approaches is usually indicated. As Rosenfeld and Sprince[35] pointed out, the particular combination of strengths and weaknesses, of innate vulnerability, and of environmental interferences create a most unusual set of characteristics for each of these children. The usual and expectable accomplishments in maturation are not readily and reliably found in these children. The strengths and capacities that are present are too easily subject to interference and regression. As Anna Freud[13] emphasized in her paper on the "Assessment of Borderline Children," it is important to assess the *total* personality development of the child before initiating treatment and not to limit the focus to the presenting symptoms alone.

If anything, these children tend to be undertreated, or treated with an approach that is either too narrow or is carried out under the pressure of immediate exigencies and emergencies. Because these children are often capable of high-level functioning, the severity and chronicity of their personality difficulties, their vulnerability to decompensation, the serious interferences with their development, and their lack of successful mastery of developmental stages is underestimated. Usually, individual psychotherapy alone is not sufficient to address their multiple needs.

Therapeutic aspects will be discussed under two broad categories: (1) the important aspects of individual treatment; and (2) general approaches that should be considered when arriving at a treatment plan for the child.

A treatment plan should usually include interventions in the child's educational and social environment as well as individual therapy. It is important that there be one responsible person to coordinate and oversee the implementation of the treatment plan in all of its components. Usually, the individual therapist undertakes this role. The parents of these children are often incapable of the organization and follow-through necessary to ensure adequate coordination of the various aspects of the treatment plan.

Not infrequently, a special day program or placement in a residential treatment program is indicated. These modalities will be discussed in a separate section, as will be the use of medication.

¶ Individual Psychotherapy

For borderline children, individual psychotherapy can be extremely helpful in working out their multiple, unresolved internal conflicts and in strengthening their adaptive defenses. In working with the child in individual therapy, it is important to keep clearly in mind the severity of the child's difficulties. One must be prepared for the sudden and unpredictable shifts in functioning to very primitive levels, and the equally sudden recovery. One must be on guard *not* to encourage the child to elaborate fantasy play, which will elicit so much anxiety that the child only deteriorates further. Most of these children, when confronted with fantasies impinging on the most vulnerable areas in their mental life, have difficulty maintaining distance from the material, even when it is brought up in a displaced form such as in doll play. Reality testing is weak, and transient lapses in the relationship to reality are frequent. Often, when the child is encouraged to express fantasies freely, the child becomes more and more stimulated and eventually the ability to deal adequately with the anxiety and excitement breaks down. For example, an eleven-year-old boy in a day program was transferred to a new male therapist. The boy came to the treatment an hour early and quite appropriately began to make a tower out of plastic blocks. The therapist encouraged the boy's fantasies while he was making the tower. The play began to shift and in rapid succession the boy introduced increasingly primitive themes, his excitment mounting until it was out of control. After a few minutes of building the tower, he introduced a male and female doll, whom he had, in sequence, kissing, having intercourse, and then merging together. He said the female was pregnant, then that the male doll was being penetrated anally and was pregnant. He then quite impulsively grabbed a pencil and pretended to stick it in his own anus, pushed the dolls together, and "glued" them together with clay, saying they were now one creature. During all of this play, he was increasingly agitated, and at one point grabbed at the therapist's penis and then ran out of the room, ending the session. He was clearly unable to maintain distance or use successful displacement and repression that is typical of latency children.

One must sensitively gauge when the child has the ability to utilize anxiety to stimulate adequate defensive maneuvers or when anxiety results in a disintegrating structural regression, as in the preceding case. For this reason, the therapist must be more active in directing the play, and intervening when the child is becoming overly stimulated; the therapist must set limits on the expression of aggressive, sadistic, and blatantly sexual material. The therapist may have to intervene physically to protect the child from harming himself, or the therapist. Frequently with these children, one treads a very narrow line between a sensitive and careful uncovering of the most vital concerns on the one hand, and a strengthening of repressive, obsessive-compulsive maneuvers on the other.

As previously mentioned, Ekstein and Wallerstein[9] emphasized the fluctuating nature of the ego organization of these children. They also emphasized the fragility of the therapeutic alliance with the child and how even subtle breakdowns in the empathic communication with the child can result in a regressive retreat to psychotic functioning. Again, these are disruptions of a transient nature but they are painful to the child and perplexing to the therapist.

Another difficulty in the individual treatment of the borderline child stems from the nature of the interpersonal relationships characteristic of these children. Many of these children do not experience close relationships as safe. In fact, closeness is frequently very threatening, although at the same time greatly desired. As the therapeutic relationship progresses, the child may become frightened of the developing closeness and will retreat, frequently becoming provocative or attacking the therapist in some manner.

For the borderline child who uses more schizoid mechanisms, a retreat to compulsive "out in space" fantasy play may result from the growing closeness to the therapist. The child will allow the therapist little or no inclusion in the play; the child becomes aloof, distant, and uncommunicative. For instance, a ten-year-old boy, after a few sessions with his therapist, in which he had been gradually more interested in and warmer toward the therapist, played out, week after week, the same theme—an endless battle in outer space consisting mainly of attacks and counterattacks by spaceships with no human or "humanoid" protagonists, and, most strikingly, little emphasis on who was good and who was bad. The therapist tried frequently to join the child in the play with little result. After several months, people rather than machines appeared and the therapist was gradually included in the play. This change coincided with reports from home that the boy was beginning to make friends for the first time and was beginning to assert himself in a positive way.

One must constantly be on the alert to strengthen reality testing whenever possible. Using an approach of gently but consistently reminding the child that it is "just pretend" or "just play" when playing out some important theme in displacement can help the child resist the disintegrative pull of his fantasies. At the same time, however, the therapist must communicate to the child that his problems are important and serious and that the therapist is not treating them lightly. Frequently, the experience of these children is that their deep concerns are not acknowledged and that they as individuals have not been taken seriously by their parents. To communicate that one understands the child without simultaneously encouraging the regressive wish to merge requires considerable sensitivity and skill on the part of the therapist.

One of the major difficulties in dealing with these children is the range of feelings the child evokes in the therapist. These children, compared with others, either healthier or with more pathology, almost universally provoke the therapist, as well as others, to experience frustration, confusion, and often helplessness. These children are the cause of some of the most heated disagreements among staff members, and not uncommonly rather desperate recommendations for alternative therapeutic approaches are proposed with increasing frequency when the child fails to respond to the more usual modes of intervention. The therapist is frequently seduced into thinking the child is more able to employ age-appropriate mechanisms on a consistent basis than is the case, and so becomes frustrated. Also, these children, because of their tendency to primary identification and immediate intimacy, fool the therapist into thinking there is a more genuine relationship than is actually possible. In addition, these children can quickly, and in a raw and undefended manner, raise very primitive issues that can be threatening to the therapist. Following the shifting levels in organization of these children can be both trying and confusing for the therapist. The problems of dealing with borderline children are magnified for persons who are not psychologically sophisticated or trained in these areas and account for the very strained relationships such children have with family members, teachers, neighbors, and peers.

¶ Work with Families

There has been a striking paucity of information in the literature on the families of borderline children. In the authors' experience, the parents of these children frequently exhibit considerable pathology. Extensive contact with the family, especially in the early phases of treatment, has been found to be essential for several reasons. Frequently, the parents feel overwhelmed, frustrated, hopeless, and angry with the child who has been creating difficulties both at home and in the community. The relationship between the parent and child is on a downhill course, caught in a vicious cycle of mutual disappointments, frustrations, and hostility. Active intervention with the parents is needed to interrupt this cycle and to offer hope of improvement.

The situation is, of course, complicated since many of these parents have serious character difficulties and the child's problems partially reflect the destructive influences of parental pathology. In regard to this, it is important to assess to what extent the child's pathological state is reactive to an interfering or destructively stimulating environment (for example, where the parent is either abusing the child outright, is alternately sadistically critical and seductive, or does not help the child to adequately assess and test reality).

The authors have often found the families of borderline children very difficult to engage in therapeutic work on their own behalf, and more often than not, the therapeutic team spends much clinical energy in maintaining an alliance with the parents in order to give them directions, advice, or even admonitions about their handling of their child. It is necessary to maintain the cooperation of the parents to make plans for their child. Many parents use mechanisms of projection, denial, and impulsive action to deal with their conflicts, and resist insight into their relationship with their child. A direct, matter-of-fact, advice-giving approach seems to work best. It is often necessary to give them simple, clear tasks and goals to effect changes in their relationships to their child. This can help minimize the often unmanageable negative transference that may arise in the parents of these children.

Day or Residential Treatment

The decision to recommend a therapeutic residential or day program depends on a number of factors, most significantly an assessment of the child's home environment and the degree of disturbance in the family relationships, especially the parent-child relationship. Often a day program is the treatment of choice if the relationship between parent and child has not become overly hostile and destructive. The combination of impulsivity, easy deterioration into anger, and difficulty on the part of others in understanding these children usually leads to a breakdown in the relationship with those who could, in the context of the relationship, help the children master their developmental tasks. These children often alienate those in their environment and establish relationships that are not helpful to growth. The parents' personality difficulties, particularly those that prevent them from separating themselves psychologically from the child, are frequently a part of the child's problem. For these reasons, a most effective intervention with many of these children is to separate them totally or partially from the home and to provide a different environment that can address the various aspects of the child's difficulties. The very fact of the separation forces the child and the family to confront their relationship,[16] particularly its destructively symbiotic aspects, as when parent and child do not have a clear sense of their individual separateness from each other. This difficulty in establishing a sense of separateness has many important clinical ramifications. For the child it can mean the development of merging fantasies as well as fears of these fantasies and the use of projective identification. For the parents, there is often a reactivation of their unresolved conflicts with their own parents, resulting in difficulty in distinguishing their experience in the past from their child's experience in the present. For these parents, it is not merely that their child's struggle *reminds* them of their difficulties as children; it *revives* these conflicts all too vividly. Their response may give validity to the child's terrifying fantasies and beliefs. This is alarming for the child, who needs, if anything, some reassurance and security that his most frightening wishes will not come true. An interruption in this destructive parent-child situation may be very useful and can be accomplished by temporarily separating the child from the parent.

One of the chief advantages of a therapeutic milieu program is that it permits a consistent and integrated approach to those aspects of the child's life that are so often pathologically disturbed—peers, school, building self-esteem, and so forth. Especially important is the close relationship between the educational and social/interpersonal components, allowing mutual feedback, illuminating the way the intrapsychic conflicts, character deficits, and cognitive and central nervous system dysfunctions mutually affect each other. For instance, the awareness by the special education teacher that the child has specific deficits in keeping in mind a sequence of instructions can help the people dealing with the child in his social environment to give directions in a simple step by step approach. One can help the child become aware of this difficulty and how it affects his social interactions and help him devise ways to deal with it. Likewise, the child-care professionals who have a deeper understanding of the child's fears and coping mechanisms can help teachers, parents, and others to understand the behavior of the child and to gain the perspective and distance that are so essential in dealing with these children.

This close coordination and feedback applies to the treatment of all children, but with borderline children it is of even more importance because their behavior is often so difficult and threatening, because they shift so rapidly from state to state, and because the countertransference feelings become so intense that people working with them need support to maintain their objectivity.

The borderline child's problems with impulse control and his tendency to take immediate action, rather than to employ fantasy or displacement when threatened, often lead to sudden and serious threats to others or to himself. At these times, the child requires direct physical controls. Often, removing the child to a safe, secure space where there is little stimulation is necessary. Going to the "seclusion" room is a most useful intervention, especially if used sensitively and with an understanding of the child's vulnerability at the time. As mentioned earlier, Geleerd[18] observed that while most children, when in the midst of a temper tantrum, will settle down if securely and firmly confronted by an adult, some children further lose control and experience panic if such an approach is used. She found that these children needed a much warmer, supportive approach with affectionate holding to help them come out of the temper tantrum. While the authors' have found this a useful observation in handling some borderline children, others become even more panic-stricken when approached physically during a temper tantrum. It is as if during their outbursts their ego boundaries are even less well defined and the closeness of another person seriously threatens their integrity. They need distance, safety, and security in order to reintegrate. They respond well to being quickly removed to the seclusion room with as little physical handling as possible.[17] Once there, verbal and/or visual contact is maintained with the counselor, but at some distance. This helps the child maintain contact with the outside world, but at a sufficiently safe distance to permit him to recover his more mature defenses. In the therapeutic milieu, a variety of methods may be employed to help the child develop better impulse control and to increase his ability to delay gratification. A reliable and consistent system of limits and rewards administered with objectivity is most important. Perhaps the most powerful therapeutic tool is the identification with a beloved counselor who has been intimately involved with the child's everyday activities. The authors' have seen children, after treatment in an inpatient program, institute for themselves limits that had previously been set by their counselor, for instance, taking a "time out" in a chair or going to the quiet room on their own.

School as a Therapeutic Milieu

As has already been mentioned, the borderline child's cognitive development is frequently delayed or impaired. This may take the form of a classic specific learning disorder or a more subtle impairment in learning functions. He may have a learning disability that has not been diagnosed because his social and behavioral difficulties are so pronounced that subtle underlying perceptual-motor or cognitive deficits are overlooked. A complete psychometric and cognitive assessment is imperative in elucidating the nature of the child's learning difficulty. Often a specific individualized program needs to be devised to meet the child's educational needs.

The school situation confronts the child with expectations and requirements that the borderline child is frequently unable to reach at first. The problems with impulse control, difficulty in delaying gratification, and the relative ease of disintegration make the classroom an especially problematic area for the borderline child. Since many of these children have not adequately internalized controls and since their social conscience still depends on the presence of an adult, these children have special difficulties being members of a group and taking their places in a more or less democratic social setting. They frequently demand the exclusive attention of the teacher and, when that is not available, behave in such a way as to force the teacher to attend to them. The more withdrawn borderline child, when not given the constant and exclusive attention of an adult, will retreat to a schizoid reverie.

Another major factor that affects the borderline child's adjustment to school is his inability to develop and elaborate satisfactory sublimatory channels. School and the process of learning provide for the healthy child a rich variety of sublimations, which, when successful, are very gratifying to the child and promote a strong investment in learning. The capacity for successful sublimations is limited for many borderline children and, when present, is subject to great fluctuations under the pressures of internal conflicts, the breakdown of age-appropriate defenses, and frequent regressions. For instance, in the child's view, the teacher too readily becomes the parent who is viewed with marked ambivalence, and struggles ensue that are based on conflicts originating well outside the classroom setting. Therefore, the learning situation may be experienced as unsuccessful and ungratifying. In addition, many borderline children have a very fragile regulatory system for self-esteem and when confronted with academic challenges at school feel overwhelmed and incapable. They tend to give up in despair, become self-denigrating, and are especially sensitive to criticism and failure. For example, a child of nine with a superior intelligence had a remarkably inconsistent report card from school with the frequent comment that when he cannot solve problems immediately, he gives up in despair, tears up his papers, sulks, refuses to do any more work, and calls himself stupid and dumb.

Generally, the children who are not learning up to their capacity and who are management problems in a regular class do function much better in a small classroom that permits considerable individual attention, with tutoring addressed to their specific limitations. The small class allows the child an opportunity for more guidance and reassurance from an adult. Also, since many of these children are easily overstimulated and have difficulty screening out distractions, the smaller, structured setting is useful. Often, by the time these children come to professional attention, they have already fallen considerably behind in scholastic achievement. This deficit contributes further to their low self-esteem. The child's teacher often needs consultation and support since, as has already been mentioned, these children present special emotional challenges to the people working with them.

Medication

In general, the use of drugs with children is complicated, unpredictable, and still rather poorly understood. However, there are some situations in which medication can be quite useful in treating the borderline child. If, as a significant part of the clinical picture, there is evidence of attentional disorder and hyperactivity (often including specific learning disorder and evidence of central nervous system dysfunction) dextroamphetamine or methylphenidate may be given a trial.

The management of anxiety is a major problem with all borderline children, and, on occasion, antianxiety medication may be useful. The minor tranquillizers so often used with adults are of limited value with children. Some children with crippling anxiety do respond to some of the major tranquillizers. The authors have found thioridazine, chlorpromazine, trifluoperazine, and haloperidol especially useful. Generally, anything that will help increase his sense of mastery is very important to a borderline child, and if medication can help him decrease or tolerate severe anxiety and control his bodily functions it should become a part of the overall treatment plan.

Side effects, although limited and usually not serious, can be distressing to borderline children and their families, who often have rather extreme and magical fantasies about drugs and their effect on the body. One father of a ten-year-old boy who had agreed to the authors' trying thioridazine for a period of time to help the boy's extreme anxiety and subsequent wild behavior, gradually became quite paranoid about the use of the drug for his son and, without telling the authors, stopped giving the medication to the boy. This father also became much more suspicious and guarded with the staff at this time. Another borderline boy, who was eight years old, showed marked attentional disorder and hyperactivity and was given methylphenidate. He experienced a decrease in appetite, especially at noontime, and, instead of sitting at the lunch table with other children as he had done previously, isolated himself, sucking his thumb and looking extremely perplexed. It gradually became clear that the loss of appetite, although not pronounced, was confusing and frightening to this child who was overly reactive to most external and internal sensations. It is important, then, to be especially alert to the fantasies, misinterpretations, and magical beliefs of both the child and the parent when administering medication to borderline children.

In summary, the treatment of the borderline child is complex and of long duration. Individual psychotherapy is most important, but only one of many factors in the total therapeutic constellation, which touches almost every aspect of the child's life. Furthermore, the authors have found that therapy often has to be continued until the child has progressed into early adolescence in order to ensure the consolidation of therapeutic gains.

¶ Bibliography

1. ARIETI, S. *Interpretation of Schizophrenia.* 2nd ed., New York: Basic Books, 1974.
2. BENDER, L. "Childhood Schizophrenia," *Nervous Child,* 1 (1942):138–140.
3. ———. "Childhood Schizophrenia: Clinical Study of One Hundred Schizophrenic Children," *American Journal of Orthopsychiatry,* 17 (1947):40–56.
4. BRADLEY, S. J. "The Relationship of Early Maternal Separation to Borderline Personality in Children and Adolescents: A Pilot Study," *American Journal of Psychiatry,* 136 (1979):424–426.
5. CHETHIK, M. "The Borderline Child," in Noshpitz, J. D., ed., *Basic Handbook of Child Psychiatry,* vol. 2, New York: Basic Books, 1979, pp 304–321.
6. CHETHIK, M., and FAST, I. "A Function of 'Fantasy' in the Borderline Child," *American Journal of Orthopsychiatry,* 40 (1970):-756–765.
7. DESPERT, J. L., and SHERWIN, A. C. "Further Examination of Diagnostic Criteria in Schizophrenic Illness and Psychoses of Infancy and Early Childhood," *American Journal of Psychiatry,* 114 (1958): 784–790.
8. DEUTSCH, H. "Some Forms of Emotional Disturbance and Their Relationship to Schizophrenia," *Psychoanalytic Quarterly,* 11 (1942):301–321.
9. EKSTEIN, R., and WALLERSTEIN, J. "Observations on the Psychology of Borderline and Psychotic Children," in Eissler, R. S., et al., eds., *Psychoanalytic Study of the Child,* vol. 9. New York: International Universities Press, 1954, pp. 344–369.
10. ———. "Observations on the Psychotherapy of Borderline and Psychotic Children," in Eissler, R. S., et al., eds., *Psychoanalytic Study of the Child.* vol. 11. New York: International Universities Press, 1956, pp. 263–279.
11. ENGEL, M. "Psychological Testing of Borderline Psychotic Children," *Archives of General Psychiatry,* 8 (1963):426–434.
12. FAST, I., and CHETHIK, M. "Some Aspects of Object Relationships in Borderline Children," *International Journal of Psychoanalysis,* 53 (1972):479–485.
13. FREUD, A. "The Assessment of Borderline Cases," in *The Writings of Anna Freud,* vol. 5. New York: International Universities Press, 1969, pp. 301–314.
14. FRIJLING-SCHREUDER, E. C. M. "Borderline States in Children," in Eissler, R. S., et al., eds., *Psychoanalytic Study of the Child,* vol. 24. New York: International Universities Press, 1970, pp. 307–327.
15. FROSCH, J. "The Psychotic Character," *Psychiatric Quarterly,* 38 (1964):81–96.
16. GAIR, D., and SALOMON, A. "Diagnostic Aspects of Psychiatric Hospitalization in Children," *American Journal of Orthopsychiatry,* 32 (1962):445–461.
17. GAIR, D., BULLARD, D., JR., and CORWIN, J. "Residential Treatment: Seclusion of Children as a Therapeutic Ward Practice," *American Journal of Orthopsychiatry,* 35 (1965):251–252.
18. GELEERD, E. R. "Observations on Temper Tantrums in Children," *American Journal of Orthopsychiatry,* 15: (1945):238–246.
19. ———. "A Contribution to the Problem of Psychoses in Childhood," in Eissler, R. S., et al., eds., *Psychoanalytic Study of the Child.* vol. 2. New York: International Universities Press, 1946, pp 271–291.
20. ———. "Borderline States in Childhood and Adolescence," in Eissler, R.S., et al., eds., *Psychoanalytic Study of the Child.* vol. 13. New York: International Universities Press, 1958, pp. 279–295.
21. GLOVER, E. "A Psychoanalytical Approach to the Classification of Mental Disorders," *Journal of Mental Science,* 78 (1932):819–842.
22. GRINKER, R. R., WERBLE, B., and DRYE, R.

The Borderline Syndrome. New York: Basic Books, 1968.

23. GUNDERSON, J. G., and SINGER, M. T. "Defining Borderline Patients: An Overview," *American Journal of Psychiatry,* 132 (1975):1–10.
24. HOCH, P., and POLATIN, P. "Pseudoneurotic Forms of Schzophrenia," *Psychiatric Quarterly,* 23 (1949):248–276.
25. KERNBERG, O. "Borderline Personality Organization," *Journal of the American Psychoanalytic Association,* 15 (1967):641–685.
26. ———. "Treatment of Patients with Borderline Personality Organization," *International Journal of Psychoanalysis,* 49 (1968):600–619.
27. KNIGHT, R. P. "Borderline States," *Bulletin of the Menninger Clinic,* 17 (1953):1–12.
28. KOHUT, H. *The Analysis of the Self.* New York: International Universities Press, 1971.
29. MAENCHEN, A. "Notes on Early Ego Disturbances," in Eissler, R.S., et al., eds., *Psychoanalytic Study of the Child.* vol. 8. New York: International Universities Press, 1953, pp. 262–270.
30. MAHLER, M. S., ROSS, J. R., and DE FRIES, Z. "Clinical Studies in Benign and Malignant Cases of Childhood Psychosis (Schizophrenic-Like)," *American Journal of Orthopsychiatry,* 19 (1948):295–305.
31. MARENS, J. "Borderline States in Childhood," *Journal of Child Psychology and Psychiatry,* 4 (1963):207–218.
32. MASTERSON, J. F. *Treatment of the Borderline Adolescent: A Developmental Approach.* New York: Wiley-Interscience, 1972.
33. MODELL, A. "Primitive Object Relationships and the Predisposition to Schizophrenia," *International Journal of Psychoanalysis,* 44 (1963):282–292.
34. PINE, F. "On the Concept 'Borderline' in Children," in Eissler, R.S., et al., eds., *Psychoanalytic Study of the Child.* vol. 29. New York: International Universities Press, 1974, pp. 341–368.
35. ROSENFELD, S. K., and SPRINCE, M. P. "An Attempt to Formulate the Meaning of the Concept 'Borderline,'" in Eissler, R.S., et al., eds., *Psychoanalytic Study of the Child.* vol. 18. New York: International Universities Press, 1963, pp. 603–635.
36. ———. "Some Thoughts on the Technical Handling of Borderline Children," in Eissler, R.S., et al., eds., *Psychoanalytic Study of the Child.* vol. 20. New York: International Universities Press, 1965, pp. 495–516.
37. SCHMIDEBERG, M. "The Borderline Patient," in Arieti, S., ed., *American Handbook of Psychiatry,* vol. 1. New York: Basic Books, 1959, pp. 398–416.
38. WEIL, A. P. "Certain Severe Disturbances of Ego Development in Childhood," in Eissler, R. S., et al., eds., *Psychoanalytic Study of the Child.* vol. 8. New York: International Universities Press, 1953, pp. 271–287.
39. ———. "Clinical Data and Dynamic Considerations in Certain Cases of Childhood Schizophrenia," *American Journal of Orthopsychiatry,* 23 (1953):518–529.
40. ———. "Some Evidences of Deviational Development in Infancy and Early Childhood," in Eissler, R. S., et al., eds., *Psychoanalytic Study of the Child.* vol. 11. New York: International Universities Press, 1956, pp. 292–299.
41. WOLFF, S., and BARLOW, A. "Schizoid Personality in Childhood: A Comparative Study of Schizoid, Autistic, and Normal Children," *Journal of Child Psychology and Psychiatry,* 20 (1979):29–46.
42. ZETZEL, E. "A Developmental Approach to the Borderline Patient," *American Journal of Psychiatry,* 128 (1971):867–871.

CHAPTER 12

PSYCHOPHARMACOTHERAPY IN CHILDREN

John S. Werry

¶ Introduction

THE USE of psychopharmacotherapy in children dates from the introduction of the stimulants by Charles Bradley in 1937. However, the early years were quiet until newer psychotropic drugs developed in the 1950s gave a new though rather slow impetus to the field. But it was not until the 1970s that pediatric psychopharmacology finally emerged as a fully fledged therapeutic modality and research endeavor in child psychiatry. As with adults, much of the early work was hampered by the lack of double-blind and other controls, by diagnostic vagueness and heterogeneity reflected in such terms as "emotionally disturbed" or "behavior disorders," and by the absence of reliable, valid, and quantitative measures of analysis.[41,68] However, there has been conspicuous improvement in the field during the last ten years.[19]

Unlike adult psychiatry, which both discovered and initially tested most of the psychotropic drugs currently used in children, pediatric psychopharmacology is still very much an empirical clinical exercise, lacking the attractive theoretical rationale for employing the antipsychotics or the antidepressants that are used in adult psychiatry. This is mostly because the uses to which these drugs are put in children are quite different from those in adults, and also because there is a serious shortage of trained investigators in pediatric psychopharmacology, particularly those with a biomedical background. As with adult psychiatry, there have been no substantive additions to the types of drugs used in children since 1960, partly because of the lack of basic research.

The rise of pediatric psychopharmacology has been accompanied by considerable public interest and concern, which is reflected in congressional and public enquiries, in legislation, and increasingly in advocacy litigation.[62] While much of the criticism has been ill-informed and is part of a sophisticated populist anti-intellectualism afflicting western society, it has revealed practices, particularly with institutionalised retardates, that are inconsistent with good medical practice and, like overprescribing in general, reflect some of the basic ills of modern medicine. Regulation and restriction on research in psychopharmacotherapy with children both at state and national levels is now well established. Some of this has effectively stopped research, yet the right to *prescribe* such drugs in children goes unhampered. Whatever the rights and wrongs of the situation, one thing is clear: As with all areas of medical practice, public accountability is a fact of life that pediatric psychopharmacology cannot ignore.[62]

With one or two exceptions, most of the indications for the use of psychotropic drugs in children are symptomatic, cutting across diagnostic entities. The drugs are prescribed mostly for the purpose of producing social conformity at the behest of adults (usually parents, teachers, and other primary caretakers). Since clinical indications and the drug action utilized are quite different, the current adult-derived therapeutic classification of psychotropic drugs (for example, antidepressant, antipsychotic, stimulant) makes little sense and, in fact, only creates confusion in pediatric psychopharmacology. Only a classification based on neuroregulatory action—for example, anticholinergic dopamine blockers (for sedative antipsychotics like chlorpromazine) or anticholinergic adrenergic facilitators (for antidepressants like imipramine)—offers any possibility of a classification suitable for both adult and pediatric psychopharmacology.

Until recently, interest in classification or diagnosis in child psychiatry languished.[40] As a result, pediatric psychopharmacology has been severely criticized for the vagueness of such clinical indications as hyperactivity.[48] The new children's section of the *Diagnostic and Statistical Manual of Mental Disorders,* third edition (DSM-III)[5] offers a unique opportunity to change this situation, since for the first time in child psychiatry there is presented a distinct, reasonably unequivocal set of criteria by which to make a diagnosis and an investigation of the relationship between diagnosis and psychopharmacotherapy in children.

Because of the rapidity of physical and psychological changes between birth and adolescence, child psychiatry puts particular emphasis on development. All psychotropic drugs have powerful effects on the brain and on systems other than those of clinical interest, particularly the higher cortical, limbic, arousal, hypothalamic, and endocrine systems. If the notion of critical periods is valid, drugs could, in theory, disrupt critical psychological, emotional, cognitive, and biological developmental events (short-term drug effects on all of these events are readily demonstrable). This calls for a spirit of proper caution and conservatism coupled with high standards of medical and multidisciplinary assessment of drug effects.

Because of the vulnerability of the developing child and his inability to express himself fluently in the face of imposing authority figures, the problem of social control and imposed treatment take on important ethical concerns in pediatric psychopharmacology. While researchers are aware of this problem, studies to date have been remarkable for the absence of child-generated data, particularly of the subjective kind.[81,82] This may be due to the methodological difficulties involved in collecting and utilizing this kind of data.[75]

While children are now fairly well protected (some would say even overprotected) by statute and regulation as far as research in pediatric psychopharmacology is concerned,[62] a most worrisome area remains the unrestricted right of physicians to treat children with psychotropic drugs. Only the acceptance of stringent peer review by the medical profession can solve this problem, though the rather unpalatable alternative of patient or advocacy litigation is having an impact as well.[62]

If pediatric psychopharmacology is to advance, child psychiatry must be seen, particularly by academic departments of psychiatry, as an underdeveloped area requiring the highest of priorities. Only in this way will the necessary cadre of skilled and creative investigators be forthcoming.

¶ Developmental Pharmacology

There is disconcerting lack of information about the effect of age on such basic pharmacological dimensions as pharmacokinetics, toxicity, dosage, and so on, particularly with respect to the psychotropic drugs.[11] For example, there are almost no studies in pediatric psychopharmacology about an area of current interest in adult psychiatry, namely the relationship between plasma levels and therapeutic effects. While some of this is understandable due to the reluctance to submit children to venipuncture, the development of salivary methods of measuring the unbound portion of drugs in plasma offers a solution to this dilemma.[37]

Research has indicated that after the first year of life children should be able to (1) clear drugs renally like adults; (2) metabolize them more quickly; (3) maintain lower plasma levels of lipid-soluble drugs such as barbiturates and benzodiazepines because of their proportionately greater amounts of body fat; (4) show different distribution of those drugs that are concentrated in such organs as the brain and liver because of their different sizes; and (5) show greater effects from drugs distributed in this way because of decreased proportion of extracellular water. The practical implications of these extrapolations from basic physiological (rather than pharmacological) data suggest that children, in general, will be more resistant to psychotropic drugs and that dosages should be based on calculations of body surface area, since these more accurately reflect the proportion of extracellular body water. However, because of the low toxicity and relatively flat dose response curves of most psychotropic drugs, calculations based on the simpler milligram/kilogram basis will ordinarily be satisfactory.

It is likely, however, that the different development of the brain and various enzyme systems in children should produce certain differences, at least of degree, in the action of psychotropic drugs at the cellular level, which is distinct from the pharmacokinetic issues already considered. For example, it is thought that acute dystonias are more frequent with antipsychotic drugs in children, but our knowledge of such possible developmental differences is merely speculative. Another factor likely to influence the cellular action of psychotropic drugs in children is that many children given medication appear to have normally functioning brains and neurotransmission. This being so, the action of drugs would only result in distortion of normal function. Such a hypothesis, however, requires formal testing, but it does illustrate some of the possible fundamental pharmacological differences that might exist between pediatric and adult psychopharmacology.

¶ Measuring Drug Effects in Children

Perhaps nowhere else is it more evident that pediatric psychopharmacology has come of age as a technical specialty than in the way in which the effect of psychotropic drugs are evaluated.[2,38,42,74] Obviously, it is no longer a field for amateurs; it requires cooperation between properly trained investigators from social, behavioral, and biomedical sciences.

Parent and Teacher Behavior Symptom Checklists

Checklists of symptoms compiled by parents and teachers form the mainstay of both patient selection and assessment of drug effects, particularly in research in pediatric psychopharmacology. A number of reliable and valid instruments exist[19,40,51] whose greatest strength derives from the capacity of the human brain to integrate a mass of information into a small number of usable clinical judgments. For example, this information comes from a number and variety of social situations, is based on differing lengths of time the child is observed (and with teachers), and is automatically compared against the behavior of comparative groups of children of similar age and background. The variability of various mothers' judgments can be overcome by the use of measures before and after medication is commenced (repeated measures design).[74] The greatest weakness of these methods lies in the rather low level of agreement between parents and teachers[40] and in the discrepancies with measures from other sources.[74]

Child-Derived Measures

Child-derived measures have not been well studied, though a number of possible techniques to do this exist.* In the clinical situation, however, this is less likely to be a problem, provided that the physician takes the time and trouble to develop a relationship with the child and discuss all aspects of treatment and its effects with him.

Interviews

While the clinical interview must form the mainstay of clinical practice, there is good reason to doubt that its validity, reliability, or exactitude is sufficient for research purposes without further structuring.[32,51] Inasmuch as psychological testing provides such a structured situation, it is not surprising that it has proved useful in pediatric psychopharmacology.[74]

Physiological and Psychophysiological Measures

Measures of a physiological and psychophysiological nature are part of physical examinations and take into account weight and height, heart rate, blood pressure, and aspects of the neurological examination (particularly motor coordination). Other highly sophisticated methods include laboratory tests such as biochemical,[17,18,57,74] electrophysiological,[74] and psychophysiological[30] measures. Although these highly technological measures offer great promise, they are still mostly restricted to the research and laboratory situation, as they require extensive technical knowledge and equipment.

Direct Measures of Behavior

Most of the developments in direct measures of behavior have occurred using human observers, but despite the spectacular results in behavior modification, the applications to pediatric psychopharmacology have so far been limited and, to a certain extent, disappointing (though there have been some successes too).[1, 9, 74] Mechanical and electronic aids such as videotape,[59] actometers, and stabilometric seats have all been used in pediatric psychopharmacology research, particularly in the measurement of motor activity.[74] In general, these techniques are cumbersome, intrusive, and expensive, restricting them mainly to particular investigations and laboratories.

Psychological and Cognitive Tests

Almost from its earliest days, pediatric psychopharmacology has been greatly concerned with the effect of drugs upon cognitive function and learning, particularly the issue of mental dulling. As a result there are a number of widely accepted and recurringly drug-sensitive measures such as the Continuous Performance Test, Paired Associate Learning, the Porteus Mazes and, to a lesser extent, standard tests of intelligence and achievement.[2,3] Regrettably, actual learning in the classroom situation has not been studied as well,[2] since the techniques necessary to assess the process are much more difficult and still to be worked out.[84]

Clinical Global Impression

One of the most robust and drug-sensitive measures in pediatric and adult psychopharmacology is the simple but crude Clinical

*See references 24, 42, 74, 81, and 82.

Global Impressions.[34] In this test, parents, teachers, and especially physicians make a simple judgment as to whether or not the patient has improved in a global sense (or got worse and if so, to what degree). In the case of the physician scale, the rating of improvement is further interpreted against the discomfort or disability of side effects produced. It thus represents the ultimate in final clinical judgment about the cost/benefit yield of a drug for an individual patient.

Which Measure is Correct?

As noted, a remarkable feature of pediatric psychopharmacology is the rather low level of agreement between measures from different sources, such as parent, teacher, laboratory, and physician.[74] Since all observers base their judgments on different aspects of function in different situations for varying lengths of time, none is a priori more valid than any other. It is the physician's job to weigh each piece of information and integrate it into a clinical decision.[74]

¶ Individual Drugs

Stimulants

The stimulants in common use (dextroamphetamine, methylphenidate, and magnesium pemoline) belong to the much larger group of sympathomimetic amines with which they share many central and peripheral adrenergic properties.[16] While their principal action is adrenergic, they also release dopamine and possibly even acetylcholine.[56,58] In general, little data on their pharmacokinetics in children is available, though their clinical effect appears immediate and, by the standards of psychotropic drugs, relatively short-ranging (from four to about twelve hours).[16] Magnesium pemoline is said to have a somewhat longer half life and thus obviates the need for a second dose. Actually it is surprising how few children need more than a single daily dose.[16,69]

Clinical Effects

Physiological effects of stimulants are much as would be expected from any sympathomimetic amine with peripheral (tachycardia, vasoconstriction) and central actions, but there are also endocrinological effects, such as release of growth hormone, as well as hypothalamic effects (anorexia, weight loss).[16] As far as behavior, motor, and cognitive function are concerned, the effects of stimulants are similar to those seen in adults, particularly when adults are fatigued or bored:[51] reduction of motor overflow; improved vigilance, attention, motor skills; and increased zest and performance in most functions.[16,19] It is for this reason that stimulants are used in pediatric psychopharmacology; and their effect in producing overall clinical improvement as perceived by adults in hyperactive/aggressive children, at least in the short term, is compelling.[7,16,19] However, effects on learning are more controversial, particularly with regard to the question of whether there is any actual increased acquisition of new knowledge or skills rather than simply increased performance of what the child already knows.[2,3,8,16] Whether stimulants improve social interaction or just passive compliance,[9,49] and whether they influence long-term clinical outcome,[72,73] are also in dispute. Another question is whether the clinical effects upon behavior seen in hyperactive/aggressive children represent a specific pharmacological effect, often called paradoxical. Studies of normal children suggest that the clinical effects are qualitatively the same as in hyperactive children, and that rather than being paradoxical the response in hyperactive children is one of degree or is even possibly rate dependent.[45,49,78]

Side Effects

While initial mild effects such as headache, stomachache, insomnia, and anorexia are common, stimulant administration causes remarkably few serious side effects.[7,16] A short-lived slight weight loss at the commencement of treatment is common. However, there is some concern that stimulants can produce continuing and significant growth suppression and weight loss, particularly in higher doses (in excess of 1 mg/kg methylphenidate).[16,52] It now appears that the frequency, degree, and durability of this problem has been greatly exaggerated,* although proper monitoring of weight and height is a clinical necessity. About 25 percent of children using stimulants show an increase in irritability, tearfulness, and even hyperactivity.[69] While an occasional child will exhibit a true amphetamine-type psychosis and even neurological syndromes, such as a dopaminergic dyskinesia,[16] such side effects dissipate rapidly upon cessation of the drug.

Clinical Indications and Use

It is generally accepted on the basis of numerous properly controlled studies that attention deficit disorder with hyperactivity (hyperkinesis) in elementary school children is a well-established clinical indication for stimulants, though many such children will neither need nor benefit from the drug. Despite a great deal of work, it is still impossible to predict without an actual clinical trial which children will respond and which will not—though there are some indications that the more severe the disorder, the more likely a good response.[6,35,70] Attempts to find psychophysiological or biochemical predictors have to date been disappointing.[30,46,57] Though dosage has been little studied, there is some evidence to suggest that the optimum dosage level is in the region of 0.5 mg/kg of methylphenidate.[64,79] Dosage may be raised above this level providing there is careful monitoring for side effects. Preschool children are generally considered not to benefit from stimulants, and many clinicians are reluctant to institute or continue medication beyond puberty because of the still unsubstantiated risk of dependence. The use of "drug holidays" during weekends and vacations has much to commend it in helping to prevent the development of tolerance and hence the chance of dosage and metabolic problems. Most children require only one dose per day, a usage that is to be encouraged, since, like drug holidays, it helps to minimize the possibility of untoward effects.

How long a child should stay on medication is an almost completely unresearched issue,[16] though one study suggests that an annual probe using a placebo should be carried out. About 24 percent of children will be found to no longer need their medication.[61] A technique developed by Swanson and associates[69] for deciding which children will respond by submitting them to a one-day laboratory trial has promise for well-equipped centers but requires further study. Despite references in the literature to other uses of stimulants in children, there are as yet no other established indications in pediatric psychopharmacology.[16,19]

Social and Ethical Issues

While the clinical research literature on the use of stimulants is the most voluminous and scientifically robust in the field of pediatric psychopharmacology, there are a number of unresolved issues, particularly in view of the fact that stimulants have a high dependency potential in adults. So far, there is no evidence that their medical use in hyperactive children leads to dependence in later life,[16,47] but such long-term studies are difficult to do and consequently few exist. The contention that stimulants lack the necessary euphoric quality critical to the establishment of dependence in children is supported by only one formal study[45] there is much anecdotal clinical evidence to the contrary.[19] While there have been a number of sensational charges about the epidemiology of stimulant-prescribing in the United States, the facts show that stimulant use is basically conservative, commoner in children of higher socioeconomic class, and becoming less prevalent since it peaked at around 500,000 patients in 1977.[16,23,63]

Whether the clear short-term benefit is re-

*See references 1, 28, 31, 36, 50, and 54.

flected in enduring better social adjustment or learning is disputed, since there is only scanty evidence to date and much of it negative.[31,72] Thus, justification for the use of stimulants must be based on the here and now and not on the basis of long-term prevention of disability.

Conclusions

Stimulants are the best studied and most clearly established of the psychotropic drugs used in children. Their only legitimate indication is in *some* cases of attention deficit disorder with hyperactivity in elementary school children, but even there their impact on long-term adjustment and academic achievement is dubious despite impressive short-term effects. As yet there is no way of predicting which child will respond to stimulants, and the contention that the effect of stimulants is paradoxical, specific, and tied to some brain dysfunction is becoming increasingly more dubious. Despite this, the stimulants are a valuable part of the overall management plan in some cases of hyperactive children and have given by far the greatest impetus to the establishment of pediatric psychopharmacology as a field of scientific endeavor.

¶ Antipsychotics (Neuroleptics, Major Tranquilizers)

There are several key reviews of antipsychotics in pediatric psychopharmacology.* They are probably used most frequently with the mentally retarded; administration percentages range from about 4 percent of children in special classes in Illinois[23] to 50 percent in institutions throughout the United States.[62]

Pharmacology

Antipsychotics have a wide variety of pharmacological effects based upon dopamine, noradrenaline, acetylcholine, histamine, and nervous impulse blockade.[83] The extent to which different antipsychotics possess any or all of these properties varies, but the two most favored in pediatric psychopharmacology (chlorpromazine and thioridazine) have all these properties.[76] The true antipsychotic property, currently thought to be dopaminolytic, is of little importance in pediatric psychopharmacology due to the infrequency and relative refractoriness of childhood psychoses.[13,14] The half life of the antipsychotics is very long (measured in days), and biodegradation is extremely complicated (chlorpromazine having well over a hundred metabolites).[83]

Clinical Effects

As would be expected with such wide-spectrum drugs, clinical effects are multiple and varied. The ones of principal importance in pediatric psychopharmacology are sedation and suppression of tics.[19,65,83] One should be able to predict from the pharmacology of these drugs that the type of sedation would be different from that produced by traditional central nervous system depressants, such as barbiturates, in that antipsychotics should produce emotional indifference not euphoria and quietness, without behavioral disinhibition and paradoxical excitement.[76] Unfortunately, there has been little systematic study of the clinical effects of these drugs in children beyond the overall behavioral change, which is generally reported to be in the direction of psychomotor slowing or sedation and perceived clinical improvement.[4,65,83] The fact that barbiturates are held to cause paradoxical excitement in children, which has caused them to be largely abandoned in favor of the antipsychotics (and antihistamines), does offer some circumstantial evidence in support of a different type of sedation resembling that reported in adults and animals. Whether this sedation is due largely to an anticholinergic effect resembling that of hyoscine or whether it is due to some additional effect is unclear, though it is significant that the two drugs most favored (chlorpromazine and thi-

*See references 4, 13, 14, 19, 22, 23, 65, and 83.

oridazine) are both strongly anticholinergic.[76] Whatever the exact nature of the effect, there is evidence that the antipsychotics can produce reduction in hyperactive, aggressive, excited behavior independent of any specific diagnosis, conspicuously in the mentally retarded, psychotic, and attention-deficit disordered children.[4,65,83] Whether this behavioral improvement is at the expense of mental alertness or cognitive function is unclear. Laboratory studies—done mostly under extremely favorable learning conditions and in children on relatively low doses—suggest a minor degree of impairment, usually impeding performance.[4,65,83] Whether this obtains in the quite different, noisy, distracting environment of institutions for the mentally retarded, which employ considerably higher doses, is not established. Nonetheless, this has not deterred current litigation from trying to reduce both dosage and frequency of the use of medication in the institutionalized mentally retarded.[62]

The suppressant action of antipsychotics upon tics, particularly useful in Tourette's disorder,[12,18,83] is probably dopaminolytic, and hence nonanticholinergic drugs such as haloperidol tend to be favored, since acetylcholine and dopamine act antagonistically in the basal ganglia. However, recent work suggests that the suppressant effects on tics may relate to interference with noradrenaline or even serotonin.[18]

Side Effects

Most of the side effects of the antipsychotics are, of course, simply normal effects of the drugs upon neurotransmitter or systems other than those of primary therapeutic interest. The well-known and varied side effects of the antipsychotics in children have been enumerated.[83] Extrapyramidal effects seem somewhat less common in children than in adults, due probably to the preferred use of strongly anticholinergic drugs in adults. Tardive dyskinesia has been reported only rarely, though cholinergic symptoms and a curious evanescent rebound dyskinesia produced on stopping of these drugs have been reported.[29,83]

Clinical Indications and Use

Unlike the stimulants, the indications for the use of antipsychotics in children are unclear and disputed. While there is evidence to suggest that these drugs will reduce certain socially disruptive behaviors, such as overactivity, aggressive outbursts, and excitement in children (particularly the mentally retarded, the psychotic, and attention-deficit disordered), the costs to the child in terms of minor but uncomfortable side effects and in mental dulling are not well established.[4,65,83] It is therefore best to regard these drugs as strictly for short-term crisis management and not as substitutes for more personalized, humane, nonbiological programs, especially in institutions. There is some evidence to suggest that low doses (0.025–0.05 mg/kg of haloperidol, 1.5–3.0 mg/kg of chlorpromazine and thioridazine) may be as effective as the more usual higher doses.[60,83] The current widespread use of these drugs in the mentally retarded and in the management of sleep disorders in young children cannot be justified on the basis of well-conducted clinical trials.[4,44] Children with pervasive developmental disorders and schizophrenia (psychoses) may constitute a special group, though evidence for a true antipsychotic as opposed to a symptomatic effect in these drugs is lacking in prepubertal children.[13,14,83]

Stereotyped movement disorders (tics and Tourette's disorder) appear to be suppressed by adequate, and often quite high, doses of antipsychotics, though there are few properly controlled studies to support what appears clinically quite convincing.[12,18,83] However, it is important that the ability of these drugs to suppress tics should not lead to their premature use, since most children's tics are largely self-resolving and of relatively minor social significance.[75]

Social and Ethical Issues

The most important ethical issue regarding the dispensation of antipsychotics to children is the risk of dulling mental capacities in children who are already handicapped in their learning ability, especially the institutionalized mentally retarded. Since these drugs are given primarily for purposes of social control (that is, for the needs of adults), particular care should be taken with children who are least able to report their own needs or side effects.[62]

Conclusions

The only reasonably clear indication for the use of antipsychotic drugs in children is in Tourette's disorder, but even there side effects can be considerable. Their use for sedation is unestablished, and the cost in terms of uncomfortable side effects and mental dulling remains unclear. Their use for this purpose is best reserved for short-term management in crisis situations, though even here properly controlled studies are long overdue. The frequency of their use and dosage is probably in inverse relationship to the quality of care given to children, particularly those in institutions. Because of the extreme degree of handicap and difficulty presented by psychotic children, the use of antipsychotics may be more defensible; however, proper documentation is always required.

¶ Antidepressants

There is considerably less data on the use of antidepressants with children than on that of stimulants and antipsychotics, probably reflecting the fact that antidepressants are prescribed less frequently for children.[19,22,43,65]

Pharmacology

It is well known that antidepressants are divisible into two main groups: monoamine oxidase inhibitors and multicyclic antidepressants. The former are more dangerous, have few advocates, and have even less supporting data for use in children.[43] The multicyclics, or "antidepressants," are thought to act primarily by blocking the reuptake of released noradrenaline, though some may be antiserotoninergic. Like the antipsychotics from which they are derived, some antidepressants possess anticholinergic, antihistaminic, and local anesthetic properties to varying degrees. And like the antipsychotics, they are long-acting drugs.

Clinical Effects

The principal effects of antidepressants upon the behavior, motor activity, and cognitive performance of children resemble those of the stimulants, at least in attention deficit disorder where most of the acceptable studies have been executed.[43,80] These effects, as noted previously, are: reduction in exuberant deviant behavior, reduced motor activity, and improved cognitive performance. Physiological effects are similar too, including slight initial weight loss, though tachycardia is much more pronounced.[43,80] Anticholinergic sedative and peripheral autonomic side effects may conceal this stimulantlike picture, especially toward the beginning of treatment. Although their effect is basically similar to those of the stimulants in hyperactive children, their effects are generally inferior, and they cause more side effects.[43] As a result, parents tend to discontinue their use more frequently than they do stimulants in the long-term management of the children.[43] Whether the time dimension of this effect is immediate or shows a latency similar to that of the antidepressants in adult depression is unclear, though clinical opinion favors the former. A true antidepressant effect in children remains to be demonstrated and is part of the continuing controversy as to whether or not adult-type depression occurs in children.[19,39,43]

Imipramine and other tricyclics have an immediate symptomatic suppressant effect upon enuresis.[10,43] The pharmacological basis of this action is generally presumed to be anticholinergic, though alpha-adrenergic and central effects must also be involved since anticholinergic agents, which act only on the bladder, do not have the powerful effect of the tricyclics.[10,43]

Side Effects

The most important side effect of the tricyclics is cardiotoxicity (seen only in doses in excess of 5 mg/kg).[43] Since such a dosage level is unusual in children, complications should be exceedingly rare except in accidental overdose, particularly in toddlers. Apart from epileptic seizures, other side effects are minor and include atropinism, tremor, tearfulness, initial sedation, and stomachaches.[43] There is some evidence[26,53] that tricyclics may be useful in the management of separation anxiety disorders (for example, school phobia), but the doses used were high, resulting in one fatality. This use has some parallels with the treatment of phobic obsessive/compulsive disorders in adults. While such disorders do occur in children, tricyclics have not yet been tried in them.[21] Thus, the use of tricyclics in the management of anxiety disorders in children must be considered at the moment experimental and subject to all the safeguards of properly controlled clinical trials.

Conclusions

The role of antidepressants in pediatric psychopharmacology appears to be much more limited than stimulants. Their principal use is probably as an alternative though less satisfactory method in the management of attention deficit disorder with hyperactivity. They are also useful in the *symptomatic* management of enuresis, though there seems little justification for their widespread use in this disorder, since they do not influence the long-term outcome of what is, after all, a benign, self-limiting, childhood disorder.[10,43] Whether they have any other roles, such as in depressive and anxiety disorders, remains to be demonstrated.

¶ Anxiolytics and Sedatives

There are several useful reviews of these drugs in children, though all have concluded that valid data upon which to make judgments is conspicuously lacking.*

Pharmacology

Drugs commonly used as antianxiety or sedative agents in children are of four main types:

1. General central nervous system depressants such as barbiturates, alcohol, gasoline, glues, and so on
2. So-called selective depressants such as the benzodiazepines
3. Antihistamines
4. Antipsychotics

While the half life of the first group varies (though it is generally less than twenty-four hours), that of the second group, the benzodiazepines, and most other psychotropic drugs is well in excess of this.[27] The action of depressants is primarily a general one, probably on membrane excitability, though phylogenetically more recent parts of the brain such as the neopallium are more readily affected. Although extravagant claims for the selectivity of the benzodiazapines are sometimes made, the most conspicuous difference between them and the traditional sedatives lies in their very flat dose-response curve and hence in their low toxicity.[27] In theory, too, since anxiolysis differs from sedation and general anesthesia only in the degree of depression of the brain, this flat dose-response curve should allow finer tuning of the pharmacological effect.[27]

The third group, antihistamines, are qualitatively different from the other sedatives and probably act primarily through a central anticholinergic action. However, those that are closely allied to the phenothiazines, such as trimeprazine and promethazine, may have a weak ataractic effect characteristic of antipsychotics as well.[76]

See page 264 for a discussion of the fourth group, the antipsychotics.

Clinical Effects

In the case of the central nervous system depressants, the clinical effects consist pri-

*See references 19, 22, 25, 27, 44, and 77.

marily in the reduction of the level of behavior along the anxiolysis/sedation/anesthesia/coma/death continuum, depending on the dose. Greenblatt and Shader[27] have argued convincingly that the distinctive characteristic of true sedatives is disinhibition of behavior—that is, the release of what is ordinarily kept suppressed by punishment or fear of punishment. Since children are ordinarily sedated in a situation where they are anxious or upset, and where adult patience is wearing thin, the long-standing observation that barbiturates often make children more rather than less excited is exactly what would be predicted from disinhibition. Studies* of the traditional sedatives show that there are few properly controlled and constructed trials of sedatives in children. However, adverse clinical experience with phenobarbital in epileptic children,[67] and the well-established fact that anxiety in children is short-lived and highly responsive to placebo and nonspecific interventions,[77] suggest that there is little role for these central nervous system depressants in children's anxiety disorders. On the other hand, sleep disorders in very young children are of a recurrent and highly disturbing nature, and it is here that a substantial literature might have been expected. A different group of sedative drugs, the antihistamines, is widely used in this group of disorders. Since they should not cause behavioral disinhibition, their use would seem to be preferable, but again there is almost no data about their efficacy and safety.[44,76]

Side Effects

Predictably, all sedative drugs should produce some impairment of cognitive function. In addition, the central nervous system depressants run the risk of producing dependence, behavioral irritability, withdrawal or other seizures, and, with the exception of the benzodiazepines, when taken in overdose, life-threatening situations. Antihistamines may have atropinic side effects that, though minor, could be quite uncomfortable, and in theory at least should produce an atropinic type of delirium in high doses or idiosyncrasy. Finally, most sedatives distort the normal pattern of sleep, resulting in a feeling of not having slept well, hangover effects, and rebound nightmares upon withdrawal.[27,44]

*See references 19, 25, 44, 76, and 77.

Conclusions

Despite widespread use, particularly of antihistamines, the use of sedatives of any kind in children has as yet no properly demonstrated role.

¶ Miscellaneous Drugs

Anticonvulsants have yet to demonstrate a bona-fide psychotropic role in children with various emotional or behavioral problems whether epileptic or not.[19,67] Some of these drugs (including phenytoin) are potentially neurotoxic and their use for psychopharmacotherapeutic reasons should be considered *strictly experimental,* requiring all appropriate safeguards.

Caffeine, amino acids, LSD and other hallucinogens, vitamins and hormones, including thyroid substances, have as yet no established use in child psychiatry,[22,44] although they have all been tried, particularly in seriously disabled mentally retarded or psychotic children.[13,14]

Lithium must also still be regarded as an experimental drug, though there is some evidence to suggest that clinical trials—particularly in adolescents with irregular behavior and explosive outbursts, and where there is a family history of lithium-responsive manic-depressive disorder—may be worthwhile.[15,44,85] Recent reports[16,18] suggest that betablockers deserve further study, though interestingly enough not in the treatment of anxiety but in organic brain disorders and Tourette's disorders. The use of chelating substances[20] depends on establishing a connection between subclinical lead poisoning and attention-deficit disorder or other childhood psychiatric disorders. The Feingold hypothesis that there are certain substances in children's diets in advanced societies that produce neurotoxic behavioral responses primarily in the area of behavior seems to grow shakier with each new properly controlled study.[33]

Specific Disorders and Their Pharmacological Treatment

It should be obvious from this review that the diagnostic indicators for psychopharmacotherapy in children are few indeed. They are reducible to attention deficit disorder with hyperactivity (stimulants, antidepressants), Tourette's and possibly chronic motor tic disorder (antipsychotics) and postpubertal schizophrenia (antipsychotics). Possible diagnostic indicators still awaiting confirmation are separation anxiety disorder, obsessive/compulsive disorder (tricyclic antidepressants), and early, atypical manic-depressive disorder (lithium). Enuresis is a qualified diagnostic indicator, the qualification being that drugs should be regarded as a temporary suppressant and not as a definitive treatment of the disorder.

As yet, infantile autism and prepubertal schizophrenia do not appear to have any specific psychopharmacological indications, though antipsychotic drugs may be helpful in dealing with certain distressing behaviors. Because of the common confusion between attention deficit disorder and conduct disorders,[40] it is entirely possible that some of the drugs currently accepted as effective in "hyperactivity" may be shown to be efficacious in certain kinds of conduct disorders. With the possible exception of separation anxiety and obsessive/compulsive disorder, neither the anxiety disorders nor the learning disorders appear to present indications for psychopharmacotherapy.

¶ Conclusions

At present, psychopharmacotherapy in children is of limited application. Part of the difficulty lies in the lack of biogenic etiological theories for any childhood disorder, most of which appear to have no resemblance to, or continuity with, adult disorders. Without proper pathophysiological formulations along the lines suggested by Cohen and Young,[17] psychopharmacology will continue as a fumbling, empirical, hand-me-down from adult psychiatry rather than emerging as an independent branch of medicine. In addition to the lack of diagnostic solidarity with the adult area, pediatric psychopharmacology presents distinctive problems at many levels. Among these are: (1) ethical issues surrounding the child's assent to treatment and adult instigated desire to produce social compliance; (2) risks of impairment of cognitive function at a time of maximum learning; (3) possible interferences with critical emotional endocrinological and other developmental stages; (4) lack of evidence for any long-term benefit concerning adjustment, self-image, and learning; (5) ignorance of dose response and other fundamental pharmacokinetic factors; (6) absence of information about the effects of drugs on learning in naturalistic as opposed to laboratory situations; (7) the apparent greater sensitivity of cognitive function than social behavior to dosage effects; (8) the possibility of state-dependent learning; (9) the probable but unstudied impact of the meaning of giving medication for the child; (10) lack of information about drug effects on children's inner mood and comfort level; and (11) the absence of significant data on the interaction between drugs and other treatments, such as psychotherapy, behavior modification, remedial education, sensorimotor training and so on, which are usually given at the same time. None of these issues has been adequately studied, though there can be no disputing their potential importance.

While the development of pediatric psychopharmacology in the 1970s has been impressive, there is still much to do. Clinically at the moment drugs occupy only a small if significant part of the overall management of children's psychiatric disorders.

¶ Bibliography

1. ABIKOFF, H., GITTELMAN-KLEIN, R., and KLEIN, D. "Validation of a Classroom Observation Code for Hyperactive Children," *Journal of Clinical and Consulting Psychology,* 45 (1977): 772–783.
2. AMAN, M. "Drugs, Learning and the Psychotherapies," in Werry, J., ed., *Pediatric Psychopharmacology: The Use of Behavior Modifying Drugs in Children.* New York: Brunner/Mazel, 1978, pp. 79–108.
3. ———. "Psychotropic Drugs and Learning Problems—A Selective Review," *Journal of Learning Disabilities,* 13 (1980):87–96.
4. ———, and SINGH, N. "Thioridazine in Mental Retardation: Fact or Fallacy?" *American Journal of Mental Deficiency,* 84 (1980): 331–338.
5. American Psychiatric Association, *Diagnostic and Statistical Manual of Mental Disorders,* 3rd ed. (DSM-III). American Psychiatric Association, 1980.
6. BARKLEY, R. "Predicting the Response of Hyperkinetic Children to Stimulant Drugs: A Review," *Journal of Abnormal Child Psychology,* 4 (1976): 327–348.
7. ———. "A Review of Stimulant Drug Research with Hyperactive Children," *Journal of Child Psychology and Psychiatry,* 18 (1977): 1–31.
8. ———, and CUNNINGHAM, C. "Do Stimulants Improve the Academic Performance of Hyperkinetic Children? A Review of Outcome Research," *Clinical Pediatrics,* 17 (1978): 85–92.
9. ———. "The Effects of Methylphenidate on the Mother-Child Interactions of Hyperactive Children," *Archives of General Psychiatry,* 36 (1979): 201–208.
10. BLACKWELL, B., and CURRAH, J. "The Psychopharmacology of Nocturnal Enuresis," in Kolvin, I., McKeith, R., and Meadows, S., eds., *Bladder Control and Enuresis. Clinics in Developmental Medicine,* nos. 48/49. London: Heinmann, 1973, pp. 231–257.
11. BRIANT, R. "An Introduction to Clinical Pharmacology" in Werry, J., ed., *Pediatric Psychopharmacology: The Use of Behavior Modifying Drugs in Children.* New York: Brunner/Mazel, 1978, pp. 3–28.
12. BRUUN, R., et al., "A Follow-up of 78 Patients with Gilles de la Tourette's Syndrome," *American Journal of Psychiatry,* 133 (1976): 944–947.
13. CAMPBELL, M. "Biological Intervention in Psychoses of Childhood," *Journal of Autism and Childhood Schizophrenia,* 3 (1973): 347–373.
14. ———, GELLER, B., and COHEN, I. "Current Status of Drug Research in Treatment with Autistic Children," *Journal of Pediatric Psychology,* 2 (1977): 153–161.
15. CAMPBELL, M., SCHULMAN, D., and RAPOPORT, J. "The Current Status of Lithium Therapy in Child and Adolescent Psychiatry," *Journal of the American Academy of Child Psychiatry,* 17 (1978): 717–720.
16. CANTWELL, C., and CARLSON, G. "Stimulants," in Werry, J., ed., *Pediatric Psychopharmacology: The Use of Behavior Modifying in Children.* New York: Brunner/Mazel, 1978, pp. 171–207.
17. COHEN, D., and YOUNG, J. "Neurochemistry of Child Psychiatry," *Journal of the American Academy of Child Psychiatry,* 16 (1977): 353–411.
18. COHEN, D., et al. "Central Biogenic Amine Metabolism in Children with the Syndrome of Chronic Multiple Tics of Gilles de la Tourette: Norepinephrine, Serotonin, and Dopamine," *Journal of the American Academy of Child Psychiatry,* 18 (1979): 320–341.
19. CONNERS, C., and WERRY, J. "Pharmacotherapy," in Quay, H., and Werry, J., eds., *Psychopathological Disorders of Childhood,* 2nd ed. New York: John Wiley, 1979, pp. 336–386.
20. DAVID, O., et al. "Lead and Hyperactivity. Behavioral Response to Chelation: A Pilot Study" *American Journal of Psychiatry,* 133 (1976): 1155–1158.
21. ELKINS, R., RAPOPORT, J., and LIPSKY, A. "Obsessive-Compulsive Disorder of Childhood and Adolescence: A Review," *Journal of the American Academy of Child Psychiatry,* in press.
22. FREEMAN, R. "Psychopharmacology and the Retarded Child," in Menolascino, F., ed., *Psychiatric Approaches to Mental Retardation.* New York: Basic Books, 1970, pp. 294–368.
23. GADOW, T. "Prevalence of Drug Treatment for Hyperactivity and Other Childhood Behavior Disorders," in Gadow, K. and Loney, J., eds., *Psychosocial Aspects of*

Drug Treatment for Hyperactivity. Boulder, Colo.: Westview Press, forthcoming.

24. GITTELMAN-KLEIN, R. "Validity of Projective Tests for Psychodiagnosis in Children," in Spitzer, R., and Klein, D., eds., *Critical Issues in Psychiatric Diagnosis.* New York: Raven Press, 1978, pp. 141–166.
25. ———. "Psychopharmacological Treatment of Anxiety Disorders, Mood Disorders and Tic Disorders of Childhood," in Lipton, M., Dimascio, A., and Killam, K., eds., *A Review of Psychopharmacology: A Second Decade of Progress.* New York: Raven Press, 1978, pp. 1471–1480.
26. ———, and KLEIN, D. "School Phobia: Diagnostic Considerations in the Light of Imipramine Effects," *Journal of Nervous and Mental Disease,* 150 (1973): 199–215.
27. GREENBLATT, D., and SHADER, R. *Benzodiazepines in Clinical Practice.* New York: Raven Press, 1974.
28. GROSS, M. "Growth of Hyperkinetic Children Taken Methylphenidate, Dextroamphetamine, Imipramine, or Desipramine," *Pediatrics,* 58 (1976): 423–431.
29. GUALTIERI, C., et al. "Tardive Dyskinesia in Children, Part I. The Wide Range of Movements," *Journal of the American Academy of Child Psychiatry,* in press.
30. HASTINGS, J., and BARKLEY, R. "A Review of Psychophysiological Research with Hyperkinetic Children," *Journal of Abnormal Child Psychology,* 6 (1978): 413–447.
31. HECHTMAN, L., WEISS, G., and PERLMAN, T. "Growth and Cardiovascular Measures in Hyperactive Individuals as Young Adults and in Matched Controls," *Canadian Medical Association Journal,* 118 (1978): 1247–1250.
32. HERJANIC, B., et al. "Are Children Reliable Reporters?" *Journal of Abnormal Child Psychology,* 3 (1975): 41–48.
33. LEVY, F. "Dietary Treatment of Hyperkinesis," in Burrows, G., and Werry, J., eds., *Advances in Human Psychopharmacology,* vol. 1, Greenwich, Conn.: JSI Press, 1980, pp. 321–329.
34. LIPMAN, R., et al. "Sensitivity of Symptom and Non-Symptom-Focused Criteria of Outpatient Drug Efficacy," *American Journal of Psychiatry,* 122 (1965): 24–27.
35. LONEY, J., et al. "Hyperkinetic/Aggressive Boys in Treatment: Predictors of Clinical Response to Methylphenidate," *American Journal of Psychiatry,* 135 (1978): 1487–1491.
36. MCNUTT, B., et al. "The Effects of Long-Term Stimulant Medication on the Growth and Body Composition of Hyperactive Children: II. Report on Two Years," *Psychopharmacology Bulletin,* 13 (1977): 36–38.
37. PAXTON, J., AMAN, M., and WERRY, J. "Salivary Estimations of Phenytoin Levels in Children," in press.
38. PSYCHOPHARMACOLOGY BULLETIN. "Pharmacotherapy of Children," Special Issue, 1973.
39. PUIG-ANTICH, J., et al. "Prepubertal Major Depressive Disorder: A Pilot Study," *Journal of the American Academy of Child Psychiatry,* 17 (1978): 695–707.
40. QUAY, H. "Classification," in Quay, H., and Werry, J., eds., *Psychopathological Disorders of Childhood,* 2nd ed. New York: John Wiley, 1979, pp. 1–42.
41. RAE-GRANT, Q. "Psychopharmacology in Childhood Emotional and Mental Disorders," *Journal of Pediatrics,* 61 (1962): 626–637.
42. RAPOPORT, J. "Self Report, Social and Emotional Functioning Measures," in *Guidelines for Evaluation of Psychoactive Agents in Infants and Children.* Rockville, Md.: Food and Drug Administration, forthcoming.
43. ———, and MIKKELSON, E. "Antidepressants," in Werry, J., ed., *Pediatric Psychopharmacology: The Use of Behavior Modifying Drugs in Children.* New York: Brunner/Mazel, 1978, pp. 208–233.
44. ———, and WERRY, J. "Antimanic, Antianxiety, Hallucinogenic and Miscellaneous Drugs," in Werry, J. ed., *Pediatric Psychopharmacology: The Use of Behavior Modifying Drugs in Children.* New York: Brunner/Mazel, 1978, pp. 316–356.
45. RAPOPORT, J., et al. "Dextroamphetamine: Cognitive and Behavioral Effects in Normal and Hyperactive Children and Normal Adults," *Journal of Pediatrics,* in press.
46. RAPOPORT, J., et al. "Urinary Amphetamine and Catecholamine Excretion in Hyperactive and Normal Boys," *Journal of Nervous and Mental Disease,* in press.

47. REATIG, N. ed. *Proceedings of the National Institute of Mental Health Workshop on the Hyperkinetic Behavior Syndrome.* Rockville, Md.: National Institute of Mental Health, 1978.
48. RIE, H. "Hyperactivity in Children," *American Journal of Diseases of Children,* 130 (1975): 783–789.
49. ROBBINS, T., and SAHAKIAN, B. "Paradoxical Effects of Psychomotor Stimulant Drugs in Hyperactive Children from the Standpoint of Behavioural Pharmacology," *Neuropharmacology,* in press.
50. ROCHE, A., et al. "The Effects of Stimulant Medication on the Growth of Hyperkinetic Children," *Pediatrics,* in press.
51. RUTTER, M., et al. "Research Report: Isle of Wight Studies, 1964–1974," *Psychological Medicine,* 6 (1976): 313–332.
52. SAFER, D., and ALLEN, R., eds. *Hyperactive Children: Diagnosis and Management.* Baltimore: University Park Press, 1976.
53. SARAF, K., et al. "Imipramine Side Effects in Children," *Psychopharmacologia,* 37 (1974): 265–274.
54. SATTERFIELD, J., et al. "Growth of Hyperactive Children Treated with Methylphenidate," *Archives of General Psychiatry,* 36 (1979): 212–217.
55. SCHREIER, H. "Use of Propranol in the Treatment of Postencephalitic Psychosis," *American Journal of Psychiatry,* 136 (1979): 840–841.
56. SHAYWITZ, S., YAGER, R., and KLOPPER, J. "Selective Brain Dopamine Depletion in Developing Rats: An Experimental Model of Minimal Brain Dysfunction," *Science,* 191 (1976): 305–308.
57. SHEKIN, W., DEKIRMENJIAN, H., and CHAPEL, J. "Urinary MHPG Excretion in Minimal Brain Dysfunction and its Modification by d-Amphetamine," *American Journal of Psychiatry,* 136 (1979): 667–671.
58. SILBERGELD, E., and GOLDBERG, A. "Pharmacological and Neurochemical Investigations of Lead-Induced Hyperactivity," *Neuropharmacology* 14 (1975): 431–444.
59. SIMEON, J., COFFEN, C., and MARASA, J. "Videotape Techniques in Pediatric Psychopharmacology Research," in Sanker, D. S., ed., *Psychopharmacology of Childhood.* Westbury, N.Y.: PJD Publications, 1976, pp. 7–27.
60. SINGH, N., and AMAN, M. "Thioridazine in Severely Retarded Patients and The Effects of Dosage," Submitted for publication, 1979.
61. SLEATOR, E., VON NEUMANN, A., and SPRAGUE, R. "Hyperactive Children: A Continuous Long-Term Placebo-Control Follow-Up," *Journal of the American Medical Association,* 229 (1974): 316–317.
62. SPRAGUE, R. "Principles of Clinical Trials and Social, Ethical and Legal Issues of Drug Use in Children," in Werry, J., ed., *Pediatric Psychopharmacology: The Use of Behavior Modifying Drugs in Children.* New York: Brunner/Mazel, 1978, pp. 109–135.
63. ———. Personal communication, 1979.
64. ———, and SLEATOR, E. "Methylphenidate in Hyperkinetic Children: Differences in Dose Effects on Learning and Social Behavior," *Science,* 198 (1977): 1274–1276.
65. SPRAGUE, R., and WERRY, J. "Psychotropic Drugs in Handicapped Children," in Mann, L., and Sabatino, D., eds., *Second Review of Special Education,* Philadelphia: JSE Press, 1974, pp. 1–50.
66. SROUFE, L. "Drug Treatment of Children with Behavior Problems," in Horowitz, F., ed., *Review of Child Development Research,* vol. 4. Chicago: University of Chicago Press, 1975, pp. 347–407.
67. STORES, G. "Antiepileptics (Anticonvulsants)," in Werry, J., ed., *Pediatric Psychopharmacology: The Use of Behavior Modifying Drugs in Children.* New York: Brunner/Mazel, 1978, pp. 274–317.
68. SULZBACHER, S. "Psychotropic Medication with Children: An Evaluation of Procedural Biases in Results of Reported Studies," *Pediatrics,* 51 (1973): 513–517.
69. SWANSON, J., et al. "Time-Response Analysis of the Effect of Stimulant Medication on the Learning Ability of Children Referred for Hyperactivity," *Pediatrics,* 61 (1978): 21–29.
70. ULLMAN, D., BARKLEY, R., and BROWN, H. "The Behavioral Symptoms of Hyperkinetic Children who Successfully Responded to Stimulant Drug Treatment," *The American Journal of Orthopsychiatry,* 48 (1978): 425–437.
71. WEISS, B., and LATIES, V. "Enhancement of Human Performance by Caffeine and the Amphetamines," *Pharmacological Review,* 14 (1962): 1–36.

72. WEISS, G. "The Natural History of Hyperactivity in Childhood and Treatment with Stimulant Medication at Different Ages: A Summary of Research Findings," *International Journal of Mental Health,* 4 (1975): 213–226.
73. ———, et al. "Effect of Long-Term Treatment of Hyperactive Children with Methylphenidate," *Canadian Medical Association Journal,* 112 (1975): 159–165.
74. WERRY, J. "Measures in Pediatric Psychopharmacology," in Werry, J., ed., *Pediatric Psychopharmacology: The Use of Behavior Modifying Drugs in Children.* New York: Brunner/Mazel, 1978, pp. 29–78.
75. ——— "Psychosomatic Disorders, Psychogenic Symptoms and Hospitalisation," in Quay, H., and Werry, J., eds., *Psychopathological Disorders,* 2nd ed. New York: John Wiley, 1979, pp. 134–184.
76. ——— "Anticholinergic Sedatives," in Burrows, G., and Werry, J., eds., *Advances in Human Psychopharmacology,* vol. 1. Greenwich, Connecticut: JSI Press, 1980, pp. 19–42.
77. ——— "Anxiety in Children," in Burrows, G., and Davies, B., eds., *Handbook of Studies in Anxiety,* Amsterdam: Elsevier, forthcoming.
78. ———, and AMAN, M. "Methylphenidate in Hyperactive and Enuretic Children," in Shopson, B., and Greenhill, L., eds., *The Psychobiology of Childhood: Profile of Current Issues.* Jamaica, N.Y.: Spectrum Publications, forthcoming.
79. WERRY, J., and SPRAGUE, R. "Methylphenidate in Children—Effect of Dosage," *Australian and New Zealand Journal of Psychiatry* 8 (1974): 9–19.
80. WERRY, J., AMAN, M., and DIAMOND, E. "Imipramine and Methylphenidate in Hyperactive Children," *Journal of Child Psychology and Psychiatry,* 20 (1979): 27–35.
81. WHALEN, C., and HENKER, B. "Psychostimulants and Children: A Review and Analysis," *Psychological Bulletin,* 83 (1976): 1113–1130.
82. ———. *Hyperactive Children, the Social Ecology of Identification and Treatment.* New York: Academic Press, 1980.
83. WINSBERG, B., and YEPES, L. "Antipsychotics," in Werry, J., ed., *Pediatric Psychopharmacology: The Use of Behavior Modifying Drugs in Children.* New York: Brunner/Mazel, 1978, pp. 234–273.
84. WOLRAICH, M., et al. "Effects of Methylphenidate Alone and in Combination with Behavior Modification Procedures on the Behavior and Performance of Hyperactive Children," *Journal of Abnormal Child Psychology,* 6 (1978): 149–161.
85. YOUNGERMAN, J., and CANINO, I. "Lithium Carbonate Use in Children and Adolescents," *Archives of General Psychiatry,* 35 (1978): 216–224.

PART THREE

Adult Clinical Conditions and Therapeutic Techniques

CHAPTER 13

THE FAMILY OF THE SCHIZOPHRENIC AND ITS PARTICIPATION IN THE THERAPEUTIC TASK

Silvano Arieti

¶ Family Dynamics

THE CONVERGENCE of the work of Harry Stack Sullivan, who stressed the interpersonal aspect of the psyche rather than the intrapsychic, the pioneering work of Nathan Ackermann in the psychodynamics of family life, and a host of contributions by many other authors, who applied in clinical practice either their own innovations or what they had learned from others, shifted the attention of many psychiatrists from the patient to the family of the patient. Rather than the patient himself, the family became the patient to be examined, treated, cured. In addition to those already mentioned, many other authors, such as Murray Bowen, G. Bateson, D. D. Jackson, L. Wynne, T. Lidz, have expanded this field. The individual is no longer seen in isolation. Of greater significance is the interaction between the patient who is a family member and the family as a group, with laws and habits pertaining to a group per se.

It would be counterproductive and regressive to deny the value of these contributions. Nevertheless, it is now time that we reevaluate their observations and data and reconsider some basic notions, especially as they relate to certain psychiatric syndromes.

In this chapter we shall reconsider the role

attributed to the family of the schizophrenic and shall present possible modifications. These issues have not only theoretical interest, but are also of practical concern since the study of them may suggest new or different approaches to the role the family can play in the treatment and rehabilitation of the patient. This reevaluation seems an impelling necessity today, when the tendency is to avoid hospitalization or reduce hospitalization to a minimum. For the considerable number of patients who do not recover completely after the initial attack and who remain a serious problem, as far as treatment, management, and rehabilitation are concerned, we may borrow an expression used by President Truman in a different context and say that the buck stops here—in the family. Since day hospitals and half-way houses are available only for a restricted number of patients, there is no other or better place to turn than to the home, no place where enlightenment and guidance from the psychiatrist are more necessary or appreciated.

The following four basic concepts, which were considered valid by most people who practiced a psychotherapy of schizophrenia with the emphasis on the role of the family, must now be drastically reevaluated.

1. The patient became schizophrenic because of what was done to him by others.
2. Whatever was done to him and was pathogenetic stemmed from family members, especially the mother, who was labeled "schizophrenogenic mother."
3. In the psychotherapeutic attempt, unless the family members participated in family therapy, they had to be left out because it was in the family that the patient had had the original traumatic conflicts that led to his illness. It was, therefore, necessary that the patient be separated from the family, unless, as already mentioned, usual family psychotherapy was instituted and the patient participated in it.
4. The disorder came to be seen solely as the effect of what the environment or the interpersonal world did to the patient. What the patient did with what was given to him by the environment, or, in other words, how he digested, or how, with his intrapsychic apparatus, he metabolized psychologically what was offered to him, was almost totally ignored.

Before discussing these four concepts, it should be stressed again that although they now seem incorrect, they had as a whole a beneficial effect, and that when we consider them in the historical continuity of scientific progress, they must be considered positive. Without them, the patient would still be seen as suffering from an endogenous disorder, or as a metabolic freak—a pathological phenomenon unrelated to or uninfluenced by an apparently normal environment. He would probably still be seen as the outcome of an exclusively genetic deviation.

As already mentioned, many authors believe that the patient becomes schizophrenic because of what was done to him by a terrible family environment. Some authors have described the mother of the patient as malevolent, and one of them spoke of her perverse sense of motherhood.[11] From some authors, one gets the impression that the parents of the schizophrenic are inhuman, cruel, perverse creatures. Others portray them as transmitting irrationality to the patient directly, just as they would transmit to him the language they speak.

Let us take some examples. One author[10] described a girl whose mother wished her to become a good writer like Virginia Woolf, even if doing so required, by implication, committing suicide. Eventually the patient did commit suicide. In the same article the author, who wished to report typical examples of parents of schizophrenics, described a mother who, referring to her son, said to the doctor, "You must cure him—he is all of my life. When he started to become sick, I slept with him just like man and wife." Other provocative examples were offered in the same article.[10] A schizophrenic woman who was hospitalized told of having her genitalia examined by her physician father each time she returned home from a date in order to make certain she was still a virgin. Also reported is the case of a female patient who not only spilled food all over herself, but blew her nose in the napkin. The patient did not know that it was wrong to do so because her father, an eminent professor, used to blow his nose in his napkin. In another case reported in the same article, the mother of a patient told her that she was afraid the father would seduce the patient's pubescent sister. The father had confided that the mother was a lesbian and a menace to the three daughters.

Many similar examples from articles by authors who have studied the father and mother of schizophrenics could be quoted. However, the point to be made is that, although dramatic and impressive, these examples are misleading. This author does not deny that parents like those reported in the preceding examples exist (having observed them in families of both schizophrenics and nonschizophrenics); however, if articles and books on the family of the schizophrenic report exclusively, or almost exclusively parents such as those mentioned, the reader may infer that these parents are typical parents of the schizophrenic. To do so would be unjust. Let us examine more closely some of the reported examples. They are not the consequence of internalization occurring through complicated intrapsychic mechanisms. Some of them are simply a result of obedience, such as the girl who committed suicide as the mother had requested. Others are examples of pure and simple imitation, such as the girl who blew her nose with a napkin as her father had done. These are not examples of schizophrenic irrationality. What is transmitted by imitation, indoctrination, conditioning, and so forth, whether considered desirable or undesirable, is not schizophrenic per se. These transmissions occur in schizophrenia, but *much more so* in neurotics and in the general population.

Both the family and the culture in general may transmit irrationality through phenomena known as psychological habituation, indoctrination, imitation, acceptance on faith, and so forth. But with the exception of rare cases of *folie à deux,* transmitted irrationality and transmitted peculiar behavior are not schizophrenic, delusional, or regressive per se. They may be unacceptable on a moral, medical, pedagogic, or orthopsychiatric basis, but they are not directly schizophrenogenic. The schizophrenic gives his own autistic, or primary process form to whatever has previously disturbed him with nonpsychotic psychodynamic mechanisms. It is the *transformation* and not the *imitation* that constitutes the schizophrenic essence of symptoms or habits. And that transformation is implemented by primary process cognition.

In the second edition of *Interpretation of Schizophrenia,*[1] this author presented evaluations and certain conclusions concerning the findings reported by others regarding the family of the schizophrenic, as well as original findings:

1. Conflicts, tension, anxiety, hostility, detachment, instability had generally existed in the family of the patient since his formative years. However, one must be aware that these findings cannot be subjected to statistical investigation. It is often an enormous task to evaluate qualitatively or quantitatively the psychological disturbance existing in a family. One must keep in mind that some authors (for instance, Waring and Ricks[13]) have found disturbed family constellations, previously considered predisposing toward schizophrenia, less frequently among schizophrenics than in control families.

2. It is common knowledge that similar family disturbances exist even in families in which there has not been a single case of schizophrenia in the two or three generations that could be investigated.

3. It is not possible to prove that the adult schizophrenics studied during family research were potentially normal children whose lives were warped only by environmental influences.

4. The one point of agreement among most authors who have subjected schizophrenic patients to deep psychodynamic investigations is that in every case so studied, family disturbance, generally serious, was found. Unless biases have grossly distorted the judgment of the investigators, one must believe that serious disturbances did exist.

5. This conclusion is important. It indi-

cates that although family disturbance of considerable seriousness is not *sufficient* to explain schizophrenia, it is probably a *necessary precondition* of schizophrenia. To have differentiated a necessary, though not sufficient, causative factor is important enough to make this factor the object of deep consideration.

6. In the last twenty years, this author has compiled some private statistics, and although personal biases cannot be excluded and the overall figures are too small to be of definitive value, has reached conclusions different from those of other authors. In relation to sexual assault, seduction, or rape by a parent of the child, events have been found much more frequently in the history of depressed, psychopathic, and hysterical patients than in the history of schizophrenics. The author has also found that in 75 percent of cases of schizophrenia, the mother did not fit the image of the schizophrenogenic mother. Prevailing nonmaternal characteristics have been found in only about 25 percent of the mothers of schizophrenics. What percentage of mothers of nonschizophrenics have been nonmaternal is not known. The mother and father of the patient have often been found to be disturbed, anxious, or hostile and detached, but only in exceptional instances to the degree described in some psychiatric literature. In the larger majority of cases the mother was a person who had been overcome by the difficulties of life. These difficulties had seemed to her enormous not only because of her unhappy marriage, but, most of all, because of her neurosis and the neurotic defenses she had built up in interacting with her children.

7. Another important point has been neglected in the literature. These studies of the patient's mother, beginning with those of Fromm-Reichmann[6] and Rosen,[11] were made at a time in which drastic changes in the sociological role of women were in incubation. It was a period immediately preceding the women's movement era. It was the beginning of a time when a woman had to contend tacitly with her newly emerging need to assert her equality. Though no longer accepting submission, she strove to fulfill her traditional role. These social factors became involved in the intimacy of family life and complicated the parental roles of both mothers and fathers.

Furthermore this was the time when the "nuclear family," a development of urban industrial society, was most fully evolved. It consists of a small number of people who live in little space, compete for room and for material and emotional possession, and are ridden by hostility and rivalry. Often deprived of educational, vocational, and religious values as well, the nuclear family is destructive not only for the children, but also for the parents, and especially for the wife and mother.

One can thus become aware of another dimension. Not only are the negative characteristics of the mother magnified and distorted by the future patient, but the seemingly original negative characteristics of the mother are in their turn a deformation, magnification, and rejection, conscious or unconscious, of roles that she believes society has inflicted on her.

What has been discussed so far can be reformulated in different words. The importance of family disturbances in the childhood of schizophrenics cannot be disregarded. Undoubtedly in the childhood of future schizophrenics there is a deviation from what is considered a normal family environment. This deviation consists predominantly of an environment characterized by more than the usual amount of anxiety, hostility, detachment, or instability in family members. This angle of deviation might have been remedied by the regenerating and self-correcting mechanisms of the organism and of the psyche; but in the case of the future schizophrenic, other circumstances did not permit this correction. Thus, the initial deviation not only persisted but was amplified by subsequent chains of causes and effects. The circumstances may be biological or hereditary. The child may be more than usually sensitive to adverse environment and psychological pain. The time of the adverse contingencies may not permit the psyche to recuperate between one blow and the next. Finally, compensatory mechanisms, such as the presence of useful parental substitutes, might be absent.

If what has been expressed so far is correct, the reason many therapists, this author included, came to believe in the reality of the schizophrenogenic mother and, less frequently, of the schizophrenogenic father must be investigated. In the majority of cases therapists have fallen into a serious error. Schizophrenics who are at a relatively advanced stage of psychoanalytically-oriented psychotherapy often describe their parents, especially the mother, in negative terms, the terms used in part of the psychiatric literature. Therapists have believed what their patients have told them. Inasmuch as approximately 25 percent of the mothers proved to be the way they were described, it was easier to make an unwarranted generalization that all the mothers of the schizophrenics were the same way.

This is a mistake reminiscent of the one made by Freud when he came to believe that his neurotic patients had been assaulted sexually by their parents. Later Freud realized that what he had believed to be true was, in by far the majority of cases, only the product of the patient's fantasy.

The schizophrenic's mother had definite negative characteristics, but the child was particularly sensitive to them because they were the characteristics that hurt him and to which—in that particular context or because of his own biology—he responded more deeply. He was less affected by, or even ignored, the positive qualities of his mother: the giver, the helper, the assuager of hunger, thirst, cold, loneliness, immobility, and other discomfort. The child who responds mainly to the negative parts of his mother will tend to make a whole of these negative parts, and the resulting whole will be a monstruous transformation of the mother. Similar observations can be made about the self-image of the future patient. The self is not merely a mirror of reflected appraisals, because the sensitive child does not respond equally to all appraisals and roles attributed to him. Those elements that hurt him more, or that please him more, stand out and are integrated disproportionately. Thus the self, although related to the external appraisals, is not a reproduction of them but in some cases a grotesque misrepresentation. This grotesque self that the patient retains would stupefy the parents if they were aware of it.

These images—the one of the mother as the major representative of the external world and eventually of the neighbor, any others, and humankind; the other the representative of the person himself—will affect, at a conscious and an unconscious level, the patient's entire life. The images are constructed not only by external contingencies, but by the patient himself. Much of the psychodynamic literature has made the error of seeing the child, the adolescent, and the young adult as entirely molded by circumstances, without addition of the elements of his own individuality and creativity to what he receives—his contribution to his transformation.

The geneticist sees the origin of the disorder in the genetic code, hidden in the chromosomes of the patient; the family therapist sees it in the effect of the family and especially of mother and father. But geneticists and a large group of psychodynamic psychiatrists are closer than they think to one another's conceptions when they see the patient as entirely shaped by circumstances alien to his being or at the mercy of obscure forces or as a passive entity that has to accept his chromosomic or familial destiny as ineluctable forces.

Obviously the patient is very much influenced by his family, but he is not just in a state of passive receptivity. Inasmuch as every human being is strongly influenced by his environment, one must acknowledge in him a fundamental state of *receptivity.* But he is not to be defined in terms of a state of receptivity alone. Every human being, even in early childhood, has another basic function which we, following the French sociologist Lucien Goldmann,[7] may call *integrative activity.* Just as the transactions with the world not only inform but transform the individual, with his integrative activity the individual transforms these transactions and in his turn he is informed and transformed by these transformations. *No* influence is received as a direct and immutable message. Multiple processes involving interpersonal and intrapsychic dimensions move back and forth. According to the philosopher Giambattista Vico:

> the being of man cannot be enclosed within a determinate structure of possibilities . . . but it moves, rather, among *indeterminable alternatives,* and even further, but its own movement generates these alternatives [Italics mine].[5]

Thus to depict the mother of the schizophrenic as a schizophrenogenic mother is a primitive simplification. The mother becomes schizophrenogenic if her negative qualities are also processed by the future patient in a schizophrenogenic fashion.

In other words, the patient makes his own contribution to his pathology. He picks up what he receives from the family and deforms it. The person who becomes schizophrenic deforms in a different way and to a greater degree than the average person and the nonpsychotic. To use an analogy, the deformation of the patient may be compared to the deformation of a sound produced by an echo if the echo in its turn is echoed several times. The original angle of deviation that existed early in life has been increased not only by its consequences, not only by the contingencies of life, but by the patient's contributions to his own pathology because of its own special integrative activity.

¶ The Family's Role in the Patient's Rehabilitation

The second part of this chapter is devoted to the therapeutic role of the family in the psychotherapy of the schizophrenic. What has been discussed about the new psychodynamic formulation can be considered in fact an introduction to what follows. Many authors have already suggested that the family should participate in the gigantic therapeutic task. What has been described by other authors in detail will not be repeated here. The focus will be only on the differences between the new approach and the others that preceded it.[2]

The family can rehabilitate the patient but cannot give him psychotherapy. There is a big difference between these two types of help. And yet rehabilitation is often confused with psychotherapy or considered a form of psychotherapy. It is useful to stress this point even if it is already familiar to the majority of the readers. Psychotherapy helps the patient become aware of the reasons for his feelings and actions. Psychotherapy helps the patient to understand how symptoms are expressions of needs that he cannot accept and that have, therefore, become unconscious. Psychotherapy also helps the patient to discard maladaptive patterns of behavior and to correct faulty ways of thinking. It would be too much to expect the family to attempt to undertake these arduous tasks. Whereas the psychotherapist and the patient engage in a common exploration of the inner life of the patient, the family members are engaged with him in an external exploration, in rediscovering that the external world is not so terrible as it once seemed but is a place where the patient, too, can find his own niche and much more.

No theory has been formulated on how rehabilitation works (either in the family or with agencies outside the family). In reference to rehabilitation carried on outside the family, it is generally felt that it is effective when it makes available methods that facilitate the patient's relating normally to others, restore his faith in himself, and lead him to engage in fruitful activities.

Relating normally to others includes good attitudes toward neighbors, interchanges with coworkers, friendships, and search for intimacy and love. Restoring faith in oneself means an attitude of hope and promise toward one's present and future. Fruitful activities include common living, work, useful habits, and also play.

Although rehabilitation includes all this, perhaps the rehabilitation that occurs within one's family includes more. Perhaps even the word rehabilitation is not appropriate in reference to the family. If one persists in using it, one would have to add that it is a special type of rehabilitation that includes reintegration in the family, not just restoring but also improving one's role in that close milieu. It involves familiarization or refamiliarization with one's own family, fraternization with siblings, and with other relatives. The words that have just been used have a warmer affective connotation than words used in association with rehabilitation carried out by agencies.

But first of all, let us face squarely the reality of the return home of a family member to whom the diagnosis of schizophrenia has been applied. A new factor has been added, and the family atmosphere is no longer the same. To make believe that everything is just as it was is masking reality; it requires the imposition of mechanisms of denial, which are likely to cause harm. Moreover, as we shall illustrate shortly, it is inadvisable for the family not to undertake some changes. To recover from schizophrenia is not the same as recovering from mumps or measles. The development of a different family climate is not generally a bad occurrence, but one possibly propitious to a satisfactory outcome. Living with the patient day by day becomes a therapeutic task, and not an easy one even for the most cooperative family.

The first problem is to decide whether it is in the patient's best interest to live with the family. Although the decision is made with the participation of everyone involved, the main responsibility for it resides with the psychiatrist in charge. Various views on this point are expressed in psychiatric circles. In a few of them the therapeutic role of the family is not appreciated at all because it was within the context of the family that the patient's conflicts leading to the illness originated. The patient's family, the patient, and the patient's illness are seen in these psychiatric circles as constituting a unity whose abnormality led to the undesirable result. There is no doubt that in a considerable number of cases this is so. The intrafamilial conflicts exist, and the solution or even amelioration of them is so improbable that the best decision is to separate the patient from his family if possible. Even when the psychiatrist thinks the strong negative feelings the patient has for his family are unjustified and based only on his distortions, it is not advisable for him to live with the family until he views his home milieu differently.

At other times, the patient is willing to live with the family, but the psychiatrist decides against it because he feels that that particular family is not able to help a sick member. Some relatives, although well intentioned, are too involved in their own problems, difficulties, illnesses, demanding occupations, or care of young children, to participate in what is always a demanding task. When the participation of the family members is not possible, the services of a therapeutic assistant or of a psychiatric companion may be resorted to. Cautiousness in making these decisions is necessary for though rehabilitation in the family may be the best, it may also be the most risky. The family must offer to the patient not just a roof but a hearth as well, a place where suffering and joy are shared in closeness and intimacy.

¶ Introducing the Family to the Task

A larger number of patients and former patients continue to live with their families because the psychiatrist feels that the family environment is satisfactory, or the only one available. It is important for the psychiatrist to prepare the family for the task by giving a general orientation. The aim is not to transform the family members into psychiatric nurses, but to make them understand more fully the problems involved so that they can add understanding to their affection and personal concern. A family member has a great advantage over even the best nurse because to the family member the patient will always be a person and not a clinical case. The family member already knows what the patient likes and what he does not like.

In his words of general orientation to the family members, this writer starts by pointing out that we human beings have learned since our early childhood to deal with others, at least in the majority of our relations, in ways that society or our particular milieu recommend. Society criticizes, rejects, or even punishes those who do not follow acceptable attitudes toward others. Acceptable attitudes generally have been evolved by traditions of many centuries' duration and have deep emotional roots in the life of most individuals. They are maintained not only by example, imitation, teaching, but also by punishment and reward, or even by the use of power. These sociological attitudes have definite educational values, but they may have disastrous effects when they are imposed on or adopted by the schizophrenic patient or one who is recovering from schizophrenia. At least in the beginning of the convalescent status, the family must exert as little as possible those pressures that the norms of society recommend. The patient must feel accepted even if he is different and unconventional. To accept the patient as he is, does not mean, however, to accept indiscriminately his behavior, as we shall see later. He must be gradually integrated into a structured life.

Most relatives insist that they never punish a patient who has returned home and who has displayed unconventional behavior. With great sincerity they state that they recognize that the patient's behavior, even when offensive, is only the result of illness and that therefore they do not consider him accountable. The truth is that, unless they train themselves to do otherwise, they do punish the patient in subtle ways—in ways that may be unconscious to them but not to the patient, who is particularly sensitized to any unpleasant input from the environment. The family member may punish the patient by avoiding him or by staying with him as little as possible; by not talking to him or talking with brief, curt sentences; by refusing to listen to him or to give explanations; by having a condescending, patronizing, or superior attitude; by being in a hurry in every interchange with the patient; by wearing a perplexed, annoyed, bored, or disapproving expression, and at times even a look of consternation.

One main requirements of the family member is to observe not only the behavior and attitude of the patient, but also his own —especially his own.

Let us assume that the brother of the patient wants to be kind, helpful, and reassuring. Instead of being grateful, the patient who has just returned from the hospital becomes distrustful, possibly contemptuous and hostile. It is normal for the brother to react by becoming impatient toward the patient, annoyed, perhaps angry and condemnatory. In turn the patient senses that the brother has such feelings and thus his prior attitude of distrust and hostility is reinforced. The vicious circle may repeat itself. The brother must train himself to respond not in the way considered normal, but by realizing that the patient still has a great need to project onto others his inner turmoil and to blame others for it.

The example just given explains the complaint which one often hears from the members of the family in approximately these words: "I want to be genuine, authentic. Since Jean came back, I have to watch every word I say to her. I can't be spontaneous any more. But I don't know if what I'm doing is right. Maybe by being artificial I'm doing harm. I believe in being authentic."

Such doubts posed to oneself or to the psychiatrist are legitimate and worthy of full consideration. The relative must analyze further what he means by authenticity. To watch one's words before talking to Jean does not necessarily mean to be artificial. To behave as if a serious illness had not occurred to a person dear or close to us is not to live authentically. It is more authentic to realize that because of the patient's particular vulnerability and sensitivity, it is better to modify some of our ways and in talking to him to refrain from using words or sentences that may sound ambiguous to him or even threatening. Moreover, let us remember that in recognizing the areas of vulnerability and great sensitivity of the patient, we may discover where and how we have been unintentionally insensitive, and perhaps even callous. We may recognize that we have wanted to impose our ways because we have considered them more appropriate, more efficient, more in agreement with what society expects, or simply because we prefer them.

Another bad habit, which fortunately is found only in very few families, is that of totally disregarding what the patient says as utterly nonsensical and at times even as a subject for ridicule.

It must be clear to the family that remarks and even complaints made by the patient must be listened to and evaluated with respect. Fears and even delusions are real, vivid, and almost always unpleasant experiences for the patients, even if based on complicated mechanisms that only the psychiatrist understands. If the family member does not understand what the patient says, he must at least respond to his request for attention and to his desire to start a dialogue. To the extent that he is capable, the relative must influence and even guide the patient, not by suppressing his activities but by increasing his understanding of them and by clarifying difficult situations. As has already been mentioned, the cooperative family member gradually increases his sensitivity about the patient's sensitivity; he becomes more aware of what may affect the patient unfavorably. His "antennae" must be ready to discern what is disturbing; he must be on the alert, but not too solicitous or too eager; he must remain near and distant, near enough to give when the need is there, distant enough not to scare the patient who is not yet capable of accepting warmth. Following Harry Stack Sullivan's terminology, it may be said that the patient who cannot yet accept too much warmth may put into effect a malevolent transformation and interpret the offer of warmth as having ulterior motives. A family capable of tolerating the difficulties inherent in living with a convalescent schizophrenic is a very important determinant of a favorable outcome.

This general attitude of acceptance, although allowing a considerable degree of permissiveness, should not extend to an unlimited laissez faire attitude. In a warm atmosphere, which does not resort to rejection, punishment, belittling, or ridicule, the patient generally understands what kinds of actions are appropriate for him. Threatening to send him back to the hospital if he does not behave is extremely disturbing to his morale. If the problems are too difficult, if in spite of the good will of everybody interpersonal tension increases, if there is a possibility of suicide or of violence, rehospitalization must be seriously considered. It should not be presented to the patient as a form of punishment, but as a need for an environment much more programmed and structured than that of a home.

It is fair to say that often the task is too big for the family unless, in addition to the individual therapy of the patient, family therapy is resorted to.

Many authors have reported that family therapy has made relapses much less frequent, has shortened the length of therapy of the individual patient, and has ameliorated the general conditions of the family, even independently from the illness of the patient. So far the role of family therapy has not been stressed sufficiently here. This is partially due to the fact that unfortunately only a small minority of families are willing to undergo this type of therapy. At times some members are willing to accept such a proposal but not the whole family.

Although family therapy is strongly advocated when possible, the family must try in any case to become a "therapeutic milieu," and in many cases, this is possible.

¶ Specific Issues

Before describing modalities of living with a convalescent schizophrenic, it must be stressed again that each case is different, each constitutes a different situation in an environment that is not identical to any one observed before.

Specific issues that come up rather frequently in living day by day with the patient must be considered. The patient who used to be delusional may no longer be so, but he may distort many interpersonal relations, see them in a worse light than they are, and may be rather accusatory, especially in relation to his parents, whom he now considers the source of his misfortune. To a lesser degree other family members are also blamed. This position of the patient is indeed hard to accept. The best attitude is not to argue with him or to tell him that he is wrong. But it is indeed difficult for many mothers and fathers not to be defensive. Their pride is hurt; they may become incensed and want to speak up as vigorously as possible, as if they were on trial. If they yield to this temptation, the trial will go on and on, endlessly, and progress will not be made. A good attitude for the parent is to say to the patient, "Perhaps the time will come when you will see what we did and what we tried to do in a different way." At the same time the parents can reassure the patient by stating that each member will see to it that the needs and rights of everybody are satisfied as fully as possible. The future then will have a greater chance of being much better than the past.

Although the impairments and areas of sensitivity of the patient should be taken into consideration, they should not be magnified. The family members should avoid making the patient more dependent than he is or treating him as an invalid or a baby. It is true that the activities of some convalescing patients are greatly curtailed, but many of them only to a minimal extent. It is necessary to exploit fully whatever is not affected or barely touched by the illness. A main goal is to find a role for the patient within the institution of the family. Some chores must be assigned to him. This is generally easier to do with female patients, who are usually more accustomed to performing domestic duties, but a male patient, too, must assume home responsibilities. The feeling that he is a contributing member of the family will be beneficial, and the residues of pity and discouragement still felt by the family members will have more chance to dissipate.

The patient must be encouraged to take care of his room, but it is also advisable not to restrict his activities to what pertains only to him. On the contrary, it is advisable for him to engage in some activity that will benefit the whole family. (It has been noted that patients from economically poor families rehabilitate faster after their return to the family than patients from well-to-do families. Possibly the difference is due to the fact that in well-to-do families it is difficult to assign domestic chores to the patient.)

Often, especially following his return from the hospital, the patient is not able to take the initiative. The relative must be the initiator and must be provided with a great deal of patience. It is a characteristic of partially recovered patients to do things at a much slower pace than the average person. Lack of concentration, inhibitions of all sorts, intruding thoughts may interfere with any activity. Nevertheless, if he continues to work on a steady basis and is encouraged in his work, no matter how slowly he does it, he will gain a rewarding sense of satisfaction. With increased confidence in himself, the tempo of his actions will speed up.

It has been observed by many therapists that from the point of view of becoming capable again of engaging in useful activities, patients who return from the hospital to live with their wives or husbands fare much better than those who return to live with their parents. Generally spouses do not treat the patient as an overly dependent person, are less willing to accept a state of passivity, and encourage the patient to resume activities. Parents, on the other hand, are more inclined to resume the parental role and to foster excessive dependency. The therapist is often asked, "Should we push the patient to be active, or shouldn't we?" Again there is no single answer. With patients who are inclined to be passive, a little push is appropriate, but it must be in the form of a kind push, given with velvet gloves, and never by an authoritarian command. The opposite attitude is valid when the patient is willing to take steps for which he is not prepared: to go immediately back to his usual job, to look for a new position, to go back to college, to finish the semester, to go to live by himself in his own apartment, and so forth. Here a kind of delaying technique should be used. The patient should be advised to postpone these plans until he is able to meet the challenge more efficiently. By no means should he be discouraged, but only invited to reprogram his plans in phases which succeed one another more deliberately. In the meantime, he must be stimulated to exploit whatever assets may be used in the home, from simple errands for the family to complicated accounting.

In dealing with some families, other types of problems appear. Expectations may be too high for the patient. It has already been mentioned that the spouse is generally more prone than the parents to stimulate the patient into an active role. Although this attitude generally has a favorable outcome, it may be detrimental if the spouse's expectations are excessive for the patient recovering from an acute episode. A wife may expect the husband to become the provider right away; the husband may expect the wife to resume fully her maternal duties. Realization that a return to health requires a longer time will ease tension, impatience, and discouragement.

A common complaint, especially among young couples, is that the convalescing patient has become sexually inadequate. If the spouse of the patient is reassured as far as the future is concerned, he will be able to tolerate better the temporary inconvenience. Generally, lack of sexual interest is due to a variety of causes. The most frequent is the medication that the patient may still take. Several neuroleptics diminish sexual desire, especially in the male, and may even prevent ejaculation. Some psychiatrists inform the patient that this is likely to occur and reassure him that this is a transitory phenomenon which will disappear with the decrease in medication, interruption of medication, or shift to another drug. Many psychiatrists, however, neglect to inform the wife of this possible occurrence. She has to be reassured, too, that the phenomenon is not permanent.

Lack of sexual interest, of course, may be due to the fact that the patient has not been concerned at all with sexual matters and has for a long time focused his attention elsewhere, so that he has lost the desire for sex or has become used to sexual abstinence. In other cases, sexual inactivity may be due to the fact that the patient has to reappraise his relation with the spouse and feels he must know where he stands with his partner. It is advisable, of course, for the spouse to suggest that the patient discuss any insecurity, anxiety, or unresolved hostility with the therapist.

¶ Involvement and Overinvolvement

Consultations with the therapist will help the family and the patient himself to avoid the opposite dangers of being either overstimulated or understimulated, of being in an environment that offers and expects too much or too little. It is difficult at times to find the proper balance. Overstimulation obligates the patient to cope with the environment beyond his ability. If the patient is withdrawn, lackadaisical, seemingly oblivious, the well-intentioned relatives try to interest him in a thousand different ways, for instance, by taking him to movies, museums, or theaters, by talking and talking, recounting stories of the good times spent together in the past. The patient may feel overwhelmed, especially if he has just returned from a hospital where, in spite of the therapy and of the occupational activities, he felt alone. It may be very strenuous for him to try to adjust to a situation that requires overinvolvement or exposure to frequent busy talk.

Some authors have made a distinction between the subjective burden—that is, the family's estimate of the hardship imposed by the patient's presence in the home—and the "objective burden," which was the researchers' estimate. According to these researchers there was a discrepancy between the objective estimate and the subjective, in the sense that the "objective" estimate was always superior to the subjective. In other words, the burden was always greater than the relatives were willing to admit. Of course, it is arguable how objective the estimate of the researchers was. Assessing the family situation from the point of view of a person who does not have to live with a recovering schizophrenic and who retains a feeling of distance may also be subjective due to the lack of intense involvement with the patient. At any rate, the fact that the objective burden was considered by these researchers as far greater than the subjective speaks well for the family of the patient. It indicates that, contrary to common belief, most families do their best to participate in the rehabilitation of a dear one and are willing to endure the concomitant hardship.

Related to the problem of overstimulation versus understimulation, but not exactly the same, is the problem of overinvolvement. British authors, inspired especially by John Wing, who has studied this issue in depth, have reported that overinvolvement on the part of the family, including too much expression of emotion, is conducive to relapse. Brown, Birley, and Wing wrote,[3] "Fifteen hours or more a week of face-to-face contact between a schizophrenic patient and a highly involved relative carries a strong risk of further breakdown."

If closeness engenders a revamping of conflicts and a renunciation of privacy, then of course we have the picture of overinvolvement described by Brown, Birley, and Wing.[3] This overinvolvement seems to be a continuation of a situation found in some families of schizophrenics even prior to the illness. In these families, each member experiences not just a feeling of competition with the others, but an extreme sense of participation, reactivity, and special sensitivity to the actions of the others, often interpreted negatively. In these cases, the members of the family want to help each other, but because of their entanglements, anxiety, distrust, and misinterpretation, end up by hurting one another.

A morbid degree of overinvolvement, however, may not be so frequent as Brown, Birley, and Wing seem to imply. Cultural differences may play a role. Brown, Birley, and Wing have worked with patients and their families who come almost exclusively from an Anglo-Saxon environment. What is considered overinvolvement in that milieu may be the usual state of affairs in Italian and Jewish families. In other words, in evaluating these factors, the ethnic background and the prevailing family culture must be considered.

Some of the contrasting, at times even opposite, positions that have to be taken in dealing with a recovering schizophrenic have already been mentioned, and the difficulty of switching back and forth between these different directions has been stressed. A few more must be mentioned. One is the situation in which both the patient's need for companionship and for privacy are essential and must be satisfied. Time must be found for both. Another difficult balance must be made between the patient's need for freedom and for structure. The patient must experience freedom of action, and yet a structure, a routine, a schedule should be worked out with him, at least for the first few months after his return from the hospital. Although structured, his day should not become packed with things to do or be too complex. The degree of complexity has to be adjusted to his capability.

¶ Important Events and Important Decisions

At times, the family is confronted by unusual happenings in the life of the patient. Although these events are discussed at length with the therapist, the family may become involved with such matters even before the therapist, or may be the only consultant, if there is no therapist. The patient has become acquainted with a person of the opposite sex or, more seldom, of the same sex, and wants to go to live with him or her, or, in other cases, wants to become engaged or get married right away. The family has the strong feeling that the patient is not ready and yet does not want to exert so much pressure on the patient that he feels unfree or unduly controlled. A delaying technique, that tries to persuade the patient to wait for a time when he feels more at ease with the programs that are formulated, is the proper approach here. However, if the patient insists and cannot be persuaded to postpone, it is best to go along with the plans and provide as much help as possible. An attitude of open opposition is not advisable and may be counterproductive.

The same principle applies to dealing with the recovering schizophrenic who wants to become pregnant. Pregnancy and motherhood are real challenges for normal women. To cause such a complication deliberately while the patient is recovering is not recommended. This point must be clarified to avoid misunderstanding. The author is not saying that recovering schizophrenics or former schizophrenics should not become mothers. Some of them make excellent mothers. There is, however, for many patients a period of time, which varies from at least a year to as many as five years, during which, even in the cases with the best results, there still is difficulty in coping with unusual and demanding challenges, such as pregnancy, childbirth, and motherhood. If the patient is under drug therapy, she must be even more careful not to become pregnant because the safety of most drugs during pregnancy and lactation has not been established.

At times, the patient wants to do something equally drastic, but in a different way, for instance, leave the spouse and children. The spouse who is threatened with being left alone (or with the children) after having gone through the hardship of the illness and having offered loyalty and support, is often mortified. At other times the spouse of the patient is ready to accept the decision, which frequently cannot be reversed. Again, the delaying technique is best, but if the patient goes through with his plans, the family must be supportive. It must be remembered that the patient is not likely to break an important family relationship because of a whim or a capricious impulse, but only because he is not able to cope with the circumstances. If children are involved, the best arrangements must be made for their care. Although, as has been noted, some former schizophrenics or even schizophrenics are excellent mothers, it is also true that a recovering mother who still feels unable to cope with the circumstances, and this may be very disturbing to a young child. In such situations, a substitute mother must be found.

A question that comes up frequently is: should the recovering patient be told the truth when some terrible event (sudden death or diagnosis of serious disease) occurs in the family or to persons dear to the patient? Over thirty years ago, when working in a state hospital, this author was instructed by older psychiatrists to advise the family always to tell the truth. Certainly one does not want to lie to patients or anybody else. However, there is a favorable and an unfavorable timing for telling the truth. State hospital psychiatrists insist that no bad effects have ever resulted from the revelation of bad news to the patient. They were referring to a group of patients who, in addition to being ill, often lived in a state of alienation aggravated by the environment. Many of these patients were not able to express their emotions. An apparent insensitivity should not be interpreted as imperviousness. Even a catatonic schizophrenic, who seems insensitive and is as immobile as a statue, feels strongly. A volcano of emotions is often disguised by his petrified appearance.

With the recovering schizophrenic the situation is completely different. He is extremely sensitive and would not forgive the relatives for not telling him the truth. And yet knowing the truth may be detrimental when he is still unstable and struggling to recover fully his mental health. The patient has to be prepared gradually and eventually be told the truth when he has already anticipated its possibility and is able to cope with it.

¶ Concluding Remarks

In summary, living with a recovering schizophrenic is a difficult task, but not an insurmountable one. It may be rewarding not only for the patient but for everyone concerned. If one compares the hardship of living with a recovering or partially recovered schizophrenic with that of living with a severe alcoholic, a blind person, an epileptic, or a chronically ill person with some incapacitating disease, the lot of living with a recovering schizophrenic is considerably better. An atmosphere of hope prevails in many cases, and the satisfaction of seeing results at least partially due to the family's cooperative efforts confers a joyful climate of further expectation. Even in a family with little children, although the situation is further complicated, the task is not necessarily an impossible one. If the children are old enough to understand, they should be told that a member of the family is ill and requires special attention. Some of the unusual attitudes of the ill person should be explained to the child in terms of illness and in a context of serious but hopeful concern. Children generally respond well to adverse or abnormal conditions provided there are compensating circumstances. In an atmosphere of warm care and frank discussion, the presence of mental illness in a member of the family tends to remain a smaller part of the child's life than is generally assumed, and in some cases a part which promotes maturation.

¶ Bibliography

1. ARIETI, S. *Interpretation of Schizophrenia,* 2nd ed., New York: Basic Books, 1974.
2. ———. *Understanding and Helping the Schizophrenic. A Guide for the Family and Friends.* New York: Basic Books, 1979.
3. BROWN, G. W., BIRLEY, J. C., and WING, J. K. "Influence of Family Life in the Course of Schizophrenic Disorders: A Replication," *British Journal of Psychiatry,* 121 (1972): 241–258.
4. CANCRO, R., ed. *Annual Review of the Schizophrenic Syndrome.* New York: Brunner/Mazel, 1978.
5. CAPONIGRI, A. R. *Time and Idea. The Theory of History in Giambattista Vico.* Chicago: Regnery, 1953.
6. FROMM-REICHMANN, F. "Notes on the Development of Treatment of Schizophrenia by Psychoanalytic Therapy," *Psychiatry,* 11 (1948): 263–273.
7. GOLDMANN, L. *La Création Culturelle dans La Societé Moderne.* Paris: Denoël-Gonthier, 1971.
8. GUNDERSON, J. G., and MOSHER, L. R., eds. *Psychotherapy of Schizophrenia.* New York: Aronson, 1975.
9. JORSTAD, J., and UGELSTAD, E., eds. *Schizophrenia 75: Psychotherapy, Family Studies, Research.* Oslo: Universitats Forlaget, 1976.
10. Lidz, T. "The Influence of Family Studies in the Treatment of Schizophrenia," *Psychiatry,* 32 (1969): 237–251.
11. ROSEN, J. N. *The Concept of Early Maternal Environment in Direct Psychoanalysis.* Doylestown, Pa.: The Doylestown Foundation, 1963.
12. STIERLIN, H. "Perspectives on the Individual and Family Therapy of Schizophrenic Patients. An Introduction," in Jorstad, J., and Ugelstad, E., eds., *Schizophrenia 75: Psychotherapy, Family Studies, Research.* Oslo: Universitats Forlaget, 1976, pp. 295–304.
13. WARING, H., and RICKS, D. "Family Patterns of Children Who Became Adult Schizophrenics," *Journal of Nervous and Mental Disease,* 140 (1965): 351–364.

CHAPTER 14

ADVANCES IN THE DIAGNOSIS AND TREATMENT OF SCHIZOPHRENIC DISORDERS

Robert Cancro

ALL OF MEDICINE has been profoundly influenced by the work of Koch, who demonstrated the bacterial origin of tuberculosis. In 1882, Koch presented a series of elegant experiments that proved the etiologic role of the tubercle bacillus in the origin and transmission of a disease. This monumental achievement made it possible to exclude other pulmonary infections, which previously had been misdiagnosed as tuberculosis, and to include those infections that belonged there but that had previously been excluded. In this way, the category was restricted to a biologically more homogeneous population. The powerful promise of this work for meaningful classification and nosology was not lost upon Kraepelin. He recognized that the protean manifestations of tuberculosis could now be understood and classified within a single disease entity conception. Kraepelin attempted to translate and apply the work of Koch to psychiatry. Obviously, it was not possible to fulfill the requirements of Koch's postulates to obtain the bacterium and have it infect a new host, in mental disorders. Kraepelin attempted, therefore, to translate the postulates into a form more appropriate for psychiatry and to create, on the basis of clinical features, entities that shared a specific etiology, presented with a consistent picture, and followed a predictable course over time. In this way, he hoped to create diagnostic categories of mental disease that would be comparable in their clinical unity to Koch's explanation of tuberculosis.

As early as 1883, Kraepelin[25] began to use the course of illness as a classificatory variable. This attempt to use the outcome of patients suffering from mental disorders as a means to group them into different diseases achieved its fullest expression in 1899 when he separated dementia praecox from manic-depressive psychosis.[26] This separation, while enormously helpful clinically, failed to create distinct and genuine entities. In the more than eight decades that have passed since Kraepelin's historic division, the problems of categorizing schizophrenic disorders have remained both baffling and frustrating.

This chapter will attempt to summarize selectively some of the more recent developments in the conceptualization, diagnosis, and etiology of the schizophrenias. These developments do offer some direction and help guide the explorations of those interested in achieving greater familiarity with the many features and facets of the problem. No chapter can hope to be an accurate nor even an adequate Baedeker for those who strive to achieve certainty in matters psychiatric. This very lack of certainty may contribute to the excitement of the activity.

¶ Conceptualization

There have been a variety of positions taken concerning schizophrenia: It has been conceptualized as a disease of the brain to a mythology developed in the name of political repression. Griesinger's[17] theories were representative of the approach that mental disorders are diseases of the brain. This conceptualization can be characterized by the slogan: "For every twisted thought there must be a twisted molecule." Conversely, observers such as Szasz[38] have argued that there is no such thing as mental illness, only socially unacceptable behaviors. This chapter will initially utilize certain assumptions that are best made explicit, particularly since not all of them are demonstrably true. It is assumed that the schizophrenic disorders represent a syndrome. This syndrome is heterogeneous in its etiology, pathogenesis, presenting picture, course, response to treatment, and outcome. In other words, the syndrome consists of multiple disorders rather than a single entity. It will also be assumed that the schizophrenias represent illnesses and not alternative life styles. The schizophrenias are seen as pathological adaptations to a highly altered sense of inner and outer reality. They are seen as pathologic adaptations because of the psychotic features that impair the quality of the person's life. Finally, the schizophrenic disorders are diagnosed on the basis of their clinical features, independent of the route of origin of these features.

¶ Diagnosis

General Comments

The careful diagnosis of a patient is important for many reasons, although it has not always been in fashion. It is important not only for the care of the individual patient but for the progress of the field. It must be understood that any clinical population labeled schizophrenic has been arrived at through the process of diagnosis. Research studies done on such populations are not more reliable than the diagnostic criteria that produced the groups studied. There are different strategies for diagnosing the disorder. Perhaps the most basic diagnostic question is whether a single criterion or multiple criteria will be used. The single-criterion approach, of course, offers greater simplicity of application. More important, it can also increase the homogeneity of the population labeled schizophrenic.[9] Nevertheless, the multiple-symptom approach has become the dominant nosologic fashion in recent years. Some of the multiple symptoms are nonspecific, while other symptoms are treated in an egalitarian manner; that is, every symptom is equal to every other for diagnostic purposes. It does not matter in this approach which symptoms or signs the patient has from the given list as long as the patient obtains the correct number of such items. Recently, diagnostic practices have also shifted to an emphasis on symptoms and signs requiring low levels of inference. This emphasis allows for greater reliability, although it suffers from a lack of clinical sophistication. The conceptualization of schizophrenia has narrowed in the last decade, and the diagnostic emphases today are on chronicity and phenomenologic manifestations of low inference and, therefore, high reliability.

Finally, it should be made clear that there is no independent test for validating the diagnosis. In the absence of such independent validation, any diagnosis must be arbitrary. This arbitrariness does not reflect lack of care on the part of the practitioner, but rather an exclusive reliance on clinical criteria. Until the time when independent diagnostic procedures are available, the diagnosis of the schizophrenic disorders will remain a clinical activity with all of the inherent unreliability of such a practice.

Strategies

There have been a number of strategies utilized through the years in making the diagnosis of schizophrenia. Kraepelin[25,26] followed the classical clinical method. He described patients in meticulous detail. In this way, the student could learn the descriptions of typical cases and compare his patient against a mental template of the disorder. Obviously, there are many deficiencies in such an approach, not the least of which is poor reliability.

Bleuler[4] moved away from the disease-entity approach of Kraepelin and introduced the concept of schizophrenia as a syndrome, or group of disorders, that contained certain clinical features in common. He identified the essential common features that were necessary for the diagnosis of the syndrome. Bleuler deemphasized delusions and hallucinations as diagnostic criteria, and relegated them to a relatively minor or accessory role. He placed major emphasis on what he considered to be characteristic disorganizations of thinking and the loss of harmony between various mental functions, in particular thinking and affect. The diagnostic advantage of this approach was that there were specific admission criteria to the category. The patient had to demonstrate particular altered fundamental signs of the disorder before the diagnosis could be made. No pathognomonic delusions or hallucinations were sufficient to diagnose the illness. While this increased reliability considerably over the Kraepelinian approach, it still was not sufficiently reliable. Bleuler's diagnostic method required clinical judgment, which reduces reliability.

Langfeldt[28] in many ways bridged and combined the diagnostic preferences of Kraepelin and Bleuler. He recognized the multiplicity of illnesses in the category by separating true schizophrenia from schizophreniform psychosis, while continuing to utilize, as did Kraepelin, a long list of typical symptoms. In the Kraepelinian tradition, he attempted to separate the schizophrenias into those with good and poor outcomes, with only the latter representing the true disorder.

In the last decade, the diagnostic approach of Schneider[36] has received increasing attention. He developed his criteria empirically by identifying the most common signs and symptoms in patients about whom there was diagnostic consensus. These were referred to as symptoms of the first rank. They included audible thoughts, voices heard arguing, voices commenting on the person's actions, somatic passivity experiences, thought withdrawal, diffusion of thought, delusional perception, and all feelings that are experienced as a result of the influence of others. The use of Schneiderian criteria tends to create a somewhat more homogeneous population with a tendency toward a chronic form of illness.

The *Diagnostic and Statistical Manual of Mental Health,* 3rd edition (DSM–III)[2] has been profoundly influenced by the Schneiderian approach. It uses a multiaxial system of diagnosis, which includes an effort to assess personality structure, other medical conditions, and the highest level of recent social functioning. The essential requirements for a diagnosis of a schizophrenic disorder are evidence of disorganization from the previous level of daily functioning, the presence of at least one symptom from a list of six during the active phase of the illness, and at least a six-month duration of symptoms (including the prodromal, active, and residual phases of the illness) during which the symptom or symptoms necessary for making the diagnosis are present. This list of required symptoms for diagnosis of the active phase of the illness includes three that are delusional in nature and two that are hallucinatory. It is obvious, therefore, that, unlike Bleuler, DSM-III places great diagnostic significance on the presence of delusions and hallucinations. Nevertheless, it recognizes the syndrome nature of the category and acknowledges that it is a group rather than a single disorder.

The third edition of the *Diagnostic and Statistical Manual* deviates very sharply from the second,[1] and it is important that the field understand the new nosology. The multiaxial approach, while new in psychiatry, has been used previously in much of medicine. Five axes are included in the new diagnostic manual. The first is the *diagnosis of the clinical syndrome.* The second is the *diagnosis of the personality.* It is possible for the clinician to specify two or more diagnoses on either of these axes. The third axis is the *diagnosis of any coexisting physical condition.* The final two axes are the *presence of psychosocial stressors and the highest level of adaptation achieved in the past year.* Obviously, this multiaxial approach gives a much richer picture of the patient than does a single diagnostic label.

The essential diagnostic features of a schizophrenic disorder are listed in DSM-III as disorganization from a previous level of functioning, the presence of characteristic symptoms, the absence of an affective disorder, the absence of an organic brain syndrome that can explain the clinical picture, a tendency toward chronicity, onset before the age of forty-five, and a duration of continuous symptoms in excess of six months. To diagnose the active phase of the illness, it is necessary for the patient to show disorganization from previous functional levels in two or more areas of daily living. These areas include work, social relationships, self-care, and so on. At least one symptom from a list of six must be present for diagnosis of the active phase of the illness to be made. The symptoms are:

1. Bizarre delusions that are obviously absurd. (Typical delusions include the belief that one's thoughts are being broadcast so that other people can hear them, the belief that thoughts are being put into the mind by external forces, the experience of thought withdrawal, or the delusion of being controlled.)
2. The presence of religious, grandiose, nihilistic, somatic, or other delusions without persecutory or jealous content. (This group of delusions need not be bizarre, but cannot be accompanied by either persecutory or jealous ideation.)
3. Delusions with a persecutory or jealous content if they are accompanied by hallucinations of any type.
4. The presence of two or more hallucinatory voices conversing with each other, or auditory hallucinations in which a voice keeps up a running commentary on the person's behaviors or thoughts as they occur.
5. Auditory hallucinations on several occasions, which are not related to either the presence of depression or elation and which are not limited to one or two words.
6. Marked loosening of associations, incoherence, illogicality, or marked poverty of speech if associated with at least one of the following: (a) blunted, flat, or inappropriate affect; (b) delusions or hallucinations; (c) catatonic or grossly disorganized behavior.

In addition to the active phase of the illness, there may be a prodromal and/or a residual phase. The prodromal precedes the active phase and the residual follows it. The prodromal and residual phases also must not be secondary to an affective disorder. In order to diagnose either a prodromal or residual phase, the patient must show at least two symptoms from a list of eight. The symptom list for the diagnosing of prodromal and residual phases includes: (1) withdrawal or social isolation; (2) marked impairment in role functioning; (3) markedly eccentric, odd, or peculiar behavior; (4) impairment of personal hygiene and grooming; (5) blunted, flat, or inappropriate affect; (6) speech that is tangential, digressive, vague, overelaborate, circumstantial, or metaphorical; (7) odd or bizarre ideation, magical thinking, overvalued ideas, ideas of reference, or suspected delusions; and (8) unusual perceptual experiences, suspected hallucinations, and sensing the presence of a force or person not actually there.

DSM–III recognizes several phenomenologic subtypes of the schizophrenic disorders: disorganized, catatonic, paranoid, undifferentiated, and residual. These subtypes reflect cross-sectional syndromes and are not assumed to be stable over time within a given person. The subtypes are descriptive of the symptoms as they appear in an individual at a given moment in time.

The disorganized category is closest to the classical concept of hebephrenia. The major clinical features required for the diagnosis of this subtype are severe incongruence and the presence of flat, incongruous, or silly affect. The delusions and hallucinations, when present, tend to be fragmentary and not organized into extended or coherent themes. A number of features associated with this category are less central than those already cited. They include the presence of grimaces, mannerisms, hypochondriacal complaints, social withdrawal, and peculiarities of behavior. The premorbid personality tends to be withdrawn and inclined toward the schizoid end of the spectrum. Onset usually is at an early age and insidious in nature. As would be expected, these patients tend to run a chronic course with few if any significant spontaneous remissions, let alone restitution to the premorbid state of the personality.

The essential clinical feature of the catatonic subtype is the marked involvement of the motor system either in the direction of over- or underactivity. There can be a relatively rapid alternation between overactivity (excitement) and underactivity (stupor). There are a number of associated features for this subtype, which include negativism, stereotypies, mannerisms, posturing, and waxy flexibility. An interesting feature that is sometimes found is mutism. The diagnosis of this category requires that the clinical picture be dominated by one of the following symptoms during the active phase of the illness: (1) catatonic stupor or mutism; (2) catatonic rigidity; (3) catatonic excitement; or (4) catatonic posturing.

The paranoid subtype should be diagnosed when the clinical picture is dominated by a persecutory or grandiose symptom complex. These symptoms can take the form of persistent delusions and/or persistent hallucinations. The critical diagnostic issues are the nature of the symptom, that is, delusion and/or hallucination; and the nature of the content, that is, persecutory and/or grandiose. Delusions of jealousy are also acceptable as one of the diagnostic criteria. During the active period of the illness, the clinical picture must be dominated by the persistence of at least one of the following symptoms in order to diagnose this subtype: (1) persecutory delusions; (2) grandiose delusions; (3) delusions of jealousy; and (4) hallucinations with a persecutory or grandiose content. The clinical features associated with, but not diagnositc of, this subtype include anger, argumentativeness, violence, fearfulness, ideas of reference, concerns about autonomy, concerns about gender identity, and preoccupation with sexual preference. These patients may show very little social impairment, particularly in the workplace. At times the delusional concerns are relatively encapsulated and do not intrude into the person's day-today life. The age of onset of the schizophrenic disorder tends to be at a later age in this subtype, and the likelihood of cognitive deterioration is less.

The so-called undifferentiated subtype tends to be less of a subtype than a "wastebasket" category. Patients who meet the criteria for a diagnosis of a schizophrenic disorder and cannot be placed in any of the three previous classes, or who manifest criteria for more than one of those subtypes, should be categorized in this grouping.

The remaining subtype is the residual.

This category is utilized for individuals who have had a clear-cut episode of a schizophrenic illness, but whose clinical picture does not, at the time of examination, contain prominent psychotic symptoms. The person must show evidence of a persistent mental disorder, otherwise the diagnosis of no mental disorder must be entertained. The common manifestations of persistence that would require the use of this category include: emotional blunting, social withdrawal, eccentric behavior, and mild communication difficulties. Delusions or hallucinations may be present if they are not prominent and have lost their affective intensity. This subcategory is quite similar to the concept of partial clinical remission. An individual can be categorized as being in full remission if there are no signs of clinical illness with or without medication. If the person is free of medication and clinical symptoms for five or more years, the diagnosis is changed to no mental disorder.

¶ General Comments on Diagnosis

The concept of acute schizophrenia has been eliminated from DSM–III. The disorder can be chronic, which is defined as in excess of two years, or subchronic, which is defined as more than six months but less than two years. DSM–III does recognize the possibility of acute exacerbations, particularly in residual phases of the disorder. During these acute exacerbations, prominent psychotic symptoms can reemerge and dominate the picture. Nevertheless, disorders that remit in six months or less must be classified elsewhere. The diagnosis of a schizophrenic disorder requires the exclusion of both organic mental disorders and affective disorders. It is not unusual for organic mental disorders to present with symptoms such as delusions, hallucinations, incoherence, and affective changes—all of which are suggestive of a schizophrenic disorder. This is particularly true in syndromes associated with substances such as phencyclidine and amphetamines. A proper differential diagnosis frequently requires the passage of a period of time; it should not place an exclusive reliance on the cross-sectional picture at a single moment. However, it is an excellent clinical rule of thumb to suspect organic mental disorder whenever confusion, disorientation, or memory impairment are significantly present in the clinical picture. The clinician must also remember that it is possible for a person to have simultaneously both an organic mental disorder and a schizophrenic disorder.

The differential diagnosis of affective disorders from schizophrenic disorders is clinically very important and is based on the presence or absence of a full affective syndrome. If a full affective syndrome is present and the disorder of mood is a prominent and relatively persistent part of the illness, the diagnosis of an affective disorder should be made. An affective syndrome may be present in a schizophrenic disorder but it must occur after the development of the psychotic symptoms. In order to diagnose mania, it is necessary for the person to show one or more distinct periods with a predominantly elevated, expansive, or irritable mood. In order to diagnose depression, the person must show a dysphoric mood or loss of interest or pleasure in most, if not all, of his usual life activities.

¶ Etiology

Of those people who are diagnosed as having a schizophrenic disorder, there is no single personality type that is premorbidly present. While no personality type is spared being vulnerable to the disorder, those premorbid personalities that are schizoid[5] and those that show autistic tendencies[15] have a relatively poor prognosis. And just as there has been no single personality type, there has been no universal biologic pattern premorbidly, not even a genetic one. At least 90 percent of the patients diagnosed as having a schizophrenic disorder fail to show any first-degree relatives with such an illness.

The premorbid history of individuals with this diagnosis frequently shows one or more episodes of a neuroticlike illness. These episodes of illness are unlike a true neurosis in that they are transient, sudden in onset, and last only four to six weeks. The symptoms include anxiety, phobias, and obsessional preoccupations. These neuroticlike episodes do not show apparent sequelae and clear completely. These premorbid microepisodes mirror, in a nonpsychotic fashion, elements of the future psychotic decompensation. They are not always present in the history and do not appear to have any particular prognostic significance. They raise, however, an important theoretical question: Are there environmental experiences that may help to suppress or exaggerate these microepisodes? It is certainly possible that the progression from a microepisode to a full-blown decompensation is influenced by a variety of fortuitous life events, some of which may be deleterious while others are healing.

Stress

There are major methodologic problems facing investigators who wish to study the role of stress in the schizophrenic disorders. Different definitions of stress will produce different findings. There is no universal definition, and the concept itself is in many ways ambiguous. Investigators have usually chosen to rate events that are externally caused and that are experienced most often as either unpleasant or undesirable.[19] It has been shown that people at the lower end of the socioeconomic scale have more of these negative life events; for example, unemployment, job insecurity, physical illness, and inadequate housing.

The literature does not answer the question of the role of stress in the etiopathogenesis of a schizophrenic disorder. The most cited studies[3,7] do suggest a relationship between the frequency of the illness and stress. However, there is a relationship between class differences and coping abilities, such that lower-class individuals are less able to cope with stress effectively.[29] It may be, therefore, necessary to conceive of a dynamic equilibrium between stress and the adaptive resources of the individual. Despite the many limitations of the studies, there is a suggestion that an acute episode of a schizophrenic disorder is preceded by an increase in stressful events and that these events are more likely to occur among the lower socioeconomic classes. It is very difficult to study life events rigorously. The more recent of the case control studies done by Jacobs and Myers[21] found that first-admission schizophrenics reported more current life events than did controls. In addition, more of these events were classified as undesirable. These data must be interpreted cautiously, but they do suggest that a precipitating role may be played by current life events.

The classic studies on the contribution of stress as a predisposing rather than a precipitating factor were done many years ago. The Faris and Dunham[12] study found that the inner or central city, and in particular the transitional zones, had the highest rates for hospital admissions. Work done by Clark[10] showed that the highest rates of schizophrenia were in the lowest status occupations. This finding of a disproportionate concentration of schizophrenics in the lowest social class was replicated in a number of cities in the United States and abroad.[24] The relationship between social class and prevalence rate is a complex one. It is linear in cities of over one million in population. There is no relationship between social class and the prevalence rate of schizophrenia in cities of less than one hundred thousand population.[24] Cities of an intermediate size—between one hundred to five hundred thousand—do not show a linear relationship but rather a clustering of schizophrenic patients in the lowest social class.

There have been several interesting reports describing a relationship between season of birth and the schizophrenic disorders.[11,18,33] These reports, from two Scandinavian countries and Great Britian, describe a disproportionate number of schizophrenics born between January and April in the northern hemisphere. While it is possible that a winter birth may constitute some as yet undefined stress, it is more likely that there is a relationship between intrauterine development and the environmental demands of the winter.

Family Studies

The early family studies, particularly in the United States, stressed the search for noxious parental interactions in the childhood of the schizophrenic patient. A variety of such destructive family interaction patterns were described by Lidz and colleagues,[30] Wynne and colleagues,[43] Wynne and Singer,[42] and Jackson.[20] These early studies suffered from major methodologic problems including the absence of normal control families. In addition, the studies were retrospective, lacking adequate controls for the effects of early peculiarities of the child on the parenting style. However, the quality and rigor of the studies have continued to improve, and currently there is more emphasis on family communication patterns, particularly the way families solve problems and arrive at closure. It seems reasonable to conclude that the families of schizophrenic patients show more communication deviancies than do nonschizophrenic families. It also appears reasonable to conclude that the communication deviance precedes the onset of the actual illness.

Relapse rates in young schizophrenic males living in their parental home have been studied as a function of the emotional climate of that home.[8,41] It has been shown that homes in which there was a tendency for the parents to make critical comments, express hostility, and show emotional overinvolvement with the schizophrenic son were more likely to have relapses. Homes in which the parents were more tolerant of their offspring's deviance were much less likely to be associated with relapses.

Genetic Studies

The genetic studies of the past decade have focused on twin and adoptive strategies. Nevertheless, six major recent consanguinity studies found that the prevalence rate in the siblings of schizophrenic patients was 10 percent.* This does not differ significantly from the prevalence rate for other first-degree relatives and, therefore, this figure suggests that rearing practices do not have as direct an effect on the prevalence rate as do other factors.

A very important twin study was reported by Fischer[13] in 1973. She followed her population of twins for a sufficient period of time so that virtually no age correction was required in her sample. She reported a monozygotic concordance rate of 56 percent in the twins she studied. This is in many ways the best estimate of the true concordance rate because it does not rely on age correction. Fischer found that monozygotic twins who were concordant for schizophrenia were not at an increased risk when compared to discordant monozygotic twins. She studied the offspring of her twin sample and found that the children of the discordant pairs were at an equally high risk for schizophrenia, as were the children of the concordant twin pairs. Perhaps the most striking finding in her study was that the nonschizophrenic twin was as likely to produce a schizophrenic child as was the schizophrenic twin. The concordance results of Fischer are quite consistent with the better earlier twin studies, which consistently showed that the concordance rate for schizophrenia was significantly higher in monozygotes as compared by dyzygotes. If one examines the five best twin studies, the average monozygotic concordance rate is 47 percent compared to an average dyzygotic concordance rate of 15 percent.† In other words, the average monozygotic concordance rate is three times as high as the average dyzygotic concordance rate.

The initial findings of the Danish adoption studies were that the adopted offspring of schizophrenic parents were more likely to develop the illness and that the biological relatives of the schizophrenic offspring were more likely to show a schizophrenia spectrum disorder.[22,35] This sample is particularly important because the majority of the off-

*See references 6, 14, 31, 32, 37, and 40.
†See references 13, 16, 27, 34, and 39.

spring of schizophrenic parents were adopted prior to the parental psychotic episode. This fact controls for the effect of possible adoptive agency bias in placement and/or the effect of the prior knowledge of a family history of mental illness on the parenting style in the adoptive home. The more recent results obtained from this sample are based on extensive psychiatric interviews of over 90 percent of the available relatives.[23] The findings consistently support the conclusion that those who bear the child are more important for the prevalence rate of schizophrenia than those who rear the child.

Vulnerability

A complex and heterogeneous group of illnesses such as the schizophrenic disorders cannot be directly transmitted genetically. What can be transmitted is a diathesis toward the development of the necessary phenotype. The current literature tends to think of this diathesis as a vulnerability to the schizophrenic disorder. The data do not suggest that the phenotype involved in the transmission of the schizophrenic disorder predisposes to psychosis, but rather predisposes to a schizophrenic form of illness should one become psychotic. Vulnerability most likely means that a given individual has the capacity to develop specific behaviors under certain conditions, which might lead a psychiatrist to make a particular diagnosis. The transmitted vulnerability need not be abnormal nor inherently pathogenic, but rather plays a role in the etiopathogenesis of the symptoms that are used for the diagnosis of the illness.

¶ Mode of Gene Action

Even in the simplest of traits, the genotype does not immutably determine the phenotype in the absence of an environmental influence. The more steps involved between the genotype and the phenotype, the more room there is for individual variation in development. The genotype is one factor that accounts for individual differences, but it is certainly not the only factor. The gene is activated by the environment, and the phenotypic outcome is determined by the nature and timing of that gene-environment interaction. The gene-environment interaction is not a passive activation, but rather a union between the gene and the activating environment. The final phenotypic outcome is determined by the complex interplay of both genetic and environmental factors. The same gene activated in different environments will produce variations in the phenotype. Any given phenotype represents only one of the possible outcomes inherent within that particular gene. The degree of freedom within the genotype is finite, and the differences in the final phenotypic characteristic may be small. Nevertheless, the difference may be of clinical significance, including the presence or absence of a pathologic trait. The amount of variability inherent within the genotype is not a theoretical question but rather an empirical one. It can only be determined by exposing the genotype to a range of environment-evoking stimuli. Having the appropriate genes necessary for a schizophrenic illness is not sufficient to produce it, as can be seen clearly from the identical twin studies.

A knowledge of the genotype does not allow a prediction of the phenotype and, similarly, a knowledge of the phenotype does not allow an inference of the underlying genotype. Any phenotype can be arrived at from different genotypes being activated in different environments. There is enormous plasticity in biologic systems, which contributes to the diversity that characterizes all species. Each individual is a unique biologic experiment in which the gene-environment interaction will never be reproduced identically again. The phenotypes that are necessary for a schizophrenic illness need not be sufficient to cause it. It is theoretically possible to have these phenotypes without becoming ill, particularly if the phenotypes represent normal variations of the trait. The phenotypes involved in the transmission of schizophrenia need not be pathogenic or inherently pathologic. For example, if shyness were to represent a phenotypic trait, it may not be pathologically exaggerated in the premorbid personality of a person who might later become schizophrenic. But should that person become psychotic, it is highly likely that the character trait of shyness will become integrated into the symptom formation. Just as the identical twin studies clearly show that the presence of a genotype is not the single cause of a schizophrenic illness, so clinical experience demonstrates that there is as yet no identifiable phenotype that is a necessary and sufficient cause of the disorder.

¶ Conclusions

The very complexity of the etiology of the schizophrenic disorders, which has become more apparent in recent years, has significant clinical implications. There are a number of different genotypes that can be involved in the production of the syndrome. And there are a number of different evoking environments that can combine with these different genotypes to produce the phenotypic characteristics that are important in the etiopathogenesis of the illness. The pathways that lead to the production of the necessary phenotypes will differ as a function of the differences in the gene-environment interactions. The clinical disorders that are called schizophrenia are, at best, modestly homogeneous syndromes deriving from different initial conditions through different biopsychosocial pathways. This variability in etiology and pathogenesis is a reality and not a nosologic deficiency. The inherent heterogeneity results from the fact that there is no single genotype or single environmental experience that will inevitably lead to a schizophrenic disorder.

The heterogeneity manifests itself not only in etiology and pathogenesis, but in the presenting picture as well. The presenting picture varies enormously, with some patients showing rapid onsets, while others show more insidious onsets. In some patients the presenting symptoms are more florid, while in others they are not. The various clinical courses also reflect the heterogeneity inherent in this group of disorders. It follows then that the attempt to identify the correct or even preferred treatment for the schizophrenic disorders is futile. There can be no preferred treatment when the group consists of multiple illnesses. Treatment helpful for some people may be useless or even harmful for others. The range of treatments must reflect the variability inherent within the disorder. There are many schizophrenias and there must be many treatments. A wide range of useful treatments can be expected. Obviously, this insight must be tempered by judgment, lest worthless fads be inflicted upon these patients.

The goal of the treatment of the schizophrenic disorders is to attain a level of psychological rehabilitation in which the individuals afflicted with the disorder begin to feel better about themselves. It has come to be recognized that psychotic symptom reduction is not enough. The patient with a schizophrenic disorder must be able to participate in the life of his or her community as a full adult member. This includes the ability to love, to work, to assume responsibilities, and so forth. While psychopharmacologic agents are very helpful in psychotic symptom reduction, they do not have a direct effect on social or personality development. There is need for social, vocational, and intrapsychic rehabilitation as well as symptom reduction. Despite our best therapeutic efforts, a significant percentage of these patients will be left with real deficits. It remains a task of psychiatry to maximize the residual assets of these patients so that they can lead as normal a life as possible.

¶ Bibliography

1. AMERICAN PSYCHIATRIC ASSOCIATION. *Diagnostic and Statistical Manual of Mental Disorders,* 2nd ed. (DSM-II). Washington, D.C.: American Psychiatric Association, 1968.

2. AMERICAN PSYCHIATRIC ASSOCIATION. *Diagnostic and Statistical Manual of Mental Disorders,* 3rd ed. (DSM-III). Washington, D.C.: American Psychiatric Association, 1980.
3. BIRLEY, J. L. T., AND BROWN, G. W. "Crises and Life Changes Preceeding the Onset or Relapse of Acute Schizophrenia: Clinical Aspects," *British Journal of Psychiatry,* 116(1970): 327–333.
4. BLEULER, E. *Dementia Praecox or the Group of Schizophrenias.* New York: International Universities Press, 1950 (originally published in 1911).
5. BLEULER, M. *Krankheitsverlauf Persönlichkeit und Verwandtschaft Schizophrener und ihre gegenseitigen Beziehungen.* Leipzig: Thieme, 1941.
6. ———. *Die schizophrenen Geistesstörungen im Lichte langjahriger Kranken- und Familiengeschichten.* Stuttgart: Thieme, 1972.
7. BROWN, G. W., and BIRLEY, J. L. T. "Crises and Life Changes and the Onset of Schizophrenia," *Journal of Health and Social Behavior,* 9 (1968): 203–214.
8. BROWN, G. W., BIRLEY, J. L. T., AND WING, J. K. "Influence of Family Life on the Course of Schizophrenic Disorders: A Replication," *British Journal of Psychiatry,* 121 (1972): 241–258.
9. CANCRO, R. "Thought Disorder and Schizophrenia," *Diseases of the Nervous System,* 29 (1968): 846–849.
10. CLARK, R. E. "The Relationship of Schizophrenia to Occupational Income and Occupational Prestige," *American Sociology Review,* 13 (1948): 325–330.
11. DALEN, P. *Season of Birth in Schizophrenia and Other Mental Disorders.* Göteborg, Sweden: University of Göteborg, 1974.
12. FARIS, R. E. L., and DUNHAM, H. W. *Mental Disorders in Urban Areas: An Ecological Study of Schizophrenia and Other Psychoses.* Chicago: University of Chicago Press, 1939.
13. FISCHER, M. "Genetic and Environmental Factors in Schizophrenia," *Acta Psychiatrica Scandinavica,* 238 (Supplement) (1973): 9–142.
14. ———. "Development and Validity of a Computerized Method for Diagnoses of Functional Psychoses (Diax)," *Acta Psychiatrica Scandinavica,* 50 (1974): 243–288.
15. FRANK, J. "Clinical Survey and Results of 200 Cases of Prefrontal Leucotomy," *Journal of Mental Science,* 92 (1946):497–508.
16. GOTTESMAN, I. I., AND SHIELDS, J. "A Critical Review of Recent Adoption, Twin, and Family Studies of Schizophrenia: Behavioral Genetics Perspectives," *Schizophrenia Bulletin,* 2 (1976): 360–401.
17. GRIESINGER, W. *Die Pathologie und Therapie der Psychischen Krankheiten,* Braunschweig, Germany: Wreden, 1871.
18. HARE, E., PRICE, J., AND SLATER, E. "Mental Disorder and Season of Birth," *British Journal of Psychiatry,* 124 (1974): 81–86.
19. HOLMES, T. H., and RAHE, R. H. "The Social Readjustment Rating Scale," *Journal of Psychosomatic Research,* 11 (1967): 213–218.
20. JACKSON, D. D. "A Study of the Family," *Family Process,* 4 (1965): 1–20.
21. JACOBS, S., AND MYERS, J. "Recent Life Events and Acute Schizophrenic Psychosis: A Controlled Study," *Journal of Nervous and Mental Disorders,* 162 (1976): 75–87.
22. KETY, S. S., ROSENTHAL, D., WENDER, P. H., AND SCHULSINGER, F. "The Types and Prevalence of Mental Illness in the Biological and Adoptive Families of Adopted Schizophrenics," in Rosenthal, D., and Kety, S. S., eds., *The Transmission of Schizophrenia,* Oxford: Pergamon, 1968, pp. 345–362.
23. KETY, S. S., et al. "Mental Illness in the Biological and Adoptive Families of Adopted Individuals Who Have Become Schizophrenic: A Preliminary Report Based on Psychiatric Interviews," in Fieve, R. R., Rosenthal, D., and Brill, H., eds., *Genetic Research in Psychiatry,* Baltimore: Johns Hopkins University Press, 1975, pp. 147–165.
24. KOHN, M. L. "Social Class and Schizophrenia: A Critical Review and a Reformulation," *Schizophrenia Bulletin,* 7 (1973): 60–79.
25. KRAEPELIN, E. *Compendium der Psychiatrie.* Leipzig: Abel, 1883.
26. ———. *Psychiatrie. Ein Lehrbuch für Studierende und Ärzte,* 6th ed. Leipzig: Barth, 1889.
27. KRINGLEN, E. *Heredity and Environment in the Functional Psychoses.* London: Heinemann, 1967.
28. LANGFELDT, G. "The Prognosis in Schizo-

phrenia," *Acta Psychiatrica Scandinavica,* 110 (Supplement) (1956): 1–66.

29. LANGNER, T. S., and MICHAEL, S. T. *Life Stress and Mental Health.* New York: Free Press of Glencoe, 1963.
30. LIDZ, T., FLECK, S., and CORNELISON, A. *Schizophrenia and the Family.* New York: International Universities Press, 1965.
31. LINDELIUS, R. "A Study of Schizophrenia: A Clinical, Prognostic, and Family Investigation," *Acta Psychiatrica Scandinavica,* 216 (Supplement) (1970): 1–125.
32. ØDEGARD, Ø. "The Multifactorial Theory of Inheritance in Predisposition to Schizophrenia," in Kaplan, A. R., ed., *Genetic Factors in Schizophrenia,"* Springfield, Ill.: Charles C. Thomas, 1972, pp. 256–275.
33. ———. "Season of Birth: A National Sample Compared with the General Population," *British Journal of Psychiatry,* 125 (1974): 397–405.
34. POLLIN, W., et al. "Psychopathology in 15,909 Pairs of Veteran Twins," *American Journal of Psychiatry,* 126 (1969): 597–609.
35. ROSENTHAL, D., et al. "Schizophrenics' Offspring Reared in Adoptive Homes," in Rosenthal, D., and Kety, S. S., eds., *The Transmission of Schizophrenia,* Oxford: Pergamon, 1968, pp. 377–391.
36. SCHNEIDER, K. *Clinical Psychopathology.* New York: Grune & Stratton, 1959.
37. STEPHENS, D. A., et al. "Psychiatric Morbidity in Parents and Sibs of Schizophrenics and Nonschizophrenics," *British Journal of Psychiatry,* 127 (1975): 97–108.
38. SZASZ, T. *Schizophrenia: The Sacred Symbol of Psychiatry.* New York: Basic Books, 1976.
39. TIENARI, P. "Psychiatric Illnesses in Identical Twins," *Acta Psychiatrica Scandinavica,* 171 (Supplement) (1963): 1–195.
40. TSUANG, M. T., et al. "Schizophrenia Among First-Degree Relatives of Paranoid and Nonparanoid Schizophrenics," *Comprehensive Psychiatry,* 15 (1974): 295–302.
41. VAUGHN, C. E., AND LEFF, J. P. "The Influence of Family and Social Factors on the Course of Psychiatric Illness: A Comparison of Schizophrenic and Depressed Neurotic Patients," *British Journal of Psychiatry,* 129 (1976): 125–137.
42. WYNNE, L. C., AND SINGER, M. "Thought Disorder and Family Relations of Schizophrenics: II. A Classification of Forms of Thinking," *Archives of General Psychiatry,* 9 (1963): 199–206
43. WYNNE, L. C., et al. "Pseudomutuality in the Family Relations of Schizophrenics," *Psychiatry,* 21 (1958): 205–220.

CHAPTER 15

NARCISSISM

Arnold M. Cooper

¶ Introduction

FEW CONCEPTS in psychiatry have undergone as many changes in meaning as has narcissism. Perhaps the single consistent element in these changes is the reference to some aspect of concern with the self and its disturbances. The word was introduced into psychiatry by Havelock Ellis.

The myth of Narcissus, as described by Bullfinch, clearly foreshadows many of the psychological descriptions that would come to be associated with the name. Narcissus was a physically perfect young man, the object of desire among the nymphs, for whom he showed no interest. One nymph, Echo, loved him deeply and one day approached him and was rudely rejected. In her shame and grief she perished, fading away, leaving behind only her responsive voice. The gods, in deciding to grant the nymphs' wish for vengeance, contrived that Narcissus would also experience the feelings of an unreciprocated love. One day, looking into a clear mountain pool, Narcissus espied his own image and immediately fell in love, thinking he was looking at a beautiful water spirit. Unable to tear himself away from this mirror image, and unable to evoke any response from the reflection, which disappeared every time he attempted to embrace it, he gradually pined away and died. When the nymphs came to bury him, he too had disappeared, leaving in place a flower.

H. G. Nurnberg[27] has pointed out that many of the features of narcissism are present in the myth: arrogance, self-centeredness, grandiosity, lack of sympathy or empathy, uncertain body image, poorly differentiated self and object boundaries, absence of enduring object ties, and lack of psychological substance.

Attempts to understand the concept of narcissism, the role of the self, and the nature of self-esteem regulation have occupied psychoanalysts and dynamic psychiatrists for three-quarters of a century. More recently, however, the "self," as a supraordinate organizing conception, has taken a more central place in the thinking of many clinicians and theorists, effecting a high yield in knowledge and understanding. This intensified interest in narcissism and the self relates to a number of current and historical trends. Some of these, briefly described, are:

1. The thrust of analytic research for several decades has emphasized the importance of early, that is, preoedipal developmental events. Psychiatrists and psychoanalysts have become increasingly interested in issues of early dependency, self-definition, separation and individuation, identity formation, and the earliest stages of object-relations. The theoretical movements of object-relational, interpersonal, and self-psychological schools have been highly influential. The works of Jacobson, Mahler, Winnicott, Rado, Horney, Sullivan, Kohut, Kernberg, Erikson, and others have been important.*

2. There has been an increasing willingness to alter or abandon traditional metapsychological language in favor of concepts that are closer to clinical experience. For instance, such designations as "self" and "identity" are incompatible with Freud's original natural science model of psychoanalysis and cannot easily be squared with the older concepts of energic and structural points of view. The concept of the self refers to a model that is more historical, experiential, intentionalistic, and action-oriented. Roy Schafer,[33] in discussing these issues, suggested that a new conceptual model for psychoanalysis is in the process of being developed, and the work of Kohut,[18,19,20,21] Jacobson,[15] Mahler,[24] and others represents a transitional step in this development. In part, the current interest in narcissism expresses a need felt by some therapists for a psychodynamic frame of reference that accommodates the unity of human behavior in terms that are appropriate to our current psychological thinking.

3. Our present interest in the self is concordant with powerful, contemporary currents in philosophy and culture. The concerns of such cultural historians as Lionel Trilling and Quentin Anderson, of philosophers such as Sartre, Heidegger, or Wittgenstein, as well as the themes of many contemporary novels and movies, are directed toward the problem of maintaining a sense of self in an alienating modern world. Psychiatry and psychoanalysis, both in theory and practice, have always been powerfully influenced by, as well as influencing, the cultural milieu in which they exist. Many social observers, from Spiro Agnew to Christopher Lasch, have expressed the view that contemporary western civilization is characterized by an intense focus on private ambitions, a loss of concern with the needs of others, and a demand for immediate gratification—in effect, producing the "gimme" or "me first" culture. This change from an earlier sense of community and concern for one's fellow human beings is attributed to the influences of a television-dominated consumer culture, the loss of moral values, the breakdown of the stable authority-centered family, the focus on youth and beauty, the difficulty of perceiving one's valued place in society, and the uncertainity of future goals in a world of nuclear threat and political chaos.

4. In the intervening years between the early part of the century and the present, psychotherapists and psychoanalysts have perceived a change in the population presenting for therapy. Glover,[13] referring to the 1930s, and Lazar,[23] referring to the early 1970s, have discussed the scarcity of the "classical" neurotic patient described in the early psychoanalytic literature, and both have mentioned the increasing numbers of patients with characterologic disorders of some severity, especially the narcissistic character.

While it is generally believed that this population change is genuine and a consequence of the cultural changes previously mentioned, there are some who feel that the change is largely in the perception of those psychotherapists who are both more sophisticated about, and interested in, character and early development. According to this view, deeper levels of personality organization are today being routinely explored, therapeutic ambitions have increased, and diagnoses have changed more than the patients.

Whether it is because of the changing population or changing diagnostic interest, ther-

*See references 6, 14, 15, 16, 17, 18, 19, 20, 21, 24, 30, 34, and 35.

apists have been increasingly willing to undertake intensive psychotherapy or psychoanalysis with patients who previously would have been considered unsuitable because of their difficulties in forming a transference. Exploratory work with these patients has yielded new knowledge concerning narcissistic aspects of the personality.

All of these factors have played a role in engaging our interest in narcissism, and they have resulted in a greatly enriched description of the developmental and functional aspects of the self.

5. More recently, it has been the work of Heinz Kohut[18] and the publication of his *Analysis of the Self* that has kindled interest in narcissism. Without attempting to review what preceded his effort, Kohut boldly set forth an independent theory of the nature of narcissism and the therapy of narcissistic disorders. His work and its later modifications engaged the imagination of analysts and therapists, both pro and con, and has focused current discussion on the topic. (Kohut, and Kernberg, a major critic of his point of view, will be discussed separately in this chapter, and will not be included in the historical review.)

Finally, it should be emphasized that there is, today, general agreement that any concept of narcissism should include normal, as well as pathological, developmental and descriptive aspects. Current discussion emphasizes that narcissism is a universal and healthy attribute of personality, which may be disordered under particular circumstances.

¶ History

Freud

Otto Kernberg[16] has pointed out that

> psychoanalytic theory has always included the concept of the self, that is, the individual's integrated conception of himself as an experiencing, thinking, valuing and acting (or interacting) entity. In fact, Freud's starting point in describing the "I" ("das ich," so fatefully translated as "the ego" in English) was that of the conscious person whose entire intrapsychic life was powerfully influenced by dynamic, unconscious forces.

While this is undoubtedly the case, it is also true as Pulver[29] has indicated that Freud had extraordinary difficulty in conceptualizing the self within the libido theory and that this difficulty was compounded as Freud developed his structural point of view alongside the instinctual one. Because of variant historical usage, and because of the different meanings derived from different frames of reference, the term "narcissism" continues to have multiple meanings. As other workers began to take up the themes of narcissism, the concept took on even more varied meanings, dependent upon the historical period of the author's frame of reference. Pulver[29] points out that early in the psychoanalytic literature narcissism was used in at least four different ways.

> 1. Clinically, to denote a sexual perversion characterized by the treatment of one's own body as a sexual object.
> 2. Genetically, to denote a stage of development considered to be characterized by the libidinal narcissistic state.
> 3. In terms of object relationship, to denote two different phenomena:
>   a. A type of object choice in which the self in some ways plays a more important part than the real aspects of the object.
>   b. A mode of relating to the environment characterized by a relative lack of object relations.
> 4. To denote various aspects of the complex ego state of self-esteem. [p. 323]

The term "narcissism" was borrowed by Freud from Havelock Ellis, who used the Greek name to describe a form of sexual perversion in which the individual takes himself as a sexual object. Freud[10] described this as "the attitude of a person who treats his own body in the same way in which the body of a sexual object is ordinarily treated —who looks at it, that is to say, strokes it and fondles it till he obtains complete satisfaction through these activities." The term was also used by Freud[7] to describe a form of homosexual object choice in which the individual takes himself as his sexual object: "they perceive from a narcissistic basis and look for a young man who resembles them and whom *they* may love as their mother loved *them.*" In 1911, in his account of the Schreber case—a patient with paranoia—Freud[8] expanded his use of the term to refer to the normal stage of libidinal development occurring between earliest autoerotism and object-love—the period in which the individual first unifies his sexual instincts by lavishing them upon himself and his own body. At this stage the self is the libidinal object, and fixation at this time could result in later perversion.

In 1913 Freud[9] described the magical omnipotent qualities of primitive or infantile thought and feeling, and considered them to be a component of narcissism.

In his paper "On Narcissism," Freud[10] elaborated the idea of narcissism as the libidinal investment of the self and described the kinds of object choice and the relationship to objects characterized by narcissism. The narcissistic individual will tend to choose and love an object on the basis of:

(a) what he himself is (i.e. himself),
(b) what he himself was,
(c) what he himself would like to be,
(d) someone who was once part of himself. [p. 90]

He described "primary" narcissism as the original libidinal investment of self and its consequent grandiose inflation, combined with feelings of being perfect and powerful. "Secondary" narcissism was seen as the self-involvement following a frustration in object-relations, and the withdrawal of libido back into the ego.

Freud attempted to understand certain symptoms of schizophrenia in terms of the withdrawal of libido into the ego, with the special characteristic that the residua of the object-attachments have been removed from fantasy. The outward manifestations of this development include the withdrawal from objects, megalomania, and hypochondriasis—all indications of pathological excessive libidinal self-involvement.

Self-regard (self-esteem) was considered by Freud[10] to be directly proportional to the "size of the ego." "Everything a person possesses or achieves, every remnant of the primitive feeling of omnipotence which his experience has confirmed, helps to increase his self-regard." Using the libidinal economic point of view, he also came to the conclusion that self-regard is lowered by being in love (since the self is divested of libido, which is sent outwards toward the object) and raised in schizophrenia. Because clinical experience demonstrates that many persons in love experience an elevation of self-esteem and most schizophrenics suffer from damaged self-esteem, later workers thought to revise that theory.

Freud[10] also considered the "ego-ideal" and the idealizing tendencies of the ego in the formation of psychic structure. Freud at this time was concerned with the criticisms of Jung and Adler, who maintained that the psychoanalytic emphasis on sexuality offered no explanation of nonsexual libidinal or aggressive behaviors. His response was to expand the concept of narcissism to describe a variety of normal and pathological states, and to postulate the ego-libido. But while Freud continued to refine his ideas on narcissism, they remained essentially intact. Elaborations of these views contributed to an explanation of depression,[12] to understanding characterologic defiance,[11] and were the starting point for the development of ego-psychology, which dominated later psychoanalytic thinking. Reich,[32] for example, took the concept of narcissism as an essential base for his description of character: "Character is essentially a narcissistic protection mechanism . . . against dangers . . . of the threatening outer world and the instinctual impulse." Reich thus further expanded the idea of narcissism as a way of conceiving defense mechanisms.

In the development of psychoanalytic theory, then, the concept of narcissism became increasingly complex as the term was adapted to fit the changing frames of reference demanded by libido-economic, topographic, developmental, genetic, and structural points of view.

In psychoanalytic literature since the development of ego psychology, the term "narcissism" has often been used either as a synonym for self-esteem or as a general reference to "a concentration of psychological interest upon the self."[27] It has become increasingly apparent that the term is so burdened with the baggage of its past that it has perhaps outlived its usefulness. The descriptive or explanatory (genetic or dynamic) ideas behind the term are not uniformly agreed-upon, and often the word is used as if it explained a phenomenon. One consequence of this trend has been an increasing focus on the concept of the "self" in an attempt to provide clearer opportunities for clinical description and research.

¶ Theorists of the Self

While many psychodynamic theorists proposed ideas about the role of self in personality, only the work of those few whose contributions were pivotal, although not always accorded full recognition at the time, will be described.

Sullivan

Harry Stack Sullivan was among the first psychoanalysts to accord a central role to the concept of the self in a systematic view of behavior. Sullivan[34] spoke of "self-dynamism," describing dynamism as "the relatively enduring patterns of energy transformation which recurrently characterize the interpersonal relations—the functional interplay of person and personifications, personal signs, personal abstractions, and personal attributions—which make up the distinctively human sort of being."

Sullivan described three types of interpersonal experience in infancy that contribute to the formation of self-dynamism: (1) that of a reward, which leads to a personification of a "good me,"; (2) that of the occurrence of anxiety, which leads to the creation of a "bad me"; and (3) that of overwhelming and sudden anxiety, which leads to the cration of the sense of "not me." "Good me" personification organizes experiences of need satisfaction and the mother's soothing ministrations. "Bad me" personification represents experiences of the infant in which increased feelings of injury or anxiety coincide with increased tenseness and forbidding behavior on the part of the mother. Both of these experiences are communicable by the infant with relatively early development of speech capacity. The concept of the personification of "not me" relates to dream and psychotic experience and is a result of intense anxiety and dread, which in turn, results in dissociative behavior. Corresponding to the "good me" and "bad me" are personifications of a good and bad mother. These personifications of self-esteem are attempts to minimize anxiety that inevitably arises in the course of the educative process between mother and infant.

Sullivan[34] goes on to say that

> the origins of the self-system can be said to rest on the irrational character of culture or, more specifically, society. Were it not for the fact that a great many prescribed ways of doing things have to be lived up to, in order that one shall maintain workable, profitable, satisfactory relations with his fellows; or, whether prescriptions for the types of behavior in carrying on relations with one's fellows were perfectly rational—then, for all I know, there would not be evolved, in the course of becoming a person, anything like the sort of self-system that we always encounter. If the cultural prescriptions characterizing any particular society were better adapted to human life, the notions that have grown up about incorporating or introjecting a punitive, critical person would not have arisen. . . . But do not overlook the fact that the self-system comes into being because of, and can be said to have as its goal, the securing of necessary satisfaction without incurring much anxiety. [pp. 168–169]

For Sullivan, this self-system was the central dynamism of human organization, the source of resistance to change in therapy as well as the source of stability in healthy functioning. Understanding the defects in the self-dynamism provides the major therapeutic opportunity for altering the more severe pathological states.

Horney

Karen Horney[14] felt that clinical observation did not support the conclusions of libido theory, which propounded that normal self-esteem is a desexualized form of self-love, and that persons tending toward self-concern or overvaluation of the self must be expressing excessive self-love. Building on H. Nunberg's concept of the synthetic function of the ego, she decided that the nuclear conflict of neurosis was not one of instincts, but one of self-attitudes. She suggested that narcissism be confined to situations of unrealistic self-inflation.

> It means that the person loves and admires himself for values for which there is no adequate foundation. Similarly, it means that he expects love and admiration from others for qualities that he does not possess, or does not possess to as large an extent as he supposes.
>
> According to my definition, it is not narcissistic for a person to value a quality in himself which he actually possesses, or to like to be valued by others. These two tendencies—appearing unduly significant to oneself and craving undue admiration from others—cannot be separated. Both are always present, though in different types one or the other may prevail.

According to Horney this type of self-aggrandizement is always the consequence of disturbed relationships in early childhood, especially the child's alienation from others provoked by "grievances and fears." The narcissistic individual is someone whose emotional ties to others are tenuous, who suffers a loss of the capacity to love. Horney describes the loss of "the real me" as occurring under conditions of parental coercion in which the child suffers impairment of self-sufficiency, self-reliance, and initiative. Self-inflation (narcissism) is one attempt to cope with these tendencies.[14]

> He escapes the painful feeling of nothingness by molding himself in fancy into something outstanding—the more he is alienated, not only from others but also from himself, the more easily such notions acquire a psychic reality. His notions of himself become a substitute for his undermined self-esteem; they become his "real me."[pp. 92–93]

This type of self-inflation also represents an attempt to maintain some life-sustaining self-esteem under conditions of potential annihilation, as well as being a desperate effort to attain admiration as a substitute for the unavailability of love. Horney describes three pathological consequences of narcissistic self-inflation: (1) increasing unproductivity because work is not satisfying for its own sake; (2) excessive expectations as to what the world owes the individual without effort on his part; and (3) increasing impairment of human relations due to constant grievances and hostility. Persons with narcissistic pathology tend to create ever more fantastic inflated versions of the self, which, lacking reality, lead to increasingly painful humiliations, which, in a vicious circle, lead to greater distortion of the self. Horney, therefore, sharply distinguishes between self-esteem, which rests on the genuine capacities that an individual possesses (which may be high or low), and self-inflation, which is an attempt to disguise a lack of qualities by a false presentation of capacities that do not exist. "Self-esteem and self-inflation are mutually exclusive."[14] Self-esteem represents the healthy development of the appropriate monitoring of self-approved action. Narcissism, therefore, is not an expression of self-love, but of alienation from the self.

She concludes:

> In rather simplified terms, a person clings to illusions about himself because, and as far as, he has lost himself. As a consequence the correlation between love for self and love for others is not valid in the sense that Freud intends it. Nevertheless, the dualism which Freud assumes in his second theory of instinct—the dualism between narcissism and love—if divested of theoretical implications contains an old and significant truth. This is, briefly, that any kind of egocentricity detracts

from a real interest in others, that it impairs the capacity to love others.[p. 100][14]

Rado

Sandor Rado,[30] in "Hedonic Control, Action-Self, and the Depressive Spell," attempted a description of what he termed the "action-self." The action-self is intended to be the organizing principle of behavior, replacing Freud's libidinal concepts.

Let me now give a rounded summary of these features of the action-self. Of proprioceptive origin, the action-self is the pivotal integrative system of the whole organism. Guided by willed action, it separates the organism's awareness of itself from its awareness of the world about it, and completes this fundamental separation by building up the unitary entity of total organism in contrast to the total environment. It is upon these contrasting integrations that the selfhood of the organism depends, as well as its awareness of its unbroken historical continuity. In accord with these functions, the action-self plays a pivotal part in the integrative action of the awareness process. This part is enhanced by its automatized organization of conscience, which increases the fitness of the organism for peaceful cooperation with the group. By its expansion and contraction, the action-self serves as the gauge of the emotional stature of the organism, of the ups and downs of its successes and failures. In its hunger for pride, it continuously edits for the organism the thought-picture of its present, past and future. [p. 304][30]

Rado attempted a functional description of a system of self-organization that was intended to replace the instinctual frame of reference of Freud.

Winnicott

Winnicott,[35] in a paper written in 1960, described a True Self as the spontaneous, biological comfort and enthusiasm that arise in the course of development.

The True Self comes from the aliveness of the body tissues and the working body-functions, including the heart's actions and breathing. It is closely linked with the idea of the Primary Process, and is, at the beginning, essentially not reactive to external stimuli, but primary. There is little point in formulating a True Self idea except for the purpose of trying to understand the False Self, because it does no more than collect together the details for the experience of aliveness. [p. 148][35]

He went on to describe the False Self as a consequence of the failure of the not-good-enough mother to meet the omnipotent fantasy of the infant during the earliest stage of object relationships

A True Self begins to have life through the strength given to the infant's weak ego by the mother's implementation of the infant's omnipotent expressions. The mother who is not-good-enough is not able to implement the infant's omnipotence, and so she repeatedly fails to meet the infant gesture; instead she substitutes her own gesture which is to be given sense by the compliance of the infant. This compliance on the part of the infant is the earliest stage of the False Self, and belongs to the mother's inability to sense her infant's needs. [p. 145][35]

In Winnicott's theory, varying degrees of False Self are constructed in an attempt to keep intact some hidden aspects of one's True Self, while presenting a false compliance to environmental demands. In severe degrees the False Self sustains the individual against the sense of total annihilation through the loss of the True Self. Anticipating Kohut, Winnicott described the extraordinary clinical importance of recognizing the existence of a False Self, and the failure of all therapeutic measures that address only the False Self while failing to understand the hidden True Self. The analyst, however, must recognize initially that he can speak only to the False Self about the True Self. As a True Self begins to emerge, the analyst must be prepared for a period of extreme dependence, often created by degrees of acting out within the analysis. A failure on the part of the analyst to recognize this need to assume the caretaker role will destroy the opportunities for further analysis of the True Self. And finally, analysts who are not prepared to meet the heavy needs of patients who become extraordinarily dependent should be careful not to include False Self patients in their caseloads, since they will not be successful in treating them.

In psycho-analytic work it is possible to see analyses going on indefinitely because they are done on the basis of work with the False Self. In one case, a man had had a considerable amount of analysis before coming to me. My work really started with him when I made it clear to him that I recognized his non-existence. He made the remark that over the years all the good work done with him had been futile because it had been done on the basis that he existed, whereas he had only existed falsely. When I said that I recognized his non-existence he felt that he had been communicated with for the first time. What he meant was that his True Self that had been hidden away from infancy had now been in communication with his analyst in the only way which was not dangerous. This is typical of the way in which this concept affects psycho-analytic work.[p. 151][35]

While Winnicott did not attempt any rigorous definition of what a self is, his work is clearly clinically relevant to, and a precursor of, current issues in narcissism. He emphasized the importance of the early failure of the "holding environment" and the need for regression of the self in the analysis. The False Self, separated from the roots that compose the matrix of psychic structure, leads to an impoverishment of the capacities for play, creativity, and love; these qualities can be achieved only through a reestablishment of the predominance of the True Self.

Erikson

Erik Erikson, wrestling with similar questions concerning the organization of unified self-perception, self-judgment, and motivation, used the term "identity" or "ego identity." He was careful never to define his meaning with great precision, believing that the definition should grow out of its developing clinical use rather than be determined in advance by theoretical considerations. He spoke of the ego identity as

the accrued experience of the ego's ability to integrate these identifications with the vicissitudes of the libido, with the aptitudes developed out of endowment, and with the opportunities offered in social roles. The sense of ego identity, then, is the accrued confidence that the inner sameness and continuity are matched by the sameness and continuity of one's meaning for others, as evidenced in the tangible promise of a "career."[p. 228][6]

Identity for Erikson meant developing a sense of one's basic personal and interpersonal characteristics, beginning in early infancy with the advent of "basic trust" and continuing through each of the eight stages of man. Adolescence is seen by Erikson to be an especially crucial period in the formation of identity since it brings together many disparate elements of ego identity—sexual, vocational, dependent. Maturation is seen as a succession of developmental crises in which the respective optimal outcomes culminate in the achievement of an ego sense of trust, autonomy, initiative, industry, intimacy, generativity, and integrity. It is clear that self-esteem is dependent on the degree of success or failure in achieving satisfying ego images at each developmental stage.

Erikson allotted special emphasis to the interaction of biological and cultural influences in the formation of ego identity. The biological matrix, essentially that of Freud's psychosexual schema, takes on its particular psychological characteristics only through the effects of specific identifications and cultural expectations, which aid or hinder the achievement of identity goals at each developmental stage.

While Erikson did not specifically address his work to the theory of narcissism, and seems to eschew all metapsychological implications, his studies bear directly on attempts to understand the formation of stable self and object representations out of bodily perceptions, parent-child interactions, and social influence, as well as on the mechanisms of the maintenance of self-esteem. Erikson has made one of the most detailed efforts to relate the vicissitudes of the individual identity, or self, to the opportunities and disadvantages that each culture provides. In addition, he offers specific analyses of several historical phenomena and some of their psychological consequences.[6]

¶ Narcissism and Culture

There is a large popular and technical literature[3,22] that maintains with varying degrees of documentation that the typical personality met with in western culture today has been deformed by consumerism and by the atmosphere of selfishness that is fostered by a child-centered society where the welfare of the child is singled out at the expense of the welfare of the family. Furthermore, the sense of anomie and hopelessness that pervades the culture at the same time that glitter and glamour are displayed on all sides has led to a general feeling of uselessness and rage, as well as a powerful urge to possess all pleasures now, ignoring future pleasures as not worth waiting for. The high divorce rate, the loss of religion, the inability to maintain an extended family, the abandonment of the home by women who join the work force, the lack of traditional pursuits, which are valued for their own sakes rather than for the material rewards they bring—all of this and more have been cited as causes for, and evidence of, the so-called narcissistic generation. From this perspective, individuals are more than ever self-centered, incapable of self-sacrifice for another person, without deeper moral, spiritual, or emotional values, and capable of experiencing only shallow transference relationships—all of which ultimately subjects them to the perils of alienation, boredom, and insecure relationships.

Christopher Lasch,[22] in *The Culture of Narcissism,* has presented an elaborate and eloquent description of the decay of western individualistic society, in which narcissism has reached a pernicious flowering, creating a mockery of older values. According to Lasch, his book "describes a way of life that is dying—the culture of competitive individualism, which in its decadence has carried the logic of individualism to the extreme of a war against all, the pursuit of happiness to the dead end of a narcissistic pre-occupation with the self." Lasch then goes on to describe a culture in which there has been a loss of both independence and any sense of competence.

> Narcissism represents the psychological dimension of this dependence. Notwithstanding his occasional illusions of omnipotence, the narcissist depends on others to validate his self-esteem. He cannot live without an admiring audience. His apparent freedom from family ties and institutional constraints does not free him to stand alone or to glory in his individuality. On the contrary, it contributes to his insecurity, which he can overcome only by seeing his "grandiose self" reflected in the attentions of others, or by attaching himself to those who radiate celebrity, power and charisma. For the narcissist, the world is a mirror, whereas the rugged individualist saw it as an empty wilderness to be shaped in his own design. . . .
>
> Today Americans are overcome not by the sense of endless possibility but by the banality of the social order they have erected against it. Having internalized the social restraints by means of which they formerly sought to keep possibility within civilized limits, they feel themselves overwhelmed by an annihilating boredom, like animals whose instincts have withered in captivity. A reversion to savagery threatens them so little that they long precisely for a more vigorous instinctual existence. People nowadays complain of an inability to feel. They cultivate more vivid experiences, seek to beat sluggish flesh to life, attempt to revive jaded appetites. They condemn the superego and exalt the lost life of the senses. Twentieth-century peoples have erected so many psychological barriers against strong emotion, and have invested those defenses with so much of the energy derived from forbidden impulses, that they can no longer remember what it feels like to be inundated by desire. They tend, rather, to be consumed with rage, which derives from defenses against desire and gives rise in turn to new defenses against rage itself. Outwardly bland, submissive, and sociable, they seethe with an inner anger, for which a dense, overpopulated, bureaucratic society can devise few legitimate outlets.[22]

While this idea seems logical and attractive, and is the theme of many novels and movies, there is little evidence that such a change of character has in fact taken place in a society that is as multifaceted as ours. It is very difficult to assess change in something as subtle as individual character or even in group behavior. Increased divorce rate, earlier appearance of sexual activity, and decline of religion need not be aspects of the failure in our ability to love, to work, or to value life itself. There has always been the tendency to blame the youth of any era for its lack of old-fashioned virtues, and as one follows the history of pop culture one must be impressed by the rapidity with which cultural movements change; for example, in a very few years an age of conformity (the 1950s) gave way to an age of rebellion (the 1960s), which in turn became an age of narcissism (the 1970s). But if we assume that character is fairly stable and slow to change, then these outward manifestations of cultural change reveal less about character than about a society that is predicated on technological goals. Of course others might say that the rapidity of cultural change is itself the source and measure of the problem of character.

Another claim for character change comes from psychoanalysts who feel that the classical neurotic patient suffering from a conflictual transference neurosis of primarily oedipal nature is now rare and has been replaced by the patient with narcissistic and even borderline features. It is difficult to know how to evaluate this claim. In the contemporary world, advances in psychoanalytic theory quickly permeate the general culture, so that even a vice-president who would later be indicted for fraud managed to have an opinion about defects in early child-rearing practices and the deformations of narcissistic character.

¶ Narcissistic Personality Disorder

Diagnosis

Because the term narcissism involves issues of self-esteem regulation and the self-representation, aspects of narcissism will appear in all psychological functioning, and disturbances of narcissism are apt to appear as a part of all psychopathology.[5] The syndrome Narcissistic Personality Disorder has been separately defined in the third edition of the *Diagnostic and Statistical Manual of Mental Disorders:*[1]

> The essential feature is a Personality Disorder in which there are a grandiose sense of self-importance or uniqueness; preoccupation with fantasies of unlimited success; exhibitionistic need for constant attention and admiration; characteristic responses to threats to self-esteem; and characteristic disturbances in interpersonal relationships that alternate between the extremes of overidealization and devaluation, and lack of empathy.
>
> The exaggerated sense of self-importance may be manifested as extreme self-centeredness and self-absorption. Abilities and achievements tend to be unrealistically overestimated. Frequently the sense of self-importance alternates with feelings of special unworthiness. For example, a student who ordinarily expects an A and receives an A minus may at that moment express the view that he or she, more than any other student, is revealed to all as a failure.
>
> Fantasies involving unrealistic goals may involve achieving unlimited ability, power, wealth, brilliance, beauty, or ideal love. Although these fantasies frequently substitute for realistic activity, when these goals are actually pursued, it is often with a "driven," pleasureless quality, and an ambition that cannot be satisfied.
>
> Individuals with this disorder are constantly seeking admiration and attention, and are more concerned with appearances than with substance. For example, there might be more concern about being seen with the "right" people than having close friends.
>
> Self-esteem is often fragile; the individual may be preoccupied with how well he or she is doing and how well he or she is regarded by others. In response to criticism, defeat, or disappointment, there is either a cool indifference or marked feelings of rage, inferiority, shame, humiliation, or emptiness.
>
> Interpersonal relationships are invariably disturbed. A lack of empathy (inability to recognize and experience how others feel) is common. For example, annoyance and surprise may be expressed when a friend who is seriously ill has to cancel a date.
>
> Entitlement, the expectation of special favors without assuming reciprocal responsibilities, is usually present. For example, surprise and anger are felt because others will not do what is

wanted; more is expected from people than is reasonable.

Interpersonal exploitativeness, in which others are taken advantage of in order to indulge one's own desires or for self-aggrandizement, is common; and the personal integrity and rights of others are disregarded. For example, a writer might plagiarize the ideas of someone befriended for that purpose.

Relations with others lack sustained, positive regard. Close relationships tend to alternate between idealization and devaluation ("splitting"). For example, a man repeatedly becomes involved with women whom he alternately adores and despises.

Associated features. Frequently, many of the features of Histrionic, Borderline, and Antisocial Personality Disorders are present; in some cases more than one diagnosis may be warranted.

During periods of severe stress transient psychotic symptoms of insufficient severity or duration to warrant an additional diagnosis are sometimes seen.

Depressed mood is extremely common. Frequently there is painful self-consciousness, preoccupation with grooming and remaining youthful, and chronic, intense envy of others. Preoccupation with aches and pains and other physical symptoms may also be present. Personal deficits, defeats, or irresponsible behavior may be justified by rationalization, prevarication, or outright lying. Feelings may be faked in order to impress others.

Impairment. By definition, some impairment in interpersonal relations always exists. Occupational functioning may be unimpaired, or may be interfered with by depressed mood, interpersonal difficulties, or the pursuit of unrealistic goals.

Prevalence. This disorder appears to be more common recently than in the past, although this may only be due to greater professional interest in the category. [pp. 315–317][1]

The Diagnostic Criteria for Narcissistic Personality Disorders are as follows:

The following are characteristic of the individual's current and long-term functioning, are not limited to episodes of illness, and cause either significant impairment in social or occupational functioning or subjective distress:

A. Grandiose sense of self-importance or uniqueness, e.g., exaggeration of achievements and talents, focus on the special nature of one's problems.

B. Preoccupation with fantasies of unlimited success, power, brilliance, beauty, or ideal love.

C. Exhibitionism: the person requires constant attention and admiration.

D. Cool indifference or marked feelings of rage, inferiority, shame, humiliation, or emptiness in response to criticism, indifference of others, or defeat.

E. At least two of the following characteristics of disturbances in interpersonal relationships:

(1) entitlement: expectation of special favors without assuming reciprocal responsibilities, e.g., surprise and anger that people will not do what is wanted

(2) interpersonal exploitativeness: taking advantage of others to indulge own desires or for self-aggrandizement; disregard for the personal integrity and rights of others

(3) relationships that characteristically alternate between the extremes of overidealization and devaluation

(4) lack of empathy: inability to recognize how others feel, e.g., unable to appreciate the distress of someone who is seriously ill. [pp. 315–317][1]

Not all psychoanalysts would agree with all aspects of this definition, since it perhaps places excessive stress on the overt grandiose and exhibitionistic qualities of the self. In fact, many persons appropriately diagnosed as possessing narcissitic personality disorders maintain grandiose fantasies at unconscious or preconscious levels, being aware primarily only of shyness, feelings of unworthiness, fears of competition, and fears of exhibiting themselves.

A detailed description of the narcissistic personality has also been given by Otto Kernberg.[17]

On the surface, these patients may not present seriously disturbed behavior; some of them may function socially very well, and they usually have much better impulse control than the infantile personality.

These patients present an unusual degree of self-reference in their interactions with other people, a great need to be loved and admired by others, and a curious apparent contradiction between a very inflated concept of themselves and an inordinate need for tribute from others. Their emotional life is shallow. They experience little empathy for the feelings of others, they obtain very little enjoyment from life other than from the tributes they receive from others or from their own grandiose fantasies, and they feel restless and bored when external glitter wears off and no new sources feed their self-regard. They envy others, tend to idealize some people from whom they expect narcissistic supplies and to depreciate and treat with contempt those from whom they do not expect anything (often their former idols). In general, their relationships with other people are clearly exploitative and sometimes parasitic. It is as if they feel they have the right to control and possess others and to exploit them without guilt feelings—and, behind a surface which very often is charming and engaging, one senses coldness and ruthlessness. Very often such patients are considered to be dependent because they need so much tribute and adoration from others, but on a deeper level they are completely unable really to depend on anybody because of their deep distrust and depreciation of others.

Analytic exploration very often demonstrates that their haughty, grandiose, and controlling behavior is a defense against paranoid traits related to the projection of oral rage, which is central in their psychopathology. On the surface these patients appear to present a remarkable lack of object relationships; on a deeper level, their interactions reflect very intense, primitive, internalized object relationships of a frightening kind and an incapacity to depend on internalized good objects. The antisocial personality may be considered a subgroup of the narcissistic personality. Antisocial personality structures present the same general constellation of traits that I have just mentioned, in combination with additional severe superego pathology.

The main characteristics of these narcissistic personalities are grandiosity, extreme self-centeredness, and a remarkable absence of interest in and empathy for others in spite of the fact that they are so very eager to obtain admiration and approval from other people. These patients experience a remarkably intense envy of other people who simply seem to enjoy their lives. These patients not only lack emotional depth and fail to understand complex emotions in other people, but their own feelings lack differentiation, with quick flare-ups and subsequent dispersal of emotion. They are especially deficient in genuine feelings of sadness and mournful longing; their incapacity for experiencing depressive reactions is a basic feature of their personalities. When abandoned or disappointed by other people they may show what on the surface looks like depression, but which on further examination emerges as anger and resentment, loaded with revengeful wishes, rather than real sadness for the loss of a person whom they appreciated.

Some patients with narcissistic personalities present strong conscious feelings of insecurity and inferiority. At times, such feelings of inferiority and insecurity may alternate with feelings of greatness and omnipotent fantasies. At other times, and only after some period of analysis, do unconscious fantasies of omnipotence and narcissistic grandiosity come to the surface. The presence of extreme contradictions in their self concept is often the first clinical evidence of the severe pathology in the ego and superego of these patients, hidden underneath a surface of smooth and effective social functioning.[17]

The chief attributes described in Kernberg's viewpoint are the individual's lack of emotional ties to others, the lack of positive feelings about his own activities, and his inability to sustain relationships except as sources of admiration intended to bolster his own faltering self-esteem. Kernberg further suggests that beneath the surface the pathological narcissist suffers from deep feelings of destructive rage and envy toward those people upon whom he depends. He also intimates that the inner fragmentation of those narcissistic individuals with good surface functioning may result in unexpected psychotic episodes during analytic treatment. Primitive defense mechanisms of splitting, projective identification, and denial are prevalent.

Kohut,[18] describing similar patients, emphasized the lack of genuine enthusiasm and joy, the sense of deadness and boredom, and the frequency of perverse activities. It is also his view that a final decision concerning the diagnosis can be made only on the basis of the transference established in the course of psychoanalysis. For Kohut,[20] the person suffering a narcissistic personality disorder is someone who has achieved a cohesive self-organization—that is, someone who is not borderline or psychotic but whose self-organization is liable to fragmentation under conditions of stress. Typically, in analysis, they form self-object transferences of the "mirror" or "idealizing" type, and these are the hallmarks of the disorder.

Finally, it should be apparent that disturbances of a psychic structure as central as the self must have consequences for all developmental stages, as well as for other psychic structures and for content and quality of intrapsychic conflict.

Differential Diagnosis

While there is continuing disagreement about the precise criteria for diagnosis, narcissistic personality disorders must, in general, be distinguished from the borderline personalities at the sicker end of the spectrum, and from the higher level (oedipal, classical, or transference) neuroses at the other end.

The borderline personality represents a more severe failure to achieve self-integration, and is characterized by greater impulsivity, varieties of sexual acting out, shifting, intense unstable relationships, frantic refusal to be alone, psychotic manifestations under stress, evidence of severe identity disturbance ("I don't know who I am"), marked and rapid lability of mood, and tendencies toward severe self-damaging behavior, including suicidal gestures.[1] While persons suffering narcissistic personality disorders may show some of these manifestations, their functioning remains characterized by a cohesive, if defective, self-organization, while the behaviors mentioned for the borderline patients are only rarely present. Self-object differentiation has been achieved and reality testing is basically intact. Relatively high levels of functioning are possible for the narcissistic personality, although there is always the tendency to "burn out" as boredom and emptiness replace the pursuit for admiration.

At the other pole, it may be impossible initially to distinguish the patient with a narcissistic personality disorder from patients with narcissistic characterological defenses against oedipal conflict, since some disturbances of the self are present in all psychopathology.[20] It is Kohut's view that only the ongoing therapeutic effort in analysis and the clarification of the nature of the transference can clearly make the distinction between these disorders. In analysis, the "classical" transference neurosis patient will develop a full tripartite oedipal fantasy relationship with the therapist, and the nuclear oedipal conflict will become apparent as narcissistic defenses are analyzed and undermined. In the narcissistic personality the oedipal palimpsest provides an "as if" sense of interpersonal involvement that quickly collapses if narcissistic defenses are analyzed and the patient is threatened with the loss of a coherent self. Kernberg[17] further emphasizes that despite surface similarities with a variety of neurotic disorders in which narcissistic defenses for self-esteem are prominent, the narcissistic personality disorder is distinguishable by the absence of genuine warmth and concern for others.

Etiology

Disturbances of narcissism arise during the early phases of infantile development in relation to beginning separation from the mother and the clear differentiation of oneself as a separate individual. It is postulated that under optimum circumstances the very young infant enjoys some vague sense of omnipotence, autarchy, and perfect union with mother and environment, since all needs are gratified relatively quickly upon their being experienced and with no special effort on the part of the infant. The experience of hunger is followed by feeding, and the experience of bodily discomfort is followed by the soothing ministrations of the mother. This experience of satisfactory unity with the caretaking environment, usually the mother, builds in the young psyche a sense of omnipotence, a fantasy of total bliss and power. With increasing psychological development, experience, and the additional complexity of needs, the infant becomes increasingly aware of his need for the mother's care and help, an awareness that reaches one peak at the rapprochement phase (the stage in which the infant, now a toddler, increasingly separated from mother and without mother's automatic aid in achieving his wishes, experiences great anxiety and frustration and ambivalently seeks both to establish autonomy and reestablish ties to mother). It is assumed that the responses at this stage are crucial for the shaping of future narcissistic characteristics. Those infants who are able to begin gradually to delegate their own sense of omnipotence to a parent for whom they have loving feelings, and to share that omnipotence while gaining a feeling of greater effectiveness, both individually and through sharing, are likely to develop a sturdy and joyful sense of self. Those infants who respond with increasing frustration and rage to the recognition of their own helplessness in satisfying their needs, or who find that the mother on whom they are dependent is an unreliable gratifier of their needs, are likely to develop rage tinged with inadequate feelings of themselves as beings incapable of providing for their own gratification.

In a brief summary then, the development of an adequate sense of self requires a mother-child "fit" that is sufficiently gratifying to both parties, so that the mother can provide the child with: (1) a "holding environment"[25] that allows a maximum of psychological comfort, including pleasures in body sensations; (2) the phase-specific wax and wane of grandiose omnipotent fantasies of perfection; (3) identifications with idealized parent images; (4) adequate experiences of loving approval of the child's body, play, and achievements; (5) control and tolerance of the child's "badness"; (6) phase-appropriate encouragement of increasing autonomy; and (7) the sense of being empathically responded to, that is, understood in some way. Clearly all of these needs are never entirely fulfilled, and the rage and frustration that routinely occur in the mother-infant interaction as a result of failures of need gratifications and subsequent disruptions of omnipotent fantasy are a part of the normal maturational process, as are the attempts to repair these feelings of injury. While it is likely that constitutional, possibly genetic, factors contribute to certain infants' difficulty in integrating the many processes that contribute to the coherent sense of self, studies on this topic are not available.

Disturbances of the self are part of all psychological disturbances, and their treatment must be part of the treatment of the major psychopathology that is present. The narcissistic personality disorders, however, require a treatment designed to repair the primary flaws in the self-organization and the related broad disturbances of functioning that are likely to be manifested in all aspects of the personality—in stability of object relations, loss of affective capacity, diminished integrity of psychic structure, unstable self-esteem, and so forth. While outcome studies are unavailable, there is general agreement that lasting treatment effects are likely to occur only with deep intensive psychotherapy or psychoanalysis, with or without modifications. In recent years two major views concerning the nature of psychotherapy for this disorder have been developed—Kohut's and Kernberg's. They are described in the next section.

¶ Therapy

Kohut

Heinz Kohut's comprehensive theory of the development of the self and treatment of disorders of the self has been a major influence in current thinking.[19,21] While Kohut's views have gone through a lengthy evolution, in their current form they define a bipolar self composed on the one hand of tendencies toward exhibitionism and ambition, and on the other hand toward idealization of parent and self. Both of these tendencies derive from early infantile precursors. Kohut posits these inferences concerning early development primarily from the nature of transferences that occur in psychoanalytic treatment. Those aspects of what are labeled the "mirror transference" reveal primitive needs for being noticed, admired, and approved in one's grandiose aspirations. When these needs are met in the course of infantile development the normal construction of an infantile grandiose self is effected, and this is a necessary basis for healthy later development. Aspects of the "idealizing transference" reveal that the infant endows the caretakers in the environment with idealized capacities for power and omniscience with which the infant can identify and from which he can borrow strengths. One pole of narcissism thus relates to the development of ambition, strivings, and achievements, while the other pole of narcissism relates to the development of values and goals. It is Kohut's view that these developmental aspects of the self precede the development of drive and that they are the sources of coherent drive expression. Failures in the cohesive development of the self lead to drive derivative "disintegration products," expressed as pathological sexual and aggressive behaviors.

The psychopathology of the narcissistic character disorder is, in Kohut's view, one of arrest of the development of adequate psychic structure—that is, it is a deficiency disease. These failures in the development of self structure are prior to, and the source of, the apparent drive-related and conflictual materials that have been traditionally interpreted as the nucleus of neurosis. According to Kohut, the exclusive focus of traditional psychoanalysis on the conflictual aspects of the problem prevents the appearance of the significant underlying etiologic deficit. Furthermore, the objective inspectional, inferential stance of the analyst contributes to a consistent attitude of muted responsiveness, which for many narcissistic characters in analysis imposes a repetition of the deprivation circumstance—that is, the lack of empathy for the patient's need for vividness, responsiveness, and so forth, which were the original source of the developmental failure. The analyst's unavoidable periodic empathic failures in the transference situation present the possibility that these original empathic failures will be analyzed in the generally empathic treatment situation rather than repeated blindly.

In this view, the first object relations of the developing child consist of partial recognitions of the actual other person as part of one's internal monitoring of the state of one's self, and are termed by Kohut "self-objects." They are objects not yet perceived as autonomous in their own right but are internalized as aspects of the self and its own needs. In the later development of healthy narcissism, when the self is sufficiently sturdy and capable of providing its own gratifications, it then acknowledges the existence of the object as autonomous and as a source of gratifications as well as an opportunity for generous giving. The development of pathological forms of narcissism is largely dependent upon the actual failures of the environment to provide appropriate empathic responses to the infant's needs. For healthy development to occur, the mother must be empathically responsive to the infant's need for admiration ("mirroring") and to the later need to idealize the parent. Empathic failures result in a developmental arrest with fixation remaining at primitive levels of grandiosity and idealization, which leads to defensive rage and distorted sexuality. The arrest of self-development and its drive-disintegration products interfere with joyous expression and prevent the development of creativity. It is Kohut's view that while aspects of narcissistic pathology can be treated by a variety of psychotherapies, only a properly conducted psychoanalysis offers the greatest opportunity for therapeutic success.

The therapeutic task, therefore, is to permit the reconstruction within the psychoanalytic situation of the original self-strivings of the patient. The feelings of empathic failure that will arise as the analytic work periodically falters, because of real empathic failures on the analyst's part, permit a reexamination of the parents' original empathic failures and an opportunity for renewed growth as the analyst senses a new object. According to Kohut, the early phase of psychoanalysis should be devoted to allowing the fullest emergence of mirror and/or idealizing transferences. This requires care on the part of the therapist to avoid a too early interpretation of defensive secondary behaviors, since this could prevent the emergence of more basic narcissistic strivings. The patient, for example, who early in the analysis expresses rage at the analyst's inadequate attention, requires an empathic understanding of what has occurred within the analytic situation (that is, what has led him to feel unattended to) rather than an interpretation concerning the nature of his habitually excessive demands for attention. If the patient is permitted to regress in the analytic situation to the stage of fixation of self, and if the therapist does not interfere with the renewed infantile needs for mirroring and idealization, then normal growth processes will resume and a more mature self can be achieved.

The emphasis on empathy is an important aspect of Kohut's work. He stresses the necessity for the therapist consistently to maintain the empathic rather than objective stance. It is the therapist's task to imagine himself "into the skin" of the patient and to understand what each situation in the transference feels like to that patient. This is more important than the attempt, with the use of theory, to understand objectively what the situation is like in some larger or more objective context.

Kohut and his followers have made the claim that the insights and technical consequences of this new theory of the self have improved their abilities to treat the full range of narcissistic disorders by the methods already indicated, as well as enabling them to bring these patients to a level where more classical psychotherapeutic-psychoanalytic interpretive techniques will be successful. Their effort is to present the patient with comprehensive reconstructive interpretations derived from an empathic mode of observation and communication as opposed to the allegedly classical part-interpretations derived from an inferential mode of observation and communication.[28]

Critics of Kohut have maintained that his work is poorly supported by data and that the clinical data produced is adequately explained by existing theories. The plea for empathy is regarded by his critics as a return to a philosophy of gratifying the patient's neurotic needs without analyzing them. His critics also claim that he provides a "corrective emotional experience" rather than an experience of deepened understanding about the conflictual nature of the difficulty.

Kernberg

Otto Kernberg has attempted to understand the dynamics of narcissism within the structural dynamic and object-relational points of view. The works of Mahler,[24] Jacobson,[15] Reich,[31] and the British School had contributed significantly to Kernberg's conception of the self as a vital aspect of the early ego developing as an original fused self/object internalization. It is Kernberg's view that all early infantile experiences contribute to the differentiation and integration of internalized self and object representations, which consist of mixtures of affective, cognitive, and drive components. Kernberg[17] states that in the narcissistic personality disorder, stable ego boundaries are established (that is, reality testing is intact), but a refusion of already differentiated internalized self and object representations occurs as a defense against anxieties arising out of interpersonal difficulties. He postulates the creation of ideal self and object images, actual self and object images, and denigrated self and object images. Whereas the normal individual maintains a structural tension of idealized self and object images (the superego), and actual self and object images (the ego), the narcissistic character pathologically fuses ideal self, ideal object, and actual self images in the attempt to destroy the actual object. As a result, there are not only distortions of the self, but structural distortions of the superego. According to Kernberg, the narcissistic character is, in effect, saying:

> I do not need to fear that I will be rejected for not living up to the idea of myself which alone makes it possible for me to be loved by the ideal person I imagine would love me. That ideal person and my ideal image of that person and my real self are all one and better than the ideal person whom I wanted to love me, so that I do not need anybody else any more.[17]

As a result of this process, denigrated unacceptable images of the self are projected onto those external objects viewed as dangerous, depriving, and attacking. The predominant self-image is itself a denigrated, hungry, weak, enraged, fearful, hating self. Kernberg discusses the feelings of emptiness, the lack of genuine feeling for others, and the paranoid projected rage that characterize these persons. Kernberg is in partial agreement with Kohut when he says that "chronically cold parental figures with covert, but intense aggression are a very frequent feature of the background of these patients."[17] The entire defensive effort of these patients is to maintain self-admiration, to depreciate others, and to avoid dependency. Kernberg's view is that the analytic task is to enable the patient to become familiar with his primitive oral rage, his hatred of the image of the aggressive mother, and to realize that this rage is linked with unfulfilled yearnings for loving care from the mother. The failure to integrate into one representation the loving and frustrating aspects of the mother—as represented in the figure of the analyst—will occupy a major portion of the analytic work. The patient's capacity to yield his own yearning for perfection in favor of accepting the terror of intimacy and the reality of another person as genuine, though imperfect, is the goal of the treatment. If successful, a new world of internalized objects is created that admits for the first time the feelings of genuineness and creative pleasure that were previosuly absent. Curiosity and interest in other persons, especially in the analyst, may begin to manifest themselves. The recognition of the reality of the analyst as a benign and actual whole person independent from the patient is, of course, the ultimate indicator of the success of the treatment.

It is Kernberg's view that narcissistic personalities can be treated without deviation from classical methods, that one must be alert to the borderline features which are displayed in more severe cases, and that one must be on the lookout for opportunities for narcissistic gratification which often hinder the analytic task. Kernberg does not agree with Kohut as to the need for a special preinterpretation phase of treatment. It is Kernberg's view that in the narcissistic personality the processes of idealization of self and object are not arrested but are faultily developed. Because the grandiose self regularly incorporates primitive components of ideal self and object, superego formation is defective and the internalized world of object-relations deteriorates, resulting in the severe disturbances of interpersonal relationships of pathological narcissism.[17] The therapeutic task is to enable the patient to arrive at new arrangements of existing structures and to undo pathological types of idealization rather than effect the resumption of growth of archaic tendencies toward idealization. For Kernberg the idealization of the analyst, early in the analysis of the narcissistic personality, would be a defensive measure related to covering underlying feelings of rage and emptiness rather than a conflict-free phase required for the building up of an adequate self. The pathological idealization is contaminated by rage, unlike the original idealization of the infant. Interpretation therefore will be aimed at helping the patient clarify his rage and greed; it will not require a preparatory phase of uncontaminated idealization.

Kernberg differentiates three levels of functioning of narcissistic personalities. The first group maintains effective surface adaptation in important areas of their lives; the patients are troubled by limited neurotic symptoms and have little insight into the inroads that narcissism has made in their lives. These patients are probably not yet willing to tolerate the anxieties that might be aroused in psychoanalysis, and are probably best treated by short-term psychotherapy. It is likely that later life experiences will bring home to them the full damage done to their personalities and they may then be amenable to psychoanalysis.

The second group of patients with narcissistic pathology is the most common and presents with severe disturbances in object relations and complicating symptoms in many areas of functioning. The treatment of choice in these cases is psychoanalysis. A third group of patients presents with borderline features and is likely to benefit from supportive-expressive psychotherapy.

Other Views

A variant of these views has been put forth by Cooper,[4] emphasizing the intermeshing of narcissistic and masochistic pathology. In his view early frustrations of narcissistic strivings lead to reparative attempts to maintain omnipotent fantasies, despite the helpless rage experienced by the infant in the course of ordinary failures of maternal care. One of these defensive efforts involves the attempt to master feelings of rage, frustration, and helplessness by the intrapsychic shift from pride in providing one's self with satisfactions to pride in the fantasy of control over a "bad mother," one who is responsible for the frustrations. Self-esteem takes on a pathological quality when an individual begins to derive satisfaction from mastery of his own humiliations, for example, when the infant begins to experience some sense of control and satisfaction when experiencing deprivations. A significant distortion of pleasure motivations has taken place and a pattern of deriving pleasure out of displeasure has begun. This pattern provides the groundwork for the later clinical picture of what Bergler[2] referred to as the behavior of the "injustice-collector." This individual engages in the following triad: (1) provocation or misuse of reality in order to suffer an injury; (2) defensive aggression designed both to deny responsibility for the unconsciously sought-for defeat and, secondarily, to escalate the self-punishment; and (3) depression, self-pity, and feelings of being singled out for "bad luck."

Cooper suggests that these individuals are basically narcissistic-masochistic characters and that their analysis regularly reveals that narcissistic defenses of grandiosity and entitlement are used to ward off masochistic tendencies toward self-abasement and self-damage. Concurrently, masochistic tendencies are used to disguise the full extent of the damage to the grandiose self. Treatment must therefore address both sides of the equation. Interpretation of narcissistic defenses produces masochistic reactions of victimization and self-pity, while interpretation of masochistic behaviors produces feelings of narcissistic humiliation.

Countertransference

Anyone who has attempted the treatment of narcissistic character pathology has noted the exceptional difficulties that arise in trying to maintain an appropriately attentive, sympathetic, and empathic attitude. The therapist is more than likely to find himself bored, or angry, or unable to make sense of the material, or just generally uneasy with the feeling of lifelessness presented in the treatment. Examination of the therapeutic situation will usually reveal that the therapist is responding to one or several of the following:

1. The patient's failure to acknowledge the therapist's existence in emotional terms. The therapist's interventions are ignored or denigrated; there is no curiosity about him, no indications that any tie exists between the two parties.
2. The patient's unspoken, grandiose, magical demand for total attention and effort on the part of the therapist, without any sense of a reciprocal relationship. The patient's feeling of icy control and detachment can be disconcerting.
3. Denigration of all therapeutic gain or effort, and destruction of all meaning.
4. Emergence of the extent of the patient's feelings of emptiness and hollowness, communicated to the therapist.
5. The patient's primitive idealizations of the therapist, arousing narcissistic anxieties in the therapist.
6. The patient's cold grandiosity, which arouses a retaliatory anger in the therapist.

Understanding the meanings of these reactions and making suitable preparations for them can aid the therapist to tolerate these periods, to remain alert for the shifts in the emotional climate of the treatment, and to avoid excessive guilt or anger on his own part.

¶ Summary

Issues of narcissism and the self have occupied a central role in psychodynamic theory and practice from the time of Freud's earliest researches. In the past several decades, increasing investigations into the diagnosis and treatment of the narcissistic personality disorders have been implemented by: (1) newer knowledge of infant development and the stages of individuation and separation; (2) developments in psychoanalytic theory that place greater emphasis on the central role of internal self and object representations and the maintenance of self-esteem; and (3) possible changes in the culture that may have produced more frequent and more severe forms of pathological narcissism. While the treatment of these patients is difficult and challenging, significant advances have been made and worthwhile therapeutic goals can often be achieved.

¶ Bibliography

1. AMERICAN PSYCHIATRIC ASSOCIATION. *Diagnostic and Statistical Manual of Mental Disorders,* 3rd ed. (DSM-III). Washington, D.C.: American Psychiatric Association, 1980.
2. BERGLER, E. *The Superego.* New York: Grune & Stratton, 1952.
3. COLES, R. "Our Self-Centered Children: Heirs of the 'Me' Decade," *U.S. News and World Report,* February 25, 1980.
4. COOPER, A. "The Masochistic-Narcissistic Character," abstracted in *Bulletin of the Association for Psychoanalytic Medicine,* 17 (1978): 13–17.
5. EISNITZ, A. "Narcissistic Object Choice, Self Representation," *International Journal of Psychoanalysis,* 50 (1969): 15–25.
6. ERIKSON, E. *Childhood and Society,* 2nd ed. New York: Norton, 1963.
7. FREUD, S. "Leonardo Da Vinci and a Memory of His Childhood," in Strachey, J., ed., *The Standard Edition of the Complete Psychological Works of Sigmund Freud* (hereafter *The Standard Edition*), vol. 2. London: Hogarth Press, 1962, pp. 63–137 (originally published in 1910).
8. ———. "Psycho-Analytic Notes on an Autobiographical Account of a Case of Paranoia (Dementia Paranoides)," in Strachey, J., ed., *The Standard Edition,* vol. 12. London: Hogarth Press, 1958, pp. 9–82 (originally published in 1911).
9. ———. "Totem and Taboo," in Strachey, J., ed., *The Standard Edition,* vol. 13. London: Hogarth Press, 1958, pp. 1–162 (originally published in 1913).
10. ———. "On Narcissism: An Introduction," in Strachey, J., ed., *The Standard Edition,* vol. 14. London: Hogarth Press, 1957, pp. 69–102 (originally published in 1914).
11. ———. "Instincts and Their Vicissitudes," in Strachey, J., ed., *The Standard Edition,* vol. 14. London: Hogarth Press, 1957, pp. 109–140 (originally published in 1915).
12. ———. "Mourning and Melancholia," in Strachey, J., ed., *The Standard Edition,* vol. 14. London: Hogarth Press, 1957, pp. 237–258 (originally published in 1917).
13. GLOVER, E. *The Technique of Psycho-Analysis.* New York: International Universities Press, 1955.
14. HORNEY, K. *New Ways in Psychoanalysis.* New York: Norton, 1939.
15. JACOBSON, E. *The Self and the Object World.* New York: International Universities Press, 1964.
16. KERNBERG, O. "Contemporary Controversies Regarding the Concept of the Self," unpublished paper.
17. ———. *Borderline Conditions and Pathological Narcissism.* New York: Jason Aronson, 1975.
18. KOHUT, H. *The Analysis of the Self: A Systematic Approach to the Psychoanalytic Treatment of Narcissistic Personality Disorders.* New York: International Universities Press, 1971.
19. ———. *The Restoration of the Self.* New York: International Universities Press, 1977.
20. ———. *The Psychology of the Self: A Casebook,* Goldberg, A., et al. eds. New York: International Universities Press, 1978.
21. ———. *The Search for the Self: Selected Writings of Heinz Kohut: 1950–1978, 2 vol.,* in Ornstein, P. H., ed., New York: International Universities Press, 1978.

22. LASCH, C. *The Culture of Narcissism: American Life in an Age of Diminishing Expectations.* New York: Norton, 1978.
23. LAZAR, N. "Nature and Significance of Changes in Patients in a Psychoanalytic Clinic," *The Psychoanalytic Quarterly,* 42 (1973): 579–600.
24. MAHLER, M., PINE, F., and BERGMAN, A. *The Psychological Birth of the Human Infant: Symbiosis and Individuation.* New York: Basic Books, 1975.
25. MODELL, A. *Object Love and Reality.* New York: International Universities Press, 1968.
26. MOORE B. E., and FINE B. D., eds. *A Glossary of Psychoanalytic Terms and Concepts.* The American Psychoanalytic Association, 1968.
27. NURNBERG, H. G. "Narcissistic Personality Disorder: Diagnosis," *Weekly Psychiatry Update Series,* 3 (1979): Lesson 17.
28. ORNSTEIN, P., and ORNSTEIN, A. "Formulating Interpretations in Clinical Psychoanalysis." Presented at the International Psycho-Analytic Congress, July 1979.
29. PULVER, S. "Narcissism: The Term and the Concept," *Journal of the American Psychoanalytic Association,* 18:2 (1970): 319–341.
30. RADO, S. "Hedonic Control, Action Self, and the Depressive Spell," in *Psychoanalysis of Behavior: Collected Papers.* New York: Grune & Stratton, 1956, pp. 286–311.
31. REICH, A. "Pathologic Forms of Self-Esteem Regulation," in Eissler, R. S., et al., eds., *The Psychoanalytic Study of the Child,* vol. 15. New York: International Universities Press, 1960, pp. 215–232.
32. REICH, W. *Character-Analysis,* 3rd ed., WOLFE, T. P. trans. New York: Orgone Institute Press, 1949, p. 158.
33. SCHAFER, R. "Concepts of Self and Identity and the Experience of Separation-Individuation in Adolescence," *The Psychoanalytic Quarterly,* 42 (1973): 42–59.
34. SULLIVAN, H. S. *The Interpersonal Theory of Psychiatry,* PERRY, H. S., and GOWEL, M. L., eds. New York: Norton, 1953, p. 168–169.
35. WINNICOTT, D. W. *The Maturational Processes and the Facilitating Environment: Studies in the Theory of Emotional Development.* New York: International Universities Press, 1965.

CHAPTER 16

MASKED DEPRESSION AND DEPRESSIVE EQUIVALENTS

Stanley Lesse

MASKED DEPRESSION is one of the more common clinical ailments seen in western medicine and rivals overt depression in frequency. Indeed, it is the type of depression most often encountered by nonpsychiatric physicians. The subject of masked depression and depressive equivalents presents us with a paradox: In spite of the frequency of the syndrome, only a relative handful of clinicians have a meaningful awareness or understanding of it. The depressive affect and even many depressive syndromes may be so masked that a nonpsychiatric or even psychiatric physician may be unaware of the fact that a serious emotional disorder is at hand until a massive, full-blown depression erupts and dominates the clinical scene.

The term "depression," in the minds of most laymen and physicians alike, usually refers only to a mood, which in psychiatric circles is more specifically labeled as sadness, melancholy, dejection, despair, despondency, or gloominess. If this overall mood pattern is not dominant in the clinical picture, the patient is not considered depressed. This view is universal among laymen. However, this narrow concept is also held by some physicians and even by psychiatrists. The masking veneer or facade may vary depending upon many factors, including: (1) the culture, (2) age of patient, (3) socioeconomic and sociophilosophic background, (4) hereditary and congenital processes, and (5) ontogenic development.

While the masked depression syndrome, hidden behind a broad spectrum of masking processes, is broadly represented in all cultures, the relevant literature is very sparse and deals primarily with those syndromes that are essentially manifested clinically as psychosomatic disorders or hypochondriacal complaints referred to various organ systems. Masked depression has been referred to by a variety of labels, which in themselves have contributed to the confusion surrounding this syndrome. The various diagnostic labels include: (1) masked depression, (2) depression sine depression, (3) depressive equivalents, (4) affective equivalents, (5) borderline syndromes, and (6) hidden depression. In many instances, where the condition has not been detected by the physician, the term "missed depression" might be appropriate.

¶ Concepts of Masked Depression and Depressive Equivalents

In western medicine, masked depressions are most commonly hidden behind psychosomatic disorders and hypochondriacal complaints.[12] Less frequently the depressions may be hidden behind various behavioral patterns.[14,16] If the clinician will look behind the presenting symptoms, a depressive core will be evident, a core that in most instances eventually becomes overt if the patient is not treated. Therefore, this type of masking hides an active depression, which can be readily discerned by careful examination.

In other situations, a psychosomatic syndrome or hypochondriacal symptom may represent an aspect of a clinical spectrum that eventually may end in an overt depressive reaction. For example, individuals who demonstrate hypochondriacal symptoms early in life are prone to develop overt depressive reactions. Women who eventually develop postpartum or involutional depressions frequently have histories of significant hypochondriacal or phobic reactions earlier in life. This is not to say that all individuals who are hypochondriacal or phobic or who have psychosomatic disorders are destined to become depressed. However, individuals with a history of these clinical phenomena have a greater propensity to eventually develop overt depressions.

With this observation in mind, one should also note that the psychodynamic mechanisms associated with hypochondriasis, phobias, and psychosomatic disorders, as they occur in western culture, are similar to those that are observed in depressed patients. Therefore, hypochondriasis, psychomatic disorders, and some acting-out behavioral patterns may be considered either as being masks of depression or depressive equivalents. When these symptoms or syndromes are merely "covering up" an underlying depressive core, they should properly be considered as depressive masks. When these symptoms or syndromes occur in the absence of a clear-cut depressive core, and then years later manifest symptoms and signs of an underlying depression, they should be thought of as depressive equivalents.

In this second context, depressive equivalents may be viewed as part of a clinical spectrum having certain psychobiologic and psychodynamic origins with features in common. These symptoms or syndromes may be seen as separate entities or as steps in a continuum that may or may not manifest themselves as phenotypical, full-fledged, overt depressions. Many years might pass before an overt depressive reaction emerges.

These observations raise the question of differentiating between those patients who have hypochondriasis, phobic reactions, or who have psychosomatic syndromes without ever developing overt depressions, from those who manifest the same symptoms and syndromes and who have a marked propensity to become depressed. Genetic studies suggest that hereditary factors may play a role in some depressive equivalents.[21] For example, the relatives of bipolar patients have a higher prevalence of hypertension, obesity, and thyroid dysfunction than do relatives of unipolar patients. In contrast to this observation, a higher incidence of chronic alcoholism and drug dependence has been noted in families of unipolar patients.[32]

From a genetic standpoint, depressive equivalents may be thought of in two ways: (1) as a different genetic subtype of affective disorder consistent with a model of heterogenic inheritance, or (2) as part of a continuum in a homogenic model of mood disturbances.

¶ Cultural and Economic Factors Influencing the Masks of Depression

It was found that depressive episodes that are masked by hypochondriasis and psychosomatic disorders are relatively uncommon in lower socioeconomic groups in the United States.[15] For example, faciopsychomyalgia, more commonly described as atypical facial pain of psychogenic origin, is rarely seen among blacks or Puerto Ricans in lower socioeconomic levels. However, this syndrome is seen among blacks and Latinos who rise in the socioeconomic scheme and who become part of the more affluent aspect of our society.

Acting-out behavior represents the more common type of masking process among lower socioeconomic groups in western society. This parallels the observation that acting-out behavior represents the common masking process in agricultural societies. For example, in India, a developing Third World country, 85 percent of the people are engaged in agriculture. Psychosomatic disorders are relatively uncommon among the nonliterate rural groups. In contrast to this observation, psychosomatic disorders and hypochondriasis are much more frequently seen among the better educated groups living in more industrialized, westernized centers such as Bombay or New Delhi.

In general, there is an evolutionary continuum of defensive confusion, anger, and acting-out from relatively frank and direct behavior in nonliterate cultures to increasing disguise and distortion in modern societies.[26] Modern societies, with their greater sophistication, use deeper disguises and more personally damaging methods of coping with problems.

In keeping with this general observation, masked depressions occur in their least severe form in most nonliterate cultures. These milder ailments are more open to spontaneous remission or shamanistic and priestly ministrations. If, however, the nonliterate cultures were strongly influenced by the European conquerors, masked depressions of the more severe type are encountered.

Simple and open confusion is the most common masking pattern in primitive or nonliterate peoples. Among these groups, confusion may be seen as a cry for help that brings the nuclear or extended family group to seek the aid of a priest or shaman. In more modern societies, however, people are relatively reluctant to show such dependent attitudes.

Hostility is a mask of depression in all societies. The direction the hostility takes, however, depends on the degree of cultural sophistication the society has attained. In nonliterate cultures, hostility is usually directed toward groups of people. In western cultures, the chief target of hostility is usually the person who is the one closest to the hostile individual. The diffusion of the objects of hostility and anger noted in technologically more primitive cultures may be accounted for, at least in part, by the fact that nonliterate cultures have more diffuse patterns of authority in the form of an extended family system.

The diffusion of hostility differs among various nonliterate societies. Opler[26] points out that among the Arctic Eskimos and the Ute Indians of Colorado and Utah, children are often adopted out of the nuclear family by relatives. This causes diffused object relations that are associated with a broad focus of hostility. In a similar fashion, when a Malaysian runs amok, there is a very diffuse portrayal of violent aggression toward anyone who crosses the path of the attacking individual. Opler also points out that acting-out among nonliterate peoples usually occurs in the presence of relatives or neighbors. In a similar fashion, Indonesian women who display the *latah* syndrome utter obscenities in the presence of friends and relatives.[24]

Among more primitive peoples, acting-out may also be in the form of imitative or negativistic behaviors; this is what occurs in Arctic hysteria and in the imu illness of the Ainu of the island of Hokkaido in Japan.[25] A similar pattern may be seen as far south as Malaysia.

Periods of confusion, occurring either as

masks of mild depressive states or as expressions of agitated euphoria, have been noted among African patients.[4] The periods of agitated euphoria may be viewed as compensations for the underlying depressions. In general, patients in primitive societies who have masked depressions can readily be restored to "health" through the psychosocial interventions of a shaman or curing cultist when the ailment is in its early phases.

¶ Acting-Out Behavior Masking Depression in Western Culture

Masked Depression in Children

The more primitive the culture the more direct and frank the clinical manifestations masking an underlying depression. In similar fashion, acting-out behavior, which is quite direct, is the most common type of depressive expression among children and, to a gradually decreasing extent, adolescents. Indeed, when viewed in this light, one can state that almost all depressions seen in childhood are masked depressions.

The literature dealing with childhood depression is quite limited. Mosse[23] points out that childhood depressions are usually subsumed under the classification of psychoneurotic disorders. She also points out that the classic and obvious symptoms of depression as they are known in adult life occur infrequently in childhood. Children who are depressed show a very diverse symptomatology. Therefore, the depressions of children and young adolescents may not be recognized at all,[6] or are classified as minor aspects of other diagnostic entities.

The masking symptoms of depression, such as defiance, truancy, restlessness, boredom, antisocial acts, and so forth, which are so common among children and adolescents, are all too often not given appropriate attention by laymen and physicians alike.[20] A study of suicidal behavior by children and adolescents indicates that months before their suicidal attempts almost half of them showed marked and definite behavioral changes that were not recognized as serious indications of depression by their parents or their teachers.[30]

Mosse points out that in most child psychiatric studies no clear distinction is made between children and adolescents.[23] She observes that both physically and psychologically there is a qualitative, and not just a quantitative, difference between childhood and adolescence and that this change affects the character of the psychopathology that is evidenced. This difference is most significant where depression is concerned.

It is most important to appreciate that suicidal attempts have been overlooked in children and adolescents due to the erroneous concept that they do not experience depression.[30] In fact, it is only recently that the subject of childhood depression has been discussed at all. Some child psychiatrists contend that depression does not exist in children. Toolan[31] points out that a popular book on child psychiatry, and several detailed monographs dealing with clinical and research aspects of depression, do not even mention childhood depression.

The scotoma that currently exists in regard to childhood depression parallels the blindness of some psychiatrists who denied the diagnosis "childhood schizophrenia" in the late 1940s and early 1950s. Psychiatrists and psychologists of that period were still fond of stating that children did not develop schizophrenia since schizophrenia could not appear until after puberty. This is no different from the ludicrous nineteenth-century belief that men could not be hysterics since hysteria was due to a "wandering uterus."

Spitz and Wolf[29] described a severe type of developmental retardation in infants that was associated with deprivation reactions and depressive elements. They labeled this syndrome "anaclitic depression." This syndrome was noted in infants and small children who had been isolated from maternal care; it was most commonly seen in children raised in institutions. These children demonstrated physical, intellectual, and emotional retardation. Initially, they protested actively, but finally became apathetic, showed decreased mental and physical activity, and rejected all adults.

Similar findings were observed in the Pavlov Institute in Leningrad in their studies of puppies.[9] If puppies, at the time of the appearance of the "awareness reflex," receive electroshocks whenever they are fed, they will withdraw from their handlers, crawl to the back part of their cages, and even refuse all food. They lose weight and hair. No matter how the future environment is improved, these puppies do not recover. If the same experiment is performed with older puppies who had initially been treated in a very humane fashion, they too withdraw in this fashion. However, among this older group, if the environment is improved, the dogs will again begin to relate to people and the overall environment in a positive fashion.

Others have also described intellectual and social retardation in institutionalized children who were deprived of close ties with their mothers or maternal substitutes.[7] John Bowlby[2] described three stages that a child undergoes when separated from the mother: (1) protest, (2) despair, and (3) detachment. Often the stage of detachment is misinterpreted by a hospital staff as a sign that the child is beginning to adjust to his situation, whereas in reality it is evidence of a profound disturbance, which Bowlby labeled "mourning" and which Toolan described as "depression."

Among older children, sociopathic manifestations and acting out are more likely to mask depression. This may take the form of disobedience, temper tantrums, truancy, or running away. Several authors have noted that underlying depressions are responsible for so-called school phobias.[1,3]

Some children will show equivalents of depression in the form of anorexia, colitis, and various other psychosomatic disorders; they may also display accident proneness and masochistic and destructive behavior.[30] Hypochondriacal and psychosomatic disorders may take the form of headache, tics, choreiform movements, abdominal complaints, nausea and vomiting, and so forth. The parents of such patients not infrequently present a history of depression.

Meyers[22] reports on a group of eighty-two childhood schizophrenics who had extensive residential and day-care treatment during their early school years. The biannual follow-ups revealed a strikingly low incidence of depressive response when the patients reached the ages of fifteen to twenty-six years. This absence of depression was even more impressive when one noted the degree of impairment in adaptation and the failures and defeats these schizophrenic individuals faced in their attempts to attain satisfying relationships with their environments. In contrast to this group, the emergence of depressive symptoms was greater in older children and in children with greater ego development and object relatedness. Meyers also observed that in severely ill schizophrenic children the grief of the mourning reaction is usually shallow, if it occurs at all. Instead there is blandness, apathy, anger, or a variety of atypical responses.

Depression may also be masked among mentally retarded children who are very often aware of their deficiencies. This is particularly true among children who are only slightly to moderately retarded. These children are frequently rejected by their peers, by their siblings, and even by their parents. Often the depression that they evidence may be masked by irritability, rage outbursts, and destructive behavior. They tend to automatically fight authority figures, but if their rage is blocked by fear of adult punishment, it may be directed toward younger children, small animals, or inanimate objects.[3] This type of masked depressive reaction is commonly misdiagnosed and mismanaged, especially in large institutions.

Masked Depression in Adolescents

Many of the depressive facades that were described for the older child are similar to those found in early adolescence. As the adolescent approaches young adult life, depressive episodes may become more overt and the masks will more closely resemble those seen in adult life. School phobia and underachievement in school may conceal underlying depressions in younger adolescents as well as those attending high school and even college. Among the older group, depressions are frequently manifested by changing courses, failures to take final examinations, dropping out of school, or changing from full-time to part-time schooling. The threat of graduation, laden as it is with the fears of unknown responsibilities, is often associated with depressive reactions masked by acting-out behavior or hypochondriacal and psychosomatic disorders.

Among adolescents one often encounters masks of depression in the form of pervasive boredom, restlessness, frantic seeking of new activities, and a reluctance to be alone.[31] The bored teenager often complains that he or she is tired. This type of adolescent may manifest an alternation between complaints of fatigue and evidence of almost inexhaustible energy.

Complaints of feeling empty, isolated, or alienated, so often described by adolescents, may also be indicative of underlying depression. He may describe himself as being unworthy and unlovable. The depressed adolescent often evinces a paradoxic combination of resentment toward his parents coupled with overdependence upon them.

The post-World War II period, particularly the past decade and a half, has been characterized by a decreased psychosocial threshold to psychologic or physical pain and frustration. This pattern has been enhanced by a multibillion-dollar advertising industry that preaches *ad nauseam* of one's birthright to wallow in material, physical, and emotional pleasure while expending little or no effort. Among older adolescents the compulsive use of drugs and sexual acting-out has become progressively more common as masks of depression. Sexual acting-out may be seen as seeking a significant other person in an attempt to relieve feelings of aloneness and alienation. At times, a depressive propensity may be aggravated by marked guilt feelings associated with this sexual behavior. Teenage pregnancies, which have increased in frequency at an astounding rate, too often compound the problem.

Chwast[5] has pointed out that depressive reactions may be masked under the guise of delinquent behavior among adolescents from lower socioeconomic backgrounds. Among these adolescent offenders, evidences of depression are commonly hidden behind sociopathic behavior patterns in a fashion also seen among adult criminals.[27] Chwast found that in a total sample of 121 delinquents, more than 75 percent appeared at least somewhat depressed, with almost 50 percent being substantially or severely depressed. Delinquent girls were usually more depressed than delinquent boys.

Among some delinquents the sociopathic acting-out served to ward off decompensating, schizophrenic defensive mechanisms. With regard to others, Chwast felt that fighting and destructive behavior should be seen as an attempt to combat depressive manifestations that threaten to become overt. To some of these individuals, the gang was a search for "significant other persons" in an attempt to compensate for a void in meaningful attachments. A separation from the gang may cause some culturally deprived persons to have feelings of inadequacy and hopelessness and to show even overt depression.

Automobile accidents and direct suicidal attempts are the two commonest causes of death among college students.[16] Many of these adolescents and young adults had exhibited masked or overt depressions. Herschfeld and Behan[8] expressed the opinion that failures in academic or social performance, which were considered "unacceptable disabilities," were converted into "acceptable disabilities" in the form of automobile accidents.

¶ Behavioral Masks of Depression Among Adults

In the vast majority of instances, masks of depression in adults take the form of hypochondriasis and psychosomatic disorders.[13]

But depression is also frequently masked by multivariant forms of acting-out. Drug dependency is one of the more common acting-out behavioral masks of depression in adults. While public attention has been focused upon problems that are secondary to narcotics addiction, which is so commonly associated with major crime in large cities, the excessive use of alcohol remains the most commonly encountered type of drug abuse.

Chronic alcoholism frequently serves to mask depression. Feelings of hopelessness, rejection, or overwhelming retroflex rage may appear precipitately following an alcoholic debauche.[16] When the mask slips, that is, when the alcoholic sobers up, massive guilt and profound depressive feelings are uncovered. Suicidal acts have followed failures in sexual performance, a problem commonly associated with alcoholism.

Marijuana, a wide spectrum of hallucinogenic agents, cocaine, amphetamines, barbiturates, antianxiety agents, neuroleptics, antidepressant drugs, and so forth, are available on the streets of American cities in vast quantities. Psychedelic drugs are often taken to mask underlying depressive syndromes. It is well known that weeks may pass before a covertly depressed patient who has been "on a trip" suddenly manifests overt depression.

Narcotics addiction may sometimes be seen as an attempt to cope with an underlying endogenous depression. Some of the suicidal attempts made when addicts are taken off narcotics may be ascribed to the emergence of massive depressive reactions that had been masked by the addiction. Individuals dependent on amphetamines and other stimulant drugs commonly have a depressive core. Precipitous depressive reactions very often result following rapid withdrawal of amphetamines from chronic users. While amphetamines are dispensed frequently to depressed patients, they usually serve only to mask the depression if it is profound enough.

Barbiturate habituation is a massive problem. There is an overproduction of barbiturates in this country, with the excess finding its way onto the streets where it is dealt with as a highly profitable, marketable product. Adolescents and adults from lower socioeconomic groups buy their barbiturates from "street pharmacists." In addition, there are literally tens of thousands of iatrogenically created barbiturate habituates. These drugs may mask underlying depressions for long periods of time, depressions that may rapidly become overt when the drugs are withdrawn.

Anger and rage are among the most commonly observed masks of depression. Spiegel[28] has stated that "the role of the equivalence of anger needs to be understood by both the patient and the therapist; and when anger or rage is dominant, the therapist should consider a relationship to depression." This is a very cogent observation. Patients with masked depression are almost without exception extremely angry individuals.[12] The rage could either be overt or covert; in most instances, it is overt. Covert anger is more difficult to manage from a therapeutic standpoint. Covert anger arises from severe childhood trauma. It is most commonly seen in patients who have been abandoned emotionally by their parents. It also occurs when parents are so hostile, domineering, critical, and sadistic that the patient, as a child, became terrified by the aggressive, punitive atmosphere. Attempts by the child to protest were usually met with overwhelming and crushing punishment. These patients, in general, are unable to react with appropriate anger in later life, even when it is justified.

Most patients with masked depressions are overtly hostile. In this type of patient, in contrast to the patient who had developed covert rage, the domineering, critical parent did not *completely* destroy the child's compensatory rage capacities or block the patient from expressing anger. This excessive anger is a compensatory mechanism that tends to dominate the patient's personality.

In the definitive treatment of patients with masked depressions, a pointed effort is made to unfold gradually the full degree of the patient's unconscious hostility. This anger is strongly guilt-linked.

Many of the patients are afraid of the intense degree of their latent anger, which is often tied to unconscious, symbolic, murderous fantasies. Many depressed patients, particularly those with covert anger, must be taught how to express anger and must be made aware of the fact that anger can be a normal, healthy reaction to certain types of stress.[18]

The apparent states of remission in depressed suicidal patients are the most serious and at times the most complicated type of masked depression.[11] They may occur in patients who have a history of suicidal ideas or who have made suicidal attempts but in whom the drive for self-destruction appears to have been ameliorated. Too frequently, the psychiatrist or psychotherapist who treats a suicidal patient may be so relieved to record some improvement that he or she may overestimate its true degree.

The availability of multiple therapies (including electroshock, psychotropic drugs, and some psychotherapies) that may be effective for various types of depressed and suicidal patients gives some psychiatrists a false sense of security simply because they use these techniques. The suicidal impulses may be merely blunted or masked by various psychotropic drugs or with electroshock therapy, particularly if the frequency or number of treatments is inadequate.

Psychotropic drugs also may result in a similar premature relaxing of clinical vigilance. Some suicidal patients may demonstrate an apparent remarkable remission following the administration of tranquilizers or antidepressant drugs. At times this apparent change may be purely a tenuous placebo reaction with the suicidal drive being only superficially masked. Any relaxation of clinical precautions during the early phase of treatment of suicidal patients, no matter what technique is used, may result in a self-destructive act.

¶ Masked Depression in Old Age

Among geriatric patients an organic mental reaction may mask an underlying depression. Patients who demonstrate fluctuations in the intensity of an organic mental syndrome require particularly close scrutiny.[16] Organically confused patients commonly show a decrease in the intensity of a depression and even of suicidal impulses. However, as they gain insight into the nature or severity of their problem, a depressive reaction leading at times to a suicidal act may occur. Among geriatric patients, depression may also be masked behind hypochondriacal symptoms and psychosomatic disorders. Marked irritability, obsessive thinking, or a gross increase in psychomotor activity may also serve as masking processes.

There is an unfortunate tendency among both physicians and laymen to attribute all changes in elderly people to organic illnesses. Not infrequently, symptoms such as listlessness, anorexia, and insomnia may be manifestations of an underlying depressive reaction.

¶ Hypochondriasis and Psychosomatic Disorders Masking Depression

In the vast majority of instances among adults in highly industrialized western countries, masks of depression assume the form of hypochondriasis and psychosomatic disorders.[12,15,17] While this type of depressive syndrome rivals overt depressions in frequency, it is insufficiently appreciated by psychiatrists and nonpsychiatrists alike, at times with tragic consequences. Physicians without formal psychiatric training are prone to treat a patient's "physical complaints" without probing to see whether the affect associated with the symptoms is secondary to a true physical deficit or whether it is a psychological expression mimicking an organic disorder. This clinical scotoma often results in patients being exposed to unnecessary and even inappropriate treatment over long periods of time.

It is likely that from one-third to two-thirds of patients past age forty who are seen by general practitioners and even specialists have masked depressions with the depressive syndromes masked by hypochondriacal or psychosomatic disorders.[17] These patients, particularly those in the late middle and older age groups, are extremely prevalent in hospital clinics; they also occupy a sizable proportion of general hospital beds. Unfortunately, they are usually subjected to a multitude of laboratory examinations and too often are exposed to a variety of organic treatments, even surgery.

In most instances, it is only after many months or even years of examinations and multiple treatments that a psychiatric consultation is requested. By the time the patient is seen by a psychiatrist, the depressions are usually of severe proportions. This observation is documented by the fact that more than 40 percent of the patients with masked depressions have suicidal ideas or drives by the time they are first seen by a psychiatrist.[15]

The masked depression syndrome poses a sharp challenge to all physicians, psychiatrists and nonpsychiatrists. A number of clinical possibilities may occur:

1. The masked depression syndrome may occur in patients without any organic processes. On the other hand, the patient may have a masked depression superimposed upon a true organic deficit. This second situation may pose a significant diagnostic problem.
2. Too often a minor organic illness is magnified by the psychogenic overlay, and it may be misdiagnosed as being a major organic disorder. In such a situation, the physician exaggerates the importance of the organic component and fails to recognize the psychogenic aspect of the problem. This usually leads to months and years of repeated studies and organic treatments. Most of these patients develop a massive iatrogenic overlay that further complicates diagnosis and treatment.
3. In other instances, a patient may have a major organic lesion with hypochondriacal or psychosomatic complaints superimposed. If the physician or therapist becomes preoccupied with the psychogenic aspects of the problem and fails to recognize the severity of the organic lesions, serious consequences may follow.

Clinical Characteristics

Sex and Age Distribution

One study[17] of 336 patients who had depressions masked by hypochondriasis or psychosomatic disorders reported that 246, or 73.2 percent, were women. This represents a female:male ratio of 2.7:1. However, in a different study[15] of 198 patients who had a type of masked depression known as "faciopsychomyalgia" (more commonly known as "atypical facial pain of psychogenic origin"), it was found that 86 percent were women. This is an 11:1 female:male ratio.

The age distribution is also very characteristic. Two hundred and ninety-five (87.8 percent) of 336 patients with masked depressions were between thirty-six and sixty-four years of age at the onset of illness.[17] One may state, therefore, that the syndrome in which depression is masked by hypochondriasis or psychosomatic disorders is primarily an ailment of middle-aged females.

It is unusual for these patients to be seen by a neuropsychiatrist early in the course of the illness. For example, 65 percent of the 336 patients were seen only after two or more years had passed from the time of onset of symptoms to the initial consultation. More than 30 percent had been ill for five or more years prior to being correctly diagnosed.

Initial Examination

A number of general characteristics can be noted during the initial examination. Patients present their history in a very wordy, forceful manner; the term "logorrhea" would be appropriate in many instances. The descriptions are replete with medical jargon gleaned from the many physicians or dentists who had examined and treated these patients. Quite often the patients consult medical texts and bring this "knowledge" to the examination.

The clinical descriptions are vague and do not represent classic descriptions of specific organic processes. At best, they are suggestive of a more unusual organic process. In addition, patients with masked depressions are far more handicapped in their vocational and social performance than are patients with true organic illnesses. These clinical descriptions, together with a tendency to exaggerate the suffering experienced, are further colored by iatrogenic factors that are secondary to prior multiple somatic examinations and treatments.

These patients come with a fixed concept that their ailment is due to a serious organic disorder. They demonstrate marked hostility toward the psychiatrist if the diagnosis of a psychogenic process is made early in the examination.

The initial phase of the history is directed primarily to ruling out a primary organic cause for the patient's complaints. Nevertheless, even the few clinical characteristics already described should warn the examining physician of the likelihood that the patient's ailments, at least in part, represent a significant psychogenic overlay, which necessitates intensive psychiatric evaluation.

The patients have a marked emotional and economic investment in their illnesses. There is a strong secondary gain mechanism behind their symptoms. Sufficient time must be allotted for the patients to expound upon their ailments, to relate the exquisite details of their symptoms, and to demonstrate their "knowledge of medicine."

The pointedness of the psychiatrist's investigations may be slowly broadened after the patient's confidence has been won. With gentle interrogation one can gradually compose a psychiatric scenario that is applicable for almost all of the patients. Patients routinely describe an agitated state, with restlessness, floor pacing, and marked feelings of anxiety. Insomnia, anorexia, persistent fatigue (especially in the morning), difficulty with concentration, loss of interest in vocational and social activities, and "feeling low" are also typical complaints. Routine personal habits become major chores. Frequently, the patients state that they are "losing their minds" and point to a "poor memory" as justification for this opinion.

Although the patients constantly refer to their "serious physical illnesses," one can gradually obtain statements indicating that they are moderately or severely depressed. Inevitably, this admission is accompanied by the disclaimer, "I would be fine if only I was free of my physical illnesses."

Once the patient admits to being depressed, one can readily elicit the presence of feelings of hopelessness. This admission can usually be brought out by questions such as, "Do you ever feel as though you will never get better?" A question that brings a positive response in almost half of the patients is "Do you ever feel as though you would like to go to bed and not awaken the next morning?" The usual response is "If I have to suffer like this, life isn't worthwhile."

One study[17] found that more than two-thirds of the patients expressed feelings of hopelessness. Even more startling was the observation that 44.5 percent had suicidal preoccupations or drives. This is evidence of the fact that depression masked by hypochondriasis or psychosomatic disorders is usually of severe proportions by the time the patient is referred for neuropsychiatric consultation. It is crucial that the intensity and imminency of these suicidal ideas be studied carefully. A number of patients examined by the author for the first time were actively contemplating suicide.[17]

If the psychiatric or nonpsychiatric physician is patient and gentle in the history taking, a close correlation between the onset of the patient's symptoms and her or his emotional traumas can often be discerned. This requires a step-by-step account of the patient's life situation prior to, during, and following the onset of the "somatic" symptoms. In some instances, specific environmental traumas cannot be documented. However, even if this is the case, careful evaluation will elicit the fact that the patients had been under chronic and severe stress with which they had difficulty coping.

PERSONALITY PATTERNS

The patients' personality patterns are rather consistent. They are typically aggressive, perfectionist, and highly intelligent individuals who have a need to dominate their environment. Characteristically, they are rigid and inflexible in their management of everyday life. In addition, they are overbearing and verbally critical of most people. These attitudes frequently alienate those around them. It can be stated that their compulsive need to dominate their surroundings is an attempt to compensate for feelings of self-derogation and inadequacy.

Although many of these patients, most of whom are women, are leaders in their communities and claim many close friendships, most of the so-called friends are usually just working acquaintances. By the time the patients are seen in initial psychiatric consultation, they are quite seclusive and are unable to function effectively vocationally, socially, or sexually. They usually express fears of being alone, strong guilt feelings related to their inability to function, and confess to a lack of sexual desires. A marked feeling of worthlessness is a characteristic clinical observation.

Family History

The family histories are also quite characteristic. Usually one or both parents are described as being very aggressive and perfectionistic. One study[12] reported that 82 percent of these patients described their mothers as the dominant individual in the home. The mother was often characterized as being an "attentive martyr," while the father was a rather passive, dependent personality dominated by the mother. Frequently, the patient's mother was hypochondriacal, phobic, or had a history of psychosomatic disorders. In many instances, the description of the mother indicated that she had referential trends. Characteristically, there is a history of a running conflict between the patient and his or her mother; this relationship universally generated marked guilt feelings in the patient.

From a psychodynamic standpoint these patients develop feelings of inadequacy and worthlessness beginning in early childhood. This is in response to the parents' real or imagined rejection. These feelings grow in crescendo fashion and color the patient's vocational and social relationships through the years. They are plagued by the anticipation that parental surrogates and peers might have the same negative image that they have of themselves.

These patients spend their lives compensating for feelings of inadequacy by a high level of performance. There is a constant struggle for self-recognition. They usually are highly critical of others (in scapegoat fashion) in an attempt to deny their own feelings of inadequacy.

There is often a history of overreacting to even mild physical illnesses. The physical ailments are a threat to the patient's constant attempts to compensate for her feelings of inadequacy. Furthermore, if one or both of the parents had been hypochondriacal, the patient tends to mimic the parents' particular hypochondriacal complaints.

¶ Treatment

The treatment of choice for patients with masked depressions, when the underlying depression is severe, is a combination of antidepressant drug therapy and appropriately designed, psychoanalytically oriented psychotherapy.[10,18,19]

The results of treatment depend upon a number of factors, including: duration of illness, amount and nature of prior medical treatment (prior drug therapies and surgical procedures), and the organ system involved. Patients with problems associated with the head and face, mammary glands, or genitourinary system are more difficult to treat for reasons that are not entirely clear at this time.

Considered as a group, more than 75 percent of those patients in whom the depression is masked by hypochondriacal complaints or psychosomatic disorders obtain excellent results if the illness is of less than one year's duration ("excellent" meaning that their symptoms disappear, the level of psychomotor activity becomes appropriate, and they are able to function vocationally and socially with pride and pleasure). Approximately 50 percent of those patients who have been ill for less than two years and who do not have strong iatrogenic overlay secondary to surgical procedures obtain excellent or good results during the initial period of therapy.

When a patient has been ill for more than two years, particularly if he or she is plagued by marked iatrogenic complications resulting from prior drug or mechanical therapies, it is difficult to predict how successful the combined therapeutic technique will be. Overall, approximately one-third of such patients obtain excellent or good results from combined therapy. While one cannot be so certain of the results that will be obtained in more chronic patients, excellent individual responses have been obtained in some who have been ill for as long as twenty-five to thirty years.

¶ Bibliography

1. AGRAS, S. "The Relationship of School Problems to Childhood Depression," *American Journal of Psychiatry,* 116 (1959): 533–536.
2. BOWLBY, J. "Childhood Mourning and Its Implications for Psychiatry," *American Journal of Psychiatry,* 118 (1960): 481–498.
3. CAMPBELL, J. D. "Manic-Depressive Disease in Children," *Journal of the American Medical Association,* 158 (1955): 154–158.
4. CARUTHERS, J. C. "The African Mind in Health and Disease," *World Health Organization Monograph H 17.* Geneva: World Health Organization, 1953.
5. CHWAST, J. "Depressive Reactions as Manifested Among Adolescent Delinquents", *American Journal of Psychotherapy,* 21 (1967): 574–584.
6. GLASER, K. "Masked Depression in Children and Adolescents," *American Journal of Psychotherapy,* 21 (1967): 565–574.
7. GOLDFARB, W. "Effects of Psychological Deprivation in Infancy and Subsequent Stimulation." *American Journal of Psychiatry,* 102 (1946): 18–22.
8. HERSCHFELD, A. H., and BEHAN, R. C. "The Accident Process, III. Disability: Acceptable and Unacceptable," *Journal of the American Medical Association,* 197 (1966): 125–128.
9. LESSE, S. "Current Clinical and Research Trends in Soviet Psychiatry," *American Journal of Psychiatry,* 114 (1958): 1018–1022.
10. ———. "Psychotherapy Plus Drugs in Severe Depressions: Technique", *Comprehensive Psychiatry,* 7 (1966): 224–231.
11. ———. "Apparent Remissions in Depressed Suicidal Patients," *Journal of Nervous and Mental Disorders,* 144 (1967): 291–296.
12. ———. "Hypochondriasis and Psychosomatic Disorders Masking Depression," *American Journal of Psychotherapy,* 21 (1967): 607–620.
13. ———. "Masked Depression: A Diagnostic and Therapeutic Problem," *Diseases of the Nervous System,* 29 (1968): 169–173.
14. ———. "The Multivariant Masks of Depression," *American Journal of Psychiatry,* 1246 (Suppl.) (1968): 35–39.
15. ———. "Atypical Facial Pain of Psychogenic Origin: A Masked Depressive Syndrome," in Lesse, S., ed., *Masked Depression.* New York: Jason Aronson, 1974, pp. 302–317.
16. ———. "Depression Masked by Acting-Out Behavior Patterns," *American Journal of Psychotherapy,* 28 (1974): 352–361.
17. ———. "Hypochondriasis and Psychosomatic Disorders Masking Depression," in Lesse, S., ed., *Masked Depression.* New York: Jason Aronson, 1974, pp. 53–74.
18. ———. "Psychotherapy in Combination with Antidepressant Drugs in Patients with Severe Masked Depressions," *American Journal of Psychotherapy,* 31 (1977): 185–203.
19. ———. "Psychotherapy in Combination with Antidepressant Drugs in Severely Depressed Patients: Twenty-Year Evaluation," *American Journal of Psychotherapy,* 32 (1978): 48–73.
20. MATTISON, A., SEESE, L. R., and HAWKINS, J. W. "Suicidal Behavior as a Child Psychiatry Emergency," *Archives of General Psychiatry,* 20 (1969): 100–109.
21. MENDELWICZ, J. "A Genetic Contribution

Toward an Understanding of Affective Equivalents," in Lesse, S., ed., *Masked Depression.* New York: Jason Aronson, 1974, pp. 41–52.

22. MEYERS, D. I. "The Question of Depressive Equivalents in Childhood Schizophrenia," in Lesse, S., ed., *Masked Depression.* New York: Jason Aronson, 1974, pp. 165–173.
23. MOSSE, H. "The Psychotherapeutic Management of Children with Masked Depression," in Lesse, S., ed., *Masked Depression.* New York: Jason Aronson, 1974, pp. 174–201.
24. ———. *Culture and Social Psychiatry.* New York: Aldine-Atherton, 1967.
25. ———. "The Social and Cultural Nature of Mental Illness and Its Treatment," in Lesse, S., ed., *An Evaluation of the Results of the Psychotherapies.* Springfield, Ill.: Charles C Thomas, 1968, pp. 280–291.
26. OPLER, M. K. "Cultural Variations of Depression: Past and Present," in Lesse, S., ed., *Masked Depression.* New York: Jason Aronson, 1974, pp. 24–40.
27. SCHMIDEBERG, M. "The Psychological Treatment of Adult Criminals," *Probation,* 25 (1946): 45–53.
28. SPIEGEL, R. "Anger and Acting Out: Masks of Depression," *American Journal of Psychotherapy,* 21 (1967): 597–606.
29. SPITZ, R., and WOLF, K. M. "Anaclitic Depressions: An Inquiry into the Genesis of Psychiatric Conditions in Early Childhood," in Eissler, R. S., et al, eds., *The Psychoanalytic Study of the Child.* New York: International Universities Press, 1946, p. 373.
30. TOOLAN, J. M. "Suicide and Suicidal Attempts in Children and Adolescents," *American Journal of Psychiatry,* 118 (1962): 719–724.
31. ———. "Masked Depression in Children and Adolescents," in Lesse, S., ed., *Masked Depression.* New York: Jason Aronson, 1974, pp. 141–164.
32. WINOKUR, G., and PITTS, F. N. "Affective Disorder: VI. A Family History Study of Prevalence, Sex Differences and Possible Genetic Factors," *Journal of Psychiatric Research,* 3 (1965): 113–118.

CHAPTER 17

BEHAVIORAL MEDICINE AND BIOFEEDBACK*

Redford B. Williams, Jr.

Why is so little attention paid to behavioral research in the treatment and cure of disease? While infectious diseases used to be the most burdensome illnesses, we now see cardiovascular disease, cancer, lung disease, accidents, homicide, and violence as the major threats to life and health. These afflictions have strong behavioral components.[33]

¶ Introduction

IN a recent address at the National Institutes of Health (NIH), Senator Edward Kennedy made a pointed reference to a growing trend in American medicine, namely, the emergence of a field of research and clinical endeavor that is now widely identified by the term "behavioral medicine." This trend is a historical fact, as documented by several recent events, including the founding of an Academy of Behavioral Medicine Research and a Society of Behavioral Medicine; the establishment of an experimental Behavioral Medicine Review Group within the Division of Research Grants of the NIH; the establishment of a Behavioral Medicine Branch within the Division of Heart and Vascular Diseases of the National Heart, Lung and Blood Institute; and a recent flurry of program announcements from the NIH in such areas as chronic pain and psychological aspects of cancer.

It is timely, therefore, to undertake a review of the emerging field of behavioral medicine. This chapter shall first consider the historical and conceptual contexts within which these recent developments have occurred, with particular emphasis on differential conceptual orientations between behavioral medicine and psychosomatic medicine, which is also concerned with emotions and disease. Then will follow a review of the substantive research contributions from studies carried out with a behavioral-medicine conceptual orientation. This review will con-

*Supported by a Research Scientist Development Award (MH70482-06) from the NIMH and by grants (HL 18589-04 and HL22740-02) from the NHLBI.

sider those aspects of behavioral medicine research relevant to etiology and pathogenesis of disease, to modification of lifestyles associated with increased risk of disease, and to direct treatment of disease. A concluding section on future directions will address the need for the integration of more psychodynamically oriented approaches (derived from psychosomatic medicine and consultation-liaison psychiatry) with the more behaviorally oriented approaches of behavioral medicine.

¶ Historical and Conceptual Context

To understand the forces behind the historical development of behavioral medicine, it is necessary to focus our attention on several recent trends. First is the realization that most patients with physical disease have not been shown to benefit from the application of "talking therapies," which focus upon verbal productions and free associations with the goal of achieving insight into neurotic conflicts. Perhaps this realization has been responsible for what both George Engel[23] and David Graham[29] have perceived to be psychosomatic medicine's lack of any substantial impact upon medicine in general in the past three decades. A second and probably more important impetus for the recent emergence of behavioral medicine as a distinct field has been the discovery in large-scale epidemiological studies that certain behaviors or lifestyles are "risk factors" for such major medical disorders as cancer and coronary heart disease. The failure of public education approaches to achieve the hoped-for dramatic changes in such risk-factor behaviors as cigarette smoking and nonadherence to antihypertensive regimens has convinced some leaders[16] in clinical medicine that behavior modification approaches that have proved effective in changing behaviors associated with mental disorders might also prove useful in attempts to modify behaviors that increase risk of major physical disorders. In addition to this realization that techniques of behavior modification might help to reduce risk-factor behaviors that presumably lead to disease, there has been the demonstration in a growing body of clinical research that behavioral treatment approaches aimed at changing physiology directly (for example, biofeedback and autogenic training) are effective[65] in the actual treatment of such physical disorders as headache, chronic pain, and Raynaud's disease—all disorders that heretofore had proven unusually resistant to the traditional biomedical approaches. The apparent ease with which these behavioral treatment approaches have found acceptance in the medical community may be because they share the traditional rationale for the use of pharmacologic agents in clinical medicine—direct modification of pathophysiological processes.

To the extent that Engel[23] and Graham[29] are correct in their observations that psychosomatic medicine has not had the desired impact upon medicine in general, it must be considered whether there is anything new or different about behavioral medicine. A brief historical digression might help to put the differences between behavioral medicine and psychosomatic medicine in perspective. In the first issue of *Psychosomatic Medicine,* which appeared in 1939, the editors defined psychosomatic medicine as the study of the "interrelation of the psychological and physiological aspects of all normal and abnormal bodily functions and thus [to achieve the integration of] somatic therapy and psychotherapy."[22] Nine years later, however, Carl Binger found it necessary to comment in an editorial: "the content of psychosomatic medicine should be greatly widened . . . beyond the ulcer, hypertension, asthma round. We should try to get away from too exclusive an emphasis on etiology and so-called psychogenesis."[8] Again, in 1958, Binger was moved to comment in another editorial that "We are not primarily interested in etiology. . . ."[9] More recently, in commenting on the definitions of psychosomatic medicine, Weiner has observed that "its aim has always been to contribute to a comprehensive account of the etiology and pathogenesis of disease."[61]

It becomes clear from these observations that psychosomatic medicine has from the beginning been concerned primarily with the role of psychological factors in the etiology and pathogenesis of disease. While initially the focus was upon seven disorders that were considered to be "psychosomatic" (duodenal ulcer, asthma, Graves' disease, essential hypertension, ulcerative colitis, neurodermatitis, and rheumatoid arthritis), the emphasis in recent years has shifted to the role of psychosocial factors in all diseases and to the underlying physiological mechanisms whereby these roles are mediated. Another feature through the years has been a primary focus upon specific personality traits that, it is felt, lead to neurotic conflicts that are in some way responsible for pathophysiologic alterations which, in turn, lead to the emergence of the disease in question. This focus upon intrapsychic conflicts determined that therapeutic interventions should focus upon the spontaneous verbal productions of the patient, with the goal of achieving insights that would result in the resolution of the neurotic conflict and, hopefully, of the disease process as well. Unfortunately, as Graham[29] noted in his recent Presidential Address before the American Psychosomatic Society, this hope has not been widely fulfilled.

Following an earlier exploratory meeting at Yale University in January 1977,[52] a group of some thirty-five behavioral and biomedical scientists was convened at the Institute of Medicine of the National Academy of Sciences in April 1978 for the purpose of forming an Academy of Behavioral Medicine Research. A direct outgrowth of that conference was the reformulation of an earlier proposed "official" definition of behavioral medicine:

> Behavioral medicine is the *interdisciplinary* field concerned with the development and *integration* of behavioral *and* biomedical science and techniques relevant to health and illness and the application of this knowledge and these techniques to prevention, diagnosis, treatment and rehabilitation.[53]

While only future events will disclose whether thirty years hence the president of the Academy of Behavioral Medicine Research will note in his or her presidential address the lack of any impact of behavioral medicine upon the practice of medicine, there are certain conceptual differences that do appear to exist between psychosomatic medicine and behavioral medicine.

In contrast to the primary focus in psychosomatic medicine upon personality factors and intrapsychic conflicts and their role in the etiology and pathogenesis of disease, behavioral medicine focuses primarily upon the overt behavior of the patient and upon modifying behavior as a means of preventing or treating the disease in question. Based upon the conditioning and learning experiments of I. P. Pavlov[45] and B. F. Skinner[55] and their extension by Joseph Wolpe[67] and A. A. Lazarus[35] into the clinical area, behavioral medicine at present does not focus upon the patient's verbal reports and free associations but rather upon the direct observation and quantification of the patient's overt behavior in real-life situations, followed by the application of the principles of learning theory and behavior modification, with the goal of changing the behaviors or pathophysiology that appear important in the initiation and maintenance of the disease process.

Even though psychosomatic medicine has focused primarily on etiological concerns, there has also been a continuing interest in issues related to treatment and intervention. Similarly, while the recent emergence of behavioral medicine stems largely from an increased awareness of, and interest in, the potential application of behavior modification techniques to prevention and treatment of physical disorders, there has also been evident a strong interest in the role of overt behaviors in the etiology and pathogenesis of physical disease. Consequently, it becomes necessary to address that area of behavioral medicine research that deals with the role of overt behavior in the etiology and pathogenesis of physical disease—with the main focus being on the relationship between the Type A behavior pattern and coronary heart disease. Bearing this in mind, it seems appropriate to review, in somewhat broader detail, the body of behavioral medicine research that deals with the use of behavior modification techniques to (1) change lifestyles and behavioral patterns that increase risk of developing physical disease and (2) modify directly pathophysiological mechanisms to treat physical disease.

¶ Review of Major Areas of Behavioral Medicine Research

Etiology and Pathogenesis

The most successful study of the role of psychosocial factors in the etiology of physical disease conducted within the psychosomatic medicine tradition was done by Herbert Weiner and his colleagues,[62] showing that among men physiologically predisposed by virtue of high serum pepsinogen levels subjected to the stress of army basic training, those who displayed a specific personality profile (high needs to be taken care of by others, frequent experience of frustration of these needs, and fear of expressing the resultant anger) were far more likely to develop active peptic ulcer disease than those not displaying the specific personality characteristics.

In contrast to this study of the role of personality in the etiology and pathogenesis of peptic ulcer disease, the research relating the Type A behavior pattern to coronary heart disease (CHD) has focused not upon the personality of research subjects but rather upon their overt behavior. As Ray Rosenman has noted: "Type A behavior pattern is . . . a style of *overt behavior* by which such individuals confront, interpret and respond to their life situations."[49] [Emphasis added.] Rosenman also notes that the appearance of such behavior depends upon the underlying personality of Type A individuals, the environmental demands with which they are confronted, and their interpretation of such demands. However, the assessment of the global Type A behavior pattern is based not upon inferences regarding such underlying personality characteristics but rather upon the "voice stylistics and psychomotor mannerisms" of the subject during the structured interview developed by Rosenman and Friedman.[48] Among the overt behaviors used to characterize subjects as Type A are rapidity of speech, explosive voice modulation, and expressions of anger or hostility.

Just as the validity of the hypothesis that the personality characteristics originally described by Franz Alexander[3] are involved in the etiology of peptic ulcer disease was confirmed in a prospective study,[62] the best evidence for the role of Type A behavior pattern in the etiology of CHD is to be found in a large-scale prospective study of over 3,000 middle-aged men by the Western Collaborative Group Study (WCGS).[50] The WCGS found that the approximately 1,500 men (free of signs of CHD at intake) who were Type A exhibited 2.37 times the rate of new CHD over an eight-and-one-half-year follow-up period as compared to their non-Type A, or Type B, counterparts. This increased CHD risk among Type A men remained highly significant even when the traditional risk factors (serum cholesterol level, cigarette smoking, and blood pressure) were statistically controlled.[14] Subsequent research has extended these findings to show that Type A patients have significantly more severe coronary atherosclerosis on arteriography even with control for traditional CHD risk factors.[11,27,68] Another line of research has shown that normal subjects exhibiting Type A behavioral characteristics also show heightened physiological[18] and neuroendocrine[28] responsivity when challenged behaviorally.

Those interested in learning more of the details of this body of research relating Type A behavior pattern to coronary heart disease are referred to a recent review volume edited by Theodore Dembroski and associates.[19] This body of research was recently reviewed by a diverse group of distinguished biomedical and behavioral scientists who were convened by the National Heart, Lung and Blood Institute (NHLBI), and who had not been directly involved in any of the Type A-related research themselves. The utility of the focus in behavioral medicine research upon overt behavior as it relates to etiology and pathogenesis was highlighted by one of the conclusions of the Review Panel:

> The Review Panel accepts the available body of scientific evidence as demonstrating that Type A behavior . . . is associated with an increased risk of clinically apparent coronary heart disease in employed, middle-aged U. S. citizens. This increased risk is over and above that imposed by age, systolic blood pressure, serum cholesterol and smoking and appears to be of the same order of magnitude as the relative risk associated with any of these other factors.[17]

The distinction between the conceptualization underlying the research on Type A behavior pattern and the primary focus upon intrapsychic factors in traditional psychosomatic medicine approaches to the study of etiology and pathogenesis was highlighted at the Timberline Conference on Psychophysiologic Aspects of Cardiovascular Disease in 1964.[58] Ray Rosenman's presentation of the then available data pertaining to the relationship between Type A behavior and coronary disease was strongly criticized as inadequate because it did not attempt to define "important traits in a personality" that could be related to disease processes.[58] Although it was to be a full decade before the first indications of the emergence of a distinct field of behavioral medicine, Rosenman's response to these criticisms was quite consistent with behavioral medicine's conceptual orientation as being primarily concerned with overt, observable behavior:

> We have not concerned ourselves with factors of motivation but only with determining the presence or absence of the *overt* pattern A, and I suspect that [those objecting to absence of concern with personality traits] are upset at this as well as our seeming oversimplification of inexact factors that are difficult to assess and even more difficult to quantitate. . . . [However] it is possible . . . to study different aspects of men that are tall and men that are short . . . without determining why they are tall or short. [p. 502][58] [Emphasis added.]

Prevention and Treatment

The almost explosive recent growth of interest in behavioral medicine stems from the hope that application of behavior modification approaches can be helpful in modifying risk-factor behaviors and lifestyles, which have been shown to increase the likelihood of developing diseases that are major public health problems in the United States today. An additional impetus for the recent emergence of behavioral medicine was the demonstration in the mid-1960s by Neal Miller and his coworkers[43] that the application of instrumental conditioning techniques, or biofeedback, could be successful in directly modifying physiological functions that previously had been thought to be beyond voluntary control. While there continues to be much controversy regarding the precise nature of the mechanisms whereby such control is achieved by human subjects, Miller's initial demonstration has spawned a new clinical specialty whereby biofeedback and other behavior therapy techniques are employed[65] on an ever-increasing scale in the direct treatment of a wide range of physical disorders that had previously proven to be quite resistant to traditional treatment approaches in clinical medicine. Representative research findings in these two areas of "applied" behavioral medicine will now be reviewed.

Behavioral Medicine and Prevention of Disease

There is probably no better case to be made for the need for better means of helping people to change risk-conferring behaviors than the fact that more than 50 million Americans continue to smoke[44] despite the massive public education campaigns that fairly shout the well-known increased risk of both lung cancer and coronary heart disease among cigarette smokers.[2] Even more disturbing is the observation that, although among adults the rate of smoking has decreased, there has been less of a decline in smoking rates among teenagers and even an increase among female teenagers.[59] Moreover, the onset of smoking is occurring at an earlier age[24] and the number of cigarettes smoked is increasing among those who do smoke.[26] This suggests that the introduction of low-tar and -nicotine brands has resulted in increased consumption among addicted smokers to compensate for the loss of nicotine.

As it became clear that educational efforts alone, even when incorporating techniques of fear arousal (for example, gory slides of emphysematous lungs) were ineffective,[7] behavioral scientists began to devote more effort toward the study of smoking behavior. It is noteworthy that this attention focused not on personality characteristics that predisposed people to smoke but rather on the various consequences of smoking and the role of social forces influencing the adoption of smoking behavior. Thus, these efforts fall clearly within the definition of behavioral medicine. One observation has been that while the long-term outcomes of smoking (lung cancer, heart attack) are clearly bad, the short-term benefits can be quite positive (stress reduction, good taste after a meal, facilitating communication and togetherness with others, and so on). This leads to a greater "subjective expected utility"[40] in continuing the addiction rather than quitting and going through the unpleasantness of breaking the habit. Numerous studies[24] have also demonstrated the importance of social influence in terms of the modeling of smoking behavior by peers, parents, siblings, and significant others perceived by children as role models (for example, teachers and celebrities).

In evaluating various early attempts of the application of behavior modification approaches (for example, electrical aversion, counterconditioning, loss of a cash deposit) to the problem of smoking cessation, R. M. McFall and C. L. Hammen[41] found that while any program could achieve abstinence in all subjects at the end of the treatment program, after six to twelve months only an average of 13 percent of participants were not smoking. More recently, the "rapid smoking" technique developed by E. Lichtenstein[36]—having the subject smoke at a rate of one inhalation every six seconds in a darkened unpleasant room—has been reported to have much higher abstinence rates on long-term follow-up than the disappointing 13 percent found in earlier efforts. Although the rapid smoking technique has obvious limitations insofar as application to the most important target populations for smoking cessation is concerned (those with lung and heart disease), it continues to be one of the most seriously considered behavioral approaches for those already addicted to smoking.[25]

In his extensive review of behavioral medicine research related to cigarette smoking, Richard Evans[25] notes a number of problems in the extensive efforts that have been made to demonstrate the efficacy of behavior modification approaches to smoking cessation in the addicted smoker. First, the early high-success rates reported for the rapid smoking technique as well as for other innovative behavior modification approaches (including operant paradigms involving self-monitoring of smoking behavior, stimulus control, and systems of self-reward and punishments) have not been confirmed in subsequent studies, particularly those where longer follow-ups have been included. Thus, the "true" rate of long-term abstinence with the best of these programs appears to be in the range of 25 to 30 percent, clearly better than the early results, but not as promising as was hoped initially. Other problems with the controlled outcome studies relate to the fact that all subjects in them are volunteers and, hence, not representative of those in the addicted smoking population who do not volunteer. Furthermore, the practice of including in the success rates only those subjects who complete the program inflates the success rates by ignoring the dropouts. Based upon his review of the results of the smoking cessation literature, Evans concludes that, "For the health professional who is asked to recommend or even judge a program there is very little basis for favoring one program over another. In fact, it would be difficult with any degree of confidence to recommend any program at all."[25]

These disappointing results have led to a general conclusion that since the resources available for the solution of the cigarette smoking problem are limited, greater efforts could be more profitably directed toward "influencing pre-addictive smokers to curtail the incidence of smoking before they become addicted or nicotine-dependent, or to focus on preventing individuals from beginning to smoke in the first place."[25] Research in this area clearly demonstrates the inadequacy of depending only on educational efforts and fear arousal; teenagers and preteens must be taught social skills that will enable them to cope with the many pressures to smoke to which they are subjected by the media, peers, family, and role models. Whether this approach, based on principles of applied social psychology, will achieve the goal of reducing over the next decades the proportion of preteens and teenagers who smoke is something that only time will tell. Preliminary studies suggest that such a "social inoculation" approach can be utilized effectively to deter the onset of addictive smoking in junior high school students.[24]

While not intended to be exhaustive, this review of behavioral medicine research related to smoking cessation does permit us to make several generalizations concerning the core characteristics of the behavioral medicine approach to a typical public health problem. The first key ingredient is a *behavioral assessment* that attempts to identify the environmental influences that play a role in the initiation, promotion, and maintenance of the behavior in question. Second is the *application of behavioral science knowledge and techniques* (in this case, derived from behavior therapy and social psychology) in attempts to prevent or reduce the behavior in question. Finally, and equally important, is a data-based *evaluation of the outcome* of the intervention employed, on both a short- and long-term basis. When this evaluation suggests that one approach (for example, behavior modification) is not having the desired effect, then further evaluation of the reasons for such failure is carried out in an attempt to identify other approaches that may offer a greater chance of success (for example, inoculation strategies to cope with peer pressures). In each case it is overt, manifest behavior that is the focus of the assessment, the intervention, and the evaluation of that intervention—thus placing this type of endeavor squarely within the mainstream of behavioral medicine.

Similar reviews could be presented here concerning the behavioral medicine approach to a variety of other risk-inducing lifestyles and behaviors, but this would not add to the general conclusions or illustrative impact of what has already been presented. The interested reader is referred to several excellent recent reviews of the behavioral medicine approach to problems of eating behavior,[56] compliance with therapeutic regimens,[30] and coronary-prone (Type A) behavior.[19]

Behavioral Medicine and Treatment of Disease

The initial demonstrations provided by basic psychophysiological research that so-called autonomic functions could be brought under voluntary control through the use of biofeedback techniques has led to the widespread use of biofeedback and other behavioral techniques in the direct treatment of a wide variety of medical disorders. As with the behavioral medicine approaches to lifestyle modification, an exhaustive review of the literature in this area of behavioral treatment approaches would far exceed the scope of this chapter. Several recent comprehensive reviews have appeared describing the research in this area.[46,65] For purposes of illustration, the present review will focus on behavioral medicine approaches to (1) neuromuscular reeducation, (2) treatment of muscle contraction headaches, (3) treatment of Raynaud's disease, and (4) treatment of idiopathic insomnia.

Before proceeding with this review, some general introduction is necessary to place the behavioral medicine approach to treatment in proper perspective. As previously noted, the main stimulus for attempting the direct treatment of various disorders was the exciting early work showing the effectiveness of biofeedback techniques in achieving changes in physiologic functions previously thought to be beyond the control of instrumental learning techniques. Much paper and ink (not to mention laboratory time) has been expended in the past decade describing studies comparing biofeedback with other, noninstrumented techniques in the treatment of various disorders. Almost without exception these studies[10] have found that various relaxation approaches, including autogenic training,[51] progressive muscle relaxation,[31] and various forms of meditation,[6] are as effective in reducing symptoms as biofeedback. This suggests that general relaxation, rather than some specific therapeutic mechanism, is a key ingredient in the therapeutic efficacy of these various techniques. It should be noted that in virtually all these studies of clinical efficacy, the behavioral treatments employed, whether biofeedback or some form of relaxation exercise, have been found more effective in reducing symptoms than has a waiting-list control group or a group given some form of attention placebo control treatment. For purposes of this review of clinical applications of behavioral treatment of physical disorders, the issue of whether biofeedback or some form of relaxation training is better is not particularly relevant. The data reviewed to date[10] indicate that where controlled-outcome studies have been conducted, both biofeedback and relaxation approaches are about equally effective and better than no treatment at all. These studies also suggest strongly that whatever means are employed to reduce muscle tension or autonomic activity, regular home practice is essential for the realization of maximum symptomatic improvement. Finally, there is some indication that where neurotic conflicts are present, they can interfere with attempts to modify physiological responses that are responsible for maintaining symptoms. There are, in fact, arguments for dealing with such underlying conflicts through the use of more dynamic interpretations in order to achieve symptom relief with the behavioral approaches.[47,63,64] With these qualifications in mind, we may now turn to the more detailed review of the disorders where strong evidence exists for the efficacy of behavioral treatment approaches.

Neuromuscular Reeducation

In contrast to the other disorder types to be reviewed here, the area of neuromuscular reeducation appears to represent an example of electromyograph (EMG) biofeedback as a *specific* treatment. Following the early case reports of Marinacci and Horande,[39] a number of systematic studies have convincingly documented the therapeutic efficacy of EMG biofeedback training in the rehabilitation of patients with a wide variety of disorders of neuromuscular function, including upper extremity paralysis,[4] lower extremity paralysis,[5,15,32] and spasmodic torticollis.[15] This has led Blanchard to conclude that: "Overall, it seems well established that EMG biofeedback can be a very useful adjunct to standard rehabilitation therapy with many neuromuscular disorders.[10]

Muscle Contraction Headache

There are now in the literature numerous reports of controlled outcome studies[63] showing that EMG biofeedback, some form of relaxation training, or a combination of the two are effective in reducing the frequency and severity of muscle contraction headaches. This represents one of the most gratifying treatment areas for the behavioral medicine clinician, since even patients with a long history of debilitating headaches requiring daily narcotic treatment can be brought to a relatively headache-free state within two to three weeks of treatment using any of the techniques typically employed to achieve reduced muscle tension. Two studies[1,42] have reported long-term follow-up data on patients with muscle contraction headaches or with mixed migraine and muscle contraction headaches. Both studies suggest that even after periods ranging in length from six months to five years, anywhere from 34 to 60 percent of patients treated behaviorally still show significant clinical improvement.

RAYNAUD'S DISEASE

Compared to the other disorders covered in this brief overview, Raynaud's disease probably accounts for much less overall suffering and financial cost on the part of those who have it. Nevertheless it represents an excellent model for evaluating behavioral treatment approaches in that (1) patients do experience discomfort and, hence, should be motivated to comply to behavioral regimens; (2) the attacks are circumscribed with respect to stimulus situations (cold weather) that elicit them and determine their frequency and severity; and (3) a logical mechanism can be postulated (try to decrease sympathetic nervous activity) whereby behavioral approaches can be employed to decrease the activity of the disease process.

Based upon this reasoning, Richard Surwit and colleagues[57] have conducted a well-designed controlled outcome study evaluating the response of disease activity (frequency and severity of vasospastic episodes) to a combination of autogenic training and finger-temperature biofeedback with frequent home practice of handwarming strategies among patients with Raynaud's disease during a severe northern winter season. The design of this study included several important features. First, the physiologic mechanisms underlying the disease and its response to the treatment were evaluated by exposing subjects to a controlled cold stress and determining the amount of decrease in finger temperature produced by this maneuver before and after treatment. Second, both an active treatment group and a wait-list control group were evaluated in terms of both disease activity and vasomotor response to the standard cold-room stress exposure. Compared to pretreatment levels, the active treatment group showed a significant decrease in frequency of attacks and a nearly significant decrease in severity of attacks following treatment. Documenting a physiologic mechanism for this symptomatic improvement, Surwit found that the active treatment group were able to maintain as high as 4°C warmer finger temperature during the posttreatment cold stress compared to the pretest levels. In contrast, repeat testing of the wait-list controls showed no difference in their finger-temperature response. After a course of active handwarming training, however, the wait-list controls were able to maintain a significantly warmer finger temperature on cold exposure following training, comparable to that of the first active treatment group. Thus, the symptomatic improvement in Raynaud's disease activity was paralleled by observable improvement in the subjects' ability to maintain a warmer hand temperature on cold exposure. By showing a demonstrable modification in pathophysiologic responsivity in association with behavioral training, Surwit's group has taken an important step forward in providing evidence for the scientific basis of the behavioral treatment approach in the clinical improvement of disease activity. In further studies employing this paradigm, Surwit has found preliminary evidence of neuroendocrine correlates of reduction in disease activity with behavioral treatment, as well as having evaluated the joint effect on disease activity of both the behavioral treatment and a pharmacologic intervention (intraarterial reserpine).* The application of this approach to the evaluation of behavioral treatment for other disorders should help considerably to establish the scientific basis and credibility of behavioral medicine approaches to treatment of physical disorders.

*Personal communication.

INSOMNIA

Sleep-onset insomnia is a very common complaint in general medical practice and probably accounts for a substantial proportion of the very high level of prescriptions for minor tranquilizers and soporifics, not to mention the record sales of over-the-counter sleep aids. Paradoxically, the chronic use of sleep medications is probably the most common cause of chronic sleep-onset insomnia.[20] Thus, the use of medications whose action is sleep induction probably has no place in the long-term management of the patient who experiences difficulty in falling asleep. (It is important to note that the following discussion is confined only to idiopathic sleep-onset insomnia.) Where there is some underlying biologic cause of the sleep disturbance, such as depression, sleep apnea, or rhythmic leg twitches,[21] the appropriate intervention must first address the underlying biologic problem. Where recent life events are responsible for distress, which interferes with falling asleep, counseling efforts should also be addressed to helping improve the patient's coping ability, as well as to the use of appropriate behavioral approaches.

When careful initial evaluation establishes the diagnosis of idiopathic sleep-onset insomnia (complaints of difficulty falling asleep, six or less hours of sleep per night, and daytime drowsiness in the absence of any of the previously mentioned biologic causes), the two behavioral approaches of relaxation training and stimulus-control procedures have both been shown to be remarkably effective treatment.

Thomas Borkovec[13] favors the use of a modified version of Edmund Jacobsen's[31] progressive muscle relaxation training as a means to provide the patient with skills at achieving a quiet state of low arousal, which is felt to be conducive to sleep onset. Borkovec reports[13] that in six studies of over 250 sleep-disturbed subjects, who followed continued home practice and honing of relaxation skills, progressive relaxation training was found to be significantly superior to both no-treatment and several placebo control conditions in reducing sleep-onset time. Follow-up evaluation of these subjects at four months and one year showed an average reduction in sleep onset time from forty-one minutes down to only nineteen minutes—bringing the treated group within the latency reports for the majority of the normal population.

Based upon learning theory considerations, Richard Bootzin[12] made the assumption that persons with sleep-onset insomnia may well have learned to associate bed-related stimuli with sleep-incompatible responses, such as reading, watching television, worrying about the day's events, or planning the next day's activities—thus leading to the circumstance that exposure to bed-related cues increases the probability of sleep-incompatible behaviors, with a corresponding reduction in the probability of sleep. With this in mind, Bootzin has proposed the following "behavioral prescription" for the treatment of sleep-onset insomnia:

1. Lie down only when you feel sleepy.
2. Set your alarm clock for the same time and get up at that time every morning regardless of how much sleep you obtained the night before.
3. Avoid daytime naps.
4. Use the bed and bedroom only for sleeping. Do not engage in other activities (e.g., eating, studying, reading, watching tv) in your bed or bedroom.
5. If you do not fall asleep within ten minutes, leave the bedroom immediately, and return to bed only when you feel sleepy again.
6. Repeat step #5 as often as necessary during the night until rapid (i.e., within ten minutes) sleep onset occurs.[12]

It is felt that with sufficient practice of these behaviors the association between bed cues and sleep grows stronger, while a corresponding decrease in the strength of the relation between bed cues and sleep-incompatible behaviors makes the probability of sleep that much greater. Bootzin[12] reported that the use of this approach resulted in an average decrease of seventy-four minutes in sleep-onset time among insomniacs requiring over ninety minutes to fall asleep prior to treatment; 61 percent of the patients averaged less than twenty minutes to sleep onset after treatment.

In view of the high prevalence of sleep-onset insomnia as indexed by the very high rates of use of pharmacologic agents to induce sleep, it is somewhat surprising that these highly effective behavioral treatment approaches are not employed more widely. Bootzin's stimulus-control procedure is particularly well-suited for use in a primary care setting, where the practitioner can simply write out the behavioral prescription (or have it on a printed form) and hand it to the patient with a brief explanation, just as if it were a prescription for a tranquilizer or sleeping pill.

As with the review of behavioral medicine approaches to modification of risk-factor behaviors, this review of behavioral medicine approaches to treatment of physical disorders is not intended to be exhaustive. It is meant to illustrate general principles underlying behavioral approaches to medical treatment, as well as those disorders for which particularly good evidence is available regarding the efficacy of behavioral treatment methods. For a more comprehensive review of this area of treatment of physical disorders, the interested reader is encouraged to refer to any of several recent review volumes.[46,65]

¶ Behavioral Medicine: Future Directions

Behavioral medicine has a great potential to contribute significantly to the understanding of the role of psychosocial factors and behavior in the etiology and pathogenesis of disease and to the prevention, treatment, and rehabilitation of a wide variety of physical disorders. As was noted, however, with regard to behavior modification approaches to cigarette-smoking cessation, the realization of this potential is not automatically assured. To promise too much, to lead the medical community to expect miracles that we cannot deliver, would be to sow the seeds of a future disillusionment that could lead to an unfortunate delay in achieving the realistic potential contributions of behavioral medicine. As with the behavioral medicine approaches to smoking cessation, it is essential that we continue to evaluate our efforts in a scientifically rigorous fashion, and we must be ready to acknowledge our failures and to modify our approaches based on the results of such evaluations. If the full potential contributions of behavioral medicine are to be realized, it is of prime importance that we follow Neal Miller's[43] injunction to "be cautious in what we claim and bold in what we try."

Another issue to be regarded as critical for the future development of behavioral medicine is the avoidance of false dichotomies. It is often tempting for those who focus primarily on overt behaviors to view with at least a certain amount of contempt efforts to explain pathogenesis of disease in terms of such intrapsychic constructs as personality or neurotic conflicts, which can only be inferred rather than observed and quantified directly. By the same token, as was noted earlier with regard to early criticisms of the Type A behavior pattern concept, it is often hard for those who focus primarily on psychodynamically oriented issues to escape the impression that more behaviorally oriented efforts to explain pathogenesis are only scratching the surface and not getting at the real, "underlying" personality issues. Might it not be wiser, however, that workers with both orientations consider the notion that the two emphases are not in conflict with one another but rather are really complementary? Surely the needs, motivations, and other intrinsic predispositions of the individual play some role in determining the overt behavior that will be displayed in any given environmental situation. Conversely, those same intrapsychic factors will exert their influence upon disease processes through the overt behavior (not just conceptualized as psychomotor, but also as physiologic and neuroendocrine) of the individual in certain environmental situations. For example, it was found that the overt Type A behavior pattern was associated with increased levels of coronary atherosclerosis on arteriography.[11] However, it was also found that a psychometrically measured personality construct, hostility, is equally and independently predictive of coronary atherosclerosis.[66] This suggests that a realization of the complementary relationship between intrapsychic and overt behavioral phenomena, and a willingness to incorporate both orientations in our attempts to understand the role of psychosocial factors in the etiology and pathogenesis of disease, will offer the greatest chances of success. To document further the need for and desirability of integration of psychodynamically and behaviorally oriented approaches, it is necessary to consider in some detail the relevance of behavioral medicine for consultation-liaison psychiatry.

Complementarity Between Behavioral Medicine and Consultation-Liaison Psychiatry

Consultation - liaison psychiatry has emerged over the years as that branch of psychosomatic medicine that is primarily concerned with issues of treatment within the clinical medicine setting. A. J. Krakowski defines consultation-liaison psychiatry as "the services which psychiatrists render outside of the psychiatric departments in the general hospitals."[34] He goes on to note that these services are rendered to help nonpsychiatric physicians care for patients with primary psychiatric disorders or whose psychiatric problems "interfere with, complicate or stem from somatic illness." Z. J. Lipowski has reviewed the kinds of psychiatric problems that are commonly seen in the setting of physical illness.[37] In addition to this consultative role, the consultation-liaison psychiatrist also has a liaison function: to educate consulting physicians and other members of the health care team about referred patients so that they can better manage such problems themselves in the future.[34] In fulfilling these roles, the consultation-liaison psychiatrist is generally seen as one who evaluates and treats psychiatric/psychosocial disorders and problems encountered in a medical (as compared to a psychiatric) setting. Traditional psychiatric approahces are employed, including psychotherapy, psychopharmacologic agents, and working with family and other key people in the patient's environment.

In contrast to consultation-liaison psychiatry's primary focus on the evaluation and treatment of psychopathology occurring in the setting of physical illness, biofeedback and other behavioral medicine treatment techniques have generally been directed primarily toward the relief or amelioration of the physical symptoms associated with physical illness per se. David Shapiro and Richard Surwit[54] have pointed out that these behavioral medicine approaches differ from the consultation-liaison psychiatric approach in two ways. First, they focus mainly on the "specific physiological problem presented by the patient" rather than on a hypothesized psychodynamic conflict that may underlie it. Second, treatment is directed toward achieving a specific change in disturbed physiology through instrumental learning rather than by any approach to the underlying psychodynamic conflicts.

The obvious conclusion from these descriptions of consultation-liaison psychiatry and behavioral medicine approaches to patients with physical illness is not that they are in conflict with each other, or that to agree with the tenets of one is to reject the principles of the other, but rather that the two approaches are, more than anything else, different. One aims at evaluating and reducing psychological distress, with the outcome often a reduction in physical distress (that is, pain); the other aims at evaluating and reducing physical distress, with the outcome often a reduction in psychological distress (that is, anxiety and depression). The word that best describes the relation between the consultation-liaison psychiatry and behavioral medicine approaches is "complementarity." Webster's defines "complementarity" as the quality of being complementary and goes on to define complementary as "serving to fill out or complete . . . mutually supplying each other's lack."[60] Thus, not only are the approaches of behavioral medicine and consultation-liaison psychiatry different, but each has the capacity to supply something that is lacking in the other.

Even though many patients with physical symptoms related to physical illness or psychophysiologic disorder will be helped by an approach that focuses exclusively on instrumentally modifying the physiologic basis of symptoms, there is a significant proportion of patients in whom underlying neurotic conflicts are such that they will be either unable or unwilling to participate in their treatment to the extent that they fail to achieve control over their physiologic function and/or symptoms. William Rickles[47] has spoken cogently of the problem of resistance to biofeedback apparatus. Also, there are other detailed descriptions[63,64] of the various ways in which underlying psychologic conflicts can prevent patients from responding to biofeedback therapy of somatic disorder, as well as of the "psychological complications" that can surface when somatic symptoms are reduced or removed in patients with severe underlying psychodynamic conflicts. Besides the advantages of addressing issues related to those psychodynamic conflicts that may prevent or complicate patients' benefiting from biofeedback therapy, it is felt that many patients with organic pain syndromes, such as low back pain, will achieve better pain control if treated with combination phenothiazine/tricyclic antidepressant therapy and EMG biofeedback-assisted-relaxation training than they will with biofeedback training alone. Again, such psychopharmacologic approaches are more often found within the province of the consultation-liaison psychiatrist than of the typical behavioral medicine practitioner. Indeed, since many working in behavioral medicine are Ph.D. psychologists, if patients are to receive the potential benefits of the above psychopharmacologic approach it is essential to have the input of the consultation-liaison psychiatrist, or some other doctor-clinician.

If these are some of the ways in which the consultation-liaison psychiatrist's approach can complement that of the behavioral medicine practitioner, how can the latter's approach help the consultation-liaison psychiatrist? One way is by making easier the insight-oriented psychotherapist's task of overcoming resistances in patients with somatic disorders. As is well-known, such patients can be quite resistant to the notion that they have any psychological problems, and they may react to the presence of a psychiatrist with anger at the implicit threat that a (medical) doctor thinks the patient is a mental case and that the pain is all in the patient's head. Such patients will often accept (if it is presented properly) the proposition that their symptom is due to real "physical" causes. For example, patients suffering from tension headache can be told that their head and neck muscles are too tight and that in order to obtain relief they will have to learn how to reduce that excess muscle contraction with the aid of a "scientific" apparatus, the biofeedback machine. If the underlying psychological problem is sufficiently severe, it may be that the patient will have difficulty learning to reduce EMG levels, and in that context will become more receptive to interpretations of underlying emotional conflicts.[64] In fact, the patient will often spontaneously report that whenever he thinks about a certain area (for example, what his mother said to him last week) he notices that the EMG feedback signal shows an increase. Thus, biofeedback approaches can often serve to prepare patients with psychological problems to enter psychotherapy who otherwise might have continued to focus only on some physical symptom and resist entering psychotherapy until the physical symptom was gone.

Another area where behavioral medicine approaches can be of help to consultation-liaison psychiatry is in narrowing the gap that Donald Lipsett[38] has noted still exists between psychiatry and the rest of medicine, despite the advances made by consultation-liaison psychiatry in recent years. First of all, by directly treating the physical disorder itself, behavioral medicine approaches not only complement the help provided by the consultation-liaison psychiatrist for psychiatric problems occurring in association with physical disorders but at the same time cannot help but impress the nonpsychiatric physician with the scientific legitimacy of a treatment modality that successfully alleviates physical symptoms which may have been resistant to his best efforts within a strictly biomedical model. By incorporating behavioral medicine approaches, consultation-liaison psychiatry can thus enhance its position vis-à-vis the rest of medicine in terms of being able to intervene directly to affect physical symptoms. As a result, it seems not unreasonable to assume that the nonpsychiatric physician will also exhibit an increased acceptance of the more traditional focus of the consultation-liaison psychiatrist upon psychopathology. Thus, consultation-liaison psychiatry, by adding behavioral medicine approaches to its usual treatment approaches, might find that not only is it able to successfully deal with a wider variety of problems "outside of the psychiatric departments of the general hospitals,"[34] but also that the personnel in those settings are more receptive to, and appreciative of, those things the consultation-liaison psychiatrist has been doing well all along.

To illustrate how some of these mutual benefits might be achieved by behavioral medicine clinicians and consultation-liaison psychiatrists working in hospital settings, it will be helpful to briefly describe how a biofeedback treatment facility was set up at Duke University, paying attention to its evolution and to the experimenters' plans for future integration between Duke Behavioral Physiology Laboratory and the Consultation-Liaison Service.[64] It was found that many of the patients referred for biofeedback treatment had underlying psychodynamic conflicts, which either prevented their being able to benefit from biofeedback therapy or complicated their treatment despite a positive response of physical symptoms. Naturally, an attempt was made to identify and deal with these psychodynamic issues. However, since the primary orientation was toward the use of behavioral techniques and not toward long-term psychotherapy, both clinicians and patients found it helpful if treatment concentrated on patients in whom psychodynamic conflicts were readily evident (or who were not learning to control physiological function using biofeedback). Such evaluation was done by psychiatrist colleagues. This often led to a joint effort—with the Behavioral Physiology Laboratory focusing on reduction of physical symptoms and the consultation-liaison psychiatrist focusing on psychodynamic issues; this effort benefited the patient. This joint approach was so successful—in terms of both patient benefit and the clinicians' own sense of doing a better job—that the chief of the consultation-liaison service now attends Behavioral Physiology Laboratory's weekly clinical case conference. Not only do the behavioral therapists appreciate and benefit from his viewpoint in understanding patient problems, but he also gains an appreciation of behavioral approaches that he can then use to enhance his liaison activities.

There are plans to incorporate this joint approach with training activities as well as with the clinical service aspect of the programs. Psychology interns rotate through the Behavioral Physiology Laboratory and receive supervision in using biofeedback and other behavioral techniques in the evaluation and treatment of patients with physical disorders. Psychiatry residents rotate through the Consultation-Liaison Service and receive supervision in the evaluation and treatment of psychiatric problems arising in the nonpsychiatric wards of Duke University Medical Center. Additionally, there are plans to form teams, consisting of a psychology intern and a psychiatric resident, that will jointly evaluate and formulate a treatment plan for patients referred for biofeedback therapy of somatic symptoms. Through joint supervision of this clinical activity and through this joint participation in the Behavioral Physiology Laboratory clinical case conference, it is hoped that this training program will achieve a model for the integration of the behaviorally and psychodynamically oriented approaches to patients referred for treatment of physical symptoms.

It is possible that both behavioral and psychodynamic "purists" may take issue with what has been said herein, asserting that to the extent that each other's viewpont has been incorporated, this review has been a waste of the author's and the patient's time. Of course, one must realize that most clinicians, whether behaviorally or psychodynamically oriented, do not take such extreme views. But to those who might—and to emphasize the point of what has been said—it bears repeating that an integrative approach, incorporating what is useful in both approaches, will not only serve to help us all care better for our patients, but it will also serve to achieve the goals of both behavioral medicine and consultation-liaison psychiatry more rapidly and to a greater extent than will ever be possible in the absence of such an integrative team approach.

¶ Bibliography

1. ADLER, C. S., and ADLER, S. M. "Biofeedback-Psychotherapy for the Treatment of Headaches: A 5-Year Follow-Up," *Headache,* 16 (1976): 189–191.
2. ADVISORY COMMITTEE TO THE SURGEON-GENERAL. *Smoking and Health.* Washington, D.C.: U.S. Public Health Service, 1964.
3. ALEXANDER, F. "The Influence of Psychologic Factors Upon Gastrointestinal Disturbances: General Principles, Objectives and Preliminary Results," *Psychoanalytic Quarterly,* 3 (1934): 501–522.
4. ANDREWS, J. M. "Neuromuscular Re-Education of the Hemiplegic with the Aid of the Electromyograph," *Archives of Physical Medicine Rehabilitation,* 45 (1961): 530–532.
5. BASMAJIAN, J. V., et al. "Biofeedback Treatment of Footdrop Compared with Standard Rehabilitation Technique: Effects on Voluntary Control and Strength," *Archives of Physical Medicine Rehabilitation,* 56 (1975): 231–236.
6. BENSON, H. *The Relaxation Response.* New York: William Morrow, 1975.
7. BERNSTEIN, D. A., and BORKOVEC, T. D. *Progressive Relaxation Training.* Champaign, Ill.: Research Press, 1973.
8. BINGER, C. "Editorial Note—Plans and Policy," *Psychosomatic Medicine,* 10 (1948): 71–72.
9. ———. "Editorial Note," *Psychosomatic Medicine,* 20 (1958):343.
10. BLANCHARD, E. B. "A Data-Based Review of Biofeedback," in Brady, J. P., and Brodie, H. K. H., eds., *Controversy in Psychiatry.* Philadelphia: W. B. Saunders, 1978, pp. 489–517.
11. BLUMENTHAL, J. A., et al. "Type A Behavior Pattern and Coronary Atherosclerosis," *Circulation,* 58 (1978): 634–639.
12. BOOTZIN, R. "Stimulus Control of Insomnia." Paper presented at the meeting of the American Psychological Association, Montreal, Agust, 1973.
13. BORKOVEC, T. D., and BOUDEWYNS, P. A. "Treatment of Insomnia by Stimulus Control and Progressive Relaxation Procedures," in Krumboltz, J. and Thoresen, C. E., eds., *Behavioral Counseling Methods.* New York: Holt, Rinehart, and Winston, 1976, pp. 328–344.
14. BRAND, R. J., et al. "Multivariate Prediction of Coronary Heart Disease in the Western Collaborative Group Study, Compared to the Findings of the Framingham Study," *Circulation,* 53 (1976): 348–355.
15. BRUDNY, J., et al. "EMG Feedback Therapy: Review and Treatment of 114 Patients," *Archives of Physical Medicine Rehabilitation,* 57 (1976): 55–61.
16. COOPER, T. "Coming National Policy on Prevention of Heart Diseases." Speech presented to Medical Dietetic Symposium, University of Arizona, March 23, 1973.
17. "Coronary-Prone Behavior and Coronary Heart Disease: A Critical Review." Conference sponsored by the National Heart, Lung and Blood Institute, Amelia Island, Fla. December, 1978.
18. DEMBROSKI, T. M., MCDOUGALL, J. M., and SHIELDS, J. L. "Psychologic Reactions to Social Challenge in Persons Evidencing the Type A Coronary-Prone Behavior Pat-

tern," *Journal of Human Stress,* 3 (1977): 2–9.

19. DEMBROSKI, D. M., et al., eds., *Coronary-Prone Behavior.* New York: Springer-Verlag, 1978.
20. DEMENT, W. C. *Some Must Watch While Some Must Sleep.* Stanford: Stanford Alumni Association, 1972.
21. ———. "Introduction to Sleep and Sleep Disorders." Symposium paper presented at the meeting of the Association for the Advancement of Behavior Therapy, San Francisco, 1979.
22. THE EDITORS, "Introductory Statement," *Psychosomatic Medicine,* 1 (1939):3–5.
23. ENGEL, G. L. "The Need for a New Medical Model: A Challenge for Biomedicine," *Science,* 196 (1977):129–136.
24. EVANS, R. I. "Smoking in Children: Developing a Social-Psychological Strategy of Deterrence," *Journal of Preventive Medicine,* 5 (1976):122–127.
25. ———, et al. "Current Psychological, Social and Educational Programs in Control and Prevention of Smoking: A Critical Methodological Review." Paper presented at Annual Meeting, Academy of Behavioral Medicine Research, Snowbird, Utah, June 1979.
26. FISHBEIN, M. "Consumer Beliefs and Behavior with Respect to Cigarette Smoking: A Critical Analysis of the Public Literature." Report prepared for the staff of the Federal Trade Commission, 1967.
27. FRANK, K. A., et al. "Type A Behavior and Coronary Artery Disease: Angiographic Confirmation," *Journal of the American Medical Association,* 240 (1978): 761–763.
28. FRIEDMAN, M., et al. "Plasma Catecholamine Response of Coronary-Prone Subjects (Type A) to a Specific Challenge," *Metabolism,* 24 (1975):205–210.
29. GRAHAM, D. T. "What Place in Medicine for Psychosomatic Medicine?" *Psychosomatic Medicine,* 41 (1979):357–367.
30. HAYNES, R. B., TAYLOR, D. W., and SACKETT, D. L. *Compliance in Health Care.* Baltimore: The Johns Hopkins University Press, 1979.
31. JACOBSON, E. *Progressive Relaxation.* Chicago: University of Chicago Press, 1978.
32. JOHNSON, H. E., and GARTON, W. H. "Muscle Re-Education in Hemiplegia by Use of EMG Device," *Archives of Physical Medicine Rehabilitation,* 54 (1973):320–325.
33. KENNEDY, E. M. Special lecture delivered at the National Institutes of Health, Bethesda, Md., April 3, 1978.
34. KRANKOWSKI, A. J. "Consultation-Liasion Psychiatry: A Psychosomatic Service in the General Hospital," in Lipowski, Z. J., Lipsett, D. R., and Whybrow, P. C., eds., *Psychosomatic Medicine: Current Trends and Clinical Applications.* New York: Oxford University Press, 1977, pp. 564–573.
35. LAZARUS, A. A. "New Methods in Psychotherapy: A Case Study," *South African Medical Journal,* 32 (1958):660–664.
36. LICHTENSTEIN, E., et al. "Comparison of Rapid Smoking, Warm Smoky Air and Attention Placebo in the Modification of Smoking Behavior," *Journal of Consulting and Clinical Psychology,* 40 (1973): 92–98.
37. LIPOWSKI, Z. J. "Physical Illness and Psychopathology," in Lipowski, Z. J., Lipsett, D. R., and Whybrow, P. C., eds., *Psychosomatic Medicine: Current Trends and Clinical Applications.* New York: Oxford University Press, 1977, pp. 172–186.
38. LIPSETT, D. R., "Some Problems in the Teaching of Psychosomatic Medicine," *in* Lipowski, Z. J., Lipsett, D. R., and Whybros, P. C., eds., *Psychosomatic Medicine: Current Trends and Clinical Applications.* New York: Oxford University Press, 1977, pp. 599–611.
39. MCFALL, R. M., and HAMMEN, C. L. "Motivation, Structure, and Self-Monitoring: Role of Nonspecific Factors in Smoking Reduction," *Journal of Consulting and Clinical Psychology,* 37 (1971): 80–86.
40. MARINACCI, A., and HORANDE, M. "Electromyogram in Neuromuscular Re-Education," *Bulletin Los Angeles Neurological Society,* 25 (1960):57–71.
41. MAUSNER, B. "An Ecological View of Cigarette Smoking," *Journal of Abnormal Psychology,* 81 (1973):115–126.
42. MEDINA, J. L., DIAMOND, S., and FRANKLIN, M. A. "Biofeedback Therapy for Migraine," *Headache,* 16 (1976):115–118.
43. MILLER, N. E. "Learning of Visceral and Glandular Responses," *Science,* 163 (1969): 434–445.
44. OFFICE OF CANCER COMMUNICATIONS. *The Smoking Digest: Progress Report on a*

Nation Kicking the Habit. Bethesda, Md.: National Cancer Institute, 1977.
45. PAVLOV, I. P. *Conditioned Reflexes.* London: Oxford University Press, 1977.
46. POMERLEAU, O., and BRADY, J. P., eds. *Behavioral Medicine: Theory and Practice.* Baltimore: Williams & Wilkins, 1979.
47. RICKLES, W. "Symptom Substitution and Biofeedback Apparatus as an Inanimate Transference Object in Borderline Patients," *Psychiatric Annals,* in press.
48. ROSENMAN, R. H. "The Interview Method of Assessment of the Coronary-Prone Behavior Pattern," in Dembroski, T. M., et al., eds., *Coronary-Prone Behavior.* New York: Springer-Verlag, 1978, pp. 55–70.
49. ———. "Introduction," in Dembroski, T. M., et al., eds., *Coronary-Prone Behavior.* New York: Springer-Verlag, 1978, pp. xiii–xvi.
50. ———, et al. "Multivariate Prediction of Coronary Heart Disease Druing 8.5 Year Follow-Up in the Western Collaborative Group Study," *American Journal of Cardiology,* 37 (1976):902–910.
51. SCHULTZ, J. H., and LUTHE, W. *Autogenic Training: A Psychophysiological Approach in Psychotherapy.* New York: Grune & Stratton, 1959.
52. SCHWARTZ, G. E., and WEISS, S. M. "Yale Conference on Behavioral Medicine: A Proposed Definition and Statement of Goals," *Journal of Behavioral Medicine,* 1 (1978):3–12.
53. ———. "Behavioral Medicine Revisited: An Amended Definition," *Journal of Behavioral Medicine,* 1 (1978):249–251.
54. SHAPIRO, D., and SURIWT, R. S. "Operant Conditioning: A New Theoretical Approach in Psychosomatic Medicine," in Lipowski, Z. J., Lipsett, D. R., and Whybrow, P. C., eds., *Psychosomatic Medicine: Current Trends and Clinical Applications.* New York: Oxford University Press, 1977, pp. 68–78.
55. SKINNER, B. F. *The Behavior of Organisms.* New York: Appleton-Century, 1938.
56. STUNKARD, A. J. "Symposium on Obesity: Basic Mechanisms and Treatment," *Psychiatric Clinics of North America,* 1 (1978):1–273.
57. SURWIT, R. S., PILON, R. N., and FENTON, C. H. "Behavioral Treatment of Raynaud's Disease," *Journal of Behavioral Medicine,* 1 (1978):323–335.
58. "Timberline Conference on Psychophysiologic Aspects of Cardiovascular Disease," *Psychosomatic Medicine,* 26 (1964):405–541.
59. U.S. PUBLIC HEALTH SERVICE. *Teenage Smoking. National Patterns of Cigarette Smoking, Ages 12 Through 18, in 1972 and 1974.* DHEW publication no. (NIH) 76–931. U.S. Department of Health, Education & Welfare, Public Health Service, National Institute of Health, 1976.
60. *Webster's New Collegiate Dictionary.* Springfield: G. & C. Merriam, 1977.
61. WEINER, H. *Psychobiology and Human Disease.* New York: Elsevier, 1977.
62. ———, et al. "Etiology of Duodenal Ulcer: I. Relation of Specific Psychological Characteristics to Rate of Gastric Secretion (Serum Pepsinogne)," *Psychosomatic Medicine,* 19 (1957):1–10.
63. WILLIAMS, R. B. "Headache," in Williams, R. B., and Gentry, W. D., eds., *Behavioral Approaches to Medical Treatment,* Cambridge: Ballinger, 1977, pp. 41–54.
64. ———. "Biofeedback: It's Time to Look Inside the Black Box," in Brady, J. P., and Brodie, H. K. H., eds., *Controversy in Psychiatry.* Phaladelphia: W. B. Saunders, 1978, pp. 519–533.
65. ———, and GENTRY, W. D., eds. *Behavioral Approaches to Medical Treatment.* Cambridge: Ballinger, 1977.
66. WILLIAMS, R. B., et al. "Type A Behavior, Hostility and Coronary Atherosclerosis," *Psychosomatic Medicine,* 1980 (In Press).
67. WOLPE J. *Psychotherapy by Reciprocal Inhibition.* Stanford: Stanford University Press, 1958.
68. ZYZANSKI, S. J., et al. "Psychological Correlates of Coronary Angiographic Findings," *Archives of Internal Medicine,* 136: (1976) 1234–1237.

CHAPTER 18

WITHDRAWAL EFFECTS FROM PSYCHOTROPIC DRUGS

George Gardos, Jonathan O. Cole, and Daniel Tarsy

THE TERM "withdrawal effects" tends to conjure up images of drug dependent persons in the throes of severe, invariably unpleasant, and at times dangerous abstinence reactions to opiates or barbiturates. There is much less attention paid to the not so dramatic, but nonetheless quite common, phenomenon of withdrawal symptoms occurring when prescribed psychotropic drugs are abruptly discontinued. Awareness of this problem could prevent a great deal of unnecessary pharmacotherapy. For instance, a patient on maintenance drug therapy may repeatedly attempt to discontinue the drug abruptly and then conclude from the resulting withdrawal symptoms that indefinite drug therapy is needed. Physicians do not always terminate drug therapy in a manner that minimizes the likelihood of withdrawal effects.

Classes of drugs will be discussed separately in the following sections. Special mention will be given to situations in which simultaneous discontinuation of two or more drugs may pose unusual problems. This chapter covers only psychotherapeutic drugs and avoids dealing with the vast literature on the phenomena of abstinence from drugs of abuse.

¶ Antipsychotic Drug Withdrawal

Very little attention has been paid to the manner in which antipsychotic drugs are discontinued in clinical practice. As often as not, drugs are stopped abruptly with no expectation of adverse consequences other than the possibility of psychotic relapse. In fact, however, a number of autonomic, behavioral, and neurological symptoms may occur in the postwithdrawal days and weeks. The importance of recognizing withdrawal symptoms and distinguishing them from psychotic relapse cannot be overemphasized.

Somatic Withdrawal Symptoms

The literature on antipsychotic drug discontinuation shows striking variations in the incidence and severity of withdrawal phenomena. Several studies reported no withdrawal symptoms,[103,108,153] although it is entirely possible that since these studies focused on other issues, withdrawal phenomena may have been missed. When abrupt antipsychotic drug withdrawal is carried out in a carefully controlled design, statistically significant increases in withdrawal symptoms can be observed.[81] A placebo effect is unlikely to play a role in the withdrawal syndrome: Battegay[10] found no lessening in the prevalence of symptoms between patients switched to placebo and patients withdrawn from antipsychotics without placebo substitution.

Common withdrawal symptoms, as reported in relevant publications,* include: nausea, vomiting, sweating, insomnia, restlessness, dizziness, and headache. Occasionally tachycardia, faintness, "flu-like" symptoms such as feeling achy or hot and cold, rhinorrhea, abdominal pain, diarrhea, malar flushing, numbness, or nightmares have been reported. The clinical picture for the individual patient is unpredictable: any one or several symptoms may be reported with varying severity.

Typically, symptoms appear on the first, second, or third day after drug discontinuation, but may occasionally be delayed as much as one to two weeks.[10,81] Symptoms usually peak during the first week, followed by a gradual attenuation and spontaneous recovery.[15]

Age appears to be an etiological factor: older patients show high prevalence of symptoms.[10] Female patients have been found to show significantly higher rates of withdrawal symptoms than male patients.[10,32] Abrupt drug withdrawal is more likely to produce withdrawal symptoms than gradual withdrawal.[59] Dosage of the antipsychotic drug before withdrawal, however, does not appear to be related to symptom prevalence.[15,81] The antimuscarinic anticholinergic effect of the withdrawn antipsychotic often determines whether or not withdrawal symptoms will occur. Luchins and associates[85] reviewed the literature and found a highly significant association between antimuscarinic potency and the number of patients showing withdrawal symptoms. Thus, of the standard antipsychotics in the United States, thioridazine and chlorpromazine account for most of the reported somatic withdrawal symptoms.[15,59,80,81] On the other hand, antipsychotic drugs with low muscarinic potency, such as piperazine, phenothiazines, and haloperidol,[141] are less likely to induce somatic withdrawal symptoms and in fact few such cases have been published.[81,136] Antiparkinsonism drugs, by virtue of their strong antimuscarinic effects, frequently induce withdrawal symptoms. Simultaneous withdrawal of antipsychotic-antiparkinson drug combinations often result in somatic symptoms that are partly, if not wholly, a result of antiparkinson drug withdrawal.*

The neuropharmacological changes underlying somatic withdrawal manifestations appear to represent mainly a cholinergic hypersensitivity reaction. It is thought to be a rebound phenomenon resulting from prolonged treatment with drugs with strong anticholinergic properties.[55] This hypothesis is supported by studies that showed that physostigmine, a powerful cholinergic agent, can induce most of the commonly noted somatic withdrawal symptoms which, in turn, can be abolished by anticholinergics.[58] However, when antipsychotics with minimal antimuscarinic potency are withdrawn, a cholinergic rebound probably does not occur. In fact, as Luchins and associates have recently demonstrated with chronic haloperidol treated mice, the converse may be true in that haloperidol withdrawal may induce a state of cholinergic subsensitivity.[85]

The principal features of somatic withdrawal symptoms are summarized in table

*See references 10, 11, 15, 31, 32, 53, 59, 80, 99, and 136.

*See references 55, 81, 85, 135, and 136.

18–1. Of particular importance to the practitioner is the self-limited nature of these symptoms and their strong association with anticholinergic drugs such as thioridazine or chlorpromazine. Clinical management in mild cases simply requires reassurance of the patient. In more severe cases, anticholinergic compounds such as benztropine or diphenhydramine may provide specific remedies. Since the symptoms usually last only a few days and may induce secondary anxiety, diazepam or other benzodiazepines may have nonspecific utility in supporting the patient while the symptoms fade. If the somatic withdrawal effects are severe or are combined with neurological withdrawal symptoms, retreatment with the previously withdrawn antipsychotic is called for.[10] More gradual dose tapering may then be attempted at a later date.

Extrapyramidal Complications of Antipsychotic Drug Withdrawal

The clinical literature provides ample evidence for the principle that every extrapyramidal symptom that can be produced by drug administration may also be seen upon drug withdrawal.

PARKINSONISM

There is almost universal agreement among experts that drug-induced parkinsonism tends to improve following antipsychotic drug withdrawal. However, the extent and the time course of improvement remains unresolved. Depending upon the duration of postwithdrawal observation, parkinsonian signs have been found to disappear in a few weeks,[32] improve substantially within three to six months,[25,33,135] or remain unchanged sixteen weeks after antipsychotic withdrawal.[64]

Case reports in which severe acute extrapyramidal symptoms occurred following withdrawal usually involve simultaneous withdrawal of antipsychotic and antiparkinson drugs.[10,46,135,136] The more rapidly excreted antiparkinson drug leaves the antipsychotic drug free to exert its neuroleptic effect unopposed.[136] Occasionally, the extrapyramidal reaction is markedly delayed. In a sixty-four-year-old woman, a dystonic reaction occurred twenty-one days after drug withdrawal,[10] while in a nineteen-year-old man, withdrawal akinesia lasted nineteen days following low dose phenothiazine therapy of relatively brief duration.[46] Alpert and

TABLE 18–1 **Somatic Symptoms of Antipsychotic Withdrawal**

Common symptoms:	nausea, vomiting, sweating, insomnia, restlessness
Onset:	typically one to three days after drug withdrawal
Duration:	one to three weeks
Type of AP:	thioridazine or chlorpromazine, rare with drugs of low antimuscarinic potency
Pharmacological substrate:	cholinergic rebound
Treatment:	reassurance, tranquilizers, anticholinergics, rarely with resumption of antipsychotic

associates[1] described a case of paradoxical worsening of extrapyramidal signs produced by antiparkinson drugs following discontinuation of chlorpromazine, trifluoperazine, and trihexyphenidyl. These case reports suggest that the clinical course of withdrawal tremor, rigidity, akinesia, and dystonia may not be a simple function of antipsychotic drug effects and that a special vulnerability exists in certain individuals to extrapyramidal effects in the postwithdrawal weeks.[1]

DYSKINESIAS

Choreoathetotic dyskinetic movements tend to increase in intensity or may appear for the first time following antipsychotic drug withdrawal. *Withdrawal dyskinesia* is a self-limiting syndrome initially indistinguishable from tardive dyskinesia. Cases of withdrawal dyskinesia have been reported following withdrawal of chlorpromazine, fluphenazine,[148] mesoridazine,[93] and haloperidol.[23,70] In these reports, dyskinesias were noted within days of drug withdrawal and lasted from one to twenty-two weeks, but in all cases, complete resolution of the syndrome was observed.

While not all antipsychotic compounds have been shown to produce withdrawal dyskinesias, it is likely that all dopamine-blocking antipsychotics may do so. Degkwitz and associates[32] noted a sex difference: withdrawal dyskinesias tended to develop later and lasted longer in female than in male patients, in contrast to postwithdrawal parkinsonism which tended to resolve faster in female patients.

Dyskinesias that become obvious only as a consequence of drug withdrawal and persist for many weeks can be regarded as a type of tardive dyskinesia. The term "covert dyskinesia"[55] is sometimes applied to this syndrome to emphasize that dopamine-blocking antipsychotics often mask an underlying dyskinesia which may be uncovered by drug withdrawal. The prevalence of covert dyskinesia varies greatly, but, in some studies, it has been strikingly high: Degkwitz and Wenzel[31] found 47 such patients (39 percent) in a double-blind drug withdrawal study of 119 persons. Escobar and Tuason[42] reported that eight out of nine patients who were on oral fluphenazine prior to withdrawal developed clinically significant dyskinesias. Female sex and higher prewithdrawal dosage have been reported to be contributing factors.[42]

The natural course of covert dyskinesia is difficult to establish. In some instances, the emerging dyskinesia is massive and may be life threatening,[24,74] and quick resumption of antipsychotic drug therapy is the indicated clinical course. Psychotic decompensation not infrequently disrupts the drug-withdrawal period and results in retreatment with antipsychotics.[10,32,65] Therefore, in many cases of dyskinesia following withdrawal, the resumption of drug therapy makes it impossible to establish whether the dyskinesia would have been self-limiting (withdrawal dyskinesia) or persistent (covert dyskinesia). In general, the prognosis of covert dyskinesia is quite similar to tardive dyskinesia: chronic, older patients with prolonged exposure to antipsychotics tend to develop persistent dyskinesias,[25,33,64] whereas younger patients who have undergone shorter courses of treatment often show reversible dyskinesias.[117]

An intriguing aspect of withdrawal dyskinsias is their apparent association with psychotic relapse in that patients with dyskinesias may be more prone to decompensation,[10,20,32,116] while no relationship was found between disappearance of parkinsonism and psychotic relapse.[32] At present, this is more a clinical observation than an established statistical correlation, but it raises fundamental questions about the way drugs exert antipsychotic and neuroleptic actions and will be discussed later. The extrapyramidal effects of antipsychotic withdrawal are summarized in table 18–2.

PSYCHOTIC RELAPSE

Two cases of delirium associated with antipsychotics and resembling alcohol withdrawal have appeared in the literature. A twenty-seven-year-old man developed delir-

TABLE 18–2 **Extrapyramidal Effects of Antipsychotic Drug Withdrawal**

		DYSKINESIA	
	DRUG-INDUCED PARKINSONISM	"WITHDRAWAL DYSKINESIA"	"COVERT DYSKINESIA"
Symptoms	Tremor, rigidity, dystonic reactions, akinesia	Choreoathetosis, motor restlessness	Choreoathetosis, tics, grimaces
Etiology	1. Continuation of already existing syndrome 2. Simultaneous withdrawal of antiparkinson drugs	Tends to occur after relatively brief drug exposure	Tends to occur after prolonged drug therapy
Onset	Within a few days, occasionally delayed two to three weeks	Usually within days	Usually within two weeks
Course	Slow, gradual improvement	Spontaneous recovery	Variable: may remit, persist or intensify
Management	Short-term treatment with antiparkinson drugs may be needed	Occasional patient may need sedative-hypnotic or benzodiazepine	Retreatment with antipsychotics in severe cases, no treatment for mild cases, antidyskinesia drugs for intermediate cases

ium with visual and auditory hallucinations twenty hours after abrupt discontinuation of haloperidol. Within three days, the symptoms abated spontaneously.[118] A forty-six-year-old man developed an acute brain syndrome lasting seven days two days after thiothixene withdrawal.[45] These isolated instances notwithstanding, the appearance of psychotic manifestations after antipsychotic drug withdrawal almost invariably heralds psychotic decompensation. The time lag between withdrawal and relapse is highly variable.[54] A number of controlled studies of antipsychotic drug withdrawal showed that relapse rates occurred at a constant rate during the first twelve months, declining thereafter. But, as long as two years after drug discontinuation, placebo relapse rates still exceeded relapse rates for drug-maintained schizophrenics.[27,41,66,115] Clearly the prediction of the time of onset of psychotic decompensation in the individual patient is problematical. Distinct and recognizable stages of decompensation have been described: (1) denial and anxiety; (2) depression and intensification of defense mechanisms; (3) internal chaos; and (4) subjective relief.[35] Early symptoms of decompensation such as anxiety, tension, and insomnia overlap somatic withdrawal symptoms; differentiating these two phenomena may be as difficult as it is clinically important.

Tardive Psychosis

Several recent reports have suggested that psychotic phenomena that are not simply attributable to a return of schizophrenic symptoms may occur following drug withdrawal. Sale and Kristall[127] described a twenty-one-year-old woman with obsessional symptoms and anxiety who developed what later progressed into chronic schizophrenia following withdrawal of chronically administered chlorpromazine. Forrest and Fahn[50] observed an array of psychotic and other symptoms on antipsychotic withdrawal that were distinct from the original symptoms for which patients were treated. The symptoms subsided with resumption of drug therapy. Forrest and Fahn considered these symptoms to result from the drug withdrawal phenomena and labeled the syndrome "tardive dysphrenia."

Davis and Rosenberg[28] presented evidence from animal studies for supersensitivity in mesolimbic dopamine receptors and suggested that cases of withdrawal psychosis might reflect mesolimbic dopamine supersensitivity. McCarthy[87] raised the issue of whether reversible "withdrawal psychosis" or persistent "tardive psychosis" might not occur analogous to withdrawal and tardive dyskinesias, reflecting limbic hypersensitivity to dopamine. Chouinard and Jones[19] presented ten cases of what they called "supersensitivity psychosis." These schizophrenic patients were treated with depot fluphenazine but appeared to require increasing doses for therapeutic effect and showed positive schizophrenic symptoms following decrease in dosage just before the next scheduled injection, or after missing one or two doses. Chouinard and Jones concluded that neuroleptic-induced mesolimbic dopamine supersensitivity accounted for their clinical findings. The thrust of these reports is that significant neuropharmacological changes may occur during antipsychotic drug treatment in areas other than the striatum, and that during drug withdrawal, these changes may be manifested in behavioral and occasionally even in neuroendocrine changes. For example, the association shown in two studies between lower prolactin levels and greater clinical deterioration after antipsychotic discontinuation points to the existence of a subgroup of schizophrenic patients with an overactive and labile dopaminergic system. These patient characteristics may play an important role in the production of withdrawal phenomena.[20,84]

Although the notion of "tardive psychosis" is an intriguing one and of great practical concern, it is certainly not yet a proven entity. It has been pointed out that "relapse" following antipsychotic drug withdrawal increases in a linear fashion during the first year at a rate of about 7 to 15 percent per month.[7] It is therefore likely that in any large clinic population of schizophrenic patients, a certain small proportion will show reemergence of psychosis very shortly after drug withdrawal. Whether the psychiatric characteristics of these patients can be differentiated from a new and superimposed withdrawal psychosis as has been claimed[19,50] remains to be confirmed by future studies.

Antipsychotic Withdrawal in Children

The same types of withdrawal phenomena may be observed in children as in adults. Yepes and Winsberg[157] reported cases with extensive symptomatology. The first patient was a nine-year-old boy who, following abrupt withdrawal of chlorprothixene 150 mg/d, developed restlessness and insomnia (first day); nausea and vomiting (fourth day); and hemiballismus, dystonia, and severe posturing of the arms and face (sixth day). The second patient, also a nine-year-old boy, when withdrawn from thioridazine 125 mg/d, developed irritability (first day); stomachaches (tenth day); dyskinesia (fourteenth day); and nausea and vomiting (twenty-first day). Vomiting was severe and lasted twelve days, while the extrapyramidal disorder persisted up to ninety days. Polizos and associates[112] studied the effects of abrupt antipsychotic drug withdrawal in thirty-four schizophrenic children, six to twelve years old. Fourteen children developed choreiform dyskinesias (mainly involving the extremities, trunk, and head) and ataxia appearing one to fifteen days after withdrawal. In half of the affected children, the dyskinesias remitted spontaneously within five weeks; in the other children, drugs had to be resumed because of massive psychotic relapse. The withdrawal dyskinesias of children are thought to be reversible.[40,86]

Neuropharmacological Considerations

Withdrawal symptoms tend to be mirror images of the drug-induced changes that occur during treatment. For example, administration of chlorpromazine produces sedation, fatigue, and hypokinetic extrapyramidal effects, while removal of the drug may induce insomnia, restlessness, and hyperkinetic extrapyramidal effects. In pharmacological terms, the issues of neurotransmitter blockade, tolerance, and supersensitivity may underlie the observed somatic and behavioral changes.

The interruption of synaptic transmission in either the peripheral or central nervous system may result in a state of denervation supersensitivity to administration of the blocked neurotransmitter or its agonist. Because of greater accessibility, denervation supersensitivity has been studied in more detail in the peripheral than the central nervous system. Preganglionic nerve section, ganglionic lesions, peripheral nerve section, and pharmacological interference with synaptic transmission have been utilized to effect denervation supersensitivity.[39,133,149] An interesting example of the importance of denervation supersensitivity in the peripheral autonomic nervous system occurs in patients with angina and hypertension who are treated chronically with propranolol, an antagonist at beta-adrenergic receptor sites. When propranolol is abruptly withdrawn, unstable angina, myocardial infarction, and cardiac irritability sometimes occur.[100] One proposed explanation for this is increased sensitivity of cardiac tissue to beta-receptor agonists.[56,140] The observation that the treatment of rats with propranolol for two weeks leads to a 100 percent increase in the number of beta-adrenergic receptors is compatible with the hypothesis of denervation supersensitivity.[56]

The extrapyramidal neurologic effects and possibly the therapeutic antipsychotic effects of neuroleptic drugs are believed to derive from their capacity to block dopamine mediated synaptic transmission. Evidence for the capacity of neuroleptic drugs to block dopamine receptors derives from their antagonism of behavioral and neuroendocrine effects of dopamine agonists, their antagonism of dopamine-sensitive adenylate cyclase in caudate and limbic brain tissue, and the capacity of neuroleptic drugs to interfere with the binding of radioactively labeled ligands such as tritiated dopamine, apomorphine, and haloperidol to dopamine receptor sites.[8]

Dopamine mediated projections lie in several discrete brain regions including pathways between midbrain and basal ganglia (the nigrostriatal tract), regions of the limbic forebrain, areas of temporal and prefrontal cerebral cortex closely associated with the limbic system, and the hypothalamic-pituitary system. Although unproven, it is currently considered that extrapyramidal effects of neuroleptic drugs are due to dopamine blocking effects in the basal ganglia, while antipsychotic efficacy relates to dopamine-blocking effects in limbic nuclei and/or limbic cortex. When given acutely, the effect of neuroleptic drugs on dopamine neurotransmission is to increase the firing rate of dopamine neurons and to increase the synthesis, release, and metabolic turnover of dopamine in dopaminergic neurons. These effects may be viewed as compensatory responses by an adaptive neuronal system seeking to maintain adequate dopamine neurotransmission in response to dopamine receptor blockade.[8]

When neuroleptic drugs are administered to animals on a more chronic basis, there is a gradual reduction in the neuroleptic-induced acceleration of dopamine turnover, such that, within seven days, following neuroleptic administration, dopamine turnover remains at baseline levels or is even reduced.[3,21,83] Following a very similar time course, the capacity of neuroleptic drugs to produce catalepsy or block apomorphine induced stereotyped behavior in rats also becomes diminished.[3,21,83,102] This loss of neuroleptic effect with repeated administration has been referred to as tolerance.[3,102] One possible explanation for appearance of tolerance to neuroleptic effects is the development of enhanced sensitivity of dopamine receptors to dopamine. This gains support from repeated observation that chronic treatment of mice, rats, or guinea pigs with drugs antagonistic to dopamine followed by their discontinuation produces increased behavioral responsiveness to dopamine agonists not accountable by other pharmacologic or pharmacokinetic mechanisms.[47,75,149] This effect, presumably representing the development of functional supersensitivity to dopamine in the brain, requires no more than several days of treatment with a neuroleptic drug and persists for several weeks with some evidence that the duration of this effect parallels the duration of pretreatment with neuroleptic drug.[105] Supporting this behavioral evidence for denervation supersensitivity of dopamine receptors have been changes in dopamine receptor binding which have been produced by chronic neuroleptic treatment. Several laboratories have demonstrated that following a course of neuroleptic pretreatment identical to that which produces behavioral supersensitivity, striatal and limbic dopamine receptors increase in number and display enhanced affinity for radioactively labeled ligands which bind at dopamine receptor sites.[17,104] Neurophysiologic studies have also shown that chronic treatment of rats with haloperidol produces a significant increase in the sensitivity of caudate neurons to microiontophoretically applied dopamine.[137,156] The fact that similar behavioral, biochemical, and neurophysiological alterations have been observed following surgical lesioning of dopamine neurons,[8] supports the concept that they reflect changes in receptor sensitivity brought about by interference with dopamine neurotransmission. On the basis of the aforementioned studies, it is suggested that chronic neuroleptic treatment results in a compensatory increase in affinity and numbers of striatal dopamine receptors which offsets the effects of dopamine receptor blockade.

Since it has been the general clinical impression that patients do not become tolerant to the antipsychotic efficacy of neuroleptic drugs, it has been assumed that tolerance and supersensitivity phenomena were restricted to the nigrostriatal system. Observations that drug-induced parkinsonism and acute dystonia tend to occur relatively early in treatment and become less frequent and severe with continued drug exposure, while transient withdrawal dyskinesias and persistent tardive dyskinesia appear later in treatment, support this concept.

In the case of the hypothalamic dopamine system, there is evidence that in both animals and man tolerance and dopamine supersensitivity fail to develop with chronic neuroleptic treatment.[51,97,146] Evidence concerning tolerance and supersensitivity in dopaminergic limbic nuclei and cortex has been less consistent, however. In early studies, chronic neuroleptic treatment produced a persistent increase in dopamine turnover in mesolimbic and mesocortical dopamine projections,[13,130] suggesting absence of tolerance in this system. However, more recent studies[14,129] have indicated that with chronic treatment, the neuroleptic-induced increase in dopamine turnover does subside in limbic nuclei, indicating a tolerance for this effect similar to that which occurs in the striatum. Muller and Seeman[105] reported an increase of dopamine binding sites in both striatal and mesolimbic regions following chronic neuroleptic treatment, while in two other studies, long-term neuroleptic treatment was found to increase the locomotor response of dopamine injected directly into the nucleus accumbens but not the striatum.[29,69]

In humans, cerebrospinal fluid homovanillic acid (HVA), a metabolite of dopamine, is increased following neuroleptic treatment but returns toward normal after three weeks of continued drug exposure.[113] Since cerebrospinal fluid HVA is derived from periventricular structures, its concentration may not reflect levels in other brain regions. Further studies in rodents, nonhuman primates, and also in man, all suggest, in fact, that tolerance to the effects of neuroleptic drugs on brain HVA concentration develops in midline nuclei such as the caudate nucleus and deeper limbic nuclei, but not in cortical regions such as cingulate, temporal, dorsal frontal, and orbital frontal cortex.[5,6] Because of this sustained biochemical change in cortical dopamine metabolism with evidence of tolerance, it has been concluded that it may be in these brain regions that antipsychotic drugs produce their therapeutic effect.

Prevention of Withdrawal Symptoms

Antipsychotic drug withdrawal symptoms may be highly unpleasant, and, in rare instances, a severe withdrawal dyskinesia may be serious and life threatening. Furthermore, withdrawal symptoms may obscure signs of early relapse, and, conversely, they may be mistaken for signs of psychotic decompensation. Avoidance of withdrawal symptoms, therefore, becomes an important goal. The following guidelines spell out the technique of withdrawal that may minimize the risk of such symptoms.

1. Withdrawal should be gradual rather than abrupt. Step-wise, gradual dose reduction will probably circumvent most withdrawal symptoms. In chronic drug-treated schizophrenics, the process of drug withdrawal may be spread over several months, delaying the onset and probably reducing the risk of psychotic relapse.

2. Antiparkinson drugs should be continued. Patients withdrawn from antipsychotics may develop a transient hypercholinergic state which produces somatic withdrawal symptoms. Continuation of antiparkinson drugs in patients on antipsychotic-antiparkinson combinations for one-two weeks beyond antipsychotic withdrawal may eliminate somatic symptoms as well as the recurrence of drug-induced parkinsonism.

¶ Antiparkinson Drugs

The major therapeutic indication for antiparkinson drugs (APK) in psychiatry is the prevention or control of the extrapyramidal side effects of antipsychotic compounds. The most frequently used APK in the United States are anticholinergics (trihexyphenidyl, procyclidine, biperiden), antimuscarinic antihistamines (benztropine, diphenhydramine), and dopamine agonists (amantadine). These drugs are rarely, if ever, administered to psychiatric patients without antipsychotic drugs, and when the latter are discontinued, APK are usually withdrawn as well. As stated in the previous section, the somatic and parkinsonian symptoms following withdrawal of high potency antipsychotic drugs are usually due to the simultaneous withdrawal of antiparkinson drugs.

Withdrawing antiparkinson drugs alone while continuing antipsychotic drug treatment is frequently attempted in clinical practice in order to ascertain whether APK are still required. Surprisingly often, however, one finds that patients are most reluctant to part with their APK.[71,96] In some cases, the desire to continue is undoubtedly related to the abuse potential of some APK, particularly trihexyphenidyl.* Adverse behavioral, neurological, and mood changes may also occur on APK withdrawal and probably explain why some patients would rather stop their clearly essential antipsychotic drug than the supposedly unnecessary APK.

Specific withdrawal effects have been investigated in open as well as double-blind placebo-controlled studies of APK discontinuation. Most studies have focused on extrapyramidal symptoms while only a few have included assessment of psychopathology or mood.

Reappearance of Drug-induced Parkinsonism

There is a wide divergence of research results with regard to the frequency with which extrapyramidal symptoms reappear after APK withdrawal. Relapse rates (that is percentage of patients developing symptoms of parkinsonism) range from 8 to 80 percent in published studies of APK withdrawal.† The most common symptoms of parkinsonism, namely tremor and rigidity, were usually focused on and were reported accurately. Akinesia, however, tended to be overlooked in some studies and to be underreported. Rifkin and associates[122] found that akinesia occurred in 27 percent of patients following procyclidine withdrawal. Other factors that may account for the wide variation in relapse rates include variations in the populations studied, variations in the tolerance for milder extrapyramidal symptoms (that is, differing criteria for "relapse"), and drug type and dosage differences of both antipsychotics and APK.

The onset of parkinsonism was usually within two weeks and nearly always within four weeks of APK discontinuation. In a carefully documented study, Pecknold and associates[109] found that symptoms first appeared an average of twelve and three-tenths days after APK withdrawal. Trihexyphenidyl or biperiden were found to produce symptoms sooner than benztropine withdrawal, reflecting the slower metabolism of benztropine.[72] No consistent differences were observed between placebo-controlled double-blind studies and open trials. Two studies by Roy and associates,[125] in which both placebo and no APK control groups were employed, produced conflicting results. No consistent relationship was demonstrated between antipsychotic dosage and extrapyramidal symptoms following APK withdrawal.[76,109] Older age was associated with reemergent parkinsonism in one study.[76] Longer prewithdrawal APK treatment was found to be correlated with lower relapse rates in three studies.[48,106,109] Previous occurrence of extrapyramidal symptoms was found to predict postwithdrawal parkinsonian symptoms;[82,109] thus therapeutic use of APK is more likely to lead to parkinsonism after withdrawal than prophylactic drug administration. No obvious differences were noted between the various antiparkinson drugs in their extrapyramidal withdrawal effects.

Somatic Withdrawal Symptoms

The frequent occurrence of somatic symptoms such as nausea, vomiting, and insomnia following withdrawal of antipsychotic drugs with marked anticholinergic effects was discussed previously. Since these symptoms are believed to reflect a cholinergic rebound, they may also be expected to result from the

*See references 57, 71, 89, 95, and 126.

†See references 12, 18, 34, 44, 48, 76, 82, 91, 92, 106, 109, 122, 125, and 143.

removal of anticholinergic antiparkinson drugs. Kruse[79] was probably the first investigator to document somatic withdrawal reactions from APK. Specific withdrawal symptoms described in the literature include restlessness,[72,76] nausea, dizziness,[122] aches and pains,[72] agitation, and stiff joints.[106] Somatic withdrawal symptoms have also been documented following the simultaneous withdrawal of butaperazine and benztropine.[136] The time course and treatment of the somatic withdrawal symptoms from APK are broadly the same as after antipsychotic withdrawal.

Mood Changes

The appearance of dysphoric symptoms is at times a striking effect of APK withdrawal.[12,71] When specifically looked for, dysphoric symptoms turn out to be quite common. In a double-blind placebo-controlled study, Jellinek and associates[72] found that out of twenty-four APK withdrawn patients, seven developed anxiety, two complained of fatigue, and one became depressed. In the often quoted study by Orlov and associates,[106] the authors found ten out of seventy-eight patients to have complained about and resisted APK withdrawal. The authors attributed this to psychological dependence, however, these patients may have experienced genuine dysphoria. Depression has been noted by some authors to occur after APK withdrawal, particularly in connection with akinesia.[12,71,122] The association of these two conditions has led to the concept of akinetic depression. It derives some support from the strong statistical association between parkinsonism and depression[123] and suggests similarities in the underlying pathophysiology. The adverse mood changes from APK withdrawal suggest that APK may possess psychotropic properties, possibly antidepressant effects, at least in some patients.

Improvement in Dyskinesia

On withdrawal of APK, the characteristic oro-facial movements of tardive dyskinesia are at times observed to remit.[43,44] In patients where parkinsonism and tardive dyskinesia coexist, changes in dyskinetic movements are often reciprocal to the changes in parkinsonism: APK withdrawal may benefit the former and aggravate the latter.[44]

The Pros and Cons of APK Withdrawal

The a foregoing review clearly shows that a considerable number of patients on antipsychotic drugs develop adverse effects from APK withdrawal. On the other hand, a great many patients are apparently totally unaffected by APK withdrawal. Some of these patients may have had prewithdrawal blood levels below therapeutic range on usual oral doses, as Tune and Coyle[150] demonstrated for benztropine. It may be assumed that such interindividual variability in blood level exists for every APK and therefore some patients who show parkinsonism during APK administration (that is, cases of treatment-resistant parkinsonism) are not on therapeutic doses and may not change following APK withdrawal.

The risk-benefit ratio of continuous antiparkinson therapy remains a matter of controversy. Some authors[36,76] stress the disadvantages, such as anticholinergic side effects, possible lowering of antipsychotic blood level, and aggravation of tardive dyskinesia. They also note the element of cost and regard prophylactic and maintenance therapy as at best unnecessary and at worst, harmful. Other researchers[72,122] consider most of these risks of APK therapy to be greatly exaggerated or largely theoretical, while they view the benefits of continuous APK treatment, such as control of subtle extrapyramidal effects and possibly the contribution of a psychotherapeutic effect, as benefits that outweigh the potential hazards.

In the current state of knowledge, it would be considered good clinical practice to attempt to withdraw prophylactic APK after ninety days of administration since acute extrapyramidal symptoms are unlikely to develop beyond this time. Careful attention to withdrawal effects requires periodic follow-up examination of APK withdrawn patients for signs of extrapyramidal disturbance as well as behavioral and mood changes. The optimal technique of APK withdrawal to minimize withdrawal effects is gradual tapering rather than abrupt discontinuation.[88] However, as demonstrated in a recent study, even when gradual APK withdrawal is instituted, about one-third of the patients still appear, after two years, to require APK.[96]

¶ Antidepressants

Withdrawal syndromes from two drug classes used in affective illness appear to be almost nonexistent. Monoamine oxidase inhibitors can be stopped abruptly with no consequences other than a possible reemergence of depressive symptoms within days or weeks. This seems most reasonable since the enzyme inhibition produced by these drugs fades very gradually over a two- or three-week period. There is one recent report describing withdrawal effects from phenelzine in which after one day two patients developed flu-like symptoms lasting for one week.[111]

Lithium also does not cause withdrawal syndromes, a fact supported by one formal study[121] and a large body of informal clinical experience. There are, however, rare occasions in which lithium toxicity may be mistaken for withdrawal. Rosser and Herxheimer[124] reported two cases of nausea and vomiting following chlorpromazine withdrawal in patients who were also on lithium. The differential diagnostic possibilities for the vomiting included: (1) removal of the antiemetic effect of chlorpromazine exposing lithium side effects; (2) risk in serum lithium level brought about by chlorpromazine withdrawal;[107] and (3) chlorpromazine withdrawal effects. Occasionally, a patient will begin to show signs of lithium toxicity that will then worsen for a couple of days after the drug is stopped. The probable mechanism is a rising serum lithium level after drug discontinuation due to dehydration and sodium loss with resulting lithium retention.

Tricyclic antidepressant drugs elicit withdrawal symptoms when dosage is abruptly or rapidly terminated. The literature on withdrawal from antidepressants is rather sparse, however, and deals almost exclusively with imipramine. Kramer and associates[78] studied withdrawal symptoms by means of interviews and nursing notes concerning forty-five patients withdrawn from imipramine after stabilization at dosages of about 300 mg/d. Twenty-two of twenty-six patients treated for over two months experienced clear withdrawal symptoms, while only three of nineteen patients on the medication for less than two months did so. Withdrawal symptoms lasted two weeks or less. Kramer and associates[78] recommended that two to four weeks be allowed for gradually reducing imipramine dosage. They noted that withdrawal symptoms (nausea, vomiting, headache, giddiness, coryza, chills, weakness, fatigue, or muscle pain) responded to a stat dose of 50 mg imipramine followed by more gradual tapering. Shatan[134] described severe withdrawal symptoms in a patient, occurring twenty-four hours after abrupt termination of imipramine (200 mg/d). Symptoms included those noted by Kramer,[78] plus cold sweat, gooseflesh, abdominal cramps, hunger, diarrhea, irritability, insomnia, and restlessness. Symptoms peaked at forty-eight hours and gradually subsided without treatment after a week off medication. Sathananthan and Gershon described three patients who abruptly stopped daily dosages of imipramine of 300 to 450 mg. Within twenty-four hours all three developed anxiety, restlessness, and forced pacing. The syndrome resembled the akathisia caused by neuroleptics. Again a stat dose of imipramine relieved the symptoms in all three cases within two hours. Andersen and Kristiansen[2] observed withdrawal symptoms in fifteen out of eighty-five patients following both gradual and acute termination of drug treatment with imipramine. Symptoms included sleep disturbances, subjective and objective restlessness and attacks of perspiration, nausea, and vomiting.

Withdrawal symptoms from other tricyclics have also been reported. Gualtieri and Staye[60] described withdrawal symptoms (nausea, vomiting, abdominal cramps, and diarrhea leading to severe dehydration) in an eight-year-old boy who had been on amitriptyline (50 mg/d) for seven months. Kraft[77] found withdrawal symptoms in seven female patients after abrupt discontinuation of clomipramine. These included dizziness, faintness, nausea, feeling "electrically charged," malaise, stomachaches, and anxiety dreams. Two of the patients reported symptoms after placebo substitution. Symptoms tended to be worse after longer term clomipramine therapy. Brown and associates[16] described withdrawal effects in a forty-one-year-old man after twelve years of record dosage of desipramine (1,000 mg/d). During the seventeen-day gradual withdrawal period, only mild symptoms were noted: sore shoulders, insomnia, vivid dreams, and "nerves," all of which faded away without treatment. Other changes during drug withdrawal were an improvement in EEG irregularities and EKG abnormalities and an increase in delta sleep and REM rebound.

In clinical practice, mild withdrawal symptoms are not uncommon if tricyclics are stopped too rapidly and in the all-too-common situation when a patient runs out of medication. Patients experiencing mild symptoms usually respond well to a dose of the interrupted antidepressant. For instance, Stern and Mendels[142] have reported two cases in which apparent imipramine withdrawal symptoms (nausea, malaise, sweating, salivation, dizziness) came on within twelve hours after the patients failed to take their nightly imipramine dosage, that is thirty-six hours after the last dose. These symptoms were relieved in both cases by resuming imipramine. Withdrawal symptoms probably occur with all tricyclics although not all have been formally studied.

If, similar to phenothiazine withdrawal, tricyclic withdrawal symptoms represent cholinergic rebound due to discontinuation of anticholinergic drugs, desipramine cessation might be less likely to elicit such reactions and newer nonanticholinergic antidepressants, such as mianserin or trazodone, might be free of withdrawal symptoms. Nevertheless, it is a good practice to taper antidepressants slowly, dropping the dose by about one-quarter per week, while watching for withdrawal effects as well as reemergence of the original depression. Shatan[134] makes the argument that tricyclics are "addictive" because they produce withdrawal symptoms and the patient becomes dependent on them for a feeling of well-being. It seems more reasonable to interpret the reemergence of depressive dysphoria as the continued presence of a chronic depression, but the existence of "rebound" depression cannot be ruled out.

¶ Antianxiety Drugs

Abrupt withdrawal of barbiturates in addicts produces a characteristic abstinence syndrome. The time course, symptom characteristics, and outcome were carefully delineated in studies on addict volunteers.[68] Barbiturate-type abstinence phenomena have since been shown to occur with a host of chemically different sedative-hypnotics, such as glutethimide, ethchlorvynol, paraldehyde,[154] and methaqualone.[145] Meprobamate, the most popular tranquilizer prior to the introduction of the benzodiazepines, has been found to produce marked withdrawal phenomena. Haizlip and Ewing[61] found that forty-four out of forty-seven patients developed insomnia, vomiting, tremors, muscle twitching, anxiety, anorexia, or ataxia on abrupt withdrawal after forty days on high doses of meprobamate. Eight patients showed hallucinosis and tremors resembling delirium tremens, while three patients developed grand mal seizures. Crawford Little[26] described a case of a thirty-one-year-old nurse who developed a psychotic state with excitement, hostility, and paranoia when coming off 6.4 g/d of meprobamate. Swanson and Okada[144] reported the death of one patient after withdrawal from very high doses of meprobamate.

Benzodiazepines have all but replaced the aforementioned compounds as standard tranquilizers and hypnotic agents. Benzodiazepines are more effective and safer, and because of their slower breakdown and elimination than the previously described drugs, they may be less likely to induce severe withdrawal reactions.[132] Nevertheless, there is extensive literature verifying that all degrees of severity of withdrawal symptoms may follow abrupt discontinuation of the use of any benzodiazepine.[4] In this discussion of benzodiazepine withdrawal, the classification offered by Wikler[154] into "minor" and "major" phenomena will be retained.

"Minor" Abstinence Phenomena

Maletzky and Klotter[90] found the following symptom prevalence in twenty-four patients after abrupt diazepam withdrawal: anxiety (95 percent), agitation (75 percent), insomnia (58 percent), tremor (42 percent), diaphoresis (29 percent), pain (25 percent), depression (17 percent), and nightmares (17 percent). Decreased appetite, nausea, muscle twitching, tachycardia, and dizziness have also been observed during diazepam withdrawal. Similar symptoms have been reported from chlordiazepoxide,[22,67,138] and from oxazepam.[62,98]

The time course of the withdrawal reaction has been established by careful daily observation of patients in a hospital setting. Pevnick and associates[110] observed a thirty-seven-year-old man with a documented daily intake of 30 to 45 mg diazepam over twenty months. Withdrawal was done abruptly under single-blind conditions with placebo substitution. During the first five postwithdrawal days, minimal changes were observed: pulse rate and tremor increased and the patient felt "nervous." During the fifth night, an abstinence syndrome emerged: there was loss of body weight, increased tremor, twitches, muscle cramps, and facial numbness. The syndrome peaked during the sixth night and seventh day at which time the patient stated that he was "kicking" and requested "Valium." He became uncooperative and also reported generalized numbness, blurred vision, and decreased appetite for cigarettes. On day eight, the symptoms began to recede and by days nine and ten, he felt "normal." Body weight returned to normal at day fourteen. A somewhat different course was observed in a thirty-two-year-old man who had taken 15 mg/d diazepam for six years.[155] The first day of placebo substitution was marked by mild symptoms. On the second day, he began to complain of more severe symptoms: anxiety, dizziness, blurred vision, tinnitus, constipation, and palpitations. His condition deteriorated further during the next two days. Readministration of a single dose of 5 mg diazepam produced a remarkable, almost euphoric effect. The duration of marked abstinence symptoms was fifteen days, followed by gradual improvement. The onset of physiological and emotional distress coincided with the plasma diazepam level dropping to 50 percent of baseline level. Hollister and associates[67] found chlordiazepoxide withdrawal symptoms appearing mostly between the fourth and seventh days when plasma levels of the drug were 25 and 10 percent of the original levels, respectively. Oxazepam withdrawal symptoms may begin as early as the first day,[98] which is almost certainly due to the faster metabolic breakdown of this compound.

Rebound insomnia is a withdrawal effect peculiar to benzodiazepines with short or intermediate half-lives. Sleep laboratory studies by Kales and associates[73] showed that discontinuation of flunitrazepam, nitrazepam, and triazolam made sleep more difficult, while flurazepam and diazepam, which have longer half-lives, did not.

"Major" Abstinence Reactions

"Major" abstinence reactions include psychotic manifestations and seizures and closely resemble similar syndromes occurring after barbiturate withdrawal. The clinical picture may be dominated by hallucinations, particularly of the visual type,[49] an organic or paranoid psychosis,[114] seizures,[120] hyperthermia,[119] or a gradual progression from minor symptoms to a full-blown toxic psychosis with delirium, paranoid delusions, seizures, and occasionally, Korsakoff's syndrome.[9,101]

Seizures from diazepam start an average of eight days after abrupt withdrawal,[30] but may occur as early as forty-eight hours.[152] Chlordiazepoxide withdrawal seizures likewise tend to occur after seven to eight days.[67] Case reports of seizures from lorazepam place the onset at three to five days,[38,151] corresponding to its faster metabolism. Instances of seizures from clorazepate and oxazepam withdrawal have also been observed.[38] Prolonged coma for up to six hours has been reported as an unusual complication of diazepam withdrawal seizures.[30]

The time course of psychotic syndromes shows great variability. The onset may range from three to fourteen days after withdrawal.[52] If left untreated, the syndrome tends to remit spontaneously in one to two weeks, but treatment is frequently required because of the severity of the condition. Phenothiazines are generally not helpful[52] and may aggravate the problem by lowering seizure threshold,[114] although occasionally haloperidol[101] or chlorpromazine[9] have been found to be effective. The usual treatment of major abstinence reactions is retreatment with benzodiazepines or the use of pentobarbital[114,131] or phenobarbital.[139]

The salient features of benzodiazepine withdrawal are summarized in table 18–3.

Clinical Considerations

While the minor withdrawal phenomena are not uncommon, the major abstinence reactions are exceedingly rare. Marks[94] estimated their incidence to be less than 1 case per 50 million months of therapeutic benzodiazepine use. The risk of seizures and psychoses is proportional to the length of drug exposure and to dosage. However, uncomfortable symptoms have been reported after withdrawal from as little as 15 mg/d diazepam[63,155] and seizures have been reported after 30 mg/d diazepam, administered for only three months.[120] Previous or concurrent use of other sedative-hypnotics tends to increase the prevalence of withdrawal phenomena,[22,52,90] and in many case reports, the severity of withdrawal reactions was aggravated by the recent or simultaneous withdrawal of other psychotropic drugs or alcohol.[9,37,114,151]

The incidence of withdrawal reactions can

TABLE 18–3 **Benzodiazepine Withdrawal Effects**

	MINOR	MAJOR
Common symptoms	Anxiety, insomnia tremor, diaphoresis	Psychosis, delirium hyperthermia, seizures
Onset	1 to 7 days	3 to 12 days
Approximate duration	1 to 10 days	7 to 21 days
Precipitating factors	Abrupt withdrawal, high dosage, prolonged administration, other sedative-hypnotics, alcohol	
Treatment	Retreatment with benzodiazepines is sometimes required	Retreatment with benzodiazepines, pentobarbital, occasionally antipsychotics

be markedly reduced through gradual tapering rather than abrupt withdrawal. When a patient is on several compounds that are crosstolerant with benzodiazepines, such as sedatives or hypnotics, slow tapering of each compound, one at a time, is likely to minimize the risk of withdrawal effects.

¶ Bibliography

1. ALPERT, M. F., DIAMOND, E. M., and LASKI, E. M. "Anticholinergic Exacerbation of Phenothiazine-induced Extrapyramidal Syndrome," *American Journal of Psychiatry,* 133 (1976): 1073–1075.
2. ANDERSEN, H., and KRISTIANSEN, E. S. "Tofranil Treatment of Endogenous Depressions," *Acta Psychologica et Neurologica Scandinavia,* 34 (1959): 387–397.
3. ASPER, H., et al. "Tolerance Phenomena with Neuroleptics," *European Journal of Pharmacology,* 22 (1973): 287–294.
4. AYD, F. J., JR. "Benzodiazepines: Dependence and Withdrawal," *Journal of the American Medical Association,* 242 (1979): 1401–1402.
5. BACOPOULOS, N. C., et al. "Regional Sensitivity of Primate Brain Dopaminergic Neurons to Haloperidol: Alterations Following Chronic Treatment," *Brain Research,* 157 (1978): 396–401.
6. BACOPOULOS, N. C., et al. "Antipsychotic Drug Action in Schizophrenic Patients: Effect on Cortical Dopamine Metabolism after Long-term Treatment," *Science,* 205 (1979): 1405–1407.
7. BALDESSARINI, R. J., et al. "Tardive Dyskinesia: A Task Force Report of the American Psychiatric Association," in press.
8. BALDESSARINI, R. J., and TARSY, D. "Relationship of the Actions of Neuroleptic Drugs to the Pathophysiology of Tardive Dyskinesia," *International Review of Neurobiology,* 21 (1979): 1–45.
9. BARTEN, H. H. "Toxic Psychosis with Transient Dysmnestic Syndrome Following Withdrawal from Valium," *American Journal of Psychiatry,* 121 (1965): 1210–1211.
10. BATTEGAY, R. "Entziehungs Erscheinungen nach abruptem Absetzen von Neuroleptica als Kriterien zu ihrer Differenzierung," *Nervenarzt,* 37 (1966): 552–556.
11. BENNETT, J. L., and KOOI, K. H. "Five Phenothiazine Derivatives: Evaluation and Toxicity Studies," *Archives of General Psychiatry,* 4 (1961): 413–418.
12. BOURGEOIS, M., and BOUEY, P. "L'antagonisme entre Correcteurs Anti-parkinsoniens et Neuroleptiques. A Propos de Diverses Experiences de Sevrage dont une Personnelle (2e partie)," *Annales de Médico-Psychologie (Paris),* 2 (1976): 699–707.
13. BOWERS, M. B., and ROZITIS, A. "Regional Differences in Homovanillic Acid Concentration After Acute and Chronic Administration of Antipsychotic Drugs," *Journal of Pharmacy and Pharmacology,* 26 (1974): 743–745.
14. ———. "Brain Homovanillic Acid: Regional Changes over Time with Antipsychotic Drugs," *European Journal of Pharmacology,* 39 (1976): 109–115.
15. BROOKS, G. W. "Withdrawal from Neuroleptic Drugs," *American Journal of Psychiatry,* 115 (1959): 931–932.
16. BROWN, G., et al. "Withdrawal from Long-term High-dose Desipramine Therapy: Clinical and Biological Changes," *Archives of General Psychiatry,* 35 (1978): 1261–1264.
17. BURT, D. R., CREESE, I., and SNYDER, S. H. "Antischizophrenic Drugs: Chronic Treatment Elevates Dopamine Receptor Binding in Brain," *Science,* 196 (1977): 326–328.
18. CAHAN, R. B., and PARRISH, D. D. "Reversibility of Drug-induced Parkinsonism," *American Journal of Psychiatry,* 116 (1960): 1022–1023.
19. CHOUINARD, G., and JONES, B. D. "Neuroleptic-induced Supersensitivity Psychosis: Clinical and Pharmacologic Characteristics," *American Journal of Psychiatry,* 137 (1980): 16–21.
20. ———, and ANNABLE, L. "Neuroleptic-induced supersensitivity Psychosis," *American Journal of Psychiatry,* 135 (1978): 1409–1410.
21. CLOW, A., et al. "Striatal Dopamine Receptors Become Supersensitive While Rats Are Given Trifluoperazine for Six Months," *Nature,* 278 (1979): 59–61.
22. COVI, L., et al. "Length of Treatment with Anxiolytic Sedatives and Response to

Their Sudden Withdrawal," *Acta Psychiatrica Scandinavia,* 49 (1973): 51–64.

23. CRANE, G. E. "Rapid Reversal of Tardive Dyskinesia (ltr to ed.)," *American Journal of Psychiatry,* 130 (1973): 1159.
24. ———. "Tardive Dyskinesia and Drug Research," *Psychopharmacology Bulletin,* 9 (1973): 33.
25. ———, and NARANJO, E. R. "Motor Disorders Induced by Neuroleptics," *Archives of General Psychiatry,* 24 (1971): 179–184.
26. CRAWFORD LITTLE, J. "A Case of Primary Addiction to Meprobamate," *British Medical Journal Memoranda,* 2 (1963): 794.
27. DAVIS, J. M. "Overview: Maintenance Therapy in Psychiatry. I. Schizophrenica," *American Journal of Psychiatry,* 132 (1975): 1237–1245.
28. DAVIS, K. L., and ROSENBERG, G. S. "Is There A Limbic System Equivalent of Tardive Dyskinesia?", *Biological Psychiatry,* 14 (1979): 699–703.
29. DAVIS, K. L., HOLLISTER, L. E., and FRITZ, W. C. "Induction of Dopaminergic Mesolimbic Receptor Supersensitivity by Haloperidol," *Life Sciences,* 23 (1978): 1543–1548.
30. DE BARD, M. L. "Diazepam Withdrawal Syndrome: A Case With Psychosis, Seizure and Coma," *American Journal of Psychiatry,* 136 (1979): 104–105.
31. DEGKWITZ, R., and WENZEL, W. "Persistent Extrapyramidal Side Effects After Long-term Application of Neuroleptics," in Brill, H., ed., *Neuropsychopharmacology,* International Congress Series 124, New York: Excerpta Medica Foundation, 1967, pp. 608–615.
32. DEGKWITZ, R., et al. "Der zeitliche Zusammenhang zwischen dem Auftreten persistierender extrapyramidaler Hyperkinesen und Psychose-recidiven nach abrupter Unterbrechung langfristiger neuroleptischer Behandlung chronisch schizophrener kranken," *Arzneim Forsch,* 20 (1970): 890–893.
33. DEMARS, J. C. A. "Neuromuscular Effects of Long-term Phenothiazine Medication, Electroconvulsive Therapy and Leucotomy," *Journal of Nervous and Mental Disease,* 143 (1966): 73–79.
34. DI MASCIO, A., and DEMIRGIAN, E. "Antiparkinson Drug Overuse," *Psychosomatics,* 2 (1970): 596–601.
35. DONLON, P. T., and BLACKER, K. H. "Stages of Schizophrenic Decompensation and Reintegration," *Journal of Nervous and Mental Disease,* 157 (1973): 200–209.
36. DONLON, P. T., and STENSON, R. L. "Neuroleptic Induced Extrapyramidal Symptoms," *Diseases of the Nervous System,* 37 (1976): 629–635.
37. DYSKEN, M. W., and CHAN, C. H. "Diazepam Withdrawal Psychosis: A Case Report," *American Journal of Psychiatry,* 134 (1977): 153.
38. EINARSON, T. R. "Lorazepam Withdrawal Seizures," *Lancet,* 1 (1980): 151.
39. EMMELIN, N. "Supersensitivity Following 'Pharmacological Denervation'," *Pharmacological Review,* 13 (1961): 17–37.
40. ENGELHARDT, D. M., and POLIZOS, P. "Adverse Effects of Pharmacotherapy in Childhood Psychosis," in Lipton, M. A., Di Mascio, A., and Killam, K. F., eds., *Psychopharmacology: A Generation of Progress.* New York: Raven Press, 1978, pp. 1463–1469.
41. ENGELHARDT, D. M., et al. "Phenothiazines in Prevention of Psychiatric Hospitalization. IV. Delay or Prevention of Hospitalization—A Reevaluation," *Archives of General Psychiatry,* 16 (1967): 98–101.
42. ESCOBAR, J. I., and TUASON, V. B. "Neuroleptic Withdrawal Dyskinesia," *Psychopharmacology Bulletin,* 15 (1979): 71–74.
43. FAHN, S., and DAVID, E. "Oral-facial-lingual Dyskinesia Due to Anticholinergic Medication," *Transactions of the American Neurological Association,* 97 (1972): 277–279.
44. FANN, W. E., and LAKE, C. R. "On the Coexistence of Parkinsonism and Tardive Dyskinesia," *Diseases of the Nervous System,* 35 (1974): 324–326.
45. FERHOLT, J. B., and STONE, W. N. "Severe Delirium After Abrupt Withdrawal of Thiothixene in a Chronic Schizophrenic Patient," *Journal of Nervous and Mental Disease,* 150 (1970): 400–403.
46. FINK, E. B. "Unexpected and Prolonged Akinesia. A Case Report," *Journal of Clinical Psychiatry,* 39 (1978): 817–818.
47. FJALLAND, B., and MOLLER-NIELSEN, I. "Enchancement of Methylphenidate-induced Stereotypes by Repeated Administration of Neuroleptics," *Psychopharmacologia,* 34 (1974): 105–109.

48. FLEISCHHAUER, J. "Open Withdrawal of Antiparkinson Drugs in The Neuroleptic-induced Parkinson Syndrome," *International Pharmacopsychiatry,* 10 (1975): 222–229.
49. FLOYD, J. B., JR., and MURPHY, C. M. "Hallucinations Following Withdrawal of Valium," *Kentucky Medical Association Journal,* (1976): 549–550.
50. FORREST, D. V., and FAHN, S. "Tardive Dysphrenia and Subjective Akathisia," (ltr. to editor) *Journal of Clinical Psychiatry,* 40 (1979): 87.
51. FRIEND, W. C., et al. "Effect of Haloperidol and Apomorphine Treatment on Dopamine Receptors in Pituitary and Striatum," *American Journal of Psychiatry,* 135 (1978): 839–841.
52. FRUENSGAARD, K. "Withdrawal Psychosis: A Study of 30 Consecutive Cases," *Acta Psychiatrica Scandinavia,* 53 (1976): 105–118.
53. GALLANT, D. M., et al. "Withdrawal Symptoms After Abrupt Cessation of Antipsychotic Compounds: Clinical Confirmation in Chronic Schizophrenics," *American Journal of Psychiatry,* 121 (1964): 491–493.
54. GARDOS, G., and COLE, J. O. "Maintenance Antipsychotic Therapy: Is the Cure Worse than the Disease?" *American Journal of Psychiatry,* 133 (1976): 32–36.
55. ———, and TARSY, D. "Withdrawal Syndromes Associated with Antipsychotic Drugs," *American Journal of Psychiatry,* 135 (1978): 1321–1324.
56. GLAUBIGER, G., and LEFKOWITZ, R. J. "Elevated Beta-adrenergic Receptor Number After Chronic Propranolol Treatment," *Biochemical Biophysical Res. Communication,* 78 (1977): 720–725.
57. GOGGIN, D. A., and SOLOMON, G. F. "Trihexyphenidyl Abuse for Euphorigenic Effect," *American Journal of Psychiatry,* 136 (1979): 459–460.
58. GRANACHER, R. P., and BALDESSARINI, R. J. "The Usefulness of Physostigmine in Neurology and Psychiatry," in Klawans, H. L., ed., *Clinical Neuropharmacology* vol. 1. New York: Raven Press, 1976, pp. 63–79.
59. GREENBERG, L. M., and ROTH, S. "Differential Effects of Abrupt Versus Gradual Withdrawal of Chlorpromazine in Hospitalized Chronic Schizophrenic Patients," *American Journal of Psychiatry,* 123 (1966): 221–226.
60. GUALTIERI, C., and STAYE, J. "Withdrawal Symptoms After Abrupt Cessation of Amitriptyline in an Eight-year-old Boy," *American Journal of Psychiatry,* 136 (1979): 457–458.
61. HAIZLIP, T. M., and EWING, J. A. "Meprobamate Habituation. A Controlled Clinical Study", *New England Journal of Medicine,* 258 (1958): 1181–1186.
62. HANNA, S. M. "A Case of Oxazepam (Serenid D) Dependence," *British Journal of Psychiatry,* 120 (1972): 443–445.
63. HASKELL, D. "Withdrawal of Diazepam," (Ltr. to Editor) *Journal of the American Medical Association,* 233 (1975): 135.
64. HERSHON, H. I., KENNEDY, P. F. and MCGUIRE, R. J. "Persistence of Extrapyramidal Disorders and Psychiatric Relapse After Withdrawal of Long-term Phenothiazine Therapy," *British Journal of Psychiatry,* 120 (1972): 41–50.
65. HOFF, H., and HOFFMAN, G. "Der Persistierende Extrapyramidale Syndrom bei Neuroleptikatherapie," *Wiener Medizinische Wochenschrift,* 117 (1967): 14–17.
66. HOGARTY, G. E., GOLDBERG, S. C. and SCHOOLER, N. R. "Drug and Sociotherapy in the Aftercare of Schizophrenic Patients. II. Two-year relapse rates," *Archives of General Psychiatry,* 31 (1974): 603–608.
67. HOLLISTER, L. E., MOTZENBECKER, F. P., and DEGAN, R. O. "Withdrawal Reactions from Chlordiazepoxide ('Librium')," *Psychopharmacologia,* 2 (1961): 63–68.
68. ISBELL, H., and WHITE, W. M. "Clinical Characteristics of Addictions," *American Journal of Medicine,* 14 (1953): 558–565.
69. JACKSON, D. M., et al. "The Effect of Long-term Penfluridol Treatment on the Sensitivity of the Dopamine Receptors in the Nucleus Accumbens and the Corpus Striatum," *Psychopharmacologia,* 45 (1975): 151–155.
70. JACOBSON, G., BALDESSARINI, R. J., and MANSCHREK, T. "Tardive and Withdrawal Dyskinesia Associated with Haloperidol," *American Journal of Psychiatry,* 131 (1974): 910–913.
71. JELLINEK, T. "Mood Elevating Effect of Trihexyphenidyl and Biperiden in Individuals Taking Antipsychotic Medication,"

Diseases of the Nervous System, 38 (1977): 353–355.

72. ———, GARDOS, G. and COLE, J. O. "Adverse Effects of Antiparkinson Drug Withdrawal," paper presented at 133rd Annual Meeting of the American Psychiatric Association, San Francisco, Calif., May 5–9, 1980.
73. KALES, A., et al. "Rebound Insomnia. A Potential Hazard Following Withdrawal of Certain Benzodiazepines," *Journal of the American Medical Association,* 241 (1979): 1692–1695.
74. KENNEDY, P. F. "Chorea and the Phenothiazines," *British Journal of Psychiatry,* 115 (1969): 103–104.
75. KLAWANS, H. L., and RUBOVITS, R. "An Experimental Model of Tardive Dyskinesia," *Journal of Neurological Transmission,* 33 (1972): 235–246.
76. KLETT, C. J., and CAFFEY, E., JR. "Evaluating the Long-term Need for Antiparkinson Drugs by Chronic Schizophrenics," *Archives of General Psychiatry,* 26 (1972): 374–379.
77. KRAFT, T. B. "Ernstige Abstinentieverschigiselen na het Gebruik van Clomipramine," *Med. T. Geneesk.* 121 (1977): 1293.
78. KRAMER J., KLEIN, D., and FINK, M. "Withdrawal Symptoms Following Discontinuation of Imipramine Therapy," *American Journal of Psychiatry,* 118 (1961): 549–550.
79. KRUSE, W. "Treatment of Drug-induced Extrapyramidal Symptoms (A Comparative Study of Three Antiparkinson Agents)," *Diseases of the Nervous System,* 21 (1960): 79–81.
80. KUMAR, B. B. "Thioridazine, Drug Holidays, and Incidence of Vomiting," (Letter to the Editor), *Journal of the American Medical Association,* 239 (1978): 25.
81. LACOURSIERE, R. B., SPOHN, H. E., and THOMPSON, K. "Medical Effects of Abrupt Neuroleptic Withdrawal," *Comprehensive Psychiatry,* 17 (1976): 285–294.
82. LA POLLA, A., and NASH, L. R. "Treatment of Phenothiazine-induced Parkinsonism with Biperiden," *Current Therapeutic Research,* 7 (1961): 536–541.
83. LERNER, P., et al. "Haloperidol: Effect of Long-term Treatment on Rat Striatal Dopamine Synthesis and Turnover," *Science,* 197 (1977): 181–183.
84. LONGHREN, T. P., BROWN, W. A., and WILLIAMS, B. W. "Serum Prolactin and Clinical State During Neuroleptic Treatment and Withdrawal," *American Journal of Psychiatry,* 136 (1979): 108–110.
85. LUCHINS, D. J., FREED, W. J., and WYATT, R. J. "The Role of Cholinergic Supersensitivity in the Medical Symptoms of Antipsychotic Withdrawal," *American Journal of Psychiatry,* in press.
86. MCANDREW, J. B., CASE, Q., and TREFFERT, D. A. "Effects of Prolonged Phenothiazine Intake on Psychotic and Other Hospitalized Children," *Journal of Autism and Childhood Schizophrenia,* 2 (1972): 75–91.
87. MCCARTHY, J. J. "Tardive Psychosis," *American Journal of Psychiatry,* (Letter to the Editor), 135 (1978): 625–626.
88. MCLELLAND, H. A., BLESSED, G., and BHATE, S. "Abrupt Withdrawal of Antiparkinsonian Drugs in Chronic Schizophrenic Patients," *British Journal of Psychiatry,* 125 (1974): 514–516.
89. MACVICAR, K. "Abuse of Antiparkinson Drugs by Psychiatric Patients," *American Journal of Psychiatry,* 134 (1977): 809–811.
90. MALETZKY, B. M., and KLOTTER, J. "Addiction to Diazepam," *The International Journal of the Addictions,* 11 (1976): 95–115.
91. MANDEL, W., CLAFFEY, B., and MARGOLIS, L. H. "Recurrent Thioperazine-induced Extrapyramidal Reaction Following Placebo Substitution for Maintenance Antiparkinson Drug," *American Journal of Psychiatry,* 118 (1961): 351–352.
92. MANDEL, W., and OLIVER, W. A. "Withdrawal of Maintenance Antiparkinson Drug in the Phenothiazine-induced Extrapyramidal Reaction," *American Journal of Psychiatry,* 118 (1961): 350–351.
93. MARCOTTE, D. B. "Neuroleptics and Neurological Reactions," *Southern Medical Journal,* 66 (1973): 321–324.
94. MARKS, J. "The Benzodiazepines," Baltimore: University Park Press, 1978.
95. MARRIOTT, P. "Dependence on Antiparkinsonian Drugs," *British Medical Journal,* (Letter to the Editor), 1 (1976): 152.
96. MARRIOTT, P., and HEIP, A. "Drug Monitoring at an Australian Depot Phenothiazine Clinic," *Journal of Clinical Psychiatry,* 39 (1978): 206–212.
97. MELTZER, H. Y., and FANG, V. S. "The

Effect of Neuroleptics on Serum Prolactin in Schizophrenic Patients," *Archives of General Psychiatry,* 33 (1976): 279–286.

98. MENDELSON, G. "Withdrawal Reactions After Oxazepam," *The Lancet,* (letter to the editor), 11 March, 1978, p. 565.

99. MILLER, R. J., and HILEY, C. R. "Anti-muscarinic Properties of Neuroleptics and Drug-induced Parkinsonism," *Nature,* 248 (1974): 596–597.

100. MILLER, R. R., et al. "Propranolol-withdrawal Rebound Phenomenon," *New England Journal of Medicine,* 293 (1975): 416–418.

101. MINTER, R., and MURRAY, G. B. "Diazepam Withdrawal: A Current Problem in Recognition," *Journal of Family Practice,* 7 (1978): 1233–1235.

102. MOLLER-NIELSEN, I., et al. "Pharmacology of Neuroleptics Upon Repeated Administration," *Psychopharmacologia,* 34 (1974): 95–104.

103. MORTON, M. R. "A Study of the Withdrawal of Chlorpromazine or Trifluoperazine in Chronic Schizophrenia," *American Journal of Psychiatry,* 124 (1968): 143–146.

104. MULLER, P., and SEEMAN, P. "Brain Neurotransmitter Receptors After Long-term Haloperidol: Dopamine, Acetylcholine, serotonin, 2-noradrenergic and Naloxone Receptors," *Life Sciences,* 21 (1977): 1751–1758.

105. ———. "Dopaminergic Supersensitivity After Neuroleptics: Time Course and Specificity," *Psychopharmacology,* 60 (1978): 1–11.

106. ORLOV, P., et al. "Withdrawal of Antiparkinson Drugs," *Archives of General Psychiatry,* 25 (1971): 410–412.

107. PAKES, G. E. "Lithium Toxicity with Phenothiazine Withdrawal," *The Lancet,* ii (1979): 701.

108. PAUL, G. L., TOBIAS, L. L., and HOLLY, B. L. "Maintenance Psychotropic Drugs in the Presence of Active Treatment Programs," *Archives of General Psychiatry,* 27 (1972): 106–115.

109. PECKNOLD, J. C., et al. "Lack of Indication for Use of Antiparkinson Medication," *Diseases of the Nervous System,* 32 (1971): 538–541.

110. PEVNICK, J. S., JASINSKI, D. R., and HAERTZEN, C. A. "Abrupt Withdrawal from Therapeutically Administered Diazepam. Report of a Case," *Archives of General Psychiatry,* 35 (1978): 995–998.

111. PITT, B. "Withdrawal Symptoms After Stopping Phenelzine?" *British Medical Journal,* 2 (1974): 332–333.

112. POLIZOS, P., et al. "Neurological Consequences of Psychotropic Drug Withdrawal in Schizophrenic Children," *Journal of Autism and Childhood Schizophrenia,* 3 (1973): 247–253.

113. POST, R. M., and GOODWIN, F. K. "Time Dependent Effects of Phenothiazines on Dopamine Turnover in Psychiatric Patients," *Science,* 190 (1975): 488–489.

114. PRESKORN, S. J., and DENNER, L. J. "Benzodiazepines and Withdrawal Psychosis. Report of Three Cases." *Journal of the American Medical Association,* 237 (1977): 36–38.

115. PRIEN, R. F., and KLETT, C. J. "An Appraisal of the Long-term Use of Tranquilizing Medication with Hospitalized Chronic Schizophrenics," *Schizophrenia Bulletin,* 5 (1972): 64–73.

116. PRYCE, I. E., and EDWARDS, H. "Persistent Oral Dyskinesia in Female Mental Hospital Patients," *British Journal of Psychiatry,* 112 (1966): 983–987.

117. QUITKIN, F., et al. "Tardive Dyskinesia: Are First Signs Reversible?" *American Journal of Psychiatry,* 134 (1977): 84–87.

118. REIMER, F. "Das Absetzungs—Delir," *Nervenarzt,* 36 (1965): 446–447.

119. RELKIN, R. "Death Following Withdrawal of Diazepam," *New York State Journal of Medicine,* 66 (1966), 1770–1772.

120. RIFKIN, A., KLEIN, D. F., and QUITKIN, F. "Withdrawal from Diazepam," *Journal of the American Medical Association,* letter to editor, 238 (1977): 306.

121. RIFKIN, A., et al. "A Study of Abrupt Lithium Withdrawal," *Psychopharmacologia,* (Berl.) 44 (1975): 157–158.

122. RIFKIN, A., et al. "Are Prophylactic Antiparkinson Drugs Necessary?" *Archives of General Psychiatry,* 35 (1978), 483–489.

123. ROBINS, A. H. "Depression in Patients with Parkinsonism," *British Journal of Psychiatry,* 128 (1976): 144–145.

124. ROSSER, R., and HERXHEIMER, A. "Chlorpromazine, Lithium and Metoclopramide: Unrecognized Synergistic and Antagonistic Effects," *Lancet,* 2 (1979): 97–98.

125. ROY, P., et al. "Studies with Antiparkin-

sonian Drugs with Chronic Psychiatric Patients," *International Journal of Neuropsychiatry,* 2 (1966): 65–69.

126. RUBINSTEIN, J. S. "Abuse of Antiparkinson Drugs," *Journal of the American Medical Association,* 239 (1978): 2365–2366.

127. SALE, I., and KRISTALL, H. "Schizophrenia Following Withdrawal from Chronic Phenothiazine Administration: A Case Report," *Australian and New Zealand Journal of Psychiatry,* 12 (1978): 73–75.

128. SATHANANTHAN, G., and GERSHON, S. "Imipramine Withdrawal: An Akathisia-like Syndrome," *American Journal of Psychiatry,* 130 (1973): 1286–1287.

129. SCATTON, B. "Differential Regional Development of Tolerance to Increase in Dopamine Turnover Upon Repeated Neuroleptic Administration," *European Journal of Pharmacology,* 46 (1977): 363–369.

130. SCATTON, B., GLOWINSKI, J., and JOULON, L. "Dopamine Metabolism in the Mesolimbic and Mesocortical Dopaminergic Systems After Single or Repeated Administrations of Neuroleptics," *Brain Research,* 109 (1976): 184–189.

131. SELIG, J. W., JR. "A Possible Oxazepam Abstinence Syndrome," *Journal of the American Medical Association,* 198 (1966): 279–280.

132. SELLERS, E. M. "Clinical Pharmacology and Therapeutics of Benzodiazepines," *Canadian Medical Association Journal,* 118 (1978): 1533–1538.

133. SHARPLESS, S. K. "Reorganization of Function in the Nervous System—Use and Disuse," *Annual Review of Physiology,* 26 (1964): 357–388.

134. SHATAN, C. "Withdrawal Symptoms After Abrupt Termination of Imipramine," *Canadian Psychiatric Association Journal,* 11 (suppl.) (1966): 150–158.

135. SIMPSON, G. M., and KUNZ BARTHOLINI, E. "Relationship of Individual Tolerance, Behavior, and Phenothiazine Produced Extrapyramidal System Disturbance," *Diseases of the Nervous System,* 29 (1968): 269–274.

136. SIMPSON, G. M., AMIN, M., and KUNZ, E. "Withdrawal Effects of Phenothiazines," *Comprehensive Psychiatry,* 6 (1965): 347–351.

137. SKIRBOLL, L. R., and BUNNEY, B. S. "The Effects of Acute and Chronic Haloperidol Treatment on Spontaneously Firing Neurons in the Caudate Nucleus of the Rat," *Life Sciences,* 25 (1979): 1419–1434.

138. SLATER, J. "Suspected Dependence on Chlordiazepoxide Hydrochloride (Librium)," *Canadian Medical Association Journal,* (Letter to editor), 95 (1966): 416–417.

139. SMITH, E. E., and WESSON, D. R. "A New Method for Treatment of Barbiturate Dependence," *Journal of the American Medical Association,* 213 (1970): 294–295.

140. SNYDER, S. H. "Receptors, Neurotransmitters and Drug Responses," *New England Journal of Medicine,* 300 (1979): 465–472.

141. SNYDER, S. H., GREENBERG, D., and YAMAMURA, H. I. "Antischizophrenic Drugs and Brain Cholinergic Receptors," *Archives of General Psychiatry,* 31 (1974): 58–61.

142. STERN, S., and MENDELS, J. "Withdrawal Symptoms During the Course of Imipramine Therapy," *Journal of Clinical Psychiatry,* 41 (1980): 66–69.

143. STRATAS, N. E., et al. "A Study of Drug Induced Parkinsonism," *Diseases of the Nervous System,* 24 (1963): 180.

144. SWANSON, L. A., and OKADA, T. "Death After Withdrawal From Meprobamate," *Journal of the Ameican Medical Association,* 184 (1963): 780–781.

145. SWARTZBURG, M., LIEB, J., and SCHWARTZ, A. H. "Methaqualone Withdrawal," *Archives of General Psychiatry,* 29 (1973): 46–47.

146. TAMMINGA, C. A., et al. "A Neuroendocrine Study of Supersensitivity in Tardive Dyskinesia," *Archives of General Psychiatry,* 34 (1977): 1199–1203.

147. TARSY, D., and BALDESSARINI, R. J. "Behavioral Supersensitivity to Apomorphine Following Chronic Treatment with Drugs Which Interfere with the Synaptic Function of Catecholamines." *Neuropharmacology,* 13 (1974): 927–940.

148. THORNTON, E. W., and THORNTON, B. P. "Tardive Dyskinesias from the Major Tranquilizers," *Journal of the Florida Medical Association,* 60 (1973): 24–26.

149. TRENDELENBERG, U. "Mechanisms of Supersensitivity and Subsensitivity to Sympathomimetic Amines," *Pharmacology Review,* 18 (1966): 629–640.

150. TUNE, L. E., and COYLE, J. T. "Serum Anti-

cholinergics and Extrapyramidal Symptoms," presented at the 132nd Annual Meeting of the American Psychiatric Association, Chicago, 1979.

151. TYRER, P. Reply to Einarson, T.R.: "Lorazepam Withdrawal Seizures," *The Lancet,* 1, (1980): 151.

152. VYAS, I., and CARNEY, M. W. P. "Diazepam Withdrawal Fits," *British Medical Journal,* 4 (October 1975): 44.

153. WHITTAKER, C. B., and HOY, R. M. "Withdrawal of Perphenazine in Chronic Schizophrenia," *British Journal of Psychiatry,* 109 (1963): 422–427.

154. WIKLER, A. "Diagnosis and Treatment of Drug Dependence of the Barbiturate Type," *American Journal of Psychiatry,* 125 (1968): 758–765.

155. WINOKUR, A., et al. "Withdrawal Reaction from Long-term, Low-dosage Administration of Diazepam," *Archives of General Psychiatry,* 37 (1980): 101–105.

156. YARBROUGH, G. G. "Supersensitivity of Caudate Neurons After Repeated Administration of Haloperidol," *European Journal of Pharmacology,* 31 (1975): 367–369.

157. YEPES, L. E., and WINSBERG, B. G. "Vomiting During Neuroleptic Withdrawal in Children," *American Journal of Psychiatry* 134 (1977): 574.

CHAPTER 19

THE CHRONIC MENTALLY ILL

John A. Talbott

¶ Historical Background

THE CARE and treatment of the chronic mentally ill date back to the founding of America. From the early Colonial period, there are reports of families who were distressed by the conduct and behavior of their kin and requested the court's permission to build outhouse-like cells in which to house their mentally ill relatives.[13] Later, jails, workhouses, and almshouses housed the chronic mentally ill.[49] An early precursor of "dumping" occurred in some Massachusetts communities when mentally ill persons were transported over county or town lines in order to shift the responsibility for their care to another community.[13]

In the mid 1800s, Dorothea Dix, appalled by the shabby treatment and dismal surroundings which localities provided for the severely and chronically mentally ill, successfully crusaded to have the states assume the burden for their care and treatment. These institutions were intended to provide the best available treatment and care of the mentally ill, known at that time as "moral treatment." Moral treatment followed Pinel's example in Paris of "striking off the chains" and was also modeled on Tuke's establishment, the York Retreat in England, which emphasized a humane, familial-like atmosphere and pleasant, open settings, with a minimum of physical restraint and a maximum of structured activity.[13,49]

The numbers of the chronic mentally ill in America soon grew larger than the institutions' capacity to provide humane housing. This situation was due to increasing immigration from Europe, the impersonality of the industrial era, and the "aging-into" chronicity of the seriously mentally ill.[11] Ultimately, state hospitals provoked scandals similar to those that motivated Dorothea Dix to agitate for elimination of local community responsibility for care of the mentally ill.[16]

As America entered the twentieth century, society was ripe for experimentation with new methods in the care and treatment of the mentally ill. The combined efforts of dedicated individuals, such as Clifford Beers,[6] accompanied by the establishment of alternative treatment settings (psychopathic hospitals, child guidance clinics, outpatient clinics, general psychiatric units) facilitated the shift from a single-sited service system into a pluralistic one. But inherent in this development was one of the major problems of the chronic mentally ill—the fragmentation of responsibility for providing the services needed by these patients. While the state hospital was overcrowded, understaffed, and inhumane, it did provide the services needed by the severely and chronically mentally ill, such as psychiatric and medical treatment, social and vocational rehabilitation, and custodial services, including food and lodging. However, with the establishment of multiple institutions and services, there was no longer any single institution responsible for the severely and chronically mentally ill population.

The foundation blocks for the movement known as community psychiatry began to be laid during the first half of the twentieth century. These included: preventive psychiatry; group, family, and systems treatment; home care; walk-in clinics; and emergency room services.[64] In addition, the philosophic basis of community psychiatry began to be articulated—that it was better to treat persons in the community than in hospitals, that community care was cheaper, and that communities would respond to the challenge to provide the necessary services.[23]

With the introduction of phenothiazines in the mid-1950s, American psychiatrists finally had the technological tool that enabled them to begin to move severely and chronically mentally ill persons from institutional to community settings, and to control psychotic symptoms in newly discovered cases, thus obviating long-term hospitalization and resultant institutionalization.[8,21]

Two other developments hastened this shift from the institution to the community. First, the federal government's assumption of funding for the poor and aged under Medicaid and Medicare, as well as funding for daily needs of the disabled indigent under Supplementary Security Income (SSI); and second, the pressure from judicial, legislative, and regulatory bodies to treat patients in the "least restrictive setting," to make involuntary admissions to mental hospitals increasingly difficult, to broaden the concept of informed consent, and to ensure the patient's right to refuse treatment. These economic and legal pressures, combined with the previously mentioned technological and philosophic developments, led to the beginning of the movement now known as deinstitutionalization.[61]

Deinstitutionalization is commonly defined as having two parts: first, shifting the locus of care from institutions to community settings; and second, blocking the admission of new patients into institutions.[2] The result of this movement was awesome. In 1955, with the patient population at an all time high in the nation, state hospitals housed 560,000 patients. By 1978, this figure had dropped to less than 150,000.[66]

Where did all these people go? Many went to live in shabby welfare hotels, flophouses, and single-room occupancy dwellings, wandering the nearby streets during the day. Many died, both in the hospital and in transition from hospital to community, and, in the earliest days of deinstitutionalization, many returned home.

While it is generally recognized that state hospitals have reduced their populations by over two-thirds and that numerous chronically mentally ill persons now walk aimlessly in America's cities, it is disconcerting to learn that the percentage of Americans housed in institutions has not changed at all since 1950.[28] While state hospitals have shrunk by two-thirds and tuberculosis sanitaria have disappeared, nursing home populations have tripled. Thus the movement is more accurately described as transinstitutionalization than as deinstitutionalization.

¶ Definitions

Up to this point, I have used the term "chronic mentally ill" without defining it. The conceptualization of this segment of the population did not occur until after the deinstitutionalization movement had begun. Before that, we referred either to the mentally ill as a single entity or to specific diagnostic groups (schizophrenics, neurotics, and the like), whether acutely or chronically ill. The deinstitutionalization movement, however, demonstrated that we needed to be concerned with a subset of the mentally ill—individuals with many diagnoses and levels of disability. The features that bound them together were their inability to survive unassisted in the community, their tendency toward episodic or chronic mental illness, and the fact that in earlier times they would probably have been housed in state mental hospitals. It was this feature that prompted Bachrach to define the population as "those individuals who are, have been, or might have been, but for the deinstitutionalization movement, on the rolls of long-term mental institutions, especially state hospitals."[2] Such persons may or may not currently be in mental hospitals. They are of all ages, including children, and their illnesses have received a variety of diagnoses, but primarily they suffer from the major psychoses (for example, chronic schizophrenia), chronic recurrent affective disorders, and severe character disorders. Such patients must be distinguished from those persons who may be in long-term psychotherapy, but who do not have chronic disability.

Chronic is a modifier that is usually defined as occurring for "a long time" or over "a long duration."[67] Most states define chronicity of illness as one or two years of hospitalization. The chronic disabled, however, can be easily defined as those meeting federal eligibility standards for Supplemental Security Income (SSI).[43]

The terms "chronic mental illness," "chronic mentally ill," and "chronic mental patient" are stigmatizing and simplistic and are, unfortunately, interpreted by some as synonymous with hopelessness, deterioration, and regression. While "long-term patient"[36] and "those in need of long-term continuing care and rehabilitation"[40] have been suggested as preferred alternatives, the terms "chronic mentally ill" and "chronic mental patient" will be used here because they are descriptive, universally understood, and generally employed by those in the field.[47,63,64]

¶ Characteristics of the Patient Population

Bachrach[3] has demonstrated that there are five subgroups among the chronic mentally ill. In the community, there are those who have been hospitalized in state hospitals and there are those who have never been institutionalized. In the hospital, there are those who are long-stay patients (almost half of state hospital residents have been there for more than five years); those who are recent admissions and who will soon be released to the community; and the new long-stay patients (some 10 to 15 percent of new admissions) who will continue to need some kind of "highly structured care." How many persons are encompassed in these five groups?

An epidemiological reconstruction must be attempted to acertain how many patients would be in state hospitals were it not for the deinstitutionalization movement. Minkoff[43] estimates that there are a total of 1,100,000 schizophrenics in the United States, of whom, 900,000 are in the community and 200,000 are in institutions (primarily nursing homes, state hospitals, and prisons). He also suggests that there are between 600,000 and 800,000 depressives, almost all of whom are in the community. Further, almost 1,000,000 elderly persons are in institutions (nursing homes and state hospitals) and between 600,-000 and 1,250,000 of the elderly living in the community are psychotic. Finally, assessing the numbers of mentally disabled individuals, Minkoff states that there are 1,762,000 institutionalized severely disabled (those who have been in an institution more than thirty days) and 225,000 severely disabled individuals living in the community (who receive SSI for mental illness). Thus, while the groups designated are not always separate (for example, there are elderly schizophrenics who are severely mentally disabled), at a minimum there are about 1 to 2 million, at a maximum 5 to 7 million chronically mentally ill in the United States.

The problems and needs of this population are complex. As articulated by the Conference on the Chronic Mental Patient,[1] their problems include: "extreme dependency needs, high vulnerability to stress, and difficulty coping with the demands of everyday living, resulting in difficulty securing adequate income and housing and holding down a job." The needs of the chronic mentally ill include medical and psychiatric treatment, social and vocational rehabilitation, and the components of everyday survival (housing, food, clothes, heat, and the like). As was stated earlier, in previous eras all these needs were filled, however poorly, by the state hospital; now they must be met by various sources in the community, through a method described by some as a "scrounging system."

Where the chronic mentally ill are housed is now moderately well known. In the early days of deinstitutionalization, 70 to 80 percent returned to their own or relatives' homes, and only 20 to 25 percent left institutions to live alone or in boarding homes. Now, however, the percentages are reversed, and only 20 to 23 percent return home; the rest go to suboptimal locations (38 percent to hotels, 11 percent to nursing homes, 28 percent to undetermined locations).[43] In one study in California, almost 50 percent of patients discharged five years earlier were now residing in board and care homes.[32]

There are considerable differences among the states in the degree of deinstitutionalization. While all states have experienced a reduction in their state hospital censuses, the speed with which the process took place has varied. In retrospect, Minkoff concluded that states that did better jobs of caring for the chronic mentally ill, had a "moderate population density, and availability of resources and living accommodations for the discharged patients."[43]

How well the chronic mentally ill function in the community is dependent on several variables: readmission due to relapse, symptomatology, vocational history, socialization, and aftercare involvement.[43] The single best predictor of readmission is the number of previous hospital admissions. Prior to deinstitutionalization, fewer than 25 percent of admissions were readmissions, whereas today the rate exceeds 60 percent in many states. Both continuation on medication and the number of aftercare visits can help to decrease readmission rates. However, a major problem results from the lack of provision of both these services to the chronic mentally ill. Studies have shown that 38 percent of discharged patients receive no aftercare referral and only 35 percent of schizophrenics applying for treatment receive it. In addition, compliance is a barrier to continuity of treatment—in one study, only 10 percent of patients from a state hospital that closed contacted the outpatient facility. Minkoff[43] has concluded that fewer than 25 percent of discharged patients continue in regular aftercare programs and fewer than 50 percent continue to take their medications.

It is not surprising, therefore, to discover that two-thirds of the chronic mentally ill living in the community have mild to moderate symptomatology, while only one-third are asymptomatic. The socialization of the chronically ill is similar—only 25 percent are fairly socially active—approximately three-fourths live isolated lives.[43]

Finally, the work history of discharged patients reveals the true ravages of serious and chronic mental illness. Only 30 to 50 percent of discharged patients are employed within the first six months following discharge, and this percentage drops to 20 to 30 percent at the end of a year. While previous employment (as with hospitalization and all other functional predictors) is the best predictor of future employability, the fact that half of those working before admission do not return to work and that 70 percent of those who do work, return to less-skilled jobs, indicates the effects of these illnesses in marketplace terms.[43]

¶ Economic Issues Regarding the Chronically Ill

The economic factors relevant to the care and treatment of the chronic mentally ill have an enormous impact on individual patient care as well as on deinstitutionalization policies. Sharfstein, Turner, and Clark[54] have analyzed two of the most important economic issues: the costs of chronic mental illness and the cost-benefit analyses of treatment and care in institutional versus community settings.

In fiscal year 1977, health care in this country accounted for 9 percent of the gross national product (GNP). Of that, 15 percent was allocated to mental health care. Forty percent of the nation's health care is now paid for with federal dollars, through Medicaid and Medicare, and over a billion dollars is spent by the federal government on public assistance to the chronic mentally ill through SSI.[54]

The direct costs of mental illness, those incurred by provision of psychiatric services, were 14.5 billion dollars in 1974—over 1 percent of the gross national product. Over one-half of these monies went toward institutional services—30 percent to nursing homes and 23 percent to state and county mental hospitals. Those sectors of the mental health system providing the care and treatment for the vast majority of the severely and chronically mentally ill consume much smaller portions of our expenditures: general hospitals, 12 percent; private psychiatrists, 9 percent; drugs, 5 percent; freestanding Out Patient Clinics (OPDs), 5 percent; community mental health centers, 4 percent; general medical services and halfway houses, 3 percent; private mental hospitals, 3 percent; and private psychologists, 1 percent.[53] It is abundantly clear from this, that money is not going where the patients are. Current funding patterns direct monies toward institutional facilities rather than toward community resources, and the problem of underfunding may be more correctly described as maldistribution rather than inadequacy of funds.

The indirect costs of mental illness in 1974 were estimated to be even greater than the direct costs—almost 20 billion dollars. These costs are allocated for loss of labor and loss of production due to mental illness. In all, the mental health bill in 1974 is estimated to have approached 37 billion dollars—of which 87 percent can be earmarked as having been spent on the care and treatment of the chronic mentally ill.[54] Thus the myth that America spends its mental health dollars on the "walking well" and on long-term intensive psychotherapy seems to be refuted. Again, with 32 billion dollars going toward the care and treatment of the chronically ill, the issue does not seem to be that insufficient funds are being expended on this population, but that they are not being allocated to the right institutions, services, and programs.

Several recently published studies have explored the issue of whether it is cheaper to treat patients in community or in institutional settings. Because state hospitals provide a variety of services (housing, food, heat, medical and psychiatric treatment, and so forth) under one roof, it has been easier to calculate the cost of institutional services. When these services are provided in the community, they are both more difficult for patients to acquire and more difficult to cost out. Despite the lack of accurate cost comparisons, state policy makers and community mental health advocates promoted the concept of deinstitutionalization in part because it would save money. Is this claim true?

Sharfstein and Nafziger[53] figured the costs of treatment for one patient over a fifteen-year period, comparing institutional to community care. They found that the costs were roughly the same for the first three years, but that after that, community care cost 10 percent less than hospital treatment.

Murphy and Datel[44] studied fifty-two patients and found that for their twelve stratified groups, all but one cost more in the hospital than in the community. On an average, the yearly saving to the U.S. economy for community care was 25 percent over that of institutional care. However, it should be noted that all persons whose costs were calculated were judged to be "successfully" deinstitutionalized, that one-third were mentally retarded, that recidivists were dropped from the study, and that the data were extrapolated from figures used after persons had been deinstitutionalized an average of only 8.5 months. For the state government, however, the saving in deinstitutionalizing these more intact patients was between 65 and 80 percent, since care in the community becomes a higher federal expense.

In another cost-benefit analysis, Weisbrod, Test, and Stein[69] examined the cost of their program in community care versus hospitalization and traditional aftercare. They found that the cost of both exceeded $7,200 per patient a year, although less than half of this expense could be attributed to direct treatment services. While the community program had both additional costs and additional benefits for the patients, it was shown that community care resulted in a 5 percent savings for the total package of care and treatment. It should be noted that the increased costs of community care derived from the added treatment services provided and the added benefits were paid for by the outpatients, whose income was double that of the hospital patients.

In sum, these controlled studies demonstrate a 5 to 10 percent cost-effectiveness advantage for community care; a 25 percent cost reduction for selected populations of healthier patients and mentally retarded persons; and a 65 to 80 percent saving for the state resulting from the discharge of more intact persons previously housed in institutions. The latter finding, however, points to one of the problems our funding system has encouraged. By allowing state governments to effect substantial savings by deinstitutionalizing patients, the federal government covertly encourages deinstitutionalization, and in the absence of adequate community services and funding of programs, poorer care and treatment frequently result.

In addition, while it may be cheaper for the United States to have an exclusively community-based care system, the institutional system is still in place at the present time, and adequate funding of both systems has become less feasible during periods of economic recession.

¶ The Problems of the Chronic Mentally Ill

The principal concern regarding the chronic mentally ill is their continued poor care and treatment. Most professionals would agree, that in the twenty years since deinstitutionalization began, whether in community or institutional settings, the quality of the patient's care has not improved. The media have described vividly the consequences of moving hundreds of thousands of discharged mental patients into communities where they are ill-equipped to survive. The media also continue to publicize the poor conditions found in most of our public mental hospitals. For the public, the problems of the chronically ill appear in the form of shabbily dressed, bizarre, or demented-looking expatients wandering the cities' streets or crowded into mental patient ghettos; for professionals the problems continue to center on the inadequacy of care and treatment for this population. There are many reasons for this publicly visible problem and for this professional concern. To assess the opinions of psychiatrists about the problems posed by and facing the chronic mentally ill, Talbott[59] conducted a survey in 1975 and found the following:

The most commonly mentioned problems were:

- The failure of deinstitutionalization to provide patients in either hospital or community settings with good treatment and care
- The inadequacy, maldistribution and discrimination in funding for this population
- The absence of a continuum of community care and housing facilities
- The lack of a model service system

- The inadequate number of housing and job opportunities, as well as rehabilitation services
- The negative attitudes about the chronic mentally ill held by legislators, the public, and mental health professionals
- The lack of definition of the role of psychiatrists and others in caring for this group
- The lack of adequately trained professionals to treat the chronic mentally ill
- The lack of involvement of families in the treatment of the population
- The lack of adequate descriptions of effective programs for the chronically ill
- The lack of continuity of care
- The paucity of community care facilities
- The problems of long-term use of psychopharmacological agents
- The lack of knowledge as to which patients should be treated at what level of care and in what facilities
- The lack of responsibility for coordination of the delivery of care
- The absence of a single, responsible person to make sure care is provided to those in need

From this list, it is apparent that several different but related areas contribute to the basic problem of inadequate care and treatment of the chronically mentally ill. These can be grouped into problems in treatment, community care, governmental responsibility, and societal and professional attitudes.

Treatment

It is clear that despite the introduction of phenothiazines in the 1950s, the treatment of the chronic mentally ill in either institutional or community settings has not met with success. The quality of state mental hospitals seems to have fallen, and their original goal of providing humane custodial care in asylum settings seems unachievable today. The resistance of patients, families, and third-party reimbursers to insure continuing long-term care is also a significant problem, as is the lack of respite facilities and emergency care for the chronically ill.

Too few professionals have received appropriate training in the care and treatment of this population and there is too little interchange between academic settings, wherein much expertise lies, and public sectors, in which most chronic patients receive their care. The roles of all mental health professionals, especially those of psychiatrists and non-psychiatric physicians, need clarification. In addition, research should be accelerated into what works, for whom, and in what setting; what prevents, maintains, or encourages chronicity; and what constitutes effective service delivery.

Community Care

Supportive services that enable patients to survive in the community are woefully lacking. There is a need for adequate housing, employment, transportation, socialization, vocational rehabilitation, and social services, as well as for an appropriate range of each type of service. The current options of a state hospital, nursing home, or aftercare clinic for the care and treatment of the chronically ill are clearly inadequate. In addition, the lack of continuity of care, aggressive outreach, and vigilant monitoring present formidable barriers to effective care and treatment.

Government Responsibility

Dozens of federal, state, and local governmental agencies have programs and funding for the chronic mentally ill. At the federal level, many of these (Medicaid, Medicare, SSI), are based in the Department of Health, Education and Welfare (HEW), but others are in the departments of Housing and Urban Development (HUD), Labor, Transportation, and so forth. Each program has its own standards, target populations, eligibility for funding, and regulations. As a result, patients do not have ready access to these funds to provide for their needs. In addition, there is a continuing discrimination against long-term care, less restrictive alternatives to hospitalization, and chronic illness. Indeed, there is not only no effective service system for delivering care and treatment to the chronic mentally ill, there seems to be no mental health system at all.

ATTITUDES

There are significant negative attitudes about the chronic mentally ill on the part of patients, families, legislators, and mental health professionals. In addition to the stigma suffered by all mentally ill persons, the chronically ill, with few articulate family members to advocate for them, fewer ego assets, and little empathy-evoking abilities, have less political clout, less lobbying ability, and less social presence than any group of have-nots now in need of services.

¶ Treatment of the Chronic Mentally Ill

Treatment of chronic mentally ill patients is complex and involves many modes of intervention. One must not only take into consideration the medical and psychiatric elements of treatment, but the social and rehabilitative components (housing, socialization, vocational rehabilitation, and so forth) as well. Since, as May,[41] and Hogarty and associates[26] have amply demonstrated, the additive effects of medication and talking therapy are considerable, one must not rely on only one modality of treatment (drugs or psychotherapy). In this section, the individual elements in the care and treatment of the chronic mentally ill will be reviewed, with the caution that the reader must assume that no one mode stands alone in the provision of effective treatment.

Medication

Experts have considered psychopharmacological agents as the single most important ingredient in the care and treatment of the chronic mentally ill. The only consistent finding among the studies of relapse among the chronically ill is that medication is the best preventor of relapse, in both schizophrenia and the affective disorders.[14,15] The fact that only 50 percent of patients discharged from mental hospitals continue to take their medication, makes patient cooperation and compliance a critical issue for this population.[34] Much of the discontinuity in treatment occurs at interfaces in the system, and there is a high rate of patient dropout and lack of follow through after hospital discharge. A recent study[57] demonstrated that only 22 percent of discharged patients follow through with aftercare if simply told to call the mental health center when problems arise, but that the percentage goes up to 68 percent if a specific appointment is made, and to 75 percent if both an appointment and predischarge contact is made. Recidivism rates are decreased 50 percent by such simple measures.

Hansell[25] has spelled out a comprehensive yet common-sense approach to consideration of pharmacotherapy with the chronic mentally ill. He lists five steps: efficacy, necessity, surveillance, cooperation, and emergency. With patients suffering from either schizophrenia or the major affective disorders, he cautions the physician to review whether the drug is indeed efficacious and necessary—suggesting a methodology for reducing the medication to a minimum; providing drug holidays; and instructing the patient in how to recognize effectiveness, side-effects, and early warning signs of recurrence. He cautions the physician to be vigilant, but with the patient's full understanding and cooperation. Finally, he suggests that the ability to respond to emergencies and exacerbations is facilitated by the therapeutic relationship, counseling, and crisis management.

Johansen,[27] a psychiatric pharmacist, has further stressed the requirement of drug monitoring and its importance with the chronically ill. She advocates a sophisticated, comprehensive program, involving the collaboration of both the patient and his physician, which includes drug education, historical review of drug use and response, and drug monitoring (especially regarding side effects which may prompt the patient to discontinue the medication). The knowledge that patients who neither possess insight about nor perceive benefits from their medication will not continue that medication, demands that the physician ensure that patients do see the value of their medication.[38]

Several authors have raised the point that while antipsychotic drugs are of inestimable value in the treatment of acute schizophrenia, their usefulness in the treatment of the chronic state is less certain.[25,52] Hansell insists, therefore, that the physician be sure that he is dealing with chronicity and not continue medication when it is not needed.[25] Segal and Aviram[52] go further. In their recent study, they demonstrate the antitherapeutic effects of medication when prescribed to less disturbed patients who were attempting to reenter society, resocialize, and so forth. The same caution about ensuring the existence of a chronic or episodic process applies to the affective disorders. In view of the increasing evidence of the detrimental effects of lithium on the kidney, consideration should always be given to prescribing tricyclics rather than lithium to those patients requiring continuing medication.

Finally, several treatment issues require reemphasis. First, there are few indications for the prescription of multiple medications (polypharmacy) for the chronic mentally ill. Second, seeking consent of the patient is sensible when prescribing long-term medications that have possible detrimental effects (such as tardive dyskinesia or renal damage). And third, nonresponders may not be absorbing adequate medication and determination of blood levels should be attempted.

Psychotherapy

Most recent discussion of the use of psychotherapy for the chronic mentally ill focuses not so much on psychoanalytically-oriented, individual psychotherapy, as that of a broader amalgam of supportive and directive therapy. In fact, much effective psychotherapy with this population involves groups, activities, and what many psychiatrists may regard as social or vocational rehabilitation.[24]

May[42] reviewed the studies on the effectiveness of psychotherapy with the chronically ill and noted that seven studies demonstrated its effectiveness. He concluded that outpatient group therapy was probably more effective than individual psychotherapy, that medication alone was insufficient to prevent relapses, and that while psychotherapy added little to drug treatment of inpatients, it was helpful with outpatients.

Lamb[30] suggests that there are several critical elements in the psychotherapy of the chronically ill that merit special attention. He lists these as: increasing the patient's sense of mastery; focusing on the healthy part of the patient's life and personality; problem-solving in the here-and-now; insight into the patient's symptoms rather than psychodynamics; taking sides with the patient against his harsh superego; and "putting the family in perspective."

Talbott[60] recently reviewed many aspects of the psychotherapy of the chronic mentally ill and enumerated several critical areas: therapeutic issues, therapeutic techniques, therapist attitudes and behavior, and content of therapy. Among the therapeutic issues that must be addressed by both patient and physician, he listed: establishing a correct working diagnosis; collaborating in setting the treatment plan; setting of clear goals; establishing a working relationship, involving the patient's family and social system; understanding the patient's communications, behavior, and thinking; titering the patient's affect so that it is not overwhelming; attending to nonverbal cues; utilizing psychodynamics without necessarily interpreting them; focusing on the here-and-now; and avoiding regression.

The therapeutic techniques employed are

1. those used with sicker patients and at the commencement of treatment (support, advice-giving, and establishment of the "real" relationship);
2. those employed much of the time (labeling, reality testing, and problem solving);
3. those used in times of relative health (exploration) and regression (suppression); and
4. those used relatively infrequently, or when the patient becomes more "neurotic" (clarification, interpretation, and abreaction)

The attitudes and behaviors of the therapist are highly important to the success of the therapeutic relationship with the chronically ill. While no one psychiatrist can embody all the features he considers important, his knowledge of what is critical may help guide his behavior. The attitudes and behavior considered important include a positive attitude about the patient and chronic mental illness, honesty without being pseudohonest or cruel, flexibility and resilience, a caring attitude without patronization, tolerance of uncertainty and dependency, an active role in therapy, a gift for intuitive thinking, and patience, persistence, dependability, and consistency.

Finally, Talbott[60] enumerated several issues in the content of treatment which he felt must be addressed during the treatment of each chronic mentally ill patient. These were: identifying the precipitating stress; delineating the defenses and coping patterns; exploring the patient's fear of intimacy, his harsh conscience, and flooding by impulses; defining his ability to function in the real world; and understanding and alleviating the patient's emotional dyscontrol. He also concluded that the attitude toward termination must be open-ended, with the understanding that the patient may return at times of regression, crisis, or destabilization.

Housing

Residential care programs constitute an essential ingredient in the total program of care and treatment of the chronic mentally ill. Optimally, there should be an adequate range of graded facilities offering an individual patient the opportunity to progress step by step from total dependence (hospitalization) to total independence (independent living), while also allowing him the option of staying for indefinite periods at any one stage when his maximum level of functioning has been achieved. Budson[10] has written a comprehensive guide to community residential care programs, and the range of options he describes will be summarized here starting from the most structured and dividing the facilities into those with staffing, those offering independent living, and finally those operated under proprietary ownership.

Those facilities that have staffing include nursing homes, quarter-way settings, halfway houses, Fairweather Lodges, and long-term group residences. Nursing homes are the most highly structured, highly staffed, and restrictive settings.[18] Because they were designed for medically ill patients, those with physical or multiple handicaps, and the elderly, nursing homes may prevent growth, be antitherapeutic, and not allow younger, chronic mentally ill persons to progress to higher levels of functioning. Quarter-way settings tend to be wards, with fewer staff than normal, established in mental hospitals to encourage skills in everyday living (such as cooking, shopping, laundering clothes, banking) and promote return to more independent settings. Halfway houses are one step further removed from the hospital.[19] They are located in the community, have on-site staff and programs, and serve a younger, more vocationally able patient who has retained his ties with the community.[10] Despite the hope that hundreds of such houses would be instituted across the country, as of 1973, there were only 200 in the United States with a capacity of 9,000 persons.[46] Their success rate in preventing recidivism, however, is admirable. Rog and Raush[48] showed that only 20 percent of patients seen in halfway houses are readmitted, while 58 percent are living independently, and 55 percent are employed or in school. Fairweather Lodges* represent a more comprehensive attempt to move persons from hospital to community settings.[17] Their basic premise is that groups of patients, taught the skills of everyday living and survival in an interdependent fashion, can move from hospital to lodge to group living. Integral to the program is the utilization of sheltered work.

*Fairweather Lodges, developed by George Fairweather, are intermediary facilities between hospital settings and community living. They are group residences, where the staff is available for training patients in the skills of everyday living and where patients can learn to live inter-dependently.

A forty-month follow-up of persons using the lodge program demonstrated that they spent only 20 percent of their time in hospital while controls spent 80 percent. The last type of staffed facility is the long-term group residence. Staffed more richly than halfway houses, these units are best suited for persons who may stay longer than a year, as opposed to those who use halfway facilities as transitional living arrangements.

Those facilities that do not have onsite staff range from semi-independent to totally independent ones. Cooperative apartments represent the first step in this series. Compared to halfway house residents, patients in these settings tend to be older, more chronic, and have fewer vocational skills and community ties.[10] Cooperative apartments are either leased by the mental health agency or a related nonprofit corporation, and/or their leases shared with the patients, who live in groups of two to five. They pay their own rent, share in household chores, and have professional supervision with crisis intervention readily available.[51] Work camps in rural environments serve patients who cannot yet survive in independent rural housing, but do not need the structure offered by an institution.[10] They function as a working farm, with daily chores and work activities, and supervision and care is available as needed. Foster care represents another concept altogether. Based on a centuries-old program in Gheel, Belgium, such placements are run by a nonclinical caretaker, who provides housing for one to four persons, often elderly or mentally retarded. Other necessary services are provided in the nearby community. A problem with foster care, despite its attractiveness, is that the caretaker may offer nothing more stimulating than custodial care, and no growth for the patient is possible if he does not successfully pursue outside programs.

Totally independent living, in individual apartments or homes, is provided by some programs. While such satellite housing or apartments may be found or initially leased by the program, the ex-patient assumes responsibility for the lease and upkeep of the housing, as well as the attendance at social programs and psychiatric/medical services. One other innovative service should be described, despite the fact that to date there has been only one of its kind and apparently it no longer functions. It is the crisis hostel—a community residence located near a psychiatric facility, where patients go for respite housing as an alternative to inpatient admission. The hostel is supervised by a nonclinical houseparent, and patients receive medication from a nurse and participate in programs at the hospital.[9]

The last large category of housing options is that of the private (proprietary) facilities. Whether designed as hotels (welfare or single-room occupancy) or homes for adults (PPHAs), they usually provide neither relief from social isolation, despite their size, nor appropriate rehabilitation or treatment. Board and care homes, California's proliferating alternative to hospitalization, carry the same hazard of fostering regression and inhibiting growth.[33]

There are several programs that offer a range of treatment options under one umbrella. Transitional Services, Inc. (TSI)[12] has a four-step program: assessment centers which are medium-sized group homes with maximum supervision; learning centers which have six to twelve units with moderate supervision; semi-independent apartments with minimal supervision; and independent apartments where services are not provided unless asked for. TSI is not a mental health service, but views itself as a housing/living agency that tries to build continuation of aftercare into ex-patients' lives.

Test and Stein[65] have provided a conceptualization of the decision-making process in selection of the correct housing alternative for each patient. They suggest that the appropriate environment should ensure that the patient's needs are met, but not meet needs that the individual can manage by himself.

Despite the growth of housing options in the past few decades, there is still a dearth of adequate alternatives. This is due to several factors. First, community resistance to establishment of such services continues to be considerable. This forces patients and staffs to seek housing in rundown or undesirable areas, creating ex-patient ghettos. Second, funding to develop these alternatives continues to be extremely limited. Under the 1978 HUD initiative, 18 million dollars were allocated for housing for the chronic mentally ill. Part of this fund was to subsidize loans for group homes, apartments, and halfway houses and part to provide rental subsidies to allow for repayment of the loans. But 18 million dollars represents only a start. Third, regulations for obtaining funding and establishing housing alternatives remain exceedingly cumbersome. At one time, in New York State, there were more than forty-seven steps in the process, and the need to gain approval from more and more community bodies extends the length of the process.[61,62] And lastly, as deinstitutionalization proceeds, there is no provision for those who will continue to need an asylum setting that provides adequate custodial care. The development of community asylum settings or domiciliary care institutions is imperative.

Social Rehabilitation

While social and vocational rehabilitation are usually conceptualized, referred to, and indeed delivered together, they really constitute two separate modes of intervention for psychiatric patients and will be presented separately.[20] In addition, while some consider treatment and rehabilitation to be separate and some consider them synonymous, others, including the author, think that rehabilitation is one portion of the total treatment. In any case, the boundaries are not sharp. However, rehabilitation does stress disability rather than disease, focuses on assets not deficits, and aims at restoration of functioning rather than relief of symptoms.

Skills in everyday living are critical to survival in the community for the chronic mentally ill. One glaring fault of deinstitutionalization, especially in its earliest days, was its failure to anticipate the degree of social ineptness of discharged patients. While the armed forces have elaborate programs to reorient servicemen, who retire after twenty to thirty years from a peculiar form of dependent living, state hospitals apparently expected patients hospitalized for decades and without such preparation to somehow negotiate complex and ill-marked transit systems, maintain budgets and handle money wisely, and select and prepare nutritious food. In retrospect, this failure to teach skills necessary to survive in our complex, urbanized, twentieth-century American environment was a catastrophic omission.

The skills of everyday living—banking, money management, and budget-planning; shopping, cooking, and serving food; learning and using transportation systems; grooming, buying clothes, and dressing appropriately; attending to personal hygiene; and using leisure time wisely—are critical to every patient's life. Any program of merit for the chronically ill now stresses the teaching of these skills. An exemplary program in Philadelphia, the Enablers Program, uses indigenous members from the patient's community to teach these survival skills others take for granted. This approach increases a patient's chance of returning to full functioning in his particular community.[68]

In addition, it is critical to provide persons with chronic mental illness the opportunity to gain or regain the ability to get along with others, to participate in social activities, and to communicate effectively and directly their thoughts and needs to others. Structured programs, such as activity groups, therapeutic groups, therapeutic milieus, individual counseling, and training, all contribute to acquisition of these abilities.[7]

In the 1930s, Joshua Bierer recognized the need for ex-patients to participate in many different social activities, structured and unstructured, and to benefit from professional guidance and planning. In his Marlborough Experiment, he pioneered in the establishment of patient-run social clubs; day, night, and weekend hospitals; rehabilitative aftercare; community hostels; and patient self-help groups (Neurotics Nomine).[7]

More recently in this country, we have seen the gradual emergence of numerous psychosocial rehabilitation centers, which combine housing, socialization, and vocational rehabilitation. Programs such as Fountain House, Horizon House, Thresholds, and so forth, concentrate on putting together in one setting all the ingredients necessary for successful rehabilitation to full community life.[20,22]

Given the early onset of some mental illness (for example, schizophrenia), habilitation—the learning of new skills to enable persons to reach levels of functioning they have never achieved before—may be more pertinent than rehabilitation. Another aspect, mentioned by Lamb,[36] is the thin line that separates patients' high but realistic expectations from unrealistic goals. Patients who attempt to attain unrealistic levels of functioning become frustrated and often regress, while patients with low-level goals may stagnate.

Vocational Rehabilitation

The primary goal of vocational rehabilitation is to evaluate the past and current work capacity of the patient and return him to his highest possible level of functioning. Again, in many cases habilitation, not rehabilitation, is meant, and the level of functioning that is aimed for may actually be higher than that previously achieved by the patient. It is important that the physician not assume that because of the severity of either the patient's diagnosis or illness, his work capacity is similarly deficient. Many chronic schizophrenics work successfully throughout their lifetimes in special settings (such as the post office), while other neurotic patients (for example, with severe compulsions) cannot work at all.

Vocational rehabilitation consists of several discrete subentities: prevocational assessment, counseling, and remuneration in a variety of settings; sheltered workshops; transitional placements; placement services; and long-term or terminal vocational settings. Prevocational assessment and counseling consists of a functional assessment of the patient's particular work skills and his ability to work in various settings, under various conditions. Also included is an evaluation of his attitudes, behaviors, and expectations about work. These assessments are usually conducted in health settings (hospitals, community agencies, or psychosocial rehabilitation centers) by use of formal interest and aptitude testing and work samples, but some agencies (for example, Council House in Pittsburgh, Pennsylvania) conduct assessments at the job site.[20]

Sheltered workshops are work settings where real work is performed by patients, side by side with staff, with professionally trained supervisors or foremen as well as clinically trained counselors. Because most employers feel that they can train motivated employees to do a particular type of work, prevocational programs and sheltered workshops often teach patients about work (interest, neatness, responsibility); behavior related to work (promptness, care, courteousness); and expectations about work (as demeaning or grandiose, offering rapid advancement without a track record, and the like) rather than the particular skills (typing, lens-grinding, and so forth). Rather than impart specialized skills, they attempt to alter maladaptive behavior so the patient can succeed in the marketplace.[29,50]

Transitional placement consists of placements in industry where a certain number of patients can try out genuine employment. Patients in transitional placements are able to begin regular work, have a sense of reality of the work world, and assume permanent employment when they feel comfortable in doing so. Many programs, private and governmental, offer placement services to facilitate the patient's search for regular employment. One other service that is a necessary element in the range of vocational services needed by the chronic mentally ill is that of long-term or terminal vocational settings or sheltered workshops. Partly because of government's reluctance to continue a long-term maintenance situation without the guarantee of "cure" or "rehabilitation," and partly because of society and psychiatry's reluctance to admit that certain patients will never return to full community life—we have too few of these placements and too little funding for existing programs.

There are several programs connected with vocational rehabilitation that are worth noting. First, given a national unemployment rate in recent years of 5 to 12 percent and regional variations of these figures (over 25 percent for young blacks), the availability of jobs for marginal persons is often insufficient. Second, some patients are unwilling to work, and their resistance and reluctance may frustrate the well-meaning professional. Third, SSI has an inherent disincentive built into the resumption of paid employment—one which must be addressed from a policy standpoint if we intend to return many of our chronic mentally ill to the work force.[35] Finally, there is a great need for part-time employment opportunities for those chronic mentally ill who cannot return to full-time employment.

Case Management

Two problems have become glaringly apparent as the process of deinstitutionalization has progressed. First, ensuring that discharged patients have access to and obtain all the services they did in an institution (food, shelter, socialization, medical and psychiatric treatment) is a difficult task. Second, finding a single point of responsibility for the total care and treatment of the chronic mentally ill has proved difficult. Case management seeks to remedy these problems by making sure that one person, team, or agency is responsible for the patient to ensure that he receives all the services he needs and that at least one person is in contact with all elements in the complex system of care and treatment.[39,58]

The tasks of case management have been defined as including assessment of needs, planning for service provision, linking the patient up with needed services, monitoring the provision of services, advocating for services, reviewing and updating the treatment plan, developing additional resources, providing direct services (from psychotherapy to escort services), integrating services, and expediting them.[29,58]

Who should perform these various tasks has been a matter of dispute. Ideally, an intact family can often be the best case management resource, but many blood relatives of the severely and chronically mentally ill are themselves impaired. Traditionally, social workers have performed many of these tasks, but in the past few decades, the profession of social work has turned away from case work and direct service provision toward the practice of psychotherapy. In truth, most physicians perform the role without knowing it, and may indeed be the most capable of doing so, as long as they recognize the need for a broader array of services, comprehend the necessity of a community support approach, and have the time and interest. The paucity of intact available family members, the decreasing numbers of interested social workers, and the insufficient number of psychiatrists in the country, have, however, led some planners to call for a new profession or paraprofession of case managers to assume this role.[45] Whether this new paraprofession is the answer is problematic, given the history of other paraprofessional efforts in the mental health field.

Certainly, there remain several critical unanswered questions about case management. Is it best performed by the person who has the most information about the patient, such as the psychotherapist?[31] Is it best performed by a team rather than an individual?[37] Is it best performed by a professional or paraprofessional? These and other questions will have to be answered in the next few years if these case management services are to be provided more expeditiously.

¶ Effective Program and Service Systems

In the 200 years of organized mental health care of the chronic mentally ill, neither hospital nor community programs have demonstrated impressive effectiveness in treating the severely and chronically mentally ill.

However, recently there have been a number of program and service systems that combine the treatment and care described in the preceding sections. While still isolated and experimental, such systems have been described in detail and scientifically based evaluation studies have been conducted, allowing others to attempt their replication.

Programs for the chronic mentally ill are basically of two types: institutional programs attempting to return patients successfully to community life; and community programs that try to keep chronic mentally ill persons functioning at an effective level in their communities. The best description of the former is by Paul and Lentz.[47] They described their four-and-a-half-year program, which teaches state hospital patients skills necessary to everyday living through a resocialization-relearning program oriented along behavioral modification lines. On outcome, their study group had a readmission rate of less than 10 percent over the year-and-a-half period following discharge, while the control group treated with milieu therapy had a readmission rate of 30 percent, and those treated with "traditional" state hospital treatment, 50 percent.

The program that best exemplifies the community approach to treatment of the chronic mentally ill is that described by Stein and Test.[56] Their Training in Community Living Program (TCL) takes randomly selected patients who otherwise would have been admitted to a state hospital. Employing retrained staff from a state hospital ward and concentrating on skills in everyday living, providing linkage with essential services, and ensuring medical and psychiatric treatment, their program also sharply decreased the amount of time spent in hospital compared to their hospital-treated controls. In addition, they discovered that the group they treated had less symptomatology than their controls, the burden to the patients' families and communities was no greater than that of the controls, their self-esteem levels were the same, and the quality of life of both groups was the same.

Two recent publications provide further descriptions of apparently successful programs for the chronic mentally ill.[55,63] While often touted as "model" programs, these descriptions do not so much offer models as examples of programs that seem to work in certain areas (for example, rural, urban, nonindustrial), with certain populations (firstbreak schizophrenics, multiply admitted chronic patients, children, the aged), from certain institutional bases (nursing homes; state, general, and private hospitals; CMHCs; and freestanding community agencies). As Bachrach[4] has noted, rather than replicable models, they offer examples to others who must then adapt their own programs to local exigencies and conditions.

There are certain characteristics that seem to typify the newer and more successful programs for the chronic mentally ill. Barter[5] notes that such programs have leaders who understand chronic illness and know what they are doing; have effective outreach, advocacy and monitoring systems; provide highly individualized treatment; show appreciation of small progress in their patient populations; have sound vertical and horizontal administration; and have effective accountability and responsibility systems. Bachrach[4] also notes that they assign top priority to the most severely impaired, link their programs with other resources, provide patients with a range of needed services, individually tailor treatment, adapt programs to meet local needs, utilize specially trained staff, have accessible liaisons with hospital units, and provide good internal evaluation of their operations. Talbott[61] concludes that there are several more important characteristics of successful programs: their ability to conceptualize and win necessary community support; their concentration on the factors known to help survival in the community, such as medication and skills in everyday living; and their effective use of a case management approach.

It should be emphasized, however, that despite impressive progress in beginning to understand what works and what doesn't for the chronic mentally ill, there is a long way to go. The programs described require continued appraisal, replication, and comparison. The process of teasing out exactly what it is in each program that works needs to be continued.

When it comes to looking at the larger picture—that of the state, regional, or national mental health systems—the situation is even more tentative. While several states have begun to deal with the problem of the chronically ill on a statewide or systems level,[63] and while the National Institute of Mental Health has initiated a community-support program to fund new programs for the chronically ill, nationally there remain a huge number of formidable tasks to be accomplished if effective care and treatment of the chronic mentally ill is ever to be realized.

¶ The Future

In its "Call to Action on The Chronic Mental Patient," the American Psychiatric Association[1] spelled out the magnitude of the problem and the multiplicity of solutions needed if the current situation were to be remedied. These included changes in:

1. the funding of services to the chronic mentally ill, so that discrimination against long-term care and maintenance in the community would be eliminated and funding could follow patients from institutional to community settings;
2. the attitudes all Americans hold about the chronic mentally ill;
3. the role of the psychiatrist (who has been leaving the field of public psychiatry), making him more involved in planning, implementation, and monitoring of services to the chronically ill;
4. the current administrative structure, which fragments funding, program planning, and service provision;
5. research efforts to provide increased understanding of epidemiology, prevention of chronicity, evaluation of what treatment works for which patients, and the like;
6. training priorities so that medical students, psychiatric residents, and practicing physicians, who all receive too little training in the care and treatment of chronic patients, understand more about effective modern programs; and
7. legislation that discriminates against the chronically ill.

It is clear that no single or simple solution will solve the multitude of problems posed by and encountered by the chronic mentally ill. Many small steps must be taken in research, training, and service delivery if the current deplorable situation is to change.

¶ Bibliography

1. AMERICAN PSYCHIATRIC ASSOCIATION. "A Call to Action for the Chronic Mental Patient," *American Journal of Psychiatry,* 136 (1979): 748–752.
2. BACHRACH, L. L. *Deinstitutionalization: An Analytical Review and Sociological Perspective.* Rockville, Md: U. S. Department of Health, Education and Welfare, 1976.
3. ———. "A Conceptual Approach to Deinstitutionalization," *Hospital and Community Psychiatry,* 29 (1978): 573–578.
4. ———. "Model Programs for Chronic Mental Patients: What Was the Question?" *American Journal of Psychiatry,* in press.
5. BARTER, J. T. "Successful Community Programming for the Chronic Mental Patient: Principles and Practices," in Talbott, J. A. ed. *The Chronic Mental Patient: Problems, Solutions and Recommendations for a Public Policy.* Washington, D.C.: American Psychiatric Association, 1978, pp. 87–95.
6. BEERS, C. W. *A Mind That Found Itself.* Garden City, N. Y.: Doubleday, 1960.
7. BIERER, J. "The Marlborough Experiment," in Bellak, L. ed., *Handbook of Community Psychiatry and Community Mental Health.* New York: Grune & Stratton, 1964, pp. 221–247.
8. BRILL, H., and PATTON, R. E. "Analysis of Population Reduction in New York State Mental Hospitals during the First Four Years of Large-Scale Therapy With Psychotropic Drugs," *American Journal of Psychiatry,* 116 (1959): 495–500.
9. BROOK, B. D. "Crisis Hostel: An Alternative

to Psychiatric Hospitalization for Emergency Patients," *Hospital and Community Psychiatry,* 24 (1973): 621–624.

10. BUDSON, R. D. "Community Residential Care," in Talbott, J. A. ed., *The Chronic Mentally Ill: Treatment, Programs, Systems.* New York: Human Sciences Press, forthcoming.
11. CAPLAN, R. B. *Psychiatry and the Community in Nineteenth Century America.* New York: Basic Books, 1969.
12. CURRENTS. "Pennsylvania Programs for Independent Living." Spring (1979) 7–13.
13. DAIN, N. "From Colonial American to Bicentennial America: Two Centuries of Vicissitudes in the Institutional Care of Mental Patients," *Bulletin of the New York Academy of Medicine,* 52 (1976): 1179–1196.
14. DAVIS, J. M. "Overview: Maintenance Therapy in Psychiatry: I. Schizophrenia," *American Journal of Psychiatry,* 132 (1975): 1237–1245.
15. ———. "Overview: Maintenance Therapy in Psychiatry: II. Affective Disorders." *American Journal of Psychiatry,* 133 (1976): 1–13.
16. DEUTSCH, A. *The Mentally Ill in America: A History of Their Care and Treatment from Colonial Times.* New York: Columbia University Press, 1949.
17. FAIRWEATHER, G. W., et al. *Community Life for the Mentally Ill.* Chicago: Aldine, 1969.
18. GLASSCOTE, R., et al. *Old Folks at Homes.* Washington, D.C.: Joint Information Service, 1976.
19. ———, GUDEMAN, J. E., and ELPERS, R. *Halfway House for The Mentally Ill.* Washington, D.C.: Joint Information Service, 1971.
20. GLASSCOTE, R. M., et al. *Rehabilitating the Mentally Ill in the Community.* Washington, D.C.: Joint Information Service, 1971.
21. GOFFMAN, E. "On the Characteristics of Total Institutions," in Goffman, E. ed., *Asylums.* Garden City, N.Y.: Doubleday, 1961, pp. 1–124.
22. GROB, S. "Socialization," in Talbott, J. A. ed., *The Chronic Mentally Ill: Treatment, Programs, Systems.* New York: Human Sciences Press, forthcoming.
23. GROUP FOR THE ADVANCEMENT OF PSYCHIATRY. *A Re-examination of the Community Psychiatry Movement.* New York: Group for the Advancement of Psychiatry, 1980, forthcoming.
24. GUNDERSON, J. G., and MOSHER, L. R. *Psychotherapy of Schizophrenia.* New York: Jason Aronson, 1975.
25. HANSELL, N. "Medication," in Talbott, J. A., ed., *The Chronic Mentally Ill: Treatment, Programs, Systems.* New York: Human Sciences Press, forthcoming.
26. HOGARTY, G. E., et al. "Drug and Sociotherapy in the Aftercare of Schizophrenic Patients. III: Two-year Relapse Rates," *Archives of General Psychiatry,* 31 (1974): 603–608.
27. JOHANSEN, C. "Drug Monitoring," in Talbott, J. A. ed., *The Chronic Mentally Ill: Treatment, Programs, Systems.* New York: Human Sciences Press, forthcoming.
28. KRAMER, M. *Psychiatric Services and the Changing Institutional Scene.* Bethesda, Md.: National Institute of Mental Health, 1975.
29. LAMB, H. R. "Rehabilitation in Community Mental Health," *Community Mental Health Review,* 2 (1977): 1–8.
30. ———. "Individual Psychotherapy," in Talbott, J. A., ed., *The Chronic Mentally Ill: Treatment, Programs, Systems.* New York: Human Sciences Press, forthcoming.
31. ———. "Therapist-Case Managers: More Than Just Brokers of Service," *Hospital and Community Psychiatry,* in press.
32. ———, and GOERTZEL, V. "The Demise of the State Hospital—A Premature Obituary?" *Archives of General Psychiatry,* 26 (1972): 489–495.
33. ———. "Discharged Mental Patients—Are They Really in the Community?" *Archives of General Psychiatry,* 24 (1973): 621–624.
34. ———. "The Long-Term Patient in the Era of Community Treatment," *Archives of General Psychiatry,* 34 (1977): 679–682.
35. LAMB, H. R., and ROGAWSKI, A. S. "Supplemental Security Income and the Sick Role," *American Journal of Psychiatry,* 135 (1978): 1221–1224.
36. LAMB, H. R., et al. *Community Survival for Long-Term Patients.* San Francisco: Jossey-Bass. 1976.
37. LANGSLEY, D. G. and KAPLAN, D. M. *The Treatment of Families in Crisis.* New York: Grune & Stratton, 1968.
38. LIN, I. F., SPIGA, R., and FORTSCH, W. "In-

sight and Adherence to Medication in Chronic Schizophrenics," *Journal of Clinical Psychiatry,* 40 (1979): 430–432.

39. LURIE, N. V. "Case Management," in Talbott, J. A. ed., *The Chronic Mental Patient: Problems, Solutions and Recommendations for a Public Policy.* Washington, D.C.: American Psychiatric Association, 1978, pp. 159–164.
40. MASSERMAN, J. Personal communication, 1978.
41. MAY, P. R. A. *Treatment of Schizophrenia.* New York: Science House, 1968.
42. ———. "Schizophrenia: An Overview of Treatment Methods," in Freedman, A., Kaplan, H. I., and Sadock, B. J., eds., *Comprehensive Textbook of Psychiatry* vol 2. Baltimore: Williams & Wilkins, 1975, pp. 923–938.
43. MINKOFF, K. "A Map of Chronic Mental Patients," in Talbott, J. A., ed., *The Chronic Mental Patient: Problems, Solutions and Recommendations for a Public Policy.* Washington, D.C.: American Psychiatric Association, 1978, pp. 11–37.
44. MURPHY, J. G., and DATEL, W. E. "A Cost-Benefit Analysis of Community Versus Institutional Living," *Hospital and Community Psychiatry,* 27 (1976): 101–103.
45. OZARIN, L. D. "The Pros and Cons of Case Management," in Talbott, J. A., ed., *The Chronic Mental Patient: Problems, Solutions and Recommendations for a Public Policy.* Washington, D.C.: American Psychiatric Association, 1978, pp. 165–170.
46. OZARIN, L. D., and WITKIN, M. J. "Halfway Houses for the Mentally Ill and Alcoholics," *Hospital and Community Psychiatry,* 26 (1975): 101–103.
47. PAUL, G., and LENTZ, R. J. *Psychosocial Treatment of Chronic Mental Patients: Milieu vs. Social Learning Programs.* Cambridge: Harvard University Press, 1978.
48. ROG, D. J., and RAUSH, H. L. "The Psychiatric Halfway House: How Is It Measuring Up?" *Community Mental Health Journal,* 11 (1975): 155–162.
49. ROTHMAN, D. J. *The Discovery of the Asylum: Social Order and Disorder in the New Republic.* Boston: Little, Brown, 1971.
50. SALKIND, I. "The Rehabilitation Workshop," in Lamb, H. R., ed., *Rehabilitation in Community Mental Health,* San Francisco: Jossey-Bass, 1971, pp. 50–70.
51. SANDALL, H., HAWLEY, T. T., and GORDON, G. C. "The St. Louis Community Homes Program: Graduated Support for Long-Term Care," *American Journal of Psychiatry,* 132 (1975): 617–622.
52. SEGAL, S. P., and AVIRAM, U. *The Mentally Ill in Community-Based Sheltered Care.* New York: John Wiley, 1978.
53. SHARFSTEIN, S. and NAFZIGER, J. C. "Community Care: Costs and Benefits for a Chronic Patient," *Hospital and Community Psychiatry,* 27 (1976): 170–173.
54. SHARFSTEIN, S., TURNER, J. E., and CLARK, H. W. "Financing Issues in the Delivery of Services to the Chronically Mentally Ill and Disabled," in Talbott, J. A., ed., *The Chronic Mental Patient: Problems, Solutions and Recommendations for a Public Policy.* Washington, D.C.: American Psychiatric Association, 1978, pp. 137–150.
55. STEIN, L., and TEST, M. A., eds. *Alternatives to Mental Hospital Treatment.* New York: Plenum, 1978.
56. ———. "Alternative to Mental Hospital Treatment: I. Conceptual Model, Treatment Program, and Clinical Evaluation," *Archives of General Psychiatry,* 37 (1980): 392–397.
57. STICKNEY, S. K., HALL, R. C. W., and GARDNER, E. R. "The Effect of Referral Procedures on Aftercare Compliance," *Hospital-Community Psychiatry,* 31 (1980): 567–569.
58. SULLIVAN, J. P. "Case Management," in Talbott, J. A., ed., *The Chronic Mentally Ill: Treatment, Programs, Systems.* New York: Human Sciences Press, 1980.
59. ———. "What Are the Problems of Chronic Mental Patients—A Report of a Survey of Psychiatrists' Concerns," in Talbott, J. A., ed., *The Chronic Mental Patient: Problems, Solutions and Recommendations for A Public Policy.* Washington, D.C.: American Psychiatric Association, 1978, pp. 1–7.
60. ———. "Psychotherapy with the Chronic Mentally Ill," in Karush, A., ed., *Psychotherapy and the Primary Care Physician.* Menlo Park, Calif.: Addison-Wesley, 1980.
61. ———. "Towards a Public Policy on the Chronic Mentally Ill Patient," *American Journal of Orthopsychiatry,* 50 (1980): 43–53.

62. ———. "The Historical Background of Community Psychiatry." Unpublished manuscript.

63. ———, ed. *The Chronic Mentally Ill: Treatment, Programs, Systems.* New York City: Human Sciences Press, (forthcoming).

64. TALBOTT, J. A., ed. *The Chronic Mental Patient: Problems, Solutions and Recommendations for a Public Policy.* Washington, D.C.: American Psychiatric Association, 1978.

65. TEST, M. A., and STEIN, L. I. "Special Living Arrangements: A Model for Decision-making," *Hospital and Community Psychiatry,* 28 (1977): 608–610.

66. TRAUBE, C. Personal communication, 1980.

67. *Webster's Third New International Dictionary.* Springfield, Mass.: Merriam, 1971.

68. WEINMAN, B., et al. "Community Based Treatment of the Chronic Psychotic," *Community Mental Health Journal,* 6 (1970): 13–21.

69. WEISBROD, B. A., TEST, M. A., and STEIN, L. I. "An Alternative to Mental Hospital Treatment: III. Economic Benefit-Cost Analysis," *Archives of General Psychiatry,* in press.

CHAPTER 20

APPROACHES TO FAMILY THERAPY*

Ira D. Glick, David R. Kessler, and John F. Clarkin

¶ Definition

MARITAL and family treatment can be defined as a professionally organized attempt to produce beneficial changes in a disturbed marital or family unit by using essentially interactional, nonpharmacological methods. Its aim is the establishment of more satisfying ways of living for the entire family and for individual family members.

Family therapy is distinguished from other psychotherapies by its conceptual focus on the family system as a whole. Major emphasis is placed on acknowledging that individual behavior patterns arise from, and inevitably feed back into, the complicated matrix of the general family system. Beneficial alterations in the larger marital and family unit will therefore have positive consequences for the individual members, as well as for the larger systems themselves. The major emphasis is placed upon understanding and intervening in the family system's current patterns of interaction; the origins and development of these patterns of interaction usually receive only secondary interest.

In many families, some member or members may be "selected" as "symptom bearers." Such individuals will then be described in a variety of ways that will amount to their being labeled "bad," "sick," "stupid," or "crazy." Depending on what sort of label such individuals carry, they, together with their families, may be treated in any one of several types of helping facility—for example, psychiatric, correctional, or medical.

But there may not always be an *identified patient.* Occasionally a marital or family unit presents itself as being in trouble without singling out any one member. For example, a couple may realize that their marriage is in trouble and that the cause of their problems

*Parts of this chapter were abstracted from: Glick, I.D., and Kessler, D. R. *Marital and Family Therapy,* 2nd ed. New York: Grune & Stratton, 1980, with the permission of the publisher.

stems from interaction with each other and not from an individual partner.

The intrapsychic system, the interactional family system, and the sociocultural system can be viewed as a continuum. However, different conceptual frameworks are utilized when dealing with each of these systems. A therapist may choose to emphasize any of the points on this continuum, but the family therapist is especially sensitive to, and trained in, those aspects relating specifically to the family system; he is aware of both its individual characteristics and the larger social matrix.

Family therapy is not necessarily synonymous with *conjoint family therapy* (in which the entire family meets together consistently for therapy sessions). For example, instead of having regular sessions with the entire family, one of the clinical and theoretical pioneers[15] in the family field has in recent years been experimenting with the almost exclusive use of the healthiest member of the family system as the therapeutic agent for change in the family unit. This therapist has also reported on his use of somewhat indirect means (for example, provocative letters to family members) as an imaginative way to conduct family therapy; that is, to bring about change or movement in a family system.

There are instances in which a family may be seen together while the therapist's frame of reference remains limited to that of individual psychotherapy. Family members in such a setting may be treated as relatively isolated individual entities. In effect then, such a therapist may be practicing conventional individual psychotherapy in a family therapy setting.

Family therapy might broadly be thought of as any type of psychosocial intervention utilizing a conceptual framework that gives primary emphasis to the family system and that, in its therapeutic strategies, aims for an impact on the entire family structure. Thus, any psychotherapeutic approach that attempts to understand or to intervene in an organically viewed family system might fittingly be called "family therapy." This is a very broad definition and admits various points of view, both in theory and in therapy.

¶ Historical Overview of the Differing Approaches

Current approaches have been synthesized from such fields as psychology, psychiatry, sociology, psychoanalysis, game theory, communication theory, Gestalt therapy, and the like. How did this state of affairs come about? The significance attributed to the family's role in relation to the psychic and social distress of any of its members has waxed and waned over the centuries. The important role of the family with regard to individual problems was mentioned by Confucius in his writings, as well as by the Greeks in their myths. The early Hawaiians would meet as a family to discuss solutions to an individual's problem. For a long time in our own culture, however, what we now call mental illness and other forms of interpersonal distress were ascribed to magical, religious, physical, or exclusively intrapsychic factors. It was not until the turn of this century that individual psychodynamics were postulated by Freud to be the determinants of human behavior. Although he stressed the major role of the family in the development of individual symptoms, he believed that the most effective technique for dealing with such individual psychopathology was treatment on a one-to-one basis.[49] At about this same time, others working with the mentally ill began to suggest that families with a sick member should be seen together, and that the mentally ill should not be viewed "as individuals removed from family relationships."[56] Eventually, psychiatric social workers in child guidance clinics, who often saw parents individually or together, began to recognize the importance of dealing with the entire family unit.

In the 1930s a psychoanalyst reported his experience in treating a marital pair.[45] And in the 1940s, Frieda Fromm-Reichmann postulated that a pathologic mother (called the "schizophrenogenic mother") could induce schizophrenia in a "vulnerable" child.[18] This speculation led other psychoanalysts to study the role of the father.[35] Their work suggested that the father also plays an important role in the development of psychopathology. At the same time, Bela Mittelman began to see a series of marital partners in simultaneous, but separate, psychoanalyses.[42] This approach was quite innovative because psychoanalysts had previously believed that this method of treatment would hinder the therapist from helping his patient, since it was thought that neither spouse would trust the same therapist and consequently would withhold important material. Therefore, the other marital partner was usually referred to a colleague.

Outside the field of psychiatry proper, marital counselors, ministers, and others have been interviewing spouses together for some time. In the early 1950s the first consistent use of family therapy in modern psychotherapeutic practice in the United States was reported by several different workers.[5,42] Nathan Ackerman[1] began utilizing family interviews in his work with children and adolescents; and Theodore Lidz and associates,[34] as well as Murray Bowen,[8] began a more extensive series of investigations of family interactions and schizophrenia. Gregory Bateson and associates,[3] and Lyman Wynne and associates[62] then embarked on the more intensive study of family communication patterns in the families of schizophrenic patients compared to families in which the patient had another psychiatric disorder.

It was not until the early 1960s, however, that these ideas were integrated into a general theory of family interrelationships, thereby shaping the modern field of family therapy.[1,50] Various schools of thought developed and journals such as *Family Process* were established. Many people became interested in learning about family therapy and in utilizing its techniques. As a matter of fact, a recent poll taken of California psychologists[63] showed that 90 percent (as expected) practiced individual therapy, but *now* more than 60 percent also practiced family therapy, while only 30 percent were doing group therapy. These statistics illustrate the rise in the growth of the family therapy field in just two decades. During the 1970s the use of family therapy was expanded to include the application of a "broad range of psychiatric problems with families differing widely in socioeconomic origin." The results, however, remained poor until crisis-oriented and short-term methods were developed to meet the needs of these families.[64]

¶ Evaluation Using Differing Approaches

There are several points of view regarding the type and quantity of the evaluative data to be gathered. Some family therapists begin with a specific and detailed longitudinal history of the family unit and its constituent members that may perhaps span three or more generations. This procedure has the advantage of permitting the family and the therapist to go over together the complex background of the present situation. The therapist will begin to understand unresolved past and present issues, will usually gain a sense of rapport and identification with the family and its members, and may then feel more comfortable in defining problem areas and in planning strategy. The family, for its part, may benefit by reviewing together the source and evolution of its current condition—a clarifying, empathy-building process that was not previously engaged in by its members. The good and the bad are brought into focus, and the immediate distress is placed in a broader perspective. Sometimes a family in crisis, however, is too impatient to tolerate exhaustive history gathering, and in acute situations lengthy data gathering must be curtailed.

Other therapists do not appear to rely heavily on the longitudinal approach, attempting instead to delineate the situation that has led the family to seek treatment and to obtain a cross-sectional view of its present functioning. This procedure has the advantage of starting with the problems with which the family is most concerned, and it will not be as potentially time consuming or as seemingly remote from the present realities as the preceding method. The therapist, however, may not emerge with as sharp a focus on important family patterns, and much of the discussion may be negatively tinged because of the family's preoccupation with its current difficulty.

More experienced (and often more courageous) therapists may severely curtail past history gathering and may also minimize formalized discussions of the family's current situation. They may begin, instead, by dealing from the onset with the family's important characteristic patterns of interaction as they are manifested in the interview setting. They may tend to utilize primarily, or exclusively, the immediate "here and now" observable family transactions. The therapist, understanding these transactions to be characteristic of the family, will clarify and comment on them, and intervene in a variety of ways. This approach has the advantage of initiating treatment right from the outset, without the delay of history gathering. There is often a heightened sense of emotional involvement, and more rapid changes may occur. Sometimes families are overwhelmed by such an approach, however, feeling threatened and defensive. Also, when specific information and patterns are allowed to emerge in this random fashion, the therapist does not always have the same degree of certainty as to whether the emerging family patterns are indeed relevant and important.

To a considerable extent these differences in technique may mirror differences in the therapist's training, theoretical beliefs, and temperament. Most therapists, however, probably use combinations of these approaches as the situation warrants, for there is no evidence of one technique being superior to the others.

Obviously, there is more than one way to evaluate a family—each way potentially useful—depending on the situation. The procedure offered in the evaluation outline depicted in table 20–1 combines, in a somewhat condensed manner, useful aspects of the first two approaches already discussed. It offers a practical alternative to gathering an exhaustive history or to plunging into the middle of the family interaction.

Formulating the Family Problem Areas

When meeting with the family, the therapist experiences its patterns of interaction and uses the data obtained in order to begin formulating a concept of the family problem.[38] Data for these formulations may come from historical material, but just as important will be what the therapist has observed in personal contact with the family. This will help to form a basis for hypotheses and therapeutic strategies. The data gathered from the outline provided (see table 20–1) should permit the family therapist to pinpoint particular areas or aspects of the family that may require attention. In addition, the data assist in laying out a priority system, so that the therapist can decide which areas of the family problem should be dealt with first. The data also clarify therapeutic strategy and the tactics indicated for the particular phases and goals of treatment.

Family Patterns of Communicating Thoughts and Feelings

Depending on the approach, some of the areas of communication to be assessed include: (1) expressions of affection, empathy, and mutual support; (2) areas of sexual satisfaction and dissatisfaction; (3) daily interaction, including the sharing of activities; (4) flexibility of roles, rivalry and competition, and the balance of power; (5) major conflicts in the marital relationship, including development intensity and means of resolving conflict; and (6) relationships to family, including children and friends. To what extent does the family group engage in meaningful and goal-directed negotiations, rather than being engulfed in incoherent, aimless talk?

TABLE 20–1 **Outline for Family Evaluation***

- I. Current Phase of Family Life Cycle
- II. Explicit Interview Data
  - A. What is the current family problem?
  - B. Why does the family come for treatment at this time?
  - C. What is the background of the family problem?
    1. Composition and characteristics of nuclear and extended family, e.g. age, sex, occupation, financial status, medical problems, etc.
    2. Developmental history and patterns of each family member
    3. Developmental history and patterns of the nuclear family unit
    4. Current family interactional patterns (internal and external)
  - D. What is the history of past treatment attempts or other attempts at problem solving in the family?
  - E. What are the family's goals and expectations of the treatment? What are their motivations and resistances?
- III. Formulating the Family Problem Areas
  - A. Family patterns of communicating thoughts and feelings
  - B. Family roles and coalitions
  - C. Operative family myths
  - D. Family style or typology
- IV. Planning and Therapeutic Approach and Establishing the Treatment Contract

*SOURCES: M. Gill, R. Newman, and F. Redlich. *The Initial Interviews in Psychiatric Practice.* New York: International Universities Press, 1954; and Group for the Advancement of Psychiatry. *The Case History Method in the Study of Family Process.* Report no. 76. New York: Group for the Advancement of Psychiatry, 1970.

Other factors of communication assessment include the general feeling tone of the family and of individual members, dyads, and triads, together with appropriateness, degree of variability, intensity, and flexibility. Questions to be considered might include: To what extent does the family appear to be emotionally "dead" rather than expressive, empathic, and spontaneous? What is the level of enjoyment, energy, and humor? To what extent does there appear to be an emotional divorce between the marital partners? To what extent is the family system skewed around the particular mood state or reaction pattern of one of its members?

Family Roles and Coalitions

Differing models of evaluation ask to what extent does the family seem fragmented and disjointed, as though made up of isolated individuals. Or does the family appear rather to be one relatively undifferentiated "ego mass"? To what extent is the marital coalition the most functional and successful one in the family system? To what extent are there cross-sectional dyadic coalitions that are stronger than the marital dyad? How successfully are power and leadership issues resolved? To what extent is this a schismatic family in which there are two or more alliances seemingly in conflict with one another?

Operative Family Myths

Some individuals in families are "selected" to be "bad," "sick," "stupid," or "crazy," and often these roles constitute a kind of self-fulfilling prophecy. Families as well as individuals function with a set of largely unexamined fundamental attitudes that have been termed "myths." These markedly influence the family's manner of looking at and coping with itself and the world.

Family Style or Typology

There is no one way of classifying, or making universally applicable the complexity of marital and family life styles. Some of the characterizations presented in table 20–2 may be found helpful, depending on the circumstances and the therapist's approach.[6]

After the evaluation data have been gathered and formulated into hypotheses and goals regarding important problem areas, the therapist is ready to consider what therapeutic strategies will be appropriate. The type of strategy is sometimes made explicit by the therapist; in other instances it remains covert; but irrespective of whether a therapist specializes in one approach or is eclectic, some hypotheses will be formed about the nature of the family's difficulty and the preferable approach to adopt.

¶ Differing Approaches to Treatment

Elements of Psychotherapy and Their Relationship to Family Therapy

Regardless of approach, most kinds of psychotherapy have in common at least eight elements:[39]

1. A good patient-therapist relationship.
2. Release of emotional tension.
3. Cognitive learning.
4. Operant reconditioning of the patient toward more adaptive behavior patterns by explicit or implicit approval-disapproval cues and by a corrective emotional relationship with the therapist.
5. Suggestion and persuasion.
6. Identification with the therapist.
7. Repeated reality testing or practicing of new adaptive techniques in the context of implicit or explicit emotional therapeutic support.

Family therapy, too, may use all eight of these elements to improve the overall functioning of the entire family. The particular mix of the elements will vary with the specific needs of the family. There is hardly any specific technique of individual or group therapy that could not in some way or another be adapted for use in family therapy.

Currently, there are a number of differing approaches for treating families. Each may emphasize different assumptions and types of interventions. Some therapists prefer to operate with one strategy in most cases; others may intermix strategies depending on the type of case and the phase of treatment.

TABLE 20–2 **Family Style or Typology**

Classification 1: Based on rules for defining power

1. The symmetrical relationship
2. The complementary relationship
3. The parallel relationship

In symmetrical relationships, both people exhibit the same types of behavior (which minimize the differences between them), role definitions are similar, and problems tend to stem from competition.

In complementary relationships, the two people exhibit different types of behavior, and this is found most often in the so-called traditional marriage. This form maximizes differences and tends to be less competitive and often highly workable. Unless role definitions are agreed on, however, serious problems can result.

In parallel relationships, the spouses alternate between symmetrical and complementary relationships in response to changing situations.

Classification 2: By Parental Stage

The move from the dyadic marital configuration to the larger, more complex one involving children tends inevitably to bring with it the potential for increased activities. Possible subcategories under this classification are as follows:

1. Before children
2. Early childhood
3. Latency and adolescent children
4. After the children have left home (empty-nest syndrome)

Classification 3: By Level of Intimacy

1. The conflict-habituated marriage is characterized by severe conflicts, but unpleasant as it is, the partners are held together by fear of alternatives.
2. The devitalized marriage has less overt expressions of dissatisfaction, with the marital partners conducting separate lives in many areas. This interaction is characterized by numbness and apathy and seems to be held together principally by legal and moral bonds and by the children.
3. The passive-congenial marriage is "pleasant" and there is a sharing of interests without any great intensity of interaction. The partners' level of expectation from the relationship is not very high, and they derive some genuine satisfaction from it.
4. The vital marriage is intensely satisfying to the spouses in at least one major area, and the partners are able to work together.
5. The total marriage, which is very rare in the investigators' findings, is characterized by similarity to the vital marriage except that the former is more intense and satisfying in the whole range of marital activities.

Classification 4: Personality Style

1. The obsessive-compulsive husband and the hysterical wife. Conflicts of intimacy often become of major importance.
2. The passive-dependent husband and the dominant wife. Power is the central theme of this system.
3. The paranoid husband and the depression-prone wife.
4. The depression-prone husband and the paranoid wife.
5. The neurotic wife and the omnipotent husband. Power is the primary conflict area. The wife's resentment is expressed through depression and a variety of other symptoms.

These material styles often work very satisfactorily if the needs of the two partners are met and if they are not overly inflexible in their application. Problems arise only when the cost of keeping the system going is too high—when one spouse changes, thereby upsetting the system, or when one partner indicates the desire to change the "rules."

Classification 5: Descriptions of Families in Treatment

No overall concept or model underlies the following six clusters; they are descriptive in nature. Because they were derived from families referred for treatment, the clusters imply a generally maladaptive tendency.

1. Constricted. Characterized by excessive restriction of a major aspect of family emotional life, such as expression of anger, negative affect, or ambivalence. These emotions become internalized into anxiety, depression, and somatic complaints. The presenting patient is often a passive, depressed child or young adult.
2. Internalized ("enmeshed"). Characterized by a fearful, pessimistic, hostile, threatening view of the world, leading to a constant state of vigilance. Such a family has a well-defined role structure, high family loyalty, and a pseudomutual bond between parents.
3. Object-focused. Characterized by overemphasis on the children ("child centered"), the outside community, or the self ("narcissistic"). Motivation for treatment depends on the willingness of the marital couple to form an effective coalition.
4. Impulsive. Characterized by an adolescent or young adult acting out anger toward a parent onto the community or expressing his or her parents' difficulties in a socially unacceptable way.
5. Childlike. Characterized by spouses who have remained dependent on their own families or on the community, based on either inadequacy or immaturity.
6. Chaotic. Characterized by disintegration. Lack of structure, chronic psychosis and delinquency, and low commitment to the family unit.

Various schools of family therapy may differ on where they place their emphasis on the following major treatment dimensions:[37]

1. Past versus present orientation.
2. Verbal interpretations versus action.
3. Growth model versus problem model.
4. General method versus specific plan for each problem.
5. Therapeutic focus on one individual versus focus on two or more individuals.
6. Equality versus hierarchy in therapeutic relationship.
7. Analogical versus digital thinking (Digital thinking concentrates on individual "bits" of behavior; the analogical view is more concerned with multiple levels and contexts of behavior).

Some therapists emphasize reconstruction of past events, whereas others choose to deal only with current behavior as manifested during the therapy session. Some therapists favor verbal exploration and interpretation, whereas others favor utilizing an action or experiential mode of treatment, either in the session itself or by requiring new behavior outside the interview. Some therapists think in terms of problems and symptoms and attempt to decode or understand possible symbolic meanings of symptomatology. Other therapists may focus on the potentials for growth and differentiation that are not being fulfilled. Some therapists utilize one or a limited number of methods in dealing with a whole range of "problems"; others are more eclectic and attempt to tailor the treatment techniques to what they consider the specific requirements of the situation.

With the therapeutic focus on one person, the emphasis is often on the individual's perceptions, reactions, feelings, and on the equality of status between the individual and the therapist. When two people constitute the operative system, attention is directed to interactions and relationships. Therapists who think in terms of a unit of three people look at coalitions, structures, and hierarchies of status and power. The number of people actually involved in the interviews may not be as important as *how many people are involved in the therapist's way of thinking about the problem.*

Three Basic Strategies of Family Therapy

Elsewhere, the authors[19] have singled out three major strategies that are especially useful for beginners.

1. Those that facilitate communication of thoughts and feelings.
2. Those that shift disturbed, inflexible roles and coalitions.
3. Those that aid family role-assumption, education, and demythologizing.

These three strategies are not necessarily mutually exclusive and may overlap. To some extent they represent different frames of reference for understanding and dealing with the same family phenomena. Nevertheless, each strategy seems to offer something unique in its concepts and techniques. In a clinical situation the therapist will be hard put to remain a purist. A therapist's efforts to clarify communication may produce shifts in family coalitions or initiate an exploration of family myths that may lead to a considerable outpouring of previously concealed affect.

Although some specific therapeutic strategies are listed above, there is no one magical phrase or technique that will "cure" the family. Instead, interventions are a series of repetitive maneuvers designed to change feelings, attitudes, and behavior. If the overall goals and strategy are kept in mind, specific interventions will suggest themselves and be modified by the particular circumstances as well as the therapist's own style.

What is unique in family therapy is not so much the specific approach but rather the overall focus and strategy that aims to evaluate and produce a beneficial change in the entire family system.

Models of Family Therapy

The different models of family therapy can be distinguished by their data base, goals, treatment techniques, selection criteria, explanatory concepts, and the stance of the therapist. Over time there has evolved a differentiation of three distinct but overlapping orientations to family treatment, with each one closely related to a different theoretical orientation. These are:[21]

1. *The insight-awareness approach:* Observation, clarification, and interpretations are used to foster understanding (and presumably change).
2. *The structural-behavioral approach:* Manipulations are devised to alter family structure and conduct.
3. *The experiential approach:* Emotional experience is designed to change the way family members see and presumably react to one another.

INSIGHT-AWARENESS MODEL

The insight-awareness orientation has also been known as the "historical," the "psychodynamic," or the "psychoanalytic school." In a real sense, this is the oldest school of family therapy since it grew naturally out of the psychoanalytic tradition. One of the earliest family therapists was a child analyst, Nathan Ackerman, who utilized his analytic background to inform and lend substance to his approach and understanding of families. One has only to read the transcripts of his sessions to appreciate the influence that analytic thinking and techniques had upon his work with families and couples.[1]

By changing the transference distortions, correcting the projective identifications, and infusing insight and new understanding into the arena of interpersonal turmoil and conflict, this school of family therapy attempts to change the functioning and interrelationships of the various members of the family or marital system. The data base is derived from historical material of the current and past generations, from transference/countertransference phenomena, unconscious derivatives, and resistances. A basic assumption is that intrapsychic conflict, interpersonal problem foci, and defensive and coping mechanisms are modeled and taught within the family system. Portions of the data base that are of paramount interest to the practitioners of this model are dream and fantasy material, fantasies and projections about other family members, and transference distortions about other family members and the therapist. Understanding the history and mutations over time of these dynamics is considered crucial to understanding current dysfunction. A broad use of the terms "transference" and "countertransference" is used here. Such phenomena can be understood in terms of transference on at least five levels: (1) man to woman; (2) woman to man; (3) woman to therapist; (4) man to therapist; (5) couple to therapist. Just as there are multiple transference reactions, there are multiple countertransference responses.[43] It is assumed that understanding of unconscious derivatives and their resistances is usually necessary to effect change.

The theoretical underpinnings of this model are the familiar ones of psychoanalytic thinking, including especially topographical concepts of conscious, preconscious, and unconscious; constructs of the id, ego, and superego; and concepts that focus on the interaction of individuals, such as secondary gain, transference, and projective identification. While early analysts, who opted for a more interactional model (for example, Nathan Ackerman[1]), criticized this model for its lack of attention to, and language for, interactional data, more recent authors in the analytic tradition of object relations (for example, Henry Dicks[15]) have applied these concepts to the understanding and analytic treatment of these interactional problems.

The major techniques of this model include clarification, interpretation, exploration of intrapsychic as well as interpersonal dynamics, and development of insight and empathy. Using such analytic techniques with an individual in the presence of a spouse or other family member represents a unique development. At the very least, it is possible that while the therapist is addressing an interpretation to one individual, the other members of the family can utilize the interpretative method to explore their own conflicts and difficulties, as well as to begin to

TABLE 20–3 **Models of Family Therapy**

MODEL (APPROACH OR SCHOOL)	REPRESENTATIVE THERAPISTS	PRIMARILY DERIVED FROM A MODEL OF LEARNING &/OR CHANGING	DATA BASE	GOALS	TECHNIQUES	STANCE OF THERAPIST
1. Insight-Awareness (also known as Historical, Psychodynamic, or Psychoanalytic)	Ackerman; Nagy & Spark; Paul; Nadelson;	Understanding	1. History 2. Unconscious derivatives 3. Transference/countertransference	Foster understanding and insight to effect change	1. Observation 2. Clarification 3. Interpretation	1. Listener 2. "Therapeutic distance" 3. Tries to form a therapeutic alliance
2. Strategic-behavioral (also known as Systems, Communications, or Structural)	Palo Alto Group (Jackson, Bateson, Haley, Satir); Sluzki; Bowen; Minuchin; M. Erickson; Palazzoli	Observing	1. Sequences 2. Communication 3. Rules 4. History 5. Behavior	Change family structure, communication pattern, and roles, which changes perception and behavior	1. Strategies desensitize and/or condition patient to alter family structure and behavior 2. Observe and transform using directives	Therapist observes and moves in and out of process
3. Experiential (also known as Existential)	Whitaker; Bowen; Nagy	Imitation (via the experience) and identification	1. Observed 2. Shared feelings (including the therapist's feelings)	Change ways family members experience and presumably react to each other	1. Therapist designs and/or participates with family in the emotional experience 2. Empathy	Therapist offers himself for interaction to minimize distance between family and himself

understand the family member more fully with the input from the therapist.

The goal is to foster understanding and insight in order to effect change in individuals as well as in the family unit.

THE STRATEGIC-BEHAVIORAL MODEL

This school of family therapy emphasizes understanding sequences, patterns, and structure, with the emphasis on manipulation as a technique to effect change. The data base is derived from elements of the structural school, systems theory, behavioral theory, and communications theory.

The orientation that is unique to the family movement and that has given it impetus is the systems understanding of family function and dysfunction. The family systems approach uses the metaphor of communication theorists such as Gregory Bateson and the computer metaphors of our modern age, such as feedback loops, communication exchange, and so forth. These explanatory concepts lead the practitioner to focus on a data base composed of repetitive interchanges between individuals in a marriage or family that occur in the here and now (as opposed to the interest in the history of the relationship in the psychoanalytic model), and that define, limit, and structure the behavior and experience of the individuals in the system. While the assumption of the psychoanalytic model is that individuals in a family group can perceive and distort the working of the group, the assumption in this model is that the group is greater than the individuals who compose it, and that the individual is governed and regulated by this greater entity (the family or the marital dyad). This model works on the assumption that there is no such thing as nonbehavior—even silence is a means of communication. There is a focus on syntax (ways of communicating), semantics (the meanings of communicative acts), and pragmatics (the effects of communication).

The treatment techniques (or strategies) in this model include: (1) changing family transactional patterns; (2) marking boundaries; (3) escalating stress; (4) assigning tasks both within and outside of the therapy session; (5) focusing on, exaggerating, deemphasizing, or relabeling symptoms; (6) reframing; (7) manipulating mood; (8) clarifying communication; (9) interrupting repetitive interactional patterns; and (10) prescribing paradoxical injunctions.[23,40,55] In one way or another, the techniques are intended to interrupt current repetitive interpersonal behaviors and introduce new interactional patterns that will result in the creation of new and more mature, or less symptomatic, interactions and inner states.

The behavioral component of the model grows out of the behavioral orientation that has historically flourished parallel, and in reaction, to the psychoanalytic tradition. The data base for this orientation is quantifiable, measurable behavior, whether internal (thoughts) or external (actions). Explanatory concepts are those of the behavioristic tradition; for example, stimulus, response, and concepts of learning theory such as classical conditioning, operant conditioning, schedules of reinforcment, and so forth. With the behavioral model becoming increasingly applied to interactional systems such as the family, other concepts have been introduced to expand the model into the interpersonal sphere. Perhaps the most influential has been the behavior exchange model of John Thibaut and Harold Kelley,[60] in which it is postulated that the benefit and cost ratio for each individual is an exchange situation (for example, marriage) has a major influence on the course and outcome of that relationship. The goal in this model is to effect change in discrete, observable, measurable behaviors that are considered problematic by the individuals seeking assistance. As opposed to the psychoanalytic model, which often seems to have more ambitious goals of character change and "insight," this model tends to focus more on discrete problem areas defined by clear behavior patterns. Thus, treatment in this model tends to be briefer and more circumscribed. Emphasis is placed not on pathology but upon behavioral deficits and excesses that are to be changed. If undesired behaviors are eliminated (for example, husband hitting wife), it is not assumed that more social behaviors will necessarily spontaneously emerge, but rather that the therapist might be required to teach new and more adaptive behaviors to the spouses or family members.

Techniques include helping family members learn means of effecting the desired behavior in another member. Some of the major tactics utilized are behavioral contracting based on good faith or *quid pro quo* agreements (if you do this, I will do that), training in communication skills, training in effective problem solving, and combining positive reinforcement with a decrease in destructive interchanges.

The stance of the therapist is quite active since he sees his job as a means to introduce behavior change into the repertoire of the family members. While in the past behaviorists have written little about how they handle individual and family resistance to their suggestions, making them sometimes appear somewhat naive to the experienced clinician, they are increasingly paying more attention to resistance and at times suggest the use of paradoxical injunctions more typically enunciated by the systems theorists. Indeed, it has been suggested[46] that there is a congruence, if not a growing similarity, between the techniques of the behaviorists and the systems interventionists.

This school of family therapy avoids the traditional techniques of individual therapy. It does not concentrate on eliciting historical material, it is not particularly interested in fostering increased awareness or expression of buried feelings, and it does not engage in interpreting psychodynamics. It does not consider understanding and insight to be important or essential in producing change. Instead, this group of family therapists manipulates variables such as the participants and rules of therapy by active suggestion and direction. They may utilize paradoxical commands and clearly attempt to alter the arrangement and intensity of family coalitions.[11]

The ultimate goal of this technique is not so much to foster understanding and insight as to change family structures, the communication patterns, and roles—changes that will presumably change perception of the individuals in the family unit and ultimately change behavior.

Experiential-Existential Model

This school believes that it is vital for the therapist to be aware of and take into account not only the experience of each member of the family, but also the therapist's experience as an outsider entering the family.[21] Empathy is the key here—that is, the ability of the therapist to experience what a particular family member feels at any given moment in the context of the family. The data base is not only what the therapist sees, but what he and/or the family *feel.*

The data base is derived from situations that the therapist designs, permitting him to participate with the family in an emotional experience. If a subject cannot be discussed, the therapist brings it up. His use of empathy enables him not only to understand what an individual family might be feeling but allows him sometimes to serve as a model for identification. He may be a role model for a family or an advocate and help the family achieve something. If a family is starving, the family therapist helps change the family by going with them to get food stamps.

Thus, in this form of therapy the therapist offers himself as a real person in order to minimize the distance between himself and the family. He is always on the side of the family, but his behavior is different from any other family member and is designed to promote more functional behavior.

The goal is to change the way family members experience and presumably react to each other. A secondary goal is growth and differentiation of family members.

Relationship of Theory to Therapy

It is important and interesting to note that there is often not a one-to-one match between a practitioner's theoretical position and his therapeutic techniques and performance. This is true, of course, for all fields of therapy and not just family therapy. However, the field of family therapy has been noted for its vigor rather than rigor, as well as for its energetic deployment in all kinds of human problems. There has been less attention to the development of a theoretical understanding of family functioning and family pathology. It follows, then, that while we can isolate three theoretical schools of family therapy, the practitioners and what they do in a family session can be classified from points of view other than that of their theoretical metaphors or concepts. For example, in one of the earliest and still useful classifications of family therapists, Chris Beels and Andrew Ferber[4] talk of conductors (therapists who take charge and direct the family sessions) and reactors (therapists who wait for stimuli to arise from the family interaction before beginning to shape those interactions with subtle and increasingly direct intervention. In this schema, Nathan Ackerman, Norman Paul, and Murray Bowen can be seen as conductors, while Carl Whitaker and Ivan Nagy are reactors, and Jay Haley and Don Jackson are described as reactors who are also systems purists, a reference once again to the theoretical stance. Both points of view —the theoretical stance of the practitioner and a more descriptive term relating to his style while working with families—are useful in the conceptualization and teaching of this field.

Other Approaches

Certain approaches are associated with innovative therapists. Don Jackson and others of the Palo Alto group focused much of their attention on pathologic communication. Together with Gregory Bateson and others, Jackson wrote about the "double bind" as playing a prominent part in family difficulties.[3] Jay Haley,[25] originally a member of the same group, has recently become more interested in the paradoxical intervention approach of Milton Erickson.[24] Virginia Satir, long interested in the communication frame of reference, has recently moved into the area of family growth enhancement.[51] A current member of the Palo Alto group, Paul Watzlawick,[61] believes that "illogical," unreasonable action can produce the desired change. He borrows from the concepts of communications, double bind, and action-oriented techniques of problem resolution.

Murray Bowen,[9] a family therapy pioneer, concerned himself on one hand with ideas about the lack of differentiation of various family subsystems and, on the other hand, with the extreme disengagement of many individuals in families—individuals who hardly seem to participate meaningfully in the family. He has used a variety of techniques over the years, including seeing in therapy only the healthiest family member and using that person as the agent, or model, of family change. He has used letters written from one family member to another as an indirect method for stirring up change in family systems.

Carl Whitaker[44] believes in the technique of paradoxical intention and manipulates the family members into believing the therapist has to overpower them, as if they are all engaged in a battle. He believes that the therapist has to deprogram himself and advance his own growing edge in order to conduct effective family therapy. In many ways Carl Whitaker appears to see family sessions in experiential terms with a good deal of emphasis on "feeling states" during the session and during immediate feedback.

Jerry Lewis, Robert Beavers, and their coworkers[32] have found that well-functioning families may have particular attributes that are lacking in dysfunctional families. A family therapist can apply this knowledge in his work as follows:

1. *The therapist works as collaborator,* implying respect for the other's subjective world view. He demonstrates with the treatment family an affiliative rather than an oppositional attitude, and a commitment to negotiation as the basis for treatment.
2. *The therapist conveys a belief in complex motivations,* especially with respect to the reciprocal nature of human interactions, paying special attention to (and including

himself as being involved in) the interactions that occur in the family-treatment setting.

3. *The therapist needs to be a model of clarity, spontaneity, nonintrusiveness, and permissiveness with respect to the expression of all types of affects,* frequently expressing empathic comments, acknowledging other people's views, and demonstrating a caring attitude.

An authoritarian style of treatment, in which the therapist sets himself up as the all-knowing and all-powerful manipulator of the family's interactions and goals, may at times prove to be helpful, even though it contradicts the model just outlined. People in some sociocultural settings are accustomed to a directive style and may be left confused or unconvinced by an invitation to share authority and responsibility. This directive style may be needed at times with the most dysfunctional families, aiding them in becoming less chaotic and enabling them to move to an intermediate "adequate" stage.

We believe it is wisest to employ a pluralistic approach; that is, tailoring the approach to the problem, rather than using a single approach in all situations.

Specific Techniques

Many specific techniques (in addition to the basic three that have already been described) can be employed during the course of treatment.

FAMILY TASKS

Family therapy focuses on behavioral change. Accordingly, many family therapists routinely prescribe various tasks for the family to perform during the session and, more commonly, between sessions. The rationale for this is to have the family work out and repeat behavior patterns outside the session. The therapist (rather than the family) takes control of the symptom or problem and helps realign coalitions. For example, family members that have not had any recreational activity together in several years may be asked to take a vacation together, or a husband and wife may be instructed to discuss a family secret.

Special techniques have been devised for helping lower socioeconomic class families, ghetto families, and highly disorganized families. The work of Salvador Minuchin[41] and others[36,48] indicate that it is both necessary and possible to help these families deal with some of their basic needs by using indigenous populations as family advocates when dealing with social agencies; by mobilizing the most constructive forces in the family system; and by providing training in basic task performance. Such tasks might involve getting the family registered with a housing agency. This serves to train and strengthen the family unit's ability to handle its problems in concrete terms; it also helps to solidify the often shaky and inadequate manner in which the family provides for its elemental needs. In this way the family can gain the experience of accomplishing something meaningful for itself during its daily struggle for existence and stability. These methods may be more useful than the more symbolic, attitudinal, psychological techniques appropriate for middle- and upper-class families.

PRESCRIBING THE SYMPTOMS*

Don Jackson, John Weakland,[27] and Jay Haley[23] have written about a therapeutic technique in which the therapist "prescribes the symptoms." After the therapist "orders" the family members (or individual) to intensify effect and the frequency of the symptoms, the symptoms begin to lose their autonomy, mystery, and power. Whereas they previously seemed to have been out of control, they now appear to come under the therapist's control. The participants in the behavior become more conscious of them, and often the disruptive behavior lessens or disappears. A marital couple that has engaged in nonproductive arguing may be told to continue fighting and even to increase it; for example, the couple may be ordered to

*Also called "paradoxical prescription [or] intention," "symptom scheduling," "negative practice," or "reverse psychology."

fight about the menu *before* dinner, so that they can enjoy the food. This injunction jars the continuing process, and they may rebel against the outsider's orders (which is often a necessary step to change).

The therapist is obligated to follow through to make sure that the directions have been followed in the way that was intended. The therapist does this by seeing the family in his office on an ongoing basis, by asking more than one family member what changes have taken place, or by visiting the family at home.

Some therapists write a family prescription after the initial session, telling the members they will receive a message about what the therapist thinks is wrong with the family and what needs to change.[52] This gives the therapist time that is not available in the heat of the session and creates an opportunity for a more accurate formulation of the family's problems. The prescription is a typewritten letter sent separately to each family member. It may describe what is happening in the family and ask each member to continue his course of action. For example, the therapist agrees with Joe (the identified patient) that he should not move out of the house at present. His parents, however, are told that they should continue to vacillate by alternately supporting his moving out and undermining it. This prescription[26] was sent to a family with a thirty-two-year-old son who kept "messing up" each time he left home, so that he was always forced to return to the family. The prescription had the effect of making Joe angry, of shifting him out of the house, and of identifying what his parents were doing. For some families there is something quite powerful about a well-thought-out message that is "official" and to the point.

Family Reconstruction

An in-depth exploration of family background is believed to improve the therapeutic outcome. All family participants explore their own life histories, learning about themselves and one another in the process. Such techniques as *role playing* and *psychodrama* can be used to bring out significant past events in the lives of families. A "family map" or genogram is used to diagram the family of origin.

Humor and Banter

With the technique of humor and banter, the therapist intentionally makes humorous comments in order to ease a tense moment or to highlight a problem area in the family. The therapist exaggerates aspects of an individual's or a family's behavior. Prerequisites include, but are not limited to, the family and the therapist having a sense of humor and being able to maintain a good rapport.

Including the Family of Origin

James Framo[16] believes that involvement of the family of origin is one of the most effective techniques in family therapy. It is based on the accepted rationale that current family problems are grounded in part on reenactments of previous problems that the husband and wife have had with their own families of origin. The therapist routinely has at least one session with each marriage partner, together with that partner's own family of origin. The individual's spouse and children are not present in order to minimize emotionalism. This enables the therapist to discuss the here-and-now "corrections" with the aim of increasing present relatedness to the spouse's current family. There is usually a great deal of resistance to this technique.

Coaching[2,25,32]

With the coaching technique, the therapist acts like a coach in helping the family member make changes. For example, the therapist may explain concepts and theories, give examples, draw diagrams, ask questions, make predictions, or suggest alternatives. The therapist can get up from a chair and stand behind family members whispering instructions or a supervisor can phone instructions into a specially equipped room.

Mourning and Empathy[47]

With the technique of mourning and empathy, the therapist elicits unresolved grief for a parent, child, or relative in order to effect change. This technique is borrowed in part from Gestalt therapy, in which there is an attempt to release long-hidden feelings, expectations, and emotions.

VISITS

With the voyages technique (modeled on the home-visit techniques) the therapist travels wherever necessary to bring leverage to the family problem. For instance, he may go to schools, homes, housing projects, churches, clinics, bars, hospitals, and so forth. He meets with individuals or agencies that influence the family (for example, the principal of a school).

SELF-DISCLOSURE

In individual psychotherapy the therapist usually does not reveal much information about himself. The therapy is focused on the patient's problems, feelings, and behavior rather than on those of the therapist's. Some family therapists prefer, however, to use themselves as a major instrument in changing the family by revealing material about themselves, their nuclear and extended families, job philosophy, conflicts, goals, and plans.

This technique has not been accepted for most training programs because of the belief that it may create more problems than it solves.

GUIDED FANTASY[17]

With the technique of guided fantasy, the therapist helps the individual share his internal system of fantasies and thoughts with other family members. The rationale is that "daydreaming" can provide people with a powerful tool for their growth and problem solving. It is important to have each member share his or her inner thoughts with the rest of the family, so that they can be empathic in helping the individual grow.

FAMILY SCULPTURE[54]

Family sculpture is a technique in which the therapist asks one or all of the members of the family to create at a given moment a physical representation of their relationships by arranging their bodies in space. Alliances and estrangement can be concretized by such an exercise. The technique can be used as part of the diagnostic workup to generate hypotheses or to represent a concept being worked on concretely during the course of therapy. Both the content of "the sculpture" and the way "the sculptor" (that is, family member) uses mass and form are examined. It is an excellent technique for nonverbal families.

MULTIPLE-FAMILY GROUP THERAPY*

The technique of multiple-group therapy brings together several family units into a group just as traditional group therapy brings together individuals. These groups may meet on a regular basis. The size of such groups may vary from three to eight families at any one time. Groups can include infants and those old enough to be living independently, as well as significant others such as grandparents, in-laws, and fiance(e)s. The duration of the treatment on an outpatient basis can be from three months to one year, whereas on an inpatient basis the family might participate in treatment usually only as long as the identified patient remains hospitalized.

This technique seems to work best when there is a good balance between the families who have been in the group previously (that is, more experienced families) and those that have not. A balance between interfamily and intrafamily interaction is also desirable.

As part of the process, there is a great deal of mutual disclosure and sharing, as well as peer review and evaluation of what has gone on. Socializing between these families outside of the group formal sessions has been used but with uncertain results.

Gould has summarized the process of such groups as follows:[20]

1. They are sharing and interactional rather than analytical.
2. They are fast moving, experiential, often hectic, and very much in the here-and-now.

*See references 7, 11, 12, 29, and 30.

3. There is a great deal of crucial interfamily contact that makes it possible for families to learn from one another.

It has been claimed that multiple-family therapy groups, in contrast to individual family treatment sessions, have very few dropouts. They are thought to be especially useful when the family expresses a great deal of denial.

Network Therapy

Ross Speck and his associates[57,58,59] have described a novel approach to help the identified patient. Members of the kinship system, friends of the family, and all significant others who bear on the problem, are brought together to work on the problem. This adds healthier voices to the mix. These groups meet for three to six biweekly sessions lasting about four hours. The meetings are held usually in the identified patient's home, and thirty to forty-five people can be involved.

Psychodrama and Role Playing

Psychodrama and role playing techniques have also been used to help families enact family problems and work out new patterns. They are especially useful in nonverbal families.[36] In role playing or reverse role playing, one partner either plays himself in a hypothetical situation or takes on the role of his partner, often switching roles back and forth and commenting on the observations, feelings, and behavior elicited. Role reversal is believed to be useful for developing empathy in family members.

Gestalt Therapy[31]

In adapting Gestalt therapy to family problems, the therapist stresses that the only real time is the present, and he does not rehash the past. He stresses that each individual is responsible for his or her own behavior (countering the familiar resistance, "I did it only because he or she made me do it"). He stresses that symptoms and conflicts are the here-and-now expressions of unresolved situations of childhood that can be finished in treatment. Significant attention is paid to nonverbal behavior.

Weekend Family Marathons

Weekend family marathons have been reported in which one or several entire family units get together for extended periods (anywhere from eight to twelve hours or longer) with leaders ("facilitators") for a variety of intensive types of encounters, usually including affect catharsis and nonverbal experience.[29]

Behavioral Approaches

Behavioral approaches deal with the means by which certain behaviors can be learned, reinforced, or extinguished, irrespective of the original causes for such behavior. Even relatively minor changes in the behavior of one family member, or in a dyad, may bring about a significant alteration in the behavior and feelings of other family members. External behavioral change may precede internal attitudinal change.

Techniques of behavioral therapy include assertiveness training, operant conditioning, relaxation and desensitization, contingency reinforcement, and cognitive behavior modifications. Family members can be utilized as cotherapists in various behavior modification exercises that are rehearsed initially in the therapist's office and are assigned for practice at home.

Videotape

Therapists often find it beneficial to review their sessions and to have a record of an entire course of therapy. Immediate playback of a videotape helps families attain some psychological distance, makes them increasingly self-aware, helps to correct distortions or conflicts about communication, and is invaluable in revealing the important nonverbal aspects of interactions that might otherwise be lost.[53] Families often comment constructively while viewing their own videotaped interactions and see things that they might deny when a therapist points them out.

AUDIOTAPE

Audiotape has also been used as an adjunct to family therapy.[13] A tape of a session can be made and the family can take it home and listen to it, or a tape can be made at home and then played at the session.

ONE-WAY MIRROR[41]

The family meets in a room equipped with a one-way mirror. The therapist can leave the family alone and observe its members through the one-way mirror or have one or more selected members, including an estranged member, observe the interactions. The family member comes out of the "heat of battle" and is presumably then able both to distance himself from what is going on and to change the unwanted behavior in the family system.

Some cotherapists find it useful for one therapist of the team to be in the room with the family while the other therapist observes (alone or with selected family members). The therapy can be interrupted at any time, so that the cotherapists can confer and plan. The therapist who functions primarily as an observer can be very objective, and this will facilitate treatment.

¶ Conclusion

Family therapy is an approach rather than a single technique. It is a group of therapeutic interventions, all focusing on the family, but directed toward a variety of specific therapeutic goals. Therefore the relative importance of a particular guideline depends in large part on the extent to which the therapist uses the family model. For instance, if the therapist treats *all* problems with family therapy, then guidelines are not important. Conversely, when different problems are treated in different ways, then guidelines become crucial.

Until about ten years ago, indications and contraindications for family therapy had been based on ingenious hunches regarding treatment efficacy in a specific situation and on clinical experience (a term once defined as "making the same mistake for thirty years"). More recently there have been some controlled outcome data that define situations in which family or marital therapy might be the treatment of choice. These situations include:[22]

1. Marital therapy for a marital problem.
2. Marital therapy for sexual dysfunction.
3. Family therapy for certain childhood and adolescent behavior problems.
4. Family therapy for the "chronic patient" (i.e., those in need of long-term continuing care and rehabilitation).

Any final authoritative pronouncement as to when and which family therapy approach should be used must be withheld until more controlled data are available comparing family therapy with other types of treatment.

¶ Bibliography

1. ACKERMAN, N. W. *Treating the Troubled Family.* New York: Basic Books, 1966.
2. BACH, G. R., and WYDEN, P. *Intimate Enemy: How to Fight Fair in Love and Marriage.* New York: William Morrow, 1969.
3. BATESON, G., et al. "Towards a Theory of Schizophrenia," *Behavioral Science,* (1956) 1:251–264.
4. BEELS, C., and FERBER, A. "What Family Therapists Do," in Ferber, A., Mendelsohn, M., and Napier, A., eds, *The Book of Family Therapy.* Boston: Houghton Mifflin, 1973, pp. 176–194.
5. BELL, J. E. *Family Group Therapy.* Public Health Monograph No. 64. Washington, D.C.: Department of Health, Education and Welfare, Public Health Service, 1961.
6. BERMAN, E. M. and LIEF, H. I. "Marital Therapy from a Psychiatric Perspective: An Overview," *American Journal of Psychiatry,* 132 (1975): 583–592.
7. BLINDER, M., et al. "MCFT: Simultaneous Treatment of Several Families," *American Journal of Psychotherapy,* 19 (1965): 559–569.

8. BOWEN, M. "A Family Concept of Schizophrenia," in Jackson, D. D., ed., *The Etiology of Schizophrenia.* New York: Basic Books, 1960, pp. 346–372.
9. ———. *Family Therapy in Clinical Practice.* New York: Jason Aronson, 1978.
10. CAMP, H. "Structured Family Therapy: An Outsider's Perspective," *Family Process,* 12 (1973): 269–277.
11. CHAZAN, R. "A Group Family Therapy Approach to Schizophrenia," *The Israel Annals of Psychiatry and Related Disciplines,* 12 (1974): 177–193.
12. CURRY, A. "Therapeutic Management of a Multiple Family Group," *International Journal of Group Psychotherapy* 15 (1965): 90–96.
13. DAVID, A. "Using Audiotape as an Adjunct to Family Therapy: Three Case Reports," *Psychotherapy,* 7 (1970): 28–32.
14. DICKS H. V. *Marital Tensions.* New York: Basic Books, 1967.
15. FRAMO, J. L. Towards the differentiation of a self in one's own family. *Family Interaction: A Dialogue between Family Researchers and Family Therapists.* New York: Springer Pub., 1972, pp. 111–166.
16. ———. "Personal Reflections of a Family Therapist," *Journal Marriage & Family Counsel,* 1 (1975): 15–28.
17. FRIEDMAN, P. H. "Outline (alphabet) of 26 Techniques of Family and Marital Therapy: A through Z," *Psychotherapy: Theory, Research and Practice,* 11 (1974): 259–264.
18. FROMM-REICHMANN, F. "Notes on the Development of Schizophrenia by Psychoanalytic Psychotherapy," *Psychiatry,* 11 (1948): 267–277.
19. GLICK, I. D., and KESSLER, D. R. *Marital and Family Therapy,* 2nd ed. New York: Grune & Stratton, 1980.
20. GOULD, E. "Self-Help Aspects of Multi-Family Group Therapy." Unpublished paper, p. 2.
21. GRUNEBAUM, H., and CHASIN, R., "Thinking Like a Family Therapist." Paper presented at the Downstate Medical Center's Symposium on Family Therapy, Training of Child Psychiatrists, New York, December, 1978.
22. GURMAN, A. S., and KNISKERN, D. P. "Research on Marital and Family Therapy: Progress, Perspective and Prospect," in Garfield, S. L., and Bergin, A. E., ed., *Handbook of Psychotherapy and Behavior Change: An Empirical Analysis,* 2nd ed. New York: Wiley, 1978. pp. 817–901.
23. HALEY, J. *Strategies of Psychotherapy.* New York: Grune & Stratton, 1963.
24. ———. *Uncommon Therapy: The Psychiatric Techniques of Milton H. Erickson, M.D.* New York: Norton, 1973.
25. HALEY, J., and HOFFMAN, L. *Techniques of Family Therapy.* New York: Basic Books, 1967.
26. HARRIS, L. "Analysis of a Paradoxical Logic: A Case Study," *Family Process,* 19 (1980): 19–33.
27. JACKSON, D., and WEAKLAND, J. "Conjoint Family Therapy: Some Considerations on Theory, Technique and Results," in Haley, J., ed., *Changing Families.* New York: Grune and Stratton, 1971, pp. 13–35.
28. LANDES J., and WINTER, W. "A New Strategy for Treating Disintegrating Families," *Family Process,* 5 (1966): 1–20.
29. LAQUEUR, H. P., WELLS, C., and AGRESTI, M. "Multiple-family Therapy in a State Hospital," *Hospital & Community Psychiatry,* 20 (1969): 13–20.
30. LAQUEUR, H. P. "Multiple Family Therapy and General Systems Theory," in Ackerman, N. W., ed., *Family Therapy in Transition.* Boston: Little, Brown & Co., 1970, pp. 82–93.
31. LEVETON, A. "Elizabeth is Frightened," *Voices,* 8 (1972): 4–13.
32. LEVETON, E. *Psychodrama for the Timid Clinician.* New York: Springer Pub., 1977.
33. LEWIS, J. M., et al. *No Single Thread: Psychological Health in Family Systems.* New York: Brunner Mazel, 1976.
34. LIDZ, R., and LIDZ, T. "The Family Environment of Schizophrenic Patients," *American Journal of Psychiatry,* 106 (1949):322–345.
35. LIDZ, T., et al. "Intrafamilial Environment of the Schizophrenic Patient. VI. The Transmission of Irrationality," *Archives of Neurology and Psychiatry,* 79 (1958): 305–316.
36. MCKINNEY, J. "Adapting Family Therapy to Multi-deficit Families," *Social Casework,* 51 (1970): 327–333.
37. MADANES, C., and HALEY, J. "Dimensions of Family Therapy," *Journal of Nervous & Mental Disease* 165 (1977): 88–98.
38. MANDELBAUM, A. "Diagnosis in Family

Treatment," *Bulletin of the Menninger Clinic,* 40 (1976): 497–504.

39. MARMOR, J. "Marmor Lecture," *Psychiatric News,* 5 November 1975, p. 1.
40. MINUCHIN, S. *Families and Family Therapy.* Cambridge, Mass.: Harvard University Press, 1974.
41. ———., et al. *Families of the Slums: An Exploration of Their Structure and Treatment.* New York: Basic Books, 1967.
42. MITTELMAN, B. "The Concurrent Analysis of Married Couples," *Psychoanalytic Quarterly,* 17 (1948): 182–197.
43. NADELSON, C. C. "Marital Therapy from a Psychoanalytic Perspective," in Paolino, T. J., and McCrady, B. S., ed., *Marriage and Marital Therapy.* New York: Brunner Mazel, 1978, p. 123.
44. NAPIER, A. Y., and WHITAKER, C. A. *The Family Crucible.* New York: Harper & Row, 1978.
45. OBERNDORF, C. P. "Folie a deaux," *International Journal of Psychoanalysis,* 15 (1934): 14–24.
46. PAOLINO, T. J., and MCCRADY, B. S., eds. *Marriage and Marital Therapy.* New York: Brunner Mazel, 1978.
47. PAUL, N. L. "The Role of Mourning and Empathy in Conjoint Marital Therapy," in Zuk, G. H., and Boszormenyi-Nagy, I., eds., *Family Therapy and Disturbed Families.* Palo Alto, Ca: Science and Behavior Books, 1967, pp. 186–205.
48. SAGER, C., BRAYBOY, T., and WAXENBERG, B. *Black Ghetto Family in Therapy: A Laboratory Experience.* New York: Grove Press, 1970.
49. SANDER, F. M. "Marriage and the Family in Freud's Writing," *Journal of the American Academy of Psychoanalysis,* 6 (1978): 157–174.
50. SATIR, V. M. *Conjoint Family Therapy: A Guide to Theory and Technique.* Palo Alto, Ca.: Science and Behavior Books, 1964.
51. ———. *Peoplemaking.* Cupertino, Ca.: Science and Behavior Books, 1972.
52. SELVINI, P. M., et al. *Paradox and Counterparadox. A New Model in the Therapy of the Family Schizophrenic Transaction.* New York: Jason Aronson, 1978.
53. SILK, S. "The Use of Videotape in Brief Joint Marital Therapy," *American Journal of Psychotherapy,* 26 (1972): 417–424.
54. SIMON, R. "Sculpting the Family," *Family Process,* 11 (1972): 49–58.
55. SLUZKI, C. E. "Marital Therapy from a Systems Theory Perspective in New York," in Paolino, T. J., and McCrady, B. S., ed., *Marriage and Marital Therapy.* New York: Brunner Mazel, 1978, pp. 366–394.
56. SMITH, Z. E. *Discussion on Charity Organizations.* Proceedings of the National Conference on Charities and Correction, 1890, p. 377.
57. SPECK, R. V., and ATTNEAVE, C. L. "Network Therapy," in Haley, J., ed., *Changing Families.* New York: Grune & Stratton, 1971, pp. 312–332.
58. ———. *Family Networks.* New York: Pantheon Books, 1973.
59. SPECK, R. V., and RUEVENI, U. "Network Therapy: A Developing Concept," *Family Process,* 8 (1969): 182–191.
60. THIBAUT, J. W., and KELLY, H. H. *The Social Psychology of Groups.* New York: John Wiley, 1959.
61. WATZLAWICK, P. J., BEAVIN, H., and JACKSON, D. D. *Pragmatics of Human Communication. A Study of Interactional Patterns, Pathologies, and Paradoxes.* New York: Norton, 1967.
62. WYNNE, L., et al. "Pseudo-Mutuality in the Family Relations of Schizophrenics," *Psychiatry,* 21 (1958): 205–220.
63. ZIMET, C. N. "NIMH Backs up on its Forward Plan and the National Register Survey of Licensed/Certified Psychologists," *Psychotherapy Bulletin,* 10 (1977): 1–3.
64. ZUK, G. H. "Editor's Introduction: The Three Crises in Family Therapy," *International Journal of Family Therapy,* 1 (1979): 3–8.

CHAPTER 21

CURRENT CONCEPTS OF HUMAN SEXUALITY

Helen Singer Kaplan

THE LAST DECADE has witnessed intense activity and rapid development in the area of human sexuality. The intellectual and moral climate regarding sex has become increasingly rational, humane, and nonjudgmental to the point where the human sexual response has become a legitimate topic for scientific study. As evidence, people all over the world are openly seeking help for their sexual difficulties. This has led to the accumulation of extensive clinical experience, substantially changing traditional concepts about sexuality and sexual disorders. At the same time, clinical techniques have been greatly improved and extended.

¶ The Triphasic Concept of Human Sexuality and Sexual Dysfunction

In the past, it was believed that human sexual response was a single entity beginning with lust and ending with a climax. But this monistic view is wrong and has hampered the quest for theoretical clarity and for the effective treatment of sexual problems.

Probably the single most important recent theoretical advance in this field has been the emergence of the *triphasic concept* of the sexual response, which is based on the discovery that the male and female sexual response is made up of three interlocking, but neurophysiologically separate, phases: desire, excitement, and orgasm. This separation of the three phases, in addition to its theoretical importance, is also of great practical significance. Each of the phases can be impaired separately, and inhibitions of the specific phases produce the clinical dysfunction syndromes of orgasm phase, excitement phase, and desire phase dysfunction. Each of these phase dysfunctions is associated with related, but distinct and different, psychopathological patterns, and each responds to related, but specific treatment strategies.[10] The new classification of sexual disorders as described in the DSM-III is organized according to this concept.[5]

¶ The Physiology of the Three Phases and the New Classification of Psychosexual Dysfunctions

The male excitement and orgasm phases were first clinically differentiated by James Seaman. In 1956, he developed a specific treatment for premature ejaculation, a condition that had previously been located, under the label of impotence, in the undifferentiated mass of male sexual difficulties. This distinction substantially improved the outcome for premature ejaculation, which had, until this time, a very poor prognosis. A similar distinction between the female excitement and orgasm phase disorders was proposed by the author in 1971.[13] This met with similar results. When anorgasmia is differentiated from other forms of "frigidity," the effectiveness of treatment is significantly improved. Recently, a third and separate sexual phase has been recognized—the phase of sexual desire. The separation of desire phase disorders from the disorders of the phases of the sexual response has clarified some of the conceptual confusion between libido and genital reflexes; it also shows promise of contributing to clinical advances of comparable significance.[11,12,15]

Desire

Desire in both genders is the experiential concomitant of specific neural activity in the brains "sex circuits." The generation of desire is analagous to the process that governs the other drives serving biological purposes; for example, the need for food. We feel hungry only when the "hunger circuits" of the brain are activated. The usual trigger for this is the need for nutrients. If these centers are not active; that is, if they are inhibited by satiety or by psychic conflict, or biologically by appetite suppressant medication, there is a loss of appetite or anorexia.

The sex centers of males and females normally are activated rhythmically by hormones, or they may be stimulated by an attractive sexual opportunity. When such neural activation occurs, we feel, in popular terminology, "sexy," "horny," or "hot," that is, sexually hungry.

If the sex centers are inhibited by psychic events or suppressed biologically, we are not sexually receptive. We lose our "sexual appetite" or suffer from "sexual anorexia."

The anatomy and physiology of the sex circuits of the brain have not yet been delineated with precision. However, enough is known to permit inferences regarding the general model of their structure and function. It is known that sexual circuits are located in the limbic brain and have important way stations or nuclei in the hypothalmic and preoptic regions. Both genders require adequate amounts of testosterone for the proper functioning of the sex centers and therefore for the experience of libido. Testosterone is the "the libido hormone" for both men and women. It is also known that the neurotransmitters serotonin (5HTP) and dopamine act in an inhibitory and excitatory capacity respectively.[5,6,11]

Sexual desire may be inhibited by psychological and physical factors, presumably because of their inhibitory effect on the sex centers. In clinical practice, anger, depression, stress, psychological conflicts about sex, and certain illnesses and drugs are commonly associated with low libido states.

A person who suffers from a low level of sexual desire has little or no spontaneous interest in sex, and cannot easily be aroused. He does not masturbate, has no fantasies, and when approached sexually experiences little pleasure. He may react with extreme avoidance, repulsion, boredom, or merely with tolerance and a mechanical genital response. Clinically, inhibited sexual desire (ISD) may be described along two axes, producing four clinical subtypes.[10,11] ISD may be *primary,* or a life-long condition, in which a patient has never experienced much of a sex drive or interest in sex, or it may be *secondary,* occurring after a period of normal desire later on in life. The parameters of sexual desire can also be described as being *global,* refering to a complete loss of libido, or *situational,* meaning a loss of libido only in specific situations; for example, the absence of desire for a spouse or lover. The former is often seen in depression, while situational ISD is typical when the inhibition is based on psychological conflict.[2,11]

Emergency emotions such as fear or anger have biological priority over the procreative drives. In physiologic terms, it may be inferred that the activity of the sexual centers and circuits of the brain (upon which the experience of sexual desire depends) is suppressed when brain mechanisms that subserve the "emergency" emotions and that insure personal survival are activated. In other words, fear and rage of sufficient intensity are incompatible with sexual desire, and attention to a dangerous situation will take priority over an amorous one. On a clinical level this means that when a person is sufficiently angry, anxious, in conflict, depressed, or under stress, his libido will be diminished. This can occur regardless of whether the source of stress is sexually related.

The paraphilias may be regarded as special instances of situational ISD. Traditionally, the emphasis in these disorders has been placed on the variant object or aim, or on the content of the fantasy or fetish that arouses desire in these individuals.[11] But perhaps the essential point involves the pattern of *inhibition* and not the nature of the arousing stimulus. For such patients only mature heterosexual expression represents a "danger," whereas the variant situation is symbolically "safe." Thus desire is inhibited only in the threatening heterosexual situations. But in "safe" circumstances, such as those created by the fantasy, fetish, or other paraphilic devices, which symbolically "circumvent" the threat of heterosexuality, the genital reflexes can function normally, and the conflicted individual is free to experience erotic desire and pleasure.

Excitement

Physiologically, excitement is caused in both genders by the reflex vasodilatation of the genital blood vessels. This produces a swelling which changes the shape of the genital organs to prepare them for their reproductive functions.

In the male the vasodilatation of sexual excitement is marked by erection. More specifically, erection is produced when the reflex vasodilatation of penile arteries increases the flow of blood to the penis, while concomitant reflex constriction of the penile veins impedes the blood's outflow. This traps blood under relatively high pressure in the corpora cavernosii of the penis, distending that organ and making it erect and hard and capable of penetration.

In the female, analagous reflex vasodilatation of the genital organs occurs during the excitement phase. The female genital anatomy is simpler. Women do not possess specialized valves that trap the blood in the area, nor do they have special distensible caverns that produce erection. Therefore, vasodilatation of the female genitals produces a diffuse swelling around the vaginal and labial areas, which has been termed the *orgasmic platform.* This genital vasodilatation is also responsible for the characteristic deep coloration of the labia during excitement. In addition, transudate from the engorged tissues escapes into the vaginal cavity, producing lubrication.[8]

The genital reflexes that subserve excitement (genital vasodilatation) and those that control orgasm (genital myotonic contractions) are controlled by separate spiral reflex centers. These are richly connected and receive inflow from various levels of the central nervous system (CNS). The multiple neural contractions provide the biological infrastructure that make it possible for erection and orgasm to be enhanced or inhibited by a wide range of experiences and emotions.

Penile erection in the male, and by analogy genital lubrication-swelling in the female, are primarily parasympathetic responses. The neural centers and circuits that govern penile erection are distributed on all levels of the CNS. All these influences finally impinge on two lower reflex centers that are located in the thoracolumbar portion of the spinal cord. One of these centers mediates psychogenic erection, while the other mediates erection attained on the basis of local tactile stimulation.[2,27]

The vasodilatory reflexes that produce the genital swelling of the excitement phase of males and females are subject to inhibitory influences from the higher brain. If the individual should become frightened or inhibited by conflict or even mildly startled, the penile blood vessels immediately constrict while the polsters and venus valves, which have retained the blood within the cavernous sinuses, relax. These reflex physiological responses to fear result in the instant drainage of blood from the penis, producing detumescence. When this occurs on a regular basis in appropriate sexual circumstances, the clinical syndrome of *impotence* (excitement phase dysfunction) results.

The patient's specific pattern of impotence will depend on which aspect of the sexual experience evokes anxiety of sufficient intensity to produce detumescence. Thus, some men have difficulty attaining an erection in the presence of a partner, some lose it at the moment of penetration, while others become anxious and consequently impotent inside the vagina before they ejaculate.

Excitement phase inhibition of the female produces a clinical syndrome that in the past has been mislabeled *frigidity.* Such women feel sexual desire and may be orgastic, but they are unable to feel pleasurable sensations or show the physiological response of lubrication and swelling. Impotence is common, but excitement phase inhibition of females is a rather rare clinical syndrome.

Orgasm

Orgasm in both the male and female is produced by the reflex contraction (myotonia) of certain genital muscles. The experiential concomitant of these muscle contractions constitutes the pleasurable sensations of orgasm.

The male orgasm consists of two subphases: emission and ejaculation. After rhythmic stimulation of the shaft of the penis, which produces excitement, the orgastic threshold is reached. This triggers a discharge of the sympathic nerves to the internal male reproductive organs (which are of Wolffian origin): the vasa differentia, the seminal vesicles, and the prostate gland. This discharge stimulates the smooth muscles of the organ to contract and causes a bolus of semen to be deposited in the posterior urethra. This response, which is mediated by the sympathetic nerves, is called emission. Emission is perceived by the male as a signal that ejaculation is about to occur. This phase of the orgasm response is not accompanied by pleasurable orgastic sensations. In the normal male, emission is followed a split second later by the 0.8 per second rhythmic contractions of the striated bulbo and ischio cavernosi muscles at the base of the penis. This contraction propels the semen through the urethra out of the penis in squirts. This phase is termed "ejaculation," and it is accompanied in the unconflicted and healthy male by pleasurable orgastic sensations. The male refractory period is associated with the first phase, emission; this means that after emission occurs a period of time must elapse before the male is responsive again. The second phase, the contractile or ejaculatory phase, does not seem to be associated with a significant refractory period.

Female orgasm lacks the first, or emission phase, because the female does not possess the Wolffian-derived internal male reproductive organs. However, female orgasm is clearly analagous to the second phase of the male orgasm. Upon tactile stimulation of the clitoral area, which is the anatomic and neurophysiologic analogue of the tip and shaft of the penis, the orgastic threshold is reached. This causes a discharge of nerve impulses to the striated muscles in the female perineum (the ischio and bulbo cavernosi muscles) producing rhythmic 0.8 per second contractions.[8,21,23] In the normal female these contractions are experienced as the pleasurable sensations of orgasm.

The understanding of the physiology of female orgasm has made it clear that (1) female and male orgasms are highly analagous, and (2) there is no dichotomy between clitoral and vaginal orgasm. The affector arm of the female orgasm reflex is located in the sensory nerve endings of the clitoral area, while its effector expression is executed by contractions of the muscles surrounding the vagina. Thus all female orgasms are both "clitoral" *and* vaginal. The lower centers, which constitute the final common pathway for orgasm, are located in the sacral portion of the cord near the reflex centers that govern defecation and urination. They are close to, but definitely separate from, the nerve pathways and nuclei serving sexual excitement.

Again, the orgasm reflex is subject to multiple inhibitory and facilitory influences from higher neural centers. Thus, when a person is in a negative emotional state, is frightened, angry, or ambivalent, the orgasm reflex may become blocked, or, conversely, escape from voluntary control.

Inhibition of the orgasm reflex is the most prevalent clinical complaint of females. *Anorgasmia* may be a part of general sexual inhibition or it can occur in women who feel strong sexual desire and who lubricate and function well during the excitement phase. Inhibition of the orgasm phase can occur with various degrees of severity. It is estimated that 8 to 10 percent of women in the United States have never had an orgasm at all.[8,10] At the other end of the spectrum of orgastic threshold are those women who are easily orgastic during coitus and who do not require direct clitoral stimulation. Between these extremes on the orgasm continuum lie various degrees of stimulatory intensity or thresholds for reaching orgasm. There is some controversy about where the clinical demarcation between the normal and the pathologically blocked female response should be drawn. Some authorities—particularly those influenced by Freudian concepts about vaginal versus clitoral orgasm—will define any woman who requires direct clitoral stimulation as suffering from an inhibition. More recent concepts and attitudes regard a woman as normal as long as she can experience a pleasurable orgasm together with her partner regardless of whether this requires direct clitoral stimulation.[2]

This concept makes sense when it is recognized that coitus is not as intensly stimulating in the physical sense as is the direct stimulation of the clitoris. The orgasm threshold is multiply determined. Both psychic and physiological elements contribute to the final threshold. According to the classical analytic view, all woman who are free of sexual conflict should be able to reach a climax in response to the stimulation provided by coital thrusting alone. The need for clitoral stimulation is taken as evidence that the threshold has been elevated by inhibition. An alternate view holds that while psychic conflict can raise the threshold of the orgasm reflex, many unconflicted and normal women also need clitoral stimulation in order to experience orgasm; in other words, this pattern of orgastic release is a normal variant of the female sexual response.

There are two orgasm phase dysfunctions of the male, *premature* and *retarded ejaculation.* The syndrome of premature ejaculation, or *inadequate ejaculatory control,* is produced by a man's failure to learn voluntary control over his ejaculatory reflex,[8] with the result that he climaxes reflexively and rapidly as soon as he reaches a high level of sexual excitement. Inadequate orgastic control is also occasionally seen in females, but does not represent a clinical problem.

The opposite orgasmic syndrome of males, *retarded ejaculation,* is analagous to female orgasm inhibition in that the ability to release the orgasm is inhibited. With males, as with women, the syndrome may occur with various degrees of severity. Men with mild retardation simply require an unusually long period of stimulation before they can ejaculate. Men who are somewhat more inhibited can reach orgasm on manual or oral stimulation but not within the vagina. Still more severely inhibited patients can only ejaculate upon masturbation by themselves, and in the most severe forms of this disorder males may not be able to ejaculate at all.

An interesting subvariety of retarded ejaculation is seen in men who experience emission but not the expulsive phase of the ejaculatory response, which is specifically inhibited. Clinically, such patients experience a pleasureless seepage of semen, but not the 0.8 per second contraction of the perineal muscles nor the attendant pleasurable sensations.[8]

Other Sexual Syndromes

Sexual phobias and psychophysiologic sexual problems associated with painful spasms of genital muscles are not related to the impairment of one of the three phases of the sexual response. They are included here because they share a similar etiology and respond to similar treatment strategies with regard to the psychosexual dysfunctions just described. Specifically all these disorders are associated with sexual anxiety and all are amenable, at least in some cases, to sex therapy methods.

DISORDERS ASSOCIATED WITH GENITAL MUSCLE SPASMS

In the female, the reflex spasm of the muscles surrounding the vaginal introitus, when severe, produces a syndrome called *vaginismus.* This condition accounts for many unconsummated marriages because penetration is impossible while the muscles are in spasm. Lesser degrees of vaginal muscle spasm produce various intensities of *dyspareunia* or pain on sexual intercourse.

A certain kind of *male dyspareunia* is also associated with spasms of the genital musculature. These patients experience acute ejaculatory pains, which may be caused by painful spasm of the cremasteric and perineal muscles or of the muscles of the internal reproductive organs at the point of ejaculation or a moment later.

SEXUAL PHOBIAS

The previously described syndromes fall into the category of psychophysiologic disorders in that they represent disturbances of physiologic processes caused by psychological determinants. But often sexual problems are not psychophysiological; instead they are the products of sexual phobias and sexual avoidances. Patients may suffer from phobias of the entire sexual experience or of its various aspects. Kissing, touching, erotic feelings, intimacy, penetration, the genitals, semen, vaginal secretion, oral and anal sex, and so forth, may evoke anxiety and an attendant avoidance response. Instead of pleasure, such patients experience discomfort and panic in the sexual situation. The sexually phobic patient requires specific treatment. When the avoidance is based on a simple phobia, sex therapy can be modified to fit the requirements of in vivo desensitization. When sexual avoidance is a manifestation of an underlying panic or phobic-anxiety syndrome, appropriate medication combined with sexual therapy may be indicated.[11,27]

¶ Etiology

All the sexual symptoms described in this chapter, with the exception of the sexual phobias, can be caused by psychological conflict, or they can all result in part or entirely from depression, severe stress, certain medical illnesses, drugs, or substance abuse. Impotence in the male is especially likely to be caused by organic determinants such as the vascular and neurogenic pathology associated with diabetes, arteriosclerotic vascular disease of the pelvis, endocrine deficiencies, antihypertensive medication, and alcoholism. Such medical factors may be subtle and difficult to diagnose except with specialized procedures, which are not yet widely available. Consequently, the specific organicity may not be detected on a routine medical or urologic examination. Organicity must, of course, be ruled out before psychiatric intervention is planned. Also, sexual dysfunctions, especially the low libido states, are frequently secondary manifestations of primary psychopathological states such as depression and severe stress reactions. In such cases, intervention on the level of the underlying psychiatric condition is indicated, and not sex therapy. The following discussion of etiology applies only to primary psychogenic sexual disorders and not to sexual disorders secondary to medical factors or to primary psychopathologic states.

Past beliefs regarding the etiology of psychosexual disorders contained two major errors: (1) that all psychogenic sexual disorders are variants of a single psychopathologic entity, and (2) that these have only one cause: serious and unconscious sexual conflict that was acquired during early childhood, usually by the fifth year of life. This view, a derivative of psychoanalytic theory, had been until recently widely accepted in the field. Accordingly, the conflict responsible for sexual problems of males centers around the unconscious fear of injury or castration. Unconscious fear of injury was also thought to play an etiological role in female sexual problems, as was unconscious competition with males and "penis envy." It was believed that conflicts acquired early in childhood were reactivated in the adult sexual situation, impairing the patient's sexual adequacy.

It followed from this theoretical position that the treatment of choice for sexual disorders of all kinds—prematurity, anorgasmia, the paraphilias, libido disorders, impotence, and so on—was held to be psychotherapy, which has the capability to foster insight into and resolve the presumed underlying sexual conflicts.

In the last decade this view of a serious and specific universal cause of sexual disorders has been challenged by the work of Masters and Johnson[21] and also by some behaviorists. These clinicians demonstrated that sexual problems do not invariably originate in deeply unconscious neurotic conflict but are, in fact, frequently the product of superficial anxieties such as the simple fear of sexual failure; that is, performance anxiety. This "minor etiologies hypothesis" was supported by the high success rate of the brief Masters and Johnson treatment regime, which was designed essentially to modify the relatively minor forms of sexual anxiety.

The clinical experience of the past decade suggests that both points of view have validity. It appears that there is no specific etiologic agent associated with sexual dysfunction. Multiple factors can produce the same sexual problems. Sexual conflict or sexual anxiety may be considered the ultimate cause of all psychosexual disorders. Anxiety evoked by sex is the "final common pathway" that leads to all psychosexual disorders. But the intensity and source of the sexual anxiety vary widely, even in those patients who display identical symptoms.

Clinical experience suggests that at this time the largest proportion of sexually dysfunctional patients suffer from the kinds of minor, superficial, or mild anxieties described by Masters and Johnson.[22] These patients respond well to the brief sex therapies. But at the other end of the causal continuum there are patients, a smaller but substantial group, whose sexual symptoms originate in the more complex—the major or "deeper"—and usually unconscious kind of psychopathology postulated by psychoanalytic theory. Such patients, although their presenting symptoms may be identical to those of performance-anxiety sufferers, usually require more intensive and more psychodynamically oriented therapy for the relief of their problems.

A Psychosomatic Concept of Sexual Dysfunctions

The observation that the same sexual symptom can be associated with a broad spectrum of causal factors makes sense when sexual dysfunctions are conceptualized as psychophysiological or psychosomatic disorders. Such disorders are produced when the normal physiological processes are disrupted by the physiological concomitants of emotional arousal, by fear, and by rage.

MULTIPLE LEVELS OF CAUSES

The disruptive effects of adversive emotion on the reflexes that comprise the sexual response are independent of their source or cause. In other words, the physiological concomitants of sexually disruptive anxiety are identical whether this anxiety is caused by a simple anticipation of failure to perform sexually or by an unconscious identification of the sexual partner with the mother. In either case, the penile arterioles will constrict, the polsters that impede the outflow of blood from the corpora cavarnosii will open, and the patient will experience an involuntary and unwelcome detumescence. According to a psychosomatic conception the same symptom can result from a variety of causes as long as the anxiety reaches an intensity that is sufficient to impair the reflex in question.[11] In some cases, this will result from simple consciously recognized stimuli, while in others deep unconscious conflicts may be responsible.

Specificity

The psychosomatic concept of sexual disorders does not answer the question of specificity or symptom choice, or why one patient with sexually related anxiety fails to learn control over ejaculations while another loses his libido. While the question of specificity remains in many respects a mystery, some interesting hypotheses have been advanced.[7,11]

Psychoanalytic theory postulates that unconscious psychodynamic conflicts related to disturbances of specific developmental periods produce specific symptoms. For example, premature ejaculation has been related to the disturbances of the urethral substage of development, and retarded ejaculation to unconscious hostility toward women deriving from oedipal and preoedipal conflicts. However, experimental and clinical support for these interesting hypotheses is lacking.

From a physiological perspective one determinant of symptom choice is "physiologic response specificity."[7] This hypothesis holds that people have a highly individualistic physiologic response to stress. A person's characteristic physiologic response pattern is evident from infancy. Accordingly, some vascular responses in an individual are particularly reactive to emotional arousal. These individuals will tend to develop hypertension or impotence or vascular headaches in response to various stresses. Persons with a different pattern might respond to the identical stressor with anorgasmia, another with a loss of libido, and still another will develop peptic ulcer or an allergy. Experimental and clinical evidence supports this hypothesis; and it is likely that such individual response patterns constitute one determinant of specificity.

Observations of the specific and immediate experiences of patients who suffer from sexual dysfunctions have yielded some additional insights regarding the determinants of the various specific sexual symptoms.[11] More specifically, three variables can play a role in the "choice" of the sexual symptom: (1) *intensity* or depth of the underlying anxiety; (2) the precise *time* at which this anxiety is evoked during the lovemaking experience; and (3) the *specific adaptation* to or *defense* against this anxiety employed by this individual.

Quality of Conflict

It is difficult to define the intensity or depth or quality of anxiety in a clinical situation with any degree of precision. However, despite this unfortunate ambiguity, there is consensus among clinicians who have experience with sexually dysfunctional patients that their underlying anxiety encompasses a wide range of intensities or complexities. Some patients suffer from minor anxieties that can be diminished, to a point at which sexual functioning will resume, by simple reassurance and training in sexual and communication skills. Others suffer from severe and major marital and/or intrapsychic problems that require extensive and complex therapeutic intervention.

The clinical study of the sexually dysfunctional patient population suggests that the severity of sex-related anxiety is not evenly distributed among the various sexual syndromes.[11] As a group, but with many exceptions, patients who suffer from orgasm phase problems (that is, those who have desire and erections but whose orgasms are either blocked or uncontrolled) tend to suffer from the mildest and most easily modifiable sexual anxiety. Such patients, as a group, have the best prognosis with the brief sex therapy methods. They often have no other discernible psychological problems, enjoy good marital relationships, and frequently improve without gaining significant insight into underlying problems, if, in fact, these exist.

By contrast, dysfunctional patients with the severest and most tenacious types of sexual conflicts and relationship problems tend, again with individual exceptions, to develop ISD.[11] These patients have the poorest prognosis with brief therapy, and treatment tends to be stormy, with resistances arising even when the outcome is eventually successful. Severely conflicted patients generally do not improve merely in response to counseling or educational and behavioral methods. Usually they must gain at least some measure of insight into underlying conscious intrapsychic and/or relationship difficulties before improvement occurs. Such patients usually require longer and more insight-oriented therapy than is provided by the standard fourteen-session, behaviorally focused treatment that is the traditional Masters and Johnson mode.[21]

Impotent patients (that is, those with excitement phase problems) as a group fall in between. Some, who suffer from secondary impotence, are clinically like the premature ejaculators and anorgastic females in that they harbor minor or superficial, and often consciously recognized, performance anxiety. Such patients have an excellent prognosis with rapid sex therapy. Others, usually those with life-long potency disturbances, are more like the typical ISD patient in the complexity of their underlying sexual conflicts. They will have a poorer prognosis and more difficult course of treatment.

It should be emphasized that the preceding conceptualization refers only to trends and not to individual patients. There are many anorgastic patients who suffer from severe and complex intrapsychic and/or relationship problems; while some low libido patients need only relief of performance anxiety or improved communication with their partner in order to regain their desire.

TIMING

The time at which anxieties are evoked during lovemaking also influences the nature of the symptom. When anxiety occurs near the end, after considerable pleasure has already been experienced, orgasm-phase problems tend to occur. When anxiety disrupts the sexual experience during the excitement phase, the patient is likely to develop impotence. When negative affect is evoked early on, at the time when lovemaking is just contemplated or initiated, then desire tends to become suppressed.

SPECIFIC DEFENSES

It has been emphasized that the source of the sex-related anxiety that produces sexual symptoms is not specific.[11] Such anxiety may derive from intrapsychic and/or relationship difficulties, and it may be "minor" or "major." Its genesis could be found in any or all stages of early development or else it might be the product of current stress. The patient or couple may be perfectly aware of what is upsetting him or her in the sexual situation, or the sexual conflict may operate entirely on an unconscious, deeply defended level. In this sense, the remote sources of sexual dysfunctions are not specific. On the other hand, the *immediate antecedents* of sexual symptoms appear to be highly specific. The study of the immediate and specific experiences of sexually dysfunctional patients suggests that the different dysfunctional syndromes are associated with specific defenses against, or adaptations to, the anxiety that emerges during the sexual experience.[11] And it is the interaction of the specific defenses or adaptations to anxiety with the physiologic process of the sexual response that produces the specific sexual symptom.

More specifically, anorgastic women and men who suffer from retarded ejaculations become anxious at high levels of sexual excitement. At the moment when abandonment is appropriate they tend to "put on the brake." Such patients deal with their anxiety with an obsessive form of self-observation:

"Will I come?" or "Is it taking too long?" In this manner the orgasm reflex becomes inhibited, just as any reflex that is under voluntary control can become inhibited when scrutinized obsessively. Again, the underlying source of the anxiety that arises as the patient makes love is highly variable and may involve unconscious hostility toward the partner, guilt, fear of loss of control, oedipal transferences, fear of rejection deriving from preoedipal issues, unrealistic expectations about sexual functioning, and so on. And if the patient deals with his anxiety obsessively (what Masters and Johnson call "spectatoring"), the orgasm reflex will be inhibited and the syndromes of retarded ejaculation or anorgasmia will result.

Similar specific antecedents may be observed with the other psychosexual disorders. Vaginismus and the functional dyspareunias of males and females are associated with painful spasms of the genital muscles. Again, this involuntary spasm is the specific final cause of the syndrome and represents the response to the sexual anxiety, which may have a variety of origins.

The syndrome of premature ejaculation is produced when a male becomes anxious at high levels of sexual pleasure to the point of total distraction or denial of these pleasurable sensations. This distraction is a perceptual defense against erotic pleasure. It interferes with the normal process of sensory integration, the process by which voluntary control over all biological reflexes that are subject to such control is learned.

Excitement phase disorders, or impotence, represent the inhibitory effect of undefended anxiety on the delicate parasympathetic genital vascular reflexes. The source of this anxiety is highly variable. Frequently it involves the anticipation of sexual failure, "performance anxiety." Some impotent males will unconsciously conjure up such performance fears in the service of deeper and unconscious conflict about sexual adequacy; for others, performance anxiety is "pure," that is, not associated with deeper sexual difficulties.

The specific and immediate antecedents of inhibited sexual desire are not yet completely clear. The study of the specific and immediate sexual experiences of such patients suggests that many, if not most, ISD patients unconsciously and involuntarily focus on depressing, frightening, and frustrating thoughts and images when they are in a potentially erotic situation.[11] These "antierotic" mental processes may include focusing on negative aspects of the partner, recalling stressful events in external life, or evoking derogatory and depreciatory thoughts about themselves. In this manner, such persons upset themselves to a point where sex is avoided. They tap into the natural physiological mechanisms that suppress sexual desire in an adaptive manner, much as a person does who finds himself in a dangerous or sexually inappropriate situation. Such patients typically have no insight into their active sexual self-sabotage.

Multiple Levels of Etiology

In some clinical situations no discernible pathology underlies the immediate causes just discussed. For example, a woman has simply acquired for no discoverable reason a condition spasm of her perivaginal musculature. No underlying trauma or sexual conflict can be detected. When the spasm is desensitized or extinguished, the patient functions normally, or is cured. In other clinical situations, however, a variety of underlying or remote causes are clearly operative. For example, the woman has acquired her vaginal spasm because the prospect of coitus evokes unconscious guilt about pleasure, or because of oedipal taboos, or because penetration has acquired the infuriating symbolic meaning of submission to the male. Where such remote causes exist, resistances to the removal of the symptom tend to be mobilized during rapid treatment. These have to be circumvented or resolved for a successful clinical outcome.

Among the remote etiological elements commonly seen with sexually dysfunctional patients are guilt, or superego conflicts regarding sex, competition, and pleasure. Unconscious fears of romantic success and intimacy are also highly prevalent in these patients, as is unconscious rage at the partner. Finally, severe psychopathology and serious difficulties in the relationship may also produce sexual dysfunctions. This group of patients does, in fact, suffer from severe problems. These are the deeply neurotic patients whose early presexual (preoedipal) and sexual (oedipal) development was substantially pathological. Couples engaged in severe contractual misunderstandings and power struggles, especially when these have their genesis in primitive parental transferences, are also in this severe or major group. Sometimes these deeper problems can be bypassed with behavioral means that modify only the immediate antecedents of the symptom. In such cases the patient is left with his major psychological problems, but his sexual functioning is improved. Frequently, however, such severe underlying problems give rise to tenacious resistances to behavior modification and are not amenable to brief treatment methods.

¶ Current Concepts of Treatment

Historical Perspective

In the past, insight therapy, that is, psychoanalysis (or one of the modifications that falls under the rubric of psychoanalytically oriented psychotherapy) was considered the appropriate modality for sexual disorders. This judgment was based on the traditional premise that sexual problems are reflections of unconscious conflict, and it was expected that resolution of this conflict would result in the cure of the sexual symptom.

Marital or couples therapy approaches have also been employed for treating sexual problems. Conjoint treatment presumes that conflicts and problems in the marital system are responsible for some sexual problems. It is thought that both partners participate in the genesis and maintenance of these relationship problems and that, to some extent at least, the dynamics of the struggle is usually beyond either partner's conscious awareness. By fostering insight into, and hopefully resolving, such interactional issues, the therapist attempts to improve the couple's sexual experience.

Patients undergoing individual insight therapy often appear to derive considerable benefits in terms of insight, growth, and improvement in other aspects of life.[25,26] But even when significant unconscious conflict is resolved, the actual sexual symptom is cured in only a relatively small portion of patients.

Although systematic outcome studies are lacking, the same is probably true regarding the outcome of conjoint treatment. With the resolution of contractual disappointment, power struggles, communication problems, mutual transference, and other difficulties in the couple's interactions, the relationship often improves, but it does not automatically follow that concomitant sexual improvement will also occur. Often the couple gets along much better, but he still climaxes too rapidly and she does not climax at all.

Sexual problems have also been treated by various behavioral methods, based on the concept that sexual symptoms and sexual anxiety are learned and can be unlearned. Accordingly, it is held that if the contingencies are constructed so as to extinguish sexual anxiety and to foster the learning of sexual skills and improved attitudes, sexual problems can be cured.

Some successes have been reported with the behavioral treatment approaches, but these have been limited. There have been no reports of large-scale successes comparable to the Masters and Johnson[21] outcome study with behavior therapy. It is my impression that strictly behavioral approaches to sexual symptom are successful for only a limited patient population.

The most impressive results thus far reported are those of Masters and Johnson,[21] who reported an 80 percent cure rate. Extensive clinical experience, accumulated since their study, has by and large confirmed the original finding, and has made it clear that sex therapy represents a genuine and significant advance in the treatment of sexual problems. Therefore sex therapy, either in its classic form or in one of the modifications in current use, is widely regarded as the treatment of choice for psychosexual dysfunctions, with the possible exception of severe disorders of sexual desire. Of course, sex therapy is indicated for the psychosexual dysfunctions only when no contraindications exist in the form of organic etiologic factors, substance abuse, depression, the presence of significant stress or crisis, active and serious psychopathology in either partner, or severe problems in the relationship.

Sex Therapy

It is interesting to speculate on the possible reasons for the efficacy of sex therapy as compared to the other forms of treatment. Possibly the limited success of the insight therapies that rely exclusively on verbal interchange between the therapist and the patient or couple is due, at least in part, to the neglect of these modalities as the specific antecedent or immediate cause of the sexual symptom, in favor of focusing on the underlying deeper or remote causes. Most psychoanalysts and couples' therapists typically do not clarify the details of the patient's or couple's sexual interaction or immediate erotic experience, nor will they give suggestions regarding sexual corrective experiences. Instead, the therapist will work to clarify and resolve the underlying intrapsychic and/or interactional conflict in the hope that such insight might be curative. However, clinical experience suggests that the resolution of unresolved unconscious intrapsychic or marital conflict, will frequently not *per se* result in relief of the sexual symptom.[10] In many clinical situations, specific sexual difficulties persist even after excellent insight has been attained and will only improve after the patient or couple's sexual experiences are specifically and adequately modified in addition to insight.

Behavioral techniques have an obvious advantage in treating sexual dysfunction because these methods are exquisitely designed to illuminate and modify the immediate and specific contingencies that ultimately produce the sexual symptom. But these techniques are helpless in dealing with resistances that typically appear during the rapid modification of sexual behavior, except in those patients where the underlying causes are minor. Since the concept of unconscious motivation has no place in behavioral theories, no behavioral techniques exist for dealing with unconscious sexual conflict and with the tenacious resistances to which they give rise.

Perhaps the great advantage of sex therapy lies in the fact that it is *integrated.* It intervenes on both levels: at the level of the immediate sexual symptom, and, when it is necessary, at the level of deeper psychopathology. Integrated techniques employ an amalgam of behavioral and psychodynamic modes. Sexual tasks designed to modify the sexual symptom are integrated with the psychotherapeutic sessions, permitting the therapist to deal with resistances and with the underlying emotional conflict from which these arise.

Recent Modifications of Sex Therapy

The original sex therapy procedure that was described by Masters and Johnson[21] in 1970 has been extensively modified by many of the clinicians who have been working in the field in recent years.[10] The purposes of these modifications have been to streamline treatment, that is, to eliminate those features that are not necessary for therapeutic efficacy, and to extend the range of therapeutic effectiveness to a greater and more troubled patient population, that is, those who suffer from deeper conflicts and resistances and who tend to fail to respond to classic sex therapy.

All modifications retain the two essential features of sex therapy: (1) The focus on relief of the sexual symptom to the exclusion of other problems, which are only dealt with to the extent that is necessary to improve sexual functioning, and (2) The employment of an integrated combination of corrective sexual tasks and psychotherapeutic sessions that are usually conducted conjointly with both partners.

PROCEDURAL MODIFICATIONS

Masters and Johnson originally felt that that the use of a dual gender cotherapy team was essential to the success of their method. However, no solid evidence has been presented attesting to the superiority of cotherapists over a single therapist. Many therapists are currently working alone, and clinical experience suggests that a single well-trained therapist, who is sensitive to the sexual experiences of both genders, is as effective as a team.

The original Masters and Johnson format required that the couple isolate themselves from the stresses and distractions of ordinary life, check into a motel, and see the therapist intensely for daily sessions for a period of two weeks. After this time, the patients were discharged regardless of treatment outcome, with the assumption that if cure had not taken place by this time, the couple was not amenable to the method. Although some clinicians and programs still operate according to this model, most have modified this procedure so that the couple remains at home and visits the therapist in their own community.

The time frame of treatment has also become more flexible in most programs. Masters and Johnson employed a fourteen-session treatment format for all their patients—once a day for two weeks. In other programs, patients are typically seen once or twice a week and length of treatment varies from three sessions to as many as thirty. Clinical experience suggests that these modifications have a more positive outcome for a wider range of patients.[10]

TECHNICAL MODIFICATIONS

Masters and Johnson's[21] original treatment program encouraged the exchange of sensuous pleasure between the partners, deemphasized performance pressures during sex, fostered open and authentic communication, promoted attitudes of mutual respect and caring, and provided training in sexual skills. This same treatment format was used for all the dysfunctions and for all couples.

The recognition of three physiologically separate phases of the sexual response, which ultimately differentiates between the three types of dysfunctional syndromes, has lead to the recognition that each syndrome is associated with specific psychopathologic antecedents. Within this theoretical framework, more specific and rational treatment strategies have been developed. Thus many clinicians now employ different therapeutic techniques for each specific syndrome. Descriptions of these are available in literature.[9,10,11]

Basically, it is assumed that in order to cure the sexual symptom, it is necessary to modify the immediate antecedents that gave rise to it. Thus, the behavioral aspect of treatment is designed to modify the specific antecedents or immediate causes of the sexual symptom. Since these are different for the various syndromes, the "sexual exercises" or behavioral tasks assigned to the couple are different and specific. For example, in anorgasmia of females and retarded ejaculation in males,[8] the exercises are designed to maximize genital stimulation and at the same time reduce the obsessive self-observation by which orgasm is retarded. Focus on erotic fantasy or on physical attributes of the partner may be used in the service of this intervention. By contrast, premature ejaculation is treated with techniques that help the patient focus on the sensations premonitory to orgasm and to accept the pleasurable sensations concomitant with high level of sexual arousal. Both the "squeeze" and the "stop-start" methods have proved effective in accomplishing this objective. By these means it is hoped that the patient's distraction, or tendency toward perceptual defense against erotic pleasure, will be modified to the extent that he can attain voluntary control of his ejaculatory reflex. Potency disorders, which are frequently associated with performance anxiety, are often successfully dealt with by structuring the couple's erotic interaction so that it is reassuring and free from performance pressures. In this manner the fear of sexual failure can often be diminished. The improvement of desire disorders generally requires that the patient gain some insight into his or her tendency to evoke or focus on negative images or "anti-fantasies" when in an erotic situation. This is the hypothetical immediate cause of ISD. To this end the tasks are structured: to help the patient confront his or her defenses against sexual pleasure and the tendency to turn off.

Tasks may be also modified in order to meet the specific and individual dynamic requirements of the patient or the couple. Also, the pace of treatment can be adjusted to accommodate the particular level of anxiety and the patient's tolerance for anxiety. Within such a flexible therapeutic framework, sensitivity and creativity must be used to devise the specific tasks. These should be structured so that they will enable the patient to make progress, while at the same time these experiences must not evoke anxiety of sufficient intensity to mobilize counterproductive resistances to treatment.

Thus, for example, a highly anxious couple may not be able to tolerate the rapid tempo of standard treatment, and tasks can be assigned at a slower and more gradual pace without sacrificing an eventual favorable outcome. Or, if it appears that a specific sexual experience has acquired some threatening symbolic meaning to the patient (and for this reason evokes resistances), alternative tasks may be employed that will "bypass" this obstacle. For example, if the female superior position, which is usually employed during the treatment of premature ejaculation, female excitement phase disorders, and some potency disturbances, is threatening on some symbolic level (perhaps evoking fears of passivity and helplessness in the male, or concerns with aggression or sexual responsibility on the the part of the female), the therapist has two choices. He may attempt to resolve the unconscious threat evoked by this position during the therapy sessions, or he may decide to substitute a side-by-side position or some other one that will still enable the patient to carry out the appropriate and crucial exercise without having to deal explicitly with the anxiety generated by the unconscious meaning of these exercises.

Psychodynamic Emphasis

Masters and Johnson[20,21,22] did not make reference to unconscious conflict or motivation in the description of their treatment procedure. They relied on education, training in sexual and communication skills, in vivo desensitization of sexual anxiety, and on the construction of stimulating and reassuring sexual tasks. These methods have helped numerous patients, and many competent therapists still adhere to this essentially nondynamic therapeutic model. However, others, including this author, feel that sexual pathology gains clarity and that the therapeutic process becomes effective for a greater range of patients when sexuality and therapy are conceptualized in psychodynamic terms with reference to multiple levels of etiology, including those that occur on an unconscious level.[9,10] This is especially true when dealing with patients whose symptoms have deeper roots, symptoms that are the product of more serious conflict than those produced by minor performance anxieties and lack of communication. With minor problems such as the fear of sexual failure (particularly when this disturbs only the orgasm phase), cure can often be obtained, without insight, on the basis of behavioral modification alone. But this is not true of more complex problems such as those seen with the more serious potency disorders and with the majority of desire-phase problems. In such cases, more emphasis needs to be placed on fostering insight.

The insight-promoting aspects of sexual therapy are emphasized during the therapy sessions. These sessions may be conducted conjointly or with either partner, as resistances arise.

In the attempt to develop more effective treatment procedures for the more difficult dysfunctions, a variety of treatment styles or variations are currently in use.[10,11] In general, these can be thought of as modifications of brief active crisis intervention forms of psychotherapy that have been especially adapted for working with sexual problems. This author's own method employs a fine balance between support of the patient's pleasure functions and active confrontation with his resistances to the development of sexual adequacy. These new approaches are promising and make sense from a theoretical perspective. However, they still need to be objectively evaluated and compared to other treatment approaches.

¶ Bibliography

1. ABEL, G. C. "Women's Sexual Response During REM Sleep." Paper presented at the 131st American Medical Association meeting, Atlanta, Ga., 1978.
2. American Psychiatric Association, *Diagnostic and Statistical Manual of Mental Disorders* (DSM-III). Washington D.C.: American Psychiatric Association, 1980.
3. ARAOZ, D. L. "Hypnosis in Treating Hypoactive Sexual Desire," *American Journal of Clinical Hypnosis,* (1977).
4. CONTI, G. "L'Erection du Penis Human et ses Bases Morpho-Vascularis," *Acta Atomica,* 14 (1952): 217–262.
5. DAVIDSON, J. M. "Neurohormonal Bases of Male Sexual Behavior." *International Review of Physiology,* vol. 13. Reprinted in Greep, R. O., ed., *Reproductive Physiology.* Baltimore: University Park Press, 1977.
6. GORSH, R. A. "The Neuroendocrine Regulation of Sexual Behavior," in Newton, G., and Riesen, A. H., eds., *Advances in Psychobiology,* vol. 2. New York: Wiley, 1974.
7. KAPLAN, H. S. "Current Concepts of Psychosomatic Medicine," In Friedman, A. M. and Kaplan, H. S., eds., *Comprehensive Textbook of Psychiatry.* Baltimore: Williams & Wilkins, 1967.
8. ———. *The New Sex Therapy.* New York: Brunner Mazel, 1974.
9. ———. *The Illustrated Manual of Sex Therapy.* New York: Quadrangle, 1975.
10. ———. "Hypoactive Sexual Desire," *Journal of Sex and Marital Therapy,* 3 (1977): 3–9.
11. ———. *The New Sex Therapy, vol. II: Disorders of Sexual Desire.* New York: Brunner Mazel, 1979.
12. ———. "A New Classification of Female Sexual Disorders," *Journal of Sex and Marital Therapy.*
13. KERNBERG, O. F. "Boundaries and Structure of Love Relations," *Journal of the American Psychoanalytic Association,* 25 (1976): 81–114.
14. KOLODUY, R. C., et al. *Textbook of Sexual Medicine.* Boston: Little, Brown, 1979.
15. LEVIN, R. J., and WAGNER, G. "Hemodynamic Changes of the Human Vagina during Sexual Arousal Assessed by Heated Oxygen Electrode."
16. LIEF, H. "What's New in Sex Research? Inhibited Sexual Desire," *Medical Aspects of Human Sexuality,* 2 (1977): 94–95.
17. LOPICCOLO, J., and LOPICCOLO L., eds. *Handbook of Sex Therapy.* New York: Plenum, 1978.
18. MACLEAN, P. D., and PHLOOG, D. M. "Cerebral Representation of Penile Erection," *Journal of Neurophysiology,* 25 (1952):19–55.
19. MARBERGER, H. *The Mechanism of Ejaculation. Basic Life Sciences,* vol. 4, Hollander, A., Ed. "Part B: Physiology and Genetics of Reproduction." New York: Plenum, 1974.
20. MASTERS, W., and JOHNSON V. *The Human Sexual Response.* Boston: Little, Brown, 1966.
21. ———. *Human Sexual Inadequacy.* Boston: Little, Brown, 1971.
22. ———. *Homosexuality in Perspective.* Boston: Little, Brown, 1979.
23. MAURUS, M., MITRA, J., and PHLOOG, D. M. "Cerebral Localization of the Clitoris in Ovariectomized Squirrel Monkeys," *Experimental Neurology,* 13: (1965) 283–288.
24. MONEY, J., and MUSTAPH, H. *The Handbook of Sexology.* Amsterdam: Excerpta Medica, 1977.
25. O'CONNOR, J. F., and STERN, L. O. "Results of Treatment in Functional Sexual Disorders," *New York State Journal of Medicine,* 72, (1972): 1927–1934.
26. SPITZER, R. L., and KLEIN, D. F., eds. *Evaluation of Psychological Therapies.* Baltimore: Johns Hopkins Press, 1976.
27. WEISS, H. D. "The Physiology of Human Erection," *American Journal of Internal Medicine,* 76 (1972): 793–799.

CHAPTER 22

SLEEP DISORDERS: EVALUATION AND MANAGEMENT IN THE OFFICE SETTING

Anthony Kales, Constantin R. Soldatos, and Joyce D. Kales

¶ Introduction

THE evaluation and treatment of sleep disorders constitute an important area of psychiatric practice. Psychological factors are prominent in the etiology of insomnia, certain cases of hypersomnia, secondary enuresis, and in cases of adults who suffer from sleepwalking, night terrors, and nightmares. In childhood the development of sleepwalking, night terrors, and nightmares is usually related to maturational factors; psychiatric disturbances are occasionally primary. In other sleep disorders such as narcolepsy, primary enuresis, and sleep apnea, psychological factors are rarely causative. However, since these disorders often have extensive psychosocial consequences, they are frequently the cause of psychological disturbances.

Disturbance of sleep is a common symptom of such psychiatric conditions as depression, mania, schizophrenia, or anxiety. Even when disturbed sleep is associated with a medical illness, psychological factors such as anxiety, depression, insecurity, lowered self-esteem, and fear of a more permanent invalidism or death are often causative.

The prevalence of sleep disorders in the general population is quite striking. In a survey[17] of representative households in the Los

Angeles metropolitan area, the overall prevalence of current or past sleep disorders in adults was 52.1 percent. The prevalence of specific sleep problems was as follows: insomnia, 42.5 percent; nightmares, 11.2 percent; excessive sleep, 7.1 percent; and sleepwalking, 2.5 percent. In many cases, these conditions were chronic, having started early in life. The prevalence of insomnia, nightmares, and hypersomnia was higher among individuals with a lower socioeconomic status or educational attainment. In addition, the presence of these disorders was correlated with more frequent complaints of general physical and mental health problems. Previous studies have reported that mental disturbances occur more frequently in groups with limited education and lower income.[88,223] Consequently, it is not surprising that the development and persistence of sleep disorders in adults is more prevalent among individuals of lower socioeconomic and educational status.

Sleep disorders are frequently encountered in general medical practice. A survey[15] of nearly 5,000 physicians determined the prevalence of sleep disorders seen within each major medical specialty. Physicians reported that an average of 17 percent of their patients had insomnia; the highest prevalence of insomnia, 32.4 percent, was reported by psychiatrists. The estimated prevalence of the other sleep disorders was as follows: nightmares, 4.3 percent; hypersomnia, 2.9 percent; enuresis, 2.2 percent; night terrors, 1.2 percent; narcolepsy, 0.6 percent; and somnambulism, 0.6 percent. Insomnia, nightmares, and hypersomnia were reported most frequently by psychiatrists and child psychiatrists; narcolepsy was encountered most often by neurologists; and enuresis, somnambulism, and night terrors were most often seen by child psychiatrists and pediatricians. Areas of high-population density had a greater prevalence of insomnia, hypersomnia, night terrors, and nightmares.[15] These data show that physicians, particularly psychiatrists, need to keep current with the considerable clinical advances that are being made in the etiology, assessment, and management of sleep disorders.

At the present time there is no official diagnostic classification for sleep disorders. The proposed classification of sleep disorders, which is included as an appendix in the *Diagnostic and Statistical Manual of Mental Disorders* (DSM-III),[5] has a number of major shortcomings that seriously limit its usefulness. There are about seventy diagnoses, many of which are either unsubstantiated or based on electrophysiologic criteria that have relatively little clinical significance. Also, this classification does not follow the DSM-III format. While the DSM-III encourages the use of multiple, separate diagnoses within a multiaxial system, the sleep classification consistently combines diagnoses—an approach that is confusing and misleading. It is probably for these reasons that this "sleep appendix" has limited applicability from a clinical standpoint and is therefore uncoded.

The International Classification of Diseases, 9th Revision, Clinical Modification (ICD-9-CM),[251] on the other hand, provides succinct and practical diagnoses for sleep disorders, particularly in the section dealing with sleep disorders of nonorganic origin. This chapter basically follows the format and most of the individual diagnoses included in the ICD-9-CM. Included are discussions of insomnia, disorders of excessive sleep (narcolepsy, the hypersomnias, and sleep apnea), and episodic sleep disturbances (sleepwalking, night terrors, nightmares, and enuresis).

¶ Insomnia

Insomnia is the most frequently encountered sleep disorder. In several national or regional surveys, the prevalence of insomnia has been estimated at 30 to 35 percent.[11,17,146,179] Although the term "insomnia" literally denotes a complete lack of sleep, it is used to indicate a relative inability to sleep, that is, difficulty in falling asleep, difficulty staying asleep, early final awakening, or combinations of these complaints.

Sleep laboratory studies demonstrate a positive correlation between subjective complaints of insomnia and the objective measurements of sleep disturbance.[16,22,100] Compared with age-matched controls, insomniac subjects show increased objective values for sleep latency, wake time after sleep onset, and total wake time. However, the subjective estimates of their sleep difficulty are usually exaggerated. Thus, the clinical complaint of insomnia, although usually valid, is often overestimated. When compared with controls in terms of sleep stages, insomniacs have less stage 4 sleep, and the amount of rapid-eye-movement (REM) sleep they obtain is either similar or slightly less.[97,174] In addition, insomniacs vary considerably from night to night in terms of sleep latency, wake time after sleep onset, stage REM, and stage 4 sleep.

The sleep of insomniacs is different qualitatively as well as quantitatively from the sleep of normal subjects. Poor sleepers have higher levels of physiological arousal both prior to and during sleep; heart rate, peripheral vasoconstriction, and rectal temperature are elevated both before and during sleep, and they have more body movements during sleep.[174] Also, insomniacs show a degree of physiological instability in the production of sleep spindles from night to night.[218]

Causative Factors

In most cases, chronic insomnia is secondary to psychological disturbances. When medical conditions and aging are excluded, about 80 percent of patients with chronic insomnia have one or more elevated Minnesota Multiphasic Personality Inventory (MMPI) scales.[127] These patients usually fall within the diagnostic categories of the neuroses or personality disorders, and have a history of chronic anxiety or depression, or both. The most frequently elevated MMPI scales are depression, conversion hysteria, psychopathic deviate, and psychasthenia.

Analysis of the MMPI patterns of insomnia patients strongly suggests that many insomniacs characteristically internalize their emotions rather than express them outwardly. The internalization of psychological conflicts may lead to chronic emotional arousal, which in turn leads to physiological arousal and, consequently, insomnia.[127] The fact that poor sleepers have higher levels of physiological arousal during sleep[174] supports this hypothesis. During the day, the insomniac typically tries to deny and repress conflicts. At night, however, as external stimulation and distractions wane, attention is focused internally, the individual relaxes, and regression occurs. Feelings of anger, aggression, and sadness threaten to break through into consciousness and, as the insomniac fights to ward off preconscious thoughts, sleeplessness worsens. Thus, in addition to the underlying psychological disturbances, a fear of sleeplessness, which is independent of the primary psychological causes, soon develops in insomnia patients. Also, a chronic pattern of disturbed sleep eventually conditions the patient to expect insomnia. The ultimate result is a vicious circle of continued sleep disturbance, with an escalation of psychological conflict, physiological arousal, sleeplessness, fear of sleeplessness, further psychological arousal and still further sleeplessness.[127]

Sleep disturbance is an extremely common complaint of patients who have significant psychiatric problems; 70 to 75 percent of any given outpatient or inpatient psychiatric population have some type of sleep disturbance.[41,247] For example, sleep disturbance is one of the most consistent symptoms of depressive illness. During the acute manic phase of manic-depressive illness, the typical patient has a marked reduction in total sleep, though he may not complain of a lack of sleep.[74] In schizophrenic patients, the degree of sleep disturbance varies considerably, depending upon whether the process is acute or chronic. Severe sleep disturbances are frequent in acute schizophrenic episodes and often reach the point of complete sleeplessness.[154]

Insomnia can also be caused by situational circumstances, medical conditions, or pharmacological agents.[104,218] Situational insomnia is most often transient and related to major life events, such as difficulties with work or family, or personal loss. It is a universal but highly individualized phenomenon; almost everyone has experienced situational insomnia at one time or another. Insomnia is frequently a symptom in medical conditions that are accompanied by pain, physical discomfort, anxiety, or depression. Sleep difficulties can also result from the use of various pharmacological agents, such as amphetamines or other stimulants, steroids, central adrenergic blockers, bronchodilators, or the caffeine contained in colas and coffee.[218] Also, an intense degree of insomnia, which is termed "rebound insomnia," may follow the withdrawal of certain short- or intermediate-acting benzodiazepine hypnotics, even when they are given in single nightly doses for short periods.[110] "Drug-withdrawal insomnia" is produced by the abrupt withdrawal of large nightly doses of nonbenzodiazepine hypnotics after prolonged administration.[123]

The elderly, especially women, frequently complain of insomnia.* Older people not only sleep less than younger persons and have many nocturnal awakenings, but the amount of stages 3 and 4 sleep they obtain is markedly reduced and may even be absent.[47,48] The aging process itself may contribute to sleep difficulty; the elderly often have medical illnesses, in which pain and discomfort cause frequent sleep disturbance. Also, as the older person faces the realities of declining function and inevitable death, he may become anxious, fearful, and depressed. Fear of death and uncertainty about the "after death" situation are more prominent during the night, when sleep is perceived as a transient deathlike state. Consequently, the process of falling asleep involves varying degrees of emotional regression.[218]

Diagnostic Considerations and Procedures

Failure to take an adequate history is one of the most common shortcomings in the management of insomnia.[124] A complete history includes a general sleep history, a psychological history, and a drug history.[112,218] It is essential to thoroughly describe the sleep problem, including the type(s) of insomnia, the nature of the onset of the problem and its duration (that is, acute or chronic), the severity and frequency of the sleep difficulty, and the circumstances that precipitate or accentuate the insomnia. Critical information is obtained from the bed partner to rule out the presence of sleep apnea or myoclonus nocturnus. Assessment of the sleep problem should not only focus on an eight-hour, nightly period, but also cover the twenty-four-hour sleep-wakefulness pattern.

A complete psychiatric history is essential and can usually point out a relationship between the onset of sleep difficulties and psychological conflicts or major life-stress events.[218] In taking a psychiatric history, the physician should be aware that insomniac patients tend to focus on symptoms that are ego-syntonic and socially acceptable, such as the complaint of insomnia itself, as opposed to such ego-dystonic symptoms as depression and impotence. The psychiatric history should include a thorough sex history and should assess for symptoms suggestive of endogenous depression or a psychoneurotic disorder.

The drug history should evaluate the patient's consumption of coffee, colas, or alcohol.[218] Excessive caffeine may result in difficulty in falling asleep,[147] whereas alcohol causes difficulty in staying asleep. The drug history also assesses previous pharmacological treatment for insomnia or other disorders. Therapy with a tricyclic antidepressant is often unsuccessful because the daily dosage has been insufficient or excessive daytime sedation has been caused by daytime administration of the medication.[124] Overtreatment is frequently a problem with hypnotic drugs and is in large part caused by the relative ineffectiveness of these drugs over intermediate- or long-term use.[129] The loss of efficacy may cause the patient to increase the dose. Under such conditions, drug-withdrawal insomnia may occur when abrupt withdrawal is attempted.

*See references 11, 17, 71, 146, 170, 179, and 233.

Therapy

Nonpharmacologic Management and Psychotherapy

Since transient insomnia is most often a reaction to physical or psychological stress, it usually subsides when the individual adapts through his own coping mechanisms. If elimination of the stress-generating situation is impossible or impractical, the main therapeutic role of the physician in these cases is to help the patient strengthen these mechanisms. In treating chronic insomnia, however, the physician must be aware that the disorder is multidimensional: Any approach that is directed to only one of the factors involved will usually be inadequate or unsuccessful.[218] In general, the most effective treatment for the patient with insomnia combines the following elements: (1) general therapeutic measures; (2) supportive, insight-oriented and behavioral psychotherapeutic techniques; and (3) appropriate adjunctive use of antidepressant or hypnotic medication.

General therapeutic measures may be applied in most cases of insomnia.[104] The patient should be encouraged to increase his physical activity and exercise during the day, but not close to bedtime since exercise at that time may heighten physiologic arousal. Complex mental activity, such as studying, especially late in the evening, may aggravate insomnia; thus, patients are instructed to avoid mentally stimulating situations and engage in relaxing mental activities prior to bedtime. In addition, patients should regulate their daily schedules and establish a regular bedtime hour, although the patient's sleep schedule should remain flexible so that he does not become unduly obsessive about the schedule itself. The insomniac should go to bed only when he is sleepy. If he is unable to sleep, he should get up, leave the bedroom, and engage in some kind of relaxing activity. In this way, the patient learns to associate the bedroom with sleep rather than with obsessive thoughts and concerns. Naps during the day should be discouraged. Finally, cigarette smoking[219] and drinking beverages containing caffeine[147] should also be discouraged, particularly close to bedtime.

Support, education, and reassurance often help to minimize the patient's fear of sleeplessness and the consequences of insufficient sleep. Patients are often obsessed with the fear that their insomnia will severely affect their physical health and even cause their death. It can be very helpful to explain to the patient how his own anxiety can become part of the vicious circle that exacerbates and maintains insomnia.

Behavioral treatment approaches that minimize rumination and help the patient to focus his attention can be beneficial.[176] In this regard, relaxation training combined with suggested pleasant imagery is therapeutic, since it focuses the thoughts of the insomniac patient on a positive or neutral theme, helping him to avoid rumination by shifting his attention from the internal to the external. Accordingly, attention and patterned thought are substituted for the ruminative concerns that maintain the insomniac's high levels of cognitive arousal. One advantage of behavioral therapy is that the individual's active participation reduces his feelings of passivity and helplessness. The effectiveness of behavioral therapies in treating insomnia is generally limited, however, to patients who have difficulty falling asleep. In addition, behavioral approaches require a much greater degree of cooperation and compliance by the patient.

In treating the psychological aspects of insomnia, the psychiatrist needs to be active and direct in exploring conflict areas rather than using a gradual, uncovering approach.[104,218] Insomniacs deny the psychologic conflicts that underlie their sleep disorder. Instead, they focus on the somatic aspects of their problem and are mainly interested in symptomatic relief. If areas of psychological conflict are delineated early in treatment, these patients are more likely to become actively involved in therapy.

Psychodynamic techniques allow for delineating and resolving the original conflict that underlies the insomnia and its development.[104,218] A chronic insomniac patient may fear going to bed because of suppressed memories of traumatic events and experiences in childhood or adolescence that were associated with sleep or bedtime (for example, parental drunkenness, domestic violence, or incest). Insomniac patients often have difficulty expressing and controlling their aggressive feelings. Going to sleep represents a loss of control, and insomnia is a defense against this fear.

Insight-oriented psychotherapy thus provides a means for dealing with current, unexpressed, psychological conflicts and emotions that predispose the patient to emotional and physiological arousal at night. For the insomniac, getting in touch with anger, overcoming binding inhibitions and apprehensions, and restoring the balance of outwardly expressed versus self-restrained aggression are important benefits from psychotherapy. This, of course, minimizes the potential for emotional release and arousal at bedtime.

Sexual difficulties, such as avoidance of sexual relations and unsatisfactory sexual relations, are frequent in patients with insomnia.[104] These problems may be related to general interpersonal difficulties with the spouse, to fear of aggression, or attempts to control or manipulate the spouse through the sexual aspects of their relationship.

DRUG TREATMENT

When patients with transient insomnia are severely anxious, prescribing anxiolytic medication as an adjunctive treatment for a short time can be helpful. In treating chronic insomnia, the use of hypnotic medication should be only an adjunct to the main therapeutic procedures. It is particularly helpful in breaking the vicious circle of insomnia, fear of sleeplessness, emotional and physiologic arousal, and further insomnia. The issues involved in the pharmacological treatment of transient and chronic insomnia are similar, except that the lack of continued effectiveness of the medication is an important additional consideration when treating chronic insomnia.

Barbiturates should generally be excluded from the adjunctive pharmacotherapy of insomnia. They have many shortcomings, including interaction with anticoagulants,[64] and a high potential for drug-withdrawal insomnia[123] and drug dependence. Most important, the potential for a lethal overdose is very high with barbiturates. The benzodiazepine hypnotics are preferred over barbiturates because they have a very wide therapeutic window and can be used with relative safety. Most of them are effective for a period of one to two weeks, that is, within the usual treatment period of transient insomnia. Nevertheless, all benzodiazepines are not alike. Those with shorter half lives have been shown to produce rebound insomnia even after brief periods of nightly administration of single, therapeutic doses.[110] On the other hand, long-acting benzodiazepine hypnotics may reduce performance during the day following their bedtime administration.[26,189] Certain benzodiazepines may also cause anterograde amnesia.[18]

The clinician needs to be aware of the side effects of the benzodiazepines he chooses to prescribe, and he should instruct his patient to use the medication cautiously. With shorter-acting benzodiazepines, tapering of the dose before complete withdrawal is necessary to minimize difficulty with rebound insomnia. The potential for impairment of memory or daytime performance should be dealt with by educating the patient about the possibility of their occurrence with certain benzodiazepines and cautioning him to avoid tasks that necessitate heightened memory or performance capabilities.

Although L-tryptophan has been proposed as a sleep-inducing agent, the hypnotic efficacy of this compound has not been proven. This failure to demonstrate effectiveness may be due to methodological problems, since most of the existing studies to date involve noninsomniacs.[76] Similarly, the effectiveness of over-the-counter "hypnotics" has not been established, and these compounds should not be used. Sleep-laboratory assessment of these drugs suggests that they are ineffective.[136] Moreover, over-the-counter drugs containing scopolamine can be hazardous since even in the recommended dose range they may precipitate acute glaucoma, especially in elderly patients who have a narrow corneal-iris angle. Also, a dose of two to three times the recommended dosage may produce transient disorientation and hallucinations.[14]

To date, only two hypnotics, nitrazepam[2] and flurazepam,[125,126] have been found to be effective for more than a two-week period. However, only flurazepam is available in this country. With most patients, treatment should be initiated with a dosage of 15 milligrams (mg) of flurazepam at bedtime. If after one to two weeks the 15 mg dose does not sufficiently improve sleep, the dosage should be increased to 30 mg.[104,218] When administering flurazepam, the physician should alert the patient to possible decrements in his daytime performance.[26,189] Also, the physician should avoid increasing the dose in the elderly and in individuals with impaired metabolic systems, since active metabolites of the drug are more likely to accumulate in these individuals.[132,211]

Withdrawal of flurazepam is facilitated by its carry-over effectiveness. On the first and second nights following withdrawal of flurazepam, sleep is still significantly improved,[126] whereas upon withdrawal of most hypnotic drugs, sleep difficulty returns immediately to predrug levels and, in some cases, rebound insomnia occurs.

When anxiolytics or antidepressants are properly used in the pharmacotherapy of insomnia, they reduce the anxiety or depression underlying the insomnia, and their sedative side effects help to ameliorate the sleeplessness itself.[218] When insomnia is secondary to anxiety states, the use of anxiolytic benzodiazepine drugs (for example, clorazepate, chlordiazepoxide, diazepam, or lorazepam) is indicated for both the daytime anxiety and, in an increased bedtime dose, the sleep difficulty. Insomnia associated with agitated depression should be treated with an antidepressant (for example, amitriptyline or doxepin). Administration of most or even all of the daily dose of the sedative antidepressant at bedtime not only alleviates sleeplessness, but also reduces daytime sleepiness.

Insomnia associated with retarded depression may be treated with an energizing tricyclic antidepressant, such as imipramine, which may be used in divided doses during the day together with a benzodiazepine hypnotic, such as flurazepam, at bedtime.[104] When the therapeutic effects of the tricyclic have been established, which takes about two to three weeks, flurazepam can be discontinued, since relief of insomnia usually accompanies improvement of the underlying depression.

A sufficient dose level is critical to the effectiveness of antidepressant medications. A major problem with the tricyclic antidepressants that have sedating side effects is that many patients are given too much of their medication during the day rather than at bedtime.[124] The sedative side effects of the tricyclic antidepressants are immediate, while the antidepressant effect is more likely to occur after about ten days. Confusing the side effects of these drugs with their basic action may account for much of the difficulty in attaining a sufficient total daily dosage, and also in appropriately adjusting the daytime-to-bedtime dose ratio.

On the other hand, the administration of a large bedtime dose of a sedative tricyclic antidepressant should be implemented gradually. In this manner, the psychiatrist can avoid severe decrements in performance the following day, as is often the case when a sedative tricyclic is initiated in a bedtime dose of 75 to 100 milligrams.

For the treatment of insomnia associated with schizophrenic psychosis or the manic episodes of manic-depressive psychosis, neuroleptics with sedative properties, such as chlorpromazine and haloperidol, are indicated.[104,218] Since a large bedtime dose is usually effective in controlling the patient's sleeplessness, administration of hypnotics can be avoided.

¶ Narcolepsy

The disorders of excessive sleepiness are narcolepsy and the various hypersomnias, including those associated with sleep apnea.

About 7 percent of the general adult population has had a complaint of excessive sleep at some time in their lives.[17]

Recognition of the narcolepsy syndrome has important psychiatric implications since a lack of a definite diagnosis and therapy may have serious psychosocial consequences for the patient. In addition, problems arising from prolonged administration of stimulant drugs when treating narcolepsy are of special interest to psychiatrists. Daytime sleep attacks, the primary component of the narcoleptic tetrad, may be associated with one or more of three auxiliary symptoms: cataplexy, sleep paralysis, and hypnagogic hallucinations.* Narcoleptic sleep attacks may be as short as a few seconds in duration or as long as thirty minutes. In milder forms of narcolepsy, sleep attacks are precipitated by boring, monotonous, and sedentary situations; whereas in severe narcolepsy, sleep attacks may occur under circumstances which are quite stimulating or exciting, for example, work or sports activities and even sexual intercourse. Cataplexy, a sudden loss of muscle tone without loss of consciousness, may be complete, causing the patient to fall as if fainting. In less pronounced situations, the cataplexy may be partial, with simply a buckling of the knees or sagging of the facial musculature. Cataplectic attacks are generally quite brief, lasting from seconds to one or two minutes. The attacks are characteristically precipitated by strong emotions such as laughter, anger, or surprise. The remaining auxiliary symptoms of sleep paralysis and hypnagogic hallucinations occur during the transition between wakefulness and sleep. Episodes of sleep paralysis last from a few seconds to one or two minutes and may occur in normal individuals, leaving the individual fully conscious but unable to move. The hypnagogic hallucinations of narcolepsy are visual, auditory, or tactile perceptions that, in general, are more vivid, emotionally charged, and unpleasant than those that sometimes occur in normal individuals.

Narcolepsy is not a rare disorder. The prevalence of narcolepsy is estimated to be from 0.02 to 0.07 percent of the general population.[37,180] The actual prevalence is most likely higher than reported, however, since the disorder is frequently misdiagnosed. The incidence in men and women is about equal. Narcolepsy usually appears between the ages of ten and thirty, and in 90 percent of the cases, onset is prior to age twenty-five.[32,33,40]

The narcoleptic tetrad, especially the auxiliary symptoms, corresponds to manifestations of REM sleep. Narcoleptic patients who also have cataplexy have REM periods at or close to sleep onset, rather than after the normal seventy to ninety minutes of non-REM (NREM) sleep.* Also, sleep attacks are about as long as REM periods, that is, fifteen to thirty minutes. Furthermore, the loss of muscle tone in cataplexy is a reflection of the motor inhibition of REM sleep, and hypnagogic hallucinations correspond to the dream activity associated with REM sleep. In those narcoleptic patients who do not have associated auxiliary symptoms (independent narcolepsy), sleep-onset REM periods do not usually occur.[207]

Disturbance of nocturnal sleep is common in narcolepsy.[255] The sleep disturbances may consist of generally restless sleep, frequent nocturnal awakenings, or less total sleep time. Since sleep attacks are reported to be more frequent after a night of poor sleep, the physician needs to be concerned about the patient's nocturnal sleep and especially how it may be adversely affected by stimulant medication taken late in the day or in the evening. Excessive nocturnal sleep is not characteristic of narcolepsy.

Causative Factors

Genetic factors are important in the development of narcolepsy; narcoleptic patients frequently have positive family histories either for narcolepsy or hypersomnia.[149,157,180,206] In a study of narcoleptic probands, it was found that the prevalence of narcolepsy among relatives was 2.5 percent,

*See references 32, 33, 40, 197, 206, 217, 253, and 255.

*See references 39, 69, 85, 86, 201, 207, and 234.

a rate about sixty times greater than the estimated prevalence in the general population.[149]

The exact pathophysiology of narcolepsy/cataplexy has not been determined; however, it is clear from the absence of epileptic discharges in the electroencephalogram (EEG) that narcolepsy is not related to epilepsy. Narcoleptic patients show a high degree of psychological conflicts, but these generally appear to be consequences of the disorder rather than causative factors.

Diagnostic Considerations and Procedures

A thorough history is important in establishing the diagnosis of narcolepsy and initiating treatment.[112,217] The physician must determine the age of onset for narcolepsy, the frequency and duration of episodes, the time of their occurrence, the circumstances precipitating the symptoms, and the general daytime behavior and adjustment. If cataplexy and other auxiliary symptoms are present, the diagnosis is easily made on a clinical basis. If sleep attacks are the only symptom, an all-night hypnopolygram and a daytime-nap EEG recording may confirm the diagnosis by detecting a sleep-onset REM period. If the hypnopolygram and the nap EEG are negative, the physician again proceeds on the basis of the clinical symptomatology.

A thorough history is usually sufficient to differentiate narcolepsy from other disorders. Narcolepsy is frequently misdiagnosed as hypersomnia, hypothyroidism, hypoglycemia, epilepsy, myasthenia gravis, or multiple sclerosis.[33,255] The characteristics that differentiate narcolepsy from hypersomnia are the presence of auxiliary symptoms (cataplexy, sleep paralysis, or hypnagogic hallucinations), usually shorter daytime sleep episodes, an absence of postsleep confusion, and an absence of prolonged nocturnal sleep. Narcolepsy is differentiated from epilepsy by the presence of auxiliary symptoms, the absence of bladder or bowel incontinence and tongue biting, and the lack of clinical EEG abnormalities characteristic of epilepsy.

Therapy

NONPHARMACOLOGIC MANAGEMENT AND PSYCHOTHERAPY

Narcoleptics often are considered by their family, friends, fellow employees, and employers as being lazy, malingering, or psychologically disturbed.[197,217] The physician should inform the patient and the people who comprise his family, social, and occupational lives that the sleep attacks and other symptoms are beyond the patient's control. Since it is not uncommon for narcoleptic patients to be terminated from employment, it is sometimes helpful to suggest to the employer that he allow the patient one or two short naps every day. Such naps often significantly improve work performance.

Narcoleptic patients should be advised of the potential dangers of long-distance driving or other activities that would expose them to danger if a sleep attack or cataplectic episode were to occur. The physician has to be judicious whenever advising restriction of activities to the narcoleptic patient. Warnings should not exceed those warranted by the clinical course of the illness before, and especially after, treatment.

Psychotherapy alone, of course, is not effective in treating narcolepsy itself, since accompanying psychological difficulties are secondary rather than primary. Supportive psychotherapy is helpful, however, as an adjunct to the pharmacological treatment of narcolepsy, since alleviation of the serious psychosocial consequences of the disorder facilitates the overall treatment.

DRUG TREATMENT

The sleep attacks of narcolepsy can be effectively treated with stimulant medication such as the amphetamines and methylphenidate.[33,217,253] Methylphenidate has been recommended as the drug of choice for the management of narcoleptic sleep attacks because of its prompt action and the relatively low incidence of side effects.[33,253] Since this drug has a short duration of action and is effectively absorbed only from an empty stomach, patients should take the medication at least forty-five minutes before, or more than one hour after, eating. Initially, patients take 5 mg upon awakening in the morning, at midday, and again at 4 P.M. Depending upon the patient's initial response and the time of day when sleepiness is worst, one or all of the three doses may be increased as high as 30 to 40 mg.

Methamphetamine is the most tolerated of the amphetamines since it causes fewer sympathomimetic side effects.[33] Initially, patients may be given 10 mg upon arising in the morning. They are then assessed so that the effectiveness and approximate duration of action of a single dose can be determined. Subsequently, this dose can be gradually increased until the appropriate dose level is reached, at which point a regimen is developed to ensure alertness throughout the day. Many patients require only a single dose, others may need two doses, and a few may require three doses a day. The total daily dose of methamphetamine may be as low as 20 mg for mild narcolepsy, or higher than 100 mg for severe narcolepsy.

Treatment of narcolepsy with amphetamines, and to a lesser degree with methylphenidate, is complicated by the many problems that are associated with these drugs. For example, psychiatric syndromes may complicate amphetamine treatment; a syndrome of irritability, paranoid tendencies, and even psychosis may develop with prolonged amphetamine use.[212,217] The incidence of these side effects, however, even with high doses of amphetamines, is lower in narcoleptic patients than it is in other individuals taking these drugs. Stimulant drugs may also disturb nocturnal sleep, particularly if they are taken in the late afternoon or early evening. This may cause the patient to take more drug in the daytime in order to ward off sleep attacks, which occur more frequently when disturbed nocturnal sleep is present.

Recent data suggest that propranolol in a dose range of 240 to 480 mg may be effective in treating the sleepiness and sleep attacks of narcolepsy.[131] In most cases, however, the beneficial effects do not extend beyond several months.

When the auxiliary symptoms of narcolepsy are present, the drugs of choice are antidepressant medications such as the tricyclic drugs. The most frequently used drug for treating the auxiliary symptoms has been imipramine.* In managing cataplexy (the most frequent and disturbing auxiliary symptom), imipramine has a rapid onset of action (within hours in some cases) and is effective at lower doses than those used to treat depression. While it is extremely effective in relieving auxiliary symptoms, imipramine is only minimally effective in decreasing sleep attacks.

When the patient complains of both sleep attacks and auxiliary symptoms, treatment with the analeptics and imipramine may be combined. Since this combination may produce serious side effects such as hypertension, dose increments must be carefully titrated and monitored. As suggested by the manufacturer, methylphenidate may inhibit the metabolism of tricyclic antidepressants so that downward dosage adjustment of imipramine may be necessary.[72]

¶ Hypersomnia

Hypersomnia is generally characterized by periods of excessive daytime sleepiness and sleep attacks that usually are longer than narcoleptic sleep attacks; episodes of hypersomnia may last from one to several hours.† Nocturnal sleep is often prolonged in hypersomnia; otherwise it is not usually disturbed. Hypersomnia usually begins later in life, often between thirty to forty years of age. Approximately one-third of patients with hypersomnia experience sleep drunkenness, which consists primarily of difficulty in fully awakening in the morning and lasts from fifteen minutes to an hour.[197,208]

*See references 4, 87, 197, 217, and 232.
†See references 21, 197, 206, 217, and 255.

Causative Factors

Hypersomnia may be idiopathic, psychogenic, periodic, associated with sleep apnea, or secondary to other organic conditions.* In the idiopathic type there is often a positive family history.[180] About 40 percent of the primary probands among hypersomniac patients have a family history of hypersomnia or narcolepsy, suggesting that there is an autosomal dominant mode of genetic transmission for this disorder.

In the sleep laboratory, hypersomniacs average close to nine hours of total sleep time.[197,199] The sleep-stage distribution is similar to normal, and although a decrease in autonomic functioning would be expected in these patients, their heart rate is actually higher than that of controls.

Psychological factors are quite important in the hypersomnias, specifically in the psychogenic type.[42,155,221] Psychogenic hypersomnia is usually secondary to a depressive disorder and is characterized by nocturnal sleep that is not only prolonged but also restless and generally disturbed. These patients also feel worse upon arising in the morning, but their symptomatology is distinct from that of sleep drunkenness. In psychogenic hypersomnia, the underlying depression may be "atypical" and therefore difficult to diagnose. More often, it may represent the depressive phase of a bipolar affective disorder.[42,154]

The Kleine-Levin syndrome, which is the most frequent type of periodic hypersomnia, is nevertheless an extremely rare condition.[31,158] The excessive sleepiness and sleep usually last two or three days, occur once or several times a year for several years, and then terminate. This condition is usually associated with excessive appetite and is most common in adolescent males.

Sleep apnea is frequently associated with excessive daytime sleepiness or hypersomnolence.† In patients with sleep apnea, there may be sleep attacks, which are often indistinguishable from narcoleptic sleep attacks.

*See references 21, 68, 163, 197, 206, 221, and 255.
†See references 38, 60, 68, 160, 162, and 163.

There are three types of sleep apnea: central, peripheral, and mixed. Central apnea is a cessation of air flow through the nares and mouth accompanied by a cessation of thoracic and abdominal respiratory efforts. In peripheral or obstructive sleep apnea, no air flow is recorded at the entrance to the upper airway in spite of persistent thoracic and abdominal respiratory movements. In mixed apnea, a brief period of central apnea is quickly followed by a longer period of obstructive apnea.

When hypersomnia is related to sleep apnea, the polygraphic sleep recording typically shows repetitive respiratory pauses in both REM and NREM sleep.[68,163] The apneic episodes are usually of the peripheral type, indicating obstructive rather than central sleep apnea. The exact pathophysiologic mechanism of the upper airway obstruction associated with sleep apnea is unclear.

Organic hypersomnia is usually secondary to tumors, vascular lesions, or trauma involving the mesodiencephalic area of the brain stem, resulting in a disturbance of the reticular activating system.[21] Other causes of organic hypersomnia include infectious and toxic encephalopathies as well as several endocrine and metabolic disorders.

Diagnostic Considerations and Procedures

The history from the hypersomnia patient should include the age of onset, the frequency and duration of episodes, their time of occurrence, precipitating circumstances, the ability to resist sleepiness, the presence or absence of auxiliary symptoms and sleep drunkenness, the duration and quality of nocturnal sleep, the presence or absence of snoring or interrupted breathing during the night, and the patient's general daytime behavior and adjustment.[112]

Questioning the spouse, parent, or other family members for the presence of snoring with interrupted nocturnal breathing is important in diagnosing sleep apnea. The physician determines if there are frequent periods of interrupted nocturnal breathing associated with snoring, gasping, gurgling, choking, periodic loud snorting, or morning headache.[68] The snoring associated with sleep apnea is unique:[165] There are ten- to forty-second periods of suspended respiration followed by very loud and abrupt snorting sounds that are two to four seconds in duration. The common type of snoring in non-apneic individuals is not as loud, fluctuates in intensity, and is continuous without any gaps of appreciable duration.

It is important to differentiate hypersomnia from narcolepsy. In hypersomnia, there are no auxiliary symptoms, the daytime sleepiness usually begins later in life, there are generally fewer but longer episodes per day, and nocturnal sleep is often prolonged but not disturbed.[217,255] A detailed psychiatric history is necessary in order to rule out depression as a causative factor. Also, whenever hypersomnia is suspected, it is important to rule out thyroid disease, hypoglycemia, and diabetes mellitus.

Therapy

NONPHARMACOLOGIC MANAGEMENT AND PSYCHOTHERAPY

The physician should educate the hypersomniac patient and his family regarding the symptomatology of hypersomnia and clarify any misconceptions they may have about the disorder. Supportive, and less frequently, insight-oriented psychotherapy may be therapeutic for certain patients with hypersomnia. For example, when the patient sleeps excessively in an attempt to deny and escape from life stresses, the physician might advise him to adopt a more assertive life style. If psychological difficulties arise from the patient's family relations, family therapy may be more appropriate. In hypersomnia secondary to depressive illness, psychotherapy may be applied as an adjunct to the use of antidepressant medication.

In overweight patients with sleep apnea, weight reduction is first indicated and, at times, results in improvement. However, many patients will not maintain their diet. Patients with obstructive apnea may require tracheostomy, which is almost invariably effective in alleviating both the sleep apnea and the excessive daytime sleepiness.[29,164,210,231] At present, an effective treatment of central sleep apnea has not been well established.

DRUG TREATMENT

Stimulant drugs are indicated for treating the excessive daytime sleepiness, the prolonged nocturnal sleep, and sleep drunkenness of idiopathic hypersomnia.[197,208] Methylphenidate is the drug of choice, primarily because of its prompt action and fewer side effects. For sleep drunkenness, the drug is either given at bedtime, or the hypersomniac patient is awakened and given the medication thirty minutes before the desired time of awakening. For excessive daytime sleepiness, drug administration is scheduled according to the patient's needs for full awareness.

In psychogenic hypersomnia, when endogenous or characterological depression is the primary difficulty, nonsedating tricyclic antidepressants such as protriptyline are indicated. Lithium can be used to treat the hypersomniac patient who is in a depressive phase of a bipolar disorder. The patient with an "atypical" depression, however, may respond more favorably to a mono-amine oxidase inhibitor. The sedating tricyclics are contraindicated in all cases.

¶ Sleepwalking

Sleepwalking, or somnambulism, is not uncommon in childhood, but it is rather infrequent in adults.[17,150,230] It is a state of dissociated consciousness in which phenomena of the sleeping and waking states are combined. The episodes occur out of stages 3 and 4 sleep,[59,89,114] when dream recall is less frequent, rather than out of REM sleep, when dreaming is most likely. Sleepwalking is therefore not the acting out of a dream, as was once commonly believed.

Somnambulism commonly begins in childhood and less often in early adolescence. When sleepwalking is outgrown, its onset was almost invariably before age ten, whereas individuals who continue to sleepwalk as adults have a significantly later age of onset. In addition, those who do not outgrow sleepwalking have more frequent events, experience episodes earlier in the night, and have more intense clinical manifestations of their events.[134]

The typical somnambulistic episode occurs within the first three hours of sleep; the patient sits up, rises out of bed, and moves about slowly in a poorly coordinated, automatic manner. Sometimes there is more complex activity, which is clearly out of context and reflects the patient's lack of awareness. Effort is usually required to awaken the patient. Unless fully aroused the patient is usually amnesic toward the episode.[59,89,114]

Causative Factors

Sleepwalking is a disorder of impaired arousal.[20] In addition to the somnambulist's general confusion and lack of responsiveness during an episode, reactions to even strong stimuli are impaired.

Etiological factors in sleepwalking appear to include both physical and psychological components. An increased familial occurrence of sleepwalking has been identified.[1,9,135] Families of sleepwalking probands also show a higher prevalence of night terrors. These data appear to fit a two-threshold, multifactorial model of inheritance in which sleepwalking is the more prevalent condition.[135]

Maturational/developmental delay is considered a significant etiological factor in sleepwalking because most sleepwalkers outgrow the disorder before adulthood. Furthermore, it has been found that an "immaturity factor" in the sleep EEG (that is, sudden, rhythmical, high voltage, slow activity) persists to a later age in child somnambulists than in normal children.[114] Child somnambulists, however, do not show appreciably more psychopathology compared with normal children.

Adult sleepwalkers may have been affected by a delay in central nervous system (CNS) maturation, but psychological factors also appear to play an important role in the condition's development and, particularly, in its persistence.[134] Sleepwalkers who do not outgrow their disorder show much more psychopathology compared with both past sleepwalkers and normal controls. Their MMPI profiles are consistent with active, outwardly directed, aggressive behavioral patterns. Individuals with this type of profile commonly struggle to deal with life frustrations and become extremely angry when frustrated. Excessive anger in response to frustration, failure, or loss of self-esteem is outwardly directed rather than being self-directed, as is the case in depression.[134]

Diagnostic Considerations and Procedures

Evaluation of the child or adult who presents with a complaint of sleepwalking should begin with the taking of a thorough history.[112,134] This includes determining age of onset, frequency of episodes, time of occurrence, behavior during the episode, length of episode, degree of recall of mental activity, general daytime behavior and adjustment, major life-stress events associated with onset, and the effects of stressful situations on frequency of episodes.

Much of this information has to be obtained from the parents or, in the case of adult somnambulists, from the spouse or roommate.[134] A detailed history will often help to determine whether the disorder is due to underlying organic factors or psychopathology. Underlying psychopathology is suggested when sleepwalking begins after about age ten; when it occurs frequently over a period of six months to a year; when there is a negative family history of sleepwalking; when daytime symptoms suggest a functional disorder; and when the onset of sleepwalking appears to be related to major life events or when mental stress increases the frequency of episodes.

On occasion, it may be necessary to differentiate sleepwalking from other conditions, such as dissociative states or nocturnal wanderings.[134] In contrast to sleepwalkers, individuals in dissociative states are much more alert and are capable of performing more complex activities. Their episodes may last for periods of hours to days. A complaint of sleepwalking in the elderly may actually represent a nocturnal wandering episode, which can be easily differentiated from sleepwalking by the history. Nocturnal wandering episodes most frequently occur in patients who have organic brain syndromes with associated intellectual impairment, and they may occur at any time during the night.[46]

Drug-induced, somnambulist-like episodes have been observed to be relatively frequent in psychiatric patients who have taken lithium together with neuroleptics.[25] These patients did not have a prior history of somnambulism. Also, febrile illness has been found to induce somnambulistic episodes.[138] When the episodes persist long after the febrile illness, there is usually a positive family history for sleepwalking or night terrors.

Therapy

NONPHARMACOLOGIC MANAGEMENT AND PSYCHOTHERAPY

An important aspect of the management of a sleepwalker is to recommend appropriate safety measures.[105] These may include sleeping on the ground floor and providing special bolts or locks for windows and doors. The sleepwalker generally returns to his room on his own, but if he does not, he usually can be gently led back. The sleepwalker's episodes should not be interrupted, however, if it is known that such intervention may lead to more confusion and fright.

Any groups or agencies, such as summer camps, that are responsible for childrens' safety on overnight outings away from home should be made aware of a patient's sleepwalking and the need for safety measures.[134]

Sleepwalking poses special problems in the armed forces; thus a number of studies have been conducted in this setting.[193,194,209,222] In general, the nocturnal behavior of the sleepwalker is inconsistent with military standards of discipline. The sleep of fellow soldiers may be disturbed by the sleepwalker's activity, there may be an uneasy feeling that there is an "odd" comrade in the barracks, and there is a risk that the somnambulist may be shot by failing to halt at a guard's command. In the navy, a sleepwalker may fall overboard during an episode.

Most children grow out of the sleepwalking condition, particularly if it has an early age of onset. Thus, the parents of children who sleepwalk need to be reassured that, in most cases, sleepwalking does not reflect a serious psychological disorder.[105] On the other hand, adults who sleepwalk often have significant psychiatric disturbances that require psychotherapeutic intervention.[134]

Since adult sleepwalkers typically cope with frustration in an externally directed manner rather than struggling with internalized distress, insight-oriented psychotherapy involves developing the patient's ability to constructively react to stressful life events. Thus, a major therapeutic goal is to help the patient identify and then discharge his feelings of frustration, anger, and aggression, and thereby relieve the need for aggressive responses.[134] The sleepwalker can be taught to think of a sleepwalking episode as a cue or indication of an important, current life frustration with which he is not dealing appropriately. He might say, "I sleepwalked last night. What is it that is so frustrating as to provoke an episode, but is also so troublesome and distressing that I do not want to cope with it?"

DRUG TREATMENT

The physician needs to be very cautious in prescribing psychotropic medication for children who sleepwalk. It should be used only when absolutely necessary and for a very limited period, since the long-term effects of psychotropic drugs on childrens' development are not known. The use of stage 4 suppressant drugs in treating adult somnambulism is being investigated. Certain drugs, primarily benzodiazepines such as diazepam and flurazepam, markedly suppress stages 3 and 4 sleep.[118,120] However, these drugs have not yet been shown to result in a clear-cut decrease in the frequency of sleepwalking episodes.

The effectiveness of imipramine in reducing the frequency of sleepwalking episodes has been suggested but has not been demonstrated in controlled studies.[192,229] The mechanism for this effect of the drug is not related to its psychotropic action in alleviating depression, since depression is not a primary problem in patients with sleepwalking. In addition, this effect of the drug is not mediated through any changes in stage 4 sleep, since imipramine does not significantly alter this sleep phase.[128] It may be that imipramine's effectiveness in reducing sleepwalking or night-terror events is related to the fact that the drug has increased the arousal levels of patients with sleepwalking and night terrors. In this way, the drug would minimize the arousal disorder that would ordinarily occur out of stages 3 and 4 sleep.

¶ Night Terrors

Night terrors are nocturnal episodes of extreme terror and panic that are associated with intense vocalization and motility and high levels of autonomic discharge. The episodes are of relatively short duration, lasting from one to several minutes. During a typical episode, the patient is confused and disoriented and has little or no recall of the event either immediately following it or the next morning.[141]

While night terrors are not as frequent as nightmares, they are not rare. In one survey of approximately 1,000 children, the incidence of night terrors was reported to be 2.9 percent.[156] In another survey of 1,000 children seen consecutively in a child psychiatry clinic, the incidence was 1.5 percent.[213]

In most cases, night terrors can be easily differentiated from nightmares by their clinical characteristics. The nightmare is accompanied by much less anxiety, vocalization, motility, and autonomic discharge than the night terror. In addition, the sleeper is more easily aroused, and there is usually vivid and detailed dream recall. Most night terrors occur in the first few hours of sleep and arise out of stages 3 and 4 sleep. In contrast, nightmares are related to REM sleep, which is more predominant during the later part of the night.*

Causative Factors

Night terror events are often accompanied by sleepwalking activity. A patient may actually present with both conditions, or he may initially have somnambulism and later develop or switch to night terrors.[140] The two disorders share many of the same clinical and psychological characteristics. In addition, patients with either disorder frequently have a family history for either sleepwalking or night terrors. Pedigrees studied appear to fit a two-threshold multifactorial model of inheritance, in which night terrors are the least prevalent condition.[135] Other similarities are that both sleepwalking and night terrors start in childhood and are usually outgrown before adulthood, and in both cases, episodes occur out of slow-wave sleep and are typically associated with impaired arousal. Further, both disorders are usually due to a maturational/developmental lag. However, when they persist or start in adulthood, psychological factors are more prominent. Thus, the two conditions have been considered as two different manifestations of the same pathophysiological substrate.[134,140]

In addition to the etiological role that genetic and developmental factors may play, psychological factors also contribute considerably to the development and/or persistence of night terrors in adulthood.[140] Night

*See references 20, 51, 54, 59, and 105.

terrors are more likely to persist into adulthood when their age of onset is after age twelve, when the frequency of the episodes is high, and when there is a major life-stress event at the time of onset. The primary role of psychological factors is underscored by the high levels of psychopathology in adult night-terror sufferers. Their MMPI profiles show an inhibition of outward expression of aggression, a predominance of anxiety, depression, and phobicness, and a secondary "schizoid" self-negativity in the absence of overt psychoticism. The inability to express aggression in night-terror patients may lead to the aggression being directed inward, thereby precipitating night terror events marked by extreme defensiveness and fighting behavior, which further frighten the patient.[140]

Diagnostic Considerations and Procedures

Complete evaluation of the child or adult who appears to have night terrors begins with a thorough history.[112,140] In addition to a carefully taken family history, important data to obtain include the age of onset, frequency of episodes, time of occurrence, behavior during the actual episode, length of the episode, degree of recall of mental activity, general daytime behavior and adjustment, and the relation of the onset of night terrors to any major life stress event. Since the patient is usually amnesic for the episodes, much of this information is obtained from the parents, or in the case of adults, from the spouse or roommate.

A thorough history is essential in evaluating the contribution of organic factors, psychopathology, or both, in the development and persistence of the condition.[140] It is quite unusual for night terrors to begin during or after middle age; if they do, underlying organic pathology such as a brain tumor is suggested. The possibility of organic pathology should therefore be carefully evaluated.

It may be necessary to differentiate night terrors from other disorders. They are easily differentiated from hysterical dissociative states, such as amnesia, fugue states, and multiple personality, since the individual in a dissociative state is more alert, more capable of complex behavior, and has longer episodes with much less violent or agitated activity. It may be necessary occasionally to distinguish the bizarre and explosive behavior of night-terror episodes from that associated with schizophrenia, which is characterized by disturbed daytime behavior.

Contrary to previous reports,[156] there does not appear to be a relationship between night terrors and epilepsy.[220,228] In differentiating night terrors from epileptic disorders, the latter are characterized by more repetitive and stereotyped behavior. Also, the duration of epileptic events is briefer, often lasting only a period of seconds. Night terrors may be confused with temporal lobe epilepsy, but it is rare for this condition to occur only at night. In a patient who is thought to have night terrors, a clinical sleep EEG should be obtained if there are similar daytime attacks or other episodic behavior.

Therapy

NONPHARMACOLOGIC MANAGEMENT AND PSYCHOTHERAPY

In the management of the patient with night terrors, safety measures are often necessary.[105] Such measures include sleeping on the ground floor and providing special bolts for windows and doors. The spouse, bed partner, or roommate should understand that forcibly interrupting a night-terror episode may aggravate the patient's violent behavior. As with sleepwalking, those groups and agencies that take responsibility for the safety of children with night terrors when they are away from home should give special consideration to the safety of these patients.[140]

Since adult patients with night terrors are often psychologically disturbed, psychotherapy is usually indicated. Night-terror patients commonly inhibit expression of their feelings of aggression and anger and may as a result direct these feelings inward. Such internalization may precipitate night-terror events that consist of extreme defensiveness and fighting behavior that further frightens the patient.[140] Thus, while the sleepwalker is conditioned to respond to stress outwardly both while awake and during sleep, the night-terror sufferer is conditioned to react with fear and apprehension. As a consequence, insight-oriented psychotherapy, by actively exploring fears of failure and hostility as well as of the night-terror event itself, can be beneficial. It is also helpful to strengthen the patient's self-assertiveness. This helps to counteract fears of hostility as well as anxieties over how others would react if the patient were to experience failures in life.

Assuming that night terrors are a type of discharge of accumulated anxiety and fear, the patient should appreciate the importance of dealing with the anxiety, however distressing, in psychotherapy rather than discharging it during sleep. If the fear involves a specific phobia, then behavior therapy using desensitization or implosive techniques may be useful.

Drug Treatment

Benzodiazepine drugs (diazepam) that suppress slow-wave sleep[118] are effective in reducing the frequency of night-terror events, but drug withdrawal usually leads to a relapse.[53] It is not clear whether diazepam's effectiveness in reducing night-terror events is related to its suppression of stage 4 sleep or to its general anxiolytic properties.

As previously discussed, effectiveness of imipramine has been suggested for reducing the frequency of sleepwalking and night terrors,[192,229] but this effectiveness has not been demonstrated in controlled studies.

Psychotropic drugs should be used only when a reduction in the frequency of night terrors is absolutely necessary; that is, when special consideration is needed for the safety of the night-terror patient. The use of such drugs in children, however, should be discouraged.

¶ Nightmares

Nightmares occur frequently in childhood, particularly in children who are between the ages of three and eight.[166,213] Nightmares are also fairly common in adults. In the general adult population of Los Angeles, 11.2 percent of those interviewed indicated that they had difficulty with nightmares at some time in their lives; 5.3 percent reported a current complaint of nightmares and 5.9 percent had only a past complaint.[17] In other studies, frequent nightmares (at least one per week) were reported as 5 percent in college undergraduates and 7 percent in a psychiatric population.[50,83]

Nightmares are nocturnal episodes of intense anxiety and fear associated with a vivid and emotionally charged dream experience.[133] Nightmares are often accompanied by some vocalization and increased autonomic activity. They can be a serious cause of sleep disruption, since the events are most often followed by lengthy periods of arousal. Typically, the individual has detailed recall of the nightmare. The patient usually awakens easily, often spontaneously, from a nightmare, is lucid following arousal, and usually becomes quickly oriented to the environment, although it may take some time for him to separate nightmare dream content from reality.

In most cases, nightmares, which occur in REM sleep, can be easily differentiated from night terrors, which occur in slow-wave sleep, because of their general clinical characteristics. The nightmare is accompanied by much less anxiety, vocalization, motility, and autonomic discharge, and there is much greater recall of the event than in the night terror. Also, night terrors usually occur during the first two to three hours of sleep, whereas nightmares may occur at any time of the night.*

Causative Factors

Childhood nightmares are most often related to specific phases of growth, being most common during the preschool and early school years; they are related both to ego development and psychological conflicts.[166] Because the child has an active fantasy life during this period of development, fears of imaginary figures and misperceptions of shadows and objects frequently trouble him while he is preparing to go to sleep. Consequently, the child often develops a number of vague, imaginary fears or phobias that may be frightening during the night, such as a fear of death, of the dark, of being alone, or of monsters and ghosts.

Nightmares are also frequently reported during febrile illness, especially in children. Since febrile conditions appear to supress REM sleep, nightmares may be more likely to occur after the fever subsides, when there is a REM rebound consisting of longer REM periods.[145] The confused and semidelerious state that often accompanies febrile illness could contribute to the child's inability to distinguish between dream material, fantasy, and reality, and thus compound his anxiety and fear.

The onset of nightmares after adolescence and their persistence into adulthood appear to be correlated with the presence of significant psychopathology.[133] MMPI data show that adults with chronic nightmares have considerable psychopathology and specific personality characteristics. They are typically distrustful, alienated, estranged, oversensitive, overreacting, and egocentric. Their distrustfulness is not paranoid, however, since it is diffuse and generalized rather than directed to a specific object. Chronic sufferers may present with an underlying chronic schizoid pattern of adjustment, although in general they are not overtly psychotic.

Chronic nightmares can be secondary to long-term difficulties in dealing with interpersonal hostility and resentment.[133] When the excessive hostility resulting from intensely neurotic object relations is not entirely discharged in everyday life, it may be carried over and released in the nightmare. The nightmare serves not only as a vehicle for discharging unreleased hostility, but also for the extinction of unfinished anger and generally negative emotions. When these emotions are finally expressed in the nightmare, they become anxiety-provoking.

The physiological basis for the occurrence of nightmares is not entirely known. It is well documented, however, that nightmares occur out of stage REM sleep and that an excess of REM sleep, as in REM rebound following drug withdrawal, is associated with a temporarily increased incidence of intense dreaming and nightmares.[118,119]

Diagnostic Considerations and Procedures

In evaluating the patient complaining of chronic nightmares, the physician needs to determine the age at onset, frequency of episodes, time of occurrence, behavior associated with the episode, length of the episode, degree of recall of mental activity, general daytime behavior and adjustment, and relation of the onset or recurrence of nightmares to major life-stress events.[112,133] The age of the patient is important in deciding whether the nightmares are related to a phase of development or whether they are due to psychological conflicts. Underlying psychopathology is suggested if the nightmares begin after about ten years of age, if they persist for several months without decreasing in frequency, if daytime symptoms suggest a functional disorder, or if there is a correlation between the onset or recurrence of nightmares with major life-stress events.

It is also helpful to obtain a general sleep history from the patient, since individuals with frequent nightmares tend to have a higher incidence of other sleep disturbances, particularly insomnia.[23,81,83,133] Similarly, patients with a major complaint of insomnia

*See references 20, 51, 54, 59, and 105.

frequently report difficulty with nightmares dating back to childhood.[82]

Since many drugs, including alcohol, induce marked changes in both the frequency and intensity of dreaming, a complete drug history is essential. Drug-withdrawal nightmares secondary to increases in REM sleep occur when REM-suppressant drugs are abruptly withdrawn,[118,119,186] but they can also occur on the same night that the drug is given.[118,119] Nightmares during nights when drugs are taken are probably rare but may occur when short-acting hypnotics are used, since REM sleep increases late in the night as the action of the drug diminishes. Nightmares may also occur on drug nights when an individual sleeps past the duration of action of the drug (nine to ten hours), as on a weekend.

Although most drug-induced nightmares are associated with REM rebound following the withdrawal of REM suppressant drugs, the actual administration of certain drugs may increase both the intensity and frequency of dreaming and REM sleep. For example, reserpine,[73] thioridazine,[122] and mesothioridazine[188] all produce increases in REM sleep and may intensify dreaming.

Therapy

NONPHARMACOLOGICAL MANAGEMENT AND PSYCHOTHERAPY

Education and reassurance are often necessary, since both the patient and his family not infrequently consider nightmares as a psychotic symptom. Parents of children with nightmares should know that children frequently experience nightmares as part of their normal development and usually "outgrow" the disorder within a relatively short time. Preventive measures are also very helpful. Parents should know that terrifying experiences such as watching violent television programs or listening to frightening bedtime stories may be harmful. Other than eliminating these specific types of potentially traumatic experiences, however, parents should allow the child a normal range of activities and experiences.

If there is considerable psychopathology, as is commonly the case in adults who have nightmares, the nature of the intrapsychic conflicts is explored in psychodynamic psychotherapy.[133] For example, the physician may determine that early life experiences are causing concern and a preoccupation with death. Monosymptomatic psychiatric conditions, such as chronic fears or phobias, may be responsible for nightmares in a limited number of cases. Most often, however, a more global psychological disturbance underlies the presence of nightmares in adulthood.

In the few instances in which nightmares appear to be an isolated problem rather than symptomatic of a basic psychiatric disturbance, behavioral therapies, such as systematic desensitization, have been successful.[24,62,214] Nevertheless, since the underlying psychiatric condition is seldom monosymptomatic, behavioral treatment has only limited application; insight-oriented psychotherapies are more likely to be effective in most cases.[133]

The psychiatrist should be aware that the nightmare sufferer often feels distrustful and alienated, and that he may consequently suddenly terminate treatment.[133] Thus, a major consideration at the beginning of therapy is to develop a sound therapeutic relationship. A subsequent therapeutic task is to have the patient achieve a better understanding of his emotions, particularly his anger. This enables him to more efficiently cope with his emotions, rather than become excessively frustrated and, as a consequence, discharge incompletely expressed resentments in his nightmares. Finally, the psychiatrist should be alert to the presence of depression, especially in men who have nightmares,[77] since the nightmare patient tends to avoid treatment and suicidal potential may go unnoticed for lack of follow-up.

DRUG TREATMENT

The physician should be aware of the schizoid adjustment in the chronic nightmare sufferer's life style, but he should also understand that the patient's coping mechanisms are neurotic in nature. When overt psychotic behavior is associated with the presence of nightmares, however, the physician should proceed with antipsychotic treatment. Antidepressant medication should be used whenever depression is identified. Administration of REM-suppressant drugs to decrease the nightmare frequency should be avoided, since their withdrawal may lead to a REM rebound associated with an increase in dream intensity and possibly nightmares.

¶ Enuresis

Attainment of complete bladder control depends upon sociocultural and developmental factors, personality characteristics of the child, general emotional climate in the home, and parental attitudes toward toilet training.[177,178] Enuresis is the term most often used to refer to bedwetting after control of the urinary bladder should have been acquired. Most children begin to develop urinary control by eighteen months of age and achieve dryness between ages two and three; some do not attain dryness until four to six years of age.[141,182]

There are two types of enuresis: primary and secondary. In primary enuresis, the child has never been dry for more than one or two weeks; in secondary enuresis, he may be dry for several weeks, months, or years before enuresis begins. Primary enuresis is also called functional or persistent enuresis, and secondary enuresis may be referred to as acquired, regressed, or onset enuresis.

After age three, about 10 to 15 percent of all children are enuretic;[19,45] at age twelve about 3 percent are enuretic.[182] Studies of military recruits indicate that in young adulthood, the prevalence of enuresis may be as high as 1 to 3 percent.[159]

Causative Factors

Enuresis appears to be related to time of night, not to sleep stages. About two-thirds of all enuretic episodes occur in the first third of the night, and most take place during NREM sleep.[20,59,128,203] However, the enuretic episodes are distributed across the various sleep stages in proportion to their rate of occurrence during each third of the night.[128] Thus, since NREM sleep predominates in the first third of the night, it is more likely for the episodes to occur in NREM sleep. If the child is not awakened and changed after wetting the bed, the sensation of wetness may be incorporated into a dream during the next period of REM sleep and result in dreams of wetting or being wet. This phenomenon accounts for the common misconception that enuresis occurs while the child is dreaming.[59]

In primary enuresis, genetic[10,70,195] and developmental / maturational factors[44,224] underlie the disorder, while in secondary enuresis, psychological factors are the most important. From a developmental standpoint, children with primary enuresis have a smaller-than-normal functional bladder capacity.[44,224] Secondary enuresis is most often psychogenic. Psychological factors related to secondary enuresis may reflect a need to regress or a need for excessive attention, for example, when a new sibling is born.[141] It should also be noted that bedwetting may be a symptom of diabetes mellitus, diabetes insipidus, nocturnal epilepsy, severe mental retardation, neurologic disorders, or other systemic conditions.

Diagnostic Considerations and Procedures

Since there are multiple causes and contributing factors in enuresis, it is critical for the physician to thoroughly evaluate and treat each patient individually rather than follow any single, stereotyped approach to management. The evaluation of the enuretic patient includes a thorough history, with specific attention paid to the history and course of bedwetting.[112] A physical examination and urinalysis are also important.

The physician first determines whether the enuresis is primary or secondary; if secondary, he then determines the age of onset and its relation to major life-stress events.

Other factors to consider are: the family history for enuresis, the child's general sleep habits, the presence of other sleep disorders, the attitude of the child and parents toward enuresis, and the degree of psychopathology in the child and family.

The physician should be careful not to misinterpret any psychopathology as the cause rather than the effect of enuresis. This is particularly true for children with primary enuresis who are commonly mishandled by their parents, sometimes with devastating psychological consequences. On the other hand, in children with secondary enuresis, the physician needs to distinguish between psychological difficulties that cause bedwetting and the psychological problems the child may develop because of his enuresis.

When conducting the physical examination, the physician directs special attention to the child's general growth and development. The urinary stream is observed to see if it is full and forceful or if there is deviation, narrowing, or dribbling. In the urinalysis, a urinary specific gravity of 1.024 or less excludes renal disease. A culture of the urine should be made when an infection is suspected. The baseline functional bladder capacity is also determined by having the child refrain from voiding as long as possible and then measuring the volume of the urine voided.[224,225]

Therapy

Nonpharmacologic Management and Psychotherapy

The type of treatment to be used for enuresis depends mostly upon whether the disorder is primary or secondary. Education and reassurance for the parents and the child are an important aspect of treating both types of enuresis. Parents of children with primary enuresis should be counseled to be tolerant, patient, and understanding of their child's disorder. They should be aware that overreacting with harsh and punitive behavior may cause their child to feel guilty, angry, and anxious, and that their mishandling of the situation may result in serious psychological complications.[105]

Because children with primary enuresis often have a small functional bladder capacity, bladder-training exercises are often used to expand the bladder's capacity.[178,224] These exercises consist of three major elements: (1) drinking fluids in unrestricted amounts during the daytime, (2) voluntarily withholding urination as long as possible at least once a day, and (3) recording the daily amount of urine passed after maximum holding. These exercises give the child a feeling of mastery and, by stretching the bladder, increase its capacity.

The treatment of primary enuresis in adolescents and adults starts with bladder-training exercises that help achieve initial mastery of bladder functioning. Subsequently, insight-oriented psychotherapy may help the patient to attain permanent dryness.

Since psychological difficulties usually underlie secondary enuresis, various psychotherapies are often employed in its treatment. The proper assessment of the psychological basis of the problem determines the therapeutic approach. Family therapy, seeing the parents and child separately in psychodynamic treatment, behavioral therapy, or simply educating and instructing the parents, are all helpful if appropriately applied by the psychiatrist. When treating adults with secondary enuresis that is unrelated to any organic factors, insight-oriented psychotherapy is most effective.

Drug Therapy

Imipramine has been proven effective in reducing enuretic frequency;[172,196] however, the relapse rate is high after the drug is withdrawn.[172] Use of the drug is recommended only in older children or adolescents, and then for only limited periods of time and special situations.

The efficacy of imipramine in treating enuresis apparently is not related to its antidepressant effect since there is no evidence to suggest that enuretic children are more depressed than nonenuretic children. Also, sleep laboratory studies have shown that the reduction in enuretic frequency with imipramine is not related to any sleep stage alterations produced by the drug. It was found that imipramine produces a significant increase in wakefulness during the night.[128] The hypothesis is that these heightened levels of arousal during the child's sleep allow a more conscious control of micturition. This factor, together with a decrease in bladder excitability, results in greater bladder capacity and fewer enuretic events, particularly early in the night when enuresis is most frequent. This allows the child to continue to sleep without micturition, and later in the night, when sleep is lighter, to be more aware of stimuli from a full bladder. Children who are being treated with imipramine, as well as their parents, should be advised of this effect of the drug. In this way, the child is encouraged to get up and urinate when he experiences wakefulness during the latter part of the night, rather than remain in bed.[128]

Imipramine should not be prescribed on a long-term basis since the effects of long-term administration of psychotropic medication in children have not been determined. Treatment with imipramine is initiated with a dose of 25 mg one or two hours before bedtime, and this dose is then gradually increased to a therapeutic level. The FDA-approved dose for treating enuresis is 1 to 2.5 mg per kilogram per day, or 25 to 75 mg daily.[80]

¶ Conclusion

Sleep disorders are frequently encountered in medical practice. Although the general physician is capable of diagnosing and treating most cases of sleep disturbance, patients who have disorders that are chronic or that present special difficulties are usually referred to specialists, most often psychiatrists and, less frequently, neurologists. Psychiatrists will most often be presented with cases of intractable insomnia, sleepwalking, night terrors, nightmares, and secondary enuresis, while neurologists are more likely to see patients with narcolepsy, the hypersomnias, and occasionally night terrors.

The clinical indications for sleep-laboratory diagnostic studies are limited.[137,139] Insomnia, enuresis, sleepwalking, night terrors, and nightmares rarely require sleep laboratory evaluations because the diagnosis nearly always can be established by taking a thorough history. In most cases, the psychiatrist's skills are far more useful in diagnosing sleep disorders than are time-consuming and expensive sleep-laboratory procedures.

There are certain specific and limited situations in which nocturnal electroencephalographic recordings or other sleep laboratory procedures can aid the psychiatrist in formulating a diagnosis. In narcoleptics who do not have cataplexy, diagnostic nap recordings to detect sleep-onset REM periods may be helpful; such studies can be performed in a clinical EEG laboratory. If sleep apnea is present or suspected, sleep-laboratory diagnostic studies are indicated to quantify the number and duration of episodes, changes in blood gases, and the degree of sleep disturbance. All-night sleep recordings may also be useful in diagnosing nocturnal epilepsy when daytime clinical sleep EEG recordings are not sufficient. Finally, psychogenic impotence can be differentiated from organic sexual difficulty by recording the occurrence of penile erections in relation to REM periods.

¶ Bibliography

1. ABE, K., and SHIMAKAWA, M. "Predisposition to Sleepwalking," *Psychiatric Neurologie,* 152 (1966): 306–312.
2. ADAM, K., et al. "Nitrazepam: Lastingly Effective but Trouble on Withdrawal," *British Medical Journal,* 1 (1976): 1558–1560.
3. AGNEW, H. W., WEBB, W. B., and WILLIAMS, R. L. "Sleep Patterns in Late Middle Age Males: An EEG Study," *Elec-*

troencephalography and Clinical Neurophysiology, 23 (1967): 168–171.

4. AKIMOTO, H., HONDA, Y., and TAKAHASHI, Y. "Pharmacotherapy in Narcolepsy," *Diseases of the Nervous System,* 21 (1960): 704–706.
5. AMERICAN PSYCHIATRIC ASSOCIATION. *Diagnostic and Statistical Manual of Mental Disorders,* 3rd ed. Washington, D.C.: American Psychiatric Association, 1980.
6. ARMSTRONG, R. H., et al. "Dreams and Gastric Secretions in Duodenal Ulcer Patients," *New Physician,* 14 (1965): 241–243.
7. ASERINSKY, E., and KLEITMAN, N. "Regularly Occurring Periods of Eye Motility, and Concomitant Phenomena During Sleep," *Science,* 118 (1953): 273–274.
8. ———. "Two Types of Ocular Motility Occurring in Sleep," *Journal of Applied Physiology,* 8 (1955): 1–10.
9. BAKWIN, H. "Sleepwalking in Twins," *Lancet,* 2 (1970): 446–447.
10. ———. "Enuresis in Twins," *American Journal of Diseases of Children,* 121 (1971): 222–225.
11. BALTER, M. B., and BAUER, M. L. "Patterns of Prescribing and Use of Hypnotic Drugs in the United States," in Clift, A. D., ed., *Sleep Disturbances and Hypnotic Drug Dependence,* New York: Excerpta Medica, 1975, pp. 261–293.
12. BERGER, R. J., and OSWALD, I. "Effects of Sleep Deprivation on Behavior, Subsequent Sleep, and Dreaming," *Journal of Mental Science,* 108 (1962): 457–465.
13. BERGER, R. J., OLLEY, P., and OSWALD, I. "The EEG, Eye Movements and Dreams of the Blind," *Quarterly Journal of Experimental Psychology,* 14 (1962): 183–186.
14. BERNSTEIN, S., and LEFF, R. "Toxic Psychosis from Sleeping Medicine Containing Scopolamine," *New England Journal of Medicine,* 277 (1967): 638–639.
15. BIXLER, E. O., KALES, A., and SOLDATOS, C. R. "Sleep Disorders Encountered in Medical Practice: A National Survey of Physicians," *Behavioral Medicine,* 6 (1979): 1–6.
16. BIXLER, E. O., et al. "Subjective Clinical Characteristics of Insomniac Patients," *Sleep Research,* 7 (1978): 212.
17. BIXLER, E. O., et al. "Prevalence of Sleep Disorders in the Los Angeles Metropolitan Area," *American Journal of Psychiatry,* 136 (1979): 1257–1262.
18. BIXLER, E. O., et al. "Effects of Hypnotic Drugs on Memory," *Life Sciences,* 25 (1979): 1379–1388.
19. BLOMFIELD, J. M., and DOUGLAS, J. W. B. "Bedwetting: Prevalence among Children Aged 4–7 Years," *Lancet,* 1 (1956): 850–852.
20. BROUGHTON, R. J. "Sleep Disorders: Disorders of Arousal?" *Science,* 159 (1968): 1070–1078.
21. ———. "Neurology and Sleep Research," *Canadian Psychiatric Association Journal,* 16 (1971): 283–293.
22. CARSKADON, M. A., et al. "Self-Reports Versus Sleep Laboratory Findings in 122 Drug-Free Subjects with Complaints of Chronic Insomnia," *American Journal of Psychiatry,* 133 (1976): 1382–1388.
23. CASON, H. "The Nightmare Dream," *Psychological Monographs,* XLVI (1935): 1–49.
24. CAUTELA, J. R. "Behavior Therapy and the Need for Behavioral Assessment," *Psychotherapy,* 5 (1968): 175–179.
25. CHARNEY, D. S., et al. "Somnambulistic-like Episodes Secondary to Combined Lithium-Neuroleptic Treatment," *British Journal of Psychiatry,* 135 (1979): 418–424.
26. CHURCH, M. W., and JOHNSON, L. C. "Mood and Performance of Poor Sleepers During Repeated Use of Flurazepam," *Psychopharmacology,* 61 (1979): 309–316.
27. CLEMENTE, C. D., and STERMAN, M. B. "Limbic and Other Forebrain Mechanisms in Sleep Induction and Behavioral Inhibition," *Progress in Brain Research,* 27 (1967): 34–47.
28. COCCAGNA, G., and LUGARESI, E. "Insomnia in the Restless Legs Syndrome,", in Gastaut, H., et al., eds., *The Abnormalities of Sleep in Man,* Bologna: Aulo Gaggi Editore, 1968, pp. 139–144.
29. COCCAGNA, G., et al. "Tracheostomy in Hypersomnia with Periodic Breathing," *Bulletin de Physio-Pathologie Respiratoire,* 8 (1972): 1217–1227.
30. COURSEY, R. D., BUCHSBAUM, M., and FRANKEL, B. L. "Personality Measures and Evoked Responses in Chronic Insomniacs," *Journal of Abnormal Psychology,* 84 (1975): 239–249.
31. CRITCHLEY, M. "Periodic Hypersomnia and Megaphagia in Adolescent Males," *Brain,* 85 (1962): 627–656.
32. DALY, D. D., and YOSS, R. E. "Pathologic

Sleep," *International Journal of Neurology,* 5 (1965): 195–206.

33. ———. "Narcolepsy," in Vinken, P. J., and Bruyn, G. W., eds., *Handbook of Clinical Neurology,* vol. 15, *The Epilepsies.* Amsterdam: North-Holland Publishing, 1974, pp. 836–852.

34. DEMENT, W. "The Effect of Dream Deprivation," *Science,* 131 (1960): 1705–1707.

35. ———. *Some Must Watch While Some Must Sleep.* Stanford, Calif.: Stanford Alumni Association, 1972.

36. DEMENT, W., and KLEITMAN, N. "Cyclic Variations in EEG During Sleep and Their Relation to Eye Movements, Body Motility and Dreaming," *Electroencephalography and Clinical Neurophysiology,* 9 (1957): 673–690.

37. DEMENT, W. C., CARSKADON, M., and LEY, R. "The Prevalence of Narcolepsy II," *Sleep Research,* 2 (1973): 147.

38. ———, and RICHARDSON, G. "Excessive Daytime Sleepiness in the Sleep Apnea Syndrome," in Guilleminault, C., and Dement, W. C., eds., *Sleep Apnea Syndromes,* New York: Alan R. Liss, 1978, pp. 23–46.

39. DEMENT, W. C., RECHTSCHAFFEN, A., and GULEVITCH, G. "The Nature of the Narcoleptic Sleep Attack," *Neurology,* 16 (1966): 18–33.

40. DEMENT, W. C., et al. "Narcolepsy: Diagnosis and Treatment," *Primary Care,* 3 (1976): 609–623.

41. DETRE, T. "Sleep Disorder and Psychosis," *Canadian Psychiatric Association Journal,* 2 (1966): 177.

42. DETRE, T., et al. "Hypersomnia and Manic-Depressive Disease," *American Journal of Psychiatry,* 128 (1972): 123–125.

43. DUNLEAVY D. L. F., "Hyperthyroidism, Sleep and Growth Hormone," *Electroencephalography and Clinical Neurophysiology,* 36 (1974): 259–263.

44. ESPERANCA, M., and GERRARD, J. W. "Nocturnal Enuresis: Studies in Bladder Function in Normal Children and Enuretics," *Canadian Medical Association Journal,* 101 (1969): 324–327.

45. ESSEN, J., and PECKHAM, C. "Nocturnal Enuresis in Childhood," *Developmental Medicine and Child Neurology,* 18 (1976): 577–589.

46. FEINBERG, I. "Sleep in Organic Brain Conditions," in Kales, A., ed., *Sleep Physiology and Pathology,* Philadelphia: J.B. Lippincott, 1969, pp. 131–147.

47. FEINBERG, I., KORESKO, R. L., and HELLER, N. "EEG Sleep Patterns as a Function of Normal and Pathological Aging in Man," *Journal of Psychiatric Research,* 5 (1967): 107–144.

48. FEINBERG, I., KORESKO, R. L., and SCHAFFNER, I. R. "Sleep Electroencephalographic and Eye Movement Patterns in Patients with Chronic Brain Syndrome," *Journal of Psychiatric Research,* 3 (1965): 11–26.

49. FEINBERG, I., et al. "Sleep Electroencephalographic and Eye Movement Patterns in Schizophrenic Patients," *Comprehensive Psychiatry,* 5 (1964): 44–53.

50. FELDMAN, M. J., and HERSEN, M. "Attitudes Toward Death in Nightmare Subjects," *Journal of Abnormal Psychology,* 72 (1967): 421–425.

51. FISHER, C., et al. "A Psychophysiological Study of the Nightmare," *Journal of the American Psychoanalytic Association,* 18 (1970): 747–782.

52. FISHER, C., et al. "A Psychophysiological Study of Nightmares and Night Terrors I. Physiological Aspects of the Stage 4 Night Terror," *Journal of Nervous and Mental Disease,* 157 (1973): 75–98.

53. ———. "A Psychophysiological Study of Nightmares and Night Terrors: The Suppression of Stage 4 Night Terrors with Diazepam," *Archives of General Psychiatry,* 28 (1973): 252–259.

54. ———. "A Psychophysiological Study of Nightmares and Night Terrors," *Psychoanalysis and Contemporary Science,* 3 (1974): 317–398.

55. FISHER, C., et al. "Evaluation of Nocturnal Penile Tumescence in the Differential Diagnosis of Sexual Impotence: A Quantitative Study," *Archives of General Psychiatry,* 36 (1979): 431–437.

56. FOULKES, D. *The Psychology of Sleep.* New York: Scribner's, 1966.

57. FRANKEL, B. L., PATTEN, B. N., and GILLIN, J. C. "Restless Legs Syndrome. Sleep Electroencephalographic and Neurologic Findings," *Journal of the American Medical Association,* 230 (1974): 1302–1303.

58. FREEMON, F. R. *Sleep Research: A Critical Review,* Springfield, Ill.: Charles C Thomas, 1972.

59. GASTAUT, H., and BROUGHTON, R. "A Clin-

ical and Polygraphic Study of Episodic Phenomena During Sleep," *Biological Psychiatry,* 7 (1964): 197–221.

60. GASTAUT, H., TASSINARI, C. A., and DURON, B. "Polygraphic Study of the Episodic Diurnal and Nocturnal (Hypnic and Respiratory) Manifestations of the Pickwick Syndrome," *Brain Research,* 2 (1966): 167–186.
61. GASTAUT, H., et al., eds. *The Abnormalities of Sleep in Man.* Bologna: Aulo Gaggi Editore, 1968.
62. GEER, J. H., and SILVERMAN, I. "Treatment of a Recurrent Nightmare by Behavior-Modification Procedures: A Case Study," *Journal of Abnormal Psychology,* 72 (1967): 188–190.
63. GILLIN, J. C., et al. "An EEG Sleep Study of Bipolar (Manic-Depressive) Patients with a Nocturnal Switch Process," *Biological Psychiatry,* 12 (1977): 711–718.
64. GREENBLATT, D. J., and SHADER, R. I. "Drug Interactions in Psychopharmacology," in Shader, R. I., ed., *Manual of Psychiatric Therapeutics,* Boston: Little, Brown, 1975, pp. 267–279.
65. GUILLEMINAULT, C., and DEMENT, W. C., eds. *Sleep Apnea Syndromes.* New York: Alan R. Liss, 1978.
66. GUILLEMINAULT, C., DEMENT, W. C., and PASSOUANT, P., eds. *Advances in Sleep Research,* vol. 3, *Narcolepsy.* New York: Spectrum Publications, 1976.
67. GUILLEMINAULT, C., CARSKADON, M., and DEMENT, W. C. "On the Treatment of Rapid Eye Movement Narcolepsy," *Archives of Neurology,* 30 (1974): 90–93.
68. GUILLEMINAULT, C., VAN DEN HOED, J., and MITLER, M. M. "Clinical Overview of the Sleep Apnea Syndromes," in Guilleminault, C., and Dement, W. C. eds., *Sleep Apnea Syndromes,* New York: Alan R. Liss, 1978, pp. 1–12.
69. GUILLEMINAULT, C., WILSON, R. A., and DEMENT, W. C. "A Study on Cataplexy," *Archives of Neurology,* 31 (1974): 255–261.
70. HALLGREN, B. "Nocturnal Enuresis in Twins," *Acta Psychiatrica Scandinavica,* 35 (1960): 73–90.
71. HAMMOND, E. C. "Some Preliminary Findings on Physical Complaints From a Prospective Study of 1,064,004 Men and Women," *American Journal of Public Health,* 54 (1964): 11–23.
72. HARASZTI, J. S., and DAVIS, J. M. "Psychotropic Drug Interactions," in Clark, W. G., and del Guidice, J., eds. *Principles of Psychopharmacology.* New York: Academic Press, 1978, pp. 495–510.
73. HARTMANN, E. "Reserpine: Its Effect on the Sleep-Dream Cycle in Man," *Psychopharmacologia* 9 (1966): 242–247.
74. ———. "Longitudinal Studies of Sleep and Dream Patterns in Manic-Depressive Patients," *Archives of General Psychiatry,* 19 (1968): 312–329.
75. ———. *The Functions of Sleep.* New Haven: Yale University Press, 1974.
76. ———. "L-Tryptophan: A Rational Hypnotic With Clinical Potential," *American Journal of Psychiatry,* 134 (1977): 366–370.
77. HARTMANN, E., and RUSS, D. "Frequent Nightmares and the Vulnerability to Schizophrenia: The Personality of the Nightmare Sufferer," *Psychopharmacology Bulletin,* 15 (1979): 10–12.
78. HAURI, P., et al. "Sleep of Depressed Patients in Remission," *Archives of General Psychiatry,* 31 (1974): 386–391.
79. HAWKINS, D. R., and MENDELS, J. "Sleep Disturbances in Depressive Syndromes," *American Journal of Psychiatry,* 123 (1966): 682–690.
80. HAYES, T. A., PANITCH, M. L., and BARKER, E. "Imipramine Dosage in Children: A Comment on 'Imipramine and Electrocardiographic Abnormalities in Hyperactive Children,'" *American Journal of Psychiatry,* 132 (1975): 546–547.
81. HAYNES, S. N., and MOONEY, D. K. "Nightmares: Etiological, Theoretical, and Behavioral Treatment Considerations," *The Psychological Record,* 25 (1975): 225–236.
82. HEALEY, S. "Onset of Chronic Insomnia and Life-stress Events," *Sleep Research,* 5 (1976): 174.
83. HERSEN, M. "Personality Characteristics of Nightmare Sufferers," *Journal of Nervous and Mental Disease,* 153 (1971): 27–31.
84. HINTON, J. M. "Patterns of Insomnia in Depressive States," *Journal of Neurology, Neurosurgery and Psychiatry,* 26 (1963): 184–189.
85. HISHIKAWA, Y., and KANEKO, Z. "Electroencephalographic Study on Narcolepsy," *Electroencephalography and Clinical Neurophysiology,* 18 (1965): 249–259.
86. HISHIKAWA, Y., et al. "H-reflex and EMG of the Mental and Hyoid Muscles During

Sleep, with Special Reference to Narcolepsy," *Electroencephalography and Clinical Neurophysiology,* 18 (1965): 487–492.

87. HISHIKAWA, Y., et al. "Treatment of Narcolepsy with Imipramine (Tofranil) and Desmethylimipramine (Pertofran)," *Journal of the Neurological Sciences,* 3 (1966): 453–561.
88. HOLLINGSHEAD, A, and REDLICH, R. C. *Social Class and Mental Illness.* New York: John Wiley, 1958.
89. JACOBSON, A., et al. "Somnambulism: All-Night Electroencephalographic Studies," *Science,* 148 (1965): 975–977.
90. JOHNSON, L. C., BURDICK, A., and SMITH, J. "Sleep During Alcohol Intake and Withdrawal in the Chronic Alcoholic," *Archives of General Psychiatry,* 22 (1970): 406–418.
91. JOHNSON, L. C., SLYE, E. S., and DEMENT, W. "Electroencephalographic and Autonomic Activity During and After Prolonged Sleep Deprivation," *Psychosomatic Medicine,* 27 (1965): 415–423.
92. JOUVET, M. "Telencephalic and Rhombencephalic Sleep in Cat," in Wolstenholme, G. E. W., and O'Connor, M., eds., *Nature of Sleep,* Boston: Little, Brown, 1961, pp. 188–208.
93. ———. "Biogenic Amines and the States of Sleep," *Science,* 163 (1969): 32–41.
94. ———. "Neurophysiological and Biochemical Mechanisms of Sleep," in Kales, A., ed., *Sleep Physiology & Pathology,* Philadelphia: J. B. Lippincott, 1969, pp. 89–100.
95. KALES, A., ed. "Sleep and Dreams—Recent Research on Clinical Aspects," *Annals of Internal Medicine,* 68 (1968): 1078–1104.
96. KALES, A., ed. *Sleep: Physiology and Pathology.* Philadelphia: J. B. Lippincott, 1969.
97. KALES, A. "Psychophysiological Studies of Insomnia," *Annals of Internal Medicine,* 71 (1969): 625–629.
98. KALES, A. "Treating Sleep Disorders," *American Family Physician,* 8 (1973): 158–168.
99. KALES, A., and BERGER, R. J. "Psychopathology of Sleep," in Costello, C. G., ed., *Symptoms of Psychopathology.* New York: John Wiley, 1970.
100. KALES, A., and BIXLER, E. "Sleep Profiles of Insomnia & Hypnotic Drug Effectiveness," in Burch, N. and Altschuler, H., eds., *Behavior & Brain Electrical Activity,* New York: Plenum Press, 1975, pp. 81–92.
101. KALES, A., and KALES, J. D. "Sleep Laboratory Evaluation of Psychoactive Drugs," *Pharmacology for Physicians,* 9 (1970): 1–6.
102. ———. "Evaluation, Diagnosis and Treatment of Clinical Conditions Related to Sleep," *Journal of the American Medical Association,* 213 (1970): 2229–2235.
103. ———. "Recent Advances in the Diagnosis and Treatment of Sleep Disorders," in Usdin, G., ed., *Sleep Research and Clinical Practice,* New York: Brunner/Mazel, 1973, pp. 59–94.
104. ———, and BIXLER, E. O. "Insomnia: An Approach to Management and Treatment," *Psychiatric Annals,* 4 (1974): 28–44.
105. ———. "Sleep Disorders: Recent Findings in the Diagnosis and Treatment of Disturbed Sleep," *New England Journal of Medicine,* 290 (1974): 487–499.
106. ———. "Shortcomings in the Evaluation and Promotion of Hypnotic Drugs," *New England Journal of Medicine,* 293 (1975): 826–827.
107. KALES, A. and KALES, J. D. "Nocturnal Psychophysiological Correlates of Somatic Conditions and Sleep Disorders," *International Journal of Psychiatry in Medicine,* 6 (1975): 43–62.
108. KALES, A., ADAMS, G. L., and PEARLMAN, J. T. "Rapid Eye Movement (REM) Sleep in Opthalmic Patients," *American Journal of Opthalmology,* 69 (1970): 615–622.
109. KALES, A., KALES, J. D., and HUMPHREY, F. "Sleep and Dreams," in Freedman, M. A., Kaplan, H. I., and Sadock, B. J. eds., *Comprehensive Textbook of Psychiatry II,* Baltimore: Williams & Williams, 1975, pp. 114–127.
110. KALES, A., SCHARF, M. B., and KALES, J. D. "Rebound Insomnia: A New Clinical Syndrome," *Science,* 201 (1978): 1039–1041.
111. KALES, A., SOLDATOS, C. R., and KALES, J. D. "Childhood Sleep Disorders," in Gellis, S., and Kagan, B., eds., *Current Pediatric Therapy,* Philadelphia: W. B. Saunders, 1980, pp. 28–30.
112. ———. "Taking a Sleep History," *American Family Physician* 22 (1980): 101–107.
113. KALES, A., et al. "Dream Deprivation: An Experimental Reappraisal," *Nature,* 204 (1964): 1337–1338.

114. KALES, A., et al. "Somnambulism: Psychophysiological Correlates I. All-night EEG Studies," *Archives of General Psychiatry,* 14 (1966): 586–594.
115. KALES, A., et al. "Somnambulism: Psychophysiological Correlates II. Psychiatric Interviews, Psychological Testing and Discussion," *Archives of General Psychiatry,* 14 (1966): 595–604.
116. KALES, A., et al. "All-night Sleep Studies in Hypothyroid Patients, Before and After Treatment," *Journal of Clinical Endocrinology and Metabolism,* 27 (1967): 1593–1599.
117. KALES, A., et al. "Sleep Studies in Asthmatic Adults: Relationship of Attacks to Sleep Stage and Time of Night," *Journal of Allergy,* 41 (1968): 164–173.
118. KALES, A., et al. "Psychophysiological and Biochemical Changes Following Use and Withdrawal of Hypnotics," in Kales, A., ed., *Sleep: Physiology and Pathology,* Philadelphia: J.B. Lippincott, 1969, pp. 331–343.
119. KALES, A., et al. "Hypnotics and Altered Sleep and Dream Patterns: I. All-night EEG Studies of Glutethimide, Methprylon and Pentobarbital," *Archives of General Psychiatry,* 23 (1970): 211–218.
120. KALES, A., et al. "Hypnotic Drugs and Their Effectiveness: All-night EEG Studies of Insomniac Subjects," *Archives of General Psychiatry,* 23 (1970): 226–232.
121. KALES, A., et al. "Sleep Patterns Following 205 Hours of Sleep Deprivation," *Psychosomatic Medicine,* 32 (1970): 189–200.
122. KALES, A., et al. "Sleep Laboratory Drug Evaluation: Thioridazine (Mellaril), A REM-Enhancing Drug," *Sleep Research,* 3 (1974): 55.
123. KALES, A., et al. "Chronic Hypnotic Use: Ineffectiveness, Drug Withdrawal Insomnia, and Dependence," *Journal of the American Medical Association,* 227 (1974): 513–517.
124. KALES, A., et al. "Common Shortcomings in the Evaluation and Treatment of Insomnia," in Kagan, F., et al., eds., *Hypnotics: Methods of Development and Evaluation,* New York: Spectrum Publications, 1975, pp. 29–40.
125. KALES, A., et al. "Effectiveness of Hypnotic Drugs with Prolonged Use: Flurazepam and Pentobarbital," *Clinical Pharmacology and Therapeutics,* 18 (1975): 356–363.
126. KALES, A., et al. "Sleep Laboratory Studies of Flurazepam: A Model for Evaluating Hypnotic Drugs," *Clinical Pharmacology and Therapeutics,* 19 (1976): 576–583.
127. KALES, A., et al. "Personality Patterns in Insomnia: Theoretical Implications," *Archives of General Psychiatry,* 33 (1976): 1128–1134.
128. KALES, A., et al. "Effects of Imipramine on Enuretic Frequency and Sleep Stages," *Pediatrics,* 60 (1977): 431–436.
129. KALES, A., et al. "Comparative Effectiveness of Nine Hypnotic Drugs: Sleep Laboratory Studies," *Journal of Clinical Pharmacology,* 17 (1977): 207–213.
130. KALES, A., et al. "Clinical Evaluation of Hypnotic Drugs: Contributions from Sleep Laboratory Studies," *Journal of Clinical Pharmacology,* 19 (1979): 329–336.
131. KALES, A., et al. "Propranolol in the Treatment of Narcolepsy," *Annals of Internal Medicine,* 91 (1979): 742–743.
132. KALES, A., et al. "Hypnotic Drugs," in Buchwald, C., Cohen, S., and Solomon, J., eds., *Frequently Prescribed and Abused Drugs: Their Indications, Efficacy, and Rational Prescribing,* Medical Monograph Series, vol. 2, No. 1, New York: Career Teaching Center, State University of New York, 1980, pp. 57–71.
133. KALES, A., et al. "Nightmares: Clinical Characteristics and Personality Patterns," *American Journal of Psychiatry* 137 (1980): 1197–1201.
134. KALES, A., et al. "Somnambulism: Clinical Characteristics and Personality Patterns," *Archives of General Psychiatry* 37 (1980): 1406–1410.
135. KALES, A., et al. "Hereditary Factors in Sleepwalking and Night Terrors," *British Journal of Psychiatry* 137 (1980): 111–118.
136. KALES, J. D., et al. "Are Over-the-Counter Sleep Medications Effective? All-night EEG Studies," *Current Therapeutic Research,* 13 (1971): 143–151.
137. KALES, J. D., et al. "Resource for Managing Sleep Disorders," *Journal of the American Medical Association,* 241 (1979): 2413–2416.
138. KALES, J. D., et al. "Sleepwalking and Night Terrors Related to Febrile Illness," *American Journal of Psychiatry,* 136 (1979): 1214–1215.

139. KALES, J. D., et al. "Sleep Disorders: What the Primary Care Physician Needs to Know," *Postgraduate Medicine,* 67 (1980): 213–220.
140. KALES, J.D., et al. "Night Terrors: Clinical Characteristics and Personality Patterns," *Archives of General Psychiatry* 37 (1980): 1413–1417.
141. KANNER, L., ed. *Child Psychiatry,* 4th ed., Springfield, Ill.: Charles C Thomas, 1972.
142. KARACAN, I. "Painful Nocturnal Penile Erections," *Journal of the American Medical Association,* 215 (1971): 1831.
143. KARACAN, I., SALIS, P. J., and WILLIAMS, R. L. "The Role of the Sleep Laboratory in Diagnosis and Treatment of Impotence," in Williams, R. L., and Karacan, I., eds., *Sleep Disorders: Diagnosis and Treatment,* New York: John Wiley, 1978, pp. 353–382.
144. KARACAN, I., et al. "Erection Cycle During Sleep in Relation to Dream Anxiety," *Archives of General Psychiatry,* 15 (1966): 183–189.
145. KARACAN, I., et al. "The Effects of Fever on Sleep and Dream Patterns," *Psychosomatics,* 9, (1968): 331–339.
146. KARACAN, I., et al. "Prevalence of Sleep Disturbance in a Primarily Urban Florida County," *Social Science and Medicine,* 10 (1976): 239–244.
147. KARACAN, I., et al. "Dose-Related Sleep Disturbances Induced by Coffee and Caffeine," *Clinical Pharmacology and Therapeutics,* 20 (1976): 682–689.
148. KILOH, L. G., and GARSIDE, R. F. "The Independence of Neurotic Depression and Endogenous Depression," *British Journal of Psychiatry,* 109 (1963): 451–463.
149. KESSLER, S., GUILLENMINAULT, C., and DEMENT, W. "A Family Study of 50 REM Narcoleptics," *Acta Neurologica Scandinavica,* 50 (1974): 503–512.
150. KLEITMAN, N. *Sleep and Wakefulness.* Chicago: University of Chicago Press, 1963.
151. KOELLA, W. P. *Sleep: Its Nature and Physiological Organization.* Springfield, Ill.: Charles C Thomas, 1967.
152. KORESKO, R. L., SNYDER, F., and FEINBERG, I. " 'Dream Time' in Hallucinating and Non-Hallucinating Schizophrenic Patients," *Nature,* 199 (1963): 1118–1119.
153. KRIPKE, D. F., et al. "Short and Long Sleep and Sleeping Pills: Is Increased Mortality Associated?" *Archives of General Psychiatry,* 36 (1979): 103–116.
154. KUPFER, D. J., et al. "Sleep Disturbance in Acute Schizophrenic Patients," *American Journal of Psychiatry,* 126 (1970): 47–57.
155. KUPFER, D. J., et al. "Hypersomnia in Manic-Depressive Disease," *Diseases of the Nervous System,* 33 (1972): 720–724.
156. KURTH, V. E., GOHLER, I., and KNAAPE, H. H. "Untersuchengen uber den Pavor nocturnus bei Kindern," *Psychiatrie, Neurologie und Medizinische Psychologie,* 17 (1965): 1–7.
157. LECKMAN, J. F., and GERSHON, E. S. "A Genetic Model of Narcolepsy," *British Journal of Psychiatry,* 128 (1976): 276–279.
158. LEVIN, M. "Periodic Somnolence and Morbid Hunger: A New Syndrome," *Brain,* 59 (1936): 494–515.
159. LEVINE, A. "Enuresis in the Navy," *American Journal of Psychiatry,* 100 (1943): 320–325.
160. LUGARESI, E., COCCAGNA, G., and MANTOVANI, M. "Hypersomnia with Periodic Apneas," *Advances in Sleep Research, vol. 4.* New York: Spectrum Publications, 1978.
161. LUGARESI, E., et al. "Symond's Nocturnal Mycolonus," *Electroencephalography and Clinical Neurophysiology,* 23 (1967): 289.
162. LUGARESI, E., et al. "Il Disturbo del Sonno e del Respiro Nella Sindrome Pickwickiana," *Sistema Nervosa,* 20 (1968): 38–50.
163. LUGARESI, E., et al. "Hypersomnia with Periodic Breathing: Periodic Apneas and Alveolar Hypoventilation During Sleep," *Bulletin de Physio-Pathologie Respiratoire,* 8 (1972): 1103–1113.
164. LUGARESI, E., et al. "Effects of Tracheostomy in Two Cases of Hypersomnia with Periodic Breathing," *Journal of Neurology, Neurosurgery, and Psychiatry,* 36 (1973): 15–26.
165. LUGARESI, E., et al. "Snoring," *Electroencephalography and Clinical Neurophysiology,* 39 (1975): 59–64.
166. MACK, J. E. "Nightmares, Conflict, and Ego Development in Childhood," *International Journal of Psycho-Analysis,* 46 (1965): 403–428.
167. MANDELL, M. P., et al. "Activation of the Pituitary-Adrenal Axis During Rapid Eye Movement Sleep in Man," *Life Sciences,* 5 (1966): 583–587.
168. MARKS, P. A., and MONROE, L. J. "Correlates of Adolescent Poor Sleepers," *Journal*

of Abnormal Psychology, 85 (1976): 243–246.

169. MARSHALL, J. R. "The Treatment of Night Terrors Associated with the Posttraumatic Syndrome," *American Journal of Psychiatry,* 132 (1975): 293–295.
170. MCGHIE, A., and RUSSELL, S. M. "The Subjective Assessment of Normal Sleep Patterns," *Journal of Mental Science,* 108 (1962): 642–654.
171. MENDELSON, W. B., GILLIN, J. C., and WYATT, R. J. *Human Sleep and Its Disorders.* New York: Plenum Press, 1977.
172. MILLER, P. R., CHAMPELLI, J. W., and DINELLO, F. A. "Imipramine in the Treatment of Enuretic Schoolchildren: A Double-Blind Study," *American Journal of Diseases of Children,* 115 (1968): 17–20.
173. MITLER, M., et al. "Narcolepsy-Cataplexy in a Female Dog," *Experimental Neurology,* 45 (1974): 332–340.
174. MONROE, L. J. "Psychological and Physiological Differences Between Good and Poor Sleepers," *Journal of Abnormal Psychology,* 72 (1967): 255–264.
175. MONROE, L. J., and MARKS, P. A. "Psychotherapists' Descriptions of Emotionally Disturbed Adolescent Poor and Good Sleepers," *Journal of Clinical Psychology,* 33 (1977): 263–269.
176. MONTGOMERY, I., PERKIN, G., and WISE, D., "A Review of Behavioral Treatments for Insomnia," *Journal of Behavioral Therapy and Experimental Psychiatry,* 6 (1975): 93–100.
177. MUELLNER, S. R. "Development of Urinary Control in Children," *Journal of the American Medical Association,* 172 (1960): 1257–1261.
178. ———. "Development of Urinary Control in Children: A New Concept in Cause, Prevention and Treatment of Primary Enuresis," *Journal of Urology,* 84 (1960): 714–716.
179. NATIONAL CENTER FOR HEALTH STATISTICS. "Selected Symptoms of Psychological Distress." U. S. Public Health Service Publication 1000, Series 11, No. 37. Washington, D.C.: U. S. Department of Health, Education, and Welfare, 1970.
180. NEVSIMALOVA-BRUHOVA, S., and ROTH, B. "Heredofamilial Aspects of Narcolepsy and Hypersomnia," *Archives Suisses de Neurologie, Neurochirurgie et de Psychiatrie,* 110 (1972): 45–54.
181. NOWLIN, J. B., et al. "The Association of Nocturnal Angina Pectoris with Dreaming," *Annals of Internal Medicine,* 63 (1965): 1040–1046.
182. OPPEL, W. C., HARPER, P. A., and RIDER, R. V. "The Age of Attaining Bladder Control," *Pediatrics,* 42 (1968): 614–626.
183. ORR, W. C., ROBINSON, M. G., and JOHNSON, L. F., "Acid Clearing During Sleep in Patients with Esophagitis and Controls," *Gastroenterology,* 76 (1979): 1213.
184. OSWALD, I. *Sleeping and Waking,* Amsterdam: Elsevier, 1962.
185. ———. "Sleep and Dependence on Amphetamine and Other Drugs," in Kales, A., ed., *Sleep: Physiology and Pathology.* Philadelphia: J. B. Lippincott, 1969, pp. 317–330.
186. OSWALD, I., and PRIEST, R. G. "Five Weeks to Escape the Sleeping Pill Habit," *British Medical Journal,* 2 (1965): 1093–1095.
187. OSWALD, I., and THACORE, V. R. "Amphetamine and Phenmetrazine Addiction: Physiological Abnormalities in the Abstinence Syndrome," *British Medical Journal,* 2 (1963): 427–431.
188. OSWALD, I., et al. "Alpha Adrenergic Blocker, Thymoxamine and Mesoridazine Both Increase Human REM Sleep Duration," *Sleep Research,* 3 (1974): 62.
189. OSWALD, I., et al. "The Effects of Two Hypnotics on Sleep, Subjective Feelings, and Skilled Performance," in Passouant, P., and Oswald, I., eds., *Pharmacology of the States of Alertness.* New York: Pergamon Press, 1979, pp. 51–63.
190. PARMELEE, A. H., JR., WENNER, W. H., and SCHULZ, H. R. "Infant Sleep Patterns from Birth to 16 Weeks of Age," *Journal of Pediatrics,* 65 (1964): 576–582.
191. PARMELEE, A. H., et al. "Sleep States in Premature Infants," *Developmental Medicine and Child Neurology,* 9 (1967): 70–77.
192. PESIKOFF, R. B., and DAVIS, P. C. "Treatment of Pavor Nocturnus and Somnambulism in Children," *American Journal of Psychiatry,* 128 (1971): 134–137.
193. PIERCE, C. M., and LIPCON, H. H. "Somnambulism: Psychiatric Interview Studies," *U. S. Armed Forces Medical Journal,* 7 (1956): 1143–1153.
194. ———. "Somnambulism: Electroencephalographic Studies and Related Findings," *U. S. Armed Forces Medical Journal,* 7 (1956): 1419–1426.

195. PIERCE, C. M., et al. "Enuresis: Clinical, Laboratory and Electroencephalographic Studies," *U. S. Armed Forces Medical Journal,* 7 (1956): 208–219.
196. POUSSAINT, A. F., and DITMAN, K. S. "A Controlled Study of Imipramine (Tofranil) in the Treatment of Childhood Enuresis," *Journal of Pediatrics,* 67 (1965): 283–290.
197. RECHTSCHAFFEN, A., and DEMENT, W. C. "Narcolepsy and Hypersomnia," in Kales, A., ed., *Sleep: Physiology & Pathology.* Philadelphia: J.B. Lippincott, 1969, pp. 119–130.
198. RECHTSCHAFFEN, A., and KALES, A., eds. *A Manual of Standardized Terminology, Techniques, and Scoring System for Sleep Stages of Human Subjects,"* National Institute of Health Publication No. 204. Washington, D.C.: U.S. Government Printing Office, 1968.
199. RECHTSCHAFFEN, A., and ROTH, B. "Nocturnal Sleep of Hypersomniacs," *Activitas Nervosa Superior,* 11 (1969): 229–233.
200. RECHTSCHAFFEN, A., SCHULSINGER, F., and MEDNICK, S. A. "Schizophrenia and Physiological Indices of Dreaming," *Archives of General Psychiatry,* 10 (1964): 89–93.
201. RECHTSCHAFFEN, A., et al. "Nocturnal Sleep of Narcoleptics," *Electroencephalography and Clinical Neurophysiology,* 15 (1963): 599–609.
202. REDING, G., et al. "Nocturnal Teeth-Grinding: All-Night Psychophysiology Studies," *Journal of Dental Research,* 47 (1968): 786–797.
203. RITVO, E. R., et al. "Arousal and Nonarousal Enuretic Events," *American Journal of Psychiatry,* 126 (1969): 115–122.
204. ROFFWARG, H. P., DEMENT, W. C., and FISHER, C. "Preliminary Observations of the Sleep-Dream Patterns in Neonates, Infants, Children and Adults," in Harms, E., ed., *Problems of Sleep and Dream in Children.* New York: Pergamon Press, 1964, pp. 60–72.
205. ROFFWARG, H. P., MUZIO, J. N., and DEMENT, W. C. "Ontogenetic Development of the Human Sleep-Dream Cycle," *Science,* 152 (1966): 604–619.
206. ROTH, B. "Narcolepsy and Hypersomnia: Review and Classification of 642 Personally Observed Cases," *Archives Suisses de Neurologie, Neurochirurgie et de Psychiatrie,* 119 (1976): 31–41.
207. ———, BRUHOVA, S., and LEHOVSKY, M. "REM Sleep and NREM Sleep in Narcolepsy and Hypersomnia," *Electroencephalography and Clinical Neurophysiology,* 26 (1969): 176–182.
208. ROTH, B., NEVSIMALOVA, S., and RECHTSCHAFFEN, A. "Hypersomnia With Sleep Drunkenness," *Archives of General Psychiatry,* 26 (1972): 456–462.
209. ROTTERSMAN, W. "The Selectee and His Complaints," *American Journal of Psychiatry,* 103 (1946): 79–86.
210. SACKNER, M. A., et al. "Periodic Sleep Apnea: Chronic Sleep Deprivation Related to Intermittent Upper Airway Obstruction and Central Nervous System Disturbance," *Chest,* 67 (1975): 164–171.
211. SALZMAN, C., VAN DER KOLK, B., and SHADER, R. I. "Psychopharmacology and the Geriatric Patient," in Shader, R. I., ed., *Manual of Psychiatric Therapeutics.* Boston: Little, Brown, 1975, pp. 171–184.
212. SEGAL, D. S., and JANOWSKY, D. S. "Psychostimulant-Induced Behavioral Effects: Possible Models of Schizophrenia," in Lipton, M. A., DiMascio, A., and Killam, K. F., eds., *Psychopharmacology: A Generation of Progress.* New York: Raven Press, 1978, pp. 1113–1123.
213. SHIRLEY, H. F., and KAHN, J. P. "Sleep Disturbances in Children," *Pediatric Clinics of North America,* 5 (1958): 629–643.
214. SILVERMAN, I., and GEER, J. H. "The Elimination of a Recurrent Nightmare by Desensitization of a Related Phobia," *Behavior Research and Therapy,* 6 (1968): 109–111.
215. SNYDER, F. "The New Biology of Dreaming, *Archives of General Psychiatry,* 8 (1963): 381–391.
216. ———. "Sleep Disturbance in Relation to Acute Psychosis," in Kales, A., ed., *Sleep: Physiology and Pathology.* Philadelphia: J. B. Lippincott, 1969, pp. 170–182.
217. SOLDATOS, C. R., KALES, A., and CADIEUX, R. "Narcolepsy: Evaluation and Treatment," in Seymour, R. B., et al., eds., *Amphetamine Use, Misuse, and Abuse.* Boston: G.K. Hall, 1979, pp. 128–140.
218. SOLDATOS, C. R., KALES, A., and KALES, J. D. "Management of Insomnia," *Annual Review of Medicine,* 30 (1979): 301–312.
219. SOLDATOS, C. R., et al. "Cigarette Smoking

Associated with Sleep Difficulty," *Science*, 207 (1980): 551–553.

220. SOLDATOS, C. R., et al. "Sleepwalking and Night Terrors in Adulthood: Clinical EEG Findings," *Clinical Electroencephalography* 11 (1980): 136–139.

221. SOURS, J. A. "Narcolepsy and Other Disturbances in the Sleep-Waking Rhythm: A Study of 115 Cases with Review of the Literature," *Journal of Nervous and Mental Disease*, 137 (1963): 525–542.

222. SOURS, J. A., FRUMKIN, P., and INDERMILL, R. R. "Somnambulism: Its Clinical Significance and Dynamic Meaning in Late Adolescence and Childhood," *Archives of General Psychiatry*, 9 (1963): 112–125.

223. SROLE, L., et al. *Mental Health in the Metropolis: The Midtown Manhattan Study.* New York: McGraw-Hill, 1962.

224. STARFIELD, B. "Functional Bladder Capacity in Enuretic and Nonenuretic Children," *Journal of Pediatrics*, 70 (1967): 777–781.

225. ———. "Enuresis: Its Pathogenesis and Management," *Clinical Pediatrics*, 11 (1972): 343–350.

226. STERMAN, M. B., et al. "Circadian Sleep and Waking Patterns in the Laboratory Cat," *Electroencephalography and Clinical Neurophysiology*, 19 (1965): 509–517.

227. SYMONDS, C. P. "Nocturnal Myoclonus," *Journal of Neurology, Neurosurgery and Psychiatry*, 16 (1953): 166–171.

228. TASSINARI, C. A., et al. "Pavor Nocturnus of Non-Epileptic Nature in Epileptic Children," *Electroencephalography and Clinical Neurophysiology*, 33 (1972): 603–607.

229. TEC, L. "Imipramine for Nightmares," *Journal of the American Medical Association*, 228 (1974): 978.

230. THOMAS, C. B., and PEDERSON, L. A. "Psychobiological Studies II: Sleep Habits of Healthy Young Adults with Observations on Levels of Cholesterol and Circulating Eosinophils," *Journal of Chronic Disease*, 16 (1963): 1099–1114.

231. TILKIAN, A. G., et al. "Hemodynamics in Sleep-Induced Apnea," *Annals of Internal Medicine*, 85 (1976): 714–719.

232. TOYODA, J. "The Effects of Chlorpromazine and Imipramine on Human Nocturnal Sleep Electroencephalogram," *Folia Psychiatrica et Neurologica Japonica*, 18 (1964): 198–221.

233. TUNE, G. S. "Sleep and Wakefulness in Normal Human Adults," *British Medical Journal*, 2 (1968): 269–271.

234. VOGEL, G. W. "Studies in Psychophysiology of Dreams: III. The Dream of Narcolepsy," *Archives of General Psychiatry*, 3 (1960): 421–428.

235. VOGEL, G. W., and TRAUB, A. C. "REM Deprivation. I. The Effect on Schizophrenic Patients," *Archives of General Psychiatry*, 18 (1968): 287–300.

236. VOGEL, G. W., et al. "REM Deprivation. II. The Effects on Depressed Patients," *Archives of General Psychiatry*, 18 (1968): 301–311.

237. VOGEL, G. W., et al. "The Effect of REM Deprivation on Depression," *Psychosomatics*, 14 (1973): 104–107.

238. WALSH, J. K., STOCK, C. G., and TEPAS, D. I. "The EEG Sleep of Workers Frequently Changing Shifts," *Sleep Research*, 7 (1968): 314.

239. WEBB, W. "Twenty-four-hour Sleep Cycling," in Kales, A., ed., *Sleep: Physiology and Pathology*, Philadelphia: J.P. Lippincott, 1969, pp. 53–65.

240. WEBB, W., ed. *Sleep: An Active Process—Research and Commentary.* Glenview, Ill.: Scott, Foresman, 1973.

241. WEBB, W. B., and FRIEL, J. "Sleep Stage and Personality Characteristics of 'Natural Long and Short Sleepers,'" *Science*, 171 (1971): 587–588.

242. WEITZMAN, E. D., ed. *Advances in Sleep Research*, vol. 1. New York: Spectrum Publications, 1974.

243. WEITZMAN, E. D., SCHAUMBURG, H., and FISHBEIN, W., "Plasma 17-Hydroxycorticosteroid Levels During Sleep in Man," *Journal of Clinical Endocrinology*, 26 (1966): 121–127.

244. WEITZMAN, E. D., et al. "Acute Reversal of the Sleep-Waking Cycle in Man," *Archives of Neurology*, 22 (1970): 483–489.

245. WEITZMAN, E. D., et al. "The Relationship of Sleep and Sleep Stages to Neuroendocrine Secretion and Biological Rhythms in Man," *Recent Progress in Hormone Research*, 31 (1975): 399–446.

246. WEITZMAN, E. D., et al. "Delayed Sleep Phase Syndrome: A Biological Rhythm Disorder," *Sleep Research*, 8 (1979): 221.

247. WEISS, H. R., KASINOFF, B. H., and BAILEY, M. A. "An Exploration of Reported Sleep

Disturbance," *Journal of Nervous and Mental Disease,* 134 (1962): 528–534.
248. WILLIAMS, R. L., AGNEW, H. W., JR., and WEBB, W. B. "Sleep Patterns in Young Adults: An EEG Study," *Electroencephalography and Clinical Neurophysiology,* 17 (1964): 376–381.
249. WILLIAMS, R. L., and KARACAN, I. "Sleep Disorders and Disordered Sleep," in Reiser, M. F., ed., *American Handbook of Psychiatry,* vol. 4, *Organic Disorders and Psychosomatic Medicine,* New York: Basic Books, 1975, pp. 854–904.
250. ———, eds., *Sleep Disorders: Diagnosis and Treatment.* New York: John Wiley, 1978.
251. WORLD HEALTH ORGANIZATION. *International Classification of Diseases, 9th Revision, Clinical Modification (ICD-9-CM).* WHO Center for Classification of Diseases for North America, National Center for Health Statistics. Ann Arbor, Mich.: Edwards Brothers, 1978.
252. YOSS, R. E., and DALY, D. D. "Criteria for the Diagnosis of the Narcoleptic Syndrome," *Proceedings of the Staff Meetings of the Mayo Clinic,* 32 (1957): 320–328.
253. ———. "Treatment of Narcolepsy with Ritalin," *Neurology,* 9 (1959): 171–173.
254. YULES, R. B., FREEDMAN, D. X., and CHANDLER, K. A. "The Effect of Ethyl Alcohol on Man's Electroencephalographic Sleep Cycle," *Electroencephalography and Clinical Neurophysiology,* 20 (1966): 109–111.
255. ZARCONE, V. "Narcolepsy," *New England Journal of Medicine,* 288 (1973): 1156–1166.

CHAPTER 23

OBESITY

Albert J. Stunkard

OBESITY is a condition characterized by excessive accumulations of fat in the body. By convention, obesity is said to be present when body weight exceeds by 20 percent the standard weight listed in the usual height-weight tables.[9] This index of obesity, however, is only an approximate one at lesser degrees of overweight, since bone and muscle can make a substantial contribution to overweight. In the future, diagnosis will probably be based upon newer and more accurate methods of estimating. Skin-fold calipers have already gained acceptance because of their convenience and because half of body fat is localized in subcutaneous tissue.[61] But for most clinical purposes the eyeball test is still the most reasonable: If a person looks fat he is fat.

¶ Epidemiology

Strikingly little information is available about the prevalence of obesity. Since most good diagnostic methods are too cumbersome for use in large-scale studies, much of our information is derived from height and weight data of poor quality, averaged over populations, and subjected to the criterion of 20 percent over standard weight. Present data suggest that prevalence of obesity reaches a peak at age forty when 35 percent of men and 40 percent of women can be designated as obese.[9,80]

There have been studies of more limited populations utilizing more reliable data that permit more valid inferences. Unfortunately, these studies differ in their criteria of obesity, making their data difficult or impossible to use for comparisons with other studies. These studies show a striking effect of age, with a monotonic increase in the prevalence of obesity between childhood and age fifty, and a twofold increase between ages twenty and fifty.[53] At age fifty, prevalence falls sharply, presumably because of the very high mortality of the obese from cardiovascular disease in the older age groups. Since these studies use the height-weight criterion, and since the fat content of the body increases per unit weight with age, these studies almost certainly underestimate the prevalence of obesity in older persons. The increasing use of skin-fold calipers should soon provide far more satisfactory data.

All studies comparing the sexes report a higher prevalence of obesity among women; this discrepancy is particularly pronounced after age fifty because of the higher mortality rate among obese men in this age group.

Social factors exert a powerful influence on the prevalence of obesity. In many countries undernutrition limits the development of obesity. Where there is no shortage of food, as in an affluent American society, many ethnic groups show a marked increase in the prevalence of obesity in the first generation. Thereafter, a variety of social influences combine to radically reduce the prevalence of obesity. One study reported a fall from 24 to 5 percent between the first and fourth generations in this county.[29]

The most striking antiobesity influence is that of socioeconomic status. Figure 23–1 shows that obesity is six times as common among women of low status as among those of high status in New York City.[29] A similar, though weaker, relationship was found among men. Two findings suggest that a causal relationship underlies these correlations. First, as figure 23–1 shows, the social class of one's parents is almost as closely linked to obesity as is the subject's own social class. Although obesity could conceivably influence a person's own social class, his obesity can hardly have influenced the social class of his parents. Furthermore, obesity is far more prevalent among lower-class children than it is among upper-class children; highly significant differences are already apparent by age six.[79] Similar analyses have shown that social mobility, ethnic factors, and generational status in the United States also influence the prevalence of obesity.[29]

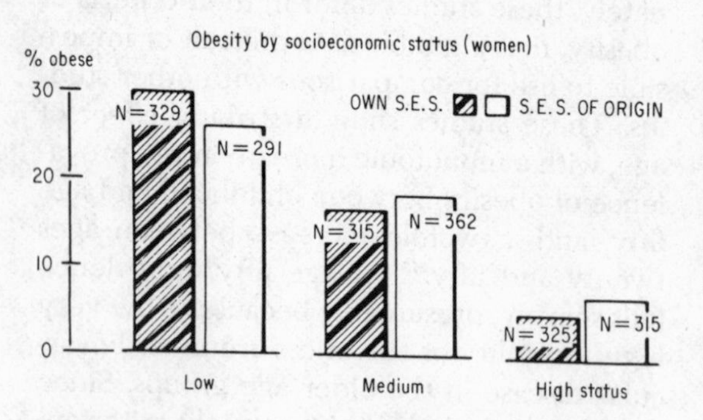

Figure 23–1. Decreasing prevalence of obesity with increase in socioeconomic status (SES). (From Goldblatt, P. B., et al. By permission of the *Journal of the American Medical Association,* 192 (1965):1039–1044.)

¶ Genetics

The existence of numerous forms of inherited obesity in animals and the ease with which adiposity can be produced by selective breeding make it clear that genetic factors can play a determining role in obesity. For years, textbooks of biology and medicine have included brief accounts implying that we know a good deal about the topic and that genetic factors play an important part in human obesity. It therefore comes as a surprise to realize the extent of our ignorance about the heritability of human obesity. Not much research has been done and we know very little about it.

Animal models of obesity have helped to put into perspective the either-or controversy of nature versus nurture; these models have indicated how several possible channels of influence can each have its own particular interaction of genetic and environmental variables, including interactions among and between genes. Genetic influences, for example, may be expressed in determining fat cell number in response to different diets at different periods of development, in regulating the efficiency of metabolic processes, and in establishing the sensitivity of different parts of the nervous system to nutrient depletion and repletion.[26] Already a few such interactions have been studied.

One of the most instructive examples of gene-gene interaction was effected by Mayer's[47] breeding the "waltzing gene" into genetically obese (ob/ob) mice. The increased physical activity of these mice prevented the development of obesity. A similarly interesting example of gene-environmental interaction is the different responses to a high-fat diet observed in two different strains of mice.[26] The C3H or A strain became fat while the C57BL or I Strain remained thin or lost weight.

With problems of human obesity, there are probably examples of single-gene obesity, comparable in some ways to the genetic obesity of the ob/ob mouse or the Zucker obese rat. But such single-gene obesity in humans, if it exists, seems confined to such rare conditions as the Lawrence-Moon-Biedl and Prader-Willi syndromes.[26] Most human obesity is probably of polygenic origin and quite possibly encompasses a large number of different conditions. Efforts to study this problem have utilized three methods: the study of familial resemblance, of twins, and of adoptees.

Examining familial resemblance has provided strong support for the familiar belief that obesity runs in families. But little progress has been made in determining whether this phenomenon is the result of genetic or environmental influences, and no progress has been made at all in determining genetic-environmental interactions. Garn's[27] quantitative studies have been the most revealing. He has established that there is a significant correlation between the skin-fold thicknesses of those parents and children who share common genes and environment ($r=0.30$). However, the correlation coefficient between the skin-fold thickness of marital partners, who share only their environment, is almost as high ($r=0.25$). These data have led Garn to propose that human obesity is primarily of environmental origin.

The study of twins, the second method of investigating human obesity, accords primary importance to genetic factors. Three studies utilized estimates of heritability based on the difference between the intrapair resemblances of monozygotic and dizygotic twins. All were high: 0.74 for 200 English twin pairs (meaning that 74 percent of the variance was accounted for by heredity), 0.88 for 100 Swedish twin pairs, and 0.78 and 0.77 for 4,000 American twin pairs (studied by the author at the time of induction into the Army and again twenty-five years later).[26]

The results of the three major adoption studies, the third method of investigating human obesity, are in complete disagreement with each other. Withers[84] reported findings that were originally interpreted as supporting a genetic origin of human obesity. In fact, however, these results are probably more supportive of an environmental origin. More recently, Garn[27] posited a purely environmental origin on the basis of his finding that the correlation of skin-fold thickness between parents and adopted children was the same as that between parents and biological children. Biron,[4] on the other hand, proposed that obesity is of genetic origin. He reported zero correlation between measures of obesity of parents and adopted children and highly significant correlations between parents and their biological children.

Etiology

What causes obesity? In one sense, the answer is simple: Fat is accumulated when more calories are taken in as food than are expended as energy. In another sense, the answer still eludes us. The regulation of body weight in normal-weight organisms is understood only poorly, and in obese organisms it is understood even less well. It does appear, however, that body weight is regulated even by many obese organisms. This idea that body weight may be regulated (even in obesity) rather than being the result of a number of unconnected influences is relatively new. The evidence supporting it, however, is strong.[35,39]

¶ The Regulation of Body Weight

It has been known for some time that the body weight of animals of normal weight is regulated. After the body weight of most experimental animals has been altered—lowered by starvation or raised by forced-feeding—it returns promptly to baseline. Only recently, however, has it become clear that animals suffering from a variety of forms of experimental obesity possess the same capacity for regulation.[39] Thus, in animals, at least, obesity need not be due to a disorder in the regulation of body weight, as had been believed in the past. Instead, it can be due to an elevation in the level about which the regulation occurs, a level that has been viewed by some as a regulatory "set point."

We know far less about the regulation of body weight in humans. Their weight tends to be the same year after year, despite the exchange of vast amounts of energy. For example, the average nonobese man consumes approximatlely one million calories a year; his body fat stores, however, remain unchanged during this time because he expends an equal number of calories. An error of no more than 10 percent either in intake or output would lead to a thirty-pound change in body weight within a year. There are only two studies of perturbation of this system in humans, and each supports the idea of regulation. Sims[62] found that normal-weight volunteers who were fattened by overfeeding and underactivity returned to their normal body weight without any special effort soon after they resumed their usual patterns of eating and activity. Keys's[41] classic study of experimental semistarvation showed that when subjects were permitted free access to food, their body weight also rapidly returned to normal. There are no similar studies of obese persons and we do not know whether they regulate body weight. The evidence, however, suggests that they do not.

Lifetime weight histories of obese persons rarely show a level about which body weight appears to be regulated. How can we explain this curious phenomenon? Why should obese people seem to be the only organisms that do not regulate body weight?

Nisbett has proposed an ingenious explanation for this apparent failure of regulation.[54] According to his theory, obese people may well have the capacity to regulate body weight. However, the set point about which their weight would be regulated, if only biological pressures existed, is higher than that which is tolerated by the society in which they live. As a result, such people go on reducing diets. And even if their weight does not fall to normal levels according to the height/weight charts, it still falls below what would be biologically normal for them. The result is the paradox of people who are statistically overweight and biologically underweight. Nisbett has described seven ways in which the biology and the behavior of obese people resemble that of people whose usually normal body weight has been reduced by starvation or other caloric restriction. In brief, they act as if they are hungry.

Nisbett's theory, attractive as it is, poses a major problem—it cannot be tested, at least not directly. Such a test would require, first, that obese people gain to their putative body-weight set point and, second, raise or lower their weight to see if it would return to this (elevated) baseline. This test is unfeasible on theoretical as well as on ethical grounds, since the putative set point cannot be estimated.

Although Nisbett's theory cannot be tested directly, indirect tests are possible. One such test, which will be described, supports the theory at least as far as *some* obese people are concerned. It was the results reported by Björntorp[5] of the treatment of obese people who possess such an excess number of fat cells that they can reduce to a statistically normal weight only by reducing the lipid content of their individual fat cells below normal values.[63]

The second test of Nisbett's theory has been carried out by Herman,[34] who has shown that some obese people (and some nonobese people) who habitually exercise restraint in the amount they eat share psychological characteristics that distinguish them from persons who do not restrain their food intake. Such "restrained eaters," for example, may show "counterregulation" of food intake and eat to excess when their habitual restraint is disinhibited. The range of such disinhibiters is impressive and includes dysphoric emotions such as anxiety and depression, alcohol, and even a high-calorie preload in a taste-testing experiment. Indeed, Herman has proposed that most of the behaviors attributed to the "externality" of obese persons, as is suggested by Schachter's popular theory,[59] is actually due to disinhibition of restrained eaters, who are more common among obese than among nonobese populations. Since this restraint may prevent obesity in persons of normal weight and mitigate its severity in those already obese, it suggests that the body weight of these people is below a biological set point.

A number of theories of the regulation of body weight have been proposed. Most ascribe this regulation to the regulation of a single nutrient. They start with the common-sense view that a person stops eating at the end of a meal because of the repletion of some nutrient that had been depleted. And one becomes hungry again when the nutrient, which had been restored by the meal, is once again depleted. It seems reasonable that some metabolic signal, derived from food that has been abosrbed, is carried by the blood to the brain, where it activates receptor cells, probably in the hypothalamus, to produce satiety. Hunger is the consequence of the decreasing strength of this metabolic signal, secondary to the depletion of the critical nutrient.

Four classical theories[10,40,48,50] of the regulation of body weight have been based upon this argument. They differ from each other only in the signal to which they ascribe primary importance. The thermostatic theory, for example, proposes that postprandial increases in hypothalamic temperature mediate satiety, with hunger resulting from a decrease in temperature at this site. Lipostatic, aminostatic, and glucostatic theories each assign the critical regulatory role to blood-borne metabolites of fat, protein, or carbohydrate, respectively.

Although each of these theories explains some of the many phenomena involved in the control of food intake, Mayer's[48] glucostatic theory has had the greatest influence on the field. According to this theory, depletion of carbohydrate stores decreases the amount of "available glucose" in the circulating blood; a fall in available glucose, signaled to hypothalamic glucoreceptors, becomes the stimulus for hunger. An increase in available glucose, with carbohydrate repletion, activates hypothalamic satiety areas and terminates eating. For more than twenty years this theory has exerted great organizational and heuristic power, and evidence for it continues to accumulate.

Despite the attractiveness of the glucostatic theory, it shares with all single-factor theories the general difficulty of encompassing the many events that are involved in the regulation of body weight. In addition to this difficulty, single-factor theories encounter two specific problems.

1. How can a mechanism of short-term, meal-to-meal control of food intake account for the remarkable stability of body weight over long periods of time and in the face of frequently marked short-term fluctuations? As a result of the problems of single-factor theories in modeling such stability, the newer theories of the regulation of body weight are multifactorial ones.[7,8] They are achieving increasing success in predicting food intake under a wide variety of conditions.

2. The second specific problem of single-factor, primarily physiological theories is how they can explain the function of satiety. For satiety occurs so soon after the beginning of a meal that only a small proportion of the total caloric content of the meal can have been absorbed. If satiety were based solely upon the limited information about food intake available at that time, it could contribute little or nothing to the regulation of food intake.

If humoral factors do not terminate eating, what does? A "full stomach" may be the answer. In addition to common sense and personal experience, Jordan[37] has added experimental evidence indicating that gastric filling, quite irrespective of the nutritive value of the meal, is the major determinant of satiety in single-meal experiments.

Although the nutritional value of meals plays little part in satiety in single-meal experiments, humans learn, as do other animals, to change food intake and even meal size in response to changes both in energy expenditure and in the nutritive value of the food. Booth[6] and Stunkard[70] have proposed that this learning is a special case of Pavlovian, or classical, conditioning. In this theory, oral, gastric, and perhaps duodenal factors serve as conditioned stimuli; humoral factors absorbed from the gastrointestinal tract serve as the later, unconditioned stimuli. This sequence accounts both for the termination of eating early in the process of food absorption from the intestine and for the long-term adjustment of meal size to changing caloric needs. Booth[7] has extended this theory to encompass "conditioned hunger" as well as conditioned satiety.

Until recently, it was believed that classical conditioning could not occur when the interval between conditioned and unconditioned stimuli was more than a few seconds. CS-US intervals longer than these have been reported only in the special case of taste aversions. But taste aversions may simply be special cases of a more general "alimentary learning." It has been proposed[70] that this alimentary learning serves as the bridge between the long-term physiological control of body weight based on humoral factors and the short-term control of hunger and satiety based upon conditioning. If this view is even approximately correct, impaired alimentary learning may lead to obesity and, more important, therapies based upon classical conditioning may become feasible.

¶ General Determinants of Obesity

There are at least six known determinants of obesity. Social determinants have been discussed under "Epidemiology," and genetic determinants under "Genetics." The other four determinants are: developmental, physical activity, brain damage, and emotional.

Developmental Determinants

A key to the understanding of obesity is provided by our growing knowledge of the anatomy of adipose tissue. It has become clear that the increased adipose tissue mass in obesity can result either from an increase in fat cell size ("hypertrophic obesity"), from an increase in fat cell number ("hyperplastic obesity"), or from an increase in both size and number ("hypertrophic-hyperplastic obesity"). Johnson and Hirsch's study[36] of six forms of experimental obesity in rodents reveals that most are either hypertrophic or hypertrophic-hyperplastic, and it appears that obese humans also usually fall into one of these two categories. These findings have important implications for prevention and treatment.

Most people whose obesity began in adult life suffer from hypertrophic obesity. When they lose weight it is solely by a decrease in the size of their fat cells; fat-cell number does not change. Salans and colleagues[53] have elegantly illustrated the dynamics of hypertrophic obesity in their study of human experimental obesity. When normal-weight men were induced to gain forty to sixty pounds, they did so solely through an increase in fat-cell size; when they lost the weight it was solely by a decrease in fat-cell size. Fat-cell number remained constant.

Persons whose obesity began in childhood are more likely to suffer from hyperplastic obesity, usually of the combined hypertrophic-hyperplastic type. They may have up to five times as many fat cells as persons of normal weight or those suffering from pure hypertrophic obesity. We still do not know all of the reasons for the elevation in fat-cell number, and at the present time the field is wracked with controversy. Recent research has challenged the old orthodoxy that fat-cell number cannot increase after early childhood and that events during a relatively brief "critical period" early in life is largely responsible for adipose tissue hyperplasia.[63] Instead, it has been proposed that fat-cell hypertrophy is a major stimulus for fat-cell hyperplasia and that this circumstance, and not a critical period, accounts for the common association of hyperplastic adipose tissue and juvenile-onset obesity.[63]

This view is compatible with Stern and Johnson's[65] review that describes at least two periods when cellular proliferation is enhanced in normal-weight children. One is before the age of two years and one is between the ages of ten and fourteen. In obese, presumably overfed children, however, the period of cell proliferation may extend well past two years of age, with consequent development of pronounced hypercellularity even early in life. Whether overnutrition alone can account for this prolonged period of cellular proliferation is not known; a genetic predisposition may also be required. Johnson and Hirsch's[36] studies of the genetically obese Zucker rat suggest the intriguing possibility that genetic factors may exert their influence by extending the period during which proliferation is particularly susceptible to the influence of overnutrition. The hypercellularity of the adipose tissue of these animals results from an ability to produce new fat cells well beyond the period of regulated proliferation found in their nonobese litter mates. Whatever the final outcome of this research, enough is already known to single out the early years of life as particularly important in the genesis of hyperplastic obesity. The public health implications are clear. To the compelling psychological reasons for the prevention of childhood obesity must now be added these compelling anatomical ones.

Our growing understanding of adipose tissue has clinical implications of equal importance. They are highlighted in a report by Björntorp and colleagues[5] on twenty-six outpatients who lost thirty-three pounds on a dietary regimen. Body weight and body fat content at the end of treatment varied widely and had reached normal limits in only ten of the subjects. Individual fat-cell size, however, was quite similar in all patients at the end of treatment and had fallen to normal in twenty-three of them. Most patients stopped treatment just at the point when further weight loss could be achieved only by the reduction of their fat cells to subnormal size. It was as if fat-cell size (perhaps particularly certain events at the cell membrane) had set a biological limit to weight reduction. If this is the case, it would explain the difficulty that hyperplastic obese persons experience in reducing to normal body weight and their proclivity to regain the weight that they have lost. More speculatively, it also suggests that reduction to a normal body weight may not be as important a health measure for these patients as we had believed. There is evidence that increased cell size, and not increased body fat alone or increased cell number, is responsible for the malignant metabolic sequence of insulin resistance, hyperinsulinemia, and lipid derangement.[58]

Physical Activity

The only component on the energy-expenditure side of the caloric ledger that both fluctuates and is under voluntary control is physical activity. As such, it is a vital factor in the regulation of body weight. Indeed, the marked decrease in physical activity in affluent societies seems to be the major factor in the recent rise of obesity as a public health problem. Obesity is a rarity in most underdeveloped nations, and not solely because of malnutrition. In some rural areas, a high level of physical activity is at least as important in preventing obesity. Such levels of physical activity are the exception in this country. If the trend exemplified by automatic can openers and mechanized swizzle sticks continues, we may succeed in reducing our energy expenditure to near basal levels. Among many obese women, the trend is already far advanced.

Figure 23–2 shows marked reduction of physical activity in a group of Philadelphia housewives; this reduction is so great as to account almost entirely for their excess weight.[18] But such low levels of physical ac-

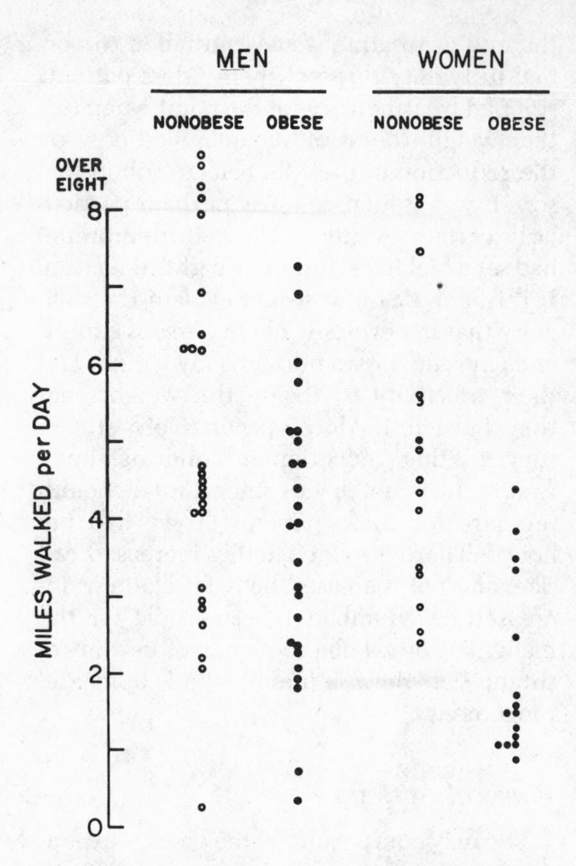

Figure 23–2. Comparison of the physical activity of obese and nonobese men and women. Each point represents the average distance walked each day by the subjects, as measured by a mechanical pedometer. Most obese women walked shorter distances than nonobese women. Among men, there is less difference in the distances walked. (From Chirico, A. M., and Stunkard, A. J. By permission of the *New England Journal of Medicine,* 263 (1960):935–946.)

tivity are not present among all obese persons. Figure 23–2 shows that the differences in physical activity among the men were so small that the additional energy expended by obese subjects in moving their heavier bodies produced a caloric expenditure equal to that of nonobese men.

Until recently, physical inactivity was considered to cause obesity primarily through its restriction of energy expenditure. There is now good evidence that inactivity may contribute also to an increased food intake. Although food intake increases with increasing energy expenditure over a wide range of energy demands, intake does not decrease proportionately when physical activity falls below a certain minimum level,[49] as shown in figure 23–3. In fact, restricting physical activity may actually increase food intake. Conversely, when sedentary organisms increase physical activity, their food intake may decrease. The importance of this phenomenon is probably even greater than was realized when it was first demonstrated by Mayer in Sprague-Dawley rats of normal weight.[49] Quite recently, studies of three forms of experimental obesity in rodents have shown that activity controls food intake even more powerfully in obese animals than in those of normal weight. The mechanism involved in this intriguing control is still unclear, but its great therapeutic potential makes it worthy of careful study.[12]

Brain Damage

Brain damage can lead to obesity, although it probably does so only rarely in humans. Nevertheless, brain damage is of great theoretical interest in understanding obesity.

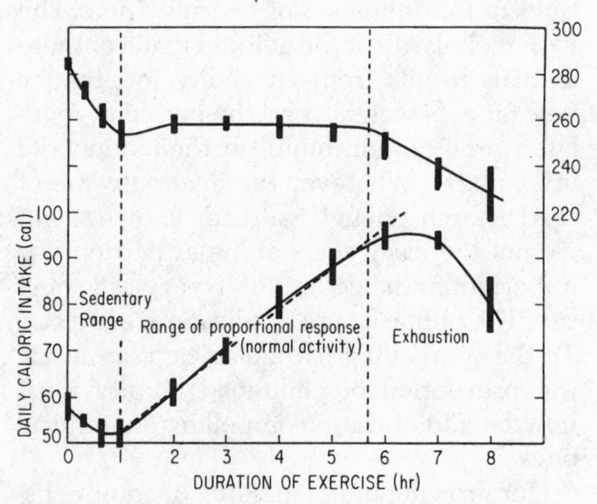

Figure 23–3. Calorie intake and body weight as functions of duration of exercise in adult rats. Within the range of normal physical activity, food intake increases with increasing physical activity and body weight remains stable. In the sedentary range of activity, however, decreasing physical activity is associated with *increased* food intake and an increase in body weight. (From Mayer, J., and Thomas, D. W. By permission of the American Association for the Advancement of Science, 156 (1967):328–337.)

This discovery, during the 1940s, that destruction of the ventromedial hypothalamus could produce obesity initiated the modern investigation of the condition. Subsequent work has delineated two broad anatomical systems mediating hunger and satiety—the former with special representation in the lateral hypothalamus, the latter in the ventromedial hypothalamus.

The way obesity is produced by ventromedial hypothalamic damage is particularly relevant to the thesis that obesity results from the level of regulation of body weight rather than from a disorder of regulation itself. Animals with such damage regulate body weight, but at a higher level. In the course of maintaining their new fat stores and body weight, such animals demonstrate interesting and potentially highly significant changes in behavior. Many of the features of this behavior in obese rats with hypothalamic lesions were described years ago by Miller and coworkers in their report on the paradox of "Decreased Hunger and Increased Food Intake in Hypothalamic Obese Rats."[47]

The cardinal feature of the rats' behavior was that they overate when food was freely available; but when an impediment was placed in the way of their eating, they not only decreased their food intake but actually decreased it to a far lower level than that of control rats without hypothalamic lesions. Furthermore, it seemed to make little difference what kind of impediment was used; motivation to work for food was impaired in every manner of task that could be devised. These rats seemed to be relatively unresponsive to all physiological cues concerning their nutritional state, and they responded imperfectly to signals both of satiety and of deprivation.

Nonetheless, the obese rats seemed hyperresponsive to the taste of food and to its availability. They increased their overeating when fat and sweet substances were added to their diet and radically restricted intake when the palatability of their food was decreased by the addition of quinine. Similar eating patterns have been reported in a wide variety of animals when they became obese for natural reasons—for example, in the genetically determined yellow obese mouse, in the rat when it becomes obese with aging, and even in the dormouse during the hyperphagia that precedes its hibernation. Experimental obesity of various types in animals thus seems to possess some common behavioral correlates. Clinical research has revealed intriguing parallels to the behavior of obese humans.

In the exceptional case, obesity in humans results from hypothalamic damage from a strategically placed tumor or vascular lesion. Usually, however, the cause of the impaired satiety exhibited by many obese persons remains unknown. Such persons characteristically complain that it is difficult to stop eating; it is the unusual obese person who reports being driven by hunger or who eats in a ravenous manner. Instead, obese persons seem particularly susceptible to the palatability of foods and find it difficult to keep from eating if food is available.

Studying the problems of obese persons without brain damage, Bruch[14,15] has described their misperception of important visceral events. Some obese persons, who are also neurotic, have difficulty in identifying hunger and satiety. They frequently seem unable to distinguish between hunger and other kinds of dysphoria. Bruch has linked this "conceptual confusion"[14] to severe deficits in identity and to feelings of personal ineffectiveness. She has convincingly described the need on the part of these patients for external signals to tell them when to eat and when to stop eating. Support for Bruch's position has come from studies that show that neurotic obese persons have a strong response bias that impairs their perception of gastric motility.[31] Unfortunately, correction of the bias did not result in weight loss.

On the basis of a long series of experiments, Schachter[59] has proposed a theory of obesity compatible with these ideas, which has achieved wide popularity. According to this theory, obese people are unusually susceptible to all kinds of "external" stimuli to eating, while remaining relatively unresponsive to the usual "internal" or physiological signals of hunger. At the present time this theory is under revision. Although externality as a personality trait seems well established, its relationship to obesity remains inconclusive.[56]

Emotional Determinants

Many obese persons report that they often overeat and gain weight when they are emotionally upset. But it has proved singularly difficult to proceed from this provocative observation to an understanding of the precise relationship between emotional factors and obesity. The most clear-cut evidence of how emotional factors influence obesity has come from two small subgroups of obese persons, each characterized by an abnormal and stereotyped pattern of food intake.[68] About 10 percent of obese persons, most commonly women, manifest a "night-eating syndrome" characterized by morning anorexia and evening hyperphagia with insomnia. This syndrome seems to be precipitated by stressful life circumstances and, once present, tends to recur daily until the stress is alleviated. Attempts at weight reduction when the syndrome is present have an unusually poor outcome and may even precipitate a more severe psychological disturbance.

Bulimia, found in fewer than 5 percent of obese persons, is one of the rare exceptions to the pattern of impaired satiety.[68] It is characterized by the sudden, compulsive ingestion of very large amounts of food in a very short time, usually with great subsequent agitation and self-condemnation. It, too, appears to represent a reaction to stress. But in contrast to the night-eating syndrome, these bouts of overeating are not periodic and they are far more often linked to specific precipitating circumstances. Binge eaters can sometimes lose large amounts of weight by adhering to rigid and unrealistic diets, but such efforts are almost always interrupted by a resumption of eating binges.

¶ Complications

Troublesome as obesity may be from a cosmetic standpoint, it is the health hazards associated with it that have caused it to be described as the nation's greatest *preventable* cause of death.

Effects on Mortality

Obesity has a strong adverse effect on morbidity and mortality rates. The death rate from several diseases is significantly higher among obese persons, and the rate increases in proportion to the severity of the obesity.[9] The most serious consequence of obesity is its impact upon the cardiovascular diseases that now cause more than half of all deaths in this country. The Framingham[30] and Chicago Peoples Gas Company[23] studies have shown a very strong relationship between obesity and coronary artery disease.

This evidence of the direct effect of obesity on mortality is matched by evidence of its indirect effect. Two of the most potent risk factors for coronary artery disease—adult-onset diabetes and hypertension—are also highly correlated with obesity. Weight reduction has a powerful effect: 75 percent of adult-onset diabetics may discontinue medication, and the blood pressure of 60 percent of hypertensives returns to normal levels after significant weight loss.[3,17]

In recent years, arguments against the importance of obesity in coronary disease have been raised, citing evidence that it is at best a weak independent risk factor.[45] Its powerful indirect effects, via diabetes, hyperlipidemias, and hypertension, however, detract from the strength of this argument.

Physical and Laboratory Abnormalities

The most serious physical[1] manifestation of obesity, and the only one that is life-threatening, albeit very rarely, is the encircling of the thorax with fatty tissue, together with pressure on the diaphragm from below due to intraabdominal accumulations of fat. The result is reduced respiratory excursion, with dyspnea from even minimal exertion. In very obese persons, this condition may progress to the so-called "Pickwickian syndrome," characterized by hypoventilation with consequent hypercapnia, hypoxia, and finally somnolence.[16]

Severe obesity leads to a variety of orthopedic disturbances, including low-back pain, aggravation of osteoarthritis, particularly of the knees and ankles, and often enormous calluses over the feet and heels. Even mild degrees of obesity are associated with amenorrhea and other menstrual disturbances. Subcutaneous fat is an excellent heat insulator, and the skin of obese persons is often warm and sweaty, particularly after meals. Hyperhidrosis leads to intertrigo in the pendulous folds of tissue, making itching and skin disorders common. Mild to moderate edema of the feet and ankles often occurs, probably due to venous obstruction; diuretics are not indicated. What is most notable about all of these complications is the ease with which they can be controlled and eliminated by weight reduction, often of only a moderate degree.

Blood-pressure elevations are frequently found in obese persons, often due to an artifact, that is, the presence of masses of subcutaneous tissue between the blood-pressure cuff and the brachial artery. This problem can usually be overcome by using a wider blood-pressure cuff.

Hyperuricemia is sometimes found in obesity, and it may reach a significant degree in persons who fast intermittently. When obesity has produced respiratory distress, hypercapnia may develop along with a respiratory acidosis.

A particular problem in the laboratory evaluation of obesity is the impaired glucose tolerance, and even the presence of fasting hyperglycemia, that occurs in many obese persons without a family history of diabetes. The high insulin levels in the fasting state and after a glucose load, usually associated with obesity, are related to the presence of muscle and adipose tissue resistance to carbohydrate metabolism.[58] The precise relationship between tissue resistance and insulin levels is not clear. It may be that tissue resistance signals the pancreas to produce more insulin, or that a high-calorie diet may increase insulin production, with tissue resistance a secondary phenomenon. However these questions are finally resolved, the response to weight reduction is highly gratifying. Most such abnormalities disappear completely unless the patient is truly diabetic.

Plasma-lipid levels are often moderately elevated in the obese. Weight reduction decreases both total cholesterol and triglycerides, and it is one of the few measures that produces elevations (among men, at least) in the protective high-density lipoprotein cholesterol.

Emotional Disturbances

Reports on emotional disturbances among obese people have flooded the literature. The better the study the less the evidence for distinctive psychological features and disabilities. The two most careful studies have shown little differences in psychopathology between obese and nonobese people. Moore and colleagues[53] reported slightly higher levels of psychopathology among obese people; Crisp and McGuiness[20] reported slightly lower levels. Even massively obese people do not seem to suffer undue psychiatric disability. The view that obese persons have a specific personality pattern is no longer held.

Although the differences in psychopathology are relatively small for the obese population as a whole, they may be quite significant for certain subgroups. Prominent among these are young women of upper and middle socioeconomic status. The reasons for the special vulnerability of these groups are of interest. Both obesity and emotional disturbances are common among persons of lower socioeconomic status;[29,52] any association between the two conditions in persons in this stratum is apt to be coincidental. Higher up on the socioeconomic ladder, however, obesity is far less prevalent, and the sanctions against it are far stronger. There is also far less emotional disturbance at this level. As a result, when obesity and emotional disturbance coexist in this group, the likelihood that they are associated is far greater. Among young, upper-class women, obesity is usually linked to neurosis. What is the nature of this linkage?

Of the various emotional disturbances to which obese persons are subject, only two are specifically related to their obesity. One is overeating, which has been discussed; the other is disparagement of the body image.[76] Persons suffering from disparagement of the body image characteristically feel that their bodies are grotesque and loathesome and that others view them with hostility and contempt. This feeling is closely associated with self-consciousness and impaired social functioning. While it may seem reasonable to suppose that all obese persons have derogatory feelings about their bodies, such is not the case. Emotionally healthy obese persons have no body-image disturbances and, in fact, only a minority of neurotic obese persons have such disturbances. The disorder is confined to those who have been obese since childhood; and even among these juvenile-onset obese, less than half suffer from it. But in the group with body-image disturbances, neurosis is closely related to obesity, and this group contains a majority of obese persons with specific eating disorders.

The extent and severity of complications following weight-reduction programs have been the subject of controversy in recent years. It now appears that as many as half of the patients routinely treated for obesity by family physicians may develop mild anxiety and depression.[78] An even higher incidence of emotional disturbance has been reported among morbidly obese persons undergoing long-term treatment by fasting or severe caloric restriction even when carried out in the hospital. These complications should be balanced against the likelihood of a decrease in anxiety and depression among those who diet successfully. Such newer treatments as behavior modification and by-pass surgery carry far less risk of emotional disturbances.

Obese persons with extensive psychopathology, those with a history of emotional disturbance during dieting, or those in the midst of a life crisis should attempt weight reduction, if at all, cautiously and under careful supervision. For others, the possibility of complications need not preclude treatment when it is indicated.

General Considerations

Weight reduction confers such great benefits on obese persons and is apparently so simple that we might expect it to be a common occurrence. Perhaps the large number of women who try to reduce without medical assistance (following diets and advice from the women's magazines) have more success. But "most obese persons will not enter outpatient treatment of obesity; of those who do, most will not lose a significant amount of weight, and of those who do lose weight, most will regain it."[69] Furthermore, these results are poor not because of failure to implement any simple therapy of known effectiveness, but because no simple or generally effective treatment exists. Obesity is a chronic condition, resistant to treatment and prone to relapse.

The basis of weight reduction is utterly simple—establish a caloric deficit by bringing intake below output. All of the many treatment regimens have this simple task as their goal. The simplest way to reduce caloric intake is by means of a low-calorie diet. The best long-term effects are achieved with a balanced diet that contains readily available foods. For most people, the most satisfactory reducing diet consists of their usual foods in amounts determined by tables of food values available in standard works. Such a diet gives the best chance of long-term maintenance of the weight lost during dieting. But it is precisely the most difficult kind of diet to follow during the period of weight reduction.

Many obese persons find it easier to use a novel or even bizarre diet, of which there has been a profusion in recent years. Whatever effectiveness these diets may have is due, in large part, to monotony—almost everyone gets tired of almost any food if that is all he or she gets to eat. As a consequence, when one ends the diet and returns to the usual fare, the incentives to overeat are multiplied.

Fasting has had considerable vogue as a treatment of obesity in the recent past, but it is now rarely used. Its importance lies primarily in what it has taught us about radical dietary restriction in the treatment of obesity. The great virtue of fasting is the rapid weight loss it engenders, often with relatively little discomfort. After two or three days without food, hunger largely disappears and patients get along well as long as they remain in an undemanding environment. The main problem with fasting as a treatment is the failure to maintain weight loss: Most patients regain most of the weight they have lost.[21] Furthermore, fasting, although surprisingly safe for such a radical procedure, is not without complications and deaths have been reported.

The most important consequence of the experience with therapeutic fasting has been the interest it has aroused in the idea that one can take advantage of the benefits (rapid weight loss) while avoiding the complications. These effects can be achieved by the administration of very small amounts of protein. The so-called "protein-sparing-modified-fast" can effectively maintain nitrogen balance by amounts of protein small enough to have only a negligible effect upon the rapid weight loss. The diet of Genuth and others,[28] for example, contains no more than 320 calories (45 grams of egg albumen and 30 grams of glucose). Vitamin and mineral supplements are necessary. These diets appear to be safe and have the merit of being able to be carried out on an outpatient basis.

The largest series of patients—1,200—treated by protein-sparing-modified-fasts has been reported by Genuth and others.[28] Seventy-five percent lost more than 18 kilograms, and blood pressure was reduced to normal in 67 percent of hypertensives. Few complications were reported: mild hyperuricemia and occasional mild orthostatic hypotension, cold intolerance, and anemia. The major problem of the protein-sparing-modified-fast is that of all conservative treatments for obesity: failure to maintain weight loss. The limited data on long-term follow-up are not encouraging.

An important distinction must be made between the carefully studied protein-sparing-modified-fast and the spate of commercially exploited "liquid protein diets" that have appeared in recent years. The best known of these is Linn's "Last Chance Diet."[43] Generally composed of hydrolysates of cowhide, collagen, and gelatin, these "liquid proteins" are of low biological quality and do not contain an adequate balance of essential amino acids. Not unexpectedly, complications began to be reported soon after these diets were introduced, and already a number of deaths have been reported.[64] "Liquid protein" diets have no place in the treatment of obesity.

Pharmacological treatment of obesity has been greatly altered by recent directives of the Drug Enforcement Administration, which has progressively restricted the use of amphetamines as appetite suppressants. A variety of agents is taking their place: diethylpropion (Tenuate), fenfluramine (Pondimin), and mazindol (Sanorex) are the common examples.[42] The efficacy and side effects of these agents are similar and their potential for abuse is limited.

Pharmacotherapy of obesity is currently out of favor. Nevertheless, its efficacy in weight reduction has been underestimated. In a recent study, weight loss with fenfluramine was increased from thirteen to thirty-two pounds simply by changing the circumstances of its administration from a traditional doctor's office format to a weekly group meeting.[19] Patients regained weight rapidly after medication was stopped. This finding suggests that tolerance to the effects of fenfluramine did not develop and that this is not a reason for restricting its use. It seems likely that pharmacotherapy of obesity has been prematurely written off; new medications and new circumstances of administration, however, may restore its popularity.

An interesting new treatment for bulimia (with or without obesity) has been reported by Wermuth and coworkers.[83] They administered phenytoin (Dilantin) to nineteen obese and nonobese persons who suffered from eating binges at least three times a week. Six patients reported a marked decrease in eating binges during the double-blind trial, and two of the four who continued treatment with phenytoin reported no binges in the next eighteen months. This report of such a simple treatment of a distressing disorder is most promising, and replication is sorely needed.

Thyroid or thyroid analogues are indicated for the occasional obese person with hypothyroidism, but should be discouraged otherwise. Bulk producers may help control the constipation that follows decreased food intake, but their effectiveness in weight reduction is doubtful. Four controlled studies of chorionic gonadotropin have found it to be ineffective.

Increased physical activity is frequently recommended as a part of weight-reduction regimens, but its usefulness has probably been underestimated, even by many of its proponents.[12] Since caloric expenditure in most forms of physical activity is directly proportional to body weight, obese persons expend more calories with the same amount of activity than do people of normal weight. Furthermore, increased physical activity may actually cause a *decrease* in the food intake of sedentary persons. This combination of increased caloric expenditure with probable decreased food intake makes an increase in physical activity a highly desirable feature of any weight-reduction program.

Group methods, propagated by the burgeoning self-help movement, are being used by increasing numbers of obese people.[67] The two largest organizations are the nonprofit Take Off Pounds Sensibly (TOPS)[71] with over 300,000 members, and the profit-making Weight Watchers, which is even larger. Costs for attending meetings are small, and many people report that the group support and frequent weighings are quite helpful. Objective assessment of these organizations, however, has lagged. A recent report makes it clear that drop out rates are very high. This problem is partly compensated for by high reentry rates, but we know very little about the weight losses of representative samples of participants. These organizations are unique in the weight-reduction field in providing economical and readily accessible assistance, in making few demands upon the participants, and in permitting them to leave and rejoin without penalties. As such they should have an important place in an overall approach to obesity.

Surgical treatments for obesity are relatively new but are highly significant for that small fraction of people who suffer from "morbid" obesity, that is, 100 percent over ideal weight. Four factors have made surgery the treatment of choice for many such people: (1) recent demonstration of the severity of the various physical complications often with profound psychosocial disability, resulting in a twelvefold increase in mortality among younger persons,[22] (2) the inefficacy of conventional treatments; (3) the continuing development of newer surgical measures; and (4) many health benefits at acceptable levels of risk.

Although the prevalence of morbid obesity is very low—less than 1 percent—over half a million Americans suffer from this condition and many of them seek psychiatric help at some time in their life.[22] It is, therefore, worth describing the newer surgical procedures and some of their surprisingly favorable behavioral sequelae. Two operations currently dominate the treatment of morbid obesity—jejunoileal bypass and gastric bypass.[24,46] In the jejunoileal bypass operation, fourteen inches of proximal jejunum is anastomized to four inches of terminal ileum, bypassing the remaining bowel and radically reducing the absorptive surface. In the gastric bypass operation, a stomach pouch of 50 ml capacity is constructed with a 1.2 cm outlet to the proximal jejunum, radically reducing the amount of food that can be consumed at one time. The original rationale for the jejunoileal bypass was to decrease intestinal absorption of nutrients. Although malabsorption occurs, most of the weight loss following jejunoileal bypass (and all of the weight loss following gastric bypass) is due to voluntary restriction of food intake. Weight loss following both procedures occurs at a decelerating rate for twelve to eighteen months, during which time at least 50 percent of excess weight is lost, although there is, of course, considerable individual variability. This weight loss is accompanied by a number of significant benefits; among them, relief of the Pickwickian syndrome, reduction of elevated blood pressure and blood glucose to normal levels among most hypertensives and diabetics, and correction of a wide variety of the mechanical ill effects of excessive weight. One of the most gratifying results is marked amelioration of the psychosocial disabilities that afflict most morbidly obese persons. Against these benefits must be considered the risks of bypass surgery. Mortality in the operative and postoperative period is below 3 percent for both procedures, if they are carried out as they should be by skilled interdisciplinary teams able to provide continuing supervision. Postoperative results, however, favor gastric bypass. Jejunoileal bypass is often followed by severe diarrhea, fluid and electrolyte disturbances over the short-term, and liver disease that is fatal in 2 percent of patients over the longer term. Hyperoxaluria leads to nephrolithiasis in 10 percent of patients and to serious focal nephritis induced by oxalate crystals in an undetermined number of other patients. Complications of jejunoileal bypass do not seem to decrease over the years and the long-term adverse effects have been serious enough to cause surgeons to turn increasingly to gastric bypass. The only common complications of this surgery are epigastric distress and vomiting; these are readily controlled as the patient learns new eating habits.

Some unexpected behavioral consequences of both operations are of interest for both theoretical and practical reasons. Five studies have reported that bypass surgery is followed by unusually benign behavioral consequences, but four of these studies have understated the benefits of this surgery. The latter studies used what was probably an inappropriate control period. For comparison with the emotional status after the surgery period, they used as a control the time just before surgery. A more appropriate control period would seem to be the time when the patient was attempting to lose weight (usually with far less success) without surgery. During such times a majority of these patients had experienced depression, anxiety, and a variety of depressive affects. By contrast, during the weight loss that follows surgery patients rarely experience such affects and, on the contrary, usually report enhanced feelings of well-being.[32,52] Furthermore, the restriction of food intake, which plays the largest part in the weight loss, is achieved without particular effort and is accompanied by a striking normalization of eating patterns. There is a marked decrease in snacking, nighteating, and binge eating; the ability to stop eating is also enhanced. Even more surprising is the effect upon the large percentage of persons who until then did not eat breakfast: during a period of rapidly falling weight, most of these people began to eat breakfast.

These phenomena are striking enough to suggest that far from merely altering the mechanics of the bowel, bypass surgery brings about a major change in the biology of the organism. It has been proposed that this surgery has the effect of lowering the set point at which the body weight of obese people may be regulated.[72] According to this view, the lack of dysphoric reactions to weight loss and the normalization of eating patterns result from the body adjusting to this new, lower set point, since it no longer has to struggle to reduce against the pressures of a higher set point.

Specialized Psychotherapeutic Techniques

Information about reducing diets is so widely available that only those who have already failed to lose weight on their own come to the doctor's office. And only those who have failed with medical treatment seek out the psychiatrist. This process of selection makes it understandable why there has not been, until recently, a systematic study of the effects of psychoanalysis upon obesity and why such an approach has fallen from favor in treating obesity. No more than 6 percent of persons entering psychoanalysis do so for treatment of their obesity, and analysts themselves have been skeptical of their ability to deal with this problem.

PSYCHOANALYSIS

A recent study by Rand and Stunkard[55] suggests that a more optimistic view of the influence of psychoanalysis upon obesity may be justified. The weight losses of a sample of eighty-four obese men and women treated by seventy-two psychoanalysts compared favorably with those achieved by other conservative methods. Thus, mean weight loss was 9.5 kg during treatment that averaged forty-two months in duration. Furthermore, 47 percent of patients lost more than 9 kg, and 19 percent lost more than 19 kg during this time. Analysts reported a striking decrease in severe body-image disparagement in their patients. At the beginning of treatment 40 percent reported severe disparagement; at its termination this figure had fallen to 14 percent. Obese patients were generally more difficult to treat than nonobese patients. For example, more obese than nonobese patients terminated treatment prematurely, and those who remained in treatment showed less improvement in psychological functioning than did nonobese patients.

This study may reawaken interest in psychoanalytic psychotherapy of obese persons. Some general observations are in order. First, there is no evidence that uncovering putative unconscious causes of overeating can alter the symptom choice of obese people who overeat in response to stress. Years after successful psychotherapy and successful weight reduction, persons who overate under stress continue to do so. Second, many obese people seem inordinately vulnerable to the overdependency on the therapist and to the severe regression that can occur in psychoanalytic therapy. Bruch[15] has provided excellent descriptions of measures designed to minimize such regression, to cope with the "conceptual confusion" described earlier, and to increase the patient's often seriously inadequate sense of personal effectiveness.

Although psychoanalysis and psychoanalytic therapy are very expensive ways to lose weight, they may be indicated for persons suffering from severe disparagement of the body image. This condition has not been influenced by other forms of treatment, even those which effect weight reduction. Psychoanalysis and psychoanalytic therapy may also be indicated for treatment of bulimia, another particularly resistant condition. Furthermore, obese people may seek psychotherapy for reasons other than their obesity; helping them to cope with their obesity may help them to resolve other problems. We have noted that many obese people overeat under stress. If psychotherapy helps them to live less stressful and more satisfying lives, then they are less likely to overeat. As a result, they may reduce and stay reduced. These benefits are not less significant for being nonspecific results of treatment.

BEHAVIOR THERAPY

Behavior therapy was introduced into the treatment of obesity a decade ago, and within five years the topic had achieved a popularity bordering on faddism. The next five years, however, saw the appearance of over fifty controlled clinical trials, and it is now possible to ascertain what has and has not been accomplished by this vast expenditure of effort.

It is clear that behavior therapy represents an improvement over traditional outpatient treatments for mild and moderate obesity, and it is the treatment of choice for these conditions. It is also clear that its great early promise has been only imperfectly realized.[74] There is consensus on six issues:

1. Dropouts from outpatient treatment have been greatly reduced, from figures as high as 25 to 75 percent to not more than 10 percent.

2. Emotional complications of behavioral weight-reduction regimens are uncommon, in contrast to rates as high as 50 percent

among persons in traditional outpatient regimens.

3. Weight losses, although greater than those achieved by alternate treatments in controlled clinical trials, have been modest and of limited clinical significance.[77] There are many reasons for these limitations: Most of the programs were short-term, many involved patients who were only mildly overweight, and a large number were conducted by inexperienced therapists. Nevertheless, mean weight loss exceeded fifteen pounds in only a minority of clinical trials.

4. There is great variability in weight losses and still no way to predict which patients will do well and which ones will not.

5. Weight lost during behavioral treatment tends to be regained. However, weight losses are probably better maintained than they are with other forms of treatment.

6. The most important aspect of behavior therapy may be the fact that its procedures can be so clearly specified and so readily taught. Detailed instructions for the conduct of behavioral weight-reduction programs for small groups and for individuals have been provided by Ferguson,[25] Jordan and colleagues,[38] the Mahoneys,[44] and Stuart.[66] Brownell's recent *Partnership Diet Program*[10] describes an original self-help program for couples.

There is a growing trend for the behavior therapy of obesity to be carried out by persons with less and less formal training, and these services are increasingly delivered by nonprofessionals. The most ambitious of these efforts is that of Weight Watchers, which enrolls 400,000 persons a week in classes taught entirely by lay persons.[67] The program is outlined in a series of "modules" that contain a brief written summary of the behavioral tasks to be accomplished during the next two weeks, along with forms for recording progress in this endeavor. Within the medical profession the lead in behavioral treatment has been taken by the Society of Bariatric Physicians, many of whose members carry out behavioral treatment programs, often with the aid of their office nurses.

¶ Obesity in Childhood

The obesity of persons who were obese in childhood—the so-called "juvenile-onset obese"—differs from that of persons who became obese as adults. Juvenile-onset obesity tends to be more severe, more resistant to treatment, and more likely to be associated with emotional disturbances.

Obesity that begins in childhood shows a very strong tendency to persist. Long-term prospective studies in Hagerstown, Maryland, have revealed the remarkable degree to which obese children become obese adults. In the first such study,[1] 86 percent of a group of overweight boys became overweight men, as compared to only 42 percent of boys of average weight. Even more striking differences in adult weight status were gound among girls: 80 percent of overweight girls became overweight women, as compared to only 18 percent of average-weight girls. A later study[75] showed that the few overweight children who reduced successfully had done so by the end of adolescence. The odds against an overweight child becoming a normal-weight adult, which were 4:1 at age twelve, rose to 28:1 for those who did not reduce during adolescence. An even more recent study,[2] which used a longer interval (thirty-five years) and, unfortunately, different (more rigid) criteria for obesity, found the difference in adult weight status continuing to grow: 63 percent of obese boys became obese men, as compared to only 10 percent of average-weight boys.

It is widely believed that obese children are very inactive and that their inactivity plays a major part in the development and maintenance of their obesity. Recent research suggests that excessive food intake is a far more important factor. At least four studies[76] that used objective measures of physical activity failed to reveal significant differences between obese and nonobese children, while only one study reported that obese children were less active.[82]

A recent intensive study of energy intake and expenditure suggests that excessive food intake and not decreased physical activity maintains childhood obesity and may even produce it. Waxman and Stunkard[76] directly measured food intake and energy expenditure in four families at meals and at play in three different settings. In each family there was one obese boy and a nonobese brother whose ages were within two years of each other. The subjects' oxygen consumption at four levels of activity was measured in the laboratory to permit calculation of energy expenditure from time-sampled measures of observed activity. The study showed that the obese boys consumed far more calories than did their nonobese brothers. Furthermore, their levels of physical activity did not differ greatly from those of the nonobese boys. When measures of activity were converted into calories, it was found that the obese boys actually had higher levels of energy expenditure than did their nonobese controls. These findings need confirmation with larger samples and with studies of girls. Nevertheless, the evidence that obese boys overeat is so striking that it justifies directing treatment at this problem.

Reduction of food intake has been the major focus of behavioral treatments of obesity in children. Although behavioral treatment of childhood obesity has lagged behind the application of comparable methods to obese adults, it is attracting increasing attention. Nine studies have already been reported and more are currently underway.[14] The results to date warrant a cautious optimism; behavior therapy of children may well prove as effective with children as it has been with adults.

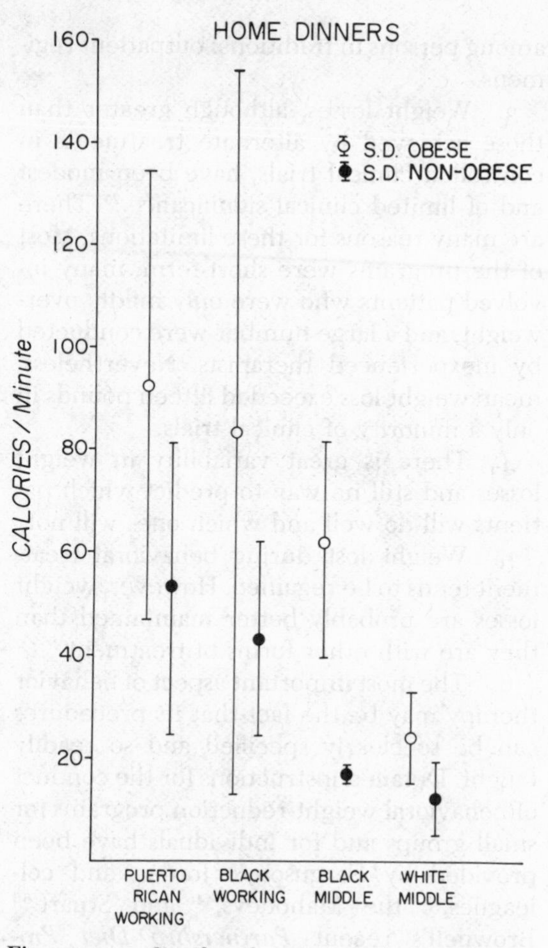

Figure 23-4. Significantly greater calorie intake by obese boys at home dinners compared with that of their nonobese brothers. (From Waxman, M, and Stunkard, A. J. By permission of the *Journal of Pediatrics,* 96 (1980):187-193.)

¶ Conclusion

Obesity, a condition characterized by excessive accumulations of body fat, is widely distributed within the population, affecting one-third of adult Americans. It has recently been viewed, not as the result of a disturbance in the regulation of body weight, but rather as the result of an elevation of the set point at about which body weight is regulated. At least six factors may affect this regulation: genetic and developmental, social and emotional, physical (inactivity) and neural (impaired brain function). The relative importance of these different factors probably varies among different obese persons. Recent studies[34] suggest that the behavior of many obese persons is affected by their efforts to restrain their natural inclinations to eat in order to maintain socially approved levels of body weight.

Obesity adversely affects morbidity and mortality, and the rates increase in direct proportion to the severity of the obesity. Although obesity is not a strong independent risk factor for cardiovascular disease, it is a major determinant of hypertension and insulin-independent diabetes, and both these conditions are markedly improved by weight reduction. Partly for health benefits and partly for cosmetic reasons, large numbers of obese people try to lose weight. Most are unsuccessful. The poor results of weight-reduction efforts are not due to failure to implement any therapy of known effectiveness but to the fact that no simple or generally effective therapy exists. Obesity is a chronic condition, resistant to treatment and prone to relapse.

The unfavorable therapeutic outlook for obesity has been brightening in recent years with the development of new treatments and renewed interest in old ones. New pharmacological treatments together with new circumstances of administration show promise of improving the results of pharmacotherapy; phenytoin (Dilantin®) has been found to be effective in the management of some cases of bulimia. Jejunoileal bypass surgery and, more recently, gastric bypass surgery have begun to bring hope to some morbidly obese persons. Psychiatric studies have shown favorable emotional concomitants of the large weight losses achieved by surgical means.[32,52] A recent study[55] indicates that psychoanalysis may have unsuspected merits in reducing body weight as well as in lessening body image disparagement of some obese persons. A very large effort has been expended in the development of behavioral therapies for obesity, making these measures the treatment of choice for mild and perhaps also moderate obesity, as well as providing a useful model for psychotherapy research. Furthermore, the ease with which behavioral measures can be taught has resulted in the increasing delivery of weight-reduction services by nonprofessionals. Lay groups are providing valuable vehicles for the introduction of behavior therapy to large numbers of people. But the main hope for the control of obesity lies in a better understanding of the factors that regulate body weight. Fortunately research into this problem is proving increasingly fruitful.

¶ Bibliography

1. ABRAHAM, S., and NORDSIECK, M. "Relationship of Excess Weight in Children and Adults," *Public Health Reports,* 75 (1960): 263–273.
2. ABRAHAM, S., COLLINS, G., and NORDSIECK, M. "Relationship of Childhood Weight Status to Morbidity in Adults," *HSMHA Health Reports,* 86 (1971): 273–284.
3. BIERMAN, E. L., BAGDADE, J. D., and PORTE, D., Jr. "Obesity and Diabetes: The Odd Couple," *American Journal of Clinical Nutrition,* 21 (1968): 1434–1437.
4. BIRON, P., MONGEAU, J. G., and BERTRAND, D. "Familial Resemblance of Body Weight and Weight/Height in 374 Homes with Adopted Children," *Journal of Pediatrics,* 91 (1977): 555–558.
5. BJÖRNTORP, P., et al. "Effect of an Energy-Reduced Dietary Regimen in Relation to Adipose Tissue Cellularity in Obese Women," *American Journal of Clinical Nutrition,* 28 (1975): 445–452.
6. BOOTH, D. A. "Satiety and Appetite are Conditioned Reactions," *Psychosomatic Medicine,* 39 (1977): 76–81.
7. ———. "Acquired Behavior Controlling Energy Input and Output," in Stunkard, A. J., ed., *Obesity.* Philadelphia: W. B. Saunders Co., 1980, pp. 101–143.
8. ———, ed. *Hunger Models: Computable Theory of Feeding Control.* New York: Academic Press, 1978.
9. BRAY, G. *The Obese Patient.* Philadelphia: W. B. Saunders Co., 1976.
10. BROBECK, J. "Neural Control of Hunger, Appetite and Satiety," *Yale Journal of Biological Medicine,* 29 (1957): 565–574.
11. BROWNELL, K. D. *The Partnership Diet Program.* New York: Rawson-Wade Publishers, 1980.
12. ———, and STUNKARD, A. J. "Physical Activity in the Control of Obesity," in Stunkard, A. J., ed., *Obesity.* Philadelphia: W. B. Saunders Co., 1980, pp. 300–324.
13. ———. "Behavioral Treatment for Obese

Children and Adolescents," Stunkard, A. J., ed., *Obesity.* Philadelphia: W. B. Saunders Co., 1980 pp. 415–437.

14. BRUCH, H. "Conceptual Confusion in Eating Disorders," *Journal of Nervous and Mental Disease,* 133 (1961): 46–54.
15. ———. *Eating Disorders: Obesity, Anorexia Nervosa, and the Person Within.* New York: Basic Books, 1973.
16. BURWELL, S. "Extreme Obesity Associated with Alveolar Hypoventilation—A Pickwickian Syndrome," *American Journal of Medicine,* 21 (1956): 811–818.
17. CHIANG, B. N., PERLMAN, L. V., and EPSTEIN, F. H. "Overweight and Hypertension: A Review," *Circulation,* 39 (1969): 403–421.
18. CHIRICO, A. M., and STUNKARD, A. J. "Physical Activity and Human Obesity," *New England Journal of Medicine,* 263 (1960): 935–946.
19. CRAIGHEAD, L. W., STUNKARD, A. J., and O'BRIEN, R. M. "Behavior Therapy and Pharmacotherapy of Obesity," *Archives of General Psychiatry,* in press.
20. CRISP, A. H., and MCGUINESS, B. "Jolly Fat: Relation Between Obesity and Psychoneurosis in General Population," *British Medical Journal,* 1 (1975): 7–9.
21. DRENICK, E. J., and JOHNSON, D. "Weight Reduction by Fasting and Semi-starvation in Morbid Obesity: Long-term Follow-up," *International Journal of Obesity,* 2 (1978): 123–132.
22. DRENICK, E. J., et al. "Excessive Mortality and Causes of Death in Morbidly Obese Men," *Journal of American Medical Association,* 243 (1980): 443–445.
23. DYER, A. R., et al. "Relationship of Relative Weight and Body Mass Index to 14-year Mortality in the Chicago Peoples Gas Company Study," *Journal of Chronic Diseases,* 28 (1975): 109–123.
24. FALLOON, W. W., ed. "Symposium on Jejunoileostomy for Obesity," *American Journal of Clinical Nutrition,* 30 (1977): 1–129.
25. FERGUSON, J. M. *Learning to Eat: Behavior Modification for Weight Control.* Palo Alto, California: Bull Publishing Co., 1975.
26. FOCH, T. T., and MCLEARN, G. E. "Genetics, Body Weight and Obesity," in Stunkard, A. J., ed., *Obesity.* Philadelphia: W. B. Saunders Co., 1980, pp. 48–71.
27. GARN, S. M., and CLARK, D. C. "Trends in Fatness and the Origins of Obesity," *Pediatrics,* 57 (1976): 443–456.
28. GENUTH, S. M., VERTES, V., and HAZLETON, J. "Supplemented Fasting in the Treatment of Obesity," in BRAY, G., ed., *Recent Advances in Obesity Research II,* London: Newman, 1978, pp. 370–378.
29. GOLDBLATT, P. B., MOORE, M. E., and STUNKARD, A. J. "Social Factors in Obesity," *Journal of American Medical Association,* 192 (1965): 1039–1044.
30. GORDON, T., and KANNEL, W. B. "The Effects of Overweight on Cardiovascular Disease," *Geriatrics,* 28 (1973): 80–88.
31. GRIGGS, R. C., and STUNKARD, A. J. "The Interpretation of Gastric Motility, II. Sensitivity and Bias in the Perception of Gastric Motility," *Archives of General Psychiatry,* 11 (1964): 82–89.
32. HALMI, K. A., STUNKARD, A. J., and MASON, E. E. "Emotional Responses to Weight Reduction by Three Methods: Diet, Jejunoileal Bypass, and Gastric Bypass," *American Journal of Clinical Nutrition,* 33 (1980): 446–451.
33. HALMI, K. A., et al. "Psychiatric Diagnosis of Morbidly Obese Gastric Bypass Patients," *American Journal of Psychiatry,* 137 (1980): 470–472.
34. HERMAN, C. P. and POLIVY, J. "Restrained Eating," in Stunkard, A. J., ed., *Obesity,* Philadelphia: W. B. Saunders Co., 1980, pp. 208–225.
35. HOEBEL, B. "Neural Control of Food Intake," *Annual Review of Physiology,* 33 (1971): 533–568.
36. JOHNSON, P. R., and HIRSCH, J. "Cellularity of Adipose Depots in Six Strains of Genetically Obese Mice," *Journal of Lipid Research,* 13 (1972): 2–11.
37. JORDAN, H. A. "Voluntary Intragastric Feeding: Oral and Gastric Contributions to Food Intake and Hunger in Man," *Journal of Comparative Physiology and Psychology,* 68 (1969): 498–506.
38. ———, LEVITZ, L. S., and KIMBRELL, G. M. *Eating Is Okay.* New York: Rawson Associates, 1976.
39. KEESEY, R. E. "A set Point Analysis of the Regulation of Body Weight," in Stunkard, A. J., ed., *Obesity.* Philadelphia: W. B. Saunders Co., 1980, pp. 144–165.
40. KENNEDY, G. C. "The Role of Depot Fat in

the Hypothalamic Control of Food Intake in the Rat," *Proceedings of the Royal Society,* 140 (1953): 578–592.

41. KEYS, A., et al. *The Biology of Human Starvation,* 2 vols. Minneapolis: University of Minnesota Press, 1950.
42. LASAGNA, L. "Drugs in the Treatment of Obesity," in Stunkard, A. J., ed., *Obesity.* Philadelphia: W. B. Saunders Co., 1980 pp. 292–299.
43. LINN, K. *The Last Chance Diet.* New York: Bantam Books, 1976.
44. MAHONEY, M. J., and MAHONEY, K. *Permanent Weight Control: A Total Solution to the Dieter's Dilemma.* New York: Norton, 1976.
45. MANN, G. V. "The Influence of Obesity on Health," *New England Journal of Medicine,* 291 (1974): 178–185.
46. MASON, E. E., et al. "Gastric Bypass for Obesity After Ten Years Experience," *International Journal of Obesity,* 2 (1978): 197–206.
47. MAYER, J. "Decreased Activity and Energy Balance in the Hereditary Obesity-Diabetes Syndrome of Mice," *Science* 117 (1953): 504–505.
48. ———. "Some Aspects of the Problem of Regulation of Food Intake and Obesity," *New England Journal of Medicine,* 274 (1966): 610–616, 662–673, 722–731.
49. ———, and THOMAS, D. W. "Regulation of Food Intake and Obesity," *Science,* 156 (1967): 328–337.
50. MELLINKOFF, S. M., et al. "Relationship Between Serum Amino Acid Concentration and Fluctuations on Appetite," *Journal of Applied Physiology,* 8 (1956): 535–538.
51. MILLER, N. E., BAILEY, C. J., and STEVENSON, J. A. F. "Decreased Hunger and Increased Food Intake in Hypothalamic Obese Rats," *Science,* 112 (1950): 256–259.
52. MILLS, M. J., and STUNKARD, A. J. "Behavioral Changes Following Surgery for Obesity," *American Journal of Psychiatry,* 133 (1976): 527–531.
53. MOORE, M. E., STUNKARD, A. J., and SROLE, L. "Obesity, Social Class and Mental Illness," *Journal of American Medical Association,* 181 (1962): 962–966.
54. NISBETT, R. E. "Hunger, Obesity and the Ventromedial Hypothalamus," *Psychological Review,* 79 (1972): 433–453.
55. RAND, C. S. W., and STUNKARD, A. J. "Obesity and Psychoanalysis," *American Journal of Psychiatry,* 135 (1978): 547–551.
56. RODIN, J. "The Externality Theory Today," in Stunkard, A. J., ed., *Obesity.* Philadelphia: W. B. Saunders Co., 1980, pp. 226–239.
57. SALANS, L. B., HORTON, E. S., and SIMS, E. A. H. "Experimental Obesity in Man: The Cellular Character of the Adipose Tissue," *Journal of Clinical Investigation,* 50 (1971): 1005–1011.
58. SALANS, L. B., KNITTLE, J. L., and HIRSCH, J. "The Role of Adipose Cell Size and Adipose Tissue Insulin Sensitivity in the Carbohydrate Intolerance of Human Obesity," *Journal of Clinical Investigation,* 47 (1968): 153–165.
59. SCHACHTER, S. "Some Extraordinary Facts about Obese Humans," *American Psychologist,* 26 (1971): 129–144.
60. SCROLE, L., et al. *Mental Health in the Metropolis: The Midtown Manhattan Study,* New York: McGraw-Hill, 1962.
61. SELTZER, C. C., and MAYER, J. "A Simple Criterion of Obesity," *Postgraduate Medicine,* 38 (1965): 101–107.
62. SIMS, E. A. H. "Experimental Obesity, Diet-induced Thermogenesis and Their Clinical Implications," *Clinics in Endocrinology & Metabolism,* 5 (1976): 377–395.
63. SJÖSTRÖM, L. "Fat Cells and Body Weight," in Stunkard, A. J., ed., *Obesity.* Philadelphia: W. B. Saunders Co., 1980, pp. 72–100.
64. SORUS, H. E., et al. "Sudden Death Associated with Protein-Supplemented Weight Reduction Regimens," *American Journal of Clinical Nutrition,* in press.
65. STERN, J. S., and JOHNSON, P. R. "Size and Number of Adipocytes and Their Implications," in Katzer, H., and Mahler, R., eds., *Advances in Modern Nutrition,* vol II. New York: John Wiley, 1978, pp. 303–340.
66. STUART, R. S. *Act Thin, Stay Thin.* New York: Norton, 1978.
67. ———. "How to Use Self-help Groups," in Stunkard, A. J., ed., *Obesity.* Philadelphia: W. B. Saunders Co., forthcoming.
68. STUNKARD, A. J. "Eating Patterns and Obesity," *Psychiatric Quarterly,* 33 (1959): 284–294.
69. ———. "From Explanation to Action in Psychosomatic Medicine: The Case of Obesity," *Psychosomatic Medicine,* 37 (1975): 195–236.

70. ———. "Satiety Is a Conditioned Reflex," *Psychosomatic Medicine,* 37 (1975): 383–387.

71. ———. "Studies on TOPS: A Self-help Group for Obesity," in Bray, G., ed., *Obesity in Perspective.* Rockville, Md.: National Institute of Health, 1976, pp. 75–708, 387–391.

72. ———, ed. *Obesity.* Philadelphia: W. B. Saunders Co., 1980.

73. ———. "Eating, Affects and the Regulation of Body Weight," in Weiner, H., Hofer, M., and Stunkard, A. J., eds., *Brain, Behavior and Bodily Disease.* New York: Raven Press, forthcoming.

74. ———. "Obesity," in Hersen, M., Bellack, A. S., and Kazdin, A. E., eds., *International Handbook of Behavior Modification and Therapy.* New York: Plenum Press, forthcoming.

75. ———, and BURT, V. "Obesity and the Body Image II. Age at Onset of Disturbances in the Body Image," *American Journal of Psychiatry,* 123 (1967): 1443–1447.

76. STUNKARD, A. J., and MENDELSON, M. "Obesity and the Body Image I. Characteristics of Disturbance in the Body Image of Some Obese Persons," *American Journal of Psychiatry,* 123 (1967): 1296–1300.

77. STUNKARD, A. J., and PENICK, S. B. "Behavior Modification in the Treatment of Obesity: The Problem of Maintaining Weight Loss," *Archives of General Psychiatry,* 36 (1979): 801–806.

78. STUNKARD, A. J., and RUSH, A. J. "Dieting and Depression Reexamined: A Critical Review of Reports of Untoward Responses During Weight Reduction for Obesity," *Annals of Internal Medicine,* 81 (1974): 526–533.

79. STUNKARD, A. J., et al. "The Influence of Social Class on Obesity and Thinness in Children," *Journal of American Medical Association* 221 (1972): 579–584.

80. VAN ITALLIE, T. B. "Obesity: Prevalence and Pathogenesis," U.S. Congress, Senate, *Congressional Record,* 95th Cong., 1st sess., 1 and 2 February, 1977, pt. 2, pp. 44–64.

81. VOLKMAR, F. R., et al. "High Attrition Rates in Commercial Weight Reduction Programs," *Archives of International Medicine,* in press.

82. WAXMAN, M., and STUNKARD, A. J. "Caloric Intake and Expenditure of Obese Boys," *Journal of Pediatrics,* 96 (1980): 187–193.

83. WERMUTH, B. M., et al. "Phenytoin Treatment of the Binge-eating Syndrome," *American Journal of Psychiatry,* 134 (1977): 1249–1254.

84. WITHERS, R. F. J. "Problems in the Genetics of Human Obesity," *Eugenics Review,* 56 (1964): 81–90.

CHAPTER 24

SOCIAL MALADJUSTMENTS

Thomas Kreilkamp

SOCIAL MALADJUSTMENT was one of several specific diagnostic categories under the general heading of "Conditions without Manifest Psychiatric Disorder" in the *Diagnostic and Statistical Manual of Mental Disorders* (DSM-II).[1] Some of the other categories under this heading were occupational and marital maladjustment. The mere existence of this heading is already indication of the difficulties faced by anyone who wants to provide a diagnostic manual that will be of use to practitioners. Clinicians see a wide variety of people, for a wide variety of difficulties, some of which are not psychiatric in any simple sense. This occurs for many complicated reasons, but two important ones include (1) the role of the psychiatrist in our society, which is vaguely defined and which permits attempts to treat almost any manner of problem in living; and (2) the ever-broadening notions of what constitutes mental illness, which in turn are based on that form of psychiatric theory that points up the continuity between illness and health, or that, as in some forms of Freudian theory, even erodes the distinction between illness and health altogether.

"Social maladjustment" does not appear in the same form in the new edition of the *Diagnostic and Statistical Manual of Mental Disorders* (DSM-III).[2] There is instead a section called "Codes for Conditions Not Attributable to a Known Mental Disorder that Are a Focus of Attention or Treatment." This includes, for example, malingering, childhood or adolescent antisocial behavior, marital problems, parent-child problems, and a residual category of "other interpersonal" problems. In addition, there is a section called "Adjustment Disorders." In describing this section, the new manual states that "The essential feature is a maladaptive reaction to an identifiable psychosocial stressor . . . It is assumed that the disturbance will eventually remit after the stressor ceases. . . ."[56] This attempt at adumbrating the essential features of the "Adjustment Disorders" condition is a good place to begin consideration of social maladjustment.

First of all, it is clear that the emphasis here is on a circumscribed maladaptive reaction. That is, it is not a chronic condition that is being talked about, but instead importance of the life event that is assumed to precipitate the maladaptive reaction. DSM-III[2] goes on to make explicit that this category of adjustment disorder must be distinguished both from "normal" adjustment reactions (such as "Simple Bereavement") and from more chronic life problems (such as would be included under "Conditions Not Attributable to a Known Mental Disorder"). Furthermore, the draft goes on to make explicit that yet other diagnostic categories must take precedence, such as "identity disorder" and "emancipation disorder," both of which would presumably refer to difficulties typical of those adolescents who are attempting to solidify a sense of themselves as separate individuals, and who are engaged in moving away from the family of origin toward some more independent status in the world.

Second, we can see that this category of "adjustment disorders" is not meant to include conditions that can reasonably be seen as exacerbations of already existing mental disorders. This sounds perhaps an easier discrimination than in fact it is. Sorting out various maladjusted reactions into different groups based on whether those manifesting them were or were not previously afflicted with a mental disorder is always a hazardous enterprise, since it depends both on one's theoretical stance with regard to all psychiatric disorders and on an accurate knowledge of the past, which is not always obtainable. A discussion of this difficulty will help to explore the implications of the preliminary definitions of adjustment disorders.

DSM-III[2] is overly inclusive on purpose, according to those who have created it, and the emphasis is not on either etiology or treatment, but rather on descriptive completeness. Regardless of its attempt to eschew etiology for description, any psychiatric practitioner, in order to be able to operate effectively, must have some kind of theoretical orientation toward psychiatric disorders, though it need not be one that specifies etiology for any or all diseases. An example of a general orientation common in psychiatry is that which sees mental disorders as characterizing individuals. The individual is the site of the "disease" and the field within which it makes itself known. However, psychiatrists must of necessity acquaint themselves with the larger social field within which the afflicted individual lives, in order to determine whether or not, for example, a depression is reactive or chronic. If the psychiatrist finds that the patient's spouse just died, then the diagnosis is more likely to be simple bereavement than depression. Similarly, in all psychiatric diagnostic endeavors, one needs to find out what has been happening in the patient's life in the recent and sometimes in the more remote past.

This necessity is especially acute with regard to the category of social maladjustment. In order even to consider such a diagnosis, one must know what the patient might be reacting to, and what social world he might be having difficulty coming into satisfying relation with; this necessity immediately broadens the horizons of the psychiatrist. That is, the psychiatrist is no longer simply examining a patient, but is assessing the nature of the fit between an individual and a social realm with which the psychiatrist may or may not be familiar. The less familiar he is with that social world, the harder it will be, of course, to assess the nature of the difficulty the patient is having in coming to terms with that world. Here the psychiatrist will want to consider the nature of the patient's coping mechanisms in general: the way in which past difficulties have been coped with and the variations, if any, in the nature of the patient's social world (what changes have occurred, such as moves, new jobs, additions to or subtractions from the family). The need to consider all these matters makes the psychiatrist's task a very broad one, and this breadth is sometimes overwhelming. Take for example the so-called "mental status" exam. An accurate carrying-out of this exam requires, minimally, some cooperation from the patient. Some patients do not cooperate, or do so in such a way as to cast doubt on the validity and/or reliability of their answers. That is, they may neither be saying what they think nor answering the same way today they might tomorrow, and this lack of cooperation may or may not be part of a disease process.

Ascertaining this probability requires immense skill, and considerable familiarity with the patient being examined and with the circumstances that have brought him to the psychiatrist. An angry adolescent brought in by an equally angry parent is less likely to offer cooperation to the psychiatrist (who is perceived as the parent's ally) than a self-referred adult who is bothered by some intrapsychic stress and wants some relief.

In addition to considering such matters in doing a mental status exam, the psychiatrist will need enough flexibility and scope to be able to recognize the possibility that there may be a suspected mismatch between a given individual, who happens to be presenting as a patient, and a given social world. He will then have to consider whether that person is better off not trying to adjust to the environment. This would be the case if adjustment to the situation is not the best outcome in the long run. The psychiatrist might make this assessment if the patient appears to be only temporarily in his present situation. For example, a patient who finds he cannot keep up with his workload may in fact be overworked by standards other than those of his present situation. He might be best off considering the possibility of changing jobs and thus reducing his workload, rather than trying to transform his character to the degree necessary to find that heavy workload tolerable. Or in another related example, there may be no good solution to a person who presents severe anxiety that appears to arise out of a conflict between his work and family demands. The problem here may not be so much social maladjustment—an inability to deal with the kinds of conflicts and pressures that other people appear to deal with—but rather a culturally determined conflict between changing definitions of what a man's role ought to include. Formerly, he would have been expected to earn a living and make progress in a job or career; now he is expected to do that in addition to spending more time with his children and helping at home. Here there is indeed social maladjustment, but it is not between an individual and society but between various facets of a complex and changing social environment.

Another case in which the psychiatrist may want to consider the possibility that "adjustment" between an individual and his environment is not possible, and where the problem is not entirely that of the patient, is when the patient is a child whose family is not able to take care of him. In such a case, the psychiatrist may want to explore the possibility of finding another home for the child, perhaps temporarily with a relative or friend or perhaps on a more permanent basis.

The diagnostic category of social maladjustment, then, demands of the psychiatist not only a familiarity with conventional diagnostic nosology, with psychopathology as it is conventionally construed, but an awareness of social developments as well. In some cases the psychiatrist will want to recognize the degree of stress in the patient, but instead of recommending therapy for the patient, may want to recommend therapy for a larger social system (a marital couple or a family) or may encourage the patient to think about rearranging his social world so as to change either one or another aspect of it, or to alter the degree of his involvement in a social sector that is coming into conflict with another important social sector.

¶ Coping and Adaptation

Maladjustment is clearly in part a function of the complexity and possibly conflictual nature of one's social world; it is also, however, related to one's own means of coping with stress. When one examines the intrapsychic (and interpsychic) measures that people develop for coping with stress, one can begin to learn something about how people ordinarily manage the sorts of conflicts and crises that are thought to be more or less routine. Ordinary management of stress involves not just intrapsychic maneuvers that are conventionally referred to as defense mechanisms, but other skills as well. As Engel[8] has pointed out,

"In recent years there has been a marked shift in the study of human adaptation from concern with intrapsychic defense mechanisms to much greater emphasis on the skills and supports required to meet typical life challenges." Traditionally this has not been studied as intensively as neurosis or difficulties in living, but there are several studies worth noting. In Robert White's *The Study of Lives,*[32] there is a section by Theodore Kroeber called "Coping Functions of the Ego Mechanisms" that attempts to differentiate between the defensive (or unhealthy) and the coping (or healthy) aspects of various ego functions. For example, he says of the ego function of discrimination (the ability to separate one idea from another, or a feeling from an idea) that this has a healthy form when it involves objectivity, and an unhealthy form when it involves isolation. Both objectivity and isolation involve separation of feeling from idea, but the former is perhaps more voluntary, more flexible, more in the service of adaptation to some requirement of external reality.

A different approach, but one that is equally indebted to Freudian ideas of ego functions and what are referred to as defense mechanisms, is espoused by George Vaillant.[30] He argues that one can arrange ego mechanisms in a hierarchy, with the more mature ones at one end and the more pathological at the other. Examples of those at the least healthy level include denial and delusional projection. Examples of the healthiest mechanisms include sublimation, humor, anticipation, and altruism. There are, of course, an array of others between the two extremes.

Vaillant proposes that the more mature mechanisms are found more frequently in the more satisfied men in his study and also in those who are better adjusted to their social worlds (and on the average, more successful). He does not seem to be saying that mature ego functions necessarily lead to more successful adaptation to one's society, only that in his sample they happen to do so. This is a very difficult argument to defend, since the data are very difficult to untangle. But his general point of view is not uncommon in psychiatry: Maturity and health, as reflected in psychological functioning, accompany successful adjustment to one's social world. Vaillant does not have a simpleminded view of adjustment; rather he argues that success has objective and subjective aspects, the former connected with worldly success, and the latter with happiness or fulfillment; and he makes it clear throughout that his ultimate support is his data, which are the lives of the men in the study. Empirically, the men who scored best on the psychological measures (and the scorers worked without knowledge of the social status and success of the subjects) were also, on average, those who did better in worldly terms. Thus for Vaillant, social adjustment appears to be connected with psychological health.

This, of course, is the conventional view upon which the psychiatric nosology including the diagnostic category of social maladjustment is based. However, there are alternative ways of looking at the whole issue, ways that are equally rooted in Freudian and dynamic psychology. For example, Richard Coan[6] attempts to analyze the notion of "optimal" personality functioning. He first provides an interesting review of the various concepts that have been described by different writers as characterizing the healthy or mature or normal person. He then argues that certain of these characteristics are incompatible, so that for a given individual to rank high on one of them, he would necessarily rank less high on another. To illustrate, one can examine a characteristic that many psychologists regard favorably: being open to experience, allowing one to experience the richness or fullness of events. Coan argues that this capacity cannot increase indefinitely without adversely affecting another highly thought of characteristic, namely the kind of stability of personality organization that provides freedom from distress. Similarly, he argues that other variables stand in opposition to one another: (1) an orientation toward harmony; (2) toward relatedness; (3) toward unity with other people and the world versus a sense of clear differentiation from others; (4) striving for autonomy, self-adequacy, mastery and individual achievement; and (5) an optimistic confident attitude toward the world versus a realistic appraisal of world conditions.

Coan's argument raises an interesting question. Perhaps the adjustment concept (referring to a condition of being where one achieves a smooth existence, gets along well, and experiences a state of well-being in which negative emotional states occur infrequently) is not the same as the "healthy" or "optimum" concept. This in turn creates difficulties in our conception of social maladjustment, and we are then led, perhaps, to consider the possibility that a person might be socially maladjusted and yet, in one way or several ways, still moving toward a healthy or "optimum" state of existence. Or perhaps social maladjustment (the psychiatric nosological category) is compatible with achievement in any of several realms: in art, in finance, even in the psychological richness of one's life.

Part of the issue here is that of ascertaining what is the best dimension on which to array examples of social maladjustment. This whole issue would be clearer if psychiatry as a discipline were more certain about whether there is an illness-health dimension or whether what is called mental illness can be shown to be the opposite of what we might call health. Other possibly relevant dimensions are those of good and bad or even happiness and unhappiness. Until psychiatry has a more coherent view of how these different aspects of experience and behavior are related, questions concerning the definition of social maladjustment will seem bewildering.

Another aspect of the same problem comes into view when one realizes that people have greater or lesser abilities to find appropriate social niches for themselves. If one can locate the appropriate social niche, in which one's deficits become advantages, then the question of maladjustment will never arise. In our pluralistic world, there are numerous social levels, any one of which might provide a comfortable habitation for a particular individual. We each have considerable choice about where we live, what work we do, whom we marry, which friends we have. These choices are never so numerous as perhaps they may appear, nor so varied as perhaps we wish, but nonetheless there are always real choices.

There is an interesting question that psychiatrists do not often ask but they well might consider: How do individuals manage to find a social world for themselves that allows them comfortable adaptation? This would be another vantage point on social maladjustment. Maladjustment can only arise, after all, when an individual has not made a successful choice of habitation. Psychiatrists are prone to think only in terms of coping styles, defense mechanisms, and personality structure, as though all of these functioned independently of social context once development has occurred. The prevailing assumption is that "development" occurs when a person is young, but when the person is "mature" he carries his personality around with him, applying his coping style to whatever social situation arises. Social psychologists and sociologists can contribute to psychiatry in this matter, since they are more likely to invest energy in articulating ways in which particular social networks provide support for, and give sustenance to, particular personality constellations (while helping to discourage the expression of others). Sometimes psychiatrists recognize this mode of thought when they discuss stress and concede that almost anyone is vulnerable to certain forms and degrees of stress. Implicit in this view is the notion that stress is less in some situations and greater in others. Not much further thought is required to recognize the ways in which certain people inhabit realms that generate less stress than others, or even to recognize that what appears as "healthy" well-adjusted functioning in one context may not appear so in another. Psychiatrists who work with pairs or groups of people (for example, with families) are more ready to recognize this, since they often have rich clinical experience that forcefully brings home the fact that adjustment is always within a context, and that the context (of the marriage, or the family, for example) implicitly provides support for certain modes of functioning. Psychiatrists who work with marital couples always have vividly before them the fact that two individuals may have chosen each other mistakenly, that each individual might do better in a different marriage. But at the same time there is a pervasive trend in psychiatry toward insisting that a person who has difficulty in one marriage (or job, or social world) is likely to have difficulty in another. This is the strand of psychiatric thinking that overemphasizes the extent to which the individual carries his personality around with him, using it equally well in one situation or another. If psychiatrists stayed in closer touch with sociology they would be less prone to make this mistake in emphasis. The difficulty is not, of course, that psychiatrists are mistaken, but simply that they do not give enough credence to another, different point of view.

¶ Sociological Contributions

The concept of social maladjustments can be approached from a clinical or from a sociological point of view. A clinical approach would first consider, in a given patient, the degree of psychological discomfort. The patient comes into the psychiatrist's purview in part because of some form of psychic pain. This pain may be connected with cognitive inefficiency, with disturbances of bodily functionings, or with a vague kind of anhedonia that is not specifiable. But in the case of patients who are likely candidates for the category of social maladjustment, there is in addition some deviation in behavior from social norms. This deviation may be only apparent to the patient (and may, in fact, be illusory), or it may be so noticeable that various community representatives become involved in the referral, bringing the patient to the attention of the psychiatrist. If the deviations from social norms and mores are particularly flagrant, they may lead representatives of the community (whether they be people who live in close proximity to the identified patient or are rather more remote cohabitants of the patient's social world) to bring the patient to the attention of the psychiatrist against the patient's will, or at least against his conviction about what is best for him. In such a situation, the psychiatrist may come to feel that he is, to a greater extent than usual, upholding the public interest, rather than (or in addition to) ministering to a particular mind diseased.

Thus, contemplating the use of the diagnostic category of social maladjustment may quickly lead the psychiatrist to a consideration of the social fabric and his part in maintaining it. Some sociologists argue that categories such as social maladjustment are in fact mainly attempts on the part of most of us to maintain some benchmarks against which we will compare ourselves favorably. If we have no standards about what is proper and improper behavior, then we are adrift in a sea of social relativity. But if we have some standards, which we enforce against others, then we gain a sense of solidity and substance. As Albert Cohen[7] puts it, "each generation establishes benchmarks for measuring wickedness . . . and one determinant of where those marks are placed is interest in finding unfinished moral work that might provide opportunities for earning moral credit." For a psychiatrist, this may seem an unconventional point of view. But one of the disadvantages of the inclusiveness of the DSM-III is that it attempts to provide a category for everything a psychiatrist is likely to see in the office. This necessarily means, in some cases, that a psychiatrist will be attempting to assess cases where the social dimension of the difficulty is preeminent. Social maladjustment is the category that refers to such cases, and in order even to consider using such a category, the psychiatrist must be assessing not just a patient's own mental state but the ways in which the patient's behavior impact upon the lives of those around him, and, further, the ways in which those around the patient view the patient's behavior. This involves assessment of social norm violation, and insofar as psychiatrists are involved in trying to change the behavior of their patients, they may well become involved in attempts to uphold social norms.

Now most psychiatrists will not, ordinarily, describe themselves in such terms. They will routinely and conventionally react with disapproval to suggestions that in some countries psychiatry may become an arm of the police establishment, used occasionally to suppress political dissent. And yet since the revolution in psychiatry wrought by Freud and his followers (a revolution that in part collapsed the distinctions between normal and pathological), all of psychiatry has become vulnerable to being embroiled in very similar situations.

In order for a psychiatrist to gain as clear a view as possible of this situation, some acquaintance with several aspects of sociological thought is desirable. One trend in current sociological theory that is particularly pertinent is referred to as labeling theory.

Labeling Theory

One specialty within sociology is the study of deviance. And within that specialty, one theoretical point of view that has been particularly influential in the past twenty years or so is the labeling theory, or the labeling perspective.[12,17] This perspective is used, of course, to discuss various forms of deviance, including the form of deviance ordinarily referred to as mental illness.[26,27] From this perspective, what is seen as crucial is not the act committed by the patient, but rather the label that is applied to the act and then, by extension, to the actor (the patient). The interest focuses on how the label comes to be applied to some people and not to others. There is great attention paid to the fact that only a small proportion (the exact ratio is unknown) of deviant acts falling into any particular category come under the scrutiny of professionals in that field. In psychiatry, for example, only a small proportion of people who are unhappy or who do bizarre things come to see psychiatrists. Why do these few come, rather than others? How are they selected? What makes them, if you will, more open to the mental health professionals than others, more susceptible to having the mental illness label put on them?

These are the sorts of questions asked by those espousing the labeling perspective. Since psychiatrists are crucial in the process of making the "mentally ill" label stick, the practice of psychiatry is of particular interest to sociologists who find this point of view reasonable. However, many of the studies of mental illness from this perspective emphasize the hospitalization process, which, although of considerable importance, is far from being seen by psychiatrists as the action that provides meaning to their endeavors. Although psychiatrists do think some people could benefit from hospitalization, they do not see committing them as their main justification for existence. In fact, the influential psychiatrists who write articles and books, who develop points of view that in turn win adherents, who help run training programs, are less interested in the process of hospitalization than in understanding what is going on with patients who are presented for treatment. And many of these patients are outpatients, not in a mental hospital or planning to spend time in one.

But even when hospitalization procedures are not the focus, labeling theorists still have ideas of potential interest to psychiatrists, since they consider the processes through which people first come to define themselves as having a mental illness, and because they often see therapy with a professional as a reasonable route to pursue. Thus, the data they gather inevitably becomes of interest to psychiatrists. Indeed, psychiatrists should take an interest in using the same methods generated by the more sociologically oriented researchers in order to study the question of how people who have entered therapy come to see themselves as either having benefited or as having remained the same. Rather than assuming, as we too often do, that there are patients with diseases to whom we apply various methods of therapy, which then either work (and produce cure) or do not work, we might instead become more sociological and reflect on how all of these nouns (patient, illness, therapy, cure) reflect very complicated social interactions. Too often, perhaps, we tend to see them as so complicated that they cannot be studied, but the intricate studies that labeling theorists have carried out make clear that this is not the case.

Social Construction of Reality

Because psychiatrists are engaged in upholding ordinary reality, they need to become sensitive to their role in maintaining social reality, especially when they find themselves using diagnostic categories such as that of social maladjustment. In order to comprehend the intricacy of the whole process of defining social reality, some acquaintance with the tradition in sociology that explores this question is desirable.[4,16] Too often psychiatry relies on a simple-minded notion of reality. The very concept of "reality-testing," which is presumably assessed in any psychiatric evaluation, reflects this sort of simple-mindedness. If the reality we each want to be in touch with were there in any simple sense, the entire question of whether we are indeed in touch with it would not be so vexing. Patients who fall into some of the residual categories in DSM-II[1] or DSM-III[2] force us to think more deeply about our ideas of reality, since often they are not psychotically impaired in their thinking but rather are engaged in some form of deviant reality that may or may not warrant psychiatric intervention. In practice, of course, psychiatrists recognize this, and they will usually only treat those who want to be treated, who want therefore to change. But psychiatrists do become implicated in working with nonvoluntary populations, and whenever that happens—whether with prison inmates, or with children, or with people who are being forced in more subtle ways to seek out psychiatric consultation by their families, relatives, or job associates—they run headlong into the dilemmas being discussing here. An acquaintance with the sociological literature on what is called the "social construction of reality" would at least serve to sensitize psychiatrists to many of these issues, issues that are relevant to assessment of patients who might fall into the category of social maladjustment.

¶ Social Traps

Ever since Darwin, the term "adaptation" has accrued a variety of meanings within an evolutionary framework. That is, adaptive behavior tends toward survival, either of the species or of the individual. Diagnosing survival in the physical sense has never been acutely problematic for psychiatrists (though, in fact, defining "death" is a difficult matter for doctors in some situations). But ascertaining what psychological survival might be is not so easy.

Survival, of course, is a rather stark goal. Psychiatrists do not ordinarily see themselves as trying to ensure bare physical survival. Rather, they view themselves as promoting something more elusive, connected perhaps with happiness or fulfillment or authenticity (no good word exists for this elusive state). However, in any attempt to assess the degree of social maladjustment present in any given individual, it would behoove psychiatrists to be at least aware of a developing field of study that intersects other more traditional fields. One landmark in this newly developing field is an article by Garrett Hardin called "The Tragedy of the Commons."[13] Actually, his argument is not unknown within the somewhat esoteric branch of social psychology known as game theory (as outlined for example in the works of Anatol Rapaport[22] and Thomas Schelling).[28] What Hardin points out, in compelling terms, is that sometimes individuals pursue what they assume to be their own best self-interest, only to find that, in the long run, they become less and less happy, or even die (for lack of food). Much of the controversy generated by the field of ecology is fueled at least in part by an awareness of this dilemma. The example Hardin uses is that of a common grazing land to which each individual member of a community has free access for his animals. As long as not too many animals graze on the land, there is plenty of grass, which keeps on growing and replenishing itself without much effort on the part of the farmers. But if too many animals are allowed to graze on the land, then the grass will be eaten up and be unable to regenerate. This tragedy—whereby each individual citizen loses an entire flock for lack of grazing fodder—can be prevented only if it is foreseen and if some form of mutual regulation is arrived at, according to which, for example, each person is only allowed to pasture a limited number of animals on the commons.

Much of human society today is predicated on the existence of such control mechanisms. There is a limit to many supplies or goods, so that rationing the goods becomes, sooner or later, a problem. At present, this problem may not be so acute as to threaten humanity so far as food, oil, water, space, and air are concerned, but many people now foresee a time when all these resources will be in such scarce supply that some control mechanisms will have to exist for their allocation.

What is the relevance of all this to psychiatry? John Platt[21] makes the connection, emphasizing that a social trap exists in the fact that "each individual . . . continues to do something for his individual advantage that collectively is damaging to the group as a whole." What psychiatrists need to keep in mind is the possibility that what looks like a coping mechanism in a given individual, insofar as it enables him to thrive in his own current social sphere, may not in the long run turn out to be optimal with regard to ensuring survival of either the individual or of the species. For example, the ability to commit oneself to both a family and a career may lead quickly to having children and to doing work that, while well rewarded in terms of money, is not truly productive in the sense of helping to create some worthwhile product that will in the long run help to ensure the survival of the species. Much work is not very productive (Paul Goodman's polemic of more than twenty years ago, *Growing Up Absurd,*[11] probably puts the case as well as any). And there may be too many children already.

Obviously, a psychiatrist is no better suited by virtue of his training than anyone else to evaluate these issues. But having some awareness of them would probably help broaden a psychiatrist's perspective and at least make him more open to considering the possibility that a given course of action, while not adaptive in the sense of not being congruent with the social mores of the world the patient lives in, might in some other context be more beneficial, more valuable, more worthwhile. Making this judgment would not be easy, but being aware that such judgments are always being made would, at least, reduce the vulnerability of psychiatry to the charge that it is a conformist institution that specializes in simply adjusting people to a society that might not be worth adjusting to.

Psychiatrists do not often consider openly whether there is such a thing as overadjustment. But sociologists raise this possibility in *The Organization Man*[33] and *The Lonely Crowd,*[23] and literary and cultural critics have also explored this possibility.[31] Psychiatrically inclined writers who explore this usually do so only in their less technical writings (as Erich Fromm did in his famous book *Escape from Freedom*).[9]

The whole issue is brought into clear-cut relief when one considers a diagnostic category such as that of social maladjustment. For example, why is there no opposite category (social overadjustment)? The fact that there are so many categories for people who cause troubles, and so few for people who get along without making any fuss, is in itself a partial indictment of psychiatric nosology. The issue is not whether psychiatrists think people who get along are healthy; clearly no such opinion exists, and any immersion in the psychiatric literature will quickly bear this out. Psychiatrists, in fact, are more prone than most to see troubles everywhere, and whenever they take the time to work closely with ordinary people who have not sought out help, they conclude that even in such populations there is substantial illness or difficulty, well hidden perhaps but nonetheless real. And yet, psychiatric nosology does not take this into account. Instead, the diagnostic manuals seem to restrict themselves almost entirely to consideration of what enters through the psychiatrist's door. And while this approach is convenient from the point of view of enabling a psychiatrist to find a category for most of the people he sees professionally, it does not encourage psychiatrists to think in terms of such problems as "overadjustment."

¶ Social Change

The whole question of social maladjustment becomes more complicated when one considers not only the question of an individual's adjustment to his social world, but also the phenomena of social change (more rapid now than ever before) and the fact that historically some societies have been known to thrive and then disappear. The whole complicated question of what enables a society to survive is far from clear, though of course there are many speculations about this topic. For the psychiatrist who is engaged in daily work with patients, most of these speculations are not directly relevant. But a psychiatrist might do well to keep the matter in mind. The implication of social change and even the disappearance of a society is that any trait that helps a given individual survive today in his own society (traits that appear socially adaptive) may not help him tomorrow or in the next decade, and that further, it may be helping to create a situation that will eventually lead to the demise of the society.

The fact of rapid social change has frequent impact on the work that psychiatrists do, since they occasionally see patients who are having difficulties adjusting to a change in their environment. Some of these changes are common in certain life histories; for example, the change from having no children to having children, and then from having children to living without them. But even these normal changes have different meanings today than they did fifty years ago. For example, when children leave the parental home, they are more likely today to go farther away, given the ease of transportation and the desire for job mobility. Thus, the skills a parent needs in order to promote a smooth and healthy separation for his children may vary from one time period to another. Similarly, the pressures on women today are different from those that prevailed even as little as thirty years ago. Psychiatrists need to be aware of these changes if they are to assess accurately the nature of the difficulties a given patient is having with adjusting to some life change.

The other question of whether a given society is adaptively organized in the long run is more difficult to ascertain. There is no way to know at present whether our society is doing well or poorly, whether it is on a rising curve or a falling one. But it is worth keeping in mind that adaptation today may not mean adaptation tomorrow. Population geneticists stress the value of diversity in the gene pool for enhancing the ability of a given species to respond adaptively to environmental changes. Perhaps there is a metaphor here that is useful for considering social evolution. That is, perhaps a certain amount of diversity in social adjustment patterns is valuable, not because the diversity necessarily means that each individual is regarded as equally successful by his peers, but because having diversity present in the society makes it more likely that if and when conditions change, there may be individuals around who have developed patterns of living and thinking which will make them more capable of working out new ways of living, which in turn can be taught to their more conventional fellows. Such individuals are often labeled "geniuses." Being a genius—that is, being able to see things in different ways and to do things that turn out to be "better" than the things others do—is often accompanied by considerable eccentricity. The word "eccentricity" is an interesting one since it identifies a form of behavior that is clearly at variance with accepted social norms, and yet its connotations are not entirely unfavorable. There is a note of forgiving acceptance in our use of the term, which is probably altogether reasonable. For some people should be allowed, even encouraged, to be different from their fellows; and psychiatrists need to be more attuned to this form of "adjustment" and to do more to promote it.

The essential questions to keep in mind, when addressing a given patient's desires for change, are why does this person want to change, and is change really a good idea in this situation? Asking such questions, for example, would have helped avoid some of the controversy surrounding the American Psychiatric Association's change of position with regard to homosexuality. Psychiatrists might have thought to themselves, in considering homosexuality, whether this kind of behavior really is harmful, and if so, to whom. They might have considered its adaptive characteristics; for example, it does not lead to reproduction, and thus acts as a population control. This is admittedly an unconventional way to think about homosexuality, but almost any deviation from the norm that might be labeled "maladaptive" may have beneficial aspects. Psychiatrists would do well to be more attuned to this possibility when considering the patients they are evaluating.

Psychiatrists could broaden their scope by studying sociological writing that emphasizes the adaptation required for anyone to work out any coherent orientation to life, even if it be called a "deviant" orientation. Many, though not all, "deviants" have adapted quite successfully to a subculture that happens to be tangential to, or even in opposition to, the mainstream culture. Thus, what looks on the surface like maladaptive behavior (in terms of the mainstream culture) may, in fact, be adaptation in terms of a smaller culture. Psychiatrists working with children are often forced to recognize that a given child, who is identified by a school as "being in trouble," may on closer examination turn out to be rather well adapted to a peculiar family, which in turn makes smooth and easy adaptation to the school (with its different standards) quite difficult.

Second, psychiatrists could benefit from a familiarity with the thinking of those who are interested in the relationships between basic personality structure and social organization. Certain societies, especially more homogeneous ones than our own, often systematically reward certain kinds of personalities (the kind identified as "well adjusted"). But this is not to say that other personality types may not exist within that culture or that they could not make contributions to the larger culture by virtue of their "outsider" position. Keeping these matters in mind may not make the nosological task of the psychiatrist any easier, but it will help to ensure his sensitivity and flexibility in the use of the diagnostic categories.

A central difficulty is that adaptation is judged by success, and success is never a permanent fact but a contingent one. What appears to be success today may turn out to be failure tomorrow. In a larger context, a cultural group or even a nation may be overrun by another cultural group or nation. Thus, the attributes that helped the tribe or nation prosper during one era may, in fact, one day contribute to its defeat in another. In our own case, if a nuclear disaster occurs, future historians (if there are any) may well speculate about how those traits that enabled western civilization to be so successful for centuries led to its demise. There is, of course, a certain vein of contemporary writing, some of it psychological and psychiatric, that assumes that our society is "mad" and that asserts that fitting into such a society is not a desirable goal. "Thus, a feeling of lack of fit between self and social structure is no longer perceived as pathological or even accidental. Such contemporary works hypothesize that what is positive in the self can never fit with society, which is, by its very nature, mad."[3] This position is certainly more extreme than what most psychiatrists would be comfortable with, but it is representative of one strand of contemporary thought about the relation between the individual and society. And such considerations are germane when considering the use of the diagnostic category of "social maladjustment."

¶ Developmental Psychology

We might legitimately wonder how it is that people who are raised by others manage to turn out in ways that are not consistent with the larger society. Part of the answer lies in the fact of social complexity; not every family that raises children according to its own lights has beliefs and values congruent with those of the larger society. Another part of the answer is inherent in the human's potential for adaptation, the phenomenal plasticity that exists at birth. Animals whose behavioral repertoire is more clearly dictated by genetic factors have less capacity for maladaptation, since they are born with the equipment necessary for carrying out their lives. People, on the other hand, are born with less of what they need, and must acquire many skills in order to be able to survive in their social and cultural context. Our phenomenal immaturity at birth is, of course, the reason we are able to learn so much after we are born. In a sense, immaturity is the capacity for adaptation.[5] From the point of view of evolution, this is a tremendous strength, but there are costs; one of them is our capacity to develop in ways not directly encouraged by the society. Because we are so flexible at birth, we can learn any of a multitude of languages, depending on what is being spoken around us in our formative years. But by the same token, we are prey to various disorders that have communication and language difficulties at their source. Thus, adaptation and maladaptation are two sides of the same coin.

Within developmental psychology there is a wealth of information about how people develop. Much of this information illuminates some of the possible sources of maladaptation. Research conducted, for example, by Louis Sander and associates[24,25] makes clear the exquisite delicacy of the mutual adaptation that occurs between parent and child during the early weeks and months of life. In this process of mutual regulation, there are many possibilities for maladaptation. One of them arises when the parent's tempo does not match that of the child. This kind of mismatch between the capacities of the child and those of the parent may be more common than we think, and may in turn lead to more complicated kinds of maladaptation later.

Another rich source of ideas regarding maladaptation is the literature about marriages, since it often contains specific detailed examples of interactions between husband and wife, which illustrate the varied ways in which two people may work out adjustments to one another. All of these sources of data make clear that adaptation is a dynamic process that occurs either between individuals or between one person and a small group or a society.

At present this point of view is not easily accommodated within the psychiatric nosological system, but the existence of a category such as social maladaptation makes clear the necessity of considering such interconnections. Keeping this point of view in mind will help the psychiatrist assess the value of alternative therapeutic strategies. In some cases, individual therapy for the designated patient who is "maladapted" may be less appropriate than an attempt to change the nature of the system within which the patient functions. Thus, a psychiatrist might consider, in dealing with a younger patient, the possible value of working with the family as well, or he might alternatively consider the value of advising the parents of the patient how to adjust themselves to their maladjusted child. A psychologist working in a school setting might consider some alteration in the school program. Such a change could provide a better environment for the needs of a particular child and thus eliminate the maladaptation that previously had appeared to be in the child but that was actually a feature of the relationship between a child and the school system.

¶ Conclusion

The term "maladjustment" is such a general one that it might, in truth, be applied to nearly all the patients a typical psychiatrist is likely to see. Most people who are neurotic, character disordered, and psychotic are likely to have a degree of maladjustment in some general sense. In addition, many of those who do not consult a psychiatrist could be considered maladjusted, in the sense that they are not perfectly adjusted either to the outer society or to their inner natures. If adjustment means no conflict, then maladjustment is everywhere.

This chapter has focused on a more limited form of maladjustment, one implied by the use of the diagnostic category of social maladjustment. However, the distinction between a category of social maladaptation and other categories of problems in living or illness is artificial. Making the distinctions necessary to use this category in practice is difficult. However, there are several issues that appear relevant to the kinds of evaluations a psychiatrist would have to carry out in order to arrive at a diagnosis of social maladaptation. Many of these considerations involve nonpsychiatric data and theoretical orientations and approaches. That is as it should be, since the category of social maladjustment arises when there is a clear lack of harmony between an individual and society; and if the psychiatrist is to consider the matter from both ends—from the individual or intrapsychic end, and from the sociological or interpersonal end—then some familiarity with what are ordinarily considered sociological issues is necessary.

However, it bears remembering that a familiarity with such sociological notions as labeling theory and the social construction of reality does not necessarily make the psychiatrist's task an easy one. But in spite of the difficulties, it is possible to broaden one's scope and deepen one's imagination by being aware of larger existential problems, and this cannot but aid the therapist in his work.

¶ Bibliography

1. AMERICAN PSYCHIATRIC ASSOCIATION. *Diagnostic and Statistical Manual of Mental Disorders,* 2nd ed. (DSM-II). Washington, D.C.: American Psychiatric Association, 1968.
2. ———. *Diagnostic and Statistical Manual of Mental Disorders,* 3rd ed. (DSM-III). Washington, D.C.: American Psychiatric Association, 1980.
3. BERGEN, B., and ROSENBERG, S. "The New Neo-Freudians: Psychoanalytic Dimensions of Social Change," *Psychiatry,* 34 (1971): 19–37.
4. BERGER, P., and LUCKMAN, T. *The Social Construction of Reality.* New York: Doubleday, 1966.
5. BRUNER, J. "The Uses of Immaturity," in Coelho, G., and Rubinstein, E., eds., *Social Change and Human Behavior,* Rockville, Md.: National Institute of Mental Health, DHEW No. (HSM) 72–9122, pp. 3–20.
6. COAN, R. *The Optimal Personality.* London: Routledge and Kegan Paul, 1974.
7. COHEN, A. *The Elasticity of Evil.* Oxford, Eng.: Basil Blackwell, 1974.
8. ENGEL, G. "The Need for a New Medical Model: A Challenge for Biomedicine," *Science,* 196 (1977): 129–136.
9. FROMM, E. *Escape from Freedom.* New York: Farrar and Rinehart, 1941.
10. GOODE, E. *Deviant Behavior.* Englewood Cliffs, N.J.: Prentice-Hall, 1978.
11. GOODMAN, P. *Growing Up Absurd.* New York: Random House, 1960.
12. GOVE, W. *The Labelling of Deviance.* New York: John Wiley, 1975.
13. HARDIN, G. "The Tragedy of the Commons," *Science,* 162 (1968): 1243–1248.
14. LAZARUS, R. *Personality and Adjustment.* Englewood Cliffs, N.J.: Prentice-Hall, 1963.
15. LEMERT, E. *Human Deviance.* Englewood Cliffs, N.J.: Prentice-Hall, 1967.
16. MANNHEIM, K. *Ideology and Utopia.* New York: Harcourt, 1936.
17. MATZA, D. *Becoming Deviant.* Englewood Cliffs, N.J.: Prentice-Hall, 1969.
18. MECHANIC, D. *Mental Health and Social Policy.* Englewood Cliffs, N.J.: Prentice-Hall, 1969.
19. ———. *Medical Sociology.* Glencoe, Ill.: Free Press, 1978.
20. PHILLIPS, L. *Human Adaptation and Its*

Failures. New York: Academic Press, 1968.

21. PLATT, J. "Social Traps," *American Psychologist,* 28 (1973): 641–651.
22. RAPOPORT, A. *Fights, Games and Debates.* Ann Arbor, Mich.: University of Michigan Press, 1960.
23. RIESMAN, D. *The Lonely Crowd.* New Haven: Yale University Press, 1950.
24. SANDER, L. "Issues in Early Mother-Child Interaction," *Journal of the American Academy of Child Psychiatry,* 1 (1962): 141–166.
25. ———. "Regulation and Organization in the Early Infant-Caretaker System," in Robinson, R., ed., *Brain and Early Behavior.* London: Academic Press, 1969, pp. 311–332.
26. SCHEFF, T. *Being Mentally-Ill.* Chicago: Aldine, 1966.
27. ———. *Mental Illness and Social Processes.* New York: Harper & Row, 1967.
28. SCHELLING, T. *The Strategy of Conflict.* Cambridge, Mass.: Harvard University Press, 1960.
29. STRAUSS, A., et al. *Psychiatric Ideologies and Institutions.* New York: Free Press, 1964.
30. VAILLANT, G. *Adaptation to Life.* Boston: Little, Brown, 1977.
31. VIERICK, P. *The Unadjusted Man.* New York: Capricorn Books, 1962.
32. WHITE, R. *The Study of Lives.* New York: Atherton Press, 1964.
33. WHYTE, W. *The Organization Man.* New York: Simon and Schuster, 1956.

CHAPTER 25

ADULT PLAY: ITS ROLE IN MENTAL HEALTH

Norman Tabachnick

THE PHRASE "adult play" is almost incongruous. While we know that many adults do play, playing seems inappropriate, even slightly sinful, for grown-ups. Playing properly belongs to childhood. Sometimes it even seems to be the main business of childhood. Yet the word "business" suggests an ambiguity. For "play" evokes visions of fun while "business" implies sustained effort directed toward important goals. So perhaps play is more serious for children than is at first apparent, and perhaps the notion that play is not important in adult life should be reconsidered.

There are several questions dealing with the development and role of "play" in adulthood:

1. Does play disappear in adult life? If so, what are the reasons for it?
2. If play continues—to what degree does it persist in adult life?
3. Does play undergo certain developments or transformations in adult life? Should we develop an epigenesis of play?
4. Finally, can play be revived in adult life? What value might result from such an achievement?

Even the definition of play contains problems, for "play" is difficult to encompass within one set of concepts or specific formulas. To play is to experiment and create—yet it is not as serious as to be an Experimenter or Creator. For to play implies that the participants are not to be held to account either for what happens during the play or for the product of the play. To play means to enjoy one's self by acting whimsically. One *need* not play according to some set pattern. (Games with rules may be different from play or they may be a special form of play.) But one can *choose* to repeat certain roles and this can also constitute play.

One can play alone (yet strongly experience one's relationship to others). And one can play with others (and although that *interplay* is intense, one can at the same time create highly personal fantasies). One can play to learn societal roles (such as mother and baby, cops and robbers, doctor and patient), to grow up faster and farther (for example, as Buck Rogers in the twenty-fifth century), or to experience nonhuman life ("Let's be flowers," or "Let's be lions.") There are endless possibilities and this too is essential to play.

¶ Characteristics of Play

Activity called "play" has been described as taking place among many animal species. The possibility has even been raised that nonpurposeful behaviors of one-celled animals may be play. However, as might be expected, the qualities of play and the purposes postulated for it in these different species have been varied. The multiplicity of theories for the purposes of play exists within a given group as well. When we focus on primates, and indeed man, we find a number of different explanations.

Variability and vagueness in defining play, accompanied by the multiplicity of explanations for play, have actually retarded serious exploration of the topic. According to Bruner:

> The behavioral sciences tend to be rather sober disciplines, tough-minded not only in procedures but in choice of topics as well. These must be scientifically manageable. No surprise then that when scientists began extending their investigations into the realm of early human development they stayed clear of so frivolous a phenomenon as play. For even as recently as a few decades ago, Harold Schlosberg (1947) of Brown University, a highly respected psychological critic, published a carefully reasoned paper concluding sternly that since play could not even be properly defined it could scarcely be a manageable topic for experimental research. His conclusion was not without foundation, for the phenomena of play cannot be impeccably framed into a single operational definition. How, indeed, can one encompass so motley a set of capers as childish punning, cowboys-and-Indians and the construction of a tower of bricks into a single, or even a sober, dictionary entry? [p. 13][1]

Bruner further suggests that what overcame the inertia in the study of play was an increasing number of reports on subhuman primates. These stressed the significance of play for education and adaptation among the young. His view may be extreme since in fact a small number of studies concerning play were appearing regularly. However, the general point—that the difficulty in defining the concept of play has held back more intense study of it—is probably valid.

What then are some of the defining characteristics of play?

1. *Human play is largely an activity of the young.* Play is an activity of children. Most of the published work on play deals with children. (In fact, we personally have only encountered one study[15] of adult play —so designated.) Whatever play is, it decreases radically as children grow up. The end of puberty is pretty much the end of human play. Gilmore, in discussing Piaget's theory of play, says, "Speaking generally, Piaget sees play as the product of a stage of thinking through which the child must pass in developing from an original egocentric and phenomenalistic viewpoint to an adult subjective and rationalistic outlook" [p. 316].[6]

And Bruner, in evaluating the place of play in the animal's life cycle, states, "But perhaps most important its role during immaturity appears to be more and more central as one moves up the living primate series from Old World monkeys through great apes to man" [pp 13–14].[1]

However, limiting play to puberty and before may be inaccurate. Most people believe that there is some play by all adult humans and that some adults play a good deal of the time.

2. *Play is spontaneous.* Much play is unplanned and seems to be created "on the spot." Although general topics ("Let's play house") are often preselected, the details are not. Indeed, the essential nature of much play seems to be spontaneous entrance into

action and interaction. Children playing together appear to be continually improvising.

3. *Play explores new issues.* Often play is used to try out new or prospective social roles. Sometimes these are close to what that child will certainly grow into (as an older child or adult), sometimes the selected roles are just a possibility (fireman, doctor). At the same time as personal roles (and the abilities, talents, and fantasies associated with them) are played at, adaptation to societal mores simultaneously occurs.

Although spontaneity and exploration are central to this quality of play, a certain amount of repetition takes place. This may be indicative of a "practicing" characteristic. All those new issues (in the self, in the society) that are being discovered are also being repeated so that the child can learn to perform more effectively. All this, of course, suggests that play is crucially linked to learning and is perhaps one of the most important educative modes.

4. *One is not held to account for play.* In many kinds of learning, one is held to account for the excellence of one's performance and for the product—what one achieved or gained. A characteristic of play is that this does *not* occur. The anthropological observation is that this not holding to account is a development of higher primates. Bruner[2] points out that in lower primates atypical behavior is punished by the dominant male. In the great apes, however, dominance is more relaxed and interrelation among a group of apes occurs without fighting. Here the young learn through play. There is "an enormous amount of observation of adult behavior by the young with incorporation of what has been learned into a pattern of play" [pp. 28–60].[2]

5. *Pleasure and fun are part of play.* For humans, pleasure and fun are usually implicit or explicit in activities called play, and whether it is a matter of the human observer's projection of his feelings onto "lower" animals or not, we usually believe that they *are* having fun when they play. Even when the attitude of children during play seems to be serious (as in cops and robbers), the evaluation, "That was fun," seems right.

A fascinating but unsatisfactorily answered question is, "Whence comes the fun of play?" It has been suggested that the pleasure is the pleasure of creativity. However, since play itself may be a type of creativity, that explanation may not advance our knowledge. It has also been suggested that the pleasure is related to exploring new situations and/or achieving mastery of them. Perhaps exploring, discovering, and mastering without being held to account always yields pleasure.

6. *Much play is behavior in the simulative mode.*[12] The player(s) acts at being or doing something, but it is evident that he is simulating. There are important differences between the play situation and the real situation. Consider a game of cops and robbers. All the players know it is only a game. If there are onlookers, they also understand this. If any of them did not recognize it as play, it would quickly be changed from play. Participants would genuinely defend themselves. Onlookers would flee, enter the fray, or call the police.

¶ Theories of Play

Play has several purposes, but they are probably not all present in all instances. Some play may focus on the *exercise* of newly discovered or newly developing abilities. Other kinds of play may focus on *discovering* new skills, fantasies that are anticipations of emerging new abilities or other new possibilities. Still other forms of play may emphasize *learning from others*—learning how to do things or what the rules of a particular culture are. Some play may intricately weave all these strands together. Finally, since play involves learning cultural rules, it is associated with an ability of great significance—symbol development. According to Bruner, "The evolution of play might be a major precursor to the emergence of language and symbolic behavior in higher primates and man" [p. 21].[1]

¶ The Play of Adults

To begin, let us repeat a few earlier ideas. Most people believe that play is predominantly an activity of childhood, and that view is certainly supported by the small number of scientific articles on adult play. Play is generally considered to take place in the years prior to adolescence. Let us examine this concept critically. Is it indeed true that play mostly belongs to childhood?

First, we shall consider how prevailing values might bias our opinion. This is important not only because societal values often lead to incorrect evaluations, but also because in the instance of play there are (to our knowledge) no statistical surveys that evaluate the relative quantities of child and adult play. If there is any doubt that societal values affect what we think about play (its general value and its specific value in adult life), that doubt can quickly be dispelled by perusal of certain scientific articles. There is much literature detailing differences in the quality of play and the value attributed to play by different cultures. A number of these studies are reported in Millar's book, *The Psychology of Play.*[11] Furthermore, within a given society, values concerning play undergo change with the passage of time.

In a fascinating sociohistorical study, Stone[15] lays the groundwork for some important changes in attitudes toward, and the actual nature of play. He relates that in western society before the seventeenth century there was no real social distinction between adults and children. Their dress did not differentiate them and expectations about them were not age specific. Play was present in all parts of society. Adults and children played the same games.

However, important changes occurred with the rise of the Protestant ethic. Again, according to Stone[15] in the seventeenth century and later, the Catholic church attempted to suppress play. Complete suppression was impossible, Stone believes, because industrialization had not yet occurred. In fact, it was the rise of an entrepreneurial class, which put work at the center of social arrangements, that helped establish childhood as a separate social period. Later, with industrialization and the rise of Protestantism in England and America, an attempt was made to suppress play among all age groups. However, it is obvious that this effort did not result in complete suppression. The impact of the Protestant ethic (which states that work is good and that play is wasteful, if not sinful) is still widespread today. This attitude, although not the only prevalent one, may bias some of us toward believing that there is relatively little play in adult life.

Play, however, is valuable—this is true not only in childhood but in adult life as well. With a group of colleagues, this author has for some years been studying the effects of play in adult life and in adult psychotherapy. Although far from providing a conclusive answer, this study suggests that play, and certain special adult forms of it do exist, is valuable, and could be of great value to human beings.

Keeping the biases in mind, let us evaluate the actualities of play in adult life. Is there less of it than in childhood?

First, it seems clear that play, particularly if we look for identical characteristics, occurs much less in adult life than in childhood. In our culture at least, there are important responsibilities associated with survival and a good deal of effort must be expended on routine, but essential, tasks. Adults for the most part are involved with these responsibilities and tasks.

Yet play is far from absent among adults. Its forms are somewhat different, but if we recall the defining characteristics of play, we can find a number of varieties of it in adult life. Joking, teasing, high-spirited "fooling around," or horseplay are frequent occurrences. For many people, sex play, with its opportunities for fantasy, spontaneity, and variations, is an important creative field. Arts and crafts offer another opportunity for playful involvement. Whether one produces a work of art or only appreciates art, many of the criteria of play are fulfilled. (In regard to art, as well as some other areas, however, it is important to distinguish the spontaneous "creative" component—which is the playful one—from others—such as making a living from the activity—that are not playful.)

The spontaneous and free-flowing direction of much scientific work is playful. Einstein[3] wrote of his creative work as play, and he was not the only scientist to feel this way. Writing, dancing, acting—all can be playful, as can a free-flowing discussion, an approach to a task, or an approach to life.

Thus, the potentiality for play continues in adults. The amount of actual play differs among individuals, and there are some life-styles (the person of spontaneous ingenuity) and career styles (the artist) that contain more play than others.

Next, there are the factors that account for changes in the quantity of play as human beings age. The theories of child's play suggest that play should disappear in adult life for the following reasons:

1. Because play is necessary for the development of immature body organs (muscle, neural tissue, and the like), and since these organs attain full or almost full growth during childhood, adults have no further need to play.
2. Likewise, some theories of play focus on learning about and adapting to one's culture. As one accomplishes those tasks, the need to play may decrease.
3. Some play is based on the need to exercise newly discovered or newly adopted modes of thinking or doing. As time goes on, this need decreases.

From these standpoints, then, perhaps there is relatively little function for play after childhood. Of course, there may be a few minor problems of adjustment, and play is fun—so to the degree that we value pleasure we may see that play has value. But for the most part—after we have used play in childhood to learn what we must do—adult life is the time to *do* it.

Let us, however, reconsider the issue of when play stops. First, in the just-stated reasons why play ends with childhood there is an important assumption. It is that the development of human intelligence—the establishment of essential structures (psychological and organic) and essential processes—is almost entirely completed by adolescence. After that time, one learns more but the basic tools of learning are already established.

However, that assumption, as well as the one that implies that the need for most play is over by the time of adolescence, may be questioned for the following reasons:

1. Many students of play theory agree that there is much to learn about even the basic principles of the subject. (In other words, our present state of knowledge does not give us the right to draw broad conclusions, such as play is over by the end of childhood.)
2. One argument offered for the absence of the necessity for play is that after adolescence there is no further growth of organic structures (brain, muscle, and the like). As far as we now know, this is true. However, it is not true that *organic changes* stop with adolescence. Deteriorations or limitations in organic function begin with birth. (Some are quite apparent and well known. Consider the decrease in muscular power which begins in young adulthood.) In addition, although new types of cells are not formed after adolescence, new growths or developments (for example, in size or firmness of muscle cells) do occur. Thus, the need for adaptation, education, and play persists after adolescence if they are involved in *changes* in organic function.
3. Similarly, and even more obviously, changes in intrapersonal processes, interpersonal ones, and relationships with constantly altering cultural influences continue throughout life.

So perhaps those individuals whose playful life-styles or playful occupations mark them as somewhat unusual are not really different types of human beings. Perhaps they just display more markedly the manifestations of a possibility inherent in all people. This possibility is that change and the capacity for dealing with it by playful learning are present throughout life.

¶ Toward an Epigenesis of Play

Among the number of factors that influence and are influenced by play are: (1) somatic factors—the unfolding of genetically coded growth structures and patterns in body organs; (2) intrapersonal forces; (3) interpersonal forces; and (4) cultural forces. Molded by play and other educational modes, the human personality develops. If we accept the belief that any factor in one of the preceding four groups may influence factors in the others, we become aware of the possibility of complex interrelationships. For example, a cultural trend that favors intense physical effort (such as might exist in a hunting community) would tend to favor early and intense neuromuscular development and influence the kinds of play in that community.

As we presently conceive of human psychology, it is apparent that individual peculiarities, connected on the one hand with genetic coding and on the other with a broad variety of interpersonal and cultural forces, result in many different patterns of function. However, at least from the time of Freud, scientists have constructed epigenetic outlines of man's development. (An epigenetic outline is one that describes successive stages of development of individuals within cultures.)

These outlines describe the general (as opposed to idiosyncratic) characteristics of individuals as they move through a sequence of developmental stages. It is possible to conceive of an epigenesis of play, and we contend that it would be valuable to construct one. Such an outline would yield many kinds of knowledge concerning the dynamics and potentialities of play.

Most of the adult play observed by the author's research group took place in drama workshops.[16] As the work continued, the importance of setting up special conditions for adult play was noted. There needed to be a set time and a special place. The place needed to be the same one each time, private and altered especially for play purposes. It was preferable to have the same playmates at each meeting. It was important that there be a group leader. This leader had to have special qualities including: (1) the ability to assume responsibility for final decisions in choosing exercises, sequence of exercises, and related dramatic choices; (2) knowledge of and interest in communication of acting skills; and (3) a supportive attitude toward individuals and the project as a whole. It was also important that the leader always be the same person. In effect, constant and (in relation to the play of children) special conditions make play possible or at least more acceptable to adults.

This is an early observation and needs more detailed study; nevertheless it seems a valid one. What would explain it? Why can children get together and play spontaneously while adults need special conditions? Could it be that adults must do something additional to establish that play is important at their stage of life? Do they need to isolate their play from other humans and provide special supports (for example, a mother-teacher-encourager)? In what ways do these features differ from childhood play? At what age do they begin to manifest themselves, and so on?

As time goes on, answers to such questions will tell us more about the meaning and evolution of play in human life.

A specific issue in the epigenesis of play deals with that aspect of adult life that is focused on decision making. Throughout life, problems arise that call for decisions. However, children—roughly up to the completion of adolescence—do not take complete responsibility for their choices. It is the young adults who, for the first time, may have "no one to turn to." These young adults must make important decisions about a life partner, about the kind of work that will be chosen, and about the type and quantity of play they will allow in their lives.

¶ The Protestant Ethic, Power, and the Suppression of Play

There is general agreement that the quantity of play decreases in adult life. It is also possible that the value of play decreases. Widely accepted theories of play suggest that play is most significant in childhood. Children have a great deal to learn and must develop *methods* of learning. Adults, having mostly achieved those goals, have less need to play. However, it is possible that other considerations may be important in understanding the decrease of play and the value of play in adult life.

Some of the decrease in the quantity and valuation of adult play is related to the influence of the "Protestant ethic" in our culture. This influence is encouraged by different groups that have a variety of motivations, through a number of methods.

Recalling Stone's[15] report that prior to industrialization children were not a special class and that the play of adults and children was much the same, we can also posit that in the eighteenth century industrial power, (primarily measured by wealth) became highly valued. Those who could achieve control of that power would wish to maintain and extend it. Among the methods used to implement this wish were the discouragement and disparagement of play, for play meant time spent other than in the specific goal-directed efforts that constitutes industrial labor.

What of the laborers? What might explain their concurrence in this societal trend? First, the necessity to survive—in a developing industrial society this would mean being willing to perform routine industrial labor. From this standpoint, it was to the laborers' advantage to favor work over play.

Second, it made sense for the owners of industry to attempt to extract more labor from the labor force. Thus in a variety of ways they attempted to encourage work and discourage play. Tangible rewards—often money—would be given to good workers (and poor players). In addition, explicitly or implicitly, a philosophy of the value of work and the wastefulness of play (the Protestant ethic) began to develop.

The effort to implement this trend took diverse forms in dealing with the various trade-offs that had to be negotiated. As one example, consider the issue of pleasure. One of the important ingredients of play is pleasure. If people are going to sacrifice the pleasure in play in order to work, what will they substitute for it? Money and the opportunity to survive is the first answer. But that may not be enough. Other pleasures (which are relatively distant from play) may be brought in as substitutes for play. The characteristics of these pleasures should be that they can produce a maximum of gratification with a minimum of playful, *active* involvement. We emphasize active because play's pleasure is associated with activity. If a designated pleasure calls for highly involved, spontaneous minds and bodies in action, then it approaches play.

But since there has been an effort to discourage play after the Industrial Revolution, other modalities of a more passive nature are sought. Marx[10] spoke of religion as "the opiate of the people." We suggest, in addition, that opiates *are* the opiate of the people. Drugs that give pleasure without activity fill the need of a society that wishes to suppress play. Woody Allen, in his movie *Sleeper*, imagined an orgasmatron, or sex machine, which made it unnecessary for two people to act with each other (or even individually) to produce an orgasm; the machine did it all.

One could, of course, deal with this issue from the standpoint of ethics, but for now it is sufficient to note that choices have been made. People have decided that work is good and reciprocally that play is bad. They have done this in the pursuit of certain goals—goals valued in an industrial society.

But we can make different choices. We can move toward other goals or other combinations of goals and means. We might disavow industrialism. We might decide that although industrialism and the kinds of effort associated with it are valuable, play is not valueless.

From one standpoint, we see that human beings make choices that influence their fate. From another (which does not exclude the first), we are dealing with evolutionary issues. As an active mode of learning, the utilization of play may have great survival value for the human race.

¶ Play in Adult Psychotherapy

Having noted the cultural suppression and inhibition of play, it may be asked, How can one start to play again in adult life? We have found that most adults play a little, and some are very playful. But for those who do not play very much, are there paths toward increased play?

In our culture, many individuals with problems about living enter psychotherapy. Some focus on specific distressing symptoms such as depression or anxiety. Others are concerned about general approaches to life; they wish to find meaningful goals and valuable modes of living. Although play is not usually associated with adult psychotherapy, it often happens that in psychotherapy an individual becomes more playful. Thus, as an unintended result of psychotherapy, certain people learn that more play and/or a more playful approach to life is of value. Let us examine the ways in which psychotherapy helps bring this about.

For many years psychotherapeutic approaches have utilized playfulness. The degree to which playfulness is considered therapeutic varies in the different approaches. Also, the therapists involved hold different opinions as to how much play actually occurs. But in general, it is probably accurate to say that there are a group of psychological therapies based on play. They include dance therapy, art and drama therapy, swimming and various "physical" therapies, and a number of others.

In addition, if one bears in mind our definition of play, one can see the playful quality in certain other techniques. Gestalt therapy, for example, with its spontaneity, encouragement of innovation, and nonserious approach, often contains much play.[4]

Also, when one considers the general tone of psychotherapy, one notes that certain psychotherapists (using any type of psychotherapy) are relatively playful. They contrast with others who, utilizing the same techniques, create an aura of profundity, seriousness, and deep responsibility.

This evaluation, however, has not been the focus of much previous scientific discussion. Few psychoanalytic writers, for example, have dealt with this issue. This is particularly interesting because "play therapy" for children is, for the most part, a psychoanalytic invention. Melanie Klein and Anna Freud are the two most important pioneers of the technique. The fact remains, however, that most of the writers on psychoanalytic "play therapy" focus on the content of play; they are not interested in play as a process it itself.

Nonetheless, at least one psychoanalyst believes that play has important value. D. W. Winnicott[18] writes:

> It is play that is universal and that belongs to health; playing facilitates growth and therefore health; playing leads into group relationships; playing can be a form of communication in psychotherapy. . . . Psychotherapy is done in the overlap of two play areas, that of the patient and that of the therapist. [p. 53]

In 1975, a group of psychoanalysts and analytically-oriented psychotherapists, including this writer, began utilizing improvisational theater techniques aimed at achieving psythotherapeutic benefits. We named our approach "Experiential Theater." There are certain activities and attitudes that characterize this approach. They include encouragement of spontaneous expression; discovering of unknown feelings, attitudes, and approaches to one's self and others; a sense of transcendent accomplishment as an accompaniment of new expression; and feelings of pervasive pleasure.[14] In attempting to understand the mechanisms of change in the experience, we came to feel that play was a key conception.

As the work continued, we tested the technique in a variety of therapeutic situations. These have included drama workshops alone, workshops combined with other psychotherapeutic approaches, workshops at the beginning of psychotherapy, and workshops for specific therapeutic problems. It is too early to present conclusions concerning the ultimate value of the approach. However, we have been impressed by its general acceptance by patients and its apparent value in different situations.

For the most part, the entire group of play-oriented psychotherapies have been auxiliaries or at least part of more comprehensive psychological treatment programs. Of course, this is not always so. A playful Gestalt therapy is on occasion a chief therapeutic effort, and if a particular psychotherapist uses a play-oriented approach, it is possible that the playfulness is the essential element of therapeutic change.

But the central question remains: How are playful approaches in psychotherapy valuable? There are some indications. In dance and movement therapies, the evaluators (both patients and therapists) may say that the muscular "loosening-up" which accompanies the therapy is reflected in a general "loosening-up" (that is, an increased flexibility) in the patient's adaptational processes. From one standpoint, it seems that the muscular tightness and/or awkwardness symbolizes a general rigidity of the personality.

In experiential theater, a participant's problem that has been worked at directly or indirectly in a workshop moves quickly toward elaboration and solution in general life. For example, inhibitions of anger are frequently expressed in improvisations. As a series of improvisations continue, new ways to express and/or deal with anger are developed. This "working out" of anger then continues in real life.

In an oft-quoted vignette, Freud[5] demonstrated the significance of play in overcoming unpleasurable experiences in the case of an eighteen-month-old boy. The infant threw away a wooden reel attached to a piece of string so that it vanished. Then, pulling the string, he made it reappear. In this way he dealt with a psychic trauma—the temporary absence of his mother. Freud felt that he changed a passively endured threat into an experience over which he had control. In addition, it can be postulated that the boy was learning that what disappears will return. The child's play helped "work out," or even better, "play out" the difficulty.

Of course, there are many problems that arise in evaluating such examples. One question is this: Suppose we accept that something valuable has occurred in the foregoing instances. What is it about a playful approach that makes an unusually valuable contribution? After all, many nonplayful therapies lead to similar results.

Mahler and her coworkers[9] have described the separation-individuation of human infants. In brief, this process describes, through psychological and psychoanalytic interpretation, the process by which infants develop a sense of identity. After an initial phase (autism) in which the infant senses (without a concept of himself as an individual) that he is all-powerful, he feels himself to be intimately fused with his mother. In subsequent stages, the infant separates from his mother, simultaneously developing a sense of individuality. These processes begin at birth and continue until the age of approximately two-and-a-half years. We can assume that these processes have some important concomitants. They include:

1. The sense of self is established.
2. Communication becomes a *self* conscious process. (Communication has gone on from birth, but only with the establishment of a self could it be sensed as occurring between two individuals.)
3. A universe of many objects is conceived.
4. The The symbolic function is established. (An early achievement of this phase is the sense of self as "me" and the sense of others

as "not me." These are probably the earliest symbols.)

With all of the preceding, there is an important need to explore, understand, and integrate aspects of "what is inside me" and "what is outside me."

As separation-individuation occurs, the role of illusion and what Winnicott[17] calls transitional objects, becomes important. First, let us understand how Winnicott uses the concept "illusion."

From the standpoint of the adult observer, the autistic infant (who believes he can achieve anything that he needs or wants) is experiencing repeated illusions. As an example, consider his need for food and comfort from the breast. He feels this need and metaphorically creates or imagines the breast as a means of satisfying it. Because he has an adequate mother who provides a breast at just the right time, that illusion is reinforced. It seems to be a reality to the infant. But the outside observer knows that the infant's concept of the breast as being under his control is an illusion stemming from the coincidental gratification of the infant's needs.

As the infant moves into the period of separation-individuation, the good mother will no longer wish to satisfy all needs immediately. In this phase of mother-child interaction it is important for her increasingly to frustrate her child because separation-individuation is a valuable developmental step that the mother wishes to foster. The child's frustration is an important stimulus toward further separation-individuation.

Incidentally, it is at this time that the illusory nature of the infant's concept of the "controlled" breast becomes clear. In the earlier phase the illusion was maintained because the mother supplied the breast just at the time it was desired. But now, when the breast is desired, it is often not there.

At this point in the infant's development, Winnicott postulates the appearance of transitional phenomena and transitional objects. A transitional object is one that first occurs between the period of complete fusion with the mother (primary narcissism in Mahler's terms) and the period when more mature object relations exist (that is, relations with entities perceived as separate from one's self). The transitional object exists in a particular and unique psychological space. It belongs neither totally in the world of one's self and external objects nor totally in the primary narcissistic world of no objects (the world in which the infant and mother are fused and, indeed, all of existence is fused). This is an intermediate "transitional" world that has elements of both of the others.

Consider the infant and his teddy bear (often an early transitional object). The teddy bear is separate from the infant and does many things that the infant wishes it to do. It is partially an external object. One can see this demonstrated at a later stage when the infant speaks of his teddy bear. "My teddy went on a trip today. He packed up his picnic lunch and went on a lovely walk into the forest. There he met another teddy bear and they danced and played together. . . ." But much earlier, before the infant can speak, he sucks his thumb and holds his teddy bear (or blanket or whatever) next to his cheek. The transitional object (Winnicott[17] calls it the first "not me") stands partly for the breast.

There are many aspects of transitional phenomena that remind us of play. Transitional objects are the carriers of illusions. Infants are not held to account for the results of their interaction with them. The illusions are accepted by normal parents. When a child tells of his fantasies about teddy, they are accepted. The interchange with transitional objects undergoes alteration and development. In those games (the word "game" suggests that the child is now playing), the infant learns how to utilize aspects of himself as well as aspects of the object, and he learns how to fit those two kinds of aspects together.

Thus it seems that transitional activities are the first play activities, and transitional objects, the first toys. These phenomena mark developments of the illusionary potential in human beings. They are an important way for humans to learn about themselves and the world of objects outside of themselves and of the relationship between the two. They are a form of trying out and practicing and communicating. All of these activities are strongly connected with the development of a sense of self.

We have focused up to now on the first appearance of transitional phenomena. However, these phenomena go on throughout life. They are related to play, religion, esthetics, and dreaming. They are related to all activities in which illusion plays a significant part.

If the therapist appreciates the significance of these phenomena, therapy can present a further opportunity to experience and utilize them. For therapy, then, becomes a later and special version of the early relationship between the developing child (patient) and the mother (therapist). From this standpoint, psychotherapy is a special modality developed and sanctioned by the culture, the purpose of which is to utilize transitional phenomena.

¶ Play in Psychodynamic Psychotherapy

Dramatic techniques—as opposed to other psychotherapeutic techniques which are also play-oriented—can be employed effectively in psychotherapy. (It should be said that these expositions are limited in their descriptions of presenting problems and histories of the patients and also in the accompanying theoretical discussion. This is because the emphasis here is on demonstrating techniques and how they are utilized.)

A sixty-year-old physician had been in psychoanalysis for lifelong episodes of depression. These were first noted at the age of thirty, at which time he sought psychiatric treatment. He tried many therapists and a number of different therapeutic approaches: supportive psychotherapy, reconstructive analytic therapy, family therapy, pharmacotherapy, and others. However, about two episodes of depression occurred each year. Thirty years ago, each episode was about one month in duration. The time (although variable) increased to three to six months for each depression. The depression was moderate to severe in intensity. The affective changes were the usual ones. Very prominent were doubts (about doing and thinking many things) and related inhibitions in activity.

His most recent analyst had been working with him for three years. The therapy had utilized psychoanalytic theory but had been flexible in technique. During the three years of therapy there had been no significant change in the occurrence of the depressions or the patient's attitude. After several weeks of one depressive episode, the therapist decided to use dramatic techniques. The patient had been limited in emotion and expression. Feeling frustrated, the therapist decided to experiment with a few dramatic exercises of the sort used by actors. The patient was willing to try. First some physical warm ups were performed (stretching, bending, walking around). Then some sounds were tried. The therapist and patient "passed" sounds between themselves—one would repeat the sound of the other, change it, then "give" it to the other, who would then repeat the new sound, change it, and so forth.

Then the therapist gave instructions for a simple solo improvisation. "Lie on your back on the floor. Relax and let your mind go blank. Then let an animal come into your mind. When and if it becomes comfortable, start to *be* that animal. If it doesn't work, doesn't feel comfortable, let a different animal come in. Start to move your body a little to see what that animal feels like, how it moves its parts. If you feel like it, let the animal make some sounds. When you are ready, have the animal start moving around and exploring its world."

The patient easily, comfortably, and in an interested way entered into the activities. He became a bear. Perambulating on all fours he came upon the seated therapist. He started to sniff and rub against the therapist's legs, shuffled away, and then returned. After about ten minutes the bear was finished with his walk. Then there was a discussion of the episode. Some time was spent on the specific meanings of the play occurrences, particularly the interest shown by the bear in the therapist. There was also some discussion of the patient's feelings about acting. He thought it was both interesting and fun.

This was the first of a number of "acting in therapy" experiences. The patient subsequently participated in two ten-week dramatic workshops. The first dramatic exercise was followed by the beginning of a termination of his depressive episode. During the subsequent eighteen months, the rate and intensity of depressive episodes markedly decreased. For the first time in his work with the present therapist, the patient became involved in meaningful and fruitful analytic work.

¶ The Method of "Experiential Theater"

In "Experiential Theater," six to nine student participants and one or two instructors form a workshop. Most workshops meet for ten to twenty sessions. Each is two to two and one-half hours in length. The first one and one-half hours are spent on the exercises. Discussion of meaning or reaction to the exercises is left to the final thirty to sixty minutes. There may be additional discussion when participants meet outside of the scheduled times or if they meet privately with their psychotherapists. This additional discussion, however, is neither suggested nor discouraged.

The specific exercises utilized at a particular session are chosen from a repertoire which is roughly divided into beginning, intermediate, and advanced sections. The instructors choose exercises, not in predetermined sequence, but according to what they sense might work for the group as a whole. In addition, they take into account reactions of workshop members. As in analytic therapy, there are a number of techniques that generally are appropriate for different phases (introductory, middle, or end) of the therapy, but adaptations in the form and timing of techniques will depend on a number of considerations in each specific case.

The purpose of the workshop is to "make theater." The instructors indicate that they wish participants to "have fun," "to feel loose," and to move toward pleasurable, authentic expression. The expression will be of different attitudes, emotions, and fantasies. Sometimes it will consist mainly of something *within* an individual; sometimes it will deal with his reactions toward other workshop members. Although members may wish to persevere, no one is expected to continue if he prefers to stop. The instructors are noncritical. Their efforts are directed toward making suggestions that may improve the work or to selecting new exercises that allow for richer or freer expression.

At the beginning of each session there are "loosening up" exercises. These include body stretching, isolated use of various muscle groups, and other movements designed to acquaint the participant with the body, its tension areas, and weaknesses and strengths. Then follows some relatively spontaneous movements designed to encourage playful interaction. One example is a game of "tag." Then there are exercises that allow for spontaneous creativity. For example, a sound or an imaginery object is passed around a circle composed of workshop members. First it is imitated by each member, but later each person changes it. As the activity proceeds, careful attention is paid to the (imagined) shape, weight, size, texture, and smell of the object.

When the group is ready for further work, creative improvisations begin. At first, these are structured and nonhuman. For example, in "machine," the first participant is asked to make a movement with a related sound, which represents some part of a complex machine. The next participant creates a second unit for the machine connecting it to the first. Then one by one the other workshop members add to the "machine."

Next come "human improvisations" in which two or more group members interact. Sometimes fairly specific instructions are provided: "Decide who you are and where you are. Then pick a situation in which one of you wishes to escape, and the other wishes to block that escape."

At other times the instructions are more general. The whole group may be told, "Half of you are going to be in or at a specific place. The other half are going to join them. Split into two groups. Agree on what the place is. Then as each person senses a role, let him or her begin to act it."

These are only a small number of the exercises. Their general purpose is to increase the flow of spontaneous and authentic expression of fantasies. In addition, an attempt is made to encourage interaction and cooperation among workshop members. The emphasis is away from talking out or acting out personal problems. However, most participants realize that the fantasies they are acting may well be related to personal issues. The main criteria for group approval is the spontaneity and feeling of conviction (or sense of authenticity) that may emerge through an exercise.

A final example of the dramatic method demonstrates how painful *ego,* alien, and conflictual impulses may be expressed in the group setting and how an opportunity exists to test out some of the significant implications of these impulses.

A humorous, poignant, and absolutely enthralling exercise had two men acting out a thwarted homosexual seduction. (Neither of these men were homosexuals or had dealt with homosexual conflicts in their therapy.) They agreed that their exercise would center around a fishing and nature-exploring trip in a canoe. As the second one entered the canoe, he "accidentally" touched the shoulder of the first. This stimulated him to believe that he was sexually attracted to his partner. He then subtly began to turn their conversation toward the pleasure of bodily contact. From time to time he would make a more overt (but not quite direct) proposal to the first man. The latter acted the role of someone becoming slowly aware of what was happening, being disconcerted by it, and wanting to fend off the proposition in a polite way. The improvisation went on for over half an hour and was enthusiastically applauded by the group.

An opportunity to understand some deeper implications of this scene soon occurred. Two days later, the second man (the one who had initiated the exercise and tried to seduce the other man) became severely depressed. Discussion with him revealed the following. His young son, who was "the most important thing in my life," lived with the patient's estranged wife. About a week before, she had vanished, apparently taking the son with her. The patient was concerned, but since the wife had acted in this way previously and had then returned, he was not too worried. But during the next few days, he received reports that made him believe that this time the wife might be permanently gone—and with her, the son. It was then that the depression became manifest. It is possible that the improvisation facilitated the patient's "getting in touch" with his fear of losing his son. In the exercise, he had reached out to touch another man. However his approach to the desired man was unsuccessful. Perhaps the experience of losing the man in the workshop was important in preparing him for the trauma of the ensuing loss of his son.

¶ Life as Transitional Phenomenon

If psychotherapy is an opportunity for growth through the use of transitional phenomena, can life in general provide a similar opportunity?

Life has many purposes. Human beings must exist; they try to satisfy needs, further their selected goals, and do many other things. One purpose is to continually define themselves. As new challenges present themselves, we learn how to deal with them, and in the process we learn what we can do and who we are. There are more active and less active ways of responding to challenge. Less active ones include allowing external issues to settle themselves and giving one's self over to a leader. A more active one is to allow one's self to form illusions, to play with the problems. This is more anxiety provoking perhaps, but it can also be more fun, and it provides the transcendant gratification of creativity and mastery.

The last word on play is not yet available. But what does seem true is that play is an important mode of learning that starts early in life. It is pleasurable because for most people it starts in extremely pleasant conditions. When a good mother, who has symbolized life and the power of the universe, begins to let her child go, to let her child learn about the world and how to handle it, the child begins to participate in his own creation as an individual human being. The mother is pleased because she participates in that creation. She continues the wonderful and crucial role of giving birth. This is the beginning of play, and this mode of learning and living, creating and being created, can continue throughout life. It changes and assumes special forms at different times. But its essence always remains—to give us pleasure in learning more about new worlds and more about ourselves.

¶ Acknowledgments

Daniel Tabachnick contributed to this article through bibliographic research, suggestions, and much valuable discussion.

The members of the seminar on movement at the Southern California Psychoanalytic Institute provided a general setting in which many of the ideas in the article took form.

¶ Bibliography

1. BRUNER, J. S. "Introduction," in Bruner, J. S., Jolly, A., and Sylva, K., eds., *Play: Its Role in Development and Evolution.* New York: Basic Books, 1976, pp. 13–24.
2. ———. "The Nature and Uses of Immaturity," reprinted in Bruner, J. S., Jolly, A., and Sylva, K., eds., *Play: Its Role in Development and Evolution.* New York: Basic Books, 1976, pp. 28–60. (Originally in *The American Psychologist,* 27: 8 (1972): 1–64.)
3. EINSTEIN, A. "A Mathematician's Mind," in *Ideas and Opinions.* New York: Crown, 1954, pp. 25–26.
4. FAGAN, J., and SHEPPARD, E. L., eds. *Gestalt Therapy Now.* Palo Alto, Calif.: Science and Behavior Books, 1970.
5. FREUD, S. "Beyond the Pleasure Principle," in Strachey, J., ed., *The Standard Edition of the Complete Psychological Works of Sigmund Freud,* vol. 18. London: Hogarth Press, 1955, pp. 3–64.
6. GILMORE, B. J. "Play, A Special Behavior," in Herron, R. E., and Sutton-Smith, B., eds., *Child's Play.* New York: John Wiley, 1971, pp. 311–375.
7. GROOS, C. *The Play of Animals.* New York: D. Appleton, 1898.
8. HALL, G. S. *Youth: Its Education, Regimen and Hygiene.* New York: Appleton, 1907.
9. MAHLER, M. S., PINE, F., and BERGMAN, A. *The Psychological Birth of the Human Infant.* New York: Basic Books, 1975.
10. MARX, K. "Contributions to the Critique of Hegel's Philosophy of Right," in Marx, K., and Engels, F., eds., *On Religion.* Moscow: Foreign Languages Publishing House, 2nd Impression, Deutsch-Französische Jahrbücher, 1844, 41–58.
11. MILLAR, S. *The Psychology of Play.* New York: Jason Aaronson, 1974.
12. REYNOLDS, P. "Play, Language and Human Evolution," in Bruner, J. S., Jolly, A., and Silva, K., eds., *Play: Its Role in Development and Evolution.* New York: Basic Books, 1976, pp. 621–633.
13. SPENCER, H. *Principles of Psychology.* New York: Appleton, 1873.
14. SPOLIN, V. *Improvisations for the Theater.* Evanston, Ill.: Northwestern University Press, 1963.
15. STONE, G. "The Play of Little Children," in Herron, R. E., and Sutton-Smith, B., eds., *Child's Play.* New York: John Wiley, 1971, pp. 4–17.
16. TABACHNICK, N., and OSMAN, M. "Acting, Play and Human Growth." in press.
17. WINNICOTT, D. W. "Transitional Objects and Transitional Phenomena," *The International Journal of Psycho-Analysis,* 34: 2 (1953): 89–97.
18. ———. *Playing and Reality.* New York: Basic Books, 1971.

CHAPTER 26

THERAPY OF MENTAL ILLNESS IN LATE LIFE

Ewald W. Busse

PRIOR to the selection and implementation of a therapeutic plan for the treatment of mental illness, it is necessary to establish the proper diagnosis and to have an understanding of the disease. The clinician should know as much as possible about the disease process, the prognosis, the various biological and environmental factors that influence the course of the disorder, and the effectiveness of treatment. This is consistent with the multiarial approach of DSM-III.[58] There are no mental disorders that occur exclusively in late life. There are, however, a number that are more common to the latter part of the life span, after the age of sixty-five. These include organic mental disorders and affective disorders, namely depression, paranoia, hypochondriasis, and sleep disturbances.

¶ Demography

The percentage of the population in the United States of persons over the age of sixty-five has increased steadily, from 4.1 percent in 1900 to 8.1 percent in 1950; 9.8 percent in 1970; 10.3 percent in 1975; and 11 percent in 1980. If current trends continue, the percentage of the older population will be 11.7 percent in the year 2000 and 16.1 percent in 2050.[7]

Computed from death rates in 1977, the average life expectancy at birth for both sexes combined was 73.2 years. For males, it was 69.3 years and for females, 77.1 years, a difference of 7.8 years. At age sixty-five the remaining expected years of life for women are 18.3 years and for men 13.9 years, a difference of 4.4 years. If recent decreases in death rates continue, especially from cardiovascular conditions, life expectancies will increase even further. This may not, however, improve the quality of life, as it is possible that more individuals will develop senile dementia or related organic mental diseases.[6]

Statisticians and epidemiologists frequently refer to persons sixty-five years of age or over as "the older population." However, from a health standpoint those sixty-five to seventy-five years (61.8 percent of the older population) are remarkably different from those seventy-five years of age and older (38.2 percent).[6,7] The majority of those sixty-five to seventy-five years are relatively healthy, although they may have some incapacities. Their health as a group is more similar to those between the ages of fifty-five and sixty-five. Consequently, those seventy-five years of age and over constitute the target population for many of the medical and supportive services that are provided by both public and private programs.

Based on data for 1978, the average American fifty-five to sixty-four years of age spends 1.9 days per year in a short hospital stay. This increases to an average of 3.2 days for persons age sixty-five to seventy-four and to 6.0 days for those seventy-five years plus. Nursing home utilization increases rapidly with age. In 1976, on the average, a person aged fifty-five to sixty-four spent a fraction of a day per year in a nursing home. Between sixty-five and seventy-four years of age, this jumped to 4.4 days; between seventy-five and eighty-four years, to 21.5 days; and for those over eighty-five years of age, to 86.4 days per year.

In 1977, there were 1.1 million older people in nursing homes. Fifteen percent were between the ages of sixty-five and seventy-four; 41 percent were seventy-five to eighty-four; and 40 percent were 85 years and over. In the nursing home population, 74 percent were women, 69 percent were widowed, 14 percent were single, and 12 percent were married. Ninety-three percent of occupants of nursing homes are white.[6]

Older people represent 11 percent of the total population of the United States but account for 29 percent (41.3 billion dollars) of total personal health care expenditures.

¶ The Aging Brain

With the passage of time there are numerous biological changes in the human body and brain. So-called normal aging is accompanied by gradual loss of muscle cells and of the number of neurons within the brain and nervous system. Physiological changes also transpire that alter the way the brain functions while awake and asleep. Such changes have an impact on the way the nervous system responds. However, this does not necessarily mean that the older person who is physically healthy and mentally active inevitably suffers intellectual losses and mental incapacity. When serious losses of intellectual capacity develop and when disruptive patterns of behavior appear, these are the result of one of several diseases which are referred to as organic brain disorders. Included in the organic brain diseases are senile dementia, Alzheimer's disease, and multi-infarct diseases (cerebral vascular disease). All of these incapacitating brain disorders are receiving increasing attention by biological and behavioral scientists. Epidemiological studies throughout the world indicate that these disorders are widespread and occur in 4 to 6 percent of the population age sixty-five and over. So-called Alzheimer's disease is characterized by relatively early onset, that is, sixty years of age or earlier, while senile dementia is likely to make an appearance after the age of seventy. The cause of these diseases is unknown, but it appears that a genetic predisposition plays a role.

¶ Metabolism and Drugs in Late Life

Drugs are metabolized differently by older people than they are by younger adults. The metabolic conditions affected by age include absorption, distribution, destruction, excretion, kinetics of drug binding, and alterations in biological rhythms.[53]

All of these changes can be exacerbated by disease and trauma. Some of the major age changes are the loss or decline of efficiency of renal function. Between the ages of forty and eighty-nine, there is a 55 percent reduction in renal clearance.[54] There is a redistribution of body content, with a decline in protein and an increase in fat. It is estimated that the basic metabolic rate during the adult years declines 16 percent from age thirty to seventy years, while the caloric requirement drops approximately one-third because of decreased metabolism and exercise.[54] The loss of nonreplaceable cells plays an important role in aging physiological changes. The loss of brain cells may not only alter important aspects of body metabolism but also may make the brain more sensitive to certain drugs. Striated musculature diminishes to about one-half by approximately eighty years of age.[54] As these muscle cells disappear, they are replaced by fat cells in fibrous connective tissue. Hence, the storage capacity for those drugs that are stored in fat cells is clearly increased. Drugs given orally may not be absorbed as quickly, as the blood flow to the upper gastrointestinal tract is decreased. Aging produces a decrease in heart cells, and there is a decline in cardiac output. In some elderly people, the loss of P cells results in a dysrhythmia or other conductive disturbance of the cardiac rhythm. The aging changes that affect neural transmitters may contribute to causing a number of mental disorders in the elderly patient. There is a decline in concentration and a decrease in certain substances within the brain, including dopamine, norepinephrine, serotonin, tyrosine hydroxylase, and cholinesterase.[44] The activity of monoamine oxidase increases with age.[47] This would contribute to a decline in norepinephrine, dopamine, and serotonin.

¶ The Health Status of the Older Population

Chronic conditions frequently limit the activity of older Americans. For males age sixty-five and over, two in five report restricted activity, and one in four indicates an inability to carry on some major activity. For adult males between the ages of forty-five and sixty-four, only one in ten reports restrictions in amount or kind of major activity, and only one in sixteen is unable to carry out a major activity. Again, it should be noted that the accumulated data on the older population are significantly influenced by those seventy-five years of age and older.[55]

Medicare did make a difference in the reported use of physicians by older persons between 1962 and 1975. In 1962, 35 percent of older persons had not seen a physician for a year or more. In 1975, this figure dropped to 19 percent. The more severely handicapped older people are far more likely to see a physician now than before Medicare.

As to psychiatric disorders in old age, it appears that 1 percent of persons over the age of sixty-five are in private or public mental institutions. Of the 3 to 4 percent of old people in nursing homes, homes for the aged, geriatric, or chronic disease hospitals, it is estimated that more than one-half suffer from significant psychiatric disturbances. Therefore, it can be concluded that at least 60 percent of all the old people who have been institutionalized are suffering from psychiatric illnesses.

There is no doubt that elderly people have underutilized outpatient psychiatric and mental health clinics and facilities. This has led to the requirement that all federally supported community mental health centers give special attention to the problems of the elderly. As to the mental health status of the elderly living in the community, a number of surveys conducted in this country and abroad reveal that approximately 4 to 5 percent are either psychotic or else have serious psychopathological symptoms. Another 10 percent have mental problems of moderate extent. There is considerable variation, ranging from an additional 15 to 40 percent, in those considered to have mild disturbances. There is little doubt that there is a deficiency of outpatient care provided for the elderly. Projections by the National Institute of Mental Health indicate that at least 80 percent of the elderly with mental health problems are being neglected in this area.[5]

¶ Organic Mental Disorders

The term *dementia* was first utilized by Philippe Pinel in the early 1800s and gained widespread attention as the result of a paper published by Benjamin Rush in 1812. Today, it is estimated that 4.4 percent of the elderly population of the United States, that is, 1,000,000 people, have some degree of organic mental impairment. It is estimated that more than 50 percent of Americans in nursing homes are there because of mental impairment. The cost of maintaining residents in nursing homes has gone up sharply. Over $10 billion was spent for nursing home care in 1976, $12 billion in 1977, and $15 billion in 1978.[17]

Considerable scientific effort has been expended to determine if senile dementia can be differentiated from presenile dementia (formerly called Alzheimer's disease). Neuropathologists generally hold that they are the same disorder, as the histopathologic findings are sufficiently similar that they cannot be differentiated. The clinical course of the disorders is similar, but, as defined in DSM-III,[3] the age of onset is the differentiating criterion. Some epidemiologists, behavioral geneticists, and others feel the two disorders are separable. There does appear to be a gap between the usual age of onset of presenile and senile dementia, and familial influence is more evident in presenile dementia than in senile dementia. Two presenile dementias, Pick's disease and Creutzfeldt-Jakob disease, do not enter into this controversy, as the neuropathological findings are characteristic, and the possible cause of these two disorders is better understood.

DSM-III has been revised and undoubtedly will continue to be revised. Most changes are concerned with name or diagnostic criteria and a distinction is made between organic brain syndromes and organic mental disorders.[3] (The DSM-III classification is given in parentheses when different from traditional diagnosis.) The essential feature of organic mental disorders is the presence of a biologic (organic) defect that permanently or transiently impairs the functioning of the brain. This biologic or physiologic defect may be primary or secondary; that is, it may be located exclusively within the brain or the brain dysfunction may be the secondary result of a disease process located elsewhere in the body. Clearly, the location is a major determinant in planning treatment. The alterations of brain functions that are usually observed include perceptions and interpretation of perceptions, learning, memory, orientation, decision making, speech, and behavior.

Primary Degenerative Dementia, Senile

Early recognition of dementia or any type of organic brain disease is facilitated by observing the sequence of onset of signs and symptoms. Subjective and objective signs and symptoms may not have a consistent relationship. In general, in senile dementia the earliest symptom is short-term memory loss. This is followed by impairment in decision making, which precipitates depression, anxiety, and fear of impending loss of independence. Often clear evidence of emotional instability follows, with reactions that are inappropriate to the situation. Fluctuation in alertness and level of awareness, a decline in attention span, and speech and sleep disturbances can often be seen throughout the course of the disease. Defects of orientation to time, place, and person then become more evident, followed by global defects in intellectual activities and behavioral changes.

In 1971, Wang and Busse[63] reported in some detail observations from the Duke Longitudinal Study regarding dementia in old age. In the Duke study, the term *brain impairment* was utilized to designate measures of loss of brain function based upon a number of laboratory procedures, including EEG, cerebral blood flow, cerebral metabolism, and so forth. The manifestations of possible brain impairment were determined by person-to-person observations, such as psychological tests, and clinical measures both of intellectual performance and emotional variations.

It was found that there is often a poor correlation between these two types of evaluation, that is, laboratory procedures versus qualified person-to-person observations. Of particular concern is the discrepancy—as high as 25 percent—found in subjects with a precipitous decline in clinically observable mental signs and symptoms which was not paralleled by evidence of physiologically measured brain changes. Careful consideration of factors such as general physical health, economic status, social environment, and previous living habits led to the conclusion that dementia in late life is a sociopsychosomatic disorder.

The technique of assessment for organic brain syndrome in Longitudinal Study I was a systematized mental status examination which included two sections; the first specifically rated the presence or absence and degree of organic signs and symptoms, and the second was referred to in data as the "Q-SUM"–a six-point global Organic Brain Syndrome (OBS) rating scale.

Longitudinal analysis reveals that approximately one-half of the subjects at some time after the age of sixty years received mental impaired ratings, yet many in a subsequent year (or years) were not impaired. These findings point to a high variability in the course of organic brain syndrome.[13]

The factors contributing to the appearance of organic brain signs and symptoms have been studied in this longitudinal population of subjects. The major contributors appeared to be decompensated heart disease, low socioeconomic status, and decreased physical and mental activity.

The relationship to cardiovascular pulmonary disease has been considered very carefully. Mild elevation of blood pressure was positively correlated with preserved brain function. It is our speculation that the relatively high blood pressure may be necessary to maintain sufficient blood supply to the brain for adequate cerebral function.

Therefore, we have concluded that the loss of mental ability in late life is not usually the result of a pathologic brain change alone. Incapacitating dementia is not an inevitable consequence of old age. The loss of mental ability (what one does) and capacity (what is possible) is influenced by changes within the brain, the health status of the entire body, behavioral patterns, and socioeconomic determinants. Consequently, the clinical course is often found to be one of periodic exacerbation and remission, or recovery and recurrence.

Differential diagnosis is extremely important, particularly in recognizing those treatable conditions that are masquerading as senile dementia. There are a number of pathological conditions that secondarily affect the brain, including heart failure, infection, uremia, hypothyroidism, vitamin deficiency, and toxicities from a variety of medical drugs as well as alcohol. All of these conditions can result in temporary or permanent mental impairment, and often in memory difficulties and confusion. Treating the medical problem or withdrawing the toxication can often clear the patient's sensorium and return the patient to a reasonable state of functioning. In addition, depressive illnesses are frequently mistaken for organic brain disease. It is unfortunate to miss these cases as they do respond to antidepressive drugs and, if necessary in severe cases, to electroconvulsive therapy.

Diagnostic Procedures

The clinician who is uncertain of a diagnosis of organic mental disorder will often request the assistance of a psychological laboratory to determine the extent or existence of organic mental impairment. The Weschler Adult Intelligence Scale (WAIS) has been widely used for over thirty years. The WAIS continues to be one of the best measures of intelligence for use with the aged.[56] It is important to understand the composition of the test and "normal" age changes. The WAIS is organized into two major components: verbal and performance intelligence, with eleven subtests, six verbal and five nonverbal. Siegler[56] concludes that verbal intelligence tends to increase until the sixties and then falls off gradually. Performance intelligence increases until the forties with a gradual decline until the sixties and a sharp decline thereafter. Clearly, verbal abilities are maintained considerably longer than nonverbal abilities. The abilities which require speed for optimal performance are most affected by the aging process and decrements appear early in life, between the ages of thirty and fifty. However, there are always some individuals who apparently maintain many skills into late life.

The estimate of organic impairment utilizing the Weschler Adult Intelligence Scale is usually based on a discrepancy between the verbal scale and the performance scale. Obviously, because of the vagaries of the aging process among the elderly, this discrepancy may not be as significant as when found in the younger adult. In addition to intelligence testing, psychologists have attempted to develop standardized procedures to assess specific organic impairment.

A test familiar to most clinicians is the Bender-Gestalt.[62] This procedure is reported to be more sensitive in discriminating organic brain damage from functional problems when augmented by the Background Interference Procedure.[19]

The Halstead-Reitan test[62] is a rather time-consuming procedure which requires a reasonable degree of cooperation on the part of the patient. When properly carried out, the examination has been useful, but it has obvious limitations.

The most readily available laboratory diagnostic procedures are plain radiographs of the skull. This noninvasive technique is well established as a routine screening procedure. It is particularly useful in revealing the presence of intercranial space-occupying mass that may be an expanding lesion. It is also useful in identifying the presence and extent of a skull fracture. Pneumoencephalography and ventriculography are relatively complicated procedures that have been utilized for years. They have been rapidly replaced, however, by computerized tomography of the brain. Other diagnostic procedures that are giving way to computerized tomography (CT) include cerebral angiography, radioisotope brain scan, and radioisotope-cisternography. Wang,[62] in reviewing all of these diagnostic procedures, concludes that there are three noninvasive procedures that are most useful for diagnostic purposes. These include electroencephalography, computerized tomography, and the determination of reasonable cerebral blood flow utilizing the xenon-133 inhalation method. This latter procedure is likely to be available in only a few medical centers.

Treatment

The treatment of primary degenerative dementia must be selected with the emphasis upon the most serious symptoms. The techniques can be separated into certain categories including psychosocial therapy (interpersonal and social skills), ergotherapy (work), ludotherapy (playing games), kinesitherapy (movement and exercise), and group therapy. All of these approaches have merit and should be part of a well-organized comprehensive program. A single approach cannot be expected to produce good results. Each of these techniques must be adapted not only to the condition of the patient but to the strengths and weaknesses of the social and physical environment. Reality orientation is a technique that is particularly applicable to hospitalized elderly or brain-damaged persons with moderate symptoms including periods of confusion and disorientation. One of the leading proponents of this early phase rehabilitation believes that the approach is based upon the recognition that the patient with brain damage is likely to withdraw from reality and in general avoid contact with the environment.[32] As the term reality orientation implies, the process utilizes continuing stimulation. There is repetition of such basic material as the patient's name, the physical location, the day of the week, the month, the year, what meal comes next, and so forth. The success of this approach is heavily dependent upon the dedication and competency of personnel. Reality orientation deals not only with the restoration of information; it can only be successfully implemented if confusing elements in the environment and the routine of the hospital are reduced to a minimum and communication becomes consistent, clear and nonthreatening.

Milieu therapy is particularly useful in the treatment of dementia patients in a residential setting. It places a large amount of responsibility on the individual patient for his or her own therapeutic program. It includes a structured series of meaningful behaviors. Positive reinforcement for appropriate behavior is emphasized.[22]

Attitude therapy has been added to the regime of techniques for the institutionalized elderly. In brief, attitude therapy contributes to the consistency of how a particular patient is approached. All of the members of the team follow a predetermined pattern in dealing with the patient. This particular therapeutic approach is often incorporated in both milieu and reality orientation therapy.

The choice of an antipsychotic medication is to a large extent determined by the possible side effects of a particular drug weighted against the behavioral and medical status of the individual patient. Two of the most commonly used antipsychotic agents used in the treatment of the elderly are thioridazine and haloperidol. Haloperidol has minimal anticholinergic and alpha adrenergic blocking properties, but thioridazine may be more effective.[25] The geriatric patient should receive 0.5 mg. daily as a starting dose to lessen the chances of extrapyramidal side effects. An average daily dose of 2 mg. is effective for a variety of symptoms.

¶ Multi-infarct Dementia

Multi-infarct dementia was previously labeled organic brain syndrome with cerebral arteriosclerosis. It is sometimes referred to as repeated infarct disease. This diagnosis of the disorder has been confirmed by autopsy. The clinical course is remittent and fluctuating, with episodes of confusion. Consequently, the intellectual deficits tend to be "patchy." Not infrequently multi-infarct dementia occurs in patients with hypertension. Hence, treatment of this cardiovascular disorder may influence the progression of the mental disorder. Repeated infarct dementia is not nearly as common as senile dementia, and it is generally believed to be more common in men than in women. The age of onset is often in the late sixties, between the usual age of onset of presenile dementia and senile dementia. However, epidemiological studies are far from satisfactory. In addition to the signs and symptoms of mental deterioration, transient or persistent focal neurological signs and symptoms are associated with the disorder. These signs and symptoms include unilateral exaggeration of deep tendon reflexes, dysarthria, balance, and gate disturbances.

Cardiac Arrhythmias in Dementia

Although brain dysfunction can produce cardiac arrhythmias, it appears that cardiac arrhythmias more frequently contribute to brain dysfunction. Cardiac arrhythmias often lower cardiac output so that cerebral blood flow is impaired. It is also possible that arrhythmias predispose to atrial thrombi resulting in multiple cerebral emboli.[51]

Among the cardiac arrhythmias, auricular fibrillation, auricular flutter, and ventricular arrhythmias are very likely to impair cardiac output and therefore to produce signs and symptoms of cerebral insufficiency.

Patients with tachycardia and bradycardia not infrequently experience cerebral ischemic attacks.[59] To establish the relationship of a cardiac dysrhythmia to cerebral changes, it may be necessary to conduct twenty-four-hour monitoring. Cardiac monitoring during periods of sleep may reveal that serious dysrhythmias are occurring and that there are sufficient indications to justify implanting a cardiac pacemaker.[46]

The Decline in the Incidence of Stroke

During the last twenty-five years there has been a decline in the incidence of stroke (cerebral infarction and intracerebral hemorrhage). The explanation for this apparent reduction is not clear. There is no doubt that mortality from stroke in recent years has altered. The reduction appears to have occurred in two phases—one from 1945 to 1959 when the overall average annual decline in the incidence rate adjusted for age and sex was 3.1 per 100,000. A plateau was reached for the next five years, followed by an acceleration in decline. The average annual decreases were 4.8 and 5.3 per 100,000 for the quinquennial periods 1965 to 1969 and 1970 to 1974, respectively. Since effective antihypertensive therapy was not readily available until the early 1950s, it is hard to explain the 5 percent decline between 1945 and 1949 and 1950 to 1954 on the basis of such treatment.[33]

¶ Transient Ischemic Attacks

Episodic cerebral disturbances commonly referred to as transient ischemic attacks (TIA) are not infrequent in the older population. The majority of these episodes are probably attributable to thrombi and emboli affecting areas where restoration of adequate blood flow is possible largely because of collateral circulation. If, however, adequate cerebral blood flow to the affected areas cannot be reestablished, the disease merges into multiple infarct disease.

The carotid bifurcation has a remarkable predilection for atherosclerotic change. In fact, it is reported that one-third of the people beyond the age of sixty-five have an advanced atherosclerotic plaque at the carotid bifurcation.[37] Consequently, palpation and auscultation of the neck are essential, and a bruit at the carotid bifurcation should never be ignored. Carotid endarterectomy is often indicated.

Treatment of Cerebral Vascular Disease

The inhalation of CO_2 in normal individuals has demonstrated a significant increase in cerebral blood flow. Unfortunately many patients with vascular disease fail to show the expected increase in cerebral blood flow following CO_2 inhalation. CO_2 is a selective cerebral vasodilator, but most drugs do not have this selective capability. Many so-called vasodilators result in a transient increase of cerebral blood flow followed by a peripheral vasodilation, causing postural hypotension and a decrease in cerebral blood flow. Consequently, there are serious limitations in the utilization of many vasodilator drugs. There appear to be emerging two types of so-called vasodilators. The first are primary vasodilators, and the second are mixed action, that is, both vasodilators and metabolic stimulants of the brain. The primary vasodilators include cyclandelate (Cyclospasmol), papaverine hydrochloride (Pavabid), and isoxuprine (Vasodilan). Despite the fact that these drugs continue to be considered vasodilators, there is little evidence that this is the mechanism by which they influence the physiological status of the brain. The mixed action drugs are predominantly the dihydrogenated ergot alkaloids (Hydergine) and deapril-ST. In Europe there are a number of other similar drugs such as naftidrofuryl (Praxilene).[65] One theory explaining the mixed action or stimulating drug is that it improves the mechanism of cerebral ganglion cells; that is, it increases their uptake of water, glucose, and oxygen which in turn allows astrocytes and capillaries, particularly on the arterial side, to return to normal dimensions.[21]

¶ Normal Pressure Hydrocephalus

The syndrome of normal pressure hydrocephalus was first defined about fifteen years ago.[1] It is a diagnostic category not found in DSM-III. Typically this disease is manifest by a gait disturbance, incontinence, and intellectual impairment. The diagnosis is established by demonstrating enlarged ventricles attributed to a communicating hydrocephalus in the presence of normal cerebral spinal fluid pressure. Katzman[40] describes two forms of hydrocephalus that must be distinguished. In one group the disorder appears to be secondary to previous head trauma, subarachnoid hemorrhage, or meningitis, while in the other group, the cause is not apparent. Therefore, it is designated idiopathic normal pressure hydrocephalus. In this type, the onset may be insidious, but the course is progressive. It appears that idiopathic normal pressure hydrocephalus is a disease of the presenium. The usual age of onset is between age fifty-five and sixty-five. After surgical intervention, 60 percent show definite improvement.

A major clinical problem is the differential diagnosis of idiopathic normal pressure hydrocephalus from presenile dementia of the Alzheimer's disease type. Although the usual diagnostic procedures should be carried out, particularly computerized axial tomography, the clinical presentation becomes of utmost importance. Gait impairment, which is common in normal pressure hydrocephalus, is rare in Alzheimer's disease. The clinician must keep in mind the classical triad of gait disturbance, incontinence, and dementia.

As to clinical management, consideration of a neurosurgical shunt procedure is important.[45] The percentage of positive responders appears to be much higher in the secondary forms of normal pressure (65 percent) hydrocephalus than in the idiopathic form (40 percent).

¶ Transient Global Amnesia

In 1964, Fisher and Adams[31] reported a particularly interesting variety of transient amnesia which they labeled transient global amnesia. Since that initial observation, similar cases have been observed. The clinical picture is striking. The victim, usually sixty years of age or older, has a sudden onset episode of retrograde amnesia. The amnesia may last for hours and then may gradually clear. The victim has no recollection of the amnesic episode, but the cognitive facilities during the amnesic period are not disturbed, although the person is aware of the memory loss and can become very anxious about the disability. Episodes of transient global amnesia are sometimes repeated, but usually they are a single occurrence. Various etiological explanations have been offered, including vascular disease and localized disturbance of blood flow, hysteria, seizures, and postictal reactions. Joynt[38] observed six cases in which there were prominent electroencephalographic abnormalities in the temporal lobe. It is likely that this is significant. It is possible, however, that the temporal lobe disturbance is the same as that frequently encountered in elderly people.

¶ Pick's Disease

Pick's disease, named after A. Pick, who first described its symptoms in 1892, is a rare disorder.[50] Pick's original purpose was to illustrate the different types of aphasic manifestations that may occur in senile brain diseases. He did not recognize it as a distinct pathological entity. The work of others was necessary to establish the disease as a distinct heredodegenerative process. It is more frequent in females, occurring in a ratio 2 to 1. Clinical criteria to diagnose Pick's disease are not uniformly accepted, although behavioral changes seem to precede memory defects.[39] In a few cases, it is said to be clinically distinguishable from Alzheimer's disease because its symptomatology is related to maximum atrophy of the orbitofrontal and temporal areas, particularly on the left.

Microscopically the nerve cell loss and the replacement gliosis are obvious in the supragranular layers. Sometimes the changes associated with Alzheimer's disease are present. Hence, diagnosis is dependent on the recognition of the "Pick cell." These cells have large agyrophilic inclusion bodies, nuclear eccentricity, and a distorted cell contour.[26] Etiology remains undetermined and the treatment is merely symptom modification.

¶ Creutzfeldt-Jakob Disease

There are several progressive degenerative diseases resulting in dementia which are believed to be caused by a slow virus. Creutzfeldt-Jakob disease is one that occurs in late middle life and is accompanied by the usual symptoms of a dementia, including memory and cognitive changes, visual difficulties, and behavioral alterations. Hallucinations may occur. Other symptoms include myoclonus, hyperesthesia, ataxia, and dysarthria. This disorder was described by Creutzfeldt in 1920.[24] It was not until 1968, however, that it became associated with a slow virus. At that time, Gibbs and his coworkers[34] reported the transmission of the disease to a monkey from a patient diagnosed as having Creutzfeldt-Jakob disease. Since that time it has been observed that it can be transmitted from one human to another by direct tissue contact, for example, corneal transplant. A slow virus is also the cause of kuru, a somewhat similar disease that was first recognized in New Guinea and was attributed to their practice of cannibalism. The major effects of kuru involve the cerebellum.

¶ Depressive Episodes

Evidence indicates that depressive episodes increase in frequency and depth in the advanced years of life. Elderly subjects are aware of these more frequent and more annoying depressive periods, and they report that during such episodes they feel discouraged, worried, and troubled, and often see no reason to continue their existence.[10] However, only a small number admit entertaining suicidal ideas; a larger percentage state that during such depressive episodes they would welcome a painless death. During such periods, the elderly are more or less incapacitated, but they rarely seek medical help. This type of reaction must be distinguished from a major depressive illness with persistent biologic signs and symptoms, as a major depression requires pharmacologic treatment and often hospitalization.

The observation that elderly subjects were aware that they were experiencing more frequent and more annoying depressive episodes is based upon a study made some years ago[16] and confirmed by more recent longitudinal studies.[11] Observations indicate that there is a difference in the process leading to depressive episodes in the elderly as compared with middle-aged or young adults. Guilt and the turning inward of unconscious impulses (interjection) are common mechanisms in the depressions of young adults. This is not the case with elderly subjects. Depressive episodes can be readily linked with the loss of so-called narcissistic supplies. The older subject becomes depressed when he cannot find ways of gratifying his needs; that is, when the social environment changes or the decreased efficiency of his body prevents him from meeting his needs and reducing his tensions, he is likely to have a loss of self-esteem. Hence, he feels depressed.

There is clear evidence that the frequency of depressive episodes is influenced by the life situation. For example, three groups of subjects reported mood disturbances occurring at least once a month and lasting from a few hours to a few days.[17] The highest number of subjects (48 percent) reporting mood disturbances were persons over the age of sixty, unable to work, attending an outpatient clinic for various physical disorders, and suffering financial hardships. Depressive spells occurred in 44 percent of those who were retired, in good health, and in acceptable financial condition. Only 25 percent of subjects continuing to work past the usual age of retirement reported such experiences. Most of the subjects in the three groups denied that they had experienced depressive spells of similar frequency or duration earlier in life.

To appreciate fully the factors that are

important to depressive episodes in the elderly, particular attention must be given to attitudes toward chronic disease, disability, and death. When studied longitudinally, the importance of physical health as a determinant of depressive feelings becomes increasingly evident. It appears that the aged person can tolerate the loss of love objects and prestige better than a decline in health, as physical disability often disrupts mobility and results in partial isolation. Hence, the opportunities for restoration of self-esteem are reduced.[57]

Important factors that contribute to depressive feelings of elderly persons are often conscious, as approximately 85 percent of elderly subjects are able to identify the specific event or stimulus that precipitated the feelings of depression. Many depressive episodes in the elderly therefore are a realistic grief response to a loss and not primarily influenced by unconscious mechanisms. The symptom is relieved when the actual loss or threat is removed or compensated for.

A recent review of life change events and the onset of major depression in adults indicates a significant relationship between frequency of upsetting experiences and the onset of depression. This suggests that there exist common precipitating factors in both depressive episodes and serious depressions.[52]

Major Depressive Disorders

DSM-III[3] attempts to separate episodic depressions from major depressive episodes. The distinction is based on the presence of a dysphoric mood of at least two weeks' duration and the existence of at least four of eight symptoms which have persisted and are of a significant degree. These include alteration in appetite with weight loss or weight gain, sleep changes and insomnia or hypersomnia, a loss of energy, psychomotor agitation or retardation, a loss of interest or pleasure in usual activities, a decrease in sexual drive, feelings of self-reproach or inappropriate guilt, decreased ability to think, indecisiveness, and lastly, suicide ideation.

Sleep, EEG Changes, and Depression

In recent years there has been increasing evidence that patients with primary depressions have a number of EEG changes including a reduction in total sleep time and a short rapid eye movement (REM) latency (that is, the time between sleep onset and the first REM). In turn, it appears that patients with depression secondary to medical disorders also have characteristic EEG changes. It appears that EEG sleep changes have a relatively high predictive value in determining those who have primary depressions. These studies are summarized in a recent report by B. J. Carroll.[20] It is possible that these EEG changes associated with depressions in young and middle life may not be as accurate a predictor in late life because of the complication of EEG changes accompanying old age.

Biological Measures and the Differential Diagnosis of Depression

Although EEG recordings during nocturnal sleep are of value in differentiating types of depression, for young and middle-aged adults other procedures are emerging which appear to be less time-consuming, equally effective, and applicable to late life. Growth hormone (GH) stimulation tests require a half day of the patient's time and can be carried out on outpatients as well as inpatients. For several years, studies of pituitary growth hormonal regulation have revealed suggestive evidence that depressed patients secrete less growth hormone than normal in response to a variety of stimuli.[43] Several substances that may be used in diagnosis include amphetamine, clonidine, and desipramine. Utilizing desipramine, it has been demonstrated that endogenous depressives had low growth hormone responses, while the neurotic depressives have exaggerated responses. Such diagnostic studies should be viewed with caution until they are adequately repeated utilizing groups of elderly persons. Although human pituitary *content* of growth hormone is relatively constant with age, the loss of a few hypothalamic cells that influence the *release* of pituitary hormones may be a confusing factor in the diagnosis of those in late life.

¶ The Treatment of Depression

There are several major classes of pharmacological agents that are utilized for treatment of depression in the elderly. These include the tricyclic compounds, the monoamine oxidase inhibitors, the stimulant drugs, and the benzodiazepines. The latter are used primarily for anxious patients with neurotic depressions. Lithium is sometimes used as a prophylactic against recurrent manic attacks and depressions and to modify bipolar mood swings (A first manic attack is rare in late life). The tricyclic compounds are frequently utilized to treat elderly patients with biological signs of depression. Although the tricyclic drugs have a common structure, they do differ chemically from each other. Consequently, these differences influence their clinical effects as well as their undesirable side effects.

The tricyclic compounds are usually administered orally and are rapidly absorbed. However, the drug is found in higher levels in an active unbound form in the blood stream of the elderly. This presence of unbound tricyclic drugs is related to the higher incidence of side effects. Furthermore, the rate of metabolism for the tricyclic agents decreases with advancing age. Consequently, patients in late life, despite receiving a lower daily dose of the drugs, tend to have a higher blood level, and the plasma level tends to be unstable. Elimination of the drug in older patients is also prolonged. For the reasons given, the determination of tricyclic plasma levels is particularly useful in the elderly where a relatively small change in the dose may produce a marked alteration in the plasma level and alter the therapeutic effect or expose the older person to the undesirable side effects. Blood collection for plasma levels should be done before the morning dose, no anticoagulant should be used, and caffeine should not be ingested twelve hours before the sample is taken.

Walker and Brodie[61] believe that there is a curvilinear relationship between the plasma levels of the secondary amine tricyclic antidepressants and the therapeutic effect, while the tertiary means have a linear or a sigmoid relationship. Task[60] originally reported this observation.

The tricyclic drugs can be separated into two groups: The tertiary amines that block the reuptake of inactivation of biogenic amines, principally serotonin, at the synaptic junction, while the secondary amines block reuptake inactivation of norepinephrine. The tertiary amines include imipramine, amitriptyline, and doxepin. The secondary amines are desipramine, nortriptyline, and protriptyline. Tricyclics are powerful anticholinergic agents. The most common side effects are due to their anticholinergic properties and include dry mouth, sweating, blurred vision, urinary retention, and paralytic ileus.

Monoamine oxidase inhibitors (MAO) are rarely used as the first drug of choice. The MAO inhibitors increase the amounts of norepinephrine, dopamine, and serotonin in the brain. Generally, they are not believed to be as effective as the tricyclic and, in addition, are associated with hypertensive reactions when the patient ingests foods containing tyramine such as cheese and wine.

Amphetamine and similar stimulating drugs are not effective as antidepressant medications.[4]

When the depressive reaction is associated with considerable anxiety, the benzodiazepines are believed to have limited value.

¶ Suicide Rates

The suicide rate in the United States is approximately 12.7 per 100,000 population.[5] This means that just over 1/100th of 1 percent of the total population commits suicide in a single year. Although persons over the age of sixty-five make up only 11 percent of the total population, 16.4 percent of all suicides are persons sixty-five years and over. This age-related phenomenon is influenced by the high suicide rate among elderly males. In 1977, suicide by females was more likely between the ages of forty-five to fifty-four, but even at this age the rate per 100,000 of male suicides is almost double that of females. Throughout adulthood, and abruptly increasing at age seventy, the risk of suicide for the male is much higher. Suicide attempts, however, are much more frequent among females than among males, but the male's attempt is much more likely to be lethal. Between 1968 and 1977, there was a sharp increase, almost a doubling, of the suicide rate for all males between fifteen and thirty years of age. The older nonwhite male, although more likely to commit suicide than the nonwhite female, does not display the sharp rise in suicide of the older white male. Suicide rates among the nonwhite population are consistently lower than those found among the white population.

The onset of what appears to be organic brain disease is now more frequently associated with suicide than prior studies indicated. This, too, is complicated, as severe depressions are frequently very difficult to distinguish from organic brain disease. There is little doubt that the presence of an incapacitating physical illness is a factor in suicide among the elderly.[15] At the present time, it is most difficult to detect and predict those who are suicidal. This is in contrast to the young would-be suicide for whom the attempt is frequently an overt but disguised cry for help.

Marital status has an influence on suicide rates. The highest suicide rates are seen among men who are divorced, followed by the widowed, and then those who never married. Suicide rates are lowest among persons with intact marriages. Other characteristics of persons who commit suicide in old age are lack of employment or rewarding social roles, unsatisfactory living arrangements, and, as previously noted, a serious concern regarding physical and mental decline.[7]

¶ Paranoid Disorders and Reactions

A fine line separates an attitude or behavior that can be considered within normal limits and one that is considered excessive or pathological. A satisfactory adjustment in late life appears to be related to the individual's ability to maintain social activities as well as health and physical activity and other factors that have been described by Palmore and Maddox.[49] The elderly need to recognize that because some people with whom they must come into contact may be inconsiderate and self-centered, they must assert their rights, to avoid being ignored by the very systems (including government, private service agencies, church, and family) intended to provide them with assistance. If the recognition of defects in others cannot be kept in perspective, an older person is likely to distort events and relationships, becoming suspicious and paranoid. For the majority of individuals, one of the major functions of social contacts is to maintain this precarious balance of evaluating events and relationships to others.

Eisdorfer[28] emphasizes that this lack of precision regarding paranoid behavior among the aged has made it most difficult to ascertain the prevalence and incidence of pathological paranoid reactions. In order to determine the existence of maladaptive suspiciousness, the examiner must determine the pervasive scope of the symptom focus; that is, is the symptom consistently maladaptive, and, particularly, has it become so widespread that it interferes with the individual's ability to function within the total environment? Eisdorfer has more or less arbitrarily divided paranoid ideation into four degrees of severity: (1) suspiciousness; (2) transient paranoid reaction; (3) paraphrenia (late onset paranoia without evidence of schizophrenic illness); and (4) paranoia associated with schizophrenia of late onset.

There are a number of predisposing and contributing factors to the development of excessive suspiciousness and paranoia. There are some individuals who have elected to cope with the complexities of life by adapting to a life-style of partial isolation. Such individuals can make an acceptable adaptation.[42] This is not the normal life style, since isolation is maladaptive. The clinician must constantly keep in mind that late life is constantly influenced by losses that impact upon the individual's capacity to maintain self-esteem. These major losses include the possible disappearance of satisfactions derived from work and a meaningful social role; the loss of friends, spouse, and family; economic instability; and the need to move from a familiar home and neighborhood. The biological losses that seem to have the greatest impact are concerned with the decline in perceptual skills, primarily auditory and visual. The loss of auditory acuity has been associated with paranoia in many stages in the life span, including the aged.[23,27] Hearing loss is a very frequent problem in late life. Consequently, the elderly person is vulnerable to misinterpretation of communication, and this failure to interpret properly is easily projected onto others.

In dealing with the increasing suspiciousness of an elderly person, the astute clinician will utilize a number of relatively simple techniques that are of considerable value. Obviously, the use of a hearing aid is of importance, but, in addition, when communicating with a person with a hearing defect, it is essential that both the speaker and the listener have eye contact. This can be encouraged by touching the elderly person when he is addressed so that he turns toward the speaker. The subject of the statement should be clearly stated, for example, "I want to talk with you about your daughter." The speaker must carefully observe the person for what he appears to misunderstand. The speaker should not hesitate to repeat without apology. In the presence of an elderly person, the speaker should avoid making side remarks in a low voice. Also, excessive background noise can severely interfere with the ability of the older person to understand. Perception defects are not the only problems for the suspicious older person. If an older person is moved from one living area or residence to another, it is advisable to attempt to arrange the bedroom in a manner that is as similar as possible to that previously occupied. An older person with loss of some cognitive skills should be assisted in becoming familiar with the location and characteristics of the toilet and kitchen facilities. Attention to this type of procedure can reduce considerably anxiety and paranoid ideation.

The loss of ability to convert short-term memory to longer-term memory and a decline in level of alertness are often associated with difficulty in locating objects and the subsequent suspicion that the object was intentionally moved or taken by someone. In such a situation, if possible, the family or staff should remain with the older person to assist in the search for the lost object. When the object is found, the person will often say, "Now I remember where I left it." Seeing the object in place in the proper surroundings is important to such recall. If this is not done, the person may continue to be suspicious.

Transient paranoid reactions may be associated with hallucinations, both visual and auditory, and although paranoia is widespread, the reaction can be frequently precipitated by a major change in life-style. It appears that this reaction is more likely to occur in the older person who has preferred relative isolation but finds this pattern suddenly disrupted. The therapeutic approach may include pharmacological treatment but centers upon the reinstitution of the life-style preferred by the individual.

Paraphrenia, that is, late onset paranoia without evidence of prior paranoid reactions, and paranoid schizophrenia of late onset are difficult to distinguish. Kay and Roth[41] believe they can be distinguished, particularly by identifying the presence of those attributes of schizophrenia which interfere with cognitive skills. Regardless of the difficulty of separating these two possible paranoid conditions, the prognosis is not particularly encouraging in spite of the utilization of a protective environment and psychopharmacological agents.[35]

¶ Hypochondriasis

Hypochondriasis is one of five diagnostic entities that are included under the category Somatoform Disorders. These DSM III disorders have in common a symptom or symptoms that suggest organic physical illness, but for which there is no discernible organic explanation. In addition, there is evidence that the symptoms are linked to psychological factors. The symptom formation is largely unconscious, which distinguishes it from factitious disorders. According to the 1980 revision of DSM-III[3] there are four diagnostic criteria for hypochondriasis:

A) The predominant disturbance is an unrealistic interpretation of physical signs or sensations as abnormal, leading to preoccupation with the fear or belief of having a disease.
B) Thorough physical examination does not suggest the diagnosis of any physical disease that accounts for the physical signs or symptoms.
C) The unrealistic fears or beliefs of having a disease persist despite medical reassurance and cause impairment in social, occupational, or recreational functioning.
D) The hypochondriacal preoccupation is not due to schizophrenia, affective disorder, somatization disorder, or anxiety disorder.

Two closely related diagnostic categories are Psychogenic Pain Disorder and Atypical Somatoform Disorder. The former is associated with severe and prolonged pain. Although primarily attributable to psychological factors, it is recognized that there are incidences "in which there is some related organic pathology. The complaint of pain is grossly in excess of what should be expected with such physical finds."[3] The existence of organic pathology is particularly pertinent to the hypochondriacal reaction in the elderly. This will be elaborated. Atypical Somatoform Disorder is considered to be "a residual category."[3] An example of a case that would fit this classification is an individual who is preoccupied with some imagined defect in physical appearance. Such atypical disorders are rare.

Clinical experience and a number of studies indicate that hypochondriasis is prevalent in late life, is more frequent among older women, and is associated in many instances with depression.[2,14] Hypochondriasis in late life is complicated by the fact that many elderly patients do have evidence of chronic physical disabilities, and although their complaints appear to be grossly exaggerated, the actual existence of organic problems cannot be dismissed. Of particular importance to the clinician is the recognition that the hypochondriacal individual who is given attention on an outpatient basis is much more likely to respond to therapy than a patient who has been hospitalized because of persistent physical complaints. It appears that hospitalization increases resistance, as the patient is convinced that an organic explanation must exist and that the actual cause of the disorder has been missed by the examiners.

It is believed that an older person is particularly vulnerable to hypochondriasis not only because of chronic disease and biological age changes but also because of economic insecurity, the loss of a meaningful social role, and a fear of decline in mental functioning. Consequently, although escape from a personal failure or threatening circumstance into "the sick role" is available at all ages, it seems to be particularly frequent in elderly persons. The escape into "the sick role" by the hypochondriacal elderly person can be successful for varying periods of time; that is, the elderly person is permitted to become more dependent, expectations are decreased, and the previous loss of self-esteem is restored to some degree. Unfortunately this status quo does not persist indefinitely, as family members and associates usually recognize that an organic illness is questionable or nonexistent, and their attitudes towards the "sick person" begin to change. Most normal individuals at some time during their lives have "played sick" to avoid trouble. Most well-adjusted people, however, consider this type of defense an immature one, and when it is recognized in others, tolerance of such a person decreases. Recognition that the excuse of illness is physically unjustified makes many people, including physicians, feel that they are being exploited. Four psychological mechanisms play a major role in the dynamics of hypochondriasis: (1) withdrawal of psychic interest from other persons or objects and a redirection of this interest on one's self, one's body and its functioning; (2) a shift of anxiety from a specific psychic area to a less threatening concern with bodily disease; (3) use of a physical symptom as a means of self-punishment and atonement for unacceptable, hostile, or vengeful feelings towards persons close to the individual; and (4) an explanation for the failure to meet personal and social expectations. These primary mechanisms are reinforced by a secondary gain, and that is that the person for varying periods of time receives increased attention and sympathy from friends and health care providers. An awareness of these mechanisms makes the patient's complaints more understandable and contributes to the development of any meaningfully designed treatment approach. For example, retirement may produce a loss of a meaningful social role. This partial isolation permits the individual to focus increased attention upon normal bodily functions. Such a person can easily become preoccupied with gastrointestinal functioning.

Because hypochondriasis is influenced, if not precipitated, by social stress, it is important to realize that hypochondriasis, as well as other psychoneurotic reactions of the elderly, is not infrequently fortuitously alleviated by changes in the environment. For this reason, longitudinal studies confirm the fact that psychoneurotic signs and symptoms can come and go over a period of time. The exacerbations and remissions are largely determined by an identifiable constellation of life events. Furthermore, some individuals tend to react to stress in a habitual manner. There are some individuals in whom the hypochondriacal pattern dominates, while in others a depressive attitude is the major factor. In general, the hypochondriacal elderly person is more likely to be a female of low socioeconomic status with little change in her work role and with patterns of social activity that are not conducive to a good adjustment. Such an individual, because of a number of factors, is placed in a situation where criticism is the rule, and appreciation and work satisfactions are absent. This is compounded by the loss of rewards from the restricted social activity.[12]

In contrast to the hypochondriacal elderly person, there are those individuals who utilize a neurotic mechanism of denial; that is, they fail to deal realistically with important physical diseases. This type of person, a persistent optimist, is more likely to be a male. The physician should not confuse denial with courage, as the courageous person does have a realistic appraisal of the situation. The older male who is likely to utilize denial is often a lifetime achiever from a higher economic status who is not burdened with financial losses but has utilized the work role as a major, if not the sole, source of self-esteem. This person is vulnerable in that at some point the denial mechanism breaks down and the existence of a serious physical illness can no longer be ignored. At that point, the person can become seriously depressed.

The Treatment of Hypochondriasis

The treatment approach which will be described in some detail was originally developed by Busse and coworkers in a special clinic for such patients in a university medical center.[9] In the years since the development of this treatment approach, it is of particular interest to note that the precipitating social stresses have changed, and it is possible that other factors such as the law prohibiting mandatory retirement will have impact upon this reaction. Experience continues to demonstrate that the therapeutic approach has considerable merit. Beginning these therapeutic techniques prior to clear establishment of the existence of hypochondriacal reaction is in no way detrimental. In fact, it usually strengthens the patient-physician relationship.

There are a few techniques utilized by health professionals which are of doubtful value in dealing with the hypochondriac. For example, it is usually believed that a patient has the right to have a full explanation of his medical condition. The truth of this cannot be denied, but one also has to recognize that it must be approached with considerable skill. For example, if the physician finds that no organic explanation for the patient's complaints can be found, the patient has the right to know this. But if the explanation is stopped at this point, the patient is suddenly deprived of a psychological defense that was necessary for maintaining self-esteem. Patients may react quickly and hostilely. Therefore, it is essential that the physician combine the explanation with a reassuring and supportive statement, such as, "I realize that you have considerable discomfort, and I am willing to continue to work with you."

For the hypochondriacal person to continue to live with his family and in society, his psychological defenses must be maintained until more reasonable defense mechanisms can be put in place. Initially it is particularly important for the physician to convey to the patient that he is considered to be sick and that he deserves medical attention. In the early therapeutic contacts, it is most unusual for a hypochondriacal patient to be capable of dealing with personal adjustment problems that may appear simple to the physician. The patient is only able to deal effectively with emotionally charged problems after the therapeutic relationship is established.

The hypochondriacal person usually wants relief from his complaints, and he expects the physician to provide it. If the physician does not do so, he will turn to other sources for relief. Therefore the physician often must comply with this expectation. The patient may be given actual medication or a placebo. The physician must be careful to avoid utilizing any medication that is likely to produce side effects since this would only complicate an already confused picture. Particular attention must be given to the avoidance of drugs that have been used previously by the patient without success. The drug or placebo must be given in an assured manner since hypochondriacal patients are alert for any expression of doubt on the part of a physician. There is a degree of deception in utilizing medication or a placebo but its value cannot be ignored. The placebo technique has symbolic value and it contributes to a good patient-physician relationship. It also is possible that in certain hypochondriacal elderly patients who are exaggerating an organic disorder, appropriate medication may reduce the disrupting stimuli. A hypochondriacal complaint is a distress signal, and the patient's anxiety may be reduced and his self-esteem increased by knowing that a highly regarded professional person is "taking care" of him. Thus, the placebo (Latin for, "I will please") can symbolically represent security and satisfaction to the patient. Although the use of medication can to a limited degree be useful, surgical procedures should be avoided. The physician cannot afford to give in to the hypochondriacal patients who request exploratory surgery. Experience has proven that postoperatively the patient is likely to be worse rather than better, and the resistance to psychological insight is dramatically increased.

¶ Sleep

Normal Sleep in the Aged

Obrist[48] believes that changes in all night EEG sleep patterns are among the most sensitive age-related physiological variables. Although there is considerable individual variation, overall sleep becomes more fragmented in the elderly, and awakenings during the night are longer and more frequent. This is associated with a marked reduction in stage 4 (high amplitude slow waves), and a moderate decrease in the amount of time occupied by rapid eye movement (REM) sleep. In addition, there is a significant decline in the number of 12–14 cycles per second spindle bursts, which are replaced by lower frequency spindle-like rhythms. Feinberg[30] reported that the amount of REM sleep correlates well with performance scores on the Wechsler Adult Intelligence Scale in both normal elderly adults and in groups with evidence of organic brain disease. Later Feinberg reported that intelligence also correlated with the reduction in the number of spindles.[29]

The sharp decline in stage 4 requires further study. Stage 4 is somewhat like REM sleep in that in normal young subjects following stage 4 sleep deprivation there is a compensatory increase in subsequent sleep. Hence, it is assumed to have a biologic function that is altered by the aging process.[48]

Sleep Disorders

According to a government report, 50 million Americans have trouble sleeping during any given year.[36] Furthermore, in one year, 10 million Americans are sufficiently concerned regarding their sleep that they consult a physician. Of these, 5 million get sleeping pill prescriptions (33 million sleeping pill prescriptions are given out per year). Although the majority of patients with sleep disorders experience other difficulties such as pain or insomnia, the government report suggests that 15 percent of the 5 million regular sleeping pill takers are chronic insomniacs with no apparent underlying disorder. Hence, the insomnia is the primary problem.[16]

Busse and coworkers[17] found that in a group of subjects over the age of sixty, 7 to 10 percent use sleeping pills habitually. Although the subjects were apparently well adjusted and living in the community, 20 to 40 percent use sleeping pills on occasion. Pain, particularly that from arthritis, is a common contributor to insomnia. In elderly subjects who were free of physical pain, those who used sleeping pills excessively were found to have many other neurotic complaints and to be poorly adjusted socially.

Sleep requirements and sleep patterns of the elderly are different from those found in early and middle adult life. On an average, the elderly need less sleep. The requirement for sleep in the human appears to decrease gradually over the entire life cycle. Many elderly report that they require less than seven hours of sleep per night. It is possible, however, that there is an alteration in sleep distribution, as some healthy older people will nap fifteen minutes or more several times during the day. The sleep of the elderly is lighter and is associated with more frequent awakenings during the night. Nocturia from prostatic hypertrophy in males contributes to increased awakenings. In normal elderly people, sleep awakenings are increased and prolonged. Also early morning awakening is not unusual. Many elderly people do not understand this physiological alteration in their sleep and become very concerned that they are sleeping poorly and that the condition will lead to serious illnesses. Such lack of information can be easily corrected and this correction will be of considerable help to many of the elderly.

Treatment of Sleep Disorders

A program of good sleep hygiene is a first step in dealing with sleep disorders in late life. Exercise in the afternoon or early evening appears to aid sleep, but exercise two hours before retiring should be avoided. A cool rather than a warm room is often conducive to sleep as is a light bedtime snack. Regularity in retiring as well as arising in the morning appears to strengthen the normal sleep cycle. More than one brief nap in the early afternoon is to be avoided. The prolonged use of sedatives and hypnotics can result in many complications. Many of these drugs lose their potency, particularly the bar-

Karasu, T., eds., *Geriatric Psychiatry.* New York: Grune & Stratton, 1976.

40. KATZMAN, R. "Normal Pressure Hydrocephalus," in Katzman, R., Terry, R. D., and Bick, K. L., eds., *Alzheimer's Disease: Senile Dementia and Related Aging Disorders.* New York: Raven Press, 1978, pp. 115–124.
41. KAY, D., and ROTH, M. "Schizophrenias of Old Age," in Tibbetts, C., and Donahue W., eds., *Processes of Aging.* New York: Basic Books, 1963, pp. 402–448.
42. LOWENTHAL, M. F. "Social Stress and Adaptation: Toward a Life Course Perspective," in Eisdorfer, C., and Lawton, M. P., eds., *The Psychology of Adult Development and Aging,* Washington, D.C.: American Psychological Association, 1973; pp. 281–310.
43. MATUSSEK, N. "Neuroendokrinologische Untersuchungen bei Depressiven Syndromen," *Nervenarzt,* 49 (1978): 569–575.
44. MEIER-RUGE, W., et al. "Experimental Pathology in the Aging Brain," in Gerson, S., and Raskin, A., eds., *Aging.* New York: Raven Press, 1975, pp. 55–126.
45. MILHORZT, T. H. "Surgical Treatment of Hydrocephalus," in Tower, D. B. ed., *The Nervous System.* New York: Raven Press, 1975, pp. 395–406.
46. MOSS, A. J. "Pacemakers in the Elderly," in Reichel, W., ed., *Clinical Aspects of Aging.* Baltimore: Williams & Wilkins, 1978, pp. 65–68.
47. NIES, A., et al. "Changes in Monoamine Oxidase with Aging," in Eisdorfer, C., and Fann, W., eds., *Psychopharmacology and Aging.* New York: Plenum Press, 1973, pp. 41–54.
48. OBRIST, W. D. "Cerebral Blood Flow and EEG Changes Associated with Aging and Dementia," in Busse, E. W., and Blazer, D. B., eds., *Handbook of Geriatric Psychiatry.* New York: Van Nostrand Reinhold, 1980, pp. 83–101.
49. PALMORE, E., and MADDOX, G. "Sociological Aspects of Aging," in Busse, E. W., and Pfeiffer, E., eds., *Behavior and Adaptation in Late Life,* 2nd ed. Boston: Little, Brown, 1977, pp. 31–58.
50. PICK, A. "Über die Beziehungen der Senilen Hirnatrophie zur Aphasie," *Mendicinishche Wochenschrift* 17:16 (1892): pp. 165–167.
51. PLUM, F. "Cardiac Arrhythmias and Neurological Dysfunction," in Busse, E. W., ed., *Cerebral Manifestations of Episodic Cardiac Dysrhythmias.* Princeton: Excerpta Medica, 1979, pp. 11–17.
52. RAHE, R. J. "Life Change Events and Mental Illness: A Review," *Journal of Human Stress,* 5 (1979): 2–10.
53. REINBERG, A. "Chronopharmacology in Man," in Aschoff, J., Cersa, F., and Halberg, F., eds., *Chronobiological Aspects of Endocrinology,* New York: F. K. Schattauer Verlag, 1974, pp. 305–333.
54. ROSSMAN, I. "Bodily Changes with Aging," in Busse, E. W., and Blazer, D., eds., *Handbook of Geriatric Psychiatry.* New York: Van Nostrand Reinhold, 1980, pp. 123–146.
55. SHANAS, E., and MADDOX, G. "Aging, Health and the Organization of Health Resources," in Binstock, R., and Shanas, E., eds., *Handbook of Aging and the Social Sciences.* New York: Van Nostrand Reinhold, 1976, pp. 592–616.
56. SIEGLER, I. C. "The Psychology of Adult Development and Aging," in Busse, E. W., and Blazer, D., eds., *Handbook of Geriatric Psychiatry,* New York: Van Nostrand Reinhold, 1980, pp. 169–221.
57. SIMON, A. *Background Paper,* White House Conference on Aging, Washington, D.C.: U.S. Government Printing Office, 1971.
58. SPITZER, R. L. "Introduction," in American Psychiatric Association, *Diagnostic and Statistical Manual of Mental Disorders,* 3rd ed. (DSM-III). Washington, D.C.: American Psychiatric Association, 1980, pp. 1–12.
59. STERN, S., and LAVY, S. "Frequency and Type of Cerebral Ischemic Attacks in Patients with Tachyarrhythmias and Bradyarrhythmias," in Busse, E. W., ed., *Cerebral Manifestations of Episodic Cardiac Dysrhythmias.* Princeton: Excerpta Medica, 1979, pp. 67–80.
60. TASK, R. "Clinical Laboratory Aids in the Treatment of Depression; Tricyclic Antidepressants Plasma Levels and Urinary MHPG," *Comprehensive Psychiatry,* 3 (1979): 12–20.
61. WALKER, I., and BRODIE, K. "Neuropharmacology of the Aging," in Busse, E. W., and Blazer, D. G., eds. *Handbook of Geriatric Psychiatry.* New York: Van Nostrand Reinhold, 1980, pp. 102–124.

62. WANG, H. S. "Diagnostic Procedures," in Busse, E. W., and Blazer, D., eds., *Handbook of Geriatric Psychiatry.* New York: Van Nostrand Reinhold, 1980, pp. 285–304.
63. WANG, H. S., and BUSSE, E. W. "Dementia in Old Age," in Wells, C. E., ed., *Dementia.* Philadelphia: F. A. Davis, 1971, pp. 152–162.
64. WEISSERT, W. *Effects and Costs of Day Care and Homemaking Services for the Chronically Ill; A Randomized Experiment,* Hyattsville, Md.: Office of Health Research, Statistics and Technology, National Center for Health Services Research, 1980.
65. YESAVAGE, J. A., et al. "Vasodilators in Senile Dementia," *Archives of General Psychiatry,* 36 (1979): 220–222.

CHAPTER 27

ART THERAPY

Carolyn Refsnes Kniazzeh

IN ITS DEVELOPMENT over the past four decades, the field of art therapy can be likened to a spectrum of colors, with each color representing a different facet of the discipline in its relationship to mental health and medical settings, to education, and to community programs. There are also spectra within spectra: for example, different philosophical and psychological orientations; various methods geared toward adults, adolescents, children, the elderly, the emotionally disturbed, mentally retarded, and physically handicapped; and applications with children with exceptional needs and learning disabilities in public and special schools.

Because the full spectrum of art therapy is too vast to be fully covered in one chapter, only dynamically oriented art therapy will be presented here (other approaches will be summarized in the conclusion). Dynamically oriented art therapy is the oldest movement in, and the source of many of the directions of, art therapy. It should be of interest to readers in psychiatry and related professions because it originated in both psychiatric and educational settings. It offers versatile methods that can be adapted to all ages and populations, to people with most types and degrees of disturbance and disability, and to a great variety of settings and purposes.

The essential feature of dynamically oriented art therapy is free art expression. The role of the art therapist and his methods are designed to cultivate individual expression in art and to turn this to therapeutic advantage through the healthy, ego-building experience inherent in the artistic process and through the communication, both nonverbal and verbal, that art inspires. One or both aspects of such therapy may be emphasized in work with groups or individuals.

Art therapy is at once a profound and practical form of treatment. It can be done in simple ways that are cost effective in many kinds of settings, both private and public. It provides opportunities for deep therapy or evaluation in one or few sessions, when other intensive therapies may be limited or not feasible, as is increasingly the case in many institutions. In settings that emphasize intensive treatments, art therapy contributes to evaluation and treatment in ways that enhance the work of all the staff.

¶ Art in Related Disciplines

The literature in psychology, psychoanalysis, psychiatry, education, art, and philosophy reveals that interest in art as clinical phenomena began in the early 1900s and continues to flourish. Sigmund Freud's analysis of movement in Michelangelo's sculptured figure of Moses[24] and his subsequent speculations on the unconscious fantasies and early childhood experiences in the paintings of Leonardo da Vinci[22] are prototypes of the psychoanalytic approach to art. Fundamental theories of the unconscious, the mechanisms of primary process and dream work, and the methods of associative interpretation[23] have fashioned a plethora of studies and approaches to art and creativity, all of which have greatly influenced art therapy. In 1914, Freud expressed both the fascination and puzzlement that art poses:

> Precisely some of the grandest and most overwhelming creations of art are still unsolved riddles to our understanding. We admire them, but we are unable to say what they represent to us. . . . In my opinion, it can only be the artist's intention, insofar as he has succeeded in expressing it in his work and in conveying it to us, that grips us so powerfully. I realize that it cannot be merely a matter of intellectual comprehension; what he aims at is to awaken in us the same emotional attitude, the same mental constellation as that which in him produced the impetus to create. [pp. 257–258][24]

Carl Jung used art from many eras as cross-cultural evidence for his theoretical formulations on symbolism.[46,47] Hans Prinzhorn first published his remarkable collection of art by the mentally disturbed in 1922, describing and classfying the styles along phenomenological lines.[96] At the same time, Oskar Pfister explored expressionism in art, employing it as an integral part of the treatment process.[93]

In the 1940s, psychologists Anne Anastasi and John Foley collected a vast number of art works and data from some 200 hospitals across the United States in one of the earliest and largest research projects to classify and compare characteristics of form and content in the art of patients.[2,3] In 1947, Rose Alschuler and LaBerta Hattwick compiled a two-volume study of style and motivation in children's art.[1]

In the 1950s, contributions of psychoanalytic ego psychology, especially by Ernst Kris,[59] Heinz Hartmann,[39] Felix Deutsch,[14,15,16] and others explored the role of instinctual drives in sublimation, as well as studying unconscious symbolic processes and facets of motivation that are more pertinent to art therapy than to verbal therapies.

In the 1960s and early 1970s, D. W. Winnicott brought into focus still another valuable theoretical dimension for art therapy by showing the relation of creative abilities to fantasy and play and how these develop in childhood in response to early object relations.[137]

In 1976, Silvano Arieti brought a major review and synthesis of psychoanalytic thought to the understanding of creativity and creative individuals that is especially useful to art therapists.[5]

In the past fifty years there have been countless case studies using art. Works by Marguerite Sechehaye,[112,113] Gustav Bychowski,[12] Wilfred C. Hulse,[44] and Harry B. Lee[69] are only a few examples. Marion Milner's *The Hands of the Living God*[81] presents a case of twenty years duration in which art was an integral part of the psychoanalysis of a schizophrenic woman, whose improvement was carefully documented.

Projective psychology has developed concurrently from the early 1900s through the present, utilizing visual stimuli to elicit projections of personality organization. Such data was often interpreted in a psychoanalytic context. Some methods pertinent to art therapy are: Rorschach (early 1900s),[107] Henry A. Murray's Thematic Apperception Test (1938),[82] J. N. Buck's House-Tree-Person Test (1948),[11] Karen Machover's Draw-A-Person Test (1949),[77] and Emanuel F. Hammer (1958)[36] and E. M. Koppitz on the evaluation of children's drawings (1968).[52] Because of its systematic research methods, projective analysis offers valuable contributions to graphic interpretation. Art in art therapy should not be viewed simply as another projective technique. It is not done in the highly controlled manner necessary to testing or research; consequently artwork in art therapy is often a more highly developed and richer form of expression.

¶ Pioneers of Art Therapy

During the past four decades the founders of art therapy have drawn on diverse sources as they molded the discipline. In the 1940s, Margaret Naumburg began developing theories and methods for using spontaneous art as a modality in psychoanalytically oriented art therapy, first with behavior-problem children at the New York Psychiatric Institute[88] and later in individual art therapy with adults. Her premise was that spontaneity in art elicited feelings and fantasies not so readily expressed in words. The meaning of art was explored, as in psychotherapy, using the interpretations and associations of the patient. The important aspect of the art was its content and the insight it permitted into unconscious dynamics. An extensive review of the psychoanalytic literature of art appears in her book *Schizophrenic Art*[84] and an extensive bibliography in her *Psychoneurotic Art.*[85] *Principles of Dynamically-Oriented Art Therapy*[87] sums up decades of her work and considers many professional issues of the field.

Naumburg first employed spontaneous art as a means of progressive art education in the early 1900s at the Walden School in New York. Her sister, Florence Cane,[13] an artist and art educator, developed a method of art teaching based on freedom of expression. Some of her major contributions were the development of rhythmic exercises to overcome tensions and inhibitions, and the scribble technique to release the imagination and spark creative experimentation in images and art forms, both of which have become frequently used procedures. It was Naumburg, however, who first recognized and rigorously applied spontaneous art as the projection of unconscious imagery in psychotherapy, a standard method in the field today.

Like the opposite side of a coin, another fundamental approach in art therapy emphasizes the activity of art as well as its content. Edith Kramer,[53,54] during her twenty-five years of working with disturbed and delinquent children, developed a method of art therapy for groups and individuals utilizing the therapeutic benefits of creative activity, both self-expression and sublimation. Because art is a technique to aid communication with children whose ability to talk is generally limited, Kramer emphasized freedom of expression and the development of artistic skills, using the illuminations of the pictures in diagnosis and treatment. At the same time, she developed as a part of treatment the ego-building elements of creative activity, so important to growing children. She stressed the integration process in creating art through which the child finds new outlets for feelings, impulses, and tensions, and which offers new ways of resolving emotional conflicts through catharsis and sublimation.

Elinor Ulman, as a practitioner of art therapy for ten years in a large municipal hospital, integrated features of Naumburg's and Kramer's approaches in a method for adult patients.[124] As founder and editor of *The Bulletin of Art Therapy* in 1961, the first professional journal in the field (now called the *American Journal of Art Therapy),* she has been a major force in delineating and maintaining the identity of art therapy as an emerging profession in the midst of more established helping professions. She has defined the arts as "a way of bringing order out of chaos—chaotic feelings and impulses within, the bewildering mass of impression from without . . . a means to discover both the self and the world, and to establish a relation between the two." She defines therapy as "procedures . . . designed to assist favorable changes in personality or in living that will outlast the session itself."[129]

Ulman, together with psychologist Bernard I. Levy, have done basic research in art therapy in the assessment of graphic form and expression.[72,130] These are among the few studies in art therapy research that try to take into account the expansive meaning of art while meeting the requirements for controls and objectivity in science. Ulman has devised a brief and effective method for assessment of personality using a series of unstructured and semistructured (scribble) drawings.[125]

Hanna Yaxa Kwiatkowska was the innovator of highly specialized techniques of evaluation and therapy in the use of art with families. This work was part of a large, ten-year research project in family treatment in the 1960s at the National Institute of Mental Health.[61]

¶ Art in Therapy and Evaluation

From the beginning art therapy has been a synthesis of disparate disciplines in both the arts and various psychologies. Like psychiatry, psychoanalysis, and clinical psychology, it was born in response to troubled people who needed help, help derived from and dependent on communication. Art, an age-old form of expression, thus became a natural alternative or supplement to communication in words. Those who initiated the uses of art therapy inevitably worked in the context of the already developing disciplines for understanding and treating the array of human troubles.

The uniqueness of art therapy lies in the nature of art and in its concern with the artwork and the art making as the source of therapy. Of the many definitions of art, Susanne Langer's assessment comes closest to suggesting the intrinsic goal of art therapy. She says, "All art is the creation of perceptible forms expressive of human feeling."[68] Art therapy, however, is not concerned with art in its aesthetic dimension, neither in traditional nor more recent modes. "Art" has always been a generic term that includes efforts by the great masters as well as the art of the general populace—that done by amateurs, children, or naive practitioners. While persons in art therapy occasionally produce art of a high order, the art therapist is concerned with the art maker's cultivating personal expressiveness rather than perfecting artistic techniques or skill.

Even the most simple, primitive work, if it is expressive, has something interesting, possibly pleasing, compelling, something of art about it. If it conveys feeling or inner experience that cannot be expressed in other ways, it has meaning; it may have impact and power while expressing uniqueness and personality. The distinction between expressive and aesthetic power can be left to the aestheticians.

Art therapy embraces the continuum from the simple or limited to the great in art. It does not preclude the aesthetic that may occur and may even be promoted through the emphasis on expressiveness. But it is the many other elements in art that are the key; and the artistic at times may be sacrificed for the stereotyped, the defensive, even the unexpressive, if considered more conducive to well-being.

The deemphasis of the aesthetic aspect notwithstanding, the cultivation of art is the primary concern in art therapy. For the purposes of art therapy the pursuit of art means the cultivation of expression that springs from a unique inner need or desire within the art maker. It takes a particular form special to each individual and it may exist in each and every person, generally going unrecognized.

The pursuit of art is not therapy in itself, although art is gratifying and its pursuit can be turned to therapeutic purpose. In art therapy, many elements of art are used to help people. For example, art can restore and develop: (1) basic functions of expression, communication, and understanding; (2) mental abilities and manual skills involved in organization and integration inherent in artistic processes; and (3) ego processes important to mastering feelings, impulses, and conflicts. All of these therapeutic features of art are contingent upon expressiveness.

The role of the art therapist and the basic art methods are designed to elicit individual expression. Art methods can be simple. The most basic and easiest ways of drawing, painting, collage, and clay sculpture will suffice. Style—whether traditional, representational, or abstract—may be left to the art maker's choice. Creativeness or expressiveness is more likely to be brought about through indirect means than by direct attack. If the untrained person lets himself go a little and just plays with the paint, he may discover himself experimenting with an image or a design uniquely his, or, if not unique, then a common image done in his own way. If it comes from within himself, it will not be beyond his skill, it will not be frustrating. Inspiration starts as a moment of small invention and grows when nurtured. Encourage the art maker to tinker with his own devices, to doodle, to play with scribbles or whatever he wants, provide structure if needed, and he will find what interests him in imagery or design. He will discover his own style. When his artwork is accepted on his own terms and not in relation to some grander external expectation of art or therapy, he will come to accept himself and find the real gratification of creative work. The role of the art therapist involves positive expectation and usually nondirective instruction to generate the art maker's initiative. How the art therapist does this is in itself an artistic and intuitive skill prerequisite to the profession. It is as important and as difficult to communicate as empathy.

A variety of media should be available for the art maker's choice. He should have access to pastels, poster paints, acrylics, charcoal, clay, colored papers and fabrics for collage, and magazine photographs for montage. There should always be concern for the danger of media and techniques becoming empty gimmicks, such as hasty scribbles or squiggles that are not developed into formed expression. Techniques such as cork or potato prints provide quick and slick results. They do not yield themselves to personal expression but force the art maker into the mold of the medium. Materials are best that require the individual to make his imprint on them and with them, that yield to the impulses of the art maker even before he knows clearly what his inclinations are.

The art therapist with such media fosters experimentation, which then elicits images and ideas before they are thought out. This is the essence of spontaneity in art; it allows the untrained and unskilled person to come quickly into expressive artwork. Spontaneity and inventiveness can be encouraged in structured as well as unstructured methods. Structuring may entail suggesting the medium, subject, or intent, while an unstructured approach usually implies a greater degree of freedom for the art maker. Some people respond more readily to structured modes, while others flourish in the opportunity for autonomy. Each approach has its place, and the selection depends first on the needs of the art maker, then on the therapeutic style of the art therapist and on the goals of the art therapy.

Art has already been spoken of as the expression and integration of inner experience in visual form. Going further: Art is a visual phenomenon that has both content and formal properties or style. Content refers to *what* is expressed and style refers to the *means* of expression. These are not really separable entities but two facets of the artwork.

Content in art has its roots in feeling and fantasy, both conscious and unconscious. Art derives its symbolizing power from the capacity of the image to combine, condense, and substitute disparate feelings, ideas, qualities, time, space, and objects. These attributes permit the simultaneous expression of multiple, sometimes contradictory, meanings, which creates the ambiguity essential for the visual image both to reveal and conceal. It is this paradoxical process of revealing and concealing that permits the art to bring to light the hidden or unacceptable feelings that are not so readily accessible in words.

The province of fantasy and dreams has already been explored by psychoanalysis, psychology, and literature, providing art therapy with many suitable methods to approach the meaning of feelings, fantasy, conflict, and psychodynamics in the imagery of art. For purposes of evaluation or therapy, exploration can be done in a limited way with individuals in the context of the art group, or more extensively in occasional individual sessions (see cases 1, 2, and 3), or with the entire group together using techniques of group process. Intensive exploration of art is done using psychotherapeutic techniques in individual art therapy (see cases 4 and 5). Case illustrations show how the content of art is approached through the associations and interpretations of the art maker, his history and present circumstances, both in groups and individual art therapy.

Compared to content, the province of form, or the means of expression, is relatively uncharted territory in psychology and psychoanalysis. It may be because form is more difficult to interpret than content, since the elements in form—line, color, tonalities, space, rhythm—have meaning only in abstract terms of structure; for example, as tones in music. Given the interdisciplinary nature of art therapy, its foundation in art and psychology, the art therapist has a dual perspective: he combines the eye of the artist, art historian, art critic, the trained objective eye to perceive form with the eye of the psychodiagnostician. Such combination gives the art therapist a unique cross-referencing of viewpoints that yields new insights into the art and the person. In looking for the determinants in style of visual forms, three general areas emerge: (1) the form of art reflecting style in behavior and personality (see cases 1, 2, and 3); (2) the relation of artistic style to inner impulses, including the modes of expressing impulses (see cases 2 and 3) and modes of containing or defending against impulses (see case 1); and (3) the reflection of mood or psychodynamics in form (see cases 1, 2, and 3). Sometimes the art maker can shed light on this, but the inquiry should not become an intellectual exercise, or spontaneity will evaporate.

The issue of interpretation is always fundamental in art therapy, whether dealing with content or form. There are two ways to approach the meaning of art: first, from the point of view of the art maker, using his interpretations; and second, from the point of view of the art therapist, the reactions of the viewer. The first approach, the interpretation of the art maker, has the advantage of eliciting documentation from the originator of the art and avoiding the pitfalls of projections and speculations of the art therapist. Some art therapists consider the projections of the art maker as pitfalls to be circumvented and prefer the educated interpretation of the art therapist. In either case, one of the unique values of art is its function as nonverbal communication. It conveys things the art maker cannot speak about. This means that the second approach, the trained, empathic response of the art therapist/viewer is essential, whether using the art maker's interpretation or not. Difficulties here involve understanding the seemingly inscrutable nature of the abstract formal elements of art and documenting the relationship between the manifest expression in the art and the vast reservoir of meaning to the person who made it. As in dream interpretation, sound inference and intuition are paramount, reinforced with rigorous questioning. One must always keep in mind Freud's observation on the nature of art: The art maker's aim is to awaken in the viewer the same emotional attitude, the same mental constellation that moved him to create.[24]

Interpretation serves a vital function in art therapy beyond providing the understanding of the art. The process of interpretation, the give and take between art maker and therapist, forms the very structure of the therapeutic alliance through which understanding and help transpire. This involves both the transference (the art maker's feelings projected onto the art therapist, which may or may not be interpreted) and the realities of the relationship (see individual art therapy case 4). The way the art therapist relates—chooses how and what to say—determines the acceptance of and response to the art on all levels, explicit and implicit (see cases 4 and 5). In summary, the two main concerns in art therapy are the cultivation of the art maker's individual expression in his art and the recognition and response to it by the art therapist.

Both the artwork and the art making are sources of therapy, and one or both avenues may be emphasized. One therapeutic facet is the personal imagery in art that functions as another medium of communication, enabling the art maker to express deeper levels of experience in art than in words. Exploration leading to insight involves both the art maker's and the art therapist's interpretations. The art maker may simply talk about his pictures, his troubles, or explain the meaning of his images. The art plays a crucial role in helping the art maker work on difficult issues. How this is done is always a highly individual matter, as will be seen throughout the case vignettes.

Another therapeutic facet of art is the creative activity of art itself, which is beneficial, partly through catharsis and the complex process of sublimation. Catharsis involves the direct expression of impulses in art finding relief through this ventilation and through sharing the feelings with the art therapist. This is thought to be an ego-building experience in that it aids the art maker in relieving tensions and restoring balance, all in the context of making the gratifying artwork (see case 3). The process may also work when the art is not gratifying.

Sublimation is the more complex therapeutic aspect of creative activity in art therapy. The dictionary defines "sublimation" as "the complex process of expressing socially unacceptable impulses and biological drives in socially acceptable forms such as art."[134] Because of the essential relation of artistic expression to deep inner experience and inner forces, art is an age-old function leading to restoration and health. In the literature of therapy using art, cases reveal how art serves as an appropriate outlet for all kinds of unacceptable, intolerable, unspeakable feelings and impulses, allowing direct expression, as in catharsis, or providing mitigating forms of disguise and substitution through sublimation. Creative activity itself channels and transforms destructive and other impulses into constructive expression and form. The integration inherent in art aids in resolving conflicts, reintegrating feelings, and restoring and expanding ego function (see cases 1, 2, and 3). Both through symbolic transformation (see case 1) and in the transmutation of impulses (see case 3), disruptive feelings are expressed, contained, reassimilated, reintegrated—allowing the art maker to resolve and move beyond his trouble. This can happen whether verbally acknowledged or not. Powers of the ego are asserted and developed as the art maker experiences new ways of expressing and coping with his conflicts. Often the artwork appears as an island of healthy activity amid a sea of trouble (see case 3). All this accounts for the joy and deep gratification of creative work, which is a therapeutic balm in itself.

As in other therapies, changes of deep nature may appear to occur in art therapy and may or may not last, or may not even be known. Deep-seated and complex changes may be difficult to document. The artwork serves as a lasting document both for the art maker and for the art therapist, illuminating the issues in therapy and the changes broached. Art and art therapy provide many avenues and means for change and for therapeutic work of a deep nature outside the traditional verbal therapies.

¶ Case Illustrations: Art Therapy Groups with Adults

Several examples illustrate different features of art in therapy in groups at a large psychiatric hospital and in the acute inpatient service of a general hospital. People with different kinds and degrees of problems are included and are described in terms of human problems and conflict rather than in terms of diagnostic categories.

The art groups met once or twice a week for two hours over a number of years. People came by referral or at their own request. They worked individually in the context of the group where they could choose to work off to one side by themselves or to join a small group around a large table. There was opportunity and encouragement to talk with the art therapist and among themselves, but there was no regular group discussion. Art therapy groups can emphasize group process, but these groups did not. Individual interviews were available to provide confidentiality for more extensive exploration than was possible within the group. The need for silence was always respected. It was found that there could be too much emphasis on talk and this could shift the energies from the creative work and the art to verbal exploration, which then could interfere with the therapeutic purposes of the art.

CASE 1: ABSTRACT SYMBOLIC IMAGERY

This brilliantly colored pastel painting, titled "Thread of Life,"* was done by a woman in her early forties who was a mother and an able laboratory technician, and had been hospitalized twice. Some years before, Mrs. Roberts† had been a practicing artist, and she was now able to resume her artwork in the art group. She developed an entirely new style in which she translated her fantasies into elaborate geometric designs with remarkable spaciousness and beauty. Each color and shape had a specific symbolic meaning that she described in a poem. Green stood for the "earth mother," drawn here in the form of the "bridge of life," stretching across the top to the cluster of "eggs and embryos." Black is the color of "death." She pointed out how black outlines the bridge and shoots into the eggs and embryos, conveying the paradox of death involved in life-giving processes. Positive elements, she said, are represented by red as "the red river of life," which has to do with the "life blood" and "sexual feeling," and by yellow as "dynamic energy," which is the powerful spirit of certain people. In the midst of all this a large, "pink embryo" falls down from the "bridge mother" into the "flat, neutral world," implying the danger and tenuousness of the "Thread of Life."

It was thought that sharing these feelings with the art therapist through the symbolic medium of the picture and the poem were steps that eventually helped Mrs. Roberts to approach these issues in talking with her psychotherapist. She was able later to discuss the intense ambivalence, her feelings and fears of dependence and abandonment, which troubled her relationships with her mother, her family, and her therapist. She returned in a few weeks to her home and full-time work. Art therapy continued for several weeks. She continued psychotherapy some time longer and she was able to maintain her artwork on her own.[98]

A high degree of sublimation occurred in

Figure 27–1. *Thread of Life* (pastel, 24″ × 18″)
NOTE: Refsnes, C. C. *A Presentation of Art Therapy.* The Exhibition of Psychopathological Art, IV World Congress of Psychiatry, Madrid. Catalog published by Boston State Hospital and the Harvard Psychiatry Service of Boston City Hospital, 1966, p. 3.

*Titles and words in quotation marks are the art maker's words.

†All names are fictitious.

this picture, both in the energy manifested in this highly aesthetic abstract design and in the transformation of inner experience and fantasy into the abstract symbolic image. This style of imagery can be termed an "abstract symbol," just as the "Thread of Life" may be seen as a diagram of conflicts at the root of lack of trust. The abstracting served both to express the feelings and to distill or disguise them; that is, to make them remote and, therefore, perhaps more tolerable. This example demonstrates the profound level of expression that can occur simply and quickly in the practical setting of the art class.

Case 2: Expressionistic Style of a Man with Neurotic Depression

The artwork of a man hospitalized for depression over his impending divorce provides a penetrating view of conflicts at a neurotic level and shows the influence of psychodynamics in imagery and style in art. Mr. Farmer had painted from photographs as a hobby. In the art class it was suggested that he try drawing from his imagination by using the scribble approach, making a scribble and drawing the images suggested by the random lines. He responded by inventing a reversal of the usual procedure. With an idea in mind, he sketched by scribbling lightly, "feeling out" by trial and error how to form his images. This process enabled him to discover many unique compositions and resulted in a free, vigorously expressionistic style.

This man, in his mid thirties, a college-educated father of four, was a good provider, although his passive-aggressive conflicts had long interfered with his work and marriage. In speaking about "The Family Portrait" (figure 27–2a), he was openly critical and antagonistic toward his wife, who was expecting their fifth child. The picture suggested his underlying dependency needs as he portrayed himself grouped with all the smiling children behind the mother holding the baby. The resemblance of the infant with its old-looking face to his own portrait suggests his own infantile wishes, while the broadly smiling faces of everyone appear to contradict the actual family friction. The picture clarified his ambivalence by uncovering the positive yearnings behind his overt hostility for his wife.[98]

Figure 27–2a. *The Family Portrait* (charcoal, 18″ × 24″)
NOTE: Refsnes, C. C. *A Presentation of Art Therapy.* The Exhibition of Psychopathological Art, IV World Congress of Psychiatry, Madrid. Catalog published by Boston State Hospital and the Harvard Psychiatry Service of Boston City Hospital, 1966, p. 7.

The imposing "Portrait of Father" (figure 27–2b) displays the sensitivity of Farmer's style and conveys a mixture of feelings for his father, whom he outwardly venerated. He called him a "good man, honest, salt of the earth," and he felt he could not live up to him. How a picture can indicate suppressed

Figure 27–2b. *Portrait of Father* (charcoal, 18″ × 24″)
NOTE: Refsnes, C. C. *A Presentation of Art Therapy.* The Exhibition of Psychopathological Art, IV World Congress of Psychiatry, Madrid. Catalog published by Boston State Hospital and the Harvard Psychiatry Service of Boston City Hospital, 1966, p. 7.

areas of concern is shown in the obscurity of the drawing of the father's arm. It looks like a raised fist and at the same time resembles the form of the baby in mother's arms (figure 27–2a). While he did not mention in art therapy the beatings he endured from his father, Farmer was known to inflict strict treatment and spankings on his own children. The ambiguity of the arm and its similarity to the form of the baby suggested not only his early fears but also his wishes for his father's love. The exaggerated, enigmatic smile masks the more threatening aspects of the father. The picture both reveals and conceals the roots of Farmer's conflicts.

Intense feelings and impulses are released directly in this vigorous mode of drawing, providing Farmer much cathartic benefit. At the same time, a profound and complex integration of past and present experience occurred through the processes of sublimation in these images. Alongside the gratification of artistic work, this man found therapeutic value through the communication of his art and the opportunities to talk about his pictures in several individual sessions.[98]

Case 3: Literal and Expressionistic Qualities in Style of a Man with Severe Character Disorder

The last example from the art group demonstrates the efficacy of art with a young man who suffered severe tendencies to slash himself. Mr. Bogard was in his early twenties, married, and was hospitalized after the death of one of his children. He had a long history of delinquency, drug and alcohol problems, and scars bore witness to frequent episodes of cutting his arms and throat. In the hospital he was able to restrain these impulses, and his remarkable performance in the art group mirrored his great if transient capacity for improved behavior on the ward.

Mr. Bogard said that he had no previous experience in art, but immediately displayed a talent for design and use of color. He worked with ardor, producing a great variety of artwork, beautiful abstracts and images that dramatically revealed the dynamics of his conflicts.

One of his first pictures, "Razor Blade and Hand" (figure 27–3a), resembled his abstractions, but on closer scrutiny the outline of the giant razor blade, the red hand (center), slash lines, sutured wounds and flames—all of which he pointed out—can be discerned. He said the long black lines were "just a screen to hide the meaning," and then he said they were "slashes." He said the red across the bottom stood for "blood and anger," the blue in the background for "sadness" and the "peace and calmness of the Madonna." Then he talked of his cutting episodes, about the tension preceding his slashing outbursts and the calmness afterward, and how the only way to find relief was by cutting himself. Throughout the weeks of art therapy and other treatment, his behavior greatly improved; he had a job in the hospital and helped out on the ward. He drew as if he were channeling his destructive impulses into the art by literally acting out the cutting with chalk on paper, finding relief through catharsis and sublimation. It is tempting to view his intense art activity as a constructive substitute for his self-mutilating impulses.[92,98]

While figure 27–3b, "Dear Mother," may be considered less artistic than his other pictures, the slashing lines shifting between the figure of mother and his initials (he assumed the name of actor Dirk Bogard) dramatically illuminate the source of his destructive and

Figure 27–3a. *Razor Blade and Hand* (pastel, 18″ × 24″)
NOTE: Patch, V. D., and Refsnes, C. C. "An Art Class in a Psychiatric Ward," *Bulletin of Art Therapy*, October 1968, p. 21.

Figure 27–3b. *Dear Mother* (pastel, 18″ × 24″)
NOTE: Patch, V. D., and Refsnes, C. C. "An Art Class in a Psychiatric Ward," *Bulletin of Art Therapy*, October 1968, p. 22.

self-destructive impulses in his conflicts over his mother who neglected him and his several siblings and finally abandoned them when the patient was in his teens. His tone, "Dear Mother," was sarcastic, but it was obvious that he blamed her for his troubles. The picture demonstrates the capacity of art to condense layers of meaning and combine both the conscious and unconscious, the present and past, in imagery. The image indicates the role of the fantasy of the all bad, threatening, devouring mother in his episodes of self mutilation.

Bogard's last picture, "The Suffering Christ" (figure 27–3c), revealed the full scope of Bogard's untrained artistic talent in both expressionistic style and symbolic imagery. He spoke of conflicts over guilt and innocence, saying, "Through religion, there is some good in me yet, some hope." Juxtaposed against the image of "Dear Mother," the all bad, is the opposite extreme, the image of Christ, the all good, symbolizing, for Bogard, a positive resolution. The role of the fantasy of redemption through punishment, death, and resurrection can be seen in Bogard's self-inflicted martyrdom. Expressed in all these pictures are the dynamics of frustrated dependence, anger, punishment, and the wish for forgiveness—all reflecting early issues of separation and fusion. These dynamics influence the style as well as the content of Bogard's art and are seen in the bleeding

Figure 27–3c. *The Suffering Christ* (pastel, 24″ × 18″)
NOTE: Vernon, V. D., and Refsnes, C. C. "An Art Class in a Psychiatric Ward," *Bulletin of Art Therapy*, October, 1968, p. 22.

head with its crown of slashlike thorns against a fiery sky. The pictures enable Bogard to reveal the depths of the intolerable in his inner world and to communicate it in the vivid and constructive form of art. The communication made possible through art gave him considerable gratification and some insight, but the greater gratification was in the artistic work itself and in the recognition by others and himself of his real achievements. His newfound benefits in art and in the hospital job did not last long, however, as Bogard abruptly left one night and did not return.[92,98]

¶ Case Vignettes: Individual Art Therapy

Only in individual art therapy can art as a tool in psychotherapy be fully utilized. Individual art therapy may be done in a mental health setting, on an outpatient basis, or in private practice. In the latter, it is important to establish the availability of medical or psychiatric consultation in case it is needed. Additional individual supervision is recommended beyond the master level training in art therapy.

CASE 4: INDIVIDUAL ART THERAPY AS PSYCHOTHERAPY

Spontaneous art introduces a new dimension in the traditional verbal therapeutic process. It brings into play two means of communication instead of one. In one case it facilitated psychotherapy with a patient whose ability to talk was limited by her need to avoid unacceptable, dangerous feelings. Individual art therapy was the only form of psychotherapy used with a young mother who suffered three postpartem psychotic reactions. Art therapy began when she was hospitalized after the birth of her second child and continued through her third pregnancy, delivery, rehospitalization, and recovery. A synopsis of four pictures out of a total of 175 highlights her conflicts over hostility, traces their derivation in prolonged maternal deprivation, and shows its influence in her attitudes toward pregnancy and care of the baby.

The first picture portrays Mrs. Janson's "Negative Reactions" (figure 27–4a) upon learning of her third pregnancy. While she claimed she would "want the baby and love it," she said the figure was "odd looking" and "too big." As she drew, she remembered her first baby and her fears of doing "wrong things" such as "buying the baby clothes too big," as depicted in the picture. Drawing the bulging brown apron, she complained she will have been changing diapers for six years. At the time, she commented that baby carriages must be deep enough so that the baby cannot fall out, but she did not recognize the more hostile aspects of the shoe kicking the carriage. At this early phase of art therapy, talk was about her feelings of inadequacy as a mother. She had recently recovered from her second postpartem psychotic episode when she entered art therapy.

Figure 27–4a. *Negative Reactions* (pastel, 12″ × 18″)
NOTE: Refsnes, C. C., and Arsenian, J. "Spontaneous Art as a Tool of Psychotherapy," *Proceedings, IV World Congress of Psychiatry, Madrid: Exerpta Medica,* 1966, p. 763.

The drawing of "The Bottles with Plant and Snake" (figure 27–4b) occurred early in the second trimester. The "balloon with a plant inside" was turned into a "bottle" and filled in with "soil." Plants frequently represented her two children. Referring to the snake, she said ". . . something bad. I'll put a lid on so he can't get out and hurt anyone." In the plethora of associations were fears of a recent exposure to tuberculosis, memories of her mother's death from this disease when the patient was twelve years old, and talk of

Figure 27–4b. *The Bottles with Plants and Snake* (pastel, 12″ × 18″)
NOTE: Refsnes, C. C., and Arsenian, J. "Spontaneous Art as a Tool in Psychotherapy," Proceedings, *IV World Congress of Psychiatry, Madrid: Exerpta Medica,* 1966, p. 764.

the cause of her own mental illness in her childhood and broken family. The images symbolized many aspects of her confusion about these things: What is inside? What kind of mother is she? Why is she sick? These wonderings included her childhood conceptions that something inside her was actually mother's tuberculosis and how this now influenced her present attitudes toward her pregnancy. The art in twice-weekly art therapy provided opportunity to work on these complex issues.

These themes culminated in the third trimester in "The Two M's" (figure 27–4c), which she said "stands for myself and my mother." With this picture, the patient realized how much she had missed her mother after her death and she now wept for her for the first time. Shortly after, she made arrangements for her mother's gravestone, neglected thirteen years. The art therapist observed the arrow pointing to the "bird," another frequent symbol for her children, but Mrs. Janson did not recognize the implications about herself or the impending birth of the baby. Other pictures equated feelings for her mother, her husband, her sister, and her father with her feelings for the art therapist. The transference was not interpreted since the art therapist preferred to clarify the realities of the patient's feelings and her family relationships.

After delivery of a healthy baby, Mrs. Janson suffered a milder psychotic reaction than before, and the art therapy continued through her rehospitalization. She recovered within a month. Figure 27–4d, "The Baby with a Propeller on His Rear," completed as she was preparing to leave the hospital, represented her first drawing of the baby. The fragmented or incomplete ground, the strangeness of the structure of the swings and seesaw and the primitive quality of the drawing reflect her disturbed state. After the mourning for her mother and the recent working through of many psychotic conflicts, she was finally able to confront her hostile feelings directly. Her associations to this picture included references to a woman who actually killed her children, and complaints that she was plagued with thoughts and fears of hurting the baby. About "The Baby with a Propeller" she said, "He will be like that in a while, all over the place and into everything." She did not recognize the possible underlying wish that he could fly away or the resemblance to herself and her wish to escape. A previous picture and interpretation had made clear to her how she kept herself apart from the children in order to protect them from her anger. Her associations at this time led her to recall the resentments she felt when she had lived with her alcoholic

Figure 27–4c. *The Two M's* (pastel, 12″ × 18″)
NOTE: Refsnes, C. C., and Arsenian, J. "Spontaneous Art as a Tool in Psychotherapy," *Proceedings, IV World Congress of Psychiatry, Madrid: Exerpta Medica,* 1966, p. 765.

Figure 27–4d. *The Baby with a Propeller on His Rear* (pastel, 12″ × 18″)
NOTE: Refsnes, C. C., and Arsenian, J. "Spontaneous Art as a Tool in Psychotherapy," *Proceedings, IV World Congress of Psychiatry, Madrid: Exerpta Medica,* 1966, p. 768.

mother and then with her father after her mother's death. She remembered being so angry that she wished he would die and how she thought it was better that her mother had died. It was possible at this time to clarify for her that she was so frightened of her feelings because her mother had died at a time when she had had such angry feelings, but that her death was not *because* of her feelings. The clarification that no feelings are so powerful brought great relief to the patient.

Gradually, over several months and many pictures, the individual art therapy enabled Mrs. Janson to sort out and work through some of these crippling feelings and to understand how they interfered with caring for her children. This helped her in resuming her family responsibilities and care of the baby. A sign of change was her increasing ability to understand and meet the needs of her children. For example, an older child's bedwetting problem cleared up quickly when she perceived the child's jealousy and need for attention. Mrs. Janson terminated individual art therapy, but resumed some months later when her father was dying. She said she asked for the art therapist again because what she had said about her mother was true. At this time she chose not to draw but only to talk in therapy. It is not uncommon that therapy with art eventually leads to verbal psychotherapy or an alternation of verbal and art therapy.[102]

CASE 5: INDIVIDUAL ART THERAPY AS ADJUNCT TO PSYCHOTHERAPY

Individual art therapy is effective as an adjunct to psychotherapy. Cooperation between two individual therapists—for example, an art therapist and a psychiatrist—can be a vital factor in amplifying and quickening the work of traditional psychotherapy. In one case, joint supervision with the same psychoanalytic supervisor enhanced the coordination of the two therapies.

Three pictures demonstrate how individual art therapy elicited feelings and conflicts too threatening for the patient to talk about directly in therapy. A young woman, Mary, dropped out of college because of severe phobias first experienced in her earliest years of school. She did over 600 drawings and paintings in sixteen months of art therapy. Art became an ardent daily activity in which she translated dreams into pictures, pouring into them all kinds of feelings, fantasies, and preoccupations. This case demonstrated that art serves a unique function, allowing the art maker to work constructively alone and bring the art for work in the sessions. The art can later be carried on after termination of treatment, permitting the art maker to continue the therapeutic work alone.

An early picture, "My Fear of Going into the Street" (figure 27–5a), vividly conveys, better than words, the intensity of Mary's phobic experience. Previous pictures reflected many aspects of the swarming cars: There were drawings of her fantasies and dreams about sperm being everywhere; the resemblance of cars to footprints; and a picture of her grandfather's shoe with memories of her warm relationship with him and of his death, a traumatic event when she was four years old. During treatment, the patient used her phobias to persuade her father to take her to and from the hospital and elsewhere. All of these things were eventually brought into discussion in her psychotherapy in conjunction with the art therapy.[103]

Figure 27–5a. *My Fear of Going into the Street* (pastel, 18" × 24")
NOTE: Refsnes, C. C., and Gallagher, F. P. "Art Therapy as Adjunct to Long-Term Psychotherapy," *Bulletin of Art Therapy*, 7:2 (1968): p. 69.

Figure 27–5b, "The Family Scene," demonstrated how art therapy elicited Mary's perceptions of her family relationships and brought many levels of these dynamics into both therapies. The patient portrayed herself in the coffin in the center of the picture. She described positive and negative aspects about her mother shown facing the table, with her back turned to the sink. About the stick figure under the duncelike cap, she said:

that when she was with her mother she felt "like in a shroud, as though a shroud is coming down over me. Mother makes me feel that way she is so miserable herself . . ."

Father appears in the corner of the page looking stern. "Father," she said, "is so awful to her. Cruel, he is cold and critical like a rat." As she drew the image at the top of the picture she talked about mother: "She is so tense, like a dam. I'm afraid she will burst and go crazy. . . . She was independent and happy before she married. I wonder why she stays . . . why she doesn't leave and make a life for herself." When the art therapist interjected, "How about yourself?" Mary burst into tears and said, "There is nothing outside, at least there is some life with them. . . ." "My mother may crack, she can only take so much; like a net with a stone, it may break through. . . ." The patient pictures this simile, portraying herself as the infant with the stone tied to her neck, who may have broken through but still dangles by the cord. [p. 71][103]

The picture clarified the patient's conflicts over her own identity and her dependence, and her difficulty in separating from mother. The patient began to talk more freely about these issues in twice-weekly psychotherapy sessions with the psychiatrist.

The last picture in this case vignette, "The Dog/Mother, Cat/Father" (figure 27–5c), presented symbolic expressions of Mary's perceptions of her parents and herself.

The focus . . . is on the mother and child, with mother caricatured as a dog looming large in the center and looking daggers at the tiny child below. Father is off to the side, portrayed as a dark gray cat, fitting the patient's description of him as "cool, aloof, and independent." One wonders how often he evades some family fray saying, "Who, me? Don't be silly."

It is father who is wearing the crown but the real power appears to be invested in the dog-mother. . . . The red color and glaring eyes add to the fierce, threatening expression of the beast but it is the enormous bared teeth which reveal the patient's unconscious fantasy and fear of being devoured by her mother.

The image conveys the impact of the message the patient receives from the banal phrase, "Mother knows best." We see how it is that she feels overwhelmed and writes, "Foolish Mary, trying to win an argument or assert herself." At the same time she is enraged and blames the threatening, domineering mother, writing, "I hate you

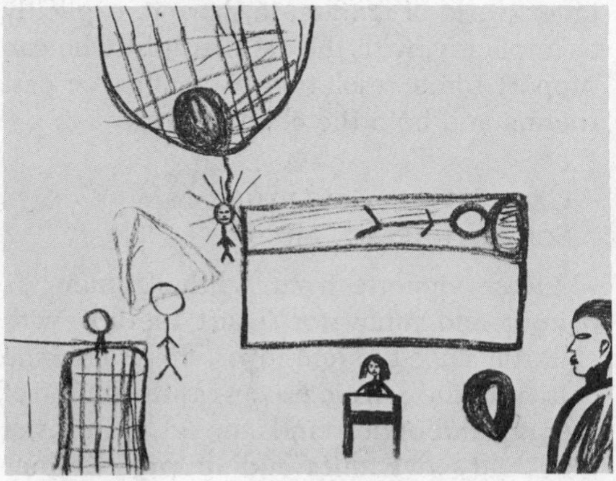

Figure 27–5b. *The Family Scene* (charcoal, 12″ × 18″)
NOTE: Refsnes, C. C., and Gallagher, F. P. "Art Therapy as Adjunct to Long-Term Psychotherapy," *Bulletin of Art Therapy,* 7:2 (1968): p. 71.

Figure 27–5c. *The Dog/Mother, Cat/Father* (pastel, 12″ × 18″)
NOTE: Refsnes, C. C., and Gallagher, F. P. "Art Therapy as Adjunct to Long-Term Psychotherapy," *Bulletin of Art Therapy,* 7:2 (1968): p. 74.

for making me this way." The picture reveals the primitive core of the hostile, dependent relationship between mother and patient. [p. 74][103]

In this picture, Mary wrote her associations in lieu of talking, and demonstrated how the art offered a daily relief as an outlet for the intolerable feelings with which she constantly struggled. All the pictures helped to clarify the source of her conflicts in the family dynamics and to facilitate the process of working through in her therapies.

While this synopsis of pictures focused on her severe, crippling feelings, many pictures and much work was done involving the positive transference with both therapists. It was thought the double transference permitted the patient to share feelings too threatening to share with only one therapist. The welter of artwork indicates that this patient was overwhelmed by the profusion and intensity of her feelings. The art provided her with a constructive outlet and new way to deal with her feelings on a daily basis. Individual art therapy continued on an outpatient basis for sixteen months as an adjunct to psychotherapy. The patient found and was able to maintain a clerical position for more than a year when she considered resumption of her college study. Psychotherapy continued for several years.[103]

While art therapy may not be for each and every individual, the kinds of people who respond and benefit are virtually limitless—all ages, diagnostic categories, the talented as well as the artistically unskilled. Art can be especially useful with people whose verbal capacities are severely limited or with people whose verbal facility is itself an obstacle in therapy. Art therapy can be adapted to all kinds of settings and many purposes. These few examples give an impression of how art works as therapy and also how it contributes to evaluation and diagnosis.

Art therapy provides an enrichment for the entire milieu, for the staff as well as for the patients. The artwork forms primary documents that can be used to teach psychodynamics and to give fresh assessments of patients on a day-to-day basis. Regular showing and discussion of the art in staff meetings may act as a catalyst, enhancing the understanding and interaction of all the staff. As the art maker resolves something of his own inner and outer experience through his art, so each staff member can see and utilize this in his work with the patient.

Art therapists should be full participants in the treatment team. Art therapists work best independently of other disciplines, such as occupational or recreational therapies, and when they are responsible directly to the people in charge of treatment or research. To flourish, art therapists need access to supervision by qualified art therapists or other professionals of choice.

¶ Art Therapy with Children

Applications of art therapy with children are as diverse as with adults and include group and individual art therapy in psychiatric and medical hospitals; in residential treatment centers for the disturbed, delinquent, or retarded; in special schools; and in public education systems. Whether the emphasis is on art in psychotherapy or the therapy inherent in artistic process, the art always involves an element of play. It engages the child in his unique way and at his own level of development and permits him to externalize his inner world of fantasy, to share it, explicitly or implicitly, with the art therapist who can support some resolution to conflict or past trauma and help the child to grow.

CASE 6: CLAY SCULPTURE OF A SCHIZOPHRENIC CHILD

A case vignette from Edith Kramer,[57] a painter and innovator in art therapy with children since the mid 1950s, shows how she cultivates the artistic process as the source of therapy and understands the art in terms of the child's dynamics and history. Kramer offers the most thorough integration of psychoanalytic theory with art therapy practice. Her method attains a delicate balance between the roles of art teacher and art therapist in her emphasis on art as therapy.

Ten-year-old Jasper used clay-modeling in . . . his task of restoring the absent [mother]. . . . When he was 7 years old, his mother had died of cancer, after a long illness during which a leg had been amputated. The father . . . once when . . . drunk . . . overturned mother's wheelchair. After her death, Jasper developed the idea that by causing mother's fall father had been responsible for her death. Jasper was hospitalized suffering from depression and nightly hallucinations when he saw mother's ghost flying across the bedroom and heard her voice calling him.

Jasper liked art from the beginning. When his symptoms had abated after several months of treatment he . . . made up his mind to make a sculpture of his mother's head. He said he could not remember her face very well, but he did recall the shape of her hairdo and that her hair had been grey . . . he spoke of mother's kindness to him. He stressed how she had refrained from meting out well deserved punishment, saying that she had "spared him" many beatings. He spoke of no positive acts of giving or of care.

It was not easy to help Jasper produce a satisfactory image in the face of his inexperience as a sculptor, the vagueness of his memories, and his unconscious ambivalence toward mother. A mother who had become progressively unable to fulfill her functions, had been horribly maimed, and had ultimately abandoned him through her death, could not fail to arouse intense aggression that conflicted with Jasper's equally powerful yearnings.

Figure 27–6a. *Sculpture of Mother's Head* (clay)
NOTE: Kramer, E. *Childhood and Art Therapy*. New York: Schocken Books, 1979, p. 254.

The conflict became apparent when Jasper, feeling quite helpless about recapturing mother's features at all, declared he would change the head into a devil and impulsively placed two horns on it. At this point I intervened. Even though mother's demonic qualities would have to be dealt with if Jasper was to be cured of his hallucinations and depression, it seemed important to help him first to restore the benign aspects if it was at all possible.

I removed the horns and told him that we should not give up so easily. I made him describe mother's face to me as best he could and together we managed to construct an image that could conceivably stand for her. . . . The completed work bears evidence of Jasper's ambivalence. The blackened eyes and aggressively incised eyelashes in particular seem to tell of mother's sinister, persecuting aspects. Her benign qualities nevertheless outweigh the malign ones. Jasper, as well as the other children on the ward, treated the head as an object that inspired respect mixed with tenderness, not fear. . . . Jasper's mother's ghost had frightened the whole ward so that the heavy, unghostlike head helped bind much communal fear. Eventually, Jasper gave the head to father to install in their apartment and luckily father was able to appreciate the reconciliatory gesture.

Did I do right when I prevented Jasper's aggression against mother from gaining the upper hand while he made the head? His being able to complete it and his later treatment of it as a precious possession indicate that my intuition was sound. Had his aggression been too powerful the sculpture would have either turned into a devil in spite of my intercession or else come to a bad end in some other way.

Having been able to restore a positive image of mother, Jasper was in a better position for working on the fear and aggression her death had aroused without being overwhelmed by guilt. . . . He constructed a scene [in clay] of mother lying in state in an open coffin beneath a large golden cross [figure 27–6b]. While he worked on it he spoke about the funeral and how frightening it had been when he was made to kiss her goodbye. He had feared that he might catch her disease when he touched her.

Children can only gradually assimilate a parent's death. In his future life Jasper was undoubtedly destined to work at coming to terms with his mother's death in many ways. For the time being, having restored mother's image and having laid her body to rest, Jasper was free to pursue more ordinary childlike interests in his art. His last sculpture before his discharge was a large horse, a subject that was at the time fashionable among the ward's children. [pp. 254–257][57]

While Edith Kramer emphasizes the artistic process in therapy with children in groups, Mala Betensky,[10] psychologist and art therapist, uses art in individual psychotherapy with children and adolescents. She provides extensive case studies of individual art therapy, exemplifying sound clinical practice within an eclectic orientation.

A different psychoanalytic approach to art therapy with children is the highly versatile one presented by Judith Rubin,[108] an art therapist and art educator in psychoanalytic training, who utilizes several creative arts in therapy—drama, movement, and music, as well as the visual arts. She emphasizes art as means of communication and interaction in therapy with groups and individuals. She describes applications of art therapy with the emotionally disturbed, the handicapped, school children, groups of mothers and children, and family groups. While integrating several arts, Rubin upholds the integrity and emphasis of the visual arts, a difficult but important balance to maintain in art therapy.

Figure 27–6b. *Mother Lying in State* (clay)

¶ Family Art Therapy and Evaluation

Art can yield rapid and penetrating insights into complex family relations both for evaluation and family treatment. Adequate space and staff is necessary for this highly specialized application of art therapy. A recorder, observer, or cotherapist to the art therapist is recommended in treatment of families using art. However, the art evaluation procedure can be easily done by the art therapist alone in a single session in most any kind of program. An example from the innovative work of Hanna Yaxa Kwiatkowska[61] from the ten-year family research project at the National Institute of Mental Health reveals a young schizophrenic man's perception of the role of his domineering mother in his illness.

CASE 7: FAMILY EVALUATION THROUGH ART BY A SCHIZOPHRENIC

[Figure 27–7] is another family portrait. It was produced by a twenty-two-year-old male schizophrenic patient during an acute psychotic episode. In family psychotherapy sessions as well as during the family art evaluation, he constantly attacked his mother. Here we see a paranoid view of the mother's malignant influence on all family members. They are all tied to her; she directs them by means of electrodes implanted in their brains, reducing them to marionettes on strings. The patient, on the lower right, is the one who

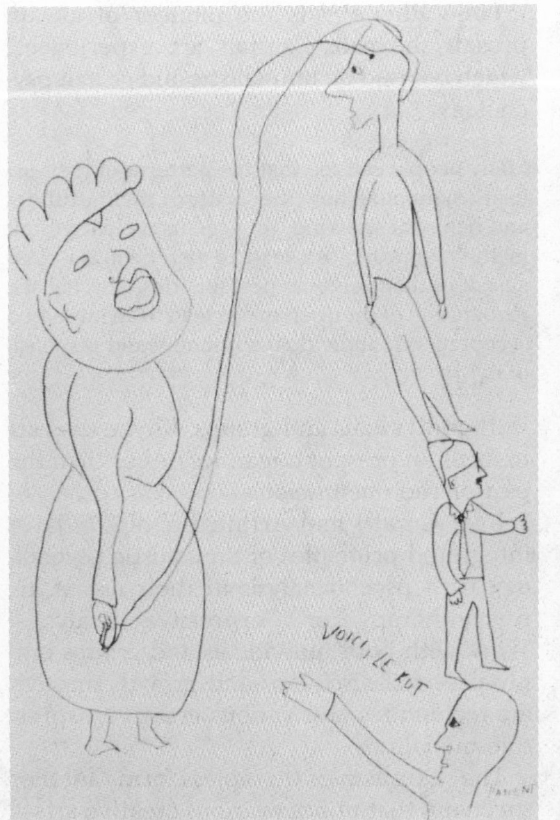

Figure 27–7. *Family Portrait*
NOTE: Kwiatkowska, H. Y. *Family Therapy and Evaluation through Art.* Springfield, Ill.: Charles C Thomas, 1978, p. 100.

accuses; in a mixture of French and schizophrenese, he designates her as the evil one. [pp. 99–100][61]

In the difficult work of evaluating and treating families, the art by each family member serves to cut through the complexity and confusion inevitable in the family group, and can isolate and bring into focus essential issues. A noncommunicative member may find a new way to be seen or heard and may be helped to evolve a healthier relationship within his family.

¶ Art in Therapy and Evaluation

Case illustrations from art therapy with groups, individuals, children, and families reveal the fundamental therapeutic elements inherent in the nature of art. The artist Wassily Kandinsky wrote in 1914 (the year of Freud's similar observation on art):

> He [the artist] must realize his every deed, feeling, and thought are raw but sure material from which his work is to arise. . . . Form, in the narrow sense, is nothing but the separating line between surfaces of colour. That is its outer meaning. But it has also an inner meaning, of varying intensity, and properly speaking, *form is the outward expression of this inner meaning.* [p. 29][49]

By means of the integration of inner experience and the response to outer realities, art offers the art maker a unique rapprochement between his inner and outer worlds, and conveys his vision to the viewers of his art. Some professionals express concern that art poses a threat of intolerable confrontation that may exacerbate the art maker's disturbance. But in years of art therapy the visions of art were found not to disturb but to heal. Art heals for many reasons. The great investment of feeling and energy gives validity and value to the art. The profundity and genuineness of expression in art generates self-confidence in the art maker. The outlet for unacceptable impulses and fantasies in safe forms provides resolution, acceptance, and relief. All of these factors allow the intolerable to be balanced by the gratification inherent in the making of art.

The artistic process may involve a regression in the confrontation with the intolerable, but art making at the same time primarily entails a restoration and development of ego function both in mastery of feelings and mastery of artistic expression. The ability of the art maker to do this, even in limited ways, involves the organization of mental functions and emotions in a profound way. Greater self-autonomy occurs as the art maker forges his art out of his inner life and out of his response to his external world. The severely disturbed and disabled can do this, at some level, in their own way, as well as can the healthy functioning person. All of these are reasons why art is a remarkable therapeutic medium for so many different people.

Art permits therapeutic work at a deep

level independent of talk and the transference relationship. It documents the opportunity for and the nature of change—change, with or without insight, not dependent on verbalization and overt acknowledgment. The artwork stands self-evident when the art maker does it and when the art therapist/art viewer is open to recognizing and responding to it.

¶ Diverse Approaches in Art Therapy

During the past forty years, art therapy has proliferated along many lines, based on various philosophical or theoretical orientations. Dynamically oriented approaches to art therapy predominate. In addition to approaches reviewed here, methods of art psychotherapy based on the medical model,[64,65,70,71] and on Jungian,[135] gestalt,[104] and expressive therapies[105] are becoming widespread.

Two pieces of scholarship in research and theoretical exploration indicate recent trends in dynamically oriented art therapy. In "Theory and Practice of Art Therapy with the Mentally Retarded,"[136] Laurie Wilson describes a two-year case study with a severely retarded woman in which she traces the derivation of repeated imagery, a circle with radial lines, to a cluster of bells (transitional object) and eventually to the breast (original object). She correlates the developing body image in the art with improvements in ego function and behavior, which are also linked to relinquishing the fixation on the bells and the radial image. Mildred Lachman-Chapin[63] explores the implications for art therapy of Kohut's psychoanalytic theories on the self and early narcissism. She sees these early formative issues mirrored in the healthy functions of artistic process and suggests art therapy can use art to deal directly with preoedipal deficits in ways that traditional verbal interpretation of imagery and conflict cannot.

Janie Rhyne[104] is the pioneer of an approach she calls "gestalt art experience," which is based on humanistic and gestalt psychology.

> Many people can see that the patterns of their art forms symbolize how they pattern their attitudes and behavior in living; thus seeing a clear gestalt in their artwork can lead to perceiving a clear gestalt of themselves as personalities. This holistic recognition of themselves can lead to an increased acceptance of individual autonomy and responsibility. [p. 157]

With individuals and groups, Rhyne uses art to focus on present dynamics rather than the past or the unconscious.

Josef Garai[27] and Arthur Robbins[105] have integrated principles of humanistic psychology with psychoanalysis in their use of art psychotherapy or "expressive analysis." Work with both individuals and groups emphasizes self-discovery and growth through art techniques and various creative/expressive modalities.

The expressive therapies form another spectrum that utilizes various creative arts—dance/movement, music, psychodrama, poetry, as well as the visual arts—in therapy, emphasizing either group interaction or directed toward art psychotherapy. Sandra Kagan and Vija Lusebrink in "The Expressive Therapies Continuum"[48] draw on psychoanalysis, humanistic, Jungian, and developmental psychologies to define modes and levels of expression in an attempt to delineate common denominators pervasive in all the arts. Their categories are Kinesthetic/Sensory, Perceptual/Affective, Cognitive/Symbolic and the Creative. These categories pertain to the interaction with media and to the effort involved in specifying the degree and healing value of creative involvement—a difficult task whether dealing with one complex art or several.

Mildred Lachman-Chapin,[62] artist, dancer, and art therapist, has clarified the roles of art and movement (or dance) in art therapy, a distinction vital to the understanding and use of various modalities. The work of philosopher Susanne Langer[67] is also useful to art

18. EHRENZWEIG, A. *The Hidden Order of Art.* London: Weidenfeld and Nicholson, 1967.
19. ERIKSON, E. H. *Childhood and Society.* New York: Norton, 1950.
20. ERIKSON, J. M. *Activity, Recovery, Growth: The Communal Role of Planned Activities.* New York: Norton, 1976.
21. FREUD, A. *The Ego and the Mechanisms of Defense.* New York: International Universities Press, 1936.
22. FREUD, S. *Leonardo da Vinci, a Study in Psychosexuality.* New York: Modern Library, Random House, 1947.
23. ———. *The Interpretation of Dreams.* New York: Modern Library, Random House, 1950.
24. ———. "The Moses of Michelangelo," in Jones, E., ed., *Collected Papers,* vol. 4, New York: Basic Books, 1959.
25. ———. *On Creativity and the Unconscious.* New York: Harper Torchbooks, n.d.
26. *GANTT, L., and SCHMAL, M., eds. *Art Therapy: A Bibliography.* Rockville, Md.: National Institutes of Mental Health, 1974.
27. GARAI, J. "The Humanistic Approach to Art Therapy and Creativity Development." Paper presented at the meeting of the American Art Therapy Association, Milwaukee, 1971
28. GARDNER, H. *The Arts and Human Development.* New York: John Wiley, 1973.
29. GHISELIN, B., ed. *The Creative Process.* Berkeley: University of California Press, 1952.
30. GIEDION, S. *The Eternal Present: The Beginnings of Art.* New York: Bollingen Series, Pantheon, 1962.
31. GOLOMB, C. *Young Children's Sculpture and Drawing.* Cambridge, Mass.: Harvard University Press, 1974.
32. GOMBRICH, E. H. *Meditations on a Hobby Horse.* London: Phaidon, 1963.
33. ———. *Art and Illusion.* New York: Bollingen Series, Pantheon, 1965.
34. GREENACRE, P. "The Childhood of the Artist: Libidinal Phase Development and Giftedness," in Eissler, R. S., et al., eds., *The Psychoanalytic Study of the Child,* vol. 12. New York: International Universities Press, 1957.
35. ———. "Play in Relation to Creative Imagination," in Eissler, R. S., et al., eds., *The Psychoanalytic Study of the Child,* vol 14. New York: International Universities Press, 1959.
36. HAMMER, E. F. *Clinical Applications of Projective Drawings.* Springfield, Ill.: Charles C Thomas, 1958.
37. HARMS, E. "Awakening into Consciousness of Subconsciousness of Subconscious Collective Symbolism as a Therapeutic Procedure," *Journal of Child Psychiatry,* 1 (1948): 208–238.
38. HARRIS, J., and JOSEPH, C. *Murals of the Mind: Image of a Psychiatric Community.* New York: International Universities Press, 1973.
39. HARTMANN, H. "Notes on the Theory of Sublimation," in Eissler, R. S., et al., eds., *The Psychoanalytic Study of the Child,* vol. 10. New York: International Universities Press, 1955.
40. HATTERER, L. J. *The Artist in Society.* New York: Grove Press, 1965.
41. HILL, A. *Art Versus Illness.* London: George Allen and Unwin, 1945.
42. ———. *Painting Out Illness.* London: George Allen and Unwin, 1951.
43. *HOROWITZ, M. J. *Image Formation and Cognition.* New York: Appleton-Century-Crofts, 1970.
44. HULSE, W. C. "Symbolic Painting in Psychotherapy," *American Journal of Psychotherapy,* 3 (1949): 559–84.
45. JAKAB, I., ed. *Psychiatry and Art.* New York: S. Karger, vol. I (1968), vol. II (1970), vol. III (1971), vol. IV (1975).
46. JUNG, C. G. *Symbols of Transformation.* 2 vols. New York: Harper Torchbooks, 1962.
47. ———. *Man and His Symbols.* New York: Doubleday, 1964.
48. KAGAN, S., and LUSEBRINK, V. B. "The Expressive Therapies Continuum," *Art Psychotherapy,* 5 (1978): 171–80.
49. KANDINSKY, W. *Concerning the Spiritual in Art.* New York: Wittenborn, 1947.
50. KELLOGG, R. *Analyzing Children's Art.* Palo Alto, Cal.: National Press Books, 1969.
51. *KIELL, N., ed. *Psychiatry and Psychology in the Visual Arts and Aesthetics: A Bibliography.* Madison, Wis.: University of Wisconsin Press, 1965.
52. *KOPPITZ, E. M. *Psychological Evaluation of Children's Human Figure Drawings.* New York: Grune & Stratton, 1968.
53. *KRAMER, E. *Art Therapy in a Children's Community.* Springfield, Ill.: Charles C Thomas, 1958.
54. ———. *Art as Therapy with Children.* New York: Schocken Books, 1971.

55. ———. "Art and Craft," in Ulman, E., and Dachinger, P., eds., *Art Therapy in Theory and Practice.* New York: Schocken Books, 1975, pp. 106–109.

56. ———. "The Problem of Quality in Art," in Ulman, E., and Dachinger, P., eds., *Art Therapy in Theory and Practice.* New York: Schocken Books, 1975, pp. 43–59.

57. ———. *Childhood and Art Therapy.* New York: Schocken Books, 1979.

58. KRIS, E. "Neutralization and Sublimation, Observations on Young Children," in Eissler, R. S., et al., eds., *The Psychoanalytic Study of the Child,* vol. 10. New York: International Universities Press, 1955, pp. 30–46.

59. *———. *Psychoanalytic Explorations in Art.* New York: International Universities Press, 1952.

60. KUBIE, L. *Neurotic Distortion in the Creative Process.* n.p. Noonday Press (paperback), 1961.

61. KWIATKOWSKA, H. Y. *Family Therapy and Evaluation through Art.* Springfield, Ill.: Charles C Thomas, 1978.

62. LACHMAN-CHAPIN, M. "The Use of Movement in Art Therapy," *American Journal of Art Therapy,* 13:1 (1973): 22–34.

63. ———. "Self Psychology: Kohut's Theories on Narcissism and Their Meaning for Art Therapy," *American Journal of Art Therapy,* 19:1 (1979): 3–9.

64. LANDGARTEN, H. "Adult Art Psychotherapy," *International Journal of Art Psychotherapy,* 2:1 (1975).

65. ———. "Art Therapy as a Primary Mode of Treatment for an Elective Mute," *American Journal of Art Therapy,* 14:4 (1975): 121–25.

66. LANGER, S. K. *Philosophy in a New Key.* New York: Mentor Books, 1948.

67. *———. *Feeling and Form.* New York: Scribner's, 1953.

68. ———. *Problems of Art.* New York: Scribner's, 1957.

69. LEE, H. B. "The Creative Imagination," *Psychoanalytic Quarterly,* 18 (1948): 351–60.

70. LEVICK, M. "Art in Psychotherapy" in Masserman, J., ed., *Current Psychiatric Therapies.* New York: Grune & Stratton, 1975, pp. 93–99.

71. ———. "Transference and Countertransference as Manifested in Graphic Productions," *Art Psychotherapy,* 2 (1975): 203–215.

72. LEVY, B. I., and ULMAN, E. "Judging Psychopathology from Paintings," *Journal of Abnormal Psychology,* 72 (1967): 182–187.

73. LEVY, B., et al. "Symposium: Integration of Divergent Points of View in Art Therapy," *American Journal of Art Therapy,* 14:1 (1974): 13–17.

74. LOMBROSO, C. *The Man of Genius.* London: Scott, 1895.

75. *LOWENFELD, V. *Creative and Mental Growth.* New York: Macmillan, 1957.

76. LYDDIATT, E. M. *Spontaneous Painting and Modelling.* London: Constable and Co., 1971.

77. MACHOVER, K. *Personality Projection in the Drawing of the Human Figure.* Springfield, Ill.: Charles C Thomas, 1949.

78. MEARES, A. *The Door of Serenity.* Springfield, Ill.: Charles C Thomas, 1958.

79. ———. *Shapes of Sanity.* Springfield, Ill.: Charles C Thomas, 1960.

80. MILNER, M. *On Not Being Able to Paint.* New York: International Universities Press, 1957.

81. *———. *The Hands of the Living God.* New York: International Universities Press, 1969.

82. MURRAY, H. A. *Explorations in Personality.* New York: Oxford University Press, 1938.

83. NAUMBURG, M. *The Child and the World.* New York: Harcourt, 1928.

84. ———. *Schizophrenic Art, Its Meaning in Psychotherapy.* New York: Grune & Stratton, 1950.

85. *———. *Psychoneurotic Art, Its Function in Psychotherapy.* New York: Grune & Stratton, 1953.

86. ———. "Expanding Non-Verbal Aspects of Art Education on the University Level," *Journal of Aesthetics and Art Criticism,* 14:4 (1961): 439–51.

87. ———. *Dynamically-Oriented Art Therapy, Its Principles and Practice.* New York: Grune & Stratton, 1966.

88. ———. *An Introduction to Art Therapy: Studies of the "Free" Art Expression of Behavior Problem Children.* New York: Teachers College Press, 1973.

89. ———. "Spontaneous Art in Education and Psychotherapy," in Ulman, E., and Dachinger, P., eds., *Art Therapy in Theory and*

CHAPTER 28

POETRY THERAPY

Owen E. Heninger

POETRY may be used as a device to involve patients in the therapeutic process. It is the speech of interiority, the "language of being," both intense and compact, and gets at the true utterance in the therapeutic endeavor. Some therapists[28] maintain that all poetry is therapy, that it is one of the techniques man has developed to cope with inner turmoil. Poetry may serve as therapy either by the resolution of mental anguish or by providing a way for man to face his conflicts more honestly. It is a means to engage in meaningful personal communication, it is a link to a greater awareness of unspoken thoughts and a way of sharing deep feelings.

¶ Introduction

Throughout history poetry has been a means to express the emotional forces that act upon men and women. The poet, bard, storyteller, like a shaman or doctor, has always been accorded a respect that transcends political disputes and national boundaries. Poetry was the original language of religion, and verse was a means by which ancient races developed a national character. It was left to Aristotle (c. 330 B.C.) to formally delineate the influence that literature had on the psyche, specifically the purifying or purgative effect of tragedy. This process he called *katharsis* (catharsis).[14] Not only did poetry afford its readers (viewers) pleasure in the aesthetic realm, it also acted upon them in a restorative manner. One of the better documented instances of poetry's therapeutic properties comes from the experience of John Stuart Mill, the English philosopher and economist. Mill noted in his autobiography that reading Wordsworth's poetry had been instrumental in his recovery from a nervous breakdown.[53]

While poetry in the form of narrative, plays, and poems has doubtlessly contributed to the well being of people for almost three thousand years, it was not until the twentieth century that men of science began to see in it a nascent discipline suited for therapeutic purposes. In 1908, Freud[22] expounded upon the possible relation between poetry and psychology in *The Relation of the Poet to Day-Dreaming.* In 1922,[63] Frederick Prescott published *The Poetic Mind,* in which he makes connections between Freud's theories and those of John Keble—an Anglican clergyman who regarded poetry as a veiled or indirect representation of the poet's true feelings, which may help stave off mental disease. Also, at around this time, in New York, Eli Greifer[42] began to organize poetry groups at a hospital and clinic; in fact, he is credited with giving "poetry therapy" its name.

The past fifty years have continued to witness an influx of therapists who have come to realize that poetry therapy is indeed a serious and viable method by which to effect mental health. And as the beneficial aspects of poetry therapy have come more and more to the attention of therapists and patients alike, a concomitant increase in the literature, in academic courses, in panels, and in workshops has occurred.

¶ Place in Therapies

As an Adjunct to Psychotherapy

Poetry therapy is not a "school" of therapy, but rather works with other therapies to help effect additional insight and emotional release.[1] Using poetry in therapy can have a catalytic function in moving psychotherapy along, stimulating thinking, and bringing out emotionally significant relationships. Poetry can work synergistically to facilitate the psychotherapeutic process.

As an Expressive Variable

Poetry therapy can help patients reveal more about themselves. Sometimes revelations may be triggered by a poem in which the poet has expressed what the patient has not been able to. Patients may read or write poems that articulate their own feelings. They can be guided by the therapist or a trained poetry therapy facilitator. (That is, paraprofessionals who have been specially trained in the technique of using poetry in group work).[15]

To Achieve Insight

The use of poetry in therapy helps both the therapist and the patient to get a better look at the psychological makeup of the patient. Lerner[45] notes that poetry may serve to explore more deeply the inner workings of the mind. By implementing observation and presenting a sharper, clearer exposure, poetry therapy contributes to insightful psychotherapy.

Exhortative Function

Poetry therapy can function in an authoritative or directive manner, exhorting the suppression of fears or the acceptance of certain values. Poetry may be used in an attempt to regulate, inspire, or support the patient. The exhortative approach has close ties with some forms of bibliotherapy where the therapist may assign the reading of articles, pamphlets, or books in order to motivate the patient. Poems that have clear exhortative messages have been collected and published by Blanton.[7] Leedy[41] notes that poems that are in harmony with the patient's particular mood or psychic condition have a supportive role in the patient's therapy.

As Creative Therapy

A patient reading or writing poetry is being creative as well as taking part in an organizing process. In poetry therapy this creativity is encouraged and supported. Heninger[30] has pointed out that this act of organizing often leads to unique and original poems that bring into existence a new order.

Settings

Poetry therapy has been utilized in a wide variety of settings; for example in a correctional setting, in the private practice of psychotherapy, in mental hospitals; with geriatric populations, with adolescents, and with drug abusers. It has been used in group therapy and in mental health centers. Molly Harrower[28] has given an elaborate account of her personal growth experience and how she worked out her problems by writing poetry.

¶ Theory

Poetry therapy works through the multidimensional experience of any or all of the psychological avenues of ventilation and catharsis, exploration, support, active mastery and understanding. It produces results because, "in poetry we experience the most effective, the most concentrated and emotionally textured communication man has, as yet, devised."[26] It should also be noted that there is a good deal of safety for the patient who uses poetry in psychotherapy.

VENTILATION AND CATHARSIS

Poetry therapy offers an opportunity for the patient to openly express his innermost thoughts (ventilation). Often this open expression of ideas is accompanied by an appropriate emotional reaction; that is, the discharge of strong feelings and unconscious tension, followed by a significant sense of relief (catharsis). Using poetry in therapy allows patients to pour out their emotional venom and discharge strong sentiments, thus freeing themselves by cleansing their emotional systems.[54,74] This is akin to lifting a cover from a deep well of emotions, which can then be discussed, dealt with, and integrated.[14] Poetry, used in the therapeutic encounter, may pry loose elements of thought that would not otherwise surface, and this process itself may lead to a reduction in symptoms.[32] Prescott[63] has observed that the ordered flow of poetry is like the eruption that prevents an earthquake. He likens the use of poetry to putting a veil over one's emotions in order to wear them in public. According to him, poetry is a spontaneous overflow of powerful feelings and a way to purge long-standing anger, which relieves the overburdened mind.

Patients are able to find poems that express unacceptable thoughts in a manner that gains both acceptance and relief. Reilly[65] recognized that poems can be containers for the unacceptable. Rothenberg[67] noted that the feverish activity of writing a poem indicates an anxiety that requires discharge. He observed that poets may put their personal conflicts into poetry in a way that unearths preconscious and unconscious material, thereby relieving tension. Through the use of poetry, rage can be channeled, anger and gloom dispelled, and melancholy purged.[8,17] Chaliff[12] related how Emily Dickinson wrote poetry as a kind of catharsis. She also notes how Dickinson's poetry can be employed therapeutically to help others and quotes the poet herself on the purgative effects of poetry: "We tell a hurt to cool it."

Poetry can be used for confessional purposes.[13] Patients may find themselves capable of revealing their most private selves unashamedly, giving voice to the deepest sense of their being. A significant example occurs where poetry is used to express individual responses to illness,[31] or where it expresses the essence of an illness.[16] Patients can sometimes express poetically what they are afraid of or unable to articulate. Here the ego is searching for a safe outlet to reveal forbidden impulses. Poetry offers a safe method of discharge that is free from guilt and humiliation. As Robert Jones put it: "The artful writer can put us into a position in which we can enjoy our own daydreams without reproach or shame."[35]

EXPLORATION

The process of psychotherapeutic exploration is aided by using poetry. Prescott[63] noted, "Poets have been disposed to introspection and self analysis; and where they have been so disposed they have far surpassed ordinary men in subtlety of discrimination and acuteness and depth of insight . . . they have exceptional powers of observation." The legacy of poetry is further recognized by Wolberg,[78] who observed that, "poets have revealed themselves and have analyzed man's condition long before human behavior was conceptualized as a science." Robinson[66] explained that poets skillfully use poetry "to express rather than to conceal. Poets use fantasy to elicit meaning in a fresh context; they distort in order to clarify; they symbolize in an effort to illuminate." Poetry therapy aims at using the patient's powers of observation, introspection and self-analysis.*

Poetry may function as a tool to effect self-discovery.[45,68,74] Reik[64] suggested that the poet's (patient's) creation gives embodiment to "the life of the mind." Meerloo[51] saw poetry as assisting in "self-recollection." Poetry that is read, written, or reacted to by patients gives an indication of where they are in their individual growth process.[5,27,61] Chaliff[12] related how poetry can prod slumbering personal issues into full wakefulness and allow one to view the underlying psychic structure. By circumventing repressive barriers, poetry creates a means by which to explore the patients unconscious.†

Poetry can be used to penetrate and illuminate the depths of psychic life. It offers a new way of looking at one's self.[1] It can provide a sharp picture of one's self (insight).‡ Poetry can reveal a "self-reality"[38] or get at what Meerloo[51] called, "the essential 'me' ". Andrews[2] saw poetry as a tool for unearthing hidden stores from within the self. Edgar[19] viewed poetry as a "reservoir for the expression of human feelings." When poetry is used in psychotherapy, it opens up "new vistas,"[4] increasing self-awareness, self-confrontation, and self-exposure.[68,72] The use of poetry can make one aware of patterns in one's own psychic functioning.[56] It can get psychological forces out in the light to be observed.[14]

Because poetry itself seeks for the most honest and truthful way to express human thought and feeling, by using words as "truth seekers,"[71] it follows that an honest and personal truth comes with using poetry in psychotherapy.[9,71,74] The poet/patient shows his conflicts honestly and what he writes is an honest reflection of how he lives.[54]

*See references 3, 19, 27, 41, and 59.
†See references 30, 58, 61, 63, 72, and 74.
‡See references 6, 30, 55, 68, and 77.

SUPPORT

Poetry can be used in psychotherapy to support patients. For example, Blanton[7] has a collection of poems that can be given to patients when courage is needed. Koch[36] uses poetry to instill confidence and promote communication and concentration. By using poetry in therapy it is possible to console patients.[38] Poetry can help patients see that their problems are similar to the problems of others (universalization).[6] They can relate to others and share their thoughts and feelings. This sharing may stimulate improved interpersonal relations, even if it's despair that is shared.[41] Through poetry patients discover psychological kinships and overcome interpersonal barriers.[57]

There are various ways by which poetry can rescue patients. They may borrow or lean on the ego strength of the poet and his manner of facing and/or handling conflicts.[14] Poetry can be used to give them guidelines concerning special attitudes or modes of behavior.[14] Well-known poetry may be used to give patients a psychological mooring.[28] Murphy[54] points out that patients even use poetry as a means of requesting help. It can help patients validate their feelings and reduce their doubts and uncertainty.[14] Poetry is often used for its self-sustaining power.[28] Greenberg[23] notes that it can even be used as a sermon to the self. When patients put their conflicts into poetry, they are usually praised. They feel they share respect with poets and develop a sense of pride and self-esteem. They may even attract praise from their community.[72]

AS ACTIVE MASTERY

"It is primarily through language, an extremely important tool for active mastery of the environment, that the poets (patients) achieve identification."[50] The language of poetry helps one to create order where chaos existed.[6] Poetry provides order and authority, channeling crude instinct.[21] It allows one to organize unconscious forces and refine what might be called "raw emotional spillage."[49] Poetry offers a means of transfering the emotionally unbearable into something that can be faced.[2] The poet/patient uses words to disguise, muffle, and transform the deep-seated thoughts and feelings that imperiously demand issue.[17] He puts these thoughts and feelings into words that can be managed, or actively mastered.[28] Poetic forms can be used to reorganize mood and perception.[57] Poetry can be ego building[9] and can allow the writer/reader to rearrange the world to be more self-pleasing.[22] Poets/patients may use poetry to escape the "miseries of the mind," to sever themselves from griefs and woes, and to channel their rage.[17]

Poetry can help in handling feelings. It is an artful device providing the writer/reader with a process to turn the frightening and unacceptable into something acceptable.[65] Through a kind of psychological metamorphosis it allows one to face formidable or forbidden subjects and reduce them to a manageable form.[60] The making of poetry can blunt the original raw emotion and objectionable ideational content and make it tolerable and even pleasurable.[69] Taking something that is ugly and turning it into something esthetic is one way poetry mollifies discomfort. Putting unappealing thoughts and feelings into appealing poetry has been likened to the process whereby an oyster takes an irritating grain and creates a beautiful pearl around it.[12] The writer/reader of poetry may use it to find his own shape and rationale for the "monsters of the mind" and may even use poetry to harness them.[60] As the patient masters his problems he may become a poet and his poems can then be seen as evidence of the therapeutic results.

Poetry can both exorcise and neutralize trauma and suffering.[50] Poetry has been seen as a form of art that is molded by the poet's/patient's inner struggle against defect.[57] Creating poetry out of one's vulnerability may even become a special strength. Emily Dickinson, it is said,[50] was able to come to terms with her problems by facing them squarely in her poetry. "Art [poetry] gives a way of reshaping the pain of experience into wholeness, health and harmony."[17] The poet/patient makes his poetry out of the deepest sense of his powerlessness.[17] There is relief when one can safely review a disaster; reading or writing poetry allows one to exorcise disaster by reviewing it. The poets/patients may have their poetry issue from their own suffering; they can create, out of transient pain, a universal statement of the human condition.[12]

The use of poetry in psychotherapy can aid in self-healing. Otherwise passive prisoners have been seen to write poetry to actively help themselves. For example, Merloo[51] has described how concentration camp inmates utilized poetry to relieve their situation. Poems may accomplish the resolution of a conflict, proclaiming the poet/patient as self-therapist.[23] Aspects of the self provide substance for what is written (or read) into poetry.[3] Poems may be connected with persons as transitional objects, and people can use poems to assist in the process of merging and separation. Poetry may be utilized to circumvent the barriers of repression,[18] permitting the observing ego to reflect on the material exposed. Revealing important psychological forces (repressed and forgotten material) and experiencing one's feelings and thoughts under the domain of the self-regulatory capacities make assimilation of these materials possible. Poetry offers a preliminary exercise for the active functions of life and can be used to study a part before playing it;[63,77] that is, reading or writing poetry provides a means to rehearse for life's activities, such as preparing for an emotional encounter with another person. Through the reading or writing of poetry a more beautiful or healthier balance can be obtained.[30] Poetry works toward harmony and is geared to bring about a more integrated personality.

The reading and writing of poetry represents a maturity of effort and a growth-producing experience.[66] It can strengthen a personal sense of identity. Poetry can be used to allow the self to bloom[23] and assists the patient's innerself to stretch, breathe, look about, and grow.[26] The inception of a poem is the beginning of a movement toward psychological freedom.[67,74] Poetry can intrigue,[52] reawaken, and rejuvenate.[30,54] It can capture a moment of intense emotion and, like amber catching a fossil, keep it protected against the ravages of time.[61] One might even call poetry, "ambered experience." Thus, the capturing, preserving, or "ambering" of deeply experienced thoughts and emotions in poetry allows them, over time, to be more readily examined, handled, and actively mastered.

UNDERSTANDING

One of the hallmarks of literature (poetry) is its ability to enrich, broaden, and deepen the reader's experience and understanding. It also establishes a bridge between the self and the world.[5] The writer's attempts to understand the world become vehicles by which the reader may also find self-understanding and self-help.[12,45] Poems are successful when they can create an empathetic state in the reader, allowing him to participate in the emotion that is so vividly depicted.[12] Through poetry we can get a good deal of understanding about a poet's (patient's) life.[13] Poetry is a means of understanding without pressure.[18] It can help in understanding the poet's development,[28] and may lead to self-recognition.[30]

SAFETY

Poetry may indirectly express thoughts, attitudes, and feelings that can not be expressed in a direct manner because of the fear of retaliation.[9] With poetry, there can be a catharsis without total psychological nudity. The poet/patient can use poetry to both hide and reveal his innermost thoughts and feelings. Poetry gratifies simultaneously both the impulse for expression and for control (concealment).[63] One may face very disquieting conflicts in the disguise of poetry.[59] The writer/patient may encode his message in a poem,[11] or, he may hide his identity behind a pseudonym and confess safely. He may even attribute his writing to inspiration from the muses. Chaliff[12] notes how Emily Dickinson used the protective disguise of her poetry to meet the dread of revealing herself fully. "Tell all the Truth but tell it slant/Success in Circuit lies."

Patients/writers talk about themselves indirectly as a safe way of sharing feelings that have been smoldering within.[19,20] Freud[22] noted that the direct expression of repressed wishes repels and leaves one cold. However, if these wishes are given expression through literary skill, which softens their egotistical character by disguising and using the esthetic pleasure as a cover, these same wishes are then presented without reproach or shame. Poetry provides an acceptable camouflage for the ventilation of unconscious material.[32] It can be used to express emotion in a socially acceptable form.[24] By encapsulating a hurt in verse,[28] poetry protects one from the violence of passion. The veil of poetry allows many otherwise unacceptable ideas and feelings to be aired in safety.

¶ Techniques

"Poetry helps the individual see where he is and where he is going. The therapist's problem is to apply poetic principles in the most productive manner possible."[33] The therapist is primarily concerned with poetry as a vehicle for the patient to explore and express himself. It is important in poetry therapy that the poetry is not judged as a work of art. It is no more appropriate to find fault with a patient's poetry than to criticize his associations in psychotherapy or his drawing in art therapy. The poems used in poetry therapy are judged by how well they share feelings rather than by their literary merit.[19,41,73]

An established poet's verse may be introduced in poetry therapy to encourage patients to disclose more about themselves. Generally, it is best to read poems aloud and maybe even to repeat the reading. At times, patients are helped even further by memorizing poems.[7] It is wise to avoid presenting poems too rapidly since it may take some time for a poem to penetrate into the mind. Sometimes a poem's influence will surface days, weeks, or even decades later. It is also wise to avoid introducing poems of such complexity that it takes special academic study to understand them.

Individual Therapy

Poetry can be used in an adjunctive way in individual psychotherapy. Perhaps the best way to begin is to ask patients about their past interests in reading or writing poetry.[61] If they bring poetry with them, it is analyzed in the process of psychotherapy.[25,30,58] At a time of strong emotion it can be helpful to encourage patients to write a poem so that the images of the moment are not lost.[30] Another technique is to select emotionally laden words or statements from the patient's speech and suggest that the patient write these words in poetic form.[26] A therapist may introduce and/or prescribe poems for patients to read.[7,14,41] Patients are then encouraged to relate the ideas and emotions that come to them on reading or hearing a poem.[55,58,61] At times, poems introduced by the therapist seem to sustain patients even in the therapist's absence.[7,61]

Group Therapy

Poetry therapy has a dynamic influence on the process of group therapy (group workshop). Using poetry shortens the warm-up period.[19] Patients who share poetry in a group experience empathy and catharsis within a framework of human kinship.[61] Reading poetry brings the members of the group closer together by starting an immediate group relationship and accentuating the ideas and feelings people have in common.[54,61] Poetry encourages feelings of equality in the group, although the sharing of personal information between group members is optional.[61]

It is often helpful to have copies of poems available for all group members; with ongoing groups filing folders are useful. Group meetings are often started with a well-known poem, although group members are encouraged to write and read their own poetry or lesser known poetry that speaks for them.[9,20,55] It is beneficial to investigate what the writer was trying to say, how other members react to the poem, and to expand on the feelings expressed.[61] Although group participants are encouraged to talk about the feelings the poems evoke or express, they may steer away from personal problems. A patient should be engaged at a depth that the therapist deems safe. It may be useful to examine the condition a patient was in when he wrote or read a particular poem. With the encouragement of the expression of feelings, group meetings become free flowing.

The therapist should provide a stable and healthy image that can control the group rather than get too involved with group members on a personal level.[61] The therapist should be spontaneous and empathetic but should avoid personal confessions and displays of strong emotion directed toward patients. The therapist should not insist that group members call him by his first name; a cotherapist or facilitator, however, may be more egalitarian.[61] The therapist must accept pauses, give time for poems to be understood, and keep track of and examine the responses the poems may elicit.

Indications

Who are the patients who should be considered for possible benefit from poetry therapy as an adjunct to psychotherapy? Good candidates are patients who cannot express themselves well through the usual channels but who can write out what they cannot say.[30,37] Those who are lonely, withdrawn, alienated or stiff, rigid and inflexible are usually good responders to poetry therapy.[37] One can usually use poetry therapy to good advantage with patients who already write verse or speak in poetic phrases.[30] At times, poetry therapy can be used to break an impasse in psychotherapy.[37] Many people without serious psychopathology may be helped toward more individual growth with poetry. Poetry therapy is also indicated in patients who need additional help in gaining insight.[1]

Contraindications

It can be dangerous to send poems to patients whom the therapist does not see regularly. It is also unwise to use poetry with those who cannot tolerate getting deeper into their own thoughts and feelings. Poetry may hit such a tender spot that it sets off a near uncontrollable emotional storm. When there are pressing problems such as a serious brain syndrome, sociopathy, acute psychosis, or addiction, poetry therapy is inadvisable.[37] It is wise to avoid using poetry with patients who markedly misunderstand it or who take poetic images too literally.[21] Poetry therapy is not indicated in patients who see it as a panacea or get fanatic about its use.[37] At times, certain initiates to poetry therapy become so excited and euphoric over their personal contact and response that they mistake these emotional responses for a cure.[37,46] Lauer[37] notes that poetry therapy is not for those who think "anyone can do it." Much the same contraindication applies to those who have the "Sorcerer's Apprentice"* syndrome. Some therapists learn the magic words (of poetry) that turn people on, but they do not know how to handle them afterwards.

Training

Special training is required for those who would be safe and efficient using poetry in psychotherapy. Training in both poetry and psychotherapy is important. Not all therapists necessarily know how to use poetry in the therapeutic setting and those familiar with poetry may not know how to do psychotherapy.[46] Training in psychotherapy would include courses in psychology and/or psychiatry. The training in poetry should include extensive courses in literature with emphasis on poetry. The academic courses should include semantics, language as behavior, and literary criticism.[18] A course in oral interpretation would be a definite help in the field.[29] A therapist using poetry in therapy should be acquainted with a large variety of poems and learn which poems are most effective in therapy.[45] There should be lengthy supervised experience in the clinical practice of poetry therapy. Emphasis should be given in relating poetry to the behavioral sciences. The goal of training in poetry therapy is to prepare one for clinical practice, teaching, research, writing, and advancing the field.[62]

*This refers to Goethe's famous ballad of a sorcerer's apprentice who made more magic than he could control and was unable to stop an enchanted broomstick from carrying water for him.

Research

There are many paths for the use of poetry therapy that have yet to be explored and documented. For example, how are poetry and psychotherapy combined in other countries and cultures? Is there a significant difference in therapeutic effect when poetry of high artistic merit is used as compared with verse of lesser merit? Should the poetry used in poetry therapy change along with a patient's growth in therapy? How can familiarity with poetry therapy become a part of psychiatric or psychological training? Does poetry therapy have a place in peer counseling, marriage counseling, or counseling by the clergy? Where else can poetry therapy be used in psychotherapy? How can poetry therapy be combined with the other art therapies? What is the best way to teach poetry therapy? Is there a prophylactic power against mental illness available in poetry therapy? Whatever the answers to these questions, poetry therapy is finding ever greater acceptance and usefulness in the practice of psychotherapy.

¶ Bibliography

1. ABRAMS, A. S. "Poetry Therapy in the Psychiatric Hospital," in Lerner, A., ed., *Poetry in the Therapeutic Experience.* New York: Pergamon Press, 1978, pp. 63–71.
2. ANDREWS, M. "Poetry Programs in Mental Hospitals," *Perspectives in Psychiatric Care,* 8:1 (1975): 17–18.

3. BALDWIN, N. "The Therapeutic Implications of Poetry Writing: A Methodology," *Journal of Psychedelic Drugs,* 8:4 (1976): 307–312.
4. BARKLEY, B. "Poetry in a Cage: Therapy in a Correctional Setting," in Leedy, J.J., ed., *Poetry the Healer.* Philadelphia: Lippincott, 1973, pp. 1–16.
5. BARRON, J. "Poetry and Therapeutic Communication: Nature and Meaning of Poetry," *Psychotherapy: Theory Research and Practice,* 11:1 (1974): 87–92.
6. BERGER, M. M. "Poetry as Therapy—and Therapy as Poetry," in Leedy, J. J. ed., *Poetry Therapy.* Philadelphia: Lippincott, 1969, pp. 75–87.
7. BLANTON, S. *The Healing Power of Poetry.* New York: Thomas Y. Crowell, 1960.
8. BLINDERMAN, A. "Shamans, Witch Doctors, Medicine Men and Poetry," in Leedy, J.J., ed., *Poetry the Healer.* Philadelphia: Lippincott, 1973, pp. 127–141.
9. BUCK, L. A., and KRAMER, A. "Poetry as a Means of Group Facilitation," *Journal of Humanistic Psychology,* 14:1 (1974): 57–71.
10. ——— "Creative Potential in Schizophrenia," *Psychiatry,* 4 (1977): 146–162.
11. BURKE, K. "Thoughts on the Poets' Corner," in Leedy, J. J., ed., *Poetry Therapy.* Philadelphia: Lippincott, 1969, pp. 104–110.
12. CHALIFF, C. "Emily Dickenson and Poetry Therapy: The Art of Peace," in Leedy, J. J., ed., *Poetry the Healer.* Philadelphia: Lippincott, 1973, pp. 24–49.
13. CHESSICK, R. D. "What Can Modern Psychotherapists Learn From Modern Poets?" *Current Concepts in Psychiatry,* 4:6 (1978): 2–8.
14. CROOTOF, C. "Poetry Therapy for Psychoneurotics in a Mental Health Center," in Leedy, J. J., ed., *Poetry Therapy.* Philadelphia: Lippincott, 1969, pp. 38–51.
15. DAVIS, L. "The Paraprofessional and Poetry Therapy," in Lerner, A., ed., *Poetry in the Therapeutic Experience.* New York: Pergamon Press, 1978, pp. 108–113.
16. DICKEY, J. "Diabetes," in *The Eye-Beaters, Blood, Victory, Madness, Buckhead and Mercy.* Garden City, N.Y.: Doubleday, 1970, pp. 7–9.
17. EDEL, L. "The Madness of Art," *American Journal of Psychiatry,* 132:10 (1975): 1005–1012.
18. EDGAR, F., and HAZLEY, R. "A Curriculum Proposal for Training Poetry Therapists," in Leedy, J.J., ed., *Poetry Therapy.* Philadelphia: Lippincott, 1969, pp. 260–268.
19. ———, and LEVIT, H. I. "Poetry Therapy with Hospitalized Schizophrenics," in Leedy, J.J., ed., *Poetry Therapy.* Philadelphia: Lippincott, 1969, pp. 29–37.
20. ERICKSON, C. R., and LE JUENE, R. "Poetry as a Subtle Therapy," *Hospital and Community Psychiatry,* Feb. 1972: 40–41.
21. FORREST, D. V. "The Patient's Sense of the Poem: Affinities and Ambiguities," in Leedy, J. J., ed., *Poetry Therapy.* Philadelphia: Lippincott, 1969, pp. 231–259.
22. FREUD, S. "The Relation of the Poet to Day-Dreaming," in Jones, E., ed., *Sigmund Freud Collected Papers.* London: Hogarth Press, 1949, pp. 173–183.
23. GREENBERG, S. A. "Poetry Therapy in a Self-Help Group AFTLI and/or Poetry Therapy," in Leedy, J.J., ed., *Poetry Therapy.* Philadelphia: Lippincott, 1969, pp. 212–222.
24. GREENWALD, H. "Poetry as Communication in Psychotherapy," in Leedy, J.J., ed., *Poetry Therapy.* Philadelphia: Lippincott, 1969, pp. 142–154.
25. HAMILTON, J. W. "Gender Rejection as a Reaction to Early Sexual Trauma and Its Partial Expression in Verse," *British Journal of Medical Psychology,* 41 (1968): 405–410.
26. HAMMER, E. F. "Interpretations Couched in the Poetic Style," *International Journal of Psychoanalytic Psychotherapy,* 7 (1978–79): 240–253.
27. HARROWER, M. "Poems Emerging From the Therapeutic Experience," *Journal of Nervous and Mental Disease,* 149:2 (1969): 213–233.
28. ———. *The Therapy of Poetry.* Springfield, Ill.: Charles C Thomas, 1972.
29. HAYAKAWA, S. I. "Postscript: Metamessages and Self-Discovery," in Leedy, J. J., ed., *Poetry Therapy.* Philadelphia: Lippincott, 1969, pp. 269–272.
30. HENINGER, O. E. "Poetry Therapy: Exploration of a Creative Righting Maneuver," *Art Psychotherapy: an International Journal,* 4:1 (1977): 39–40.
31. ———. "Iritis II," *Newsletter—American Physicians Poetry Association,* 1:5 (1978): 15.
32. ———. "Poetry Therapy in Private Prac-

tice: An Odyssey into the Healing Power of Poetry," in Lerner, A., ed., *Poetry in the Therapeutic Experience.* New York: Pergamon Press, 1978, pp. 56–62.

33. HENINGER, O. E. Personal communication, Nov. 15, 1978.
34. JONES, R. E. "Treatment of a Psychotic Patient by Poetry Therapy with a Historical Note," in Leedy, J. J., ed., *Poetry Therapy.* Philadelphia: Lippincott, 1969, pp. 19–28.
35. ———. "The Double Door Poetry Therapy for Adolescents," in Leedy, J. J. ed., *Poetry Therapy.* Philadelphia: Lippincott, 1969, pp. 223–230.
36. KOCH, K. "Teaching Poetry Writing to the Old and the Ill," *Milbank Memorial Fund Quarterly/Health and Society,* 56:1 (1978): 113–126.
37. LAUER, R. "Abuses of Poetry Therapy," in Lerner, A., ed., *Poetry in the Therapeutic Experience.* New York: Pergamon Press, 1978, pp. 72–79.
38. LAWLER, J. G. "Poetry Therapy?" *Psychiatry,* 35 (1972): 227–237.
39. LEEDY, J. J. ed., *Poetry Therapy.* Philadelphia: Lippincott, 1969.
40. ———. "Introduction to Poetry Therapy," in *Poetry Therapy.* Philadelphia: Lippincott, 1969, pp. 11–13.
41. ———. "Principles of Poetry Therapy," in *Poetry Therapy.* Philadelphia: Lippincott, 1969, pp. 67–74.
42. ———. "In Memoriam Eli Greifer 1902–1966," in *Poetry Therapy.* Philadelphia: Lippincott, 1969, pp. 273–275.
43. ———, ed., *Poetry the Healer.* Philadelphia: Lippincott, 1973.
44. ———, and RAPP, E. "Poetry Therapy and Some Links to Art Therapy," *Art Psychotherapy: an International Journal* 1 (1973): 145–151.
45. LERNER, A. "Poetry Therapy," *American Journal of Nursing,* 73: 8 (1973): 1336–1338.
46. ———. "A Look at Poetry Therapy," *Art Psychotherapy: an International Journal,* 3: 1 (1976): i–ii.
47. ———, ed. *Poetry in the Therapeutic Experience.* New York: Pergamon Press, 1978.
48. ———. Personal communication.
49. LIVINGSTON, M. C. personal communication.
50. LORD, M. M., and STONE, C. "Fathers and Daughters—A Study of Three Poems," *Contemporary Psychoanalysis,* 9 (1973): 526–539.
51. MEERLOO, J. A. "The Universal Language of Rhythm," in Leedy, J. J., ed., *Poetry Therapy.* Philadelphia: Lippincott, 1969, pp. 52–66.
52. MORRISON, M. R. "Poetry Therapy with Disturbed Adolescents—Bright Arrows on a Dark River," in Leedy, J. J., ed., *Poetry Therapy.* Philadelphia: Lippincott, 1969, pp. 88–103.
53. ———. "A Defense of Poetry Therapy," in Leedy, J. J., ed., *Poetry the Healer.* Philadelphia: Lippincott, 1973, pp. 77–90.
54. MURPHY, J. M. "Forward" in Leedy, J. J., ed., *Poetry the Healer.* Philadelphia: Lippincott, 1973, pp. ix–xvi.
55. ———. "The Therapeutic Use of Poetry," in Masserman, J. H., ed., *Current Psychiatric Therapies,* vol. 18. New York: Grune & Stratton, 1979, pp. 65–71.
56. NEMIAH, J. C. "The Art of Deep Thinking: Reflections on Poetry and Psychotherapy," *Seminars in Psychiatry,* 5:3 (1973): 301–311.
57. PARKER, R. S. "Poetry as a Therapeutic Art in the Resolution of Resistance in Psychotherapy," in Leedy, J. J., ed., *Poetry Therapy.* Philadelphia: Lippincott, 1969, pp. 155–170.
58. PATTISON, E. M., "The Psychodynamics of Poetry by Patients," in Leedy, J. J., ed., *Poetry the Healer.* Philadelphia: Lippincott, 1973, pp. 197–214.
59. PIETROPINTO, A. "Exploring the Unconscious Through Nonsense Poetry," in Leedy, J. J., ed., *Poetry the Healer.* Philadelphia: Lippincott, 1973, pp. 50–76.
60. ———. "Monsters of the Mind: Nonsense Poetry and Art Psychotherapy," *Art Psychotherapy an International Journal,* 2 (1975): 45–54.
61. ———. "Poetry Therapy in Groups," in Masserman, J. H., ed., *Current Psychiatric Therapies,* vol. 15. New York: Grune & Stratton, 1975, pp. 221–232.
62. "Poetry Therapy," in *The International College Catalog,* 1979–1980. Los Angeles, Cal.: The International College Catalog, p. 76.
63. PRESCOTT, F. C. *The Poetic Mind.* Ithaca, N. Y.: Cornell University Press, 1959.
64. REIK, T. "Forward" in Leedy, J. J., ed., *Poetry Therapy.* Philadelphia: Lippincott, 1969, pp. 5–7.

65. REILLY, E. "Sylvia Plath: Talented Poet, Tortured Woman," *Perspectives in Psychiatric Care,* 16:3 (1978): 129–136.
66. ROBINSON, S. S., and MOWBRAY, J. K. "Why Poetry?," in Leedy, J. J., ed., *Poetry Therapy.* Philadelphia: Lippincott, 1969, pp. 188–199.
67. ROTHENBERG, A. "Poetic Process and Psychotherapy," *Psychiatry,* 35 (1972): 238–254.
68. ———. "Poetry and Psychotherapy: Kinships and Contrasts," in Leedy, J. J., ed., *Poetry the Healer.* Philadelphia: Lippincott, 1973, pp. 91–126.
69. ———, and HAUSMAN, C. eds. *The Creativity Question.* Durham, N. C.: Duke University Press, 1976.
70. ———, and GREENBERG, B. *The Index of Scientific Writings on Creativity.* Hamden, Conn: Shoe String Press, 1976.
71. SCHECHTER, R. L. "Poetry: A Therapeutic Tool in the Treatment of Drug Abuse," in Leedy, J. J., ed., *Poetry the Healer.* Philadelphia: Lippincott, 1973, pp. 17–23.
72. SCHLOSS, G. A. *Psychopoetry.* New York: Grosset & Dunlap, 1976.
73. ———, and GRUNDY, D. E. "Action Techniques in Psychopoetry," in Lerner, A., ed., *Poetry in the Therapeutic Experience.* New York: Pergamon Press, 1978, pp. 81–96.
74. SILVERMAN, H. L. "Creativeness and Creativity in Poetry as a Therapeutic Process," *Art Psychotherapy: An International Journal,* 4:1 (1977): 19–28.
75. STONE, A. A. "Preface," in Leedy, J. J. ed., *Poetry the Healer.* Philadelphia: Lippincott, 1973, pp. xvii–xviii.
76. STONE, C. "Three Mother-Daughter Poems: The Struggle for Separation," *Contemporary Psychoanalysis,* 2:2 (1975): 227–239.
77. WATERMAN, A. S., KOHUTIS, E., and PULONE, J. "The Role of Expressive Writing in Ego Identity Formation," *Developmental Psychology,* 13:3 (1977): 286–287.
78. WOLBERG, L. R. "Preface: the Vacuum," in Leedy, J. J., ed., *Poetry Therapy.* Philadelphia: Lippincott, 1969, pp. 9–10.
79. WOOD, J. C. "An Experience in Poetry Therapy," *JPN and Mental Health Services* (Jan–Feb 1975): 27–31.

CHAPTER 29

MUSIC THERAPY

Karen D. Goodman

¶ Historical Use of Music in Healing

FROM ancient times music has played an essential role in the humane treatment of individuals with mental, physical, and emotional illnesses. The design of this treatment is fascinating because it not only parallels the evolution of civilization but also the evolving integration of "magical" and scientific healing.

The primordial imitation of, and subsequent identification with, certain sounds in the environment (wind, rain, waves, trees, animals) gave primitive man access to nonverbal communication with his "invisible world"—a "supernatural, magical" world which, he was convinced, controlled his well being. Healing rites, consisting of music, rhythms, songs, and dances, were led by a "magician" who was conversant with the formulas for communicating with the "spirits."

Eventually, ancient man moved from magical healing via music to religious healing via music—still utilizing music as a communicative language with the supposed sources of disease rather than recognizing the specific ability of music to affect man's psyche and soma. As it was written in the Bible: "Seek out a man who is a cunning player on a harp [David]: and it shall come to pass, when the evil spirit of God is upon thee, that he shall play with his hand and thou shalt be well." (Samuel I, Chapter 16, verse 16).

Practices found in the medical tradition of ancient Greece represented a similar divergence of attitudes. The Greeks used music and musical instruments, believed to be gifts from the gods, to propitiate deities they had created in their own image. Incantations, songs, and music were a standard part of Greek and, later, of Christian ritual. Also available were environmental healing treatments involving supplications for help from the gods. At some 420 temples of Aesclepius the basic treatment consisted of ritual purification, special diet, and sleep-inducing drugs, with a musical background as an integral component of the dreaming/sleeping period.

At the same time, rational and scientific ideas about music and medicine were being developed by philosophers. Cassiodorus, Plato, and Aristotle attributed certain emotional and ethical characteristics to various musical modes. Socrates referred to the subliminal action of intoxicating music, and Aristotle recognized the cathartic power of music. Plato introduced the idea of specific harmonic modes, instruments, and rhythms affecting the human psyche (*Republic,* III) and suggests specific musical "recipes" for such afflictions as Korybantism and insomnia.

These elementary Greek concepts of music therapy, "a kind of psychotherapy that affected the body through the median of the soul,"[4] were not isolated phenomena. In scrutinizing the history of psychiatry, it is striking to note that almost all of the humane treatment movements have included music as either a specific or nonspecific therapy.[102] The tradition appeared in Baghdad (eighth century, A.D.), Damascus (ninth century), Leleppo, Kalaoma, Cairo, and Fez as part of the Arabic-Hebraic medical tradition (thirteenth century). As early as 1100 B.C. travelers returning from the Middle East to Europe reported "humane" psychiatric treatments that carried on the Aesclepian tradition. An essential part of treatment was the concerts in which the instruments were tuned in a special way so that they would not jar the patients' nerves—an example of the gentleness and humanity of these ancient institutions.

While this Arab and Judaic tradition had some impact on the Christian practice of medicine developing in Europe in the Middle Ages (for example, the Salerno Medical School founded by Constantinus Africanus), it was overshadowed by the prevalent attitude that the mentally ill, labeled as evil, were to be punished rather than helped.

The emergence of the Renaissance revitalized the use of music therapy in terms of physiology and psychology. Moreover, the possibility of man's self-actualization via the arts was recognized in that the patient was encouraged to express himself through a musical experience. Obviously this experience was still not fully understood in relation to the therapeutic growth of the psyche, soma, and/or interpersonal relationship between therapist and client. The part the musicians themselves played in music therapy was to remain largely empirical until shortly after World War II.

¶ Music Therapy as a Profession

As a distinct discipline, music therapy contributed to the holistic point of view prevalent in the treatment of disease after World War II. Before and during the war, musicians had been encouraged to use their talents purely for morale purposes. However, the need to assess the influence of music on behavior, its relationship to psyche and soma, and its potential as a specific treatment process required validation in the scientific age. Music therapy developed into a profession with recognized institutions (National Association of Music Therapy established in 1950, American Association of Music Therapy established in 1969), a degree (B.A., M.A.), internships, registrations (Registered Music Therapist, Certified Music Therapist), textbooks, and journals (*The Journal of Music Therapy*).

Many of the initial music therapists were music educators with training and experience in public school music. Musicians from other backgrounds joined them, and they and representatives of other professions—primarily representatives of schools of psychiatric and psychoanalytic thought, clinical psychology,[83,84] milieu and attitude therapy (Menninger Foundation)—contributed to the field of music therapy. Continuing research cast doubts on beliefs on which earlier clinical practice had been based. Several popular concepts—that specific music will predictably stimulate or sedate patients, that background music increases verbalization in therapy groups, or that learning of other subject matter by retardates can be enhanced by utilizing their specific musical abilities—were not supported by research.[17,87,110]

The effect of music therapy research on clinical practice led to diversification as some practitioners altered older methods while others turned to new ones. While the mainstream of music therapy in the fifties and sixties remained only supportive, pioneers in the field developed a profession with additional depth. In Europe, Nordoff and Robbins,[80] beginning their innovative improvisation work with multiply handicapped children, made a great impact in America by establishing a humanist tradition as well as a musical therapeutic rapport between client and therapist. Such innovators as Florence Tyson[111,112] and Mary Priestley[91] began to translate these concepts into clinical practice, embodying the analytic model suggested previously,[83,84] as well as contributing a wealth of new material. Work with the adult psychiatric population was now beginning to take shape. The suggestion that musical imagery was as valid as dream imagery[83] was utilized in clinical practice by Dr. Helen Bonny who opened the Institute for Music Consciousness and Research in Baltimore, Maryland. In the 1970s, clinicians realized the need for specialized approaches in music therapy. Specialized music therapy now existed for the hearing-impaired,[10,96] aphasic,[101] autistic,[79,82] severely/profoundly handicapped preschooler,[35] and criminal offender.[90] Indeed, specialized music therapy offered a variety of clinical orientations including psychoeducational,[79] behavior modification[64] analytic,[91,111,112] and eclectic.[4,6] Tremendous variety of age (infancy to geriatric[20]) and handicap of clientele, as well as influences from shifts in institutional treatment philosophies and methods, have provided a plethora of approaches to music therapy.

Most music therapy was and is practiced in institutions for psychiatric and retarded patients, although settings presently include educational institutions, nursing homes, private practice in conjunction with psychiatric practice, alienated subsystems in communities (such as addiction treatment centers and prisons), and outpatient and inpatient clinics for treatment of the physically handicapped.

The field of music therapy is a "changing elastic diversity of clinical activities and points of view, carried out by a polyglot inclusive group whose bond remains the use of music for the benefits of the sick and disabled."[105]

¶ Music as a Tool

Music[1,11] is the art of sound in time, expressing ideas and emotions in significant forms through the elements of melody, harmony, and color. Tones or sounds occur in single line (melody) or multiple lines (harmony) and are sounded by voice(s) and instrument(s). Appreciation of or responsiveness to musical sounds or harmonies is inherent in the concept of music.

There is no doubt that a medium that affects the emotions, endocrines, circulation, respiration, blood pressure, mood, association, and imagery of man[2] can and must be used to meet specific needs in clinical practice. With this in mind, a body of music therapists, psychologists, psychiatrists, neurologists, pediatricians, physiologists, speech pathologists, and numerous others continues to examine the ways in which music affects individual physiological and psychological processes in man.[4]

One vital aspect of the biological foundation of music is the separate roles of the right and left hemispheres in music perception. The investigation of hemispheric roles and dominance, and the implication for education, culture, and creativity is nothing new. However, this investigation takes on added significance in regard to neurologically impaired or suspected split-brain individuals who express themselves well through music. Music may prove a means of transfer between left and right hemispheric functioning. Left and right hemispheres are employed differently in the musical process. The right "metaphoric" hemisphere, notably intact in many neurologically impaired clients, is responsible for major aspects of musical perception and musical behaviors; that is, the recognition of pitch, a gestalt sense of melody, rhythm, style, and musical memory.[94]

The commonalities between the components of speech and music[38,79] are a basis for the perceptual processes of the right hemisphere, which influences language functions and behavior.[44] The left hemisphere is predominantly involved with analytic, logical thinking, especially in verbal and mathematical functions. With someone who is musically sophisticated the left hemisphere

> dissects its passages in a manner analogous to the feature-detecting capacity of the left hemispheric visual fields. In other words, the right hemisphere, in effect, thinks, "ah, yes, Silent Night," two Columbia psychologists report, but the educated left hemisphere thinks, "two sequences, the first a literal repetition, the second a repetition at different pitch levels—ah, yes, Silent Night by Franz Gruber, typical pastorale folk style." [p. 7][94]

This assumes the processing involved in score-reading and score notation as well as mediating temporal order and acuity.[12]

The neurological potential for music processing to relate to the emotional life of any individual is enormous. Altschular says:

> The thalmus is a main relay station of emotions, sensations, and feelings. It is believed that even aesthetic feelings are relayed by the thalmus to the master brain. The thalmus is connected to the master brain by nerve pathways, and the stimulation of the thalmus almost simultaneously arouses the master brain. Once the brain is aroused, it sends impulses back to the thalmus and thus a reverberating circuit is set in motion. We can conclude then that music arouses our emotions. [p. 29][3]

As an affective response, evidenced physiologically, music leads to alterations of blood pressure; changes in respiration and pulse, the cardiovascular system, and galvanic skin response; lowering of the thresholds of sensitivity to other forms of stimulation; and delay of the onset of muscular fatigue.[19] Why music has this power, what exactly it is about music that is responsible for its effects, and how one can effectively tap music's power to utilize it therapeutically still remain open questions although a wealth of information on the subject is available. Generalities of human affective responses exist but only to the degree to which individuality of physical, organic, psychological, and experimental differences come into play.

As parts making up the whole, melody, rhythm, and harmony each have a vital, purposeful origin in the history of man and a psychological significance in the form or lack of form they present.

Melody appears to have its origin in primitive man's sound language—the sounds with which he reacted to external stimuli and expressed affect. As the only instrument belonging to man, voice, able to produce both melody and rhythm, best represents the "hidden person, his individuality, his uniqueness."[4] As a primitive response, melody appears as the first "informal" music experience of the child when his voice assumes differentiated crying and then vocal contagion/babbling.[85] The infant's attendance to the timbre of the mother's voice is an initial melodic and affective experience: "With its constant rising and falling and many inflections, it [melody] approximates the affective component of the endocept. The 'breathing' of melody approximates some of the physiological concomitants of emotion."[23]

As primitive and childish melodies (feeble, poorly organized, lacking in definite affective significance and nuance[72]) take "shape," they logically feed into definite tonal relationships (intervals) which resolve on a key note. Modes, the selection of tones within a scale,[7] will vary from culture to culture. Most common in this country are the major and minor scales. The timbre or quality of sound offered to the child by the human voice may prove either pacifying or disquieting.

The choice of tone relations in a melody provides many possibilities for psychological tension and relief—the frequency of and distance between consonant and dissonant intervals; the direction of movement toward consonance; the size of the intervals (small intervals, except for semitones relieving tension); the movement/direction of the intervals (ascending tones being generally more tension laden); the fulfillment of tonal expectation based on past association;[69] and resolution to the lower tone of a successively played interval and/or the lowest point of a descending melodic line.[55] Through our experience with probabilities in sound, our expectations concerning music develop.[69]

Rudolph Steiner relates the choice and directionality of intervals to dance movement in the art of eurhythmy, in which particular sounds and intervals assume physical postures and cathartic emotional value.[106]

Once perceived as a Gestalt, the familiar movement of the melody is repeated in the song and may be perceived consciously or unconsciously.

Rhythm, the feeling of movement in time, with its implication of both regularity and differentiation,[7] is, like melody, primordial. Rhythm is of biological origin, beginning prenatally when the fetus produces rhythmic heartbeats, chest movements, and limb and body movements.[85] Sounds transmitted by vibrations are responded to by the fetus in the third trimester. The information suggesting that rhythmicity is at least "hard wired" into the circuits of our brains is well founded: witness the seasonal or monthly cycles or even diurnal rhythms in depressives; the basic rhythm of the rapid eye movement of the sleep cycle;[45] the obvious microrhythms of heart rate, respiration, and brain wave activity; the various cries that change in rhythm at different periods of life;[118] the predictable rhythms of spontaneous neonatal mouthing and sucking activity;[119] the mannerisms and stereotypies of neonatal motor patterns;[120] and the movement patterns and rhythms found in each dynamically invested phase of development.[53] These and related studies may be potential diagnostic tools not only for music therapists but for the entire medical profession.

Gaston[43] said, "without rhythm, there would be no music, for its unique potential to energize and bring order is seen as music's most influential factor." According to Gaston, "rhythm is the primitive and driving factor of music." The manner in which rhythm is indicated or expressed by an individual determines to a large degree the amount of energy invested in responding to music.[1] Rhythmic patterns between mother and infant impart to some individuals an emotional message.[1,121] Later, rhythmic interaction with another person can become a dynamic medium for self-expression reflecting the earlier experiences as well as being a means of relating to another person.[1]

The choice of rhythms in a given piece of music, whether it be a listening or performing experience (precomposed or improvised), again lays bare psychological possibilities for tension and relief. The predictable accented interplay between basic beat (meter) and melodic rhythms, the stability and consistency of basic beat, and the gradual evolution of tempi changes (much faster or slower tempi can be tension laden) are all indicative of a need for order, control, and security.[55] This pleasure in order and control is an orientative Gestalt for the mind as well as the body: "the ultimate foundation of rhythm is to be found in mental activity."[69] This supposition concerning mental activity and organization of rhythmic impulses led the music therapy pioneers, Paul Nordoff and Clive Robbins, to suggest relationships between rhythmic response and emotional and/or physical difficulties in hundreds of handicapped children.[80]

As previously mentioned, music is fundamentally melodic, and so it is not surprising that the possibility of simultaneous combinations of tones, that is, harmony, was only gradually appreciated and understood.[72] As a direct qualitative experience, a chord possesses its own "color." There may be tension and/or relief in an emotional reaction to one chord. However, it is more common to perceive a succession of harmonic devices as tension or relief provoking. The Western listener, unacquainted with alternative means of harmonic movement, expects to hear the consonant root progressions inherent in music of the European tradition. Various listeners and cultures have retained subjective definitions of consonance and dissonance in spite of numerous attempts to provide an objective definition.[7] As consonance is thought to produce "an agreeable effect" and dissonance "a disagreeable effect,"[7] the composer arouses tension by delaying the consonant and/or presenting the dissonant. "Either too much tension or too much relaxation can be fatal. Excessive consonances produce stagnation, while too many dissonances often result in 'irritation', undue restlessness. Dissonances in music furnish an element of motion or progress, and keep the mind and the imagination of the hearer aroused."[71]

Initial contact with sound occurs at birth, assuming different qualities and producing different affective states contributing to the satisfaction of bodily functions.[1,73] As the infant develops, different sounds and auditory impressions become associated with different responses and create definite emotional reactions within the child.[1,73] According to Nass, the ego has simultaneous adaptive functions, and the "early listening and hearing experiences may serve to master the outside world."[73] Thus, the ego can function to re-experience early phenomena, as well as the present experience of listening during a therapeutic situation.[1] For some, the recreation of a way of "being with another person" based on early sound[1,114] experiences, comes through music—melody, harmony, and rhythm.

¶ Music as a Therapeutic Process

In therapy, the music session is the ongoing context within which the relationship between client and therapist is formed and the individual personality elucidated. Both active and passive involvement with music provide therapeutic opportunities that have been translated into clinical practice within the last three decades.

Listening to music is essentially a selective response in which some elements of the musical experience become dominant while others are subordinated.[72] This attendance to certain elements of music varies from individual to individual and is, in most cases, not conscious. According to Mursell, extrinsic factors in listening include: (1) the general mood or affective set of the listener, which is reinforced and/or prevailed upon by the music (only when the listener's affective state is very definite, strong, and in sharp conflict with the prevailing mood of the music being presented, does it constitute a disturbing factor); (2) the flow of association and the arousal of imagery usually based on the mood and emotions elicited by the music; and (3) the visual stimulation provided by the musician. The intrinsic factors in listening, in order of their priority, include shifting volume and quantity of tonal content; the Gestalt of the melody, rhythm, harmony, and the general architectonic design of the music.

With the great majority of listeners, attendance and emotional reactions to intrinsic factors of listening are not controlled and usually exist on a level that is apparently not conscious. Meyer[69] attempted to clarify the role of mental organization of rhythmic and melodic stimuli, expectation and learning, and psychological "norms" of necessary repetition, resolution, good continuation, completion, and balance of tension and relief. The ability to recognize and respond to these components of the intrinsic listening experience may prove to be a key diagnostic factor in work with mentally disorganized clients.[46]

As music "activates tendencies, inhibits them and provides meaningful and relevant resolutions,"[69] it will effect tension or relief. Any state of suspense in music, resulting in expectation, can provide positive excitement in leading up to a resolution and/or negative upset in further delaying or blocking a resolution. It has been speculated that the greater the familiarity of a person with a certain piece of music or the higher his musical training in general, the more he will enjoy musical sequences of greater unexpectedness and complexity. In addition, the effects of repetition and familiarity will provide an internal sense either of more mastery over the music or of eventual boredom since there no longer exists any tension. How increased tension can become pleasurable is a problem that has puzzled psychoanalysts for years. Miller[70] provided us with an analytic interpretation of tension and relief via the music experience: Increased tension is the result of the musical elements symbolically disturbing the listener's degree of ego-mastery. According to Meyer,[69] the ego-mastery is upset when control over the predicted resolution of the music is temporarily or permanently lost. According to Miller,[70] "Ego-mastery is symbolically disturbed during music by the listener being reminded of previously disturbing situations such as environmental noises (frightening and unexpected sounds), forbidden id impulses (sexual and aggressive sounds), ego failures (complex and difficult sounds), and super-ego pressures (unconforming and inartistic sounds)." Increased tension can become pleasurable when the listener realizes emotionally that the music is only "make-believe" and not threatening to the self; the eventual regaining of mastery over the music situation leads to decreased tension. "Ego mastery is attained in music when sufficient defensive energy is invested to understand, overcome and re-master the symbols disturbing situations."[70]

The ability of music to portray multiple, even opposite, emotions simultaneously makes it possible for music to become deeply meaningful to the listener in both positive and negative ways. The connection for these feelings may be the world of imagery and associations. In many cases, the submerged memory of a particular event and/or feeling is triggered by a song or instrumental composition associated with the memory. The therapeutic value of reawakening feeling in order to reexperience and elucidate it is a key feature of music therapy. Frances Hannett,[49] reviewing the significant themes of American popular songs during the early twentieth century, pinpointed a primary (69 percent) thematic reference to an unfulfilled love relationship (that is, mother-child). She also pointed out the value of a patient spontaneously singing, whistling, or hearing a song as a means of conveying emotions and feelings not being directly expressed.

Musical daydreaming, attention to the suggestions and associations of music and to its "meaning beyond itself," may be viewed as the symbolic expression of unconscious feelings and contents.[72] As dreams are valuable interpretative material for the analyst, so musical daydream, arising from the emotional reactions a piece of music elicits, can prove valuable interpretative material for the music therapist. It is on this basis that Guided Imagery in Music (GIM), a process involving "listening to music in a relaxed state for the purpose of allowing imagery, symbols and deep feelings to arise from the inner self and then be used for therapeutic intervention or self-understanding," was developed by Helen Bonny, R.M.T. at the Maryland Psychiatric Research Center in Baltimore, Maryland.[16]

The reader will keep in mind that although the direct effect of a music listening experience may be to provide pleasure in and of itself, this superficially therapeutic "music to sooth the savage beast" notion does not at all fulfill the real purpose of therapy—to promote behavioral change of long-standing value.

¶ Performing Music

Certainly our greatest concern is to encourage personal expression by the music therapy client by means of either formal (precomposed) or informal (improvisational) active music experience. Whether it be vocal, instrumental, or involving movement, the perceptual, cognitive, motor, and emotional processes incorporated into music performance make it an ideal therapeutic opportunity:

> The voice represents the hidden person, his individuality, his uniqueness. To be born means to become sounding, to have a voice means to be something which has its own growth, its own development. . . . The primitive or modern player has always identified himself with his instrument, which is a prolongation of his body and transforms into sound his psycho-motor impulses and liberates them. The manipulation of an instrument demands also conscious control of movement in time and space and obedience to certain laws of acous-

resentation? On what level does the translation process of the unconscious content into tonal structure take place? What is the nature of the motive or desire to find expression and gratification in music?[84]

Perhaps part of the answer to the last question can be found by recognizing the roles of the id, ego, and superego in musical expression. From the point of view of the developmental hierarchy of psychological stages, the meaning of music ". . . is derived from its capacity to allow subtle regression via extraverbal modes of psychic function."[54] The id is served by music since catharsis of primitive impulses is transformed into an emotional experience. The ego is served by music since organization and hence mastery of sound impulses is achieved; music is, in this sense, a form of play. The superego is served by music since an expression of rules to which one submits becomes a task to be fulfilled as well as an aesthetic experience. The elements of music themselves serve to satisfy the libido. In summary, rhythmic repetition and emphasis are a pleasurable means of discharged energy.[28] Melody, produced under muscular tension in the tones of the human voice, also produces this effect; and harmony provides an additional and increased enjoyment since a number of items concur in any simultaneous expression of pleasure.

Further evidence to satisfy the question of motives in finding expression and gratification in music is found in the writings of Maslow[66,67] and Jung.[91] Maslow's hierarchy of human needs—physiological, safety, belonging, esteem, and self-actualization—are all within the capacity of musical experience. Jung's four functions of the psyche—thinking, feeling, sensation, and intuition—are also integral components of the music experience.

The key to the music therapy work, which calls itself analytic in orientation, is recognizing that unconscious contents of the music experience and the approach to the music experience must become conscious to the extent that the client can achieve greater self-awareness. In some cases a musically expressive way of relating may be linked to the rest of the behavioral personality and help the client achieve greater personality integration. In "pure" music therapy—therapy where the music experience itself is sufficient to promote behavioral change outside the music room—verbalization regarding the music experience is not necessary. Music therapy in this "pure" sense has been achieved most often with children. With adults, particularly the neurotic, it has usually been necessary for the therapist and client to verbally "process" the musical experience. The client is often guided by the therapist to reflect back and clarify what pertinent dynamics the musical expression has offered. This process may elucidate other behavioral patterns in the client's life.

As the client is allowed to "work through" his conflicts musically and verbally by being offered, and personally identifying with, an enriching musical experience and a supportive client-therapist relationship, alternative means of approaching and integrating the musical experience into the self become possible. This emotional awareness frequently effects extramusical change.

Music Therapy and Cultural Diversity

In concluding the argument that affective and cognitive structures are related to musical development and must be clarified in forthcoming years to enhance music therapy expertise, it is necessary to include the subject of cultural diversity with regard to human development and musical expression. It is dangerous to extend any findings beyond the population (in our case, Western) in which they have been established. As musicians,[93] musicologists,[63] and psychologists[42] have demonstrated, the music of one's culture may be influenced by such factors as relationship to religion and art, "vocabulary" (that is, musical mode), and different emphasis (for example, greater emphasis on rhythm in primitive cultures). Lomax,[63] investigating the universal relationships between folk music styles and cultural characteristics of fairly primitive societies, recognized the relevance of such findings to music therapy practice.[104] For example, the finding that unison instrumental folk music is more common in societies whose political and social interaction patterns are simply organized while counterpoint occurs more frequently in societies characterized by complex, specifically defined and differentiated patterns of political and social interaction, is somewhat analagous to the observation, made frequently by music therapists, that appropriate individual participation in a contrapuntal musical texture is associated with a fairly high degree of individual behavioral organization and social competence. It has also been found that recorded contrapuntal musical textures are responded to by some psychotic patients as though they were social interactions and that patients' acceptance or avoidance of contrapuntal music can provide clinical implications about their tolerance of interactive, interpersonal situations.[103] These and other findings suggest a strong relationship between social structure and group musical performance not only within a given society but within any social subculture.

¶ Basic Purpose and Structure of Music Therapy

The intention of music therapy is to encourage, structure, and develop musical activity that will serve to incorporate the affective, cognitive, communicative, perceptual, and motor needs of the client. The demonstration of enriching change and growth through musical, verbal, and nonverbal communication and interaction with the therapist is directly related to the extramusical goals the therapy was intended to meet. In making a referral to a music therapist another clinician (psychiatrist, psychologist, occupational therapist, physical therapist, speech pathologist, special education teacher, social worker, or the like) realizes that successful methods will vary for each individual. The strength of music as a therapeutic tool is that it motivates a client to work toward difficult goals in a personally meaningful, expressive way. According to I Ching:

> Music has power to ease tension within the heart and to loosen the grip of obscure emotions. The enthusiasm of the heart expresses itself in a burst of song, in dance and rhythmic movements of the body. From immemorial times, the inspiring effects of the invisible sound that moves all hearts and draws them together has mystified mankind.

As in psychotherapy,[116] music therapy can and does proceed on three levels:[115]

1. supportive therapy—strengthening of existing defenses, development of new and better mechanisms to maintain control, and restoration to an adaptive equilibrium;
2. insight therapy with reeducative goals—insight into the more conscious conflicts with deliberate efforts at readjustment, goal modification, and living up to existing creative potentialities
3. insight therapy with reconstructive goals—insight into unconscious conflicts with efforts to achieve extensive alterations of character structure and expansion of personality growth with development of new adaptive potentialities.

Due to the tremendous diversity of age (preschool to geriatric) and disability (developmentally disabled, emotionally handicapped, physically impaired, multiply handicapped, substance abusers, criminal offenders), musical activity is structured and developed differently for each client. Depending on the training and orientation of the music therapist, the capacity of the client to be involved in therapy, and the interdisciplinary goals set for that client, the extent to which musical activity is developed to incorporate client needs differs dramatically (see "Specific Applications of Music Therapy"). In those cases where the client will undergo short-term therapy and may be chronically ill and/or highly resistant to therapy, supportive music therapy provides immediate access to the benefits of music (catharsis, self-organization, socialization, time-ordering, mood-setting) without necessitating an investigation of the dynamics of the activity to provide insight and change. Pleasurable activities emphasizing interpersonal relationships, task achievement, and social behavior in groups are common approaches to supportive music therapy.

Many cases of insight therapy with reeducative goals are represented in the literature of music therapy with handicapped children—particularly those children who are developmentally delayed. Most of these children are in an educational setting in which interdisciplinary team members work toward psychoeducational goals. Reeducative goals for geriatric clients are commonly approached through music. In cases of stroke or deterioration of the brain leading to disabled motor, communication and/or thought processing, music, perceived through the right hemisphere, can be a basis for motivating clients and "teaching" the left hemisphere to be operative again.

Insight therapy with reconstructive goals, the most in-depth treatment, can be brought about with music creating an environment in which a patient can explore intrapsychic phenomena, experience affect, develop mastery, direct himself toward resolution of conflicts, and risk change.

In all cases, music is used because it evokes various moods or affective states[16] and demands reality-oriented behavior to those stimuli—aural, visual, kinesthetic, musical, and verbal—built on the time-ordered necessities of a given musical situation.[98] Through involvment with the musical process, one can develop awareness of affectual states while maintaining the reality of the moment.[1] Music also provides time-ordering, self-organization, and experience in relating to others[98] or to the self.[46]

Since music can be very much a formal art, the capacity for entry level music involvment will vary tremendously. Many clients may simply have an emotional attachment to music and the desire (whether directly or indirectly expressed) to become more involved. Others will have had previous amateur and/or professional experience and emotional attachment (whether positive, negative, or ambivalent) to their training in voice, an instrument, composition, or the like. In all cases, the capacity of music to influence behavior is clear to the music therapist.

Recognizing the current capacity of the client to become involved in music, and acting to provide support, guidance, and organization, the therapist structures the client's musical experience via listening, improvisation (movement, personalized songs, simple instrumental activity, musical drama), and/or preplanned song, movement, and instrumental activity. As in normal development, the individual's dependency upon the guidance of the therapist should decrease as a repertoire of skills is built up, and the client is able to initiate choices. A feeling of autonomy and self-respect develops as the client identifies with the musical activity, incorporates more and more of himself into it, and increases a sense of mutuality with the therapist or group in the music-making experience.

Other music therapy sessions may involve the gradual intervention of the music therapist in dealing with a client's rigidly bound patterns of relating to an instrument. Assuming the client has a stake in preserving these patterns, the therapist's job is to listen, to help clarify the problem, and to present opportunities for the client to try out alternate ways of relating and musically communicating.

As has been implied all along, the relationship between client and therapist in the music therapy process is of paramount importance. The development of basic trust, dependency, and then the emotional recognition of the autonomy of the two individuals (or members of the group) is essential. In music therapy, transference and countertransference come about through verbal, nonverbal, and musical situations:

> The loving feelings of the music therapy positive transference do not have the same frustrating aim-inhibited quality as they do in analysis. To play music together and, even partially, to relieve physical tensions in this way, can be an unconscious symbolic equation for various basic impulses such as feeding, making love or even kill-

ing. Music therapy transferences are therefore deep but more manageable, both in their positive and negative aspects. [p. 243][91]

The amount and nature of verbal "processing" between client and therapist once again depends upon the orientation of the therapy as well as on the training of the therapist. A therapist incorporating reeducative and/or reconstructive goals into therapy will provide as much structure and direction as the client needs and will verbally support and "process" feelings evoked by the music while possibly relating them to extramusical patterns of behavior.

Music therapy structure depends on the suitability of individual or small group therapy, and necessarily involves assessment of client needs, musical materials based on these needs, and evaluation of client response.

¶ Specific Applications of Music Therapy

The Multiply Handicapped Child

Today the profession of music therapy is probably most often associated with handicapped children. These children have many needs in relation to their primary difficulties and secondary delays. All of these developmental delays (social-emotional, cognitive, communicative, motor, perceptual) may be revealed in musical experience and the child-therapist relationship. Common goals include speech initiation and language development, ego organization and personality development, behavior modification, motor control, lengthened attention span, and memory development:

Often a handicapped child lives in a state of continuous tension, and is unable to express his emotions in an orderly fashion. Frequently he is unable to assimilate his life's experiences. He may be confused because he fails to interpret them. He has little or no faith in his own capacities. He feels rejected and he rejects himself. Often he is unable to communicate his needs, unable to control his inner conflict. The results are well-known manifestations of unacceptable behavior. [p. 143][71]

The music therapist must communicate with the child at the level of the child's developmental capacity for musical experience. In some cases the therapist may begin this communication with a rhythmic tapping on the soles of the child's feet—"Hel-lo." The child can appreciate this as a basis for further vocal and/or movement feedback to the therapist via differentiated tonal cries and rhythmic movement. Following a reflexive level of vocal and movement response, a child has the capacity to move on to a tonal and/or rhythmic response that bears a relationship to a therapist's musical piano improvisation. The expansion of the child's world is made possible by the therapist musically "joining" the mood of the child as well as by attempting to incorporate the rhythms and tonalities of even primitive behaviors—"stereotypic" rhythmic movements and/or undifferentiated crying.[82] The idea of reinforcing the present mood of the client is based on the age old "Iso principle" that "like acts on like."[4] Clinical music therapy improvisation techniques incorporate the tonal and/or rhythmic behavior of a child as well as predictable music structures by which the child may "join" the therapist. The therapist works for a "sustaining of directed response-impulses setting up musical communication"[82] as a basis for the "music child's" further involvment in individualized and, eventually, group activity. Self-expressive musical confidence, enthusiasm for musical creativity, and free functioning and communication of musical intelligence and skills are demonstrated in more demanding, often highly structured, play songs,[59,60,75,78] or orchestrations for instruments such as drums,[77] resonator bells,[61,81] and assorted percussive and tonal instruments, such as reed horns, glockenspiels, claves, wood blocks, bird whistles, cymbals, adapted string instruments, autoharps, and zithers, designed for developmental needs.[59,60,79] The child able to engage in more abstract musical role-playing is encouraged to do so in musical plays designed for voice and simple rhythmic/tonal instrumentation.[74,76]

As in verbal therapy, there is often resistance to transition:

> For months he [eight-and-one-half-year-old, severely retarded, nonverbal, psychotic child] responded only by rocking happily to simple, conventionally consonant, rhythmic music—and only to this. Any time the therapist introduced dissonances, added words to her singing, or tried to structure his musical experience by repeating patterns, introducing a song, or the like, he would start to cry with a tone of misery and anger. He showed no interest when his name was sung. Any communication had to be entirely on his conditions. [p. 11][82]

The exploratory efforts of the therapist to engage the child musically in conjunction with an understanding of the dynamics of the child's personality led to further growth: "But gradually, he began to accept as part of his music a Tyrolean waltz played bitonally; this highly dissonant music, which still contained some of his 'old, uncomplicated' music, engaged him and provoked some vocal responses."[82] The ability to sing longer tonal phrases, the subsequent exploration of vowel sounds, and finally holophrastic language allowed this child an opportunity to sing about himself and his environment ("What's That?"[77] "Roll Call Song,"[75] "Goodbye Song"[75]) while playing with the therapist.

The playing-out of a child's emotions through improvised or precomposed music should lead to the child's eventual control of the musical situation; in this regard the ability to perceive and play a regular beating pattern is a measure of inner control. The twelve catagories of rhythmic beating responses observed by Nordoff and Robbins[79] in their more than twenty years of clinical work all have psychological and neurological correlates.[79,82] One example of this is the compulsive beater: "What a compulsive beat signifies is an enigma. It is almost always associated with children generally described as autistic. Perhaps its most salient characteristic is its apparent meaninglessness; no variety, no mobility, no expression, remoteness. It is unrelated to the environment.[79]

The therapist employs techniques to create a comprehension of the music rather than the ritualistic action of repeated beating:

> Gradually the boy [nine-year-old, brain-damaged, aphasoid, autistic] is attracted to two songs sung by the therapists; one is about him and contains his name. His emotional response can initially come no further than facial expression and deeper breathing. His stiff posture relaxes to the warmth of the melodies and harmonies. The songs alternate with rhythmic work, the therapist leading from one to the other as the child's responses indicate. Subjected to syncopation and dissonance in the improvisation his compulsive beat begins to break down. His beating now acquires some meaning and relates to the songs (basic beat, melodic rhythm). Gradually he begins to sing fragments of the songs with the therapists. As he works his way into musical expression through a series of new experiences, he activates parts of his nervous system he has not used before.[79]

The planning and implementation of musical involvement for the handicapped child develops musical skills as a secondary benefit to developing behavioral growth. As already mentioned, composed and improvised music therapy materials are highly personalized in terms of the client's personality and developmental needs. Composed music therapy materials, based on speech and language development, follow the inflections, rhythms, and phrasing of speech, as well as reflecting the array of genuine emotional experiences children undergo.[79] Special motor needs for physically handicapped children,[25] asthmatic children,[68] and developmentally delayed children[59,79] are met in music therapy composition and instruments designed. The music therapy profession continues to develop specific programming for such populations as the autistic,[6,79,82] hearing-impaired,[96] severely/profoundly handicapped,[35] communication disordered,[39] cerebral palsied[13] learning disabled,[47] and mentally retarded.[62]

The Neurotic

With neurotic clients in individual and/or small group therapy (see "Group Work with Emotionally Disturbed Adults"), music therapy goals include awareness of and feelings about one's musical expression, extension and integration of music expression, and spontaneous musicality and interaction. The awareness of inner sensations and freedom to respond to these is essential. In neuroses, the patient is frequently blocked due to particular or diffuse anxiety. All music therapy goals bear a specific relationship to the client's particular psychopathology, which is, whenever possible, verbally processed between client and therapist. "As the modern analyst resolves the patient's resistances to putting his thoughts and feelings into words (the music therapist can) resolve resistances to putting them into music. This process is begun by learning as much as possible about the client and discovering with him what specific needs exist."[14] An ostensible "music lesson" may really be a continuation of the learning process "with an exploration of the many possibilities within the music and within the person in terms of what he might want to say through his body and his own decoding of the composer's message on the printed page."[14]

Depending on the level of anxiety, warm-up relaxation techniques utilizing rhythmic movement, deep breathing, appropriate vocal sounding, and simple melodic piano improvisations may help the client begin the session. By his own selected musical modality and materials the client provides the therapist with a self-portrait of his emotional and cognitive styles/patterns toward music making. These emotional and cognitive styles may reflect on the patient's emotional and cognitive styles outside of the music session. Through the guidance of the music therapist, the client is encouraged to process his feelings about musical experience in relation to self. The client is encouraged to attempt alternate means of dealing with the music situation, as he would with extramusical situations. Such experiences may lead to new understanding and possible alternatives in behavior.

Often the highly verbal neurotic will gain more insight from the emotionally-laden music experience than from incessant talking:

> The client, preoccupied with conflict about satisfying sexual desires, entered the music room as usual wanting to discuss her problem. Encouraged to sing her chosen selection, Jerome Kern's "Make-Believe" (from "Showboat") she was vocally supported by the therapist. Following the first stanza in which the client sang with strained serious voice, the therapist chose a jazzy harmonically sensual backdrop for the piano accompaniment of the second stanza. Going along with this "sexy" piano accompaniment, the therapist modeled an uninhibited vocal timbre, rather breathy at times and suggestively inflected. The client, vocally supported by the therapist, apparently found the "sexy" style of singing spontaneously contagious and attempted to try it herself. With difficulty she spontaneously shook her arms and hips, snapped her fingers and, when not laughing, "dared" to sing sultry, low-pitched sounds. These were in direct contrast to her usual shrill, strained voice. Her obvious delight in singing was accompanied by embarrassed laughter and a blushing face. She claimed she "didn't like singing this way." When the therapist pointed out her physical reactions in relation to the experience, she began to understand her gut reactions to sexual experience. It was gratifying for her to realize she could express another "more submerged" side of herself safely through the music.[46]

Priestley[91] wrote that "as a reference point for working through the psychic needs and tension—musical forms provide a safe environment and structure for personal expression." Furthermore, the therapist helps the client by "providing a musical container to receive and complement the expression of her painful emotions and makes it safe for her to talk about them and the memories they evoked afterwards." This "musical container" can be precomposed or spontaneous; it is always shared—either directly or indirectly—by the therapist. It can be a listening experience in guided imagery[16] or a written composition or song offered by the client. In many instances the improvisatory effort between client and therapist results in dramatic casting of ideas or roles.[91]

Among her various techniques of "analytical" improvisation, Priestley[91] suggested: (1) "holding," that is, a means of allowing the client fully to follow an emotional experience through to a climax in order to diminish the fear of disintegration under high emotional stress, thus giving the emotion the chance to be expressed while expending enough bound energy to allow the client to think more and feel less about the subject; and (2) "splitting-useful," that is, where the client has projected part of himself onto another character. Therapist and client take on different roles and then switch musical feelings of "characters" in order to further clarify identities. This technique is also useful in conflict situations where all the energy is being held in maintaining the status quo. After the client describes her feelings about both sides of the conflict in word pictures, the music therapist starts off in the character of one person or idea:

> A university student, Eva, wanted to explore with me her unequal relationship experiences. I started by being "Doormat" while she played "Dominant" but she was not concentrating on her own expression but trying to provoke me all the time. She said that she thought that if I did last out then she would have to be "Doormat." Next I was "Doormat" but I left spaces for her to answer back musically but she never did. Next I made a long decrescendo (gradual lowering of volume) to see when she would dare to reverse roles. At my "pp" (pianissimo—soft level of volume) she was "ppp" (even softer) and then when I reached "ppp" suddenly, right at the end, she did assert herself. Following this, she felt able to resist her tutor's efforts to make her take up work at a "suitable" school whose principles she disbelieved in, and to risk looking for a position in which she would feel happy and honest. [p. 125][91]

The Psychotic

With the psychotic in individual or small group therapy, basic goals (as with the neurotic) include awareness of one's musical expression, extension and integration of musical expression, and spontaneous expression and interaction. The need for more structure is necessary to bring the patient into closer touch with external reality. Of paramount value for the psychotic is the formation of an attachment, the expression of self within a structure, and the further channeling and structuring of emotional effort when musical communication becomes established as the ongoing basis for attachment.

Personal integration of emotion and meaning, whether it is apparent to the therapist or not, is often a result of music therapy experience. The schizophrenic patient usually does not express himself verbally, but may manifest the organization of his feelings via the music. Experiencing a moment that is relevant and real, the patient confirms his attachment within an ongoing process; feeling is brought out of chaos into a structure; there is a delineation of experience:

> Using his musical ear to pick out the notes on the keyboard, Arthur [seventeen-year-old schizophrenic] very much wanted to recreate the musical composition but was defeating himself by mechanically trying to reproduce the melody. Since he had already learned the accompanying chords of the composition while supported melodically by the therapist, he was guided to recognize that the melody was composed of the same notes as the accompanying three-note chord. He joyfully experimented with the concept while the therapist aided him. Given a meaningful structure within which to work, it was easier for him to grasp the melodic meaning of the composition. The rhythm had to be structured and modeled for Arthur by the therapist as well; his ability to organize himself musically is tremendously impaired although he is quite gifted musically.[46]

Cognitive organization is inherent in recognizing and enjoying the structural parts of the music (melody, variations of melody, rhythms in relation to melody, harmonies in relation to melody, and so forth) in and of themselves, as well as in relation to a whole musical composition. One musical idea follows another as the scheme unfolds. The joy in taking apart the puzzle and then putting it back together in a meaningful way helps give the client a feeling of control and mastery.

The recognition of the psychotic client's present capacity to participate in and/or create music is essential. As in visual arts experience,[8] it is suggested that levels of musical participation may correspond to a certain stage of pathology—particularly in the case of the schizophrenic client:[46]

Marie was consistent in her creation of very literal music materials. Harmonies were traditionally consonant and always repetitive in terms of three basic Western chord structures. Melody was confined to a five-note range and repeated twice. Rhythm was 4/4—an easily grasped structure which always corresponded directly to melodic rhythm. She was a conservatory graduate of Juilliard and had received considerable musical training in piano and theory and composition. The usual musical directions of accent, phrasing, dynamics, and tempi markings were completely absent from her composition. Only as she became more trusting of the therapist and more revealing of her emotions in her dearly loved classical music was she able to begin to compose with syncopated rhythms, dissonant harmonies, larger tonal range, and an emotional content related to the title of the music. In addition she spontaneously began to include musical markings and to point out the "new" aspects of her work.[46]

If psychotic patients resort to thought disorder as a means of defense against stress,[109] then their complete disorganization and need for structure in music may "progress" into a defensive, literal, tightly structured musical interpretation as an alternative defense. This hypothesis is supported by the observation that psychotic patients sometimes produce melodies in constricted ranges with even patter rhythms,[24,46,104] and that this constriction of range and lack of rhythmic differentiation were found to function defensively. Often, patients whose behavior is overcontrolled also speak in pinched, narrow, or nasal tones. Severely restrictive, constrictive behavior usually indicates much anxiety, and severe constriction of vocal tones in patients may also be indicative of conflict.[108] Association between remission of these musical traits and remission of a psychotic thought process has also been observed.[99]

Adult Group Music Therapy

The awareness that one can confront conflict and grow not only by being one's self but by participating with other selves may be achieved by participation in musical structures.[1] Ain described some of the roles music serves in a group situation. The group as a whole perceives the sound (as do the individuals) and reacts.[1,4] The need to channel cathartic and play activities is met by acceptable adult expression—instrumentation, songs, verbal expressions, and the association accompanying the musical process and expression. As previously mentioned, these activities and forms provide psychic distance[1,56] from threatening affect; the manner in which the person expresses himself musically may represent his psychic constellation.[1] The acting out of feelings provides a possible basis for increased reality testing when feelings, expressions, and mastery are overtly expressed within the group: "The musical stimuli can act as a rehabilitative vector in revealing conflicting situations and manifesting catharsis."[1] This structured reality simultaneously creates a means to explore one's affective states in relation to other people.[1] Music is universally capable of arousing affective responses and modifying existing moods in individuals with only minimal musical talent: "Everyone understands music and it is a communication form which no one can withdraw from . . . therefore . . . it is a social force which unites people during therapy."[1] One example of this is the use of rhythm since it provides a "bond for joining people together because rhythm 'persuades' individuals to act with other individuals. Thus we can give in to rhythmic impulses and simultaneously maintain form and control in expression."[1] It has been suggested that imitation and initiation of group rhythmic patterns not only can facilitate and sustain the group's attention span and develop awareness of other members' sounds but also can be the basis for a group rhythmic sound. Likewise, the creation of a group melody to which each member adds his pitch (tone) at a specified or spontaneous time can be a basis

a Music Therapy Program," *Journal of Music Therapy,* 6 (1969): 12–14.

25. CHADWICK, D., and CLARK, C. *Clinically Adapted Instruments for the Multiply Handicapped.* Westford, Mass.: Modulations, Inc., 1979.
26. CHASE, M., in Chauklen, H., ed. *Marion Chase: Her Papers.* Washington, D.C.: American Dance Therapy Association, 1975, pp. 52–53.
27. COOKE, D. *The Language of Music.* London: Oxford University Press, 1959.
28. COPLAND, A. *Music and Imagination, The Charles Eliot Norton Lectures.* Cambridge: Harvard University Press, 1952.
29. CORIAT, F. "Some Aspects of a Psychoanalytic Interpretation of Music," *Psychoanalytic Review,* 32 (1945): 408–418.
30. DAVIES, J. B. *The Psychology of Music.* Stanford: Stanford University Press, 1978.
31. DOBBS, J. P. B. *The Slow Learner and Music.* London: Oxford University Press, 1966.
32. DREIKURS, R. "Psychiatric Concepts of Music Therapy for Children," *Proceedings of the National Association for Music Therapy, Inc.* Lawrence, Kan., 1953, pp. 81–84.
33. EAGLE, C., ed. *Music Therapy Index,* vol. 1. Lawrence, Kan.: National Association for Music Therapy, 1976.
34. ———. *Music Psychology Index,* vol. 2. Denton, Tex.: Institute for Therapeutics Research, 1978.
35. EWING, S. R. "Piagetean Approach to Music Therapy with the Severely/Profoundly Handicapped Child," Conference of the National Association for Music Therapy, Atlanta, Georgia; November 1979.
36. FRANCES, A., and SCHIFF, M. "Popular Music as a Catalyst in the Induction of Therapy Groups for Teenagers," *International Journal of Group Psychotherapy,* 26 (1976): 393–398.
37. FROMM-REICHMANN, F. *Principles of Intensive Psychotherapy.* Chicago: The University of Chicago Press, 1950.
38. GALLOWAY, H. "Music and the Speech Handicapped," in Graham, R., ed., *Music for the Exceptional Child.* National Conference of Music Educators, Reston, Va., 1972, pp. 15–48.
39. ———. "A Comprehensive Bibliography of Music Referential to Communicative Developmental Processing, Disorders and Remediation," *Journal of Music Therapy,* 12:4 (1975): 164–196.
40. GARDNER, H. *The Arts and Human Development: A Psychological Study of the Artistic Process.* New York: John Wiley, 1973.
41. ———. *The Shattered Mind.* New York: Vintage Books, 1976.
42. ———. *Developmental Psychology: An Introduction.* Boston: Little, Brown, 1978.
43. GASTON, E. T., ed. *Music in Therapy.* New York: Macmillan, 1968.
44. GESCHWIND, N. "Language and the Brain," *Scientific American,* April 1972, pp. 76–83.
45. GLOBUS, G. G. "Quantification of the REM Sleep Cycle as a Rhythm," *Psychophysiology,* 7:2 (1970): 248–253.
46. GOODMAN, K. D. Unpublished Papers, 1979.
47. GRAHAM, R., ed. *Music for the Exceptional Child.* National Conference of Music Educators, Reston, Va.; 1975.
48. HAMEL, P. M. *Through Music to the Self.* Boulder, Col.: Shambula Publications, 1978.
49. HANNETT, F. "The Haunting Lyric: The Personal and Social Significance of American Popular Songs," *Psychoanalytic Quarterly,* 33 (1964): 226–229.
50. HEIMLICH, E. P. "An Auditory-Motor Percussion Test for Differential Diagnosis of Children with Communication Difficulties," *Perceptual and Motor Skills,* 40 (1975): 839–845.
51. HURWITZ, I., WOLFF, P., and KOTAS, K. "Nonmusical Effects of the Kodaly Music Curriculum in Primary Grade Children," *Journal of Learning Disabilities,* 8:3 (1975): 167–173.
52. KASLOFF, L., ed. *Holistic Dimensions in Healing.* New York: Doubleday, 1978.
53. KESTENBERG, J. S. "Role of Movement Patterns in Development," *Psychological Quarterly,* 34 (1965): 1–36.
54. KOHUT, H., and LEVARIE, S. "Observations on the Psychological Functions of Music," *Journal of American Psychoanalysis,* 5 (1957): 389–407.
55. KREITLER, H., and KREITLER, S. *Psychology of the Arts.* Durham, N.C.: Duke University Press, 1972.
56. KRIS, E. *Psychoanalytic Explorations in Art.* New York: International Universities Press, 1952.

57. LANGER, S. *Philosophy in a New Key.* Cambridge: Harvard University Press, 1951.
58. ———. *Feeling and Form: A Theory of Art.* New York: Charles Scribner's Sons, 1950.
59. LEVIN, H. *Learning Through Music.* Boston: Teaching Resources, Inc., 1975.
60. ———. *Learning Through Song.* Boston: Teaching Resources, Inc., 1977.
61. ———. *A Garden of Bellflowers.* Bryn Mawr, Pa.: Theodore Presser, Inc., 1978.
62. ———, and LEVIN, H. "Instrumental Music—A Great Ally in Promoting Self-Image," *Music Educators Journal,* 58:8 (1972): pp. 31–34.
63. LOMAX, A. *Folk Song Style and Culture.* Washington, D.C.: American Association for the Advancement of Science, Publication no. 88, 1968.
64. MADSEN, C. K., and MADSEN, C. H. "Music as a Behavior Modification Technique with a Juvenile Delinquent," *Journal of Music Therapy,* 5:3 (1968): 72–76.
65. MAHLER, M. *Psychological Birth of the Infant.* New York: International Universities Press, 1975.
66. MASLOW, A. H. "Music Education and the Peak Experience," *Music Educators Journal,* 54:6 (1968): 73–75.
67. ———. *The Farther Reaches of Human Nature.* New York: Viking Press, 1972.
68. MARKS, M. "Musical Wind Instruments in Rehabilitation of Asthmatic Children," *Annals of Allergy,* 33:6 (1974): 313–319.
69. MEYER, L. B. *Emotion and Meaning in Music.* Chicago, Ill.: University of Chicago Press, 1956.
70. MILLER, M. D. "Music and Tension," *Psychiatric Review,* 54:1 (1967): 141–155.
71. MORETTI, V. "Music in Special Education," *Academic Therapies,* (1971): 143–148.
72. MURSELL, J. *The Psychology of Music.* New York: W. W. Norton, 1937.
73. NASS, M. L. "Some Considerations of a Psychoanalytic Interpretation of Music," *Psychoanalytic Quarterly,* 40:2 (1971): 303–316.
74. NORDOFF, P., and ROBBINS, C. *Pif-paf-poultrie.* Bryn Mawr, Pa.: Theodore Presser, 1961.
75. ———. *Playsongs-Book I.* Bryn Mawr, Pa.: Theodore Presser, 1962.
76. ———. *The Three Bears.* Bryn Mawr, Pa.: Theodore Presser, 1966.
77. ———. *Fun for Four Drums.* Bryn Mawr, Pa.: Theodore Presser, 1968.
78. ———. *Playsongs-Book II.* Bryn Mawr, Pa.: Theodore Presser, 1968.
79. ———. *Music Therapy in Special Education.* New York: John Day, 1971.
80. ———. *Therapy in Music for Handicapped Children.* New York: St. Martin's Press, 1971.
81. ———. *Folksongs with Resonator Bells.* Bryn Mawr, Pa.: Theodore Presser, 1977.
82. ———. *Creative Music Therapy.* New York: St. Martin's Press, 1978.
83. NOY, P. "The Psychodynamic Meaning of Music," *Journal of Music Therapy,* 3 (1966): 126–134.
84. ———. "The Psychodynamic Meaning of Music," *Journal of Music Therapy,* 4 (1967): 7–23, 25–51, 81–94, 117–125.
85. OSTWALD, P. "Musical Behavior in Early Childhood," *Developmental Medical Child Neurology,* 15 (1973): 367–375.
86. OSTWALD, P., and PELTZMAN, P. "The Cry of the Human Infant," *Scientific American,* 230:3 (1974): 84–90.
87. PETERS, M. L. "A Comparison of the Musical Sensitivity of Mongoloid and Normal Children," *Journal of Music Therapy,* 7 (1970): 113–123.
88. PFLEDERER, M. "The Responses of Children to Musical Tasks Embodying Piaget's Principle of Conservation," *Journal of Research in Music Education,* XLL (1964): 251–268.
89. PRIESTLEY, M. "Counter-transference in Analytical Music Therapy," *British Journal of Music Therapy,* 9:3 (1978): 2–5.
90. ———. "Music, Freud and Recidivism," *British Journal of Music Therapy,* (Autumn 1977): 10–13.
91. ———. *Music Therapy in Action.* New York: St. Martin's Press, 1975.
92. RACKER, H. "Contribution to Psychoanalysis of Music," *American Imago,* 8 (1951): 129–163.
93. RECK, D. *Music of the Whole Earth.* New York: Charles Scribner's Sons, 1977.
94. REGELSKI, T. A. *Arts Education and Brain Research.* National Conference of Music Educators, Reston, Va., 1978.
95. REID, C. *Voice: Psyche and Soma.* New York: Joseph Patelson Music House, 1975.
96. ROBBINS, C., and ROBBINS, C. M. *Music*

with the Deaf: A Resource Manual and Curriculum Guide. forthcoming.

97. ROSENBAUM, J. B. "Songs of the Transference," *American Imago,* 20 (1963): 257–269.
98. SEARS, W. W. "Processes in Music Therapy," in Gaston, E. T., ed., *Music in Therapy.* New York: Macmillan, 1968, pp. 30–47.
99. SLAUGHTER, F. "Clinical Practices: Approaches to the Use of Music Therapy," in Gaston, E. T., ed., *Music in Therapy.* New York: Macmillan, 1968, pp. 238–244.
100. SOIBELMAN, D. *Therapeutic and Industrial Uses of Music.* New York: Columbia University Press, 1948.
101. SPARKS, R., and HOLLAND, A. "Method: Melodic Intonation Therapy for Aphasia," *Journal of Speech and Hearing Disorders,* XLI (1976): 287–297.
102. SPIRO, H. R. "The Role of Music in Mental Health Treatment Programs," Address to the Conference of the National Association of Music Therapy, Milwaukee, Wisconsin, September 1977.
103. STEIN, J. "Crazy Music: Theory," *Psychotherapy,* 8 (1971): 137–145.
104. ———. "Musicology for Music Therapists: The Lomax Study," *Journal of Music Therapy,* 10 (1973): 46–51.
105. ———, and EUPER, J. "Advances in Music Therapy," in Masserman, J., ed., *Current Psychiatric Therapies.* New York: Grune & Stratton, 1974, pp. 107–113.
106. STEINER, R. *Eurhythmics as Visible Music.* London: Rudolf Steiner Press, 1977.
107. STERBA, R. "Toward the Problem of the Musical Process," *Psychoanalytic Review,* 33 (1946): 37–43.
108. STEVENSON, I. "The Psychiatric Interview," in Arieti, S., ed., *American Handbook of Psychiatry,* vol. 1. New York: Basic Books, 1959, pp. 197 ff.
109. SULLIVAN, H. S. "The Onset of Schizophrenia," *American Journal of Psychiatry,* 84 (1927–28): 105. Reprinted in Sullivan, H. S., *Schizophrenia as a Human Process.* New York: W. W. Norton, 1962.
110. TAYLOR, D. B. "Subject Responses to Precategorized Stimulative and Sedative Music," *Journal of Music Therapy,* 10 (1973): 86–94.
111. TYSON, F. "Therapeutic Elements in Out-Patient Music Therapy," *The Psychiatric Quarterly,* 39 (1965): 1–13.
112. ———. "Music Therapy Practice in the Community: Three Case Studies," *Psychoanalytic Quarterly Supplement,* part 1 (1966): 1–20.
113. ———. "Music Therapy Expands to Include Outpatients in Private Practice," *Roche Report: Frontiers of Psychiatry,* 1:1 (1971): 1–2.
114. VISCOTT, D. S. "A Musical Idiot Savant—Psychodynamic Study and Some Speculations of the Creative Process," *Psychiatry,* 33:4 (1970) 494–515.
115. WHEELER, B. Verbal communication, November 1979.
116. WOLBERG, L. R. *The Technique of Psychotherapy.* New York: Grune & Stratton, 1954.
117. WOLFF, P. "The Natural History of Crying and Other Vocalizations in Early Infancy," in Foss, B. M., ed., *Determinants of Infant Behavior.* London: Methuen, 1968, pp. 81–109.
118. ———. "Role of Biological Rhythms in Early Psychological Development," in Chess, S., and Thomas, A. eds., *Annual Progress in Child Psychiatry and Child Development.* New York: Brunner/Mazel, 1968, pp. 1–21.
119. ———. "The Serial Organization of Sucking in the Young Infant," *Pediatrics,* 42:61 (December, 1968): 943–956.
120. ———. "Sucking Patterns of Infant Mammals," *Brain Behavior,* 6 (1968): 354–367.
121. YATES, S. "Some Aspects of Time Difficulties and Their Relation to Music," *International Journal of Psychoanalysis,* 16 (1935): 341–354.

CHAPTER 30

THE EFFICACY OF INDIVIDUAL PSYCHOTHERAPY: A PERSPECTIVE AND REVIEW EMPHASIZING CONTROLLED OUTCOME STUDIES*

Douglas W. Heinrichs and William T. Carpenter, Jr.

¶ Introduction

FOR three decades the great debate concerning the value of psychotherapy has achieved no clear consensus. Recently, this debate has moved from the arena of scholarly interest to that of public attention, becoming a matter of major practical importance to all the mental health professions. While a great deal of professional time and interest is devoted to individual psychotherapy, an ever broadening range of efficacious pharmacologic treatments now provides the clinician with alternatives to traditional psychotherapy in treating many psychopathologic states. A crucial sociopolitical factor is now relevant. With the growing reliance on third-party financing of health care and

*Dr. Morris Parloff, chief, Psychiatry and Behavioral Intervention Section, Clinical Research Branch, National Institute of Mental Health, was generous in making available unpublished material, critically reviewing an earlier draft of this manuscript, and offering suggestions to the authors.

the anticipation of national health insurance, treatment modalities are receiving closer and more public scrutiny. Demonstration of efficacy must satisfy not only clinicians, but increasingly policy makers and the public at large as well. Treatments not fitting a narrow biomedical approach to therapeutics are especially suspect. It is imperative that clinicians and clinical trainees become experts not only in the theory and application of psychotherapy, but also about research bearing on the efficacy question and the theoretical and methodological issues relevant to this debate. This chapter provides a review of the data base for judging psychotherapy efficacy, and related concepts and special problems are also discussed.

In the hands of trained clinicians psychotherapy has taken many forms involving innovative experiments, new theory, and applications to an ever-increasing range of problems. New approaches that have been judged irresponsible by the mainstream of the mental health professions are not uncommon. Of greater concern has been the willingness of nonprofessionals and persons with no rigorous training and scant clinical backgrounds to become popular advocates of a plethora of psychosocial techniques offered as psychotherapy. No conceptual definition of psychotherapy can clearly differentiate "proper" psychotherapy from all other interpersonal strategies designed to be therapeutic. Nor are sufficient standards for education, training, experience, and performance available for readily distinguishing between a socially sanctioned expert and a self-appointed mental health care provider. We intend to provide a perspective for assessing treatment efficacy and to review results of relevant studies. This is best accomplished by avoiding the wide border between psychotherapy and pseudopsychotherapy, and between the mainstream professional and self-appointed clinician. While not denying the legitimacy of many activities outside the core tradition of psychotherapy, the purposes of this chapter are best served by a focus on dyadic psychotherapy used by highly trained clinicians in the treatment of common psychiatric and psychosomatic diseases in adults, specifically schizophrenia, affective illnesses, psychoneurotic and personality disorders, and psychosomatic illnesses.

The definition of psychotherapy used in this chapter will be clarified by the selection of studies for inclusion. As a general definition, we view psychotherapy as a treatment in which the relationship between patient and therapist provides a context for understanding the psychological components of illness and the psychosocial matrix of its development. The therapist should be prepared to use the data generated in this setting to enable the patient to increase his self-understanding in the belief that insight into one's own psychology and psychopathology may induce therapeutic change. While there are major cognitive or intellectual components in developing self-understanding, most forms of psychotherapy presume that the emotional components of the therapeutic relationship are indispensable to accumulating information and assimilating insight. Hence, psychotherapeutic goals range from the in-depth understanding and intrapsychic restructuring that are the goals of psychoanalytic and psychodynamic treatments to the clarification and articulation of more observable processes (for example, behavioral, affective, and interpersonal patterns) that result in the maladaptive consequences for the patient that are the focus of more time-limited psychotherapeutic strategies.

The technique of psychotherapy may vary from nondirective and interpretive to a more direct and advice-giving mode. The critical ingredient is the use of the therapeutic relationship to encourage a cognitive/affective reappraisal by the patient of himself and his situation as a prerequisite to therapeutic change. Thus, treatments based on other types of interpersonal techniques designed to alter behavior directly, such as behavioral therapy, social skills training, provision of a reassuring relationship, and social case work, are not reviewed as core psychotherapy.

Nevertheless, insofar as many of these other approaches incidentally involve a shift in the patient's understanding of important issues, the delineation of traditional psychotherapy is often imprecise. This is especially true since identification with the therapist may be an important (if unwitting) ingredient in the effectiveness of all these modalities, and each procedure waxes and wanes in its "purity," often containing aspects of other modalities.

Only controlled outcome studies will be reviewed here in detail, but the entire range of relevant information will be noted. Since the issues are complex and the data confusing, a context has been developed for weighing the validity of the various data bearing on this subject. To this end, it is essential to consider the nature of medicine and, in particular, of psychiatry.

¶ Psychiatry's Scientific Mode and the Nature of Evidence

The empirical sciences gather information predominantly in one of two modes (or a mixture of both). The experimental mode is applicable to disciplines for which the relevant objects can be manipulated for scientific study. This mode provides an unequaled degree of precision and clarity; as a result those disciplines that can rely heavily on experimentation have become the prototype of valid scientific methodology. For many scientific disciplines, however, the objects of investigation do not lend themselves to such manipulation. In such circumstances, science must rely on careful and critical observations. In the case of astronomy, for instance, where the size of objects limits experimental manipulation, most laws derive from the recognition of stable and repetitive patterns. Rarely is any scientific discipline purely experimental or observational, although one modality often dominates. Experimental fields usually look to the observational domain to define important questions and hypotheses to be pursued experimentally. In the case of observational sciences, certain aspects of the relevant objects of study can frequently be subjected to experimental manipulations in a limited way or in the context of another discipline.

The location of a science on the observational-experimental continuum largely determines the range of information deemed relevant and the methods with which it is approached. Medicine is primarily an observational science that in certain areas has been able to rely on a body of precise experimental work for supplementation. The limitation of experimental manipulation in medicine and human psychology comes from the complexity and adaptability of the human organism and from ethical considerations. The study of neuropathology is a prime example. In the case of humans, it is unethical to cause experimental lesions to observe functional consequences. Knowledge in neuropathology accrues from the careful and meticulous observation of various naturally-occurring lesions and their functional correlates. Where animal systems are sufficiently similar, experimental methods may be applicable and the knowledge gained inferentially applied to human neuropathology. In the case of psychiatry, however, where the primary concern is with higher and distinctively human functioning, animal models are more restricted in their applicability. It is due to these considerations that the great tradition of medicine has largely been one of careful observation. In the case of psychiatry, there is a second limitation that has a pervasive effect. This is the fact that the objects being studied, human beings, are self-conscious and form opinions, judgments, and reactions about the manipulations to which they are subjected. As a result, each subject's response is in part dependent on expectations rather than simply reflecting the consequences of the manipulation (for example, the placebo effect). Furthermore, human subjects may refuse to participate, and many informative manipulations in psychiatry are impractical because of the limited number of people (scientists or subjects) willing to participate.

It is important to fully appreciate the fact

that medicine (including psychiatry) is primarily an observational science. Recent preoccupation with a narrow biomedical model of disease has caused some critics to decry data not derived in a strict experimental mode. This position ignores the scientific base of modern medicine—clinical observation. It also fails to acknowledge the tremendous gap between a demonstrated function in isolation (for example, a single neuron preparation) and the overall harmony and interplay among human systems (such as social, psychological, and biological). With this understanding of medicine's fundamental position, clinical sciences will utilize the experimental mode wherever possible, and the interplay between hypothesis generation from observation and hypothesis testing in experimentation will be continuous.

Therapeutics often provides opportunities to mix scientific modes in clinical tests of efficacy. Psychopharmacologic investigations have been strikingly productive in this regard. In spite of the inherent limitations of medical research, rigorous efficacy tests of many drugs have been executed, resulting in an impressive body of information. Study paradigms rely on both manipulable experimental attributes (drug dose, placebo, random assignment, repeated trials, external replication) and clinical observations (subject selection, assessment of behavioral change, clinical context for conduct of the experiment). A number of factors make such research more difficult in the case of psychotherapy. Difficult, however, does not mean impossible, and we will comment briefly on both necessary and unnecessary impediments to a more definitive testing of psychotherapeutic efficacy.

The following factors illustrate unavoidable problems in designing mixed modal experiments to assess the efficacy of psychotherapy:

1. Double-blind conditions for clinical trials have proven of inestimable value in tests of drug efficacy, but there is no apparent method for having patient and psychotherapist unaware of what treatment (if any) is being provided. Single-blind conditions are imaginable (that is, the therapist, but not the patient, knows what therapeutic modality is being used). However, the patient not being specifically informed and being actually uninformed are two different matters. Furthermore, single-blind studies have not proven superior to open studies (both patient and doctor know the treatment) in psychopharmacologic studies, since the clinician unwittingly and perhaps nonverbally communicates his expectations.[18] At present, the best compensation for this problem is reliance on objective measures of change and independent (blind, if possible) raters of change. Objective measures are especially difficult where the aims of treatment are to alter subjective symptoms.

2. Placebo-controlled assessment of efficacy is a problem in psychotherapy since no one has been clever enough to create the trappings of psychotherapy while making the essential therapeutic ingredients inert. The two alternatives are informative but limited. Comparing psychotherapy to no therapy (for example, waiting list) can show efficacy but cannot determine the mechanism (that is, placebo or psychotherapeutic effect). Comparing different forms of interpersonal clinical contacts may reveal comparative merits without determining the absolute extent of effect or the placebo contribution to effect. Here, we are using placebo to refer to any factors contributing to beneficial change other than those purported to be the therapeutic modality. Placebo traditionally means that the treatment is inert only with respect to the mechanism of action being studied—usually chemical. Placebo may be quite active, but by psychological mechanisms. In psychotherapy research, the experimental treatment is also assumed to depend on psychological mechanisms, so the distinction between active treatment and placebo is clouded.

3. Many forms of psychotherapy require months, with most benefits being realized late in treatment. Furthermore, individuals respond at remarkably different rates. It is difficult to justify withholding treatment from a control group of sick patients for long

periods. Many aspects of a patient's status change with time, so prolonged clinical trials have more spontaneous variances (noise) and hence require larger numbers of subjects. The time required of clinicians to do psychotherapy, the length of a clinical trial, and the number of subjects required converge to make these studies expensive and logistically complicated.

4. Experimental designs undeniably alter to some extent the clinical circumstances in which a treatment is received, often improving care by bringing more resources, better follow-up, innovative treatment, and an aura of expectation. Changes in the therapeutic setting may, however, introduce anti-therapeutic artifacts in psychotherapy. Many argue that informed consent procedures, random assignments, research criteria for patient selection, explicit interest in special variables as reflecting change, and other routine considerations in clinical studies undermine the uniqueness and complexity of psychotherapy as clinically practiced. Crucial factors such as patient motivation, patient-therapist matching, and a strong belief shared by patient and therapist that the psychotherapy being done is the most desirable treatment option are obvious problems affected by research design.

5. Psychotherapy and therapist cannot be standardized and monitored in detail. One can imagine the problems to be encountered in drug studies if one could not determine the amount of drug per capsule, could not be sure whether other active drugs were mixed with the treatment drug, or be confident that the placebo was chemically inert. Problems such as compliance and metabolic variance do complicate drug research, but psychotherapy research cannot easily solve the "pre-packaging" problem.

6. Some experts such as Strupp[101] contend that psychotherapy efficacy may be minimally dependent on the precise mode of treatment and maximally dependent on innate qualities of the therapists. If so, psychotherapy efficacy research will be dependent on defining the innately gifted therapists. The inability to assure the quality of the therapist will consistently weaken the measured effect of therapy.

These formidable problems in the scientific testing of psychotherapeutic efficacy must be recognized. However, it is also important to emphasize that factors not intrinsic to research design have unfortunately hindered the scientific assessment of psychotherapy efficacy.

1. There has been a long tradition in psychiatry of appeals to authority, most obviously in the psychoanalytic movement. The preeminence of psychoanalysis in American academic psychiatry during the 1950s and 1960s created circumstances that substituted authority for science in validating theory. Less abstract is the fact that so long as everyone "knows" a treatment is good, there is little urgency in pursuing its evaluation. While psychodynamic psychiatrists may have been restrained by theoretical blinders, the problem is certainly not unique to them as the many charismatic proponents of alternative models amply demonstrate. (The authors have discussed the role of arrogance as an impediment to clinical science elsewhere.[11])

2. The claim that psychotherapy is far too complex and the necessary observations too numerous to permit scientific scrutiny has often been made without a realistic appreciation that the nature of experimentation is to extract and simplify from the complexity of the natural experience and, in a controlled way, to examine the relationship between a few of the many variables involved. As such, experimentation neither threatens or supplants clinical observation, but may allow a more rigorous examination or confirmation of some limited aspects of the larger clinical experience. Such claims are also offered without the mathematical background to assess current techniques for statistical evaluation of multivariate designs.

3. The theoretically-oriented clinician tends to underestimate the extent that his presuppositions influence his clinical observations. Inference and observation have not always been differentiated, hence the field has collected a great deal of verifying "obser-

vation" without realizing how inferential and perhaps unreliable these clinical findings were.

4. An understandable human tendency exists on the part of the psychotherapist to shun experimental research because of the inherent threat of negative findings. If a pharmacotherapist finds, as a result of controlled trials, that a particular drug is not efficacious, he does not feel that his person has been indicted or his career jeopardized. He simply searches for an alternative pharmacologic intervention. In the case of the psychotherapist, however, who has all too often been trained exclusively in one modality, the prospect of finding the sort of treatment he offers to be of little or no value has major personal and professional ramifications. In this regard, Frank[27] quoted Confucius as saying that "a wise man does not examine the source of his well-being."

¶ Observational Data on Psychotherapy

The case report is the most common form of observation relating to psychotherapy outcome, particularly in the psychiatric literature. Its strength is in the detailed presentation of clinical material, and thus, it is more effective in illustrating an approach and generating hypotheses than in confirming them. Surveys are an extension of this sort of observation. The larger number of cases adds to the persuasive power. However, the lack of controls makes it impossible to assure that changes can be attributed to the treatment. This is particularly true given the fact that the natural history of psychiatric illnesses is highly variable and poorly specified.

The meaning of the survey data of psychotherapy outcome has been hotly debated for several decades. The challenge was first articulated by Eysenck[19] and further argued by Eysenck[20,21,22,23] and Rachman.[88,89] These studies reported that approximately two-thirds of all neurotics improved substantially over a two-year period, irrespective of whether or not they received psychotherapy. This conclusion has been challenged by a number of investigators.* A careful review of this debate is provided by Bergin and Lambert.[6] Several issues are involved. The first is the spontaneous remission rate of untreated patients. Eysenck and Rachman maintain there is a two-thirds improvement rate over two years. Bergin and Lambert review a number of studies and find a median spontaneous remission rate of 43 percent. They note, however, a high variability between the studies reviewed, ranging from 18 percent to 67 percent. Among other factors determining this variability is diagnosis. Although the literature does not provide definitive data, anxiety and depressive neuroses have the highest spontaneous recovery rates with lower recovery rates for hysterical, phobic, obsessive compulsive, and hypochondriacal disorders. A second problem in reviewing the survey literature is that different investigators use different criteria for improvement. Such inconsistency may have a major impact on results. Bergin and Lambert illustrate this by repeatedly calculating the collective improvement rate in five surveys of psychoanalytic treatment using several different sets of criteria, each of which seems reasonable. Yet, the overall percentage of improvement ranged from 44 percent to 83 percent, depending on the criteria used.

A third confounding factor is the possibility that psychotherapy may make some of the patient population worse (deterioration effect) and that this factor would interfere with discovering a significant improvement due to psychotherapy in other patients. Some data suggest such an effect.[6,58,103]

The many sources of variability in survey data make it impossible to draw indisputable conclusions. However, it is fair to say that most recent assessments of the literature, such as Meltzoff and Kornreich,[76] draw considerably more positive inferences about the effectiveness of psychotherapy than the ear-

*See references 5, 7, 12, 57, 62, 66, 76, 99, 100, and 102.

lier and widely publicized reviews of Eysenck.[19,22] These recent reviews conclude that most studies (80 percent according to Meltzoff and Kornreich) show moderate positive results to a greater degree than would be expected by chance. These reviews generally include group and family, as well as individual, modalities. At the same time, they note limitations in interpreting such data and express the hope that there will be a decline in this type of broad study.

¶ The Controlled Study

The controlled experiment has the greatest potential for persuasively demonstrating the efficacy of therapy, particularly when built upon a firm foundation of careful observations. Controlled outcome studies vary immensely, however, in the quality of their design and execution, and the degree of confidence in the findings must vary with the adequacy of design.

Studies are improved to the extent that patients are homogeneous with respect to diagnosis, prognosis, duration and severity of illness, premorbid functioning, and prior treatment experience. Control patients should be comparable on these and other relevant variables and the control experience should minimize contaminating and biasing elements. Studies are further strengthened to the extent that the psychotherapy itself is highly specified, administered by experienced therapists, and of a duration and intensity likely to maximize therapeutic change. Given the lack of firm correlation between the array of outcome criteria used, it is desirable to collect a broad range of data. Various sources should be utilized—patient self-reports, therapist reports, and ratings by independent evaluators. Information should also relate to a range of outcome dimensions—for example, symptoms, social and occupational functioning, contentment and satisfaction, personality change, and treatment utilization. It is also important that the patients chosen be acceptable candidates for the treatment in question and that the outcome dimensions examined include those most likely to be affected by the therapy. Although space does not permit a systematic presentation of the degree to which the reviewed studies meet each of these standards, these factors are considered in assessing the literature and are noted in cases where they are particularly critical.*

The investigations reviewed here include controlled outcome studies of individual psychotherapy in the core tradition for adults seeking treatment for the common diseases noted earlier. Excluded are studies that do not make a serious effort to use a control group of comparable subjects or that use subjects as their own controls (since the variable natural history of these illnesses and the order of treatments confound such efforts). Also excluded are studies that use individual treatment as a minimum contact control to test some other intervention.

A computerized literature search covering the last three years was conducted. Earlier work was identified from previous reviews.† Of these several hundred potentially relevant reports, only the following met the preceding criteria for inclusion in this discussion.

Controlled Outcome Studies of Schizophrenia

The most influential work in the area is that by May, Tuma, and coworkers.‡ Over 200 hospitalized schizophrenics in the mid-prognostic range participated in these studies. Treatment was administered by psychiatric residents or psychiatrists without extensive experience and it consisted of one of five modalities—psychotherapy alone, psychotherapy plus neuroleptics, neurolep-

*Some reports fail to mention important aspects of the research design and its execution. This is reflected in the absence of key information in the reviews of some of the following studies.

†See references 5, 6, 7, 10, 13, 14, 16, 17, 25, 26, 42, 47, 52, 53, 63, 64, 65, 69, 70, 76, 79, 84, 92, 98, 103, and 105.

‡See references 68, 71, 72, 74, and 75.

tics alone, electroconvulsive therapy(ECT), and a control group with only general treatment in a psychiatric ward. Although the therapists were inexperienced, all psychotherapy was supervised by experienced psychoanalysts, who strongly believed in the efficacy of the treatment they were supervising. The therapy was primarily ego-supportive in nature with an emphasis on defining reality. There was a minimum of depth interpretations and a substantial focus on current problems and confronting the patient with the reality of his own behavior, as well as the clarification of perceptual distortions. The therapists were seen as acting as suitable models for interjection. Therapy was to be given on an average of not less than two hours weekly, with an absolute minimum of one hour. For nonpsychotherapy cohorts, a serious attempt to minimize time spent with the doctor was effective, resulting in considerably less doctor contact than experienced by patients in psychotherapy. Treatment continued for one year or until discharge from the hospital, although treatment could be ended after six months, if the treating physician and supervisor agreed that a given case was a treatment failure with little likelihood of responding for the duration of the study. Patients were followed for up to five years. This investigation is impressive in its attempt to assess patients on a wide range of outcome variables. Rating scales, such as the Menninger Health/Sickness Scale and the Cammarilo Assessment Scale were used to evaluate affective contact, anxiety, ego strengths, insight, motivation, object relations, identity, and sexual adjustment. Behavioral ratings were made by nursing staff. Ratings were made on idiosyncratic symptoms for each patient. Other ratings and psychological tests assessed cognitive functioning, thought disorder, and affective state. Duration of the index hospitalization, as well as number of days in the hospital from the first admission or from the index discharge over the subsequent five-year period were assessed. Antipsychotic drugs proved significantly superior to psychotherapy, which was, in general, no more effective than the treatment given to the control group. On many variables ECT was intermediately effective. Drugs plus psychotherapy worked slightly better than drugs alone. This work is impressive evidence that psychotherapy administered in the hospital by inexperienced therapists to midprognostic-range schizophrenics (not selected for psychotherapy suitability) is not effective. The generalizability of this work has been challenged in the belief that a more selective group of patients or more experienced therapists would have made a difference. Nevertheless, this investigation was a telling assessment of what was then the most common psychotherapeutic experience available to the hospitalized schizophrenic patient.

Karon and Vandenbos[51] randomized thirty-six hospitalized schizophrenics to psychotherapy alone, psychotherapy plus drugs, or drugs alone. The patients were primarily poor inner-city blacks, two-thirds of whom never had been previously hospitalized. The therapy-without-medication group received an active psychoanalytic therapy stressing oral dynamics and utilizing "direct interpretations." Sessions were held five days a week until discharge and usually once per week thereafter. The group receiving both therapy and drugs was given a psychoanalytic therapy "of an ego-analytic variety" conducted three times per week and eventually reduced to once weekly. The third group was hospitalized in a public institution in which phenothiazines were used as the primary treatment. Treatment was available for all groups for twenty months, and the therapy groups received an average of approximately seventy sessions. Of the twelve psychotherapists participating in the study, four were regarded as experienced and eight as inexperienced. Outcome variables included a clinical status interview, projective tests, the Porteus maze, and tests for vocabulary and intelligence. The two psychotherapy groups had significantly shorter hospital stays and performed significantly better on the clinical status interview as well as on a number of performance tests. Differences relative to the experience level of the therapists demonstrated some advantages for the more experienced therapists. These results have been challenged on methodological and statistical grounds by May and Tuma,[73] to which the authors have provided a rebuttal.[50] In addition to the small sample size and statistical issues, the most telling inadequacy is the difference in the hospital experience of the three groups (that is, nonpsychotherapy groups were treated in state hospitals). The problems are sufficient to weaken the merits of this study, and a larger scale replication attempt is warranted.

Messier and coworkers[77] reported a follow-up of an earlier study of hospitalized schizophrenics conducted by Grinspoon and coworkers.[39,40] The original study compared thioridazine and placebo in twenty patients, all of whom received twice-weekly analytically-oriented psychotherapy for two years from senior staff psychiatrists in an active milieu. A vast superiority for the thioridazine group was demonstrated. All patients received psychotherapy, but its efficacy could not be assessed. The follow-up study attempted to evaluate the impact of psychotherapy by comparing the twenty patients in the original study with twenty-one other patients chosen at the same time, but assigned to stay in the state hospital where neuroleptics were the main treatment modality and psychotherapy was uncommon. Outcome criteria included psychotic symptoms, employment, recreational functioning, living situation, and capacity to live outside of the hospital. There were no significant differences between the state hospital group on the one hand and the psychotherapy alone, psychotherapy plus thioridazine, or the combined psychotherapy groups on the other. There were several serious methodologic problems in this study. The no-therapy controls, in fact, had an extremely different hospital experience than did the psychotherapy patients. They had been treated in a state hospital, whereas the psychotherapy patients were treated in a special research ward with an active therapeutic milieu. Furthermore, patients were not assigned to groups in a strictly random manner, in that patients transferred to the research ward were only those who consented to participate in the research project and whose families agreed to be involved in the treatment. If patient or family refused, they were assigned to the state hospital control group. Furthermore, all the patients were chronic, having been hospitalized for three or more years. Hence, they are hardly representative of patients most likely to demonstrate benefits from psychotherapy. For these reasons, the results of this study are severely compromised and difficult to interpret.

Rogers and coworkers[90] studied thirty-two hospitalized schizophrenics, half of whom had been hospitalized over eight months. Patients were randomized to psychotherapy or no psychotherapy conditions, and an attempt was made to minimize the use of medication in therapy patients. The experience and orientation of the therapist was highly variable and poorly controlled, but approximated Rogers's client-centered therapeutic approach. Therapy lasted from four months to two and one-half years with sessions held, on the average, twice a week. Outcome variables included symptomatology, work behavior, hospitalization status, the Minnesota Multiphasic Personality Inventory (MMPI), and the Q-sort. There was no significant difference between client-centered therapy patients and controls, but a few trends favored the former. This research was designed to study process variables and the mechanisms of change in psychotherapy, and the assessment of outcome was ancillary. The use of medication and assignment of psychotherapists were poorly controlled, hence the results of this study relevant to efficacy are compromised.

Bookhammer and associates[8] compared fourteen hospitalized schizophrenics treated with Rosen's "direct analysis" with thirty-seven controls. All patients were suffering from their first attack of overt psychotic symptoms at the time of the study. Control patients received a wide range of treatments (probably including interpersonal therapies) in various facilities, with no attempt to standardize or define their treatment experience. All patients were evaluated periodically for five years by the investigative team, who judged signs and symptoms, patient's attitude toward himself, interpersonal relationships, contact with reality, useful work, and the amount of time spent out of the hospital. No significant differences were found. One may conclude that "direct analysis" did not prove superior to a hodgepodge of other treatments, but one may not judge whether all forms of treatment were equally effective or equally ineffective.

Marks and associates[67] compared psychotherapy and token economy in twenty-two chronic hospitalized schizophrenics, all of whom received both treatments in a crossover design. Patients were evaluated with respect to work, social, and conceptual competence, word association tests, symbolic literal meaning test, and several tests of speed and maintenance of work set. The two treatments had similar effects, showing significant improvement on twelve of eighteen variables. The authors then compared thirteen subjects whose medication was held constant, prior to and during the study, with patients participating in a drug study conducted a short time before this investigation. Both studies covered a similar period of time, used a crossover design, and had similar behavioral assessment (especially of ward behavior). The drug study revealed no difference between drug and placebo treatment in these chronic patients. Comparing the results of the two studies on the nine measures common to both, patients receiving token economy treatment or psychotherapy showed significantly more improvement on eight of the nine measures than patients in the drug/placebo study conditions. The drug study included eleven patients who participated in the later therapy study as well and thus were serving as their own controls. The post hoc nature of this comparison cannot answer questions as to the adequacy of randomization and comparability of the control conditions between studies. Thus, these results are hardly persuasive. The aforementioned studies evaluate psychotherapy in hospitalized patients and do not relate to the role of psychotherapy in the outpatient setting.

Only two[51,67] of the six studies reviewed purport to demonstrate efficacy for individual psychotherapy with hospitalized schizophrenics, and serious methodologic flaws in these leave the verdict unsettled. The work of May and associates* in the 1960s was an exceptionally accomplished initiative, but whether the negative findings are applicable to more experienced therapists and schizophrenic patients judged suitable for psychotherapy is not yet determined. No controlled study of individual psychotherapy in the outpatient context has been reported to date. The most relevant study to mention, therefore, was conducted by Hogarty and coworkers.[44,45,46] They examined the role of individual social casework and vocational counseling, termed "major role therapy," in the aftercare of schizophrenic patients. Although not fitting even the broad definition of psychotherapy used in this review, their work demonstrates several points important in conceptualizing and designing studies of psychotherapy in the outpatient setting. After randomizing 374 newly discharged schizophrenics to chlorpromazine or placebo aftercare, each group was further randomized to either no psychological treatment or "major role therapy" (MRT). Patients were treated for two years or until relapse. MRT had no demonstrable value during the first six months, but for the seven to twenty-four-month period it significantly reduced the relapse rate independent of drugs. At eighteen and twenty-four months, a significant interaction appeared between MRT and drugs on measures of symptoms, social and occupational adjustment, and overall functioning among the subgroup of patients completing the study without relapse. For medicated patients, MRT improved functioning, especially in interpersonal relations and overall functioning. Unmedicated patients, however, did better without MRT! These findings were achieved despite the small difference—less than one social work

*See references 68, 72, 73, 74 and 75.

contact per month—between MRT and non-MRT groups. Two major implications are: (1) long duration of treatment may be necessary to demonstrate benefits from some interpersonal treatments; and (2) beneficial effects of psychological therapy may only be apparent in patients receiving medication. False negatives (type II error) may be obtained in studies of too brief a duration or where psychotherapy is not evaluated in combination with drug treatment. This latter point has been reinforced by the recent demonstrations by Goldstein and coworkers of a drug-family therapy interaction.[37]

Regarding schizophrenia, the authors find the reports of skilled clinicians working intimately with their patients extremely informative as to the phenomenology of schizophrenia and, to a lesser extent (since it is necessary to allow for theoretically based bias), informative regarding the intrapsychic and psychodynamic components of schizophrenia. The focus of this review is efficacy of psychotherapy as treatment, not as a clinical method for observation. Here the survey data contribute little, and only a half dozen controlled studies of dyadic psychotherapy have been reported. Considering the complexity of psychotherapy and the heterogeneity of schizophrenia, these few studies could not be definitive even if results were consistent and methods without serious flaw. These contrast with twenty-nine controlled studies of antipsychotic drug therapy in 3,519 outpatients[15] and scores of such studies on inpatient units. The results of psychotherapy on schizophrenia to date have not been consistent, but the modest benefits noted in several studies[51,67] are outweighed by the negative results in the others.* Also of note is the fact that in the negative studies more patients have been studied by better methods.

Since some mental health professionals consider it axiomatic that "talking therapies" do not favorably alter the course of schizophrenia, it is worth noting several recent reviews[42,79,80] of a broader range of interpersonal treatment techniques for schizophrenia, including milieu group and family therapy. These reports find stronger evidence for treatment efficacy than do the studies reviewed in this chapter. Many of these broader studies mix individual psychotherapy with other psychological treatments, and drugs are less likely to be excluded as a component of treatment.

Finally, the clinician judging which treatment modalities may be attempted with schizophrenic patients is cognizant of the limited effects of all present treatments (including pharmacotherapy) and the formidable morbidity endured for decades by those patients with chronic forms of this illness. While some treatment effects may seem modest or even trivial, the humane and financial benefits that accrue to patients who become slightly more able to maintain relationships, slightly more likely to hold a job, and slightly more likely to recognize and avoid pathogenic stresses, are enormous when illness begins in young adulthood and may last sixty years. The monetary savings that would be associated with reducing unemployment in discharged schizophrenics from 67 percent to 60 percent are so vast that even the cost containment expert for third-party payers should be eager to avoid prematurely closing the door on rationally derived, potentially beneficial therapeutic techniques. In the absence of definitive answers from controlled studies, the clinician weighs all available data with judgment and intuition. It is worth keeping in mind that only 10 to 20 percent of all medical therapeutics have been proven effective in controlled studies.[43]

Controlled Outcome Studies of Affective Disorders

There are no controlled outcome studies of individual psychotherapy with manic patients. There are, however, three well-designed investigations of psychotherapy efficacy in depressive disorders.

The Boston-New Haven Collaborative De-

*See references 68, 72, 73, 74, 75, 77, and 90.

pression Project[54,106,107] studied 150 depressive female outpatients, randomized to either high- or low-contact groups. Each group was further randomly assigned to amitriptyline, placebo, or no medication. Most of these neurotically depressed women had one previous depressive episode, and only 5 percent had bipolar affective illness. Each patient had an acute depressive episode of significant severity but had responded to four to six weeks of amitriptyline therapy prior to inclusion into the study. The study focused on the aftercare phase of treatment. The high contact group received therapy consisting of at least one hourly session per week with an experienced social worker. It focused on identifying current maladaptive patterns of interpersonal functioning and altering them. There was little attempt to reconstruct early experiences in the patients' lives. All patients were seen for a monthly fifteen-minute visit with a psychiatrist to assess clinical status and to adjust medication. This was the only clinical session for the low-contact group. Duration of treatment was eight months. This design permits evaluating drug effects in aftercare (drug versus placebo), psychotherapy effects (high versus low contact), the effectiveness of each treatment group versus no treatment, and drug/psychotherapy interactions. Patients receiving amitriptyline had less depressive symptomatology early in treatment and fewer relapses into depressive episodes. A tendency to fewer relapses in psychotherapy patients was not statistically significant, but measures of occupational and interpersonal functioning revealed significant benefit from the fifth to the eighth month of psychotherapy. Since amitriptyline did not affect these variables, this study neatly demonstrates that pharmacotherapy and psychotherapy may have their major effects on different aspects of psychopathology, a demonstration that affirms expectations based on common sense and uncontrolled clinical observation. Treatment was not controlled following the eight-month trial, and six and twelve months follow-up did not find persisting group differences.

Weissman and coworkers[108] report on a study of eighty-one acute unipolar, nonpsychotic depressed patients, randomized to individual psychotherapy alone, psychotherapy plus amitriptyline, amitriptyline alone, and nonscheduled treatment with a maximum of one visit per month. Treatment lasted sixteen weeks. Psychotherapy was similar to the Boston-New Haven Project, but differed in two respects. First, it was administered by psychiatrists rather than social workers. Second, by a careful examination of what actually occurred in psychotherapy sessions in the Boston-New Haven Project, a manual was developed that prescribed the psychotherapy used with a degree of specificity uncommon in this sort of research. Again, the therapy focused on the social context of the depression and the identification of maladaptive patterns of interpersonal functioning. This time, psychotherapy was as effective as drugs in reducing depressive symptoms and relapse—both to a significantly greater degree than nonscheduled treatment. There was a trend favoring the combination of therapy and drugs. This work suggests an efficacy of psychotherapy in treatment of acute depressive symptomatology equal to that of medication, over and above any effect on interpersonal adjustment, which was not assessed in this report.

Rush and coworkers[91] randomized forty-one significantly depressed outpatients to either cognitive therapy or treatment with imipramine for twelve weeks. All patients were at least moderately depressed on the Beck Depression Inventory, most had multiple prior depressive episodes and reported suicidal ideation. Over one-third had been depressed more than one year and nearly one-quarter had previous psychiatric hospitalizations. All patients were unipolar. The technique for cognitive therapy was highly specified and elaborated in a treatment manual, and averaged one and one-half sessions per week. The focus of the therapy involved altering negative and pessimistic cognitive attitudes of the patient toward himself and the environment. Cognitive psychotherapy was significantly more effective than imipramine in reducing depressive symptoms, as judged by the patient, the therapist, or an independent clinical evaluator. This difference was maintained at three months follow-up and persisted as a trend at six months. Furthermore, 68 percent of the drug group reentered treatment for depression during the follow-up period as compared to only 16 percent for the cognitive therapy group.

These three studies were especially well designed to test efficacy of special forms of psychotherapy and to contrast these effects with those of an established effective treatment. The investigators assured that both pharmacotherapy and psychotherapy were conducted according to standards, and both therapeutic approaches were superior to minimal or no treatment. Psychotherapy showed a beneficial effect on psychosocial aspects of course of illness, but also rivaled or surpassed antidepressant medication on symptom and relapse measures in two of the studies. Future studies are required to determine if these findings are generalizable to more severely ill patients, to mildly depressed patients, or to patients with bipolar affective disorder. Also, whether other forms of psychotherapy are effective in treating moderately to moderately-severe depressed patients awaits demonstration.

Although only three studies can be cited, the evidence strongly affirms the efficacy of special forms of psychotherapy in outpatient depressives. These studies illustrate the applicability of carefully designed, controlled studies of psychotherapy efficacy and should prove as influential as they have proven informative. Investigations of depression have an advantage in being able to select relatively homogeneous patient cohorts for study, to focus on a more limited range of change criteria, and to use briefer periods of treatment than seem plausible in schizophrenia, where heterogeneity of patients and pervasiveness and chronicity of psychopathology create a greater challenge. Nonetheless, these study paradigms may be fruitfully applied in testing treatment effects in other psychiatric disorders.

Controlled Outcome Studies of Psychosomatic Disorders and Psychological Sequelae of Physical Disease

The following studies relate to the use of individual psychotherapy to treat either illnesses traditionally seen as psychosomatic in nature or as the adverse consequences of physical illness.

Grace and coworkers[38] studied the impact of a form of "superficial psychotherapy" designed to alleviate the stress of patients suffering from chronic ulcerative colitis. Two groups of thirty-four patients each were matched with respect to age, sex, severity of ulcerative colitis, duration of illness prior to therapy, age of onset, and X-ray changes. Patients ranged in age from fifteen to fifty-four years; 60 percent were classified as severely ill. The duration of illness ranged from one month to over ten years. One group of patients received psychotherapy of unspecified intensity and duration. The second group was treated medically, with an emphasis on diet and antispasmodic agents. All patients were observed for at least two years. All psychotherapy patients were treated by the senior author. Outcome was assessed in terms of deaths, operations performed, symptoms of colitis, complications, time spent in hospitals, visits to physicians, and X-ray changes. Although outcome evaluations were performed by the authors, who knew the treatment assignments, the potential for bias was somewhat mitigated by the fact that several outcome measures were primarily objective criteria requiring a minimum of interpretation and, in the case of X-ray changes, the X-rays were read by radiologists unaware that a study was being conducted. Although no tests of statistical significance were performed on the data, large differences favoring the psychotherapy group were found for nearly all of the outcome measures. An additional group of the 109 patients with ulcerative colitis treated at the same hospital but not included in the study was also examined. They received standard medical treatment, and although significantly less ill than the study patients, their outcome more closely resembled the control group than the psychotherapy group on most measures.

O'Connor and associates[82] studied 114 patients suffering from ulcerative colitis of at least five years' duration. The psychotherapy group consisted of 57 patients referred for psychiatric treatment. The majority had a diagnosis of personality disorder, although one-third of this group were diagnosed as schizophrenic (criteria unspecified). The psychotherapy ranged from formal psychoanalysis for six patients to short-term therapy of less than twenty sessions directed at current conflicts for 13 patients. The remaining portion of the sample received psychoanalytically-oriented therapy twice weekly for one to two years. The control group was matched with the therapy patients for severity of ulcerative colitis, sex, age of onset, and use of steroids. It is important to note that the groups were not matched for psychopathology, given the fact that the therapy group were all referred for psychiatric treatment and the control patients were not. It is not surprising that psychopathology was markedly more severe in the therapy group. The patients were followed for at least seven years after the initiation of therapy. Outcome measures included periodic protoscopic examinations and ratings of bowel symptoms. Psychological criteria were derived from ratings in occupational functioning, sexual adjustment, family relationship, and self-esteem. Also evaluated were hospitalizations, amount of steroid therapy, amount of surgery, and mortality rate. It was found that patients with a schizophrenic diagnosis did very poorly regardless of treatment and clearly worsened over the course of the study. When the schizophrenic patients were removed from the analysis, the psychotherapy patients were found to improve over the entire course of follow-up, while the control patients worsened. In spite of the symptomatic advantage for psychotherapy patients in both somatic and psychologic domains, the mortality and surgical rates were approximately equal in treated and untreated groups. While this study suggests an advantage for psychotherapy in patients with ulcerative colitis, there are a number of significant methodologic flaws. No tests of statistical significance were performed on the data, the length of the follow-up period was not uniform, and, most importantly, the patients were not assigned in a random manner. If psychopathology and ulcerative colitis interact negatively, this design may underestimate psychotherapy benefits.

Glen[35] studied forty-five patients with confirmed diagnoses of duodenal ulcers. The therapy group received once weekly psychotherapy based on the method of Alexander,[1] concentrating initially on disturbing life situations and later on events of early life and dreams. The control group received standard medical treatment consisting of advice on diet and alkali use. All patients were treated for approximately six months and evaluated for a two-year period. Unfortunately, the only outcome criterion reported was histamine-induced maximal acid output. The result was in favor of the psychotherapy group, but not significantly so. In contrast to the studies of O'Connor and associates[82] and Grace and associates[38], where a wide range of relevant outcome criteria were evaluated, this study demonstrates the loss of a large body of potentially valuable information when assessment is limited to a single variable.

Schonecke and Schuffel[93] demonstrated no benefit for psychotherapy combined with either bromazepam, placebo over bromazepam, or placebo alone in the treatment of functional abdominal disorders. Outcome focused on abdominal symptoms, depression, anxiety, and a personality inventory. Although both groups improved on a number of outcome measures, there was no significant advantage for either group. However, the design of this study is grossly inadequate for evaluating the potential benefits of psychotherapy in that the therapy consisted of a total of only 60 minutes over a six week period. Few clinicians would anticipate tangible results from such minimal contact.

Although there are methodologic flaws in each of these studies, it is illuminating to note that the two studies that evaluated outcome with a broad range of clinically relevant measures demonstrated an advantage for psychotherapy. Conversely, the studies that were limited to a single narrow outcome measure or used an unreasonably brief trial of psychotherapy had negative results. While it is difficult to draw firm conclusions from so few studies, the methodologically adequate studies do demonstrate a value for psychotherapy in at least some psychosomatic disorders.

Two studies examined the efficacy of individual psychotherapy in managing the psychological sequelae of physical illnesses. Gruen[41] studied seventy patients in an intensive care unit following their first myocardial infarction. The therapy group was seen for thirty-minute sessions five or six days a week thoughout the hospitalization. The initial phase of therapy consisted of a nonprobing discussion of the patient's feelings and reactions to the hospital, during which time the therapist assessed the patient's strength and coping mechanisms. In a context of empathic concern and reassurance, the therapist then began to help the patient explore his fears and anxieties and to clarify unrealistic attitudes toward his illness and his future. The patient was encouraged to articulate and resolve conflicts, develop his coping strategies, and utilize existing resources. Measures of outcome included time spent in the hospital, in the intensive care unit, and on the monitor. Other somatic measures included the amount of angina, arrhythmias, and heart failure. In addition, physicians and nurses made ratings of depressive behavior, nervous and anxious behavior, refusals of treatment, violations of orders, and weak and exhausted behavior. Affects were also evaluated with a number of psychological tests. Follow-up interviews were carried out approximately four months after the infarction, usually in the patient's home. At follow-up, the patient's physician was also asked to assess the patient's functioning. The follow-up interviews were rated for level of anxiety and the degree to which the patient had resumed a normal life compared with the physician's judgment of the patient's capabilities. The results demonstrated a wide range of significant advantages for the psychotherapy group. These included less time in the hospital, in the intensive care unit, and on the monitor; fewer patients with evidence of congestive heart failure and supraventricular arrhythmias; less evidence of weakness and depression in the nurses' ratings; and less depression in the physicians' ratings. Several ratings from psychological tests also showed a significant benefit for the treatment group. At follow-up, the psychotherapy group showed significant advantages both in anxiety ratings and level of activity.*

Godbole and Verinis[36] studied sixty-one inpatients at a rehabilitation hospital, who were referred for psychiatric consultation. The patients were predominantly older women, widowed or divorced, with an average age of sixty-nine years. All had major physical disabilities and the majority had multiple physical diagnoses. The psychiatric diagnosis was either reactive depression or life situational reaction. Patients were randomly assigned to either no therapy or one of two therapy conditions. Both forms of therapy consisted of ten- to fifteen-minute sessions three times a week for two to four weeks. One type was characterized as brief supportive psychotherapy, the second was brief psychotherapy utilizing a confrontation statement according to the method of Garner.[32] Both therapists and nurses responsible for the care of the patients completed a series of rating scales describing aspects of the patients' behavior, including psychiatric and physical symptomatology, personal interaction with others, and self-care. Patients completed scales measuring depression and self-concept. In addition, a record was kept of the discharge plans for each patient with the assumption that those patients returning home were more improved than those who had to be rehospitalized or continued in aftercare facilities. On the basis of ratings by

*While the severity of the infarction may be a factor, there was no specific matching for this variable. Presumably, it was handled by the random factor. This probably reflects the fact that it is difficult to initially rate the severity of the infarction accurately.

both the nurses and the therapists, patients in either form of brief psychotherapy improved significantly more than the no therapy group, and the confrontation approach was significantly more effective than the other two methods in improving the patients' ratings of depression and self-concept. Both types of psychotherapy were significantly more effective in returning patients to their homes than no therapy, the brief supportive psychotherapy being most effective in this regard.

Thus, in addition to the studies of psychosomatic illnesses per se, these two investigations support the efficacy of psychotherapy as part of the overall management of medical illnesses and their sequelae.

Controlled Outcome Studies of Psychoneuroses and Diagnostically Mixed Groups

Most of these studies suffer from the lack of homogeneous or well-specified patient groups, but diagnoses of psychoneuroses or personality disorders predominate. Some studies include psychotic patients.

Frank and his colleagues are pioneers in this area. In a series of reports,* they compare patients receiving individual or group psychotherapy with low-contact controls. Individual therapy was one hour weekly and group therapy was one and one-half hours per week. The minimal contact group saw a psychiatrist for no more than one-half hour every two weeks. The minimal treatment condition was intended as an alternative to a pure no treatment control, which was regarded by these investigators as difficult to implement and ethically questionable. All patients were diagnosed as psychoneurotic or suffering personality disorder other than antisocial personality. Alcoholism and organic brain disease were exclusion criteria. There were eighteen patients in each of the three groups. Twenty-three patients who dropped out of therapy before the fourth meeting were replaced in their original groups and were analyzed as a separate cohort. Treatment was offered for six months, with 89 percent of the patients having at least four months of treatment. Both the individual and group therapy focused on current interpersonal difficulties and the feelings they aroused. Two outcome measures were used: (1) a discomfort scale consisting of the patient's self-rating of forty-one common complaints; and (2) a social ineffectiveness scale consisting of fifteen categories of behavior involving interpersonal relationships rated by trained observers following interviews with the patient and a relative. Evaluations were made at six months, one year, two years, five years, and ten years. At the end of the six-month experimental period, the three groups and dropout group all showed a similar and significant decrease on the discomfort scale. This improvement occurred very early in the six-month period and was interpreted as a nonspecific response to any offer of treatment. On the social ineffectiveness scale, however, there was significantly greater mean improvement, and a higher percentage of patients improved in the individual and group therapy cohorts compared with the minimal treatment or dropout groups at six months. By the time of the five-year follow-up, all three groups demonstrated progressive, negatively accelerated improvement. As a result, at five years, there were no significant differences between treatment conditions on either outcome measure. By the time of the ten-year follow-up, there was again a significant advantage in social effectiveness for the individual and group therapy patients over the minimal contact controls. This seems to be the result of a return of the minimal contact group to levels approaching their scores immediately following the treatment; whereas the other two groups maintained their improvement at a steady level between five and ten years. Follow-up evaluations were completed on 50 to 65 percent of the original cohorts. This work suggests that social adequacy may be specifically responsive to psychotherapy, as opposed to subjective discomfort. This again demonstrates the importance of assessing

*See references 28, 29, 30, 31, 48, and 49.

more than one dimension of outcome in evaluating the efficacy of psychotherapy. Combined with the results of the Boston-New Haven Collaboration Depression Project, it suggests that interpersonal functioning may be a dimension of outcome particularly sensitive to psychotherapeutic interventions. Finally, this work also demonstrates the value of long-term follow-up assessment in determining the full impact of psychotherapy on patients' lives.

Sloane and associates[96,97] studied 94 outpatients, randomized to either behavioral therapy, psychotherapy, or waiting list status (no treatment). All persons eighteen to forty-five years of age applying for treatment at a university psychiatric outpatient clinic were considered for the study. Patients were excluded if they seemed "too mildly ill"; were too seriously disturbed to risk a waiting period; evidenced signs of psychosis, mental retardation, or organic brain disease; or were judged to be primarily in need of drug therapy. Patients were also excluded if psychotherapy was not considered to be the treatment of choice. Of a total of 119 patients, 98 met the criteria and were accepted for this study. They were predominantly white women in their early twenties, roughly two-thirds of whom suffered from a psychoneurosis and the other third from a personality disorder. Treatment consisted of hour-long sessions on a weekly basis for four months. A list of stipulative definitions of each therapy was developed indicating procedures common to both treatments and those that were allowable only within one or the other modality. Thus, for example, elements characteristic of the psychotherapy included infrequent direct advice, interpretation of transference and resistance, the use of dreams, the interpretation of symptoms, and the eliciting of childhood memories. Behavioral therapy was characterized by the lack of these elements, plus the use of specified behavioral techniques.

All therapists participating in the study were highly experienced in the modality of treatment they were providing. Outcome measures included a list of three target symptoms developed individually for each patient and a structured interview assessing general level of functioning and overall improvement. Several personality tests were also administered. The patients and independent evaluators also rated work, social, sexual, and overall adjustment. Patients were evaluated initially, at the end of the four-month treatment period, and after one year.

At the end of treatment, both the psychotherapy and behavioral therapy groups improved significantly more on the target symptoms than did the waiting list group. The psychotherapy and behavior therapy groups did not significantly differ from the waiting list group in work or social adjustment. The behavior therapy group showed a significant advantage on the global measure. Results of the one-year follow-up are difficult to interpret since well over half of the waiting list patients subsequently received psychotherapy, and many of the patients assigned to a treatment group received varying amounts of additional therapy after the four-month period.

This study is outstanding in many ways, such as, the use of highly experienced therapists, selection of patients judged to be good candidates for psychotherapy, careful characterization of treatment modalities, careful implementation of random assignment, and inclusion of numerous clinically relevant outcome criteria including some specifically tailored for the individual patient. This study demonstrates that psychotherapy is helpful in improving target symptoms of particular importance to specific patients. The lack of effect on more general adjustment measures is surprising, but may reflect the brevity of the therapy. Investigations that do provide strong evidence for such a generalized effect, such as the work by Frank and his coworkers* and the Boston-New Haven Collaborative Depression Project,[54,106,107] continued active treatment for six months or more.

Koegler, Brill and associates[9,55,56] studied 299 patients drawn from applicants to a psychiatric outpatient clinic. All patients were

*See references 27, 28, 29, 30, 46, and 70.

white females between twenty and forty years of age. Exclusion criteria included psychosis, severe depression, disabling physical illness, and sociopathic disorders. The most common diagnoses were personality disorders, psychoneuroses, psychosomatic disturbances, and borderline schizophrenic states. Patients were randomly assigned to one of six conditions: individual psychotherapy, meprobamate, prochlorperazine, phenobarbital, placebo, or waiting list (no treatment). Psychotherapy consisted of a fifty-minute session at least once a week for an average of five months. The treatment was primarily psychoanalytically oriented and generally nondirective. Patients in each of the three drug groups were seen for fifteen-minute visits either weekly, biweekly, or monthly. All treatment was administered by psychiatric residents. Patients were evaluated initially, after five and ten weeks of treatment, at the end of treatment, and at follow-up averaging twenty-one months posttreatment. Outcome measures included a symptom check list and rating of change completed by therapists and the patients' ratings of change on twelve dimensions, including symptoms, self-satisfaction, and social and occupational functioning. At termination, a close relative also rated change over the course of therapy with respect to symptoms and overall functioning. In addition, the MMPI was administered initially and at termination. Finally, a social worker rated several aspects of general adjustment and work adjustment based on written reports about the patient. Although no differences between groups were apparent at five and ten weeks, by the end of treatment patients receiving either psychotherapy or meprobamate were significantly more improved on self-ratings than the other groups. Meprobamate and psychotherapy patients also showed a significant advantage in social work evaluations and some aspects of the MMPI. In addition, there were numerous nonsignificant trends in the data that favored psychotherapy and meprobamate patients over patients in the other groups. The impression from the data is that meprobamate and psychotherapy seemed superior to the other treatments. There was overall improvement in all treated groups, contrasting with a lack of improvement in the waiting list group. At the time of follow-up, there were no longer significant differences between groups, but a tendency remained for the psychotherapy group to be the most improved.

Lorr and coworkers[61] studied 150 male patients applying for treatment at Veterans' Administration clinics who were judged suitable for intensive individual psychotherapy. Exclusion criteria included psychiatric hospitalization or psychotherapy during the previous three months, history of neurologic disorder or alcohol addiction, patients who could not discontinue current medication for the study, and patients over fifty-five years of age. Patients were randomly assigned to either psychotherapy or no psychotherapy. Each of these groups was further divided into one of three medication conditions, either chlordiazepoxide (Librium), placebo, or no pill. The no psychotherapy-no pill group was placed on a waiting list. Psychotherapy consisted of fifty-minute interviews once a week. All patients, except the waiting list group, also saw a separate clinician for regulation of medications. All treatments were continued for four weeks. Therapists, which included staff members and trainees, had widely different levels of experience. Ratings were collected from the patients, the therapists, and the physicians on a range of outcome dimensions, including degree of discomfort, level of symptoms, feelings and attitudes, patient self-assessment of social and psychological change, and global ratings of improvement. Both physicians and patients rated active drug treatment significantly more helpful than placebo. For all groups receiving either active medication or placebo, there was no indication that psychotherapy improved outcome. However, the psychotherapy only group did better than the waiting list group, and the improvement in the psychotherapy only patients was in a pattern indistinguishable from patients also receiving active drugs. The results of this study are consistent with the hypothesis of

Frank and associates[30,31] that the early effects of treatment are rather nonspecific and occur in a similar fashion with any serious offer of help made to the patient. Thus all forms of treatment—active medication, psychotherapy, and placebo showed significant benefits to the patient as compared to the waiting list condition. The length of treatment, four weeks, is better suited to demonstrate specific drug effects than specific psychotherapy effects.

Morton[78] studied forty subjects referred by vocational counselors at a university center. Although psychotic patients were excluded, the personal and social adjustment of these subjects was judged by the counselors to be significantly impaired. All subjects were seen for an initial diagnostic interview that systematically explored fourteen areas of adjustment. Subjects were matched, based upon the results of this interview, an incomplete sentence test, and a problem checklist. One member of each matched pair was randomly assigned to psychotherapy, the other to a waiting list control condition. The psychotherapy consisted of three sessions conducted within a three-week period. The therapy utilized the Thematic Apperception Test (TAT) to elicit and elaborate areas of conflict and maladaption. Following the therapy, both experimental and control subjects were reinterviewed by the experimenter and the vocational counselor who had made the original referral. The subjects also completed an incomplete sentence test and a problem checklist. The final interview by the vocational counselor and the experimenter was essentially a survey interview designed to elicit the subject's awareness of any change that had taken place since the initial interview. Initial and terminal interviews were tape recorded. Outcome measures included the incomplete sentence test, the problem checklist, a global rating of adjustment by the vocational counselor, a similar rating by the experimenter, and global ratings made by three independent raters based upon the tape recorded initial and terminal interviews. A significant advantage for the psychotherapy group was found on the pooled global ratings of change and the incomplete sentence test, with a trend favoring psychotherapy on the problem checklist. The vocational counselors judged some improvement in 93 percent of the experimental group, but in only 47 percent of the control subjects.

Fairweather and coworkers[24] investigated the effect of individual and group psychotherapy as components of therapeutic inpatient programs. The ninety-four patients who participated were equally divided among long-term psychotics (over one year previous hospitalization), short-term psychotics (less than one year previous hospitalization), and nonpsychotics. Each of these three diagnostic groups was equally divided into four experimental conditions. Group C was provided an individual work assignment and a plan for posthospital living by a rehabilitation team. Group I had the same treatment as Group C, plus individual psychotherapy two to four hours per week. Group G received the same treatment as Group C, plus group psychotherapy twice per week. Group GG received group psychotherapy and participated in group work situations in the context of a group living environment. Psychotherapy was described in this study as "psychoanalytically oriented." Both experienced and inexperienced therapists were used. Evaluations were made shortly following a patient's transfer to the experimental ward and a second time shortly before he left the ward, either at time of discharge or after six months of treatment. Instruments used included a Ward Behavioral Scale, the MMPI, the Q-Sort, the Holland Vocational Preference Inventory, and the TAT. A follow-up questionnaire was completed six months after the patient left the experimental program either by a person with whom the patient was living or, if the patient was hospitalized, by a staff psychologist. This questionnaire assessed the amount of time employed, the amount of time in the hospital, alcohol use, antisocial behavior, number of friends, verbal communication, general adjustment, problem behavior, and degree of illness.

Patients receiving individual psychotherapy remained hospitalized significantly longer than patients belonging to the other three groups. There was little difference between groups on the MMPI, but interactions between treatment and diagnosis were apparent on some scales. In general, these suggested that long-term psychotic patients did better without psychotherapy, while nonpsychotic and short-term psychotic patients benefited from psychotherapy. There were no significant differences between groups on the Ward Behavior Scale, the Vocational Preference Inventory, and the Q-Sort. The TAT showed no significant differences overall, but again interactions between treatment and diagnosis were present. Group therapy methods were advantageous with nonpsychotic patients, and all psychotherapy approaches benefited short-term psychotics. When long-term psychotics were treated, however, individual psychotherapy showed differential benefit. On the follow-up questionnaire, only the amount of employment significantly differentiated treatment groups. All psychotherapy groups showed a higher percentage of employment than the control condition. This was most pronounced with the individual psychotherapy and the group living conditions.

When all measures are considered, there is the suggestion that nonpsychotic and short-term psychotic patients show more adaptive change with psychotherapy, while long-term psychotic patients change more adaptively without it. The authors use differences in the variance of outcome measures among groups to argue that psychotherapeutic approaches result in more change, both positive and negative, than control treatment. However, the validity of this line of reasoning is highly debatable. With the exception of the follow-up questionnaire, there is a heavy reliance on various psychological tests as outcome criteria. Scores on these tests are not easily translated into meaningful clinical change, thus limiting the value of this study for assessing clinical efficacy of psychotherapy. This investigation illustrates the importance of diagnosis to the impact of psychotherapy, and the need for well-defined patient groups in clinical trials.

The following studies are sufficiently flawed methodologically or have such narrow outcome criteria of questionable clinical relevance that they receive only brief comment. Argyle and coworkers[2] failed to demonstrate a benefit for psychotherapy or social skills training on a social skills rating scale. In this case, the outcome measure is extremely narrow and oriented toward specific goals and theoretical assumptions more appropriate for social skills training than for psychotherapy. Furthermore, its relation to other aspects of patient functioning was undetermined. Shlien and associates[94] demonstrate an advantage for psychotherapy on a Q-Sort procedure. In addition to limiting outcome to a single measure far removed from daily life, this study leaves many important methodological considerations unspecified, including the manner of assignment to treatment condition and comparability of patients in each group. Barron and Leary[4] and Levis and Carrera[60] found no advantage for psychotherapy on MMPI scores. Again, outcome is limited to a test score without assessing the clinical status of patients. Furthermore, the former study made no attempt to use random assignment, whereas the latter did not specify how patients were assigned to groups and provided no characterization of the patient group.

Another series of studies has compared different forms of psychotherapy with one another or with behavioral therapy. Since neither treatment in these studies has established efficacy, and control groups not receiving treatment were not used, they add little to the question of psychotherapy's value. Furthermore, most of them have not persuasively demonstrated an advantage of one approach over the other—behavioral therapy versus traditional psychoehtrapy,* reflective versus leading therapy,[3] and cathartic versus traditional therapy.[81] The one exception in this regard is Siassi,[95] who found a persistent and significant advantage for psychodynamic

*See references 33, 34, 59, 85, 87, and 88.

psychotherapy (in a group of patients judged to be good candidates for psychodynamic treatment) when compared to a range of eclectic, reality-oriented therapies.

Each of the reasonably adequate studies on predominantly psychoneurotic outpatient populations demonstrates some significant benefits for individual psychotherapy. The nature of that gain differs somewhat among studies and does not always persist at follow-up. Furthermore, these studies may underestimate the potential for psychotherapy since therapy is usually of a relatively brief duration (from a few weeks to six months), often used with extremely heterogeneous populations (substantial portions of whom may be ill suited for therapy), and frequently implemented by inexperienced therapists or trainees.

Many process studies (not reviewed here) have begun the critical task of identifying the attributes of patients, therapists, and the treatment process that maximize therapeutic benefits and minimize ineffective or harmful results. These issues are far from resolved. At present, there is persuasive evidence that psychotherapy is efficacious in a substantial number of psychoneurotic and personality disordered patients. The next generation of studies may better define which psychotherapies are most promising with which patients and determine the limits of therapeutic generalizability.

¶ Discussion

Despite widespread interest in dyadic psychotherapy, few carefully designed and controlled studies of its efficacy have been undertaken. This reflects not only the very considerable complexity of the task, but also a prolonged willingness of psychotherapy advocates to ignore the scientific requirement to demonstrate results. Times have changed, and with an ever increasing public and professional assumption that little or no evidence supports psychotherapy as a treatment, the social, professional, and financial base for interpersonal treatments is in jeopardy. Old arguments against scientific scrutiny of psychotherapy are giving way to careful study of some aspects of treatment for some patient types.

The authors have emphasized that medicine is a predominantly observational science, hence the experimental results reviewed in this chapter must be clarified and extended by the rich body of case reports and surveys. Concerning several classes of illness (affective, psychosomatic, psychoneurotic, and personality disorders), there is a confluence of evidence supporting the efficacy of psychotherapy. The many reports of patients being helped by psychotherapy are supported by the majority of methodologically adequate studies examining the same issues under the controlled circumstances these reports lack. The most important present questions are: (1) which subgroup of patients in these categories are most likely to benefit, and which (if any) may suffer a detrimental effect; (2) precisely which psychotherapies are effective, and what training, experience, and setting are required for their optimal administration; (3) under what circumstances is psychotherapy superior to alternative treatments, additive and/or synergistic with other treatments, and inferior to them; (4) what are the cogent patient/therapist matching considerations; and (5) upon what aspects or mechanisms of psychotherapy is benefit dependent. There is considerable evidence that psychotherapy works, but why it works is still argued from doctrine rather than data.

The efficacy of dyadic psychotherapy in schizophrenia has been more difficult to establish, and both the observational and experimental data are inconsistent. Some anecdotal reports describe considerable success with psychotherapeutic strategies, but many clinicians are unimpressed with their own experiences treating schizophrenics with psychotherapy. The controlled studies have been in hospital settings and have failed to demonstrate any strong effect. One exception, in addition to observational data, suggests that if such therapy can be helpful it requires selected patients and experienced therapists. The term schizophrenia covers an extraordinarily broad range of psychopathology comprising many aspects of human functioning. Many psychotherapeutic innovations will require study before the field can securely define what is therapeutic for which aspect of psychopathology in what phase of illness, and so forth. The vigorous pursuit of these studies is mandated by the following considerations: (1) the potentials of interpersonal therapeutic approaches for schizophrenic patients have barely been touched by scientific study (hence any closure on the issue is premature); (2) no alternative treatment provides definitive relief to the schizophrenic patient, especially in the long-term deterioration of interpersonal and intrapsychic functioning; (3) while antipsychotic drugs are clearly effective in reducing positive symptoms and relapse rates, it is urgent to determine if combining psychotherapeutic approaches with drug treatment can enhance benefits and reduce drug exposure and the risk of tardive dyskinesia; and (4) schizophrenia is so common and severe an illness that the broadest range of etiologic and treatment research must be encouraged.

If the case for psychotherapy can be made with at least moderate persuasive power for many diagnoses, why is it currently under such severe attack? Medicine is an applied science. Consequently, patients seeking relief from their suffering require treatment based on the best available information even in the absence of certainty. The requirement for scientific rigor is a laudable development in psychiatry, but it can result in a distorted perspective that precludes reasonable action based upon significant but imperfect knowledge. In spite of its limitations, the data on psychotherapy indicate benefits for many patients, often where there is no alternative treatment. In medicine, many treatments are accepted despite the lack of rigorous experimental validation because they possess an intuitive rationale persuasive to professionals and laymen alike. Taking a thorough medical history and coronary bypass surgery are examples. Other maneuvers may be so well grounded on indisputable facts of anatomy, physiology, or chemistry that they are persuasive, at least to those educated in the field. Only 10 to 20 percent of medical procedures are based on evidence from controlled studies.[43] Psychotherapy is handicapped in this regard. Its mode of action often seems improbable and nearly mystical to many. Lacking a well-established basic science of mental and behavioral phenomena, psychotherapeutic strategies are all too often justified by appeals to highly abstract and debatable models that are acceptable only to some subset of professionals. While some of these difficulties are intrinsic and unavoidable, many psychotherapeutic interventions can be related to less abstract and relatively atheoretical concepts. Thus, it seems reasonable that a person with impaired patterns of relating to others can be helped by ongoing interaction with an individual trained to recognize those patterns, help the patients appreciate their nature and consequences, and assist in developing more adaptive modes. Such formulations not only make psychotherapy more intuitively acceptable but can articulate clinically tangible hypotheses about how therapy operates that can be tested by workers from various theoretical backgrounds.

While some experts may argue that only rigorously validated treatments should be employed, the authors believe that the intuitive rationale and the danger and invasiveness of a treatment must be considered. Hemodialysis, for example, has no intuitive relationship to schizophrenic functioning, and it is a highly invasive procedure with considerable risk. Thus, it is reasonable to require experimental demonstration of efficacy before hemodialysis becomes an accepted treatment for schizophrenia. Psychotherapy, on the other hand, is not invasive in the same sense, is relatively safe, and can be intuitively related to many problems of the mentally ill. The authors believe clinicians are justified in using such a treatment until subsequent research definitively resolves the question of its efficacy.

Even when careful research in psychotherapy is carried out, a major impediment to confidence in the results is that the personality of the therapist is extremely important in ways that are difficult to define. Most medical treatments can be specified in terms of a technique with little reference to the provider beyond an assurance of a basic level of competence. For example, the ability to perform a spleenectomy requires a set of skills that can be rather clearly specified. Thus, it is relatively easy to determine that a residency in surgery provides adequate training in those skills. Consequently, although all surgeons are not equally skilled, the public is justified in assuming that physicians completing designated training programs have a basic competence in the surgical treatment of illness. In assessing the efficacy of spleenectomy, one is able to examine the outcome of cases treated by a group of trained surgeons and have reasonable confidence that the results were generalizable to the surgical profession at large. For psychotherapy, these crisp distinctions are lacking. Not only is it difficult to specify a precise "correct" technique for a given patient, but it is generally assumed that the personal attributes of the therapist are extremely important in determining outcome quite apart from technique. Thus, in contrast to the example of surgery, there are three important gaps in our knowledge. First, it is not known what aspects of technique are specifically therapeutic for a given patient. The future generation of process studies should help clarify this issue. Second, there are no criteria for determining the personal attributes a person must have to be an effective therapist for a specific patient. And third, it is consequently difficult to know if a given training program provides the necessary education and experience to ensure that its graduates are competent in psychotherapy. The resulting problem for research is that a given study may say more about the efficacy of psychotherapy as practiced by a specific person(s) than about psychotherapy as a generalizable technique of treatment.

The professional psychotherapist finds himself surrounded by lesser trained and untrained self-appointed experts. The public must have some consistent basis for placing trust in the health professions, and third-party payers justly demand guidelines for determining when psychotherapy is a legitimate treatment of illness by qualified health professionals. While not offering a detailed proposal, the authors suggest the following guidelines regarding this issue.

1. As health care practitioners, it is perilous to claim expertise in areas that cannot be integrated within the framework of a broad medical model encompassing sociologic, psychologic, and biologic data relevant to health and disease. The scope of this chapter was limited to psychosocial intervention in disease states, but psychotherapy within a medical model would also include prevention. The authors do not consider the plethora of human interventions into normal functioning (troubled or not), whether these be traditionally valued (for example, academic counseling, pastoral counseling) or the self-actualizing and personal growth movements whose popularity rose during the 1970s, to be the legitimate domain of the health professions.

2. Expertise in psychotherapy is often assumed of graduates of programs in medicine, psychology, and social work. Sociology and psychology are taught in these fields, but programs vary tremendously and little or no training in psychotherapy may be provided. The public deserves some cross-disciplinary collaboration in establishing minimal requirements for the training of those practicing psychotherapy.

3. To the extent that psychotherapy is offered to patients (persons with a diagnosable illness), it should be provided in a medical framework. It is no longer justifiable to treat with a limited repertoire those with mental illness. The depressed patient may need psychotherapy (not necessarily by a physician), but clinical judgment concerning pharmacotherapy, ECT, genetic counseling, family or group therapy, occupational guidance, and other considerations must be continuously integrated in clinical care. Treatment

of psychiatric illness within a medical framework can no longer permit exclusive and doctrinaire treatment approaches, and collaboration between physician and psychotherapist should not merely be "medical screening."

The confusion as to the medical legitimacy of mental health efforts has resulted in an unacceptable compromise of partial financial support for a diffuse array of alleged treatments. To counterbalance the possibility that at times payments are made for psychological interventions of questionable legitimacy, the truly mentally ill suffer from inhumane restrictions on the health care they receive. If the mental health profession fulfills its responsibility to clearly demarcate the domain of appropriate treatments for legitimate mental illnesses, there is no scientific rationale for excluding psychiatric patients from full access to health care. The poor prognosis cancer patient is not offered thirty days maximum inpatient care and a limited number of outpatient visits, but the poor prognosis schizophrenic patient is forced out of the public's consciousness and pocketbook as quickly as possible. A pragmatic as well as a humanitarian rationale applies here. Health care planners have been preoccupied with direct costs and have ignored the social and financial burdens associated with unemployment, inability to manage a household, and so forth. If the financiers of health care paid social security disability, welfare, and unemployment benefits, they might be more eager to assure the adequacy of inpatient and outpatient treatment for mental illness. Various psychotherapies seem particularly likely to improve functioning in these areas of "hidden" health costs.

¶ Conclusions

Considering the full range of evidence, there is a strong case that the core tradition of psychotherapy for adults suffering from common disorders offers clinical benefits for many patients. With the lack of comprehensive alternatives, this justifies continued support for psychotherapy at this time. Yet, there is no room for complacency. Support in the future requires extensive efforts to determine what elements of therapy are effective and with which patients. Process studies can provide useful hypotheses to help focus and specify future outcome studies. Since controlled studies are most meaningful when imposed on an extensive body of clinical observation, the traditional core of psychotherapy with its wealth of information should be the top priority of such focused investigation. In addition, this is the most prevalent type of psychotherapy currently being practiced. Innovative techniques are certainly to be encouraged, but validating nearly one hundred years of clinical effort deserves our most urgent attention.

¶ Bibliography

1. ALEXANDER, F., *Psychosomatic Medicine: Its Principles and Applications.* New York: W.W. Norton, 1950.
2. ARGYLE, M., et al. "Social Skills Training and Psychotherapy: A Comparative Study," *Psychological Medicine,* 4 (1974): 435–443.
3. ASHBY, J.D., et al. "Effects on Clients of a Reflective and a Leading Type of Psychotherapy," *Psychology Monograph,* 71, no. 453: 1957.
4. BARRON, F., and LEARY, T. F. "Changes in Psychoneurotic Patients With and Without Psychotherapy," *Journal of Consulting Psychology,* 19 (1955): 239–245.
5. BERGIN, A. E. "The Evaluation of Therapeutic Outcomes," in Bergin, A. E. and Garfield, S. L., eds., *Handbook of Psychotherapy and Behavior Change.* New York: John Wiley, 1971, pp. 217–270.
6. ———, and LAMBERT, M. J. "The Evaluation of Therapeutic Outcomes," in Garfield, S. L. and Bergin, A. E., eds., *Handbook of Psychotherapy and Behavioral Change: An Empirical Analysis.* 2nd ed., New York: John Wiley, 1978, pp. 139–190.

7. BERGIN, A. E., and SUINN, R. M. "Individual Psychotherapy and Behavior Therapy," *Annual Review of Psychology,* 26 (1975): 509–556.
8. BOOKHAMMER, R. S. "A Five-year Clinical Follow-up of Schizophrenics Treated by Rosen's Direct Analysis," *American Journal of Psychiatry,* 123 (1966): 602–604.
9. BRILL, N. Q., et al. "Controlled Study of Psychiatric Outpatient Treatment," *Archives of General Psychiatry,* 10 (1964): 581–595.
10. BUTCHER, J. N., and KOSS, M. P. "Research on Brief and Crisis-Oriented Psychotherapies," in Garfield, S. L. and Bergin, A. E., eds., *Handbook of Psychotherapy and Behavioral Change: An Empirical Analysis.* 2nd ed., New York: John Wiley, 1978, pp. 725–768.
11. CARPENTER, W. T., and HEINRICHS, D. W. "The Role for Psychodynamic Psychiatry in the Treatment of Schizophrenic Patients," in Strauss, J., ed., *Psychotherapy of Schizophrenia.* New York: Plenum Press, 1980, pp. 231–247.
12. CARTWRIGHT, D. S. "Effectiveness of Psychotherapy: A Critique of the Spontaneous Remission Argument," *Journal of Counseling Psychology,* 20 (1956): 403–404.
13. ———. "Annotated Bibliography of Research and Theory Construction in Client-Centered Therapy," *Journal of Counseling Psychology,* 4 (1957): 82–100.
14. CRISTOL, A. "Studies of Outcome of Psychotherapy," *Comprehensive Psychiatry,* 13 (1972): 189–200.
15. DAVIS, J. M., et al. "Important Issues in the Drug Treatment of Schizophrenia," *Schizophrenia Bulletin,* 6 (1980): 70–87.
16. DI GIUSEPPE, R., et al. "A Review of Rational Emotive Psychotherapy Outcome Studies," *Counseling Psychologist,* 7 (1977): 64–72.
17. DITTMANN, A. T. "Psychotherapeutic Processes," in Farnsworth, P. R., McNemar, O., and McNemar, Q., eds., *Annual Review of Psychology,* 16. Palo Alto: Annual Reviews, 1966, pp. 51–78.
18. EVANS, F. J. "The Placebo Control of Pain: A Paradigm for Investigating Non-specific Effects in Psychotherapy," in Brady, J. P., et al. eds., *Psychiatry: Areas of Promise and Advancement.* New York: Spectrum, 1977, pp. 129–136.
19. EYSENCK, H. J. "The Effects of Psychotherapy: An Evaluation," *Journal of Consulting Psychology,* 16 (1952): 319–324.
20. ———. *Handbook of Abnormal Psychology.* London: Pitman, 1960.
21. ———. "The Effects of Psychotherapy," *International Journal of Psychiatry,* 1 (1965): 97–178.
22. ———. *The Effects of Psychotherapy.* New York: International Science Press, 1966.
23. ———. "The Non-Professional Psychotherapist," *International Journal of Psychiatry,* 3 (1967): 150–153.
24. FAIRWEATHER, G. W., et al. "Relative Effectiveness of Psychotherapeutic Programs," *Psychology Monograph,* 74, no. 492: 1960.
25. FEINSILVER, D. B., and GUNDERSON, J. G. "Psychotherapy for Schizophrenics: Is It Indicated? A Review of the Relevant Literature," *Schizophrenia Bulletin,* 6 (1972): 11–23.
26. FISKE, D. "Findings on Outcome in Psychotherapy. The Shaky Evidence is Slowly Put Together," *Journal of Consulting and Clinical Psychology,* 37 (1971): 314–315.
27. FRANK, J. D. "What We Don't Know About Psychotherapy," *Today in Psychiatry.* Abbott Laboratories.
28. ———. *Persuasion and Healing: A Comparative Study of Psychotherapy.* New York: Schocken, 1973.
29. ———. "Therapeutic Components of Psychotherapy: A 25 Year Progress Report of Research," *Journal of Nervous and Mental Disease,* 159 (1974): 325–342.
30. ———, et al. "Why Patients Leave Psychotherapy," *Archives of Neurology and Psychiatry,* 77 (1957): 283–299.
31. FRANK, J. D., et al. "Patients' Expectancies and Relearning as Factors Determining Improvement in Psychotherapy," *American Journal of Psychiatry,* 115 (1959): 961–968.
32. GARNER, H. H. "Psychotherapy Applied to Comprehensive Medical Practice," *Illinois Medical Journal,* 135 (1969): 289–295.
33. GELDER, M. G., and MARKS, I. M. "Severe Agoraphobia: A Controlled Prospective Trial of Behaviour Therapy," *British Journal of Psychiatry,* 112 (1966): 309–319.
34. GELDER, M. G., et al. "Desensitization and Psychotherapy in the Treatment of Phobic States: A Controlled Inquiry," *British Journal of Psychiatry,* 113 (1967): 53–73.

35. GLEN, A. I. M. "Psychotherapy and Medical Treatment for Duodenal Ulcer Compared Using the Augmented Histamine Test," *Journal of Psychosomatic Research,* 12 (1968): 163–169.
36. GODBOLE, A., and VERINIS, J. S. "Brief Psychotherapy in the Treatment of Emotional Disorders in Physically Ill Geriatric Patients," *Gerontologist,* 14 (1974): 143–148.
37. GOLDSTEIN, M. J., et al. "Drug and Family Therapy in the Aftercare of Acute Schizophrenics," *Archives of General Psychiatry,* 35 (1978): 1169–1177.
38. GRACE, W. J., et al. "The Treatment of Ulcerative Colitis," *Gastroenterology,* 26 (1954): 462–468.
39. GRINSPOON, L., et al. "Long-Term Treatment of Chronic Schizophrenia," *International Journal of Psychiatry,* 4 (1967): 116–128.
40. GRINSPOON, L., et al. *Schizophrenia: Pharmacotherapy and Psychotherapy.* Baltimore: Williams & Wilkins, 1972.
41. GRUEN, W. "Effects of Brief Psychotherapy During the Hospitalization Period on the Recovery Process in Heart Attacks," *Journal of Consulting and Clinical Psychology,* 43 (1975): 223–232.
42. GUNDERSON, J. G. *Drugs and Psychosocial Treatment of Schizophrenia Revisited: The Effects of Psychosocial Variables on Outcome of Schizophrenia.* Fifth International Symposium on Psychotherapy. Oslo, Norway, August 13–17, 1975.
43. HERRINGTON, B. S. "Klerman Weighs Role in Therapy Efficacy Study," *Psychiatric News,* 15 (1980): 10.
44. HOGARTY, G. E., and GOLDBERG, S. C. "Drug and Sociotherapy in the Aftercare of Schizophrenic Patients," *Archives of General Psychiatry,* 28 (1973): 54–64.
45. HOGARTY, G. E., et al. "Drug and Sociotherapy in the Aftercare of Schizophrenic Patients. II. Two-Year Relapse Rates," *Archives of General Psychiatry,* 31 (1974): 603–608.
46. HOGARTY, G. E., et al. "Drug and Sociotherapy in the Aftercare of Schizophrenic patients. III. Adjustment of Nonrelapsed Patients," *Archives of General Psychiatry,* 31 (1974): 609–618.
47. HOLLON, S., and BECK, A. T. "Psychotherapy and Drug Therapy: Comparisons and Combinations," in Garfield, S. L. and Bergin, A. E., eds., *Handbook of Psychotherapy and Behavioral Change: An Empirical Analysis,* 2nd ed., New York: John Wiley, 1978, pp. 437–490.
48. IMBER, S. D., et al. "Improvement and Amount of Therapeutic Contact: An Alternative to the Use of No-Treatment Controls in Psychotherapy," *Journal of Consulting Psychology,* 21 (1957): 309–315.
49. IMBER, S.D., et al. "A Ten-Year Follow-up Study of Treated Psychiatric Outpatients," in Less, S., ed., *An Evaluation of the Results of the Psychotherapies.* Springfield, Ill.: Charles C Thomas, 1968, pp. 70–81.
50. KARON, B. P. and VANDENBOS, G. R. "Experience, Medication and the Effectiveness of Psychotherapy with Schizophrenics: A Note on Drs. May & Tuma's Conclusion," *British Journal of Psychiatry,* 116 (1970): 427–428.
51. ———. "The Consequences of Psychotherapy for Schizophrenic Patients," *Psychotherapy: Theory, Research, and Practice,* 9 (1972): 111–119.
52. KELLNER, R. "The Evidence in Favour of Psychotherapy," *British Journal of Medical Psychology,* 40 (1967): 341–358.
53. ———. "Psychotherapy in Psychosomatic Disorders: A Survey of Controlled Outcome Studies," *Archives of General Psychiatry,* 32 (1975): 1021–1028.
54. KLERMAN, G. L., et al. "Treatment of Depression by Drugs and Psychotherapy," *American Journal of Psychiatry,* 131 (1974): 186–191.
55. KOEGLER, R. R. "Brief-contact Therapy and Drugs in Outpatient Treatment," in Wayne, G. J. and Koegler, R. R., eds., *Emergency Psychiatry and Brief Therapy.* Boston: Little, Brown, 1966, pp. 139–154.
56. KOEGLER, R. R., and BRILL, N. Q. *Treatment of Psychiatric Outpatients.* New York: Appleton-Century-Crofts, 1967.
57. LAMBERT, M. J. "Spontaneous Remission in Adult Neurotic Disorders: A Revision and Summary," *Psychological Bulletin,* 83 (1976): 107–119.
58. LAMBERT, M. J., BERGIN, A. E., and COLLINS, J. L. "Therapist-induced Deterioration in Psychotherapy," in Gurman, A. S., and Razin, A. M., eds., *Effective Psychotherapy: A Handbook of Research.* New York: Pergamon Press, 1977, pp. 452–481.
59. LEVENE, H., et al. "A Training and Research Program in Brief Psychotherapy,"

American Journal of Psychotherapy, 26 (1972): 90–100.

60. LEVIS, D. J., and CARRERA, R. "Effects of Ten Hours of Implosive Therapy in the Treatment of Outpatients," *Journal of Abnormal Psychology,* 72 (1967): 504–508.
61. LORR, M., et al. "Early Effects of Chlordiazepoxide (Librium) Used With Psychotherapy," *Journal of Psychiatric Research,* 1 (1963): 257–270.
62. LUBORSKY, L. "Another Reply to Eysenck," *Psychological Bulletin,* 78 (1972): 406–408.
63. LUBORSKY, L., and SPENCE, D. P. "Quantitative Research on Psychoanalytic Therapy," in Garfield, S. L. and Bergin, A. E., eds., *Handbook of Psychotherapy and Behavioral Change: An Empirical Analysis,* 2nd ed. New York: John Wiley, 1978, pp. 331–368.
64. LUBORSKY, L., et al. "Comparative Studies of Psychotherapies: Is It True That 'Everybody Has Won and All Must Have Prizes'?" *Archives of General Psychiatry,* 32 (1975): 995–1008.
65. MALAN, D. H. "The Outcome Problem in Psychotherapy Research: A Historical Review," *Archives of General Psychiatry,* 29 (1973): 719–729.
66. MALAN, D. H., et al. "Psychodynamic Changes in Untreated Neurotic Patients. II. Apparently Genuine Improvements," *Archives of General Psychiatry,* 32 (1975): 110–126.
67. MARKS, J., et al. "Reinforcement versus Relationship Therapy for Schizophrenics," *Journal of Abnormal Psychology,* 73 (1968): 397–402.
68. MAY, P. R. A. *Treatment of Schizophrenia.* New York: Science House, 1968.
69. ———. "Psychotherapy and Ataraxic Drugs," in Bergin, A. E. and Garfield, S. L., eds., *Handbook of Psychotherapy and Behavior Change.* New York: John Wiley, 1971, pp. 495–540.
70. ———. "Schizophrenia: Overview of Treatment Methods," in Friedman, A. M., Kaplan, H. I., and Sadock, B. J., eds., *Comprehensive Textbook of Psychiatry,* vol. 2. Baltimore: Williams & Wilkins, 1975, pp. 923–938.
71. ———, and TUMA, A. H. "The Effect of Psychotherapy and Stelazine on Length of Hospital Stay, Release Rate and Supplemental Treatment of Schizophrenic Patients," *Journal of Nervous and Mental Disorders,* 139 (1964): 362–369.
72. ———. "Treatment of Schizophrenia: An Experimental Study of Five Treatment Methods," *British Journal of Psychiatry,* 111 (1965): 503–510.
73. ———. "Methodological Problems in Psychotherapy Research: Observations on the Karon-VandenBos Study of Psychotherapy and Drugs in Schizophrenia," *British Journal of Psychiatry,* 117 (1970): 569–570.
74. MAY, P. R. A., et al. "Schizophrenia—A Follow-up Study of Results of Treatment," *Archives of General Psychiatry,* 33 (1976): 474–478.
75. MAY, P. R. A., et al. "Schizophrenia—A Follow-up Study of Results of Treatment: II. Hospital Stay Over Two to Five Years," *Archives of General Psychiatry,* 33 (1976): 481–486.
76. MELTZOFF, J., and KORNREICH, M. *Research in Psychotherapy.* New York: Atherton Press, 1970.
77. MESSIER, M., et al. "A Follow-Up Study of Intensively Treated Chronic Schizophrenic Patients," *American Journal of Psychiatry,* 125 (1969): 1123–1127.
78. MORTON, R. B. "An Experiment in Brief Psychotherapy," *Psychology Monograph,* 69, no. 386: 1955.
79. MOSHER, L. R., and KEITH, S. J. "Research on the Psychosocial Treatment of Schizophrenia: A Summary Report," *American Journal of Psychiatry,* 136 (1979): 623–631.
80. ———. "Psychosocial Treatment: Individual, Group, Family, and Community Support Approaches," *Schizophrenia Bulletin,* 6 (1980): 10–41.
81. NICHOLS, M. "Outcome of Brief Cathartic Psychotherapy," *Journal of Consulting and Clinical Psychology,* 42 (1974): 403–410.
82. O'CONNOR, J. F., et al. "An Evaluation of the Effectiveness of Psychotherapy in the Treatment of Ulcerative Colitis," *Annals of Internal Medicine,* 60 (1964): 587–602.
83. ORLINSKY, D. E., and HOWARD, K. I., "The Relation of Process to Outcome in Psychotherapy," in Garfield, S., and Bergin, A., eds., *Handbook of Psychotherapy and Behavior Change: An Empirical Analysis,* 2nd ed. New York: John Wiley, 1978, pp. 283–330.
84. PARLOFF, M., et al. *Assessment of Psychoso-*

cial Treatment of Mental Disorders: Current Status and Prospects. National Technical Information Service (PB287640), Washington, D.C.: Government Printing Office, 1980.

85. PATTERSON, V., et al. "Treatment and Training Outcomes With Two Time-Limited Therapies," *Archives of General Psychiatry,* 25 (1971): 161–167.
86. PATTERSON, V., et al. "A One-Year Follow-Up of Two Forms of Brief Psychotherapy," *American Journal of Psychotherapy,* 31 (1977): 76–82.
87. PIERLOTT, R., and VINCK, J. "Differential Outcome of Short-Term Dynamic Psychotherapy and Systematic Desensitization in the Treatment of Anxious Out-Patients. A Preliminary Report," *Psychologica Belgica,* 18 (1978): 87–98.
88. RACHMAN, S. *The Effects of Psychotherapy.* Oxford: Pergamon Press, 1971.
89. ———. "The Effects of Psychological Treatment," in Eysenck, H., ed., *Handbook of Abnormal Psychology.* New York: Basic Books, 1973, pp. 805–861.
90. ROGERS, C. R., et al. *The Therapeutic Relationship and Its Impact.* Madison: University of Wisconsin Press, 1967.
91. RUSH, A. J., et al. "Comparative Efficacy of Cognitive Therapy and Pharmacotherapy in the Treatment of Depressed Outpatients," *Cognitive Therapy and Research,* 1 (1977): 17–37.
92. RUSH, A. J., and BECK, A. "Adults with Affective Disorders," in Hersen, M., ed., *Behavior Therapy in the Psychiatric Setting.* Baltimore: Williams & Wilkins, 1978, pp. 286–330.
93. SCHONECKE, O., and SCHUFFEL, W. "Evaluation of Combined Pharmacological and Psychotherapeutic Treatment in Patients with Functional Abdominal Disorders," *Psychotherapy and Psychosomatics,* 26 (1975): 86–92.
94. SHLIEN, J. M. "Time-Limited Psychotherapy: An Experimental Investigation of Practical Values and Theoretical Implications," *Journal of Counseling Psychology,* 4 (1975): 318–322.
95. SIASSI, I. "A Comparison of Open-Ended Psychoanalytically-Oriented Psychotherapy with Other Therapies," *Journal of Clinical Psychiatry,* 40 (1979): 25–32.
96. SLOANE, R. B., et al. *Psychotherapy versus Behavior Therapy.* Cambridge: Harvard University Press, 1975.
97. SLOANE, R. B., et al. "Short-Term Analytically Oriented Psychotherapy versus Behavior Therapy," *American Journal of Psychiatry,* 132 (1975): 373–377.
98. SMALL, L. *The Briefer Psychotherapies.* New York: Bruner/Mazel, 1971.
99. STRUPP, H. H. "The Outcome Problem in Psychotherapy Revisited," *Psychotherapy,* 1 (1963): 1–13.
100. ———. "The Outcome Problem in Psychotherapy: A Rejoinder," *Psychotherapy,* 1 (1964): 101.
101. ———. "Specific vs. Nonspecific Factors in Psychotherapy and the Problem of Control," in Strupp, H. H., ed., *Psychotherapy: Clinical, Research and Theoretical Issues.* New York: Jason Aronson, 1973, pp. 103–121.
102. ———, and BERGIN, A. E. "Some Empirical and Conceptual Bases for Coordinated Research in Psychotherapy: A Critical Review of Issues, Trends and Evidence," *International Journal of Psychiatry,* 7 (1969): 18–90.
103. STRUPP, H. H., HADLEY, S. W., and GOMES-SCHWARTZ, B. *Psychotherapy for Better or Worse: An Analysis of the Problem of Negative Effects.* New York: Jason Aronson, 1977.
104. UHLENHUTH, E. H., et al. "Combined Pharmacotherapy and Psychotherapy: Controlled Studies," *Journal of Nervous and Mental Disorders,* 148 (1969): 52–64.
105. WEISSMAN, M. "The Psychological Treatment of Depression. Evidence for the Efficacy of Psychotherapy Alone, in Comparison with, and in Combination with Pharmacotherapy," *Archives of General Psychiatry,* 36 (1979): 1261–1269.
106. ———, et al. "Treatment Effects on the Social Adjustment of Depressed Patients," *Archives of General Psychiatry,* 30 (1974): 771–778.
107. WEISSMAN, M., et al. "Follow-Up of Depressed Women after Maintenance Treatment," *American Journal of Psychiatry,* 133 (1976): 757–760.
108. WEISSMAN, M., et al. "The Efficacy of Drugs and Psychotherapy in the Treatment of Acute Depressive Episodes," *American Journal of Psychiatry,* 136 (1979): 555–558.

CHAPTER 31

ADVANCES IN THE PREVENTION AND TREATMENT OF MENTAL RETARDATION

Frank J. Menolascino and Fred D. Strider

IN THIS CHAPTER current approaches to the prevention and treatment of mental retardation will be reviewed with a focus on specific psychiatric treatment and intervention appropriate to the combined syndromes of mental retardation and mental illness. In a concluding section, the authors will indicate their views regarding the essential elements of comprehensive approaches and treatment.

¶ Introduction

The model of *primary* (prevention of the appearance of a disorder), *secondary* (very early diagnosis, effective treatment, and return of the person to a normative state), and *tertiary* (minimization of the remaining handicaps and return of the person to as high a level of functioning as possible) prevention will be utilized to review the currently available and possible future preventive approaches in each of these three dimensions. Although this three-step approach may seem simple, it is actually quite complex because there are over 350 causes of mental retardation. Major prevention programs have been successful in certain states: for example, Illinois has been successful in the area of screening and prevention of lead poisoning; California has energetically encouraged public education concerning mental retardation; Connecticut has mounted an excellent program to prevent Rh blood incompatibility; and Massachusetts has a well-established program for discerning a number of preventable forms of inborn errors of metabolism. The interconnections between different levels of prevention and ongoing service delivery, and possible linkages between the different human service systems that serve the mentally retarded will also be discussed. This section will examine the continuity between prevention and clinical treatment of the physical and psychiatric symptoms of mental retardation.

¶ Current Status of Prevention Efforts

A recent national report entitled "Preventing Mental Retardation: More Can Be Done"[20] noted that, at the national level, the current "state of the art" in the prevention of mental retardation is fragmented and, while promising much, delivers little. There are always major lags between the evolution of new knowledge of mental retardation and its direct application in the field. The gap is currently being narrowed by advocacy, public information programs, and such exceptional programs as the regionalization and expansion of metabolic screening of newborns for the inborn errors of metabolism. However, federal funds have frequently been withdrawn from excellent pilot programs in prevention, and the states have not always been able to maintain enough programs even though these preventive efforts are cost-effective and the lack of money "up front" results in increased costs and unnecessary human suffering later on.

In the United States, there are eight areas of activity in mental retardation in which preventive efforts are currently being implemented—albeit with some major gaps: (1) comprehensive prenatal care (including recent increased attention to "high risk" pregnancies and prenatal nutrition); (2) infectious diseases (both prenatal and postnatal); (3) chromosome disorders; (4) metabolic diseases (such as the inborn errors of metabolism); (5) internal (for example, Rh blood incompatibility) and external (for example, lead poisoning) intoxications; (6) adverse early childhood experiences within the family; (7) childhood accidents; and (8) other approaches such as screening for neural tube disorders. Each of these areas will be briefly reviewed.

Comprehensive Prenatal Care

The dearth of comprehensive prenatal services is a significant national problem. Low birth weight and prematurity are often associated with the symptoms of mental retardation, epilepsy, and cerebral palsy. A major issue in prenatal care is malnutrition during pregnancy, which may cause infant death or permanent brain impairment.[8] Excessive maternal alcohol consumption or drug use raise the risk of having an abnormal infant to almost 44 percent.[29]

After birth, the vulnerability of the developing brain to permanent damage as a result of inadequate nutrition has been clearly implicated.[43] This susceptibility has been observed in both experimental animals and human beings.[23] Nutritional supplement programs must include pregnant and lactating women, infants, and preschool children. Since the brain is still very rapidly developing during the first six years of life, one of the great advances in national health care in the last quarter century has resulted from improved general nutrition; this advancement must be vigorously extended to the poor.[20]

National and state statistics[20] indicate that many women still receive insufficient prenatal care, even though federal and state programs have been established to reach persons in economically depressed areas who might not otherwise receive such services. The extent of need for additional services is unknown because neither federal nor state agencies have adequately analyzed the extent of current prenatal care needs, the areas of greatest needs, or the effect of existing prenatal programs.

Following conception, comprehensive prenatal care can help prevent low birth weight and prematurity, which are in turn directly associated with mental retardation.

This is the point at which early detection of and direct treatment intervention for chronic conditions (for example, diabetes mellitus, essential hypertension, hypothyroidism,) in the mother are critical. The identification of "high risk" pregnancy, in conjunction with the rapid initiation of specialized obstetrical care, can save lives and minds, and can also save dollars in long-term care. The Vermont Association for Retarded Citizens[52] recently completed a study that concluded that a comprehensive perinatal program should be composed of the elements (noted in figure 31–1) that delineate the mechanics of dealing with "at-risk" pregnancies.

The components indicated by this flow chart are neither new nor are they expensive to initiate. This approach encompasses putting together the already known basic components of a successful prenatal program. As to linkages, the Vermont plan for perinatal care recommended that the perinatal program be regionally based throughout the state via a strong university and department of health collaboration. Although Vermont is a small state, it is interesting that it felt that its projected perinatal program could build on current resources by adding—over a four-year period—the following components to bring the program to fruition: (1) family planning and counseling services; (2) genetic screening and counseling; (3) maternal/fetal transportation; (4) high-risk obstetrical unit; (5) intensive care of the newborn; and (6) program evaluation

A greater level of awareness by family physicians and family planning centers of the importance of high-risk pregnancy identification is needed. A key factor is the availability of the mother's obstetrical history, which can yield the necessary information for a prompt referral to regional diagnostic programs for the further detection of, and direct dealing with, high-risk pregnancies. Increased funding and resources will be needed to carry out these prenatal efforts in our country, and the system (or individual programs) to be utilized can be a series of specialized units in regional general hospitals or a medical service system that is integrated into current generic health services.

•

TEENAGE PREGNANCIES

The alarming number (20 percent of all live births and 26 percent of low-birth-weight infants) of teenage pregnancies directly contributes to increased numbers of premature infants and infants who are gravely "at risk" to develop mental retardation. Adolescent pregnancies increase health risks, such as higher incidences of toxemia, anemia, prolonged labor, and, for the infant, increased incidences of low birth weight and related signs and symptoms of mental retardation.

Teenage pregnancy is a pressing social problem compounded by ignorance, immaturity, illness, illegitimacy, and poverty. It costs billions of dollars each year in welfare support and medical costs. It has been estimated that about half of the women currently on the welfare roles had their first child during adolescence. A significant number of teenagers are not adequately informed about, or disregard, contraception. Every year, more than one million teenage girls, most of them single, become pregnant; many are younger than fifteen. These pregnancies result in about 600,000 live births a year. The others are terminated by miscarriages (100,000) and abortions (300,000).[49] Teenage mothers give birth to a disproportionate number of premature babies born with low birth weight (under 2,500 grams): 16.5 percent for women under age fifteen; 10.1 percent for women age fifteen to nineteen. Among very small premature babies (under 1,500 grams), the incidence of mental retardation in those who survive is about 26.3 percent compared to only 1.6 percent for full-term babies.

Teenage motherhood brings many complications. Many young girls are ill informed of medical care needs and do not receive adequate prenatal care or nutrition. Others, especially those in their early teens, ages eleven to fourteen, are not physically mature and are often unable to bear the stress of

Figure 31-1 Perinatal Program Flow Chart

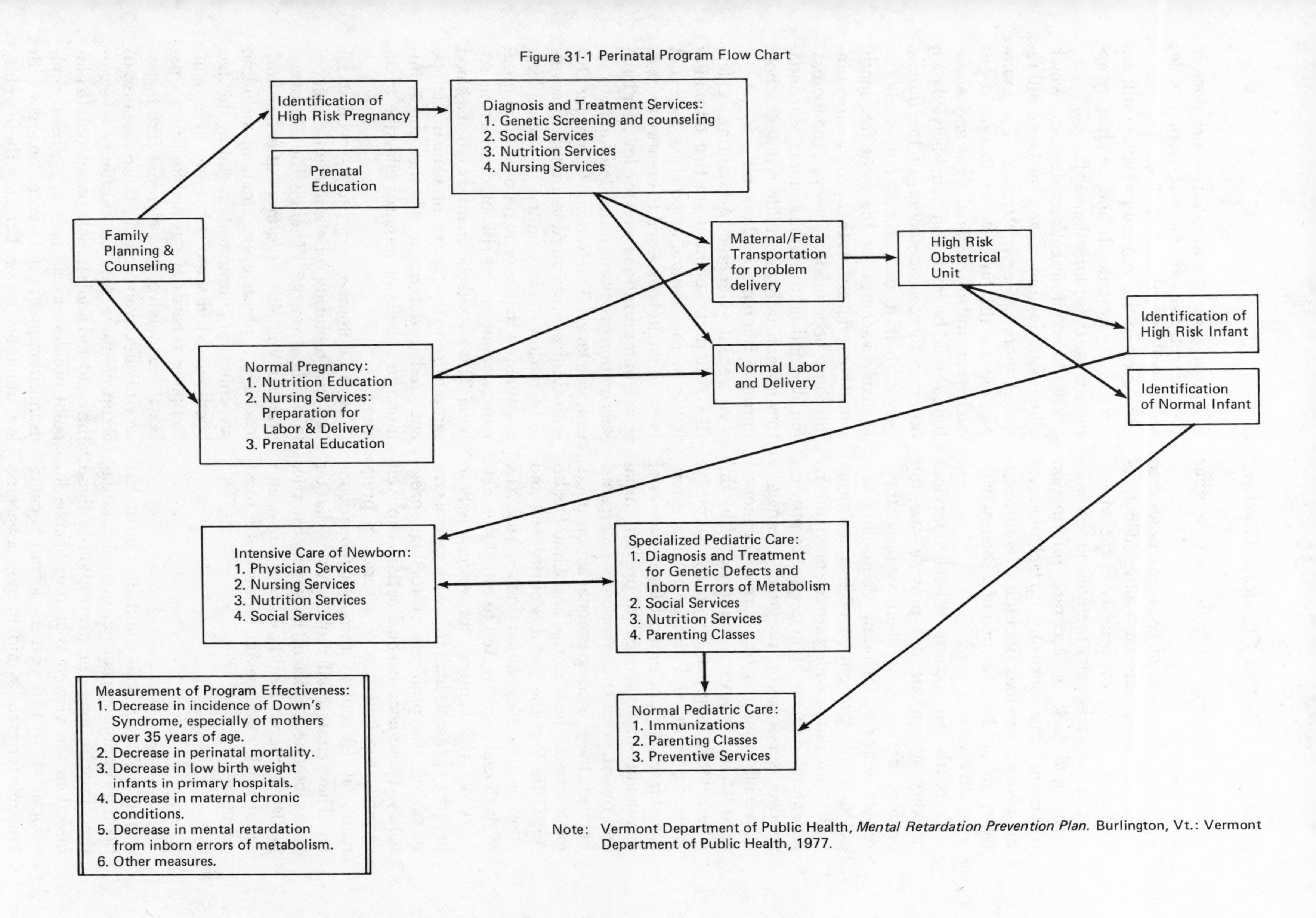

Note: Vermont Department of Public Health, *Mental Retardation Prevention Plan.* Burlington, Vt.: Vermont Department of Public Health, 1977.

having a baby without high risk to mother and child.

Beyond the medical concerns related to adolescent pregnancies are some difficult social issues: (1) the low priority of the adolescent in our society; (2) the conspiracy of silence and lack of accurate information regarding human sexuality; (3) lack of access to medical services, including contraceptives; and (4) the influence of a society with a negative view of sex.[33]

Currently, our country's health services delivery systems do not provide one-stop centers offering comprehensive pregnancy-related services, including counseling oriented to the prevention of teenage pregnancy. Also not provided are the mechanisms for extended follow-up necessary to provide needed services to pregnant teenagers whose offspring may be "at risk" for mental retardation. However, a bill enacted by the Ninety-fifth Congress (S. 910 by Kennedy, Williams, Javits, and Hathaway) has established federal grant support for networks of community-based services to prevent initial and repeat pregnancies among adolescents. Further, there is congressional action pending to amend the Maternal and Child Health Act (Title V of the Social Security Act) and other pertinent legislation (such as Title XIX of the Social Security Act) to establish a network of "risk centers" for women whose pregancy presents high risk, with special emphasis on the problems of teenage pregnancy. These congressional actions can well result in the initiation of a series of pregnancy "at-risk centers" throughout our country. These centers will require a close liaison (and subsequent linkages) between obstetric, pediatric, mental health, visiting nurse, and social work resources in order to be most effective.

Infection

Infection, both before and after birth, has been a continuing cause of mental retardation. Prior to 1963, the prevention of these infections was viewed as an improbable development. In that year a vaccine became available for measles and, in 1965, a vaccine for rubella was produced. The incidence of mental retardation resulting from measles and rubella has plummeted, but the immunity levels in the general public are still unnecessarily low; death and residual symptoms from these infections still occur.

The topic of infectious diseases, which affect both the unborn and the born, still requires more research attention. For example, menningitis strikes over 20,000 citizens each year and its cardinal signs and symptoms should be made common knowledge in order to increase detection of the disease very early in its course. Few field studies of available vaccines for the most frequently noted bacterial agents that produce menningitis have been made. However, the increasing demand (and establishment) of mandatory immunization of children before school entrance is a nationwide trend.

A recent General Accounting Office (GAO) Report[20] pointed out that immunity levels in the general population have not been adequately assessed to identify areas in which immunization levels are low, nor have vaccination programs raised immunity to acceptable levels. The National Center for Disease Control still considers immunity levels to be too low in our country. For example, it is estimated that over 10 percent of all pregnant women are at-risk of contracting rubella. This is a sad commentary when a blood sample from the pregnant woman can be analyzed to determine whether she is immune to rubella or requires effective treatment.

As of September 1977, forty-seven states had laws requiring rubella immunization before school entry.[20] Though the Headstart program guidelines require that each enrollee have a physical examination, including screening for immunization status, this has been waived in some states and so only incomplete records are available. From a national perspective, two elements are necessary to effectively combat these diseases: (1) comprehensive data on immunity levels in the child and adult female population to pinpoint problem areas; and (2) aggressive immunization programs targeted on areas with low immunity levels. Utilizing these guidelines, at least 95 percent of all children would be immunized before they entered school and rubella titers would be a part of all family planning services. Private physicians must be educated to be highly conscious of the medical-legal implications of providing this protection to women of childbearing age.

As with preschoolers' immunizations, rubella titer evaluation and follow-up immunization before entering high school should be made mandatory. Since vaccination must precede conception by two months, at the time of marriage a reevaluation should be required and a vaccination or booster given if the antibody titer is not adequate.[2] This vital preventive step could be legislated by the federal government as a prerequisite for obtaining a marriage license.

Certain federal programs, in particular Headstart and family planning programs, can improve surveillance data and possibly raise immunity levels. These programs are all important linkages to other components of a state's approach to prevention. Mandatory vaccination, before entering school, against diphtheria, pertussis, tetanus, measles, rubella, and poliomyelitis is a current trend that has been firmly supported by recent federal government statements. The Department of Health, Education and Welfare (HEW) committed an additional 3 million dollars for childhood immunization programs in 1979; and former HEW Secretary Califano made mandatory immunizations an integral part of the Child Health Assessment Program,[35] which focuses on the health status of young children.

The normal childhood diseases (mumps, measles, chicken pox) can occasionally result in encephalitis with subsequent mental retardation as a residual. Supportive treatment of these "normal" childhood diseases, which infrequently have "nonnormal" outcomes, must be actively sought.

Chromosome Disorder

Chromosome abnormalities are estimated to account for about 16 percent of institutionalized mentally retarded citizens. Amniocentesis makes possible the prenatal diagnosis of an increasing number of developmental disorders such as: chromosome abnormalities (Down's syndrome), neural tube defects (meningomyelocele, anencephaly), inborn errors of metabolism, and sex-linked genetic disorders.[16] The need for an amniocentesis evaluation is appropriate when the following factors are present: increased maternal age (over thirty-five years), maternal carrier of X-linked diseases, previous trisomic offspring, previous neural tube defect, maternal carrier of an established chromosome translocation, parental carriers of inborn errors of metabolism, and high-risk pregnancies with repeated miscarriages. Amniocentesis may also be indicated when there is distinct evidence of genetically-based instances of severe mental retardation in the family. An obstetrician and/or genetic counselor may recommend that amniocentesis be performed between the fifteenth and eighteenth weeks of pregnancy. About 5 percent of mid-term pregnancies (250,000 yearly) are considered "high risk" and would benefit from the amniocentesis procedure.

Contrary to some current opinions, many children's lives are saved by genetic counseling and amniocentesis evaluation. For example, if the amniocentesis procedure showed the presence of a normal child, then recourse to abortion is unnecessary. If a Down's syndrome infant is detected by amniocentesis, then the parents can make specific plans for early physical intervention, psychosocial stimulation early in life, and similar treatment-management approaches that can serve to ameliorate the developmental delay. Continuing and increased availability of amniocentesis, in the authors' opinion, will also accelerate the development of methods for very early treatment *in utero* that may ameliorate any prenatally diagnosed conditions.

Currently, genetic screening and counseling services are only readily available to the higher income groups, while the majority of people who would benefit from such services are not receiving them. Also, an insufficient number of at-risk mothers are not receiving these services because the needed pool of trained manpower does not exist. However, current research in selectively detecting fetal cells that have intermingled with the maternal bloodstream during early pregnancy, holds promise for replacing the current amniocentesis technique with a relatively simple peripheral blood sampling test —thus enhancing the availability of prenatal diagnostic services.

The incidence of neural tube disorders is related to genetic factors and the need to screen for these disorders. Yet screening for neural tube disorders (that is, alpha-fetoprotein levels in peripheral blood and amniotic fluid, via amniocentesis) should become a routine aspect of all prenatal care programs.

A concise review of the indicators for chromosomal disorders in pregnant women should be routinely utilized in prenatal care in all health programs. Outreach into the lower socioeconomic groups is needed. Because family planning programs contact so many women, these programs are ideal as sites for outreach activity. The referral consciousness of the family practitioner in regard to potential chromosome disorders in at-risk mothers must also be raised.

Metabolic Disorders

Metabolic disorders and chromosome abnormalities are separated because the nature of the diseases and methods of diagnosis are, at times, very different. Many hereditary metabolic disorders can be diagnosed in the newborn and a few (methylmalonic acidemia) have been detected and treated *in utero.* The classical illustration of the detection of an inborn error of metabolism in the newborn period is phenylketonuria (PKU). Disorders such as hypothyroidism (twice as common as PKU) are also detected by means of newborn screening, and treatment is available.

A number of model state laws have appeared in the last two decades mandating the screening of newborn infants for metabolic disorders that are associated with mental retardation. These screening techniques have been automated to make them more cost-efficient.[24] Expansion of metabolic screening to include many other treatable disorders is feasible; with automation the cost of multiple screening may be little more than that of one routine PKU test. For example, the following disorders can be detected early and rapidly by mass screening: (1) PKU; (2) hypothyroidism; (3) galactosemia; (4) valinemia; (5) homocystinuria; and (6) histidinemia.[24]

Yet, much remains to be accomplished. A national survey of newborn screening[34] to which forty-three states responded, noted: Six states were not conducting PKU screening programs; thirteen states were screening for galactosemia; seventeen states were screening for hypothyroidism; and fifteen states were carrying out additional tests, some of which included alanine deaminase deficiency, histidinemia, lead poisoning, sickle cell disease, thalassemia, leucine deficiency, (such as, maple syrup urine disease), methionine deficiency, Tay-Sachs disease, and toxoplasmosis. In the twenty-one state questionnaire returns that were usable, a composite 95 percent coverage for PKU was found. Two states have programs to detect maternal phenylketonuria early in pregnancy so that a reduced phenylalanine level diet can be instituted to protect the baby *in utero.* Screening requirements are, to some extent, specific for different populations, and transient neonatal phenomena are responsible for many abnormal results. It is important to have sound methods for confirmation and follow-up.

PKU, which occurs in one in 10,000 newborns, has become the classic model for illustrating the prevention of mental retardation. In the past, 1 percent of the incidence of severe retardation was secondary to PKU. Now these tragic consequences can be avoided by a phenylalanine controlled diet.

Six states attempt to detect maternal PKU by assessing elevated phenylalanine blood levels present in the neonate. Maternal phenylketonuria is an emerging problem, since the symptoms of mental retardation are a consequence of the infant being literally saturated *in utero* by high concentrations of phenylalanine in the mother's bloodstream. Routine screening of expectant mothers for PKU should be followed by treatment of the maternal carriers with a reduced phenylalanine diet. The cost-benefit ratio for prevention of PKU is clearly in society's favor.[20]

Newborn screening for thyroid deficiency has been actively pursued in recent years.[27] Since the incidence of congenital hypothyroidism has been reported to be from one in 3,000 to one in 7,000 live births, it is clear that this disorder is a frequent metabolic cause of mental retardation. The clinical diagnosis of hypothyroidism in an infant is difficult, and the screening technique becomes both a crucial diagnostic and preventive tool. The initial mass screening is accomplished by a filter paper technique (the thyroid-4-component), and the positive samples are validated by further testing (the thyroid stimulating hormone component). This procedure has recently become operational in Michigan where the State Health Department coordinates both the screening (by young volunteers from the Illinois Association for Retarded Citizens) and treatment coordination (by the medical staff of general hospitals throughout the state).[40] Their program is an excellent example of putting current knowledge about the prevention of mental retardation into direct service for young citizens. The condition of these infants would otherwise probably not have been detected within the first three months—the critical period of intervention. If treated with exogenous thyroid hormone within the first three months, the majority of the afflicted infants will develop normally.

Some maintain that "meddlesome" intervention with our genetic endowment, by permitting genetically abnormal people to survive and to reproduce, will be detrimental to our "national genetic balance" (that is, the gene pool). However, it has been estimated that by allowing PKU patients to reproduce it would require over 100 generations for the gene frequency to double.[21] It is only since the infant mortality rates started to fall that we have been able to differentiate more clearly genetic diseases from acquired diseases. The process of human evolution has been altered far more rapidly by antibiotics than by the advent of PKU or hypothyroid testing.

With automation of the assessment of direct blood specimens (via the punch-index machine), newborn screening tests for treatable inborn metabolic error diseases such as homocystinuria, tyrosinemia, galactosemia, maple syrup urine disease, histidinemia, and valinemia can be added to PKU testing without excessive addition of laboratory personnel. If the laboratory processes a minimum of 25,000 newborn specimens a year, the cost increase over PKU screening alone is minimal. It is most frequently recommended that a regional program be based in a university department of pediatrics in close alliance with the state department of health. This linkage is an essential one if the operational problems noted in the recent past in attempting to implement neonatal screening programs are to be avoided. Guthrie[24] noted that for most states, the legacy of the 1960's controversy concerning PKU remained in the 1970's in the form of three problems: (1) lack of liaison between the medical centers and the screening programs; (2) many states with laws requiring PKU testing simply do not have sufficient population to make multiple testing of newborn infants practical; and (3) many states with large populations allow each private laboratory to perform small numbers of PKU tests for a profit, thus causing the same problem that exists in small states. In May 1975, a regional model of linkage, the Oregon Neonatal Metabolic Screening Program, was established in Portland. It receives screening test specimens from Oregon, Alaska, and Montana. In January of 1976 the Massachusetts Metabolic Disorders Detection Program in Boston started receiving screening test specimens from Massachusetts, Rhode Island, and Maine.

Although an increasing number of the inborn errors of metabolism can be tested on a regional basis and the cost is only about $2.50 per complete testing, the follow-up of high-risk infants should be assured. This follow-up care should be available at special treatment-management regional centers (for example, local crippled children facilities, medical centers, or in conjunction with regional treatment centers) such as those established in California (California Assembly Bill #45, 1977). Experiences in England, Sweden, and France have clearly documented that over 50 percent of the motor and special sensory handicaps noted in developmentally at-risk newborn children can be completely reversed, to the great developmental enhancement of these children.[54] Accordingly, early secondary prevention efforts can inadvertently become tertiary preventive challenges if follow-up services are not available after early diagnosis has been accomplished.

Intoxications

LEAD POISONING

Lead intoxication and Rh blood incompatibility disease are fine examples of primary preventive accomplishments in the field of mental retardation. Recognition of childhood lead poisoning is relatively new—it was first described in children in the early part of the twentieth century. The cause was assumed to be the ingestion of paint, but the role of air-borne lead in general exposure—primarily from automobile emissions—has recently received increasing attention.

There has been a clear recognition of lead's ubiquity, and intervention has taken place on the federal legislative level with the "Lead Paint Act," and planned reductions in the content of lead in gasoline. Young children up to the age of six years are at great risk, especially from peeling paint and chipping plaster in dilapidated housing. There is a strong association between pica and lead poisoning. Programs have been instituted to eliminate asymptomatic and symptomatic lead poisoning. The free erythrocyte protoprophyrin (FEP) test, which can be conducted from a drop of blood on filter paper, has made wide-scale testing feasible. Spinoffs of the FEP screening test include detection of iron deficiency anemia and the hemoglobinopathies.

Lead poisoning is not just an inner-city problem, but a widespread one that justifies more than a high-risk target area approach. This insidious and cumulative metal affects American children in epidemic proportions; over 600,000 bear blood levels above 40 mg/100 cc of blood.[49] In 1970, the U.S. Surgeon General recommended that 40 mg/100 cc blood be the borderline at which the child is in potential danger of clinical lead poisoning; in 1975 the revised borderline was lowered to 30 mg/100 cc of blood.[20] Ages one through six years of life are at the greatest risk for lead poisoning, with the peak years being from one to three. Theoretically, even very low levels of lead are associated with toxicity, and lead is known to cross the placenta during pregnancy.

Does a threshold for lead toxicity exist? On both theoretical and empirical grounds, there are several reasons to believe that if there is a threshold, many children have exceeded it.[48] First, there is no known metabolic function for lead. Secondly, global environmental levels of lead have increased since the industrial revolution—most markedly since the invention of the automobile—and, in the evolutionary scale of time, the human organism has had only a brief opportunity to adapt to contemporary environmental levels. Third, no toxicologist would accept a margin of safety of 50 percent, yet thousands of children carry blood-lead levels of 40 ug/dl. Lastly, the symptoms of lead toxicity are vague and easily missed. Headache, lethargy, colic, or clumsiness are not readily identifiable as symptoms of lead poisoning. Perino and Ernhart [46] reported that black preschoolers with blood-lead levels above 50 ug/dl had significant impairment on the McCarthy Scales of Mental Development when compared to children matched for race and controlled for other variables with blood-lead levels below 30 ug/dl. Undoubtedly, many children with these symptoms are misdiagnosed by both parents and physicians.

Of considerable interest is the report of Beattie and coworkers.[4] They identified seventy-seven children with mental retardation of unknown cause and matched them with normal children on age, socioeconomic status, and geographic residence. The place of residence of the mother was identified and a sample of the drinking water was analyzed for lead. No normal children came from homes with excess lead, while eleven mothers of retarded children lived in homes with high levels of lead in the water during the time they carried the child. The authors concluded that the risk of retardation is increased by a factor of seven by living in a home which has a high lead level in the water during the mother's pregnancy.

Mass screening for lead poisoning utilizes the free erythrocyte protoporphyrin test (FEP), which consists of a dried blood sample collected on filter paper. Volunteers can be quickly trained to collect the dried blood specimens, and the samples are forwarded to a state (or regional) laboratory. The Illinois Association for Retarded Citizens has been particularly effective in organizing volunteers to screen large regions of the state of Illinois. Their testing technique (noted previously) and the more recent zinc-protoporphyrin test are inexpensive. Positive tests are further evaluated via blood level determinations. The family physician carries out treatment with British Anti-Lewisite Factor (BAL) when necessary.

An effective program against lead poisoning must concern itself with a number of issues. Parents and the general public must be educated as to sources of lead, its effects, and what can be done. Screening with an inexpensive test (such as the free erythrocyte protoporphyrin test) is necessary because signs and symptoms of toxicity are too variable and covert. Following detection at the screening level, confirmation, and the institution of therapy, the child should be periodically monitored as to the effectiveness of the treatment—this may sometimes be necessary for up to two years. Finally, the environment of the child must be corrected to prevent re-exposure.[8]

In summary, lead poisoning is directly linked to mental retardation, epilepsy, and cerebral palsy; it is a widespread problem, is frequently asymptomatic, and often presents with generalized or vague symptoms. The new free erythrocyte protoporphyrin test has made widespread screening for lead poisoning feasible. Although the total elimination of lead from our environment is probably not possible as a relatively short-term goal, the elimination of leaded gasoline is nearing reality.

Rh Incompatability Disease

Hemolytic disease of newborns secondary to Rh incompatibility, is now a preventable disease—only ten years after the introduction of routine postbirth administration of hyperimmune globulin (i.e., RhoGAM) to Rh-negative mothers. Careful adherence to clinical and preventive guidelines can help ensure that every susceptible woman is treated, so that the disease will be eliminated.

The latest available information (1974)[20] reveals that there are over 7,000 infants born each year in our country who have sequellae of Rh-blood type incompatibility. Yet, only five states have Rh-blood type registries or education of the general public and professionals about the use of RhoGAM for preventing this disorder. An example of an excellent state-wide educational program on the use of RhoGAM is that carried out in Connecticut.

Immunoglobulin (RhoGAM) was licensed for human treatment in 1968. Although it can protect children from Rh incompatibility, it is currently underutilized. Problems exist because of its lack of availability in rural areas and failure to administer the agent to women after abortions as well as after births. Public and professional awareness must be raised concerning its vital preventive role. Prevention of Rh incompatibility includes the following issues: (1) early detection of all Rh-negative mothers; (2) availability of this information during pregnancy and delivery; and (3) prompt utilization of immunoglobulin after the delivery of each Rh-positive child to an Rh-negative mother. Yet, there is a lack of complete national information on the incidence of Rh-negative and the accompanying immunization by immunoglobulin. For example, the recent GAO Report[20] noted that only five states had mechanisms for fully monitoring Rh hemolytic disease (Connecticut, California, Illinois, New Jersey, and Colorado); only seven states required by law either premarital or prenatal blood typing; and only six states had special programs for immunoglobulin utilization, as of the date of the report. One would think that the private physicians would routinely check for Rh incompatibility, but the current Rh incidence rate indicates that this is not true. The consequences of not checking for Rh-negative incompatibility might lead to tragic consequences for the afflicted child.

Interestingly, Connecticut has had a complete Rh prevention program in operation for the last ten years. A recent report from the Center for Disease Control shows that Connecticut has had a 96 percent decline since 1970 in maternal sensitization and Rh blood disease in the newborn. Their prevention program includes routine premarital or prenatal blood typing. Comprehensive data are compiled on immunoglobulin usage at the time of birth, abortion, or miscarriage; Rh incidence and mortality (fetal and infant); the number of sensitized women; and the overall effectiveness of each of these preventive efforts are regularly monitored. State law in Connecticut requires that the Rh typing be listed on the marriage licenses, and each hospital is mandated to report information concerning Rh determination, the use of immunoglobulin, and follow-up results.

Mercury Poisoning

The hazardous situation occurring through large amounts of mercury being discharged into waterways was first recognized in 1970 when eating fish that had been contaminated was established as a definite link to human toxicity. There are two types of mercury poisoning, acute and chronic. The acute situation arises from the ingestion of soluble mercuric salts (mercuric chloride) and produces serious renal tubular damage. The chronic form develops from inhalation of mercury vapor or ingestion of small quantities of mercuric nitrate or other salts; it clinically presents with psychiatric and gastrointestinal symptoms.[5]

At an increasing rate, alkyl mercury compounds have become significant environmental problems. The compounds ethyl and methyl mercury are soluble in organic solvents, and the covalent carbon/mercury bond is not biologically degradable. Intestinal absorption is nearly complete, and it is difficult for the body to rid itself of these compounds because of the continuous interchange of the gastrointestinal-liver blood circulation. Seed grain which has been treated with alkyl mercury compounds (such as antifungal agents) has been diverted into food, resulting in epidemics in Japan and several other countries. Methyl mercury is especially found in tuna and swordfish taken from waters contaminated with mercury. In humans, this compound freely passes the placental barrier, accumulating in the fetus, with resultant cerebral palsy and mental retardation in the newborn.[5]

It has been found that n-acetyl-DO-penicillamine, in a dose of 500 mg four times a day, accelerates the excretion of mercury and has been used successfully in the therapy of chronic mercury poisoning. However, this treatment is not effective in methyl mercury poisoning, in which case the use of ion-absorbing resin to interrupt the body circulation of toxic levels of mercury shows promise as a method of treatment.

The prevention of mercury poisoning will necessitate the shift of industries away from mercuric chloride, calomel, and mercury ointments—for all of which safe substitutes are available. This would eliminate acute poisoning and many instances of chronic poisoning. This shift to alternative compounds has already occurred to some degree. Mercuric nitrate has been eliminated from the felt-processing industry. Mercury salts for fingerprinting by police departments have been replaced by barium, zinc, or bismuth salts. Silver has replaced mercury in the manufactory of mirrors. This shift to alternatives, however, should be accelerated.

Early Childhood Experiences

The noxious developmental effects of adverse childhood experiences have, unfortunately, only recently been directly addressed by our society. Child abuse (physical, sexual, and emotional), secondary effects of severe childhood mental illness (childhood depression and the psychoses), minimal opportunities for developmental stimulation secondary to poor parental modeling, racism, and the "culture of poverty" all represent major current prevention frontiers. Children born and reared in urban ghettos or impoverished rural areas are fifteen times more likely to be diagnosed mentally retarded than children from middle-class, suburban environments.[7]

The Milwaukee Project[18] was a research undertaking in which a target group of women from socioeconomically blighted census tracts in the metropolitan area of Milwaukee, Wisconsin, were studied. A very high frequency of mental retardation had previously been clearly documented in these census tracts. The research focused on two major approaches: (1) intensive training programs for pregnant mothers (for example, counseling and vocational training, instruction in mothering, child care, and homemaking); and (2) direct assistance to the families in stimulating their child by providing enriched early developmental experiences. The results clearly showed that the study group of children born to these mothers developed intelligence quotients averaging 25 points higher than the control group of children who were not provided these mother-child special training/stimulation experiences. This study indicates what can be done to alter socio-cultural causes of mental retardation by intensive environmental stimulation programs.

Unfortunately, most of the research in this area has not utilized the rigorous methodology employed in the Milwaukee Project. Indeed, definitive evaluations of the bulk of current environmental enrichment programs—and the people they are reaching—still needs to be determined. For example, the GAO Report[20] notes that the University of Connecticut completed a national survey to determine what developmentally enriching programs for children under three years of age were operating or proposed. Information was solicited from several sources, including the state departments of education, state offices of child development, and early education program directors. A total of fifty-three ongoing and proposed programs were identified as operating at 116 sites and involving about 19,000 children and their families. Additionally, eight universities and twenty-three community colleges were involved in infant and toddler research and service programs. No clear set of program guidelines (or results) was discernible. The availability of these development enrichment programs is far below the demonstrated need for them. The national-state evaluation picture has changed little since this 1974 study.

The prevention of psychosocial retardation requires education, improved childrearing practices, and environmental correction for developmental high-risk infants. Ideally, programs would start early and continue until the child enters the formal school system, concentrating on definitive developmental corrections and on increasing verbal-social skills. Attitudinal changes in the homes (and in school) will probably have the most long-lasting effect, since the intellectual/educational atmosphere of the home is an important determinant of future development. Regional and state programs must work aggressively to train mothers of at-risk children (for example, low-income adolescent mothers). Unfortunately, it is very difficult to clearly delineate the linkages necessary in order to accomplish this goal. The complexity of this issue which literally cuts across virtually all areas of the concept of prevention is noted in table 31–1.

Accordingly, it appears that improvements in national policies, commitments to enhanced living standards, and increased human service provision can result in a Gestalten that acts favorably on the future lives of these psychosocially at-risk individuals. It is not the authors' intention to belittle the efforts of specific prevention programs,

TABLE 31–1 **Relationship between Poverty and School Failure**

I. Inadequate Intrauterine Environment due to:
  A. Poor maternal growth
  B. Poor maternal nutrition
  C. Poor maternal health and health care
  D. Too many pregnancies too close together
  E. Extremes of maternal age (too young or too old)
  F. Poor obstetrical care
II. Poverty:
  A. Increased infant mortality
  B. Increased infant morbidity
  C. Elevated family size
  D. Poor nutrition
  E. Increased illness
  F. Absence of adequate health care
  G. Inadequate physical surroundings
  H. Inadequate home learning environment
  I. Inadequate home emotional environment
  J. High level of social stress
III. Effects in childhood and adulthood
  A. Increased chance of school failures and under-achievement
  B. Adult unemployment and under-employment

SOURCE: Birch, H.G., and Gussow, J.D. *Disadvantaged Children: Health, Nutrition, and School Failure.* New York: Grune & Stratton, 1970, p. 268.

which can appreciably affect the noted key issues (for example, child abuse prevention projects), but to stress that the scope of this prevention challenge is a massive one since it eventually addresses key issues such as the persistence of "illiteracy."

Childhood Accidents and Postnatal Injuries

Childhood accidents and allied postnatal instances of head injury continue to be a major cause of mental retardation. Similarly, accidental poisoning also remains difficult to prevent.

Accidents are the number one cause of death among infants and children. Although postnatal head injury seldom results in mental retardation, there is an increased incidence of residual behavioral disturbances and neurological handicaps. A recent Canadian report on this topic[17] noted that preschoolers, with their energy and lack of judgment, are at greatest risk. Other predisposing factors are hyperactivity, aggressiveness, hunger, and tiredness. Accidents that can result in severe injury to the brain include auto accidents, falls, sports injuries, near suffocation or drowning, and extensive body burns. The American Academy of Pediatrics[2] recently stated that approximately 90 percent of these childhood accidents are preventable.

Prevention strategies must include an increased public education program, promoting safety in the home and in sports, packaging of drugs in child-proof containers, and special hazards protection, such as auto seat restraints. Similarly, there should be stringent enforcement of child abuse and "driving while intoxicated" laws. There also needs to be an increase of correct information for parents about the signs and symptoms in a child that require immediate medical attention. The utilization of information services (for example, poison control centers) and similar public information can help in dispersing this information.

Other Disorders

There are many disorders for which the causes are not always known, but the disorders themselves can be ameliorated or prevented. For example, the role of nutritional supplements, both during pregnancy and in early life, has been shown to improve the physical integrity of infants.[1] Accordingly, states such as California have passed legislation that automatically provides these supplements to low-income pregnant mothers, their infants, and their young children. This prevention program emphasizes the importance of giving all children an optimal chance for full physical development—even though the specific role of early nutritional intervention is heatedly debated among scientists.[50]

Similarly, even though the exact causes of hydrocephaly and spinal cord disorders such as meningomyelocele (herniation of the spinal cord through its overlying bone and skin covering in the lower lumbar back region) are not known, therapeutic alternatives are available. Specifically, mothers who have had a previous child afflicted with one of these disorders can now be screened during subsequent pregnancies via a peripheral blood test for the alpha-feto-protein (a protein that leaks out of the spinal fluid system of the fetus into the surrounding amniotic fluid, and hence into the maternal circulation). If the test is positive, it can be confirmed via amniocentesis, and the parents may be given the option of a therapeutic abortion.[41] Interestingly, Dr. Robert Cooke[12] has recently suggested that abortion may not be necessary if we further explore the evolving research work on fetal surgery *(in utero)* on monkeys. This work provides a direct parallel for future interventions in humans—interventions that would both avoid abortion and provide the future infant with an intact central nervous system. The current hope for prevention of these disorders is best provided by the previously mentioned regional chromosomal screening/genetic study centers. Last, an enhanced awareness of this prenatal screening test and a rapid increase in genetic counselors will be needed to offer this alpha-feto-protein assessment to all potential cases.

¶ Priorities and Linkages

The prevention of mental retardation will have its greatest impact if attention is focused on the following areas: (1) optimizing preconception conditions; (2) improving pregnancy outcomes; and (3) optimizing infant growth and development. The activities necessary in each of these three areas constitute an intervention strategy for the prevention of mental retardation. This involves a Cycle of Continuing Intervention (see figure 31–2), which indicates that persistent and broad-based interventions are necessary to

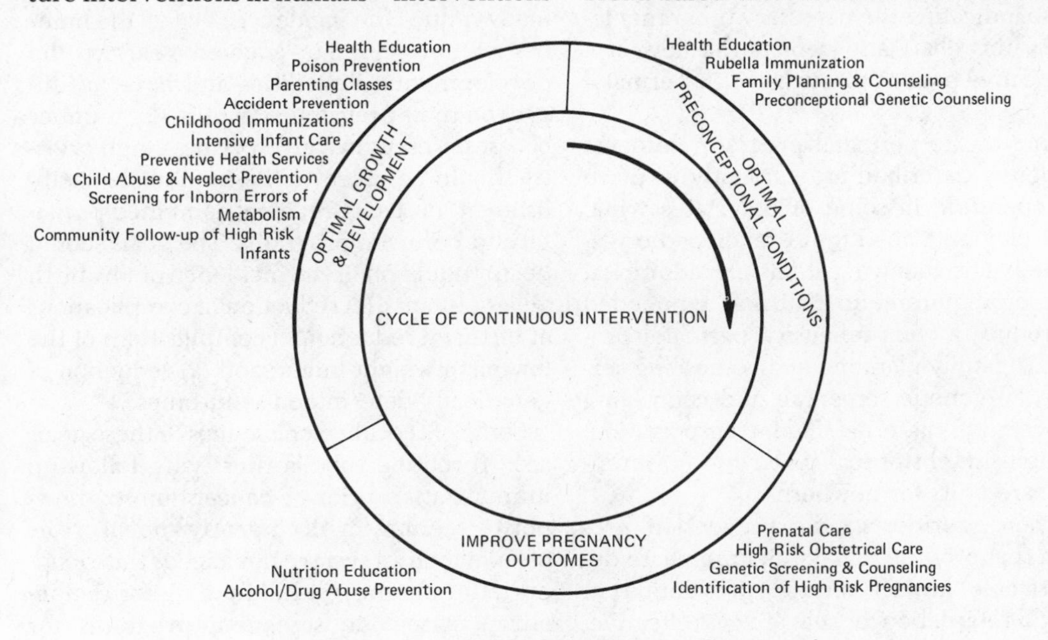

Figure 31–2. **Cycle of Continuous Intervention.**
NOTE: Vermont Department of Public Health. *Mental Retardation Prevention Plan,* Burlington, Vt.: Vermont Department of Public Health, 1977.

prevent mental retardation in future generations.

A fully operational program for the prevention of mental retardation would contain all of the elements listed in the "Cycle of Continuing Intervention." Since the availability of financial resources may preclude the unilateral implementation of the prevention program, priorities should be established. One criterion for establishing priorities may be to choose those program elements that will have an immediate impact upon the prevention of mental retardation. Another criterion would be to choose those program elements that produce the greatest impact for the dollars expended. Together, these criteria indicate that the perinatal program be selected as a high priority item because of its ability to influence a number of factors that cause mental retardation. For example, the activities of a comprehensive perinatal program should produce a decrease in the number of low birth weight infants, chronic maternal conditions, and genetically determined abnormal infants. The elements of such a comprehensive perinatal program were previously reviewed in the "Perinatal Program Flow Chart" (figure 31–1). In addition to indicating the necessary program elements, this chart also depicts the flow of a prospective patient through such a perinatal program.

Some of the perinatal program elements previously described are currently in place and operating in some of our states while other elements need to be developed or expanded. The following items are additional resource requirements that may be needed to produce a comprehensive perinatal program:(1) family planning and counseling services; (2) genetic screening and counseling services; (3) maternal/fetal transportation; (4) high-risk obstetrical units; and (5) intensive care units for newborns.

Other priorities in the prevention program that promise to have an immediate demonstrable impact on the prevention of mental retardation, or that are cost effective, include regional screening of all newborns for the seven inborn errors previously reviewed and for hypothyroidism and lead poisoning. Provisions must also be in place (via linkages with regional centers) for intervention services such as the PKU diet, further assessment of thyroid status (T-4 testing), and treatment, (for example, BAL treatment for lead poisoning), and so forth. A measles and rubella immunization program, an Rh identification program with subsequent steps included to ensure that desensitization is provided (availability and administration of RhoGAM), and a vigorous child abuse and neglect prevention program should be implemented.

It should be remembered, however, that the implementation of these priority activities will not entirely prevent the occurrence of mental retardation. With our current incomplete records and knowledge, however, it is important to pursue these activities until all feasible preventive measures are known and/or enacted.

Throughout this presentation, linkages have been suggested between early diagnostic screening services and how or where follow-up treatment can be provided. Priorities must be established so that appropriate intervention can take place. In order to maximally reduce the incidence (that is, the number of new cases in a given year) of the developmental disabilities and have an impact on their prevalence (that is, the number of cases present in a given year), a high priority should be given to the immediate establishment of strong regional combined prenatal and perinatal programs. The goals should be: (1) reduction of the incidence of low birth weight infants; (2) reduction in complications at birth; (3) reduction in complications of the low birth weight infant; and (4) reduction of genetically determined syndromes.

Some of the allied challenges to these goals are: (1) routine rubella titers with follow-up immunization prior to conception; (2) newborn screening for all currently known errors of metabolism via regionalized state supported programs; (3) full funding for regionalized state-wide screening programs for lead poisoning, Rh incompatibility, hypothyroidism, and mercury screening programs; and (4) adequate immunization clinics in each school.

All of these services could be provided by neonatal and perinatal risk centers, located in the major population centers. The rapid availability of genetic counseling (on a regular basis) is an associated necessity. Intervention techniques (alpha-feto protein, amniocentesis and, in the future, intrauterine fetal surgery) should also be made available to all who need them.

Further, a high priority must be established for widespread public education for responsible parenthood. To achieve this priority goal, there should be appropriate family planning and counseling available to all persons of childbearing age. The necessary skills and responsibilities of parenthood must be emphasized.

Other services that could be provided by neonatal and perinatal risk centers are:

1. development and support of a state program for family planning under the department of health;
2. increased support of planned parenthood and other women's "walk-in" health clinics to provide contraceptive information and counseling regarding responsible sexual behavior; and
3. acknowledgment that sexuality and family planning education are part of health education and should be actively supplemented by the educational and health care system.

There are a number of other priority areas where a lesser impact can be made in order to decrease the incidence of mental retardation:

1. rapid implementation of an extensive health education curriculum throughout all schools (starting in the first or second grade);
2. programs to prevent child abuse and neglect, including: (a) identification of the family at risk when the mother is pregnant and close follow-up (that is, a maternal child-care nurse who is psychologically sophisticated, can provide a nurturing experience for the parents, serve as a resource person in helping them with childrearing, and foster optimal child development); and (b) use of the nurse (or other developmentally-oriented health care professional) as a liaison with other social-educational-medical service support systems to maximize the services offered to the family—a variety of well-coordinated supportive services over a prolonged period of time do significantly help a large portion of actual or potential abusers;
3. a nutritional education program for all pregnant women; and
4. assessment of the status of catastrophic insurance needs (Major Medical provisions in third-party insurance contracts), so that the outcome of pregnancy is not complicated by unbearable financial burdens to the family.

First Steps

A professional who is committed to advocating prevention-oriented programs in mental retardation should focus on the following "first steps" to effectively aid the citizens in his area:

1. Assure immunization for all infants and preschoolers.
2. Expand program of Rh screening and mandatory marriage license recording of blood typing as soon as possible and include ensurance of RhoGAM availability and utilization.
3. Initiate a lead poisoning prevention program by rapidly screening a mass population with the FEP test, treating those currently ill, and spurring the development of rigid lead control measures.
4. Establish at least one regional intensive care unit for at-risk newborns.
5. Establish at least one regional genetic and/or metabolic laboratory to screen more extensively for the inborn errors of metabolism, lead poisoning, and hypothyroidism.
6. Initiate a teenage pregnancy risk center, utilizing currently available federal funds.
7. Activate and extend local public education programs, especially for primary and secondary school pupils, in the prevention of mental retardation.
8. Extend early intervention education

programs (such as Project Homestart and Headstart) to broaden understanding of the psychosocial causes of mental retardation. Support local programs to combat child abuse and neglect.

The current state of the art in prevention of mental retardation has been reviewed. The issues surrounding the establishment of a comprehensive approach to this challenge have been noted. A system for initiating state-wide or regional programs to meet this challenge, including priorities and linkages, has been presented.

Our country has gone a long way toward helping to prevent mental retardation—those disorders that have blighted the lives of millions of its citizens and their families. It can do more. The authors' hope is that this chapter will aid in bringing about a brighter future of full developmental opportunity for all of our country's citizens.

¶ Treatment

Current treatment strategies utilized in the field of mental retardation can be subsumed under the general subtitles of "Cure for a Few," (for example, early dietary management in phenylketonuria, neurosurgical intervention in craniostenosis), "Treatment For Many" (for example, specific correction of speech and hearing problems, seizure management, psychotherapy for concomitant emotional illnesses), and "Habilitation for All" (for example, special education, vocational training). This section will review the current state of the art by describing possible treatment of the various etiological groupings of the retarded. The following overview of suggested treatments is presented in greater detail by Menolascino and Egger.[39]

Genetic and Chromosomal Disorders

In the chromosomal disorders, cure, as we understand the concept, is not possible because no specific treatment interventions are currently available. At present, for example, strategies of treatment for Down's syndrome consist of: (1) meticulous medical care during the newborn period; (2) close observation for other associated anomalies; (3) early adjunctive treatment for associated disorders (for example, respiratory infections, which frequently require energetic treatment because of anatomical problems in the palate, nose, and sinuses; and (4) a high general vitamin regimen with optional doses of vitamin B and C in particular because of proneness to early and repeated infectious diseases. Down's syndrome patients should also be medically evaluated early in life to establish cardiac status because of the high prevalence (18 percent) of congenital heart disorders among these patients. During the past ten years advances in techniques of cardiac assessment and surgical intervention have been especially helpful in correcting the cardiac defects in Down's syndrome. These and other treatment approaches can be of great assistance in minimizing the secondary signs of this chromosomal syndrome.

Habilitation strategies can accomplish a great deal in all of the chromosomal disorders (including the cri du chat, trisomy D, and trisomy E syndromes) to maximize the development of these youngsters early in life. These modern habilitation strategies are most clearly observed in recent approaches to Down's syndrome. For example, at the University of Washington, groups of Down's syndrome patients one year of age and older have been taught at an advanced rate a wide variety of perceptual and motor skills (such as tactile and visual-motor differentiation and color discrimination). Progress reports of this eight-year study[25] indicate that most of the Down's syndrome patients in the study are functioning in the mild to borderline range of intellectual ability (rather than the usual severe to moderate intellectional levels seen in nonstimulated Down's syndrome patients)—showing that early habilitation efforts do have a profound impact on the developmental prognosis of patients with major chromosomal disorders. In all the chromosomal disorders, as a general rule, habilitation represents a major resource for amelioration of the effects of these disorders even though treatment and curative procedures are not currently available.

Disorders associated with sex chromosome abnormalities may or may not also be associated with mental retardation. In both Turner's and Klinefelter's syndromes, endocrine treatment can be very helpful, as can counseling these youngsters and their parents about problems associated with sex differentiation. The disfiguring signs and symptoms of the ears and neck in Turner's syndrome can be corrected by plastic surgery so that these patients appear as normal individuals. Similarly, individuals with Klinefelter's syndrome tend to have a high incidence of cardiac difficulties, and can profit from close attention to factors such as hypertension management and alternation in the level of their physical activity. Although a cure is not available for either Turner's or Klinefelter's syndromes, treatment for physical remediation and supportive psychotherapy may be of considerable benefit by increasing the patient's ability to lead a normal life in the community.

In the congenital neurocutaneous syndromes of childhood (for example, tuberous sclerosis, neurofibromatosis, Sturge-Weber syndrome, etc.) curative strategies of treatment are also not available. However, treatment approaches may be helpful. In tuberous sclerosis, recent work suggests the presence of isolated tubers (firm connective tissue masses) in the brain. Since these tubers seem to be the foci for seizures, their surgical removal has, in selective instances, dramatically reduced the frequency of seizures. As in all seizure disorders, ongoing neurological consultation and continued care are necessary because the organic basis for the continuing seizure disorder (the tuber) makes these individuals prone to seizure phenomena unless meticulously followed to monitor anticonvulsant status.

Neurofibromatosis patients frequently have disfiguration of the face and neck secondary to the neurofibroma soft tumor masses sited in the meyelin sheaths of the cranial nerves of the face and neck. Cosmetic surgery may aid by removing the neurofibromas, and similar surgical treatment can remove the disfiguring facial-skull hemangiomas associated with the Sturge-Weber syndrome. Further, recent treatment advances in the Sturge-Weber syndrome have developed from the observation that calcification of the brain occurs on the same side (hemisphere) as the large papillary hemangioma of the face. In selected instances partial hemispherectomies that remove this calcified brain tissue have significantly decreased the frequency of seizures.

There have been dramatic and significant advances in the cure, active treatment, and habilitation of the mentally retarded associated with the inborn errors of metabolism. Phenylketonuria is the classical example in this area. A child with phenylketonuria whose disease is diagnosed very early and who is placed on a therapeutic diet consisting of synthetic foods specially formulated to be very low in phenylalanine, will tend to develop normally both physically and intellectually. Recent work[6] suggests strongly that if a mother who has had a previous phenylketonuric child is placed on the PKU diet during her next pregnancy, she will produce a child who has not been excessively exposed to the toxic effects of high phenylalanine levels during gestation. This approach, when coupled with early postnatal dietary treatment, further reduces the risk of damage to the child during the middle years of childhood.

Early diagnosis and treatment, coupled with an appropriate dietary regime, can also ameliorate the effects of maple syrup urine disease. If the diagnosis is made at a sufficiently early age, similar treatment strategies are effective in Wilson's disease. Later in life, treatment of Wilson's disease includes the use of BAL as an aid in precipitating the excess copper out of the system. Similarly, homocystinuria can be effectively treated with dietary supplements of methionine and dosages of vitamins B and C. More active treatment approaches are developing rapidly and, even when such approaches are not curative, they are frequently effective in slowing down the rate of the effects of the inborn error of metabolism.

Both galactosemia and fructosuria are examples of defects that can be cured. After early diagnosis, simply switching the child from cow's milk to goat's milk (or a different type of nondairy component) eliminates the deleterious effects of both of these metabolic defects.

The lipid metabolism disorders (for example, Tay-Sachs disease, Spielmeyer-Vogt disease, Niemann-Pick disease) remain an area of concern and active research. Currently there are no available treatment regimens leading to cure. The same can be said in general of the connective tissue disorders such as the Ellis-van Creveld syndrome, Ehlers-Danlos syndrome, and Marfan's syndrome. Here one treats the associated symptoms and uses appropriate supportive habilitation procedures. In the complex inborn errors of sugar metabolism, there is currently a good deal of argument as to whether Hurler's syndrome is curable or not. The enzyme defect in the disorder is known, and some early experimental diets (almost like the early PKU diets) appear to offer the possibility of a cure.

Recent work has shown that youngsters suffering from the Lesch-Nyhan syndrome have a gross abnormality of uric acid metabolism. The high uric acid blood level is associated with major behavioral problems in which they aggressively bite their fingers and lips. Medication similar to that used in treating gout returns the uric acid blood levels to normal and tends to abolish dramatically the behavioral symptoms. Whether these medications will change the long-term course of the disease is not known at this time.

In summary, more and more is being learned about the disorders caused by genetic factors and more and more about specific enzyme defects; this knowledge has provided the medical community with procedures that permit intervention at a specific biochemical level or dietary level, and similar approaches that actually slow down, if not reverse, the disease process. Many treatment approaches for associated handicaps can appreciably alter the life of the afflicted individuals and greatly assist in ongoing habilitation efforts.

Metabolic Diseases and Intoxications

Metabolic diseases and intoxications leading to the symptom of mental retardation represent, as a general group, disorders that can be controlled and, in selected instances, definitively treated. These disorders represent the result of abnormal levels of enzymes, metabolic substitutes, or derangements of endocrine substances. Primary hypoglycemia, diabetes mellitus, and hypothyroidism are the metabolic diseases most frequently noted during the early years. In the past, without specific treatment, such conditions produced serious medical disease with accompanying brain damage. Now, with therapeutic agents such as insulin, control of the blood sugar level can prevent this disorder from ravaging the brain or other organ systems. Similarly, the early diagnosis and treatment of hypothyroidism (depending on the nature of the enzyme-organ problem involved) is a dramatic medical advance.

Drug toxicity raises the issue of the side effects of particular compounds, such as the risk of deafness associated with streptomycin and the risk of kidney toxicity secondary to the use of sulfa compounds. The essential strategy in these disorders is to obviate the use of potentially harmful pharmaceuticals when there are other less toxic agents available. For example, returning to the treatment of thyroid disorders, radioactive iodide has been found to be extremely toxic, and perhaps carcinogenic; whereas propylthiouracil can be rather safely utilized. The Thalidomide experience is a clear example of the need for fully understanding the potential for drug teratogenesis and avoiding such pharmaceuticals in favor of less toxic agents. Preventive postures should also govern the use of the phenothiazines and barbiturates; when such drugs are utilized, the lowest dosage possible should be administered in a fashion that minimally impairs the autonomy of the individual.

There is no specific treatment (and probably never will be) for mental retardation caused by exposure of the fetus to radiation. Prevention is the only effective means of control.

The alkylating agents (such as those for the excretion of lead and mercury) are excellent treatment interventions for individuals exposed to excessive amounts of lead or mercury compounds. The energetic use of BAL factor for decreasing the lead levels in the body can be considered curative for associated symptoms such as the anemia that is produced by this intoxication.

In reducing mental retardation related to maternal diabetes, the optimal treatment approach is meticulous clinical management of the diabetes mellitus in the pregnant mother. It is essential to be on the alert to identify the fragile juvenile diabetic who becomes pregnant and will require close clinical monitoring of her metabolic status.

As a cause of mental retardation, toxemias of pregnancy have been decreasing in frequency as a result of mass screening for hypertension. Treatment is supportive during the acute toxemia state. Toxemias of pregnancy are best approached by preventive measures and meticulous obstetrical care of the mother.

Prenatal, Perinatal, and Postnatal Factors

This is a very complex area since it bridges such diverse disease entities as toxic states, infections, blood incompatibility, and improper diet. Treatment of the multiple factors leading to mental retardation associated with the prenatal, perinatal, and postnatal periods is often a secondary consideration. There are a significant number of general preventive strategies, such as assuring excellent prenatal care, focusing on maternal nutrition, and providing general social and financial support for the mother and her family. These strategies, however, transcend the resources of the usual health care system. They include social programs that foster the well-being of all individuals, such as the recent federally funded program to establish centers for pregnant teenagers, not only to give help while they are pregnant, but also to provide counseling and redirection of their life styles, in hopes of preventing future unwanted pregnancies.

Prenatal Factors

Among the specific prenatal causes of mental retardation, are congenital syphilis and rubella. With syphilis, the issue of public education and early diagnosis with specific laboratory testing during the newborn period arises again. If there is an active infection phase in the newborn child, a cure is now possible via penicillin treatment. Treatment of syphilis takes on more specificity in the secondary and tertiary stages of this infection.

Rubella is a stellar example of what can be accomplished in the primary prevention of mental retardation. Recent national public health efforts have produced population immunization levels in the United States of between 92 to 95 percent of females under eighteen years of age. This figure provides great hope that the recurrent epidemics of rubella—and its tragic aftermath for pregnant women—will no longer occur in our country. Yet, even when primary prevention efforts have not been utilized, there is still much that can be done therapeutically after the birth of a child whose mother had rubella during the first trimester of her pregnancy, and there is ample opportunity for active tertiary prevention programs for the child. The most common residual handicap of congenital rubella, a high frequency hearing loss, can be effectively ameliorated by special programs for the hearing-impaired. Other associated handicaps include mental retardation, microencephaly, and blindness. Programs for the visually impaired can be helpful in dealing with the child's visual problems. These children also have special needs for intensified contact with people, especially early mother-child interactions. Accordingly, a very energetic program to promote personal interaction is needed in the first year of life. Without this psychosocial development intervention, the youngster is likely to develop a fixed autistic orientation to the outside world.[10]

Infectious diseases of both prenatal and postnatal origin continue to present major treatment challenges. For example, though toxoplasmosis continues to be a relatively frequent prenatal syndrome which can produce mental retardation, no specific treatment exists for it at this time. Repeated attempts at a cure have been made with anti-parasitic drugs, but it continues to be extremely resistant to definitive treatment. Antiviral agents have been utilized in another prenatal infection, cytomegalic inclusion body disease, but the findings as to their efficacy are still equivocal.[39] These and similar prenatal disorders continue to present a treatment challenge, whether *in utero* or in the newborn and early infancy periods. Of the childhood diseases, rubella in particular presents a special challenge. Ironically, many professionals continue to view these disorders as being "usual," "common," or "expected" in childhood. Accompanying this attitude is the opinion that these disorders are not dangerous even though according to clear data, one-third of all children who display clinical signs of rubella tend to have abnormal electroencephalograms during the febrile phase of this disease.[10] They are extremely at risk to develop encephalitis secondary to this systemic viral disorder. Such considerations underscore the need for meticulous medical care during the acute period of infection. Utilization of supportive medical care should be addressed to the use of anticonvulsants and/or antibiotics for the secondary symptoms and infections of the primary generalized rubella infection. This contemporary approach to the active management of rubella can be helpful in preventing secondary manifestations or resultant handicaps.

PERINATAL FACTORS

The perinatal factors that increase the likelihood that a mother will have a retarded child are well known: prolonged labor, toxemia, prematurity, difficult birth, perinatal hypoxia, maternal syphilis, maternal pelvic radiation, maternal diabetes, prolonged maternal infertility, low socioeconomic status, previous birth of a defective child, maternal age (very early adolescence or over age thirty-five years), and high parity (five or more pregnancies, especially if closely spaced). In all of these factors, preventive treatment should focus on: (1) full awareness and documentation of the fact that the mother is at risk; and (2) maximum prenatal care for the mother. Modern prenatal care represents the best possible means of providing the newborn child as great a chance as possible to develop normally. After birth, the child of the at-risk mother should be fully evaluated and closely observed, with close attention to supportive physical care. For example, physical factors that present major treatment challenges are frequently noted in infants of low birth weight with associated congenital anomalies of the limbs. Special sensory defects of these infants require energetic treatment. From across the world, the findings of the last decade strongly indicate that if early treatment is provided, there is a sharp decrease in both the number and extent of symptoms. For example, infants who have experienced a physically traumatic birth may develop symptoms of cerebral palsy. Treatment techniques, such as developmentally oriented physical therapy, which focuses upon increasing the range of motion of the child's limbs and decreasing spasticity, may well result in the elimination of 75 to 80 percent of cerebral palsy symptoms. These children should be closely monitored by follow-up observation and allied interventions. These treatment strategies also foster early development and habilitation.

POSTNATAL AND EARLY CHILDHOOD DISEASES

Postnatal infections continue to be a major cause of mental retardation. The two major groups of these disorders are the infection processes subsumed under the terms meningitis and encephalitis. It is of crucial importance to remember that these two disorders are entirely different, tend to have distinct infectious agents, and subsequently may require different treatment approaches. Meningitis is an infectious disorder that primarily involves the meninges of the brain, is bacterial in nature, and, if diagnosed early, can be effectively treated by a broad range of antibiotics. Encephalitis, on the other hand, does not specifically involve the meninges. It directly attacks brain substance itself, tends to be caused by viral agents, and is typically not responsive to currently available antibiotic therapy. Effective treatment for meningitis requires: (1) rapid diagnosis; (2) laboratory identification of the bacterial agent involved; and (3) specific treatment interventions which may involve exceeding the usual dosage range of antibiotics. Delayed or ineffective treatment may leave the child with seizures, coma, and mental retardation. Clearly, these meningeal infection disorders are curable. Even those children who are brought in too late for early definitive treatment, usually have far fewer residual handicaps if vigorous treatment and supportive general measures are initiated.

The effective treatment of the infections that cause encephalitis is, at the present time, a most difficult and discouraging problem. There are only a few effective antiviral agents that may be helpful in selected instances of encephalitis. Current treatment emphasizes general physical support during the period of acute infection to prevent the body from being overwhelmed (for example, cortisone may be needed to prevent adrenal exhaustion), and youngsters who develop seizures during the acute phase need active treatment with anticonvulsants. Active treatment, albeit supportive, increases the rate of cure in encephalitis, but, more importantly, it significantly decreases the number of residual effects of the infectious disease process. Special rehabilitation treatment resources are especially needed for those youngsters who have regressed in their development as a result of a meningitis or encephalitis episode.

Tuberculosis, especially the pulmonary form with brain foci, formerly was endemic in certain parts of the United States. It is rarely seen today. Treatment involves: (1) specific and early diagnosis; (2) specific and early antibiotic therapy; and (3) persistence in the treatment. This treatment regimen has made tuberculosis a minor problem in our country since it is a curable infection. However, if early diagnosis and specific antibiotic agents are not utilized (or not utilized in appropriate amounts), youngsters who have suffered pulmonary tuberculosis may have residual brain damage, special sensory impairments, and mental retardation.

New Growth and Tumors

New growths or neoplasms, which result in the symptom of mental retardation, encompass disorders that are essentially restricted to the central nervous system. Benign and malignant brain tumors are the most frequent causative agents. Treatment centers around: (1) early diagnosis before the tumor has grown so far that it has destroyed adjacent brain tissue; (2) attempts to remove neurosurgically the abnormal tissue to determine whether or not it is malignant; and (3) support for the child whose world has changed as a result of the tumor, whether it is malignant or not. Accordingly, the major treatment approach is definitive surgery or surgical techniques that focus on aiding the child to recoup as much intact function or ability as possible.

Hypothyroidism

It has become clear, during the last five years in particular, that hypothyroidism in the newborn is probably as frequent as phenylketonuria (that is, it occurs in one in 6 thousand to one in 10 thousand births). Mass screening and laboratory confirmation have made possible accurate checks for hypothyroidism in the newborn nursery and subsequent definitive treatments. Although the spectrum of causology can range from enzyme deficiency to tissue anomalies of the thyroid, it can be said that, if detected early, congenital hypothyroidism is curable; if the diagnosis is not confirmed within the first year of life, it is no longer curable. In this instance, the word "curable" indicates reversal of all the skin, blood, and body changes, and of mental retardation. It is the latter aspect, development of intelligence, that is nonresponsive to treatment intervention after one year. Even so, one must still inaugurate treatment and management to provide specific metabolic support to the individual (for example, thyroid compounds and/or iodine) so as to salvage as much developmental function as possible.

Today, primary prevention strategies do exist for detecting hypothyroidism. In some states, in which newborn screening for hypothyroidism is mandatory, the disorder is no longer present. Universal application of these primary preventive approaches should cause hypothyroidism to become a disease of the past.

General

Disorders that reflect brain or skull malfunction as a cause of mental retardation encompass brain malformations and specific syndromes of the face and skull. In this area, cure, or primary prevention, is currently not a viable treatment goal for the majority of individuals suffering from these disorders. However, a small number of these disorders are amenable to surgical intervention. For example, in the area of spina bifida with associated disorders such as meningomyelocele, surgery in the first week or two of life allows the physician to open the herniated sac in the posterior sacral area of the spine, carefully resect the tissue around the area of the spinal cord, replace the spinal cord in the vertebral column, and cover it over with a bone graft. If this particular surgery is not performed early in life, the spinal cord will become adhered to the overlying sacral area tissues and will stay fixed at the point where it is herniated out of the spinal column. Later brain-spinal column growth will literally pull the brain down onto the upper cervical vertebrae and produce serious compression of the brain stem. In a number of these instances, surgical intervention can be a curative procedure. In other instances, resultant bilateral paralysis (distally from the umbilicus) may be secondary to a spinal lesion and, in these cases, a general rehabilitative treatment program similar to those utilized in cerebral palsy is needed.

Craniostenosis usually implies that the sutures of the human skull, which are normally about 80 percent closed by two to three years of age, begin instead to close at nine to ten months of age. At this developmental time (under normal conditions) the growing brain is literally pushing the overlying, and still very mobile, skull to ever greater expansion. If the sutures close prematurely, the rapidly enlarging brain begins to fill a closed cranial vault with resulting secondary pressure and atrophy of the underlying brain. The treatment of choice is early diagnosis by means of radiographic studies of the skull. Neurosurgical intervention involves crushing the suture lines, placing strips of platinum or polyethylene in the suture lines, and allowing this material to remain in place until the child is almost three years of age. For many children with craniostenosis this neurosurgical procedure is curative.

Over the past thirty years, great strides have been made in the treatment of hydrocephalus. This condition is caused by production of an excess amount of cerebrospinal fluid in the midline structures of the brain. This excess fluid balloons out the inside of the brain and pushes the skull out to form a very large external configuration. Following radiographic studies to locate and categorize the nature and/or cause of the individual case, a Holtner valve can neurosurgically be placed in one of the lateral ventricles of the hydrocephalic brain and connected to surgical tubing that exits out the adjacent mastoid or is threaded under the skin to the heart or one of the kidneys. The Holtner valve is arranged so that when a certain pressure gradient of the cerebrospinal fluid is reached inside the skull, the valve opens and discharges excess fluid into the mastoid, heart, or kidney for excretion. The use of the Holtner valve and associated surgery can be regarded, in many instances, as curative for hydrocephalus.

Other disorders, such as the first arch syndrome, Rubinstein-Taybi syndrome, and the Klippel-Feil syndrome, are not curable or effectively treatable at this time. However, as with many other disorders, treatment can provide general physical support. Habilitation efforts for the child should be started very early in life.

¶ Psychiatric Disorders

Emotional disturbances in the retarded continue to be a frequent cause for their institutionalization. A study of emotional disturbance in a sample of institutionalized individuals with Down's syndrome noted that whereas only 37 percent of the total sample were emotionally disturbed at the time of the study, 56 percent had displayed significant symptoms of emotional disturbance at the time of their initial admission to the institution.[38] One might ask why most of these people who were emotionally upset were sent to an institution for the retarded rather than to a mental health facility. In contrast, a brief period of inpatient or outpatient psychiatric care, coupled with revised expectations of the parent and increased support in the community, could frequently have prevented institutionalization.[36]

Compared with the incidence of emotional illness in the general population, the retarded are more at risk. Early prevalance studies (before 1960) were carried out primarily in institutional or hospital settings, and frequency rates reported ranged from 16 to 40 percent.[28,31] A recent series of reports on mentally retarded persons living with their primary families or in community-based settings at the time of study have appeared during the last fifteen to twenty years. These studies, especially those focusing on retarded children under the age of twelve years, have rather consistently reported a 20 to 35 percent frequency of emotional disturbances.[9,37,47] These findings are especially important when one notes that epidemiological studies on mental illness, for example, the Mid-Town Manhattan and Sterling County studies,[26] in the general population reveal an incidence of 14 to 20 percent.

In the broadest sense, one finds that the mentally retarded develop essentially the same types of emotional illness that befall persons of normal intellectual ability.[11] One finds in the retarded the full range of psychoses, neuroses, personality disorders, behavior disorders, psychophysiologic disorders, and transient situational disturbances that are noted in the "normal" population.

In community-based psychiatric programs that treat the retarded (community mental health centers), it is not unusual to note combined diagnoses, such as childhood schizophrenia and moderate mental retardation, or unsocialized aggressive reaction of adolescence and mild mental retardation. Practically speaking, certain diagnostic categories, such as the neuroses, tend to be underrepresented in the retarded, while others are seen with relative frequency (for example, schizophrenia, the various behavioral reactions, and transient situational disturbances). The following section reviews the diagnostic entities seen most frequently and those that present special problems of diagnosis and treatment.

Psychotic reactions of childhood have presented a major challenge to the clinicians since they were recognized as distinct in the first decade of this century.[14] Delineation of types and etiologies has been delayed, in part, by the fact that the psychotic child frequently functions at a mentally retarded level, and early observers believed that all psychotic children deteriorated. In 1943, early infantile autism was described and became the focus of much interest, including speculation as to whether it represented the earliest form of childhood schizophrenia.[30] The term "autism" is frequently employed in the differential diagnosis of severe emotional disturbances in infancy and early childhood. Yet, to label a child autistic presents some formidable problems with regard to definition of the term, specific etiological-diagnostic implications, and treatment considerations.[44] All too often, the word is used as if it were a diagnosis, a synonym for childhood schizophrenia, or an abbreviation for early infantile autism. Such usage obviously is imprecise and contributes further to the diagnostic confusion that has abounded in the literature concerning childhood psychosis.

Interestingly, today there is not the degree of fervor over diagnosis, treatment, and differential outcome concerning the functional childhood psychoses and their interrelationships to mental retardation that existed ten to fifteen years ago. A number of follow-up studies,[38] the rediscovery of the wide variety of primitive behavioral repertoires in the retarded (the same behavior that had been termed "psychotic" in the past), and a lack of relative differences in treatment modalities and corresponding responses to them, have all tended to mute the intensity of this earlier clinical debate. For example, an excellent review of the past relationships between emotional disturbance and mental retardation by Garfield and Shakespeare[19] addressed almost a third of its content to the relationships between emotional disturbance and mental retardation. Indeed, as Creak[13] and Penrose[45] have noted, the most common challenge is to ascertain not whether the patient is retarded or psychotic, but how much of his condition is attributable to retardation and how much to psychosis.

Earlier it was noted that the psychoses of childhood intensified the study of the interrelationships between mental retardation and the psychoses of childhood. Now in the 1980s, the issue has been clarified, and it is becoming apparent that the number of functional etiologies of infantile autism and childhood schizophrenia is limited. The reported evidence for central nervous system pathology in the psychoses of childhood is the most frequent trend noted in the past ten years.[51]

In summary, these clinical reports have shown clearly that: (1) the psychoses of childhood, particularly autism, are strongly associated with dysfunction of the central nervous system; (2) the appearance of psychotic behavior (and/or autistic behavior) and mental retardation in young, nonverbal children indicates both common etiology and a diminished capacity to tolerate stress; (3) retarded patients may show stereotyped, self-stimulating behavior that resembles autism; and (4) relief of the psychotic condition in "autistic" children far more commonly produces a retarded child who is able to interact with others rather than a child of normal intelligences.

Personality disorders are characterized by chronically maladaptive patterns of behavior (for example, antisocial personality, passive-aggressive personality, and so forth) that are qualitatively different from psychotic or neurotic disorders. Studies reported in the earlier history of retardation tended indiscriminately to see antisocial behavior as an expected behavioral accompaniment of mental retardation. Indeed, the much discussed earlier reports on the relationship between retardation and personality disorders —especially the antisocial personality—were couched in moralistic-legal terms rather than containing definitive descriptive criteria.[3] The antisocial personality designation continues to receive much attention, and it is frequently overrepresented in references to borderline and mildly retarded individuals. It would appear that, for a variety of reasons, behavioral problems of an antisocial nature are more frequently seen in this group. The same poverty of interpersonal relationships during childhood that leads to retardation associated with psychosocial deprivation can also lead to impaired object relations and poorly internalized behavioral and emotional controls. Also, the diminished coping skills of this group often lead them to perform deviant acts simply to exist. Finally, this group is most likely to be released from institutional settings in young adulthood, and their behavior illustrates graphically the effects of institutionalization on personality structure. It is interesting to note that other personality disorders (for example, schizoid personality) have been reported only rarely in the retarded. Indeed, the only other personality disorder in the retarded that has received much attention is the "inadequate personality," even though the application of exact diagnostic criteria would exclude this disorder as a primary diagnosis in mental retardation.

In summary, although personality disorders do occur in the mentally retarded, they are based primarily on extrinsic factors, have no distinct etiological relationships to mental retardation, and, despite persistent folklore, are not abnormally frequent in the noninstitutionalized retarded population.

Psychoneurotic disorders were rarely included in discussion of the frequency and types of emotional disturbances in the mentally retarded before 1950.

In 1970, a study[9] disputed the concept of incompatibility between neurosis and retardation. Neurotic phenomena tend to be associated with atypical developmental patterns in conjunction with disturbed family functioning. Psychoneurotic disorders of retarded children tend to be linked to symptoms of anxiety, fear of failure, insecurity, and to exogenous factors such as chronic frustration, unrealistic family expectations, and deprivations. Psychoneurotic disorders are more common in children functioning in the high-moderate and mild ranges of mental retardation. This finding has prompted speculation as to whether the relative complexity of psychoneurotic transactions is beyond the adaptive limits of the severely retarded.[53]

Transient situational disturbances are a rather large category of minor emotional disturbances that are perhaps underutilized in the clinical assessment of emotional disturbances in the retarded population. The authors think that this underutilization is one of the major drawbacks of descriptive approaches to the retarded. It is defined as a category reserved for transient disorders of any severity (including those of psychotic proportions) that occur in individuals who have no apparent underlying mental disorders and that represent an acute reaction to overwhelming environmental stress. The transient nature of these disorders is their paramount feature, and the assessment of adaptive capacity poses a recurrent dilemma when one works with a retarded population. If the clinican thinks that retarded people have poor adaptive capacities and, therefore, expects little resolution, treatment intervention is less than energetic, and other diagnostic categories are often utilized. Furthermore, if the mentally retarded are considered to be excessively prone to emotional disturbances, "transient" is then viewed as the beginning of a chronic emotional disturbance that has emerged to accompany the retardation.

In the authors' experience, a great number of emotional and behavioral problems in the retarded are transient in that they are frequently caused by inappropriate expectations or rapid changes in life patterns. These problems often respond rapidly to environmental adjustments. In summary, even though it is not possible to use the transient situational diagnosis if one follows the letter of current diagnostic guidelines, it is hoped that professionals will conceptualize disorders in the retarded in this manner when it is appropriate.

¶ Problems Associated with Different Levels of Retardation

The severely retarded are characterized by gross central nervous system impairment, multiple physical signs and symptoms, and a high frequency of multiple handicaps (in particular, special sensory and seizure disorders). Such severe problems directly impair their ability to assess and effectively participate in ongoing interpersonal-social transactions. Clinically, these patients manifest primitive behaviors and gross delays in their developmental repertoires. Studies by Chess, Korn, and Fernandez[10] of severely retarded children with the rubella syndrome and by Grunwald[22] on the multiply handicapped-severely retarded clearly document the high vulnerability of these children to psychiatric disorders.

From a diagnostic viewpoint, the very primitiveness of the severely retarded person's overall behavior, in conjunction with much stereotyping and negativism, may be misleading. For example, when under minimal stress in an interpersonal setting, mentally retarded children frequently exhibit negativism and out-of-contact behavior. This behavioral response may initially suggest a psychotic disorder of childhood. However, these children do make eye contact and will interact with the examiner quite readily, despite their minimal behavioral repertoire.

The authors have been impressed by the extent of personality development the severely retarded can attain if early and energetic behavioral, educational, and family counseling interventions are initiated and maintained. True, they remain severely handicapped in their cognitive and social-adaptive characteristics, but there is a world of difference between the severely retarded child with many self-help skills who graduates from a standing table to a wheelchair and the untrained, severely retarded one who tries to withdraw from, or is aloof to, interpersonal contacts and who is totally lacking in self-help skills. Interestingly, these youngsters tend to be accepted by their parental support systems and peer groups (if adequate evaluations and anticipatory counseling are accomplished), perhaps reflecting empathy for the obvious handicaps they display.

The moderately retarded encompass some of the same etiological dimensions noted previously accompanied by a wide variety and high frequency of associated handicaps. The children's slow rate of development and their specific problems with language elaboration and concrete approaches to problem-solving situations present both unique and marked vulnerabilities. In an outstanding study, Webster[53] viewed these personality vulnerabilities as stemming from the characteristic postures moderately retarded children tend to use in their interpersonal transactions: more autism (selective isolation), inflexibility, repetitiousness, passivity, and a simplicity of emotional life. This simplicity of emotional life, a cardinal characteristic of the moderately retarded, reflects their undifferentiated ego structures and poses the clinical challenge of attempting to modulate their tendency toward direct expression of basic feelings and wants, as noted in their obstinacy, difficulties in parallel play situations, and so forth. The limited repertoire of personality defenses, coupled with their concrete approaches, tends to be fertile ground for overreaction to minimal stresses in the external world. Proneness to hyperactivity and impulsivity, rapid mood swings and temporary regression to primitive self-stimulatory activities are characteristic of their fragile personality structures. Unlike the severely retarded, these youngsters tend to be rejected by their parents and peers. Their significant attempts to approximate developmental expectations, coupled with the aforementioned behavioral traits, appear to alienate them from those very interpersonal contacts they so desperately need.

The mildly retarded have given rise to debate as to whether they represent the statistical expression of the polygenetic basis of the symptom of mental retardation, or whether they are the untutored "have-nots" of a society that tolerates only minor deviations from the norm.[15] Emotional disturbances of the mildly retarded reflect the well-known residuals of a person who is labeled deviant and then becomes caught in the dynamic interplay of disturbed family transactions. The typical delay in establishing that these youngsters have a distinct learning disability (usually not confirmed until six to nine years of age) presents the mildly retarded individual with a constant source of anxiety about his inability to integrate the normal developmental sequences at the appropriate time in his life. Usually, during the latent period of psychosexual personality integration, mildly retarded children have considerable difficulty in understanding the symbolic abstractions of schoolwork and the complexities of social-adaptive expectations from both family and peer group. Often at this stage they gain some insight into their limitations and, by early adolescence, have established an identity that incorporates both retardation and deviance. Frequently the vulnerabilities of the mildly retarded are not buffered or redirected by loved ones into new interpersonal coping styles to help correct earlier misconceptions about the self. Without some community support and direction, the mildly retarded are at high risk for failure in society —especially urban society. In the past, if they managed to avoid an institutional setting for the retarded, it was not unusual to find them, eventually, in other types of institutions, such as correctional facilities or state-supported psychiatric hospitals.

In summary, it appears that mildly mentally retarded individuals are very likely *not* to be readily identified as handicapped and needing support. Rather, they are seen as society's misfits who, if not simply ridiculed, are apt to be taken advantage of in far more serious ways because of their lack of judgment and limited coping skills.

¶ Psychiatric Problems Associated with Different Models of Care

Another way to conceptualize the problems of the retarded, in addition to types of emotional disturbance and levels of retardation, is by problems that appear to be related to different models of care. Providing treatment guidelines for optional care for the retarded at home, in the community, or in an institutional setting is extremely difficult since there is no such thing as an "average" retarded child. In a general way, they can be grouped by overall abilities; but one of the most striking things about the retarded is the great variation in abilities that may be seen in each individual. This variability, plus the great difficulty caregivers are likely to experience in attempting to understand fully the retarded person's abilities and disabilities, appears to be the basis of a number of the psychiatric problems seen in the retarded. The most common type of error in the care of the retarded is a result of the caregiver's expecting either too little or too much. Too few expectations, combined with too little effort on the part of the care providers, is often the lot of formerly institutionalized retarded individuals. They tend to show a pattern of underachievement and a detachment syndrome that is typical of people reared in barren institutional environments. One common characteristic is a profound and often indiscriminately expressed affect hunger. Because they often have not had experience with significant or meaningful object relations and have been accustomed to living amid large numbers of minimally involved people, their indiscriminate approach to strangers may lead to serious problems. Another variant is the situation in which the caretaker actually does too much instead of too little—an overprotective model. Parents sometimes feel that the only acceptable solution is to keep their retarded child or adult at home, and too often they are assigned to an isolated part of the home away from the bulk of family or external social contacts. Here, the family caters to their retarded relative's every need and, in doing so, increases his dependence and almost totally eliminates his ability to develope effective social-adaptive functions.

In summary, while the detached, mildly or moderately retarded person is at risk of becoming the counterpart of a person of normal intelligence with a character disorder, the overprotected retarded person is likely to show symptoms of inflexibility, autistic thinking, and situational anxiety.

At the other end of the spectrum, there are retarded individuals who show clear evidence of their caretakers' having expected too much from them. One of the most common problems in very young, moderately retarded children who do not have physical stigmata is a failure by the parents to recognize their children's intellectual limitations before the normal time for language acquisition. It would appear that one common cause of autistic-like psychoses is the placement of a sensitive, intelligent-appearing, but nevertheless retarded, child in a situation in which conscientious parents are doing all the "right" things during the second year of life to facilitate language skills. Verbal demands often cause the moderately retarded child with a language disability to react with increasing anxiety and a variety of avoidance behaviors that result from his lack of pleasure in verbal interactions. Similar examples of excessive expectations are occasionally seen in innovative institutional or community programs in which children who are severely retarded may be urged into overintense efforts to maximize their capabilities. In some cases this results in more frequent seizures, and in others produces a pattern of autistic withdrawal similar to that noted previously. One of the most distressing problems with older children in this group is outbursts of violent behavior when excessive expectations have been maintained too long. All too often such children are placed on high doses of medication in an effort to control aggression that is actually reactive in nature and not a symptom of major emotional disturbance.

¶ Comprehensive Approach to Treatment

A comprehensive-treatment approach to mentally retarded children with associated emotional disturbances should include the following basic principles.

Open-minded Approach

The diagnosis and treatment of children who are both mentally retarded and emotionally disturbed necessitate an open-minded approach. This is the first basic requirement for the clinician who plans treatment for these youngsters, and it is important to maintain this approach throughout treatment. Periodic reevaluation often reveals developmental surprises that underscore the need for a flexible diagnostic-prognostic attitude.

Active Family Participation

The second principle in planning treatment for children with both mental retardation and emotional disturbance is to engage the family in active participation as early as possible. The family is the key to any effective treatment program. The clinician's attitudes and level of interest frequently determine the success of this endeavor; thus, future cooperation with the family (or lack of it) may reveal his unspoken, as well as spoken, attitudes at the time of initial contact. The therapist should convey to the family his willingness to share the facts of the case—not at the end of treatment, but as part of the first step. Treatment plans must be a cooperative process that parents and clinician work out during the course of treatment.

It is valuable to indicate at an early contact that treatment planning rarely results in a single recommendation; it may shift in focus and alter its course as the child grows and develops. Diagnostic and treatment flexibility in the early stages help a clinician to view the total child and refer to other special sources of help as required.

Much has been written about the grief reactions of families with handicapped children. Such a reaction frequently occurs in parents of mentally retarded children. The clinician evaluating these children must be alert to this grief reaction when first offering parents an interpretation of their child's condition and in subsequent interviews.

Assessment of family interactions and strengths is a necessary part of the total evaluation, since these assets are essential to planning a comprehensive treatment program. Conversely, some of the family psychopathology may reactivate the difficulties of the child being treated. Several interviews may be necessary to determine the nature of family transactions.

Early Diagnosis and Treatment

A third principle of the comprehensive treatment approach is early descriptive diagnosis and early treatment. This includes clarification not only of what needs treatment but also of what can and cannot be actively treated. Full discussion of therapeutic goals will assist families in establishing realistic treatment expectations so that mutual frustration is reduced and fewer secondary psychiatric problems are encountered. In this sense, prevention becomes part of the ongoing work with the child and his family. This total approach requires continued follow-up of the patient. Periodic reevaluations must be done, and appropriate shifts in treatment and expectations carried out.

Initial Contact

The fourth principle is to accept each child as he is at the time of initial contact. He needs acceptance for what he is, not what he might have been without his problem or if therapy had been undertaken sooner. A corollary of this principle is awareness of the family's feelings and acceptance of them as they are. Increasing the parents' guilt feelings is rarely, if ever, desirable in attempting to motivate them toward therapy.

Maximization of Developmental Potential

The fifth principle requires focusing on the maximization of developmental potential. It calls for a different type of goal setting from the usual treatment expectation, since the focus often must be on what the child can do rather than on an anticipated cure. The goal then becomes one of trying to provide the child with the necessary opportunity and support to develop as fully as possible and to reduce obstacles to a minimum.

Coordination of Services

The sixth principle is to coordinate the services needed by the child. This requires awareness of the various services available in a given community and an attitude that permits collaboration. It necessitates sharing the overall treatment plan with the child (when his or her overall coping ability permits) and the parents. As with other groups of dependent psychiatric patients, child or geriatric, the clinician's efforts must frequently be directed as much toward assessing the strengths and weaknesses of the family and coordinating the available community support systems, as they are toward care of the individual patient. Many retarded people with acute emotional problems are institutionalized unnecessarily. In nearly all cases, the retarded can be maintained in the community with the help of mental health personnel who are willing to provide short-term and long-term care for them.

¶ Bibliography

1. ACOSTA, P. "Prenatal Supplements and Breast Feeding," in Koch, J., ed., *International Summit on Prevention of Mental Retardation from Biomedical Causes, President's Committee on Mental Retardation,* Washington, D.C.: Department of Health, Education and Welfare, Pub. No. HDS 78-21023, 1978, pp. 111–135.
2. AMERICAN ACADEMY OF PEDIATRICS. *Report of the Committee on Infectious Diseases,* 18th ed. Evanston, Ill.: American Academy of Pediatrics, 1977.
3. BARR, M. W. *Mental Defectives, Their History, Treatment and Training.* Philadelphia: P. Blakiston, 1904.
4. BEATTIE, A. D., et al. "Role of Chronic Low Level Lead Exposure in the Etiology of Mental Retardation," *Lancet,* 1: 7907 (1975): 589–592.
5. BEESON, P. B., and MCDERMOTT, W., eds. *Textbook of Medicine,* 14th ed. Philadelphia: W. B. Saunders, 1975.
6. BERRY, H. K., et al. "New Approaches to Treatment of Phenylketonuria," in Mittler, P., ed., *Research to Practice in Mental Retardation,* vol. 3. Baltimore: University Park Press, 1977, pp. 229–239.
7. BIRCH, H. G., and GUSSOW, J. D. *Disadvantaged Children: Health, Nutrition, and School Failure.* New York: Grune & Stratton, 1970.
8. CALIFONIA ASSOCIATION FOR THE RETARDED. *Prevention: An Agenda for Action.* Sacramento: California Association for the Retarded, 1977.
9. CHESS, S. "Emotional Problems in Mentally Retarded Children," in Menolascino, F. M. ed., *Psychiatric Approaches to Mental*

Retardation. New York: Basic Books, 1970, pp. 55–67.

10. CHESS, S., KORN, S., and FERNANDEZ, P. B. *Psychiatric Disorders of Children with Congenital Rubella.* New York: Brunner/Mazel, 1971.
11. COLLINS, D. T. "Head-banging: Its Meaning and Management in the Severely Retarded Adult," in Menolascino, F. J., ed., *Psychiatric Aspects of the Diagnosis and Treatment of Mental Retardation.* Seattle: Special Child Publications, 1970.
12. COOKE, R. Personal communication, November 16, 1978.
13. CREAK, E. M. "Childhood Psychosis: A Review of 100 Cases," *British Journal of Psychiatry,* 109 (1963): 84–89.
14. DESANCTIS, S. "Sopra Alcune Varieta Della Demenza Precoce," *Rivista Sperimentale di Freniatria e di Medicina Legale in Relazione con l'Antropologia e le Scienze Gueridiche et Sociale,* 32 (1906), 141–165.
15. EISENBERG, L. "Caste, Class, and Intelligence," in Murray, R. F., and Rossner, P. L., eds., *The Genetic, Metabolic, and Developmental Aspects of Mental Retardation.* Springfield, Ill.: Charles C Thomas, 1972, pp. 185–198.
16. FEINGOLD, M. "Amniocentesis: For Whom, By Whom?" *Patient Care,* 11:20 (1977): 16–61.
17. FOTHERINGHAM, J. B., and MORRISON, M. *Prevention of Mental Retardation.* Toronto: National Institute of Mental Retardation, 1976.
18. GARBER, H., and HEBER, F. R. "The Milwaukee Project—Indications of the Effectiveness of Early Intervention in Preventing Mental Retardation," in Mittler, P., ed., *Research to Practice in Mental Retardation,* vol. 1. Baltimore, Md.: University Park Press, 1977.
19. GARFIELD, A., and SHAKESPEARE, R. "A Psychological and Developmental Study of Mentally Retarded Children with Cerebral Palsy," *Developmental Medicine and Child Neurology,* 6 (1964): 485–89.
20. GENERAL ACCOUNTING OFFICE. *GAO Plans Federal Effort to Prevent Retardation.* Washington, D.C.: General Accounting Office, 1977.
21. GHADIMI, H. "Aminoacidopathies and Mental Retardation," in Murry, R. F., and Rossner, P. L., eds., *The Genetic, Metabolic, and Developmental Aspects of Mental Retardation.* Springfield, Ill.: Charles C Thomas, 1972, pp. 45–57.
22. GRUNWALD, K. "International Trends in the Care of the Severely and Profoundly Retarded and Multiply Handicapped," in Menolascino, F. J., and Pearson, P. H. eds., *Beyond the Limits: Innovations in Services for the Severely and Profoundly Retarded.* Seattle: Special Child Publications, 1974.
23. GUTHRIE, R. "A Position Paper: Undernutrition in the Infant as a Cause of Mental Retardation," *Mental Retardation News,* 19:10 (1970): 2.
24. ———. "Current and Newer Aspects of Newborn Screening," in Mittler, P., ed., *Research to Practice in Mental Retardation,* vol. 3. Baltimore: University Park Press, 1977, pp. 13–19.
25. HAYDEN, A. H., and HERING, N. G. "The Acceleration and Maintenance of Developmental Gains in Down's Syndrome School-age Children," in Mittler, P., ed., *Research to Prevention in Mental Retardation,* vol. 1. Baltimore: University Park Press, 1977.
26. HOLLINGSHEAD, A. B., and REDLICH, F. C. *Social Class and Mental Illness.* New York: John Wiley, 1958.
27. HUSEMAN, C. A. "Neonatal Thyroid in Screening for the State of Nebraska," *Nebraska Medical Journal,* 63:8 (1978): 278–280.
28. JERVIS, G. "The Mental Deficiencies," in Arieti, S. ed., *The American Handbook of Psychiatry,* vol. 1. New York: Basic Books, 1959, pp. 1289–1316.
29. JONES, K. L., SMITH, D. W., and HANSON, J. W. "The Fetal Alcohol Syndrome: Clinical Delineation," *Annals of the New York Academy of Science,* 273 (1976): 130–139.
30. KANNER, L. "Autistic Disturbances of Affective Contact," *Journal of the Nervous Child.,* 2 (1943): 217–50.
31. ———. "Feeblemindedness: Absolute, Relative and Apparent," *Journal of the Nervous Child,* 7 (1948): 363–97.
32. KOCH, J., ed. *International Summit on Prevention of Mental Retardation from Biomedical Causes, President's Committee on Mental Retardation,* Washington, D.C.: Department of Health, Education and Welfare, Pub. No. HDS 78-21023, 1978.
33. MCCOY, K., "Report on the Wingspread

Conference on Adolescent Sexuality and Health Care," *Journal of Clinical Child Psychology,* 3:3 (1974): 18–22.

34. MACCREADY, R. A. "Rh Hemolytic Disease Prevention in the United States," Paper presented at Meeting of the American Association on Mental Deficiency, Chicago, Ill., May 3, 1977.

35. MARAVILLA, A. "The Early and Periodic Screening, Diagnosis, and Training Program," *Clinical Pediatrics,* 16 (1975): 173–178.

36. MENOLASCINO, F. J. *Challenges in Mental Retardation: Progressive Ideology and Services.* New York: Human Sciences Press, 1977.

37. ———, and BERNSTEIN, N. R. "Psychiatric Assessment of the Mentally Retarded Child," in Bernstein, N. R., ed., *Diminished People.* Boston: Little, Brown, 1970, pp. 201–222.

38. MENOLASCINO, F. J., and EATON, L. "Psychoses of Childhood: A Five-year Follow-up Study of Experiences in a Mental Retardation Clinic," *American Journal of Mental Deficiency,* 72 (1967): 370–380.

39. MENOLASCINO, F. J., and EGGER, M. J. *Medical Dimensions of Mental Retardation.* Lincoln, Neb.: University of Nebraska Press, 1978.

40. Michigan Department of Public Health. *Hypothyroid Mass Screening.* Lansing, Mich.: Michigan Department of Public Health, 1978.

41. MILUNSKY, A. *Know Your Genes.* Boston: Houghton, Mifflin, 1977.

42. MITTLER, P., ed. *Research to Practice in Mental Retardation,* vol. 3. Baltimore: University Park Press, 1977.

43. National Academy of Sciences. *Maternal Nutrition and the Course of Pregnancy.* Washington, D.C.: National Academy of Sciences, 1970.

44. ORNITZ, E. M., and RITVO, E. R. "The Syndrome of Autism: A Critical Review," *American Journal of Psychiatry,* 133 (1976): 609–621.

45. PENROSE, L. S. "The Contribution of Mental Deficiency Research to Psychiatry," *British Journal of Psychiatry,* 112 (1966): 747–755.

46. PERINO, J., and ERNHART, C. "The Relation of Sub-clinical Lead Level to Cognitive and Sensorimotor Impairment in Black Preschoolers," *Journal of Learning Disabilities,* 7:10 (1974): 616–620.

47. PHILLIPS, I., and WILLIAMS, N. "Psychopathology: A Study of 100 Children," *American Journal of Psychiatry,* 132 (1975): 1265–1273.

48. PIOMELLI, S. "A Model Program for Lead Screening," in *International Summit on Prevention of Mental Retardation from Biomedical Causes.* Washington, D.C.: Department of Health, Education and Welfare, 1979.

49. President's Committee on Mental Retardation. *Islands of Excellence,* Washington, D.C.: U.S. Government Printing Office, 1973.

50. ROSSO, P. "Nutrition, Brain Growth, and Prevention of Mental Retardation," in Koch, J., ed., *International Summit on Prevention of Mental Retardation from Biomedical Causes, President's Committee on Mental Retardation,* Washington, D.C.: Department of Health, Education and Welfare, Pub. No. HDS 78-21023, 1978, pp. 98–110.

51. RUTTER, M., and SCHOPLER, E., eds. *Autism: A Re-appraisal of Concepts and Treatment.* New York: Plenum Press, 1978.

52. Vermont Department of Public Health. *Mental Retardation Prevention Plan.* Burlington, Vt: Vermont Department of Public Health, 1977.

53. WEBSTER, T. G. "Unique Aspects of Emotional Development in Mentally Retarded Children," in Menolascino, F. J., ed. *Psychiatric Approaches to Mental Retardation.* New York: Basic Books, 1970, pp. 3–54.

54. WYNN, M., and WYNN, A. *Prevention of Handicaps of Perinatal Origin.* London: Foundation for Education and Research in Child Bearing, 1976.

PART FOUR

Hospital, Administrative, and Social Psychiatry

CHAPTER 32

THE NOSOLOGY OF PSYCHIATRY

Robert J. Campbell

NOSOLOGY (from *nosos,* "disease") is the study of diseases from the point of view of their grouping, ordering, and relationship to one another; it includes the classification of diseases as well as the formulation of principles for differentiating one disease from another.

Diagnosis (from *dia,* "through, dividing into parts," and *gnosis,* "knowledge, recognition") is the process of distinguishing or recognizing the whole from its manifestations, of detecting the presence of disease from its symptoms. The process of diagnosis affirms that a disease is present; it defines the nature or character of that disease at the greatest level of specificity possible; and it provides a summary statement of what was discovered. Diagnosis is therefore both a process and a statement of the conclusion to which that process leads.

Nomenclature (from *nomen,* "name," and *calare,* "to call") is the agreed-upon label or wording that is used to communicate the results of the diagnostic process. Nomenclature is the shorthand name for the disease that has been identified, but in addition it implies that there is some reason for preferring one name to another.

Classification is the grouping of diseases into classes or orders, a logical scheme for organizing and categorizing so that different types of diseases can be distinguished and assigned their proper places.

All four terms—diagnosis, nomenclature, classification, and nosology—refer to various aspects of the conceptualization of disease. Because they are overlapping and interdependent, rather than mutually exclusive, it is not surprising that usage has tended to blur the distinctions between them. In itself, that is of little matter; what is unfortunate is that the vagueness and uncertainty that surround their use have spread as well over the assumptions on which they are based. Often lost sight of is that each of them reflects current speculation and hypotheses about the conditions to which they are applied and not only "hard" knowledge or scientific "fact."

Expanding knowledge dispels far fewer hypotheses than it generates, and the diagnostic process and classificatory scheme must accommodate themselves both to facts as they are established and to the speculative models and innovative guesses that guide the research of the day. General paresis, for example, was well described as a clinical entity, or syndrome, by Haslan in 1798, by Bayle in 1822, and again by Esquirol in 1826. Its relationship to syphilis was suggested by Esmarch and Jessen in 1857 and again by Krafft-Ebing later in the century. But it was not until the discovery of the spirochete by Schaudinn and Hoffmann in 1905, the development of the Bordet-Wassermann reaction over the period 1901 to 1907, and Noguchi's demonstration in 1911 of spirochetes in the brain that it became possible to identify positively as syphilitic many conditions whose etiology previously had been merely speculative.

Psychiatric nosology, like all medical nosology, is thus ever changing, not only to correct the demonstrable errors and misconceptions of the past but also to provide a proper place for the discoveries that current technology promises.

Thomas Sydenham (1624–1689) is credited with the founding of nosology. He differentiated between symptom, syndrome, and disease and defined a disease syndrome as a group of symptoms, intercorrelated and not each a separate illness, differentiable from other syndromes, and with a characteristic pattern over time. In psychiatry, the greatest systematist was Emil Kraepelin (1856–1926). He introduced the prognostic approach into the classification of psychiatric disorders and thereby separated the endogenous psychoses with good prognosis (manic-depressive psychosis) from those with poor prognosis (dementia praecox, the group that Bleuler would later call the schizophrenias). Perhaps more than any other classification scheme proposed since Kraepelin's, the American Psychiatric Association's *Diagnostic and Statistical Manual of Mental Disorders* 3rd ed. (DSM-III)[5] reaffirms the Kraepelin and Sydenham position that nosology includes both a phenomenologic description of disorders and also a prediction of their outcome. Because the etiology and pathophysiology of most psychiatric disorders are unknown, they are not readily identifiable or separable on the basis of causative agent. Yet some of them disappear or abate while others do not, and a system that groups them according to what their response to time or treatment will be provides an invaluable tool to the clinician.

Change always provokes some degree of resistance, and the architects of DSM-III have had to contend with a substantial number of objections to their proposals. Those objections were met head on, and they often resulted in major changes in the classification. They also pointed up the difficulties that should be recognized by any who would construct their own or discard another's system of classification.

¶ What Is Illness?

As already noted, an essential part of the diagnostic process is saying that there is or is not a disease. But what is a disease? Often it is defined as any deviation from normal form and function, but, particularly in the area of human behavior and emotional reactions, such a definition is likely to be unsatisfactory. Who defines what is normal, and by what standards? Normality is often defined on the basis of statistical criteria, but how is the dividing line determined? One standard is that 95 percent of the population measured fall in the normal range. Yet any number of variants whose incidence is less than 5 percent of the population can be cited, and despite their rarity, they would hardly be considered disorders. How, for example, should one regard the basketball star? He is seven feet tall, far outside the normal range, and has cardiac hypertrophy, also a statistical abnormality. It is well known that cardiac hypertrophy is a significant factor in many types of heart disease. It is also well known that athletes' hearts respond to the demands of repetitive strenuous activity by enlarging. Cardiac hypertrophy in athletes is thus considered an appropriate, desirable, and even necessary response to the functional demand.

Functioning, then, and not mere counting or measuring, may be the answer to the question of what is disorder or disease. So long as a person functions well and is able to meet the ordinary stresses of life, he is to be considered healthy or normal and free from disorder. Such a definition, while it may be an improvement over a purely statistical approach to disease, has three major failings: it ignores latent disorders and conditions that may manifest themselves only under special conditions or after a period of development; it gives little due to the subjective or complaint aspect of disorder; and it avoids the issue of how a judgment of adequate functioning will be reached.

Think for a moment of the light-complexioned person in the days before chemical sun screens were invented. As long as he lived in a northern clime, worked mainly indoors, and moved about outside only briefly or after dark, no measures would detect deviation from the norm. Yet if forced to work under an equatorial sun, he would in a matter of hours be acutely ill. His "disease" could not accurately be described only in terms of overexposure to the sun, for its development depended at least as much on his inherent susceptibility or sensitivity. At what point does that person have an illness—only when the burn appears, at the first moment of exposure, or when his potentiality for developing the reaction is recognized?

As will be seen, the situation is relevant to the controversy surrounding the diagnostic criteria for schizophrenia. For any chronic progressive condition, the possibility that it might be prevented completely, or that its downhill course might be halted, would favor broad and overinclusive criteria so that the disorder could be recognized early enough to permit application of preventive efforts while there is still a chance they might be effective. On the other hand, if a syndrome is a final common pathway for the symptomatic expression of a group of diverse disorders, no completely rational approach to their prevention can be derived until they are identified and separated into homogenous entities. In order to do that it is necessary to apply rigidly exclusive criteria in order to prevent extraneous factors from contaminating the sample.

Another problem in defining disorder is that disease and illness are not the same. The sun-sensitive person described previously may carry his "pigment disorder" for years without knowing it, but when he is exposed to the sun becomes acutely aware of the fact that he is ill. The diabetic has a disease, to be sure, but so long as it is well-controlled by insulin, he is no more ill than the sun-sensitive person who screens out the harmful rays with para-aminobenzoic acid.

Diabetes illustrates many of the problems of defining disease. It consists of a cluster of symptoms, including polyuria, polydipsia, bulimia, weight loss, weakness, malaise, dehydration, and coma. It also includes various somatic and biochemical abnormalities, including glycosuria, hyperglycemia, abnormal glucose tolerance curve, and abnormal plasma insulin response. In a few cases, a clear cause can be identified, but not in most, although there is reason to believe that in a significant proportion of those in whom the cause cannot be identified the disease is a manifestation of genetic defect. In those cases, the disease must, of course, be present from the moment of conception, yet it will not become clinically apparent for many years. In the *latent* phase, it is not even possible to detect a biochemical abnormality by the techniques currently available. In the *preclinical* phase, biochemical abnormalities can be demonstrated, but the affected person remains symptom free and has no clinically apparent disorder of carbohydrate metabolism. It is only when the patient enters the *clinical* phase that signs and symptoms develop.

When is the diagnosis of "diabetes" justified? All would agree that by the time the patient is ill with symptoms the diagnosis is warranted. Its typical symptoms were clearly described by the first century. Sweetness of diabetics' urine was described by the fifth century, and by the late eighteenth century it was known that the sweetness was due to sugar in the urine. Would it have been correct to include all persons with glycosuria under the term "diabetes," even though they had not developed all the symptoms of the syndrome? With the development of the glucose tolerance test and, more recently, the plasma insulin response, it became possible to detect many "chemical" diabetics—some of whom would not develop clinical manifestations for many years, others never would except, perhaps, under special conditions. It is to be expected that future research will make it possible to devise tests of biochemical action that are even closer to the gene level than are plasma insulin levels—that is, within ten years it is likely that more "potential" or "chemical" diabetics will be uncovered. At what point along that inadequately charted course from genetic defect to biochemical abnormalities to early symptom formation to full syndrome development does one apply the diagnostic label?

Another level of difficulty is posed by a group of disorders due to a deficiency of the enzyme hypoxanthine-guanine phosphoribosyltransferase (HGPRT). When the enzyme is present at only a 0.005 percent level of normal activity, the affected subject develops a severe neuromuscular disorder with involuntary choreoathetoid movements, mental retardation, biting of the lips and fingertips, and a severe gouty arthritis because of high uric acid levels. To that cluster of symptoms the name "Lesch-Nyhan syndrome" has been given. In some members of the families of patients with Lesch-Nyhan syndrome, the enzyme is deficient, but not to the same degree. In those where the activity level is between 0.01 percent and 0.5 percent of normal, spinocerebellar syndromes of variable severity will develop; but if the enzyme level is as high as 1.0 percent of normal, the resultant syndrome is gout.[21] All three syndromes, as well as the group of conditions termed diabetes, underscore the fact that categorization of disease, even though it changes, is not wholly arbitrary; rather, it reflects developing knowledge, at increasingly discrete levels, of the process of pathogenesis.

It should be clear then that one can be diseased, even for many years, without being ill. Is it also possible that one can have an illness and not have a disease? To some extent, the question invites circular semantic debates, but at the same time it emphasizes the significance of the symptom or complaint level that any classificatory scheme must take into account. At what point, for example, does "obesity" become the appropriate designation for a subject's body weight? Supposing that a physician and patient can agree on the latter's ideal body weight, how many pounds need be added before the doctor is justified in prescribing a weight-loss regimen? The number might be very different for a middle-aged factory foreman and a twenty-year-old starlet or cover girl, for whom five extra pounds of weight might constitute a disastrous illness. A tiny pimple beneath the hairline on the base of the neck may be of no consequence; yet the same-sized pimple on the eyelid can be an excruciatingly painful illness as well as a significant disease.

A related difficulty in nosology is illustrated by Baker's example of cultural/national/racial influences on the designation of disease.[26] The axillary sweat of whites and blacks is "smelly" and offensive by Japanese standards. But about 10 percent of the Japanese people (mainly those of Ainu ancestry) also have "smelly" armpits. Their condition (osmidrosis axillae) is recognized as a disorder of enough significance to warrant exemption from military service, and there exist in Japan physicians who specialize in its treatment. In the United States and Europe, by contrast, if it is admitted at all that axillary odor is offensive, it is combated by an array of antiperspirants and deodorants that are consigned to the realm of the cosmetologist. It seems highly unlikely that it will be given disease status in the United States classificatory system, or that its management will be transferred to the dermatologist, the pathologist, or the psychiatrist.

It can be seen that, even outside the nebulous realm of psychiatry, the determination of what is disease is not an easy one. The definition is often man made, based on cultural or philosophic biases rather than objective phenomena, and reflective of personal and idiosyncratic value judgments rather than scientific data. It is probably useless, then, to try to differentiate between disease, disorder, derangement, ailment, malady, sickness, and illness. It might be better to accept Feinstein's dictum, ". . . that the only workable definition of disease is that it represents whatever the doctors of a particular era have defined as disease."[12]

¶ Classification and Nomenclature

Classification is a systematic arrangement, in this case, of disorders into classes so that different orders or levels can be distinguished from each other. Taxonomy is the theory of how classificatory systems should be structured and formalized, but even so abstract a level of operation cannot be divorced from the conceptualization of what it is that is being classified.

Every classification reflects the purpose(s) for which it was constructed in the first place. If the main purpose is to provide access to information that is not readily at one's fingertips, an index might be the most appropriate system, that is, a classification based on an alphabetical listing of names that are likely to be recognized by the user. Thus a "Directory of Mental Health Services" might list clinic, community mental health center, electroconvulsive treatment, hospitalization, insurance coverage, outpatient department, payment mechanisms, pharmacotherapy, psychotherapy, somatic treatments, and so forth without regard for where each might fall within a hierarchical ranking and without concern for the fact that the system is over inclusive and duplicative. Similarly, an "Index of Diagnostic Terms" might list in alphabetical order every name within the nomenclature, including names that are obsolete or not preferred (for example, "mongolian idiocy") as well as the more acceptable terms ("Down's syndrome" or "trisomy 21") and class names as well as genus and species names ("developmental disorder," "childhood onset pervasive developmental disorder"). The clinician who uses such an index, however, would expect that the page to which he is referred will place the disorder named within some logical frame and will indicate that language disorder is a specific developmental disorder of childhood. If it were only a so-called *key classification* that used "language disorder" as the single characteristic that would divide one subject from all others, the clinician would find himself in a hodgepodge of aged aphasics with cerebrovascular disease, alcohol or barbiturate abusers whose dysarthria betokens cerebellar involvement, bright children who stutter, anencephalics with no language whatever, catatonic patients with verbigeration, and a host of others.

Most biological classifications have progressed to a *natural classification,* grouping together forms that seem to share fundamental and significant characteristics. Such a classification provides not only conciseness, by reducing the number of separate elements that have to be examined, but also the prospect of efficient storage of the information obtained. In addition, it provides some degree of predictability, in that the new "case" with some of the characteristics of the group is likely to share other characteristics even though they are not as yet obvious. At the same time it must be recognized that how the grouping is made in the first place depends on the state of the art or science that decreed such and such characteristics to be fundamental and significant. Most natural classifications describe groups in somewhat exaggerated terms of what is believed to be significant, the prototypical case. Ignored are the innumerable factors that are believed to be insignificant or secondary. Particularly when one is dealing with clusters of symptoms (syndromes) rather than with well-defined diseases and when those symptoms are expressed in thoughts, feelings, or social relationships, rather than as more discrete and localized variations in a well-defined organ system, what may seem to be unimportant or irrelevant, or what is not seen at all because no technique has evolved to measure it, may in the long run turn out to be the most essential feature of a group. The schizophreniform episodes of acute intermittent porphyria, for example, appeared "naturally" to fall within the schizophrenic group. It was not until the Watson-Schwartz and glycine loading tests were devised that it was possible to demonstrate that such episodes were accompanied by increased excretion of δ—aminolevulinic acid and porphobilinogen in the urine. Only then could such patients be grouped correctly, within disorders of porphyrin metabolism.[18,20]

Behavior is a final common pathway for many disparate processes, which converge upon the only outlet available. Such *functional convergence*, as it is now commonly termed, has profound implications for both treatment and research. If altered behavior (be it hallucinations, melancholia, avoidance, aggressiveness, or withdrawal) were the only basis for defining a disease category, all "patients" with the same behavior would be given the same treatment. But the "schizophrenia" of one patient may be due to an inborn metabolic error, while the "schizophrenia" of another may reflect intrafamilial conflict, and it is unlikely that the same treatment will be optimal for both. Similarly in research, if behavior alone were used to make the groupings, the truly discriminant abnormalities (such as porphyrinuria) would be ignored because they would appear to be statistically insignificant within a large heterogeneous group. Many studies of schizophrenic populations have found that when defined behaviorally, the group showed no consistent abnormalities in any number of physiologic and biologic measurements. The comment was often made that greater variability was the only characteristic, when in fact the extremes had been averaged out by researchers, blotting out the differences between the distinct subpopulations with abnormally high and abnormally low scores.

Once the inadequacies of a single level approach were generally recognized, classification moved toward a polythetic approach. Variations in behavior (including thinking, feeling, and interpersonal relationships) were no longer the sole determinants of classificatory groupings; they were to be supplemented by as many measurements as possible, from as many levels as possible—physiologic, metabolic, previous history, course of illness, response to treatment, family history, and so forth. Syndromes could then be described in terms of clusters of measurements from all levels, rather than clusters only of symptoms. *Numerical taxonomy* provides a computerized system for quantifying the various measures, subjecting them to multivariate analysis, and thereby deriving objective, operational classification schemes.[13]

Yet even a computer classification is not without its pitfalls. The decision as to what is a disorder or what is undesirable is not the computer's, it is the investigator's. The decision, accordingly, will reflect the investigator's bias (or that of his culture) as to what is diseased, what should be changed, and what should be abolished. The decision as to which measures are relevant, which measure the same function (thus artifically exaggerating its importance by counting it over and over again in the clustering process), and which will give the broadest possible range of information also remains with the investigator. In order to gain maximal usefulness from any set of measures, Corning and Steffy[9] recommend the following criteria for their selection:

1. Use standardized procedures with known standard errors of measurement within population groupings;
2. Use tests with high construct validity, that is, tests whose items measure what they are intended to measure and not some subordinate or related function; a test of arithmetic ability, for example, should test a broad range of arithmetic tasks, and each item of the test should measure only arithmetic and not reading ability;
3. Select tests that have already demonstrated their sensitivity to psychopathology;
4. Give preference to indicators of vulnerabil-

ity or outcome, to tests that measure deficits over a period of time rather than to cross-sectional assessments of acute episodes of decompensation;
5. Use tests that are known to highlight or elicit distinctive abnormalities in the group(s) under study.

The final test of any new groupings thus obtained, of course, is their applicability to the patients the clinician sees, and particularly their predictive value relative to outcome. It is to be hoped that increasingly homogeneous diagnostic groups can be differentiated until the ideal is finally reached: each group identified has a common etiology, pathogenesis, and epidemiology. While we are far from that ideal now, present-day research promises to provide increasing refinement of the classificatory system with its insistence on a multi–axial, polythetic description: clinical characteristics, physical and neurologic factors, familial distribution of psychiatric illness, natural history, and biological indices (such as rapid eye movement latency, dexamethasone suppression, pharmacological responsiveness, and so forth).[1] Throughout the United States, there is growing emphasis on pragmatic, operational criteria for diagnosis, as free as possible of theoretical speculations about etiologic and pathogenetic mechanisms. Indeed, in the introduction to their book on psychiatric diagnosis, the St. Louis group[11] states: "There are few explanations in this book. This is because for most psychiatric conditions there are no explanations."[27]

¶ ICD-9, ICD-9-CM, and DSM-III

The *International Classification of Diseases* (ICD) is a product of the World Health Organization (WHO) and was designed originally for the classification of morbidity and mortality information for statistical purposes. Later its use was extended to include the indexing of hospital records by disease and operation for data storage and retrieval. Because the original classification was used to indicate causes of death, mental disorders were not included in the ICD until the fifth revision (1938), when they were included in the section on "Diseases of the Nervous System and Sense Organs." The sixth revision (1948) had a separate section for mental disorders, but many psychiatrists throughout the world felt that the classification did not reflect satisfactorily the expanding amount of knowledge within the field. WHO subsequently revised the classification of mental disorders, and when the eighth revision was adopted in 1965, it included a "Glossary of Mental Disorders" as a guide to more uniform usage of the principal diagnostic terms. In preparing the section on mental disorders for the ninth revision (ICD-9,[28] 1977), WHO convened a series of international seminars devoted to a consideration of recognized problem areas in psychiatric diagnosis. Those deliberations, by psychiatrists from more than forty countries, led to a recasting of the classification of many disorders in ICD-9 as well as the introduction of several new categories. The section on mental disorders again includes a glossary, this time as an integral part of the text. The glossary provides a common frame of reference for diagnoses that are ordinarily based upon descriptions of behavior and feelings rather than on independent, confirmatory, laboratory data, and for terms that otherwise might be used with markedly different meanings by different clinicians or statisticians.

ICD-9-CM,[8] the Clinical Modification of ICD-9, was adopted for use in the United States to provide greater specificity than was possible with ICD-9. ICD-9 is a three-digit system; thus, the numbers from 001 through 999 must contain all recognized diseases. They are subdivided into seventeen major groups (for example, infectious diseases, diseases of the circulatory system and so forth), and each is allocated a specific set of numbers. Mental disorders have been allocated thirty numbers, from 290 through 319; that is, every recognized mental disorder must in some way be incorporated within that span of numbers, and the only greater specificity possible in ICD-9 is a maximum of ten subtypes for each number, provided by adding a fourth digit after the decimal point. ICD-9-CM, striving for more precise clinical groupings rather than mere statistical groupings or trend analysis, adds a fifth digit and thereby makes possible a refinement ten times greater than can be achieved with ICD-9.

EXAMPLE

In ICD-9, the number 290 identifies "Senile and Presenile Organic Psychotic Conditions"; Subtypes within that group are

290.0 Senile Dementia, simple type
290.1 Presenile Dementia
290.2 Senile Dementia, depressed or paranoid type
290.3 Senile Dementia with acute confusional state
290.4 Arteriosclerotic Dementia
290.8 Other
290.9 Unspecified

ICD-9-CM, while remaining compatible with the parent system, nonetheless provides much greater specificity with the addition of a fifth digit. The same number 290 can be further subdivided into

290.0 Senile Dementia, uncomplicated
290.1 Presenile Dementia
  290.10 Presenile Dementia, uncomplicated
  290.11 Presenile Dementia with delirium
  290.12 Presenile Dementia with delusional features
  290.13 Presenile Dementia with depressive features
290.2 Senile Dementia with delusional or depressive features
  290.20 Senile Dementia with delusional features
  290.21 Senile Dementia with depressive features
290.3 Senile Dementia with delirium
290.4 Arteriosclerotic Dementia
  290.40 Arteriosclerotic Dementia, uncomplicated
  290.41 Arteriosclerotic Dementia with delirium
  290.42 Arteriosclerotic Dementia with delusional features
  290.43 Arteriosclerotic Dementia with depressive features
290.8 Other specificed senile psychotic conditions
290.9 Unspecifed senile psychotic condition.

While ICD-9-CM was being prepared, the American Psychiatric Association's Task Force on Nomenclature and Statistics was working on the third (1980) edition of the Diagnostic and Statistical Manual of Mental Disorders, (DSM-III).[5] Among several features new to DSM-III as compared with DSM-I (1952)[3] and DSM-II (1968)[4] were the addition of some categories, the deletion of others, the coinage of names for new categories, and, on occasion, for older categories whose names seemed inappropriate or misleading. All the terms in DSM-III are included in ICD-9-CM as recommended terms or as inclusion terms (that is, acceptable as alternatives to the recommended terms). For example, ICD-9-CM, under 290.4 "Arteriosclerotic Dementia," lists as an inclusion term the DSM-III name, "Multi-infarct Dementia or Psychosis." Thus DSM-III and ICD-9-CM are compatible in that the latter contains the diagnostic terms of DSM-III; the reverse, however does not hold, for many ICD-9-CM codes and terms do not appear in DSM-III.

DSM-III represents an attempt to reflect the current state of knowledge about mental disorders. In some instances, the name of a disorder, or its placement within the classificatory scheme, or indication of what should be excluded from or included within the boundaries of the disorder appear to be a radical departure from earlier classifications. The clinician familiar with any of those other systems may not at first be comfortable with the innovations of DSM-III, but the elaborate field testing that the manual has already undergone suggests that it will quickly be recognized as coming closer to clinical reality than many other systems which were based more on theory than on fact—and that by and large the advantages of the new approach far outweigh the disadvantages.

The major innovations of DSM-III are the following:

1. The descriptive approach used es-

chews theory in favor of reporting objective clinical data—behavior, symptoms, signs, test results, and so forth. Assumptions about how those manifestations came into being—that is, assumptions about etiology, pathophysiology, or psychopathologic mechanisms—are avoided. Different disorders are grouped according to the degree to which they share such objective clinical features, and not according to a theory of what kind of hypothesized conversion, displacement, or substitution mechanism might be operating unconsciously. Disorders that were in the class of "Neuroses" in DSM-II, for example, are scattered throughout several classes in DSM-III. "Dysthymic Disorder" ("Depressive Neurosis" of old) is a "Specific Affective Disorder"; "Obsessive Compulsive Neurosis" is a type of "Anxiety State," along with "Panic Disorder" and "Generalized Anxiety Disorder." All those "Anxiety States," together with "Post-traumatic Stress Disorder" and "Phobic Disorders," make up the major group of "Anxiety Disorders." The conversion type of hysterical neurosis is a "Somatoform Disorder," while the dissociative type constitutes the group of "Dissociative Disorders."

2. Operational criteria are given for each diagnostic category—*inclusion criteria* for the clinical features that support or warrant the diagnosis, and *exclusion criteria* for features that are incompatible with the diagnosis. The development of such guidelines is an outgrowth of the work of the St. Louis group,[11] whose original description of diagnostic criteria for fifteen conditions was later expanded for a group of twenty-three disorders by Spitzer and his colleagues[23] into the *Research Diagnostic Criteria* (RDC). The operational criteria for DSM–III categories were developed by fourteen advisory committees and numerous consultants in the various subgroups of the clinical field.

3. As already mentioned, field trials of the system were made during its development. The field trials provided continuing feedback about the applicability of the criteria, their clinical relevance, and the degree of reliability that characterized their use. Not only did the trials lead to many modifications of the original drafts of DSM-III, but they also presented evidence that more than 800 clinicians were able to use DSM-III in diverse settings with relative ease.

4. Explicit principles of classification, including a definition of mental disorder are provided. While the definition does not claim to draw a sharp line between "normal" and "disordered," it nonetheless faces squarely the issue of mislabeling social deviance by classifying it as a disorder. The definition emphasizes that disorder occurs within the individual; that is, it does not aim to make a group or social diagnosis. It consists of behavioral or psychologic manifestations that are clinically significant, typically because they include either a distressing symptom or some degree of impairment in one or more important areas of functioning. Finally, it is presumed that the disturbance reflects some biologic, behavioral, or psychologic dysfunction and is not only a disturbance in the relationship between the individual and society. Spitzer and associates[24] point out that such a definition clarifies the position assigned homosexuality in DSM-III. Clinicians can agree that sexual functioning is an "important" area of functioning; there is disagreement, though, as to whether the function must be exercised only in heterosexuality. DSM-III takes the position that it is the patient's decision as to whether or not an inability to function heterosexually is a significant impairment. Consequently, homosexual activity is not a mental disorder—no matter how society may view it—unless the person who engages in such activity is persistently distressed by it or by the fact that while heterosexual arousal is desired, it cannot be attained. In that case, the diagnosis of "Ego-Dystonic Homosexuality" is warranted.

DSM-III recognizes three levels of conceptualization of disorders: (1) symptom or sign, without reference to the context within which it occurs, such as "sadness" or "forgetfulness"; (2) syndrome, a distinctive clinical picture produced by a clustering or grouping of signs or symptoms, such as sadness expressed as feelings of painful dejection with loss of self-esteem, psychomotor retardation, and difficulty in thinking (the syndrome of clinical depression)—or forgetfulness expressed as marked impairment of immediate and recent memory, impaired judgment, concretistic thinking, difficulty in abstract conceptualization, dressing apraxia, and nominal aphasia (the syndrome of dementia); and (3) disease, wherein a specific etiology or pathophysiology is known to account for the distinctive picture, such as dementia in a sixty-seven-year-old hypertensive man with a history of intellectual deterioration that occured in irregular spurts over the preceding four years, eventually complicated by dysarthria, small-step gait, and funduscopic changes suggestive of arteriosclerosis ("Multi-infarct Dementia"). Disorders are grouped on the basis of the symptoms or signs they have in common and, in general, are arranged in a hierarchy with those at the top having the wider range of symptoms. Disorders high on the list, in other words, may have symptoms that disorders below them also have, but the lower groups do not have the additional symptoms that are found in diseases listed above them. Such an arrangement allows the branching of a series of decision "trees" for differential diagnosis in major symptom areas.

5. Extensive descriptions of each disorder are given—essential and associated features, age at onset, usual course, degree of impairment, complications, predisposing factors, prevalence, sex ratio, family pattern, and differential diagnosis—to the extent that such factors are known at the present time.

6. A system of multiaxial evaluation includes five axes for recording information. Axis I includes all the mental disorders except for specific developmental disorders in children and personality disorders in adults, which fall into Axis II. The reason for the separation into two axes is to focus attention on an underlying personality disorder, for example, which is of significance in treatment planning and prediction of outcome but is often overlooked when the Axis I disorder occupies the foreground of the clinical picture. Axis I also provides for coding of conditions that are a focus of clinical attention even though they may not constitute a mental disorder (for example, marital problem, academic problem, antisocial behavior). Axis III records concomitant physical disorders of significance to the overall management of the patient, whether etiologic (such as hypothyroidism in a patient with "myxedema madness") or otherwise relevant (such as glaucoma in a patient whose depression would ordinarily be treated with a tricyclic antidepressant). Axes I, II, and III are necessary for the full diagnostic assessment, and multiple diagnoses can be recorded on each of them. Axes IV and V are supplemental recordings for use in research and other special settings. Axis IV notes the severity of any psychosocial stressor that has been identified as contributory to the development of the present illness (coded as none, minimal, mild, moderate, severe, extreme, or catastrophic). Axis V is used to indicate the highest level of adaptive functioning maintained for at least a few months during the past year. As defined in DSM-III, adaptation is a composite of functioning in three areas—social relations, occupation, and use of leisure time.

¶ The Major Changes in Current Classifications

In ICD-8, "Psychoses" occupied categories 290 through 299. In ICD-9, "Organic Psychotic Conditions" are 290–294; "Other Psychoses" are 295–299, but 299 is used for a new category, "Psychoses with Origin Specific to Childhood."

In ICD-8, "Neuroses, Personality Disorders, and Other Nonpsychotic Mental Disorders" were 300–309; in ICD-9, that group is expanded (300–316) by the addition of several new categories: "Nondependent Abuse of Drugs"; "Acute Reaction to Stress"; and "Adjustment Reaction" (replacing the single category previously called "Transient Situational Disturbances"); "Depressive Disorder," not elsewhere classified; "Disturbance of Conduct," not elsewhere classified, "Disturbance of Emotions Specific to Childhood and Adolescence," "Hyperkinetic Syndrome of Childhood," and "Specific Delays in Development" (all four replacing the single previous category of "Behavior Disorders of Childhood"); and "Psychic Factors Associated with Diseases Classified Elsewhere."

In ICD-9, "Mental Retardation" has been reduced from six to three categories: "Mild Mental Retardation" (317), "Other Specified Mental Retardation" (318), and "Unspecified Mental Retardation" (319).

ICD-9-CM is compatible with ICD-9. The contents and the sequence of the three-digit categories are retained, with the exception of "Affective Psychoses" (296), and further specificity is gained through the addition of a fifth digit. DSM-III is compatible with ICD-9-CM to the extent that the latter contains all the terms of the former and that DSM-III keeps the same *numbers* for diagnoses as ICD-9-CM. The arrangement within groups is different, however, so that the numbers in DSM-III do not always follow in sequential order. The DSM-III group, "Somatoform Disorders," for example, does not occur as such in ICD-9-CM, and the members of the group occur in different categories of ICD-9-CM.*

ICD-9-CM begins with "Organic Psychotic Conditions." In contrast, DSM-III begins with "Disorders Usually First Evident in Infancy, Childhood or Adolescence," subdivided into five groups on the basis of the area of predominant disturbance:

I. Intellectual—Mental Retardation (317–319)

The fifth digit in these categories is used to indicate that other behavioral symptoms are present that require clinical attention (such as aggressive behavior that is not part of another codable disorder). In both DSM-III and ICD-9-CM, 318.1 signifies "Severe Mental Retardation"; in DSM-III the fifth digit provides more clinical specificity. Thus 318.11 indicates that the severely retarded child has other significant behavioral symptoms; 318.10 indicates that the retardation is not complicated by such symptoms.

II. Overt Behavior
 Attention Deficit Disorder
  with Hyperactivity (314.01)
  without Hyperactivity (314.00)

 Conduct Disorders (312.00)

DSM-III subdivides "Conduct Disorders" into four types, depending upon whether behavior is predominantly aggressive or nonaggressive, *and* socialized or undersocialized (referring to the ability or inability to establish adequate social bonds, empathy, affection for others, and so forth). Included herein are the DSM-II categories of "Runaway Reaction," "Unsocialized Aggressive Reaction," and "Group Delinquent Reaction." ICD-9-CM, incidentally, includes within "Conduct Disorders" various disorders of impulse control; these appear much later in DSM-III because they are disorders of adulthood rather than of infancy or childhood.

*This table is not presented as a reproduction of the DSM-III classification, but rather to indicate the ways in which DSM-III differs from ICD-9-CM even while retaining compatibility with it.

TABLE 32–1 **Somatoform Disorders**

DSM-III LISTING	CODE	ICD-9-CM LISTING
Somatization Disorder	300.81	Other Neurotic Disorders, Somatization Disorder
Conversion Disorder	300.11	Hysteria, Conversion Disorder
Psychogenic Pain Disorder	307.80	Special Symptom—Psychalgia–Psychogenic pain, site unspecified
Hypochondriasis	300.70	Neurotic Disorder, Hypochondriasis
Atypical Somatoform Disorder	300.70	Neurotic Disorder, Hypochondriasis

III. Emotions
  Anxiety Disorders
    Separation Anxiety Disorder (309.21)
    Avoidant Disorder (313.21)
    Overanxious Disorder (313.00)
  Other Disorders of Infancy, Childhood or Adolescence
    Reactive Attachment Disorder (313.89)
    Schizoid Disorder (313.22)
    Elective Mutism (313.23)
    Oppositional Disorder (313.81)
    Identity Disorder (313.82)

The order of listing in this section once again highlights the emphasis of DSM-III upon overt and objective clinical manifestations. ICD-9-CM places "Separation Anxiety Disorder" under "Adjustment Reactions"; DSM-III places it with "Avoidant Disorder" and "Overanxious Disorder" under "Anxiety Disorders of Childhood or Adolescence" because anxiety is the predominant clinical feature of all of them. The distinction between "Avoidant Disorder" and "Schizoid Disorder of Childhood or Adolescence" reflects the clinical judgment that the child who is afraid of strangers but at the same time wants to make contact with them is probably very different from the child who has no desire or capacity for emotional involvement.

IV. Physical (307.00)
  Eating Disorders (including anorexia nervosa, bulimia, pica, rumination disorder)
  Stereotyped Movement Disorders (including transient and chronic motor tic disorders, Tourette's disorder)
  Other (stuttering, functional enuresis, functional encopresis, sleepwalking disorder, sleep terror disorder)

In ICD-9-CM, all of the preceding fall within the category of "Special Symptoms" or "Syndromes, not elsewhere classified." DSM-III subdivides them into specific disorders when that is warranted by different clinical features, course, and treatment implications. Also included as a "Special Symptom" in ICD-9-CM is "Psychalgia," which is grouped within "Somatoform Disorders" in DSM-III.

V. Developmental
  Pervasive Developmental Disorders
    Infantile Autism (299.0x)
    Childhood Onset (299.9x)

Within this group, the fifth digit is used to indicate whether the full syndrome is present (x = 0), or whether the full syndrome was present in the past but that only residual symptoms are currently evident (x = 1). In ICD-9-CM, category 299, "Psychoses with Origin Specific to Childhood" contains a disorder termed "Disintegrative Psychosis." It is not included in DSM-III because of the evidence that it is a nonspecific organic brain syndrome, which belongs more properly among the dementias.

"Specific Developmental Disorders" (315) includes reading, arithmetic, language, and articulation disorders. All of these are coded on Axis II, inviting full attention to the developmental disorder(s) as well as to any other disorder(s) that may coexist.

The next major group in DSM-III is "Organic Mental Disorders." This section is subdivided into two sections—those organic mental disorders in which the etiology or pathophysiology is known (specifically, disorders related either to aging of the brain or to drug/substance intake), and a group of organic brain syndromes whose etiology is either unknown or related to a disease that is coded outside the mental disorders section (and noted on Axis III). The organic syndromes are differentiated on the basis of clinical symptoms alone; unlike many other classifications, DSM-III does not subdivide on the basis of acute versus chronic, or psychotic versus nonpsychotic, or reversible versus irreversible. Each is described as are the other disorders in DSM-III, and the description of clinical features, course, and complications is followed by operational diagnostic criteria.

The nine organic brain syndromes are grouped into six categories:

1. "Delirium" and "Dementia," with relatively global cognitive impairment;
2. "Amnestic Syndrome" and "Organic (or Drug) Hallucinosis," with relatively selective cognitive impairment;
3. "Organic Delusional Syndrome" and "Organic Affective Syndrome," with features that mimic schizophrenic or affective disorders;
4. "Organic Personality Syndrome," with

changes in attitudes, traits, and the general style of relating to the environment;

5. "Intoxication" and "Withdrawal," related to intake of or abstinence from a substance, when the symptoms do not meet the criteria for inclusion in any of the foregoing syndromes;
6. Atypical or mixed.

Within the "Organic Mental Disorders" are dementias related to aging and substance-induced disorders. The former include "Primary Degenerative Dementia" of senile or presenile onset, and "Multi-Infarct Dementia" (formerly called cerebral arteriosclerosis and, renamed because of evidence that the dementia is due to repeated infarcts rather than to arteriosclerosis per se). The fifth digit is used to indicate if the dementia is uncomplicated, or if delirium, delusional features, or clinical depression complicate the picture.

Terminology and the sequence of codes in the DSM-III categories of substance-induced "Organic Mental Disorders" often vary considerably from ICD-9-CM. In DSM-III, intoxication is recognized as a specific syndrome due to the direct effect of the substance in question on the central nervous system. Except for those substances in which no such syndrome occurs (hallucinogens and tobacco), intoxication is always listed first among the brain syndromes induced by the substance. In ICD-9-CM, in contrast, intoxication is subsumed under nondependent abuse of drugs rather than under the organic mental disorders.

In the interest of more accurate description, some of the alcohol-related disorders have been renamed. "Delirium Tremens" becomes "Alcohol Withdrawal Delirium"; "Pathologic Intoxication" becomes "Alcohol Idiosyncratic Intoxication"; "Korsakoff's Psychosis" becomes "Alcohol Amnestic Disorder"; and "Alcoholic Deterioration" or "Alcoholic Dementia" is termed "Dementia Associated With Alcoholism," in view of the doubt that alcohol is the etiologic agent in such cases. In this last category, the severity of dementia is indicated in the fifth digit. The "Alcohol Paranoid State" in DSM-II has been eliminated, as has been the "Alcoholic Jealousy" in ICD-9-CM, because there is no convincing evidence that either exists as a distinct entity. "Alcohol Withdrawal" is separated from "Alcohol Withdrawal Delirium," giving the clinician the opportunity to specify the condition he is ordinarily treating when he places a patient on an alcohol detoxification regimen. In both DSM-II and ICD-9-CM, the clinician was forced to label such a patient either "Acute Intoxication" (even though it was absence of the "poison" rather than too much of it that produced the condition), or "Delirium Tremens" (even though that was the condition that treatment aimed to prevent).

For the other substances in this section, DSM-III also offers greater specificity than was possible in previous classifications. Nine classes of drugs are specified—the ones that, in addition to alcohol, are most commonly used nonmedically to alter mood or behavior. Within each class, the specific brain syndromes known to be produced by the drugs are listed. In ICD-9-CM, code numbers are assigned to conform to the less discriminant classification, but an additional coding is available to reflect the unique specificity of DSM-III. This was made possible by using a number assigned to but not currently used by the section on "Diseases of the Nervous System" (327) for substance-induced mental disorders other than alcohol. The fourth digit is used to indicate the class of drugs (for example, barbiturate, opioid, hallucinogen) and the fifth digit to indicate the syndrome (for example, intoxication, withdrawal, delusional disorder). Thus "Barbiturate Amnestic Disorder" is coded 327.04 (0 = barbiturate; 4 = amnestic disorder), while "Amphetamine Delusional Disorder" is 327.35 (3 = amphetamine; 5 = delusional disorder).

Syndromes induced by barbiturates (and similar sedatives and hypnotics) include intoxication, withdrawal, withdrawal delirium, and amnestic disorder. Under opioids, intoxication and withdrawal are specified, but under cocaine only intoxication is specified, since no withdrawal syndrome has been consistently described. Syndromes produced by amphetamines (and similarly acting sympathomimetics) are intoxication, delirium, delusional disorder, and withdrawal. Phencyclidine (PCP) and similarly acting arylcyclohexylamines produce intoxication, delirium, and mixed organic mental disorder. In addition to hallucinosis, hallucinogens may also produce delusional disorder and affective disorder. Cannabis syndromes include intoxication and delusional disorder, although some doubt that the latter is a separate entity since it disappears by the time symptoms of ordinary cannabis intoxication abate. "Tobacco Withdrawal" is recognized as an entity, as is "Caffeine Intoxication" (caffeinism).

"Substance Use Disorders" are the next section in DSM-III, so placed because many patients who fall into this category will at times also develop intoxication, withdrawal or some other organic mental disorder induced by the substance they are abusing. For most substances, there are two major patterns of pathologic use—abuse and dependence—both of which are differentiated from nonpathologic use for recreational or medicinal purposes. It should be noted that three of the substance classes—cocaine, phencyclidine, and hallucinogen—are not known to be associated with a pattern of physiologic dependence in that there is no evidence of tolerance or withdrawal; thus only the "abuse" pattern is coded for them. Also unusual is tobacco, for which only a pattern of "dependence" is clinically significant, appearing as an inability to stop and/or development of "Tobacco Withdrawal" (327.71), an "Organic Mental Disorder." For all the "Substance Use Disorders," the fifth digit is used to indicate course (that is, continuous, episodic, in remission, or unspecified).

Mention has already been made of the likelihood that any classification will tend either to be overly exclusive or overly inclusive, and the reasons for erring in either direction were discussed. The "Schizophrenic Disorders" section of DSM-III will seem overly exclusive to many clinicians in the United States whose aim has been to identify these disorders as early as possible in the hope that early intervention might halt or retard their progression. The approach of DSM-III is to narrow this group considerably in an attempt to achieve more homogeneous subgroupings and greater reliability of diagnoses made by clinicians in widely different settings. DSM-III attaches the label of schizophrenia only if there has been a period of active psychosis (for example, delusions, hallucinations, loosening of associations, or other disturbances of the form of thought, altered psychomotor behavior, changes in affect and/or relationships to the external world, disturbance in goal-directed activity), deterioration from a previous level of functioning, onset before the age of forty-five years, and duration of at least six months. Such requirements eliminate the older category of "Acute Schizophrenic Episode" (which would now be termed "Schizophreniform Disorder"), as well as the many categories that included illnesses without psychotic manifestations, such as the latent, borderline, pseudoneurotic, and even the classic simple forms (most of which would fall into the "Personality Disorders," coded on Axis II). The schizoaffective type has also been eliminated; many so diagnosed in the past would now be placed within the "Affective Disorders," while a few would be labeled "Schizoaffective Disorder" within the new group, "Psychotic Disorders Not Elsewhere Classified."

What remain are five subtypes: disorganized (the hebephrenic of other classifications), catatonic, paranoid, undifferentiated, and residual (with previous episode of schizophrenia but currently without prominent psychotic symptoms). ICD-9-CM retains the simple, latent, and schizoaffective subtypes, as well as acute schizophrenic episode, so the clinician who wishes to use those non-DSM-III diagnoses may continue to do so. Because their identifying fourth digits do not appear in DSM-III, they should not be any source of difficulty for researchers or record room librarians. The serious drawback to their use is that it will perpetuate the uncertainties of the older systems and simultaneously prolong the time needed for adequate testing of the new one.

There is general agreement that since the time of Kraepelin efforts to refine the classification of psychiatric disorders have not been notably successful. In the case of the schizophrenic group, efforts have proceeded in two directions. Eugen Bleuler broadened Kraepelin's concept of dementia praecox at the expense of the manic-depressive group. His diagnostic criteria leaned heavily on clinical judgment rather than on empirically derived factors, with a high degree of interrater reliability. At least in parts of the United States, his concepts were extended to the degree that even a "trace" of schizophrenia was enough to establish the presence of the disorder, no matter what the intensity or number of accompanying affective symptoms.

Kraepelin's successor, Kurt Schneider, took a different direction. Like Bleuler, he employed a broader concept of schizophrenia than Kraepelin, but he paid less attention to the course of the disorders and focused to a greater extent on symptoms. Finally, he developed his set of "first-rank symptoms," whose presence or absence could be established with relative ease. According to the standards held by many American clinicians, the Schneiderian criteria were too heavily weighted in the direction of nuclear, process, poor-prognosis, or far-advanced schizophrenia. Yet the criteria espoused by those clinicians were notoriously unreliable and to many they seemed so vague and impressionistic as to be almost mystical. The growth of computer technology and the possibility of its application to the masses of data generated by the psychiatric interview seemed to offer a way back to a more solid, objective and pragmatic science. The operational criteria set out in DSM-III for the diagnosis of schizophrenic disorders are a contemporary reaffirmation of the Schneiderian approach and a refinement by numerical taxonomy and other statistical methods that ensure maximal reliability.

Field trials have already demonstrated that the reliability achieved with the DSM-III operational criteria probably surpasses that of any other classificatory system in wide use. To achieve that reliability within the schizophrenic group, however, a great deal of what was previously admissible to the group is now excluded.[19] Reliability alone does not secure validity, nor does lack of reliability mean nonexistence. An overemphasis on reliability deifies counting and measuring, but counting the trees may not be the best way to appreciate the intricacies of the forest. Perhaps DSM-III has reduced schizophrenia to its proper size, but where will the unreliable rejects be placed?

There is a growing tendency to put all those with any degree of affect disturbance into the manic-depressive or affective disorders. It can only be hoped that this swing of the pendulum will not finally so adulterate the affective group that the next generation of classifiers will find them as hopelessly heterogeneous and overinclusive as they regard the schizophrenic group now. The other rejects from the schizophrenic categories—mainly the acute and episodic psychoses in which affect disturbance does not predominate—end up in a no-man's land of uncertain, atypical, or unclassified disorders. While this is bound to be discomfiting to the clinician, the declaration of uncertainty should in the long run be preferable to unwarranted assignment to a specific category of affective disorder. It will remain for future studies—ideally broad enough to include qualitative parameters along with measurements and scales—to establish the validity of DSM-III's proposed groupings as well as to point the way for classification of the now uncertain and atypical cases.[17]

"Paranoid Disorders" follow the schizophrenic group and include "Paranoia," "Shared Paranoid Disorder" (*"folie à deux"* or "Psychosis of Association" in other classifications), and "Acute Paranoid Disorder." The difficulty in differentiating these disorders from severe "Paranoid Personality Disorder" and the paranoid type of schizophrenia are acknowledged, but the operational criteria provide some guidelines in this regard. Involutional paranoid state and paraphrenia are not described as separate entities.

The next section, "Psychotic Disorders Not Elsewhere Classified," is to some extent the "wastebasket" category that every system of classification needs. Unlike other systems, however, in DSM-III this category is likely to contain a significant proportion of the entire group of mental disorders because of the exclusionary provisions of diagnostic criteria. By "psychotic" is meant behavior indicative of gross impairment in reality testing, such as delusions, hallucinations, or marked disorganization of speech or other activity. The specific categories included are "Schizophreniform Disorder" (where most previously diagnosed "Acute Schizophrenia" will fall), "Brief Reactive Psychosis" (of less than two weeks, following an identifiable psychosocial stressor), "Schizoaffective Disorder" (to be used when the clinician cannot make a differential diagnosis between "Schizophrenia" and "Affective Disorder"), and "Atypical Psychosis" (for all other unclassifiable psychoses).

"Affective Disorders" in DSM-III include entities that fall into the "Affect Psychoses," "Neuroses," and "Personality Disorders" of other classifications. They are grouped on the basis of the degree of expression of an affective syndrome, rather than in terms of psychotic versus neurotic, endogenous versus reactive, and so forth. The terminology follows that suggested by Leonhard,[15] who divided affective disorders into bipolar (with episodes of both mania and depression) and monopolar types. "Unipolar" has since replaced "monopolar" on etymologic grounds, and on the basis of clinical evidence it has come to be limited to cases manifesting only depressive episodes. Mania thus is the deciding factor in this particular dichotomy; cases with a full manic syndrome, whether or not depression has also been manifested, are bipolar. Affective syndromes without mania are unipolar depressions; in DSM-III these are termed "Major Depressive Disorders," to emphasize that a full affective syndrome is required for grouping within the category of "Major Affective Disorders."

ICD-9-CM departs markedly from ICD-9 in its classification of "Affective Psychoses," and while it moves in the direction of DSM-III, it is not totally consistent with that classification either, as table 32–2 indicates.

The fifth digit in DSM-III is used for further subdivision according to clinical features. For depressive episodes, 6 = in remission; 4 = with psychotic features (if not specified, it is assumed that such features are mood-congruent and consistent with the themes of inadequacy, unworthiness, death, or need for expiation; mood-incongruity should be specified and may be indicated by using 7 instead of 4 for the fifth digit); 3 = with melancholia ("vegetative signs" in others' terminology); 2 = without melancholia; and 0 = unspecified.

For manic episodes, in the fifth digit 6 = in remission; 4 = with psychotic features (as with depressive episodes, to be specified as to whether such features are congruent or incongruent with the mood); 2 = without psychotic features; and 0 = unspecified.

"Other Specific Affective Disorders" are those in which only a partial affective syndrome develops; to be included here however, the disturbance must be present for at least two years. "Cyclothymic Disorder" corresponds to the "Cyclothymic Personality" in DSM-II, and "Dysthymic Disorder" includes both the "Depressive Neurosis" and "Depressive (or Asthenic) Personality" of other classifications. Affective disorders that cannot be placed in either "Major Affective Disorders" or "Other Specific Affective Disorders" are classified as "Atypical."

As already indicated, one effect of shrinking the group of schizophrenias may be to transfer uncertain or doubtful cases into the "Affective Disorders." DSM-III strives for the same level of certainty within the affective group as with the schizophrenic group by using exclusion and inclusion criteria for diagnosis. But even at this early date, reports in the literature suggest that more and more clinical material will come within the affective sweep. Indeed, if the mere presence of an affective syndrome is significant enough to call into question a large number of patients previously considered clinically to be schizophrenic, will it not cast equal doubt on a diagnosis of personality disorder?

Despite the many recent contributions to our understanding of the genetic, develop-

TABLE 32–2. **Affect Disorders—Classification (Affective Psychoses)**

ICD-9		ICD-9-CM		DSM-III
296.0	Manic-depressive psychosis, manic	296.0	Manic Disorder, single episode	Major Affective Disorders Bipolar Disorder 296.6x mixed 296.4x manic 296.5x depressed
296.1	Manic-depressive psychosis, depressed	296.1	Manic Disorder, recurrent episode	
296.2	Manic-depressive psychosis, circular; currently manic	296.2	Major Depressive Disorder, single episode	Major Depression 296.2x single episode 296.3x recurrent
296.3	Manic-depressive psychosis, circular; currently depressive	296.3	Major Depressive Disorder, recurrent episode	
296.4	Manic-depressive psychosis, circular; currently mixed	296.4	Bipolar Affective Disorder, manic	Other Specific Affective Disorders 301.13 Cyclothymic Disorder 300.40 Dysthymic Disorder (or depressive neurosis)
296.5	Manic-depressive psychosis, circular; unspecified	296.5	Bipolar Affective Disorder, depressed	
296.6	Manic-depressive psychosis, other and unspecified	296.6	Bipolar Affective Disorder, mixed	Atypical Affective Disorders 296.70 Atypical Bipolar Disorder 296.82 Atypical Depression
		296.7	Bipolar Affective Disorder, unspecified	
296.8	Other	296.8	Manic-depressive psychosis, other and unspecified	
296.9	Unspecified	296.9	Other and unspecified Affective Psychoses	

mental, neurohormonal, biochemical, and cognitive factors that may contribute to depressive disorders, in the final analysis they remain as heterogeneous a group as the schizophrenias. The thrust of many investigations is to define more homogeneous subgroups, and to that end Dunner, Fleiss, and Fieve,[10] for example, suggest that bipolars be subdivided into Type I (the patient has been hospitalized for mania) and Type II (the patient has been hospitalized for depression but not for mania, even though a history of manic symptoms removes him from the unipolar category). Akiskal and associates[2] propose a subtyping of Type II bipolar disorders into (1) recurrent clinical depressions with occasional hypomanic periods; (2) unipolars who develop mania only when treated with antidepressant agents; and (3) cyclothymics who develop a clinical depression (on the basis of data supporting consideration of cyclothymia as a phenotypical variant of full-blown bipolar affective psychosis).

The major depressive disorders also remain troublesome for the classifier. Does a subdivision only into single episode and recurrent types do justice to the large group? Should the dichotomies of other systems be applied as a means of subgrouping major depressions? If one does accept a division into primary and secondary types, would it be reasonable to keep all the secondary depressions together? Or would it be more reasonable to subclassify them on the basis of their antecedent conditions and separate depressions arising in obsessive-compulsive patients, for example, from those in antisocial personalities and from depressions that are complications of alcoholism?

There is no general agreement on the answers to these questions, nor is there unanimity of opinion as to the subtyping of primary unipolar depressions. Some differentiate "pure depressive disease" (PPD), which has neither mania nor alcoholism in the family of the index case, from "depressive spectrum disease," familial in that a first degree relative is alcoholic and there may also be a relative with depression. But examination of the predepression history revealed that many of both types were in fact secondary depressions, with significant antecedent psychopathology. In contrast, only a small number of "sporadic (nonfamilial) depressions" were secondary. So it could be argued that "sporadic depression" is a purer disease than familial "pure depressive disease" and may even be a distinctly separate entity.[6] Attempts to define more clearly the validity of these clinical groupings with the use of biologic or pharmacologic measures—such as urinary 3-methoxy-4-hydroxy-phenylglycol assays, dexamethasone suppression test, response to monoamine oxidase inhibitors or tricyclic antidepressants, or electroconvulsive treatment—have verified only that not all depressions are the same. The subgroupings elicited by any one method are not wholly consistent with those generated by other approaches, and it will remain for future investigations to tell how far off the mark DSM-III is, whether it should be supplanted, or how it should be refined. It may well be that strict attention to the fifth digit subdivisions of DSM-III will be as clinically useful a typology of depression as any other; only consistent use of its diagnostic criteria over time can establish its true worth.

Not to be lost sight of is a fundamental axiom of classification theory—the first step in defining a clinical syndrome is to demonstrate that it is a recognizable entity that can be discriminated from other disorders. Only when that is done can one begin to worry about what larger class it may be a part of. It has been suggested, for example, that the "Borderline Syndrome"[25] and "Hysteroid Dysphoria,"[16] and perhaps other mixed states fit more properly into the affective group than elsewhere. Just as with the typology of the major depressive disorders, it is essential to gain more clinical experience to establish the validity of such arguments. To push every entity into the affective stream because it has a trace of dysphoria would merely churn further the already muddied water. Such an approach, indeed, would be to regress to a key classification, likely to compound rather than resolve the existing confusion.

The next major group in DSM-III is "Anxiety Disorders." While this group includes many of the neuroses of other classifications, not all are placed here. Directly experienced anxiety is the essential inclusion factor in this group, which is subdivided into "Phobic Disorders" (or "Phobic Neuroses"), "Anxiety States" (or "Anxiety Neuroses"), and "Post-traumatic Stress Disorder." Three types of "Phobic Disorders" are differentiated, on the basis of differences in response to treatment as well as in the clinical picture—"Agoraphobia" (with or without panic attacks), "Social Phobia" (fear of situations in which one might be scrutinized by others and in which one might behave in a way that would be humiliating or embarrassing), and "Simple Phobia" (all other types).

Within the "Anxiety States" are "Panic Disorder," "Generalized Anxiety Disorder," and "Obsessive Compulsive Disorder." As with other groupings in DSM-III, designation of subtypes reflects current knowledge, which suggests that "Panic Disorder" is a discrete entity when treatment response is considered as one of the descriptive parameters. "Obsessive Compulsive Disorder" is considered a type of "Anxiety State" because if the obsession or compulsion is resisted, anxiety is experienced directly.

Two forms of "Post-traumatic Stress Disorder" are distinguished, because evidence suggests that they differ in outcome. The acute form is manifested within six months of the trauma, and the symptoms last no more than six months. The chronic or delayed form has a more malignant course; onset may be months or even years after the trauma, and symptoms persist for a long time.

"Somatoform Disorders" are characterized by physical symptoms that mimic organic or physical disorders and an absence of organic findings or physiologic abnormalities that might explain them. Included are several forms that are recognizable as parts of the formerly referred to conversion hysteria: "Somatization Disorder" (in other systems, "Hysteria" or "Briquet's Syndrome"); "Conversion Disorder" (or "Hysterical Neurosis, Conversion Type"); and "Psychogenic Pain Disorder." A fourth type of "Somatoform Disorder" is "Hypochondriasis" (hypochondriacal neurosis). The suggested guidelines for assignment to the residual category within this group, called "Atypical Somatoform Disorder," illustrate the reliance that DSM-III places on objective and verifiable data. "Dysmorphophobia," a symptom (or syndrome) that some clinicians would consider a delusion and/or schizophrenic-like impairment in reality testing, consists of preoccupation with some imagined defect in physical appearance. If dysmorphophobia is the only symptomatic expression of disorder, it would be labeled "Atypical Somatoform Disorder" in DSM-III because it fulfills the inclusion criteria: physical complaint not explicable by demonstrable organic findings and apparently related to psychologic factors.

"Dissociative Disorders (Hysterical Neuroses, Dissociative Type)" is a separate category, subdivided on the basis of different clinical manifestations, predisposing factors, and course. Included are "Psychogenic Amnesia," "Psychogenic Fugue," "Multiple Personality," and "Depersonalization Disorder" (or "Depersonalization Neurosis"). "Sleepwalking Disorder" is classified within the group of "Childhood Disorders" because of its usual time of onset; otherwise, on symptomatic grounds, it would belong here.

Another major category is "Psychosexual Disorders," greatly expanded in comparison to previous classifications and ICD-9 to reflect major advances in knowledge about human sexuality. Not only are more subtypes differentiated with the use of the fourth and fifth digits, but the ordering of the categories indicates how DSM-III is based on a higher level conceptualization of disorder than mere counting of variations from an assumed "normal" (table 32–3).

The fifth digit in the code for "Transsexualism" is used to indicate the predominant prior sexual history: 1 = asexual; 2 = homosexual; 3 = heterosexual; 0 = unspecified.

"Factitious Disorders," the next major group, are disorders in which the production of symptoms appears to be under the per-

TABLE 32–3 **Classification of Psychosexual Disorders in ICD-9 and DSM-III**

ICD-9	DSM-III
302 SEXUAL DEVIATIONS AND DISORDERS	PSYCHOSEXUAL DISORDERS
.0 Homosexuality	*Gender Identity Disorders*
.1 Bestiality	302.5x Transsexualism
.2 Paedophilia	302.60 Gender Identity Disorder of Childhood
.3 Transvestism	302.85 Atypical Gender Identity Disorder
.4 Exhibitionism	
.5 Transsexualism	*Paraphilias*
.6 Disorders of psychosexual identity	302.81 Fetishism
.7 Frigidity and impotence	302.30 Transvestism
.8 Other	302.10 Zoophilia
.9 Unspecified	302.20 Pedophilia
	302.40 Exhibitionism
	302.82 Voyeurism
	302.83 Sexual Masochism
	302.84 Sexual Sadism
	302.90 Atypical Paraphilia
	Psychosexual Dysfunctions
	302.71 Inhibited Sexual Desire
	302.72 Inhibited Sexual Excitement
	302.73 Inhibited Female Orgasm
	302.74 Inhibited Male Orgasm
	302.75 Premature Ejaculation
	302.76 Functional Dyspareunia
	306.51 Functional Vaginismus
	302.70 Atypical Psychosexual Dysfunction
	Other Psychosexual Disorders
	302.00 Ego-dystonic Homosexuality
	302.89 Psychosexual Disorder not elsewhere classified

son's voluntary control; unlike malingering, where the goal of seeming to be sick is clearly related to conscious wishes, "Factitious Disorders" reflect no such needs but appear instead to be related to intrapsychic needs that are understandable only in terms of the subject's psychologic makeup. "Factitious Disorder with Psychological Symptoms" includes what has been called the "Ganser Syndrome," "Pseudopsychosis," or "Pseudodementia." "Chronic Factitious Disorder with Physical Symptoms" corresponds to the "Münchhausen syndrome." Most people with any of the factitious disorders also have a personality disorder of significant dimensions; that disorder is coded on Axis II.

"Disorders of Impulse Control Not Elsewhere Classified" include "Pathological Gambling," "Kleptomania," "Pyromania," "Intermittent Explosive Disorder," and "Isolated Explosive Disorder."

"Adjustment Disorder" replaces "Transient Situational Disorder" of DSM-II, and it is subdivided on the basis of predominant symptoms rather than on the age at which these symptoms appeared. The diagnosis is made only for reactions to identifiable stressors that develop within three months of the stressful event; disturbances of psychotic proportion are excluded.

"Psychological Factors Affecting Physical Condition" is another major category. Although the name is cumbersome, it avoids the mind/body dualism of the "psychosomatic" and "psychophysiologic" terms it replaced. Further, it requires that the physical disorder be coded on Axis III, the Axis I diagnosis being used to emphasize the role of psychologic factors in exacerbating, precipitating, or maintaining the underlying physical disorder.

The final diagnostic category consists of the "Personality Disorders," which are coded on Axis II. As noted previously, they will often coexist with disorders that will be coded on Axis I. In addition, it is possible to give more than a single coding of personality disorder to the same subject, a welcome relief for the clinician whom other classifications forced into an uncomfortable choice that could only inadequately suggest the condition of the patient. The personality disorders are clustered to conform with the way they generally present themselves to an outside observer: (1) the person affected often seems eccentric or odd, as in the paranoid, schizoid, and schizotypal personality disorders; (2) the person is erratic, overemotional, or dramatic, as in the histrionic, narcissistic, antisocial, and borderline personality disorders; and (3) the person is generally anxious or fearful, as in the avoidant, dependent, compulsive, and passive-aggressive personality disorders.

It should be noted that the "Cyclothymic Personality" of DSM-II has been placed within the "Affective Disorders" of DSM-III, that "Asthenic Personality" is subsumed under "Dysthymic Disorder" (depressive neurosis), and that "Explosive Personality" does not appear as a personality disorder since by definition it is not a part of the characteristic, typical, and usual behavior of the subject. "Schizoid Personality" remains, but more careful delineation of clinical features has expanded the DSM-II concept into three forms—"Schizoid," "Schizotypal," and "Avoidant Personality Disorders." New to the grouping in DSM-III are "Borderline Personality Disorder" and "Narcissistic Personality Disorder."

Finally, DSM-III provides a set of V codes, for conditions that are the focus of clinical attention even though they are not attributable to a mental disorder.

¶ Summary

No classification of human disorders is perfect, and no classification can anticipate the multiple uses to which it will be put nor the countless theories it will be strained to encompass. Some of the difficulties in classification have been outlined, and DSM-III has been reviewed with those difficulties in mind. Some potential pitfalls in its use have been identified, not as a way of saying, "Do not use," but rather to caution, "Learn to use correctly." No matter what defects of DSM-III may turn out to be, it represents the most serious attempt in this century to provide a classification of mental disorders that is based on a minimum of theory and a maximum of established knowledge. It admits of gaps in that knowledge, it warns the user when it has been forced to make an educated guess because objective data were lacking, and it invites continuing validation of its applicability. The classifiers who will use it as a base for the next edition of the *Diagnostic and Statistical Manual of Mental Disorders* could ask no more.

¶ Bibliography

1. AKISKAL, H. S., ed. "Affective Disorders: Special Clinical Forms," *Psychiatric Clinics of North America,* vol. 2, no. 3. Philadelphia: W.B. Saunders, 1979.
2. AKISKAL, H. S., KHANI, M. K., and SCOTT-STRAUSS, A. "Cyclothymic Tempermental Disorders," in Akiskal, H. S. ed., "Affective Disorders: Special Clinical Forms," *Psychiatric Clinics of North America,* vol. 2, no. 3. Philadelphia: W.B. Saunders, 1979, pp. 527–554.
3. AMERICAN PSYCHIATRIC ASSOCIATION. *Diagnostic and Statistical Manual of Mental Disorders,* 1st ed. (DSM–I). Washington, D.C.: American Psychiatric Association, 1952.
4. AMERICAN PSYCHIATRIC ASSOCIATION. *Diagnostic and Statistical Manual of Mental Disorders,* 2nd ed. (DSM–II). Washington, D.C.: American Psychiatric Association, 1968.
5. AMERICAN PSYCHIATRIC ASSOCIATION. *Diagnostic and Statistical Manual of Mental Disorders,* 3rd ed. (DSM–III). Washington, D.C.: American Psychiatric Association, 1980.
6. ANDREASEN, N. C., and WINOKUR, G. "Secondary Depression: Familial, Clinical, and Research Perspectives," *American Journal of Psychiatry,* 136 (1979): 62–66.
7. CARPENTER, W. T., and STRAUSS, J. S. "Diagnostic Issues in Schizophrenia," in Bellak, L. ed., *Disorders of the Schizophrenic Syndrome.* New York: Basic Books, 1979, pp. 291–319.
8. COMMISSION ON PROFESSIONAL AND HOSPITAL ACTIVITIES. *International Classification of Diseases, 9th Revision, Clinical Modification,* (ICD-9-CM). Ann Arbor: Commission on Professional and Hospital Activities, 1978.
9. CORNING, W. C., and STEFFY, R. A. "Taximetric Strategies Applied to Psychiatric Classification," *Schizophrenia Bulletin,* 5 (1979): 294–305.
10. DUNNER, D. L., FLEISS, J. L., and FIEVE, R. R. "The Course of Development of Mania in Patients with Recurrent Depression," *American Journal of Psychiatry,* 133 (1976): 905–908.
11. FEIGHNER, J. P., et al. "Diagnostic Criteria for Use in Psychiatric Research, "*Archives of General Psychiatry,* 26 (1972): 57–63.
12. FEINSTEIN, A. R. "A Critical Overview of Diagnosis in Psychiatry," in Rakoff, V. M., Stancer, H. C., and Kedward, H. B., eds., *Psychiatric Diagnosis,* New York: Brunner/Mazel, 1977, pp. 189–206.
13. JOHN, E. R., et al. "Neurometrics," *Science,* 196 (1977): 1393–1410.
14. KROLL, J. "Philosophical Foundations of French and U.S. Nosology," *American Journal of Psychiatry,* 136 (1979): 1135–1138.
15. LEONHARD, K. *Aufteilung der endogenen Psychosen.* Berlin: Akademie Verlag, 1959.
16. LIEBOWITZ, M. R., and KLEIN, D. F. "Hysteroid Dysphoria," in Akiskal, H. S., ed., "Affective Disorders: Special Clinical Forms," *Psychiatric Clinics of North America,* vol. 2, no. 3. Philadelphia: W.B. Saunders, 1979, pp. 555–575.
17. MCGLASHAN, T. H., and CARPENTER, W. T. JR. "Affective Symptoms and the Diagnosis of Schizophrenia," *Schizophrenia Bulletin,* 5 (1979): 547–553.
18. MELBY, J. C., and WATSON, C. J. "Disorders of Porphyrin Metabolism," *Medical Science,* (1960): 821–837.
19. OVERALL, J. E., and HOLLISTER, L. E. "Comparative Evaluation of Research Diagnostic Criteria for Schizophrenia," *Archives of General Psychiatry,* 36 (1979): 1198–1205.
20. RICHARDS, F., and BRINTON, D. "Glycine Loading Test," *Brain,* 85 (1962): 657.
21. SHIELDS, J., HESTON, L. L., and GOTTES-

MAN, I. I. "Schizophrenia and the Schizoid: The Problem for Genetic Analysis," in Fieve, R. R., Rosenthal, D., and Brill, H., eds., *Genetic Research in Psychiatry.* Baltimore: Johns Hopkins University Press, 1975, pp. 167–197.

22. SNEATH, P., and SOKOL, R. R. *Numerical Taxonomy.* San Francisco: W.H. Freeman, 1973.
23. SPITZER, R. L., ENDICOTT, J., and ROBBINS, E. "Clinical Criteria for Psychiatric Diagnosis and DSM-III," *American Journal of Psychiatry,* 132 (1975): 1187–1192.
24. SPITZER, R. L., WILLIAMS, J. B. W., and SKODOL, A. E. "DSM-III: The Major Achievements and an Overview," *American Journal of Psychiatry,* 137 (1980): 151–164.
25. STONE, M. H. "Contemporary Shift of the Borderline Concept from a Subschizophrenic Disorder to a Subaffective Disorder," in Akiskal, H. S. ed., *Affective Disorders: Special Clinical Forms.* Psychiatric Clinics of North America, vol. 2, no. 3, Philadelphia: W.B. Saunders, 1979, pp. 577–594.
26. WING, J. K. "The Limits of Standardization," in Rakoff, V. M., Stancer, H. C., and Kedward, H. B., eds., *Psychiatric Diagnosis,* New York: Brunner/Mazel, 1977, pp. 84–108.
27. WOODRUFF, R., GOODWIN, D., and GUZE, S. *Psychiatric Diagnosis.* New York: Oxford University Press, 1974.
28. WORLD HEALTH ORGANIZATION. *Manual of the International Classification of Diseases, Injuries, and Causes of Death.* Geneva, Switzerland: World Health Organization, 1977.

CHAPTER 33

LIAISON PSYCHIATRY

Maurice H. Greenhill

¶ Introduction

THE THRUST for the psychological care of the sick came from psychiatry rather than from medicine. It was part of a strenuous effort by psychiatry to gain a share of the practice of medicine. How this was done and at what expenditure of effort will be described in this presentation. But psychiatry has been so eager to be accepted as a discipline by medicine, that it seems at times to have lost sight of the patient as the primary concern of psychological care. The goal of liaison psychiatry is the biopsychosocial care of the medically ill patient, but so many obstacles to this have arisen that efforts to persuade members of the hospital power base (nonpsychiatrist physicians and administrators) to acknowledge the value of psychiatric methods have often taken precedence over patient care itself.[58] This proselytizing effort has been going on since 1929[74] and has led to the development of a specialty —liaison psychiatry.

Throughout its history, the locus of liaison psychiatry has been the general hospital. Here geographical fact and the presence of the patient in residence facilitates transactions between medical disciplines. There has been relatively little experience with liaison programs for ambulatory or home care patients. In community hospitals, psychiatric consultation systems alone may be standard; in tertiary hospitals a variety of consultative-liaison programs exist depending on the characteristics of the hospital. Their function, success, or failure depends largely on the liaison psychiatrist, whether he is a consultative psychiatrist in private practice, a full-time hospital physician, or a psychosomatic fellow. His effectiveness depends largely on the strategies he has devised to overcome the obstacles inherent in the health system. For the liaison psychiatrist, these obstacles are legion; in no other area of medicine are there so many.[58,110,134] They include functioning in clinical territories over which he has little authority and working with physicians who give little credence to psychosocial issues, have different value systems, are resistant to engaging in psychosocial transactions, and have low expectations of his effectiveness. He may be dependent upon the willingness of one man, such as a chairman of a department of medicine, to establish or maintain a liaison program. He receives little financial support from his own psychiatric department or the department he serves. He deals with patients who often do not want his services and are often not willing to pay for them. He knows that his own department of psychiatry does not place his service high on their list of priorities. He must negotiate with hospital administrators whose lukewarm interest is concerned mainly with protecting the legal and financial position of the institution. Shaping his liaison work is, therefore, a creative task dependent upon the obstacles and assets that confront him in the particular setting.

¶ Definitions

Liaison psychiatry has come to be the name of choice for identifying the *system* whereby psychiatry and other disciplines of medicine cooperate in clinical activity in order to deal with psychosocial variables in their concerns with health and disease. The liaison program in the field of liaison psychiatry refers to the organizational structure within which the delivery of mental health services to medical and surgical patients takes place. Since there is some lack of clarity in the use of synonymous terms, there is a need for definition. The name "psychosomatic medicine" has been retained over a long period of time and refers to the interrelationship of emotion and disease, or the effects of reaction to stress, life change, illness, and neuroendocrine and other biological influences on disease process. Under its umbrella have been studied the psychological characteristics of medical and surgical disorders, and recently, theories of disease. Another term connected with liaison psychiatry is "psychosocial medicine." This refers to the influence of social factors in disease through epidemiological approaches or with methodologies of psychological care within the hospital and in the wider community in the context of the growing knowledge of the social implications of health care. "General hospital psychiatry" is yet another term that has recently become more prevalent, as general hospitals expand their services in the field of mental health and as increased recognition has been given to the role of psychiatry in critical care medicine.

In the literature and in the conference rooms of general hospitals, the expression "the psychosomatic approach" is frequently used.[64] By this is meant awareness of psychological-sociological-biological interrelatedness in disease and of readiness to identify psychosocial factors and to deal with them in clinical situations. Finally, a term that tends to be all-encompassing is "psychiatric medicine."[192] Whether or not this name will find widespread acceptance is problematical, but it indicates the difficulty of crystallizing psychosocial considerations within the field of medicine. Allowing for this difficulty, for the present liaison psychiatry expresses most clearly the working relationship between psychiatry and medicine.

¶ The Nature of Liaison Psychiatry

When all is said and done, the goal of liaison psychiatry is to effect a relationship—a liaison with other departments of medicine to promote the recognition of psychosocial factors in clinical work and to ascertain that the medically ill get the benefit of complete care which requires inclusion of these factors. In order to do this, psychiatry has relied upon medical protocol, or the medical model—that the contract between physician and patient is so strictly private that accountability for decision and result depends upon the judgment of the physician. In consequence, a resource person cannot approach the situation uninvited. Within liaison psychiatry, this is honored, although it is now thought that since the patient is a member of the hospital system, the psychiatrist may contribute to direction of his care.

But this right has been slow to evolve. Liaison psychiatry began with the psychiatric consultation and was called consultation psychiatry. As interest grew in the psychological and social aspects of medical disorders, psychiatrists sought teaching and case finding opportunities in general medical units of hospitals. To indicate the conjoint clinical and teaching functions of such psychiatric programs, consultative-liaison services were organized. All are now subsumed under the name of liaison psychiatry or liaison program.

In 1959, Beigler and his coworkers[12] had called attention to two major categories of liaison psychiatry:

> 1) the consultation-type functions and 2) the specifically 'liaison' functions. The former comprise the services usually rendered by a psychiatrist summoned as a consultant; the latter constitute . . . functions of the psychiatrist as he works over an extended period of time on the various non-psychiatric divisions of a general hospital.

Lipowski,[125] in 1973, attempted to refine the meaning of these terms and what they represent when he stated that:

> Psychosomatic medicine as a scientific discipline attempts to collect a body of facts and build a unified theory about the interrelationships between man's psychological and biological attributes and functions on the one hand, and his physical and social environment on the other.

He defined consultation-liaison psychiatry as "the area of clinical psychiatry that encompasses clinical, teaching, and research activities of psychiatrists and allied mental health professionals in the non-psychiatric divisions of a general hospital." He continued:

> The designation "consultation-liaison" reflects two interrelated roles of the consultants. "Consultation" refers to the provision of expert diagnostic opinion and advice on management regarding a patient's mental state and behavior at the request of another health professional. "Liaison" connotes a linking of groups for the purpose of effective collaboration.

In 1976,[128] he included under a definition of psychosomatic medicine, "clinical activities at the interface of medicine and the behavioral sciences subsumed under the term consultation-liaison psychiatry."

Strain,[185] in 1977, described the nature of liaison psychiatry by strictly dividing the functions of liaison and consultation psychiatry:

> Although consultation in the hospital setting provides the cornerstone for the liaison effort, there are major differences between these models. . . . Briefly, in contrast to the psychiatric consultant, whose primary function is to alleviate acute psychiatric symptomatology in the individual patient, the liaison psychiatrist seeks to enhance the psychological status of all medical patients . . . In addition, the liaison psychiatrist differs from the psychiatric consultant in that he/she participates in case detection rather than awaiting referral, clarifies the status of the caretaker as well as the patient, and provides an educational program that promotes more autonomous functioning by medical, surgical, and nursing personnel with regard to handling their patients' psychological needs.

In this way, he spelled out some of the functions of a liaison program in the delivery of mental health services to the medically ill and offered an overview of the nature of liaison psychiatry. But to most liaison psychiatrists, modern consultation practices and liaison activities are inseparable.

¶ Historical Perspective

Liaison psychiatry was an outgrowth of the psychosomatic movement, which began in Germany and Austria in the second and third decades of this century and reached its apogee in the United States between 1930 and 1950. Many theoretical, research, and clinical studies of the interrelationship of the emotions and bodily functions were conducted under the influence of psychoanalytic investigators, physiologists, and clinical psychiatrists. Methods of application of concepts and vehicles of administration soon developed. Henry,[74] in 1929, published a significant paper in the *American Journal of Psychiatry.* It is noteworthy not only because it was the first exposition of the consultation model of service, but because it described many of the classical obstacles and indicated their solutions.

In 1933, the true development of liaison programs was set in motion. The Medical Sciences Division of the Rockefeller Foundation,[161] under the leadership of Dr. Alan Gregg, placed major emphasis on the development of psychiatry by providing funding for full-time teachers of psychiatry in selected American medical schools and by establishing departments of psychiatry or extensions of departments within certain university hospitals. Grants were given for these purposes to Harvard (Massachusetts General Hospital) (1934), Tulane University (1936), George Washington University in St. Louis (1938), and Duke University (1940). The foresight of Alan Gregg and his associate, Dr. Robert Lambert, set the course of psychiatry for a generation and placed psychiatry in the general hospital where it could exercise a significant influence on the rest of medicine.

The growth and influence of these new or extended departments developed at different rates but all in the course of time made striking contributions to psychiatry, medicine, and psychosomatic medicine. The first two institutions to influence liaison psychiatry were the University of Colorado Medical Center and Harvard's Massachusetts General Hospital. At the University of Colorado, the extended department was called "The Psychiatric Liaison Department" or "P.L.D." As far as can be determined, here was the origin of the term "psychiatric liaison" and it is attributed to Franklin G. Ebaugh and Edward G. Billings[17] who developed the concept there. The model established has been followed by many centers. "The department purposely had no hospital beds assigned to it and no specific niche in the outpatient clinic. Patients were examined, treated, and utilized as the focus for teaching and research wherever they might be bedded—whether in a pediatric, surgical or medical ward."[20] Billings stated that the liaison department was organized around three aims, which are still, after forty-five years, the objectives of many liaison services:

1. To sensitize the physicians and students to the opportunities offered them by every patient, no matter what complaint or ailment was present, for the utilization of a common sense psychiatric approach for the betterment of the patient's condition, and for making that patient better fitted to handle his problems—somatic or personality—determined or both.
2. To establish psychobiology as an integral working part of the professional thinking of physicians and students of all branches of medicine.
3. To instill in the minds of physicians and students the need the patient-public has for tangible and practical conceptions of personality and sociological functioning. This was to be not so much in the sense of "prevention" of mental disorders per se, but rather in the sense of preventing false thinking, misconceptions, misunderstanding, folk-lore and taboos which made it difficult for the patient to accept help or to allow the physician to be of help. [p. 30][20]

In 1948, Kaufman and Margolin[89] offered the following principles:

The organization of the psychiatric service in a general hospital at any given time depends on the level of sophistication with respect to psychology. *Therefore, no blueprint of an organization can be regarded as universally applicable* [author's italics] . . . Its structure will be sufficiently dynamic and flexible as to permit revision, in terms of shifts of emphasis and foci of activity as the level of psychological indoctrination changes.

These principles are as effective today as they were in 1948, and have been utilized in the organization of many liaison services. There is no one ideal liaison program; each is shaped to fit the potentials of the institution and its liaison psychiatrists.

Kaufman and Margolin also described the objectives of a liaison program:

The administrative set up must be built around the professional needs of the institution. The *primary needs* are always:

1. Psychiatric services: i.e., diagnoses and treatment of the hospital population, both outpatient and inpatient.

2. Teaching, which involves two aspects—one, the further training of the psychiatric staff, and two, the indoctrination and teaching of every member of the hospital staff from administration through chiefs to house staff.
3. Research. [p. 612][89]

A variety of strategies was developed to attain objectives. At the Massachusetts General Hospital the goal was to demonstrate that the scientific method was applicable to psychiatry; at Rochester the operational strategy was to merge with medicine; at Duke the aim was to reach objectives through concentrated training of the medical house officer; at Johns Hopkins there was refinement of the consultative process; at Washington University emphasis was on the comprehensive approach to clinical problems; at Cincinnati the psychosomatic ward and hospital psychosomatic conferences were emphasized; and at Einstein there was a hospital-consultation service and psychoanalytically oriented teaching. All of these were designed to influence the acculturation of the nonpsychiatrist physician toward the acceptance of psychiatry as relevant to the care of the sick and dying.

In the history of liaison psychiatry, six models of consultation liaison programs developed.

The Consultation Model

Patients are referred to the psychiatrist for evaluation and possible treatment, and/or for recommended emotional care by the consultee and/or caretaker staff. This is basic to all models, whether alone or in combinations with other models. Despite the fact that there is no solid data on the availability of psychiatric consultations in general hospitals, probably most of the general hospitals in the United States carry on psychiatric consultations, principally following the consultation model.[71,163] This is particularly true in community hospitals, and is the sole model in many teaching hospitals.

The Liaison Model

Psychiatrists and other mental health workers are assigned by a liaison division of the department of psychiatry to selected hospital units (usually in the department of medicine) to consult, case find, and teach. This design is sometimes referred to as the consultation-liaison model. It often relies on "islands of excellence" or model demonstration. This design has been the one of choice in many training centers, with the implication in some that ultimately liaison arrangements will cover the entire hospital. This has never been fully achieved.

The principal example of the liaison model is at the University of Rochester.[41,43,46,171] The program was started in 1946, has remained under the same leadership and has had the full support of both the departments of psychiatry and medicine. It has been a principal area of psychiatric undergraduate teaching and has provided training for a large number of psychosomatic fellows (109 between 1946 and 1977), 60 percent of whom became full-time medical educators.[169] Its operational strategy seems to have been to interdigitate or merge with medicine, through what has been termed the Medical Psychiatric Liaison Group. Through the years this group has consisted of five to ten full-time senior staff and six to nine fellows in training. Like leaders in several other liaison programs, the majority of the staff first received training in internal medicine and later qualified in psychiatry. The tightness of the liaison arrangement, which permits the liaison worker to act as a resource person in both psychiatry and medicine on select medical hospital units and exemplifies the role model of the internist who integrates psychosocial factors into his clinical considerations by performance, is an important characteristic of the Rochester liaison program.

The Milieu Model

An extension of the liaison model, the milieu model places emphasis upon the group aspects of patient care, group process, staff reactions and interactions, interpersonal theory, and the methods of Stanton and Schwartz.[182] Several centers combine the liaison and the milieu models.

The Critical Care Model

Mental health personnel are assigned to critical care units rather than to clinical departments. The goal is patient care with the psychiatrist a participating member of the unit team, in which he often becomes the unofficial leader. Teaching combines behavioral, biological psychiatric, and psychoanalytic theoretical models. This model was developed in the 1970s as a result of changes within clinical medicine.

There are now many centers in which psychiatric services attach psychiatrists to intensive care units in liaison arrangements. Two that can be cited as examples are the psychiatric departments of the Massachusetts General Hospital and the Hospital of the Albert Einstein College of Medicine. At the Massachusetts General Hospital active liaison psychiatry is an integrated part of clinical work in the intensive care unit, the coronary care unit, the pulmonary care unit, burn unit, and in oncology.* At the Einstein Hospital, all available psychiatrists are scheduled for priority service on critical care units, in acute psychiatry, and in oncology and terminal care rather than assigned to liaison arrangements with clinical departments.[57,59,64]

Additional liaison psychiatry centers may concentrate on the critical care model in one specific area, such as in hemodialysis units, as has been the case at the Downstate Medical Center[116,117] and at the University of Southern California.[108]

The Biological Psychiatry Model

This is a more exacting example of the critical care model with strict emphasis upon neuroscience and psychopharmacology in which the psychiatrist maintains his status as a peer scientist. He provides the psychological care by assessment of cognitive disturbance and changes in levels of awareness, and treatment at these biological levels by management with psychopharmacologic and other physical agents and by maintaining vigilance to assure an optimal environment for the patient. The Massachusetts General Hospital utilizes this model, as does Columbia,[105,106,107] the University of Southern California,[91,92] and Montefiore Hospital.[186]

*See references 28, 29, 72, 194, 196, and 197.

The Integral Model

The aforementioned models of liaison programs depend in the main upon traditional consultation with patients and staff and liaison arrangements whereby initiative for psychosocial intervention rests entirely with the nonpsychiatrist physician. As a result of the burden of critical care, medicine with its load of technocratic systems and complexities and social pressures, the integral model of liaison psychiatry is developing a new direction in medical care. It was first conceptualized and established at the Hospital of Albert Einstein College of Medicine,[58,62] while at the same time its possibilities were recognized and reported at Montefiore Hospital.[186] This model is based upon the inclusion of psychological care as a component of patient care, as the right of every sick person, and provides for the availability of the psychiatrist to function at the points of clinical and administrative need. The initiative comes from staff consensus on the need for psychiatric intervention more than from the judgment of the nonpsychiatrist physician. The integral model will be discussed at greater length later in the chapter.

¶ Liaison Consultation

Whether the psychiatrist is a member of a liaison team or functions alone, his principal contribution to the medically ill and to medical and surgical colleagues is through the psychiatric consultation. Such consultations require skilled techniques about which much has been written since midcentury.[174] We can distinguish between the "psychiatric consultation" and the "liaison consultation." The first is essentially an arrangement between the patient's attending physician and the psychiatrist, ostensibly for purposes of diagnosis, management, or treatment planning in which the patient is interviewed and a verbal or written report is given to the consultee. In the absence of a liaison program the result is apt to be a single examination, a so-called "one shot" consultation which may prove valuable as an assessment if properly done. The solitary psychiatrist, if he has undergone liaison training, will often develop the single contact into a liaison consultation.

This author has described the nature of the liaison consultation:

At the core of liaison work is the dynamic contact between the liaison psychiatrist and the key figures in the clinical field: patients, families, physicians, nurses, social workers, administrators, psychologists and others. The interaction at this point of contact is called liaison consultation, and the principal participant is the liaison psychiatrist whose task it is to serve as the resource expert on psychological and social variables in disease. The substantive knowledge, methods, and techniques of psychiatry are brought to bear on the task. The liaison psychiatrist is expected to have additional knowledge concerning the characteristics of forces at the interface of psychiatry and the other medical disciplines. The points of dynamic contact are variable so that the consultations may be with the patient alone, with the patient and consultee, with the patient and nurse, with the patient and all key persons in his clinical field, with the consultee alone, with the family alone, or with other combinations. This fluidity of consultative endeavor is one of the principal skills of the liaison psychiatrist. He is adept at changing role models and is familiar with systems and boundaries of systems. [p. 132][58]

Approaches to the Psychiatric Consultation

The preceding description serves as a background for the definition and discussion of approaches to consultative work and of models of psychiatric consultation. Although the patient is ostensibly the object of the consultation, the working orientation to the basic purpose for the consultation helps to determine the approach. There are four approaches to consultation work: (1) the patient-oriented approach, (2) the consultee-oriented approach, (3) the situation-oriented approach, and (4) the professional-oriented or supervisory approach.

The patient-oriented approach is the traditional psychiatric consultation. How to carry this out has been carefully spelled out in many publications.* There is general agreement on the method: preparation for the consultation, the setting for the examination, the approach to the patient, the interview, the consultant-patient relationship, the written report, and transactions with the consultee. The approach is always patient-centered; the objective is the assessment of the patient himself.

In 1959, Schiff and Pilot[167] introduced the concept of the consultee-oriented approach.

It is based on a point of view which is primarily consultee-oriented rather than patient-oriented, and attempts to examine carefully the manner in which the consultation is requested and the background of each situation. The assumption is made that every psychiatric consultation, if not every consultation, stems from the referring physicians concerns, of which the most cogent are frequently not explicitly stated. [p. 357]

This point of view focuses on the latent reasons for the consultee's request in terms of his position in the clinical situation and has become a frequently used component of the consultative process. Whether or not the results of the consultation will be acceptable to the consultee and be acted upon productively by him may depend on the incorporation of the consultee-oriented approach.

The situation-oriented approach has essentially a group process emphasis and was described by Greenberg[55] in 1960:

At times, the interaction of members of the clinical staff may produce an atmosphere in which certain aspects of the patient's historical behavior produce anxiety in one or more staff members, or in which covert symptoms became manifest . . . A situation-oriented approach is suggested to meet with the conditions found in some research settings, as well as with the conditions of a general hospital. This approach takes into account the interpersonal transactions of all the people involved in the direct care of the patient. [p.691]

*See references 34, 88, 137, 174, and 195.

This approach acknowledges that the medical-surgical inpatient unit is a therapeutic community, although it is usually neither organized nor monitored as such. We have here a transformation of the concepts and methods of Stanton and Schwartz and others in the psychiatric hospital setting. This approach is, of course, easier to use in the milieu liaison model, but should be considered in any consultation event.

The professional-oriented approach is essentially a supervisory one and was described by Greenhill in 1977:

> In this approach psychiatrists consult with physicians regarding patients whom the latter does not want seen or whom it may not be necessary to see, may advise them on the psychological management of patients without interviewing everyone, and may conduct psychotherapy supervision as a learning experience for the medical trainee or practicing physician without meeting the patient. Nurses and social workers subscribe to this approach frequently, and in addition, some patients are electively not seen in staff conferences held in their behalf. [p. 133][58]

Although the implication is that this approach is reserved for situations in which the patient is not to be seen, this is not always the case. The supervisory function is at the core of the professional-centered approach with the psychiatrist in the background, whether he has seen the patient or not.

Theoretical Models of Psychiatric Consultation

In concept, theoretical models are not always different from these approaches to consultation except that they seem to include unified theories on the background and meaning of the consultation and how the process is worked through. There are four consultation models described in the literature: (1) The operational group model, (2) The communications model, (3) The therapeutic consultation model, and (4) The crisis-oriented model.

The operational group model was described in 1961 by Meyer and Mendelson.[139] It is essentially a social process model. The operational group consists of four people: the patient, the internist, the consulting psychiatrist, and the nurse. By the systematic collection of data from the transaction of these four people, the identification of the problem and its solution is forthcoming. Schwab[174] interprets the concept as:

> the request for psychiatric consultation reflects a crisis within the group, usually a disruption of trust and communication between the patient and the "caring for" people. The entrance of the psychiatrist redefines the operational group, thus reducing anxiety and establishing trust and communication. The interaction between the patient and staff then becomes therapeutic.

The communication model was described in 1964 by Sandt and Leifer.[164] It represents once more the application of a social science theoretical model to medicine, in this instance, communications theory. Sandt and Leifer do not go beyond analyzing the communication (language) factors in the request for psychiatric consultation by decoding latent reasons for the request. Brosin,[24] in 1968, carried the communication concept beyond the request to the consultation process by suggesting that the consultant carry out a message system throughout the duration of the case. The successful encoding and decoding of messages determines the outcome of the specific liaison arrangement between consultee and consultant.

Weisman and Hackett's[195] therapeutic consultation model has as its objective the formulation and implementation of a management program for patients with psychological problems. They have written:

> There are four phases to the work of therapeutic consultation: Rapid evaluation, with special attention to the personal factors and the reason for consultation; psychodynamic formulation of the major conflict, predomoninant emotional patterns, ego functions, and object relationships; rational planning of a therapeutic intervention, based on the formulation; and active implementation by the psychiatrist himself.

The theoretical approach is "patient-oriented, rather than disease-oriented" in that it attempts to provide psychiatric management by focusing upon crisis, conflict, and reality testing in the patient's brief hospitalization.

The crisis-oriented model, as described by Greenhill,[58] places a greater emphasis on behavioral therapy. It considers psychosocial factors as emotional stressors that produce exacerbations of symptoms and behavioral reactions by the sick person which are presented in some form as a crisis. The exacerbation that brought him to the hospital and his course in the hospital may be marked by a series of critical events, and he may be influenced in the social setting of the medical unit by crises within the staff or of other patients. In the consultation, the liaison psychiatrist identifies the event-dysfunction sequences and the patient's communication defenses which attempt to shield him from disclosures to the staff. The consultant also screens for staff involvement in the relevant crises and for other sources of crises in the milieu. Thus a pattern of behavioral and somatic reactions to emotional stress is noted and communicated to the consultee and staff, and concomitantly a therapeutic approach to the patient is begun.

¶ The Delivery of Mental Health Services to the Medically Ill

Psychiatric Consultations

In the field of medical care, those patients in need of psychiatric intervention who are referred for consultation are "case found" in liaison programs, or remain anonymous. As we shall see, the majority remain anonymous.

There are several categories of patients with medical and surgical disorders who require mental health services, but there are a limited number of studies on frequency. From a series of 2,521 consultations in two Yale teaching hospitals, Kligerman and McKegney[102] listed diagnoses and frequency of occurrence:

1. Acute and Chronic Brain Syndromes 31.0 percent
2. Depressive Reactions 57.4 percent
3. Conversion Reactions 11.5 percent
4. Neurotic Reactions 32.2 percent
5. "Classical" Psychosomatic Disease 8.8 percent

When one considers Schmale's[168] figures on depression and separation reactions in medical patients, the incidence of depression ranges from 57.4 percent to 69.0 percent. Other studies show that the frequency is close to that range. Poe[151] found that 52 percent of 191 patients at the Peter Bent Brigham Hospital suffered depressions and West and Bastani[198] found depressive disorders in 52 percent of 1,039 patients at the University of Nebraska Medical Center. Shevitz, Silberfarb, and Lipowski[176] diagnosed depressive reactions in 53 percent of 1,000 consultations; of those uncovered 20 to 25 percent were severe enough to warrant treatment. Mood disorders are undoubtedly the major psychiatric complication of the medically ill.

Statistics on other psychiatric diagnostic categories, dependent upon the population studied, vary more than depression. Organic brain syndromes varied from 18.0 percent to 31.0 percent and neurotic reactions from 3.0 to 31.2 percent.[102,176] It appears that mood disorders, organic mental syndromes, and management problems account for the principal efforts of psychiatric consultants. When added to this is the reported fact that 47 percent to 68.2 percent of referrals show both a physical and psychiatric disorder, the purpose of liaison psychiatry is clear.[176]

But this is only a portion of those needing help. These patients are those on whom consultations had been requested. There are others. Kligerman and McKegney[102] reported that between 39 percent and 45.8 percent of the patients on the Yale-New Haven Hospital Medical Service were moderately or severely emotionally disturbed. Lipowski,[122] in 1967, cited the prevalence of psychiatric morbidity in "medical" populations in nine studies as ranging between 15 percent and 72.5 percent, depending on the study. There is general agreement in all prevalence studies carried out since then that in 20 to 50 percent of medical-surgical patients there is an associated psychiatric morbidity, with most reports citing a figure of 40 to 45 percent.

In contrast, the frequency of psychiatric consultations requested is low. Kligerman and McKegney[102] reported that only 2.94 percent of all patients at Yale-New Haven Hospital have been subjects of such consultations. At the same center, Duff and Hollingshead[36] reported there were consultations requested on only 6 percent of 161 identified psychiatric problems on a medical-surgical unit. From eight other studies over several years (reviewed in 1967), the frequency ranged from 4 to 13 percent with an average of 9 percent.[122] It is suspected that this is high, for all reports since then have shown that the consultation rate rarely goes above 3.0 percent. It is at this level at the Einstein Hospital,[57] and at 3.3 percent in the Dartmouth study.[176] Benson[13] reported consultation rates of 0.5 percent, 0.7 percent, and 2.0 percent in three additional studies. Cavanaugh and Flood,[30] in a survey of attending physicians, reported that 1 to 5 percent of their private hospitalized patients received inhospital psychiatric consultation. At any rate, only a small proportion of the medically ill who need psychiatric intervention receive it; in round figures, only one in fifteen.

The obvious questions that arise are: Does this represent a striking lack of interest and strong resistance on the part of physicians, even where there are active educational programs? Or does it indicate that the liaison system itself provides enough "know-how" on the part of nonpsychiatric personnel that the mental health needs of patients are being met and not reported? Or are there forces at work that result in the emotional needs, the "felt needs," even the psychopathology, being grossly neglected, no matter how great the zeal and efforts of liaison people? There are no evaluation studies, but the answer is probably some of all of these. All of the evidence points to the fact that after forty years and two generations of effort by liaison psychiatrists, physicians as a class do little about emotional care.

The Liaison Program

The purpose of the liaison system is to improve the delivery of mental health care and to overcome the obstacles to its effectiveness. This is achieved, first of all, by insuring the availability of psychiatrists in the units of other clinical departments of the general hospital by utilizing strategies that attach them to clinical rounds, conferences, and teaching exercises. When the psychiatrist has been assigned by negotiation with the chief of another service, he utilizes opportunities for contact and relationship with the nonpsychiatrist physicians and other staff members to insinuate and promote psychosocial considerations into patient care. This is brought about by case finding and education. In distinction to the passive position of the psychiatrist who waits to react until he is invited during the traditional psychiatric consultation, the liaison psychiatrist, if he is skilled in the liaison process and psychosomatic approach, assumes an active posture for clinical intervention. His aim is to relate to the nonpsychiatrist physician and to work through the latter's resistance to psychosocial intervention. This is best realized by education and training.

¶ Education and Training

To teach the physician to include psychosocial variables in patient care and then to deal with them himself, or to teach him to permit a psychiatrist to participate, has proven to be a special and challenging task for the psychiatric educator. He aims at an improvement through education in clinical science for the benefit of the somatically ill patient. This he undertakes to do by broadening the base of clinical considerations by enlarging the concept of the biomedical model in use throughout medicine. Adolf Meyer attempted this early in the century by implementing his concept of psychobiology.[119] Numerous others have followed with comprehensive and holistic approaches. Engel[45] the most recent and vocal advocate, emphasized the necessity for a "biopsychosocial" model as fundamental to the theory of disease. The aim is to educate the medical clinician to be a "whole" thinker, not an exclusionist or reductionist.

The logistics of the teaching situation has special characteristics. A small number of liaison teachers (one to ten), with a small number of liaison fellows (one to six) or none at all, have undertaken the task of teaching clinical faculty, medical students, psychiatric residents, attending physicians, interns, residents, and fellows in other departments, nursing students, nursing staff, nurse clinicians, and a variety of ancillary personnel, whose multidiscipline collaboration is needed. The strategies developed to allow teachers to do an effective job in these proportions depend on the needs of the institution and the interest and personality of the liaison leader. On the whole, the attempted strategies aim at (1) concentrating the teaching area in a limited geographic area, that is, one or more medical units in the department of medicine; (2) establishing a demonstration model or "island of excellence" in one hospital unit or subspecialty; (3) attempting to reach the learner while he is still a medical student; (4) enlarging the population of available liaison workers by concentrating on graduate training (fellowships); and (5) utilizing the services of available fellows to do the principal work of teaching house officers in other departments by peer effect. As a matter of fact, in many programs the task is so disproportionate that the patient may become the end point of interest, so great must be the strategy of educating caretakers.

The following are a comprehensive group of teaching objectives that were derived from a survey of several centers,[58] and that are applicable in degrees of intensity to the education of medical students, nonpsychiatrist residents, and fellows:

1. to teach those psychiatric methods and techniques that are relevant to the physically ill, including the collection and assessment of raw psychosocial data by use of interviewing technique and instruction in supportive therapy, limited-goal brief psychotherapy, and crisis intervention;
2. to present a body of substantive psychiatric knowledge the content of which is relevant to medical and surgical disorders. This would include delirium and dementia; depression, grief, and separation reactions; psychoneurotic equivalencies in somatic disease (anxiety attacks, conversion reactions), emotional stress-sensitive medical disorders (peptic ulcer, asthma, and so forth); psychological reactions to illness, to interpersonal stress, and to terminal states; and addictive reactions and borderline states;
3. to examine the administration of psychopharmacological agents to the physically ill;
4. to present the influence of social science, with the effect of social stresses and patterns on exacerbation and course of medical and surgical disorders;
5. to change the attitudes of the learner regarding psychological processes, the image of the psychiatrist, scientific dogma, the quest for certainty, and the counter values of other teachers;
6. to influence the learner's methods of communication, verbal and written, to include recognition of psychosocial phenomena and willingness to engage the patient on emotional topics;
7. to interest the learner in the problems of chronicity;
8. to offer the learner experiential involvement in the physician-patient relationship in medicine by continued case supervision, brief, or extended; and
9. by a combination of points of concentration on interviewing technique, group process, physician-patient relationship, and modification of attitudes to increase the learner's humanitarianism.

We now come to the consideration of role models in this educational process. In 1967, Engel wrote:

Of critical importance for the establishment of the program and its subsequent growth was, I believe, the fact that I and those who joined me in the early days were fully qualified as internists. This enabled us to establish ourselves as peers on the medical service and gradually to overcome the misconception that we were alien poachers on their domain.... When such programs are staffed only or predominately by psychiatrists, they never really become anything other than psychiatric consultation services. As a result students and house officers never have as a model a physician who combines in his own personal skill both the psychological and somatic aspects of illness. And without such a model the student has no alternative but to believe that it takes two specialists to deal with psychosomatic issues.[41]

Kaufman presented another point of view in 1953:

A psychiatrist is a catalyst, an integrator. He has a great deal to contribute to medicine, but his contribution must be made primarily as a psychiatrist. The writer has no patience with the type of psychiatrist who tries to smuggle himself into medicine under false colors and who feels that it behooves him to demonstrate to the surgeon or to the internist that after all he too, is a top internist or surgeon . . . The psychiatrist is a psychiatrist, just as the surgeon is a surgeon; and it is only as a psychiatrist, standing firmly based on his own discipline, that he can eventually demonstrate the value of his orientation in the understanding and treatment of patients.[88]

The stand taken by liaison psychiatrists concerning role model appears to be influenced by their own predominant identifications. As a teacher of many years, this author knows that many liaison fellows are searching for their identity in medicine; most of them find it, on one side of the fence or on the other, but always retaining that "liaison touch." In 1977, this author wrote:

I think we are historically beyond the image of the internist as the role model. The role model is a proven expert clinician in any field, including psychiatry, whose enthusiasm for clinical science is contagious, and who can demonstrate that psychosocial data and interpersonal processes are powerful factors in medicine. The internist role model may be less prepared to deal with the exigencies of social and behavioral pressures on medicine today than the psychiatrist or the professor of community medicine. [p. 152][58]

When all is said and done, if one cuts through all models, designs, and efforts in teaching and training in the psychosomatic approach in graduate and postgraduate education, the basic ingredients are *exposure* and *engagement.* Those patient care and teaching programs that encourage, by whatever means, consistent and meaningful exposure of the physician to the emotional implications of the clinical state of his patients and that assist him to engage with the consultant, the patient, the nurse, and the family in open-ended acknowledgment of such implications reflect realistic expertise.

¶ Resistance

Throughout the history of psychosomatic medicine and liaison psychiatry the obstacles encountered in patient care and teaching have been much discussed and reported. McKegney,[134] Lipsitt,[131] and Krakowski[110] in the United States wrote about these during the 1970s. Limitations of the psychosomatic approach have been caused by several factors including economics, space, and curriculum, but the most serious and prevalent are physicians themselves.

For want of a generic term, I have called these particular physician-generated obstacles, "resistances." In 1950, Greenhill and Kilgore[66] reported on types of resisters encountered among medical house staff engaged in a liaison teaching program and on how to deal with their resistances. Although the term has been used frequently through the years, it bothers some who align it with resistance in the therapeutic process and the repression of unconscious conflicts. Here, the term "resistance" is used in a wider sense to denote the efforts of physicians to avoid, withstand, deny, deter, and obstruct, by any means, conscious or unconscious, the consideration of the influence of emotion in disease. It is a well-known fact, well documented in the literature, that physicians tend to resist utilizing psychiatric service and give low priority to the emotional concomitants of clinical situations.

Resistance to psychological medicine on the part of nonpsychiatrist physicians is complicated, puzzling, and stubborn. It has taken many forms, and many methods have been employed in attempts to combat it. Conciliation, concession, internist role modeling, use of somatic language, equality of rank, emphasis on physiology and biochemistry, and attitudes indicating the validity of a psychiatric approach are among the methods that have been used by liaison psychiatrists. Early exposure of the medical student, with reinforcement at later stages in his career, is another. Causes advanced for resistance to psychological medicine include the intensive pressures of medical training, assimilation by medical students of the negative attitudes toward psychiatry and psychological medicine held by teachers in other fields, anxiety aroused by unconscious forces within physicians, and a distrust of the perceived lack of certainty of human behavior.

But the nonpsychiatrist physician is not alone in his resistance, for the psychiatrist participates in it in his own fashion. A factor that has not been thoroughly explored by liaison psychiatry is the resistance within psychiatry itself. Most psychiatrists prefer to work with the intricacies of interpersonal relations and intrapsychic forces, specific symptom groups, and psychotherapy. They are apt to be strongly individualistic or oriented toward a particular social group. It may not always be a matter of psychiatrists feeling uncomfortable in the medical situation and with the medical model, but rather that they do not find medical patients very interesting psychiatrically. Such patients seem psychopathologically superficial and their psychological aberrations, not being readily presented, must be sought out. Besides, psychiatrists in the main do not like to work with unmotivated patients.

The behavior of psychiatrists toward medicine and other physicians has been considered by many to be an important source of avoidance of the psychosomatic approach on the part of their medical colleagues. For example, Lipsitt[131] has written that (1) most psychiatrists withdraw from the medical model; (2) psychiatric residents tend to be uneasy in the liaison rotation, considering it "regressive"; (3) internists reject the psychiatrist or exhibit discomfort in his presence, and the psychiatrist helps to promote this; (4) the psychiatrist tends to misperceive the difference in style, rhythm, and demands of office practice; and (5) the psychiatrist has little understanding of how to synthesize with the "doctor's job." But the tendency to blame the interrelationship between physicians and liaison psychiatrists for this problem seems too narrow a view; many other causes are at work as well.

¶ The Role of Psychiatric Units in the General Hospital

The general hospital has become the focal point for the delivery of mental health care in the United States.[63] This has come about as a result of the evangelical campaign for deinstitutionalization of the mentally ill and the concomitant expansion of the number of psychiatric units in general hospitals. The increase in such units has been phenomenal.[84] In the decade of the 1940s, there were an estimated 40 psychiatric units in general hospitals; by 1971 there were 750.[84] Another 289 hospitals provided inpatient psychiatric treatment on their regular wards. Therefore a total of 1,039 hospitals, a minimum of 19 percent of the 5,565 community general hospitals in the United States, admitted psychiatric patients from the community.[84] In 1971, there were 542,000 patients admitted to these units, which was 43 percent of all psychiatric admissions compared to 34 percent admitted to state hospitals.[84] That percentage has held to the present time.[63] Of the general hospitals that have psychiatric units, the median unit size is twenty-eight beds.[84]

These facts are presented in order to consider the relationship of liaison psychiatry to psychiatric units in the general hospitals. In actuality, there is very little relationship. It might be presumed that having an active psychiatric unit in their midst, physicians would request more consultations, and that medical-surgical patients with psychiatric complications would be transferred to the psychiatric units for optimal care. It might also be expected that liaison divisions of the hospital department of psychiatry would be strengthened. Nothing of the kind has occurred.

Benson[13] has pointed out that the consultation rate remains constant when "It would be thought that a viable psychiatric unit alert to psychiatric problems in the rest of the general hospital population could increase the consultation rate." As for the transfer of medical-surgical patients with depressions, organic mental syndromes, psychological management problems, or other psychiatric conditions to the psychiatric units, all evidence shows that this seldom occurs.[63] One has but to study the censuses of psychiatric units to see that they have a distinct population.[84]

On the other hand, it has been reported that the psychiatric unit has a salutory effect on the general hospital.[158] It is reassuring to medical and surgical staff, no matter what their criticism of the presence of a psychiatric unit in their midst, to know there is a backup facility at hand should any patient become unmanageable. It is ironical that these backup facilities are loathe to accept transfers. The units prefer to admit only patients with psychiatric disorders—and only from the community—because the pressure from the community to take psychiatric patients is great. Further, staff on psychiatric units believe that admission of medically ill patients will contaminate the therapeutic milieu, and psychiatric nurses find it difficult to take care of medical patients and psychiatric patients at the same time.

It is a paradox that psychiatric units in general hospitals do not serve the hospital populations, particularly in view of the high incidence of psychiatric disorder within the general hospital population.[176] At one time in the history of liaison psychiatry, it appeared that this problem might be avoided by the establishment of "psychosomatic units."[58] These were sections of the hospital that accepted only combined medical-psychiatric problems and were staffed by physicians and nurses from both psychiatry and medicine. Such units existed in the 1940s and 1950s at the University of Cincinnati,[85] Mount Sinai Hospital in New York,[88] Montefiore Hospital in New York, and the University of Maryland in Baltimore. With the exception of the unit at Cincinnati, they did not survive for various reasons, but the recommendation has been made that, not withstanding the cost, general hospitals of the future should have two psychiatric units, one for psychiatric patients from the community and the second to serve the medically ill in the hospital.[64] Otherwise, the latter are neglected and the hospital department of psychiatry does not have a proper liaison with its own liaison group.

¶ Evaluation

Evaluation of the scope and effect of liaison psychiatry has been sparse. Houpt[81] attributes this to the complexity of its goals and theoretical viewpoints, the characteristics of settings in which it operates, and the variety of the organizational structures of liaison programs. To this may be added the fact that so limited were the facilities and the number of workers in the field and so meager the financial support that demands for teaching and service were all that could be handled, to say nothing of ongoing assessment. Besides, until they became required, methodologies of evaluation and accountability held little interest for psychosomatists and liaison psychiatrists. Few evaluation studies of liaison psychiatry existed before 1977. The increased number after that may well be a result of the high priority that, beginning in 1974, was given to the development and expansion of psychiatric consultation-liaison teaching services by the Psychiatric Education Branch of NIMH. Financial support has been provided for more than fifty programs each year since then, and each of the supported programs is required to have an evaluation component.[155] This last requirement may also account for the increased quality of evaluation in recent studies.

¶ The Roles of Nonmedical Disciplines

In essence, liaison psychiatry deals with multifactorial health issues through multidiscipline channels. In this pluralistic approach, any situation in the patient care setting involves not just psychiatrist and patient, nor attending physician and patient, nor pschiatrist and attending physician, nor nurse-patient-consultee-psychiatrist, but many professionals in a group effort that spans the boundaries of many disciplines. The more all disciplines dealing with the sick know about the psychosocial aspects of illness and include them in the techniques and attitudes of their clinical efforts, the more accurate the care. Nurses, social workers, psychologists, mental health paraprofessionals, physical therapists, activity therapists, dieticians, and clinical technicians are all involved. Hospital administrators must be consistently informed and educated to the liaison approach. Because an informed therapeutic community is the aim of the experienced director of liaison psychiatry, he should include all health disciplines in his planning. The goal of a complete therapeutic community is beyond achievement because the systems of the general hospital are too numerous and complex, but the thrust of an active program is always in that direction.

The nonpsychiatric physicians and psychiatrists have the major responsibility in liaison psychiatry, but most of the work is performed by medical-surgical nurses:*

*See references 15, 25, 26, 67, 79, 178, and 190.

When all is said and done, the nurse deals with the emotional care of the patient more than anyone else. As a matter of proximity alone, it falls to the nurse either to be exposed to the crises of patients or to be confronted by them through default, because there may be no one else.[58]

Nurses become involved in liaison programs as result of the recognition and enhancement of their clinical influence by the psychiatrist, by in-service training, and through the development of the concept of the psychiatric liaison nurse (PLN). The latter is usually a psychiatric nurse clinician but may be a medical nurse clinician with psychiatric liaison training. The PLN functions collaboratively with general hospital nurses in a fashion analagous to the work of the liaison psychiatrist with the nonpsychiatrist physician.

In 1971, Barton and Kelso[10] called attention to the following functions of the liaison nurse: (1) providing perspective from the viewpoint of the nursing profession; (2) gathering information about patients for the diagnostic process; (3) being involved in prevention of crises and intervening in crises; (4) providing specialized nursing care otherwise unavailable; (5) coordinating available resources by improving communication; (6) providing an educational experience for the members of the liaison team in the transactional field of the patient (nursing care); and (7) participating in research into aspects of nursing care. In 1973, Kimball[96] cited the report of Pranulis on the role of the PLN at Yale, which included not only participation in diagnostic evaluations, staff sensitivity training, problem solving regarding gaps in patient-staff communication, and a brief psychotherapeutic approach, but also acted as a triage person for referral of the patient to the most appropriate liaison team member.

From the beginning, physicians have been more comfortable in referring the psychosocial aspects of patient care to medical social workers. In liaison work, the objective is to incorporate this tendency into the mainstream and not allow it to be a factor in the physician's resistance. Social workers have considerable effect on the patient's attitudes toward convalescence and recovery in planning for the after-hospital period, are in positions to take leadership in aiding families to cope with and accommodate to illness. Lately, they have made a major contribution to hospice and terminal care process in the general hospital.

Psychologists are much needed in evaluation studies, case finding, research, and procedural planning regarding failure of coping devices and changes in levels of consciousness in critical care situations.

Attempts have been made to train and utilize paraprofessionals as mental health counselors on medical-surgical units and with cancer and dialysis patients, for case finding and triage, but these programs have been difficult to maintain within the hierarchy of the hospital structure.[57,177]

¶ The Economics of Liaison Programs

Reimbursement for consultative-liaison psychiatric services is poor. There are many reasons for this including: (1) the resistances of nonpsychiatrist physicians, attitudes of general hospital administrators, and policies of third-party carriers; (2) national problems inherent in the issue of health insurance for psychiatric disorders; (3) low-key interest on the part of organized psychiatry and departments of psychiatry in campaigning for appropriate reimbursement for consultation-liaison work; (4) slowness of psychiatry to participate in general hospital quality assurance and medical audit developments; (5) inequities of reimbursement for consultation as a problem shared by several disciplines in addition to psychiatry; and (6) the unwillingness of medically ill patients to take responsibility for payment or to press third-party carriers for it, either due to poor preparation for psychiatric intervention or because of characteristic reactions against such intervention. The attempts made to overcome this inequity have been of little or no avail, particularly in approaching third-party carriers or hospital administrators. In the face of the high costs of medical care, reimbursement for consultative-liaison psychiatry has low priority at the administrative level. Psychiatry is often put off by insistence that it must set down the criteria for diagnosis, therapy, and prognosis in order to provide a firmer basis for establishing fees.

Be that as it may, there is no third-party coverage for liaison services. As Sanders[163] has pointed out, third-party carriers place several constraints on reimbursement:

> For inpatients, payment is allowed for one consultation, but follow-up care is not a covered service for most patients . . .
>
> Simultaneous care from one service, such as the provision of a general psychiatric consultation and the services of a psychopharmacologist, will not both be reimbursable; only one caretaker per discipline is covered.
>
> Important therapeutic functions are not covered; for example the hospital is not reimbursed when a patient is granted a leave of absence.
>
> Similarly, liaison work and the provision of one-to-one supervision for the housestaff so necessary to protect the patients contract for a specific type of therapy and for confidentiality is not a recognized reimbursable service any more than is the care rendered to family members of a child who has been identified as the patient.

Within the arrangements for liaison between departments, the liaison service is usually inadequately subsidized by the department of psychiatry and little or not at all by the other departments it serves. At Rochester, with its thirty-year collaboration with medicine, in 1976 the Department of Medicine paid 10 percent of the budget of the psychiatric liaison department.[169] Infrequently, another department will subsidize a liaison fellowship.[154] It is reported that at the University of Vermont, the general medicine service contracts annually for a dollar amount of liaison service, but it is too early to measure its effects.[163] In 1972, McKegney[134] estimated that at the University of Vermont Hospital "the amount of time necessary for a staff psychiatrist to perform an adequate role on one non-psychiatric unit seems to approximate 10 hours per week" at a cost of $15,000 to $20,000 per year per inpatient unit based on an hourly rate for psychotherapy. At a contractual rate, a lower base could be reasonably expected. But almost all departments other than psychiatry contribute nothing at all. As for the hospital administration, if it budgets for a chief of psychiatry, it expects him to be responsible for the liaison work in addition to all other responsibilities.

It is well known that psychiatrists are poorly compensated for consultations. The lack of patient interest and frequently poor preparation for the consultation by the consultee contribute to this. Consultants spend at least double the time ordinarily given to a psychotherapy session and may receive half the fee or none at all.[58] However, improvement is possible where patients themselves pay for the service, particularly in well-organized liaison programs.

Guggenheim[69] indicated the outlook for the future economics in liaison psychiatry.

> Can new national health legislation be influenced to benefit consultation psychiatry? Should hospital administrations be urged to put the cost of consultation psychiatry on to the per diem rate charge for all patients? The fiscal planning of consultation work offers the challenge of developing an optimum and a minimum cost-benefit figure in consultation psychiatry. Guidelines need to be established in setting up the disbursal of a given mental health budget for a general hospital as well as for the relative evaluation of the effectiveness of different models of consultation activity. [p.178]

The claim has been made that "as yet there are insufficient data to support the premise that caretaking in the general hospital is both more effective and more efficient when consultative-liaison psychiatry is present."[163] These data will gradually be collected. Cost-benefit ratios are urgently needed in planning for the health insurance of the future. The data that exist (such as the studies demonstrating that early referral in consultation reduces long-term maladaptive responses,[104] the medical audit at the Einstein Hospital,[61] which shows that patients with organic mental syndromes who do not receive psychiatric intervention have appreciably longer hospital stays, and others) point to what will be proven to be obvious. Cost-benefit ratios will undoubtedly show that third-party carriers and hospital administrators are emotionally blind to the importance of psychological care of the sick in reducing the overall cost of medical care. If they could add one to two dollars per day to the per diem hospital reimbursement rates, cost-benefit ratios would result and much needed psychological care would be improved at the same time.

The single most positive contribution to the financial aspect of liaison psychiatry is the financial support that is given to education and training in this field by the training branch of NIMH. This may be as far-sighted as the action of the Rockefeller Foundation in 1934 that led to the birth of the liaison design. It may in time produce the leaders who will guide health care toward the objectives of liaison psychiatry.

¶ Empirical Therapeutic Approaches in Liaison Psychiatry

Liaison psychiatry is made up of three parts: a consultative process, a liaison program, and an integrated therapeutic approach to disease. That the first two parts are well recognized is demonstrated by common usage of the term consultative-liaison psychiatry. The therapeutic aspect of psychosomatic medicine has been well studied in clinical research and in the literature in connection with specific medical diseases and disorders of organ systems. With the exception of occasional overview studies of the treatment of psychosomatic disorders and a few position papers on the use of psychotherapy in treating psychosomatic diseases, few attempts have been made to integrate models of comprehensive psychosocial management with medical treatment. Gildea,[51] Alexander,[4] Hopkins and Wolff,[80] Sperling,[181] and Lipowski[124] made earlier approaches to this subject. But the place of the therapeutic or management approach as a major component in liaison psychiatry, although commonly practiced, has not been squarely addressed.

Recently, an interest in liaison therapy has

been emerging. Hill,[76] Wittkower, and Warnes,[200] Strain and Grossman,[186] Strain,[183] Karasu,[86] and Karasu and Steinmuller[87] have published works on this topic. All make some attempt to assess the role of psychotherapy in medical illness, present programs for clinical management of the medically ill, or edit a series of articles by diverse authors on different aspects of management in an eclectic potpourri.

The state of knowledge on the subject of integration of psychosocial and medical therapy needs much investigation and refinement. Yet therapy and management proceed in a growing number of clinical settings; what is being practiced is mainly empirical. There are several such empirical approaches.

Conflict-Specificity Approach

Alexander,[4] French,[50] Graham,[54] and Wolff[201] were among the proponents of the conflict-specificity approach. Alexander was the first to present this idea in what has been termed his "tripartite" theory: (1) each psychosomatic disorder has its specific psychodynamic constellation; (2) there is an onset life situation that activates the specific psychological conflict of the patient, leading to the exacerbation of the disorders; and (3) the disorder occurs only in the presence of a constitutional vulnerability of a specific tissue, organ, or system (X factor of Alexander). Psychiatrists working with medical patients continue to find this theory useful as a guideline, despite the contention of some modern theoriests, such as Kimball,[94] Reiser,[156] and Weiner,[193] that the "The Holy Seven Diseases" as they call them (hyperthyroidism, neurodermatitis, peptic ulcer, rheumatoid arthritis, essential hypertension, bronchial asthma, and ulcerative colitis) with which Alexander worked, are much too multifactorial to permit such a concept.

In general, the conflict-specificity approach identifies the type of conflict, psychodynamic constellation, and onset situation, and attempts to bring these to awareness in the patient by brief or recurrent focal psychotherapy, and by behavioristically teaching the patient to avoid, where possible, his particular critical issues.

Personality-Specificity Approach

Dunbar,[37] Reusch,[157] Rosenman and Friedman,[162] Nemiah and Sifneos,[144,145] Groves,[68] and others have emphasized the significance of the personality patterns. The concept is to approach an infantile, borderline, or hypersensitive character structure by direct and corrective management, with limitsetting and by appropriate staff planning.

Crisis-Specificity Approach

Engle,[42] Schmale,[168] Lindemann,[120] Greenhill,[56,65] and others give attention to the importance of the precipitating situation and its repetitive nature, such as separation, loss, bereavement, and other critical events, with accompanying crisis intervention. This will be discussed in more detail later.

Somatic Orientation Approach

Kiely,[91,92] Hackett,[72] Kornfeld,[105,106] McKegney,[133] Levy,[116] and others assist and support coping mechanisms, emphasize life maintenance, and attempt to correct failing somatic mechanisms utilizing psychological and cognitive disturbances for guidance. This has led to such methods as the use of psychopharmacological agents in somatic disorders; conditioning and behavior modifications; biofeedback and self-control of physiological functions; increased psychophysiological considerations in pain, and in sleep disorders; consideration of the neurobiological basis of psychopathology; and psychiatric intervention in oncology, cardiac surgery, hemodialysis, and during treatment in intensive care units. This approach is an integral component of medical care, both acute and chronic.

Psychotherapeutic Approach

Since he is dealing with psychosocial influences in medical illness, the liaison psychiatrist is continually involved in interpersonal and intrapsychic phenomena, in one context or another. Consequently he maintains a ceaseless psychotherapeutic orientation to the transactions between the patient and himself, the patient and other caretakers, and staff and psychiatrist. In this day of emphasis on biological psychiatry during which the importance of interpersonal influences has receded, efforts are being made to maintain psychotherapeutic attitudes during liaison psychiatry.[39]

It is beyond the scope of this chapter to review the details of psychotherapeutic intervention in liaison psychiatry. But the orientation of psychiatrist, nurse, social worker, and psychologist, among others, as well as the objectives in teaching nonpsychiatrist professionals are the principal concepts used empirically as guidelines in liaison psychiatry. Whatever doubts have been raised concerning their usefulness remain unproven.

¶ The Influence of the Cost of Care

The cost of health care is one of the major issues of the 1970s and 1980s, and economists have evolved an industrial concept based on their inquires into the health care system. Critical care practice contributes significantly to this cost and as a result forces within society are exerting pressure which influences patterns of care. What bearing does this have on liaison programs? The power structure within the hospital has shifted. The administrator and lay boards have greater power, supported as they are by Blue Cross reimbursement rates, hospital council edicts, governmental support, and utilization review. Admissions may be controlled and hospitalization is shortened. Liaison programs have too little time to work with hospitalized patients and are beginning to shift the site of clinical work to the outpatient department. Rotation of house officers is accelerated and this handicaps the liaison psychiatrist in his teaching. Protracted care of chronic and recurring medical disorders has been the ideal clinical situation for the teaching of psychosomatic medicine and for liaison arrangements. The tendency now is to keep many of these patients in the community rather than in the hospital, except for brief contact or in situations requiring heroic measures. The classic psychosomatic disorder either stays home, undergoes radical surgery, or receives emergency medical intervention.

The Changing Role of the Physician

The emergence of critical care medicine to change the patterns of patient care, the effects of social accountability upon the flow of patient care and its financing, and the recalcitrance of physicians to act upon the reality of a psychological and humanistic medicine is serving to reduce the influence and control on decision making of the individual physician. He appears to be relinquishing some power to medical councils, peer review boards, hospital administrators, and governmental officials. He finds himself not only in greater need of the counsel of the psychiatrist because of critical care medicine and psychopharmacology, but because societal determinants more than education, compel him to consider psychosocial influences on clinical medicine.

The principal obstacle to the delivery of mental health services to medical and surgical patients has been the resistance of the physician. There has been ample proof that he uses psychiatric consultations sparingly and gives low priority to emotional care in his clinical management. The job of reaching the physician and of securing psychological care for patients has been a difficult one for the liaison psychiatrist, who began his mission forty years ago with enthusiasm and hope, expecting that he and his disciples could make converts out of overburdened physicians. He has persisted in that hope. Yet what scanty evaluation there is demonstrates, as we have seen, that only a small fraction of physically ill patients receive the psychiatric consultation or the emotional care they need.

John Whitehorn[199] wisely wrote in 1963:

Perhaps the greatest benefit of a liberal education is to escape the tyranny of first impressions and of naive preconceptions—to learn to suspend judgment and actions, not indefinitely and vaguely, but long enough and sturdily enough for the orderly review of evidence and the weighing of probabilities and values . . . It is humanly difficult to weigh alternatives unless one can cultivate some *tolerance of uncertainty.*

His hope was that when students had more opportunities opened to them, as part of their educational program, for what he called "scientific questing," there would be "less of the phobic aversion for the uncertainties of the human being." Physicians have been erroneously taught to be secure only with the certainty of the fact that science is absolute, which it is not, and "the tolerance of uncertainty" which is required in the multivariables of human behavior is anathema to them.

The changing role of the physician in the face of the changing patterns of health care has led to a reexamination of the traditional aim of liaison psychiatry to provide psychological care for the medically ill through the education of the physician. Greenhill[58] has written:

This leads to a conclusion which psychiatry should carefully consider. Perhaps one of the principle aims of liaison programs—the conversion of physicians—has been premature. We have grasped a problem but we do not have sufficient information to make much headway with it. I would suggest that we leave the doctor alone until we have that information, that we stop proseltytizing, that we desist in our attempts to reform him.

The physician is an earnest and overworked professional carrying the load of sick and dependent human beings and making crucial decisions, often in the face of fatigue. He needs our expertise but seems to fear it or drifts with our liaison programs half-heartedly. Let us not expect of the physician that he can be part psychiatrist. The emotional care of patients does not and will not reside entirely in his hands, but also in the hands of nurses and others who enthusiastically desire that function. The task of educating the physician in psychological medicine need not be abandoned. But it is slow work, about which there is still much to be learned. It will take a long time, and the sick person should not have to wait. In the meantime, let us design programs in which Psychiatry is more direct and decisive in the care of the sick and the dying. Social forces have now given us that opportunity and the timing is right to grasp it. [p. 179]

Emerging Designs in Liaison Psychiatry

The concept of liaison psychiatry is an evolutionary step in the historical development of the psychological care of the sick and dying. It is but a stage in the process; in time it may not be a separate entity. Ultimately, there is no more need for "liaison psychiatry" than there is today for "liaison radiology" or "liaison pathology." The designs of new psychosocial programs to meet the changing patterns of health care and medical science appear to shape themselves around two forms: active psychiatric and societal intervention and the establishment of alternative approaches, such as primary care education, medical audit, bioethical monitoring, and terminal care (hospice) programs.

Active Psychiatric and Societal Intervention

It has been reported that in situations in which a liaison service functions, the percentage of patients receiving needed psychological care increases. Sanders[163] has shown that at the Massachusetts General Hospital 3 percent of private, 10 percent of general hospital, and 40 percent of cardiac-surgical admissions receive psychiatric consultation. Active intervention in the ICU and CCU accounts for the 40 percent consultation rate. Strain[185] has claimed that active case finding is a part of the work on liaison units. In 1979, Torem, Saraway, and Steinberg[189] described a controlled study in which an "active" approach increased the rate of referrals on liaison units from 2 percent using the "reactive" approach (traditional consultation model) to 20 percent.

In 1977, Greenhill[58] described the integral model of liaison psychiatry developed at the

Einstein Hospital. The integral model was based on the evidence that both the traditional medical model and liaison model were insufficient to meet the needs of the patient population of any hospital and that a new design was required. In the first place such a design must include, as the right of every sick person, psychological care as an integral component of patient care. In the second place, since it has been amply demonstrated that physicians do little about the emotional care of their patients, that care must be assured by another system.

There are five major components to the integral model.[62] The first is open access psychiatric consultation in which the psychiatrist has the freedom to see any patient who needs his care and to enter into staff relationships in any situation in which the staff needs his assistance. Other staff members in addition to the physician, such as nurses, psychologists, social workers, and physical therapists may ask for a psychiatric consultation by applying to the liaison department. Such procedures reduce the responsibilities of the physician only to the extent that they take from him the complete right to initiate formal psychological care. The key step in the free consultation procedure is that the physician is spoken to by the psychiatrist before the consultation takes place. If the physician resists in the face of the need for the consultation, it may become a matter for clinical and administrative review.

Mandatory high-risk evaluation and care of certain clinical situations in which most psychosocial oversights occur is the second component of the model. These are suicidal and homocidal risks, medical and surgical problems with psychiatric complications, open heart surgery, dialysis, and transplant patients, repeated hospital admissions, hospital stays beyond sixty days, instances of drug and alcohol abuse, repeated surgery, families with hard-core medical problems, and cancer patients. In these instances, the psychiatric service is administratively mandated to initiate consultations by the same method used in open access consultation.

A third component is the early identification of stress problems related to age, culture, and ethnicity. This monitoring is initiated by the social service department and nursing staff, beginning with the information gathered by the admitting office upon reception of the patient to the hospital unit. An open channel to psychiatry is available from the start.

A fourth and important component of the integral model is a system of triage that makes it logistically possible. It is ethical to expect that all patients should receive some psychological surveillance, but no staff is large enough to succeed at the task of providing emotional care without a screening process for focusing of effort. All patients do not have to be seen by mental health personnel; there are those that do, and those who can benefit by informal consultations between mental health personnel and physicians and nurses.

Data for triage come from consultations, informal reporting, case finding rounds at nurses' stations, and monitoring processed through the department of psychiatry. The system makes possible the assignment of clinical problems by triage workers to forms of intervention practical for the available staff, assuring that every patient identified as needing psychological help will receive it in some form.

A fifth and most recently developed component of the integral model is quality assurance monitoring. This is an outgrowth of the medical and auditing procedures correlated with utilization review, Medicaid review, and the influence of the Professional Standards Review Organization, all of which assess patient needs for greater economy and improved care. It is a built-in device that permits the refinement of psychological case finding and the evaluation of intervention in emotional reactions to illness. Since quality assurance is already practiced as an integral part of hospital medicine, the expansion of this service to include emotional care monitoring is a logical step.

The integral model is a system of active intervention which is not meant to be intrusive, authoritarian, or aggressive. On the contrary, it requires diplomacy, tact, and above all, a long preparation by an experienced liaison director to build trust in the competence of his liaison workers. The system proceeds by negotiation and education in connection with every circumstance. At the Einstein Hospital, in four years of experience with the integral model, not one physician has prevented a consultation. Emergency consultations have been reduced 75 percent, and the request for psychological care of patients from all caretakers has doubled.

At the same time that active psychiatric intervention emerges in these ways, societal forces intervene. These interventions are mainly by third-party carriers and consumers. Hospital administrations are concerned with these forces and are frequently compelled to step into liaison territory to assure reimbursement and prevent legal complications. As a result, there is an increasing connection between hospital administration and liaison departments with signs of a slow realization by the former that patients with emotional reactions to illness must be identified and treated.

The forces of consumerism are accelerating their demands that physicians and health care systems be accountable, not only for financial cost, but for a humanitarian program of health protection. Informed consent, pharmaceutical habituation, and advocacy have become areas of concern. It is but a short step to claims of patient abuse and class action suits against hospitals, in addition to malpractice suits against physicians. All of this is one aspect of emerging recognition of psychosocial considerations in the treatment of disease.

Alternative Approaches and Primary Care

Alternative means of achieving the goals of liaison psychiatry are developing in the context of changing patterns of health care. The most important is primary care and primary care education.* Eaton and his coworkers have written:[38]

> In 1974, the Psychiatry Education Branch of NIMH began to give high priority to the development and expansion of psychiatric consultation-liaison teaching services throughout the country. Among the many reasons for this emphasis was the fact that the country was moving toward a comprehensive health care system that would rely heavily on primary care physicians, who would be expected to handle preventive, diagnostic, and therapeutic tasks for which their training had not prepared them. An active consultation-liaison program would help educate non-psychiatric house officers and staff to recognize and manage the less complicated mental illness and to develop a more comfortable approach to the "problem patient." [p. 21]

They also pointed out that "there seemed to be increasing interest on the part of primary care physicians in behavioral medicine and behavioral pediatrics" and that "the consultation-liaison psychiatrist would certainly be well-equipped to teach primary care faculty and trainees an open, comfortable approach to patients."

The demographic basis for these conclusions turns out to be sound. In 1978, it was pointed out that approximately 60 percent of mental health problems in the United States were treated by primary care practitioners and only 15 percent by mental health personnel.[154] Brodie[21] has written, "we are faced with the realization that primary care providers have a major responsibility for the recognition and treatment of mental illness and that they therefore need adequate training in the psychological aspects of patient care." There is universal agreement on this point.

There has as yet been little opportunity to test the results of the training of primary care providers and although the need for psychiatric training of such physicians is frequently expressed, only a few programs are in existence. No valid data on their locations and programs have been published. Yet there is no doubt that in time the sites for liaison psychiatry will extend from general hospitals to neighborhood primary care centers and health maintenance organizations. Indeed the trend has already begun.

*See references 47, 48, 53, 159, and 179.

Liaison with Health Care Functions

In the evolution of the role of the liaison psychiatrist within the health care systems, his presence has given indirect but significant psychosocial exposures to nonpsychiatrist physicians, other caretakers, and administrators. The presence and participation of the psychiatrist at medical audit committee and bioethical committee meetings, and his input into hospital terminal care and clinical-pathological conferences, where nonpsychiatrist physicians are reluctant to relinquish their leadership, provide useful educational forums for liaison psychiatry. In these arenas, the nonpsychiatrist participants feel reasonably secure and the psychiatrists crystallize *their* competence.

Another indirect activity connected with changing patterns of health care, which may have an influence on liaison psychiatry, is modification of nomenclature by the newly revised *Diagnostic and Statistical Manual of Mental Disorders* (DSM-III).[7] Lipp, Looney, and Spitzer[130] believe that the new classification system of psychosomatic disorders will "reduce conceptual ambiguity" and "promote collaborative care rather than care by triage." The expectation is that a code number from DSM-III will be added to a diagnostic category for a physical condition now listed in the *International Classification of Diseases* which would indicate that psychological factors are important in etiology. Once such a notation becomes a requirement for hospital accreditation and third-party reimbursement, nonpsychiatrist physicians should take collaborative care more seriously.

¶ Beyond the Historical Perspective

As a branch of psychiatry, liaison psychiatry has advanced into the territories of general hospitals and health care in a manner analogous to the entry of fellow mental health professionals into legislatures and courts to improve the delivery of mental health care. They have made a large contribution to psychiatry by increasing nation-wide recognition of mental health as an important part of the national health care system.

In his editorial establishing the journal *General Hospital Psychiatry,* Lipsitt[131] wrote:

> whatever the determinants, psychiatry has clearly established its rightful role in and valid contribution to the teaching and practice of medicine. Now the potential exists for psychiatry to reach beyond the historical perspective. From its well established base in the general hospital, psychiatry can move in tandem with medicine in exploring new directions in health care.

For all of societies' concern for the sick and dying, they are the least protected against psychological vulnerability. In spite of the fear and despair expressed about the chronically psychotic, they are the most neglected. The liaison psychiatrist and the general psychiatrist have each deplored these conditions. From his firm base as a physician, the liaison psychiatrist has moved with greater ease into the councils plotting the future of general health care, while the general psychiatrist, struggling with role diffusion, is concerned with changing patterns of psychiatric care "in the community." Both directions are important, but "beyond the historical perspective" the model of the physician-psychiatrist, with his scientific medical training, has to prevail as the integrationist of forces contributing to health and relief from disease. The predictable weight of future biological advances will ensure this.

¶ Bibliography

1. ABRAM, H. S. "Psychological Aspects of Intensive Care Units," *Medical Annals of the District of Columbia,* 43 (1974): 59–62.
2. AITKEN, C. "Prospects and Services for Medical Rehabilitation," *Bibliotheca Psychiatrica,* 159 (1979): 145–154.
3. ALDRICH, C. K. "Office Psychotherapy for the Primary Care Physician," in Arieti, S.,

ed., *American Handbook of Psychiatry,* vol. 5. New York: Basic Books, 1975, pp. 739–755.

4. ALEXANDER, F. *Psychosomatic Medicine: Its Principles and Applications.* New York: W. W. Norton, 1950.
5. ———. "The Development of Psychosomatic Medicine," *Psychosomatic Medicine,* 24 (1962): 13–24.
6. ALEXANDER, F., FRENCH, T. M., and POLLACK, G. H., eds. *Psychosomatic Specificity.* Chicago: The University of Chicago Press, 1968.
7. AMERICAN PSYCHIATRIC ASSOCIATION. *Diagnostic and Statistical Manual of Mental Disorders,* 3rd ed. (DSM-III). Washington, D.C.: American Psychiatric Association, 1979.
8. BALINT, M. "Medicine and Psychosomatic Medicine—New Possibilities in Training and Practice," *Comprehensive Psychiatry,* 9:4 (1968): 267–274.
9. BALINT, M., BALL, D. H., and HARE, M. L. "Training Medical Students in Patient-Centered Medicine," *Comprehensive Psychiatry,* 1:4 (1969): 249–258.
10. BARTON, B., and KELSO, M. T. "The Nurse as a Psychiatric Consultation Team Member," *Psychiatry in Medicine,* 2 (1971): 108–115.
11. BASTIAANS, J. "Models of Teaching Psychobiological Medicine to Medical Students," *Bibliotheca Psychiatrica,* 159 (1979): 48–61.
12. BEIGLER, J. S., et al. "Report on Liaison Psychiatry at Michael Reese Hospital, 1950–1958," *American Medical Association Archives of Neurology and Psychology,* 81 (1959): 733–746.
13. BENSON, R. "The Function of a Psychiatric Unit in a General Hospital—A Five Year Experience," *Diseases of the Nervous System,* 37 (1976): 573–577.
14. BERGEN, B. J. "Psychosomatic Knowledge and The Role of the Physician: A Sociological View," *International Journal of Psychiatry in Medicine,* 5:4 (1974): 431–442.
15. BERMOSK, L. S. "Interviewing: A Key to Therapeutic Communication in Nursing Practice," *Nursing Clinics of North America,* 1:2 (1966): 205–214.
16. BERNSTEIN, S. S., SMALL, S. M., and REICH, M. J. "A Psychosomatic Unit in a General Hospital," *American Journal of Nursing,* 49:8 (1949).
17. BILLINGS, E. G. "The General Hospital: Its Psychiatric Needs and the Opportunities. It Offers for Psychiatric Teaching," *American Journal of Medical Science,* 194 (1937): 234–243.
18. ———. "Liaison Psychiatry and the Intern Instruction," *Journal of the Association of American Medical Colleges,* 14 (1939): 375–385.
19. ———. "Value of Psychiatry to the General Hospital," *Hospitals,* 15 (1941): 305–310.
20. ———. "The Psychiatric Liaison Department of the University of Colorado Medical School and Hospitals," *American Journal of Psychiatry,* 122 (1966): 28–33.
21. BRODIE, H. K. H. "Mental Health and Primary Care," *Psychosomatics,* 20:10 (1979): 658–659.
22. BRODSKY, C. M. "A Social View of the Psychiatric Consultation. The Medical View and the Social View," *Psychosomatics,* 8 (1967): 61–68.
23. ———. "Decision-making and Role Shifts as They Affect the Consultation Interface," *Archives of General Psychiatry,* 23 (1970): 559–565.
24. BROSIN, H. "Communication Systems of the Consultation Process" in Mendel, W., and Solomon, P., eds., *The Psychiatric Consultation.* New York: Grune & Stratton, 1968.
25. BURNETT, F. M., SITES, P., and GREENHILL, M. H. "Learning the Mental Health Approach Through the Chronic Medical Patient," *Public Health Nursing,* (June, 1951).
26. BURSTEN, B. "The Psychiatric Consultant and the Nurse," *Nursing Forum,* 2 (1963): 7–23.
27. CAPLAN, G., and KILLILEA, M., eds. *Support Systems and Mutual Help—Multidisciplinary Exploration.* New York: Grune & Stratton, 1976.
28. CASSEM, N. H., and HACKETT, T. P. "Psychiatric Consultation in a Coronary Care Unit," *Annals Of Internal Medicine,* 75 (1971): 9–14.
29. ———. "Psychological Aspects of Myocardial Infarction." *Medical Clinics of North America,* 61:4 (1977): 711–721.
30. CAVANAUGH, J. L., JR., and FLOOD, A. B. "Psychiatry Consultation Services in the Large General Hospital: A Review and a

New Report," *International Journal of Psychiatry in Medicine,* 7:3 (1976–7): 193–207.

31. COBB, S., and FINESINGER, J. "Psychiatric Unit at the Massachusetts General Hospital, Monthly Bulletin," *Massachusetts Society for Mental Hygiene,* 15 (1936).
32. CRAMOND, W. A., KNIGHT, P. R., and LAWRENCE, J. R., "The Psychiatric Contribution to a Renal Unit Undertaking Chronic Hemodialysis and Renal Homotransplantation," *British Journal of Psychiatry,* 113 (1967): 1201–1212.
33. CRAMOND, W. A., et al. "Psychological Aspects of the Management of Chronic Renal Failure," *British Journal of Medicine,* 1 (1968): 539.
34. CUSHING, J. G. N. "The Role of the Psychiatrist as Consultant," *American Journal of Psychiatry,* 106 (1950): 861–864.
35. DOHRENWEND, B. S., and DORENSEND, B. P. *Stressful Life Events: Their Nature and Effects.* New York: John Wiley, 1974.
36. DUFF, R. R., and HOLLINGSHEAD, A. B. *Sickness and Society.* New York: Harper & Row, 1968.
37. DUNBAR, H. F. *Psychosomatic Diagnosis,* New York: Paul B. Hoeber, 1943.
38. EATON, J. S., JR., et al. "The Educational Challenge of Consultation-Liaison Psychiatry," *American Journal of Psychiatry,* 134 (1977): 20–23.
39. EISENBERG, L. "Interface Between Medicine and Psychiatry," *Comprehensive Psychiatry,* 20:1 (1979): 1–13.
40. ENGEL, G. L. "The Concept of Psychosomatic Disorder," *Journal of Psychosomatic Research,* 11 (1967): 3–9.
41. ———. "Medical Education and the Psychosomatic Approach: A Report on the Rochester Experience 1946–1966," *Journal of Psychosomatic Research,* 11 (1967): 77–85.
42. ———. "The Psychosomatic Approach as a Prophylactic Measure in the Care of Patients with Peptic Ulcer," in Weiner, H., ed., *Advanced Psychosomatic Medicine,* vol. 5. Basel: S. Karger, 1971, pp. 186–189.
43. ———. "The Education of the Physician for Clinical Observation. The Role of the Psychosomatic (Liaison) Teacher," *Journal of Nervous and Mental Diseases,* 154:3 (1972): 159–164.
44. ———. "Is Psychiatry Failing in Its Responsibilities to Medicine?" *American Journal of Psychiatry,* 128:12 (1972): 111–114.
45. ———. "The Need for a New Medical Model: A Challenge for Biomedicine," *Science,* 196, (1977): 42–86, 129–136.
46. ———. et al. "A Graduate and Undergraduate Teaching Program on the Psychological Aspects of Medicine," *Journal of Medical Education,* 32 (1957): 859–871.
47. FINK, P. H., and OKEN, D. "The Role of Psychiatry as a Primary Care Specialty," *Archives of General Psychiatry,* 33 (1976): 998–1003.
48. FLECK, S. "Unified Health Service and Family Focused Primary Care," *International Journal of Psychiatry in Medicine,* 6:4 (1975): 501–516.
49. FOX, H. M. "Teaching Integrated Medicine—Report of a Five Year Experiment at Peter Bent Brigham Hospital," *Journal of Medical Education,* 26 (1951): 421–429.
50. FRENCH, T. M. *The Integration of Behavior,* vol. 3. Chicago: University of Chicago Press, 1958, pp. 99–102, 384–385.
51. GILDEA, E. F. "Special Features of Personality Which Are Common to Certain Psychosomatic Disorders," *Psychosomatic Medicine,* 11 (1949): 273–283.
52. GLAZER, W. M., and ASTRACHAN, B. M. "A Social Systems Approach to Consultation—Liaison Psychiatry," *Journal of Psychiatry in Medicine,* 9:1 (1978–9): 33–47.
53. GOLDBERG, R. L., et al. "Psychiatry and the Primary Care Physician," *Journal of the American Medical Association,* 236 (1976): 944–945.
54. GRAHAM, D. T., et al. "Specific Attitudes in Initial Interviews with Patients Having Different 'Psychosomatic' Diseases," *Psychosomatic Medicine,* 24 (1962): 259–266.
55. GREENBERG, I. M. "Approaches to Psychiatric Consultation in a Research Hospital Setting," *Archives of General Psychiatry,* 3 (1960): 691–697.
56. GREENHILL, M. H. "Psychotherapy within Psychosomatic Medicine," *North Carolina Medical Journal,* 4:6 (1943).
57. ———. *Annual Report on the Department of Psychiatry.* Bronx, N.Y.: Hospital of the Albert Einstein College of Medicine, Office of the Administrator, 1975, 1976.
58. ———. "The Development of Liaison Programs," in Usdin, G., ed. *Psychiatric Medi-*

cine. New York: Brunner/Mazel, 1977, pp. 115–191.

59. ———. "Current Concepts in Psychosomatic Medicine," *Weekly Psychiatry Update Series,* 2:42 (1978): 3–7.

60. ———. "Fundamentals of Liaison Psychiatry," *Weekly Psychiatric Update Series,* 2:10 (1978): 2–7.

61. ———. "Organic Mental Syndrome (OMS) as a Secondary Diagnosis," *Report of the Medical Audit Committee,* Hospital of the Albert Einstein College of Medicine, 1978.

62. ———. "Models of Liaison Programs That Address Age and Cultural Differences in Reaction to Illness," *Bibliotheca Psychiatrica,* 159 (1979): 77–81.

63. ———. Psychiatric Units in a General Hospital: 1979," *Hospital and Community Psychiatry,* 30:3 (1979): 169–182.

64. ———. "Teaching and Training of the Psychosomatic Approach," *Bibliotheca Psychiatrica,* 159 (1979): 15–22.

65. GREENHILL, M. H., and FRATER, R. M. B. "Family Inter-relationships Following Cardiac Surgery," *Archives Foundation of Thanatology,* 6:1 (1976).

66. GREENHILL, M. H., and KILGORE, S. R. "Principles of Methodology in Teaching the Psychiatric Approach to Medical House Officers," *Psychosomatic Medicine,* 12 (1950): 38–48.

67. GREGG, D. "Reassurance," in Skipper, J. K., and Leonard, R. C., eds., *Social Interaction and Patient Care.* Philadelphia: Lippincott, 1965.

68. GROVES, J. E. "Management of the Borderline Patient in a Medical or Surgical Ward: The Psychiatric Consultant's Role," *International Journal of Psychiatry in Medicine,* 6:3 (1975): 337–348.

69. GUGGENHEIM, F. G. "Marketplace Model of Consultation Psychiatry in the General Hospital," *American Journal of Psychiatry,* 135 (1978): 1380–1383.

70. GUZE, S. B. "Nature of Psychiatric Illness: Why Psychiatry is a Branch of Medicine," *Comprehensive Psychiatry,* 19:4 (1978): 295–307.

71. HACKETT, T. P. "The Psychiatrist: In the Mainstream or on the Banks of Medicine?" *American Journal of Psychiatry,* 134 (1977): 432–434.

72. HACKETT, T. P., and CASSEM, N. H. "The Psychology of Intensive Care: Problems and Their Management," in Usdin, G., ed., *Psychiatric Medicine.* New York: Brunner/Mazel, 1977, pp. 228–258.

73. HAWKINS, O. R. "The Gap Between the Psychiatrist and Other Physicians," *Psychosomatic Medicine,* 24:1 (1962): 94–102.

74. HENRY, G. W. "Some Aspects of Psychiatry in General Hospital Practice," *American Journal of Psychiatry,* 7 (1929): 481–499.

75. HIENE, R. W., ed. *The Student Physician as Psychotherapist.* Chicago: University of Chicago Press, 1962.

76. HILL. O. "The Psychological Management of Psychosomatic Diseases," *British Journal of Psychiatry,* 131 (1977): 113–126.

77. HOLMES, T. H. "Some Observations on Medical Education," *Psychosomatic Medicine,* 31:3 (1969): 269–273.

78. HOLMES, T. H., and RAHE, R. H. "The Social Readjustment Rating Scale," *Journal of Psychosomatic Research,* 11 (1967): 213–218.

79. HOLSTEIN, S., and SCHWAB, J. "A Coordination Consultation Program for Nurses and Psychiatrists," *Journal of the American Medical Association,* 194 (1965): 103–105.

80. HOPKINS, P., and WOLFF, H. H., eds. *Principles of Treatments of Psychosomatic Disorders.* London: Pergamon Press, 1965.

81. HOUPT, J. L. "Evaluating Liaison Program Effectiveness: The Use of Unobtrusive Measurement," *Journal of Psychiatry in Medicine,* 8:4 (1977–8): 361–370.

82. IAWSAKI, T. "Teaching Liaison Psychiatry and Clinical Practice of Psychosomatic Medicine in the General Hospital," *Bibliotheca Psychiatrica,* 159 (1979): 32–38.

83. JACKSON, H. A. "The Psychiatric Nurse as a Mental Health Consultant in a General Hospital," *Nursing Clinics of North America,* 4 (1969): 527–540.

84. KANNO, C., and SCHEIDEMANDEL, P. L. *Psychiatric Treatment in the Community: A National Survey of General Hospital Psychiatry and Private Psychiatric Hospitals.* Washington, D.C.: Joint Information Service, 1974.

85. KAPLAN, S. M., and CURTIS, G. C. "Reactions of Medical Patients to Discharge from a Psychosomatic Unit of a General Hospital," *Postgraduate Medicine,* 29 (1961): 358–364.

86. KARASU, T. B. "Psychotherapy of the Medi-

cally Ill," *American Journal of Psychiatry*, 136:1 (1979): 1–11.

87. KARASU, T. B., and STEINMULLER, R. I., eds. *Psychotherapeutics in Medicine.* New York: Grune & Stratton, 1979.
88. KAUFMAN, M. R. "The Role of a Psychiatrist in a General Hospital, *Psychiatric Quarterly*, 27 (1953): 367–381.
89. KAUFMAN, M. R. and MARGOLIN, S. G. "Theory and Practice of Psychosomatic Medicine in a General Hospital," *Medical Clinics of North America*, (1948): 611–616.
90. KEMPH, J. P. "Renal Failure, Artificial Kidney and Kidney Transplant," *American Journal of Psychiatry*, 122 (1966): 1270.
91. KIELY, W. F. "Coping with Severe Illness," *Advanced Psychosomatic Medicine*, 8 (1972): 105–118.
92. ———. "Psychiatric Syndromes in Critically Ill Patients," *Journal of the American Medical Association*, 235:25 (1976): 2759–2761.
93. KIMBALL, C. P. "A Predictive Study of Adjustment to Cardiac Surgery," *Journal of Thoracic Surgery*, 58 (1969): 891–896.
94. ———. "Conceptual Developments in Psychosomatic Medicine: 1939–1969," *Annals of International Medicine*, 73 (1970): 307–310.
95. ———. "A Liaison Department of Psychiatry," *Psychotherapy Psychosomatics*, 22 (1973): 219–225.
96. ———, ed. "A Report of the First Workshop in Liaison Psychiatry and Medicine," *Psychosomatic Medicine*, 35:2 (1973): 176.
97. ———. "Medical Psychotherapy: A General Systems Approach," in Arieti, S., ed., *American Handbook of Psychiatry*, vol. 5. New York: Basic Books, 1975, pp. 781–806.
98. ———. Psychosomatic Theories and Their Contributions to Chronic Illness," in Usdin, G., ed., *Psychiatric Medicine.* New York: Brunner/Mazel, 1977, pp. 259–333.
99. ———. "The Issue of Confidentiality in the Consultation-Liaison Process," *Bibliotheca Psychiatrica*, 159, (1979): 82–89.
100. ———. "Teaching Medical Students Psychosomatic Medicine: Of Substance and Approaches," *Bibliotheca Psychiatrica*, 159 (1979): 23–31.
101. KIMBALL, C. P., and KRAKOWSKI, A. J., eds. "The Teaching of Psychosomatic Medicine and Consultation-Liaison Psychiatry: Reactions to Illness," *Bibliotheca Psychiatrica*, 159 (1979).
102. KLIGERMAN, M. J., and MCKEGNEY, F. P., "Patterns of Psychiatric Consultation in Two General Hospitals," *Psychiatry in Medicine*, 2 (1971): 126.
103. KLIGERMAN, S. "A Program of Teaching a Psychodynamic Orientation to Resident Physicians in Medicine," *Psychosomatic Medicine*, 14 (1952): 277–283.
104. KORAN, L. M., et al. "Patients" Reactions to Psychiatric Consultation," *Journal of the American Medical Association*, 241:15 (1979): 1603–1606.
105. KORNFELD, D. S. "Psychiatric View of the Intensive Care Unit," *British Medical Journal*, 1 (1969): 108–110.
106. ———. "The Hospital Environment: Its Impact on the Patient," *Advanced Psychosomatic Medicine*, 8 (1972): 252–270.
107. KORNFELD, D. S., et al. "Psychiatric Complications of Open Heart Surgery," *New England Journal of Medicine*, 273:6 (1965): 287–292.
108. KORSCH, B. M., et al. "Experience with Children and Their Families during Extended Hemodialysis and Kidney Transplant," *Pediatric Clinics of North America*, 18:2 (1971).
109. KRAKOWSKI, A. J. "Consultation-Liaison Psychiatry: A Psychosomatic Service in the General Hospital," *International Journal of Psychiatry in Medicine*, 6 (1975): 283–292.
110. ———. "Psychiatric Consultation in the General Hospital: An Exploration of Resistances," *Diseases of the Nervous System*, 36:5 (1975): 242–244.
111. ———. "Liaison Psychiatry in North America in the 1970's," *Bibliotheca Psychiatrica*, 159 (1979): 4–14.
112. ———. "Psychiatric Consultation for the Geriatric Population in the General Hospital," *Bibliotheca Psychiatrica*, 159 (1979): 163–185.
113. KRUPP, N. E., and RYNEARSON, E. K. "Consultation-Liaison: A Patient-based Overview," *Psychosomatics*, 20:2 (1979): 108–117.
114. KUBLER-ROSS, E. *On Death and Dying.* New York: Macmillan, 1969.
115. LEIGH, L., and REISER, M. F. "Major Trends in Psychosomatic Medicine: The Psychiatrist's Evolving Role in Medicine,"

Annals of Internal Medicine, 87:2 (1977): 233–239.

116. LEVY, N. B. *Living or Dying: Adaptation to Hemodialysis.* Springfield, Ill.: Charles C Thomas, 1974.
117. ———. "Teaching of Liaison Psychiatry in the Hemodialysis Center," *Bibliotheca Psychiatrica,* 159 (1979): 141–144.
118. LIDZ, T., and FLECK, S. "Integration of Medical and Psychiatric Methods and Objectives on a Medical Service," *Psychosomatic Medicine,* 12 (1950): 103.
119. LIEF, A., ed. *The Commonsense Psychiatry of Adolf Meyer.* New York: McGraw-Hill, 1948.
120. LINDEMANN, E. "Psychiatric Problems in Conservative Treatment of Ulcerative Colitis," *Archives of Psychiatry and Neurology,* 53 (1945): 322–325.
121. LIPOWSKI, Z. J. "Review of Consultation Psychiatry and Psychosomatic Medicine," *Psychosomatic Medicine,* 29:3 (1967): 153–171.
122. ———. "Review of Consultation Psychiatry and Psychosomatic Medicine. II. Clinical Aspects." *Psychosomatic Medicine,* 29:3 (1967): 201–224.
123. ———. "Review of Consultation Psychiatry and Psychosomatic Medicine. III. Theoretical Issues." *Psychosomatic Medicine,* 30:4 (1968): 395–422.
124. ———, ed. "Psychosocial Aspects of Physical Illness," *Advanced Psychosomatic Medicine,* 8 (1972).
125. ———. "Psychosomatic Medicine in a Changing Society: Some Current Trends in Theory and Research," *Comprehensive Psychiatry,* 14:3 (1973): 203–215.
126. ———. "Consultation-Liaison Psychiatry: An Overview," *American Journal of Psychiatry,* 131:6 (1974): 623–629.
127. ———. "Psychiatry of Somatic Diseases: Epidemiology, Pathogenesis, Classification," *Comprehensive Psychiatry,* 16:2 (1975): 105–124.
128. ———. "Psychosomatic Medicine: An Overview," in Hill, O., ed., *Modern Trends in Psychosomatic Medicine,* 3. London: Butterworths, 1976, pp. 1–20.
129. ———. "Psychosomatic Medicine in the Seventies: An Overview," *American Journal of Psychiatry;* "Past Failures and New Opportunities," *General Hospital Psychiatry,* 1 (1979): 3–10.
130. LIPP, M. R., LOONEY, J. G., and SPITZER, R. L. "Classifying Psychophysiologic Disorders," *Psychosomatic Medicine,* 39:5 (1977): 285–287.
131. LIPSITT, D. R. "Some Problems in the Teaching of Psychosomatic Medicine," *International Journal of Psychiatry in Medicine,* 6 (1975): 317–329.
132. ———. "Psychiatry and the General Hospital: An Editorial," *General Hospital Psychiatry,* 1:1 (1979): 1–2.
133. MCKEGNEY, F. P. "The Intensive Care Syndrome," *Connecticut Medicine,* 30 (1966): 633–636.
134. ———. "Consultation-Liaison Teaching of Psychosomatic Medicine: Opportunities and Obstacles," *Journal of Nervous and Mental Diseases,* 154:3 (1972): 198–205.
135. ———. "The Teaching of Psychosomatic Medicine: Consultation-Liaison Psychiatry," in Arieti, S., ed., *American Handbook of Psychiatry,* vol. 4. New York: Basic Books, 1975, p. 910.
136. MENDEL, J. G., and KLEIN, D. F. "Utilization of the Psychiatric Consultation Service in a Large City Hospital," *Psychosomatics,* 14 (1973): 57–58.
137. MENDEL, W. M., and SOLOMON, P. *The Psychiatric Consultation.* New York: Grune & Stratton, 1968.
138. MENDELSON, M., and MEYER, E. "Countertransference Problems of the Liaison Psychiatrist," *Psychosomatic Medicine,* 23:2 (1961): 115–122.
139. MEYER, E., and MENDELSON, M. "The Psychiatric Consultation in Postgraduate Medical Teaching," *Journal of Nervous and Mental Diseases,* 130:78 (1960).
140. ———. "Psychiatric Consultations with Patients on Medical and Surgical Wards: Patterns and Processes," *Psychiatry,* 24 (1961): 197–220.
141. MILLER, W. B. "Psychiatric Consultation: Part I. A General Systems Approach," *Psychiatry in Medicine,* 4:2 (1973): 135–145.
142. ———. "Psychiatric Consultation: Part II. Conceptual and Pragmatic Issues of Formulation," *Psychiatry in Medicine,* 4:3 (1973): 251–271.
143. MOORE, G. L. "The Adult Psychiatrist in the Medical Environment," *American Journal of Psychiatry,* 135:4 (1978): 413–419.
144. NEMIAH, J. C., and SIFNEOS, P. E. "Affect

and Fantasy in Patients with Psychosomatic Disorders," in Hill, O. W., ed., *Modern Trends in Psychosomatic Medicine,* vol. 2. New York and London: Appleton-Century-Crofts, 1970, pp. 24–26.

145. NEMIAH, J. C., FREYBERGER, H., and SIFNEOS, P. E. "Alexithymia: A View of the Psychosomatic Process," in Hill, O. W., ed., *Psychosomatic Medicine,* vol. 3. London and Boston: Butterworths, 1976, pp. 430–439.

146. PASNAU, R. O., ed., *Consultation-Liaison Psychiatry.* New York, Grune & Stratton, 1972.

147. PAYSON H. E., and DAVIS, J. M. "The Psychosocial Adjustment of Medical Inpatients After Discharge: A Follow-up Study," *American Journal of Psychiatry,* 123 (1967): 1220–1225.

148. PEPLAU, H. E. "Psychiatric Nursing Skills and the General Hospital Patient," *Nursing Forum,* 3 (1964): 29–37.

149. PETERSEN, S. "The Psychiatric Nurse Specialist in a General Hospital," *Nursing Outlook,* 17 (1969): 65–58.

150. PETRICH, J., and HOLMES, T. H. "Life Change and Onset of Illness," *Medical Clinics of North America,"* 61:4 (1977): 825–838.

151. POE, R. O., LOWELL, F. M., and FOX, H. M. "Depression. Study of 100 Cases in a General Hospital," *Journal of American Medical Association,* 195, (1966): 345–350.

152. RAHE, R. H., MCKEAN, J. D., and ARTHUR, R. J. "A Longitudinal Study of Life-Change and Illness Patterns," *Journal of Psychosomatic Research,* 10 (1967): 355–366.

153. REGIER, D. A., GOLDBERG, I. D., and TAUBE, C. A. "The DeFacto Versus Mental Health Services System," *Archives of General Psychiatry,* 35 (1978): 685–693.

154. REICHSMAN, F. "Teaching Psychosomatic Medicine to Medical Students, Residents, and Postgraduate Fellows," *International Journal of Psychiatry in Medicine,* 6 (1975): 307–316.

155. REIFLER, B., and EATON, J. S., JR. "The Evaluation of Teaching and Learning by the Psychiatric Consultation and Liaison Training Programs," *Psychosomatic Medicine,* 4:2 (1978): 99–105.

156. REISER, M. F. "Changing Theoretical Concepts in Psychosomatic Medicine," in Arieti, S., ed., *American Handbook of Psychiatry.* vol. 4. New York: Basic Books, 1975, pp. 477–500.

157. REUSCH, J. "The Infantile Personality: The Core Problem of Psychosomatic Medicine," *Psychosomatic Medicine.*

158. RICHMOND, J. B. "Relationship of the Psychiatric Unit to Other Departments of the Hospital" in Kaufman, M. R., ed., *The Psychiatric Unit in a General Hospital.* New York: International Universities Press, 1965, pp. 425–437.

159. RITTELMEYER, L. F., and FLYNN, W. E. "Psychiatric Consultation in an HMO: A Model for Education in Primary Care," *American Journal of Psychiatry,* 135 (1978): 1089–1092.

160. ROBINSON, L. "Liaison Psychiatric Nursing," *Perspective Psychiatric Care,* 6 (1968): 87–91.

161. ROCKEFELLER FOUNDATION. Information Furnished by Rockefeller Foundation, Health Sciences Division and Central Reference Service, Courtesy Edith King, 1976.

162. ROSENMAN, R. H., et al. "Coronary Heart Disease in the Western Collaborative Group Study: A Follow-up Experience of 4 1/2 Years," *Journal of Chronic Diseases* 23, (1970): 173–190.

163. SANDERS, C. A. "Reflections on Psychiatry in the General Hospital Setting," *Hospital and Community Psychiatry,* 30:3 (1979): 185–189.

164. SANDT, J. J., and LEIFER, R. "Psychiatric Consultation," *Comprehensive Psychiatry,* 5 (1964): 409.

165. SASLOW, G. "An Experiment with Comprehensive Medicine," *Psychosomatic Medicine,* 10 (1948): 167–175.

166. SCANLAN, J. M. "Psychiatric Consultation in a General Hospital," *Minnesota Medicine,* 57:11 (1974): 922–924.

167. SCHIFF, S. K., and PILOT, M. L. "An Approach to Psychiatric Consultation in the General Hospital," *Archives of General Psychiatry,* 1 (1959): 349–357.

168. SCHMALE, A. H. "Relationship of Separation and Depression to Disease," *Psychosomatic Medicine,* 20 (1958): 259–277.

169. ———. Personal communication, University of Rochester, 1976.

170. SCHMALE, A. H., and ENGEL, G. L. "One Giving Up—Given Up Complex Illustrated on Film," *Archives of General Psychiatry,* 17 (1967): 135–145.

171. SCHMALE, A. H., et al. "An Established Program of Graduate Education in Psychosomatic Medicine," *Advanced Psychosomatic Medicine,* 4 (1964): 4–13.
172. SCHUBERT, D. S. P. "Obstacles to Effective Psychiatric Liaison," *Psychosomatics,* 19:5 (1978): 283–285.
173. SCHULMAN, B. M. "Group Process: An Adjunct in Liaison Consultation Psychiatry," *International Journal of Psychiatry in Medicine,* 6:4 (1975): 484–499.
174. SCHWAB, J. J. *Handbook of Psychiatric Consultation.* New York: Appleton-Century-Crofts, 1968.
175. SCHWAB, J. J., et al. "Differential Characteristics of Medical Inpatients Referred for Psychiatric Consultation: A Controlled Study," *Psychosomatic Medicine,* 27 (1965): 112–118.
176. SHEVITZ, S. A., SILBERFARB, P. M., and LIPOWSKI, Z. J. "Psychiatric Consultations in a General Hospital: A Report on 1,000 Referrals," *Diseases of the Nervous System,* 37:5 (1976): 295–300.
177. SHOCHET, B. "The Role of the Mental Health Counselor in the Psychiatric Liaison Service of the General Hospital," *International Journal of Psychiatry in Medicine,* 5:1 (1974): 1–16.
178. SKIPPER, J. K., and LEONARD, R. C. "The Importance of Communication," in *Social Interaction in Patient Care.* Philadelphia: Lippincott, 1965, pp. 51–60.
179. SLABY, A. E., POTTASH, A. L. C., and BLACK, H. R. "Utilization of Psychiatry in a Primary Care Center," *Journal of Medical Education,* 53 (1978), 752–758.
180. SOLOFF, P. H. "The Liaison Psychiatrist in Cardiovascular Rehabilitation: An Overview," *International Journal of Psychiatry in Medicine,* 8:4 (1977–8): 393–401.
181. SPERLING, M. "Psychotherapeutic Techniques in Psychosomatic Medicine," in Bychowski, G. and Despert, J. L. eds., *Specialized Techniques in Psychotherapy.* New York: Basic Books, 1952, pp. 279–301.
182. STANTON, A. H., and SCHWARTZ, M. S. *The Mental Hospital: A Study of Institutional Participation in Psychiatric Illness and Treatment.* New York: Basic Books, 1954.
183. STRAIN, J. J. *Psychological Interventions in Medical Practice.* New York: Appleton-Century-Crofts, 1975.
184. ———. "The Medical Setting: Is It Beyond the Psychiatrist?", *American Journal of Psychiatry,* 134:3 (1977): 253–256.
185. ———. "In Response to a Review," *American Journal of Psychiatry,* 134:3 (1977): 331.
186. ———., and GROSSMAN, S. *Psychological Care of the Medically Ill: A Primer in Liaison Psychiatry.* New York: Appleton-Century-Crofts, 1975.
187. STRATAS, N. E. "Training of Non-Psychiatric Physicians," *American Journal of Psychiatry,* 125 (1969): 1110.
188. STUNKARD, A. "A New Method in Medical Education," *Psychosomatic Medicine,* 22:5 (1960): 400–406.
189. TOREM, M., SARAWAY, S. M., and STEINBERG, H. "Psychiatric Liaison: Benefits of an 'Active' Approach," *Psychosomatics,* 20:9 (1979): 598–607.
190. TRYON, P. S., and LEONARD, R. C. "Giving the Patient an Active Role," in Skipper, J. K., and Leonard, R. C., eds., *Social Interaction and Patient Care.* Philadelphia: Lippincott, 1965, pp. 120–127.
191. TUCHER, G. L., and LEONARD, R. C. "Psychiatric Attitudes of Young Physicians: Implications for Teaching," *American Journal of Psychiatry,* 124:7 (1968): 146–151.
192. USDIN, G., ed. *Psychiatric Medicine.* New York: Brunner/Mazel, 1977.
193. WEINER, H. "The Illusion of Simplicity: The Medical Model Revisited," *American Journal of Psychiatry,* 135 (Suppl.) (1978): 27–33.
194. WEISMAN, A. D. "A Model for Psychosocial Phasing in Cancer," *General Hospital Psychiatry,* 1 (1979): 187–195.
195. WEISMAN, A. D., and HACKETT, H. T. "The Organization and Function of a Psychiatric Consultation Service," *International Record of Medicine,* 173, (1960): 306–311.
196. ———. "Predilection to Death: Death and Dying as a Psychiatric Problem," *Psychosomatic Medicine,* 23:3 (1961): 232–256.
197. WEISMAN, A. D., and WORDEN, J. W. "The Existential Plight in Cancer: Significance of the First 100 Days," *International Journal of Psychiatry in Medicine,* 7, (1976): 1–15.
198. WEST, N. D., and BASTANI, J. B. "The Pattern of Psychiatric Referrals in a Teaching Hospital," *Nebraska Medical Journal,* 15, (1973): 438–440.
199. WHITEHORN, J. C. "Education for Uncer-

tainty," *Perspective Biological Medicine,* 7 (1963): 118.

200. WITTKOWER, E. D., and WARNES, H., eds. *Psychosomatic Medicine: Its Clinical Applications.* Hagerstown: Harper & Row, 1977.

201. WOLFF, H. G., and GOODELL, H. *Stress and Disease.* Springfield, Ill.: Charles C Thomas, 1968.

202. ZABARENKO, R. N., and ZABARENKO, L. M. "Teaching Psychological Medicine on Hospital Rounds: A Liaison Experiment," *International Journal of Psychiatry in Medicine,* 8:4 (1977–8): 325–333.

CHAPTER 34

MEDICAL FACULTY, ORGANIZATIONAL PRESSURES, AND SUPPORT

Pamela J. Trent and H. Keith H. Brodie

¶ Introduction

MEDICAL EDUCATION in the United States has not faced such careful scrutiny since Abraham Flexner published his report in 1910.[9] Whereas that report disclosed the need to organize medical education, present studies of the issues reflect the overwhelming complexity of that organization. The recent report, *The Organization and Governance of Academic Health Centers*[14,24] makes this abundantly clear. Academic health centers (AHCs), direct descendants of post-Flexnerian efforts to improve medical schooling, are now complex and diverse institutions "consisting of a medical school, at least one other health school, and a teaching hospital (owned or affiliated)."[14] Note the historical factors leading to this complexity, as delineated in the report:

1. There has been an overwhelming increase in the operating and capital budgets for programs, staff, students, and faculty.
2. There has been a tremendous growth of new medical knowledge in the past forty years.
3. There has been a large and steady increase in federal support for health care, medical education, and medical research since 1945, adding to the administrative complexity.
4. Since the 1960s, there has been pressure to provide more and better health care to all members of society.
5. There has been a major increase in external regulation from all levels of government.
6. There has been increasing competition from nonphysician professionals who seek to improve their status, capabilities, and credibility in health-care delivery.

Note also the diverse modes for organizing and governing AHCs:

1. Some are public, others private.
2. Most are subdivisions of a "parent" university, but a large minority are autonomous.
3. Of those that are part of a "parent" university, some are on the same campus, some are on a different campus in the same city, and some are in a different city.
4. Some include a university-owned hospital while others make arrangements with affiliated hospitals.
5. The number of health schools on a campus may range from one to seven.
6. The majority have a position identified as chief administrative officer, often with the title of vice president for health affairs.

Finally, observe the various factors that induce complexity at the policy-making level:

1. The several interacting components in AHCs incorporate different modes of government.
2. The three distinct missions of the AHC—education, research, service—frequently conflict, leading to frustration and ambiguity.
3. Because AHCs must respond to diverse client groups with often contradictory expectations, it has been impossible to develop performance measures for institutional activities.
4. AHCs are part of and vulnerable to a powerful, dynamic environment vis-à-vis changing technology, consumer demands, and governmental regulation.
5. AHCs accommodate professionals with wide-ranging skills, interests, and roles, who often experience conflict between professional values and organizational expectations.
6. The power and influence in AHCs are issue-specific.

Clearly, the *Organization and Governance of Academic Health Centers* report is enlightening. The complexity of the medical education enterprise challenges the notion that the people who work in these organizations can understand their unique role, the position of their administrative component, and their relative security in the organization. Although research on academic health centers is still forthcoming,[24] we do know from classical research and theory[1,18,27] that high complexity in organizations leads to employee confusion about lines of authority, alienation from the task (or inability to make a meaningful commitment to the work), fragmentation of efforts, and overall insecurity. In short, the person gets lost in the organizational labyrinth.

This chapter is concerned with medical faculty as teachers in academic health organizations. It focuses on the teacher for three reasons. First, there is much thought and writing, especially in psychiatry, about medical education issues (for example, students, learning styles, teaching methods, and evaluation). But there is an obvious dearth of discussion and data regarding the needs of teachers in medical education. Second, teachers make choices that affect students' lives. These choices, never based purely on intellectual consideration or quantitative data, include implicit values, philosophies, personal meanings, and assumptions about life. Teachers need to heighten their awareness of these implicit underpinnings. Third, the teacher-student relationship is a model for the doctor-patient relationship, a relationship that should be based on trust and acknowledgment of a personal commitment. It is difficult for the teacher to make a commitment if his or her own needs are not addressed. This chapter delineates the pressures affecting medical faculty in general, it describes strategies used by faculty for coping with the pressures, and it then focuses on the particular needs of psychiatry faculty with a recommendation for addressing those needs.

¶ Organizational Pressures

Factors impinging on the teaching function of faculty members are external and internal to the organization. External factors considered here to be especially deleterious include economic conditions, knowledge explosion, and legal influences. Internal factors include administrative responsibilities, tenure and promotion considerations, and performance evaluation.

External Factors

Economic Conditions

Spiraling inflation, rising costs for medical care and medical education, and limited research and training funds have caused internal stress on AHCs, which saw an increasing support base annually during the 1950s and 1960s. In a climate of diminishing resources, individual faculty members must develop various fiscal responses.

Decreasing research and training funds require that clinical faculty devote more time to patient care, generating money from their clinical practice to support their own salaries and other departmental expenses. The transfer of these monies is organized in a variety of ways. In some institutions, the funds from individual practice are given directly to the medical school that, in turn, pays the clinician's salary and overhead costs. Other clinical departments are organized as partnerships, and still others allow physicians to function in solo fashion with or without constraints on total income. Regardless of how the financial arrangements are organized, some faculty receive negligible financial support from their university and many are even required to pay for their own office space, secretarial assistance, and supplies.[4]

Faculty members who must depend solely on research and training funds suffer from equally burdensome financial problems. Federal and private grants and contracts have become limited;[4,31] this results in a highly competitive market made even more competitive by fluctuating funding priorities. What was considered highly fundable in 1980 may be a low priority in 1984. This puts faculty in a precarious position as they spend time and energy developing grants, bidding on contracts, reporting about ongoing projects, and trying to predict what will be attractive to funding agencies when the priorities change. It encourages "project-hopping" rather than research built on a logical and sequential approach that is conducive to the development of a research product and career of excellence.

The financial burden on individual faculty members is augmented by the need to contain costs in an uncontrolled economy. This means restricting research and teaching, limiting growth, decreasing certain activities, and adjusting to strict allocation of resources. The situation encourages competition for scarce resources in a profession that is already highly competitive. Pressures from economic conditions are most deleterious because they chip away at the faculty's fundamental need for security (that is, a steady income and predictable job) and for commitment to a sustained effort.

Knowledge Explosion

Rapid increase in knowledge has encouraged specialization and a competency-based approach to education, placing additional pressure on faculty members.

Specialization, although necessary for organizing profuse knowledge and providing skilled care, tends to fragment the medical profession and encourages "empire building nationally, regionally, and locally."[4] For instance, specialization perpetrates a keen competition among faculty in the same institution for available time in an overcrowded curriculum. As a result, specialization builds barriers to communication among faculty who could otherwise share theories, interests, and frustrations. In similar fashion, specialization affects communication between teacher and student. Tosteson[30] comments about this:

> It is my impression that the opportunities for developing meaningful fruitful relations between students and faculty have decreased in our medical schools during the past decade even though the faculty-to-student ratio has increased. One factor leading to this situation is the growing specialization of medical education. An expert appears briefly to present his knowledge and disappears, rarely to be seen again by the students. Such brief encounters do not allow the students to know the faculty as persons. [p. 693]

Because of the overwhelming amount of knowledge and scientific information a student must learn and apply through medical training, the competency-based educational model has become popular for its ability to specify learning objectives and evaluate performance. Indeed, federal training grant proposals are given higher priority scores if they use the standards of this model. The model requires a defined competence of all students; makes explicit the knowledge, skill, and attitude objectives; evaluates achievement according to a criterion rather than a norm; and allows students to repeat certain training until they achieve competence in that area. Preliminary studies[22,29] indicate its effectiveness in medical training, and the benefits are self-evident: It provides a guide for teachers and students, organizes content, and minimizes competition since all students are encouraged to succeed.

However, there are subtle pressures inherent in the competency-based system. The amount of time and energy that must be devoted to developing a competency-based teaching "module" or program is substantial and might unduly strain an already overworked teacher who conceivably receives few rewards for teaching. A cost-benefit analysis of the model should test this observation. In addition, competency-based proponents encourage teachers to think that every iota of student learning can and should be measured. This restricts evaluation of learning to a very objective, technical meaning, one that may ignore the teacher's subjective ability to observe subtle changes in students. Psychiatrists, for instance, realize that psychiatric education is largely an artful, intuitive activity that often defies precise measurement.[26] Finally, since competency-based teaching is systematic and structured, teachers may think that students can travel through the learning experience with minimal guidance. Where this encourages self-directed learning it is a noble outcome; where it encourages the teacher to become personally removed from the educational process, acting as a barrier to communication, it becomes counterproductive for both teachers and students.

LEGAL INFLUENCES

The physician's image as expert, or unquestionable authority, has been challenged extensively in litigation regarding malpractice suits, students rights' to due process, and privacy rights of students and parents. The potential for litigation forces the physician to be constantly aware of his or her actions, adding pressures unanticipated ten years ago. In terms of malpractice suits, Rogers[25] states succinctly:

> The threat of lawsuits adds to the pressures that constrain and sometimes paralyze the physician's ability to act quickly and effectively in treatment. And the physician's response is typically and predictably one of caution, demonstrated always in more conservative decisions, more protracted tests and consultation, and more guardedness in sharing information with the patient. [p. 41]

Psychiatry, for instance, faces a skeptical public's demand for evaluation of treatment modalities and clarification of patients' right to treatment and right to refuse treatment. This forces psychiatrists into more conservative postures. Further pressure ensues when the guarded conservative approach to patient care conflicts with the need to contain costs, forcing the physician to arbitrate between these conflicting needs. The clinical teacher is required then to be duly conscious of what patients and procedures can be used for medical educating and of the patients' rights to informed consent in patient care, teaching, and research.

Due process, a legal concept expressed in the fifth and fourteenth amendments to the Constitution of the United States, provides that neither the federal nor state government shall "deprive any person of life, liberty or property, without due process of law." This is a significant issue in medical education where students serve in a professional capacity, and faculty must evaluate students' professional as well as personal judgment, ethical integrity, clinical skills, and relation to and management of patient welfare. Because evaluation must often be based on subjective interpretations, students have become increasingly concerned and vocal about right to due process. Litigation has forced interpretation of the law. Pursuant to litigation and in accordance with interpretation, the Liaison Committee on Medical Education, sponsored by the Association of American Medical Association, developed a policy as follows:[20]

> III. *Educational Program.* A medical school should develop and publicize to its faculty and students a clear definition of its procedures for the evaluation, advancement, and graduation of students. Principles of fairness and "due process" must apply when considering actions of the faculty or administration which will adversely affect the student to deprive him/her of valuable rights. [p. 3]

In addition to due process, the Privacy Rights of Parents and Students Act of 1976[32] allows parents and students direct access to students' educational records. Students are able to review professors' comments and evaluations. This law protects students from derogatory evaluation, but if taken to an extreme it allows students to challenge the professor's perception about what constitutes quality patient care. On the one hand, this law together with due process pressures the faculty to be very circumspect in evaluating student performance. On the other hand, it pressures them to be less than honest in cases where they may realistically fear litigation.

A short anecdote told by a psychiatry colleague puts this dilemma into a personal perspective. In a psychiatry curriculum committee meeting, a faculty member explained that a student requested a change in the rhetoric of an evaluation. The student challenged the psychiatry educator's use of the term "passive" in describing the student's lack of involvement in learning. The teacher changed the term, agreeing that use of psychiatric terms was inappropriate in student evaluation, and suggested to colleagues that they take greater care in describing student performance. (It is easy to phrase student evaluations with the same language used in patient summaries.) The colleagues agreed about the language but some were aghast over the student's ability to review the file and request a change. They felt threatened by this, for it undermined their perception that they were autonomous and that as medical faculty they could command an almost unquestioning respect regarding their professional judgment.

Internal Factors

Academic health centers have grown so rapidly that faculty members must adjust to increased administrative responsibilities, tougher guidelines for promotion and tenure, and systematic performance evaluation.

ADMINISTRATIVE RESPONSIBILITIES

Although health education became more organized after the Flexner report, the schools retained a simple administrative network. They were usually "administered by part-time deans with assistance of a small clerical staff, and department heads devoted the majority of their time to their own professional activities rather than departmental administration."[24] The past twenty years have seen a notable difference with administrative decision making branching out to include university and hospital boards, a university president, health center chief administrative officer, deans of medicine and other health schools, department chairmen, division and section heads, professional and nonprofessional support staff, students, and alumni.

This administrative complexity requires faculty members to become more involved with supervision of support staff and institutional service. They must devote time to work on committees for admissions, curriculum, recruitment, student affairs, governance, and human subjects research. The increased administrative responsibility is a burden to faculty because they often lack administrative skills. There is no defined reward for this work, and it absorbs time that would otherwise be given to patient care, teaching, and research.

Faculty must also adjust to the increasing administrative pressures placed on their department chairmen. In the past, chairmen could spend time interacting personally and professionally with each faculty member, knowing firsthand their concerns and aspirations. Now the increased size of departments, managerial duties, and fiscal responsibilities tend to insulate chairmen from faculty. This is frustrating for both, leaving a gap that may needlessly strain relationships and is difficult to fill.

Promotions, Tenure, and University Expectations

Financial constraints, probability of decreased growth rate, high cost of a large tenured faculty, difficulty of maintaining research productivity in the present market, and pressure to train more primary-care physicians rather than specialists have led to redefinition of university expectations and revision of promotion and tenure guidelines. Although there are advantages to reassessing the tenure and promotion systems, the major disadvantage is that this generation of faculty members must live with indecision regarding their security in medical academia. Some institutions have frozen tenure, others have changed the time line for consideration, still others give faculty nontenured appointments, using clinical and research professoriates. With very little knowledge about the long-term outcomes of these actions, it is difficult for faculty to choose the "right" career path. Choosing becomes even more difficult when university expectations remain uncertain and faculty performance criteria are vague.

Performance Evaluation

Over the past twenty years, there has been an increased need to document and evaluate faculty performance for promotion and tenure decisions, in fairness to faculty and in compliance with their rights by law.[16] Although elaborate evaluation systems have been proposed in higher education generally,[15,28] it is difficult to implement such systems in medical education because of the patient-care milieu and unique institutional constraints. Nevertheless, faculty evaluation has been a longstanding concern among administrators, students, and faculty in the medical professions,[4] and several issues emerge.

First, in terms of patient care, it is difficult to evaluate clinical competence because reaching a consensus regarding the definition of competence is almost impossible. "There are degrees of competence expected and necessary at different points in a physician's career and according to the specialty pursued, and the nature of the practice."[4] The lack of definitive criteria and measuring instruments necessitates further research into reliable and valid assessment in this area.

Second, differences among specialties and subspecialties make it difficult to evaluate the quality of a faculty member's research and publication contributions. At most institutions, research publications are counted rather than critically appraised. Furthermore, critical appraisal would require adjusting the evaluation criteria to reflect decreasing research funds, varying journal review procedures, different publication standards, and professional politics. It would also require objective peer evaluation and review, which are costly and time-consuming.

Third, as with research, institutional service is measured by counting the number of committees to which the faculty member is appointed. Quantity rather than quality becomes the explicit criterion. Where quality of contribution is considered, it is usually assessed through informal communication channels and remains an implicit criterion.

Fourth, although much research has been done on the evaluation of teaching,[15,3,12] the concept of teacher effectiveness remains almost as vaguely defined as physician clinical competence. Ordinarily, lists of teaching characteristics, which vary from institution to institution, are developed into rating scales and used for evaluating teaching performance. Ideally, the teacher should be evaluated by a variety of persons including administrators, peers, and students. But, too often, only student evaluations of instruction are used, and consequently the student ratings often reflect biases such as: students' general disposition toward instructors and instruction; teaching conditions including class size, elective versus required status of course, and subject matter; student preference for highly structured or less structured teaching styles; and student expectations and achievement.[19] Clearly, there is a need for further research in the development of teacher evaluation measures and methods.

Because of the relatively unsophisticated state of the art in performance evaluation, medical faculty must live with a certain amount of ambiguity. Although current evaluation efforts attempt to explicate the basis of performance decisions, their lack of sophistication pressures the faculty for the following reasons:

1. Faculty members rarely have direct input into developing criteria that will be used to assess their individual performance.
2. By and large, faculty are not trained to be administrators and teachers, but are nevertheless being evaluated in those functions.
3. Faculty are uncertain about how evaluation results are used in making promotion and tenure decisions.
4. Performance evaluation is used primarily to make judgments (for example, contract renewal, tenure, and promotion) rather than to facilitate professional growth and development.
5. Evaluators, be they students, administrators, peers, or outside observers, always hold implicit values and biases which they are not required to clarify in their ratings of faculty.
6. Evaluation is always threatening, especially when conducted in a competitive rather than trusting atmosphere. No one would deny the competitiveness inherent in medical training and the discomforting challenge of being observed and rated.

These organizational factors indicate the substantial pressures on faculty members in academic health centers—their financial insecurity, uncertain job futures, increased responsibilities, and confusion over institutional expectations and reward. What emerges is a picture of faculty members in need of support.

Coping Strategies

While physicians are trained to be confident and responsible, they now face severe limits and a certain helplessness vis-à-vis the complex organization. Rogers[25] describes the sense of helplessness as a "fear of impotence in effecting change or control," and points out physicians' unconstructive as well as adaptive responses to the current situation.

The unconstructive responses, similar to those used by larger groups of people experiencing greater helplessness, depend on psychological denial and escape. One such strategy is the compulsion to identify with symbols of power, which can obscure a person's real limits in decision making and control. Another strategy, most obvious in the medical profession, is simply overwork. Seriousness and dedication turn into compulsion. Faculty members refuse to accept the limits of responsibility, time, and energy. A third strategy is to surround oneself with technology, becoming insulated from interpersonal interactions that require openness and vulnerability. Although distancing is important in patient care, it is often extended to other human interactions, even those that should be "intimate." Finally, another strategy, which some claim is a conspicuous illness of our age generally, is the inability to make a commitment to one's work, family, and community. As Dyer[6] states: "Serious indeed is the erosion of public confidence that medicine has suffered, but equally grave is the loss of self-confidence of many physicians who often practice within the confines of what is expected of them rather than what they are committed to."

Adaptive responses rely on understanding the valuable components of unconstructive strategies and recognizing where overreaction begins. A constructive response requires acknowledging the real limits of time, training, disposition, and situation. It also means "differentiating real helplessness and professional limits from obsessive and worrisome forms of imagined helplessness."[25]

As teachers, faculty members build the future of the medical profession in their daily interaction with the next generation of physicians. They make choices which affect students' lives, and their relationships with students are models for students' relationships with patients. If indeed the stresses are producing, as Miller[21] describes, "a new breed of ambitious specialized professionals who are, by preference, by training, and by the requirements of the tenure and promotion system, more interested in the prestige of research and publication than in the humbler rewards of excellent teaching," then we must be concerned with how the teachers' choices affect students. And if teachers take "easy" options, ignoring ethical obligations to students while engaging in "the genteel art of cutting rival scholarship, of rejecting articles without reading them, of extorting free books from publishers, and the like,"[21] then we must be concerned with what is being modeled in the teacher-student relationship. But most important, organizations must be sensitive to the needs of the faculty, the pressures they face, and the support they have for coping with stress.

¶ A Focus on Psychiatry

Recommending ways for addressing faculty needs and establishing support systems in medical education is too ambitious a project for this chapter. Surely there are no simple answers to the complex issues. It seems more sensible therefore to choose to focus on psychiatry education with its unique problems and to suggest one possible method for faculty support. Perhaps then, as Herbert Pardes, Director of the National Institutes of Mental Health, suggested: "The close attention that psychiatry gives to the details of its professional education can serve as a model for the rest of medicine."[8]

In addition to coping with pressures that are part of the academic health organization, psychiatry faculty face unique issues. First, limited financial support is not new to psychiatry. Over the past ten years, research and training funds from the Department of Health, Education and Welfare have not kept pace with inflation and the increase in number of medical schools.[8,31] Furthermore, psychiatry is the lowest paid clinical specialty in medicine because of discriminatory third-party coverage, the amount of physician time spent in patient care, and the lack of profitable technologies.[8] Second, psychiatry suffers from a poor image both inside the profession (with an unclear status as a medical field) and outside the profession (from a skeptical public). Third, psychiatrists must adjust to working with a variety of mental health professionals such as social workers, psychiatric nurses, and clinical psychologists, and must also compete for patients with these professionals. Finally, psychiatry is experiencing dwindling residency enrollments: there has been a 28 percent decline in interest among medical school candidates; a drop in medical students entering psychiatry (from 11 percent in 1970 to 3.6 percent in 1978), and an increase in the resident dropout rate (from 7.1 percent in 1978 to 12.2 percent in 1979).[13]

To improve psychiatry's image and offset the dwindling enrollments, some have suggested that the profession pay special attention to improving its educational practices.[5] Indeed, psychiatry is perhaps the most conscientious of all medical professions in this regard, with conferences devoted to teaching, innovative education efforts encouraged by the Psychiatry Education Branch of The National Institutes of Mental Health, and the development of education evaluation methods.[23] Quite notable, in fact, was the report of the 1975 Conference on Education of Psychiatrists[26] which, among other things, outlined specific responsibilities of psychiatric teachers (see table 34–1). The responsibilities reflect psychiatry's focus on and immense concern with teaching. However, what remains to be addressed by the profession is the type of support teachers need to meet these responsibilities amid the pressures of complex organizations.

There is much to be learned about psychiatry faculty from research on professionals in academic organizations in general[1] and from reactions of people in various types of corporate structure.[17] But there is no substantial literature regarding pressures, coping mechanisms, and support systems for medical faculty and psychiatry faculty in particular. These issues obviously need to be addressed through rigorous research that can lead to appropriate organizational change and support.

TABLE 34–1 **Teaching Responsibilities in Psychiatry Education**

1. Serving as role models, demonstrating through their own example how a mature clinician should approach the diagnosis and care of patients
2. Supervising and guiding residents as they develop their own skills by providing advice, support, information, and extensive evaluative feedback
3. Conveying essential information concerning the intellectual and theoretical foundations of psychiatry through both clinical and didactic teaching
4. Serving as sensors to developments in the immediate and larger social milieu, the profession, and the relevant sciences; communicating these to residents; and shaping the educational program to keep pace not only with the present but the future
5. Continually expanding their own state of knowledge and skill so that what is preached is practiced as well
6. Acting as compassionate, perceptive guides to professional development (with special skills honed by professional training) who can be responsive to the individual needs of residents at a time of great stress and growth
7. Understanding and respecting residents sufficiently to include them in major decisions that affect their education and professional well-being and shaping the residency program, to the extent possible, to achieve a reasonable balance between the fulfillment of resident needs, professional responsibility, and the demands of the service setting
8. Serving as the legally and professionally responsible representatives of patients' best interests and exerting leadership to assure that residents, as well as faculty members, practice with those interests foremost
9. Participating in administrative decisions that affect the educational milieu; the curriculum; faculty; residents and other students in the training settings; the use of time, space, and personnel; the dominant philosophy and approaches to education; and the accommodation among departmental research, education, and service activities
10. Helping to articulate educational objectives, assess whether and how well they are being met by teachers and students, and introducing those changes needed to aid in their realization

SOURCE: Rosenfeld, A. H. *Psychiatric Education: Prologue to the 1980's.* Washington, D.C.: American Psychiatric Association, 1976, pp. 195–196.

In the meantime, based on our experience and sense of what is true, it is clear that faculty need support. And we must be willing to explore, develop, and assess alternative faculty support structures that are built into each faculty member's immediate milieu. One alternative is available: In order to facilitate the matching of individual faculty expectations with institutional goals, to help faculty clarify their personal values and professional commitments, to provide faculty with opportunities for developing teaching, research, and administrative skills that are not taught in medical training, and to enhance physicians' current capabilities, we suggest maximizing the relationship between individual faculty and faculty administrators such as section chiefs, division heads, and chairmen. Such a relationship could be characterized as an advisory system that could be established in the context of present departmental structures or in matrix-management arrangements. Matrix management, a preference in some academic health centers,[24] defines horizontal and vertical lines of authority in which "functional managers" are responsible for merit review and "project managers" supervise productivity[11] in patient care, teaching, research, and administration.

We suggest that advisory systems, based on relationships between functional managers (for example, section chiefs, division heads, chairmen) and faculty members, be implemented and studied as supportive means for faculty guidance, development, and accountability. Guidance would acknowledge personal as well as professional concerns by: (1) clarifying rights and responsibilities of faculty members; (2) providing appropriate counsel and support; (3) defining organizational opportunities as well as organizational constraints; and (4) helping faculty to prioritize their personal and professional goals. A commitment to faculty development would assure that: (1) performance evaluation be used as a developmental tool; (2) opportunities for faculty enrichment are studied thoroughly[2,10]; and (3) individual faculty members are made aware of different types of rewards for varying contributions to the institution. Accountability would be formulated through a written contractual arrangement that would delineate: (1) organizational expectations regarding productivity and salary support; (2) the means for documenting faculty performance; (3) long- and short-term goals of faculty members; (4) developmental time lines for promotion and tenure; and (5) methods for negotiating and altering the contract.

The very nature of the advisory relationship requires a fiduciary commitment; that is, that it be founded in trust and confidentiality. Not a new concept, the fiduciary standard has its roots in the historical beginnings of the medical profession and is intimately connected to the ethics of professional responsibility. As Dyer[7] explains:

> The fiduciary tradition in medical ethics is at least as old as the Oath of Hippocrates and the cults of Aesculapius, and it has endured not as a code but as a symbol of the ideals most deeply cherished by the medical profession. While not encompassing in terms of behavioral guidelines, it has yet to be replaced by anything more morally inspiring. [p. 989]

If we espouse the importance of the fiduciary commitment in patient care and teaching, then we must acknowledge its importance in the advising and care of faculty. Above all that is expected of our teachers, and all they are able to produce, they are first and foremost people who, like other people, need support and guidance.

¶ Conclusion

This chapter has focused on the complexity of academic health centers, describing the resultant pressures incurred by faculty. While the complexity affects medical education in general, faculty are singled out because teachers are medical education's greatest resource, and their needs, coping strategies, and support systems are sorely understudied. Moreover, teachers affect students' lives in immeasureable ways, and their relationships with students are echoed in the students' relationships with patients.

There are no easy answers, and no one answer, for meeting faculty needs in medical education. Developing and exploring the feasibility of alternative support systems, one of which could be an advisory system for faculty, is one way to meet this challenge. Because of psychiatry's unique knowledge and skills regarding human behavior, it appears to be a fertile field in which to cultivate research and development in this significantly important area.

¶ Bibliography

1. BALDRIDGE, J. V. "Environmental Pressure, Professional Autonomy, and Coping Strategies in Academic Organizations," *Research and Development Memorandum*

No. 78. Stanford: Stanford Center for Research and Development in Teaching, 1971.

2. BERGQUIST, W. H., and PHILLIPS, S. R. *A Handbook for Faculty Development.* Washington, D.C.: Council for the Advancement of Small Colleges, 1975.
3. COSTIN, F., GREENOUGH, W. T., and MENGES, R. "Student Rating of College Teaching: Reliability, Validity, and Usefulness," *Review of Educational Research,* 41 (1971): 511–535.
4. COUNCIL ON MEDICAL EDUCATION. Report of The Council. *Future Directions for Medical Education.* Washington, D.C.: American Medical Association, 1979.
5. DANIELS, R. S. "Premedical Education in The Medical School Admission Process: Factors Influencing Psychiatry as a Specialty Choice." Paper presented at National Conference on Psychiatric Manpower and Recruitment, San Antonio, March 1980.
6. DYER, A. R. "Medical Ethics, The Art and Science of Medicine," *Connecticut Medicine,* 40 (1976): 467–470.
7. ———. "Reflections on The Doctor-Patient Relationship," in Brady J. P., and Brodie, H. K. H., eds., *Controversy in Psychiatry,* Philadelphia: W.B. Saunders, 1978, pp. 983–996.
8. EATON, J. S. JR. "The Psychiatrist and Psychiatric Education," in Kaplan, H. I., Freedman, A. M., and Sadock, B. J., eds., *Comprehensive Textbook of Psychiatry,* vol. 3. New York: Williams & Wilkins, 1980, pp. 2926–2950.
9. FLEXNER, A. *Medical Education in The United States and Canada.* A Report to The Carnegie Foundation for The Advancement of Teaching. Bulletin no. 4. Boston: Updyke, 1910.
10. GAFF, J. G. *Toward Faculty Renewal.* San Francisco: Jossey-Bass, 1975.
11. GOODMAN, R. A. "Organization and Manpower Utilization in Research and Development," *IEEE Transactions on Engineering Management,* EM-15/4 (1968): 193–204.
12. GRASHA, A. F. *Assessing and Developing Faculty Performance: Principles and Models.* Cincinnati: Communication and Education Associates, 1977.
13. HALES, D. "Recruiting Psychiatrists Seen As Crucial Issue," *Newspaper of The American Psychiatric Association,* 4 April 1980, p. 1.
14. HASTINGS, D. A., and CRISPELL, K. R. "Policy-Making and Governance in Academic Health Centers," *Journal of Medical Education,* 55 (1980): 325–332.
15. HILDEBRAND, M., WILSON, R. C., and DIENST, E. R. *Evaluating University Teaching.* Berkeley: Center for Research and Development in Higher Education, University of California, 1971.
16. HOLLEY, W. H., and FEILD, H. S. "The Law and Performance Evaluation in Education: A Review of Court Cases and Implications for Use," *Journal of Law and Education,* 6 (1977): 427–448.
17. KANTER, R. M. *Men and Women of The Corporation.* New York: Basic Books, 1977.
18. KATZ, D., and KAHN, R. L. *The Social Psychology of Organizations.* New York: John Wiley, 1966.
19. KULICK, J. A., and MCKEACHIE, W. J. "The Evaluation of Teachers in Higher Education," in Kerlinger, F. N., ed., *Review of Research in Education.* Itasca, Illinois: F.E. Peacock, 1975, pp. 210–240.
20. LIAISON COMMITTEE ON MEDICAL EDUCATION. *Accreditation of Schools of Medicine: Policy Documents and Guidelines.* Washington, D.C.: Association of American Medical Colleges, 31 March 1976.
21. MILLER, J. W. "Punctiliousness or Petifoggery?: A Case for Commitment to Ethics in Academe," *American Association of Higher Education Bulletin,* 31 (1979): 3–4.
22. MOORE, J. T., and BOBULA, J. A. "A Conceptual Framework for Teaching Geriatrics in a Family Medicine Program," *Journal of Medical Education,* 55 (1980): 339–344.
23. MUSLIN, H. L. et al. *Evaluative Methods in Psychiatric Education.* Washington, D.C.: American Psychiatric Association, 1974.
24. ORGANIZATION AND GOVERNANCE PROJECT OF THE ASSOCIATION OF ACADEMIC HEALTH CENTERS. Report of the Project. *The Organization and Governance of Academic Health Centers: Presentation of Findings.* Washington, D.C., 1980.
25. ROGERS, W. R. "Helplessness and Agency in The Healing Process," in Rogers, W. R., and Barnard, D., eds., *Nourishing The Humanistic in Medicine: Interactions With*

The Social Sciences. Pittsburgh: University of Pittsburgh Press, 1979, pp. 25–51.

26. ROSENFELD, A. H. *Psychiatric Education: Prologue to the 1980's.* Report of the 1975 Conference on Psychiatric Education. Washington, D.C.: American Psychiatric Association, 1976.
27. SCHEIN, E. H. *Organizational Psychology,* 2nd ed. Englewood Cliffs, N.J.: Prentice-Hall, 1970.
28. SELDIN, P. *Successful Faculty Evaluation Program: A Practical Guide to Improve Faculty Performance and Promotion/Tenure Decisions.* Crugers, N.Y.: Coventry Press, 1980.
29. STILLMAN, P. L., and SABERS, D. L. "Using A Competency-Based Program to Assess Interviewing Skills of Pediatric House Staff," *Journal of Medical Education,* 53 (1978): 493–496.
30. TOSTESON, D. C. "Learning in Medicine," *The New England Journal of Medicine,* 301 (1980): 690–697.
31. TRENT, P. J., and BRODIE, H. K. H. "Psychiatric Research: A Process View," in Kaplan, H. I., Freedman, A. M., and Sadock, B. J., eds., *Comprehensive Textbook of Psychiatry,* vol. 3. New York: Williams & Wilkins, 1980, pp. 3298–3305.
32. U.S. DEPARTMENT OF HEALTH, EDUCATION AND WELFARE, Office of the Secretary. Rules and Regulations. "Privacy Rights of Parents and Students: Final Rule on Education Records," *Federal Register,* 41 (17 June 1976): 24670–24675.

CHAPTER 35

THE DELIVERY OF MENTAL HEALTH SERVICES*

Darrel A. Regier and Carl A. Taube

MENTAL HEALTH services in the United States are provided by a diverse group of professional personnel and facilities with little coordination or centralized governmental planning. As a result, a combination of public and private, general medical and specialty mental health, multispecialty group and solo practice settings constitute what has previously been described as the "de facto U.S. mental health services system."[24] Major changes in this service delivery system have been the result of multiple economic forces, sociocultural pressures, and therapeutic innovations, as well as governmental policy decisions.

This chapter will describe the principal characteristics of the current mental health services system and the evolutionary trends that have emerged over the past twenty-five years. By starting with a brief review of epidemiological data on the prevalence of mental disorders, it is possible to provide some perspective on the scope of the mental health problems that the service system is designed to address. The major sites for mental health service delivery will be identified, as will the trends that have emerged in the service delivery system since 1955. Finally, the characteristics and distribution of professional personnel involved in providing specialty mental health services will be briefly discussed.

*Sections of this chapter are taken in part from two previous publications: D. A. Regier, I. D., Goldberg and C. A. Taube "The De Facto U.S. Mental Health Services System," *Archives of General Psychiatry,* 35 (1978): 685–693, and C. A. Taube, D. A. Regier, and A. H. Rosenfeld "Mental Disorders" in *Health United States—1978,* DHEW publication No. (PHS) 78-1232, Hyattsville, Md., 1978. All statistical data have been updated to reflect the most recent available information from the National Institute of Mental Health and the National Center for Health Statistics surveys. The assistance of Michael J. Witkin, Statistician, Survey and Reports Branch, Division of Biometry and Epidemiology, NIMH, in updating these data is gratefully acknowledged.

¶ Scope of the Problem: Prevalence of Mental Disorders

A basic and seemingly simple question often asked of mental health experts is: How many people in the United States have mental disorders? Unfortunately, firm answers are hard to obtain. Epidemiologists responsible for such information have been hampered by several problems, including disagreement about the criteria for diagnosing mental disorders and difficulty in obtaining reliable case identification data when communities of untreated people are surveyed. Many of these technical problems are being overcome, but current epidemiological data reflect these long-standing problems. Still, it is possible to obtain some rough estimates of the proportion of the U.S. population affected by these disorders, either at one point in time (point prevalence) or over a given period of time (period prevalence), and to describe the rate at which new cases develop (incidence).

The best current estimate of the prevalence of mental disorders is that at least 10 percent of the U.S. population is affected by mental disorders at any given point in a year. This conclusion is based on the findings of several different studies. A 1954 survey of the noninstitutionalized population of Baltimore, Maryland, found that at any given point in the year 10 percent of the total population of all ages had a mental disorder.[19] Using different case identification criteria, a 1954 study found that among people twenty through fifty-nine years of age, 23 percent was affected by a serious psychiatric impairment at any point in time.[30] In 1967, a study in New Haven, Connecticut, found a point-prevalence rate of about 16 percent for mental disorders in the population twenty years of age and over.[32] A resurvey of the same population indicated that 15.1 percent had definite mental disorders, and an additional 2.7 percent had probable disorders; thus 17.8 percent of the population now twenty-six years of age and over exhibited some form of mental disorder.[34]

Such studies, although useful, do not specify how many people are mentally ill within a given time period (for example, annual-period prevalence). To obtain such data, one must account not only for the point prevalence but also for the rate at which new cases develop (annual incidence). Although studies of the incidence of mental disorders are extremely rare, a 5 percent annual rate of new cases for the nation can be extrapolated from the rate of new, treated cases in a community-wide psychiatric case register.[24]

Using the results of the study with the lowest point-prevalence estimate (10 percent) and adding another 5 percent for new cases during the year, the annual-period prevalence of all mental disorders in the United States is conservatively estimated to be at least 15 percent of the population. (This will be the prevalence rate referred to in later sections of this chapter.)[24] In the future, when newer and more precise case identification methods are used in large-scale population studies, such as the National Institute of Mental Health (NIMH) Epidemiologic Catchment Area Program,[8] evidence will probably mount for point-prevalence rates of at least 15 percent, and for annual-prevalence rates of more than 20 percent.

¶ Overview of the Mental Health Service System

Any attempts to define where and by whom mental health services are delivered must begin with some definition of mental health services themselves and the mental health service delivery system. An operational definition is necessary to place reasonable boundaries on the concept of mental health in a society where many nontraditional therapists offer a bewildering array of "mental health" services.[18] Hence, the definition of mental health services should aim for the best possible understanding of where persons with mental disorders receive services. To that end, four major sectors where mental health services are provided may be identified.

The specialty mental health (SMH) sector encompasses a wide range of facilities, which include state and county mental hospitals, psychiatric units of general hospitals, private psychiatric hospitals, residential treatment facilities for children, community mental health centers, freestanding outpatient psychiatric clinics, partial care and halfway houses for the mentally ill, college campus mental health clinics, and office-based practices of mental health professionals. In 1975, approximately 3 percent of the U.S. population, or one-fifth (21.5 percent) of those estimated as having mental disorders, received services from this sector.[24] Figure 35–1 shows that there is some degree of treatment overlap with the general medical sector for these patients.

No health specialty, whether medical, surgical, or mental health, can provide all necessary services for patients with disorders in its area of special expertise. Hence, there is a necessary division of patient care responsibility between the specialty and the general medical service sectors of the mental health services system. Figure 35–1 shows that approximately 58 percent of the mentally ill receive services exclusively in the general hospital inpatient/nursing home (GHI/NH) and the primary care/outpatient medical (PC/OPM) sectors collectively referred to as the general medical practice (GMP) sectors. As previously noted, some overlap or joint care occurs between the SMH and GMP sectors.

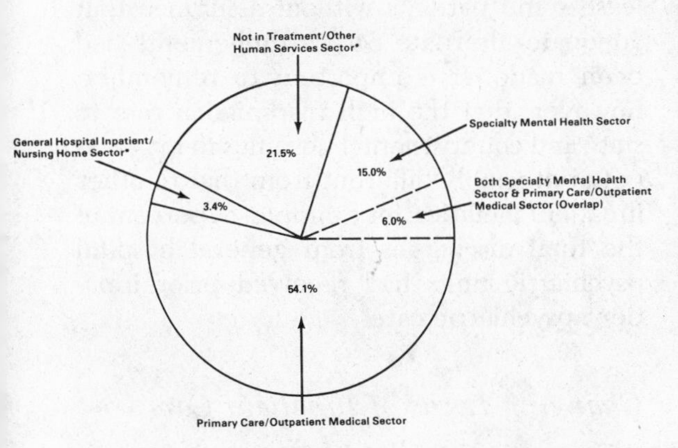

Figure 35–1 **Estimated Percent Distribution of Persons with Mental Disorder, by Treatment Setting —United States, 1975**

Note: Data relating to sectors other than the specialty mental health sector reflect the number of patients with mental disorder seen in those sectors without regard to the amount or adequacy of treatment provided.
*Excludes overlap of an unknown percent of persons also seen in other sectors.
SOURCE: Reprinted by permission from "The De Facto U.S. Mental Health Services System," by D. A. Regnier, I. D. Goldberg, and C. A. Taube, *Archives of General Psychiatry*, 35 (1978):685–693.

In addition to the general medical and specialty mental health sectors, mental health services also are provided in other human service settings, such as family service agencies and other social welfare organizations. This sector will be referred to as the other human services (OHS) sector. Although there is not yet accurate data on the amount of service provided by facilities in these sectors, it is assumed that the remaining 21 percent of persons with mental disorders are either treated in this sector or receive no mental health services at all (see figure 35–1).

This definition of mental health services includes only the direct patient treatment services of the specialty mental health sector and a limited number of general medical practice services. Data on most of these services in the SMH sector are routinely collected by the NIMH and are relatively more comprehensive than data obtained from the remaining sectors. In contrast, mental health services in the GMP sectors include only initial screening and diagnostic services to identify persons with mental disorders in their settings. Hence, information on the number of persons receiving services in these settings should be interpreted with these limitations in mind, which preclude the gathering of data on the type, amount, or quality of services provided in any of these settings.

¶ Recent Trends in the Service System

The mental health services system is an amalgam of historical trends of recent and distant origin, which continue to influence its form and functions. Understanding these trends is a prerequisite for effective planning and improvement of mental health care.

Declining Role of the State Mental Hospital

State and county mental hospitals have undergone significant changes since 1955, when the resident population in these facilities began to decline—a decline that has continued to the present. Between 1955 and 1978, the number of residents fell from an all-time high of 559,000 to 149,000. During this period, inpatient modalities of state hospitals, which had accounted for 49 percent of the total inpatient and outpatient episodes* in the country, fell to a low of 9 percent of all episodes (see table 35–1). Clearly, the locus of care had shifted.

The decline in the resident population of state mental hospitals is related to many factors, including:

1. Increased availability and use of alternate care facilities for the aged.
2. Increased availability and use of outpatient and aftercare facilities.
3. Development and use of psychoactive drug treatment.
4. Gradual reduction in the length of stay for admissions.
5. Greater use of community mental health centers and their affiliation with state mental hospitals.
6. Development of effective screening procedures to prevent inappropriate admissions.
7. Changes in state legislation regarding commitment and retention in facilities.
8. Deliberate administrative efforts to reduce the inpatient population.

These highly interrelated factors affected the rates for admission, readmission, and duration of stay, thereby affecting the number and composition of the inpatient population. While the resident population began diminishing in 1955, the annual number of additions (that is, admissions, readmissions, and returns from leave) to state mental hospitals increased yearly until 1971. Since 1971 the number of additions has declined.

The phenomenon of the "revolving door" of readmissions to state and county mental hospitals has elicited considerable concern in recent years. While the number of total admissions fell between 1972 and 1975 (in part, because of declining new admissions), the number of readmissions in 1975 was just slightly higher than the 1972 figure and remained at a high level of almost 70 percent for all admissions. The high number of readmissions might at first seem to be readily explained by the growth in the number of released mental hospital patients who constituted the population theoretically "at risk" for readmission. However, the readmission rate per 1,000 released patients rose from 174 to 197 between 1969 and 1975 (see table 35–2). Thus, other factors were involved. One factor was a shift from the use of long-term leave status to outright patient discharge, so that people needing rehospitalization were counted as readmissions rather than as returns from leave. Another possible factor, requiring further study, was a tendency to release some patients without assurance that adequate alternate care arrangements had been made. It is important to remember, however, that the high readmission rate to state and county mental hospitals in 1975 was not appreciably different from that to other inpatient facilities; for example, 61 percent of the total discharges from general hospital psychiatric units had received prior inpatient psychiatric care.

Changing Locus of Inpatient Care

The rate of total inpatient episodes per 100,000 population increased from 795 in 1955 to 842 in 1977 (see table 35–1). However, psychiatric case register data indicate that, when these episodes are unduplicated, the rate per 100,000 population of persons hospitalized has shown a slight decrease in recent years.[2] Thus, the declining role of the state mental hospital over the past two decades

*"Patient care episodes" are defined as the number of residents in inpatient facilities at the beginning of the year (or the number of persons on the rolls of noninpatient facilities) plus the total admissions to these facilities during the year (that is, new admissions, readmissions, and returns from long-term leave). This index, therefore, provides a duplicated count of persons and is not equal to a true annual-prevalence rate or the annual prevalence rate for treated mental disorder, which would require unduplicated person counts.

TABLE 35–1. **Number and Percent Distribution and Rate per 100,000 Population of Inpatient and Outpatient Care Episodes, in Selected* Mental Health Facilities, by Type of Facility—United States, 1955, 1965, 1971, 1975, and 1977.**

		INPATIENT SERVICES OF:						OUTPATIENT PSYCHIATRIC SERVICES OF:		
YEAR	TOTAL ALL* FACILITIES	ALL INPATIENT SERVICES	STATE & COUNTY MENTAL HOSPITALS	PRIVATE MENTAL† HOSPITALS	GEN. HOSP. PSYCHIATRIC SERVICE (NON-VA)	VA PSYCHIATRIC INPATIENT SERVICES	FEDERALLY ASSISTED COMM. MEN. HEALTH CEN.	ALL OUTPATIENT SERVICES	FEDERALLY ASSISTED COMM. MEN. HEALTH CEN.	Other
					Number of patient care episodes					
1977	6,392,979	1,816,613	574,226	184,189	571,725	217,507	268,966	4,576,366	1,741,729	2,834,637
1975	6,409,447	1,791,171	598,993	165,327	565,696	214,264	246,891	4,618,276	1,584,968	3,033,308
1971	4,038,143	1,721,389	745,259	126,600	542,642	176,800	130,088	2,316,754	622,906	1,693,848
1965	2,636,525	1,565,525	804,926	125,428	519,328	115,843	—	1,071,000	—	1,071,000
1955	1,675,352	1,296,352	818,832	123,231	265,934	88,355	—	379,000	—	379,000
					Percent distribution					
1977	100.0%	28.4	9.0	2.9	8.9	3.4	4.2	71.6	27.2	44.4
1975	100.0%	27.9	9.3	2.6	8.8	3.3	3.9	72.1	24.7	47.4
1971	100.0%	42.6	18.5	3.1	13.4	4.4	3.2	57.4	15.4	42.0
1965	100.0%	59.4	30.5	4.8	19.7	4.4	—	40.6	—	40.6
1955	100.0%	77.4	48.9	7.3	15.9	5.3	—	22.6	—	22.6
					Rate per 100,000 population					
1977	2,964	842	266	85	265	101	125	2,122	808	1,314
1975	3,033	847	283	78	268	101	117	2,185	750	1,435
1971	1,977	843	365	62	266	87	64	1,134	305	829
1965	1,376	817	420	65	271	60	—	559	—	559
1955	1,028	795	502	76	163	54	—	233	—	233

*In order to present trends on the same set of facilities over this interval, it has been necessary to exclude from this table the following: private psychiatric office practice; psychiatric service modes of all types in hospitals or outpatient clinics of federal agencies other than the VA (e.g., Public Health Service, Indian Health Service, Department of Defense, Bureau of Prisons, etc.); inpatient service modes of multiservice facilities not shown in this table; all partial care episodes; and outpatient episodes of VA hospitals.

†Includes estimates of episodes of care in residential treatment centers for emotionally disturbed children.

SOURCE: Unpublished provisional data from the National Institute of Mental Health (1977).

TABLE 35–2. **Readmission index and percent change for state and county mental hospitals: United States, 1969, 1972, and 1975.**

COMPONENT OF READMISSION INDEX	INDEX YEAR			PERCENT CHANGE	
	1969	1972	1975	1969–72	1972–75
Net live releases in 3 years prior to index year	995,834	1,188,104	1,179,977	19.3	−0.7
Number of readmissions during the index year	173,245	217,468	232,272	25.5	6.8
Readmission index (readmissions in index year per 1,000 net live releases in previous 3 years)	174.0	183.0	196.8	5.2	7.5

SOURCE: Unpublished data from the Division of Biometry and Epidemiology, National Institute of Mental Health.

has resulted in a shift to alternate inpatient psychiatric settings, such as general hospital psychiatric units, which have taken over inpatient care functions.

Because of the greater use of inpatient settings that have an active treatment focus, the length of inpatient care and the number of psychiatric beds have declined. Between 1971 and 1977, the total number of inpatient days decreased 40 percent, from 153 million to 91 million (see table 35–3), and the rate per 1,000 population decreased 43 percent, from 750 to 424. Between 1971 and 1977, the number of beds in inpatient psychiatric facilities declined from 471,800 to 298,783, a 37 percent decrease (see table 35–3). The corresponding rate of inpatient psychiatric beds per 100,000 population dropped 38 percent, from 225.6 to 138.9. Much of this decrease in the number of beds during this period reflects changes within state and county mental hospitals, where the number of beds decreased 49 percent between 1971 and 1977.

If changes in the number of beds in various inpatient facilities are taken as an indication of shifting loci of care, some interesting patterns can be seen. There was a net decrease in the number of psychiatric beds between 1971 and 1977 for all psychiatric facilities, largely as a result of the drop in the number of state mental hospital beds from 361,578 to 184,079. Despite this net decrease, some facilities increased the number of beds during the same period. For example, beds in private psychiatric hospitals rose from 14,412 to 16,637. Even more dramatically, nonfederal general hospital psychiatric unit beds increased from 23,308 to 29,384. These changes are but one indication of the growing role being assumed by these settings in inpatient psychiatric care.

Growth in General Hospital Psychiatry

There was a 26 percent increase in beds in psychiatric units of nonfederal short-term general hospitals between 1971 and 1977. This increase contrasts markedly with the decrease in state hospital beds and exceeds the 9 percent overall increase for general hospital beds for the same period.[1] The increase in the number of general hospital psychiatric unit beds reflects a 45 percent increase in the number of new units between 1971 and 1977.

As of January 1978, nonfederal general hospitals maintained 844 inpatient psychiatric units, 300 outpatient psychiatric services, and 166 day treatment programs for psychiatric patients. Veterans Administration general hospitals added another 100 inpatient psychiatric units, 102 outpatient psychiatric services, and 56 day treatment programs (see table 35–4). The nonfederal general hospital separate psychiatric services accounted for 20 percent of the episodes in all specialty mental health facilities in 1977.

TABLE 35-3. **Inpatient days of care and inpatient beds in mental health facilities, percent distribution and percent change according to type of facility: United States, 1971, 1973, 1975, and 1977.**

TYPE OF FACILITY*	YEAR								PERCENT CHANGE
	1971	1973	1975	1977	1971	1973	1975	1977	1971–77
	Number of inpatient days				Percent distribution of inpatient days				Inpatient days
All facilities	153,104,652	125,905,826	104,907,588	91,432,227	100.0%	100.0%	100.0%	100.0%	−40.3
Psychiatric hospitals	137,697,251	109,302,017	86,709,598	72,633,469	89.8	86.8	82.6	79.4	−47.3
State and county hospitals	119,200,126	92,210,109	70,584,014	57,206,390	77.7	73.2	67.2	62.6	−52.0
Private hospitals	4,220,216	4,107,499	4,400,522	4,791,906	2.8	3.3	4.2	5.2	13.5
VA hospitals†	14,276,909	12,984,409	11,725,062	10,635,173	9.3	10.3	11.2	11.6	−25.5
Nonfederal general hospital psychiatric units	6,826,260	6,990,253	8,349,412	8,434,691	4.5	5.6	8.0	9.2	23.6
Residential treatment centers for emotionally disturbed children	6,355,745	6,337,926	5,900,112	6,545,570	4.2	5.0	5.6	7.2	3.0
Community mental health centers	2,225,396	3,275,630	3,948,466	3,818,497	1.5	2.6	3.8	4.2	71.6
	Number of inpatient beds‡				Percent distribution of inpatient beds‡				Beds‡
All facilities	471,800	391,813	331,134	298,783	100.0%	100.0%	100.0%	100.0%	−36.7
Psychiatric hospitals	418,535	335,881	274,206	234,512	88.8	85.6	82.8	78.5	−44.0
State and county hospitals	361,578	280,277	222,202	184,079	76.7	71.4	67.1	61.6	−49.1
Private hospitals	14,412	15,369	16,091	16,637	3.1	3.9	4.9	5.6	15.4
VA hospitals†	42,545	40,235	35,913	33,796	9.0	10.3	10.8	11.3	−20.6
Nonfederal general hospital psychiatric units	23,308	24,518	28,706	29,384	4.9	6.3	8.7	9.8	26.1
Residential treatment centers for emotionally disturbed children	19,348	19,023	18,029	20,071	4.1	4.9	5.4	6.7	3.7
Community mental health centers	10,609	12,391	10,193	14,816	2.2	3.2	3.1	5.0	39.7

*Excludes multiservice mental health facilities not elsewhere classified, which represent 1 percent or less of the inpatient days and beds for each of the years.
†Includes VA neuropsychiatric hospitals and psychiatric inpatient units of VA general hospitals.
‡Counts on number of beds are obtained as of December 31 for each of the years.
SOURCE (inpatient days): Unpublished estimates from the Division of Biometry and Epidemiology, National Institute of Mental Health.
SOURCE (beds, 1971–75): National Institute of Mental Health, Division of Biometry and Epidemiology, *Statistical Notes 98, 118,* and *144*.
SOURCE (beds, 1977): Unpublished estimates from Division of Biometry and Epidemiology, National Institute of Mental Health.

TABLE 35–4. **Number of Mental Health Facilities and Service Modalities—United States, January 1978.**

TYPE OF FACILITY	NUMBER OF FACILITIES	NUMBER OF SERVICES		
		INPATIENT	OUTPATIENT	DAY TREATMENT
Total, all facilities	3,751	2,433	2,439	1,581
Nonfederal psychiatric hospitals	487	487	183	184
State and county hospitals	298	298	121	104
Private hospitals	189	189	62	80
VA psychiatric services*	137	122	128	68
Neuropsychiatric hospitals	22	22	21	9
General hospitals	110	100	102	56
Nonfederal general hospitals	925	844	300	166
Public hospitals	173	159	78	35
Nonpublic hospitals	752	685	222	131
Residential treatment centers for emotionally disturbed children	375	375	62	114
Federally funded CMHC's	563	563	563	563
Freestanding outpatient clinics	1,160	—	1,160	389
Public	397	—	397	135
Nonpublic	763	—	763	254
Other mental health facilities	104	42	43	97

*Total includes information for five VA freestanding psychiatric outpatient clinics that are not shown separately. Five of the clinics had outpatient services and three had day treatment.
SOURCE: Unpublished data from the Division of Biometry and Epidemiology, National Institute of Mental Health.

The overall role of general hospitals in providing mental health services is much larger, however, than that of their specialty psychiatric services. For example, discharges from nonfederal general hospital psychiatric units numbered 552,437 in 1977, whereas discharges with a primary psychiatric diagnosis from all hospital units numbered 1,625,000. Therefore, there were more than an additional 1 million discharges with a primary psychiatric diagnosis from general hospitals over and above those discharged from specialty psychiatric inpatient units. In addition to the 1.6 million discharges with a primary diagnosis of mental disorder, an additional 1.1 million discharges in nonfederal short-stay hospitals had a secondary psychiatric diagnosis (see table 35–5).

Organized outpatient mental health services may be categorized by their organizational location as follows:

1. Freestanding outpatient clinics that are not administratively part of or affiliated with an inpatient psychiatric facility.
2. Outpatient services affiliated with psychiatric hospitals, both public and private.
3. Outpatient psychiatric services of general hospitals.
4. Outpatient psychiatric services of other mental health facilities, such as residential treatment centers for emotionally disturbed children, outpatient services of federally funded community mental health centers, and clinics of the Veterans Administration.

Of the total 2,439 outpatient mental health services in the United States as of January 1978, approximately 183 (8 percent) were affiliated with psychiatric hospitals; 402 (16 percent) were affiliated with general hospitals (nonfederal and Veterans Administration); 1,160 (48 percent) were freestanding psychiatric services; 563 (23 percent) were affiliated with federally-funded community mental health centers; and 131 (5 percent) were affiliated with other types of mental health facilities. Dual affiliation with a general hospital and a community mental health center are counted with the latter (see table 35–4).

Additional information on the volume of services in each of the facility types is presented in table 35–6. This table reflects additions to facilities, that is, admissions, readmissions, and transfers during the year. When combined with data in table 35–1, it is possible to determine the percent of episodes beginning in the year. It is also possible to determine where the increase in outpatient mental health services has occurred in the past decade. Ninety-four percent of the absolute increase of 1,008,160 outpatient additions between 1971 and 1977 was about equally distributed between two types of outpatient settings; freestanding outpatient services

TABLE 35–5. **Distribution of Discharges, Excluding Newborns, from Nonfederal Short-stay Hospitals, According to Whether or not Primary or Secondary Diagnosis was a Mental Disorder—United States, 1977.**

DIAGNOSIS	NUMBER OF DISCHARGES
All discharges	35,902,000
Discharges with a mental disorder	2,763,000
Primary diagnosis of mental disorder	1,625,000
Primary only	1,128,000
Primary and one or more secondary	497,000
Secondary only	1,139,000

SOURCE: Unpublished data from the National Center for Health Statistics Hospital Discharge Survey.

TABLE 35–6. Admissions to Mental Health Facilities and Percent Change, According to Service Mode and Type of Facility—United States, 1971 and 1977.

TYPE OF FACILITY	SERVICE MODE								
	INPATIENT			OUTPATIENT			DAY TREATMENT		
	NUMBER OF ADMISSIONS		PERCENT CHANGE	NUMBER OF ADMISSIONS		PERCENT CHANGE	NUMBER OF ADMISSIONS		PERCENT CHANGE
	1971	1977	1971–77	1971	1977	1971–77	1971	1977	1971–77
Total, all facilities	1,269,029	1,588,964	25.2	1,378,822	2,386,982	73.1	75,545	171,118	126.5
Nonfederal psychiatric hospitals	494,640	552,854	11.8	147,383	146,797	−0.4	18,448	14,530	−21.2
State and county hospitals	407,640	414,703	1.7	129,133	107,127	−17.0	16,554	10,631	−35.8
Private hospitals	87,000	138,151	58.8	18,250	39,670	117.4	1,894	3,899	105.9
VA psychiatric services*	134,065	183,461	36.8	51,645	123,893	139.9	4,023	6,978	73.5
Nonfederal general hospital psychiatric units	519,926	552,437	6.3	282,677	225,765	−20.1	11,563	13,260	14.7
Government hospital psychiatric units	215,158	135,460	−37.0	139,077	99,543	−28.4	4,291	3,480	−18.9
Private hospital psychiatric units	304,768	416,977	36.8	143,600	126,222	−12.1	7,272	9,780	34.5
Residential treatment centers for emotionally disturbed children	11,148	15,152	35.9	10,156	18,155	78.8	994	3,147	216.6
Federally funded CMHCs	75,900	257,347	239.1	335,648	876,121	161.0	21,092	102,493	385.9
Freestanding outpatient clinics	—	—	—	484,677	889,589	83.5	10,642	21,149	98.7
Government	—	—	—	273,358	340,953	24.7	7,737	8,059	4.2
Private	—	—	—	211,319	548,636	159.6	2,905	13,090	350.6
Other mental health facilities	33,350	27,713	−16.9	66,636	106,662	60.1	8,783	9,561	8.9

*Includes Veterans Administration neuropsychiatric hospitals and Veterans Administration general hospitals with separate psychiatric modalities.
SOURCE: Unpublished data from the Division of Biometry and Epidemiology, National Institute of Mental Health.

and outpatient services of community mental health centers (see table 35–6).

Increasing Role of Nursing Homes in Care of Mentally Ill

One of the major factors contributing to the decline in the size of the state mental hospital resident population has been the growth of the nursing home industry. Under the Medicare and Medicaid programs,[11] the cost of caring for the mentally ill aged shifted from primarily state support to primarily federal support. These financing changes permitted nursing homes to flourish and assume responsibility for long-term care of many chronically mentally ill aged. Between 1954 and 1976, the number of nursing homes increased by about 210 percent, from about 6,500 to 20,185, and the number of nursing home beds grew by almost 730 percent, from 170,000 to 1,407,000. As Redick observed:

> In 1960, 615,000 or about 4 percent of persons 65 years of age and over were . . . in institutions; by the 1970 census, this number had increased to 968,000 and represented 5 percent of all persons 65 and over. At both time periods, over 90 percent of the elderly in institutions were either in mental hospitals or homes for the aged and dependent, but the proportions of elderly in each of the two types of institutions showed a significant shift over the 10-year interval. Between 1960 and 1970, the percentage of institutionalized elderly in mental hospitals decreased from about 30 percent to 12 percent, whereas the proportion in homes for the aged and dependent increased from 63 to 82 percent. [p. 1][21]

Between 1969 and 1973, the number of nursing home residents sixty-five years of age and over with a chronic mental disorder increased more than 100 percent, from 96,000 to 194,000, while the number of residents sixty-five years of age and over in all types of psychiatric hospitals decreased by 37 to 40 percent (see table 35–7). The net benefit of this trend for the mentally ill elderly has been questioned. Studies of the care provided for these individuals in nursing homes have suggested that "reinstitutionalization" rather than a deinstitutionalization to a less restrictive environment has resulted.[9] As an example of the impact of financing on the locus and quality of care, this phenomenon has important implications for national health insurance planning.

Growth in Federally Funded Community Mental Health Centers

One aspect of the growth in community-based mental health care has been the development of federally funded community

TABLE 35–7. **Resident Patients 65 Years of Age and Over in Psychiatric Hospitals or Residents 65 Years of Age and Over with Chronic Condition of Mental Disorder* in Nursing Homes and Percent Change, According to Type of Facility—United States, 1969 and 1973.**

TYPE OF FACILITY	NUMBER OF RESIDENTS		PERCENT CHANGE
	1969	1973	1969–73
State and county mental hospitals	111,420	70,615	−36.6
Private mental hospitals	2,460	1,534	−37.6
VA hospitals†	9,675	5,819	−39.9
Nursing homes‡	96,415	193,900	101.0

*Includes mental illness (psychiatric or emotional problems) and mental retardation but excludes senility.
†Includes Veterans Administration neuropsychiatric hospitals and general hospital inpatient psychiatric services.
‡Data on residents with chronic condition of mental disorder used rather than data on residents with primary diagnosis of mental disorder at last examination, since latter data were not available by age in 1969.
SOURCES: Selected publications and unpublished data from the Division of Biometry and Epidemiology, National Institute of Mental Health; A. Sirrocco, "National Center for Health Statistics: Chronic conditions and impairments of nursing home residents, United States, 1969," *Vital and Health Statistics*, Series 12, No. 22, DHEW Pub. No. (HRA) 74-1707, Health Resources Administration, Washington, D.C.: U.S. Government Printing Office, December 1973; and unpublished data.

mental health centers. The number of community mental health centers grew from 205 in 1969 to 563 in 1977 and, as noted, earlier, the outpatient services of these centers and of freestanding outpatient clinics accounted for 94 percent of the absolute increase in outpatient episodes between 1971 and 1977. In 1977, federally funded community mental health centers accounted for 31 percent of the total inpatient and outpatient episodes (see table 35–1).

The growth in the number of community mental health centers (CMHCs), which reached 763 in 1979, has resulted in a reorganization of existing facilities and an absolute increase in the number of persons served by organized mental health facilities. CMHCs generally are not newly created but rather are formed by the affiliation of existing community resources—usually general hospital psychiatric services and freestanding outpatient and day treatment programs; for example, 528 CMHCs in 1975 encompassed 2,000 affiliated facilities. General hospital psychiatric services have formed a major base for the development of CMHCs as have state- or county-operated and/or supported outpatient services. The state role in the development of CMHCs is demonstrated by the fact that 29 percent of the funding for CMHCs in 1977 was provided by state governments, an amount equal to that provided by the federal government.

In recent years, CMHCs have accounted for the major part of the growth in day treatment services, which were virtually nonexistent twenty years ago. Between January 1972 and January 1978, the number of day treatment programs increased by 60 percent. CMHCs accounted for 268 (45 percent) of the 592 new day treatment programs; freestanding outpatient psychiatric clinics accounted for 243 (41 percent); and residential treatment centers for children accounted for 54 (9 percent).

The numerical increase in day treatment programs has been greatest in CMHCs, which also sponsor the largest programs, averaging 182 annual admissions per program versus 75 annual admissions for other settings. Because of this growth, the CMHC day treatment programs now account for more than half of the annual admissions to day treatment services.

Despite dramatic increases in the numbers of day care programs and admissions to them, day treatment still remains relatively unused in the total spectrum of mental health resources. Of the 6.9 million patient-care episodes in mental health facilities during 1977, only 3.2 percent were in day treatment services.

Growth of Private Sector in Providing Mental Health Services

During the early development of mental health services, public programs were the predominant mode of service delivery. However, this dominance has been eroding at a rapid pace in recent years. (The growth in psychiatric services in general hospitals has already been noted.) Similarly, private psychiatric hospitals have grown from 151 in 1968 to 189 in 1977 and have assumed an increasing role in inpatient care. While national trend data are not available, there has probably been a significant increase in the number of people under care of private practitioners.[22] The number of people seen privately by psychiatrists and psychologists has been estimated to be almost 1.3 million, or 20 percent of the total number of people seen in 1975 in the specialty mental health sector.[24] Indeed, when the number of people seen in all private settings (both organized and private office settings are combined), the resultant number represents about half of the people under care in all specialty mental health settings during 1975.

Providing Mental Health Services in the Health Sector

Of the total number of people affected by mental disorders in 1975, about 19 million, or more than 60 percent, were estimated to have had contact with a general medical professional during the year. Only about 6 percent of the total were estimated to have been seen also in the specialty mental health sector during the year (see figure 35–1).[24] Since approximately 76 percent of the noninstitutionalized population visits a physician in one or more settings during a year,[4] this finding is not surprising. However, it does underscore the importance of the health sector as part of the treatment system for the mentally ill.

Special surveys[13,14,28] of general practitioners and internists have shown that about 15 percent of their patients are recognized as being affected by a mental disorder during periods of one month to one year, a figure reasonably consistent with the overall annual prevalence of mental illness in the population as a whole. Lower rates were found in industrial clinic settings, and somewhat higher rates were found in hospital outpatient departments.[26,27]

The rates of mental disorder found in these studies were higher than those routinely reported within the general health sector. For example, in 1975, the National Ambulatory Medical Care Survey determined that only 5 percent of visits to general practitioners, internists, and pediatricians resulted in a diagnosis of mental disorder.[23] It is believed that such underreporting results from several factors: (1) Organic illnesses are frequently the problems most presented and constitute the major focal point within nonpsychiatric office practice; (2) some nonpsychiatrist physicians are unable to recognize certain types of mental illness; and (3) many nonpsychiatrists prefer to avoid a mental disorder diagnosis whenever an alternative is available, perhaps to assure that treatment will be covered by health insurance.[25]

In a more recent study, mental health professionals examined a sample of general medical practice patients with a standardized psychiatric interview protocol. Results showed that 26.7 percent of adult patients (eighteen years and over) could be predicted to have a Research Diagnostic Criteria (RDC) diagnosis[29] of mental disorder in one year.[10] These findings indicate that higher percentages of general medical practice patients may be found to have bona fide mental disorders when more precise measures are used. However, the percent of these disorders that are mild and self-limited and that do not require specific "mental health services" has not yet been determined.

A study of general medical physicians in England[28] found that 67 percent of their patients with identified mental disorder received some form of treatment directly from the physician. Another 5 percent were referred for specialty mental health care, and 28 percent received no mental health treatment in the year. There is wide variation, however, in what is defined as "treatment" within general health care settings. Some of the U.S. general medical practice studies found that psychotropic drugs were prescribed for 60 to 80 percent of patients with identified mental disorders, and that "supportive therapy" was provided for up to 96 percent.[12,13,27] It is also obvious that some types of treatments used for patients with identified mental disorders were used for other patients as well. For example, a 1973 survey of visits to office-based physicians revealed that an antianxiety or sedative agent was prescribed in 12 percent of these visits, although only 5 percent of such visits were for mental disorder.[3]

Even if physicians in general medical practice neither recognize nor treat all of the mental disorders of their patients, it is clear that these physicians provide a substantial share of the total volume of mental health services in the United States.[7] Of all visits to office-based physicians resulting in a primary diagnosis of mental disorder, 47 percent were attributed to nonpsychiatric physicians, and 53 percent were attributed to psychiatrists.[23] Likewise—although nonpsychiatrists acknowledged use of a "psychotherapy-therapeutic listening" service in only 2 percent of their visits, compared with 73 percent of psychiatrists' visits, by sheer weight of numbers nonpsychiatrists accounted for as many as 46 percent of visits and 27 percent of the total time devoted to such therapeutic listening treatment by office-based physicians.[5]

¶ Geographic Location of Services

By almost any measure one chooses to use, specialty mental health resources are unevenly distributed geographically. Whether one looks at a national, regional, or local community level, resources tend to be clustered in certain areas, while other areas are essentially underserved or unserved. This uneven distribution results in limited or difficult access to mental health services for many who need them.

In general, mental health resources, whether facilities or personnel, tend to be clustered regionally in the Northeast and in urban rather than suburban or rural areas. Until quite recently, the location of service facilities and personnel had been planned with little consideration to local service needs and resources. The development of community mental health centers represents an effort at the federal level to complement state and local efforts to encourage more rational and equitable resource allocation and distribution, although these goals are not easily reached.

Examination of how psychiatric beds are distributed nationally will illustrate some of the current problems of resource distribution. The adequacy of a community's inpatient psychiatric care resources cannot be judged solely by its bed-to-population ratio. However, using this and other measures, it is apparent that there are vast inequities in the distribution of beds and other resources, which remain unrectified.

Psychiatric beds are distributed reasonably equally when the bed rate per 100,000 is considered by the state (see figure 35–2). However, psychiatric beds are more unevenly distributed by the state than are general

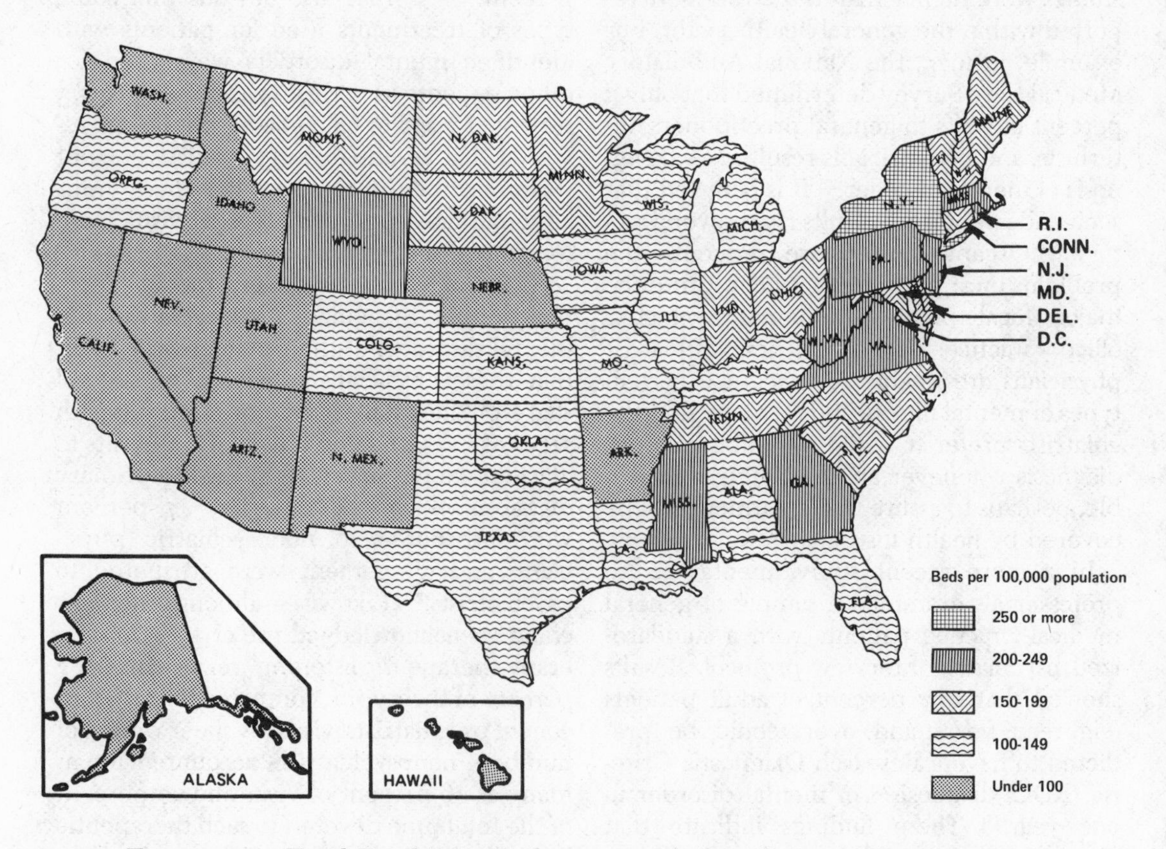

Figure 35–2. **Total Psychiatric Beds per 100,000 Population by State—United States, January 1976**

hospital beds. Particular types of psychiatric inpatient facilities show different degrees of uneven bed distribution; beds in psychiatric units in general hospitals are most evenly distributed, while beds in state and county mental hospitals are most unevenly distributed.

Compared with urban areas, rural areas and suburban areas have a relatively low rate of community-based psychiatric beds per 100,000 population. Rural psychiatric hospital bed ratios compared with urban area bed ratios are also relatively low, while psychiatric bed ratios in locales outside urban areas (but not rural) are very high, reflecting the historical tendency to locate psychiatric hospitals outside of populated areas.

One of the many objectives of the community mental health center program has been to increase the geographic accessibility of mental health care to the nation's population. In 1975, however, twelve years after passage of the community mental health care center legislation, 104 of the 1,542 geographic catchment areas in the United States still had no mental health services, 647 still had no community-based inpatient mental health service, and 334 had inpatient and outpatient mental health services but no day care or emergency services. The primary reasons for this omission are as follows:

1. Funds to support the development of CMHCs in all needed catchment areas have been limited.
2. Development of and planning for mental health services are difficult for those areas with scarce resources, and such planning may not be given highest priority by some communities.
3. Some areas are so sparsely populated that it would not be cost effective to provide a full range of services to them.

Remedies for this situation have been proposed, most recently by the President's Commission on Mental Health,[20] but these barriers to service development may not be easily overcome. Implementation of the Mental Health Systems Act, which is now being developed by Congress, would provide the greatest potential for ameliorating the current maldistribution of services.

Manpower Supply and Distribution

Concern has frequently been voiced over the adequacy of the manpower supply to meet the current and future service needs of the mentally ill. The issue becomes particularly acute considering the possibility that national health insurance, by eliminating some financial barriers, may increase the demand for services. At present, it is extremely difficult to say, except at a very general level, whether there are or are likely to be enough of the right people, with the right skills, in the right places, to respond appropriately to mental health service needs and demands. To answer these questions requires information not yet available regarding such issues as:

1. Who needs what services where?
2. What types of persons are best suited to provide various types and levels of care for particular kinds of individuals and disorders?
3. How do caregivers spend their time in various organizational settings?
4. How do various types of caregivers affect those they treat?
5. What kinds of human resources are needed (and for how long) to provide adequate treatment for various disorders?
6. What incentives can effectively alter manpower distribution patterns to make them more equitable?
7. How are the supplies of various types of manpower and other resources changing?

The issue is particularly complex because the characteristics of both the specialty mental health and the general health manpower system must be understood.

Almost two-thirds of those with mental disorders have contact with only the general health sector during a given year; it is thus critical that the need and demand for manpower in treating mental disorders be analyzed in this larger context. However, if examination is confined to the core disciplines providing mental health services (that is, psychiatry, psychology, social work, and mental health nursing), some idea of the general supply of personnel in these fields can be obtained, as well as an idea of how these individuals are distributed nationally in various service settings. Such figures, although crude, do suggest that however adequate or inadequate the current supply may be nationally, there is considerable geographical maldistribution that stands in need of correction.

There has been a substantial growth in the core disciplines during the past thirty years, as noted by Kole:[11]

> Membership of the American Psychiatric Association increased from about 12,000 in 1963 to about 23,000 in 1976; of these, 17,000 are estimated to be providing patient services in various settings, a ratio of 1:13,000 to the general population in 1976. Membership of the American Psychological Association increased from 21,000 in 1963 to 44,500 in 1977. Of these, approximately 23,000 are considered by the Association to be health care providers; approximately 81 percent of these providers have doctorate degrees and 17 percent have master's degrees, with many of the latter working toward the doctorate. The supply of social workers increased from an estimated 105,000 in 1960 to 195,000 in 1974 with perhaps 70,000 having an MSW degree or higher; about 26,000 full-time equivalent social workers were employed in mental health facilities in 1976, with 73 percent of these at the MSW level or above. In 1976, about 39,000 full-time equivalent nurses worked within organized mental health facilities; these include the entire range of training from associate degree nurses to those holding doctorate degrees. The number of mental health nurses with master's degrees or higher has increased from less than 20 in 1947 to approximately 11,000 in 1976. [p. 12]

As of January 1978, there were 469,038 filled staff positions (excluding private practitioners) in specialty mental health facilities in the United States. Of that total, 31 percent were staff not engaged in patient care. Of the professional staff, 26 percent were registered nurses, 13 percent were psychiatrists, 3 percent were other physicians, 13 percent were psychologists, 18 percent were social workers, 6 percent were physical health professionals, and 23 percent were other mental health professionals. Of the other staff engaged in patient care, 12 percent were licensed practical or vocational nurses, and 88 percent were mental health workers. Full-time staff worked an average of 39.6 hours per week, part-time staff worked 16.7 hours, and trainee staff worked 22.3 hours.

More than half of the total full-time equivalent staff of specialty mental health facilities worked in state and county mental hospitals. These hospitals deployed relatively large numbers of staff for work other than patient care and employed mental health workers with less than a bachelor's degree. Professional staff engaged in patient care in state and county mental hospitals were not as predominant, accounting for one-third of the full-time equivalent staff positions.[17]

A study of the distribution of mental health manpower in mental health facilities has reported several aspects of uneven manpower distribution.[33] First, urban areas rather than rural areas tended to attract concentrations of manpower and services. Such urban–rural manpower differences were particularly great regarding psychiatrists, social workers, and registered nurses. Although there were urban–rural disparities in the supply of psychologists, the disparities were not as great. Paraprofessionals tended to be more evenly distributed. The urban–rural manpower disparity holds even when poverty areas are compared. For psychiatrists, psychologists, social workers, and registered nurses, the highest mean number of manpower hours per 100,000 catchment area population in a poverty area was found in urban poverty areas, while the lowest manpower levels were in rural poverty areas.

From a regional perspective, the Northeast was relatively well supplied with mental health manpower, while the South, particularly the west south central and east south central regions, was poorly supplied. Certain states were outstanding either for their notably high rates of mental health manpower (for example, New York, Massachusetts, Vermont, and the District of Columbia) or for notably low rates (for example, Alabama, Alaska, and Mississippi).

Although this study was limited to manpower in mental health facilities, similar distribution patterns may exist for mental health personnel in private practice and in other care settings such as schools, industrial clinics, and other human service settings.[15]

¶ Summary

A review of the key points of this chapter provides an overview of the current mental health service system. Some trends and issues of particular importance for future planning are as follows:

1. About 15 percent of Americans are estimated to have mental disorders within any one-year period.
2. Most receive care from a variety of resources, but primarily such care comes from the general health, not the specialty mental health service system.
3. As many as 22 percent of those with mental disorders in a given year may receive no diagnostic assessment or treatment from either service system.
4. The specialty mental health service system, once largely geared toward long-term inpatient care in public facilities, is becoming increasingly oriented toward short-term and outpatient care in the private sector.
5. The length of stay in specialty mental health inpatient facilities has decreased appreciably, as has the number of inpatient beds.
6. The locus of inpatient care of the mentally ill is shifting from state and county mental hospitals to several other settings, particularly nursing homes and psychiatric inpatient units of general hospitals.
7. The growth of community mental health centers has provided new service resources and has had a profound effect on outpatient care (particularly day care), but has not yet achieved its full potential in creating more equitable geographic distribution of services and personnel.
8. Mental health personnel, like mental health facilities, are unevenly distributed geographically, with rural areas notably low in mental health service resources.

Obviously there is still much work to be done to assure that all Americans have access to appropriate, convenient, effective mental health care when it is needed. Considerable work is also required to reduce the need for mental health services through prevention. Such preventive efforts must be firmly grounded in laboratory-based and epidemiologic studies of the conditions that contribute to mental disorder; for example, risk factors. The more that is understood about the origins of mental illness and how to control it, the less reliance there will be on an extensive and expensive treatment system. Thus, future mental health planning must consider not only how to make mental health care more accessible and equitable for those with mental disorders, but also how to keep people mentally healthy.

¶ Bibliography

1. AMERICAN HOSPITAL ASSOCIATION. *Hospital Statistics, 1979.* Chicago: American Hospital Association, 1979.
2. BABIGIAN H. M. "The Impact of Community Mental Health Centers on the Utilization of Services," *Archives of General Psychiatry,* 34 (1977): 385–394.
3. BALTER, M. B., "Coping with Illness, Choices, Alternatives, and Consequences," in Helms, R. B., ed., *Drug Development and Marketing.* Washington, D.C.: American Enterprise Institute, Center for Health Research, 1974, pp. 27–45.
4. BLACK, E. R. National Center for Health Statistics. *Current Estimates from the Health Interview Survey, 1976,* Vital and Health Statistics. Series 10, No. 119. DHEW Pub No. (PHS) 78-1547. Public Health Service, Washington, D.C.: U.S. Government Printing Office, November 1977.
5. BROWN, B. S., and REGIER, D. A. "How NIMH Now Views the Primary Care Practitioner," *Practical Psychology for Physicians* 5 (1977): 12–14.
6. CHILES, C. L. "A Study of the Failure to Implement Alternate Care Recommendations for Patients in Mental Hospitals and Nursing Homes." U.S. Dept. of Commerce, National Technical Information Service, Contract Report, December 1975.
7. DELOZIER J. E. National Center for Health Statistics. *National Ambulatory Medical*

Care Survey, 1973 Summary. Vital and Health Statistics. Series 13, No. 21. DHEW Publication No. (HRA) 76-1772. Health Resources Administration. Washington, D.C.: U.S. Government Printing Office, October 1975.

8. EATON, W. W., et al. "The NIMH Epidemiologic Catchment Area Program, in Epidemiologic Community Surveys," in Weissman M. M., Myers, J. K., and Ross, K. E., ed., *Psychosocial Epidemiology Monographs.* New York: Neale Watson, forthcoming.
9. GLASSCOTE, R., et al. *Old Folks at Homes.* Washington, D.C.: American Psychiatric Association, 1976.
10. HOEPER, E. W., et al. "Estimated Prevalence of RDC Mental Disorder in Primary Medical Care," *International Journal of Mental Health,* 8 (1979): 6–15.
11. KOLE, D. M. "Report of the ADAMHA Manpower Policy Analysis Task Force," Rockville, Md.: Alcohol, Drug and Mental Health Administration, May 1978. Unpublished.
12. LOCKE, B. Z. "Patients Psychiatric Problems and Nonpsychiatrist Physicians in a Prepaid Group Practice Medical Program," *American Journal of Psychiatry,* 123 (1966): 207–210.
13. LOCKE, B. Z., and GARDNER, E. "Psychiatric Disorders Among the Patients of General Practitioners and Internists," *Public Health Reprot,* 84 (1969): 167–173.
14. LOCKE, B. Z., KRANTZ, G., and KRAMER, M. "Psychiatric Need and Demand in a Prepaid Group Practice Program," *American Journal of Public Health,* 56 (1966): 895–904.
15. MORROW, J. S. "Toward a More Normative Assessment of Maldistribution, the Gini Index," *Inquiry* 14 (1977): 278–292.
16. NATIONAL CENTER FOR HEALTH STATISTICS. "Utilization of Short-Stay Hospitals, United States, 1977, Vital and Health Statistics," Series 13, No. 41. DHEW Pub. No. (PHS) 79-1792. Washington D.C.: U.S. Government Printing Office, March 1979.
17. NATIONAL INSTITUTE OF MENTAL HEALTH. "Staffing of Mental Health Facilities, United States, 1978." Mimeographed report, Rockville, Md.: Alcohol, Drug Abuse, and Mental Health Administration, 1980.
18. PARLOFF, M. "Shopping for the Right Therapy," *Saturday Review,* 21 February 1976, pp. 14–20.
19. PASAMANICK, B., et al. "A Survey of Mental Disease in an Urban Population," *American Journal of Public Health,* 47 (1956): 923–929.
20. PRESIDENT'S COMMISSION ON MENTAL HEALTH. *Report to the President from the President's Commission on Mental Health, vol 1.* Washington, D.C.: U.S. Government Printing Office, 1978.
21. REDICK, R. W. "Patterns in Use of Nursing Homes by the Aged Mental Ill," Statistical Note 107. Rockville, Md.: Survey and Reports Branch, Division of Biometry and Epidemiology, National Institute of Mental Health, 1974.
22. REDLICH, F., and KELLERT, S. R. "Trends in American Mental Health," *American Journal of Psychiatry,* 135 (1978): 22–28.
23. REGIER, D. A., and GOLDBERG, I. D. "National Health Insurance and the Mental Health Services Equilibrium," Paper presented at the American Psychiatric Association, Miami, May 13, 1976.
24. REGIER, D. A., GOLDBERG, I. D., and TAUBE, C. A. "The De Facto U.S. Mental Health Services System," *Archives of General Psychiatry,* 35 (1978): 685–693.
25. REGIER, D. A. et al. "The Need for a Psychosocial Classification System in Primary-Care Settings," *International Journal of Mental Health,* 8 (1979): 16–29.
26. ROSEN, B. M., et al. "Identifying Emotional Disturbance in Persons Seen in Industrial Dispensaries," *Mental Hygiene,* 54 (1970): 271–279.
27. ———. "Identification of Emotional Disturbance in Patients Seen in General Medical Clinics," *Hospital and Community Psychiatry,* 23 (1972): 364–370.
28. SHEPHERD, M., et al. *Psychiatric Illness in General Practice.* London: Oxford University Press, 1966.
29. SPITZER, R. L., ENDICOTT, J., and ROBINS, E. "Research Diagnostic Criteria," *Archives of General Psychiatry,* 35 (1978): 773–782.
30. SROLE, L., et al. *Mental Health in the Metropolis: The Midtown Manhattan Study.* New York: McGraw-Hill, 1962.
31. TAUBE, C. A., REGIER, D. A., and ROSENFELD, A. H., "Mental Disorders," in *Health*

in the United States. DHEW Publication No. (PHS) 78–1232, Hyattsville, Md.: Department of Health, Education and Welfare, 1978.

32. TISCHLER, G. L., et al. "Utilization of Mental Health Services: I. Patienthood and the Prevalence of Symptomatology in the Community," *Archives of General Psychiatry,* 32 (1975): 411–418.

33. TWEED, D., KONAN, M., and LONGEST, J. W. "Distribution of Mental Health Manpower in Facilities in the United States," Contract No. 278-75-0028, College Park: University of Maryland, September 1977.

34. WEISSMAN, M. M., MYERS, J. K., and HARDING, P. S. "Psychiatric Disorders in a United States Urban Community: 1975–1976," *American Journal of Psychiatry,* 135 (1978): 454–462.

CHAPTER 36

THE PRESENT AND THE FUTURE OF THE PSYCHIATRIC HOSPITAL

Henry Brill

¶ Introduction

IN JANUARY 1976, there were an estimated 332,000 beds for the mentally ill in U.S. psychiatric facilities and an additional 425,000 in nonpsychiatric facilities, chiefly nursing homes and homes for the aged and dependent.[26,31,33,38] Commonly listed among the psychiatric facilities are the psychiatric units of general hospitals, community mental health centers (CMHC), the state hospitals (including state and county mental hospitals); Veterans Administration hospitals, residential treatment centers for emotionally disturbed children, children's psychiatric hospitals, and private psychiatric hospitals. All these treatment centers are different in many ways, but they share a common body of theory and practice, as well as a common treatment technology. And they have all been similarly affected by the social and economic changes of the past several decades. This situation is clearly described in a recent paper, "Psychiatry in New York City: Five Systems, All Overwhelmed."[32] The authors might well have added a sixth system, the nursing homes (including homes for the aged and dependent) because, nationwide, these have a larger inpatient census of psychiatric cases than all other facilities combined. All five systems face the problem of "how to deal with more patients than there are beds for, how to treat them, and how to design programs suitable for chronic patients as well as all the others who seek treatment."[32]

That a shortage of beds is common to all the various sytems of service is a matter of daily experience. It is encountered, for example, when one seeks to move a patient from a nursing home to a psychiatric hospital or from a psychiatric hospital to a nursing home. As a result, public general hospitals are often forced to keep a long-term patient in a costly acute-care bed for weeks or months, while negotiations are carried on for some type of placement. The various types of psychiatric facilities are affected by other factors as well, such as the system of multiple source funding, the requirements of continually new developments in legislative, judicial, and administrative law, and certain shifts in public attitudes toward mental patients and "former mental patients," as well as psychiatric diagnosis and treatment generally. Finally, they all face the pressures of public concern about rising costs, and, in the long run, they all compete for the same mental health funds. Thus, from one point of view the various types of psychiatric facilities may be seen to function as parts of a single system even though they are administratively heterogeneous, and the populations they serve, though they tend to be different, have a considerable overlap.

For this chapter, hospitals shall be considered as parts of a system and, for purposes of perspective, something of the history of the development of these facilities will be considered. Then the effects of advances in treatment techniques, the administrative responses to new problems and conditions, and the impact of newer fiscal and legal developments and the effect that they have had on the structure and function of the psychiatric hospitals generally will be considered. Finally, the situation with respect to several types of psychiatric hospitals will be reviewed.

¶ Inpatient Facilities Up to 1963

Early History

Hospital treatment of the mentally ill has a long and turbulent history. The classic Greek, Roman, and Arabic medical texts describe mental illness along with other forms of sickness; and to some degree, practice was in line with theory at least in the Muslim world, as space was provided for the mentally ill in Bagdad, Cairo, and Damascus general hospitals as early as the ninth century A.D. There is a story in the *Arabian Nights* about a man who was treated in such a facility and then released to the custody of his mother, after treatment with the strong deconditioning methods of that period had led to his recovery. Far more common during those times was the practice of extruding the mentally ill from society. They were left to wander as wild men in the woods[39] or set adrift at sea—as in the frequently described ships of fools.

It was the Spanish, perhaps reflecting influences from the Moorish occupation, who, in the early 1400s, first opened European asylums, and it was in the Spanish colonies that the first such facilities were established in the New World. The British followed suit not long after, perhaps as a result of the strong Spanish influence in England at the time, and by Shakespeare's day Bedlam and its wandering beggars, Toms o' Bedlam, were familiar to the public.[39]

In the eighteenth century there were asylums scattered across all of Europe. A psychiatric ward was opened at Guy's Hospital in London in the late 1720s, while in 1755 place was made for lunatics in the Pennsylvania Hospital in Philadelphia.[20]

The early asylums were run by ignorant and often brutal keepers, and conditions there soon fell below even the primitive standards of the time. This eventually triggered a period of reform, often associated with Philippe Pinel in France and William Tuke in England although many others were involved. Though this was the period of the French Revolution, the motto "Liberty, Equality and Fraternity" did not include the mentally ill and though Pinel's moral therapy was an advance, his own writings show that this was far from the golden age that has been depicted in some recent reports. Pinel described a very high mortality rate and some of the methods that he used would today be branded as brutal. Given the state of medical knowledge in his day, it could hardly have been otherwise: Paresis was rampant, untreatable, and unrecognized; anticonvulsants were unknown, the science of nutrition was still embryonic, and there was no effective way of controlling the twin asylum problems of psychotic behavior and epidemics. In the next 150 years other reform movements followed periodically; they centered in turn on such targets as privately run madhouses,[24] county-run asylums, brutal care by the community, harsh commitment practices, and generally low levels of care and treatment. Legal and administrative changes followed but they never solved the underlying problems of inadequate funding, overcrowding, and understaffing. The asylums continued to grow in population until 1955 and represented a political liability as well as a fiscal burden to whatever level of government administered them—municipal, county, or state;[16] the federal government as a matter of policy was not involved.

Compared with other branches of medicine, psychiatry was scientifically stagnant. The period after World War I saw the advent of psychoanalysis on the American scene but it was only an office practice. The prospects for improvement in the overall situation were dim, and the population of state mental hospitals continued to grow twice as fast as the general population.

The Beginnings of Community Psychiatry

Until recent years, the state mental hospitals remained the major psychiatric resource. The entire system appeared to be, unchanging but new developments were underway. The first of these was the move toward the so-called psychopathic hospitals and the psychopathic units (now called psychiatric units) in general hospitals. Deutsch[16] traced the former to the mid 1800s and stated that Pliny Earle, in an 1867 paper, first used the term "psychopathic hospital," urging separate hospitals for the "acutely insane." Early institutions of this type were often primarily receiving and observation points and performed a simple triage function for discharge or transfer to state hospitals, but gradually treatment became more important. In addition, relationships were established with medical schools, training for medical students and residents was provided, and research was undertaken. The Boston Psychopathic Hospital, the Syracuse Psychopathic, and the Colorado Psychopathic were early examples of this new type of small, locally based, academically oriented unit.

In 1902 the Mosher Memorial Unit in Albany, New York became the first modern psychiatric ward in a U.S. general hospital, and in 1933 the Rockefeller Foundation began to provide grants to create departments of psychiatry in university hospitals. Such units were set up in the Massachusetts General Hospital (1934), the University of Chicago (1935), George Washington University (1938), and Duke University (1940).[20] A similar pattern was followed in the establishment of psychiatric institutes that were autonomous but maintained relationships—with universities for example, the New York Psychiatric Institute, which was opened in 1929 as a part of the Columbia Medical School. One can see in these various facilities the emerging concept of community psychiatry based in small, short-term, treatment facilities located in the community they served. But it was only after 1963 that this became a dominant theme in American mental health policy.

Among those who prepared the ground for a shift from the state hospital to the community was H.S. Sullivan.[9] Working chiefly in the 1930s and 1940s in Washington, D.C., and building on the theories of the Chicago school of social psychology, Sullivan incorporated some of the ideas of Adolph Meyer (1866–1950), a professor of psychiatry at Johns Hopkins, and added some of his own to create an academic base for a social psychiatry, also referred to as community psychiatry. This theory brought together much of what had been developing outside of the realms of both organic and dynamic psychiatry and emphasized the importance of the social environment on the etiology, prevention, and treatment of mental illness. With this orientation, the institution came to be viewed as a noxious factor and the community as a constructive and normalizing influence. Hospitalization was to be minimized and persons were to be treated in their own homes or at least in their own communities. Community psychiatry could thus be understood as applied social psychiatry.

World War II did much to shift opinion in this direction. Support was provided by the practical experience accumulated during the war and by studies, such as that of Querido in Amsterdam, that seemed to demonstrate that emergency intervention in a civilian population could all but abolish the need for the state hospital. It was a time of confidence in social engineering as a way of correcting social problems, which was, in part, a reaction against Hitler and the Nazi regime's violations of civil and human rights. It was virtually inevitable that large mental hospitals would continue to be a target for reform, even though major medical advances had begun to affect their operations. The introduction of penicillin in the 1940s was to reduce paresis, which had accounted for 10 percent of admissions, from a fatal disease to a clinical rarity; vitamin B and advances in nutrition were wiping out pellagra and the associated psychosis that had once been a major problem in some Southern states; the use of diphenylhydantoin (Dilantin) in epilepsy was reducing admissions with epileptic psychosis to mental hospitals and would soon lead to the phasing out of the now virtually forgotten special state hospitals for epilepsy, such as New York's Craig Colony, which once had over 2,400 cases. Another major advance was in the prevention and treatment of tuberculosis. In New York state hospitals in the 1930s, this disease had a mortality rate twenty times that of the general population; the first survey in 1941 found several thousand active cases among 80,000 mental patients.[15,24] Today this figure has been reduced to several hundred and even that is maintained only by a continual inflow of "skid-row" type admissions. Similarly other types of infection have been spectacularly reduced and acrocyanosis, once so common that it was frequently considered a complication of schizophrenia, has entirely disappeared with the advent of more active programs of patient rehabilitation and better nutrition.

The introduction of somatic therapies for functional psychoses also produced important changes. The sleep therapy of J. Klaesi introduced in the 1920s left doubts but the insulin shock of Manfred Sakel, in the mid- and late 1930s, showed that a somatic therapy could produce remission of recent schizophrenia; convulsive therapy in the 1940s gave good, rapid results in an amazing proportion of depressions.[23] Brilliant individual results were now commonplace in the treatment of functional psychoses, but the effects of all advances were swallowed up in the overall picture of continued increase of mental hospital census, overcrowding, understaffing, and underfinancing, and the community facilities were still operating on only a token scale. State hospitals were still the major psychiatric resource, and in them large number of cases accumulated because they were refractory to all treatment efforts. In their back wards severe psychotic behavior was widespread and was marked by wetting, soiling, assaultiveness, chronic shouting and screaming, and destruction of clothing and furnishings. In the mid-1940s, psychosurgery appeared and was undertaken because it offered hope for amelioration of such symptoms and even remission for many intractable cases, but it remained controversial. It was abandoned with the advent of the tranquilizers in 1954 and 1955. These drugs were to produce the first overall changes in the mental hospital situation.

The Impact of Psychiatric Drug Therapy on Patients and the Public

These tranquilizers could be and were utilized on a large scale with existing resources.

The immediate effect was a radical improvement in the operating conditions of all types of mental hospitals. The distressing behavior that had become the hallmark of the psychiatric hospital rapidly faded, although it still remains embedded as a stereotype in the public mind. Within a year or so of the introduction of tranquilizers, the census of the large mental hospitals here and abroad began to decrease because these new drugs reduced or abolished delusions and hallucinations and restored social capacity on a scale never before achieved.[11] Nationally, the mental hospital population began an uninterrupted fall from a peak of 559,000 in 1955 to 504,000 by 1963,[17] a total reduction of almost 10 percent. However, the annual rate of decline was only a little over 1 percent, and soon there was growing pressure to increase the pace of population decrease once the possibility of such a decrease had been established. The mechanism for such a change was already available. Outpatient maintenance was an accomplished fact, and by the early 1970s the community hospitals could provide rapid inpatient therapy.[18] Although the service was available in the 1960s, the treatment time was cut to fourteen to thirty days in the 1970s.

The introduction of antidepressant medication in the late 1950s further extended the therapeutic potential of such facilities. In view of this, the decrease in the length of hospital stay from six months for newly admitted state hospital cases to one and one-half months was not impressive.

Public Opinion Begins to Change

For a time after the advent of the psychiatric drugs, psychiatry, as practiced privately and in both the community and the large hospitals, enjoyed a favorable press. The open hospital had become a reality after almost two centuries of frustrated hopes.[16] Voluntary admissions increased, the confidence of the general public rose, and interns competed for psychiatric residencies. But there were indications that this phenomenon was to be short-lived, and an antipsychiatry movement began to take shape. An early and rather bizarre manifestation developed in the Southwest, where a rumor was spread that left-leaning psychiatrists were planning to purge conservatives by hospitalizing them in a new facility, perhaps in Alaska. For the most part, however, the focus of attack was on involuntary hospitalization and on inadequate care and treatment in the large hospitals. Eventually the attack broadened to include the validity of psychiatric diagnosis; the effect of labeling patients,[19] and the ethics, morality, and legal status of various techniques of treatment. The somatic therapies and their adverse reactions were a special focus for criticism, but the concept of mental illness itself was questioned, as was the validity of psychiatric pronouncements in forensic matters. A large literature soon developed along these lines that was to have a major impact on the attitudes of those persons in a position to make or influence public policy.

The Joint Commission Report: Community Psychiatry Becomes Federal Policy

In 1955, the year the number of psychiatric patients in state hospitals reached its peak, Congress appointed a Joint Commission on Mental Illness and Health to redesign the U.S. mental health system. During the years of the Joint Commission's deliberations, the impact of the new drugs was already being felt. This fact was acknowledged in their history-making 1961 report,[21] although the drugs were not given a central role in the document or its proposals and predictions for the future. The document was a carefully written plan for social engineering in the mental health field, and its findings have stood up well in the light of subsequent developments. Recommendations were for a large increase in community services of all types, including general hospital units; the entrance of the federal government into the mental health field with strong financial support; a sharp reduction of dependence on state mental hospitals; and a larger participation of professionals other than psychiatrists in treatment.

The federal response was contained in President John F. Kennedy's message to Congress on February 5, 1963. He laid out a new concept of community-based services that would provide a complete spectrum of resources coordinated by community mental health centers. Institutional care and treatment was to be minimized and the large state hospital was a target of criticism. He said:

> . . . reliance on the cold mercy of custodial isolation will be supplanted by the open warmth of community concern and capability. . . . If we launch a broad new mental health program now it will be possible within a decade or two to reduce the number of patients now under custodial care by 50% or more.[25]

The proposal was quickly enacted into law and federal funds began to flow, at first for construction of the new community mental health centers (CMHC), then for staffing, and finally, more broadly, through such mechanisms as Medicaid and Medicare. Some aspects of the plan did not develop as rapidly as had been hoped. The CMHC phase was particularly slow in developing because support was time limited and the localities had be be prepared to pick up the cost or find other support. Also the integration of all services into a comprehensive system for each locality had to be left to local initiative, and this phase has posed special problems. By 1973, only 392 of the planned 2,000 centers were in actual operation[7] and as late as October 1978, it was projected that only 678 centers would soon be in operation.[30] In the meantime, the figure of 2,000 had been scaled down, and the centers themselves had come under attack on the grounds that they had merely developed more services along traditional lines and that they were not delivering the minimal services required for such facilities.

Deinstitutionalization

While the CMHC part of the Kennedy program did not develop as rapidly as hoped, the phasing out of the state hospitals did progress rapidly. The target figure was a reduction of 50 percent in a decade or two, and the actual fall was from 504,000 in 1963,[17] the date of his speech, to an estimated 175,000 in 1978.[7,8,9] In part, this was the result of a radical liberalization by the hospitals themselves of discharge policies and in part it reflected administrative policies of higher levels of state government. In at least one state, deadlines had actually been set for the discharge of certain proportions of the population. Equally important was the change in the thrust of laws, rules, regulations, and judicial decisions, which made it more difficult to admit or retain involuntary patients. Another factor was the escalation of costs, in good part the direct consequence of class-action suits and threats of such suits to upgrade hospital conditions and set minimum standards. Indeed one of the proponents of this method said that he hoped to make state mental hospital care so expensive that it would have to be abandoned—a sentiment that has been echoed by others.[7,37]

The Narrowing Definition of Mental Illness

For many generations the definition of what constitutes certifiable mental illness had been broadened by pressure of the courts and society. It had come to include an increasing proportion of the indigent population. In the 1960s and 1970s this was no longer the case; the mental disorders of old age, simple alcoholism, and drug abuse were largely ruled out. The criteria for certification were otherwise narrowed. The state hospitals were no longer willing to simply provide care for persons for whom there seemed to be no other place. And all other psychiatric hospitals moved in the same direction. As will be seen in the section on psychogeriatrics, other channels for placement began to open up during this period, especially by the operation of third-party payments such as Medicaid. But even this has not fully met the need, and there has been continued pressure for more beds in the inpatient system. A major problem that remains unsolved is finding a substitute for those public facilities that in the past served a triage function. The problem has been especially acute for those facilities serving a congested and poor metropolitan area. Old patterns of response die hard and the police, as well as social agencies and local government itself, still require that certain needs be met for homeless and/or helpless, confused, mentally incapacitated persons who cannot be left to fend for themselves. Under the circumstances, some of these community psychiatric units continue to serve reluctantly, providing a triage function that is parallel to their normal catchment area duties. They view the present situation as a transitional one during which other local resources will develop.

In the meantime, the volume of work in the entire mental health system has grown from 500,000 patient care episodes in 1950[26] to 5 or 6 million in 1979.[30] Yet the need seems to be even greater, based on estimates of the many millions who suffer from mental disabilities and disorders in the general population.[30] It is inevitable that under such pressures choices must be made at all levels, and this leads to the much-discussed situation in which "the sickest persons" are said to get the least attention. Actually it would be more accurate to say that the less treatable and more chronically disabled and those who cannot or will not cooperate with a treatment plan are displaced in favor of cases in which better results can be achieved with the same resources. This can be seen as a sort of paraphrased Gresham's Law that might state, "The more treatable and more cooperative patients will tend to displace the less treatable and less cooperative cases insofar as facilities can choose their patients." Cost is not the only factor; professional satisfaction and legal responsibilities also play a role, because cases that are maintained with difficulty and at a precarious level of adjustment may create public relations problems for the agency and the staff member who takes on the treatment responsibility. If there is a real or perceived threat to others, there are legal considerations. (See the Tarasoff case on page 745.)

In spite of all these negative factors, from all indications we have passed the most difficult phase of deinstitutionalization. In New York state, the census of state hospitals has fallen from 93,550 in 1955 to a current 25,000 (est. 1980). About half of these are aged persons admitted many years ago, and their number shrinks each year, which accounts for the current institutionalized population decrease of 4 to 5 percent per year. In California the high point was 37,000 in 1956 and an informal 1978 inquiry gave a census of 5,000. "New chronics" are still appearing but at a far slower rate than in the past, and now that it is generally recognized that certain persons will require long-term help, community support systems are beginning to arise. At one time the problems of discharged patients in the community were explained as a result of inadequate preparation for community life prior to hospital discharge, which tended to absolve everyone but the hospital of responsibility for posthospital service; today the emphasis is more realistic, especially since many of these cases are now seen only briefly in community hospitals that must discharge them after only a short-term residence. A small number of such cases are unwilling to accept assistance or to manage medication or funds by themselves. It remains to be seen how society will eventually deal with this problem.

¶ Psychogeriatrics

The care and treatment of aged persons no longer able to remain in the community because of mental disorder or disability remains a problem without satisfactory solution. The early literature on mental hospitals shows that admission for mental illness was largely limited to relatively young individuals. This appears to have continued throughout the 1800s. As late as 1914, New York state hospitals admitted only 770 cases of senile and arteriosclerotic psychosis, or 7.9 per 100,000 of the general population. But by 1955, this number had risen to 6,223, or 39.00 per 10,000 (first admissions). These older persons, two-thirds of them aged seventy-five and over, had an average duration of hospital life of less than two years in the early 1950s. Thus the practice was viewed as simply sending aged persons to the state mental hospital to die. The situation was exacerbated because it came at a time when the hospital census was rising by over 2,000 per year, and it led to a scandalous overcrowding; some geriatric wards were so crowded with beds that there was no day room or living space left for those who were ambulatory.

These overcrowded geriatric wards were among the first to feel the relief of the early decreases of hospital census, but the improvement was slow. By the end of the 1960s it had become New York state policy to limit the admission of persons over sixty-five. This policy was adopted nationwide, and between 1965 and 1972 the rate of admission of persons sixty-five and over was reduced by about two-thirds.[26] In the meantime, Medicare and Medicaid payments had become available, and the population of the homes for the aged and dependent rose from 296,783 in 1950 to 469,717 in 1960 and 927,514 in 1970.[26] The current (1980) census is estimated to be more than a million. In 1969, just over half of these residents were identified as having a mental disorder,[31] although a 1977 study showed that only 3.5 percent of 801,000 admissions gave their former residence as a mental hospital, indicating that direct transfer was unusual.[30] On the other hand, a New York state study of some 25,000 patients in adult homes showed that 29 percent had been patients in mental hospitals, which is almost three times as high as one would expect to find in a purely random selection of a group of aged persons from the general population.

These figures may be complex and even confusing, and they are not strictly comparable, but they do indicate that a very large problem of psychogeriatrics has been moved from the several hundred large mental hospitals of the United States to some 18,000 widely scattered nursing and personal care facilities. How much has been gained for the patients from this move remains to be evaluated. There have been recent scandals in New York state, in which certain of these facilities were attacked in the press and criminal charges were brought. However, this does not permit of an overall judgment, and the resulting reluctance of new entrepreneurs to enter the field has intensified the shortage of beds, which, as has been emphasized, creates a backpressure that tends to be felt throughout the hospital system.

At this time, we lack an organized and effective policy with respect to psychogeriatrics, and we may have to look to the British for a pattern. They have begun to establish special general hospital units for psychogeriatrics and are developing figures as to how many beds of this type may be needed for 100,000 of this specific population. It is of no small interest that they report a reasonable turnover of cases in such units and that the proportion of treatable cases, including depression and other problems, is considerable.[7,8]

¶ Fiscal and Administrative Issues

Action for Mental Health[21] stated that in 1954 total appropriations for state, county, and psychopathic hospital care was $568 million; this figure had risen to $854 million by 1959; the per-diem expenditure had gone up from $3.18 to $4.06 per capita. The recommendation was that expenditures for public psychiatric services be doubled in five years and tripled in the next ten. Some fifteen years later, the cost of direct care for mental illness amounted to $14,506 billion and by 1976 it was estimated at $17 billion.[30] This is only partly a result of inflation; staffing ratios of public mental hospitals were 27 per 100 patients in 1956,[21] while today the figure stands at over 100 staff per 100 patients. Total expenditure is now at about 12 percent of the health budget of the United States, which amounted to $160 billion in 1976. We seem to have reached a ceiling for health expenditures, which now amount to 8.5 percent of the gross national product. Within the health budget mental health must compete with other services, and there are indications that its competitive position may not be as strong as it was at one time.

One of the earlier inducements for deinstitutionalization was that the money saved by closing state mental hospitals wards would be more than enough to finance the brief-stay hospitals and outpatient services as well. It would require an expert in government finance to explain why this has not happened, but in retrospect, it seems overoptimistic to expect such a "plowing back" at all. In general, government fiscal policy prefers to see all savings turned back to a general fund; each new expenditure must then be judged each on its own merits. Aside from this in the present stringent fiscal climate, it seems unlikely that there will be much overall increase of mental health funds in the future, and all attempts to free funds by cutting expenditures in existing large hospitals will have to contend with class-action suits and accreditation standards that seek to raise rather than lower levels of care. In the meantime, pressure for additional funds continues from within the system, and general hospital costs are quoted as high as $300 to $400 per day in the New York area. In addition the cost of community care has proved to be far higher than was originally anticipated, and is as high or higher than in-hospital costs (in state facilities). One of the illusions of the present system of multiple source funding (federal, state, county, and third party) is the idea that somehow by shifting the cost from one agency or one level to another new money will be discovered. This technique was productive for a number of years while new sources were coming into play, and there can be no question that this system made possible the massive and revolutionary expansion of community hospital as well as outpatient services. But for the present this phase seems to have run its course. We are now encountering a highly organized and powerful move for cost containment, which comes from the large third-party interests. It is interesting that even by 1974 the care of the mentally disabled in nursing homes accounted for 29.3 percent of direct expenditures for mental health from all sources, the state and county hospitals 22.8 percent, and the general hospitals 11.7 percent,[27] and a major shift from one sector to another seems unlikely at this time. It remains to be seen how society will deal with the conflicting pressures generated by demand for more services and increased staffing on the one hand and demand for cost containment or even cost reduction on the other.

Internal Structure and Function

Revolutionary changes—many of them apparent only on inspection and they cannot be stated in quantitative terms—have taken place in all psychiatric hospitals. Furnishings are far better, the diet has improved, the doors are open almost everywhere, and patients come and go with the traditional freedom of the open hospital. Team organization has created a new informality between staff and patients and among staff members, since uniforms are now frowned upon for the most part. Nonmedical administration has been accepted at all levels except for strictly medical issues and the proportion of psychiatrist administrators of CMHC fell from 53.3 percent in 1971 to 30 percent in 1976.[4] This represents a trend in psychiatric administration generally.

The rigid sexual segregation of the past has been replaced by full integration; men and women mingle freely in the dining rooms and often in the living areas as well. Many hospitals are as fully integrated in this respect as are hotels, and only the actual sleeping rooms and the use of bathing or toilet facilities are separate. Sexual acting out does occur but it has not been a serious problem, and in at least some facilities women have access to contraceptive pills or other devices on request, just as they do in the community.

One of the most radical changes is to be found in policies with respect to admission and discharge. At one time a major complaint was that patients were admitted too readily and held too long and without adequate justification. Current complaints are in the reverse direction; that it is too difficult to gain admission, even on request, and that patients are discharged too soon. In part, this situation is due to fiscal considerations—funding is limited and cost-containment groups are vigilant. But it is also partly an expression of a general administrative pressure for deinstitutionalization. This pressure has had its major impact on the state mental hospitals, but it has influenced policy in other psychiatric facilities as well, especially the Veterans Administration and private hospitals. Compared with the general situation before 1963, a striking improvement can be seen in overall conditions due to better financing, better staffing, and to the ability of the facilities to resist being flooded with more cases than they can handle. Yet, this is not to portray the situation everywhere as ideal, because as great as the accomplishments of the past few decades have been, the expectations of the public have been even greater. It is an unusual facility that does not receive its full measure of justifiable consumer complaints. In fact, one may ask why surveys of consumer satisfaction are not a part of the regular routine of all hospital systems.

The Problem of Complexity

The complexity of mental hospital administration is by itself a major problem. The U.S. Comptroller[30] estimates that some 135 federal programs operated by eleven major departments and agencies have an impact on the mentally disabled. A New York state source states that there are some 120 state, city, county, and voluntary agencies that regulate some aspect of hospital operation. This poses a major problem of compliance, since often voluminous and detailed regulations governing fire and safety precautions, labor relations, fiscal operations, administration of medication, record keeping, and various aspects of patient care, to mention a few, are involved in running a mental hospital. The codes and rules governing each area usually have the force of law, and infractions carry the threat of a variety of sanctions. Perhaps the most important problem is that the whole regulatory structure has the characteristics of a house of cards; noncompliance with any one of the major components easily leads to withdrawal of approval by a series of other agencies, which may lead to loss of operating certificates for essential functions. This in turn may lead directly to loss of funding from the federal and state level.

A final problem is that all parts of the regulatory system tend to be in a continual state of change; a rule that is in effect one year may no longer be valid the next. The building codes are an excellent example. While a building is usually considered in compliance so long as it is not altered in structure or function, it may require costly alterations to bring it up to code again if a change is required for hospital purposes.

Often one cannot quarrel with the intent or even the effect of the regulations, as for example when it is required that a traditional system of psychiatric records be replaced by a client- or patient-oriented record. Yet this requires that the entire staff be reoriented, and for a time this detracts from the hours available for patient programs. The cumulative effect of regulation is that the amount of documentation required for patient records has increased as much as fivefold. Such increase in administrative overhead is particularly serious in large state facilities, which have already voluminous records and where the scarcity of clerical personnel is chronic. Another area of administrative concern has to do with legal issues.

Legal and Forensic Problems

In spite of all efforts to the contrary, differences still persist between the operation of a psychiatric ward or hospital and that of any other type of medical facility. Many of these, based on law but administrative and medical in nature, have to do with such issues as the admission and discharge of patients, the rights of patients to refuse treatment, hospital responsibility for acts of patients during hospitalization and afterward, competence of patients with respect to specific acts, general principles of civil rights as they have been applied to psychiatric patients, and finally the personal liabilities of staff members and the liabilities of the hospital when sued for violations in connection with such issues. It may be noticed that malpractice insurance rates are lower for psychiatry than for many other specialties, but this is misleading because the term malpractice has rather narrow medical definitions and insurance coverage may not extend to actions that can be interpreted as deprivation of civil rights, false imprisonment, or failure to observe due process.

As a result, the psychiatric facility is likely to maintain close liaison with its legal advisors to make sure that various actions are in conformity with the current climate of legal opinion. Legal guidance is thus essential in psychiatry, but its value is no more absolute than is advice of a doctor with respect to medical matters. The courts are not bound by the opinions of a lawyer, and liabilities may still be incurred even for actions that were taken in accordance with the advice of counsel, although this is far less likely than when action is taken without such advice.

One of the newer areas of the law in psychiatry has to do with constitutional guarantees of civil rights. The applicable law is Section 1 of the Civil Rights Act of 1871, a law originally passed to protect the black minority after the Civil War.[37] It reads in part:

> Every person who, under color of any statute . . . regulation . . . or usage of any state . . . subjects or causes to be subjected any citizen of the United States to the deprivation of any rights, privileges, or immunities secured by the Constitution and laws shall be liable to the party injured in an action at law. . . .

This law, whose operation has been reinforced by a Civil Rights Attorney Fees Act of 1976, has been widely used in class-action suits following the *O'Connor* v. *Donaldson* decision of 1975. This case, which was in the courts for some years, was finally settled by a cash payment from the estate of Dr. O'Connor, who had died in the interim. He was held liable for the involuntary hospitalization of patient Donaldson in a Florida state hospital because he had continued to confine, "without more" a patient who was not dangerous to himself or others and who could survive outside of the hospital with help available to him. This decision greatly increased the pressure on hospital authorities to discharge patients and made hospital personnel far more vulnerable to personal suit than they had been in the past.

Another landmark case affects a state institution and has influenced state hospital practice across the country, but its principles would seem to apply to other psychiatric facilities. This case is best known as *Wyatt* v. *Stickney.* The first major decision was in 1972 in a district federal court. As it has continued, it has set the precedent that a court may establish standards for care and treatment and make sure that constitutional rights of patients are not violated and that it may continue to monitor the results of the implementation of its orders even to the point of superseding control of the facility by the state. Personal liabilities were not invoked in this case, and the court had extensive professional assistance in the setting of standards.

The Tarasoff case (1976) in California has also had national impact, even though the decision was local. Tarasoff, a psychologist, had warning that a clinic patient of his had homicidal intentions toward a young woman. Tarasoff did not warn the woman although he did take other action; the woman was killed by the patient and liability was found by the court. This case seems to be in conflict with another major medico-legal principle, that of confidentiality, and the results are particularly difficult to apply because they may be understood to imply that the professional and the agency may be held liable for the behavior of an outpatient. One must thus choose between violation of confidentiality and the duty to warn. If all this applies to an outpatient, it might logically be held to be even truer of an inpatient.

Liability for acts of an inpatient may be taking a new form, as illustrated by a recent Long Island, New York case. This case involved a patient who murdered his wife while he was on pass from a state hospital. In this instance, indictment on charges of criminal negligence was sought against two psychiatrists. The grand jury did not return an indictment but, as always in such cases, the possibility that such charges might be brought had its own impact. A broader issue raised by the preceding case is related to the general policy regarding psychiatric patients who have criminal tendencies. At one time, New York state held several thousand such persons in special security facilities but a series of cases such as *Baxstrom* v. *Herold* (1966), in *Gault* (1967), and *Jackson* v. *Indiana* (1972) were reinforced by new laws, and New York now has specialized facilities holding only a few hundred such patients. It has been widely noted[35,36] that the reduction of the population of the old security facilities across the country did not produce any significant problems. Many of these patients could be released to the community. Nevertheless the pressure for "secure" beds has continued, and there has been an increase of admissions with a history of arrest in New York: in 1946 to 1948, 15 percent of a surveyed group of men released from New York hospitals had a previous arrest; in 1968, 32 percent had been arrested at some time before hospitalization; in 1975, the figure was 40 percent.[36] A record of arrest is an indication of antisocial tendencies, and the rising proportion of cases with such records intensifies the dilemma of responsibility for antisocial acts after hospital. On the one hand it is argued that prediction of overt acts is so unreliable that hundreds or thousands of persons would have to be incarcerated to avoid one antisocial act; on the other hand the psychiatric facility and its staff are required under severe sanctions to predict the behavior of its patients and to prevent such overt acts but without violating civil rights.[10]

Finally, the major advances in the field of patients' rights must be noted. It is now generally recognized that patients may refuse treatment, except for emergency situations, and such refusal can be overcome only by specific due process. Patients also have a right to the treatment that their condition requires, and this must be given in situations no more restrictive than necessary. Restraint and seclusion may be applied only under rigidly controlled conditions, if at all, and informed consent must be secured for various forms of specific therapy.

The preceding is but a quick glimpse of hospital psychiatry and the law, and it is indeed "A System in Transition."[36] Legal controls have evolved within the social climate and with developments in treatment technology; the overall results have been strongly positive, but much more remains to be done before the system reaches a new equilibrium.

¶ The Psychiatric Hospital Systems

As ordinarily reported, the major systems of psychiatric hospital are as follows: (1) community mental health centers; (2) psychiatric units of general hospitals; (3) private psychiatric hospitals; (4) residential treatment centers for children and psychiatric hospitals for children; (5) state and county psychiatric hospitals; (6) Veterans Administration (V.A.) hospitals. Another group, the nursing homes, have already been considered. Now the three systems that fall somewhat outside of the focus of this section, the V.A., the children's facilities, and the private hospitals, will be discussed. In January 1976 there were 35,913 psychiatric beds listed in the V.A. system. The V.A. facilities have carried on important academic and research activities in American psychiatry. The organizational and operational changes in the V.A. system have closely paralleled those in other psychiatric facilities, which have been described. Treatment centers for emotionally disturbed children numbered 331 in 1976, and they reported a population of 18,000—almost 97 percent of these patients under nongovernmental auspices. Turnover was relatively low, and there were only 12,000 additions in the year. The centers also reported 53,000 patient care episodes, of which 29,000 were inpatient. While these figures may appear large the demand for such beds far exceeds the supply and this is even truer of the twenty state and four private psychiatric hospitals for children.[40] This scarcity is not surprising in view of the untold numbers who are thought to suffer from various disabilities.[30] Finally, the private psychiatric facilities are a small but important source of treatment. Their distant past was a stormy one,[6,29] they have come to be regarded as a valuable resource. In 1975, there were 180 such facilities; they admitted 119,000 patients, ending the year with a census of some 11,500. It is interesting to note that there was an increase of 10 percent in the capacity of these facilities between 1968 and 1975, although the number of nonprofit private hospitals decreased somewhat in this period.[40]

Community Mental Health Centers

As of January 1, 1976 there were 10,193 beds listed in CMHC's, and even in 1973 they were already reporting 23 percent of all U.S. patient care episodes.[26,30] These facilities have been criticized for falling behind the program goals set in 1963, but in retrospect it seems that the goals were unrealistic and did not allow sufficiently for the problems inherent in just setting up some hundreds of new facilities, let alone having each of them integrate the resources already existing in its area. The aim was to have each of them supplement services that were lacking and insure that each area would have at least inpatient, outpatient, emergency care, partial hospitalization, and consultation and education services. It is doubtful if anyone anticipated the virtual avalanche of unmet needs that would be released by Medicaid, Medicare, and other third-party payments and by the social changes of the coming years. It is also doubtful if the complexities of coordinating a wide variety of jealously autonomous local agencies were given due consideration. Nevertheless, in spite of a slow start, the CMHC program has made important contributions. For a time it seemed that this program was in danger of being abandoned at the federal level, but the situation now looks more favorable again.

Psychiatric Units in General Hospitals

Greenhill[20] states that the general hospital has emerged as the focal point in the delivery of mental health care in the United States. There is much to support his contention. The speed with which these facilities have proliferated is impressive. While their history can be traced back for hundreds of years, an earlier official American Psychiatric Association publication[1] states that until after 1900, Bellevue was "the only such service in the land." Even as late as 1950, there were only a small number of such units in existence. Yet by 1963 almost 500 of a total of 5,400 general hospitals had psychiatric units,[20] and in 1971 this number had grown to 750 reported by 5,565 hospitals. By 1978 the figure had jumped to 1,600 units.[33] In 1963 the number of inpatient beds was 18,500;[20] it grew to 39,000 in 1975.[22] The volume of work carried on in these facilities increased correspondingly; 370,000 cases were treated in 1963, and by 1975 admissions alone numbered 543,000.[3] While these figures are not fully comparable they do show a marked trend for psychiatric units in general hospitals to play a dominant role in mental health care. In 1975 they accounted for 36 percent of all patient care episodes.[3]

As had been hoped, these facilities have indeed provided brief treatment in a community setting and on a large scale. The factor of cost is an important one in their operations, and it has been noted that the median length of stay in 1975 was related to methods of payment.[18] The overall figure was twelve days, and 84 percent of the patients were discharged within twenty-eight days, which may be related to the thirty-day limit for payment by such third parties as Blue Cross. It was also noted that median time to discharge was 6.6 days for personal payment cases and 3.3 days for no-charge cases.

These facilities have been caught in the

upward spiral of hospital costs generally, and charges of $300 to $400 per day are not unusual in the New York area. Staffing ratios are high, and it is estimated that two to three personnel are required for each bed and additional staffing is needed for those facilities which maintain a twenty-four-hour walk-in service. The general hospital units now usually try to provide overall service to a limited geographic area, but their triage function still has not been eliminated. In spite of the creation of restricted catchment areas, many patients from outside of the district are still brought to certain hospitals that once served as triage centers for large populations.

Another lingering problem for these units is posed by psychogeriatric cases. These may be admitted to medical or surgical services and then find their way to psychiatry where they may occupy a costly bed for weeks or months while alternative placement is sought. The general hospital unit may also be troubled by overall hospital staff shortages and by competing demands for emergency staffing in other parts of the facility. This may make it difficult to keep a stable staff on psychiatric wards at night or on weekends. Yet with all of these problems, persons who require hospitalization clearly prefer the psychiatric wards of the general hospital and regularly select one in their own immediate area.

State and County Mental Hospitals

These facilities were an initial target of the 1963 plan to reorganize U.S. mental health services, and they have played an important part in the changes that have taken place since then. These events have already been reviewed earlier in this paper and elsewhere.[7,8,9,11] Some of the major changes that have occurred will be recapitulated. The census of these facilities has fallen from a high of 559,000 in 1955 to an estimated 175,000 or less in 1980. Overcrowding is now a forgotten problem, and buildings or even entire hospitals are being closed. Medicaid and other third-party payments have put large sums of money into the system. Accreditation standards such as those of the Joint Commission on Accreditation of Hospitals must be met; failure to do so can and has led to loss of federal funds for some of the better known facilities in this class. In addition, class-action suits and threats of such suits have maintained a continued pressure for upgrading. Overall staff ratios, which in some facilities were once at the level of fifteen to twenty employees per one hundred patients, have now reached a level nationwide of about one hundred staff to one hundred patients. In 1975 they listed a full-time equivalent staff of 211,899 when their average census was 193,436. Their budget, which had been $568 million in 1954,[21] was $2,641,295,000 in 1974–1975.[27] With all of these improvements, however, the per-diem allowance per patient is still only 25 percent or less than that in a general hospital unit and far below the private hospital allowance. As their population continues to fall by 4 or 5 percent per year and rumors of closing circulate, morale is damaged and problems in recruiting capable staff, especially at the psychiatrist level, continue to increase.

It has been repeatedly pointed out that these hospitals could be a valuable resource, at least for the care and treatment of those who are not able to respond quickly to other approaches, and it is often said that many persons would seem to be more comfortable and even safer in state and county hospitals than they are in the slum accommodations where so many of them now are congregated. Yet there are lingering fears about creating the old system anew, and thus other solutions—such as the creation of community support systems—are being sought, and the pressures for continued deinstitutionalization continue. This author knows of no organized opposition to this aim, only demands for a more adequate alternative. In fact deinstitutionalization has already been largely accomplished; the bulk of the present state mental hospital population is composed of a highly transient short-term group plus a very large group of aged persons whose high death rate accounts for the current population decrease. There is no indication that the "new chronics" will be allowed to accumulate again in these hospitals nor that these facilities will again be opened to the heterogeneous population of mentally incapacitated. The fact that these facilities still survive at a third of their previous capacity seems to indicate that they serve an essential though reduced function. It remains to be seen whether they can or will be totally replaced.

¶ The Future

The future of the community-based psychiatric hospital as the focal point of treatment of major psychiatric disorder seems assured; the advances in treatment technology, which are now in prospect as a result of new discoveries in the basic sciences, are more likely than not to require the technological support that only such facilities can provide. It seems probable, however, that they will have to take a more active role in the treatment of drug abuse and its complications now that this condition has become so prominent in our society. The treatment of alcoholism may also have to be expanded, and there are great unmet needs in acute psychogeriatrics and in child psychiatry.

The possible future of the state mental hospital and the community support systems that now are being developed has already been discussed. Turning to the overall scene, it would seem that the future of psychiatric hospital service will be determined, as it has been in the past, by the interaction of a variety of forces, including the economic and social climate, the real cost and availability of personnel (including the professions), and the pace of scientific and technological advances. It is in this last that the major hope may lie, since there are indications that we have reached a ceiling in mental health expenditures. Only by such advances can we hope to do what is so strongly demanded, to accomplish more with less cost.

¶ Bibliography

1. AMERICAN PSYCHIATRIC ASSOCIATION. *One Hundred Years of American Psychiatry.* New York: Columbia University Press, 1944.
2. BACHRACH, L. L. *Deinstitutionalization: An Analytical Review and Sociological Perspective.* DHEW Pub. No. (ADM) 76–351, Washington, D.C.: U.S. Government Printing Office, 1976.
3. ———. "General Hospitals Taking Greater Role in Providing Services for Chronic Patients," *Hospital and Community Psychiatry,* 30 (1979): 488.
4. BASS, R. D. *CMHC Staffing: Who Minds the Store?* DHEW Pub. No. (ADM) 78–686. Washington, D.C.: U.S. Government Printing Office, 1978.
5. BEAN, P. "Psychiatrists' Assessments of Mental Illness. A Comparison of Some Aspects of Thomas Scheff's Approach to Labelling Theory," *British Journal of Psychiatry,* 135 (1979): 122–128.
6. BOND, D. *Dr. Kirkbride and His Mental Hospital.* Philadelphia: J.B. Lippincott, 1947.
7. BRILL, H. "The Future of the Mental Hospital and Its Patients," *Psychiatric Annals,* 5 (1975): 10–21.
8. ———. "The State Hospital Should be Kept, How Long?" in Talbott, J. A., ed. *State Mental Hospitals.* New York: Human Science Press, 1980, pp. 147–160.
9. ———. "Notes on the History of Social Psychiatry," *Comprehensive Psychiatry,* in press.
10. BRILL, H., and MALZBERG, B. "Statistical Report on the Arrest Record of 5,353 Male Ex-patients." Mimeographed. 1954.
11. BRILL, H., and PATTON, R. E. "Analysis of 1955–56 Population Fall in New York State Mental Hospitals in First Year of Large-Scale Use of Tranquilizing Drugs," *American Journal of Psychiatry,* 114 (1957): 509–516.
12. BROWN, B. S. "The Life of Psychiatry," *American Journal of Psychiatry,* (1976): 489–495.
13. CATHCART, R. "Issues Facing American Hospitals," *Hospital and Community Psychiatry,* 30 (1979): 193–194.
14. COMPTROLLER GENERAL OF THE U.S. *Report to the Congress: Returning the Men-*

tally Disabled to the Community. Washington, D.C.: U.S. General Accounting Office, 1977.

15. DEPARTMENT OF MENTAL HYGIENE. *Annual Report. 1946,* p. 85.
16. DEUTSCH, A. *The Mentally Ill In America.* Garden City, N.Y.: Doubleday, 1937.
17. EISENBERG, L. "The Future of Psychiatry," *The Lancet,* (1973): 1371–1375.
18. FADEN, V. B., and TAUBE, C. A. *Length of Stay of Discharges from Non-Federal General Hospital Psychiatric In-Patient Units, U.S. 1975.* Mental Health Statistical Note No. 133. Rockville, Md.: National Institute of Mental Health, May 1977.
19. GORDON, D., ALEXANDER D. A., and DIETZAN, J. "The Psychiatric Patient: A Voice to be Heard," *British Journal of Psychiatry,* 135 (1979): 115–21.
20. GREENHILL, M. H. "Psychiatric Units in General Hospitals: 1979," *Hospital and Community Psychiatry,* 39 (1979): 169–182.
21. JOINT COMMISSION ON MENTAL ILLNESS AND HEALTH. *Action for Mental Health.* New York: Basic Books, 1961.
22. JONES, R. E. "Issues Facing General Hospital Psychiatry," *Hospital and Community Psychiatry,* 30 (1979): 183–184.
23. KALINOWSKY, L., and HIPPIUS, H. *Pharmacological, Convulsive and Other Treatments in Psychiatry.* New York: Grune & Stratton, 1969.
24. KATZ, J., PLUNKETT, R., and BRILL, H. "Control of Tuberculosis Among Patients in Mental Institutions: A 10-Year Report," *Psychiatric Quarterly,* 28 (1954): 416–423.
25. KENNEDY, J. F. *Message from the President of the United States.* 88th Cong., 1st sess., Doc. 58, 5 February 1963.
26. KRAMER, M. *Psychiatric Services and the Changing Institutional Scene. 1950–1985.* DHEW Pub. No. (ADM) 77–433, Washington, D.C.: U.S. Government Printing Office, 1977.
27. LEVINE, D. S., and WILLNER, S. G. *The Cost of Mental Illness, 1974.* Mental Health Statistical Note No. 125. DHEW Pub. No. (ADM) 76–158, Washington, D.C.: U.S. Government Printing Office, 1976.
28. NEW YORK STATE DEPARTMENT OF MENTAL HYGIENE. *Annual Report 1955.* Legislative Document. Albany, N.Y., 1956.
29. PARRY-JONES, W. L. L. *The Trade in Lunacy. A Study of Private Madhouses in England in the 18th and 19th Centuries.* Toronto: University of Toronto Press, 1972.
30. PRESIDENT'S COMMISSION ON MENTAL HEALTH. *Report to the President,* vol. 1. Washington, D.C.: U.S. Government Printing Office, 1978.
31. REDICK, R. W. *Patterns in Use of Nursing Homes by the Aged Mentally Ill.* DHEW Pub. No. (ADM) 74–69, Washington, D.C.: U.S. Government Printing Office, 1974.
32. ROBBINS, E. S., et al. "Psychiatry in New York City: Five Systems, All Overwhelmed," *Psychiatric Annals,* (1979): 14–27.
33. SLABY, A. E. "News Item (in Residents' Forum)," *Psychiatric News,* 2 November 1979, p. 14.
34. SNIDER, D. A., PASCARELLI, D., and HOWARD, M. *Survey of the Needs and Problems of Adult Home Residents in New York State.* Albany, N.Y.: Welfare Research, 1979.
35. STEADMAN, H. J., and COCOZZA, J. *Careers of the Criminally Insane.* Lexington, Mass.: Lexington Books, 1974.
36. STEADMAN, H. J., COCOZZA, J. J., and MELICK, M. E. "Explaining the Increased Arrest Rate Among Mental Patients. The Changing Clientele of the State Hospitals," *American Journal of Psychiatry,* 135 (1978): 816–820.
37. STONE, A. A. *Mental Health and the Law. A System in Transition.* New York: Jason Aronson, 1976.
38. TAUBE, C. A., and WITKIN, M. J. *Staff-Patient Ratios in Selected In-Patient Mental Health Facilities: January 1974.* Mental Health Statistical Note No. 129. DHEW Pub. No. (ADM) 76–158, Washington, D.C.: U.S. Government Printing Office, 1976.
39. WELSFORD, E. *The Fool.* New York: Anchor Books, 1961.
40. WITKIN, M. J. *Residential Treatment Centers for Emotionally Disturbed Children. 1975–6.* Mental Health Statistical Note No. (ADM) 77–158. Washington, D.C.: U.S. Government Printing Office, 1977.
41. ———. *Private Psychiatric Hospitals 1974–5.* DHEW Pub. No. (ADM) 77–380. Washington, D.C.: U.S. Government Printing Office, 1977.

CHAPTER 37

COMPUTERS IN PSYCHIATRY

John H. Greist and Marjorie H. Klein

¶ Introduction

In the five years since we wrote *Computers in Psychiatry, Promises to Keep,*[74] cross currents of advances in computing technology and legislation mandating and regulating mental health services, payments and research, and professional attitudes have produced some striking successes in this field, as well as a simultaneous turbulence that makes future directions far from certain. Few clinicians, fewer patients, but perhaps more administrators, interact directly with computers. Most contacts have negative connotations: There is the obligation to complete forms to feed the computer, and the requirement of paying bills. It is thus still appropriate to ask whether computers have an important role in psychiatry beyond the standard fiscal and administrative services that can be adapted from business applications. Often, these fiscal services are purchased from Computer Service Bureaus, which also serve a wide variety of nonmedical business applications. Psychiatric computing remains at some distance from the promised land of integrated and demonstrably useful administrative and *clinical* computing services that we have been traveling toward for the past two decades.

Technologies that succeed quickly and broadly solve critical problems. Often they represent first-time solutions (for example, the Salk polio vaccine) and markedly increase the quality and/or quantity of a service in a cost-effective or cost-efficient manner (for example, banking, credit card, and airline reservation services). For computers to gain a permanent franchise in psychiatry, they must address and solve important problems. They must also enter, or sometimes create, a receptive atmosphere or else they must wait for a hostile atmosphere to change. There are those who feel that many computer applications are unjustified intrusions into the humanistic practice of psychiatry.[130] Computer applications apparently represent a paradigm shift that some professionals are incapable of making.[79]

What are the psychiatric problems for which computer applications have been proposed and tried? What attempts, accomplishments, and failures have appeared to date? What is the short- and long-term prognosis for computer applications?

Each mental health professional (whether clinician, administrator, clerical worker, teacher, or researcher) faces unsolved professional problems. Those who have no problems (no areas in which they can make improvements) are naive or dangerous or both. Solutions to the many mental health problems may or may not involve the use of computers. But as an ethical profession, psychiatry has a responsibility to address and to attempt to solve problems of prevention and treatment, in order to provide better care for patients, which, of course, is the reason that the mental health professions exist.

Where it has been possible to show that computers have solved difficult problems for psychiatry, they have sometimes been accepted and used. Prominent examples are fiscal and administrative applications, which often reduce or eliminate hand collating and tabulating, hasten the processing of accounts receivable and posting of bills, while reducing errors and lowering costs. These programs usually operate at a discrete distance from clinicians.

Mental health researchers, who have been trained in quantitative approaches to hypothesis testing and who are required to manage data somewhat akin to those of mental health administrators, have also embraced computers.

Between these two camps of computer users remain the clinicians and their patents, who are the *raison d'être* for all clinical and supportive staff. Will the clinical interaction in psychiatry remain so much an intuitive art or analogue phenomenon that digital machines (with their needs for dichotomous data) will not (and perhaps should not) be able to intrude? Is the language and logic of clinical psychiatry so idiosyncratic (even well-trained psychiatrists often disagree about diagnosis and management) that we cannot hope that computers can help?

We will briefly review the business (fiscal and administrative) and research applications that have secured a foothold in psychiatry. The remainder of this chapter will deal with clinical applications that have been developed and tested but not yet disseminated. As is the habit of those writing about fast-moving fields, we will conclude with predictions about the years ahead.

Definitions

A few concepts and definitions will assist the reader in understanding computers in general and their specific applications in the mental health field.

Hardware: Computing machinery consists of a central processing unit (CPU), memory, data storage devices (usually disc or tape), cathode ray tube or CRT (television screen and typewriter keyboard), optical scan (reads marks from paper) and printing typewriter terminals, card or keypunch for batch-oriented systems, and any equipment that connects terminals to the computer (acoustic couplers, modems, cables or telephone lines).

Central Processing Unit (CPU): This unit rapidly processes data and instructions read from memory.

Memory: There are two kinds of memory:

1. Fast, or main, memory, which feeds the CPU.
2. Slow, bulk, or mass storage memory, usually disk (similar to a record player—often with a stack of records or "platters") or tape (similar to a tape recorder).

Software: Operating systems that control communications, time sharing, and file storage; programming languages; and actual programs that command the computer to process data.

Computer sizes:

1. mainframes—large computers.
2. minicomputers—mid-size computers.
3. microcomputers—small computers.

All three sizes of machines process data in a similar manner, differing only in the speed with which they proceed and the quantities of data they can store.

Medical Information System: Lindberg[86] has provided a short, noncontroversial definition: "A set of formal arrangements by which facts concerning the health or health care of individual patients are stored and processed in computers."

Mental Health Information Systems (MHISs) are a subtype of the more general Medical Information System.

How Computers Work

While many psychiatrists now have a basic understanding of how computers work, others do not and are sometimes awed by these essentially simple machines. A computer is very much like a traditional psychiatrist's office, complete with filing cabinets (disk or tape), typewriter (terminal), telephone and interoffice communication device (intercom or some form of telecommunications equipment), and psychiatrist and secretary (central processing unit, memory, and programs).

If a psychiatrist completes a consultation and wishes to write a letter to the referring physician, he calls on the intercom (or simply asks directly) for the patient's record. The secretary remembers (with luck) where the record is stored, retrieves it from the file, and gives it to the psychiatrist. The psychiatrist (CPU) reads the record (data input), thinks about the patient, the record, and the consultation report (data processing), and creates a new data set by writing or dictating (consultation report), which can then be typed (data output).

Computers function in an analogous manner. The user commands the computer to find a record that has been put into the storage file and, if a suitable program has been written, tells the computer to process the data and type a report.

Computing Costs

Costs for computing equipment and the energy required to run computers continue to decline—a nearly unique phenomonon in our time. A comparison of one of the earliest computers with today's machines is striking:

> Today's microcomputer, at a cost of perhaps $300, has more computing capacity than the first large electronic computer ENIAC. It is twenty times faster, has a larger memory, is thousands of times more reliable, consumes the power of a light bulb rather than of a locomotive, it occupies 1/30,000 the volume and costs 1/10,000 as much. [p. 65][98]

Major cost factors are processing speed, storage capacity, and reliability of storage devices. In general, faster processing speed and larger storage capacity increase cost. However, development of memory-chip technology has permitted dramatic reductions in fast memory costs at the same time capacity has increased more than tenfold. Little else can be done to improve the reliability of the present slow disk storage devices, which remain vulnerable because they turn at many hundred revolutions per minute and have arms that must move with great speed and precision to read and write data. The advent of laser-etched disk storage systems (still several years from widespread practical use) promises a tenfold reduction in storage costs and improved reliability.[71]

Microcomputers, which are widely advertised as hobby computers, are comparatively inexpensive to purchase but suffer from two substantial limitations. Processing speed of the largest and consequently most expensive microcomputers is about one-half that of standard minicomputers. Microcomputer storage capacity on "floppy" disks is twenty to four hundred times less than that available on minicomputers. Even when the new Winchester-type fixed head disk drives with dramatically increased storage capacity are employed, storage on microcomputers usually remains smaller, and the fixed head configuration introduces a major difficulty in data backup. Standard minicomputer configurations use removable disk packs (costing $150 to $500 each), and data is stored or "dumped" nightly onto blank disk packs for several consecutive days, with additional copies kept at weekly and monthly intervals. Disk packs are reusable after their data is no longer needed for backup recovery from errors. This planned redundancy permits rapid recovery of data that might have been lost or destroyed—infrequent but possible occurrences that require careful backup in critical medical environments. Dumping from disk to disk typically requires twenty minutes to one hour and is usually done after midnight, when usage is typically low.

To achieve identical rapid backup capability with fixed head disks would require five identical disk drives (not just disk packs) costing three to eight thousand dollars each. An alternative is to dump to cartridge or to tape, although this technique is much slower and generally less satisfactory than dumping from disk to disk.

Thus, even the largest microcomputers will not support as many users, process data as rapidly, or store as conveniently as standard minicomputers. For some limited applications where powerful computing is not required (for example, interviewing), or where slow processing speed is not a problem (batch-oriented fiscal programs), microcomputers are now suitable machines for use in mental health settings. However, they cannot support in an efficient manner the broad range of useful and available mental health computing programs. We expect in a few years that microcomputers with laser-etched disk storage systems will provide computing power and convenience comparable to today's minicomputers for costs in the $15,000 to $35,000 range instead of the $45,000 to $105,000 costs for today's minicomputers.

If one has problems the computer can already help solve, it becomes a simple exercise to calculate whether waiting is cost efficient or whether savings recoverable through program use in intervening years would justify beginning now with a minicomputer. Alternatives that could reduce the cost of owning an entire minicomputer include leasing one or more ports of access on a minicomputer so that the minicomputer's power and broad range of all programs would be available. Even more inexpensive would be access to an entire machine with eight to sixty-four ports of access, or the purchase of a microcomputer to run a more limited range of programs at a slower speed, with reduced storage capacity and less convenient storage backup.

Typical monthly charges for a medium-sized community mental health center leasing access to the Multistate Information System (MSIS), which provides admission, patient census and movement, terminations and direct service reporting, average three thousand dollars for computing, five hundred and fifty to twelve hundred dollars for terminal equipment, plus all data-processing personnel and communication costs, which vary depending on the distance from MSIS headquarters in Orangeberg, New York.

A small minicomputer system can be purchased for approximately $1500 per month (for five years) to run software from the Forest Hospital, Des Plains, Illinois. Programs include evaluations of problems (Problem Severity Scale, SCL-90, Mental Status Exam, Minnesota Multiphasic Personality Inventories, treatment planning, documenting progress, and creating an aftercare plan. This package also provides summary reports for administrative purposes from the same data base. Some of these programs are also available for microcomputers.

Human Services Computing, Madison, Wisconsin, markets twenty-three hours per day access to a minicomputer for $950 per month per port, with users paying telephone communications and terminal costs. Programs include a mental health information system customized for each setting that keeps track of patients and providers over a period of time;[33] a single-encounter medical information system widely used in research studies;[30] fiscal packages; scheduling program; patient interviews (symptom change based on SCL-90);[48] mental status; suicide risk prediction;[45] general medical history; health hazard appraisal; sexual functioning;[42] Diagnostic and Statistical Manual of Mental Disorders, 3rd edition (DSM-III) and other clinical consultations;[10] bibliographical retrieval[49] and Great Paper Chase;[8] and word processing. There is no additional charge for use of any programs, and storage of 4 million characters is included in the basic monthly port charge, with additional storage available as required. Typical community mental health centers with up to twenty-five hundred active cases lease one or two ports depending largely on the number of programs they decide to use. In-house systems are also available.

Community Mental Health Systems (CMHC) of Columbus, Ohio; Systems Technology of Atlanta, Georgia; Ravenswood Mental Health Center of Chicago, Illinois; and Psych Systems of Baltimore, Maryland, also market or plan to market computing services for use in mental health settings. The appearance of vendors specializing in mental health applications is a positive sign that computing is beginning to pay off, at least commercially.

Major considerations when acquiring computing systems and services include vendor experience with computing and mental health applications, adaptability of programs to local needs, and kinds and amounts of ongoing support available. Advice obtained from independent consultants is often valuable.

These costs seldom reflect developmental costs which are often supported by research grants. Thus, the MSIS System received approximately $10,000,000 over seven years to help support development of their mental health information system. Initial development costs of nine outpatient management information systems ranged from $230,000 to $10,000,000 with continuing modification costs between $154,000 and $539,000 per annum.[55] Hospital Medical Information System (MIS) development is typically more costly with the National Data Communications/Honeywell System costing $12,000,000[93] and the Technicon MIS development priced at $25,000,000.[131]

The Place of Psychiatric Computing within Medical Computing and Computing in General

The point has been made that mental health problems are sufficiently different from other medical problems that computer programs "must usually be designed (or redesigned) to adequately meet the needs involved."[54] This opinion is not necessarily valid since programming strategies over the last decade have pointed toward "dictionary driven" programs with general features of broad-scale utility that can easily be tailored to specific needs. Individualization is accomplished by constructing a dictionary to contain items relevant to a particular user. The dictionary then "drives" all other programs (that is, entry, editing, search, report, statistics, and so forth) without further intervention by a programmer. Wide-scale use of such driver programs has been made from other medical settings; for example, CONVERSE,[10] Paper Chase,[8] WISAR,[30] and EPIC,[33] which, originally written for psychiatric settings, has also been used by medical departments as varied as ophthalmology, cardiovascular medicine, rehabilitation medicine, obstetrics, gynecology, and clinical cancer. Roberts[103] recently reported on the use of PROMIS* (a language previously employed for work in internal medicine, obstetrics, and gynecology) in performing a computerized diagnostic evaluation of a psychiatric problem. Psychiatry stands to gain appreciably to the extent that it can align itself with the rest of medicine in the development and sharing of general "driver" programs.

Medicine as a whole, however, is a very specialized area with needs and computing problems often quite different from industrial organizations with consistently higher levels of standardization than found in the medical arena. Computing giants such as International Business Machines (IBM)[86] and General Electric (GE)[1] have had difficulties with medical computing. GE's MEDINET (Medical Network) Department was established in 1966, demonstrated its Medical Information System in 1967, and announced "nationwide availability" by early 1969. MEDINET offered hookups to a centralized computer system via a nationwide telecommunications network that General Electric

*This is an acronym for Problem-Oriented Medical Information System.

used for its intracompany and commercial computer business. By late 1967 a decision was reached to emphasize business over clinical applications in an attempt to broaden the market for services. By that time, MEDINET had spent about $16,000,000 on systems development. General Electric merged its computer manufacturing business with Honeywell in 1970, and MEDINET was administratively eliminated in 1975. It appears that General Electric, having had previous experience only with banks and industrial settings with more rational, explicit, and mature management problems, seriously underestimated the complexity of the health-care delivery system. In fact, it sometimes seems inappropriate to view health-care delivery as a "system." Standard problem-solving techniques of the computer industry include a thorough system analysis, where each component of an organization is made explicit to the point that a flow diagram can specify most, if not all, organizational functions. Medicine is far less standardized, as physicians who practice in two or more hospitals quickly realize.

Even the computer languages that work well for other tasks have not been easily applied in medicine. COBOL, the standard language used for most fiscal programs, is designed for numerical manipulations where all inputs, outputs, and intermediate storage capacities must be specified to the final digit and cent. This same specificity is a marked handicap when the variables under consideration have not been completely defined. The Massachusetts Utility Multiprogramming System (MUMPS)[40] was developed in medical settings and has strengths in file handling, time sharing, and speed of programming, while yielding some ground to COBOL in "number crunching." MUMPS is one of only four languages for which national standards have been adopted, and MUMPS is now available in many medical settings.

Thus, medicine, and psychiatry as part of medicine, have been recognized in the computing industry as forming a difficult, even fractious marketplace. Health users comprise less than 3 percent of computer hardware sales.[91] Having encountered unexpected difficulties on its past forays into medical computing, computer companies will probably continue to follow their major markets, where, as Willy Sutton suggested, "the money is." Medicine can and does make use of the same machines that serve very different industries but has been and will probably remain largely on its own in the development of computing languages and applications. This fact has important implications for support of training and research in computer applications in medicine and psychiatry.

¶ Problems Facing Psychiatry for which Computers May Prove Helpful

Psychiatry and the related mental health disciplines of psychology, social work, psychiatric nursing, and psychiatric administration face enormous challenges. Administratively, there is pressure for more complete and timely data to insure accountability and quality of care. Fiscally, there are ever stricter and more complex reporting requirements to qualify for third-party payments. Clinically, many settings must deal with inadequate staffing ratios, and in other settings specialized skills are simply not available. Educationally, there is difficulty in deciding what to learn from the welter of information presented during formal academic years and residency training. Acquired skills must be maintained as well as knowledge, and it bears remembering that education is not completed upon formal training. Researchers are often submerged by data that, while potentially useful, have become largely unmanageable. Thus, across the broad spectrum of mental health professions and practice are problems that must be faced and, in time, solved.

Clearly, many of these problems can and may be solved without computers. While recognizing this possibility, one must also ask what the relative social, professional, and economic costs of using or not using computers might be. Partial answers to this kind of question are available in some areas, and reasonable extensions of these data provide some guidance about the future impacts of computing on psychiatry.

Administrative and fiscal programs are now widespread in mental health settings where several professionals work. More than 70 percent of state departments of mental health reported some use of computers in 1977,[53] and perhaps one-half of all community mental health centers use computing, most commonly through contracts with off-sight service bureaus, which provide fiscal services and sometimes other administrative reports. Several recent studies have indicated a need for additional computing services in fiscal-administrative areas.[54]

These studies have found little clinical computing in any mental health settings and little appreciation of a need for clinical computing. By clinical computing is meant all programs that directly affect patient care or care issues. Examples are: (1) direct computer interviews of patients in order to gather information or to teach, monitor, or treat; (2) teaching/consultation programs for clinicians; and (3) bibliographic retrieval programs.

Part of this reason for the acceptance of computers in the fiscal services can be explained by our growing cultural familiarity with, and acceptance of, computers in the fiscal aspects of our personal lives (banks, charge cards, travel reservations, telephone number change announcements, and so on). The lack of clinical computing in mental health settings is probably attributable to resistance on the part of some mental health clinicians to technological changes. It is interesting to note that some conscientious clinicians, concerned with confidentiality and the impact certain diagnoses might have on their patient's futures, appear to substantially distort the diagnoses they report for outpatients.[107] Payment for these "innocent" diagnoses is permitted for outpatients. Insurance coverage for inpatient treatment is often available for a more limited and generally more serious set of disorders, and the ethics of this "situation" permits more pejorative and, presumably, more accurate diagnoses. One comes to the unsavory conclusion that diagnostic reporting could be shaped by changes in disorders for which payment will be made. Hedlund[54] captures the feeling well with quotes from two other workers:

> The thesis here is that the mental health industry (particularly the clinical services segment) is essentially pre-technological, with "an emphasis on individually provided services, a minimal number of tools, a lack of standardization, and the apprenticeship system (watching and copying craftsmen) as the major way of learning the relevant vocational skills";[82] and that it "construes every technical innovation as a mechanistically insensitive encroachment on medicine's responsibility for the personal and intuitively sensitive provision of care."[105] As Rome also notes, such reverence for the art of clinical practice—sometimes as if it needed no justification other than its vaguely humanistic objective—has traditionally subordinated reliance on impersonal technical aids because it views them as inflexible, intrusive and as tantamount to exercising a lesser degree of professional skill.

Hedlund[54] has also summarized factors described by different workers to represent barriers to diffusion of computer technology in mental health:

> Lack of top-level agency support;
> Lack of adequate funding;
> The difficulties of transferring research projects to operational settings including inadequate involvement of clinical operators and administrators in conceptualizing and designing of mental health information systems;
> The piggy-backing of many clinical applications on statistical reporting systems;
> The "softness" of mental health data;
> The ambiguity of mental health goals and criteria;
> The lack of an overall guiding conceptual framework for Mental Health Information Systems;
> The uncritical acceptance of traditional mental health concepts and record's procedures as models for computer applications;

The lack of a standard clinical language and an inability to gain wide acceptance for standard or highly structured clinical forms;

Complex data collection and distribution systems which have sometimes been unable to ensure either timely or reliable information return;

Duplication of clinical reporting procedures (one manual and the other computerized);

The distrust of mental health clinicians for information that has been obtained or processed "impersonally";

Resistance related to issues of privacy and confidentiality;

The lack of clinical commitment to making computer technology work for mental health needs;

Repeated underestimation of the amount of time it takes initially to translate even relatively simple concepts into computer applications;

The lack of adequate input devices for high-volume entry from a wide variety of clinical settings;

The stereotyped nature of computer generated reports;

The frequent inaccessibility of computer stored data for special needs. [pp. 17–18]

For these reasons and perhaps because of administrators' fears of losing status and income, clinical computing has not been able to penetrate clinical practice to the extent that administrative-fiscal computing has been welcomed by administrators.

What are the justifications for introducing computer procedures into clinical psychiatry? First of all, there is a need to improve clinical services in terms of quantity, quality, cost efficiency, and effectiveness. Most clinicians acknowledge this need, but some will question the advisability of using computers to pursue these goals. Concerns expressed usually include doubts about the ability of computers to perform as well as humans, uncertainty about high costs, anxiety about possible detrimental effects of machines interacting with already disturbed patients and, as some clinicians readily admit, a global fear that computers may be too successful and displace the mental health practitioner from lucrative and satisfying work. Considerable resistance remains even though it was shown years ago in more than fifty psychiatric studies that computers consistently diagnose at least as well as clinicians;[106,111] that even when poor decision models are employed, the computer still performs well;[25] and that costs have been less than two dollars per hour on a forty hour per week basis for direct computer interviewing since 1975. Also, numerous studies have shown computer interviews to be highly acceptable to patients and sometimes preferable to interviews by clinicians.*

¶ Mental Health Information Systems

A goal of most workers in the psychiatric computing field has been the development of a comprehensive mental health information system. Hedlund[54] has defined the mental health information system as:

> A constellation of computer hardware/software and related procedures that are intended to facilitate the collection, processing, storage and/or the display (retrieval) of information relevant to the evaluation and care of mental health patients or clients. This term is used here to refer both to general information systems that attempt to integrate information about mental health care from a number of different sources in order to satisfy a wide variety of administrative and clinical needs for such data, and to special-purpose or stand-alone computer applications that play a more specific, limited role in mental health patient evaluation and care.

Such a system would integrate administrative, clinical, educational, and research functions so that redundancy in data collection would be eliminated and maximum use could be made of all data collected.

Conceptually appealing, this goal has not yet been achieved in psychiatry or in other fields of medicine. The reasons for this failure are many:

1. There is a philosophical disagreement about how to proceed to the final design.

*See references 45, 48, 63, 94, and 116.

Some advocate a total system design that is application-independent as a preliminary necessity,[58] while others favor a sequential or cumulative approach.[6] Intermediate viewpoints, promoting a general but still flexible design that accepts, accommodates, and integrates applications as they are developed and proven, have also been advanced.[97]

Total system designs are typically advocated by commercial organizations, which prefer to see similarities rather than differences between potential customers in hopes that a single design will achieve many sales. This view seems most justifiable in the realm of business applications. Advocates of a cumulative approach tend to come from settings where more innovative administrative and clinical computer applications are developed. As Lindberg[86] points out in his definitive review of medical information systems, "there has been no known case in which a business office system has ever evolved into a [comprehensive] medical information system."

2. Those who have written both fiscal-administrative and clinical computer applications quickly realize that the former are far simpler to prepare and introduce into routine use. Many computer service bureaus already provide fiscal-administrative services to mental health facilities in a cost efficient manner. By contrast, clinical applications are far more difficult to conceptualize (many times the clinical process being addressed is incompletely understood, as in diagnosis, prognosis, dosage, choice, and so forth). Integration with clinical practice patterns, which may vary widely from setting to setting and from clinician to clinician, is also a problem. An application that works well when terminals are readily available may fail completely if clinicians must walk some distance to use a terminal or if the computer running the application is overloaded and slow in responding. Underlying differences in philosophies about the practice of psychiatry can also affect program use (that is, the relative importance of somatic versus psychosocial treatments in depression and schizophrenia). Gradual progress has been made in a number of clinical areas, but a total MHIS is far from realization.

3. Many attempts at developing and integrating computer applications in psychiatry have foundered because of inadequate support that would permit full development and testing of prototypes, careful evaluations of final programs, and translation and transmission of proven products to other systems and settings. Full-scale development of a mental health information system is, at best, a time-consuming and costly endeavor.

4. Progress has also been limited at times by poor choices of hardware-software combinations. As hardware costs have tumbled, the proportion of a total mental health information system budget allocated to software development naturally rises, and some languages are inherently more economical than others. For example, a recent study comparing COBOL with MUMPS programming languages found that MUMPS programs were prepared many times more quickly than those written in COBOL.[72] Some COBOL programs could not be completed at all and others were impossible to modify. Also, COBOL consistently uses more storage than MUMPS. For both of these reasons, clinical applications written in COBOL are likely to be far more costly than those written in MUMPS. In fairness, it must be stated that COBOL fiscal programs, while more costly to write and requiring larger storage capacity than identical MUMPS programs, will run somewhat more quickly. However, fiscal programs (billing or accounts receivable) are usually run in batch mode so that run speed is not a critical factor. Overhead costs for COBOL are prohibitive for clinical applications where initial development and subsequent updating require numerous programming changes. Consequently, MUMPS, BASIC, PASCAL, and other efficient languages are gaining a dominant place in clinical computing.

The Institute of Living, in Hartford, Connecticut, has made considerable progress toward a comprehensive mental health information system. Operating on two PDP-15 minicomputers and using the MUMPS programming language, a large number of clinical and fiscal-administrative applications have been written. Long-term support has been available directly from the Institute of Living. Ongoing costs for this system averaged $2.50 per patient per day.[39]

The University of Wisconsin has moved steadily toward a comprehensive mental health information system by integrating individual programs as they are developed on a minicomputer. Work began with clinical applications in computer interviewing,[43,44,45] and added consultation,[42] bibliographic retrieval,[49] word processing, and an overtime data base.[33] Data collected from patients and clinicians by computer interview or written questionnaires can be stored in EPIC* to provide routine reports used by clinicians, trainee supervisors, clinic administrators, clerical workers, and the hospital billing office. A general search routine permits users to ask and answer a wide range of questions for which specific reports have not been prepared. Even individuals unfamiliar with computer use can perform searches, compose reports, and request statistical tests on groups of data (the statistical test will not be performed unless the data satisfy criteria for the test requested). Data stored in EPIC can be inserted into documents prepared with the word processor, and the word processor can also access references stored in a universal reference file. This resource is being used by residents, faculty, and administrators to answer questions that were previously too time-consuming to permit completion.

What evidence is there that medical information systems will produce a payoff worth the developmental costs and difficult adjustments associated with conversion from the present human systems? By analogy, it is quite clear that many business functions could not proceed in the manner we now take for granted without computer assistance. Banking, credit cards, telephone switching, travel reservations, inventory control, and computer-controlled manufacturing processes are but a few examples.

*This is a data base management system that runs in the MIIS dialect of the MUMPS programming language.

Six evaluations of specific general hospital and ambulatory information systems and two reviews of the field have shown beneficial effects attributable to the installation of medical information systems.*

Improvements typically included increased productivity of staff (up to 200 percent) and greater satisfaction with work.[34] While cost efficiency was increased, total cost did not always decrease. Most studies found high levels of acceptance among nurses, admissions officers, and pharmacy, radiology, and laboratory departments. Physicians in one study were most resistant (only 61 percent voted to retain the system after four years) but increased their approval as time passed (80 percent of the same physicians voted for system retention after five years).[86] Not all studies[12] have found beneficial results, indicating the importance of evaluations of medical information systems and their constituent parts.

A study of the Kaiser-Permanente Medical Information System demonstrated a reduction in patient morbidity when the system was in use.[126] Unfortunately, federal support for the system was withdrawn before additional studies aimed at measuring the system's effect on mortality could be completed.

In psychiatry, the work of Williams and associates[132] has provided some idea of the potential benefits of a Mental Health Information System. Their work was focused on the admissions process where patients were interviewed by computer for up to five hours, and nonphysician staff received prompts from computer terminals as they performed physical and mental status evaluations of the computerized admissions unit.

Several examinations were performed. Expert psychiatrists (faculty members from the Department of Psychiatry at the University of Utah College of Medicine), who had personally examined 195 patients and who were

*See references 7, 21, 35, 52, 86, 87, and 102.

then given admission data collected both by computer interview and routine clinical procedure, found the computer data to be statistically better organized, more complete, more readable, and *more clinically useful* than the routine clinical data. Time and total cost (including staff and computer time) required to complete the admission process were also significantly less with the computer approach. Despite these findings, there remained resistance to these procedures, and attention will undoubtedly have to focus on factors affecting staff acceptance of these programs.[70]

¶ Confidentiality

Concerns about control of access to confidential information stored in computer systems has diminished in recent years, in part because no instances of violation of medical computer systems have been discovered and in part because additional legal and computing safeguards have been established to protect medical records.[83] For example, in order to gain access to data in the EPIC Information System, one must initiate contact through a sequence of passwords, identification codes, specific user names and codes, and a unique code for each EPIC Data Base. None of this information appears on the terminal, and errors force the user to start the ten-second sequence again. Having come this far, the user is told the date and time of his or her last access, so that illicit use of an authorized user's codes could be recognized. Within EPIC, items may be identified as "confidential," and especially sensitive data may be stored in garbled format. Individual users are provided with different levels of access. Thus, most users are permitted to search and generate reports from nonconfidential data while access to confidential data is restricted to those with a legitimate need to know.

While any security system can be breached and any cryptographic code broken, given enough time and resources, the EPIC system has resisted the efforts of a programmer skilled in MUMPS, the language in which it is written. Systems that do not permit telephone access from remote sites are obviously even more secure than those that do, so "in-house" systems have that advantage at the cost of being isolated from legitimate users at, for example, satellite clinics. It is far more difficult to steal information from a well-protected computer file than to obtain a written record either by posing as a clinician or by breaking and entering the building and file cabinet containing the desired record.

It should be noted that computers have promoted access by patients to their computer records under the Privacy Act of 1974,[125] which requires that:

1. Recordkeeping systems must not be secret.
2. Individuals must be able to learn what information his/her record contains and how it is used.
3. Information obtained for one purpose cannot be used for another purpose without the individual's consent.
4. Individuals must be permitted to correct errors in their records and
5. Organizations which create, maintain, use, or disseminate records in which individuals can be identified are responsible for both the reliability and proper use of those records.

¶ Future Hardware and Software

Several trade-offs will need to be evaluated in the years ahead. To perform a reasonably broad range of computing services for mental health agencies or practicioners in an efficient manner, minicomputers are presently required. They are certainly needed for any mental health information system worthy of its title. Microcomputers are increasing in power and perhaps will be able to support moderate-size medical information systems within five years. If computing costs decrease sufficiently, there will even be cost incentives for small agencies to operate their own microcomputers. However, this course is not without problems, including a sizable capital outlay, maintenance and operation of the computer, and acquisition of improvements as they become available.

The most common form of computing today by mental health agencies involves the use of programs operating on a service bureau computer with data entered from punched cards. Most service bureau computers are minis or larger mainframes that provide a greater range of programs and faster processing than is available on microcomputers.

Communication costs (long-distance telephone charges and equipment to connect phone lines to terminals and computers [acoustic couplers/modems]) can account for a sizable part of total computing costs for online operations on a mini or mainframe computer and may account for up to one-half of the total cost if a great distance separates the computer from the service site. Communication networks are already available and promise less expensive communications in the future. The prospect of being able to obtain the best computing services from several different vendors is a reasonable near-term goal for those with the ability to identify and evaluate good programs and who can relate effectively to the vagaries of different hardware and software combinations. In the end, however, integration of all programs into a coherent Mental Health Information System running on a single computer remains the most attractive option. Whether that machine is located "in house" or at a service bureau supporting several agencies will depend largely on the size of the agency. Although not of absolute importance, economies of scale in both computing power and programming effort appear to give the edge to minicomputers and shared services for all but the largest mental health agencies. Large operations can frequently justify the cost of operating their own minicomputer at this time.

¶ Computer Interviewing

Introduction

Much of the data that mental health clinicians collect from patients for administrative and treatment purposes can be obtained directly by computers. The technology is clearly present: The hardware and software for interactive systems are available and have demonstrated the capacity of the computer to engage in dialogue ranging from rather routine multiple-choice question sets to very lively and life-like dialogues using branching and free text capabilities.[41,43] Most researchers who have developed and tested computer interviews note their advantages.

> . . . it [the computer] does not get tired, angry, or bored. It is always willing to listen and to give evidence to having heard. It can work at any time of the day or night, every day of the week, every month of the year. It does not have family problems of its own. It is never sick or hung over. Its performance does not vary from hour to hour or from day to day. It has no facial expression. It does not raise an eyebrow. It is very polite. It has a perfect memory. It need not be morally judgmental. It has no superior or social status. It does not seek money. It can provide the patient with the copy of the interview to study. It does what it is supposed to and no more (and no less). [p. 114][38]

What we find, in surveying the field, is that there are many more computer information systems that are dependent upon data provided by paper-and-pencil forms from patients and other computers than there are systems that take advantage of direct computer input from patients.* Thus, while various reviews of computer technology in mental health include sections describing computer interviews, it is clear that there are few places where this work is being done on more than an experimental basis, even though developers of computer interviews for patients consistently speak of the high levels of patient acceptance and the many advantages of these procedures. Apparently

*See references 23, 38, 54, 66, 85, and 109.

there is still widespread resistance on the part of clinicians and systems developers to the routine use of these methods.

Computer Medical Histories

Much of the early developmental work in direct computer interviewing of patients was done in the context of medical history-taking, an area where patient and physician acceptance has been excellent. The work of Slack and his colleagues,[128] first at the University of Wisconsin and later at Harvard, has done much to demonstrate the attractiveness and feasibility of the method. The first interview, which focused on allergy problems, was found to elicit more complete information than medical charts.[119] When the interview was expanded to a full medical history, patients and physicians continued to react favorably to the experience, producing higher quality information[112,116,117,118] Slack, in later work, has also explored and developed ways for the computer to process nonverbal input, such as heart rate,[113] and to facilitate actual vocal dialogue of patients with the computer program.[114] The computer was programmed to display the questions, respond to the presence and absence of sound (talk), and to encourage the continued flow of talk by responding to silence with prompts displayed on the screen. When interviews with thirty-two male volunteers, who were instructed to talk about a variety of feelings to the computer and to a doctor (in counterbalanced design), were evaluated, it was found that while subjects may have liked talking to the doctor more and spoke more words to the doctor, the content was equivalent in most respects.[114,115] When the quality of feeling-expression was judged for the two forms, it was found that while both methods elicited appropriate levels of feeling-expression, somewhat higher levels were expressed to doctors. The computer, however, was more consistent than doctors. Dialogue with doctors was more subject to experimenters and time-of-day effects.[115]

Other explorations of the potential for medical computer interviewing have tended to confirm the findings of Slack in a number of areas such as obstetrics-gynecology,[99] epilepsy,[14] dietary habits,[120] anesthesia history,[124] headache,[123] gastrointestinal complaints,[13] general medical history,[23,50,110] and health risk.[127] In general, patient acceptance and the accuracy-reliability of the information is high. Simmons and Miller[110] found that for a medical history physicians tend to miss 35 percent of history items collected by a computer. Card[13] and Lucas[88,89,90] investigated factors affecting patient reactions and reported that while patient acceptance is generally high (82 percent), and that 40 percent even prefer the computer to a doctor, acceptance is related to such aspects of the procedure as a visual display, simplicity of keyboard, choice of interframe speed, and so forth.[88,89]

Psychiatric Interviewing

Many of the same models and procedures developed for medical interviewing have been applied with great success to interviewing for psychiatric and psychological purposes. Indeed, it is in sensitive, "personal," and emotionally laden content areas that some of the advantages of computer interviewing (privacy, consistency, individualization) come to the fore. In 1969, Evan and Miller[31] reported that respondents were more open to the computer for content areas described as "highly personal and possibly disturbing" than they were when faced with impersonal and neutral questions. Similar findings have emerged from studies in several sensitive areas. When questioned about drug and alcohol abuse by computer and paper-and-pencil methods, a sample of 132 high school students gave essentially similar reports, but clearly preferred the computer.[47] In another trial of computer questions on a range of socially taboo topics, using medical outpatient volunteers, preferences tended to shift toward the computer (and away from a doctor) as topics became more sensitive.[41] This encouraged further testing that would contrast responses to a more extensive series of questions about sexual functioning versus less threatening questions about work and exercise habits. In this study, respondents were interviewed by both methods in random order. It was found that all respondents, women especially, were significantly more likely to indicate problems with sex to the computer than to a psychiatrist interviewer (even of the same sex), and women reported less embarrassment and greater preference for the computer.[41] Because of this and other experiences in questioning psychiatric patients about sensitive material, experts are developing more computer interviews for direct mental status and diagnostic assessment. While many clinicians and researchers appear to doubt the potential of computer interviews in this area, others continue to feel that the computer's promise has not yet begun to be tapped.

The considerable ambivalence about the reliability and validity of patients as informants concerning psychiatric matters is clearly reflected in the tendency of many workers to interpose a "third party" between the patient and the computer terminal. Programs for observations of psychiatric symptoms, mental status examinations, and psychological assessment at the Institute of Living,[37] and similar programs developed by Sletten,[121,122] Hedlund,[53,54] and others for a network of Missouri Mental Health agencies,[108,121,122] as well as programs in the military,[96] all exemplify this tendency to isolate the patient-respondent. Most of these procedures use trained clinicians, but some use clerical assistants or relatives as respondents.[15,29,56] This means that while various forms of psychiatric assessment by computer are now integral to some of the major mental health information systems, the potential role of patients as informants about their own symptoms is largely unexplored and undeveloped.

Assessment by Computer

Another way that clinicians often achieve a sort of "distance" from the patient in mental health evaluation is to rely on standardized test batteries instead of direct questions. This approach of assessment-by-computer has typically *not* taken advantage of some of the strengths of computer interviews. Many assessment procedures depend on computer processing and interpretation of input collected on paper-and-pencil forms; they avoid direct patient interviews altogether.[37,96] Other programs present test items directly to the patient, but do so in a standardized, lock-step fashion.[76] Only a few systems are now in operation that even begin to exploit some of the strength and advantages of the computer.[36,95]

One of the main benefits of computer testing is the potential power afforded by branching logic so that promising areas are covered and fruitless topics ignored. Kleinmuntz and McLean[75] describe an attempt to develop a branched computer version of the Minnesota Multiphasic Personality Inventory (MMPI) and report somewhat mixed reliabilities when short and full versions for various scale scores are compared. Another experience with a branched version of the SCL-90[27,28] has also found that excellent reliability exists for some scales only.[41] Much more developmental work and experimentation with branching strategies need to be done before the future of automated assessment is settled. What will stimulate growth in this area are the obvious practical advantages to busy clinicians of having test information readily available and easily integrated into data bases.[76] This further development of psychological testing will be vastly enhanced by work now being done with imaginative response formats. Kiss[73] describes a program that explores attitude structures and that makes good use of branching based on extreme values (to stimulus words). The program also records response latencies to gain both a broad and deep profile of attitude structures. Another creative aspect of computers involves Lang's[81] use of his computer "Sam," which has the ability to interact with respondents in the process of scaling of affective constructs; this is done by using the visual display as feedback to confirm the subjects' intended responses.

Psychiatric Admission Unit (PAU) at the Salt Lake City Veterans Administration

The Salt Lake City VA is one place where computer interviews have been quite extensively tested and integrated into the day-to-day activities of a mental health care system. Originally adopted in order to increase intake staff productivity in the face of heavy service demands, the system is centered on the Patient Admitting Unit, where the staff has access to thirteen cathode-ray terminals operating off of a CDC 3200 Computer.[62,63,65,132] Patients, clinical staff, and clerical workers all interact with a set of computer programs to build each patient's information data base at admission. Because the system developers felt that it was important to have an initial assessment of each patient's ability to give valid self-reports, the patient's first contact with the computer consists of a short instructional period followed by a brief questionnaire that functions as a validity screening instrument (the Q1 test).[21,67,68] For the vast majority of patients who pass the screening, there is a brief assessment and history interview carried out by the admitting staff, with results entered directly into the computer terminal along with physical examination findings. On the basis of the screening, the case coordinator selects a battery of psychometric tests and questionnaires appropriate for the patient's particular problem array. Assessments available for direct patient interaction by computer interview include: MMPI; an IQ test, which is a combination of the Shippley-Hartford, or a long-term memory test; the WAIS Arithmetic subtest; the Briggs Social History; and the Beck Depression Inventory. This material is combined with the results of a mental status exam and a detailed problem checklist completed on a computer by the case coordinator after interaction with the patient and psychiatric consultant. This assessment process, which, as a whole, generally takes several hours, yields a narrative report, a DSM-II Psychiatric Diagnosis, and a detailed problem list suitable for a problem-oriented record.[20,64]

Since this system has been in effect, a number of evaluations have been carried out in which the results of the PAU Evaluation System are compared with "traditional" assessment procedures. In one study,[77] 41 PAU reports were judged by an experienced psychiatrist to be superior overall to 37 reports obtained by traditional interview methods. In a second study,[77] where traditional and computer evaluations were directly compared for 35 patients, similar results were obtained. When the reliability of the diagnosis emerging from the PAU system was compared with traditional psychiatric diagnosis, a kappa of 0.56 was obtained, which is equivalent to the agreement of clinicians with one another.[63]

Patient acceptance and satisfaction with the PAU system is extremely high: 89 percent of 132 patients interviewed indicated that they would favor the PAU system while only 8 percent indicated dislike. Forty-six percent of the patients reported that they may have been more candid in the computer interview than they might have been with a clinical intake interviewer.[78] This high level of patient acceptance is due, in part, to the willingness of the system developers to tailor the design to patient need and reactions.[21] The system also has a definite impact on the efficiency and effectiveness of the admission process. In one evaluation, the developers reported less time from assessment to treatment for PAU patients, as well as less staff time devoted to intake and assessment activities.[77,78] Indeed, in a detailed timestudy of the admissions unit, the decrease in the percent of staff time taken up by admissions workups dropped from 13 percent to 3.5 percent. This meant that more staff time was available for treatment. On a cost basis, when assessment costs were directly compared,[63] it was estimated that the PAU Assessment costs half of that of a traditional evaluation. In an overall assessment of the efficiency of the whole unit, the investigators found that since the assessment unit had been in existence, the "annual number of patients treated nearly doubled to about 2,000 while staff size increased only 20% and the cost of delivering services dropped 8% despite inflation."[63] While these changes may not be directly attributable to the PAU system (note that the system was initiated in response to an increased caseload), it is clear that the system may make a great contribution.

Clinical decisions were also notably influenced by the PAU procedure.[78] When twenty-one PAU and eighteen physician-process-intakes were compared, it was found that the same percentage of outpatient versus inpatient referrals were made by both systems, but that independent assessors more often felt that the PAU treatment disposition was the correct decision. In another study, it was also found that the PAU leads to more appropriate medication decisions and higher goal attainment.[63] While PAU patients did not differ with respect to length of stay from traditionally evaluated patients, they did have a lower recidivism rate, presumably reflecting improved treatment.

Despite all of these advantages, clinical acceptance of the PAU system still remains the biggest stumbling block.[63,69] Of ninety-four clinicians interviewed,[63] reactions to the PAU system ranged from neutral to negative. Whether this is a result of clinician "computer anxiety" or of the more basic fear that clinicians may be replaced by computers, it is clear that system developments must concentrate on the problem of clinical acceptance. And clinical acceptance of computerized systems in mental health care can result only from diligent planning. It is important to work on a continuous basis with staff "opinion leaders" about problems pertaining to many details of the system. Because clinician acceptance is basic to the system, it must be alloted continuing attention.

Behavioral Assessment at the Duke University Medical Center

Behavioral assessment is a natural area for computer interviewing: It is difficult for clinicians to do because the guidelines are vague, and because a wide range of problems must be assessed and probed in great detail and specificity, potentially requiring a great deal of clinician time. In their work with a behavioral assessment computer program, Angle and colleagues[2,4,51] have demonstrated the superiority of the computer over the clinician in all these respects and have developed, at the Duke University Medical Center, a model behavioral assessment. Computer interview has two stages: The first, the problem screen, obtains detailed demographic information and descriptions of problem behaviors in twenty-nine life areas. Treatment motivation is also assessed. The number of questions in each area ranges from 21 to 233. Using minimal branching, most respondents get from 60 to 80 percent of the potential questions set.[3] The second stage of the interview goes into greater detail in areas considered to be relevant to treatment, which were selected by the therapist from the report of the initial problem screen. Information collected at this second stage is used both for treatment planning (that is, considerable attention is paid to eliciting information about the situational variables that control the problematic behavior) and for outcome and follow-up assessment.

This system has been used on at least 600 patients and has been extensively evaluated. Patient acceptance is high. In one study[3] of 331 clients, approximately 80 percent found the computer experience positive, despite its length, which ranged from four to eight hours per patient. Indeed, with respect to length, the authors of this study reported that no client refused to take the interview because of length, and only about 15 percent felt that the interview was too long. Another evidence of acceptance is the fact that approximately 80 percent of those interviewed expressed willingness to retake the interview, and about 60 percent reported some subjective preference for the computer over personal interview. With respect to the issue of candor, most clients felt that they would be equally truthful to either a computer or a human interviewer, but about one-third of those interviewed by the computer did state that they had been able to be somewhat more truthful under these conditions. In a direct comparison of human and computer interviews,[2,4] the investigator finds, as is to be expected, that the computer interview is vastly superior in the amount of both comprehensive problem coverage and detail elicited. Thus, in contrast to the computer interview, clinicians covered only 50 percent of relevant problem areas and provided only 6 percent of the detailed information that was found in the computer interview. In another comparison,[2] roughly the same proportion of differences between human and computer interviewers was found, and a computer was successful in identifying 76 percent more of the problems judged to be critical and relevant to the patients by independent clinical interviewers. Even when session notes were reviewed in detail over four sessions, the missing information rate for the clinician-gathered-data dropped only slightly from 75 to 62 percent.

Because of the length of the computer interview and the number of areas in which detail must be obtained, the investigators have attempted to explore the possibility of using branching strategies to reduce the number of questions presented to any one patient. Assessment in two areas—sexual problems and depression—indicate that branching on patient reports of problem frequency and intensity would yield false negatives from about 20 to 25 percent of the patients, although it is not clear whether this 25 percent are individuals who have accepted and adopted their problematic behavior to the point where they no longer experience and report it as such.[2] Presently this group is pursuing a nonbranching strategy in which most patients answer 60 to 80 percent of the total question pool. It is clear that as the question pool increases, greater reliance on branching strategies will become necessary.

As at the Salt Lake City VA, clinician reactions at Duke are much more problematic than patient reactions. When thirty-four clinicians from five treatment programs were surveyed, the clinicians overwhelmingly felt that the computer interview report was superior in comprehensiveness, detail, and extent of problem identification. Nevertheless, they questioned the computers' ability to identify patients' problems better than clinicians. Clinicians were further divided with respect to their assessments of the utility of the interview. In general, hospital psychiatrists were less favorable to the computer interview data than nonpsychiatric community mental health center staff.[2] Clearly, patient acceptance is not a sufficient condition for clinician acceptance, and the possibility remains that certain clinician groups may be particularly threatened by, and slow to accept, this new technology. In influencing clinician acceptance, the investigators found it very important to give clinicians feedback about acceptance, since clinicians are particularly concerned about experiences of dehumanization and patient reactions. Thus, once again, it was found that a computer interview data collection system is more readily accepted by patients than by clinicians.

Computing Interviewing at the University of Wisconsin

The early pioneering work of Slack and his colleagues[16,17,18,19] at the University of Wisconsin in the 1960s, focusing on medical history interviews, operated with very primitive and cumbersome hardware and software. As hardware and software became more sophisticated and complex and applicable to psychiatric interviewing, it became possible to develop an interview system that was concerned with more complex assessments and with measures of changes in patient status over time.[48] Unlike other systems, Greist and colleagues'[48] interview system has perhaps made more use of branching and of free-text possibilities (combined with multiple choice). This was found to be important both because these options make it possible to more closely model the physician-patient interchange, and because these features make it possible to develop repeated and individualized assessments.[41]

One of the first such interviews[43] developed combined individualized target symptom questions with a standardized symptom inventory.[27] The target symptom questions (up to three) encouraged the patient to "describe the most serious problems you are seeking help with today," in free text. Afterward, he was asked to make ratings on scales of frequency and intensity. Also explored, in a very limited way, was the potential of scanning these open-ended responses for key words (for example, reactions to the presence or absence of certain affect words) as a basis for additional specific questions. The SCL-90 follows in branched format, using screening questions and items presented within each cluster according to their factor loading.[41] The interview can be presented repeatedly so that patients can be asked to reevaluate their previous complaints, which are displayed on the screen. The data from this interview were compared with paper-and-pencil and personal interviews and found to be comparable. Patient reactions were mostly favorable; some patients (sixteen) expressed misgivings about the "nonhuman" method, while others felt freed by the privacy and unhurried pace.[43]

Another example of this flexibility is provided by Greist and Klein's[41] social adjustment interview, which relies heavily on branching and free text responses. Branching is particularly useful early where an initial assessment of patient demographic characteristics and patient ratings of role problems lead to appropriately detailed questions about functioning in relevant life areas. It is in these intensive inquiry sections that multiple-choice responses are interspersed with free text, so that detailed descriptions of each patient's life situation and functioning are provided for the clinical summary.

Both the target system and social adjustment interviews have been evaluated in two respects: (1) a comparison of interview summaries with clinical records indicates high reliability in most areas; and (2) a specific investigation of the effect of branching on the completeness of computer-obtained records indicates that branched lines of questionning are appropriate for most, but not all, symptom areas and aspects of life functioning. One exception to this is the finding that severely disturbed patients tended to emphasize their difficulties with leisure time activities and problems relating to their illness, while clinicians placed more stress on family problems.

Other measures currently on this computer assessment system present fairly standardized question sets in a more straightforward manner. Here the advantages of the computer administration have more to do with the availability of the immediate scoring and reports; for example, Benjamin Interpersonal Checklist[9] and suicide risk prediction.[45] The computer administration seems to make patients more comfortable and candid than does a human interviewer; for example, interviews for sexual dysfunction, and alcohol and drug use.

Other interviews (not yet published), which also make use of direct feedback to respondents, have recently been developed for a project promoting the exchange of health (wellness) information among high school students. Programs concerned with depression, nutrition and eating habits, and contraception alternate segments where knowledge or problem levels are tested with segments that appropriately teach or counsel. The enthusiasm with which the high school students responded to these interviews exemplifies the special potential of the computer to reach and engage the attention of hard-to-reach, nonpsychological-minded populations. This is reminiscent of computer-patient experiences with sex and drug abuse data. Other studies confirm these findings with respect to alcoholics,[90] working-class people,[89] blacks,[60] and other "people who are unable to participate in the usual vis-à-vis interview."[57]

¶ Bibliographic Organization and Retrieval

Computers, which are useful in storing, organizing, and retrieving data about patients and providers, may also be used for almost any sort of inventory control purposes. Clinicians and researchers are faced with an enormous and steadily growing volume of clinical and research information relevant to their work.[5] A great deal of this information is contained in journal publications that clinicians and researchers subscribe to and that are stored on bookshelves in their offices. Remembering that a particular subject in question has been discussed in a journal article is difficult enough when most articles are only skimmed upon the journal's arrival. Locating the relevant reference is an even more difficult task. Consequently, a great deal of useful information remains sequestered on bookshelves rather than assisting its owner in solving problems.

One approach to this difficulty is to use the National Library of Medicine's MEDLARS/MEDLINE computer retrieval of over 2,300 biological sciences journals. Searches can be conducted from a list of more than 14,000 different subjects. Limitations of this approach are numerous: (1) a medical record librarian is required to conduct searches; (2) lists of references obtained often number in the hundreds, leaving the recipient with another sizable search and data reduction problem; (3) MEDLARS lists arrive several days after requested; and (4) the referenced articles themselves are often from journals not readily available to the person requesting information.

Personalized library systems organized around alphabetical, numerical, or darning needle identification and retrieval systems, or involving a series of folders into which relevant articles are placed, have been the best systems available to many individual clinicians and researchers. The idiosyncratic nature of these filing systems has made them difficult for even their authors to use consistently, and the limited nature or complete absence of cross-referencing has severely restricted the completeness of retrieval. Another limitation of these personal systems is that they cover only a small subset of the available and relevant literature.

Paper Chase[8] is a program designed for direct use by clinicians and researchers that solves many of the problems presented by MEDLARS/MEDLINE and individual bibliographic systems. References (journal articles, chapters in books, books, and other documents including memoranda, letters, case notes, and so on) are quickly entered at a cathode-ray tube terminal. Written in the MIIS programming language, Paper Chase runs on Data General Eclipse, Digital Equipment, PDP 11 and PDP 15 and IBM Series 1 Computers. Entry time for each reference is approximately two minutes and includes the names of all authors, title of the referenced material, journal or publisher, specific volume, page and year of reference, and the appropriate key, index, or subject words. A modification of the program at the University of Wisconsin permits entry of abstracts or summaries. Alternatively, entire volumes of many different journals may be read into the computer from tapes obtained from the National Library of Medicine.

Once entered, references may be searched by any combination of author, title word, journal, and year and subject, and search times for registries containing several thousand references routinely average less than one minute. If the search results obtained are too general, a second search may be done immediately on the subset of references found in the first search. In this way, specific and readily available references relevant to particular subjects can be quickly found.

One specific application of Paper Chase has been the Lithium Library, the heart of the Lithium Information Center.[49] In the fifteen years after lithium's introduction into clinical psychiatry in 1949, forty-three articles on the biological uses of lithium were published. From 1964 through 1976, an additional 4,000 references appeared. For the three years from 1977 through 1979, 2,500 more publications were added to the literature. The task of keeping abreast of this burgeoning literature for clinical, research, and educational purposes quickly outstripped the best efforts of conscientious clinicians running the lithium clinic and consultation service in one state.[61] References to the lithium literature were entered into Great Paper Chase and the availability of this bibliographic resource was announced in appropriate psychiatric journals. For 1978 and 1979, over five hundred requests per year were answered by the librarian of the Lithium Information Center and more than thirty sites had on-line access to the Lithium Library from computer terminals in their institutions. All reference materials are filed at the Lithium Information Center and can be quickly sent to users who request them.

Work is now underway at the University of Wisconsin to provide synopses of information about the more than 1,400 key word or subject terms used in the Lithium Library. This approach will permit clinicians with questions about specific areas to receive summary information about that area in addition to references to the primary sources upon which the summary is based. Another program under development goes still further by integrating information from the lithium literature into a coherent consultation about a patient the clinician has described. While still in rudimentary form, these programs indicate ways in which bibliographic retrieval can play a more direct role in health-care delivery.

¶ Computer Therapy

Introduction

Simply put, computer therapy involves the use of computers to treat persons with psychiatric problems. Behind this simplistic definition lies a nascent and extremely complex field that seeks to integrate the rapid and continuing progress in computer hardware and software with prior understanding of the art of psychotherapy.

Development of the first computers made it possible to process mathematical symbols at a rapid rate. Programming languages dealing with linguistic symbols soon followed and, with steady refinement, have allowed for easy programming to process languages strings, which can express quite complex meanings. With the advent of on-line computing, immediate computer responses to user inputs became possible. Time-sharing techniques permitted single computers to interact simultaneously with many users, dramatically reducing computing costs. Harnassing the interactive computer medium to psychotherapeutic tasks seemed a natural step, and proponents prophesied widespread availability of expert and inexpensive computer therapies.

Present Status

By 1965, a program that crudely simulated Rogerian psychotherapy had been developed.[129] Colby, who has been a seminal and steadily productive worker in this field, had begun his studies,[16] Slack[119] had conducted medical interviews, which had apparent psychotherapeutic effects, and other workers[80] were beginning to apply computers to studies and treatments of psychophysiological problems. Despite this early promise, there has been neither the widespread interest in, nor extensive development of, computer therapies that many expected.

The several different techniques of proposed computer therapy in psychiatry are based on different patient problems and different conceptualizations of etiology and therapy. The hallmark of these computer therapies is a direct interaction between the patient and the computer, and it is this intimacy with, and anthropomorphization of, a machine that some find so threatening, regardless of any associated benefits.

The reactions of most psychiatric patients to computer interviews that collect past history and present symtom descriptions are strongly positive. Patients in a variety of settings have found the interview experience itself helpful and, in the words of a few, "therapeutic." In these interviews, therapeutic education, reassurance, suggestion, modeling, support, and authorization to express emotion are all possible.[114]

Colby's work has gone far beyond the simple and directly linked question-answer branching of most medical computer interviews. He has developed complex models of human thought with a capacity to evolve in different directions based on continuing patient-computer interaction. A program for autistic children who had no socially useful speech was helpful in initiating speech in thirteen of seventeen patients with whom it was tried.[17] Another Colby program has a computer simulating a paranoid patient so successfully that it becomes impossible for psychiatrists to determine whether they are interacting with a computer or a person.[18] If one can simulate a psychiatric disorder so successfully, it seems plausible that one may be able to prepare a psychotherapeutic computer program to treat psychiatric disorders.

The biofeedback field has blossomed with the availability of small computers, which can convert the patient's physiologic functions into electronic signals, which then guide the patient in modifying those very functions. Although this field has pulled back from its overly optimistic and simplistic beginnings to a more recent and strictly focused position, the on-line computer will clearly play an important role in defining the ultimate applications of biofeedback to medical problems.

Automation of Behavioral Treatment

A number of attempts have been made to automate behavioral treatment programs. Anxiety disorder treatment programs have been prescribed for snake phobia,[80] social and agoraphobia,[11,32] and flight phobia.[26] Automation has also been accomplished for avoidance and aversive treatment programs for pedophilia,[84] homosexuality,[92,104] and personal values.[104] However, these automated treatments seldom use computers directly, depending more commonly on slide or movie projection or audio or videotape presentations. Most studies found comparable benefits between these automated approaches and human therapist treatments.

The Future

One of the major problems in psychotherapeutic practice and research has been to systematically define the treatment techniques so that they may be taught to other therapists and applied in a standardized fashion to patients whose disorders may respond to a particular kind of psychotherapy. By contrast, even in the face of widely variable individual drug metabolism and incomplete compliance, the standardization of psychoactive drug therapies has permitted substantial progress in that field.

Unlike psychotherapy administered by human therapists, which often varies between different therapists and even in a single therapist's treatment of different patients with the same disorder, computer psychotherapy will have the possible advantage of holding constant computer therapy statements and interpretations across a series of similar patients. Since data about the interaction can be immediately stored in computer-processible form, program deficiencies can be quickly identified and the program changed before it is used with additional patients.

Computer technologies to understand human speech are steadily being improved. Computer-controlled speech generation devices with growing vocabularies (such as those used by the telephone company for handling changed and disconnected numbers) are already fairly flexible. Yoking of these two technologies to language-processing computer programs will ultimately provide a capability for humans (who may be patients) to speak directly to computers that speak back to them.

There has been considerable criticism of the use of computers in psychiatry in general and of data collection from patients in particular, yet clearly the use of inhuman devices in medicine is far from inhumane, since technological advances in many fields have brought substantial health benefits to patients. Critics often speak on behalf of an assumed patient constituency without directly consulting it. Whenever patient-computer interactions have been evaluated *by patients,* the reaction has been strongly positive, often to the point of preferring the computer over the doctor as an interviewer.[90]

This seems to be especially true when sensitive subject matter is being discussed, as is often the case in psychotherapy.

Too often there is an immodest overestimation of the benefits of human psychotherapy based on an absence of comparison with other treatments, and occasionally these assessments are based on simple self-interest. Compounding these deficiencies is a large public health problem that confounds, for the most part, present-day techniques.[101] Those who criticize computer therapies are often blind to large, understaffed state hospitals and community mental health centers, which do little but triage patients, often for unavailable or ineffective therapies. The ultimate computer therapy will be, in reality, several different computer treatments that have been shown to be effective for specific patient problems as well as for styles of coping and interaction. The magnitude of the specificity problem has seriously hampered progress in human psychotherapy research, and it seems reasonable to expect that computer psychotherapies, which are totally reliable and capable of systematic modification and unthinking self-scrutiny (through the process of recording, tabulating, and analyzing each class of interactions with each patient), will bring about both an acceleration of our understanding of psychotherapy and the development of more effective human and computer psychotherapies.

¶ Conclusion

Despite disappointments about the difficulties of developing high-quality mental health computer applications, a great deal has been accomplished in the past five years. Computing hardware continues to increase in speed and storage capacity while costs paradoxically decline. Programming languages have become more powerful and more efficient. These technological advances in medicine and psychiatry represent the thin edge of a wedge that can provide the leverage to dramatically improve clinical, administrative, fiscal, research, and clerical aspects of mental health services.

Harnessing this powerful potential to appropriate applications is complicated by the diversity of the mental health field and the complexity of the problems this field deals with. Many systems and problems remain poorly defined and in need of help. Computing has been most helpful where system designs can be clearly specified and where the problem progression presently favors fiscal over administrative over clerical over research over clinical areas. Clinical computing has made a good start in patient computer interviewing, but clinician and patient education and consultation is barely underway. Computer-administrated therapies are now being conceived, and some will be tested in the years just ahead.

The next five years should produce an acceleration in progress in all areas of mental health computer applications. It will become feasible for solo practitioners to use a microcomputer to advantage in many areas of their practice. However, broad-scale acceptance of this advance will probably lag several years behind its availability. As Max Planck sadly but sagely observed, "a new scientific truth does not triumph by convincing its opponents and making them see the light, but rather because its opponents eventually die, and a new generation grows up that is familiar with it."[100]

¶ Bibliography

1. AMERICAN HOSPITAL ASSOCIATION. "MEDINET Announcement." vol. 8, Press release of the American Hospital Association, Chicago, 1967.
2. ANGLE, H. V., ELLINWOOD, E. H., and CARROLL, J. "Computer Interview Problem Assessment of Psychiatric Patients," in Orthner, F. H., ed., *Proceedings: The Second Annual Symposium on Computer Application in Medical Care* Washington,

D.C.: Institute of Electrical and Electronics Engineers, 1978, pp. 137–148.

3. ANGLE, H. V., et al. "Computer-Aided Interviewing in Comprehensive Behavioral Assessment," *Behavior Therapy,* 8 (1977):-747–754.
4. ANGLE, H. V., et al. "Computer Interview Support for Clinicians," *Professional Psychology,* 10 (1979):49–57.
5. BAR-HILLEL, Y. "Is Information Retrieval Approaching a Crisis?" *American Documentation,* 14 (1963):95–97.
6. BARNETT, G. O. "Modular Hospital Information System," *Computers and Biomedical Research,* 4 (1974):243–267.
7. BARRETT, J. P. "Evaluation of the Implementation of a Medical Information System in a General Community Hospital." Final report, Battelle Columbus Laboratories.
8. BECKLEY, R. F., and BLEICH, H. L. "Paper Chase: A Computer-Based Reprint Storage and Retrieval System," *Computers and Biomedical Research.* 10 (1977):423–430.
9. BENJAMIN, L. S. "Structural Analysis of Social Behavior," *Psychology Review,* 81 (1974):392–425.
10. BLOOM, S. M., et. al. "Converse: A Means to Write, Edit, Administer, and Summarize Computer-Based Dialogue," *Computers and Biomedical Research,* 11 (1978):167–175.
11. BRANHAM, L., and KATAHN, M. "Effectiveness of Automated Desensitization with Volunteers and Phobic Patients," *Canadian Journal of Behavioral Science,* 6 (1974):234–245.
12. BROOKS, R. C., et al. *Evaluation of the Air Force Clinical Laboratory Automation System (AFCLAS) at Wright-Patterson USAF Medical Center: Summary* vol. 1. Falls Church, Va.: Analytic Services, 1977, 5:1–44.
13. CARD, W. I. et al. "A comparison of Doctor and Computer Interrogation of Patients," *International Journal of Biomedicine and Computers,* 5 (1974):175–187.
14. CHUN, R. W. M., et al. "Computer Interviewing of Patients with Epilepsy," *Epilepsia,* 17 (1976):371–375.
15. CODDINGTON, R. D., and KING, T. L. "Automated History Taking in Child Psychiatry," *American Journal of Psychiatry,* 129 (1972):276–282.
16. COLBY, K. M. "The Rationale for Computer Based Treatment of Language Difficulties in Non-Speaking Autistic Children," *Y,* 3 (1973):254–260.
17. ———. *Artificial Paranoia: A Computer Simulation of Paranoid Process.* Elmsford, N.Y.: Pergamon Press, 1975.
18. ———. "Computer Psychotherapists," *in* Sidowski, J. B., Johnson, J. H., and Williams, T. A., eds., *Technology in Mental Health Care Delivery Systems.* Norwood, N. J.: Ablex Publishing, 1980, pp. 109–118.
19. COLBY, K. M., WATT, J. P., and GILBERT, J. P. "A Computer Method of Psychotherapy: Preliminary Communication," *Journal of Nervous and Mental Disorders,* 142 (1966):148–152.
20. COLE, E. B., JOHNSON, J. H., and WILLIAMS, T. A. "Design Considerations for an On-Line Computer System for Automated Psychiatric Assessment," *Behavior Research Methods and Instrumentation,* 7 (1975):195–198.
21. ———. "When Psychiatric Patients Interact with On-Line Computer Terminals. Problems and Solutions," *Behavior Research Methods and Instrumentation,* 8 (1976):92–94.
22. COLLEN, M. F., ed. *Hospital Computer Systems.* New York: John Wiley, 1974.
23. COOMBS, G. J., MURRAY, W. R., and KRAHN, D. W. "Automated Medical Histories: Factors Determining Patient Performance," *Computers and Biomedical Research,* 3 (1970):178–181.
24. CRAWFORD, J. L., MORGAN, D. W., and GIANTURCO, D. T., eds. *Progress in Mental Health Information Systems: Computer Applications.* Cambridge, Mass.: Ballinger, 1974.
25. DAWES, R. M. "The Robust Beauty of Improper Linear Models in Decision Making," *American Psychologist,* 34 (1979): 571–582.
26. DENHOLTZ, M. S., and MANN, E. T. "An Automated Audiovisual Treatment of Phobias Administered by Non-Professionals" *Journal of Behavior Therapy and Experimental Psychiatry,* 6 (1975):111–115.
27. DEROGATIS, L. R., and CLEARY, P. A. "Confirmation of the Dimensional Structure of the SCL-90: A Study in Construct Validity," *Journal of Clinical Psychology,* 33 (1977):981–989.
28. DEROGATIS, L. R., et al. "The Hopkins

Symptom Checklist (HSCL): A Self-Report Symptom Inventory," *Behavioral Science,* 19 (1974):1–15.

29. EATON, M. E., et al. "Missouri Automated Psychiatric History for Relatives and Other Informants," *Diseases of the Nervous System,* 31 (1970):198–202.
30. ENTINE, S. M., and FRIEDMAN, R. B. "WISAR: A MUMPS Data Base System which Utilizes Non-Prime Time." *Proceedings of MUMPS Users Group Meeting,* 1978.
31. EVAN, W. M., and MILLER, J. R. "Differential Effects on Response Bias of Computer vs Conventional Administration of a Social Science Questionnaire: An Exploratory Methodological Experiment," *Behavioral Science,* 14 (1969):216–227.
32. EVANS, P. D., and KELLMAN, A. M. P. "Semi-Automated Desensitization: A Controlled Clinical Trial," *Behavior Research Methods,* 11 (1973):641–646.
33. FAULKNER, J. R., et al. "EPIC: Information Management for Mental Health Clinicians, Administrators and Researchers," in Crawford, J. L., Vitale, S., and Robinson J., eds., *Computer Applications in Mental Health.* Cambridge, Mass.: Ballinger, forthcoming.
34. FLAGLE, C. D. "Operations Research with Hospital Computer Systems," in Collen, M. F., ed., *Hospital Computer Systems.* New York: John Wiley, 1974, pp. 418–430.
35. GALL, J. E. Statement of El Camino Hospital, Mountain View, California, before the council on wage and price stability. Mountain View, Calif.: El Camino Hospital, 1976.
36. GEDYE, J. L., and MILLER, E. "The Automation of Psychological Assessment," *International Journal of Man-Machine Studies,* 1 (1969):237–262.
37. GLUECK, B. C. "Computers at the Institute of Living," in Crawford, J. L., Morgan, D. W., and Gianturco, D. T., eds., *Progress in Mental Health Information Systems: Computer Applications.* Cambridge, Mass.: Ballinger, 1974, pp. 303–316.
38. ——— and STROEBEL, C. F. "Computers and Clinical Psychiatry," in Freedman, A.M, Kaplan, H. I., and Sadock, J. B., eds., *Comprehensive Textbook of Psychiatry, vol. II.* Baltimore: Williams & Wilkins, 1975.
39. GLUECK, B. C., ERICKSON, R. P., and STROEBEL, C. F. "The Use of a Psychiatric Patient Record System," *Federation Proceedings,* 33 (1974):2379–2384.
40. GREENES, R. A., et al. "Design and Implementation of a Clinical Data Management System," *Computers and Biomedical Research,* 2 (1969):469–485.
41. GREIST, J. H., and KLEIN, M. H. "Computer Programs for Patients, Clinicians, and Researchers in Psychiatry," in Sidowski, J. B., Johnson, J. H., and Williams, T. A., eds., *Technology in Mental Health Care Delivery Systems* Norwood, N.J.: Ablex Publishing, 1980, pp. 161–182.
42. ———. and ERDMAN, H. P. "Routine On-Line Psychiatric Diagnosis by Computer," *American Journal of Psychiatry,* 133 (1976):1405–1408.
43. GREIST, J. H., KLEIN, M. H., VAN CURA, L. J. "A Computer Interview for Psychiatric Patient Target Symptoms," *Archives of General Psychiatry,* 29 (1973):247–253.
44. GREIST, J. H., VAN CURA, L. J., and KNEPPRETH, N. P. "A Computer Interview for Emergency Room Patients," *Computers and Biomedical Research,* 6 (1973):257–265.
45. GREIST, J. H. et al. "A Computer Interview for Suicide Risk Prediction," *American Journal of Psychiatry,* 130 (1973):1327–1332.
46. GREIST, J. H., et al. "Suicide Risk Prediction: A New Approach", *Life Threatening Behavior,* 4 (1974):212–223.
47. GREIST, J. H., et al. "The Computer Interview as a Medium for Collecting Questionnaire Data on Drug Use: Predicting Adolescent Drug Abuse," in Lettieri, D. J., ed., *Predicting Adolescent Drug Use: A Review of Issues, Methods and Correlates.* Washington, D.C.: U.S. Government Printing Office. 1975, pp. 147–164.
48. GREIST, J. H., et al. "Computer Measures of Patient Progress in Psychotherapy," *Psychiatric Digest,* 38 (1977):23–30.
49. GREIST, J. H., et al. "The Lithium Librarian—An International Index," *Archives of General Psychiatry,* 34 (1977): 456.
50. GROSSMAN, J. H., et al. "Evaluation of Computer-Acquired Patient Histories," *Journal of the American Medical Association,* 215 (1971):1286–1291.
51. HAY, W. M., et al. "Computerized Behavioral Assessment and the Problem-Oriented Record," *International Journal of Mental Health,* 6 (1977):49–63.

52. HEARD, M. R., and THOMAS, J. C. "A Hospital Information System, Its Impact on Costs, Personnel and Patients in one Department," in *Cost Containment, Caps and Consumerism within the Health Care Delivery System,* Proceedings of the Annual Joint Systems Conference 2, Chicago, Ill.: American Hospital Association, 1978.
53. HEDLUND, J. L., and HICKMAN, C. V. "Computers in Mental Health: A National Survey," *Journal of the Mental Health Administration,* 6 (1977):30–52.
54. HEDLUND, J. L., et al. *Mental Health Information Systems: A State of the Art Report.* Health Care Technology Center. Columbia, Mo.: University of Missouri Press, 1979.
55. HENLEY, R. R., and WIEDERHOLD, G. *An Analysis of Automated Ambulatory Record Systems: Findings,* vol. 1. University of California Medical Center Publication, 1975, p. 157.
56. HILF, F. D. "Partially Automated Psychiatric Interviewing—A Research Tool," *Journal of Nervous Mental Disorders.* 155 (1972):410–418.
57. ———. et al. "Machine-Mediated Interviewing," *Journal of Nervous Mental Disorders,* 152 (1971):278–288.
58. HODGE, M. H. "Large Scale Medical Data Systems—The Integrated Approach," *Journées D'Informatique Medicale* (symposium on medical data processing), 3 (1975):3–4.
59. JEFFERSON, J. W., and GREIST, J. H. *Primer of Lithium Therapy.* Baltimore: Williams & Wilkins, 1977.
60. JOHNSON, D. F., and MIHAL W. L. "Performance of Blacks and Whites in Computerized versus Manual Testing Environments," *American Psychologist,* 28 (1973):694–699.
61. JOHNSON, F. N., ed. *Handbook of Lithium Literature.* Lancaster, England: MTP Press, 1980, pp. 433–438.
62. JOHNSON, J. H., and WILLIAMS, T. A. "The Use of On-Line Computer Technology in a Mental Health Admitting System," *American Psychology,* 30 (1975):388–390.
63. ———. "Using On-Line Computer Technology to Improve Service Response and Decision-Making Effectiveness in a Mental Health Admitting System," in Sidowski, J. B., Johnson, J. H., and Williams, T. A., eds., *Technology in Mental Health Care Delivery Systems.* Norwood, N. J.: Ablex Publishing Corp., 1980, pp. 237–252.
64. JOHNSON, J. H., COLE, E. B., and WILLIAMS, T. A. "PROSE: A Simple User-Oriented Program for Computer Constructed Narratives," *Behavior Research Methods and Instrumentation,* 7 (1975):309–310.
65. JOHNSON, J. H., GIANNETTI, R. A., and WILLIAMS, T. A. "Real-Time Psychological Assessment and Evaluation of Psychiatric Patients," *Behavior Research Methods and Instrumentation,* 7 (1975):199–200.
66. ———. "Computers in Mental Health Care Delivery: A Review of the Evolution Toward Interventionally Relevant On-Line Processing," *Behavior Research Methods and Instrumentation,* 8 (1976):83–91.
67. JOHNSON, J. H., KLINGLER, D. E., and WILLIAMS, T. A. "An External Criterion Study of the MMPI Validity Indices," *Journal of Clinical Psychology,* 33 (1977):154–156.
68. JOHNSON, J. H., et al. "Interventional Relevance and Retrofit Programming: Concepts for the Improvement of Clinician Acceptance of Computer Generated Assessment Reports," *Behavior Research Methods and Instrumentation,* 9 (1977): 123–132.
69. JOHNSON, J. H., et al. "Strategies for the Successful Introduction of Computer Technology in a Mental Health Care Setting: The Problem of Change," in Korfhage, R. R., ed., *AFIPS Conference Proceedings: 1977 National Computer Conference,* vol. 46. Montvale, N.J.: AFIPS Press, 1977, pp. 55–58.
70. JOHNSON, J. H., et al. "Organizational Preparedness for Change: Staff Acceptance of an On-Line Computer-Assisted Assessment System," *Behavior Research Methods and Instrumentation,* 10 (1978):186–190.
71. KENNEY G. C. "Special Purpose Applications of the Optical Videodisc System, I.E.E.E.," *Transactions on Consumer Electronics,* 11 (1976):327–337.
72. KIMURA, S., et al. "Frequency Tabulation of Pathological Findings of more than 23,000 Autopsy Cases of Annual Collection through Japan." *Proceedings of MUMPS Users Group Meeting,* 1979.
73. KISS, G. R., "An Adaptive, On-Line Computer Program for the Exploration of Atti-

tude Structures in Psychiatric Patients," *International Journal of Bio-Medical Computing,* 5 (1974): 39–50.

74. KLEIN, M. H., GREIST, J. H., and VAN CURA L. J. "Computers and Psychiatry: Promises to Keep," *Archives of General Psychiatry,* 32 (1975): 837–843.
75. KLEINMUNTZ, B., and MCLEAN, R. S. "Diagnostic Interviewing by Digital Computer," *Behavior Science,* 13 (1968): 75–80.
76. KLETT, C. J., and PUMROY, D. K. "Automated Procedures in Psychological Assessment," in McReynolds, P. ed., *Advances in Psychological Assessment.* Palo Alto, Calif.: Science and Behavior Books, 1971.
77. KLINGLER, D. E., and MILLER, D. A. "Process Evaluation of an On-Line Computer-Assisted Unit for Intake Assessment of Mental Health Patients," *Behavior Research Methods and Instrumentation,* 9 (1977): 110–116.
78. KLINGLER, D. E., JOHNSON, J. H., and WILLIAMS, T. A. "Strategies in the Evaluation of an On-Line Computer-Assisted Unit for Intake Assessment of Mental Health Patients," *Behavior Research Methods and Instrumentation,* 8 (1976): 95–100.
79. KUHN, T. S. *The Structure of Scientific Revolutions.* Chicago, Ill.: University of Chicago Press, 1970.
80. LANG, P. J. "The On-Line Computer in Behavior Therapy Research," *American Psychology,* 24 (1969): 236–239.
81. ———. "Behavioral Treatment and Bio-Behavioral Assessment: Computer Applications," in Sidowski, J. B., Johnson, J. H., and Williams, T. A., eds., *Mental Health Care Delivery Systems.* Norwood, N. J.: Ablex Publishing, 1980, pp. 129–138.
82. LANYON, R. I. "Mental Health Technology," *American Psychologist,* 26 (1971): 1071–1076.
83. LASKA, E. M., and BANK, R. *Safeguarding Psychiatric Privacy: Computer Systems and Their Uses.* New York: John Wiley, 1975.
84. LAWS, D. R., and PAWLOWSKI, A. V. "An Automated Fading Procedure to Alter Sexual Responsiveness in Pedophiles," *Journal of Homosexuality,* 1 (1974): 149–163.
85. LEONARD, J. M. *Computer Assisted Instruction for the Handicapped.* Albany, N.Y.: New York State Education Department, 1970.
86. LINDBERG, D. B. *The Growth of Medical Information Systems in the United States.* Lexington, Mass.: D. C. Heath, 1979.
87. LITTLE, A. D. "Evaluation of Computer-Based Patient Monitoring Systems: Final Report, Appendix D." *A review of the MEDLAB System in Thoracic Surgery Intensive Care Unit at Latter Day Saints Hospital Rockville, Md.* Department of Health, Education and Welfare, 1973.
88. LUCAS, R. W. "A Study of Patients' Attitudes to Computer Interrogation," *International Journal of Man-Machine Studies,* 9 (1977): 69–86.
89. ———, et al. Computer Interrogation of Patients," *British Medical Journal,* 2 (1976): 623–625.
90. LUCAS, R. W., et al. "Psychiatrists and a Computer as Interrogators of Patients with Alcohol-Related Illnesses: A Comparison," *British Journal of Psychiatry,* 131 (1977): 160–167.
91. LUSTED, L. B. Computers in medicine—A Personal Perspective," *Journal of Chronic Diseases,* 19 (1966): 365–372.
92. MACCULLOCH, M. J., BIRTLES, C. J., and FELDMAN, N. P. "Anticipatory Avoidance Learning for the Treatment of Homosexuality: Recent Developments and an Automatic Aversion Therapy System," *Behavior Therapy,* 2 (1971): 151–169.
93. MATTHEWS, C. "Texas Firm Moving into Position to Control U.S. Health Care," *St. Louis Post Dispatch,* Jan. 30, 1977.
94. MAULTSBY, M. C., and SLACK, W. V. "A Computer-Based Psychiatry History System," *Archives of General Psychiatry,* 25 (1971): 570–572.
95. MILLER, E. "A Case for Automated Clinical Testing," *Bulletin of British Psychology and Sociology,* 21 (1968): 75–78.
96. MORGAN, D. W., and FRENKEL, S. I. "The Computer Support in Military Psychiatry (COMPSY)," in Crawford, J. L., Morgan, D. W., and Gianturco, D. T., eds., *Progress in Mental Health Information Systems: Computer Applications.* Cambridge, Mass.: Ballinger, 1974, pp. 253–84.
97. MORRIS, F. "General Requirements for a Medical Information System (MIS)", *Proceedings of a Conference on Medical Information Systems.* Washington, D.C.:

U.S. Department of Health, Education and Welfare, 1970, pp. 28–30.

98. NOYCE R. N. "Microelectronics," *Scientific American*, 9 (1977): 65.

99. PECKHAM, B. M., et al. "Computerized Data Collection in the Management of Uterine Cancer," *Clinical Obstetrics and Gynecology*, 10 (1967): 1003.

100. PLANCK, M. *Scientific Autobiography and Other Papers.* Gaynor, F., trans. New York: Philosophical Library, 1949, pp. 33–34.

101. President's Commission on Mental Health. *Report to the President.* Washington, D.C., U.S. Government Printing Office, 1978.

102. RICHART, R. H. "Evaluation of a hospital computer system," in Collen, M. F., ed., *Hospital Computer Systems.* New York: John Wiley, 1974, pp. 341–417:

103. ROBERTS, B. "A Computerized Diagnostic Evaluation of a Psychiatric Problem," *American Journal of Psychiatry*, 137 (1980): 12–15.

104. ROKEACH, M. "Long-Term Value Change Initiated by Computer Feedback," *Journal of Personality and Social Psychology*, 32 (1973): 467–476.

105. ROME, H. P. "Human Factors and Technical Difficulties in the Application of Computers in Psychiatry," in Kline, N. S., and Laska, E. M., eds., *Computers and Electronic Devices in Psychiatry.* New York: Grune & Stratton, 1968, pp. 37–44.

106. SAWYER, J. "Measurement and Prediction, Clinical and Actuarial," *Psychology Bulletin*, 66 (1966): 178–200.

107. SCHARFSTEIN, S. S., TOWERY, O. B., and MALOWE, I. D. "Accuracy of Diagnostic Information Submitted to an Insurance Company," *American Journal of Psychiatry*, 137 (1980): 70–73.

108. SCHOOLMAN, H. M., and BERNSTEIN, L. M. "Computer Use in Diagnosis, Prognosis, and Therapy," *Science*, 200 (1978): 926–931.

109. SIDOWSKI, J. B., JOHNSON, J. H., and WILLIAMS, T. A. *Technology in Mental Health Care Delivery Systems.* Norwood, N. J.: Ablex Publishing, 1980.

110. SIMMONS, E. M., and MILLER, O. W. "Automated Patient History-Taking," *Hospitals*, 45 (1971): 56–59.

111. SINES, J. O. "Acturial versus Clinical Prediction in Psychopathology," *British Journal of Psychiatry*, 116 (1970): 129–144.

112. SLACK, W. V. "Medical interviewing by computer," *Southern Medical Bulletin*, 57 (1969): 39–44.

113. ——— "Computer-Based Interviewing System Dealing with Nonverbal Behavior as well as Keyboard Responses," *Science*, 171 (1971): 84–87.

114. ———, and SLACK, C. W. "Patient-Computer Dialogue," *New England Journal of Medicine*, 286 (1972): 1304–1309.

115. ———. "Talking to a Computer about Emotional Problems; A Comparative Study," *Psychotherapy: Theory, Research and Practice*, 14 (1977): 156–164.

116. SLACK, W. V., and VAN CURA L. J. Patient Reaction to Computer-Based Medical Interviewing," *Computers and Biomedical Research.* 1 (1968): 527–531.

117. ———. "Computer-Based Patient Interviewing: Part 1," *Postgraduate Medicine*, 43 (1968): 68–74.

118. ———. "Computer-Based Patient Interviewing: Part 2," *Postgraduate Medicine*, 43 (1968): 115–120.

119. SLACK, W. V. et al. "A Computer Based Medical History System," *New England Journal of Medicine*, 274 (1966): 194–198.

120. SLACK, W. V., et al. "Dietary Interviewing by Computer," *Journal of the American Dietary Association*, 69 (1976): 514–517.

121. SLETTEN, I. W., and HEDLUND, J. L. "The Missouri Automated Standard System of Psychiatry: Current Status, Special Problems, and Future Plans," in Crawford, J. L., Morgan, D. W., and Gianturco, D. T., eds., *Progress in Mental Health Information Systems: Computer Applications.* Cambridge, Mass.: Ballinger, 1974, pp. 285–302.

122. SLETTEN, I. W., ERNHART, C. B., and ULETT, G. A. "The Missouri Automated Mental Status Examination: Development, Use and Reliability," *Comprehensive Psychiatry*, 11 (1970): 315–327.

123. STEAD, W. W., et al. "Computer-Assisted Interview of Patients With Functional Headache," *Archives of Internal Medicine*, 129 (1972): 950–955.

124. TOMPKINS, B. M., et al. "A computer-Assisted Preanesthesia Interview: Value of a Computer-Generated Summary of Patient's Historical Information in the

Preanesthesia Visit," *Anesthesia and Analgesia,* 59 (1980): 3–10.

125. United States Senate. Privacy Act of 1974, Public Law 93–579. 93rd Congress, 2nd Session. 1974.

126. VAN BRUNT, E. E., DAVIS, L. S., and COLLEN, M. F. "Kaiser-Permanente Hospital Computer System (San Francisco)," in Collen, M. F., ed., *Hospital Computer Systems.* New York: John Wiley, 1974, pp. 701–753.

127. VAN CURA, L. J. "A Self-Administered Health Hazard Appraisal," in Orthner, F. H. ed., *Proceedings: The Second Annual Symposium on Computer Application in Medical Care.* Piscataway, N.J.: Institute of Electrical and Electronics Engineers. 1978, pp. 147–164.

128. ———, SLACK, W. V., and FREY, S. R. "Elements of a Computer Medical Interview System," *Biomedical Sciences and Instrumentation,* 8 (1971): 33–42.

129. WEIZENBAUM, J. "ELIZA—A Computer Program for the Study of Natural Language Communication Between Man and Machine," *Communications of the Association for Computer Machinery,* 9 (1966): 36–45.

130. ———. *Computer Power and Human Reason: From Judgement to Calculation,* San Francisco: Freeman Press, 1976.

131. WHITEHEAD, E. C. "Technology Impact on Health Care Costs," Statement to United States House of Representatives Committee on Science and Technology, September 27, 1978.

132. WILLIAMS, T. A., JOHNSON, J. H., and BLISS, E. L. "A Computer-Assisted Psychiatric Assessment Unit," *American Journal of Psychiatry,* 132 (1975): 1074–1076.

CHAPTER 38

RECENT DEVELOPMENTS IN LAW AND PSYCHIATRY*

Alan A. Stone

¶ The Mentally Ill and the Civil Rights Movement

HISTORIANS who attempt to chronicle American life in the sixties will have to sort out the impact of the civil rights movement, not only as it affected the status of racial minorities, but also as it set patterns for organized advocacy on behalf of groups, such as the mentally ill, whose status seemed to bear no apparent relationship to these minorities. The civil rights movement, however, had at least three dimensions corresponding to aspects of "mental patients' liberation." The first was redress through the courts using constitutional litigation. The second was an ideological program that emphasized the dangers of paternalism and social stereotypes. Third was the development of self-help groups with a polemical orientation against the status quo. Only the first of these manifestations will be fully addressed here, but, as will become apparent, the ideological considerations are in some sense the "deep structure" of the constitutional litigation.

Central to the "deep structure" is the attack on paternalism. A political and philosophical rationale for this attack was formulated a century ago by John Stuart Mill in his famous essay, *On Liberty*:[12]

> the only purpose for which power can be rightfully exercised over any member of a civilized community against his will is to prevent harm to others. His own good, either physical or moral is not a sufficient warrant. He cannot rightfully be compelled to do or forbear because it will be better for him to do so, because it will make him happier, because in the opinion of others to do so would be wise, or even right. These are reasons for remonstrating with him, or reasoning with him, or persuading him, or entreating him, but not for compelling him, or visiting him with any evil in case he do otherwise. [p. 197]

*I would like to acknowledge the assistance of Sue Stone, J. D., who provided valuable help in the legal research for this chapter.

If one reads these sentences out of context, as is usually done, it is possible to conclude that Mill was opposed to every instance in which a citizen is coerced by law for his or her own good. That, of course, has been the traditional *parens patriae* (the state as parent) rationale invoked at the intersection between psychiatry and law. However, in the very next paragraph of the essay Mill writes, "It is perhaps hardly necessary to say that this doctrine is meant to apply only to human beings in the maturity of their faculties."[12] Furthermore, a close reading of the entire essay will make it clear that the phrase "civilized community" was meant by Mill to exclude most of the nonwestern world. However, critics of modern psychiatry, if they acknowledge Mill's exception at all, maintain that in order to exclude the insane, they must first be identified, and this they claim psychiatry is incompetent to do.

During the 1960s critics of psychiatry made their voices heard. Their most extravagant thesis was that no one is mentally ill, that madness does not exist, that reality is a matter of personal choice, and that insanity is a political invention. The apotheosis of these arguments in different versions is found in the writings of T. Szasz,[26] R. D. Laing,[8] and M. Foucault,[5] writers who had enormous impact on the marketplace of ideas toward the end of the sixties. Aspects of these extreme arguments were taken up by a growing segment of the behavioral science community and made part of "deviance theory," "the existential approach," and so forth. Many who did not question the validity of such basic diagnostic categories as schizophrenia and affective disorder nonetheless raised questions about the objectivity and reliability of psychiatric diagnoses of these conditions. Whatever the underlying arguments might be, the essential thesis was that psychiatrists are incompetent to determine who should be declared not in "the maturity of their faculties."

A second thesis follows from the first: If madness does not exist or cannot be reliably identified, then psychiatric treatment is always or almost always either brainwashing or brain damaging.

These theses found intense support among many young lawyers and civil libertarians. The civil rights of the mentally ill and the mentally retarded were seen by them as the last battlefield of the great war for civil rights. Civil libertarians, including the American Civil Liberties Union, began to challenge every aspect of the legal status of the mentally ill, and the interface between law and psychiatry took on a new political and constitutional dimension as the sixties came to an end.

¶ Applying the Precedents of Civil Rights Litigation

The Supreme Court, under Chief Justice Earl Warren, had been perhaps the most powerful liberal/progressive force in America during the fifties and early sixties. The court fashioned a variety of new constitutional rights whose impact is still being felt to this day. Many of these new constitutional rights were intended to protect the citizen against state and local government, against police brutality, against racism, and against discrimination. For example, alleged criminals who were indigent were given lawyers at government expense, and new constitutional due process safeguards were set up to protect alleged criminals against "stop and frisk," coerced confessions, and other potentially brutal intrusions by law enforcement officers.

The Bill of Rights was the mainstay of legal reform in this area. But transcending the various narrow constitutional arguments in each instance was the basic principle that loss of liberty is the most grievous penalty in a democratic society. It is historically important to recognize that much of the new criminal law handed down by the Warren Court had, in fact, racial significance. A disproportionate percentage of those charged with crimes are members of America's racial minority groups. Thus, more procedural safeguards for alleged criminals meant more protection for minorities against racially biased law enforcement. These procedural reforms of the criminal justice system can properly be considered part of the civil rights movement. Similar considerations continue to play a part in the efforts to abolish capital punishment, since it is thought that the imposition of the death penalty involves racial inequities.

Running parallel to these reforms in criminal law were many important explicit civil rights cases based on the Bill of Rights' guarantee of equal protection under the law, which stipulated that no citizen should be treated differently because of membership in a group whose members were determined on some suspect discriminatory basis. Most of these important constitutional decisions were in place by the middle of the sixties. Using these due process and equal protection arguments and the precedents that had been set by the Warren Court, the constitutional litigation on behalf of the mentally ill was packaged as part of the civil rights movement. But outstanding questions remain. Do the problems of the mentally ill really fit within that package? Do the procedural safeguards developed for the alleged criminal work when they are applied to the alleged patient? Is mental illness a suspect classification in the same sense that race is?

¶ The Supreme Court's Invitation in Jackson v. Indiana

In 1972, the United States Supreme Court made its most significant ruling regarding cases of mentally disabled individuals who faced criminal charges. In *Jackson* v. *Indiana,*[6] the Court was faced with a troubling fact situation involving the potential indefinite confinement of a deaf mute who had the intelligence of a child. Jackson had been criminally charged with the crime of handbag snatching, but had been found incompetent to stand trial and, therefore, could not under law be tried, sentenced, and processed as a criminal. This meant that he would have been indefinitely and involuntarily confined in a mental health facility. Indiana had no capacity to provide Jackson with the training in sign language that might have made him competent to stand trial. The Supreme Court rejected such an indeterminate confinement, stressing that, competent to stand trial or not, there must be some reasonable relationship between the purpose of the disabled individual's confinement and the length of that confinement. In deciding this case, the court used the occasion to comment on the legal situation of the mentally ill:

> The states have traditionally exercised broad power to commit persons found to be mentally ill. The substantive limitations on the exercise of this power and the procedures for invoking it vary drastically among the states.

Then after briefly describing the variations, the Court commented:

> Considering the number of persons affected, it is perhaps remarkable that the substantive constitutional limitations on this power have not been more frequently litigated. [p. 738][6]

Many read these words as suggesting that the Supreme Court was ready and perhaps eager to examine the constitutional implications of confining the mentally ill. It seemed the Burger Court was ready during the seventies to consider the whole panoply of civil rights arguments advanced by the critics of psychiatry and to extend the Warren Court precedents to the mentally ill.

¶ Patient's Rights and Patient's Needs

By the end of the seventies it was clear that the Burger Court was either unwilling or unready to follow the lead suggested by the Jackson decision, at least not without an opportunity to explore and define its own positions on the troublesome issues raised. On the other hand, the lower federal courts were aggressively going forward on the issues. Under the rubric of due process, the "alleged" mentally ill were given lawyers, and the medical model of civil commitment was repudiated. The psychiatrist was defined as the agent of the state and therefore as an adversary of the "alleged patient."

Mental health litigation can be divided roughly into two categories. The first emphasizes the civil rights of the patient and ignores any conflict that the exercise of these rights might have with the patient's need for treatment. A major emphasis of this litigation is to ignore potential benefits and to construe the relationship between psychiatrist and patient as analogous to policeman and criminal. Included in this category is litigation seeking to bar all medical reasons for involuntary confinement in favor of such supposedly objective legal criteria as "dangerousness." Recent litigation also seeks to give those patients who are legally confined a right to refuse treatment, particularly drugs and electroconvulsive therapy.

A second kind of mental health litigation has sought to improve the quality of care and treatment provided the mentally ill. This is the so-called "right to treatment," under which federal courts have attempted massive reforms of institutions and, in some cases, of whole state mental health systems. As one reads the judicial opinions in these cases, it is clear that the federal courts feel that state and local government have failed in their basic responsibility and that decades of legislative inertia have produced a harvest of human tragedy. Many psychiatrists would agree with judicial activists who feel their intervention is justified whatever constitutional theory is employed in the particular case. During the past two decades these increasingly progressive lower federal courts have gradually assumed a crucial new role in our society as they seek to remedy legislative abdication of responsibility. The traditional role of judges in our legal system was to resolve disputes between two parties in favor of one or the other. But the contemporary federal judge involved, for example, in a constitutional right to treatment case now takes responsibility for working out complicated solutions over long periods of time. The federal judge no longer simply lays down the law but—just as in school desegregation cases—will take it upon himself to establish an "ongoing regime" that will constantly regulate the future interactions of the parties to the suit and subject the parties to continuing judicial oversight. The judge typically becomes the de facto superintendent of the mental institution and in some cases the de facto commissioner of mental health.

¶ The Due Process Model of Civil Commitment

Due process arguments have been invoked in all mental health litigation: they have been used to attack the psychiatrist's role in civil commitment and to reject every element of the psychiatrist's discretionary authority over patients. Civil commitment in the past typically in practice gave great weight to the psychiatrist's expert opinion, and the statutes authorizing it leaned toward the medical model. Mental illness and need for treatment was an acceptable justification for civil commitment in many states.

In *Lessard* v. *Schmidt,*[9] the benchmark case, a special three-judge district court dealt with a Wisconsin civil commitment statute that embraced this medical approach and allowed involuntary confinement if the person was "mentally ill" and "a proper subject for custody and treatment." The court held that in order for this statute to be constitutional, it must be interpreted as requiring that the state bear the burden of proving "that there is an extreme likelihood that if the person is not confined he will do immediate harm to himself or others." Proof of such dangerousness must be "based upon a finding of a recent overt act, attempt or threat." Furthermore, where the state attempts to confine on the basis of these criteria, due process requires the following: (1) timely notice of the charges justifying confinement; (2) notice of right to a jury trial; (3) an initial hearing on probable cause for detention beyond two weeks; (4) a full hearing on the necessity of detention beyond two weeks; (5) aid of counsel; (6) the Fifth Amendment protection against self-incrimination; (7) proof of mental illness and dangerousness "beyond a reasonable doubt"; (8) An inquiry into less drastic alternatives before commitment for inpatient care; and (9) no treatment until the alleged patient has had a probable-cause hearing. This decision, citing the most radical critics of psychiatry as authority for the inadequacies of psychiatric diagnosis and treatment, in effect rejected the medical model and imposed all of the due process safeguards of a criminal trial on civil commitment. There is now considerable evidence that the *Lessard* approach not only creates chaos in psychiatric hospitals, which must hold patients without treatment, but also has led to more violence by psychiatric patients.[23] Thus, it fails to protect society and leads to needless suffering by psychotic patients in the name of liberty.

The federal district court decision in *Lessard* took place in 1972, the same year the Surpeme Court decided *Jackson* v. *Indiana.* It became a model for other federal courts reaching a similar result, many of which attacked the credibility of psychiatric expertise as the basis for any legal decision.

For example, in reviewing *Lessard*-type decisions (and concurring with them), one federal judge noted that along with his concern about personal freedom "a close second consideration has been that the diagnosis of mental illness leaves too much to subjective choices by less than neutral individuals."[25]

The consistent thrust of constitutional reform in the federal court is to reject the medical model, replace psychiatric opinion by "objective" evidence of dangerous behavior, and attack the *parens patriae* justification for confinement. This has been the approach adumbrated by the American Civil Liberties Union (ACLU). It adopted as a matter of constitutional right a due process model that gives short shrift to the needs of psychotic patients, the vast majority of whom are not dangerous. It is worth noting that *Lessard,* the benchmark case, was appealed to the Supreme Court, which had the chance to face these issues squarely. Instead, the court returned the case to the lower court on a technicality. The lower court met that technicality and it was appealed again. Again, the Supreme Court returned it on a technical issue. Again, the lower court stood its ground; and there it remains.

The Supreme Court's Contribution to Due Process

The Supreme Court stood silent on all these crucial issues until 1979 when it accepted the case of *Addington* v. *Texas,*[1] which asked the Supreme Court to decide only one very narrow issue: not what is an acceptable justification for civil commitment, not what procedural safeguards are required, but only what is the standard of proof —beyond a reasonable doubt, clear and convincing evidence, or preponderance of the evidence. To decide what the standard of proof is without deciding what is to be proved is an extraordinary exercise. The court had clearly been avoiding the more difficult questions it had itself raised in *Jackson,* and in *Addington* it dealt with the narrowest issue possible. The court opted for the intermediate standard: clear and convincing proof.

If there was any rationale to the court's decision in *Addington,* it was perhaps the classical medical nostrum—do no harm. Many states had already adopted the "beyond a reasonable doubt standard," and the Supreme Court's decision did not require them to reduce that standard. Almost no state had been using preponderance of the evidence, the lowest standard, and thus the Supreme Court's decision had no impact other than to signal its own caution in this area of constitutional reform. But, as already noted, the lower courts were well on their way to imposing all of the procedural due process of a criminal trial. By ignoring the issues in *Lessard* and by dealing with the narrowest issue possible in *Addington,* the Supreme Court allowed these developments to continue.

The Due Process Rights of Children

Adults can, of course, be voluntary or involuntary patients. Children, if they meet the state's statutory criteria, may be involuntary patients—but how can a child who does not have the legal capacity to consent enter the hospital as a voluntary patient? The traditional answer was that the parents or someone standing *in loco parentis* could consent for the child. The special status of parents to make decisions for their children has always been recognized in law and in constitutional interpretations of law, but where a social worker or some agent of the State Human Services' bureaucracy makes such decisions his authority is, at least in principle, more dubious. Certainly in the latter instance there is little reason to assume that the child's best interests are given full consideration. Often the state institution for children is used as a dumping ground for unwanted or troublesome children, and many times there is no prospect of adequate treatment. The plaintiffs in *Bartley* v. *Kremens*[2] challenged the constitutionality of both kinds of substituted consent, parents and those *in loco parentis.*

In the *Bartley* case, a federal court declared that the voluntary provisions for children of the Pennsylvania Mental Health and Mental Retardation Act of 1966 were unconstitutional under the due process clause of the Fourteenth Amendment. The court enjoined enforcement of the following sections of the state statutory scheme until the state legislature provided juveniles the *Lessard* type of procedural due process safeguards upon entering the hospital:

1. A probable cause hearing within seventy-two hours from the date of initial detention.
2. A full-blown postcommitment hearing within two weeks of the initial detention if probable cause was in fact found at the first hearing.
3. Juveniles were to have their own individual attorneys, and these attorneys as well as the juveniles were to receive notice forty-eight hours prior to the original probable cause hearing. If a juvenile was indigent, the state had to provide an attorney. Juveniles and their attorneys had to be provided with written reasons for the juvenile's initial admission.
4. Juveniles had the right to be present at the hearings with the caveat that the juvenile's attorney could waive this due process procedural right on behalf of the juveniles.
5. Juveniles had the right to a finding by clear and convincing proof of their need of institutionalization.
6. Juveniles had the criminal law rights of confrontation of witnesses, cross-examination, and the right to offer evidence in their own behalf.

The lower federal court in *Bartley* acknowledged the established tradition of parental authority in the rearing of children. But it emphasized that the decision to admit children to mental health facilities involved serious conflicts of interest between parent and child. The court noted that parents could not control a juvenile's right to abortion as a matter of constitutional law,[16] and it ruled that parental admission of a juvenile to a mental health or mental retardation facility should be considered a similar exception to the parental control rule. It further justified this exception to parental authority on the basis of the stigma involved, the unreliability of psychiatric diagnosis, and the loss of liberty. The sweep of the court's perspective on conflict of interests is demonstrated by the fact that it cited the respite program, which allows parents who keep retarded children at home the opportunity to place the child in a facility for a brief period, as an example of a conflict of interests that should entitle the child to a lawyer and a hearing to contest the respite program. Obviously a court willing to push that far had no trouble dismissing the authority of those acting *in loco parentis.*

The United States Supreme Court did not decide the *Bartley* case the first time it was appealed. Instead, as with *Lessard,* it avoided it on a technicality and remanded the case to the lower court.

However, on remand, the federal court did not budge. Instead, it reinstated its previous holding that plaintiffs had a liberty interest in not being institutionalized without due process of law and that interest could not be constitutionally waived by parents or guardians.

Finally, in 1979, the Supreme Court decided the *Bartley* case[21] on the merits, consolidating that appeal with a similar appeal from Parham, Georgia,[15] where a federal court had declared that Georgia's procedures for juvenile admission at the request of parents or state violated due process. The Supreme Court, although it acknowledged a due process problem, held that due process required only that a staff physician, acting as a neutral fact finder, evaluate the admission. Its description of that evaluation is not much different from what is generally accepted as good psychiatric practice. The decision maker, in addition, must have the authority to refuse to admit any child who has not satisfied medical standards for admission. After admission, a child's commitment must be reviewed periodically by an independent procedure similar to that required for initial admission; for example, a case conference review.

The Supreme Court ruled that the current Georgia and Pennsylvania statutory and administrative procedures already comported with these minimum due process requirements. The Supreme Court also refused to make any distinction between parents and those acting *in loco parentis.* Thus, the Supreme Court rejected the complex due process approach of *Lessard.* Clearly, there now exists a radical dysjunction between the Supreme Court's perspective on mental health law and that of the lower federal courts. Until some resolution of these differences is achieved, litigation will doubtless continue. Furthermore, there has been a remarkable proliferation of legislation seeking to obtain the same goals reformers have sought in the courts.

¶ The Right to Refuse Treatment

The most important current development in mental health litigation placing rights ahead of needs is the right to refuse treatment. Proponents of this right claim that it should operate even after a patient has been involuntarily confined as mentally ill and dangerous. The best example for their argument runs as follows: Let us assume that an involuntary psychiatric patient is a sincere Christian Scientist and offers that as the basis of a refusal to accept drug treatment for his schizophrenia. The federal courts in New York in *Winters* v. *Miller*[29] concluded that to impose somatic treatment over the valid religious objections of a patient is a violation of the patient's constitutional rights.

The religious issue, however, is rarely the real question in the right to refuse treatment. That kind of case simply allows us to begin to reflect on the constitutional considerations, for example, religious convictions, that judges have in mind when they examine what might justify a constitutional right of a mental patient to refuse treatment.

More radical civil libertarian arguments go much further than the clash between religious tenets and good medical practice. Such arguments urge that involuntary treatment be considered a violation of the First Amendment. Since the mind is the source of mentation and since freedom of speech originates in mentation, to influence anyone's mind against his or her will is a volation of that person's First Amendment rights. This formulation of a First Amendment right to refuse treatment was adopted by a federal court in Massachusetts in *Rogers* v. *Okin,*[19] now being appealed.

The right to refuse treatment has also been premised on the Eighth Amendment protection against cruel and unusual punishment[7] and on the constitutional right to privacy.[18] Even without rolling up the heavy artillery of constitutional argument, there has traditionally been at common law a right to refuse treatment. The crucial problem is not whether the right exists, but under what circumstances can it be waived. The state cannot send doctors into the streets to inject citizens with neuroleptic drugs against their will. But does a legally valid admission to a mental hospital, be it voluntary or involuntary, constitute a waiver of the right to refuse treatment? Advocates of the right to refuse claim that voluntary patients do not waive any such right, and they further claim that involuntary confinement, in effect, demonstrates only that the patient is committable and not that he lacks the capacity to refuse treatment. The court in *Rogers* v. *Okin* agreed with this extreme argument, ruling that involuntary confinement is not enough; the state must, except in an emergency defined narrowly by the courts, prove incompetency in a guardianship hearing, a lengthy proceeding that may take weeks to complete. The *Rogers* court has held that every person is presumptively competent to exercise individual autonomy even after being found mentally ill and dangerous. The court insisted that the burden to overcome that presumption should be on the physician or the state. There was substantial evidence in *Rogers,* ignored by the judge, that the limits set on treatment by the right to refuse had led to assaults, arson, sexual molestation, and other abuses of patients by patients, as well as self-destructive activity.[10] Other judges have suggested less cumbersome and more practical alternatives. For example, in *Rennie* v. *Klein*[18] the court decided that the objecting patient is entitled to an independent second opinion about the proposed treatment by a psychiatrist who is asked to keep in mind the balance between needs and rights.

Individual autonomy is an important value in a democratic society, and it is important to consider what values are served by allowing psychiatrists to override a patient's refusal of treatment. The most obvious value is the relief of needless suffering. There are patients, particularly schizophrenics, whose suffering can be relieved but whose mentation is so disturbed that they cannot choose to accept the treatment that will help them.

Second, although it is usually ignored in the libertarian calculus, the suffering and the behavioral manifestations of the mentally ill do have a deleterious effect on those around them, even when no physical injury occurs. The courts have long been willing to acknowledge the reality of psychic trauma. When a grossly disturbed mentally ill patient is admitted to a hospital, his or her disturbance has an impact on the other patients and their treatment. Allowing a patient to go on traumatizing other patients needlessly as a result of a refusal to accept treatment may make it impossible to treat others.

Finally, the value of freedom of mentation, freedom of choice, and privacy are not sacrificed when we impose reasonable treatment on a person who is unable to exercize his or her autonomy.

Some judges, although not unaware of these considerations, give different weight to them. For example, in *Rogers* the judge gave no weight at all to the effect of psychic trauma on other patients and the staff. He would allow emergency involuntary treatment only where there is an immediate risk of physical injury. He took that position knowing that in a mental hospital such a standard cannot be reliably applied and that considerable physical injury may in fact result. Thus, this judge in effect permits the opportunity for both physical and psychic trauma in the effort to protect the right to refuse treatment. This places such extraordinary burdens in the way of nonemergency involuntary treatment that it almost precludes all sensible clinical intervention.

Although the *Rogers* decision nowhere discusses the issues set out earlier in this chapter (for example, that mental illness is a myth and that treatment is brain washing or damaging), it is clear that the judge assumed that psychiatrists lack the ability and/or cannot be trusted with the responsibility of identifying those who are incompetent to refuse treatment. Hence, he is, in effect, condoning an elaborate legal procedure to protect patients from psychiatrists. The enthusiastic reception of decisions like this by the media must say something about the widespread perception that patients' rights are more important than their needs. Surely it also suggests a growing popular impression that psychopharmacologic treatment is coercive, punitive, abusive, and potentially more dangerous than mental illness.

¶ The Right to Treatment

Ironically, the radical criticism of psychiatry began during the 1960s when the scientific foundations of psychiatry had achieved a new respectability. New biological and psychological treatment methods had reduced the populations of state mental institutions by 50 percent, and the average length of stay of patients had been drastically shortened. All this was well underway long before mental health litigation began.

But reform initiated by the mental health profession depends on both competent leadership and substantial state support. During the sixties some states lagged behind in initiating these reforms in the care of the mentally ill, and it was in one of those states that the first right to treatment suit was brought. Alabama was last among the states in providing funds to care for mental patients. Its large state institutions were by all reports grossly inadequate by any standard of evaluation. Legal reformers, supported by many of the associations that represent the mental health professions, sought to force the state of Alabama to improve conditions in its state hospitals and retardation facilities through constitutional litigation. They brought a class action suit, *Wyatt* v. *Stickney,*[32] claiming a constitutional right to treatment on behalf of all patients in the state mental institutions in Alabama. This effort was almost simultaneous with *Lessard,* but *Lessard* focused on rights while *Wyatt* focused on needs of patients. The legal theory behind *Lessard* emphasized procedural due process whereas in *Wyatt* it was substantive due process. This difference between substantive and procedural due process is a crucial distinction in constitutional law.

If one reviews the few right-to-treatment cases prior to the *Wyatt* case, it turns out that none had been predicated on constitutional grounds. For the most part they had involved individual patients who had been diverted from the criminal justice system and confined with an implicit expectation that the state was to treat them rather than punish them. For example, like Jackson, they had been confined as incompetent to stand trial or were found not guilty by reason of insanity. Without treatment they claimed the state was in reality punishing them. Some judges had agreed, but their rulings were based on the promise implicit in the specific state statutes authorizing nonpenal confinement and not on any constitutional right.[24] The case of *Wyatt* v. *Stickney* attempted to push the right to treatment further in every respect: It was a class action on behalf of all civilly committed patients, it asked for a constitutional holding, and it finally directed attention to the plight of the large group of mentally ill and mentally disabled patients who had committed no crimes and who were confined in less than adequate hospitals.

After lengthy argument and many legal briefs, Judge Frank Johnson of the Alabama General Court held that: "To deprive any citizen of his or her liberty upon the altruistic theory that the confinement is for humane and therapeutic reasons and then fail to provide adequate treatment violates the very fundamentals of due process."[32]

Although this decision, like the decisions already discussed, emphasizes deprivation of liberty, it does not adopt a narrow due process solution for the situation of patients confined without treatment. For example, an alternative might have been a more limited holding as in *O'Connor* v. *Donaldson,*[14] where the Supreme Court ruled that all nondangerous patients who were involuntarily confined in these institutions and not getting treatment must be released. But Judge Johnson apparently realized that most of these patients were chronically disabled and, if given the legal right to leave, either had no capacity to exercise it or would be just as badly off if they did. Judge Johnson took a bold step, therefore, and asserted what was, in effect, a substantive right.

A more conservative judge would have hesitated before making such a ruling for reasons that go back to the distinction between procedural and substantive due process. It is one thing for a judge to tell a state what procedures it must employ when it confines a patient—even when those procedure are costly. This is accepted judicial practice as exemplified in *Lessard.* It is quite another thing to tell the state *how* it must take care of mental patients when it confines them, since this implicitly creates new substantive rights and imposes on the state the duty to meet those rights (needs) of patients. The latter substantive approach involves a critical legal entanglement; namely, the separation of powers. How far can the judiciary go in setting standards of institutional practice that require the legislature and the executive branch to raise new tax revenues or reorder the fiscal priorities of social needs that have been established through executive and legislative decision making? For these and other reasons a more conservative judge would have rejected the substantive due process argument. Judge Johnson, of course, did not give all of the citizens of Alabama a new substantive right; he ruled that the state must meet the treatment needs only of those patients it involuntarily confines. However, in establishing treatment standards Judge Johnson expanded somewhat the class of patients entitled to the right by making no distinction based on the legal status of the patients. He insisted that the needs of all patients in these Alabama institutions be met whether they were there voluntarily or involuntarily.

Other judges subsequently have had to confront the constitutional basis for the right to treatment more directly, looking to theories that can less easily be construed as creating new and substantive rights. One of the most unassailable approaches to a right to treatment would have been to ground it on the Eighth Amendment, which forbids cruel and unusual punishment. Stated in simplest terms, state institutions for the mentally ill and mentally retarded should not be allowed to perpetuate conditions similar to those that had already been declared unconstitutionally cruel in prisons. Other courts, building on this approach, have talked about the right not to be harmed.[13] Given the conditions in some of our state institutions, mere compliance with that standard requires enormous expenditures.

However, in the *Wyatt* case, although this kind of Eighth Amendment argument was acknowledged, Judge Johnson sought to reach a higher standard of treatment; he sought to insure adequate treatment. To accomplish this, the *Wyatt* order detailed that minimum "medical and constitutional" requirements be met with dispatch. The decree set forth requirements establishing staff-to-patient ratios, adequate floor space, sanitation, and nutrition. The court also ordered that individual treatment plans be developed, that written medication and restraint orders be filed, and that they be periodically reviewed. Outside citizen's committees were appointed to monitor enforcement of patient's rights under the order.

The *Wyatt* decree was far from a generalized array of commands arrived at arbitrarily. It was formulated from study of testimony of institutional personnel, with outside experts and representatives of national mental health organizations appearing as *amici curiae.* Most of the specifics of the order were taken from a memorandum of agreement signed by the parties. The most critical specifics—the model staffing ratio—approximate those recommended at the time by the American Psychiatric Association. However, the case proceeded without the participation of the American Psychiatric Association, although most other mental health professional groups were involved. Not surprisingly, the court's decree authorized that qualified nurses, psychologists, and social workers be allowed to take clinical responsibilities that had traditionally been limited to physicians. The dethronement of the psychiatrist as the head of the mental health team has been emulated in subsequent litigation and legislation.

The *Wyatt* decree reads like a judicial translation of the kind of document that the Joint Commission on Accreditation of Hospitals (JCAH) might be expected to promulgate. But seven years have gone by and despite Judge Johnson's continuing oversight, the decree he formulated has never been fully implemented. In 1979, the judge concluded that it was necessary to place the entire mental health system of Alabama in the hands of a receiver, a step that removed all authority from state officials.

Judge Johnson's difficulty in implementing his decree demonstrates a number of very real problems. First, just as happened with federal judicial intervention in school desegregation, recalcitrant state bureaucracies can place enormous impediments in the way of such complex institutional reforms. Second, reforming some of the worst state mental facilities in the country requires enormous resources and financial aid. Meeting the decree required Alabama to alter many of the fiscal priorities and tax strategies its legislative and executive branches had decided on. Not surprisingly, they were extremely resistant. Third, Judge Johnson's decree sought to upgrade facilities that, in these days of deinstitutionalization, might more appropriately be closed. Thus, enormous capital expenditure was being poured into outmoded facilities. Furthermore, the cheapest way to begin to approximate the mandated staff-to-patient ratios was to discharge patients without adequate aftercare and without alternative treatment settings. Judge Johnson's original decree did not clearly foresee these possibilities.

Fourth, the judge's decree deprived the mental health professions of their own flexibility in establishing independent policy and treatment strategy. This was true not only within the institutions involved but also outside, since so much money and resources had to be directed toward the judge's priorities.

Fifth, the litigation in *Wyatt* and in subsequent cases has intensified interprofessional tensions and rivalries. Obviously, litigation did not create these problems, but there are no longer clear lines of authority among the mental health professions responsible for the care of patients, and a great deal of energy is being wasted in struggles over status and control. The Oversight Committee appointed by the judge, albeit necessary, added to the disarray of authority by establishing a shadow administration. These difficulties have made it difficult to recruit good people for leadership positions. All of these problems are to be expected when a judge becomes *de facto* commissioner of mental health. But before rejecting the judge's activism, responsible mental health professionals must take a hard look at the alternative, or the situation that existed in Alabama before *Wyatt.* Obviously, Judge Johnson concluded that the Alabama situation was so bad that legal intervention to meet the needs of patients could not make matters worse.

Subsequent right to treatment litigation has been able to learn from the experience of *Wyatt.* For example, judges have attempted to get the state to negotiate with the plaintiffs so that they might set their own goals and standards rather than having the judge assume this task. Thus, a consent decree arrived at by the parties replaces the judicial decree. However, most lawyers who are experienced in this kind of litigation realize that the real struggle arises in the efforts to implement the decree, whether it has been arrived at by consent or by the judge's own findings. Although the Supreme Court has never endorsed the right to treatment, such litigation and particularly consent decrees have proliferated all over the country. This proliferation has given rise to suits even in those states that had been in the vanguard of reforming their large state institutions. As one reviews this kind of right-to-treatment litigation, one cannot help wondering if patients' real needs (as opposed to their mere rights) are being met.

It has become generally accepted psychiatric policy that state hospitals, which have a large census, which keep patients a long time, and which are often set at a distance from the community, are not desirable. Deinstitutionalization has, therefore, become the order of the day. But communities have become increasingly resistant to the opening of community-based facilities. Zoning restrictions, neighborhood protests, and political pressures have all been mounted against such needed facilities as halfway houses, sheltered living situations, and so forth. Increasingly, the "community" is the major opponent of the "community mental health approach." Furthermore, the problems of continuity of care are intensified when deinstitutionalization is compounded by revolving door policies. There is accumulating evidence that chronic patients are being lost in the shuffle and are subject to abuses at least as serious as those found in the "back wards."[17] Nonetheless, right-to-treatment litigators are demanding that the pace of deinstitutionalization be increased under the legal theory that patients are entitled to treatment in the least restrictive setting. Of course, the setting that least restricts a patient may not be the setting in which treatment needs are most effectively met. And when, as is increasingly the case, the good alternatives to total institutions are overwhelmed by applicants and the quality of the service available is suffering, the demand for the least restrictive setting will begin to place the abstract right to liberty above the concrete need for care.

¶ The Least Restrictive Alternative at the Time of Admission

"The least restrictive alternative" seems to be one of the most confused and confusing phrases in mental health litigation. The concept arose in an entirely different kind of constitutional context; it has been wrenched out of that context and applied to the mentally ill and disabled.[22] The argument asserts that the state's interest in confining the patient must be met by that treatment approach that will produce the least loss of liberty. Like the right to treatment, the least restrictive alternative, as it has been applied to the mentally ill, has not been recognized by the Supreme Court. It is another extension of the procedural due process theory, which at least potentially seems to involve substantive rights. There are many potential patients whose hospitalization could be avoided by immediate crisis intervention. Demonstration projects have, at least, suggested that a massive reduction in the need for hospitalization whether voluntary or involuntary can be achieved. A few treatment centers have the capacity to provide such crisis resources; most do not. Does the least restrictive alternative mean, with regard to civil commitment standards such as *Lessard,* that the state must provide such resources? Alternatively, many patients are committed because they have no family or anyone else to see to their needs, to supervise their taking of medication, and to keep them from wandering the streets. These "gravely disabled" might be cared for in their homes by a nurse or a housekeeper. Does the least restrictive alternative mean the state must provide such a caretaker? If answered affirmatively, these first two questions interpret the least restrictive alternative as establishing a substantive due process right requiring the state to create new services.

Or does the least restrictive alternative mean only that, given the treatment facilities the state has available, the patient must be placed in the one which is least restrictive? This interpretation is more in the nature of a procedural due process requirement. But even this latter requirement is beset by confusion. Is the patient entitled to the least restrictive alternative in light of the dangers he or she poses to self or community, or to the least restrictive alternative in which effective treatment can be provided? Few, if any, of the courts ruling on the least restrictive alternative have dispelled these confusions.

¶ The Least Restrictive Alternative in Class Action Right-to-Treatment Litigation

Where the least restrictive alternative is demanded in right-to-treatment litigation, as is increasingly the case, the plaintiffs want the state to provide new community-based facilities. But an additional goal of the litigants at times seems to be to close down the state mental hospital. For example, the plaintiffs in *Rone* v. *Fireman*[20] demanded that an eight-hundred-bed facility, which had been tentatively accredited by the JCAH for one year, be closed down and replaced by a fifty-bed unit, with all other patients transferred to less restrictive alternatives. This may be a good thing—indeed eventually all of the state hospitals should probably be closed—but a question of timing has become a central concern. Does it make sense to close down a decent state hospital facility at a time when available alternatives in the community are overwhelmed? It may be true that a good foster home is a less restrictive alternative—but is a bad foster home or a rundown welfare hotel in the inner city a less restrictive alternative? Even where the litigants have no intention of closing down the institution, the difficulties inherent in finding alternatives have made this litigation problematic. If the patients are pushed into the community without suitable alternatives, then the abstract right ignores the concrete reality of the patients' needs.

Nowhere have these problems been more intense than in litigation seeking to achieve the least restrictive alternative for the mentally retarded. A dominant treatment approach to the mentally retarded is "normalization."[30] It is a comprehensive philosophy, and its goal is to see that the mentally retarded are given the opportunity to live a life as close to the normal as possible. Most of the professionals involved in the care of the retarded see deinstitutionalization and the least restrictive alternative as essential to "normalization." Some parents and relatives of the mentally retarded object to this and not always on purely selfish grounds, as is sometimes alleged. Less restrictive alternatives in the community frequently are organized and run by activists, many of them young. Parents are concerned that this activism and idealism will wane, as has happened with other social endeavors. They believe that the young people will move on and that the mentally retarded will be lost and exploited in the community. Whatever inadequacies there may be in the large brick institutions built by the state, they do seem to promise continuity of care[11]—a continuity that will last after the parents die or become unable to look after their retarded offspring. These are deep divisions in attitude and they portend deep divisions in policy. Already there has been dispute concerning the least restrictive alternative litigation applied to the mentally retarded, with parents contesting the lawyers who claim to represent their children.[3]

The Justice Department of the United States has participated in a number of right-to-treatment suits on the side of the plaintiffs.[32] Its participation allows the FBI to become involved in the investigation phase, helping to document inadequate conditions and allegations of abuse. The Justice Department also brings to the problem the resources of a massive federally funded agency geared to litigation. The standing of the Justice Department to participate in these suits has been challenged, and some federal courts have ruled that the Justice Department does not, in fact, have the legal standing to bring such suits.[27] But many federal courts have invited the Justice Department to participate with other plaintiffs as *amicus curiae,* and the Congress has enthusiastically supported legislation that would give the Justice Department the legal standing that the courts deny it. The role of the Justice Department is crucial since it seems to have adopted the least restrictive alternative as a goal of this litigation, even where concrete needs will be sacrificed for abstract rights.

It must be emphasized that despite a great deal of rhetoric to the contrary, the least restrictive alternative has not been less expensive than institutional care when quality care is involved. Neglect is the only way to achieve real savings be it inside or outside the hospital. The appearance of savings has been achieved in effective deinstitutionalization only by budget manipulation. For example, if patients are transferred from mental hospitals to good nursing homes the mental health budget will go down, showing an apparent savings; however, the Medicaid and/or welfare budget will go up. Often the burden is merely shifted from the state to the federal government, or in some instances a new obligation is placed on cities and towns. The point to be emphasized is that the objectives to be sought in responsible deinstitutionalization do not include overall cost savings.

Rone v. *Fireman*[20] forced a federal judge to confront all the difficulties of the least restrictive alternative. In his decision, the judge described the community's resistance to the good-faith attempts of the Department of Mental Health to create alternatives to the large state hospital. He noted that even the plaintiffs acknowledged that patients would need the same kinds of services they were getting in the hospital. In light of these and other considerations, he responded to the plaintiffs' demand for the least restrictive alternative as requiring only that the patients be provided with transportation from the hospital back and forth to the city (the hospital was located some distance away and there was no public transportation). This decision, if it did nothing else, demonstrated how flexible the least restrictive alternative can be as a meaningful constitutional doctrine. But it did more—it advanced a nonpolemical analysis of the many problems that now beset any unyielding policy of deinstitutionalization.

The remarkable thing that emerges from an even cursory review of the right to treatment litigation is the expanding cast of characters whom litigation has involved in decisions affecting the situation of the mentally ill. There are the federal judges, the Justice Department, various public interest lawyers, state officials, and various bureaucracies. Others often become deeply involved; for example, the Department of Health, Education and Welfare (now the Department of Health and Human Services), the various professional associations, the Joint Commission on Accreditation of Hospitals, the American Civil Liberties Union, the National Associations for Mental Health and for Retarded Citizens, many advocacy groups, various planning and systems consultants, and many others. All these participants are in addition to the responsible mental health professionals, the relatives, and the patients themselves. It is no wonder that the state commissioner of mental health now serves on the average only eighteen months.

¶ The Supreme Court and the Right to Treatment

During the mid-seventies while the *Lessard* case was bouncing back and forth between the federal court and the Supreme Court, and while federal courts all over the country were becoming involved in right-to-treatment cases, the Supreme Court decided its first important mental health case, *O'Connor* v. *Donaldson.*[14] This was the first time in the history of the Supreme Court that it had dealt with a straightforward instance of civil commitment. But equally important, the case had involved the right to treatment at lower court levels.

For fifteen years Kenneth Donaldson was a patient at Chatahoochee State Hospital in Florida. He was diagnosed as a chronic paranoid. Dr. J. B. O'Connor was the superintendent of the institution where Mr. Donaldson was a patient. During Mr. Donaldson's stay the institution had a ratio of one doctor per eight hundred patients. Mr. Donaldson applied more than a dozen times to various state and federal courts for his release from involuntary confinement. Each time the courts rejected his plea. Finally, he turned for assistance to Morton Birnbaum, a physician and lawyer, who is considered today the father of the right-to-treatment litigation. Dr. Birnbaum initiated a right-to-treatment suit on behalf of Mr. Donaldson and other patients. Eventually, however, Bruce Ennis, a lawyer with the Mental Health Law Project, took over the case. During this time Dr. O'Connor suffered a coronary and resigned as superintendent. The new superintendent subsequently discharged Mr. Donaldson. Mr. Donaldson had never been dangerous, had refused medication, and during the fifteen years of his hospitalization never received anything that a responsible clinician would consider treatment. After Donaldson's discharge, and with Ennis's legal input, the lawsuit took a different direction. Rather than a class action right-to-treatment suit, it became a suit for damages under the Civil Rights Act.[28] Dr. O'Connor was to pay for Mr. Donaldson's loss of constitutional rights.

But as the case progressed from the lower federal courts to the Supreme Court it was not clear what constitutional right Dr. O'Connor had deprived Mr. Donaldson of. If Dr. O'Connor had violated Mr. Donaldson's right to be treated, then Mr. Donaldson indeed had such a constitutional right and so did every other involuntary patient. This was the view taken by the federal district court and the Court of Appeals.[4]

However, the Supreme Court took a much narrower view of the case. It decided only that those patients who were not dangerous to themselves or to others, who could survive outside the hospital, and who were not getting treatment within the hospital had a right to be discharged. The court was silent on the right to be treated, but in a footnote it emphasized that it was vacating the lower court's decision of a constitutional right to treatment.[14] Chief Justice Burger in a concurring opinion scathingly criticized the lower court's reasoning in reaching a right to treatment. (As of this writing, the Supreme Court has never endorsed the right to treatment, and Chief Justice Burger has in subsequent mental health cases made clear his opposition to substantive due process decisions, in which lower court judges have attempted to meet the needs of the mentally ill as opposed to their procedural due process rights. But as we have seen, the Supreme Court has not even been generous in providing procedural due process rights as is apparent in *Donaldson, Addington, Parham,* and *Bartley.*)

Donaldson was a unanimous decision. In effect, it said that if Mr. Donaldson had been dangerous, he had no right to be discharged; if he had been unable to care for himself, he had no right to be discharged; and if he was getting treatment, he had no right to be discharged. Only when there is absolutely no possible justification for confinement has a patient the right to be discharged.

Because the Supreme Court held that Mr. Donaldson had been deprived of his constitutional right to liberty, the decision was hailed as a great triumph for civil libertarians. However, *Donaldson* was in reality little more than a restatement of *Jackson.* The Supreme Court, in effect, simply announced once again that it was constitutionally impermissible to incarcerate a human being forever without some reasonable legal justification.

But the Supreme Court gave no guidance as to what reasonable legal justification would be necessary in order to justify initial confinement. Its decision was applicable only to the grounds for discharge. This is a matter of some interest. Almost all states have laws on the books that regulate civil confinement; some of them, as has been demonstrated, are very complex and set stringent standards. But once the patient is confined, the psychiatrist has almost total discretion over discharge, although many states now require periodic review. The Supreme Court's decision in *Donaldson* did impose on psychiatrists the responsibility to discharge nondangerous custodial patients who could survive outside the hospital. If psychiatrists did not comply, they risked a suit for damages under the Civil Rights Act.

But in deciding that Mr. Donaldson had been deprived of his right to be discharged, the Supreme Court, in light of its intervening opinion in *Woods* v. *Strickland,*[31] instructed the lower courts to reconsider their finding that Dr. O'Connor had been liable for damages. *Woods* had established criteria for determining the liability of persons acting under color of law, such as Dr. O'Connor, and included a criterion that, in effect, prevents retroactive liability for depriving a person of a constitutional right not yet established. The lower courts never decided this issue of liability. Instead, Mr. Donaldson settled for a minimal sum of $10,000 with a written agreement in which no admission of liability on the part of Dr. O'Connor was conceded. Perhaps the most important aspect of the *Donaldson* case is that it marked the first time that any patient had succeeded at even a lower court level with this kind of civil rights action against a psychiatrist.

Although the judgment was vacated, remanded, and settled, federal legislation makes it possible for lawyers to be compensated for their efforts in such litigation. The basic theory of such compensation is that lawyers who press for the constitutional right of citizens are functioning as a kind of private attorney general. If they win their case, even to the extent that Mr. Donaldson won his, the defendant must compensate them. Thus, the lawyer is not dependent on a contingency fee arising from damages, or on representing wealthy clients. This, of course, removes one of the major obstacles to litigation in the mental health area, since cases such as *Donaldson* entail vast sums in legal fees. Similar private attorney general statutes in some states, other provisions of pending federal legislation, and the growth of advocacy programs for mental patients guarantee that mental health litigators will be a continuing reality, shaping the practice of psychiatry in the United States.

¶ Conclusion

The rights of mental patients have been linked to the civil rights movement. This conception emphasizes due process and equal protection as the essential constitutional doctrines. When applied to the psychiatric context these doctrines have produced two kinds of litigation. One emphasizes rights even where they interfere with the patient's needs. The other emphasizes needs and attempts to express those needs as rights. The lower federal courts have taken a much more activist approach to this litigation than has the Supreme Court. Thus, there exists a profound disequilibrium and uncertainty about future judicial decisions. However, many states through legislative action have already adopted new mental health statutes that provide all of the due process safeguards sought by reformers. Therefore, whatever decisions the Supreme Court ultimately makes, we can expect that for the next few decades law and legal intervention will remain a central concern of American psychiatry.

¶ Bibliography

1. Addington v. Texas, 99 S. Ct. 1804 (1979).
2. Bartley v. Kremens, 402 F. Supp. 1039 (E.D. PA. 1975).
3. Connecticut Association for Retarded Citizens et al. v. Gareth Thorne et al. H–78–653, Docket Number.
4. Donaldson v. O'Connor, 493 f.2d 507 (1974).
5. FOUCAULT, M. *Madness and Civilization.* New York: Random House, 1965.
6. Jackson v. Indiana, 406 U.S. 715 (1972).
7. Knect v. Gillman, 488 f.2d 1136 (8th Cir. 1973).
8. LAING, R. D. *The Politics of Experience.* New York: Valentine Books, 1967.
9. Lessard v. Schmidt, 349 F. Supp. 1078 (E.D. Wis. 1972), vacated, 414 U.S. 473, reinstated 413 F. Supp. 1318 (E.D. Wis. 1974), vacated 421 U.S. 957 (1975), reinstated, 413 F. Supp. 1318 (E.D. Wis. 1976).
10. MASSACHUSETTS PSYCHIATRIC ASSOCIATION, *Brief of Amicus Curiae.* Rogers v. Okin, No. CA 75–1610–T (D. Mass. 1979); Stone, A. A., "Recent Mental Health Litigation: A Critical Perspective," 134 *American Journal of Psychiatry* 273, 278 (1977).
11. "Memorandum," *Fernald League Policy,* unpublished, p. 5.
12. MILL, J. S. "On Liberty," in Cohen, M., ed., *The Philosophy of John Stuart Mill.* New York: The Modern Library, 1961, p. 197.
13. NYARC v. Rockefeller, 357 F. Supp. 752 (E.D. N.Y., 1973).
14. O'Connor v. Donaldson, 422 U.S. 563 (1975).
15. Parham v. J.R., 99 S. Ct. 2493 (1979).
16. Planned Parenthood of Central Mo. v. Danforth, 428 U.S. 52 (1976).
17. REICH, and SIEGEL, "The Chronically Mentally Ill: Shuffle to Oblivion," *Psychiatric Annals* 35:3 (1973).

18. Rennie v. Klein, 462 F. Supp. 1331 (D. N.J. 1978).
19. Rogers v. Okin, No. CA 75–1610–T (D. Mass. 1979).
20. Rone v. Fireman, 473 F. Supp. 92 (1979)
21. SECRETARY OF PUBLIC WELFARE OF INSTITUTIONALIZED JUVENILES, 99 S. Ct. 2523 (1979).
22. Shelton v. Tucker, 364 U.S. 479 (1960).
23. SPIRO, H. R., "Affidavit," in Project Release v. Prevost, No. 78C 1467 (E.D. N.Y., 1979).
24. STONE, A. A. "Right to Treatment" in *Mental Health and Law: A System in Transition.* Washington, D.C.: U.S. Government Printing Office, 1975.
25. Suzuki v. Quisenberry, 411 F. Supp. 1113 (D. Hawaii, 1976).
26. SZASZ, T. *The Myth of Mental Illness.* New York: Dell, 1961.
27. United States v. Solomon, 563 F.2d 1121 (1977).
28. United States Code, Title 42 Section 1983.
29. Winters v. Miller, 446 F.2d 65 (2nd Cir.) Cert. denied 404 U.S. 985 (1971).
30. WOLFENSBERGER, W. *Normalization: The Principle of Normalization In Human Services.* Toronto: National Institution on Mental Retardation, 1972.
31. Woods v. Strickland, 420 U.S. 308 (1975).
32. Wyatt v. Stickney, 325 F. Supp. 781 (M.D. Ala. 1971) enforced 344 F. Supp. 387 (N.D. Ala. 1972), aff'd sub nom Wyatt v. Aderholt, 503 F.2d 1305 (5th Cir. 1974): NYARC v. Rockefeller, 357 F. Supp. 752 (E.D. N.Y. 1973); Rone v. Fireman, 473 F. Supp. 92 (1979).

CHAPTER 39

ETHICS IN PSYCHIATRY

H. Tristram Engelhardt, Jr., and
Laurence B. McCullough

¶ Introduction

WHILE reflection on medical ethical issues has been intrinsic to medicine throughout its history, it is only in the past twenty-five years that the study of medical ethics has expanded to embrace the biological and behavioral sciences —an inquiry now conducted under the rubric of bioethics. This development of a more sustained inquiry in bioethics has occurred simultaneously with the various civil and human rights movements. Like these movements, the renewed and growing interest in bioethics reflects our culture's reexamination of value commitments and the proper bounds that may be placed upon institutions that wield power and authority. The consequent convergence of intellectual and social forces has culminated in formal examinations of ethical issues in medicine.

¶ The Scope and Character of Ethics in Psychiatry

Perhaps the most prominent ethical issue in medicine has been the use of human subjects in medical research.[1,19] Multidisciplinary deliberations about the ethical dimensions of this practice achieved a public character in 1973 with the establishment of the National Commission of the Protection of Human Subjects in Biomedical and Behavioral Research in the U.S. Department of Health, Education and Welfare. In addition to general considerations of ethical issues occasioned by human research, the commission has addressed issues that bear directly on research in psychiatry; for example, research involving mentally ill subjects.[90] The commission has also considered the ethical dimensions and procedures employed for psychiatric complaints; for example, psychosurgery.[89]

This concern with ethical issues in psychiatric research did not arise apart from the broader concern with the ethical dimensions of medicine and psychiatry. In fact, the interest in the ethics of human research was pursued concurrent with, and in part gave rise to, inquiry into the rights of patients, in particular hospital patients.* The ethical dimensions of rights and of rights language also have a direct bearing on psychiatry. They

*See references 2, 9, 37, 48, and 56.

are associated with such issues as due process in the civil commitment of the mentally ill[84,99,130] and the rights to treatment of those confined to mental hospitals.[15,44,115,126] Thus, the scope of ethics in psychiatry has not been limited exclusively to research but includes inquiry into the ethical dimensions of psychiatric practice.*

There has been another change concerning the inquiry into medical ethics. Its character, as well as its scope, has altered. The discussion of ethical issues in psychiatry, and for medicine generally, has become more philosophically sophisticated. There is a growing appreciation for the importance of those basic concepts that structure ethical issues in psychiatry. This development is clearly evidenced in the burgeoning literature in bioethics, including an *Encyclopedia of Bioethics*[107] and numerous books and journals wherein the full range of ethical issues in psychiatry are addressed.†

Five Senses of Ethics in Psychiatry

There are at least five different senses of ethical reflections in psychiatry. First, one might refer to generally accepted views of proper conduct of practitioners within a particular culture. What one will discover in such an inquiry are various and often poorly examined views about what is proper in life, including sexual and other social taboos. Such informal views are often the subject of sociological or orthological study.‡ A second and similar sense of ethics in psychiatry derives from a traditional understanding of medical ethics vis-à-vis canons of professional etiquette.** In this respect, one might examine professional codes or procedures[125] that are meant to guide both professional and civil conduct. A third and more general sense of ethics consists in following legal rules and procedures.†† This sense of ethics, however, should be distinguished from the more basic ethical notion of man's right to refuse on certain justifiable occasions to act in ways that are socially or legally sanctioned. Consider the case of a psychiatrist subpoenaed to testify in court about a patient who has been in therapy for several years and is charged with consensual sodomy. Under oath, the psychiatrist is asked by the prosecuting attorney whether the patient has confided that he has engaged in the actions with which he is charged. Because the psychiatrist may regard the patient's utterances during therapy and his own written record and notes to be protected by a moral obligation of confidentiality, he may be ethically justified in refusing to answer, even though he could be in contempt of court for not answering a direct question. Thus, ethics in the sense of abiding by legal rules and procedures may generate conflicts with well-founded moral obligations and so does not by itself provide a reliable guide to proper conduct. Hence, it should not be confused with that more fundamental sense of ethics. (One might think here of the conflict that Sophocles depicts between Antigone and Creon, as well as the remarks made by Hegel on this subject.)

A fourth sense of ethics in psychiatry relates ethical conduct to various religious codes or religiously grounded views of proper conduct. This view, because it appreciates the need for critical assessment of socially and legally sanctioned conduct, is closer to the more basic understanding of ethics.[102] Its shortcoming is that, in a pluralistic society, ethics requires more general views about human values and proper conduct. This is so for two reasons. First, ethics should provide the basis for persons to inquire into issues of common interest and concern. Second, ethics should provide a common ground to critically assess socially and legally sanctioned patterns of conduct. Thus, the fifth and most fundamental manner in which to understand ethics is to perceive it as an enterprise through which we negotiate divergent moral intuitions. Such an undertaking is crucial for a professional concern such as psychiatry, since it deals directly with

*See references 17, 21, 32, 62, 76, and 106.

†See references 13, 25, 76, 78, 88, 94, 97, 120, 123, 127, 133, 134, and 137.

‡See references 42, 69, 79, 86, and 87.

**See references 3, 4, 5, 6, and 8

††See references 1, 16, 17, 20, 28, 44, and 71.

the anxieties, conflicts, and interests invoked in patients by the various axiologic dimensions of life. The goal of philosophical ethics, therefore, is not simply an ethic that is nothing more than a general impression of the good life as it is understood by certain groups in a particular society, but rather it is the development of reliable means for analyzing ethical issues and indicating how ethical disputes can be reasonably negotiated. As such, ethics applied to psychiatry should be understood more as a set of modes for analyzing problems and solutions than as a series of final answers to assorted questions.[23,55,92,103]

¶ Psychiatry in Ethics

Psychiatry has provided a number of important insights into how and why particular ethical viewpoints have developed. For example, various ethical viewpoints might be understood as different ways of coming to terms with anxieties provoked by certain conflictual situations in life.[52,54] Of course, such causal accounts of ethical systems do not impugn the intellectual validity and practical importance of ethical analysis. After all, an account of how particular personality traits lead individuals to study quantum physics would not undercut the validity and meaning of quantum mechanics. In other words, identifying hidden or unconscious motivations for ethical views does not provide an exhaustive account of ethics as an intellectual and practical enterprise. What such an inquiry into causal and, in particular, motivational forces does yield is the view that ethics is a form of intellectualization that offers a peaceful mediation of interpersonal conflicts and values. This is a useful notion, for it points to the understanding of ethics as an alternative to force, an attempt to negotiate different and sometimes conflicting moral intuitions without recourse to coercion.[70]

Characterizing philosophical ethics as an alternative to force does not mean that ethics is more efficient than open force or subtle coercion[57] in the settling of disputes concerning what choices of human conduct are proper. Rather, one faces ethical issues only when one is asking a question that is at once intellectual and practical: How, to what extent, and on what grounds, can reasonable individuals reach agreement about disputed or uncertain areas of moral conduct? Such ethical reflections face up to the problem of pluralism. Thus, in a fundamental sense, philosophical ethics demands a commitment to explore the possibilities of the logic of a pluralism of moral values.

¶ Rights, Duties, and Values

If ethics vis-à-vis psychiatry must take into account the problems inherent in a pluralistic notion of moral values, it is first necessary to understand the nature of a logic of pluralism and the application of that logic to the resolution of disputes. As with disputes in other areas of moral conduct, those in psychiatry tend to be expressed in the general philosophical language of rights, duties, and moral values. For the purpose of this essay, and in order to display more clearly what is at stake in such disputes, two senses of rights claims are distinguished: (1) those advanced as a way of enjoining the pursuit of a certain set of goods or values, and (2) those that hold independently of any interest in particular goods or values.

One sense of rights claims is consequentialist or teleological (from the Greek, *telos,* "end"). That is, rights claims can be taken as goal-oriented ways of appreciating legitimate claims of patients and the corresponding duties or obligations of professionals. For example, if one claims that psychiatrists should tell the truth to their patients, one might mean that the practice of recognizing such a duty will lead to the realization and protection of important goods and values in the conduct of psychiatric treatment and care. Leon Salzman[111] has developed this line of thought in considering Sigmund Freud's analysis of the importance of truthfulness in psychiatric treatment. Freud[54] stated that "psychoanalytic treatment is founded on truthfulness. A great part of its educative effect and its ethical value lies in this very fact." Here Freud is arguing, in effect, for a particular moral obligation—telling the truth to patients—on the grounds that fulfilling that obligation will lead to the realization of an important goal: maintaining the authority of the psychoanalyst in the context of the therapy. The realization of this good, of course, serves another: the care and treatment of the patient. These goods apply to the patients' right to truthfulness and the psychiatrist's corresponding duty to the patient in a manner characteristic of the teleological sense of rights.

The second sense in which rights claims can be understood proceeds from the recognition that one cannot understand ethics as an alternative to force without agreeing at the same time that one must respect the free choice of persons.[70] After all, respect for freedom is the single alternative to force or coercion in some form. Thus, one may hold that there are rights in the therapeutic context that exist by virtue of the very nature of a community based on neither force nor coercion. Rights based on the notion of such a moral community are not reducible to interests in goods and values.

This second justification of rights claims is deontological (from the Greek, *deon,* "obligation"). One such deontological argument has made respect for the freedom of rational persons a condition for moral conduct.[92] That is, the moral community is to be founded on mutual respect of each individual's autonomy. Insofar as psychiatry is practiced in a pluralistic society, which lacks a single, coherent view of the good life, and insofar as one views it as inappropriate to use the therapeutic relationship to impose one particular view of the good life upon another person, one is forced to acknowledge respect for the autonomy of patients as an indispensable condition for proper professional conduct. Such a deontological concern with freedom or autonomy should be familiar to psychiatrists, since it has been advocated by such individuals as Thomas Szasz,[128] who contended that psychiatry should exist in order "to liberate the patient, to support the autonomy and free choice of patient." Similarly, Peter Breggin[24] has argued that psychiatry is a form of applied ethics because of its encouragement of an ethic of autonomy.

A deontological approach to ethics in psychiatry, though, has received its sharpest focus in the context of informed consent. Because of the pluralism that characterizes our society, neither patient nor therapist should routinely assume that there exists between the two a ready-made consensus on basic values and goals. Indeed, such an assumption on the part of the therapist might lead, inadvertently, to subtle forms of coercion. This consideration points up the inadequacy of the Golden Rule as a moral maxim, especially for ethical inquiry into psychiatry. The problem here is that the injunction "Do unto others as you would have them do unto you" may inadvertently, or even advertently, become the occasion for imposing upon patients the psychiatrist's own view of the good. Respect for persons, as an alternative to coercion in even subtle forms, leads to quite a different moral maxim: "Do not do unto others what they would not have chosen to have done to themselves." It should be noted that respect for the autonomy of the patient does not require that the therapist surrender his autonomy to the patient. That is, the therapist's autonomy should be respected by the patient. Thus, the therapist need not accept or endorse every expression of a patient's values. Indeed, an interest in the patient's autonomy may justify the therapist's probing or even challenging a patient's choice of values as part of the therapeutic process.

Even so, voluntary and informed consent functions primarily to maintain patient autonomy. Indeed, the practice of obtaining free and informed consent becomes increasingly important as the likelihood of disagreement between psychiatrist and patient increases. Thus, for example, the National Commission for the Protection of Human Subjects, in its *Belmont Report,*[91] stressed that respect for autonomy grounds concern for free and informed consent by competent individuals.

Among psychiatrists there has been an understandable difficulty in interpreting the force of consent by *all* individuals who come under their care. After all, individuals under the care of psychiatrists may often be in circumstances where their competency is in question or where a choice or pattern of choices by the individual may appear to be somewhat bizarre. At the same time, however, there is a legitimate concern that lack of competence might be assumed without reliable evidence. This concern is complicated by the equivocal nature of the concept of competence. Indeed, the exact relationship between psychiatric diagnoses and levels of competence is a far from settled matter.[101] This concern is reflected in those laws regarding civil commitment that guarantee due process as the means to assure that the prerogatives of competent individuals are not abrogated.[36] A correlative concern is for those whose competence is clearly and substantially diminished, for whom the same legal procedures are meant to be a means to define and support the best interests of the individuals thus affected. These concern for protecting the patient's autonomy and best interests have been tied to arguments that the involuntary commitment of a patient ("for his own good") imposes the obligation to treat that person adequately once committed.[15]

In summary, then, it seems fair to say that issues of a genuine ethical character are an ingredient in psychiatry. These issues cannot be resolved unless one understands that they exist as points of tension concerning the best interests and free choices of individuals, of professionals, and of society. Attempts to resolve these complex conflicts will be expressed in the way in which the psychiatric profession frames particular institutions for therapy. Inevitably, resolutions will embrace practices aimed at assuring some but not other guarantees of autonomy, and they will achieve some but not other goods and values.

It is important to emphasize that these two views of ethical analysis, which are designed to address questions of autonomy and moral goods and values, are not to be taken as competing forms of ethical inquiry. Treating the two views as extremes, one of which must be chosen at the expense of the other, would create an artificial polarity that would only impede attempts to understand the complex ethical issues at stake. Instead, the complexity of the issues demands that these two approaches be understood and employed, when possible, as complementary modes of analysis. One might, by way of illustration, consider the ethical dimensions of psychiatrists engaging in sexual intercourse with their patients as a part of therapy. Leonard Riskin,[109] for example, has argued that this practice should be subjected to a study designed to determine its costs and benefits. From the perspective of philosophical ethics, this proposal amounts to an invitation to reexamine basic practices of the profession to determine whether they can be justified in light of their consequences; that is, whether the goods and values that constitute the goals of the profession will be achieved. An analysis and evaluation based on these goods and values could thus help to determine the consequences of such "therapies" for treatment and the chances for therapeutic success.

Alternatively, one might evaluate this practice in terms of respect for patient autonomy, with a view perhaps to determining whether coercion in subtle forms is or is not an inevitable feature of such "therapies." That is, by inquiring into the ethics of psychiatry we seek to display alternative ethical analyses of various practices, actual and proposed, and to critically evaluate practices from the concomitant perspective. The final goal is to weigh these perspectives, to achieve an adequate, thorough, and—where possible—coherent account of the proper bounds of professional conduct. For the example in question—sexual relations with the patient—this would require an account that gives prominence to respect for patient autonomy and to an interest in maximizing the benefit/cost ratio of therapy. At the very minimum, one would not want psychiatrists to believe that sexual intercourse with a patient would be a good therapy when it is not, either because the patient was not given an opportunity to freely consent to it or because it would, in fact, neither protect nor advance the patient's choice of values.

As this example indicates, an analysis of ethical issues in psychiatry will rarely provide one with final answers or with concrete admonitions or injunctions. Instead, ethical analysis, when it is done well, will usually suggest how the inquisitive and thoughtful practitioner might display the geography of values and the character of conflicts among diverse moral values, rights, and duties that he is likely to encounter in research and/or practice. Such analyses will allow one to identify better solutions and only rarely hit upon the best solutions. Because the human moral universe is diverse, complex, and perhaps in part incoherent, one is usually forced to choose among values. Therefore, in many important areas of concern, one is often forced to choose among several conflicting obligations. Ethical analyses therefore offer suggestions on how to approach the conflict of values at stake in psychiatric practice. It becomes necessary then to see how some of the more prominent of these concerns arise. These analyses, however, must be appreciated as attempts to suggest how the problem sketched might be understood, rather than as statements of definitive resolutions of those problems.

¶ Diagnosis and Values

The diagnosis of mental illness involves complex conceptual and normative issues.[47,83] The normative issues involved, however, are not only ethical, but include nonethical evaluations as well. For example, judging that an individual is abnormal, deviant, ill, or diseased involves, at least according to the arguments of some,[22,74] appeals to nonethical norms or ideals of psychological functions. That is, judging an individual to be mentally ill involves more than a judgment that he or she is statistically deviant.[46] The abnormality that is recognized also reflects a judgment that the individual fails to realize a minimum ideal or norm of psychological function. This recognition allows one to hold that a mental illness might be statistically the rule, though still an abnormality. It explains as well why individuals at the lower end of the distribution of IQs are considered to be abnormal in a normative sense, while those at the higher end, while equally statistically abnormal, are considered to be normal. Holding individuals to be healthy or diseased involves not only a description of facts and an explanation of their occurrence, but evaluations of them as well,[46] evaluations that often reflect the broad, transculturally recognized minimal ideals of proper psychological or behavioral function.

In addition, diagnostic labeling casts individuals into sick roles, with not only special privileges but also special obligations.[95,118] Being placed in the sick role results in limitations on an individual's liberty and ability to pursue certain goods and values. As a result, diagnosis involves the interplay of (1) nonmoral values concerning proper human behavioral and psychological capacities and function; (2) explanatory, predictive, and therapeutic interests that lead to the development of explanatory accounts (for example, notions of particular psychiatric diseases, such as schizophrenia); and (3) special social roles that are established, verified, and given concrete form through the authenticating or diagnosing role of psychiatrists. Thus, psychiatrists not only describe clinical data but join such descriptions with evaluations in explanatory models that certify individuals as falling properly within a sick role.

Now, patients can, and sometimes do, abuse and take advantage of such roles through the manipulation of psychiatrists and the health care system. Such forms of abnormal illness behavior, as described by

Izzy Pilowski,[100] afford patients various forms of secondary gain from certain sick roles. On the other hand, psychiatrists and mental health institutions can be harnessed by social groups in order to impose on others the ideals of those groups concerning proper psychological and behavioral functions. The recent debate concerning the classification of homosexuality in the *Diagnostic and Statistical Manual* (DSM-III)[7] of the American Psychiatric Association reflects such concerns about the nature of ideals of sexual function and the social power of diagnostic labeling.*

Those opposed to including homosexuality in the DSM-III may believe that terming an individual choice psychologically abnormal, deviant, or diseased is not simply descriptive; it must necessarily involve a normative interpretation of reality, one that can and does have a profound impact on an individual's autonomy and choice of values in life-style. At the same time, *not* labeling choices of values or life-style as abnormal, deviant, or a stage of arrested development (and thus a form of psychiatric disorder) also involves normative interpretations of reality. Thus, the change in the DSM-III[7] classification of homosexuality from a species of sexual deviation to ego-dystonic homosexuality may be taken to imply that this life-style is normal, healthy, and therefore good for those who choose it. In summary, various senses of mental health and well-being, as well as mental illness, abnormality, and deviance, express different views about not only ideals of function but about what pains and anxieties are to be tolerated and which are to be considered "abnormal" in the sense of being worthy of treatment.

Thus, one prominent set of ethical concerns about psychiatric diagnosis has focused on the creation of a social reality in the form of psychiatric sick roles. In addition to excusing individuals from the consequences of certain behaviors ("He can't help that he's mentally ill"), the sick role excuses one from social obligations ("He can't be expected to work, he is completely disabled due to his being mentally ill"), establishes duties ("He ought to seek treatment for his problem"), and sanctions authorities ("In fact, he should see a psychiatrist"). The normative aspect of diagnosis also establishes certain special rights ("He will receive full disability pay until he is well"). Psychiatric diagnoses can also lead to the loss of rights through the relationship between psychiatric categories of diagnoses and legal concepts of insanity ("He can't, given his illness, be responsible for his assets or write a new will"). Thus, such specially sanctioned sick roles are multifaceted. They can give special protection against criminal prosecution and can also provide grounds for civil commitment. Because of the social power that such sick roles possess, they raise opportunities for misuse[26] and therefore bring about complex ethical issues.

*See references 18, 28, 60, 64, 96, 119, 121, and 122.

Ethical Dimensions of Labeling

When nonmoral normative judgments are transmuted into performative judgments, they create social roles with socially and often legally enforced rights and duties. Being labeled a mentally diseased individual will, therefore, bring normative evaluation as well as special forces to bear on that individual. For example, a drug addict can be treated not simply as statistically deviant, but responsible for his actions and, therefore, perhaps a criminal. The addict, however, can also be regarded as diseased and hence in need of treatment to turn him aside from a self-destructive habit or life style. A more profound example of the moral import of labeling is the use of psychiatric hospitalization in the Soviet Union.[29] In this context the transmutation of political judgments into psychiatric judgments changes the political role "dissident" into the psychiatric role "insane personality," a deviant in need of treatment. Placing individuals in the sick role thus involves ethical issues concerning the protection, diminishment, and manipulation of the autonomy of individuals and of their choice of values. Such roles involve a commitment to special transfers of goods and to the sometimes profound alterations in the usual connotations of rights and duties.[99]

Ethical Dimensions of Clinical Judgment

The process of clinical judgment involves several genres of ethical issues. These turn on determination of prudent balances of likely benefits and costs in the process of working toward and then applying a diagnostic label. This process involves a determination of what is in the best interests of a patient, since any clinical judgment will expose a patient to a risk of false positive diagnoses as well as false negative ones. On the one hand, one will be inclined to hold that it is reasonable to be exposed to increased numbers of false positive diagnoses, if the treatment involved has few noxious sequelae, if there is sufficiently efficacious treatment available, if the disease is serious enough to justify the risks of diagnosis and treatment, and if the diagnostic label entails bearable social costs. On the other hand, one must try to avoid false positives, even at the risk of increased false negative diagnoses, if for instance there is not in fact a successful treatment (and the treatment has noxious side effects) or if the diagnostic label carries social risks that outweigh the benefits of treatment. In this respect, one might consider, for example, the social costs of being labeled a schizophrenic. From still another perspective, one must be concerned about false negative diagnoses if, and only if, the disease is serious and there is a sufficiently promising treatment with manageable side effects, and low enough social costs, consequent to the label involved.

Judgments about the prudent balance of benefits and costs are reflected in indications for making a diagnosis. They set the threshold of facts that ought to be established to make a diagnosis, given a risk of being wrong and therefore of needlessly exposing the patient to danger; for example, determining whether or not to recommend that a severely depressed suicidal patient be hospitalized. Even the acquisition of data to make a diagnosis involves risks of anxiety and social loss, as in studies of schizophrenia.[34] Thus, clinical judgment involves the issues of the costs of holding a particular state of affairs to be the case, even if one knows that the probability of a particular diagnosis being true is always less than 100 percent. Ethical questions arise because of differing views of which balances of benefits and risks is justified. In fact, one must ask who in the end should participate in setting such balances, and on what grounds. If it is possible to do so, should the patient be consulted? What weight should be accorded by the psychiatrist to a subsequent acceptance or refusal of a diagnosis and its label by the patient? It should be noted that the concerns expressed in these questions are similar to those raised in medicine generally, with the possible exception of the concern for labeling, for which there are only a few parallel cases in somatic medicine; for example, syphilis and leprosy.

¶ Informed Consent

In a frequently cited essay on the subject of ethics in psychiatry, Fritz Redlich and Richard Mollica[106] present the view that "informed consent is the basis of all psychiatric intervention and that without it no psychiatric intervention can be morally justified." The only exception they allow is the case of a patient who is "judged incompetent to give his informed consent," in which case consent should be sought "through proper judicial channels."

Informed Consent in Psychiatric Practice

An interesting insight into the nature of the principle of informed consent in psychiatric practice can be gleaned from a consideration of the legal understanding of informed consent as mandated by the Department of Health, Education and Welfare (now Health and Human Services)* rules for consent in research, and which has been

*See references 72, 90, 91, 92, and 110.

expressed in much of case law. Together these developments direct attention to certain obligations arising from a patient's right to informed consent. The psychiatrist should respect the freedom and integrity of patients by keeping them apprised of their diagnosis, alternative methods of treatment, the risks and benefits of each treatment option, and the prognosis under each treatment modality. At the same time, however, there has been a tendency in the law to recognize limits on a strict adherence to informed consent.[28,49] For example, there might be a need to balance the principle of informed consent with prudential judgments concerning the benefit/cost ratio of premature and thus anxiety-provoking revelations in a therapeutic context. Thus, the psychiatrist might choose to time carefully the revelation of new diagnoses of schizophrenia or latent homosexuality when the diagnosis itself might be perceived as threatening by the patient.

Both deontological and teleological aspects of the principle of informed consent must be considered. As indicated earlier, the deontological understanding of moral rights, duties, and values makes respect for persons an indispensable condition for proper moral conduct. This mode of ethical analysis captures a central feature of the principle of informed consent: respect for autonomy. Recent philosophical analyses of the concept of autonomy have distinguished its two dimensions.[41] The first of these is authenticity: that is, any person's right to self-integrity, to choose and live out whatever values one wants. The second dimension of autonomy is independence: that is, the right of any person to control the circumstances of his own life. With this understanding, the practice of informed consent can be justified as enhancing autonomy, independent of the considerations of particular goods or evils such a practice might promote. Consideration of a patient's autonomy will lead to the practice of informing patients, in a timely and routine manner, about the features of their disease and the appropriate treatment for it.

The practice of informed consent can have beneficial results for the therapeutic relationship.[27] A patient may, with the psychiatrist's assistance and guidance, begin to appreciate that he creates problems for himself because of confused or even contradictory choices. Or the patient, through a process Isaac Franck[53] terms "reflexive thought," might discover that in certain circumstances he makes compulsive choices, a feature of his life that until then was hidden. In the absence of such information about themselves, patients will remain ignorant of their unconscious motivations and thus of the full ramifications of a mental disturbance, disorder, or illness. That is, in the absence of a practice of informed consent in psychiatric treatment, certain goods and values deriving from an increased awareness of obsessions and compulsions may be lost for the patient. Informed consent, because it increases the patient's knowledge of his disorder and promotes bonds of trust between the patient and the psychiatrist, leads to increased participation by the patient in his own care and treatment. Informed consent thus comes to be appreciated as aiding the therapeutic process and, in effect, becomes a central element in that process. This is but another way of saying that a right to informed consent can be claimed as a way of securing already well-recognized goods and values in the therapeutic process.

The deontological and teleological analyses of informed consent thus converge on, and provide a justification for, the principle of informed consent for which Redlich and Mollica argue.[106] Indeed, this principle of mutual participation is at the heart of a common model of psychiatric treatment.[132] Because many forms of mental disorder or illness are best approached through this model of patient-therapist interaction, the practice of informed consent encourages respect for the patient's autonomy, and protects, defines, and advances the patient's best interests.

There are, however, limits on a principle of informed consent, because mental disorders and illnesses can often imperil autonomy or distort choices for goods and values. Indeed, it is in such terms that the very meaning of many psychiatric diagnostic categories can be understood; for example, psychoses.

The severely psychotic will not be able to have full control of the basic circumstances of their lives or to choose values in a consistent and meaningful manner. In such cases the bases for informed consent may not obtain, because the patient is unable to render an informed consent. Generally, one implication of a psychiatric diagnosis is that competence is diminished. But does the same mental illness diminish autonomy equally in all those who suffer from it? And what does diminished autonomy imply for levels of competence in making decisions? These questions suggest that the connection between psychiatric diagnostic categories and levels of competence is not of a fixed or logical nature, but is more open-textured and nuanced. Close study of the issues involved is required before reliable assessments of that connection can be offered.[101]

This is a significant undertaking because in clinical judgments concerning diminished competence what is at stake are the future freedom and best interests of the patient. That is, the patient can become an incapacitated coworker. As a consequence, one must recognize that patients in such straits are susceptible to even well-intentioned manipulations. If the psychiatrist, for example, acts on the Golden Rule, he may inadvertently choose a view of the good life for the patient that is inconsistent with what the patient might have chosen were he not incapacitated. As a counterpoise, what is required is a practice that respects what has been the patient's coherent choice of nondestructive values. That is, decision of consent to treatment should be fashioned in a manner that is maximally consistent with the patient's previous history and with a view toward avoiding the sort of choices that resulted in the present incapacitation. In short, for such patients there ought to be a procedure of substituted consent based on the goal of reestablishing competence and on the best estimation of what values the patient would choose to act upon were he fully competent to choose.

The question at this stage is: Who makes determinations of competence? And, in cases of substantially diminished competence, who should provide the substituted consent? The answers turn on a consideration of the goods and values that substituted consent is meant to protect. The patient thought to be incompetent should be shielded from those moral judgments ingredient in the process of both clinical judgment and treatment, whose consequences may not redound to his benefit or maximal future freedom. This shielding on matters of great moment can best be provided by a third party who acts as an advocate for the patient's interests. In our society an institutional practice most closely approximating a formal assessment of a patient's competence and needs by a third party would be court review. Thus, on teleological grounds, Redlich and Mollica's proposal[106] for court review of consent to treatment by the mentally incapacitated is justifiable.[35]

A clear disadvantage of adopting such a procedure is that it is both cumbersome and time-consuming. Our earlier reflections on the value dimensions of clinical diagnosis, labeling, and judgment, however, suggest that these inconveniences may prove a prudent price to pay to protect patients from potentially destructive alterations in the texture of rights and duties, which the practice of informed consent is meant to protect. Still, most concrete choices of therapy will need to remain in the hands of the therapists or the institutions that the court or guardian chooses. It will surely be inconvenient if not ridiculous to review all such choices in a formal fashion. Nor will it be justified in most circumstances to place the burden of proof on families and guardians with respect to the propriety of every choice they make on behalf of the patient. Avenues of review and protection, however, should be provided, as they are incorporated (at least in theory if not in fact) in procedures for determining an individual's level of competence.

Informed Consent in Psychiatric Research

As in other areas of medicine, psychiatry depends on research to make new discoveries and to test new therapies. Historically, the limitation on medical research was the classic "do no harm" principle: So long as harm was not done to the patient, research in medicine was permissible, even obligatory.[140] Since the revelation of Nazi medical war crimes at the Nuremberg Trials, however, the research community and the public have recognized the need for additional protection of subjects of medical research. Indeed, beginning with the Nuremberg Code[93] a consistent view has been developed: The voluntary and informed consent of the research subjects is the preeminent ethical consideration. This view is explicitly set out in the various procedural safeguards established in the Department of Health, Education and Welfare's National Commission on the Protection of Human Subjects in Biomedical and Behavioral Research, including guidelines for research on the mentally infirm,[90] as well as for the use of such techniques as psychosurgery.[81]

Here, too, both deontological and teleological analyses converge. On the one hand, a rigorous application of the principle of informed consent in psychiatric research respects the subject's autonomy. On the other hand, such a practice minimizes the chances that an individual's or group's goods and values will be sacrificed for the sake of securing the goods and values of others. Obviously, the one best situated to make such value judgments is the potential research subject, and the principle of informed consent protects his freedom to make such judgments on a voluntary and fully informed basis. At the same time, the principle of informed consent is meant to protect those who might be vulnerable to manipulation or public forms of coercion, particularly those suffering from mental disorders.[34] These two considerations have a special bearing on informed consent for research in psychiatry. The first bears on problems of deception in research, and the second on the use of institutionalized patients as subjects of research.

Research Involving Deception

In obtaining informed consent of a potential subject for psychiatric research, the person must be told of the method(s) to be employed in the research project. After all, for consent to be meaningful, it must be consent to the particular research project and not to research in general. Thus, if a research project on behavioral responses to stressful or anxiety-provoking situations will employ concealed observers to record and evaluate each subject's responses, the potential research subject should be informed of the possibility that his reactions will be monitored. The psychiatric researcher, however, may be concerned that such information is likely to render the data useless. It may well be that there is no clear-cut, final resolution of the ethical dilemma that emerges here. On the one hand, obtaining informed consent respects the subject's autonomy and enhances the likelihood that he will be a more willing and thus cooperative participant in the research project. On the other hand, deception may advance the goals of the project while sacrificing respect for autonomy, thus challenging the project's integrity and risking a cynical view on the part of others about such research when the deception is discovered, as in the much-cited studies on homosexuality by Laud Humphries.[68,139]

A similar sort of problem occurs in drug experiments matching a possible effective agent and a placebo, as in the testing of tranquilizers. In such circumstances, must potential research subjects be informed that placebos will be used? A strict application of the principle of informed consent requires that we answer "yes." But then the "placebo effect" may be diminished and the reliability of resultant data called into question. The use of double-blind trials in such cases rescues the psychiatric researcher from the dilemma. If the subject consents to the possibility of deception, the researcher at once maintains the ignorance of both the subject and the administrator of the "drugs" tested and gains the informed consent of the subject to participate.

Unfortunately, this sort of resolution to the problem of deception in research may not be possible in research of the first type, which uses concealed observers, unless a blanket permission has been given to some form of deception. There are parallel examples in ordinary life of permission to be subjected to various forms of deception, for example in the game of poker where permission is given to some, though not all, forms of deception. The clear trend in the public debates on these matters and in the National Commission's deliberations,[91] however, is to emphasize the practice of informed consent, at the expense of the possible good to be gained from research that does not easily accommodate such a practice. That is, a choice has been made to protect individual freedom and individual choices at the possible expense of a larger, common good. Like other ethical judgments involving prudential balances of risks and benefits, these should not be regarded as final and forever certain, but should be routinely subject to review and evaluations.

RESEARCH INVOLVING THE INSTITUTIONALIZED MENTALLY ILL

The use of institutionalized mentally ill patients as subjects of psychiatric research is attractive, for it maximizes a number of conditions for effective research. One is dealing with an easily identifiable and controllable population. Moreover, using this patient population for research overcomes the difficulty of securing sufficient numbers of noninstitutionalized patients to serve as subjects in research that is sometimes promising and thus felt to be important or even urgent. Finally, it is often difficult to find a sufficient number of individuals with a particular affliction except in an institutional setting. Thus, one might argue, unless we move ahead with research programs involving institutionalized subjects, we shall impede the development of possibly more effective therapies, thus harming the interests of those who may be afflicted in the future with the mental illness to be studied.

The problem, though, is that obtaining informed consent from institutionalized patients may be difficult, if not impossible. The potential subjects may be so incapacitated by their mental illness, their treatment regimen, or the institutionalization itself that they are incompetent to render an informed consent to participate in a research project. Here again, as in the case of informed consent for psychiatric treatment, substituted consent by a third party would be an appropriate practice to adopt.[106]

The issues to be considered by the psychiatric researcher can, however, be more complex than those involved in treatment decisions, which are therapeutic in nature. First, a research protocol surely may involve therapeutic measures. Second, it may be designed to determine if a treatment regimen is in fact therapeutic for the subject's condition. Third, research may be interested in more basic and sustained study of psychiatric disorders in and of themselves. The acquisition of such knowledge is not, by itself, therapeutic for the subjects of the research, though it may someday lead to benefits for others similarly affected. Those goods, one might argue, would be jeopardized by an overly strict practice of informed consent. Thus, along with respect for autonomy, important goods and values are at stake in psychiatric research.

Here, it would seem, we are faced with a conflict between respect for freedom and an interest in the goods and values of the research subject and future patients, for which there is no readily apparent, exclusive solution. On the one hand, out of respect for autonomy and a keen appreciation for the already incapacitated status of the institutionalized mentally ill, one might argue for provision of special protection or even complete immunity from research for those already at increased risk. On the other hand, a vigorous research program may increase the likelihood that research subjects or others in the future could be deinstitutionalized, brought to the point that their disorder is manageable with minimal supervision, or even cured. With such a view one might argue that such research ought to be undertaken. In short, we are faced with a classical ethical problem: how to strike a justifiable balance between assuring some level of autonomy, while still achieving certain goods and values. This conflict should be a familiar one to psychiatrists, since it mirrors a basic conflict between psychiatric roles of physician and therefore cure-giver for a particular patient, and scientist and therefore researcher.

¶ Confidentiality and Privacy

The moral issues raised by the practice of confidentiality may be understood in terms of rights to privacy, including the right to expect that confidences will be kept and that areas of privacy will not be intruded upon without consent. There are two ways in which privacy can be intruded upon. First, others can directly or indirectly disrupt one's person or circumstances. Second, information about oneself can be released by others into the public domain. It is the latter feature of violation of privacy that is especially pertinent here, since the obligation of confidentiality is designed to prevent dispersion of private information into the public arena.

One way to understand confidentiality is to perceive an analogue with the patient's personal property: The patient's autonomy is his to dispose of as he wishes. That is, issuing from autonomy as both a value and a constraint is a right to privacy: One claims the right to control information in order to protect the integrity of one's person. Thus, the burden of proof falls upon those who would use that information in ways other than those that the patient would permit. In other words, information about a patient, revealed by a patient under the assumption that it will stay between the patient and the psychiatrist, is protected by the notion of a moral community founded in respect for freedom, not force. Unauthorized disclosure of information about patients would, therefore, be a form of violence against patient autonomy. The scope of the right to confidentiality is broad, encompassing all information about patients obtained under the guarantee of confidentiality. In addition, the canons of informed consent apply: Permission by the patient to release information must be explicit, voluntary, and informed.

Now, the shortcoming of this view is that it does not apply so readily to the patient who is less than fully autonomous. It may be difficult, if not impossible, to gain permission to release information about the patient who is substantially incapacitated by a mental disorder or illness, even when it might be in the patient's own best interests. For patients whose autonomy is not wholly intact, how should psychiatrists understand their obligations of confidentiality?

In seeking an answer to this question, we should attend to the basic purposes of professions like psychiatry. We should, in part, appeal to the goods and values that shape the relationship between patient and psychiatrist. An interesting suggestion along these lines has been made by Stephen Toulmin,[135] who has argued that confidentiality is a key feature to that relationship, because it is a means to protect the patient in situations of vulnerability. Patients are vulnerable (1) to mental disorders and illnesses, (2) to the psychiatrist because of special feelings of trust and dependence, and (3) to society because of the increasingly strong interest in psychiatric patients by third parties. Thus, the patient's privacy should be accorded special protection. Moreover, choices of values may be distorted by illness, subjected to manipulation or subtle coercion by psychiatrists, or jeopardized or even sacrificed to the goods and values of others. By protecting and sustaining the patient's own best interests in these respects, in particular interests concerning privacy, confidentiality emerges as a fundamental obligation within the patient-psychiatrist relationship. Thus, interestingly, deontological and teleological lines of reasoning converge, resulting in a strong obligation to protect the patient's privacy.

There are, however, a growing number of conflicts causing concern and anxiety for psychiatrist and patient alike. On the one hand, such accepted practices as having secretaries type or transcribe notes, or presenting case histories of patients at staff conferences, can raise substantial risks to confidentiality. On the other hand, conflicts between different social institutions can raise special problems. What, for example, is the proper disposition of divergent responsibilities on the part of the psychiatrist in private practice whose patient has expressed deep and abiding hostility toward another person and now confides that he intends to physically harm or even kill that person? Should the psychiatrist in such circumstances reveal to third parties that they are in danger of harm from a patient who has expressed anger and growing hostility toward them? The *Tarasoff* case in California has described the legal conflict here: The privilege to warn endangered third parties has been replaced by a *legal duty* to warn.[10,61] But what of the moral obligations at stake here? How then can the conflict between the moral obligation to maintain confidentiality and the moral responsibility to warn third parties be resolved?

One might consider here the extent to which the psychiatric profession should guarantee confidentiality in the face of court subpoenas for information when there exists the risk of danger to third parties. The choice of a rule for practice in this regard frames a profession of a particular ethical character. What should that character be regarding confidentiality?

Consider the view that it is legitimate or even obligatory for psychiatrists to report to police authorities that a patient is likely to be dangerous to a third party. One can, on utilitarian grounds (a form of teleological ethics that holds that the right act is the one that ensures the greatest good for the greatest number) argue that such a practice is justifiable if and only if one or more of the following conditions is satisfied: (1) possibly dangerous individuals will *not* be dissuaded from seeking treatment when they know that full confidentiality will not be offered to them; (2) psychiatric treatment does not actually diminish the threat of such persons to third parties; or (3) individuals in a society would, as a rule, feel greatly ill at ease at the thought that a psychiatrist would not make such a report, even if a practice of strict confidentiality would in the long run actually diminish their risk of violence at the hands of such patients. (That is, if strict confidentiality would effectively bring individuals to treatment, it would reduce the general level of risk. In the last case, if one still did not allow strict confidentiality, a greater general value would have been assigned to the perturbation attendant to the thought of such strict confidentiality existing, than to the risk of the violence that strict confidentiality would diminish.) Such choices frame psychiatry as a profession more willing to be cognizant of the impact of certain practices on the common good, even at the expense of the goods and values of individual patients.

In contrast, one might argue that patients require at least one reliable sanctuary from conflicts with the interests of others and thus from the sanctions of society while they struggle to come to terms with their anxieties and mental conflicts. From such a view, one would urge that psychiatrists withholding patients' threats to others be accorded full privilege against criminal and civil liabilities. Thus, along the lines suggested at the beginning of this section, psychiatry can be regarded as a profession whose fundamental obligations are consistent with respect for freedom and the maintenance of the moral community, even when in some respects some larger, common goods and values might (occasionally) be diminished.

Thus, in deciding a practice of confidentiality for psychiatry, one would expect to find a conflict between two basic roles of the psychiatrist: (1) the psychiatrist as a particular patient's therapist and therefore protector of the patient's freedom and best interests; and (2) the psychiatrist as a public health officer and citizen and therefore responsible for the commonweal.[84,138] The preceding analysis shows that alternative resolutions of this conflict frame the moral character of the professions of psychiatry in starkly different ways.

Conflicts regarding confidentiality are heightened for psychiatrists who, because they are employed by someone other than the patient, have obligations, not just to their patients, but to their employers. These conflicts may be relatively minor, as in the case of a psychiatrist employed by a Health Maintenance Organization (HMO). The psychiatrist may recommend means of treatment more to maintain the cost effectiveness of the HMO than to aid the patient in the best manner possible. More serious conflicts, however, are likely to be encountered by industrial psychiatrists, military psychiatrists, school psychiatrists, and psychiatrists retained to evaluate individuals for job fitness, court proceedings, and the like. Consider the case of a psychiatrist who diagnoses a patient and discovers a condition that could prove to be very costly to his employers, as in the case of an alcoholic airline pilot or a soldier who fears combat.[77] On the one hand, the psychiatrist is bound to his patient by the obligation of confidentiality not to disclose his findings and their implications. On the other hand, he is obligated to his employers as well as to the public, and thus must report his diagnosis and its implications as to future fitness.[136] One escape from this dilemma may be found in the psychiatrist informing patients at the beginning of the relationship that he is bound by two sets of duties and that, when duties to the patient regarding confidentiality conflict with duties to their employer, the psychiatrist will take himself to be obligated to disclose certain or all of the relevant information to the employer—even if doing so might result in loss of pay or even the end of a career for the patient. The drawback of this approach, of course, is that it may result in less than fully frank disclosure by the patient and thus compromise the therapeutic process. The alternative seems to be not informing the patient of the built-in set of conflicts, which amounts to deception.

Similar conflicts arise in the context of therapy involving married couples. For example, in obtaining a sexual history a patient may relate details of an extramarital affair of which the spouse may be unaware. Such information is protected by the obligation of confidentiality, unless the patient explicitly and in advance consents to its being shared with the spouse. The advantage of such a practice is that it protects the psychiatrist from manipulation by the patient, in the form of imposing unjustifiable burdens on the psychiatrist or of drawing the psychiatrist unwillingly into a neurotic conspiracy against the other spouse. The patient is also protected from manipulation by the psychiatrist who may attempt to press one party to a level of candor with the other party to which the first (and perhaps the second) party has not consented.

¶ Patient Rights

This chapter began by noting that many moral disputes in psychiatry are expressed in the language of rights. It will close its inquiry into ethics in psychiatry by considering how rights language bears on a series of issues regarding treatment, nontreatment, and civil commitment. While some of the issues have already received a great deal of attention in the law, our interest will not be in strictly legal issues but rather in understanding how moral obligations of psychiatrists to their patients can be framed in response to rights claims (1) in treatment, (2) to treatment, (3) to refuse treatment, and (4) regarding civil commitment.

Rights in Treatment

One area in which rights claims have direct bearing on psychiatry is in rights in treatment. That is, in the therapeutic process itself respect for persons and an interest in securing certain goods frame psychiatrists' obligations to their patients. This involves: (1) therapies involving barter systems; (2) ethical dimensions of deinstitutionalization; and (3) patient access to records. As will become clear, each of these rights claims is an implication, in practice, of a more basic moral right: informed consent.

Some therapeutic styles involve a form of barter economy, especially in institutional settings where part of the goal of therapy is to encourage and sustain responsible behavior. Thus, for example, a weekend pass is offered as a kind of reward for a set of specified actions. Some patients may be best treated through the use of such barter measures. Indeed, it may be that these patients only respond well to therapy when it includes the features of a barter economy. Thus, important goods are secured for patients through the employment of this practice. At the same time, however, such practices may involve subtle and perhaps unconsented to (by the patient) forms of manipulation. Such concerns can be mitigated, though only to some extent. After all, patients who are best suited to such a therapeutic approach may already have their autonomy substantially diminished by their mental illness. And not employing barter practices will not only not help such patients but might also result in further diminishment of autonomy.

The issues at this point intertwine paternalism and informed consent. Recent philosophical analyses[40,51] of the concept of paternalism indicate (the view is not necessarily a settled one) that two criteria must be satisfied to justify coercion on paternalistic grounds: (1) the person to be coerced must be in a state of substantially diminished autonomy; and (2) the coercion must be the only means to avoid serious, far-reaching, and irreversible diminishment of autonomy in the future.

One may then argue that such barter measures will be appropriate when they are designed to restore autonomy. Thus, forms of barter economies that diminish patient autonomy, and thus amount to coercion, could not be justified on the account just given. Here, respect for patient's freedom acts as a side constraint on institutional practice and forms the basis of a right in treatment: the protection against even subtle forms of coercive therapies.

Similar rights claims can be advanced on behalf of patients who are appropriate candidates for the transition from institutionalized to deinstitutionalized care. Presumably patients become appropriate candidates for such a change because (1) deinstitutionalized care will be of greater benefit to them, and (2) they can more or less manage the new responsibilities that will devolve upon them in a noninstitutionalized setting. A number of questions should be raised about this policy change. Is deinstitutionalization indeed what the patient really wants? Are these people being moved out of institutions where, at least in some cases, they might prefer to stay? Will these people, because of the area and conditions in which they will be newly located, be subjected to greater risks without commensurate benefits? Do we sufficiently understand the benefits and costs of deinstitutionalization, and are these benefits and costs explained to candidates for deinstitutionalization or their guardians? All of these questions focus on the issue of consent by candidates for deinstitutionalization. Long-range studies and policy recommendations should take into account these questions and the issues they raise.

Given a justifiable emphasis on a practice of informed consent in psychiatric practices, questions concerning patient access to records will inevitably arise. Does the patient have the right regarding his own records and what are the justifiable limits on these rights? Before attempting an answer, it will be useful to distinguish *rights to* and *rights in* patients' records. The former set of rights can be claimed by those, such as psychiatrists and hospitals, who own the records and thus have special interest in them. The latter rights can be claimed by those who do not own the records but have participated in the production directly, or indirectly, and thus have rights *in* the record. These include the rights of consultant psychiatrists or family members whose remarks are recorded in the record but that were originally offered under a promise of confidentiality by the attending psychiatrist. In these circumstances, one's colleagues or the patient's family members can legitimately demand not to have their confidential reflections released to the patient. Thus, a patient's right to see these portions of his record may be overridden by claims of confidentiality on the part of third parties. Moreover, revelations of this material may not redound to the patient's benefit and progress in therapy, thus giving additional weight to the view that the right of access to records by patients may be a limited one.

The Rights to Treatment

Rights to treatment have been claimed principally on behalf of institutionalized psychiatric patients. Historically, people have been institutionalized without their consent for specialized care or psychiatric supervision for a variety of reasons, including: (1) public offensiveness—for example, compulsive public exposure; (2) long-term incapacity to function in our complex and demanding society, as in the case of those suffering profound mental retardation; (3) the collapse of alternative social institutions so that many in psychiatric institutions (especially public ones) are there, in part, because they have nowhere else to go; (4) dangerousness to others; and (5) dangerousness to self. It is with respect to the last of these, dangerousness to self, that the concept of a right to treatment has been articulated in legal terms. From an ethical standpoint, such a right can be articulated in a fairly straightforward manner. If patients have been institutionalized because of dangerousness to self, for example, presumably they have been treated in this way "for their own good." This phrase can be analyzed in terms of goods served by traditional goals in psychiatry: diagnosis and effective treatment to restore a person to mental well-being. Thus, in order to secure the goods and values of psychiatry for institutionalized patients, courts have claimed on their behalf: (1) a right to treatment, and (2) a right to public funding to provide adequate treatment. The right to treatment, if there is an effective treatment, secures the means to restoring autonomy. In addition, it serves the good of returning to society an intact, productive citizen. The right to funding for adequate levels of treatment is tied logically to the right to treatment, as its necessary condition. That is, a *quid pro quo* arrangement is analyzed in terms of a teleological justification for the conjoint rights to treatment and to funding to provide adequate treatment.[11,15]

The Right to Refuse Treatment

The right to refuse psychiatric[61,66] as well as medical[82] treatment is based principally on concerns for respect for autonomy. If, in the moral community, we accord persons the right to control the fundamental circumstances of their lives, then surely refusing psychiatric treatment falls within the scope of autonomy and can be recognized formally in a right to that effect. This view is coincident with the already recognized legal and moral right of competent adult citizens, once fully informed of their condition and need for medical treatment, to refuse such treatment. To coerce medical or psychiatric treatment in such cases is clearly to resort to force and thus to violate the notion of a moral community.

Again, a recurrent problem is that many patients are substantially diminished in their capacity to make free choices of values, a necessary condition for the right to refuse psychiatric treatment. Thus, the question of the right of such patients to refuse treatment is arguable. What is the conscientious psychiatrist to do when a substantially less than autonomous patient refuses treatment because of his mental illness? It is unlikely that the long-range interests of such patients will be well served by respect for their refusals. At the same time, respect for freedom cautions against the possibility of the psychiatrist imposing his own view of the good life on the patient, because commitment, by itself, may not eliminate the right to informed consent to treatment. Such considerations might lead one to argue that review of the refusal of treatment should be provided, perhaps by courts, as in the context of informed consent for treatment of the mentally incapacitated. Such a mechanism seems best suited to satisfy the ethical requirements of determining the patient's best interests, the actual absence of autonomy, and that reasonable criteria for incompetence have been fairly and impartially applied (that is, in a way that all would accept, even for themselves).

Civil Commitment

Decisions regarding involuntary commitment are made on a variety of grounds, but principally on dangerousness to self and to others.[20,30,61] The latter concerns the state's legitimate interest in providing for the safety of its citizens. The arguments made in support of such authority are traditional ones, appealing to constitutionally mandated police power and the rights of innocent third parties to protection from gratuitous acts of violence against their person or possessions. The former involves committing someone involuntarily, for his own best interests, for his "own good." This practice raises fundamental and troubling ethical issues concerning the proper bounds of psychiatric and (taken broadly) institutional authority.

The first set of issues concern the special status of psychiatric diagnoses. First, alone among physicians, psychiatrists possess a peculiar sanctioned authority: to recommend involuntary commitment. Second, there are subtle incentives—tied to risks of error on the part of the psychiatrist—to commit people involuntarily. Little or no harm comes to the psychiatrist who mistakenly commits someone. But if the psychiatrist mistakenly diagnoses someone as a noncandidate for commitment and that person subsequently harms himself or others, then the psychiatrist may be open to malpractice actions as well as social opprobrium. Thus, there may be considerable legal and social pressure to diagnosis in favor of diminished mental (and consequently moral and legal) status.

A second set of issues centers on the concept of dangerousness.[117] Is dangerousness, for example, a disease or medical concept, or is it a social concept? And how best (or at all) can dangerousness be determined?[101] Even if we could answer these questions in a reliable way, a third set of issues would emerge, for involuntary civil commitment involves justifying coercive paternalistic intervention in the choices of a citizen for his "own good." Thus, for example, on the basis of criteria for justifying such paternalism, a rational willed dangerousness to oneself in the form of suicidal intentions or actions, for example, might not warrant involuntary commitment. Recall here that respect for freedom and an interest in preserving the moral community require that each of us respect another's free choice of values. It may be unfortunate or even tragic that someone comes to view suicide rationally, deliberately, and with due consideration as the only solution, but such is the scope of freedom. Here, themes concerning manipulation of patients have bearing. An interest in the moral community and, consequently, respect for freedom together entail an obligation for the psychiatrist not to impose his view of the good life on the autonomous patient.

By contrast, a teleological approach, expressed perhaps in utilitarian terms, might take a dimmer view of a practice of so-called "rational" suicide because of likely or feared negative consequences for important social institutions such as being a parent, employer, employee, public servant, or friend. With this in mind, psychiatrists can attempt a more intrusive paternalism, a development that might give pause to those who practice psychiatry in societies committed to individual freedom. The upshot of this teleological view of matters is, of course, increased obligations on the part of psychiatrists to discourage suicidal intentions and to assist in the commitment and treatment of those who arrive at even a "rational" determination to end their lives. Perhaps, as Daniel Creson[33] has suggested, civil commitment does in fact function as an alternative social means (even if often unjustified) of dealing with deviance.

A final complication here are religious injunctions against suicide. From a number of religious perspectives, suicide is regarded as one of the most serious offenses a believer can commit against God, and hence it is almost always absolutely prohibited. Based on the notion of respect for autonomy and the analysis of defects in the Golden Rule approach, the psychiatrist should avoid imposing his own religious views on the patient, either directly by recommending commitment or indirectly in arriving at a clinical judgment of incompetence. The case of a patient whose religious views prohibit suicide, but who nonetheless is considering such an act, is more complex. Should the psychiatrist regard such intentions as contradictory and thus evidence of diminished autonomy and therefore a warrant for protection? If so, then the psychiatrist encounters the problems of the role of a moral enforcer. Or should the psychiatrist take expressions of suicidal intentions as a genuine change of belief? If so, then one must thread one's way through the difficulties involved in making such a clinical judgment in a reliable way.

On all of these issues, the moral options clearly diverge, and the profession of psychiatry is faced with a stark choice. If it moves in one direction, psychiatry will be conceived as a liberal profession, in the sense that it is committed to preserving and fostering the moral community, perhaps at some cost to particular social institutions. If it moves in the other direction, psychiatry will be conceived as a profession committed to the preservation of fundamental social institutions, perhaps at some cost to the moral community. The resolution to be found here, perhaps, is not a factual one that we could somehow discover and thus, in an empirically reliable way, settle upon. Instead, the resolution will have to be more created than discovered, on the basis of fundamental moral choices whose character and bearing on psychiatry this chapter has attempted to sketch.

¶ Conclusion

The final resolution of ethical conflicts may not always be possible. This is so because the logic of pluralism in moral values must be broad enough to embrace the full spectrum of free choices for values by moral agents. The risk of such a generous view of ethics is that it may not be able to resolve tough issues in a way that will prove satisfactory, once and for all, to everyone. That this is sometimes the case has been vividly illustrated in the issue of civil commitment. Even on more tractable issues, where diverse ethical approaches begin to assume the form of coherent analyses, difficult and fundamental questions nevertheless emerge, as in the case of confidentiality. The most urgent of these questions addresses the ethical center of the profession of psychiatry: How shall it frame its enterprise so that it at once protects and sustains the notion of a moral community not based on force, while fostering those moral goods and values to which persons freely commit themselves? It seems reasonable that the answer to this question cannot be discovered, but instead must be created as the fruit of sustained inquiry into the complex dimensions of ethics in psychiatry.

¶ Bibliography

1. AMERICAN CIVIL LIBERTIES UNION (Due Process Committee). "Proposed Policy Statement: Medical Experimentation on Human Beings," Statement presented to the Board of Directors, American Civil Liberties Union. September 20, 1973, New York.
2. AMERICAN HOSPITAL ASSOCIATION. "A Patient's Bill of Rights," in Reich, W. T., ed., *Encyclopedia of Bioethics.* New York: MacMillan, 1978, pp. 1782–1783.
3. AMERICAN MEDICAL ASSOCIATION. *Code of Medical Ethics.* New York: H. Ludwig, 1848.
4. AMERICAN MEDICAL ASSOCIATION (Judicial Council). *Opinions and Reports of the Judicial Council, Including the Principles of Medical Ethics and the Rules of the Judicial Council.* Chicago: American Medical Association, 1971.
5. AMERICAN PSYCHIATRIC ASSOCIATION. "The Principles of Medical Ethics with Annotations Especially Applicable to Psychiatry," *American Journal of Psychiatry,* 130 (1973): 1058–1064.
6. AMERICAN PSYCHIATRIC ASSOCIATION.

Ethical Standards, in Wolman, B. B., ed., *International Encyclopedia of Psychiatry, Psychology, Psychoanalysis and Neurology,* vol. 1. New York: Van Nostrand Reinhold Co., 1977, pp. 456–461.

7. AMERICAN PSYCHIATRIC ASSOCIATION. *Diagnostic and Statistical* Manual (DSM-III). Washington, D.C.: American Psychiatric Association, 1980, pp.281–283.
8. AMERICAN PSYCHOLOGICAL ASSOCIATION. *Ethical Principles in the Conduct of Research with Human Participants.* Washington, D.C.: American Psychological Association, 1973.
9. ANNAS, G. *The Rights of Hospital Patients: The Basic ACLU Guide to a Hospital Patient's Rights.* New York: Avon 1975.
10. ———. "Confidentiality and the Duty to Warn," *Hastings Center Report,* 6 (1976): 6–8.
11. ———. "O'Connor v. Donaldson, Insanity Inside Out," *Hastings Center Report,* 6 (1976): 11–12.
12. ———. "Psychosurgery: Procedural Safeguards," *Hasting Center Report,* 7 (1977): 11–13.
13. AYD, F., ed. *Medical, Moral and Legal Issues in Mental Health Care.* Baltimore: Williams & Wilkins, 1974.
14. BANCROFT, J. "Homosexuality and the Medical Profession: A Behaviorist's View," *Journal of Medical Ethics,* 1 (1975): 176–180.
15. BAZELON, D. L. "The Right to Treatment: The Court's Role," *Hospital and Community Psychiatry,* 20 (1969): 129–135.
16. BERGER, M. M. "Legal, Moral and Ethical Considerations," in Berger, M. M. ed., *Videotape Techniques in Psychiatric Training and Treatment.* New York: Brunner/Mazel, 1978, pp. 293–315.
17. BERNAL Y DEL RIO, V. "Psychiatric Ethics," in Freedman, A., Kaplan, H., and Saddock, B., eds., *Comprehensive Textbook of Psychiatry,* vol. 2, 2nd ed. Baltimore: Williams & Wilkins, 1975, pp. 2543–2552.
18. BIEBER, I. "A Discussion of Homosexuality: The Ethical Challenge." *Journal of Consulting and Clinical Psychology,* 44 (1976): 163–166.
19. BLACKSTONE, W. "The American Psychological Association Code of Ethics for Research Involving Human Participants: An Appraisal," *Southern Journal of Philosophy,* 13 (1975): 407–418.
20. BLOMQUIST, C. "From the Oath of Hippocrates to the Declaration of Hawaii," *Ethics in Science and Medicine,* 4 (1977): 139–149.
21. ———. "Some Ethical Problems in Psychiatry," *Ethics in Science and Medicine,* 6 (1979): 105–114.
22. BOORSE, C. "On the Distinction Between Disease and Illness," *Philosophy and Public Affairs,* 5 (1975): 49–68.
23. BRANDT, R. *Ethical Theory: The Problems of Normative and Critical Ethics.* Englewood Cliffs, N.J.: Prentice-Hall, 1959.
24. BREGGIN, P. "Psychotherapy as Applied Ethics," *Psychiatry,* 34 (1971): 59–74.
25. CAMPBELL, A. V., ed. *Journal of Medical Ethics.* London, Eng.: Society for the Study of Medical Ethics, Professional and Scientific Publication.
26. CARTWRIGHT, S. A. "Report on the Diseases and Physical Peculiarities of the Negro Race," *New Orleans Medical and Surgical Journal,* 7 (1850): 707–709.
27. CASSELL, E. "Informed Consent in the Therapeutic Relationship: Clinical Aspects," in Reich, W. T. ed., *Encyclopedia of Bioethics,* New York: Macmillan, 1978, pp. 767–770.
28. CASSIDY, P. "The Liability of Psychiatrists for Malpractice," *University of Pittsburgh Law Review,* 36 (1974): 108–137.
29. CHODOFF, P. "Involuntary Hospitalization of Political Dissenters in the Soviet Union," *Psychiatric Opinion,* 11 (1974): 5–19.
30. ———. "The Case for Involuntary Hospitalization of the Mentally Ill," *American Journal of Psychiatry,* 133 (1976): 496–507.
31. CLARE, A. W. "In Defence of Compulsory Psychiatric Intervention," *Lancet,* 1 (1978): 1197–1198.
32. COLES, E. "The Ethics of Psychiatry," *American Journal of Psychiatry,* 131 (1974): 231.
33. CRESON, D. L. "Function of Mental Health Codes in Relation to the Criminal Justice System," in Engelhardt, H. T., Jr., and Brody, B., eds., *Mental Illness: Law and Public Policy.* Dordrecht, Holland: D. Reidel, 1980, pp. 211–219.
34. CURRAN, W. "Ethical and Legal Consideration in High-Risk Studies of Schizophrenia," *Schizophrenia Bulletin,* 10 (1974): 74–92.
35. ———. "The Saikewicz Decision," *New Eng-*

land Journal of Medicine 298 (1978): 499–500.

36. ———. "The Supreme Court and Madness: A Middle Ground on Proof of Mental Illness for Commitment," *New England Journal of Medicine,* 301 (1979): 317–318.
37. DAVIS, J. E. "Rights of Chronic Patients," *Hospital and Community Psychiatry,* 29 (1978): 39.
38. DAVISON, G. C. "Homosexuality: The Ethical Challenge," *Journal of Consulting and Clinical Psychology,* 44 (1976): 157–162.
39. DONALDSON, K. *Insanity Inside Out.* New York: Crown, 1976.
40. DWORKIN, G. "Paternalism," *Monist,* 56 (1972): 64–84.
41. ———. "Autonomy and Behavior Control," *Hastings Center Report,* 6 (1976): 23–28.
42. EISENBERG, L. "Psychiatry and Society: A Sociobiologic Synthesis," *The New England Journal of Medicine,* 296 (1977): 903–910.
43. ENGELHARDT, H. T., JR. "Ideology and Etiology," *The Journal of Medicine and Philosophy,* 1 (1976): 256–268.
44. ENGELHARDT, H. T., JR., and BRODY, B. *Mental Illness, Law, and Public Policy.* Dordrecht, Holland: D. Reidel, 1980.
45. ENGELHARDT, H. T., JR., and MCCULLOUGH, L. "Confidentiality in the Consultant-Liaison Process: Ethical Dimensions and Conflicts," in Kimball, C., ed., *The Psychiatric Clinics of North America,* vol. 2. Philadelphia: W. B. Saunders Co., 1979, pp. 403–413.
46. ENGELHARDT, H. T., JR., and SPICKER, S., eds. *Evaluation and Explanation in the Biomedical Sciences.* Dordrecht, Holland: D. Reidel, 1975.
47. ———. *Mental Health: Philosophical Perspectives.* Dordrecht, Holland: D. Reidel, 1978.
48. ENNIS, B., and SIEGEL, L. "The Rights of Mental Patients: The Basic ACLU Guide to Patient Rights," *American Civil Liberties Handbook.* New York: Avon, 1973.
49. EPSTEIN, G. "Informed Consent and the Dyadic Relationship," *Journal of Psychiatry and Law,* 6 (1978): 359–362.
50. FANIBANDA, D. K. "Ethical Issues of Mental Health Consultation," *Professional Psychology,* 7 (1976): 547–552.
51. FEINBERG, J. "Legal Paternalism," *Canadian Journal of Philosophy,* 1 (1971): 105–124.
52. FOSTER, H. "The Conflict and Reconciliation of the Ethical Interests of Therapist and Patient," *Journal of Psychiatry and Law,* 3 (1975): 39–61.
53. FRANCK, I. "Spinoza, Freud, and Hampshire on Psychic Freedom," in Smith, J. H., ed., *Thought, Consciousness, and Reality.* New Haven: Yale University Press, 1977, pp. 257–309.
54. FREUD, S. "Moses and Monotheism," in Strachey, J., ed., *The Standard Edition of the Complete Psychological Works of Sigmund Freud,* vol. 23, London: Hogarth Press, 1964, pp. 2–127. (originally published in 1939).
55. FRIED, C. *Right and Wrong.* Cambridge, Mass.: Harvard University Press, 1978.
56. GAYLIN, W. "The Patient's Bill of Rights," *Saturday Review (of the Sciences),* 1 (1973): 22, 24.
57. ———. "On the Borders of Persuasion: A Psychoanalytic Look at Coercion," *Psychiatry,* 1 (1974): 1–9.
58. ———. "Scientific Research and Public Regulation," *Hastings Center Report,* 3 (1975): 5–7.
59. GAYLIN, W., MEISTER, J., and NEVILLE, R., eds. *Operating on the Mind: The Psychosurgery Conflict.* New York: Basic Books, 1975.
60. GREEN, R. "Homosexuality as a Mental Illness," *International Journal of Psychiatry,* 10 (1972): 77–98.
61. GRIFFITH, E. J., and GRIFFITH, E. E. H. "Duty to Third Parties, Dangerousness, and the Right to Refuse Treatment: Problematic Concepts for Psychiatrist and Lawyer," *California Western Law Review,* 14 (1978): 241–274.
62. GUREL, B. "Ethical Problems in Psychology," in Wolman, B., ed., *International Encyclopedia of Psychiatry, Psychology, Psychoanalysis, and Neurology,* vol. 4, New York: Van Nostrand Reinhold, 1977, pp. 377–380.
63. HALE, W. D. "Responsibility and Psychotherapy," *Psychotherapy: Theory, Research, and Practice,* 13 (1976): 298–302.
64. HALLECK, S. L. "Another Response to 'Homosexuality: The Ethical Challenge,'" *Journal of Consulting and Clinical Psychology,* 44 (1976): 167–170.
65. HAUENSTEIN, C. B. "Ethical Problems of Psychological Jargon," *Professional Psychology,* 9 (1978): 111–116.

66. HOFFMAN, P. B. "The Right to Refuse Psychiatric Treatment: A Clinical Perspective," *Bulletin of the American Academy of Psychiatry and the Law,* 4 (1976): 269–274.
67. HOFLING, C. "Current Issues in Psychohistory," *Comprehensive Psychiatry,* 17 (1976): 227–259.
68. HUMPHRIES, L. *Tearoom Trade: Impersonal Sex in Public Places.* Chicago: Aldine Publishing Co., 1975.
69. KAHLE, L. R., and SALES, B. D. "Attitudes of Clinical Psychologists Toward Involuntary Civil Commitment," *Professional Psychology,* 9 (1978): 428–439.
70. KANT, I. *Foundations of the Metaphysic of Morals,* Beck, L. W., trans. New York: Liberal Arts Press, 1959.
71. KASSIRER, L. B. "Behavior Modification for Patients and Prisoners: Constitutional Ramifications of Enforced Therapy," *Journal of Psychiatry and Law,* 2 (1979): 245–302.
72. KATZ, J. "Informed Consent in the Therapeutic Relationship: Legal and Ethical Aspects," in Reich, W., ed., *Encyclopedia of Bioethics,* New York: Macmillan, 1978, pp. 770–778.
73. KENNEDY, E. *Human Rights and Psychological Research: A Debate on Psychology and Ethics.* New York: T. Y. Crowell, 1975.
74. KLERMAN, G. L. "Mental Illness, The Medical Model, and Psychiatry," *Journal of Medicine and Philosophy,* 2 (1977): 220–243.
75. KOLODNY, R. "Conference Report: Ethical Guidelines for Research and Clinical Perspectives on Human Sexuality," *Newsletter on Science, Technology, and Human Values,* 24 (1978): 17–22.
76. LEVINE, M. *Psychiatry and Ethics.* New York: George Braziller, 1972.
77. LONDON, P., et al. "Fear of Flying: The Psychiatrist's Role in War," *Hastings Center Report,* 6 (1976): 20–22.
78. MCEWAN, J. M., ed. *Ethics in Science and Medicine.* Elmsfold, N.Y.: Pergamon Press.
79. MCNAMARA, R. "Socioethical Considerations in Behavior Research and Practice," *Behavior Modification,* 2 (1978): 3–24.
80. MCNAMARA, R., and WOODS, K. "Ethical Considerations in Psychological Research: A Comparative Review," *Behavior Therapy,* 8 (1977): 703–708.
81. MACKLIN, R. "Ethics, Sex Research, and Sex Therapy," *Hastings Center Report,* 6 (1976): 5–7.
82. MALMQUIST, C. P. "Can the Committed Patient Refuse Chemotherapy?" *Archives of General Psychiatry,* 36 (1979): 351–355.
83. MARGOLIS, J. "The Concepts of Disease," *The Journal of Medicine and Philosophy,* 1 (1976): 238–255.
84. MERSKY, H. "In the Service of the State: The Psychiatrist as Double Agent: A Conference on Conflicting Loyalties," *Hastings Center Report,* Special Supplement, 8 (1978): 1–24.
85. ———. "Political Neutrality and International Cooperation in Medicine," *Journal of Medical Ethics,* 4 (1978): 74–77.
86. MICHELS, R. "Professional Ethics and Social Values," *International Review of Psychoanalysis,* 3 (1976): 377–384.
87. MOORE, R. A. "Ethics in the Practice of Psychiatry—Origins, Functions, Models, and Enforcement," *American Journal of Psychiatry,* 135 (1978): 157–163.
88. MULLOOLY, J. P., ed. *Linacre Quarterly.* Elm Grove, Wisconsin: National Federation of Catholic Physicians Guild.
89. THE NATIONAL COMMISSION FOR THE PROTECTION OF HUMAN SUBJECTS OF BIOMEDICAL AND BEHAVIORAL RESEARCH. *Report and Recommendations: Psychosurgery.* Bethesda, Md.: DHEW Publications No. (05) 77–0001, 1977.
90. ———. *Report and Recommendations: Research Involving Those Institutionalized as Mentally Infirm.* Bethesda, Md.: DHEW Publication No. (05) 78–006, 1978.
91. ———. *The Belmont Report: Ethical Guidelines for the Protection of Human Subjects of Research.* Bethesda, Md.: DHEW Publication No. (05) 78–0012, 1978.
92. NOZICK, R. *Anarchy, State, and Utopia.* New York: Basic Books, 1974.
93. "The Nuremberg Code," in "Permissible Medical Experiments," *Trials of War Criminals Before the Nuremberg Military Tribunals under Control Council Law No. 10: Nuremberg, October 1946–April 1949,* vol. 2. Washington, D.C.: U. S. Government Printing Office, 1949, pp. 181–182.

94. PARKER, L. "Psychotherapy and Ethics," *Cornell Journal of Social Relations,* 9 (1974): 207–216.
95. PARSONS, T. "The Sick Role and the Role of the Physician Reconsidered," *The Milbank Memorial Fund Quarterly,* 51 (1975): 257–278.
96. PATTISON, E. M. "Confusing Concepts About the Concept of Homosexuality," *Psychiatry,* 37 (1974): 340–349.
97. PELLEGRINO, E., ed. *Journal of Medicine and Philosophy.* Dordrecht, Holland: D. Reidel.
98. PERLMAN, B. "Ethical Concerns in Community Mental Health," *American Journal of Community Psychology,* 5 (1977): 45–57.
99. PESZKE, M. "Madness, Committal and Society: Psychiatry in the Service of the Law?" *Connecticut Medicine,* 40 (1976): 705–712.
100. PILOWSKY, I. "A General Classification of Abnormal Illness Behaviors," *British Journal of Medical Psychology,* 51 (1978): 131–137.
101. PLOTKIN, R. "Limiting the Therapeutic Orgy: Mental Patients' Right to Refuse Treatment," *Northwestern University Law Review,* 72 (1977): 461–525.
102. RAMSEY, P. *The Patient as Person.* New Haven: Yale University Press, 1970.
103. RAWLS, J. *A Theory of Justice.* Cambridge, Mass.: Belknap Press of Harvard University Press, 1971.
104. REDDAWAY, P. "The Next Victims of Soviet Psychiatric Terror," *New Society,* 45 (1978): 125–127.
105. REDLICH, F. "The Ethics of Sex Therapy," in Masters, W., Johnson, V., and Kolodny, R., eds., *Ethical Issues in Sex Therapy and Research,* Boston: Little, Brown, 1977, pp. 143–181.
106. REDLICH, F., and MOLLICA, R. "Overview: Ethical Issues in Contemporary Psychiatry," *American Journal of Psychiatry,* 133 (1976): 125–156.
107. REICH, W. T., ed. *Encyclopedia of Bioethics.* New York: Macmillan, 1978.
108. RICH, V. "Heading for Honolulu," *Nature,* 268 (1977): 578–579.
109. RISKIN, L. L. "Sexual Relations Between Psychotherapists and Their Patients: Toward Research or Restraint," *California Law Review,* 67 (1979): 1000–1027.
110. ROBITSCHER, J. "Informed Consent for Psychoanalysis," *The Journal of Psychiatry and Law,* 6 (1978): 363–370.
111. SALZMAN, L. "Truth, Honesty, and the Therapeutic Process, *American Journal of Psychiatry,* 130 (1973): 1280–1282.
112. SCHMITT, J. P. "An Ethical Basis for Social Control," *Psychology,* 11 (1974): 3–10.
113. SCHOOLAR, J., and GAITZ, C., eds. *Research and the Psychiatric Patient.* New York: Brunner/Mazel, 1975.
114. SCHULTZ, L. G. "Ethical Issues in Treating Sexual Dysfunction," *Social Work,* 20 (1975): 126–128.
115. SCHWED, H. J. "Protecting the Rights of the Mentally Ill," *American Bar Association Journal,* 64 (1978): 564–567.
116. SCHWITZGEBEL, R. K. "Suggestions for the Uses of Psychological Devices in Accord with Legal and Ethical Standards," *Professional Psychology,* 9 (1978): 478–488.
117. SHAH, S. A. "Dangerousness: A Paradigm for Exploring Some Issues in Law and Psychology," *American Psychologist,* 33 (1978): 224–238.
118. SIEGLER, M., and OSMOND, H. "The Sick Role Revisited," *Hastings Center Studies* 1 (1973): 41–58.
119. SILVERSTEIN, C. "Homosexuality and the Ethics of Behavioral Intervention: Paper 2," *Journal of Homosexuality,* 2 (1977): 205–211.
120. SMITH, M. B. "Psychology and Ethics," in Kennedy, E., ed., *Human Rights and Psychological Research: A Debate on Psychology and Ethics.* New York: T. Y. Crowell, 1975, pp. 1–22.
121. SPERO, M. H. "Homosexuality: Clinical and Ethical Challenges," *Tradition,* 17 (1979): 53–73.
122. SPITZER, R. L. "The Homosexuality Decision—A Background Paper," *Psychiatric News,* 9 (1974): 11–12.
123. STEINFELS, M. O. ed. *The Hastings Center Report.* Institute of Society, Ethics, and the Life Sciences. Hastings-on-Hudson, New York.
124. STOLZ, S., et al. *Ethical Issues in Behavior Modification: Report of the American Psychological Association Commission.* San Francisco: Jossey-Bass, 1978.
125. SULLIVAN, F. "Peer Review and Professional Ethics," *American Journal of Psychiatry,* 34 (1977): 186–188.

126. SULLIVAN, P. R. "Hospital Commitment and Civil Rights," *America*, 139 (1978): 105–106.
127. SZASZ, T. *The Myth of Mental Illness.* New York: Harper & Row, 1961.
128. ———. *The Ethics of Psychoanalysis.* New York: Basic Books, 1965.
129. ———. "Psychotherapy as Applied Ethics," *Psychiatry,* 34 (1971): 59–74.
130. ———. "The Case Against Compulsory Psychiatric Interventions," *Lancet,* 1 (1978): 1035–1036.
131. ———. "The Concepts of Mental Illness: Explanation or Justification?" in Engelhardt, H. T., Jr., and Spicker, S. F., eds., *Mental Health: Philosophical Perspectives.* Dordrecht, Holland: D. Reidel, 1978, pp. 235–250.
132. SZASZ, T., and HOLLENDER, M. "A Contribution to the Philosophy of Medicine: The Basic Models of the Doctor-Patient Relationship," *Archives of Internal Medicine,* 97 (1956): 585–592.
133. TANCREDI, L., and SLABY, A. *Ethical Policy in Mental Health Care: The Goals of Psychiatric Intervention.* New York: Protist/William Heinemann Medical Books, 1977.
134. TORREY, E., ed. *Ethical Issues in Psychiatry.* Boston: Little, Brown, 1968.
135. TOULMIN, S. "The Meaning of Professionalism: Doctor's Ethics and Biomedical Science," in Engelhardt, H. T., Jr., and Callahan, D., eds., *Knowledge, Value and Belief.* Hastings-on-Hudson, N.Y.: The Hastings Center, 1977, pp. 254–278.
136. VASTYAN, E. A. "Warriors in White: Some Questions About the Nature and Mission of Military Medicine," *Texas Reports on Biology and Medicine,* 32 (1974): 327–342.
137. VAN HOOSE, W., and KOTTLER, J. *Ethical Issues in Counseling and Psychotherapy.* San Francisco: Jossey-Bass, 1977
138. VEATCH, R. "Duty to Patient and Society," in Veatch, R., ed., *Case Studies in Medical Ethics.* Cambridge, Mass.: Harvard University Press, 1977, pp. 59–88.
139. WARWICK, D. "Tearoom Trade: Means and Ends in Social Research," *Hastings Center Studies,* 1 (1973): 27–38.
140. WITHINGTON, C. F. *The Relation of Hospitals to Medical Education.* Boston: Copples, Uphman and Company, 1886, pp. 14–17.

NAME INDEX

SUBJECT INDEX

therapists in understanding the relationships and distinctions between the arts. She acknowledges the source of all the arts in the inner life and clearly defines the essential differences between the various art forms, pointing out the necessity of understanding and maintaining the integrity of each.

Finally, Joan Erikson[20] has provided a rationale for art as a healing activity. Instead of exploration of the meaning of art, she emphasizes the healthy functions involved in craft and making the created object.

¶ Art Therapy in Education

The groundwork of art therapy in education was established by Margaret Naumburg in her early work as an educator at the Walden School.[83] Years later she would continue to write about art in education and therapy.[89]

Art therapists in education look primarily to Viktor Lowenfeld's classic work, *Creative and Mental Growth,*[75] in which he promoted the natural spontaneity in the art of children and provided a full survey of content and style in children's art at different ages and stages of growth. Cognizant of Naumburg's principles, he advocated a method called "Art Education Therapy" for use in schools, and designated this special area of education as one frontier of art therapy.

In the 1960s and more in the 1970s, applications of art therapy began to appear frequently in educational settings. Elinor Ulman reviewed the history of the development of art therapy in mental health and in education in an article in the *Encyclopedia of Education.*[127] She clarified the difficult distinctions between these areas, which overlap primarily because of values inherent in the artistic process. In 1978, Judith Rubin[108] presented a review of the literature by art educators and art therapists. Sandra Pine[95] devised methods of art for large groups of school children and clarified the distinction between approaches of art in education and in psychotherapy. Robert Wolf[138] used individual art therapy in the school setting with emotional and learning problems of economically deprived children in New York City. Myer Site[115] used art as therapy with slow learners in public schools.

Frances Anderson[4] stressed the importance of developing the artistic needs of exceptional children as well as their educational needs. She emphasized the fundamentals of self-expression to develop communication and cooperation and elaborated classroom procedures and activities.

In research, Rawley Silver[114] has used sound procedures focusing on cognitive and developmental aspects of art in art therapy with school children with communication disorders and learning disabilities. She devised clear ways of perceiving and evaluating art, defined objectives and methods, and presented case studies and statistical results in a highly readable account.

Sandra Packard[90] defined categories of children's problems and of school settings ripe for development of art therapy. Packard and Frances Anderson[91] addressed the distinction between the roles of the art therapist in educational and therapeutic settings.

The potential for art therapy has only begun to be developed in educational settings for the learning disabled, the emotionally disturbed, and the physically handicapped. Defining the differences between art therapy in the mental health spectrum and the educational spectrum is a prerequisite to establishing the identity of the art therapist in school settings.

Elinor Ulman[127] has presented discerning distinctions of the issues and points of view of these two facets of the field and defines their common ground:

> The artistic process calls on the widest range of human capacities. Like maturation in general, it demands the integration of many inescapably conflicting elements, among them impulse and control, aggression and love, feeling and thinking, fantasy and reality, the unconscious and the conscious. The goal of art education is to make available to the individual resources within himself and outside himself, and the arts serve throughout life as a meeting ground for the inner and outer worlds. . . . Educators and therapists alike find

here a key to understanding the value of art education in alleviating emotional disturbance. [pp. 312–313]

Ulman's definition of the nature of the artistic process as a therapeutic agent reaffirms the thesis that art, as a manifestation of inner experience of the art maker involving all his capacities for integration and expression, is the common binding throughout the spectrum of the discipline, in all its philosophical and theoretical orientations. Through emphasis on individual expression in art, art therapy achieves its great versatility in cultivating the healing power of art, both through the artistic process and as a means of communication.

¶ Conclusion

This chapter has reviewed the pioneers of art therapy and the context of related disciplines in which they developed the field during its forty years as an independent profession. Because art is both profound and yet practical, because it is an intrinsically healthy process, art therapy fits naturally and with ease in the most diverse settings and yields rich contributions in evaluation and treatment. As the economic difficulties in provision of treatment to those in need increase, the advantages of art therapy in meeting the challenges of treatment in mental health, educational, and community settings are gradually being recognized. The field of art therapy continues to grow and spread in many directions, and its potential and benefits are becoming widely realized.

¶ Bibliography*

1. ALSCHULER, R., and HATTWICK, L. *Painting and Personality.* 2 vols. Chicago: University of Chicago Press, 1969 (originally published in 1947).

*Books with extensive bibliographies are indicated by an asterisk.

2. *ANASTASI, A., and FOLEY, J. P., "A Survey of the Literature on Artistic Behavior of the Abnormal: I. Historical and Theoretical Background," *Journal of General Psychology,* 25 (1941), 111–142; II. "Approaches and Interrelationships," *Annual of New York Academy of Science,* 42 (1941), 1–112; III. "Spontaneous Productions," *Psychological Monographs,* 52 (1940), 1–7; IV. "Experimental Investigations," *Journal of General Psychology,* 25 (1941), 187–257.
3. ———. "An Experimental Study of the Drawing Behavior of Adult Psychotics in Comparison to that of a Normal Control Group," *Journal of Experimental Psychology,* 34 (1944): 169–94.
4. ANDERSON, F. E. *Art for All the Children.* Springfield, Ill.: Charles C Thomas, 1978.
5. ARIETI, S. *Creativity: The Magic Synthesis.* New York: Basic Books, 1976.
6. ARNHEIM, R. *Art and Visual Perception.* Berkeley: University of California Press, 1954.
7. ———. *Toward a Psychology of Art.* Berkeley: University of California Press, 1966.
8. BARUCH, D. W., and MILLER, H. "Use of Spontaneous Drawings in Group Therapy," *American Journal of Psychotherapy,* 5 (1951): 45–58.
9. *BETENSKY, M. *Self-Discovery through Self-Expression.* Springfield, Ill.: Charles C Thomas, 1973.
10. ———. "Phenomenology of Self-Expression in Theory and Practice," *Confinia Psychiatrica,* 21:1–3 (1978): 31–36.
11. BUCK, J. N. "The H-T-P Test." *Journal of Clinical Psychology,* 4 (1948): 151–159.
12. BYCHOWSKI, G. "The Rebirth of a Woman: A Psychoanalytic Study of Artistic Expression and Sublimation," *Psychoanalytic Review,* 34 (1947): 32–57.
13. CANE, F. *The Artist in Each of Us.* New York: Pantheon 1950.
14. DEUTSCH, F. "The Creative Passion of the Artist and Its Synesthetic Aspects," *International Journal of Psychoanalysis,* 40 (1959): 38–51.
15. ———. "Body, Mind and Art," *Daedalus,* Winter (1960): 34–45.
16. ———. Body, Mind and Art II: Studies of the Pictographic Reflections of Body Image on the Drawings of Children," *Acta Psychotherapy,* 2 (1963): 181–192.
17. DEWEY, J. *Art as Experience.* New York: Capricorn Books, 1934.

Practice. New York: Schocken Books, 1975, pp. 221–39.

90. PACKARD, S. P. "Art for the Exceptional Child: Rapidly Expanding Job Opportunities," *Art Psychotherapy*, 3 (1976): 81–85.
91. *———, and ANDERSON, F. E. "A Shared Identity Crisis: Art Education and Art Therapy," *American Journal of Art Therapy*, 16 (1976): 21–28.
92. PATCH, V. D., and REFSNES, C. C. "An Art Class in a Psychiatric Ward," *Bulletin of Art Therapy*, October (1968): 13–24.
93. PFISTER, O. *Expressionism in Art*. New York: Dutton, 1923.
94. *PICKFORD, R. W. *Studies in Psychiatric Art*. Springfield, Ill.: Charles C Thomas, 1967.
95. PINE, S. "Fostering Growth through Art Education, Art Therapy, and Art in Psychotherapy," in Ulman, E., and Dachinger, P., eds., *Art Therapy in Theory and Practice*, New York: Schocken Books, 1975, pp. 60–94.
96. *PRINZHORN, H. *Artistry of the Mentally Ill*. New York: Springer-Verlag, 1972.
97. RANK, O. *Art and Artists*. New York: Agathon Press, 1968; ———. *Myth of the Birth of the Hero and Other Essays*. Freund, P., ed. New York: Vintage Books, 1959.
98. REFSNES, C. C. *A Presentation of Art Therapy*. The Exhibition of Psychopathological Art, 4th World Congress of Psychiatry, Madrid. Catalog published by Boston State Hospital and the Harvard Psychiatry Service of Boston City Hospital, 1966.
99. ———. "Free Art Expression in Psychotherapy of Severe Psychosis," *Aspects of Art Therapy*. Scientific Exhibits, American Psychiatric Association. Catalog published by Boston State Hospital and the Harvard Psychiatry Service of Boston City Hospital, 1968.
100. REFSNES-KNIAZZEH, C. Book review of Naumburg, M. *Introduction to Art Therapy, American Journal of Art Therapy*, 18:1 (1978): 31–33.
101. ———. "Considerations in Success and Failure in Art Therapy and in Art." Paper presented at 10th Annual Conference of the American Art Therapy Association, November 1979, Washington, D.C. (*American Journal of Art Therapy*, in press).
102. REFSNES, C. C., and ARSENIAN, J. "Spontaneous Art as a Tool in Psychotherapy," *Proceedings, IV World Congress of Psychiatry. Madrid: Exerpta Medica*, 1966, pp. 762–769.
103. REFSNES, C. C., and GALLAGHER, F. P. "Art Therapy as Adjunct to Long-Term Psychotherapy," *Bulletin of Art Therapy*, 7:2 (1968): 59–79.
104. *RHYNE, J. *The Gestalt Art Experience*. Monterey, Calif.: Brooks/Cole, 1973.
105. ROBBINS, A., and SIBLEY, L. B. *Creative Art Therapy*. New York: Brunner/Mazel, 1976.
106. ROBBINS, M. "On the Psychology of Artistic Creativity," in Eissler, R. S., et al., eds., *The Psychoanalytic Study of the Child*, vol. 24. New York: International Universities Press, 1969.
107. RORSHCACH, H. *Psychodiagnostics*. Bern: Verlag Hans Huber, 1942.
108. RUBIN, J. A. *Child Art Therapy*. New York: Van Nostrand Reinhold, 1978.
109. SACHS, H. *The Creative Unconscious*. Cambridge, Mass.: Sci-Art Publishers, 1951.
110. SCHAEFFER-SIMMERN, H. *The Unfolding of Artistic Activity*. Berkeley: University of California Press, 1961.
111. SCHILDER, P. *The Image and Appearance of the Human Body*. New York: International Universities Press, 1950.
112. SECHEHAYE, M. A. *Autobiography of a Schizophrenic Girl*. New York: Grune & Stratton, 1951.
113. ———, *Symbolic Realization*. New York: International Universities Press, 1951.
114. SILVER, R. A. *Developing Cognitive and Creative Skills Through Art*. Baltimore, Md.: University Park Press, 1978.
115. SITE, M. "Art and the Slow Learner," in Ulman, E., and Dachinger, P., eds., *Art Therapy in Theory and Practice*. New York: Schocken Books, 1975, pp. 191–207.
116. SMITH, S. R., MACHT, L. B., and REFSNES, C. C. "Recovery, Repression and Art," *Bulletin of Art Therapy*, 6:3 (1967): 103–120 (reprinted in *Art Therapy Viewpoints*. New York: Schocken Books, 1980).
117. STERN, M. "Free Painting as an Auxiliary Technique in Psychoanalysis," in Bychowski, G., and Despert, L., eds., *Specialized Techniques in Psychotherapy*. New York: Basic Books, 1961.
118. SUTHERLAND, J. D., ed. *Psychoanalysis and Contemporary Thought*. New York: Evergreen Paperback, Grove Press, 1959.

119. "Therapeutic Art Programs Around the World: Albert Einstein Medical College, New York," *Bulletin of Art Therapy,* 3 (1963): 26–27.
120. "Therapeutic Art Programs Around the World: François-Michelle School, Montreal," in Ulman, E., and Dachinger, P., eds., *Art Therapy in Theory and Practice.* New York: Schocken Books, 1975, pp. 208–212.
121. "Therapeutic Art Programs Around the World: Highland View Hospital, Cleveland, Ohio," *American Journal of Art Therapy,* 10:3 (1971): 145–152.
122. "Therapeutic Art Programs Around the World: Medical Center for Federal Prisoners, Springfield, Missouri; California Medical Facility, Vacaville," *Bulletin of Art Therapy,* 5 (1965): 21–25.
123. UHLIN, D. *Art for Exceptional Children.* Dubuque, Iowa: William C. Brown, 1972.
124. ULMAN, E. "Art Therapy in an Outpatient Clinic," *Psychiatry,* 16:1 (1953).
125. ———. "A New Use of Art in Psychiatric Diagnosis," *Bulletin of Art Therapy,* 4:3 (1965): 91–116.
126. ———. "The Power of Art in Therapy," in Jakab, I., ed., *Psychiatry and Art,* vol. 3. New York: S. Karger, 1971, pp. 93–102.
127. ———. "Art Education for Special Groups: The Emotionally Disturbed," in Deighton, L. C., ed., *The Encyclopedia of Education,* vol. 1. Riverside, N. J.: Crowell Collier, 1971, pp. 311–316.
128. ———, and DACHINGER, P., eds. *Art Therapy in Theory and Practice.* New York: Schocken Books, 1975.
129. ULMAN, E., KRAMER, E., and KWIATKOWSKA, H. Y. *Art Therapy in the United States.* Craftsbury Common, Vt.: Art Therapy Publications, 1978.
130. ULMAN, E., and LEVY, B. I. "An Experimental Approach to the Judgment of Psychopathology from Paintings," *Bulletin of Art Therapy,* 8 (1968): 3–12.
131. ULMAN, E., and LEVY, C., eds. *Art Therapy Viewpoints.* New York: Schocken Books, 1980.
132. WADESON, H. *Art Psychotherapy.* New York: John Wiley, 1980.
133. WAELDER, R. *Psychoanalytic Avenues to Art.* New York: International Universities Press, 1965.
134. WEBSTER'S NEW COLLEGIATE DICTIONARY. Springfield, Mass.: G. and C. Merriam, 1953.
135. WESTMAN, H. *The Springs of Creativity.* New York: Atheneum, 1961.
136. WILSON, L. "Theory and Practice of Art Therapy with the Mentally Retarded," *American Journal of Art Therapy,* 16:3 (1977): 87–97.
137. WINNICOTT, D. W. *Playing and Reality.* New York: Basic Books, 1971, pp. 1–25.
138. WOLF, R. "Art Therapy in a Public School," *American Journal of Art Therapy,* 12:2 (1973): 119–127.
139. ZIERER, E., STERNBERG, D., and FINN, R. "The Role of Family Creative Analysis in Family Treatment," *Bulletin of Art Therapy,* 5 (1966): 47–63, 87–104.

tics. The process has a well-known therapeutic value. [pp. 19–20][4]

Contrary to popular opinion, the music therapy candidate is not necessarily a trained musician. However, musicianship can develop. According to the "Musical Communicativeness" evaluation scale of Nordoff and Robbins,[82] vocalization on a low level of response includes "fleeting reflexive sounds that echo some parts of the music; brief sounds that have a connection with the music, tonally and/or rhythmically and/or expressively"; instrumental on a low level of response includes "[drum] beating [that] is compulsive, impulsive, disordered, or totally uncontrolled, yet . . . shows fleeting reflexive effects of the music—beating [that] is discontinuous, infantile, poorly coordinated, or sporadic, yet . . . shows some slight influence of the music"; movement on a low level of response includes "brief excited movements such as running, jumping, stamping, hand and/or head movements in response to particular musical stimuli—habitual movements show slight, irregular changes in tempo, intensity and duration; compulsive rocking or twirling patterns give way briefly." In other words, the "music child" has yet to be formed but indicates some instinct for potential music making. In addition to acquiring and/or integrating the developmental prerequisites (cognitive, motor, emotional, perceptual factors) necessary for each ensuing stage of musical communication (with the therapist and/or peers) and expressiveness, the music therapy client will spontaneously incorporate aspects of his or her personality into the music he or she chooses, physically maneuvers, interprets, and/or communicates with the therapist. Further spontaneity can proceed by improvisation with the therapist—through movement, drama, vocalization, and/or instruments. The incorporation of the performer's personality into his music may demonstrate a parallel or contradictory picture of the client's extramusical personality in need of further elucidation and healthy integration.[46] In the same way, the client's formal composition efforts—the notation and reading of musical ideas—are autobiographical. "The tonal dream awake" is a direct avenue to personal symbolic expression.[72]

¶ Relationship to a Previously Established Theoretical Framework

Music Therapy and Piaget

In an attempt to translate into clinical practice firm methodologies for music therapy with various populations, it is apparent that musical experience and development can and must ultimately fit into an affective and cognitive framework. In this way the relationship between musical experience/expression and human development is not only strikingly demonstrated but also serves the clinician in establishing relationships between developmental goals and musical objectives.

From the cognitive-structuralist viewpoint of Piaget, one may theorize that musical development evolves through the same stages as other realms of knowledge: a sensorimotor stage (from birth to eighteen months) during which children simply emit sounds and react to changes in sound; a symbolic stage (ages two to six) when sounds begin to acquire communally shared meaning; a concrete operational stage (ages seven to eleven) when children can voice to one another a set of organized patterns of sound; and a formal operational period (age twelve and up) when they can reflect on how music works, analyze a composition, and freely invent new patterns.[42]

Already, researchers have begun to confirm the effectiveness of an interrelated cognitive-musical framework approach both in music education[88] and in music therapy.[35] Ewing seems to have translated into clinical practice the even more specific stages of musical development in the first five years of life that were outlined by Peter Ostwald.[85]

While musical precocity is often attributed largely to hereditary factors[40] (and individual differences in musical ability among young children are tremendous), the stages of musical development suggested by the Piagetean framework are applicable for the general population. It is, of course, important to remember that musical precocity and, at the very least, "normal" ability are found quite frequently among children who are not intellectually outstanding and, in many cases, are retarded.[40] Autistic children, in particular, are stereotyped for their unusual musical capacities which, when developed into communicative patterns, provide the first step in remediation of a serious emotional illness.

A supplementary effort on the part of Heinz Werner is much more informative about the distinctive features of musical development.[40] Focusing on the musical aspects of children's melodies, Werner examined the ascending and descending movements of melody, the role of repetition, the emergence of cadences, and the handling of phrasing. According to Gardner[40]

> This fine-drawn analysis revealed which aspects of melodies were most salient for children at a young age, and which aspects appeared spontaneously—i.e., that resting on a lower note is a constant throughout melodic production, presumably reflecting a fundamental tendency in vocal production, but that ability to use and vary cadences waits upon the ability to repeat a simple motif.

Even with these initial studies, Werner substantiates the relation between musical development and other aspects of linguistic and cognitive development.

Music Therapy and Personality Theory: The Analytic Framework

In general, the conceptual basis for a psychoanalytic theory of the arts relies upon: (1) the libido as an energetic source; (2) the transformation of unconscious content in analogy with dreams, imagination, and humor; (3) the dominance of the sublimative mechanism; and (4) the relative flexibility of the repressive mechanism.[83] These factors must all be considered specifically in relation to music—the most formal of the arts.

According to Noy,[83] "The problem of defining the significance of forms by relating it to the analysis of the content implied in those forms presents the major challenge facing the analytical theory of art today." This problem is further compounded if one considers music "contentless" and/or "objectless."[107] However, Suzanne Langer[57] provided a convincing argument that music is indeed a symbolic language even though its content may seem unfathomable much of the time. Music lends expression to the world of feeling; it is a symbol and its symbolized object is the emotional life.

> The tonal structures we call "music" bear a close logical similarity to the forms of human feeling—forms of growth and of attenuation, flowing and stowing, conflict and resolution, speed, arrest, terrific excitement, calm or subtle activation and dreamy lapses—not joy and sorrow perhaps, but the poignancy of either and both—the greatness and brevity and eternal passing of everything vitally felt.
>
> Such is the pattern of logical form, of sentience; and the pattern of music is that same form worked out in pure, measured sound and silence. Music is a tonal analogue of emotive life. [p. 27][58]

As a language that represents emotional life rather than ideas, music is "much more liable to evade the defenses and to reach the unconscious";[29] in this way it may "safely become an object of displacement rather than language."[92]

If one does indeed accept music as a language safely lending symbolic expression to unconscious contents (although definition of these contents is difficult to fathom), then one can also accept the likelihood of perception of music through primary and secondary process. Researchers[54,84,92] have written about a musical secondary process (the tune) uncovering a deeper musical primary process (the rhythm) and have even suggested that primary process mechanisms—displacement, condensation, inversion, repetition by the opposite—are musically operative. Many questions remain however: Why does the unconscious want or need to achieve tonal rep-

for group identification: "Verbal processing is necessary for members to become aware of their affectual and cognitive responses to interpersonal relationships in the group."[1]

The use of singing to unify a group is common in music therapy. Solo, duet, and trio vocalizing are helpful "to develop autonomy and to foster peer support."[1] The nature of the song material as well as the nature of the music experience (listening versus participation) can affect the cohesion of the group. Mitchell and Zanker[1] found that active singing of folk and traditional songs, as opposed to listening to diverse classical music, increased group cohesion during music group therapy. The group members who went from a passive listening experience to active participation by singing these songs, showed further integration of their personalities during group therapy.

Frances and Schiff,[36] encouraging teenagers in group psychotherapy sessions to choose songs and recordings that appealed to them, found that it was "socially acceptable to be moved by the songs . . . [whereas] . . . it is less acceptable to share the same feelings openly in a group." During this group process "previously repressed affects became available for cognitive understanding and control." The authors did not cite any differences in growth between active and passive musical experience.

Writing lyrics to original or precomposed compositions "can be an excellent way of members exploring their attitudes related to the group."[1] Rosenbaum[97] recognized that the song may be a way to express transference feelings toward the therapist. The selection of the song can also serve as a "transitional object to gain autonomy within the group or from the leader."[1]

Group improvisation was perceived by Priestley[91] as a means for group psychotherapy. Depending on the members' musical skills, percussion and tonal instrumentation are commonly used to create nonverbal dialogues to explore affectual responses within the group: "Depending upon the group's and members' needs, individual expression is verbally processed."[1] Ain suggested that some members have not set the boundaries to control such qualities as dynamics (loud and soft), tempo (fast and slow), and pitch (high and low) in a socially appropriate way and can explore their control through the use of percussion instruments.

More active and more passive roles in the group are ongoing via musical structures, not only when group sounds are influenced (harmonically, melodically, rhythmically) by members, but also when a direct leadership role, such as conductor, is allocated or elected.

The intent of the group will vary. Ain suggested the creation of a music therapy "work" group by structuring the beginning, middle, and ending of activities. The intent of another kind of group might be to formulate attitudes toward group members and therapist through the dynamics of musical activity. Such a group would be termed by Ain[1] as an "assumption" group. "If a person asks for another to sing a song with him, this could be an indication of pairing in the group. If the patient randomly plays an instrument, this could be a flight from the group's task. If a person is silent, he has the opportunity to actively listen and, in his way, feel a part of the group."[1]

It is apparent that the verbal processing of feelings evoked by musical experience and the clarification of musical behaviors in relation to human dynamics will differ from group to group. Verbal interaction among members of the group and the subsequent relationship between verbal interaction and musical processes need further investigation.[1] Ain studied two groups (each with four members) of developmentally disabled and emotionally disturbed young adults who were exposed to maximally structured and minimally structured music therapy groups for six weeks. He concluded that "inactive leadership is contraindicated for short-term treatment with this population. The directive approach is more necessary and appropriate during short-term music therapy sessions for groups of emotionally disturbed and developmentally disabled young adults."[1]

It seems the capacity of the group members and/or the leader (therapist) to provide order through sound affects verbal processing. Ain writes:

> The leader continues to provide order through sounds and rhythms that have inherent relational patterns. Members unconsciously identify with these structures and feel ordered experiences. From this stance, the group gains courage to verbalize feelings and expression related to their intra-psychic and interpersonal functioning. An example of this process is when Yvonne [age 22, latent type schizophrenic] finishes singing and playing her rhythmic patterns she then expresses she feels less self-hatred and hopes to work in a satisfying job. [p. 104][1]

The music therapist recognizes that influential components of group process in which music is used as a tool include the musical interrelationship between group members and the therapist as well as the musical intervention and structure of activities by the therapist.[1]

¶ Bibliography

1. AIN, E. "The Relationship of Musical Structure and Verbal Interaction Process during Adult Music Therapy," (unpublished M.A. thesis, Hahnemann Medical College and Hospital, Philadelphia, Pa., 1978).
2. ALTSCHULAR, I. M. "Four Years Experience with Music as a Therapeutic Agent at Eloise Hospital," *American Journal of Psychiatry,* 100 (1944): 792–794.
3. ———. "The Past, Present and Future of Music Therapy," in Podolsky, E., ed., *Music Therapy.* New York: Philosophical Library, 1954, pp. 24–35.
4. ALVIN, J. *Music Therapy.* London: John Baker Publishers, 1966.
5. ———. *Music for the Handicapped Child,* 2nd ed., London: Oxford University Press, 1976.
6. ———. *Music Therapy for the Autistic Child.* New York: Oxford University Press, 1978.
7. APEL, W. *Harvard Dictionary of Music,* 2nd ed., Cambridge: Harvard University Press, 1972.
8. ARIETI, S. *Creativity: The Magic Synthesis.* New York: Basic Books, 1976.
9. BAILEY, P. *They Can Make Music.* London: Oxford University Press, 1973.
10. BANG, C. "Music Therapy and Musical Speech Therapy for Deaf and Multiply Handicapped Children," (unpublished paper, 1978).
11. BARNHART, C., ed. *The American College Dictionary.* New York: Random House, 1953.
12. BEATTY, K. "Cerebral Dominance in the Processing of Verbal vs. Musical Stimuli," (unpublished paper, Montclair State College, Upper Montclair, N.J., 1979).
13. BEREL, M., DILLER, L., and ORGEL, M. "Music as a Facilitator for Visual Motor Sequencing Tasks in Children with Cerebral Palsy," *Developmental Medical Child Neurology,* 13 (1971): 335–342.
14. BERSHATSKY, D. Personal communication, November 1979.
15. BONNY, H. L. "Music Therapy—Guided Imagery," in Raslof, L., ed., *Holistic Dimensions in Healing: A Resource Guide.* New York: Doubleday, 1978.
16. ———, and SAVARY, L. *Music and Your Mind.* New York: Harper & Row, 1960.
17. BONNY, H. L., et al. "Some Effects of Music on Verbal Interaction in Groups," *Journal of Music Therapy,* 2 (1965): 61–63.
18. BOXBERGER, R. "Historical Bases for the Use of Music in Therapy," *Proceedings of the National Association for Music Therapy, Inc.* Lawrence, Kan., 1966, pp. 125–166.
19. BREGENZER, D. "Affective Response to Music as Evidenced in Physiological Process," (unpublished paper, Montclair State College, Upper Montclair, N.J., 1978).
20. BRIGHT, R. *Music in Geriatric Care.* New York: St. Martin's Press, 1972.
21. BROWN, M., and SELINGER, M. "A Nontherapeutic Device for Approaching Therapy in an Institutional Setting," *International Journal of Group Psychotherapy,* 19 (1969): 88–95.
22. CAPURSO, A. *Music and Your Emotions.* New York: Liveright, 1952, pp. 56–82.
23. CASS, M. "A Note on Music," in Arieti, S., ed., *Creativity: The Magic Synthesis.* New York: Basic Books, 1976, pp. 236–241.
24. CASTELLANO, J. A. "Music Composition in

HIEBERT LIBRARY
3 6877 00070 1440